PREFACE TO THE TWENTY-FOURTH REVISED MULTICOLOURED EDITION

The author feels great pleasure in presenting to his readers, the first revised multicolour edition of **'A Textbook of Electrical Technology (Vol. IV)'.** Multicolour pictures have been added to enhance the content value and to give the students an idea of what he will be dealing in reality, and to bridge the gap between theory and practice. A notable feature is the inclusion of chapter on "Flip-Flops and Related Devices" as per latest development in the subject. Latest tutorial problems and objective type questions specially for GATE have been included at relevant places.

I am thankful to the management and the editorial team of S. Chand & Company Pvt. Ltd. for all help and support in the publication of this book.

Any suggestions for the improvement of this book will be thankfully acknowledged and incorporated in the next edition.

Dr. R.S. SEDHA

PREFACE TO THE TWENTY-THIRD REVISED EDITION

I feel pleasure in presenting our esteemed readers a thoroughly revised edition of 'A Textbook of Electrical Technology (Vol. IV)'. As before the main objective of this Volume is to present the subject of Electronic devices and circuits, both analog and digital with sufficient depth.

This edition has been updated with the addition of more examples, tutorial problems and objective test questions in almost all the chapters. The chapter on Optoelectronic Devices has been expanded to include more application examples in the area of optical fibre networks. The chapter on Regulated Power Supply carries more detailed study of fixed positive – fixed negative and adjustable-linear IC voltage regulators as well as switching voltage regulator. The topic on OP-AMPs has been separated from the chapter on Integrated Circuits. A new chapter is prepared on OP-AMPs and its Applications. The chapter on Integrated Circuits has been thoroughly revised to include advanced topics required to learn/teach about fabrication of integrated circuits. The chapter on OP-AMPs and its Applications includes OP-AMP based Oscillator circuits, active filters etc. The chapter on Logic Gates is revised to include problems related to timing waveforms. The chapter on Boolean Algebra and Logic Families has been expanded to cover more detailed treatment on Karnaugh map, simplification of Boolean expression using Boolean Algebra/Karnaugh map and logic families. The chapter on Electronic Instruments has also been expanded to include study of some more recent electronic instruments such as logic analyser, spectrum analyser, signal generators and IEEE - 488/GPIB system.

I would like to acknowledge the helpful feedback I received from many distinguished teachers and students spread all over India and abroad. I am also thankful to our publishers particularly Mr. Ravindra Kumar Gupta, M.D. and Mr. R.S. Saxena, Adviser of S. Chand & Co. Ltd. for the personal interest they took in the expeditious printing of this book. I am also thankful to the publishers for the extremely attractive cover design conceived by the team.

B.L. THERAJA
A.K. THERAJA

A TEXTBOOK OF

Electrical Technology

VOLUME IV

Electronic Devices & Circuits

IN

S.I. SYSTEM OF UNITS

(Including rationalized M.K.S.A. System)

For the Examinations of B.E. (Common Course to All Branches), B.Tech., B.Sc. (Engg), Sec. A & B of AMIE(I), A.M.I.E.E. (London), I.E.R.E. (London), Grade I.E.T.E., Diploma and other Competitive Examinations

B.L. THERAJA

A.K. THERAJA

Revised By :

Dr. R.S. SEDHA

Program Chair

B.E. (Hons.), M.Tech. (IIT Delhi).

Ph.D. (U.K.), FIETE, MIEEE

School of Engineering

Republic Polytechnic

Tanglin Campus

1 Kay Siang Road

SINGAPORE–248922

S. CHAND PUBLISHING

empowering minds

S. CHAND & COMPANY PVT. LTD.

(An ISO 9001 : 2008 Company)

RAM NAGAR, NEW DELHI - 110 055

S. CHAND & COMPANY PVT. LTD.

(An ISO 9001 : 2008 Company)

Head Office: 7361, RAM NAGAR, NEW DELHI - 110 055

Phone: 23672080-81-82, 9899107446, 9911310888; Fax: 91-11-23677446

Shop at: **schandgroup.com**; e-mail: **info@schandgroup.com**

S. CHAND PUBLISHING
empowering minds

Branches :

AHMEDABAD : 1st Floor, Heritage, Near Gujarat Vidhyapeeth, Ashram Road, **Ahmedabad** - 380 014, Ph: 27541965, 27542369, ahmedabad@schandgroup.com

BENGALURU : No. 6, Ahuja Chambers, 1st Cross, Kumara Krupa Road, **Bengaluru** - 560 001, Ph: 22268048, 22354008, bangalore@schandgroup.com

BHOPAL : Bajaj Tower, Plot No. 2&3, Lala Lajpat Rai Colony, Raisen Road, **Bhopal** - 462 011, Ph: 4274723, 4209587, bhopal@schandgroup.com

CHANDIGARH : S.C.O. 2419-20, First Floor, Sector - 22-C (Near Aroma Hotel), **Chandigarh** - 160 022, Ph: 2725443, 2725446, chandigarh@schandgroup.com

CHENNAI : No.1, Whites Road, Opposite Express Avenue, Royapettah, **Chennai** - 600014 Ph. 28410027, 28410058, chennai@schandgroup.com

COIMBATORE : 1790, Trichy Road, LGB Colony, Ramanathapuram, **Coimbatore** - 6410045, Ph: 2323620, 4217136 coimbatore@schandgroup.com **(Marketing Office)**

CUTTACK : 1st Floor, Bhartia Tower, Badambadi, **Cuttack** - 753 009, Ph: 2332580; 2332581, cuttack@schandgroup.com

DEHRADUN : 1st Floor, 20, New Road, Near Dwarka Store, **Dehradun** - 248 001, Ph: 2711101, 2710861, dehradun@schandgroup.com

GUWAHATI : Dilip Commercial (Ist floor), M.N. Road, Pan Bazar, **Guwahati** - 781 001, Ph: 2738811, 2735640 guwahati@schandgroup.com

HALDWANI : Bhatt Colony, Talli Bamori, Mukhani, **Haldwani** - 263139 **(Marketing Office)** Mob. 09452294584

HYDERABAD : Padma Plaza, H.No. 3-4-630, Opp. Ratna College, Narayanaguda, **Hyderabad** - 500 029, Ph: 27550194, 27550195, hyderabad@schandgroup.com

JAIPUR : 1st Floor, Nand Plaza, Hawa Sadak, Ajmer Road, **Jaipur** - 302 006, Ph: 2219175, 2219176, jaipur@schandgroup.com

JALANDHAR : Mai Hiran Gate, **Jalandhar** - 144 008, Ph: 2401630, 5000630, jalandhar@schandgroup.com

KOCHI : Kachapilly Square, Mullassery Canal Road, Ernakulam, **Kochi** - 682 011, Ph: 2378740, 2378207-08, cochin@schandgroup.com

KOLKATA : 285/J, Bipin Bihari Ganguli Street, **Kolkata** - 700 012, Ph: 22367459, 22373914, kolkata@schandgroup.com

LUCKNOW : 1st Floor, Arya Pratinidhi Sabha, 5-Mirabai Marg, Near Narhi Sabzi Mandi, Hazratganj, **Lucknow** - 226 001, Ph: 4065646, 4026791, 4027188, 4022948 lucknow@schandgroup.com

MUMBAI : Blackie House, IInd Floor, 103/5, Walchand Hirachand Marg, Opp. G.P.O., **Mumbai** - 400 001, Ph: 22690881, 22610885, mumbai@schandgroup.com

NAGPUR : Karnal Bagh, Near Model Mill Chowk, **Nagpur** - 440 032, Ph: 2720523, 2777666 nagpur@schandgroup.com

PATNA : 104, Citicentre Ashok, Mahima Palace , Govind Mitra Road, **Patna** - 800 004, Ph: 2300489, 2302100, patna@schandgroup.com

PUNE : Sadguru Enclave, Ground floor, Survey No. 114/3, Plot no. 8 Alandi Road , Vishrantwadi **Pune** - 411015 Ph: 64017298 pune@schandgroup.com

RAIPUR : Kailash Residency, Plot No. 4B, Bottle House Road, Shankar Nagar, **Raipur** - 492 007, Ph: 2443142,Mb. : 09981200834, raipur@schandgroup.com **(Marketing Office)**

RANCHI : Shanti Deep Tower, Opp.Hotel Maharaja, Radium Road, **Ranchi** - 834001 Mob. 09430246440 ranchi@schandgroup.com

SILIGURI : 122, Raja Ram Mohan Roy Road, East Vivekanandapally, P.O., Siliguri, **Siliguri** - 734001, Dist., Jalpaiguri, (W.B.) Ph. 0353-2520750 **(Marketing Office)** siliguri@schandgroup.com

VISAKHAPATNAM : No. 49-54-15/53/8, Plot No. 7, 1st Floor, Opp. Radhakrishna Towers, Seethammadhara North Extn., **Visakhapatnam** - 530 013, Ph-2782609 (M) 09440100555, visakhapatnam@schandgroup.com **(Marketing Office)**

First Edition 1959

Subsequent Editions and reprints 1960, 61, 62, 64, 65, 66, 67, 68, 69, 70, 71, 72, 73(Twice), 74, 75, 76, 77, 78, 79, 80, 81, 82, 83, 84 (Twice), 85, 86, 89, 91, 93, 94, 95, 96, 97, 98, 2000, 2001, 2002, Thoroughly Revised Twenty-Third Edition 2002, Reprint 2004; ***First Multicolour Edition 2006,*** *Reprint 2007, 2008, 2009, 2010, 2012, 2013, 2014*

Reprint 2015

ISBN : 978-81-219-2667-6

Code : 1010B 295

PRINTED IN INDIA

Other Parts Available

Volume I : Basic Electrical Engineering; Volume II : AC & DC Machines; Volume III : Transmission, Distribution & Utilization Combined Edition Also Available

This book is for Sale in India & other SAARC Countries only.

By Vikas Publishing House Pvt. Ltd. Plot 20/4, Site-IV, Industrial Area Sahibabad, Ghaziabad-201010 and published by S. Chand & Company Pvt. Ltd., 7361, Ram Nagar, New Delhi -110 055.

Contents

51. Semiconductor Physics ... 2017 — 2058

The Atom—Bohr's Atomic Model—Calculations concerning Bohr's Atomic Model—Normal, Excited and Ionized Atom—Electron Energy Levels in Hydrogen Atom—Orbital (or azimuthal) Quantum Number—Electron Configuration of Atoms—Orbital Magnetic Quantum Number (m_l)—Magnetic Spin Quantum Number (m_s)—Pauli's Exclusion Principle—Energy Band in Solids—Spacing Between Energy Levels of a Band—Energy Bands in Lithium and their occupancy—Valence and Conduction Bands—Insulators, Conductors and Semiconductors—Crystal Structure—Representation of Crystal Planes and Directions—Atomic Binding in Semiconductors—Types of Semiconductors—Intrinsic Semiconductors—Hole Formation in Semiconductors—Fermi Level in an Intrinsic Semiconductors—Extrinsic Semiconductors— Majority and Minority Carriers—Mobile Charge Carriers and Immobile Ions—Electron Conductivity of a Metal—Conductivity of Intrinsic Semiconductors—Conductivity of Extrinsic Semiconductors—Conductivity when Intrinsic Charge Carrier Densities are Neglected—Conductivity of Pure and P-type-Germanium—Drift—Diffusion—Combined Drift and Diffusion Currents—Relation between D and μ—Recombination—Carrier Life Time—Total Carrier Flow—P-N Junction—Formation of Depletion Layer—Junction or Barrier Voltage—Energy Band Diagram of a P-N Junction—Forward Biased P-N Junction—Forward V/ I Characteristic—Reverse Biased P-N Junction—Reverse V/ I Characteristic—Hall Effect—Objective Tests.

52. P-N Junction Diode ... 2059 — 2086

P-N Junction Diode—Derivation of Junction Resistance—Junction Breakdown—Junction Capacitance—Equivalent Circuit of P-N Junction—Diode Fabrication—Grown Junction—Alloy Junction—Diffused Junction—Epitaxial Junction—Point Contact Junction—The Ideal Diode—The Real Diode—Diode Circuits with D.C. and A.C. Voltage Sources—Diode Clipper and Clamper Circuits—Clippers—Some Clipping Circuits—Clampers—Summary of Clamping Circuits—Objective Tests.

53. Optoelectronic Devices ... 2087 — 2110

Fundamentals of Light—Light Emitting Diode (LED)—Use of LEDs in Facsimile Machines—Liquid Crystal Displays—P-N Junction Photodiode—Dust Sensor—Photoconductive Cell —Phototransistor—Photodarlington—Photo voltaic or Solar Cell — Laser Diode—Optical Disks—Read—only Optical Disk Equipment—Printers using Laser Diodes—Hologram Scanners — Laser Range Finder—Light-activated SCR(LASCR)—Optical Isolators—Optical Modulators—Optical Fibre Communication Systems—Optical Fibre Data Links—Objective Tests.

54. Special Diodes ... 2111— 2128

Zener Diode—Voltage Regulation—Zener Diode as Peak Clipper—Meter Protection—Zener Diode as a Reference Element—Tunneling Effect—Tunnel Diode—Tunnel Diode Oscillator—Varactor—PIN Diode—Schottky Diode—Step Recovery Diode—Gunn Diode—IMPATT Diode—Objective Tests.

55. D.C. Power Supplies ... 2129 — 2166

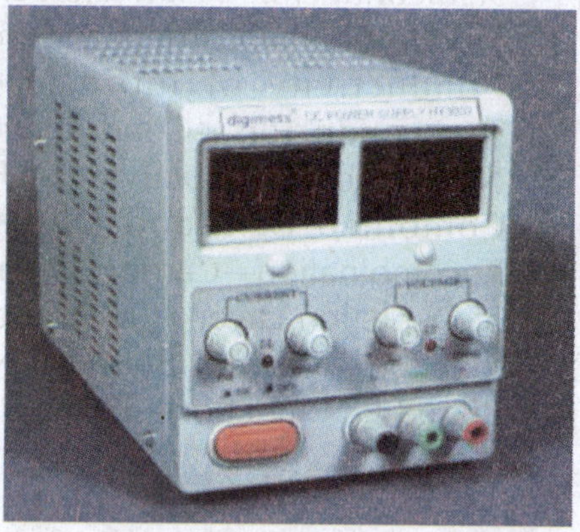

Introduction—Unregulated Power Supply—Regulated Power Supply—Rectifiers—Single-phase Half-wave Rectifier—Equivalent Circuit of a HW Rectifier—Single-phase Full-wave Rectifier—Full-wave Bridge Rectifier—Three-phase Half-wave Rectifier—Full-wave Rectification of 3-phase Currents—Six-phase Half-wave Rectifier—Three-phase Bridge Circuit—Calculations with Resistive Load—Filters—Shunt Capacitor Filter—Effect of increasing Filter Capacitance—Calculation of Shunt Capacitor Filter—Series Inductor Filter—The Choke Input of L-C Filter—The R-C Filter—The C-L-C or Pi Filter—Bleeder Resistor—Voltage Dividers—Complete Power Supply—Voltage Multipliers—Half-wave Voltage Doubler—Full-wave Voltage Doubler—Voltage Tripler and Quadrupler Circuits—Troubleshooting Power Supplies—Controlled Rectification—Output Waveform for Different Firing Angles—Output Voltage and Current Values in Controlled Rectifiers—Average Values for FW Controlled Rectifier—Silicon Controlled Rectifier—Pulse Control of SCR—90° Phase Control of SCR—180° Phase Control of SCR—SCR Controlled Load Circuit—UJT Controlled Load Circuit—Chopper—Inverters—Single Phase Inverter—Push-pull Inverter—Objective Tests

56. Regulated Power Supply ... 2167— 2186

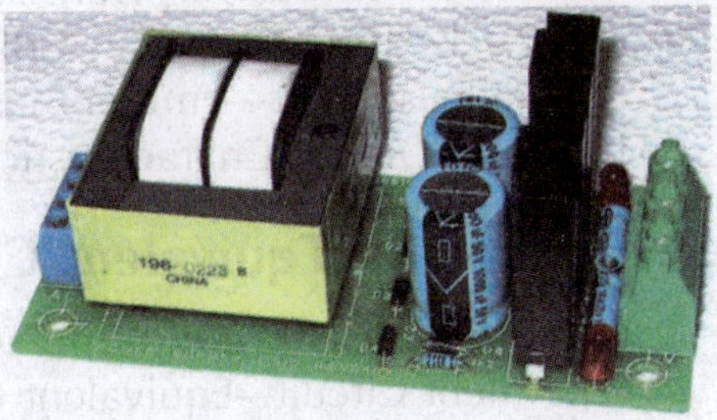

General—Voltage Regulation—Zener Diode Shunt Regulator—Transistor Series Voltage Regulator—Controlled Transistor Series Regulator—Transistor Shunt Voltage Regulator—Transistor Current Regulator—Variable Feedback Regulator—Basic OP-AMP Series Regulator—Basic OP-AMP Shunt Regulator—Switching Regulators—Step-down Switching Regulator—Step-up Switching Regulator—Inverting Switching Regulator—IC Voltage Regulators—Fixed Positive Linear Voltage Regulators—Fixed Negative Linear Voltage Regulators—Adjustable Positive Output Linear Voltage Regulators—Adjustable Negative Output Linear Voltage Regulators—Use of External Pass Transistor with Linear Voltage Regulators—Use of Linear Voltage Regulator as a Current Regulator—Switching Voltage IC Regulators—Objective Tests.

57. Bipolar Junction Transistor .... 2187— 2218

Bipolar Junction Transistor—Transistor Biasing—Important Biasing Rule—Transistor Currents—Summing up—Transistor Circuit Configurations—CB Configuration—CE Configuration —Relation between α and β—CC—Configuration—Relation between Transistor Currents—Leakage Currents in a Transistor—Thermal Runaway—Transistor Static Characteristics—Common Base Test Circuit—Common Base Static Characteristics—Common Emitter Test Circuit—Common Emitter Static Characteristics—Common Collector Static Characteristic—Different Ways of Drawing Transistor Circuits—Common Base Formulas—Common Emitter Formulas—Common Collector Formulas—The Beta Rule—Importance of V_{CE}—Cut-off and Saturation Points—BJT Operating Regions—Active Region DC Model of BJT—BJT Switches—Normal DC Voltage Transistor Indications—Transistor Fault Location—Solving Universal Stabilization Circuit—Notation for Voltages and Currents—Increase/Decrease Notation—Applying AC to a DC Biased Transistor—Transistor AC/DC Analysis—Objective Tests.

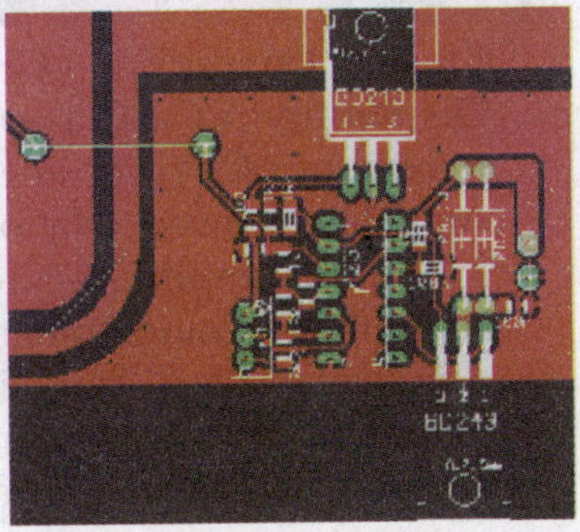

58. Load Lines and DC Biased Circuits ... 2219 — 2236

DC Load Line—Q-Point and Maximum Undistorted Output—Need for Biasing a Transistor—Factor Affecting Bias Variations—

Stability Factor—Beta Sensitivity—Stability Factor for CB and CE Circuits—Different Methods for Transistor Biasing—Base Bias—Base Bias with Emitter Feedback—Base Bias with Collector Feedback—Base Bias with Collector and Emitter Feedbacks—Emitter Bias with two Supplies—Voltage Divider Bias—Load Line and Output Characteristics—AC Load Line—Objective Tests.

59. Transistor Equivalent Circuit and Models ... 2237— 2276

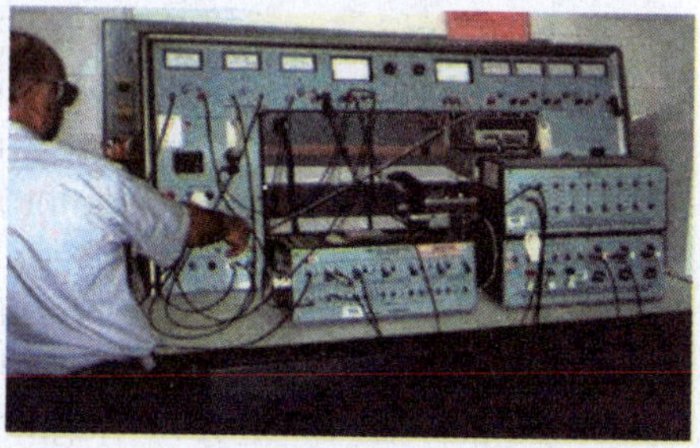

General—DC Equivalent Circuit—AC Equivalent Circuit—Equivalent Circuit of CB Amplifier—Effect of Source Resistance R_S on Voltage Gain—Equivalent Circuit of a CE Amplifier—Effect of Source Resistance R_S—Equivalent Circuit of a CC Amplifier—Small-signal Low-frequency Model or Representation—T-Model—Formulas for T-equivalent of a CB Circuit—Formulas for T-equivalent of a CE Circuit—Formulas for T-equivalent of a CC Circuit—What are h-parameters?—Input Impedance of a Two Port Network—Voltage Gain of a Two Port Network—The h-parameter Notation for Transistors—The h-parameters of an Ideal Transistor—The h-parameters of an Ideal CB Transistor—The h-parameters of an Ideal CE Transistor—Approximate Hybrid Equivalent Circuits—Transistor Amplifier Formulae using h-Parameters—Typical Values of Transistor h-parameters—Approximate Hybrid Formulas—Common Emitter h-parameter Analysis—Common Collector h-parameter Analysis—Conversion of h-parameters—Objective Tests.

60. Single-Stage Transistor Amplifiers ... 2277 — 2314

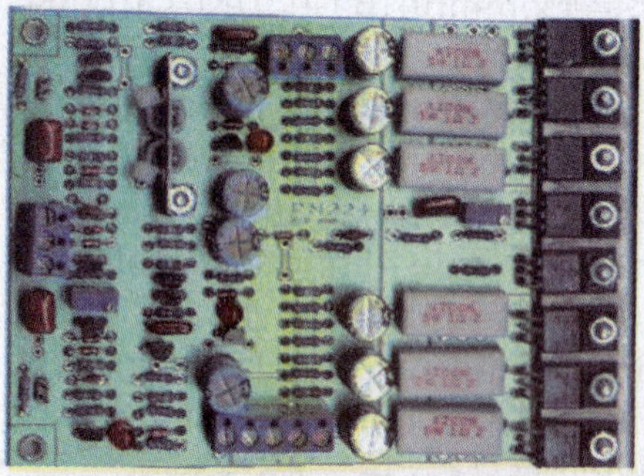

Classification of Amplifiers—Common Base (CB) Amplifier—Various Gains of a CB Amplifier—Characteristics of a CB Amplifier—Common Emitter (CE) Amplifier—Various Gains of a CE Amplifier—Characteristics of a CE Amplifier—Common Collector (CC) Amplifier—Various Gains of a CC Amplifier—Characteristics of a CC Amplifier—Comparison of Amplifier Configurations—Amplifier Classification Based on Biasing Conditions—Graphic Representation—Class-A Amplifier—Power Distribution in a Class-A Amplifier—Power Rectangle—Power Efficiency—Maximum AC Power in Load—Transformer-coupled Class-A Amplifier—Class-B Amplifier—Power Relation for Class-B Operation—Maximum Values—Class-B Push-pull Amplifier—Crossover Distortion—Power Efficiency of Push-pull Amplifiers—Complementary Symmetry Push-pull Class-B Amplifier—Class-C Amplifier—Tuned Amplifier—Distortion in Amplifiers—Non-linear Distortion—Intermodulation Distortion—Frequency Distortion—Phase or Delay Distortion—Noise—The Decibel System—Other Expressions for Power Gain—Voltage and

Current Levels—Characteristics of the Decibel System—Value of 1 dB—Zero Decibel Reference Level—Variations in Amplifier Gain with Frequency—Cause of Amplifier Gain Variations—Miller Effect—Cut-off Frequencies of Cascaded Amplifiers—Transistor Cut-off Frequencies—Alpha Cut-off Frequency—Beta Cut-off Frequency —The f_T of a Transistor—Relation between f_α, f_β and f_T—Gain-bandwidth Product— Objective Tests.

61. Multistage and Feedback Amplifiers ... 2315 — 2342

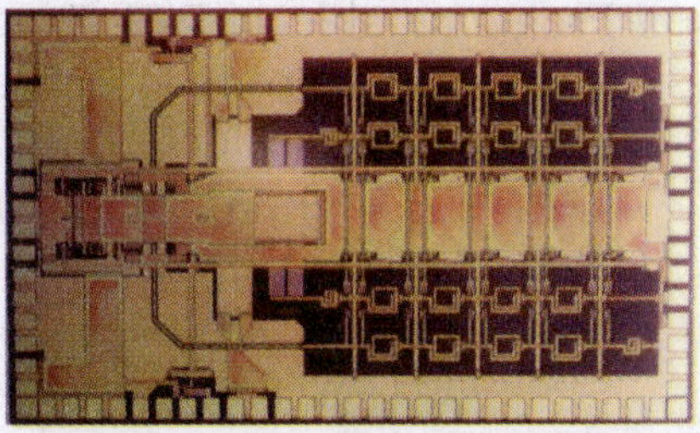

General—Amplifier Coupling—RC-coupled Two-stage Amplifier—Advantages of RC Coupling—Impedance-coupled Two-stage Amplifier—Advantages of Impedance Coupling—Transformer-coupled Two-stage Amplifier—Advantages of Transformer Coupling —Frequency Response—Applications —Direct-coupled Two-stage Amplifier using Similar Transistors—Direct-coupled Amplifier Using Complementary Symmetry of Two Transistors—Darlington Pair—Advantages of Darlington Pair—Comparison between Darlington Pair and Emitter Follower—Special Features of a Differential Amplifier—Common Code Input—Differential Amplifier—Objective Tests.

62. Feedback Amplifier ... 2343 — 2362

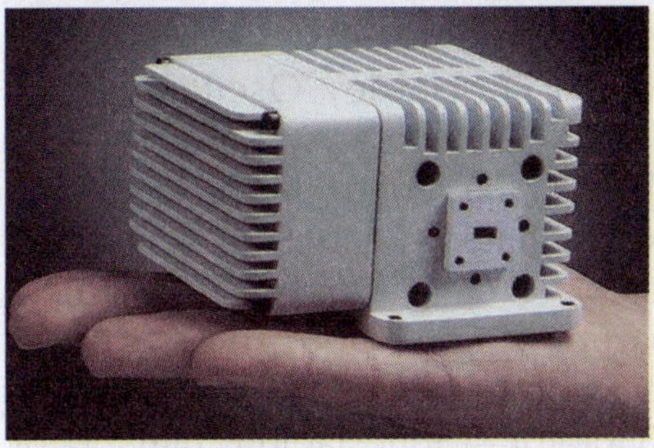

Feedback Amplifiers—Principle of Feedback Amplifiers—Advantages of Negative Feedback—Gain Stability—Decreased, Distortion—Feedback Over Several Stages—Increased Bandwidth—Forms of Negative Feedback—Shunt-derived Series-fed Voltage Feedback—Current-series Feedback Amplifier—Voltage-shunt Negative Feedback Amplifier— Current-shunt Negative Feedback Amplifier—Noninverting Op-amp with Negative Feedback—Effect of Negative Feedback on R_{in} and R_{out}—R_{in} and R_{out} of Inverting Op-amp with Negative Feedback—Objective Tests.

63. Field Effect Transistors ... 2363 — 2388

What is a FET—Junction FET (JFET)—Static Characteristics of a JFET—JFET Drain Characteristic with V_{GS} = 0—Characteristics with External Bias—Transfer Characteristic—Small Signal JFET Parameters—D.C. Biasing of a JFET—DC Load Line—Common Source JFET Amplifier—JFET Amplifier

Gains—Advantages of FETs— MOSFET or IGFET—DE MOSFET—Schematic Symbols for a DE MOSFET—Static Characteristics of a DE MOSFET—Enhancement-only N-Channel MOSFET—Biasing E-only MOSFET—FET Amplifiers—FET Applications—MOSFET Handling—Objective Tests.

64. Breakdown Devices ... **2389 — 2406**

What are Breakdown Devices?—Unijunction Transistor—UJT Relaxation Oscillator—Programmable UJT(PUT)—Silicon Controlled Rectifier—Comparison between Transistors and Thyristors—Transient Effects in an SCR—Phase Control—Theft Alarm— Emergency Lighting System—Light Activated SCR (LASCR)—The Shockley Diode—Triac—Diac—Silicon Controlled Switch (SCS)—Objective Tests.

65. Sinusoidal and Non-sinusoidal Oscillators ... **2407 — 2440**

What is an Oscillator?—Comparison between an Amplifier and an Oscillator—Classification of Oscillators—Damped and Undamped Oscillations—Oscillatory Circuit—Frequency of Oscillatory Current— Frequency Stability of an Oscillator—Essentials of a Feedback LC Oscillator—Tuned Base Oscillator—Tuned Collector Oscillator—Tuned Drain Oscillator (FET)—Hartley Oscillator—FET Hartley Oscillator—Colpitts Oscillator—Clapp Oscillator—FET Colpitts Oscillator—Crystals—Crystal Controlled Oscillator—Transistor Pierce Cystal Oscillator—FET Pierce Oscillator—Phase Shift Principle—RC Phase Shift Oscillator—Wien Bridge Oscillator—Non-sinusoidal Waveforms—Classification of Non-sinusoidal Oscillators—Pulse Definitions—Basic Requirements of a Sawtooth Generator—UJT Sawtooth Generator—Multivibrators (MV)—Uses of Multivibrators—Astable Multivibrator—Monostable Multivibrator (MMV)—Bistable Multivibrator (BMV)—Schmitt Trigger—Transistor Blocking Oscillator—Objective Tests.

66. Modulation and Demodulation ... **2441 — 2470**

Introduction—What is a Carrier Wave?—Radio Frequency Spectrum—Sound—Need for Modulation—Radio Broadcasting—Modulation—Methods of Modulation—Amplitude Modulation—Percent Modulation— Upper and Lower Side Frequencies—Upper and Lower Sidebands—Mathematical Analysis of a

Modulated Carrier Wave—Power Relation in an AM Wave—Forms of Amplitude Modulation—Generation of SSB—Methods of Amplitude Modulation—Block Diagram of an AM Transmitter—Modulating Amplifier Circuit—Frequency Modulation— Frequency Deviation and Carrier Swing—Modulation Index—Deviation Ratio—Percent Modulation—FM Sidebands—Modulation Index and Number of Sidebands—Mathematical Expressions for FM Wave—Demodulation or Detection—Essentials of AM Detection—Diode Detector for AM Signals—Transistor Detectors for AM Signals—FM Detection—Quadrature Detector—Frequency Conversion—Superheterodyne AM Receiver—Standard Superhet AM Receiver—FM Receiver—Comparison between AM and FM—The Four Fields of FM—Objective Tests.

67. Integrated Circuits ... 2471 — 2502

Introduction—What is an Integrated Circuit ? Advantages of ICs—Drawbacks of ICs—Scale of Integration—Classification of ICs by Structure—Comparison between Different ICs—Classification of ICs by Function—Linear Integrated Circuits (LICs)—Manufacturer's Designation of LICs—Digital Integrated Circuits—IC Terminology—Semiconductors used in Fabrication of ICs and Devices—How ICs are Made? — Material Preparation—Crystal Growing and Wafer Preparation—Wafer Fabrication—Oxidation—Etching—Diffusion—Ion Implantation—Photomask Generation—Photolithography—Epitaxy—Metallization and Interconnections — Testing, Bonding and Packaging—Semiconductor Devices and Integrated Circuit Formation—Comparison of ICs based on MOS and Bipolar Transistor Technology—Popular Applications of ICs—Objective Tests.

68. OP-AMP and its Applications ... 2503 — 2524

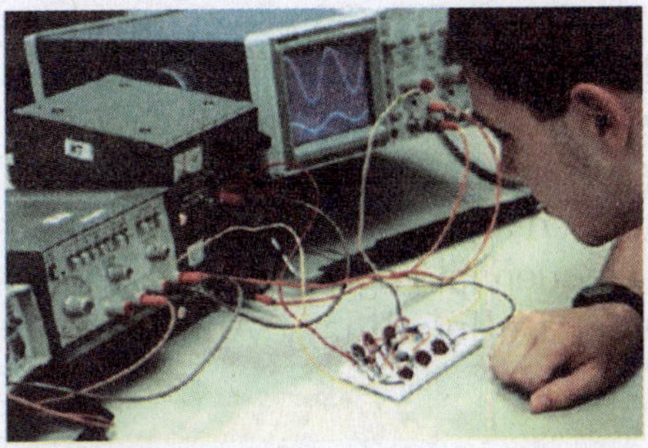

What is an OP-AMP? —OP-AMP Symbol—Polarity Conventions—Ideal Operational Amplifier—Virtual Ground and Summing Point—Why V_i is Reduced to Almost Zero? — OP-AMP Applications—Linear Amplifier—Unity Follower—Adder or Summer —Subtractor—Integrator—Differentiator—Comparator—Audio Amplifier—OP-Amp Based Oscillator Circuits—OP-Amp Based Wien Bridge Oscillator—OP-Amp Based Crystal Oscillator—A Triangular-wave Oscillator—A Voltage-controlled Sawtooth Oscillator (VCO) —A Square-wave Relaxation Oscillator—High-impedance Voltmeter—Active Filters—Low-pass Filter—High-pass Filter—Band-pass Filter—Notch Filter—Objective Tests.

69. Number Systems and Codes ... 2525 — 2554

Number Systems—The Decimal Number System—Binary Number System—Binary to Decimal Conversion—Binary Fractions—Double-Dadd Method—Decimal to Binary Conversion—Shifting the Place Point—Binary Operations—Binary Addition—Binary Subtraction—Complement of a Number—1's Complemental Subtraction—2's Complemental Subtraction—Binary Multiplication—Binary Division—Shifting a Number to Left or Right—Representation of Binary Numbers as Electrical Signals—Octal Number System—Octal to Decimal Conversion—Decimal to Octal Conversion—Binary to Octal Conversion—Octal to Binary Conversion —Usefulness of Octal Number System—Hexadecimal Number System—How to Count Beyond F in Hex Number Systems ? —Binary to hexadecimal Conversion—Hexadecimal to Binary Conversion—Decimal to hexadecimal Conversion—Hexadecimal to Decimal Conversion—Digital Coding—Binary Coded Decimal (BCD) Code—Octal Coding—Hexadecimal Coding—Excess-3 Code—Gray Code—Excess-3 Gray Code—Other Codes—ASCII Code—Objective Tests.

70. Logic Gates ... 2555 — 2584

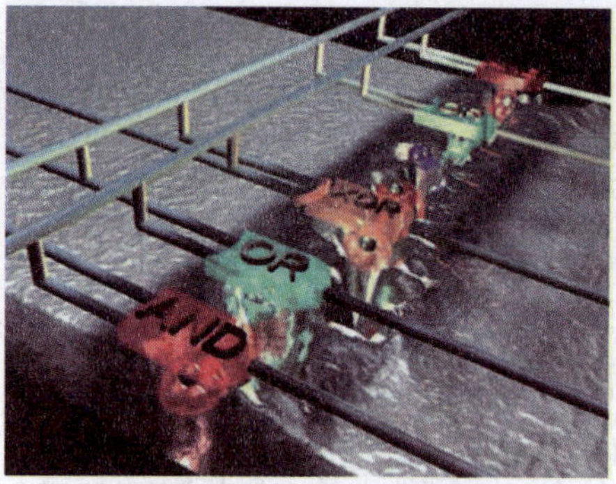

Definition of a Logic Gate—Positive and Negative Logic—The OR Gate—Equivalent Relay Circuit of an OR Gate—Diode OR Gate—Transistor OR Gate—OR Gate Symbolizes Logic Addition—Three Input OR Gate—Exclusive OR Gate—The AND Gate—Equivalent Relay Circuit of an AND Gate—Diode AND Gate—Transistor AND Circuit—AND Gate Symbolizes Logic Multiplication—The NOT Gate— Equivalent Circuits for a NOT Gate—The NOT Operation—Bubbled Gates—The NOR Gate—NOR Gate is a Universal Gate—The NAND Gate—NAND Gate is a Universal Gate—The XNOR Gate—Logic Gate at Glance—Digital Signals Applied to Logic Gates—Application of Logic Gates—Combinational Logic Circuit—Sequential Logic Circuits—Adders and Subtractors—Half Adder—Full Adder—Parallel Binary Adder—Half Subtractor—Full Subtractor—Objective Tests.

71. Boolean Algebra & Logic Families ... 2585 — 2630

Introduction—Unique Feature of Boolean Algebra—Laws of Boolean Algebra—Equivalent Switching Circuits—DeMorgan's Theorem's—Duals—Standard Forms of Boolean Expressions—The Sum-of-Products (SOP) Form—The Standard SOP

Form—The Product-of-Sums (POS) Form—The Standard POS Form—The Karnaugh Map—The Two-variable Karnaugh Map—The Three-variable Karnaugh Map—The Four-variable Karnaugh Map—Square Adjacensy in Karnaugh Map—Mapping a Standard SOP Expression on the Karnaugh Map—Mapping a Non standard SOP Expression on the Karnaugh Map—Simplification of Boolean Expression using Karnaugh Map—Mapping Directly on Karnaugh Map from a Truth Table — "Don't Care" Conditions—Main Logic Families—Saturated and Non-saturated Logic Circuits—Basic operating Characteristics and Parameters of Logic Families—DC supply voltage—TTL and CMOS Logic Levels—Noise Immunity—Noise Margin-Power Dissipation—Power Dissipation versus Frequency—Propagation Delay—Speed-Power Product—Loading and Fan-out—RTL Circuit—DTL Circuit—TTL Circuit—TTL Sub-families—ECL Circuit — I^2L Circuit— MOS family—PMOS Circuit—NMOS Circuit—CMOS Circuit—CMOS Sub-families and their Characteristics—Objective Tests.

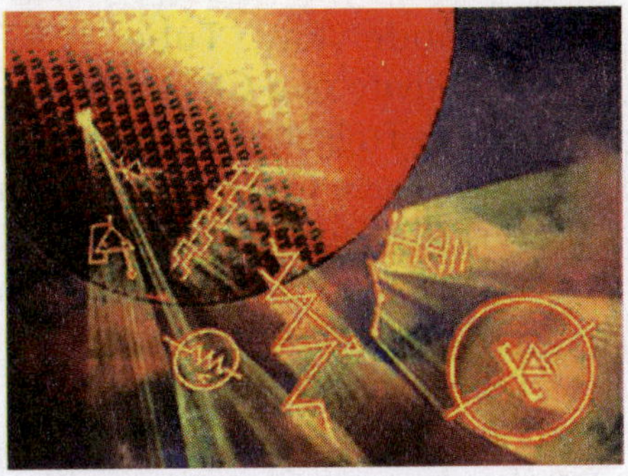

72. Flip-Flops and Related Devices ... 2631— 2694

Introduction —Flip-Flop (FF)—Latch—NAND Gate Latch—NOR Gate Latch—Clocked Signals—Some Main Ideas Common to Clocked Flip-Flops—Clocked S-C Flip-Flop—Clocked J-K Flip-Flop—Clocked D Flip-Flop—Parallel Transfer of Data Using D-Flip-Flops—D Latch (Transparent Latch)—Clocked J-K Flip-Flop with Asynchronous Inputs—Flip-Flop Timing Parameters—IC Flip-Flop Timing Values—Applications of Flip-Flop—Flip-Flop Synchronization—Parallel Data Transfer—Frequency Division—Counting—Schmitt Trigger Devices—Types of One-Shots—Retriggerable One-Shot—Actual One-Shot Devices—Clock Generator Circuits—Schmitt Trigger Oscillator—555 Timer Used as an Astable Multivibrator

73. Electronic Instruments ... 2695 — 2736

Introduction—Analog and Digital Instruments—Functions of Instruments—Electronic Vs Electrical Instruments—Essentials of an Electronic Instrument—Measurement Standards—The Basic Meter Movement—Characteristics of Moving Coil Meter Movement—Variation of Basic Meter Movement—Converting Basic Meter to DC Ammeter—Multirange Meter—Measurement of Current—Converting Basic Meter to DC Voltmeter—Multirange DC Voltmeter—Loading Effect of a Voltmeter—Ohmmeter—The Multimeter—Rectifier Type AC Meter—Electronic Voltmeters—Direct Current FET VM—Electronic Voltmeter for Alternating Currents—The Digital Voltmeter (DVM)—Cathode Ray Oscilloscope (CRO)—

Cathode Ray Tube (CRT)—Deflection Sensitivity of a CRT—Normal Operation of a CRO—Triggered and Non-triggered Scopes—Dual Trace CRO—Dual Beam CRO—Storage Oscilloscope—Sampling CRO—Digital Readout CRO—Handheld Battery Operated Oscilloscope— Lissajous Figures—Frequency Determination with Lissajous Figures—Applications of a CRO—The Q Meter—Logic Analysers—Spectrum Analysers—Signal Generators—Audio Generators—Function Generators—Pulse Generators—RF Generators—Frequency Synthesizer—Other Signal Generators—IEEE-488 General Purpose Interface Bus (GPIB) Instruments—Objective Tests.

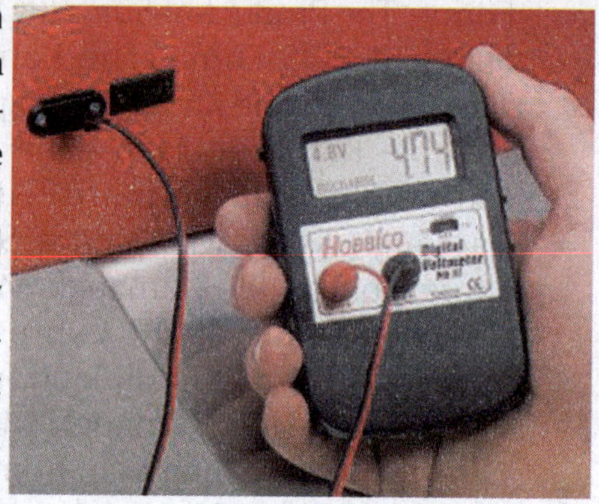

INDEX ... **(i — vi)**

VOLUME IV

ELECTRONIC DEVICES & CIRCUITS

CHAPTER 51

SEMI-CONDUCTOR PHYSICS

Learning Objectives

- The Atom
- Bohr's Atomic Model
- Electron Energy Levels in Hydrogen Atom
- Orbital (or Azimuthal) Quantum Number
- Orbital Magnetic Quantum Number (m_l)
- Magnetic Spin Quantum Number (m_s)
- Pauli's Exclusion Principle
- Energy Bands in Lithium and Their Occupancy
- Valence and Conduction Bands
- Insulators, Conductors and Semiconductors
- Crystal Structure
- Types of Semiconductors
- Mobile Charge Carriers and Immobile Ions
- Electron Conductivity of a Metal
- Combined Drift and Diffusion Currents
- Relation Between D and μ
- Carrier Life Time
- P-N Junction
- Formation of Depletion Layer
- Junction or Barrier Voltage
- Forward Biased P-N Junction
- Reverse Biased P-N Junction

A worker at a hydro-electric power station is carrying out routine maintenance on a high-voltage insulator made of a ceramic material.

51.1. The Atom

According to the model of atom proposed by Bohr in 1913, an atom is composed of a number of electrons moving in circular or elliptical orbits around a relatively heavy nucleus of protons and neutrons as shown in Fig. 51.1. Although, this simple model of an atom has been replaced by later models, yet it affords a convenient method of understanding the working of semi-conductor devices. Electron has a mass of nearly 9.1×10^{-31} kg and a charge of 1.6×10^{-19} C. The diameter of an atom is approximately 10^{-10} m and that of the nucleus about 10^{-15} m. The number of protons in the atom of an element gives its atomic number while the atomic mass number is determined by the number of protons and neutrons present in the nucleus.

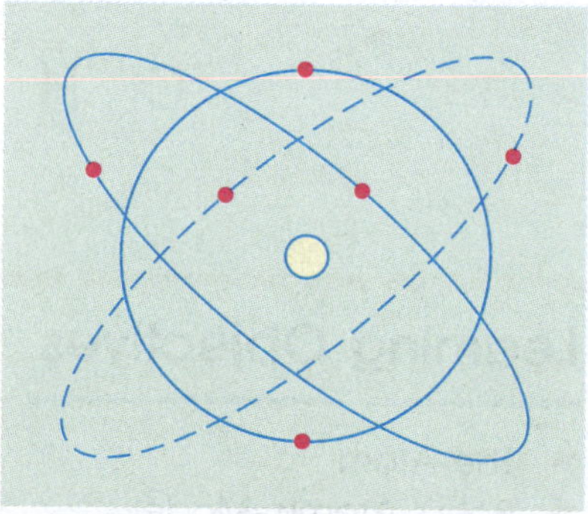

Fig. 51.1

51.2. Bohr's Atomic Model

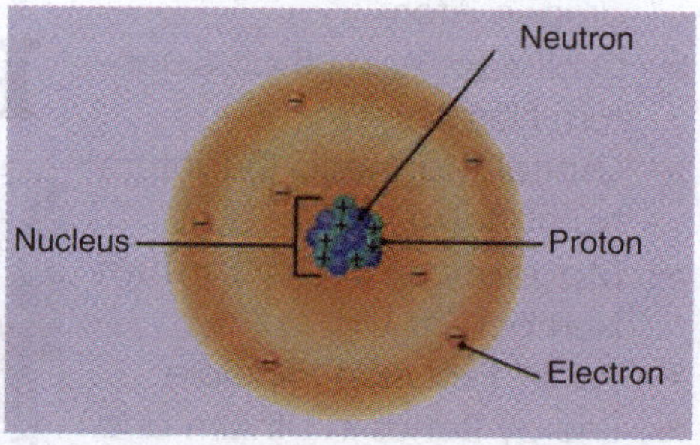

The nuclear atomic model proposed by Rutherford in 1911 was found to suffer from two serious drawbacks concerning distribution of extra-nuclear electrons and stability of the atom as a whole. It was later on superseded by atomic model proposed by Bohr in 1913. Using Planck's Quantum Theory, Bohr made the following postulates :

1. The atom has a massive positively-charged nucleus;
2. The electrons revolve round their nucleus in circular orbits, the centrifugal force being balanced by the electrostatic pull between the nucleus and electrons;
3. An electron cannot revolve round the nucleus in any arbitrary orbit but in just certain definite discrete orbits. ***Only those orbits are possible (or permitted) for which the orbital angular momentum (i.e. moment of momentum) of the electron is equal to an integral multiple of h/2π i.e. orbital angular momentum = nh/2π where n is an integer and h is Planck's constant***. Such orbits are also known as ***stationary orbits***;
4. While revolving in these permitted stationary (or stable) orbits, the electron does not radiate out any electromagnetic energy. In other words, ***the permissible orbits are non-radiating paths of the electron***;
5. The atom radiates out energy only when an electron *jumps* from one orbit to another. If E_2 and E_1 are the energies corresponding to two orbits before and *after* the jump, the frequency of the emitted photon is given by the relation

$$E_2 - E_1 = hf \qquad \text{or} \qquad \Delta E = hf$$

where f is the frequency of the emitted radiations.

Neils Bohr (1885–1962)

Explanation. If I is the moment of inertia of an electron and ω its angular velocity, then as per assumption (3) given above

$$\omega I = nh/2\pi \quad \text{or} \quad (mr^2)\,\omega = nh/2\pi \quad \text{or} \quad (mr^2)v/r = nh/2\pi$$

or $\quad mvr = nh/2\pi$

Alternatively, since the momentum of the revolving electron is mv, its moment about the nucleus is mvr (Fig. 51.2).

Hence, $mvr = nh/2\pi$

when n = 1, 2, 3 etc. for the first, second and third orbits respectively. It is called the ***principal quantum number*** and because it can take whole number values only, ***it fixes the size of the allowed orbits*** (also called Bohr's circular orbits).

Let the different permitted orbits have energies of E_1, E_2, E_3 etc. as shown in Fig. 51.3 (*a*). The electron can be raised from $n = 1$ orbit to any other higher orbit if it is given proper amount of energy.

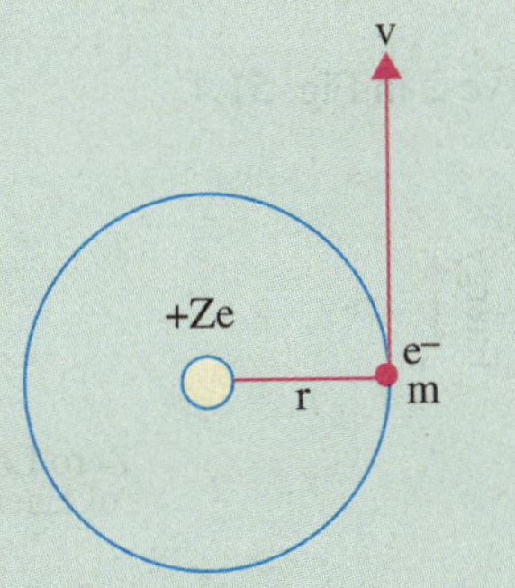

Fig. 51.2

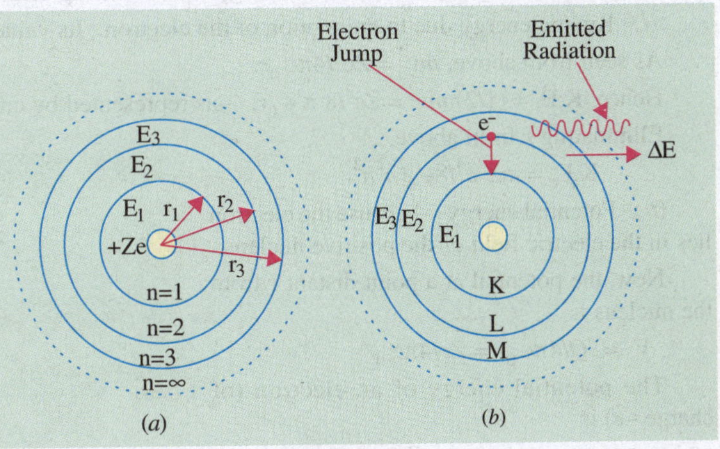

Fig. 51.3

When it drops back to $n = 1$ orbit after a short interval of time, it gives out the energy difference ΔE in the form of a radiation as shown in Fig. 51.3 (*b*). The relation between the energy released and frequency of the emitted radiation is

$$E_2 - E_1 = hf \quad \text{or} \quad \Delta E = hf$$

51.3. Calculations Concerning Bohr's Atomic Model

The above postulates concerning Bohr's atomic model can be utilized to calculate not only the radii of different electron orbits but also the velocity and orbital frequency possessed by different electrons.

Now, the stability of the atom requires that the centrifugal force acting on the revolving electron be balanced by the electrostatic pull exerted by the positively charged nucleus on the electron.

$$\therefore \quad \frac{mv^2}{r} = \frac{Ze \, . \, e}{4\pi\epsilon_0 \, r^2} \quad \text{or} \quad mv^2 r = \frac{Ze^2}{4\pi\epsilon_0}$$

Also, according to Bohr's postulates, $mvr = nh/2\pi$.

The above two equations may be used to find the radii of different Bohr's circular orbits.

(*a*) Radii of Orbits. Eliminating v from the above two equations, we get

$$r = \frac{\epsilon_0 \, n^2 h^2}{\pi \, mZe^2} \quad \text{or} \quad r_n = \frac{\epsilon_0 \, n^2 h^2}{\pi \, mZe^2}$$

It is seen that the radii of the permitted orbits vary as the square of the principal quantum number n. Also, $r_n = n^2 r_1$.

For hydrogen atom, $r_1 = 0.53 \times 10^{-10}$ m, $r_2 = 2^2 \, r_1 = 2.12 \times 10^{-10}$ m. All values between r_1 and r_2 are forbidden.

(*b*) Velocity of Revolving Electrons. The velocity of a revolving electron as found from the above equations is

$$v = Ze^2/2\epsilon_0 nh = 9 \times 10^9 . \, 2\pi Ze^2/nh.$$

It shows that velocity is inversely proportional to n *i.e.* $v \propto 1/n$. Hence, the electron in the innermost orbit has the highest velocity (nearly 1/37 of the velocity of light).

(c) Orbital Frequency. The orbital ***rotational*** frequency* of an electron is

$$f = \frac{v}{2\pi r} = \frac{mZ^2e^4}{4\epsilon_0^2 n^3 h^3}. \qquad \text{It is seen that } f \propto 1/n^3$$

(d) Electron Energy. The orbital energy of a revolving electron is of two types :

(*i*) Kinetic energy due to the motion of the electron. Its value is = (1/2) mv^2.

As seen from above, $mv^2 = Ze^2/4\pi\epsilon_0 r$.

Hence, K.E. = (1/2) $mv^2 = Ze^2/8\pi\epsilon_0 r$. —as represented by curve 2 in Fig. 51.4.

Eliminating r from above,

$$\text{K.E.} = mZ^2e^4/8\epsilon_0^2 n^2 h^2.$$

(*ii*) Potential energy—because the electron lies in the electric field of the positive nucleus.

Now, the potential at a point distant r from the nucleus is

$$V = Q/4\pi\epsilon_0 r = Ze/4\pi\epsilon_0 r$$

The potential energy of an electron (of charge $-e$) is

$$\text{P.E.} = V \times (-e) = -\frac{Ze^2}{4\pi\epsilon_0 r}$$

$$= -\frac{mZ^2e^4}{4\epsilon_0^2 n^2 h^2} \quad \text{— curve 1 in Fig. 51.4}$$

$\therefore$ Orbital energy = K.E. + P.E.

$$= \frac{Ze^2}{8\pi\epsilon_0 r} - \frac{Ze^2}{4\pi\epsilon_0 r}$$

$$= -\frac{Ze^2}{8\pi\epsilon_0 r} \quad \text{— curve 3 in Fig. 51.4}$$

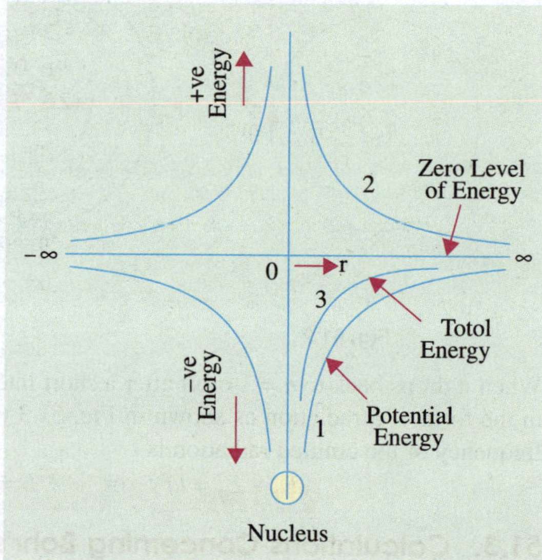

Fig. 51.4

Variations of potential energy and total energy are shown in Fig. 51.4.

If r is eliminated from the above equation, the orbital energy** is

$$E_n = -mZ^2e^4/8\epsilon_0^2 n^2 h^2$$

where n = 1, 2, 3....... etc. for the energy states that is possible for the electron to have. It will be seen that $E_n = E_1/n^2$.

51.4. Normal, Excited and Ionized Atom

Consider the case of the simplest atom *i.e.* hydrogen atom. When its only electron is in its innermost orbit (n = 1), then the atom is said to be in its normal (or unexcited) state. Generally, it is this condition in which most of the free hydrogen atoms in a gas are found to exist at normal room temperature and pressure. However, if spark is passed through hydrogen gas contained in a vessel, then high-speed electrons produced by the spark collide with hydrogen atoms and may either completely remove the $n = 1$ electron from them or raise it to higher permitted orbits having n = 2, 3, 4 etc.

When the electron is completely removed from the atom, the atom is said to be ionized. If, however, the electron is forced into an outer or higher n-value orbit, then the atom, is said to be

* It is the mechanical frequency of rotation and should not be confused with the frequency of emitted radiations.

** The negative sign only indicates that this much energy is required to remove the electron from the atom.

excited (or in an excited state). The atom does not remain in the excited state longer than 10^{-8} second because the electron under the attractive force of the nucleus jumps to the lower permitted orbit. In doing so, the electron loses the energy it had earlier gained during collision. However, the electron may return by several jumps, thereby emitting many different radiations of different frequencies.

51.5. Electron Energy Levels in Hydrogen Atom

As seen from Art. 51.3, orbital energy of an electron revolving in n_{th} orbit or shell is

$$E_n = -m Z^2 e^4 / 8\epsilon_0^2 n^2 h^2$$

In the case of hydrogen atom, $Z = 1$, hence

$$E_n = -\frac{me^4}{8\epsilon_0^2 n^2 h^2} = -\frac{me^4}{8\epsilon_0^2 h^2} \cdot \frac{1}{n^2} = -\frac{21.7 \times 10^{-19}}{n^2} \text{ joules}$$

$$= \frac{-21.7 \times 10^{-19}}{1.6 \times 10^{-19}} \cdot \frac{1}{n^2} = -\frac{13.6}{n^2} eV \qquad (\because 1\ eV = 1.6 \times 10^{-19} \text{ joules})$$

This expression gives the total energy of an electron when it occupies any one of the different orbits (or shells) of the hydrogen atom.

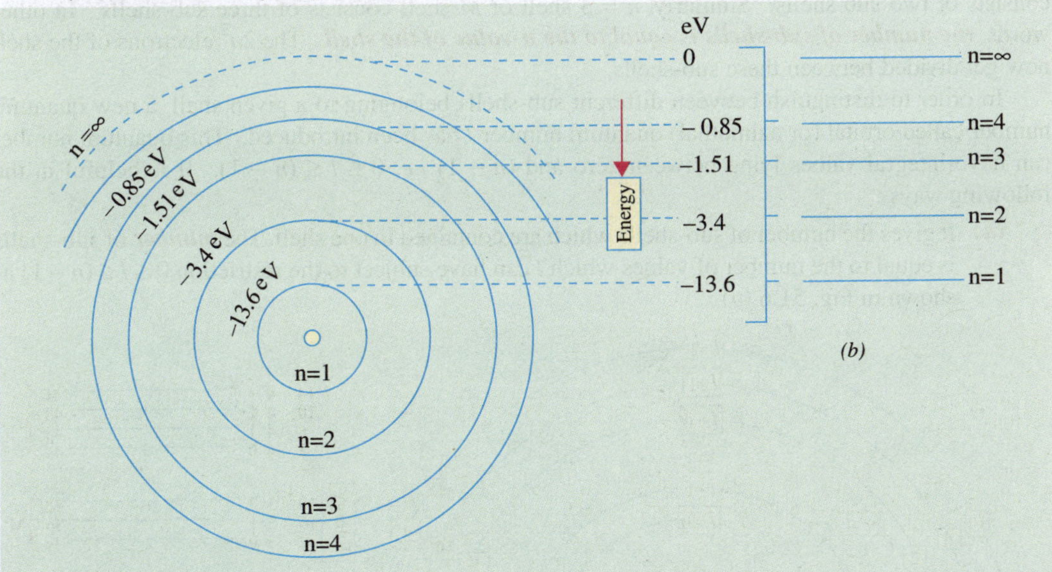

Fig. 51.5

For *K*-shell, $n = 1$, $E_1 = -13.6$ eV

For *L*-shell, $n = 2$, $E_2 = -13.6/2^2 = -3.4$ eV

For *M*-shell, $n = 3$, $E_3 = -13.6/3^2 = -1.51$ eV etc.

Instead of drawing various electron orbits to the scale of their radii as in Fig. 51.5 (*a*), it is customary to draw horizontal lines to an energy scale as shown in Fig. 51.5 (*b*) and such a diagram is called energy level diagram (*ELD*) of an atom. In this array of energies, the higher (*i.e.* less negative) energies are at the top while the lower (*i.e.* more negative) energies are at the bottom. The various electron jumps between allowed orbits now become vertical arrows between energy levels. Greater the length of the arrow, greater is the energy '*hf*' of the radiated photon.

Example 51.1. *Calculate the value of the kinetic, potential and total energy of an electron revolving in Bohr's first orbit in a hydrogen atom.*

Solution. (*i*)

$$K.E. = \frac{me^4}{8\epsilon_0^2 n^2 h^2} = \frac{9.1\times10^{-31}\times(1.6\times10^{-19})^4}{8\times(8.854\times10^{-12})^2\times1^2\times(6.625\times10^{-34})^2}$$

$$= 21.7\times10^{-19} \text{ joules} = \mathbf{13.6\ eV}$$

(*ii*)

$$P.E. = -\frac{me^4}{4\epsilon_0^2 n^2 h^2} = -43.4\times10^{-19} \text{ joules} = \mathbf{-27.2\ eV}.$$

(*iii*)

$$\text{Total energy} = K.E. + P.E. = 13.6 + (-27.2) = \mathbf{-13.6\ eV}$$

51.6. Orbital (or Azimuthal) Quantum Number

According to Bohr's postulates (Art. 51.2), there is only one orbit (and hence one energy level) corresponding to each value of the principal quantum number n. However, subsequent experimental evidence revealed that all orbits (except $n = 1$ orbit) consist of more than one orbit called sub-orbits. This group of sub-orbits is collectively known as ***shell.*** For example, $n = 1$ shell or K-shell consists of only one orbit which may also be called its own sub-orbit or sub-shell. The $n = 2$ shell or L-shell consists of two sub-shells. Similarly, $n = 3$ shell or M-shell consists of three sub-shells. In other words, ***the number of sub-shells is equal to the n-value of the shell***. The $2n^2$ electrons of the shell now get divided between these sub-shells.

In order to distinguish between different sub-shells belonging to a given shell, a new quantum number called orbital (or azimuthal) quantum number l has been introduced. This quantum number can have integral values lying between zero and $(n - 1)$ *i.e.* $0 \le l \le (n - 1)$. It is helpful in the following ways :

(*a*) It gives the number of sub-shells which are contained in one shell. The ***number*** of sub-shells is equal to the number of values which l can have subject to the restriction $0 \le l \le (n - 1)$ as shown in Fig. 51.6 (*a*).

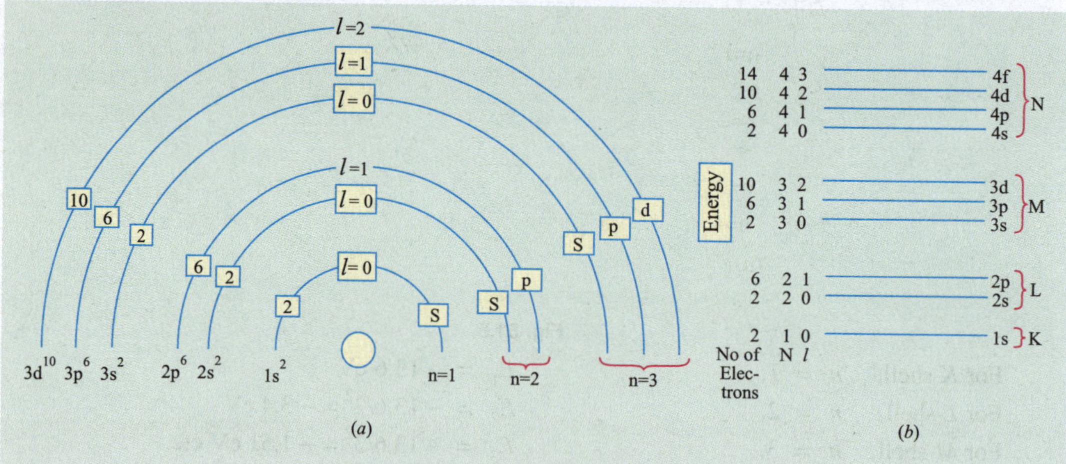

Fig. 51.6

(*b*) It helps to distinguish between different sub-shells of a shell by its different values for each sub-shell. Consider the following cases :

1. ***n* = 1 shell.** Here, l can have only one value *i.e.* zero. Hence, K-shell has only one sub-shell with two quantum numbers of $n = 1$ and $l = 0$.

2. **$n = 2$ shell.** Here, $l = 0, 1$. Hence, this shell has two sub-shells* which are distinguishable from each other by their different quantum numbers of $n = 2, l = 0$ and $n = 2, l = 1$. (Fig. 51.6).
3. **$n = 3$ shell.** Here $l = 0, 1, 2$, showing that M-shell has ***three*** sub-shells which differ in their l-values. The two quantum numbers for the three sub-shells are :
   $n = 3, l = 0; n = 3, l = 1; n = 3, l = 2$.

(c) It helps to determine the ***shape*** of different sub-shells of a given shell. Some sub-shells are circular whereas others are elliptical in shape. If a and b are semi-major and semi-minor axes respectively of an ellipse, n and l are its principal quantum number and orbital quantum number respectively, then the relation between them is

$$\frac{b}{a} = \frac{l+1}{n}$$

Consider the following cases :

1. K-shell or n = 1 shell

Here, $l = 0$, hence $b/a = (0 + 1)/1$ or $b/a = 1$ or $b = a$. In other words, the first shell (also called sub-shell or K-shell) around the nucleus of an atom is circular in shape as shown in Fig. 51.7 (a).

2. L-shell or n = 2 shell

As stated earlier and shown in Fig. 51.6, it has two more sub-shells.

First sub-shell with $l = 0$. Here, $\frac{b}{a} = \frac{0+1}{2} = \frac{1}{2}$ or $b = \frac{a}{2}$. It is elliptical in shape as shown in Fig. 51.7 (b).

Second sub-shell with $l = 1$. Here $\frac{b}{a} = \frac{1+1}{2} = 1$ or $b = a$.

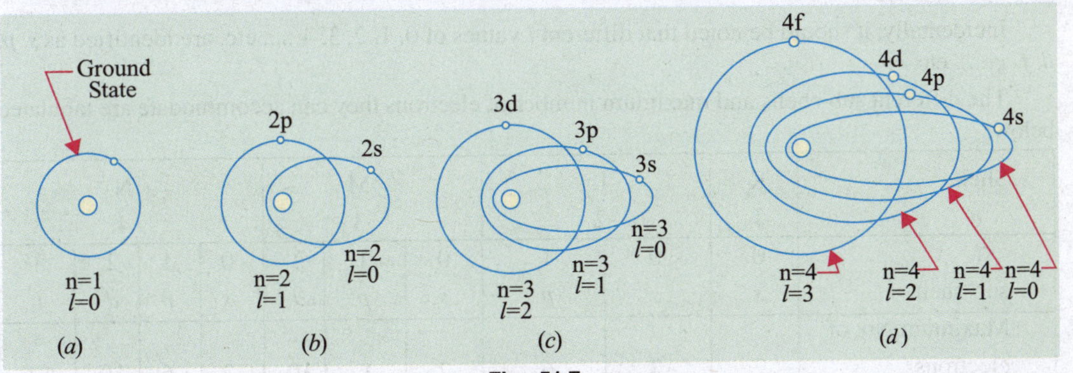

Fig. 51.7

This sub-shell has circular shape as shown in Fig. 51.7 (b).

3. M-shell or n = 3 shell

It has three sub-shells with $l = 0, 1, 2$.

First sub-shell with $l = 0$. Here, $b/a = (0 + 1)/3 = 1/3$. It is elliptical in shape.

Second sub-shell with $l = 1$. Here, $b/a = (1 + 1)/3 = 2/3$. It is also elliptical in shape.

Third sub-shell with $l = 2$. Here, $b/a = (2 + 1)/3 = 1$ or $b = a$. Obviously, this sub-shell is circular in shape.

Fig. 51.7 shows shapes of different sub-shells as per values of n from 1 to 4.

(d) It determines the distribution of electrons in various sub-shells of a shell. The maximum

---

* Different sub-shells have been shown circular for the sake of simplicity although, in practice, most of them are elliptical.

number of electrons which a sub-shell of any n-value can accommodate is = 2 (2l + 1). Consider the following :

1. **K-shell.** Here $l = 0$, hence it can have $2\,(2 \times 0 + 1) = 2$ electrons. They are designated as $1s^2$ electrons.
2. **L-shell.** First sub-shell with $l = 0$ can have $2\,(2 \times 0 + 1) = 2$ electrons which are designated as $2s^2$ electrons.
   Second sub-shell with $l = 1$ can have a maximum of $2\,(2 \times 1 + 1) = 6$ electrons which are designated as $2p^6$ electrons.
   Total number of electrons in the two sub-shells is 2 + 6 = 8 which tallies with $2n^2 = 2 \times 2^2 = 8$.
3. **M-shell.** First sub-shell with $l = 0$ has, as explained above, 2 electrons which are designated as $3s^2$ electrons.
   Second sub-shell $l = 1$ can have 6 electrons written as $3p^6$ electrons.
   Third sub-shell with $l = 2$ can accommodate a maximum of $2(2 \times 2 + 1) = 10$ electrons which are known as $3d^{10}$ electrons.
   Again, it will be seen that total number of electrons in the three sub-shells of $n = 3$ shell is = 2 + 6 + 10 = 18 which equals $2n^2 = 2 \times 3^2 = 18$.
4. **N-shell.** First sub-shell with $l = 0$ can have maximum of two electrons designated as $4s^2$ electrons. Second sub-shell can contain 6 electrons known as $4p^6$ electrons.
   Similarly, third sub-shell can have 10 electrons designated as $4d^{10}$ electrons.
   The fourth sub-shell can have a maximum of $2(2 \times 3 + 1) = 14$ electrons whose designation is $4\,f^{14}$ electrons.

The total number of electrons is = 2 + 6 + 10 + 14 = 32 which is in accordance with the number $2n^2 = 2 \times 4^2 = 32$.

Incidentally, it should be noted that different l values of 0, 1, 2, 3, 4...... etc. are identified as *s, p, d, f, g*...... etc.

The different sub-shells and maximum number of electrons they can accommodate are tabulated below.

Shell n	K 1	L 2		M 3			N 4			
l sub-shell	0 s	0 s	1 p	0 s	1 p	2 d	0 s	1 p	2 d	3 f
Maximum No. of electrons = 2 (2l + 1)	2	2	6	2	6	10	2	6	10	14
Total No. of electrons = $2n^2$	2	8		18			32			

(*e*) The orbital quantum number l also quantizes the orbital angular momentum p_l associated with each sub-shell

$$p_l = l.h/2\pi \quad \text{where} \quad l = 0, 1, 2, 3, \text{.....etc.}^*$$

However, quantum mechanical considerations indicate that $p_l \neq l.h/2\pi$ as stated above but instead $p_l = l^*.h/2\pi$ where $l^* = [l\,(l + 1)]^{1/2}$.

* According to Bohr's theory, $p_l = n\,.\,h/2\pi$.

51.7. Electron Configuration of Atoms

By electron configuration of an atom is meant the distribution of its electrons in its various sub-shells around the nucleus. Following three rules govern the electron distribution :

1. *Maximum number* of electrons a shell can have is = $2n^2$.
2. In the nth shell, there are n sub-shells having different values of l such as 0, 1, 2..... $(n - 1)$.
3. Each sub-shell can accommodate a maximum of $2(2l + 1)$ electrons.

Consider the following atoms :

(*i*) **Sodium atom, Z = 11.** It has 11 electrons. Hence, its electronic configuration is $1s^2$, $2s^2$, $2p^6$, $3s^1$. Obviously, *Na* has a single electron in its outermost sub-shell and hence is said to be monovalent. Same property is possessed by other alkali metals like Li, K, Rb and Cs. They have similar chemical properties and are, therefore, included in the same group in the periodic table.

The electron in the inner sub-shells are very tightly bound to the nucleus and cannot be easily removed. In other words, they have high ***binding energy***.

(*ii*) **Copper atom, Z = 29.** It has 29 electrons. Its electronic configuration is $1s^2$, $2s^2$, $2p^6$, $3s^2$, $3p^6$, $3d^{10}$, $4s^1$.

51.8. Orbital Magnetic Quantum Number (m_l)

It determines the spatial ***orientation*** of elliptical electron orbits with respect to an applied magnetic field. There are restrictions on the orientations of the electron orbits because of which they are said to be ***space quantized***. Whereas l determines the orbital angular momentum, m_l represents the magnitude of the component of angular momentum along the direction of the magnetic field.

Looked from a different angle, m_l determines the number of sub-subshells in a given shell.

This quantum number can have any one of the $(2l + 1)$ values ranging from $+ l$ to $- l$ including zero *i.e.* $l, (l - 1), (l - 2)$.........2, 1, 0, – 1, – 2,.......... $- (l - 2), - (l - 1), - l$.

Each sub-subshell can accommodate a maximum of 2 electrons, so that maximum number of electrons in a shell becomes $2(2l + 1)$ as stated earlier in Art. 51.6 (*d*).

51.9. Magnetic Spin Quantum Number (m_s)

It has been found that an electron spins around its own mechanical axis as it rotates in orbit around the nucleus. When the electron is subjected to a magnetic field, its spin axis orientates itself either parallel to or antiparallel to the direction of the field.

This quantum number arises out of quantization of the electron spin angular momentum. It determines the spin orientation up or down and has correspondingly two values of + 1/2 and – 1/2.

51.10. Pauli's Exclusion Principle

This principle which was enunciated by Pauli in 1925 states that in an atom, ***no two electrons can have the same set of values for its four quantum number n, l, m_l and m_s***. In other words, no two electrons can be described by an identical set of four quantum numbers. They may have at the most three numbers alike but at least one must be different. Consequently, it restricts the number of electrons an atom can have.

Consider the case of He atom which has two electrons. These electrons occupy *K*-shell ($n = 1$) and are designated as $1s^2$ electrons. Their four quantum numbers are as follows :

	n	l	m_l	m_s
1st electron :	1	0	0	+ 1/2
2nd electron :	1	0	0	– 1/2

As seen, the two electrons have different sets of four quantum numbers (1, 0, 0, +1/2) and (1, 0, 0, – 1/2) as required by Pauli's Exclusion Principle.

This principle affords a very rational theoretical justification for the electronic shell structure of the atoms deduced experimently.

Pauli's Exclusion Principle may be used to find the total number of electrons an atom can have in its various shells. Consider the following examples :

n = 1, K-shell

$l = 0$, 1s sub-shell, $m_l = 0$; $m_s = \pm 1/2$ — 2 electrons

Total — 2 electrons

n = 2, L-shell

(*i*) $l = 0$, 2 s sub-shell, $m_l = 0$; $m_s = \pm 1/2$ — 2 electrons

(*ii*) $l = 1$, 2p sub-shell, $m_l = -1, 0, +1$; $m_s = \pm 1/2$ — 6 electrons

Total — 8 electrons

n = 3, M-shell

(*i*) $l = 0$, 3s sub-shell, $m_l = 0$; $m_s = \pm 1/2$ — 2 electrons

(*ii*) $l = 1$, 3p sub-shell, $m_l = -1, 0, +1$; $m_s = \pm 1/2$ — 6 electrons

(*iii*) $l = 2$, 3d sub-shell, $m_l = -2, -1, 0, +1, +2$; $m_s = \pm 1/2$ — 10 electrons

Total — 18 electrons

n = 4, N-shell

(*i*) $l = 0$, 4s sub-shell, $m_l = 0$, $m_s = \pm 1/2$ — 2 electrons

(*ii*) $l = 1$, 4p sub-shell, $m_l = -1, 0, 1$; $m_s = \pm 1/2$ — 6 electrons

(*iii*) $l = 2$, 4d sub-shell, $m_l = -2, -1, 0, +1, +2$, $m_s = \pm 1/2$ — 10 electrons

(*iv*) $l = 3$, 4f sub-shell, $m_l = -3, -2, -1, 0, +1, +2, +3$, $m_s = \pm 1/2$ — 14 electrons

Total — 32 electrons

51.11. Energy Bands in Solids

In the case of an ***isolated single*** atom, there are ***single*** energy levels as shown for a hydrogen atom in Fig. 51.5. But there are significant changes in the energy levels when atoms are brought close together as in solids.

It is found that each of the energy levels of an atom splits into N levels of energy where N is the number of atoms in the crystal. Each original energy level becomes a ***band*** of very closely-spaced levels of slightly different energy. The individual energies within the band are so close together that, for many purposes, the energy band may be considered to be a continuous one.

Fig. 51.8 (*a*) shows the splitting of *K, L* and *M* levels as the distance between different atoms is reduced. At first, only valence level or *M*-level is affected as shown by dotted vertical line marked *A*, then as separation is reduced, inner shells also become affected as indicated by dotted vertical line *B*.

Consider the case of *Na* crystal which consists of an ordered array of many closely-packed sodium atoms usually referred to as crystal lattice. Each *Na* atom has 11 electrons arranged in different shells and sub-sheels as shown in Fig. 51.9 (*a*). As seen, the 1*s*, 2*s* and 2*p* sub-shells are filled but 3*s* sub-shell is incomplete and could hold one electron more. The electrons in each sub-shell occurs specific energy levels as shown in Fig. 51.9 (*b*). For a small sodium crystal containing 10^{20} atoms, the band formed by splitting of *s*-subshell has 2 electronic levels (one with spin up and the others with spin down). Similarly in a *p*-band there will be 6×10^{20} closely packed levels because there are 6 electrons in a filled *p* subshell.

In general, in an assembly of N atoms, the number of possible energy states is N. Since only two electrons of opposite spin can occupy the same state (as per Pauli's Exclusion Principlé discussed earlier), the maximum number of electrons which these N states can occupy is 2N.

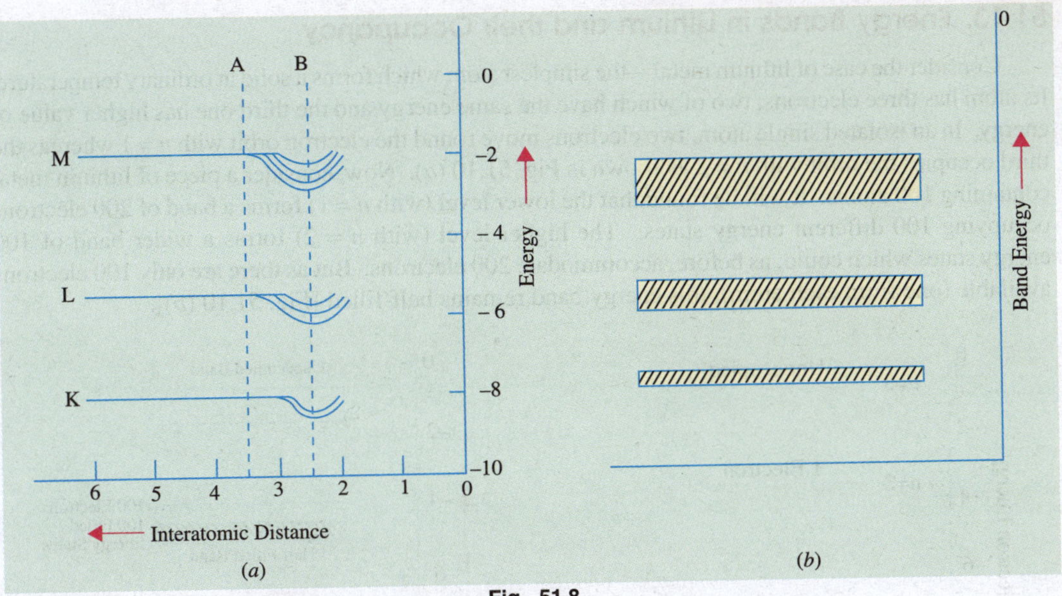

Fig. 51.8

51.12. Spacing Between Energy Levels of a Band

It will be quite interesting as well as instructive to calculate the spacing between different energy levels in an energy band. A crystal weighing one milligram contains about 10^{19} atoms. If we assume the valency band to be an s-band, it will contain 2×10^{19} levels. Suppose the width of the energy band is 2 eV. Then, it is obvious that 2×10^{19} levels per milligram are spread over an energy band width of 2 eV. Hence, spacing between different levels $= 2/(2 \times 10^{19}) = 10^{-19}$ eV. It will be appreciated that even though energy levels are discrete, the picture of a band as a continuum of energy levels is a very good approximation. This splitting of the single energy level of an ***isolated*** atom into a band of

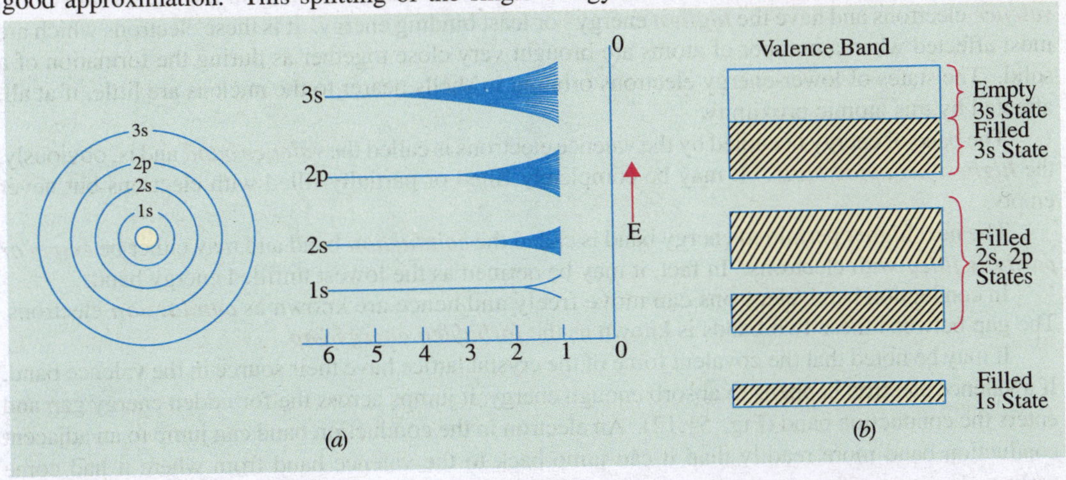

Fig. 51.9

energy in the case of a ***solid*** is responsible for most of the electrical, magnetic and optical properties of that solid. It is worth pointing out here that in gases under normal conditions of temperature and pressure, the atomic spacing is so great that there is no splitting of energy levels and hence no band formation.

51.13. Energy Bands in Lithium and their Occupancy

Consider the case of lithium metal—the simplest atom which forms a solid at ordinary temperature. Its atom has three electrons, two of which have the same energy and the third one has higher value of energy. In an isolated single atom, two electrons move round the electron orbit with $n = 1$ whereas the third occupies the orbit with $n = 2$ as shown in Fig. 51.10 (*a*). Now, consider a piece of lithium metal containing 100 atoms. It will be found that the lower level (with $n = 1$) forms a band of 200 electrons occupying 100 different energy states. The higher level (with $n = 2$) forms a wider band of 100 energy states which could, as before, accommodate 200 electrons. But as there are only 100 electrons available (one from each atom), this energy band remains half-filled [Fig. 51.10 (*b*)].

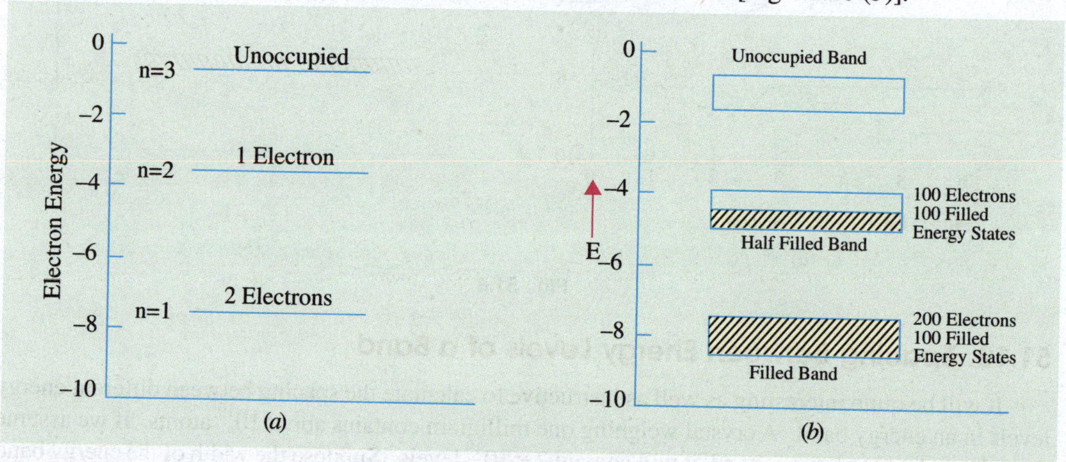

Fig. 51.10

51.14. Valence and Conduction Bands

The outermost electrons of an atom *i.e.* those in the shell furthermost from the nucleus are called ***valence*** electrons and have the ***highest*** energy* or least binding energy. It is these electrons which are most affected when a number of atoms are brought very close together as during the formation of a solid. The states of lower-energy electrons orbiting in shells nearer to the nucleus are little, if at all, affected by this atomic proximity.

The band of energy occupied by the valence electrons is called the ***valence band*** and is, obviously, the ***highest occupied band***. It may be completely filled or partially filled with electrons but never empty.

The next higher permitted energy band is called the ***conduction*** band and may either be ***empty*** *or* ***partially filled*** with electrons. In fact, it may be defined as the lowest unfilled energy band.

In conduction band, electrons can move freely and hence are known as ***conduction*** electrons. The gap between these two bands is known as the ***forbidden energy gap***.

It may be noted that the covalent force of the crystal lattice have their source in the valence band. If a valence electron happens to absorb enough energy, it jumps across the forbidden energy gap and enters the conduction band (Fig. 51.12). An electron in the conduction band can jump to an adjacent conduction band more readily than it can jump back to the valence band from where it had come earlier. However, if a conduction electron happens to radiate too much energy, it will suddenly reappear in the valence band once again.

When an electron is ejected from the valence band, a covalent bond is broken and a positively-charged hole is left behind. This hole can travel to an adjacent atom by acquiring an electron from

---

* In the algebraic sense only. Their binding energy is the least.

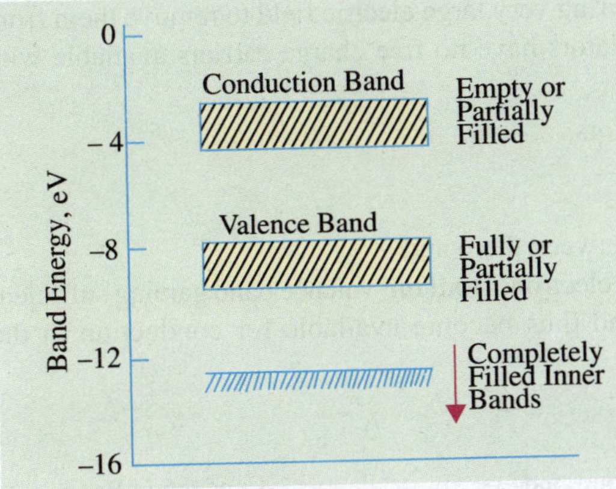

Fig. 51.11

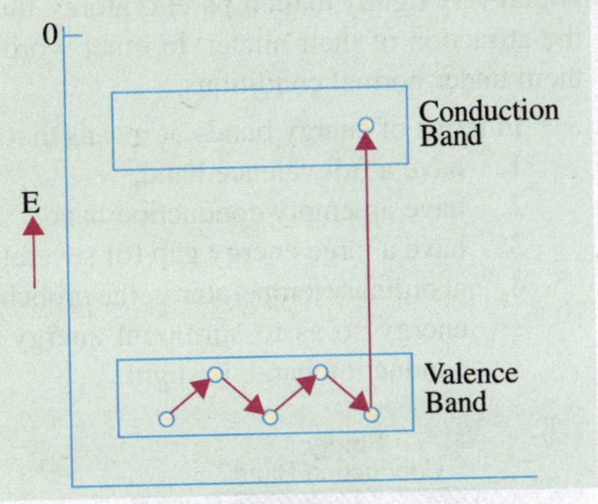

Fig. 51.12

that atom which involves breaking an existing covalent bond and then re-establishing a covalent bond by filling up the hole. It is to be noted carefully that holes are filled by electrons which move fromadjacent atoms without passing through the forbidden energy gap.

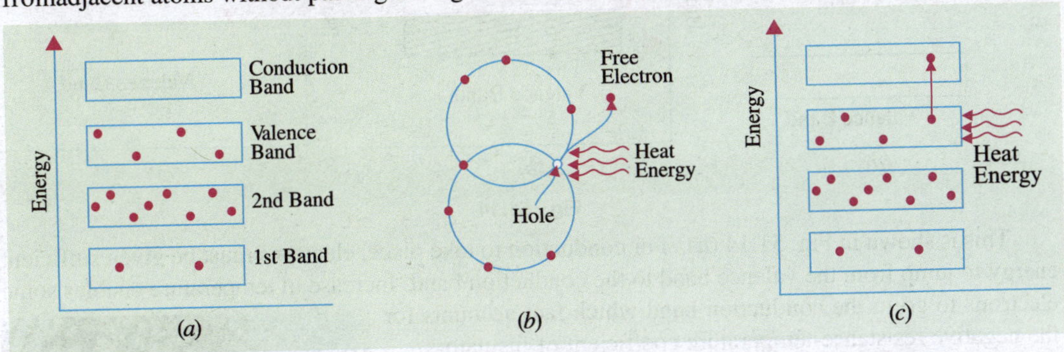

Fig. 51.13

It is simply another way of saying that conditions in the conduction band have nothing to do with the hole flow. It points to a very important distinction between the hole current and electron current— although holes flow with ease, they ***experience more opposition than electron flow in the conduction band***.

To summarize the above, it may be repeated that :

1. conduction electrons are found in and freely flow in the ***conduction*** band ;
2. holes exist in and flow in the valence band ;
3. conduction electrons move almost twice as fast as the holes.

Fig. 51.13 (*a*) shows the energy band diagram of an unexcited silicon atom ($Z = 14$) with its electronic distribution. When silicon crystal is given thermal or light energy from outside [Fig. 51.13 (*b*)], some electrons gain sufficient energy to jump the gap from the valence band into the conduction band thereby becoming free electrons [Fig. 51.13 (*c*)]. For every electron which jumps to conduction band, a hole is created in the valence band. In this way, an electron-hole pair is created.

51.15. Insulators, Conductors and Semiconductors

The electrical conduction properties of different elements and compounds can be explained in terms of the electrons having energies in the valence and conduction bands. The electrons lying in the lower energy bands, which are normally filled, play no part in the conduction process.

(*i*) **Insulators.** Stated simply, insulators are those materials in which valence electrons are bound very tightly to their parents atoms, thus requiring very large electric field to remove them from the attraction of their nuclei. In other words, insulators have no free charge carriers available with them under normal conditions.

In terms of energy bands, it means that insulators.

1. have a full valence band,
2. have an empty conduction band,
3. have a large energy gap (of several eV) between them and
4. at ordinary temperatures, the probability of electrons from full valence band gaining sufficient energy so as to surmount energy gap and thus become available for conduction in the conduction band, is slight.

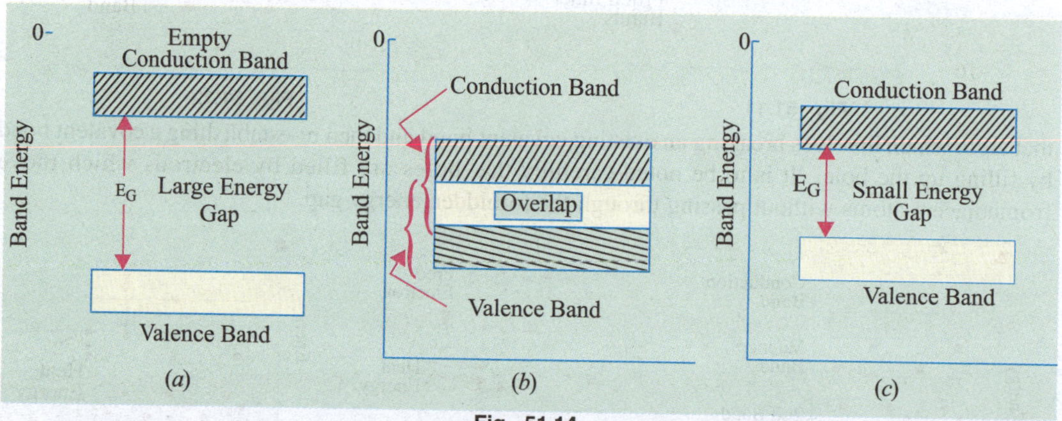

Fig. 51.14

This is shown in Fig. 51.14 (*a*). For conduction to take place, electrons must be given sufficient energy to jump from the valence band to the conduction band. Increase in temperature enables some electrons to go to the conduction band which fact accounts for the negative resistance-temperature coefficient of insulators.

Insulators

(*ii*) Conductors. Put in a simple way, conducting materials are those in which plenty of free electrons are available for electric conduction.

In terms of energy bands, it means that electrical conductors are those which have overlapping valence and conduction bands as shown in Fig. 51.14 (*b*).

In fact, there is no physical distinction ***between the two bands***. Hence, the availablity of a large number of conduction electrons.

Another point worth noting is that in the absence of forbidden energy gap in good conductors, there is no structure to establish holes. The ***total current in such conductors is simply a flow of electrons***. It is exactly for this reason that the existence of holes was not discovered until semi-conductors were studied thoroughly.

(*iii*) Semiconductors. A semiconductor material is one whose electrical properties lie in between those of insulators and conductors. Examples are : germanium and silicon.

In terms of energy bands, semiconductors can be defined as those materials which have almost an empty conduction band and almost filled valence band with a very narrow energy gap (of the order of 1 eV) separating the two.

At 0ºK, there are no electrons in the conduction band and the valence band is completely filled. However, with increase in temperature, width of the forbidden energy bands is decreased so that some of the electrons are liberated into the conduction band. In other words, conductivity of semiconductors increases with temperature. Moreover, such departing electrons leave behind positive holes in the valence band (Fig. 51.12). Hence, semiconductor current is the sum of electron and hole currents flowing in opposite directions.

51.16. Crystal Structure

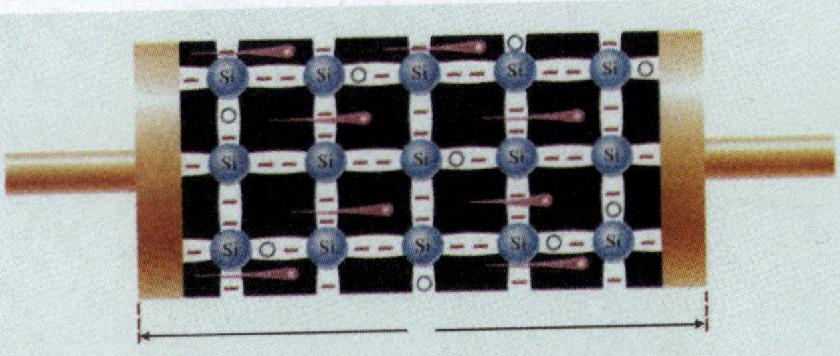

Current conduction through semiconductor

Semiconductors like germanium and silicon, have crystalline structure. That is the atoms are arranged in three-dimensional periodic fashion. The periodic arrangement of atoms in a crystal is called ***lattice***. In a crystal, an atom strays far from a single, fixed position. The thermal vibrations associated with the atom are centred about this position. For a given semiconductor (silicon or germanium), there is a ***unit cell*** that is representative of the entire lattice. By repeating the unit cell throughout the crystal, we can generate the entire lattice.

There are several different types of crystal lattice depending upon the symmetry and internal structure. One of them is the cubic crystal lattice. There are three basic types of unit cells in a cubic crystal lattice. These are simple cubic (SC), base centred cubic (BCC) and face centred cubic (FCC). Fig. 51.15 (*a*) shows a SC crystal. In this unit cell, each corner of the cubic lattice is occupied by an atom (indicated by a small sphere) that has six equidistant nearest neighbouring atoms. The dimension "*a*" is called the ***lattice constant***. Only ***podium*** is crystallised into the simple cubic lattice.

Fig. 51.15 (*b*) shows a BCC unit cell. In this unit cell, each atom has eight nearest neighbouring atoms. Crystals exhibiting the BCC lattices include those of ***sodium*** and ***tungsten***. Fig. 51.15 (*c*) shows a FCC unit cell. This unit cell has one atom at each of the six cubic faces in addition to the eight corner atoms. In an FCC lattice, each atom has 12 nearest neighbouring atoms. A large number of elements exhibit the FCC lattice form. These include ***aluminium, copper, gold*** and ***platinum***.

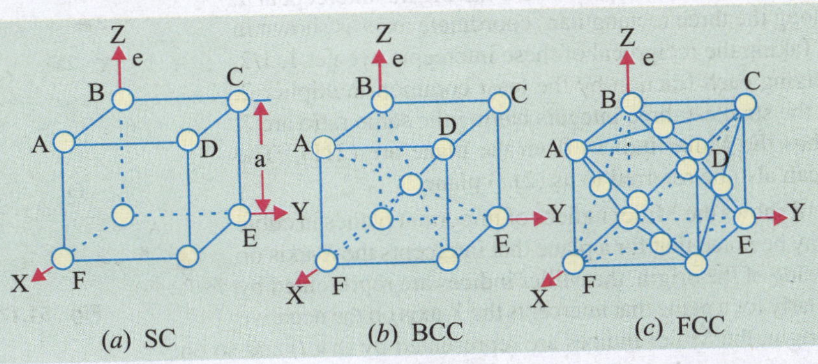

Fig. 51.15

Fig. 51.16 shows the crystal structure of silicon and germanium—the element semiconductors. This type of structure is called diamond crystal structure and it belongs to cubic crystal family. All atoms are identical in a diamond lattice. Note that each atom in diamond lattice is surrounded by four equidistant neighbours that lie at the corners of a tetrahedron.

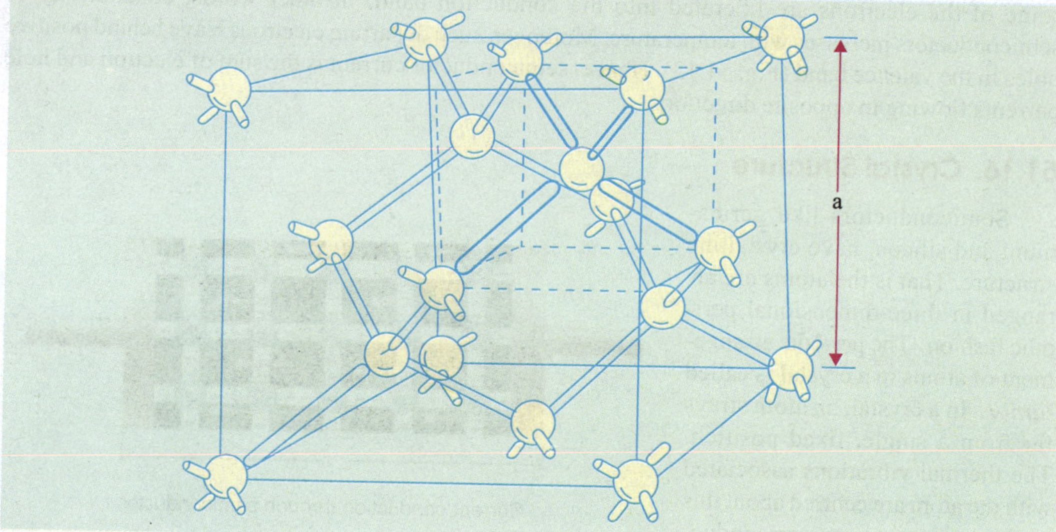

Fig. 51.16

51.17. Representation of Crystal Planes and Directions

Referring to Fig. 51.15 (*b*) again, we note that the there are four atoms in the ABCD plane and five atoms in ACEF plane (four atoms from the corners and one from the centre). Moreover, the atomic spacing are different for the two planes. Therefore the crystal properties along different planes are different. The electrical and other device characteristics are dependent on the crystal orientation. A convenient method of defining or representing the various planes in a crystal is to use ***Miller indices***. These indices are obtained using the following steps :

1. Find the intercepts of the plane on the three coordinate axes in terms of lattice constant.
2. Take the reciprocals of these numbers and reduce them to the smallest three integers having the same ratio.
3. Enclose the result in parentheses (*hk l*) as the Miller indices for a single plane.

For example, consider the plane ABC having the intercept at *a*, 2*a* and 2*a* along the three rectangular coordinate axes as shown in Fig. 51.17. Taking the reciprocal of these intercepts, we get 1, 1/2, 1/2. Multiplying each fraction by the least common multiplier 2, we find that the smallest three integers having the same ratio are 2, 1 and 1. Thus the Miller indices from the plane are (211). The plane ABC can also be referred to as (211) plane.

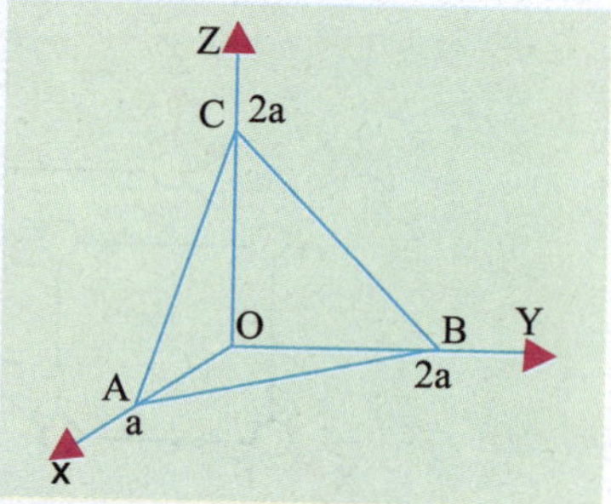

Fig. 51.17

Fig. 51.18 shows the Miller indices of important planes in cubic crystal. It may be noted that for a plane that intercepts the *X*-axis on the negative side of the origin, the Miller indices are represented by ($\bar{h}$ k *l*). Similarly for a plane that intercepts the *Y*-axis on the negative side of the origin, the Miller indices are represented by (h $\bar{k}$ *l*) and so on.

Sometimes it convenient to represent the Miller indices of the planes of equivalent symmetry by (hk*l*). For example, in cubic crystal, the planes (100), (010), (001), ($\bar{1}$00), (0$\bar{1}$0) and (00$\bar{1}$) have the same symmetry. Therefore, these planes can be represented by Miller indices as {100}. Note that we have used the curly brackets to represent a set of planes with equivalent symmetry.

The line that originates from the origin and passes through the plane at right angles to it is called ***crystal direction***. The crystal direction is indicated by enclosing the Miller indices in a square brackets

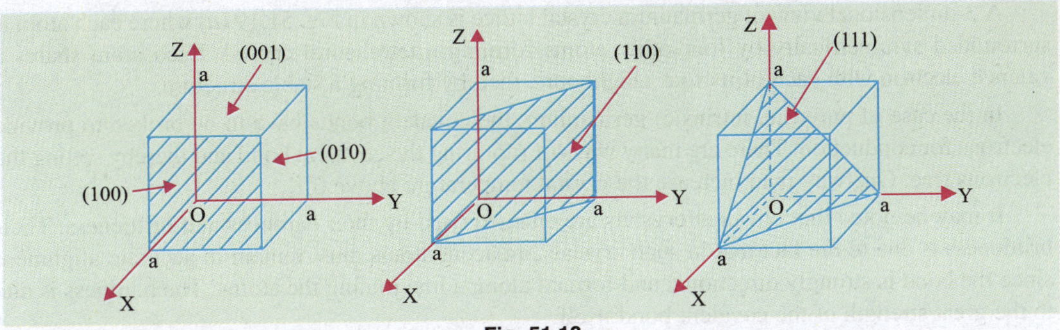

Fig. 51.18

i.e., [hk*l*]. For example, the direction idicated by [100] is a direction for the *X*-axis and it is perpendicular to (100) plane. Similarly [111] direction is perpendicular to (111) plane. It is possible to represent a set of equivalent directions by < hk*l* >. Notice the use of carat signs. Thus < 100 > represents a set of directions for [100], [010], [100], $[0\bar{1}0]$ and $[00\bar{1}]$.

51.18. Atomic Binding in Semiconductors

Bonds in semiconductors

Semiconductors like germanium* and silicon, have crystalline structure. Their atoms are arranged in an ordered array known as crystal lattice. Both these materials are tetravalent *i.e.* each has four valence electrons in its outermost shell. The neighbouring atoms form ***covalent*** bonds by sharing four electrons with each other so as to achieve inert gas structure (*i.e.* 8 electrons in the outermost orbit). A two-dimensional view of the germanium crystal lattice is shown in Fig. 51.19 (*b*) in which circles represent atom cores consisting of the nuclei and inner 28 electrons. Each ***pair*** of lines represents a covalent bond. The dots represent the valence electrons. It is seen that each atom has 8 electrons under its influence.

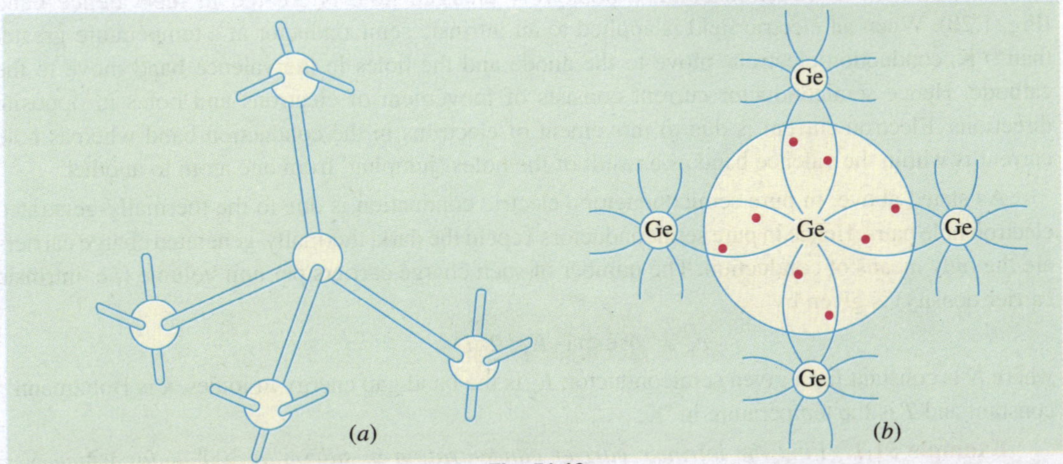

Fig. 51.19

* A single germanium atom has 32 electrons out of which only four electrons take part in electrical properties of germanium, the remaining 28 electrons being tightly bound to the nucleus. The four electrons revolve in the outermost shell and are called valence electrons.

A 3-dimensional view of germanium crystal lattice is shown in Fig. 51.19 (*a*) where each atom is surrounded symmetrically by four other atoms forming a tetrahedral crystal. Each atom shares a valence electron with each of its four neighbours, thereby forming a stable structure.

In the case of pure (*i.e.* intrinsic) germanium, the covalent bonds have to be broken to provide electrons for conduction. There are many ways of rupturing the covalent bond and thereby setting the electrons free. One way is to increase the crystal temperature above 0ºK.

It may be noted that covalent crystals are characterised by their hardness and brittleness. Their brittleness is due to the fact that in such crystals, adjacent atoms must remain in accurate alignment since the bond is strongly directional and formed along a line joining the atoms. The hardness is due to the great strength of the covalent bond itself.

51.19. Types of Semiconductors

Semiconductors can be classified as shown below :

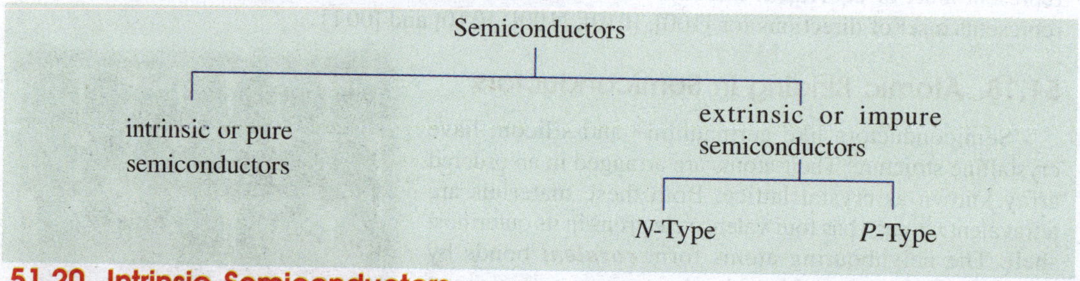

51.20. Intrinsic Semiconductors

An intrinsic semiconductor is one ***which is made of the semiconductor material in its extremely pure form***.

Common examples of such semiconductors are : pure germanium and silicon which have forbidden energy gaps of 0.72 eV and 1.1 eV respectively. The energy gap is so small that even at ordinary room temperature, there are many electrons which possess sufficient energy to jump across the small energy gap between the valence and the conduction bands. However, it is worth noting that for each electron liberated into the conduction band, a positively charged hole is created in the valence band (Fig. 1.20). When an electric field is applied to an intrinsic semiconductor at a temperature greater than 0ºK, conduction electrons move to the anode and the holes in the valence band move to the cathode. Hence semiconductor current consists of movement of electrons and holes in opposite directions. Electron current is due to movement of electrons in the conduction band whereas hole current is within the valence band as a result of the holes 'jumping' from one atom to another.

As stated above, in pure semiconductors, electric conduction is due to the thermally-generated electron hole pairs. Hence in pure semiconductors kept in the dark, thermally-generated charge carriers are the only means of conduction. The number of such charge carriers per unit volume (*i.e.* intrinsic carrier density) is given by

$$n_i = N \exp(-E_g / 2kT)$$

where N is constant for a given semiconductor, E_g is the band gap energy in joules, k is Boltzmann's constant and T is the temperature in ºK.

Example 51.2. *Find the intrinsic carrier concentration in silicon at 300º K for which* $N = 3 \times 10^{25}\ m^{-3}$, $E_g = 1.1\ eV$, $\mu_e = 0.14\ m^2/V\text{-}s$ *and* $\mu_h = 0.05\ m^2/V\text{-}s$. *Also, find the conductivity of silicon.* **(Electronics-II, Madras Univ. 1993)**

Solution. The intrinsic carrier concentration in pure silicon is given by

$$n_i = N \exp(-E_g/2kT)$$

Now, $N = 3 \times 10^{25}\ \text{m}^{-3},\ E_g = 1.1\ \text{eV} = 1.1 \times 1.6 \times 10^{-19} = 1.76 \times 10^{-19}\ \text{J}$

$$k = 1.38 \times 10^{-23}\ \text{J/K},\ T = 300°\text{K}$$

$\therefore$ $n_i = 3 \times 10^{25}\,(-1.76 \times 10^{-19})/2 \times 1.38 \times 10^{-23} \times 300) = \mathbf{1.76 \times 10^{16}\ m^{-3}}$

$$\sigma = n_i e\,(\mu_e + \mu_h) = 1.76 \times 10^{16} \times 1.6 \times 10^{-19}\,(0.14 + 0.05) = \mathbf{0.535 \times 10^{-3}\ S/m}$$

51.21. Hole Formation in Semiconductors

The formation of a hole which is a positive charge carrier is explained below :

As shown in Fig. 51.21, suppose the covalent bond is broken at *A* and the electron has moved through the crystal lattice leaving behind a hole in the covalent bond. An electron at *B* may jump into the vacant hole at *A* and later, an electron at *C* may jump into the hole at *B* and so on. In this way, by a succession of electron movements, a hole will appear at *G* and a negative charge would have moved from *G* to *A*. It would, however, be more convenient to regard positive charge to have moved from *A* to *G* and this conception gives rise to a hole as a positive charge carrier as if it were an electron with a positive charge. It should be clearly understood that these holes are due to the movement of electrons in the valence band and that each electron movement corresponds to a collision. The drift velocity of holes is, obviously, much less than the drift velocity of electrons.

Alternatively, an intrinsic semiconductor may be defined as one ***in which the number of conduction electrons is equal to the number of holes***.

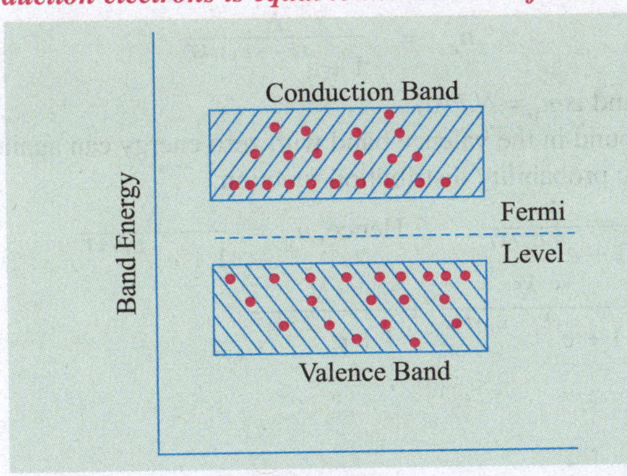

Fig. 51.20

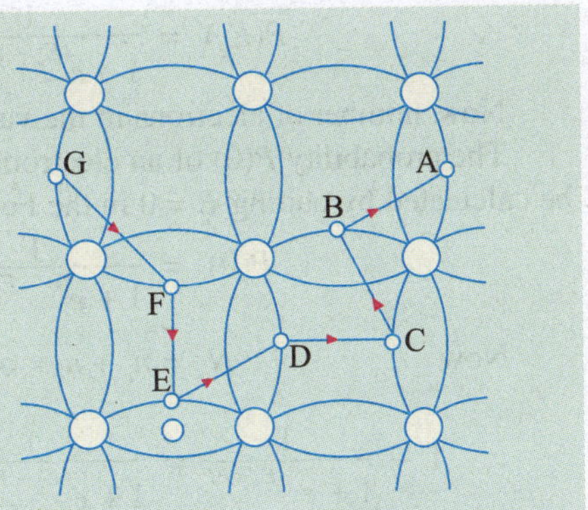

Fig. 51.21

Schematic energy band diagram of an intrinsic semiconductor at room temperature is shown in Fig. 51.20. Only two bands *i.e.* valence and conduction bands have been shown since lower filled bands are not of any consequence. Here, Fermi level* lies exactly in the middle of the forbidden energy gap.

51.22. Fermi Level in an Intrinsic Semiconductor

It can be proved that in an intrinsic semiconductor, Fermi energy level* E_F lies in the middle of the energy gap *i.e.* midway between the conduction and valence bands.

Let, at any temperature T°K

n_c = No. of electrons in the conduction band

n_v = No. of electrons in the valence band

* For the present discussion, Fermi level may be defined as the energy which corresponds to the centre of gravity of conduction electrons and holes weighted according to their energies.

$$N = n_c + n_v$$
$$= \text{No. of electrons in both bands.}$$

Further, let us make the following simplifying assumptions :

1. width of energy bands are small as compared to forbidden energy gap between them;
2. since band widths are small, all levels in a band have the same energy;
3. energies of all levels in valence band are zero as shown in Fig. 51.22;
4. energies of all levels in the conduction band are equal to E_g.

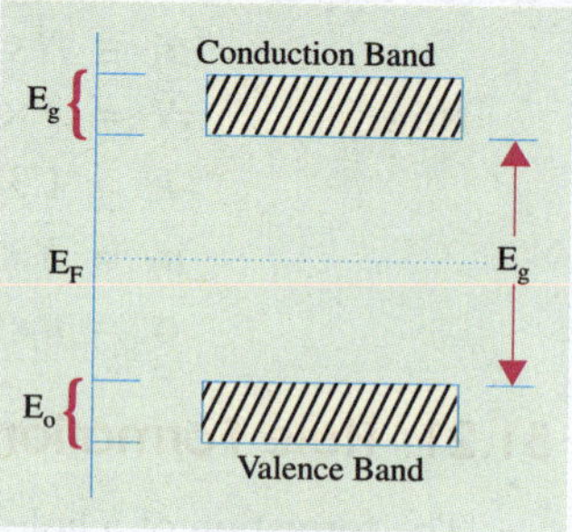

Fig. 51.22

In Fig. 51.22, the zero-energy reference level has been arbitrarily taken at the top of valence band.

Now, number of electrons in the conduction band is

$$n_c = N.P\,(E_g)$$

where $P(E_g)$ represents the probability of an electron having energy E_g. Its value can be found from Fermi-Dirac probability distribution function given by

$$P(E) = \frac{1}{1 + e^{(E - E_F)/kT}}$$

where $P\,(E)$ is the probability of finding an electron having any particular value of energy E and E_F is Fermi level.

$$\therefore \quad P(E_g) = \frac{1}{1 + e^{(E_g - E_F)/kT}} \qquad \therefore \quad n_c = \frac{N}{1 + e^{(E_g - E_F)/kT}}$$

Now, number of electrons in the valence band is $n_v = N.P(0)$

The probability $P(0)$ of an electron being found in the valence band with zero energy can again be calculated by putting $E = 0$ in the Femi-Dirac probability distribution function.

$$P(0) = \frac{1}{1 + e^{(0 - E_F)/kT}} = \frac{1}{1 + e^{-E_F/kT}}. \qquad \text{Hence, } n_V = \frac{N}{1 + e^{-E_F/kT}}$$

$$\text{Now} \qquad N = n_c + n_v \quad \text{or } N = \frac{N}{1 + e^{(E_g - E_F)/kT}} + \frac{N}{1 + e^{-E_F/kT}}$$

$$\therefore \quad 1 - \frac{1}{1 + e^{-E_F/kT}} = \frac{1}{1 + e^{(E_g - E_F)/kT}}$$

which on simplification gives $E_F = E_g/2$.

Hence, it shows that in an intrinsic semiconductor, the Fermi level lies midway between the conduction and valence bands. That this conclusion is physically possible can be best seen with the help of Fig. 51.23.

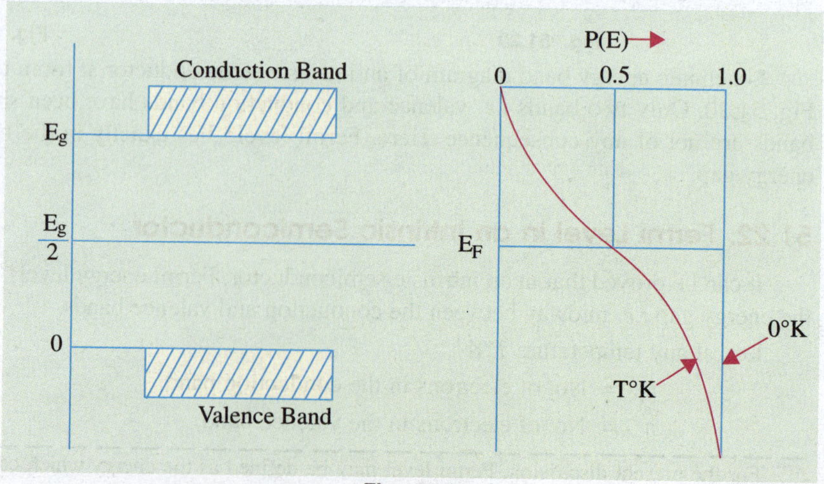

Fig. 51.23

51.23. Extrinsic Semiconductors

Those intrinsic semiconductors to which some suitable impurity or doping agent or dopant has been added in extremely small amounts (about 1 part in 10^8) are called ***extrinsic*** or impurity semiconductors.

The usual doping agents are :

1. ***pentavalent*** atoms having five valence electrons (arsenic, antimony, phosphorus) or
2. ***trivalent*** atoms having three valence electrons (gallium, indium, aluminium, boron).

Pentavalent doping atom is known as ***donor*** atom because it donates or contributes one electron to the conduction band of pure germanium. The trivalent atom, on the other hand, is called ***acceptor*** atom because it accepts one electron from the germanium atom.

Depending on the type of doping material used, extrinsic semiconductors can be sub-divided into two classes :

(*i*) *N*-type semiconductors and (*ii*) *P*-type semiconductors.

(*a*) **N-type Extrinsic Semiconductor.** This type of semiconductor is obtained when a pentavalent material like antimonty (Sb) is added to pure germanium crystal. As shown in Fig. 51.24 (*a*), each antimony atom forms covalent bonds with the surrounding four germanium atoms with the help of four of its five electrons. The fifth electron is superfluous and is loosely bound to the antimony atom. Hence, it can be easily excited from the valence band to the conduction band by the application of electric field or increase in thermal energy. Thus, practically every antimony atom introduced into the germanium lattice, contributes one conduction electron into the germanium lattice ***without creating a positive hole***. Antimony is called ***donor*** impurity and makes the pure germanium an *N*-type (*N* for negative) extrinsic semi-conductor. As an aid to memory, the student should associate the **N** in doNor with **N** in the *N*-type material and *N* in Negative charge carrier.

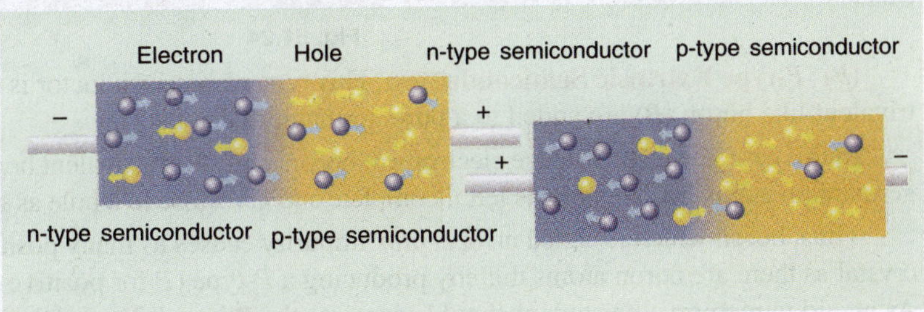

It may be noted that by giving away its one valence electron, the donor atom becomes a positively-charged *ion*. But it cannot take part in conduction because it is firmly fixed or tied into the crystal lattice. It will be seen that apart from electrons and holes ***intrinsically available in germanium***, the addition of antimony greatly increases the number of conduction electrons. Hence, concentration of electrons in the conduction band is increased and exceeds the concentration of holes in the valence band. Because of this, Fermi level shifts upwards towards the bottom of the conduction band as shown in Fig. 51.24 (*b*),* because the number of charge carriers has become more in conduction band than in valence band.

In terms of energy levels, the fifth antimony electron has an energy level (called donor level) just below the conduction band. Usually, the donor level is 0.01 eV below conduction band for germanium and 0.054 eV for silicon.

It is seen from the above description that in *N*-type semiconductors, electrons ***are the majority carriers*** while holes constitute the minority carriers. Hence, *N*-type semiconductor conducts principally by electrons in the nearly empty conduction band and the process is called '***excess***' conduction.

Another point worth noting is that even though *N*-type semiconductor has excess of electrons,

---

* Since the number of electrons as compared to the number of holes increases with temperature, the position of Fermi level also changes considereably with temperature.

still it is electrically neutral. It is so because by the addition of donor impurity, number of electrons available for conduction purposes becomes more than the number of holes available intrinsically. But the total charge of the semiconductor does not change because the donor impurity brings in as much negative charge (by way of electrons) as positive charge (by way of protons in its nucleus).

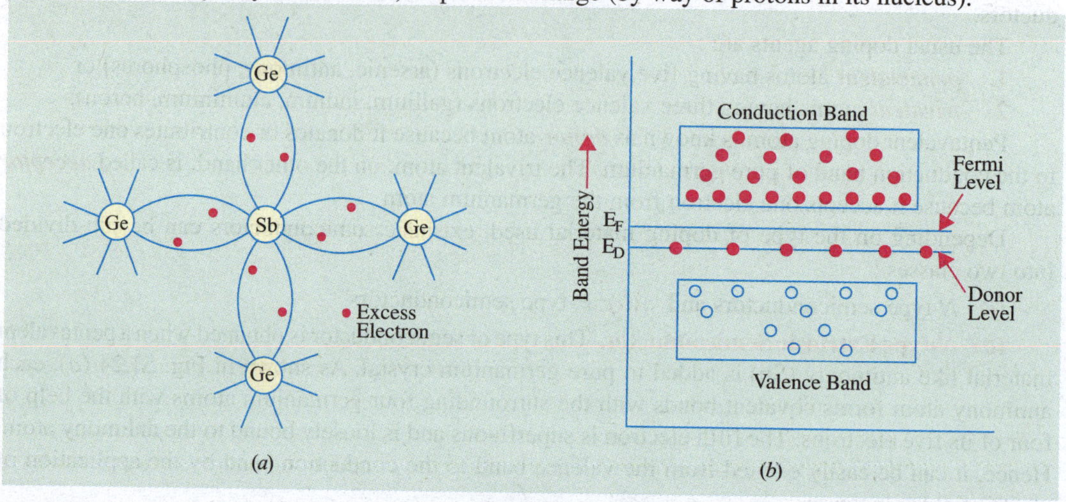

Fig. 51.24

(b) P-type Extrinsic Semiconductor. This type of semiconductor is obtained when traces of a trivalent like boron (*B*) are added to a pure germanium crystal.

In this case, the three valence electrons of boron atom form covalent bonds with four surrounding germanium atoms but one bond is left incomplete and gives rise to a hole as shown in Fig. 51.25 (*a*).

Thus, boron which is called an ***acceptor*** impurity causes as many positive holes in a germanium crystal as there are boron atoms thereby producing a *P*-type (*P* for positive) extrinsic semiconductor. As an aid to memory, the student should associate the **P** in acce**P**tor with **P** in *P*-type material and **P** with **P**ositive charge carrier.

In this type of semiconductor, conduction is by the movement of holes in the valence band. Accordingly, holes form the majority carriers ***whereas electrons constitute minority carriers***. The process of conduction is called **'deficit'** conduction.

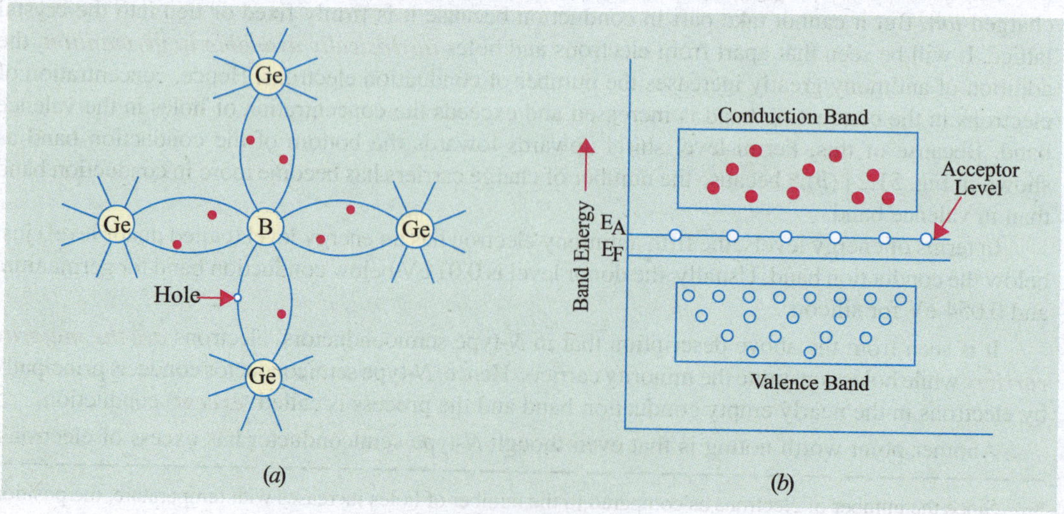

Fig. 51.25

Since concentration of holes in the valence band is more than the concentration of electrons in the conduction band, Fermi level shifts nearer to the valence band [Fig. 51.25 (*b*)]. The acceptor level lies immediately above the Fermi level. Conduction is by means of hole movement at the top of valence band, acceptor level readily accepting electrons from the valence band.

Again, it may be noted that even though *P*-type semiconductor has excess of holes for conduction purposes, on the whole it is electrically neutral for the same reasons as given above.

51.24. Majority and Minority Carriers

In a piece of ***pure*** germanium or silicon, no free charge carriers are available at 0°K. However, as its temperature is raised to room temperature, some of the covalent bonds are broken by heat energy and as a result, electron-hole pairs are produced. These are called thermally-generated charge carriers. They are also known as intrinsically-available charge carriers. Ordinarily, their number is quite small.

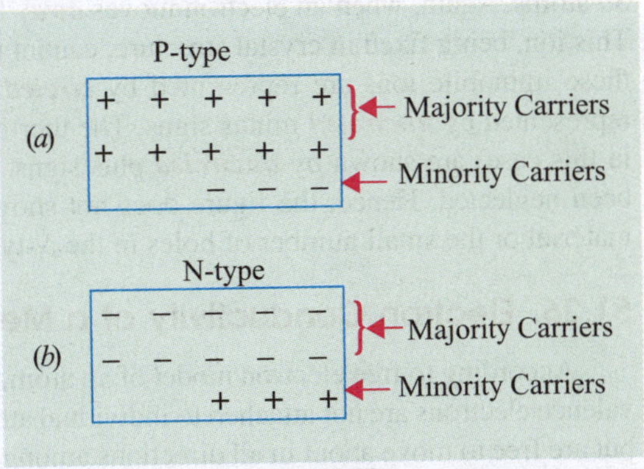

Fig. 51.26

An intrinsic of pure germanium can be converted into a *P*-type semiconductor by the addition of an acceptor impurity which adds a large number of holes to it. Hence, a *P*-type material contains following charge carriers :

(*a*) large number of positive holes—most of them being the added impurity holes with only a very small number of thermally generated ones;

(*b*) a very small number of thermally-generated electrons (the companions of the thermally generated holes mentioned above).

Obviously, in a *P*-type material, the number of holes (both added and thermally-generated) is much more than that of electrons. Hence, in such a material, holes constitute ***majority*** carriers and electrons form ***minority*** carriers as shown in Fig. 51.26 (*a*).

Similarly, in an *N*-type material, the number of electrons (both added and thermally-generated) is much larger than the number of thermally-generated holes. Hence, in such a material, electrons are majority carriers whereas holes are minority carriers as shown in Fig. 51.26 (*b*).

51.25. Mobile Charge Carriers and Immobile Ions

As discussed in Art. 51.23, *P*-type material is formed by the addition of acceptor impurity atoms like boron to the pure *Ge* or *Si* crystals. The number of holes added is equal to the number of boron atoms because each such atom contributes one hole. Now, when a hole moves ***away*** from its parent atom, the remaining atom becomes a ***negative ion***. Unlike the mobile and free-moving hole, this ion

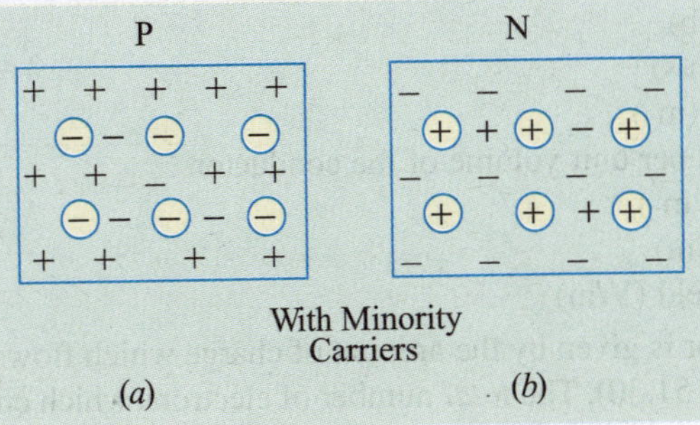

Fig. 51.27

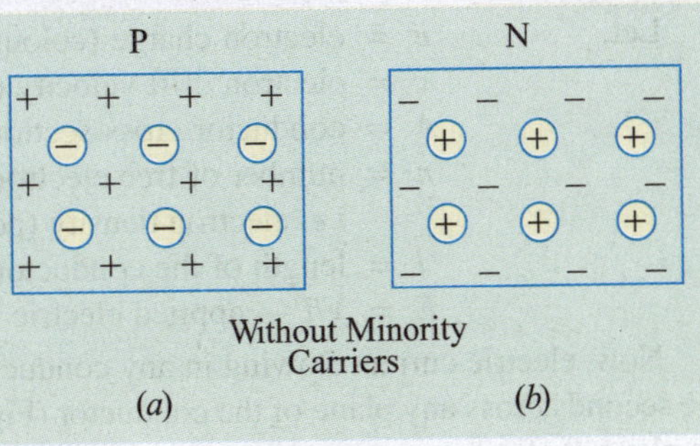

Fig. 51.28

cannot take part in conduction because it is fixed in the crystal lattice. In Fig. 51.27 (*a*), these immobile ions are shown by ***circled*** minus signs whereas free and mobile holes are shown by ***uncircled*** plus signs. Thermally-generated electrons (which form minority carriers) are shown by ***uncircled*** minus signs.

Similarly, addition of penta-valent atoms like antimony to pure *Ge* or *Si* crystal produces an *N*-type material. The number of free and mobile electrons which are added equals the number of donor *Sb* atoms. Again, when an electron moves ***away*** from its parent atom, it leaves behind a positive ion. This ion, being fixed in crystal structure, cannot take part in conduction. As shown in Fig. 51.27 (*b*); these immobile ions are represented by ***circled*** plus signs whereas free and mobile electrons are represented by ***uncircled*** minus signs. The thermally-generated holes (which form minority carriers in this case) are shown by ***uncircled*** plus signs. In Fig. 51.28, minority carriers of both types have been neglected. Hence, the figure does not show the small number of free electrons in the *P*-type material or the small number of holes in the *N*-type material.

51.26. Electron Conductivity of a Metal

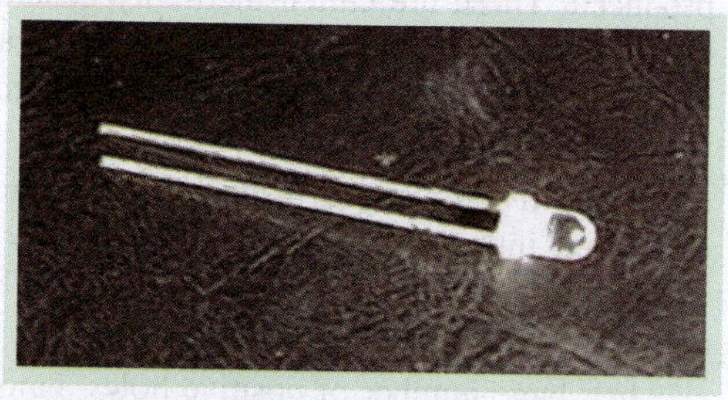

According to free electron model of an atom, the valence electrons are not attached to individual atoms but are free to move about in all directions among the atoms. These electrons are called ***conduction*** electrons and are said to form **'free electron cloud'** or **free electron 'gas'** or **the Fermi gas**. For example in copper there is one such free electron per atom, the other 28 electrons remaining bound to the copper nuclei to form positive ion cores.

When no external field is applied to the metal, the free electrons move randomly in all directions as shown in Fig. 51.29 (*a*). However, when an external electric field is applied to the metal, the free electron motion becomes directed a shown in Fig. 51.29 (*b*). This directed flow of electrons results in a net charge displacement in a definite direction. This type of motion is known as ***drift*** and the phenomenon is referred to as process of conduction by drift charge. The drift velocity (v) of the electrons is dependent upon the electron mobility (μ_e) and the applied electric field E. The actual relation is $v = \mu_e E$

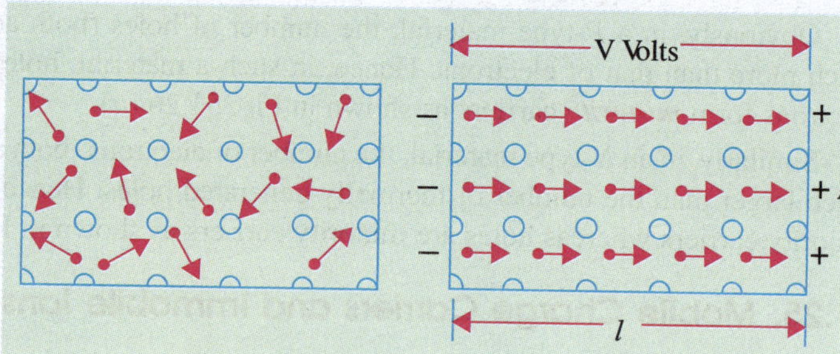

(*a*) (*b*)

Fig. 51.29

Let,
e = electron charge (coloumb)
v = electron drift velocity (m/s)
A = conductor cross-section (m^2)
n = number of free electrons per unit volume of the conductor *i.e.* electron density (per m^3)
l = length of the conductor (m)
E = V/l — applied electric field (V/m)

Now, electric current flowing in any conductor is given by the amount of charge which flows in one second across any plane of the conductor (Fig. 51.30). The ***total*** number of electrons which cross the plane P of cross-section A in one second = $n \times (v \times A)$.

Charge carried by them per second is = *envA*. Hence, $I = \nu enA$.

Substituting the value of ν, we get

$$I = neA\mu_e E = nAe\mu_e\ V/l$$

$$\therefore \quad R = \frac{V}{I} = \frac{l}{A}\left(\frac{1}{ne\mu_e}\right) = \rho\frac{l}{A}$$

∴ resistivity $\rho = 1/ne\ \mu_e$ ohm-m and conductivity $\sigma = ne\mu_e$ Siemens/m

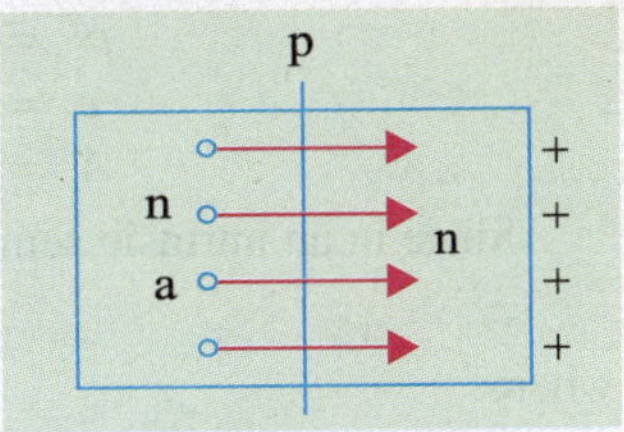

Fig. 51.30

Incidentally, it may be noted that conductivity of a semiconductor differs from that of a metal in one important respect *i.e.* in a semiconductor, charge carriers are both holes as well as electrons whereas in metals, electrons are the only charge carriers (Art. 51.27).

Example 51.3. *A copper wire of 2mm diameter with conductivity of 5.8×10^7 Siemens/m and electron mobility of 0.0032 m^2/V-s is subjected to an electric field of 20 mV/m. Find (a) the charge density of free electrons, (b) the current density, (c) current flowing in the wire, (d) the electron drift velocity.* **(U.P.S.C. Engg. Services 2002)**

Solution. $d = 2\text{mm} = 2\times10^{-3}\text{m}, \quad \sigma = 5.8\times10^7\ \text{S/m},$

$\mu_e = 0.0032\ \text{m}^2/\text{V-s}, \quad E = 20\ \text{mV/m}$

$\sigma = ne\mu_e$

(a) $$n = \frac{\sigma}{e\mu_e} = \frac{5.8\times10^7}{1.6\times10^{-19}\times0.0032} = \mathbf{1.132\times10^{29}/m^3}$$

(b) $J = \sigma E = (5.8\times10^7)\times(20\times10^{-3}) = \mathbf{1.16\times10^6\ A/m^2}$

(c) Let I = current flowing through the wire.

Area of cross-section of a wire,

$$A = \frac{\pi d^2}{4} = \frac{\pi\times\left(2\times10^{-3}\right)^2}{4} = 3.14\times10^{-6}\text{m}^2$$

$$J = \frac{I}{A} \quad \text{or} \quad I = JA = (1.16\times10^6)\times(3.14\times10^{-6}) = \mathbf{3.64A}$$

(d) $\upsilon = \mu E = 0.0032\times(20\times10^{-3}) = \mathbf{6.4\times10^{-5}\ m/s}$

51.27. Conductivity of Intrinsic Semiconductors

In their case, current flow is due to the movement of electrons and holes in opposite directions. However, since their charges are of opposite sign, the current due to each is in the same direction. Even though the number of electrons equals the number of holes, hole mobility μ_h is practically half of electron mobility μ_e.

As shown in Fig. 51.31, the total current flow which is due to the sum of electron flow and hole flow, is given by

$$I = I_e + I_h$$

Let $\quad \nu_e$ = drift velocity of electrons (m/s)

ν_h = drift velocity of holes (m/s)

n_i = density of free electrons in an intrinsic semiconductor (per m^3)

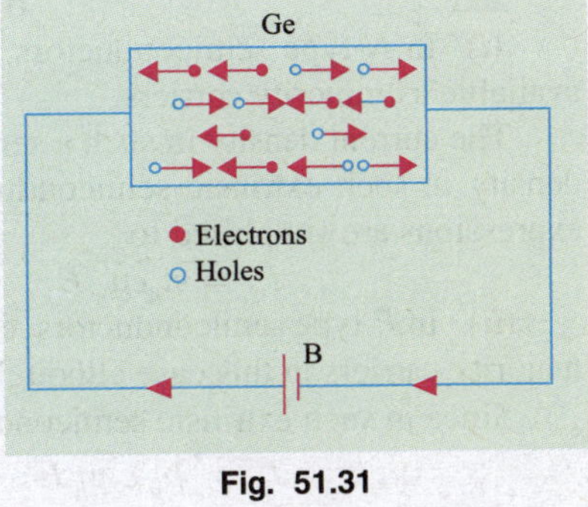

Fig. 51.31

p_i = density of holes in an intrinsic semiconductor (per m^3)
e = electron charge (coulomb)
A = cross-section of the semiconductor (m^2)

Since in an intrinsic semiconductor $n_i = p_i$

$$\therefore \quad I = n_i e\,(v_e + v_h)A = n_i e\,(\mu_h + \mu_e)\,EA$$

where μ_e = electron mobility = v_e/E

μ_h = hole mobility = v_h/E

Since $E = V/l$ where l is the length of the intrinsic semiconductor,

$$\therefore \quad I = n_i e\,(\mu_e + \mu_h)\,AV/l$$

$$\therefore \quad \frac{V}{I} = \frac{l}{A}\cdot\frac{1}{n_i e\,(\mu_e + \mu_h)} = \rho_i\,\frac{l}{A}$$

where ρ is the resistivity of the semoconductor. It is given by

$$\rho_i = \frac{1}{n_i e\,(\mu_e + \mu_h)} \text{ ohm-m}$$

The electrical conductivity which is the reciprocal of resistivity is given by

$$\sigma_i = n_i e\,(\mu_e + \mu_h) \text{ S/m}$$

Now, current density, $J = I/A$ $\quad\therefore\quad J = n_i e\,(\mu_e + \mu_h)E = \sigma_i\,E$ $\quad\therefore\quad \sigma_i = J/E$

Obviously, conductivity of semiconductors depends on two factors *(i)* number of current carriers present per unit volume and *(ii)* the mobility of the current carriers. It is found that with increase in temperature, n_i as well p_i increase and correspondingly, the conductivity of intrinsic semiconductors increases *i.e.* resistivity decreases.

51.28. Conductivity of Extrinsic Semiconductors

The general expression for current density (derived above) in the case of an extrinsic semiconductor when an electric field is employed is

$$J = (ne\mu_e + pe\mu_h)E \qquad \text{...(i)}$$

(a) If it is an *N*-type semiconductor, then the above expression becomes

$$J_n = e\,(n_n\mu_e + p_n\mu_h)E$$

where n_n and p_n represent the electron and hole densities in the *N*-type semiconductor ***after doping***.

(b) If it is a *P*-type semiconductor, then

$$J_p = e\,(n_p\mu_e + p_p\mu_h)\,E$$

where n_p and p_p represent similar quantities in a *P*-type semiconductor after doping.

The conductivity is given by $\sigma = J/E$

$$\therefore \quad \sigma = ne\mu_e + pe\mu_h$$

or $\quad \sigma_n = e\,(n_n\mu_e + p_n\mu_h)$ — for *N*-type

and $\quad \sigma_p = e\,(n_p\mu_e + p_p\mu_h)$ — for *P*-type

(i) In *N*-type semiconductors, electrons form the majority carriers although holes are also available as minority carriers.

The current density in such a semiconductor is given by Eq. (*i*) above. However, since electron density in such extrinsic semiconductors is much more than hole density *i.e.* $n_n \gg p_n$, the above expressions are simplified to

$$J_n = n_n e\mu_e\,E \quad \text{and} \quad \sigma_n = n_n\,e\,\mu_e \qquad \text{...(ii)}$$

(ii) In *P*-type semiconductors, conduction is by means of holes in the valence band which form majority carriers in this case although electrons are available as minority carriers.

Since in such extrinsic semiconductors, $n_p \ll p_p$, the above expressions become

$$J_p = p_p\,e\,\mu_h\,E \quad \text{and} \quad \sigma_p = p_p\,e\,\mu_h \qquad \text{...(iii)}$$

51.29. Conductivity when Intrinsic Charge Carrier Densities are Neglected

In case density of charge carriers available intrinsically is negligible as compared to the added impurity atoms (whether of donor or a acceptor type), then the formulae for conductivity given by Eq. (*ii*) and (*iii*) above will be changed as follows :

(*a*) For *N*-type semiconductor

As seen from Eq. (*ii*) above, the conductivity is given by $\sigma_n = n_n e\ \mu_e$ where n_n is electron density after doping.

In this relation, intrinsic hole density has already been neglected. The remaining electron density is also made up of the following two components :

1. intrinsic hole density due to holes available in a *pure* semiconductor ;
2. electron density N_d contributed by added donor impurity.

However, if we further neglect the intrinsic electron density, then $\sigma_n = N_d\ e\ \mu_e$

(*b*) For *P*-type semiconductor

As seen from Eq. (*iii*) above, the conductivity is given by $\sigma_p = p_p\ e\ \mu_h$ where p_p represents hole density. Again, in this relation, intrinsic electron density has been already neglected. This hole density further consists of the following two components :

1. intrinsic hole density due to holes available in a pure semiconductor ;
2. hole density (N_a) contributed by added acceptor impurity.

However, if we further neglect the intrinsic hole density, then $\sigma_p = N_a\ e\ \mu_h$

51.30. Conductivity of Pure and P-type Germanium

As shown in Art. 51.27, the conductivity of pure germanium is given by

$$\sigma = n_i e\ (\mu_e + \mu_h) = p_i e\ (\mu_e + \mu_h)$$

When germanium is doped by a trivalent impurity like indium, it becomes a *P*-type semiconductor. After doping, its conductivity depends on the number of charge carriers available in it. The law of Mass Action can be used for finding this number. For acceptor impurity, the law may be stated as follows :

$$n_p\, p_p = n_i\, p_i = n_i^2 = p_i^2$$

where n_p and p_p represent the 'free' electron and hole densities respectively in the semi-conductor after doping and n_i and p_i, the electron and hole densities ***before*** doping *i.e.* in an intrinsic semiconductor. Put in another way, it simply means that, at constant temperature, the product of the number of electron carriers and the number of hole carriers is independent of the density of acceptor atoms. In physical terms, it means that the introduction of *P*-type impurity fills some of the electron levels produced by thermal action.

By calculating n_p and p_p from above and knowing μ_e and μ_h, conductivity after doping can be found out as illustrated by the following example.

(*a*) Pure Germanium

$$\sigma = n_i\ e\ (\mu_e + \mu_h) = p_i\ e\ (\mu_e + \mu_h)$$

Let, $\quad n_i = p_i = 2 \times 10^{19}$ per m^3, $\quad \mu_e = 0.36$ m^2/V-s

$\mu_h = 0.17$ m^2 / V-s and $e = 1.6 \times 10^{-19}$ C

$\sigma = 2 \times 10^{19} \times 1.6 \times 10^{-19}\ (0.36 + 0.17) = 1.69$ S/m.

(*b*) *P*-type Germanium

here, $\quad \sigma_p = e\ (n_p\ \mu_e + p_p\ \mu_h)$

Suppose, we add 10^{20} atoms/m^3 of indium and that $n_i = 2 \times 10^{19}$ charge carriers (either electrons or holes) per m^3.

Then, $\quad n_p\, p_p = n_i^2 = (2 \times 10^{19})^2 = 4 \times 10^{38}$ and $p_p - n_p = 10^{20}$

$\therefore \quad p_p - 4 \times 10^{38}/p_p = 10^{20}$ or $p_p^2 - 10^{20}\ p_p - (4 \times 10^{38}) = 0$

Solving the above quadratic equation and taking positive value only.

$$p_p = 1.04 \times 10^{20} \quad \text{and} \quad n_p = 0.04 \times 10^{20}$$

$$\sigma_p = 1.6 \times 10^{-19} (0.04 \times 10^{20} \times 0.36 + 1.04 \times 10^{20} \times 0.17) = 3.1 \text{ S/m}^*$$

It is seen that conductivity is almost doubled.

Example 51.4. *What length of a round copper wire of diameter 1 mm will have a resistance of 1k Ω if copper conductivity is 60 MS/m. A cylindrical piece of silicon having a diameter of 1 mm is doped with 10^{20} m^{-3} atoms of phosphorous which are fully ionized. What length of this silicon would be required to give a resistance of 1 k Ω if electronic mobility in silicon is 0.1 m^2 / V-s ?*

(Electronic Devices & Circuits, Pune Univ. 1994)

Solution. $R = 1 \text{ k}\Omega = 1000\ \Omega, \sigma = 60 \times 10^6 \text{ S/m}, A = \pi d^2/4 = \pi \times (1 \times 10^{-3})^2/4 \text{m}^2$

$$R = l/\sigma A$$

$$l = \sigma AR = 60 \times 10^6 \times (\pi \times 10^{-6}/4) \times 1000 = 47{,}123 \text{ m} = \mathbf{47.1 \text{ km}}$$

For Silicon Wire

$$\sigma = n_i e \mu_e = 10^{20} \times 1.6 \times 10^{-19} \times 0.1 = 1.6 \text{ S/m}$$

$$l = \sigma AR = 1.6 \times (\pi \times 10^{-6}/4) \times 1000$$

$$= 1.26 \times 10^{-3} = \mathbf{1.26 \text{ mm}}$$

Example 51.5. *Calculate the intrinsic conductivity of silicon at room temperature if n = 1.41 × 10^{16} m^{-3}, μ_e = 0.145 m^2/V-s, μ_h = 0.05 m^2/V-s and e = 1.6 × 10^{-19} C. What are the individual contributions made by electrons and holes ?* **(Electronic Engg., Nagpur Univ. 1991)**

Solution. As seen from Art. 51.27, the conductivity of an intrinsic semiconductor is given by

$$\sigma_i = n_i e\, \mu_e + n_i e\, \mu_h$$

$$= 1.41 \times 10^{16} \times 1.6 \times 10^{-19} \times 0.145 + 1.41 \times 10^{16} \times 1.6 \times 10^{-19} \times 0.05$$

$$= 0.325 \times 10^{-3} + 0.112 \times 10^{-3} \text{ S/m} = \mathbf{0.437 \times 10^{-3} \text{ S/m}}$$

Contribution by electrons $= \mathbf{0.325 \times 10^{-3} \text{ S/m}}$

Contribution by holes $= \mathbf{0.112 \times 10^{-3} \text{ S/m}}$

Example 51.6. *Calculate the donor concentration in N-type germanium having resistivity of 100 Ω-m. Derive the formula you use. Take e = 1.6 × 10^{-19} C, μ_e = 0.36 m^2 V^{-1} s^{-1}.*

(Electronics ; Nagpur Univ. 1990)

Solution. As seen from Art. 51.29, $\rho_n = 1/N_d\, e\, \mu_e$

$\therefore \quad 100 = 1/N_d \times 1.6 \times 10^{-19} \times 0.36 \; ; \quad N_d = \mathbf{1.74 \times 10^{17} \text{ atoms/m}^3}$

Example 51.7. *An N-type silicon has a resistivity of 1500 Ω-m at a certain temperature. Compute the electron-hole concentration given that n_i = 1.5 × 10^{16} m^{-3}, μ_e = 0.14 m/V-s, μ_h = 0.05 m^2/V-s and e = 1.6 × 10^{-19} C.*

Solution. Being *N*-type silicon, it is assumed that $n \gg p$

$\therefore \quad \sigma = e\,(n\mu_e + p\mu_h) = en\, \mu_e$

$\therefore \quad \rho = 1/en\mu_e$ or $1500 = 1/n \times 1.6 \times 10^{-19} \times 0.14$ or $n = \mathbf{2.97 \times 10^{16} \text{ m}^{-3}}$

Now, $np = n_i^2$ or $p = n_i^2/n = (1.5 \times 10^{16})^2/(2.97 \times 10^{16}) = \mathbf{7.57 \times 10^{15} \text{ m}^{-3}}$

Example 51.8. *A specimen of pure germanium at 300° K has a density of charge carriers of 2.5 × 10^{19}/m^3. It is doped with donor impurity atoms at the rate of one impurity atom for every 10^6 atoms of germanium. All impurity atoms may be supposed to be ionized. The density of germanium atom is 4.2 × 10^{28} atoms/m^3.*

Find the resistivity of the doped germanium if electron mobility is 0.36 m^2/V-s.

Solution. Density of added impurity atoms is

$$N_d = 4.2 \times 10^{28}/10^6 = 4.2 \times 10^{22} \text{ atoms/m}^3$$

* It is the new name for the old unit of mho/m.

As seen, it is very large as compared to the intrinsic charge carrier density of $2.5 \times 10^{19}/m^3$ which will, therefore, be neglected. Now, as seen from Art. 51.29.

$$\sigma_n = N_d e \mu_e = 4.2 \times 10^{22} \times 1.6 \times 10^{-19} \times 0.36 = 2.42 \times 10^3 \text{ S/m}$$

$$\rho_n = 1/\sigma_n = 1/2.42 \times 10^3 = \mathbf{0.413 \times 10^{-3}\ \Omega\text{-m.}}$$

Example 51.9. *Compute the relative concentration of silicon atoms and electron-hole pairs at 300° K. Also, calculate intrinsic resistivity of silicon. Given Avogadro's number = 6.02×10^{23} atoms/g-atom, density = 2.33×10^6 g/m^3, atomic Wt. = 28.1, intrinsic carrier density = 1.5×10^{16} m^{-3}, $\mu_e = 0.14$ m^2/V-s, $\mu_h = 0.05$ m^2/V-s.*

Solution. n_A = Avogadro's No. × density/atomic Wt.

$$= 6.02 \times 10^{23} \times 2.33 \times 10^6/28.1 \cong 5 \times 10^{28} \text{ atoms/m}^3$$

Since intrinsic concentration n_i *i.e.*, electron-hole pairs/m^3 is 1.5×10^{16}

$$\therefore \quad \frac{n_A}{n_i} = \frac{5 \times 10^{28}}{1.5 \times 10^{16}} \cong \mathbf{3.3 \times 10^{12}}$$

It means that there are 3.3×10^{12} Si atoms for each electron-hole pair.

Since in a pure semiconductor, $n = p = n_i$, intrinsic conductivity is given by

$$\sigma_i = n_i e (\mu_e + \mu_h) = 1.5 \times 10^{16} \times 1.6 \times 10^{-19} (0.14 + 0.05) = 0.456 \times 10^{-3} \text{ S/m}$$

Hence, intrinsic resistivity is given by

$$\rho_i = 1/\sigma_i = 1/0.456 \times 10^{-3} = \mathbf{2193\ \Omega\text{-m}}$$

Example 51.10. *Silicon is doped with acceptor atoms to a density of 10^{22} m^{-3}. If it is assumed that all acceptor centres are ionized, calculate the conductivity of the extrinsic silicon. Given that intrinsic density is 1.4×10^{16} m^{-3}, $\mu_e = 0.145$ m^{-3} and $\mu_h = 0.05$ m^{-3}.*

(Electronic Devices & Circuits, Pune Univ. 1992)

Solution. The minority carrier density can be found from the equation given in Art. 51.30, *i.e* $np = n_i^2$.

Now, $\quad p = 10^{22} \quad \therefore \quad n \times 10^{22} = (1.4 \times 10^{16})^2 ; \quad n = 1.96 \times 10^{10}/m^3$

Now, $\quad \sigma = ne\,\mu_e + pe\,\mu_h$

$$= 1.96 \times 10^{10} \times 0.145 \times 1.6 \times 10^{-19} + 10^{22} \times 0.05 \times 1.6 \times 10^{-19} = \mathbf{80\ S/m}$$

51.31. Drift

Directed motion of charge carriers in semiconductors occurs through two mechanisms :

(i) charge drift under the influence of applied electric field and

(ii) diffusion of charge from a region of high charge density to one of low charge density.

Consider the drift phenomenon first. When no electric field is applied to the semiconductor which is above 0° K, the conduction electrons (as well as holes) move within the crystal with random motion and repeatedly collide with each other and the fixed ions. Due to randomness of their motion, the net average velocity of these charge carriers in any given direction is zero. Hence, no current exists in the crystal under this condition of no field.

Now, consider the case when an electric field is applied to the crystal. Under the influence of this field, the charge carriers attain a ***directed*** motion which is superimposed on their random thermal motion. This results in a net average velocity called ***drift*** velocity in the direction of the applied electric field. Of course, electrons and holes move in opposite directions but because of their opposite charges, both produce current in the same direction. In extrinsic semiconductors, this current is essentially a majority carrier flow.

The drift velocity is proportional to electric field strength E, the constant of proportionality being called mobility μ. The exact relation between the two is $v = \mu E$.

Let us find the value of drift current in a semiconductor.

(i) current density due to electron drift is $J_e = e\mu_e nE$ where μ_e is electron mobility, n is electron density and E is the electric field strength.

(ii) current density due to hole drift is $J_h = e\,\mu_h\,pE$ where p is hole density.

Total current density due to electron and hole drift is

$$J = J_e + J_h = e\mu_e nE + e\mu_h pE = e\,(n\mu_e + p\mu_h)\,E$$
$$= e\mu_h\,(p + bn)E \qquad \text{where } b = \mu_e/\mu_h$$

51.32. Diffusion

It is gradual flow of charge from a region of high density to a region of low density. It is a force-free process based on non-uniform distribution of charge carriers in a semiconductor crystal. It leads to an electric current without the benefit of an applied field. This flow or diffusion of carriers is proportional to the carrier density gradient, the constant of proportionality being called diffusion constant or diffusion coefficient D which has a unit of m^2/s.

Current density due to hole diffusion is $J_h = -\,eD_h\,dp/dx$. Similarly, current density due to electron diffusion is $J_e = eD_e\,dn/dx$.

where D_e, D_h = electron and hole diffusion constants respectively

dn/dx = density gradient of electrons

dp/dx = density gradient of holes.

It is obvious that diffusion depends on charge in homogeneity or on the presence of a space gradient of charge density. It can occur in regions free of electric field. On the other hand, drift current is a function of both electric field and charge density.

Incidentally, it may be noted that, generally, diffusion leads to redistribution of charges which further results in the development of potential difference between different parts of the semiconductor. The electric field due to this potential difference sets up drift current in opposition to diffusion current. Final equilibrium is achieved when the potential difference developed becomes sufficiently large so as to create a drift current equal and opposite to the diffusion current thus resulting in zero net flow of current.

51.33. Combined Drift and Diffusion Currents

In semiconductors, drift and diffusion processes may be present simultaneously. The expressions for total electron and hole densities become.

$$J_e = e\mu_e nE + eD_e\,dn/dx \text{ A/m}^2 \quad \text{and} \quad J_h = e\mu_h\,pE - eD_h dp/dx \text{ A/m}^2$$

51.34. Relation Between D and μ

Both diffusion constant and mobility are statistical thermodynamic phenomena and are related to each other by the following equation.

$$\mu_e = \frac{e}{kT}D_e \text{ and } \mu_h = \frac{e}{kT}D_h \quad \text{or} \quad \frac{D_e}{\mu_e} = \frac{D_h}{\mu_h} = \frac{kT}{e} = \frac{T}{11,600}$$

The relationship is known as Einstein's equation

At $t = 27°C$, $T = 300°K$, hence $D/\mu = 300/11{,}600 = 1/39$ or $\mu = 39$ D

Example 51.11. *Calculate diffusion constants for electrons and holes at 300°K in silicon if $\mu_e = 0.15\ m^2/V\text{-}s$ and $\mu_h = 0.05\ m^2/V\text{-}s$.*

Solution. According to Einstein's equation,

$$D = \mu kT/e \quad \text{or} \quad D = \mu/39\ m^2/s \quad \text{— at 300°K}$$
$$D_e = \mu_e/39 = 0.15/39 = 3.85 \times 10^{-3}\ m^2/s$$
$$D_h = \mu_h/39 = 0.05/39 = 1.28 \times 10^{-3}\ m^2/s$$

Example 51.12. *Find the diffusion coefficients of holes and electrons for germanium at 300 K. The carrier mobilities in cm^2/volt-sec at 300 K for electrons and holes are respectively 3600 and 1700. Density of carriers is 2.5×10^{13}. Boltzmann constant, $k = 1.38 \times 10^{-23}$.*

(U.P.S.C. Engg. Services, 1996)

Solution. According to the Einstein's equation,

$$D = \mu kT/e \quad \text{or} \quad D = \mu/39 \text{ m}^2/\text{s} - \text{at } 300^\circ \text{ K}$$

$$D_e = \mu_e/39 = 3600 \text{ (cm}^2/\text{V-s)}/39 = 0.36 \text{ (m}^2/\text{V-s)}/39 = \mathbf{9.2 \times 10^{-3} \text{ m}^2/\text{s}}$$

$$D_h = \mu_h/39 = 1700 \text{ (cm}^2/\text{V-s)}/39 = 0.17 \text{ (m}^2/\text{V-s)}/39 = \mathbf{4.36 \times 10^{-3} \text{ m}^2/\text{s}}$$

51.35. Recombination

Apart from drift and diffusion, a third phenomenon which occurs in semiconductors is called ***recombination*** that results from the collision of an electron with a hole.

The process is essentially the return of a free conduction electron to the valence band and is accompanied by the emission of energy. Obviously, the recombination rate is directly proportional to the carrier concentration for the simple reason that larger the number of carriers, the more likely is the occurrence of electron-hole recombination. This phenomenon is important in describing minority carrier flow.

As is well-known, in a semiconductor, thermal generation of electron-hole pairs also takes place continuously. Hence, there is ***net recombination*** rate given by the difference between the recombination and generation rates.

51.36. Carrier Life Time

It is defined as the time for which, on an average, a charge carrier will exist before recombination with a carrier of opposite charge. Its value varies from nanoseconds (10^{-9}) to hundreds of microseconds (μs) and depends on temperature and impurity concentration in the semiconductor material.

51.37. Total Carrier Flow

The total carrier flow in a semiconductor is the sum of the three flow phenomena discussed above. Each type of carrier has to be treated separately and the number of electrons or holes leaving the sample being accounted for by drift or diffusion or net recombination. The current in the semiconductor is then the sum of the electron and hole currents.

51.38. P-N Junction

It is possible to manufacture a single piece of a semiconductor material half of which is doped by P-type impurity and the other half by *N*-type impurity as shown in Fig. 51.32. The plane dividing the two zones is called ***junction.*** Theoretically, junction plane is assumed to lie where the density of donors and acceptors is equal. The *P-N* junction is fundamental to the operation of diodes, transistors and other solid-state devices.

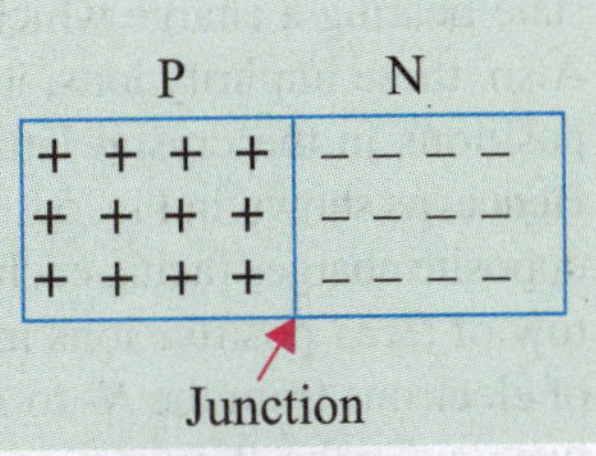

Fig. 51.32

Let us see if anything unusual happens at the junction. It is found that following three phenomena take place :

1. A thin ***depletion layer*** or region (also called space-charge region or transition region) is established on both sides of the junction and is so called because it is depleted of ***free charge carriers***. Its thickness is about 10^{-6} m.
2. A barrier potential or junction potential is developed across the junction.
3. The presence of depletion layer gives rise to junction and diffusion capacitances (Art. 52.5).

51.39. Formation of Depletion Layer

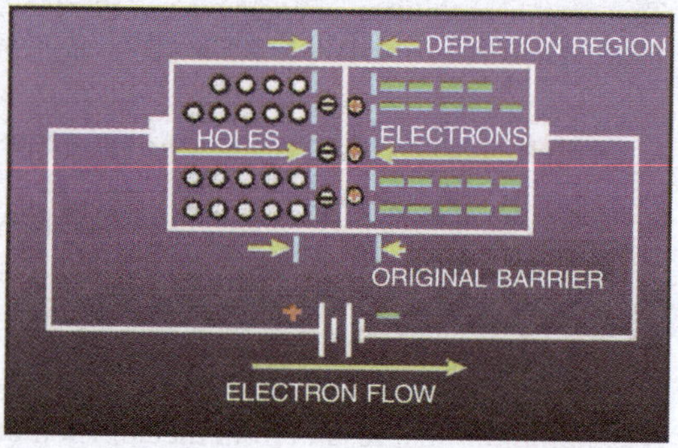

Suppose that a junction has just been formed. At that instant, holes are still in the *P*-region and electrons in the *N*-region. However, there is greater concentration of holes in *P*-region than in *N*-region (where they exist as minority carriers). Similarly, concentration of electrons is greater in *N*-region than in *P*-region (where they exist as minority carriers). This concentration differences establishes density gradient across the junction resulting in carrier diffusion. Holes diffuse from *P* to *N*-region and electrons from *N*-to *P*-region and terminate their existence by recombination [Fig. 51.33 (*a*)]. This recombination of free and mobile electrons and holes produces the narrow region at the junction called depletion layer. It is so named because this region is devoid of (or depleted of) ***free and mobile charge carriers like electrons and holes***—there being present only positive ions which are not free to move.

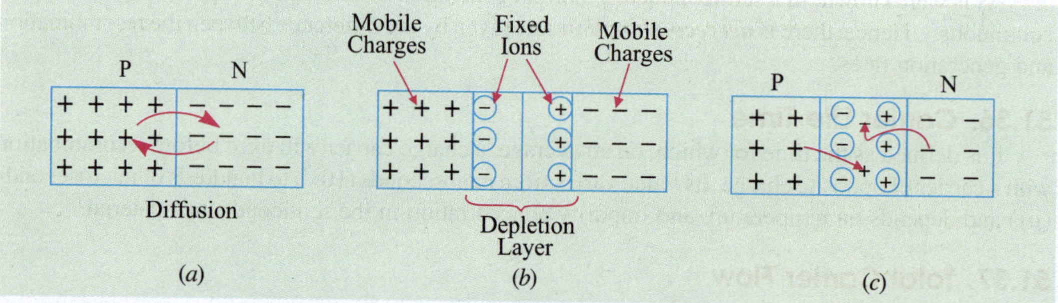

Fig. 51.33

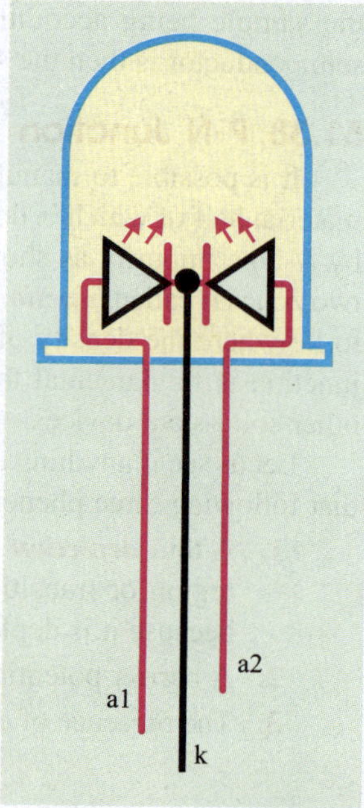

It might seem from above that eventually all the holes from the *P*-side would diffuse to the *N*-side and all the electrons from the *N*-side would diffuse to the *P*-side but this does not occur due to the formation of ions on the two sides of the junction. The impurity atoms which provide these migratory electrons and holes are left behind in an ionized state bearing a charge which is opposite to that of the departed carrier. Also, these impurity ions, just like germanium atoms, are fixed in their positions in the crystal lattice in the *P*- and *N*- regions of the diode. Hence, as shown in Fig. 51.33 (*b*), they form parallel rows or 'plates' of opposite charges facing each other across the depletion layer. Obviously, row of ***fixed*** positive ions in the *N*-region is produced by the migration of electrons from the *N*- to *P*- region. Similarly, the row of ***fixed*** negative ions in the *P*-region is produced by the migration of holes from the *P*- to *N*-region.

If a majority carrier (either an electron or a hole) tries to cross into depletion layer, it can meet either of the following two facts :

(*i*) either it can be trapped or captured by the row of fixed impurity ions of opposite sign which guard its own region. For example, a hole trying to approach the depletion layer may be neutralized by the row of fixed negative ions situated in the *P*-region itself at the edge of the depletion layer. So will be the case with the electron trying to approach the depletion layer from *N*-region [Fig. 51.33 (*c*)]; or

(*ii*) it may succeed in entering the depletion layer where it will be repelled by the row of similarly-charged impurity ions guarding the other region. But its life will be cut short by recombination with a majority carrier of opposite sign which has similarly entered the depletion layer from the other half of the diode.

Ultimately, an equilibrium condition is reached when depletion layer has widened to such an extent that no electrons or holes can cross the *P-N* junction.

51.40. Junction or Barrier Voltage

Even though depletion layer is cleared of charge carriers, it has oppositely-charged fixed rows of ions on its two sides. Because of this charge separation, an electric potential difference V_B is established across the junction even when the ***junction is externally isolated*** (Fig. 51.34). It is known as ***junction or barrier potential***. It stops further flow of carriers across the junction unless supplied by energy from an external source. At room temperature of 300ºK, V_B is about 0.3 V for Ge and 0.7 V for Si.

The value of barrier voltage is given by $V_B = V_T \log_e N_a N_d / n_i^2$ where N_a, N_d, n_i^2 and V_T have the meanings explained in Art. 51.29 and 34. The value of V_T at room temperature of 300ºK is given by

$$V_T = V_{300} = \frac{kT}{e} = \frac{1.38 \times 10^{-23} \times 300}{1.6 \times 10^{-19}} = 26 \text{ mV}$$

$$\therefore \quad V_B = 26 \log_e (N_a N_d / n_i^2) \text{ mV}$$

Barrier voltage depends on doping density, electronic charge and temperature. For a given junction, the first two factors are constant, thus making V_B dependent on temperature. With increase in temperature, more minority charge carriers are produced, leading to their increased drift across the junction. As a result, equilibrium occurs at a lower barrier potential. It is found that both for Ge and Si, V_B decreases by about 2 mV / ºC.

$$\therefore \quad \Delta V_B = -0.002 \, \Delta t \qquad \text{where } \Delta t \text{ is the rise in temperature in ºC.}$$

The strong field set up by V_B causes drift of carriers through depletion layer. As seen from lFig. 51.35 (*b*), under the influence of this field, ***holes drift from N-to P-region and electrons from P- to -N region***. This drift current must be equal and opposite to the diffusion current [Fig. 51.35 (*a*)] because under condition of equilibrium and with no external supply, net current through the crystal is zero.

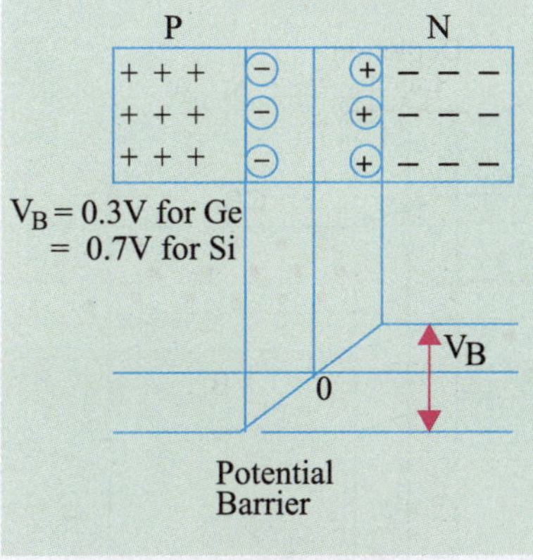

Fig. 51.34

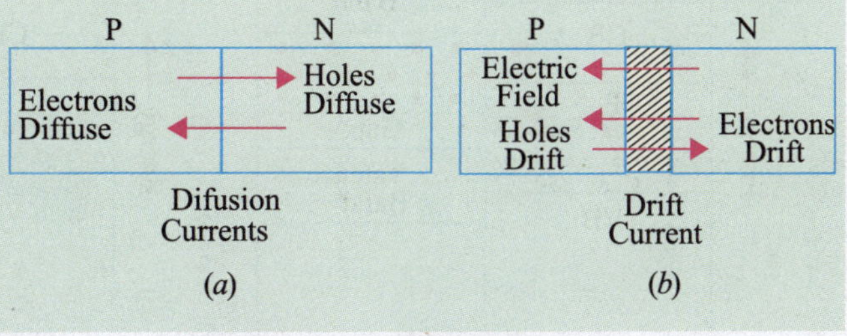

Fig. 51.35

To summarize the main points of Art. 51.39 and 51.40, we have :

1. As soon as *P-N* junction is formed, free electrons and holes start diffusing across the junction and recombining.

2. Their recombination leads to the appearance of a depletion layer across the junction which contains no ***mobile*** carriers but only immobile ions.
3. These immobile ions set up a barrier potential and hence an electric field which sets up drift current that is equal and opposite to the diffusion current when final equilibrium is reached.

Example 51.13. *Calculate the barrier potential at room temperature for P-N junction in silicon which is doped to a carrier density of 10^{21} m^{-3} on the P-side and 10^{22} m^{-3} on the N-side. The intrinsic carrier density for silicon is 1.4×10^{16} m^{-3}.* **(Electronics-I, Bangalore Univ. 1992)**

Solution. Using the relation given in Art 1.40, we have

$$V_B = 26 \log_e (N_a N_d / n_i^2) \text{ mV}$$
$$= 26 \log_e \{10^{21} \times 10^{22}/(1.4 \times 10^{16})^2\} = 641 \text{ mV} = \mathbf{0.641\ V}$$

Example 51.14. *Calculate the change in barrier potential of a P-N junction at 300°K if doping on the N-side is increased 1000 times while keeping doping on P-side unchanged.*

Solution. As seen from Art. 51.40, at 300°K

$$V_B = 26 \log_e (N_a N_d / n^2) \text{ mV}$$
$$\therefore \quad V_{B1} = 26 \log_e (N_{a1} N_{d1}/n_i^2)\ ; \quad V_{B2} = 26 \log_e (N_{a2} N_{d2} / n_i^2)$$
$$\therefore \quad V_{B2} - V_{B1} = 26 \log_e (N_{a2} N_{d2})/(N_{a1} N_{d1})$$
$$= 26 \log_e (N_{d2} / N_{d1}) = 26 \log_e 1000 = \mathbf{179.6\ mV}$$

51.41. Energy Band Diagram of a P-N Junction

Now, let us consider the operation of a *P-N* junction in terms of its energy bands. At the instant of junction formation, energy bands of the trivalent impurity atoms in the *P*-region are at a slightly higher level than those of the pentavalent impurity atoms in the *N*-region as shown in Fig. 51.36. It is so because core attraction for valence electrons (+3) in a trivalent atom is less than the core attraction for valence electrons (+5) in a pentavalent atom. Consequently, trivalent valence electrons are in slightly higher orbit and, hence, at a higher energy level. However, there is some over-lap between respective bands of the two regions. Due to this reason, some high-energy electrons near the top of *N*-region conduction band diffuse into the lower part of the *P*-region conduction band. Soon after, they recombine with the holes in the valence band as shown in Fig. 51.36 (*a*). As diffusion continues, depletion layer begins to form. Another side-effect of this electron diffusion is that energy bands in

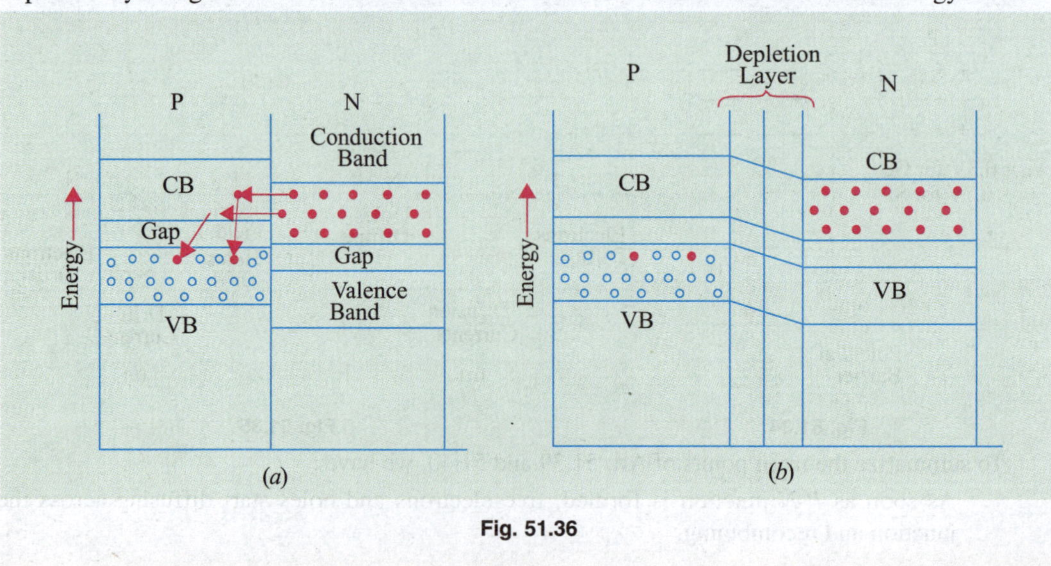

Fig. 51.36

the *N*-region shift downward due to loss of high-energy electrons. When the top of the conduction band in *N*-region reaches the same level as the bottom of the conduction band in *P*-region, further diffusion ceases and equilibrium condition is reached as shown in Fig. 51.36 (*b*). To an electron which might still try to diffuse across the junction, the path looks like a steep energy hill. It cannot climb this energy hill unless it receives energy from an external source.

51.42. Forward Biased P-N Junction

Suppose, positive battery terminal is connected to *P*-region of a semiconductor and the negative battery terminal to the *N*-region as shown in Fig. 51.37 (*a*). In that case the junction is said to be biased in the ***forward direction*** because it permits easy flow of current across the junction. This current flow may be explained in the following two ways :

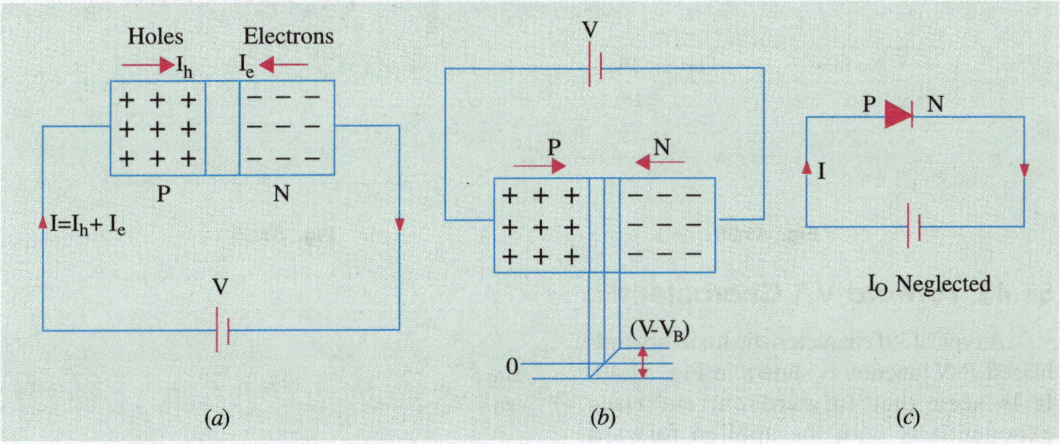

Fig. 51.37

(*i*) As soon as battery connection is made, holes are ***repelled*** by the positive battery terminal and electrons by the negative battery terminal with the result that both the electrons and the holes are driven ***towards*** the junction where they recombine. This ***en masse*** movement of electrons to the left and that of holes to the right of the junction constitutes a large current flow through the semiconductor. Obviously, the junction offers ***low resistance*** in the forward direction.

A more detailed picture of carrier flow is as under :

As free electrons move to the left, new ***free*** electrons are injected by the negative battery terminal into the *N*-region of the semiconductor. Thus, a flow of electrons is set up in the wire connected to the negative battery terminal. As holes are driven towards the junction, more holes are created in the *P*-region by the breakage of covalent bonds (Art. 51.21). These newly-created holes are driven towards the junction to keep up a continuous supply. But the electrons so produced are attracted to the left by the positive battery terminal from where they go to the negative terminal and finally to the *N*-region of the crystal.

Incidentally, it may be noted that though there is movement of both electrons and holes ***inside*** the crystal, ***only free electrons move in the external circuit i.e.*** in the battery-connecting wires.

Note. There is also present an extremely small amount of current I_0 due to minority carriers on either side of the junction. But it is negligible as compared to forward current which is due to majority carriers. This current I_0 depends on temperature but is almost independent of applied voltage.

(*ii*) Another way to explain current flow in forward direction is to say that forward bias of *V* volts lowers the barrier potential to $(V - V_B)$ which now allows more current to flow across the junction [Fig. 51.37 (*b*)].

Incidentally, it may be noted that forward bias reduces the thickness of the depletion layer as shown in Fig. 51.38.

Energy band diagram for forward bias is shown in Fig. 51.39. By comparing this figure with Fig. 51.36 it is seen that energy hill has been reduced. Because of this reduction, conduction electrons in *N*-region are able to cross over to *P*-region. After reaching there, each electron falls into a hole (path *A*) and becomes a valence electron. In this way, it is able to continue its journey towards the left end of the crystal.

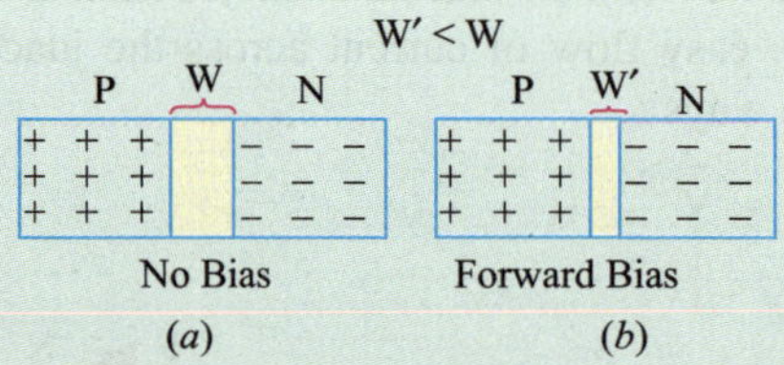

Fig. 51.38

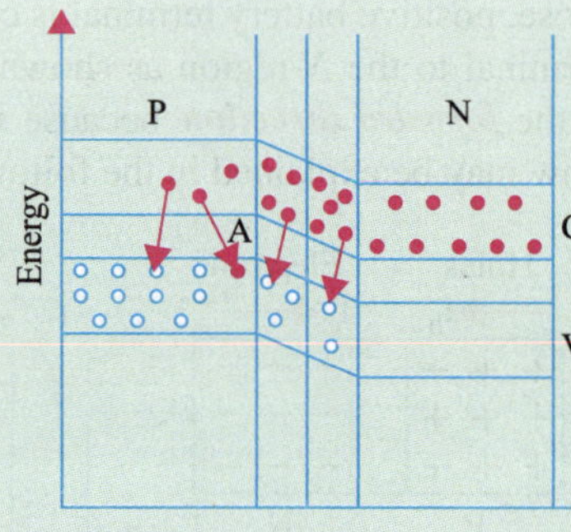

Fig. 51.39

51.43. Forward V/I Characteristic

A typical *V/I* characteristic for a forward-biased *P-N* junction is shown in Fig. 51.40. It is seen that forward current rises exponentially with the applied forward voltage. However, at ordinary room temperature, a p.d. of about 0.3 V is required before a reasonable amount of forward current starts flowing in a germanium junction. This voltage is known as ***threshold voltage*** (V_{th}) **or** ***cut-in voltage*** or ***knee voltage*** V_K. It is practically the same as barrier voltage V_B. Its value for silicon junction is about 0.7 volt. For $V < V_{th}$, current flow is negligible. But as applied voltage increases beyond the threshold value, the forward current increases sharply. If forward voltage is increased beyond a certain safe value, it will produce an extremely large current which may destroy the junction due to overheating. *Ge* devices can stand junction temperatures around 100ºC whereas *Si* units can function upto 175ºC.

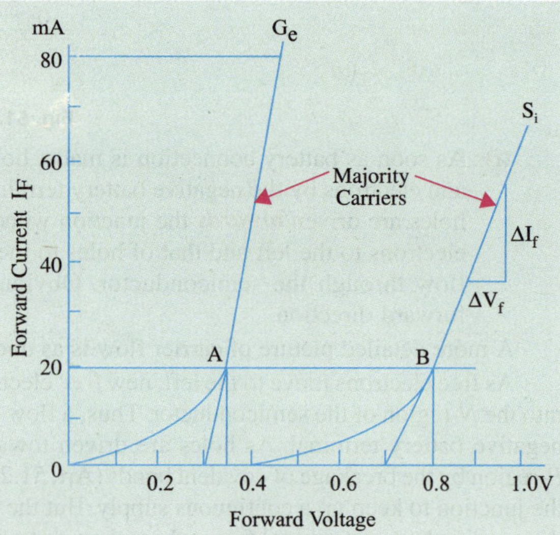

Fig. 51.40

Obviously, the ***forward-biased junction has a low resistance***. For point *B* in Fig. 51.40, the forward resistance for Si is

$$R_F = 0.8\text{V}/20 \text{ mA} = 40\ \Omega$$

Similarly, for point *A* on the *Ge* curve, $R_F = 0.36$ V/20 mA = 18 Ω

In practice, this static forward resistance is not used. Instead, the **dynamic** resistance or **incremental** resistance or **ac resistance** of the juncion is used. It is given by the reciprocal of the slope of the forward characteristic.

$$r_{ac} = \frac{1}{\Delta I_F / \Delta V_F} = \frac{\Delta V_F}{\Delta I_F}$$

Here, $\Delta V_F = 0.19$ V and $\Delta I_F = 37.6$ mA

$$\therefore \quad r_{ac} = \frac{0.19}{37.6 \times 10^{-3}} \cong 5\ \Omega$$

51.44. Reverse Biased P-N Junction

When battery connections to the semiconductor are made as shown in Fig. 51.41 (*a*), the junction is said to ***reverse-biased***. In this case, holes are attracted by the negative battery terminal and electrons by the positive terminal so that both holes and electrons move ***away*** from the junction and ***away*** from each other. Since there is no electron-hole combination, no current flows and the junction offers high resistance.

Another way of looking at the process is that in this case, the applied voltage increases the barrier potential to $(V + V_B)$, thereby blocking the flow of majority carriers.

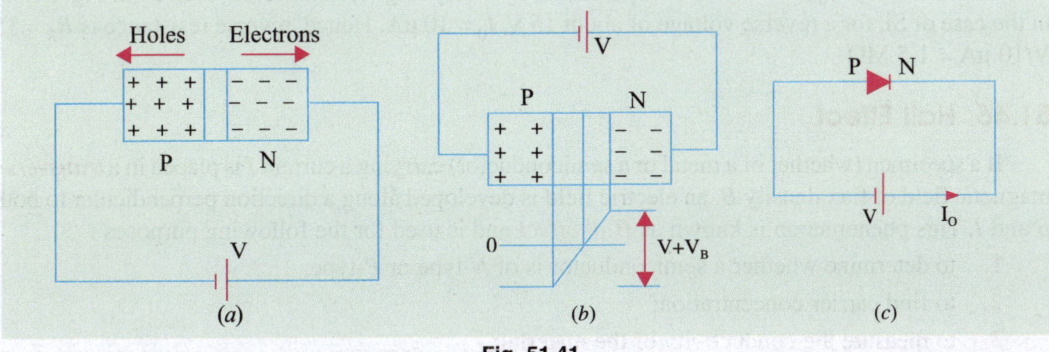

Fig. 51.41

Incidentally, it may be noted that under reverse bias condition, width of depletion layer is increased because of increased barrier potential as shown in Fig. 51.42.

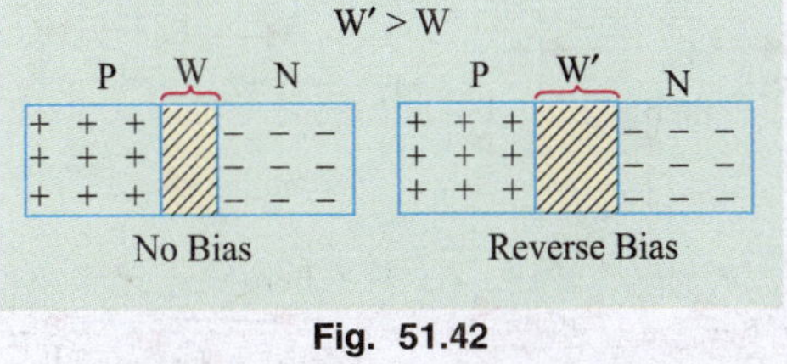

Fig. 51.42

Although, in this case, there is practically no current due to ***majority*** carriers, yet there is a small amount of current (a few μA only) due to the flow of ***minority*** carriers across the junction. As explained earlier in Art. 51.24, due to thermal energy, there are always generated some holes in the *N*-type region and some electrons in the *P*-type region of the semiconductor as shown in Fig. 51.26. The battery drives these minority carriers across the junction thereby producing a small current called ***reverse current or reverse saturation current*** I_0 **or** I_S. Since minority carriers are thermally-generated, I_0 is extremely ***temperature dependent***. For the same reason, forward current is also temperature dependent but to a much less degree because minority current forms a very small percentage of the majority current. The name saturation has been used because we cannot get minority current more than what is produced by thermal energy. In other words, ***I_S does not increase with increase in reverse bias***.

I_S is found to increase approximately 7 percent per 0°C rise in temperature both for Ge and Si. Since, $(1.07)^{10} = 2$, it means that reverse current ***doubles for every 10°C rise in temperature***. It is worth noting that reverse saturation current is also referred to as leakage current of the *P-N* junction diode.

With reverse bias, energy hill becomes too steep for majority carriers to go up the hill and cross over.

51.45. Reverse V/I Characteristic

As said earlier, the reverse saturation current is also referred to as ***leakage current*** of the *P-N* junction. Fig. 51.43 shows *V/I* characteristics of a reverse-biased *P-N* junction. It is seen that as reverse voltage is increased from zero, the reverse current quickly rises to its maximum or saturation value. Keeping temperature constant as the reverse voltage is increased, I_o is found to increase only slightly. This slight increase is due to the impurities on the surface of the semiconductor which behaves as a resistor and hence obeys Ohm's law. This gives rise to a very small current called ***surface leakage current***. Unlike the main leakage (or saturation) current, this surface leakage current is independent of temperature but depends on the magnitude of the reverse voltage.

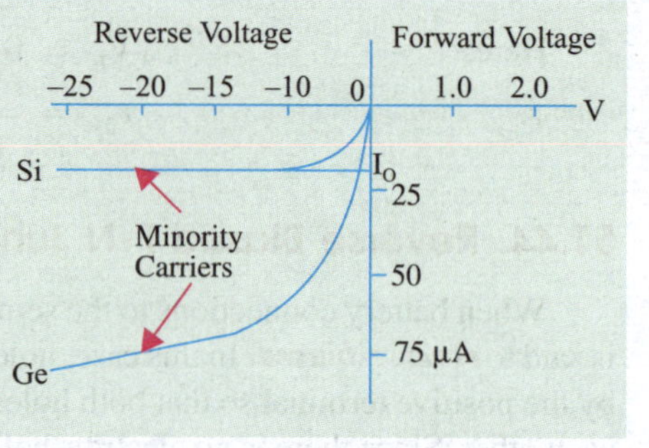

Fig. 51.43

A reverse-biased junction can be represented by a very large resistance. As seen from Fig. 51.43, in the case of Si, for a reverse voltage of about 15 V, $I_0 = 10\ \mu A$. Hence, reverse resistance is $R_R = 15\ V/10\ \mu A = 1.5\ M\Omega$

51.46. Hall Effect

If a specimen (whether of a metal or a semiconductor) carrying a current *I* is placed in a ***transverse*** magnetic field of flux density *B*, an electric field is developed along a direction perpendicular to both *B* and *I*. This phenomenon is known as ***Hall effect*** and is used for the following purposes :

1. to determine whether a semiconductor is of *N*-type or *P*-type;
2. to find carrier concentration;
3. to measure the conductivity of the material;
4. to find carrier mobility;
5. to detect and measure magnetic fields one million times smaller than that of earth with the help of Hall-effect magnetometers.

As shown in Fig. 51.44 (*a*), a current *I* is flowing through the semiconductor in the direction *MN* under the influence of an external applied electric field E_L. Obviously, electrons comprising this current move along *NM* with a velocity of v_e. Consider one such electron shown in the figure. The direction of the force exerted on it by the magnetic field *B* can be found by using Fleming's left-hand rule and is as shown*. The magnitude of force is $Be\,v_e$.

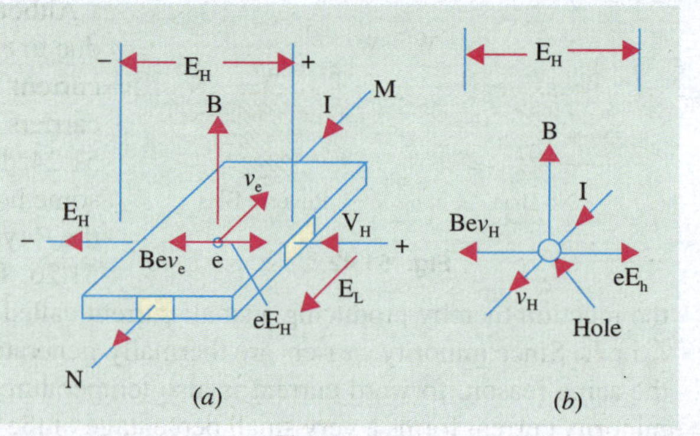

Fig. 51.44

Under the influence of this force, electrons will tend to crowd towards the left side of semiconductor. This collection of electrons to one side gives rise to electric potential difference V_H (called Hall voltage) and hence to an electric field E_H. This field begins to prevent additional electrons from arriving there. Force exerted by this field on the electrons is eE_H. Equilibrium is established when the two oppositely-directed magnetic and electric forces acting on the electron become equal in magnitude.

---

* While using the rule, the middle finger must point in the direction of conventional current *i.e.* along *MN* and not along the direction of electronic current *i.e.* along *NM*.

i.e. $Bev_e = eE_H$

or $Bv_e = E_H$ ...(*i*)

Also $v_e = \mu_e E_L$

$\therefore$ $B\mu_e E_L = E_H$ or $\mu_e B = E_H/E_L$

Hence, knowing E_L and B and measuring E_H, we can find electron mobility.

Now, current density $J = I/A = nev_e A/A = nev_e$

Substituting the value of υ_e from (*i*) above, we get

$$J = neE_H / B \quad \text{or} \quad n = JB / eE_H$$

This equation may be used to find electron density.

Now, Hall coefficient, $R_H = 1/\text{charge density} = 1/ne$ or $R_H = E_H/JB$

As shown in Art. 51.26. $\mu_e = 1/ne\ \rho = R_H/\rho$

where ρ is the resistivity of the semiconductor.

It may be noted that the above treatment is equally applicable to *P*-type semiconductors where current flow is made up of hole movement. If holes were to move in the same direction as that of the electrons in Fig. 51.44 (*a*), then polarity of Hall voltage E_H would be reversed as shown in Fig. 51.44 (*b*). In fact, it was the observation of the polarity of Hall voltage associated with *P*-type material which led to the concept of a positive hole.

Example 51.15. *The Hall-coefficient of a specimen of doped semiconductor is 3.66 × 10^{-4} m^3 C^{-1} and the resistivity of the specimen is 8.93 × 10^{-3} Ω–m. Determine the carrier mobility in $m^2 v^{-1} s^{-1}$.*

Solution. Using equation $\mu_e = R_H/\rho$

$$\mu_e = \frac{3.66\times10^{-4}}{8.93\times10^{-3}} = \mathbf{0.041\ m^2\ v^{-1}\ s^{-1}}$$

Example 51.16. *A sample of N-type semiconductor has a Hall coefficient of 160 cm^3/Coulomb. If is resistivity is 0.16 Ω-cm, estimate the electron mobility in the sample,*

Solution. Using the relation, $\mu_e = R_H/\rho$, we have

$$\mu_e = \frac{160}{0.16} = \mathbf{1000\ cm^2/Volt\text{-}Sec.}$$

Example 51.17. *A current of 50 A is passed through a metal strip, which is subjected to a magnetic flux of density 1.2 Wb/m^2. The magnetic field is directed at right angles to the current direction and the thickness of the strip in the direction of magnetic field is 0.5 mm. The Hall voltage is found to be 100 V. Calculate the number of conduction electrons per cubic metre in the metal.*

Solution. We know that $n = \dfrac{JB}{eE_H}$

Since $J = I/A = I/d.t$ where d is the distance between the two surfaces between the ends across which the elecric field is measured. The above equation can be rewritten as,

$$n = \frac{(I/A)\,B}{eE_H} = \frac{(I/d.t)\,B}{eV_H/d} = \frac{BI}{eV_H t} \qquad (\because\ E_H = V_H/d)$$

Substituting the values of V_H, t, B, I and e, we get

$$n = \frac{1.2\times50}{(1.6\times10^{-19})\times100\times(0.5\times10^{-3})} = \mathbf{7.51\times10^{21}/m^3}$$

Example 51.18. *An N-type semiconductor has a resistivity of 20 × 10^{-2} ohm-m. The mobility of electrons through a separate experiment was found to be 100 × 10^{-4} m^2 $v^{-1}s^{-1}$. Find the number of electron carriers per m^3.*

Solution. Using the equation, $\mu_e = 1/ne\ \rho$, we have

$$n = \frac{1}{e\mu_e \rho} = \frac{1}{1.6\times10^{-19}\times\left(100\times10^{-4}\right)\times\left(20\times10^{-2}\right)}$$

$$= \mathbf{3.1\times10^{21}/m^3}$$

Example 51.19. *The Hall-coefficient of a specimen of doped silicon is found to be 3.66×10^{-4} m^3C^{-1}; the resistivity of the specimen is 8.93×10^{-3} Ω-m. Find the mobility and density of charge carriers assuming single carrier conduction.* **(U.P.S.C. Engg. Services 2002)**

Solution. $R_H = 3.66\times10^{-4}\ m^3C;\ \rho = 8.93\times10^{-3}\ \Omega\text{-m}.$

$$\mu = \sigma R_H = \frac{1}{\rho} R_H = \frac{1}{8.93\times10^{-3}}\times3.66\times10^{-4} = \mathbf{0.041 m^2/V.s}$$

$$R_H = \frac{1}{ne} \text{ or } n = \frac{1}{R_H e} = \frac{1}{\left(3.66\times10^{-4}\right)\times\left(1.6\times10^{-19}\right)} = \mathbf{1.7\times10^{22}/m^3}$$

Example 51.20. *Resistivity of a sample semiconductor is 9mΩ-m. Its holes have mobility of 0.03 m^2/Vs. Calculate the value of Hall-coefficient of the sample.* **(U.P.S.C. Engg. Services 2002)**

Solution. $\rho = 9\ m\Omega\text{-m} = 9\times10^{-3}\ \Omega\text{-m},\ \mu_p = 0.03 m^2/V.s$

$$\sigma = \frac{1}{\rho} = \frac{1}{9\times10^{-3}} = \mathbf{111.1 S/m}$$

$$\mu_p = \sigma R_H$$

$$R_H = \frac{\mu_p}{\sigma} = \frac{0.03}{111.1} = \mathbf{2.7\times10^{-4}\ m^3C^{-1}}$$

Tutorial Problem No. 51.1

1. Compute the intrinsic conductivity of a specimen of pure silicon at room temperature given that $n_i = 1.4\times10^{16}\ m^{-3}$, $\mu_e = 0.145\ m^2/V\text{–}s$, $\mu_\mu = 0.05\ m^2/V\text{-}s$ and $e = 1.6\times10^{-19}$ C. Also, calculate the individual contributions from electrons and holes.
[0.437×10^{-3} S/m ; 0.325×10^{-3} S/m ; 0.112×10^{-3} S/m]

2. Find (*i*) conductivity and (*ii*) resistance of a bar of pure silicon of length 1 cm and cross-sectional area 1 mm^2 at 300°K. Given : $n_i = 1.5\times10^{16}$ per m^3, $\mu_e = 0.13\ m^2/V\text{-}s$, $\mu_h = 0.05\ m^2/V\text{-}s$ and $e = 1.6\times10^{-19}$ C. **[(*i*) 4.32×10^{-4} S/m (*ii*) 23.15 MΩ]**

3. A specimen of silicon is doped with acceptor impurity to a density of 10^{22} atoms per m^3. Given that $n_i = 1.4\times10^{16}$ per m^3, $\mu_e = 0.145\ m^2/V\text{–}s$, $\mu_h = 0.05\ m^2/V\text{-}s$, $e = 1.6\times10^{-19}$ C. All impurity atoms may be assumed to be ionized. Calculate the conductivity of extrinisic silicon. **[nearly 80 S/m]**

4. Calculate the conductivity of a specimen of pure *Si* at room temperature of 300°K for which $n_i = 1.5\times10^{16}\ m^{-3}$, $\mu_e = 0.13\ m^2/V\text{-}s$, $\mu_h = 0.05\ m^2/V\text{-}s$, $e = 1.6\times10^{-19}$ C.
The *Si* specimen is now doped 2 parts per 10^8 of a donor impurity. If there are 5×10^{28} *Si* atoms/m^3, calculate its conductivity. By what factor has the conductivity increased ?
[4.32×10^{-4} S/m ; 20.8 S/m ; ≃ 48,148]

5. Mobilities of electrons and holes in a sample of intrinsic germanium at room temperature are 0.36 m^2 /V-s and 0.17 m^2 / V-s respectively. If the electron and hole densities are each equal to 2.5×10^{19} per m^3, calculate the conductivity. **[2.12 S/m]*(Electronic-I, Bangalore Univ.)***

6. A p.d. of 10 V is applied longitudinally to a rectangular specimen of intrinsic germanium of length 2.5 cm, width 0.4 cm and thickness 0.15 cm. Calculate at room temperature

(*i*) electron and hole drift velocities;

(*ii*) the conductivity of intrinsic Ge if intrinsic carrier density is = $2.5\times10^{19}/m^3$;

(*iii*) the total current.

Given, $\mu_e = 0.38\ m^2\ V^{-1}s^{-1}$, $\mu_h = 0.18\ m^2V^{-1}s^{-1}$, $k = 1.38 \times 10^{-23}$ J-deg^{-1}, $e = 1.6 \times 10^{-19}$C

[(*i*) 152 m/s, 72 m/s (*ii*) 2.24 S/m (*iii*) 5.38 mA]

(Applied Electronics & Circuits, Grad. I.E.T.E. 1987)

7. The resistivity of a dopoed silicon material is $9 \times 10^{-3}\Omega$-m. The Hall coefficient is $3.6 \times 10^{-4}m^3$ Coulomb^{-1}. Assuring single carrier conduction, find the mobility and density of charge carriers. $e = 1.6 \times 10^{-19}$ Coulomb. **[(*i*) .04 m^2/V-S, (*ii*) 1.74 × $10^{22}/m^3$] *(U.P.S.C. Engg. Services 1994)***

8. Consider an intrinsic Ge bar with material time constant of 100 μsec across a cross-section of 1mm^2 and length 1 cm. One side of the bar is illuminated with 10^{15} photons/sec. Assume that each incident photon generate one electron hole pair and these are uniformly distributed throughout the bar. Find the bar resistance under constant excitation at room temperature.

(Electronic Devices and Circuits, Nagpur Univ. Summer, 2004)

9. Find the concentration of holes and electrons in a *p*-type germanium at 300K, if the conductivity is 100 per ohm-cm. Also find these values for *n*-type silicon, if the conductivity is 0.1 per ohm-cms. Given that

for germanium $n_i = 2.5 \times 10^{13}/cm^3$

M_n = 3800 cm^2/V–s, M_p = 1800 cm^2/V-s

for silicon, $n_i = 1.5 \times 10^{10}$ per cm^3

M_n =1300 cm^2/V-s and M_p = 500 cm^2/V–s ***(Electronics Engg; Bangalore Univ. 2004)***

10. A germanium P-N junction at 300 K has the following parameters; $N_D = 5 \times 10^{18}/cm^3$, $N_A = 6 \times 10^{16}/cm^3$, $n_i = 1.5 \times 10^{10}/cm^3$; Calculate the minority electron density in the P-region and the minority hole density in the N-region. ***(Electronics Engg. Bangalore Univ. 2003)***

OBJECTIVE TESTS – 51

1. The total energy of a revolving electron in an atom can
(*a*) have any value above zero
(*b*) never be positive
(*c*) never be negative
(*d*) not be calculated.

2. An atom is said to be ionised when any one of its orbiting electron
(*a*) Jumps from one orbit to another
(*b*) is raised to a higher orbit
(*c*) comes to the ground state
(*d*) is completely removed.

3. The maximum number of electrons which the M-shell of an atom can contain is
(*a*) 32 (*b*) 8
(*c*) 18 (*d*) 50.

4. Electronic distribution of an Si atom is
(*a*) 2, 10, 2 (*b*) 2, 8, 4
(c) 2, 7, 5 (*d*) 2, 4, 8.

5. Semiconductor materials have .... bonds.
(*a*) ionic (*b*) covalent
(*c*) mutual (*d*) metallic.

6. The maximum number of electrons which the valence shell of an atom can have is
(*a*) 6 (*b*) 8
(*c*) 18 (*d*) 2

7. Silicon has Z = 14. Its outermost orbit is
(*a*) partially filled
(*b*) half filled
(*c*) completely occupied
(*d*) empty

8. Major part of the current in an intrinsic semiconductor is due to
(*a*) conduction-band electrons
(*b*) valence-band electrons
(*c*) holes in the valence band
(*d*) thermally-generated electron.

9. Conduction electrons have more mobility than holes because they
(*a*) are lighter
(*b*) experience collisions less frequently
(*c*) have negative charge
(*d*) need less energy to move them.

10. Doping materials are called impurities because they

(a) decrease the number of charge carriers
(b) change the chemical properties of semiconductors
(c) make semiconductors less than 100 percent pure
(d) alter the crystal structures of the pure semiconductors.

11. Current flow in a semiconductor depends on the phenomenon of
(a) drift (b) diffusion
(c) recombination (d) all of the above.

12. The process of adding impurities to a pure semiconductor is called
(a) mixing (b) doping
(c) diffusing (d) refining.

13. The most widely used semiconucting material in electronic devices is
(a) germanium (b) sillicon
(c) copper (d) carbon

14. Electon-hole pairs are produced by
(a) recombination (b) thermal energy
(c) ionization (d) doping

15. Recombination takes place when
(a) an electron falls into a hole
(b) a positive and a negative ion bond together
(c) a valence electron becomes a conduction
(d) a crystal is formed

16. When a *P-N* junction is formed, diffusion current causes
(a) mixing of current carriers
(b) forward bias
(c) reverse bias
(d) barrier potential.

17. The leakage current of a *P-N* diode is caused by
(a) heat energy (b) chemical energy
(c) barrier potential (d) majority carriers.

18. Electronic components which are made of a semiconductor material are often called .... devices.
(a) solid-state (b) silicon
(c) germanium (d) intrinsic.

19. Any voltage that is connected across a *P-N* junction is called .... voltage.
(a) breakdown (b) barrier
(c) bias (d) reverse.

20. The area within a semiconductor diode where no mobile current carriers exist when it is formed is called .... region.
(a) depletion (b) saturation
(c) potential barrier (d) space charge.

21. The depletion region of a semiconductor diode is due to
(a) reverse biasing
(b) forward biasing
(c) crystal doping
(d) migration of mobile charge carriers.

22. If an intrinsic semiconductor is doped with a very small amount of boron, then in the extinsic semiconductor so formed, the number of electrons and holes will,
(a) decrease
(b) increase and decrease respectively
(c) increase
(d) decrease and increase respectively.

23. Two initially identical samples A and B of pure germanium are doped with donors to concentrations of 1×10^{20} and 3×10^{20} repectively. If the hole concentration in A is 9×10^{12}, then the hole concentration in B at the same temperature will be
(a) $3 \times 10^{12}\ m^{-3}$ (b) $7 \times 10^{12}\ m^{-3}$
(c) $11 \times 10^{12}\ m^{-3}$ (d) $27 \times 10^{12}\ m^{-3}$

24. Hall effect is observed in a specimen when it (metal or a semiconductor) is carrying current and is placed in a magnetic field. The resultant electric field inside the specimen will be in
(a) a direction normal to both current and magnetic field
(b) the direction of current
(c) a direction antiparallel to the magnetic field
(d) an arbitrary direction depending upon the conductivity of the specimen.

ANSWERS

1. (b)	2. (d)	3. (c)	4. (b)	5. (b)	6. (b)	7. (b)
8. (a)	9. (d)	10. (d)	11. (d)	12. (b)	13. (b)	14. (b)
15. (a)	16. (d)	17. (a)	18. (a)	19. (c)	20. (a)	21. (d)
22. (d)	23. (a)	24. (a)				

CHAPTER 52

P-N JUNCTION DIODE

Learning Objectives

- P-N Junction Diode
- Derivation of Junction Resistance
- Junction Breakdown
- Junction Capacitance
- Equivalent Circuit of P-N Junction
- Diode Fabrication
- Grown Junction
- Alloy Junction
- Diffused Junction
- Epitaxial Junction
- Point Contact Junction
- The Ideal Diode
- The Real Diode
- Diode Circuits with D.C. and A.C. Voltage Sources
- Diode Clipper and Clamper Circuits
- Clippers
- Some Clipping Circuits
- Clampers
- Summary of Clamping Circuits

Chemist, led the research for the molecular diode (In the semiconductor industry, called p-n junctions)

52.1. P-N Junction Diode

(a) Construction

It is a two-terminal device consisting of a P-N junction formed either in Ge or Si crystal. Its circuit symbol is shown in Fig. 52.1 (*a*). The *P*-and *N*-type regions are referred to as anode and cathode respectively. In Fig. 52.1 (*b*), arrow-head indicates the conventional direction of current flow when forward-biased. It is the same direction in which hole flow takes place.

Commercially available diodes usually have some means to indicate which lead is *P* and which lead is *N*. Standard notation consists of type numbers preceded by 'IN' such as IN 240 and IN 1250. Here, 240 and 1250 correspond to colour bands. Fig. 52.2 (*a*) shows typical diodes having a variety of physical structures whereas Fig. 52.2 (*b*) illustrates terminal identifications. Also refer to the picture of two commercial diodes shown in Fig 52.1(*c*).

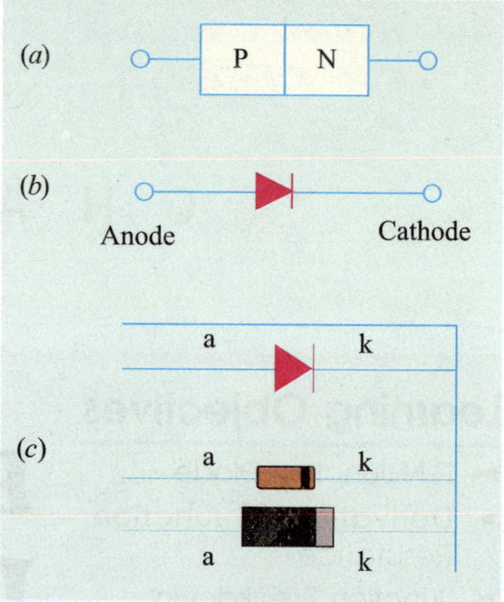

Fig. 52.1

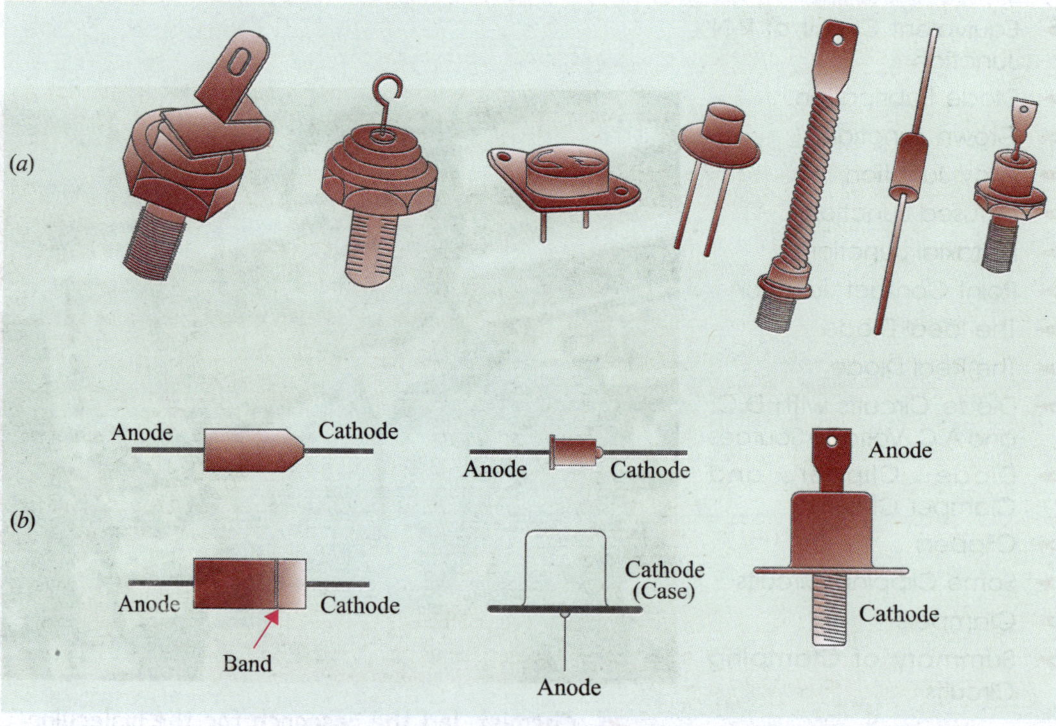

Fig. 52.2

The low-current diodes whose body is about 3 mm long can carry a forward current of about 100 mA, have saturation current of 5 μA at room temperature (25°C) and can withstand a reverse voltage of 75 V without breaking down. The medium-current diodes can pass a forward current of about 500 mA and can withstand a reverse voltage of 250 V. The high-current diodes or power diodes can pass a forward current of many amperes and can survive several hundred volts of reverse voltage.

(b) Diode Mounting

Low and medium-current diodes are usually mounted by soldering their leads to the connecting terminals. The heat generated by these diodes (when operating) is small enough to be carried away by air convection and conduction along the connecting leads. However, high-current stud-mounted diodes generate large amounts of heat for which air convection is totally inadequate. For cooling, they need heat sinks made of metals such as copper or aluminium which are good conductors of heat. The sink absorbs heat from the device and then transfers it to the surrounding air by convection and radiation since it has large surface area.

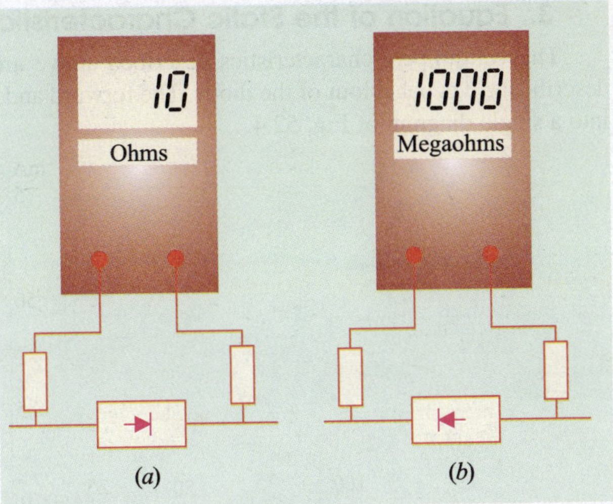

Fig. 52.3

(c) Working

A P-N junction diode is one-way device offering low resistance when forward-biased [Fig. 52.3 (*a*)] and behaving almost as an insulator when reverse-biased [Fig. 52.3 (*b*)]. Hence, such diodes are mostly used as rectifiers *i.e.* for converting alternating current into direct current.

(d) V/I Characteristic

Fig. 52.4 shows the static voltage-current characteristics for a low-power P-N junction diode.

1. Forward Characteristic

When the diode is forward-biased and the applied voltage is increased from zero, hardly any current flows through the device in the beginning. It is so because the external voltage is being opposed by the internal barrier voltage V_B whose value is 0.7 V for Si and 0.3 V for Ge. As soon as V_B is neutralized, current through the diode increases rapidly with increasing applied battery voltage. It is found that as little a voltage as 1.0 V produces a forward current of about 50 mA. A burnout is likely to occur if forward voltage is increased beyond a certain safe limit.

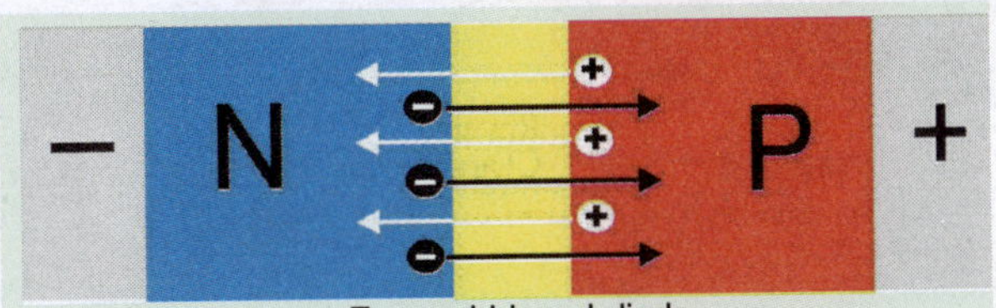

Forward biased diode.

Reverse biased diode.

2. Reverse Characteristic

When the diode is reverse-biased, majority carriers are blocked and only a small current (due to minority carriers) flows through the diode. As the reverse voltage is increased from zero, the reverse current very quickly reaches its maximum or saturation value I_0 which is also known as ***leakage current***. It is of the order of nanoamperes (nA) for Si and microamperes (μA) for Ge. The value of I_0 (or I_s) is independent of the applied reverse voltage but depends on (*a*) temperature, (*b*) degree of doping and (*c*) physical size of the junction.

As seen from Fig. 52.4, when reverse voltage exceeds a certain value called break-down voltage V_{BR} (or Zener voltage V_z), the leakage current suddenly and sharply increases, the curve indicating zero resistance at this point. Any further increase in voltage is likely to produce burnout unless protected by a current-limiting resistor.

When P-N junction diodes are employed primarily because of this breakdown property as voltage regulators, they are called Zener diodes (Art. 54.1).

3. Equation of the Static Characteristic

The volt-ampere characteristics described above are called ***static*** characteristics because they describe the d.c. behaviour of the diode. The forward and reverse characteristics have been combined into a single diagram of Fig. 52.4.

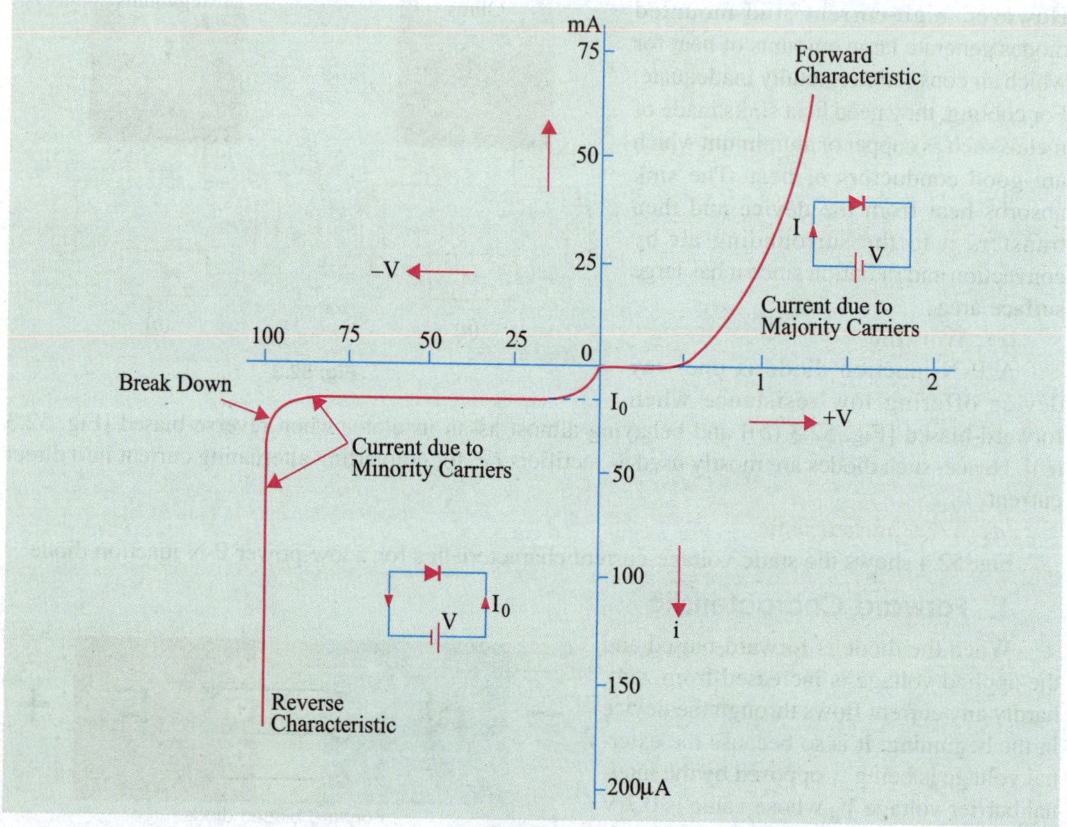

Fig. 52.4

These characteristics can be described by the analytical equation called ***Boltzmann diode equation*** given below :

$$I = I_0\left(e^{\frac{eV}{\eta kT}} - 1\right) \text{ ampere}$$

where I_0 = diode reverse saturation current

V = voltage across junction – positive for forward bias and negative for reverse bias.

k = Boltzmann constant = 1.38×10^{-23} J/°K

T = crystal temperature in °K

η = 1 – for germanium

= 2 – for silicon

Hence, the above diode equation becomes

$$I = I_0\left(e^{eV/kT} - 1\right). \quad \text{– for germanium}$$

$$I = I_0\left(e^{eV/2kT} - 1\right) \quad \text{– for silicon}$$

Now, e/k = 11,600 and putting $T/11{,}600 = V_T$, the above equation may be written as

$$I = I_0\left(e^{11{,}600V/\eta T} - 1\right) = I_0\left(e^{V/\eta V_T} - 1\right) \text{ ampere}$$

Now, at room temperature of (273 + 20) = 293°K, V_T = 293/11,600 = 0.025 V = 25 mV. Substituting the value of η, we have

$$I = I_0 (e^{40\,V} - 1) \qquad \text{– for Ge}$$
$$\cong I_0 e^{40\,V} \qquad \text{– if } V > 1 \text{ volt}$$
$$I = I_0 (e^{20\,V} - 1) \qquad \text{– for Si}$$
$$\cong I_0 e^{20\,V} \qquad \text{– if } V > 1 \text{ volt}$$

We may also write the above diode equation as under

$$I = I_0 \{e^{(V_f/\eta V_T)} - 1\} \qquad \text{– forward bias}$$
$$= I_0 \{e^{(-V_R/\eta V_T)} - 1\} \qquad \text{– reverse bias}$$

(e) Diode Parameters

The diode parameters of greatest interest are as under :

1. Bulk resistance (r_B)

It is the sum of the resistance values of the P-type and N-type semiconductor materials of which the diode is made of.

$$\therefore \qquad r_B = r_P + r_N \qquad \text{– Fig. 52.5}$$

Usually, it is very small. It is given by

$$r_B = (V_F - V_B)/I_F$$

It is the resistance offered by the diode well above the barrier voltage *i.e.* when current is large. Obviously, this resistance is offered in the forward direction.

2. Junction resistance (r_j)

Its value for forward-biased junction depends on the magnitude of forward *dc* current.

$$r_j = 25 \text{ mV}/I_F \text{ (mA)} \qquad \text{– for Ge}$$
$$= 50 \text{ mV}/I_F \text{ (mA)} \qquad \text{– for Si}$$

Obviously, it is a *variable* resistance.

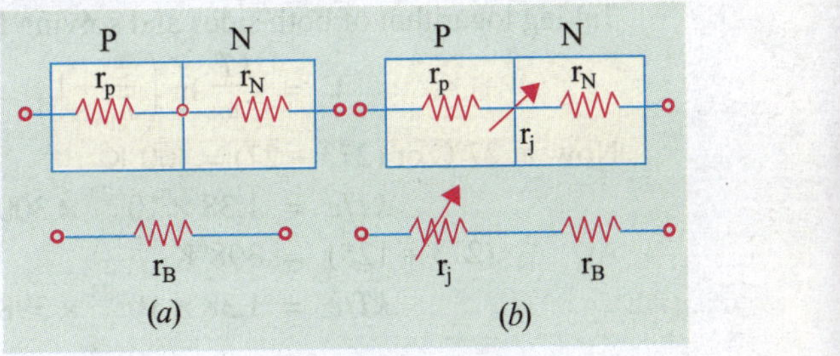

Fig. 52.5

3. Dynamic or ac resistance

$$r_{ac} \text{ or } r_d = r_B + r_j$$

For large values of forward current, r_j is negligible. Hence, $r_{ac} = r_B$. For small values of I_F, r_B is negligible as compared to r_j. Hence $r_{ac} = r_j$.

4. Forward voltage drop

It is given by the relation

$$\text{forward voltage drop} = \frac{\text{power dissipated}}{\text{forward dc current}}$$

5. Reverse saturation current (I_0). It has already been discussed in Art. 51.44.

6. Reverse breakdown voltage (V_{BR}). It is discussed in Art. 52.3.

7. Reverse dc resistance $R_R = \dfrac{\text{reverse voltage}}{\text{reverse current}}$

(f) Applications

The main applications of semiconductor diodes in modern electronic circuitry are as under :

1. As power or rectifier diodes. They convert ac current into dc current for dc power supplies of electronic circuits.
2. As signal diodes in communication circuits for modulation and demodulation of small signals.
3. As Zener diodes in voltage stabilizing circuits.
4. As varactor diodes–for use in voltage-controlled tuning circuits as may be found in radio and TV receivers. For this purpose, the diode is deliberately made to have a certain range of junction capacitance. The capacitance of the reverse-biased diode is given by $C = K/\sqrt{V_R}$ where V_R is the reverse voltage.

5. In logic circuits used in computers.

Example 52.1. *Using approximate Boltzmann's diode equation, find the change in forward bias for doubling the forward current of a germanium semiconductors at 290°K.*

(Basic Electronics, Osmania Univ. 1993)

Solution. The approximate Boltzmann's diode equation is given by $I = I_0 \exp(eV/kT)$

$\therefore \quad I_1 = I_0 \exp(eV_1/kT)$ and $I_2 = I_0 \exp(eV_2/kT)$

$$\therefore \quad \frac{I_2}{I_1} = \exp\left[\frac{e}{kT}(V_2 - V_1)\right] \qquad [\because \text{ for Ge } \eta = 1]$$

$$\text{or} \quad (V_2 - V_1) = \frac{kT}{e} \ln\left(\frac{I_2}{I_1}\right) = 25 \ln\left(\frac{I_2}{I_1}\right) \text{ mV}$$

Since $\quad I_2 = 2I_1 \quad$ or $\quad I_2 / I_1 = 2$

$\therefore \quad (V_2 - V_1) = 25 \ln 2 = 25 \times 0.693 =$ **17.3 mV**

Example 52.2. *A certain germanium P-N junction diode has a leakage current of 10^{-14} A at room temperature of 27°C and 10^{-9} A at 125°C. The diode is forward-biased with a constant-current source of 1 mA at room temperature. If current is assumed to remain constant, calculate the junction barrier voltage at room temperature and at 125°C.*

Solution. As given in Art. 52.1,

$I = I_0 \exp(eV/\eta kT - 1)$ Since, for Ge $\eta = 1 \therefore \quad (I/I_0) + 1 = \exp(eV/kT)$

Taking logarithm of both sides and solving for V, we get

$$V = \frac{kT}{e} \ln\left(\frac{I}{I_0} + 1\right)$$

Now, at 27°C or (273 + 27) = 300°K

$$kT/e = 1.38 \times 10^{-23} \times 300 / 1.6 \times 10^{-19} = 26 \text{ mV}$$

At $\quad (273 + 125) = 398°K,$

$$kT/e = 1.38 \times 10^{-23} \times 398 / (1.6 \times 10^{-19}) = 34 \text{ mV}$$

$$\text{Hence, at 27°C,} \quad V_B = 26 \ln\left(\frac{10^{-3}}{10^{-14}} + 1\right) = 660 \text{ mV} = \mathbf{0.66\ V}$$

$$\text{At 125°C,} \quad V_B = 34 \ln\left(\frac{10^{-3}}{10^{-9}} + 1\right) = 470 \text{ mV} = \mathbf{0.47\ V}$$

52.2. Derivation of Junction Resistance

Junction resistance is also known as ***incremental*** or ***dynamic resistance*** and is an important parameter particularly in connection with small-signal operations of the diode.

$$r_j = dV/dI \quad \text{or} \quad g_j = dI/dV$$

$$\text{Now,} \quad I = I_0\left(e^{V/\eta V_T} - 1\right) = I_0 e^{V/\eta V_T} - I_0$$

$$\therefore \quad g_j = \frac{dI}{dV} = \frac{I_0 e^{V/\eta V_T}}{\eta V_T} = \frac{I + I_0}{\eta V_T}$$

(*a*) Reverse bias

When reverse bias is greater than a few tenths of a volt *i.e.* when $|V/\eta V_T| \gg 1$, then g_j is extremely small so that r_j is very large. That high value is also represented by R_R.

(*b*) Forward bias

Again, for a forward bias greater than a few tenths of a volt, $I \gg I_0$, hence, $g_j = I/\eta V_T$ and $r_j = \eta V_T/I$.

Now, at room temperature of 293°K, $V_T = T/11{,}600 = 293/11{,}600 = 25$ mV. Also $\eta = 1$ for Ge and 2 for Si.

$$\therefore \quad r_j = 25 \text{ mV}/I \text{ mA} \quad \text{– for germanium}$$

$$= 50 \text{ mV}/I \text{ mA} \quad \text{– for silicon}$$

Example 52.3. *A silicon diode has a forward voltage drop of 1.2 V for a forward dc current of 100 mA. It has a reverse current of 1 μA for a reverse voltage of 10 V. Calculate*
(a) bulk and reverse resistance of the diode
(b) ac resistance at forward dc current of (i) 2.5 mA and (ii) 25 mA.

Solution. (*a*) $r_B = \dfrac{V_F - V_B}{I_F} = \dfrac{1.2\text{ V} - 0.7\text{ V}}{100\text{ mA}} = 5\ \Omega$

$R_R = V_R / I_R = 10\text{ V}/1\ \mu\text{A} = \mathbf{10\ M\Omega}$

(*b*) (*i*) $r_j = 50\text{ mV}/2.5\text{ mA} = 20\ \Omega \quad \therefore \quad r_{ac} = r_B + r_j = 5 + 20 = \mathbf{25\ \Omega}$

(*ii*) $r_j = 50\text{ mV}/25\text{ mA} = 2\ \Omega \quad \therefore \quad r_{ac} = 5 + 2 = \mathbf{7\ \Omega}$

Example 52.4. *Using analytical expression for diode current, calculate the dynamic slope resistance of a germanium diode at 290 K when forward biased at current of (i) 10 μA and (ii) 5 mA.* **(Electronics-I, Mysore Univ. 1992)**

Solution. $I = I_0 (e^{eV/\eta kT} - 1)$

For Ge $\eta = 1$

$\therefore \quad I = I_0 \{\exp (eV/kT) - 1\}$

for forward bias

$I >> I_0$

$\therefore \quad I \approx I_0 \exp (eV/kT)$

$\therefore \quad dI = \dfrac{e}{kT} I_0 [e^{eV/kT}]\, dV = \dfrac{e}{kT} I\, dV$

$\therefore \quad r_d = \dfrac{dV}{dI} = \dfrac{kT}{eI} = \dfrac{25 \times 10^{-3}}{I} - I$ in ampere

(*i*) Now, $I = 10\ \mu\text{A} = 10 \times 10^{-6} = 10^{-5}$ A

$r_d = 25 \times 10^{-3} / 10^{-5} = \mathbf{2500\ \Omega}$

(*ii*) Now, $I = 5\text{ mA} = 5 \times 10^{-3}$ A

$\therefore \quad r_d = 25 \times 10^{-3}/5 \times 10^{-3} = \mathbf{5\ \Omega}$

Example 52.5. *Find the current through the 20 Ω resistor shown in Fig. 52.6 (a). Each silicon diode has a barrier potential of 0.7 V and a dynamic resistance of 2 Ω. Use the diode equivalent circuit technique.* **(Semiconductor Devices, Gujarat BTE, 1993)**

Solution. In Fig. 52.6 (*b*) each diode has been replaced by its equivalent circuit. It is seen that diodes D_1 and D_3 are forward-biased by 5 V battery whereas D_2 and D_4 are reverse-biased. Hence, the current will flow from point *A* to *B*, then to *C* via 20 Ω resistance and then back to the negative terminal of the 5 V battery.

Fig. 52.6

The net voltage in the equivalent circuit is

$V_{net} = 5 - 0.7 - 0.7 = 3.6$ V

Total resistance seen by this net voltage is

$R_T = 2 + 20 + 2 = 24\ \Omega$

The circuit current $I = V_{net}/R_T = 3.6/24 = \mathbf{0.15\ A}$

52.3. Junction Breakdown

If the reverse bias applied to a *P-N* junction is increased, a point is reached when the junction breaks down and reverse current rises sharply to a value limited only by the external resistance connected in series with the junction (Fig. 52.7). This critical value of the voltage is known as **breakdown voltage** (V_{BR}). It is found that once breakdown has occured, very little further increase in voltage is required to increase the current to relatively high values. The junction itself offers almost zero resistance at this point.

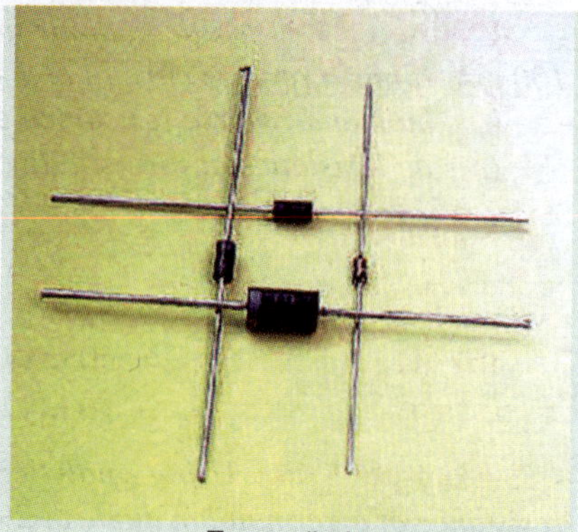

Zener diodes

The breakdown voltage depends on the width of the depletion region which, in turn, depends on the doping level.

The following two mechanism are responsible for breakdown under increasing reverse voltage:

1. Zener Breakdown

This form of breakdown occurs in junctions which, being heavily doped, have narrow depletion layers. The breakdown voltage sets up a very strong electric field (about 10^8 V/m) across this narrow layer. This field is strong enough to ***break or rupture the covalent bonds*** thereby generating electron-hole pairs. Even a small further increase in reverse voltage is capable of producing large number of current carriers. That is why the junction has very low resistance in the break-down region.

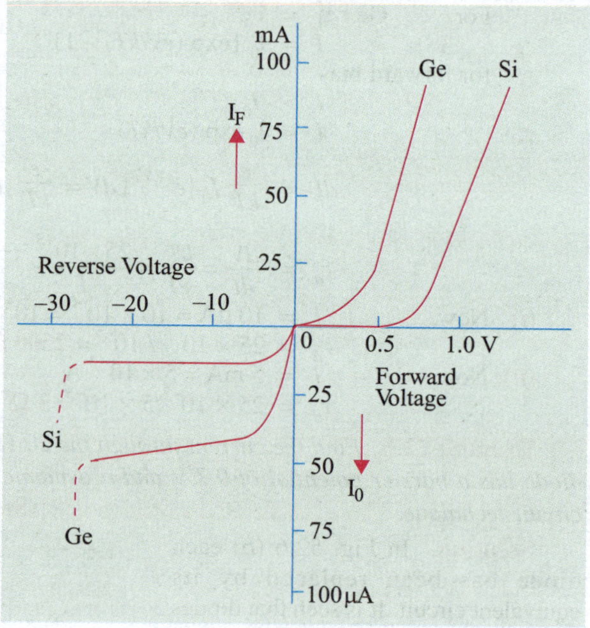

Fig. 52.7

2. Avalanche Breakdown

This form of breakdown occurs in junctions which being lightly-doped, have wide depletion layers where the electric field is not strong enough to produce Zener breakdown. Instead, the minority carriers (accelerated by this field) collide with the semiconductor atoms in the depletion region. Upon collision with valence electrons, covalent bonds are broken and electron-hole pairs are generated. These newly-generated charge carriers are also accelerated by the electric field resulting in more collisions and hence further production of charge carriers. This leads to an avalanche (or flood) of charge carriers and, consequently, to a very low reverse resistance. The two breakdown phenomena are shown in Fig. 52.7

52.4. Junction Capacitance

Capacitive effects are exhibited by *P-N* junctions when they are either forward-biased or reverse-biased.

(*a*) Transition Capacitance (C_T) or Space-charge Capacitance

When a *P-N* junction is reverse-biased, the depletion region acts like an insulator or as a dielectric material essential for making a capacitor. The *P-* type and *N*-type regions on either side have low resistance and act as the plates. We, therefore, have all the components necessary for making a parallel-plate capacitor. This junction capacitance is called ***transition or space charge*** capacitance (C_{pn} or C_T). It may be calculated by the usual formula $C = \in A/d$. Its typical value is 40 pF. Since thickness of depletion (or transition) layer depends on the amount of reverse bias, capacitance C_T can be controlled with the help of applied bias. This property of variable capacitance possessed by a reverse-biased *P-N* junction is used in the construction of a device known as ***varicap or varactor***.

This capacitance is voltage dependent as given by the relation

$$C_T = \frac{K}{(V_K + V_R)^n}$$

where V_K = knee voltage ; V_R = applied reverse voltage

K = constant depending on semiconductor material

$n = \frac{1}{2}$ – for alloy junction and $\frac{1}{3}$ –for diffused junction

The voltage-variance capacitance of a reverse-biased *P-N* junction is used in many circuits one of which is automatic frequency control (AFC) in an FM tuner. Other applications include self-balancing bridge circuits, special type of amplifiers known as ***parametric*** amplifiers and electronic tuners in TV.

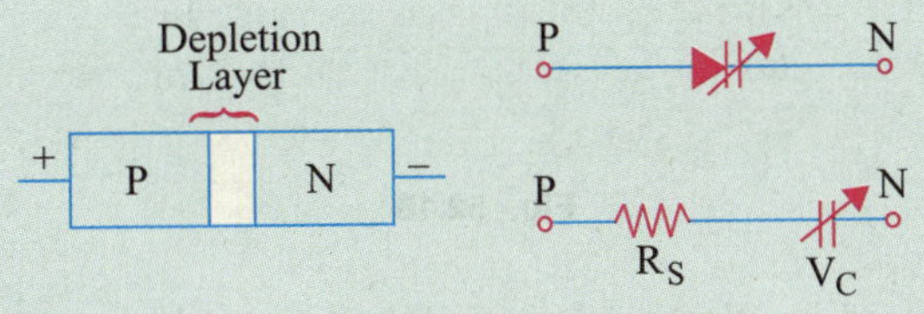

Fig. 52.8

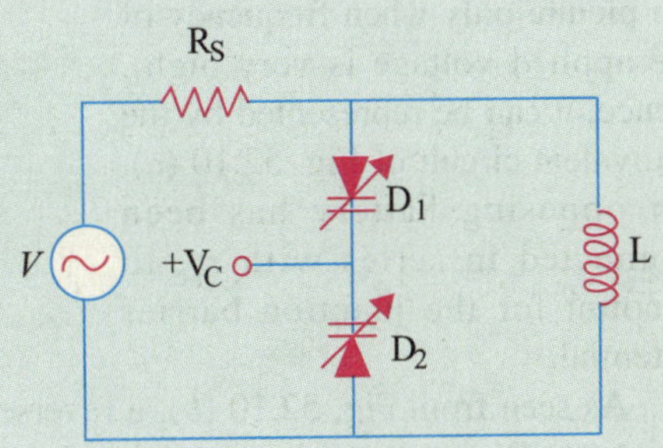

Fig. 52.9

When used in such a role, the diodes are called ***varactors, varicaps* or *voltacaps***. The symbol of these diodes when used in this role is shown in Fig. 52.8 along with its equivalent circuit.

When used in a resonant circuit, varactor acts as a variable capacitor and allows the resonant frequency to be adjusted by a variable voltage level. In Fig. 52.8, two varactors have been used to provide total variable capacitance in a parallel resonant circuit. Here V_C is a variable dc voltage that controls the reverse bias and hence the capacitance of the diodes.

(*b*) Diffusion or Storage Capacitance (C_D)

This capacitive effect is present when the junction is ***forward-biased***. It is called diffusion capacitance to account for the time delay in moving charges across the junction by diffusion process.* Due to this fact, this capacitance cannot be identified in terms of a dielectric and plates. It varies directly with the magnitude of forward current as explained below in more details.

Consider a forward-biased junction which is carrying a forward current I_F. Suppose the applied voltage is suddenly reversed, then I_F ceases suddenly but leaves lot of majority charge carriers in the depletion region. These charge carriers must get out of the region which, to their bad luck, becomes wider under the reverse bias. Hence, it is seen that when a forward-biased *P-N* junction is suddenly reverse-biased, a reverse current flows which is large initially but gradually decreases to the level of saturation current I_0. This effect can be likened to the discharging current of a capacitor and is, therefore, rightly represented by a capacitance called ***diffusion capacitance*** C_D. Since the number of charge carriers left in depletion layer is proportional to forward current, C_D is direcly proportional to I_F. Its typical value is 0.02 μF which is 500 times C_T.

The capacitance assumes great significance in the operation of devices which are required to switch rapidly from forward to reverse bias. If C_D is large, this switchover cannot be rapid. It will delay both the switch-on and the switch-off. This effect of C_D is variously known as ***recovery time*** or carrier storage.

* In the case of forward bias, the diode current is almost entirely due to diffusion (drift current being negligible).

If τ is mean lifetime of charge carriers, then a flow of charge Q yields a diode current of

$$I = \frac{Q}{\tau} = I_0\,(e^{V/\eta V_T} - 1) \cong I_0\, e^{V/\eta V_T}, \quad Q = I\tau = \tau\, I_0\, e^{V/\eta V_T}$$

$$\therefore \quad C_D = \frac{dQ}{dV} = \frac{\tau\, I_0}{\eta\, V_T}\, e^{V/\eta V_T} \cong \frac{\tau\, I}{\eta\, V_T}$$

52.5. Equivalent Circuit of a P-N Junction

We have seen from above that a forward-biased junction offers ac resistance r_{ac} and possesses diffusion capacitance C_D (which comes into the picture only when frequency of the applied voltage is very high). Hence, it can be represented by the equivalent circuit of Fig. 52.10 (*a*). An opposing battery has been connected in series with r_{ac} to account for the junction barrier potential.

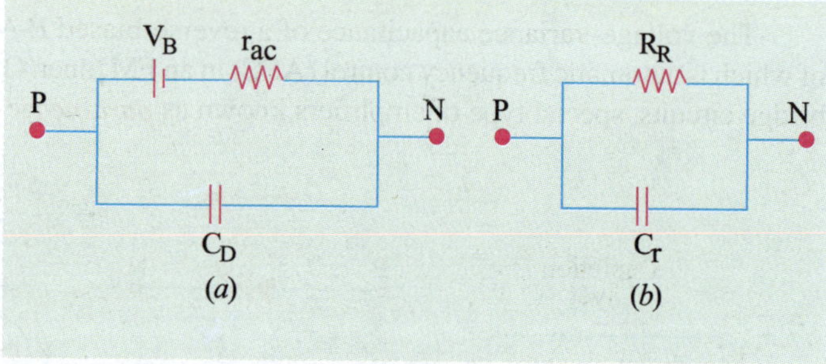

Fig. 52.10

As seen from Fig. 52.10 (*b*), a reverse-biased junction can be simply represented by a reverse resistance R_R connected in *parallel* with a capacitance C_T or C_{pn}.

Example 52.6. *Calculate the barrier potential for Si junction at (a) 100ºC and (b) 0ºC if its value at 25ºC is 0.7 V.*

Solution. As seen from Art. 51.40.

$$\Delta V = -0.002\,\Delta t = -0.002\,(t_2 - t_1)$$

(*a*) $\Delta t = (100 - 25) = 75ºC \qquad \therefore \Delta V = -0.002 \times 75 = -0.15\ V$

$\therefore \quad V_B$ at 100ºC $= 0.7 + (-0.15) =$ **0.55 V**

(*b*) $\Delta t = (0 - 25) = -25ºC \qquad \therefore \Delta V = -0.002 \times (-25) = 0.05\ V$

$\therefore \quad V_B$ at 0ºC $= 0.7 + 0.05 =$ **0.75 V**

Example 52.7. *A germanium diode draws 40 mA with a forward bias of 0.25 V. The junction is at room temperature of 293ºK. Calculate the reverse saturation current of the diode.*

Solution. $I = I_0\,(e^{40V} - 1)$ or $40 \times 10^{-3} = I_0\,(e^{40 \times 0.25} - 1)$

$\therefore \quad I_0 = 40 \times 10^{-3}/(22{,}027 - 1) =$ **1.82 μA**

Example 52.8. *Calculate forward current in Ge diode at 20ºC when forward voltage is 0.3 V. Compare this value with that after a temperature rise of 50ºC. Assume that reverse saturation current doubles for every 10ºC rise in temperature.* **(Electronics-I, Mysore Univ. 1991)**

Solution. According to Boltzmann's diode equation

$$I_{20} = I_0\,(e^{40V} - 1) = I_0\,(e^{40 \times 0.3} - 1) = I_0\,(e^{12} - 1) = 162{,}755\, I_0$$

At, $t = (20 + 50) = 70ºC$ or $T = 70 + 273 = 343ºK$

$V_T = 343/11{,}600 = 0.0296\ V \cong 0.03\ V$, hence, $V/\eta\, V_T = 0.3/1 \times 0.03 = 10$

The value of I_0 at 70ºC is $I_0' = 2^5 \times I_0 = 32\, I_0$ because it doubles for every 10ºC rise in temperature.

$\therefore \quad I_{70} = I_0'\,(e^{10} - 1) = 32\, I_0\,(22{,}026 - 1) = 704{,}800\, I_0$

It means that $I_{70} = (704{,}800 \div 162{,}755)\, I_{20} = 4.3\, I_{20}$ *i.e.* forward current has increased only 4.3 times whereas I_0 has increased 32 times for the same rise in temperature.

Example 52.9. *The capacitance of a varactor varies from 5 to 50 pF. Two such varactor diodes are used in the tuning circuit of Fig. 52.11. If L = 10 mH, determine the tuning range of the circuit.*

Solution. It should be noted that the two varactor capacitances are connected in series (Fig. 52.11).

$$C_{T(min)} = 5/2 = 2.5 \text{ pF} ; C_{T(max)} = 50/2 = 25 \text{ pF}$$

$$f_{r(max)} = 1/2\pi\sqrt{LC} = 1/2\pi\sqrt{10\times10^{-3}\times2.5\times10^{-12}}$$

$$= \textbf{1 MHz (approx)}$$

$$f_{r(min)} = 1/2\pi\sqrt{10\times10^{-3}\times25\times10^{-12}} = \textbf{318 KHz}$$

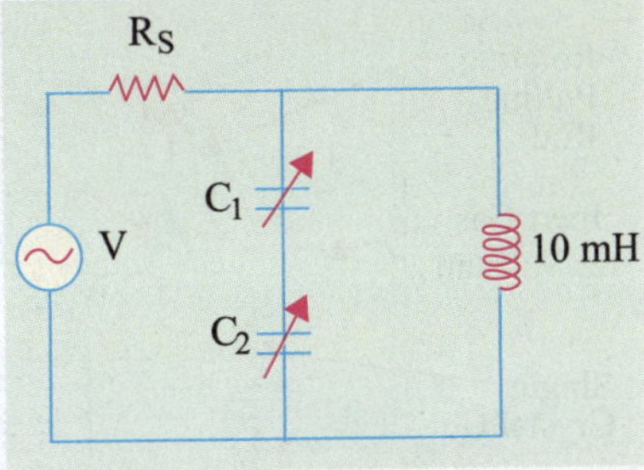

Fig. 52.11

52.6. Diode Fabrication

The electrical characteristics of a semiconductor diode depend on two factors *(i)* the material employed and *(ii)* the type of *P-N* junction used.

The two most commonly-used materials are germanium (Ge) and silicon (Si). Since Ge has higher electrical conduction than Si, it is often used in low- and medium-power diodes. On the other hand, Si has been found more suitable for high-power applications because it can be operated at higher temperatures than Ge. A new material called gallium-arsenide (GaAs) is found to combine desirable features of both Ge and Si and is finding ever-increasing use in many new applications.

Polycrystalline structures

The *P-N* junction may be produced by any one of the following methods :

1. grown junction
2. alloy junction
3. diffused junction
4. epitaxial growth
5. point contact junction.

The first step in the manufacture of any semiconductor device is to obtain the semiconductor material in an extremely pure form. The accepted impurity level is less than one part of impurity in one billion (10^9) parts of the semiconductor material. To begin with, the raw material is subjected to a series of chemical reactions and then to a zone refining process which employs induction heating to reduce the impurity level of the polycrystalline structure. Next, the Czochralski or floating zone technique is used to form single crystals of Ge or Si for fabrication of diodes. These crystals are then cut into wafers as thin as 0.025 mm (nearly one fourth the thickness of this paper).

Now, we will briefly discuss the four basic processes commonly used in the manufacture of semiconductor diodes.

52.7. Grown Junction

Such junctions are produced by employing either the Czochralski or floating zone technique. The apparatus used for Czochralski technique is shown in Fig. 52.12. A single crystal seed of the desired impurity level is immersed in the molten semiconductor material contained in a crucible. Then, it is gradually withdrawn while the shaft holding the seed is slowly turning. When crystal is being pulled out, impurities of *P-* type and *N*-type are alternately added to produce a *P-N* junction. This large area crystal is then cut into a large number of smaller-area diodes.

52.8. Alloy Junction

The alloy process produces junction diodes that have high PIV and current ratings but which have large junction capacitance due to their large junction area.

In this process, a tiny dot (or pellet) of indium (or any other *P*-type impurity) is placed on the

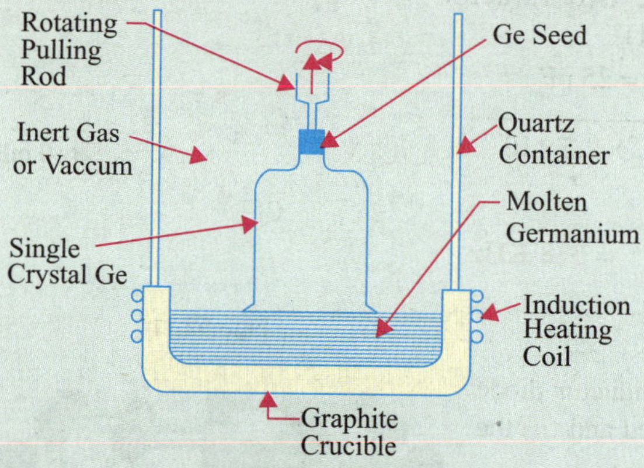

Fig. 52.12

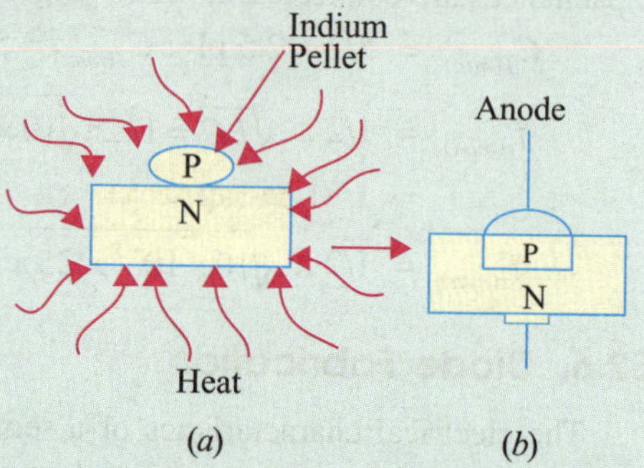

Fig. 52.13

surface of an *N*-type silicon wafer and the two are heated well above the melting temperature of indium (about 150ºC) as shown in Fig. 52.13 *(a)*. Consequently, indium melts and dissolves some of the silicon. The temperature is then lowered and silicon refreezes to form a single crystal having a *P-N* junction as shown in Fig. 52.13 *(b)*.

52.9. Diffused Junction

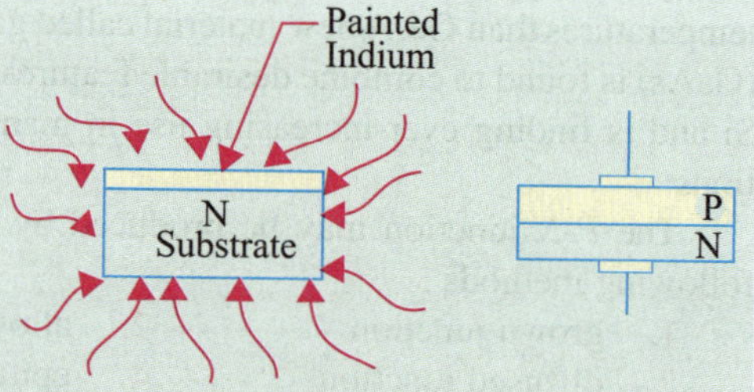

Fig. 52.14

The diffusion process employs either solid or gaseous diffusion. This process takes more time than alloy process but is relatively cheaper and more accurately controllable. In this process, particles from an area of high concentration drift to surrounding region of lesser concentration.

***(i)* Solid Diffusion**

The solid diffusion process starts with the 'painting' of a *P*-type impurity (say, indium) on an *N*-type substrate and heating the two until the impurity diffuses a short distance into the substrate to form *P*-type layer (Fig. 52.14).

***(ii)* Gaseous Diffusion**

In the gaseous diffusion process, an *N*-type material is heated in a chamber containing a high concentration of an acceptor impurity in vapour form (Fig. 52.15). Some of the acceptor atoms are diffused (or absorbed) into the *N*-type substrate to form the *P*-type layer thus creating a *P-N* junction. By exposing only part of the *N*-type material during the diffusion process (the remainder being covered by a thin coating of SiO_2), the size of the *P*-region can be controlled. Finally, metal contacts are electroplated on the surface of each region for connecting the leads.

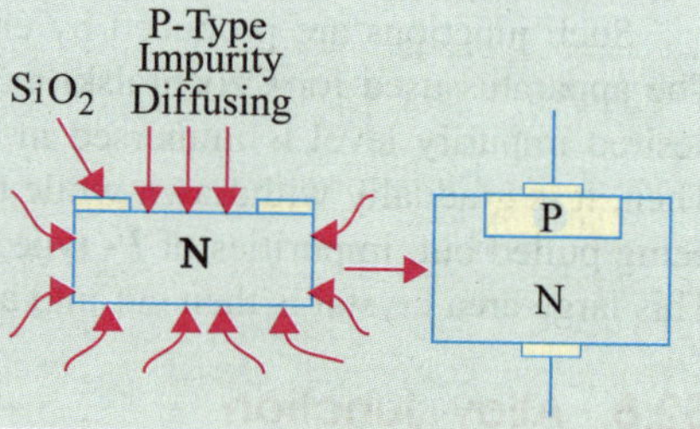

Fig. 52.15

The diffusion technique enables simultaneous fabrication of many hundreds of diodes on one small disc of a semiconductor material. That is why it is the most frequently-used technique not only for the manufacture of semi-conductor diodes but also for the production of transistors and integrated circuits etc.

52.10. Epitaxial Junction

This junction differs from the diffusion junction only in the manner in which the junction is fabricated. Such junctions are grown on top of an *N*-type wafer in a high temperature chamber. The growth proceeds atom by atom and hence is exactly similar to the crystal lattice of the wafer on which it is grown. Such junctions have the advantage of low resistance.

52.11. Point Contact Junction

It consists of an *N*-type germanium or silicon wafer about 1.25 mm square by 0.5 mm thick, one face of which is soldered to a metal base by radio-frequency heating as shown in Fig. 52.16 (*a*). The other face has a phosphor bronze (or tungsten) spring (called a cat's whisker) pressed against it. The *P-N* junction is formed by passing a large current for a second or two through the wire while the crystal face with wire point is kept positive. The heat so produced drives away some of the electrons from the atoms in the small region around the point of contact thereby leaving holes behind. This

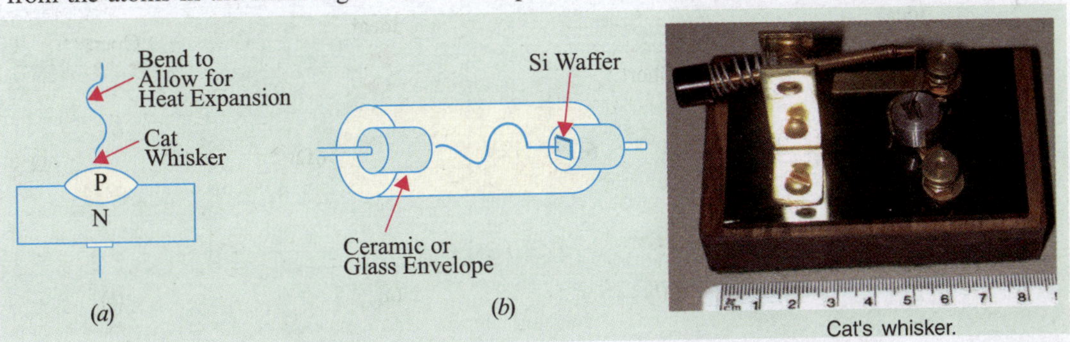

Cat's whisker.

Fig. 52.16

small region of the *N*-type material is, consequently, converted into *P*-type material as shown in Fig. 52.16 (*a*). The small area of the *P-N* junction results in very low junction capacitance as mentioned earlier.

52.12. The Ideal Diode

There is no such thing as an ***ideal*** diode or perfect diode. The existence of such a diode is visualized simply as an aid in analysing the diode circuits.

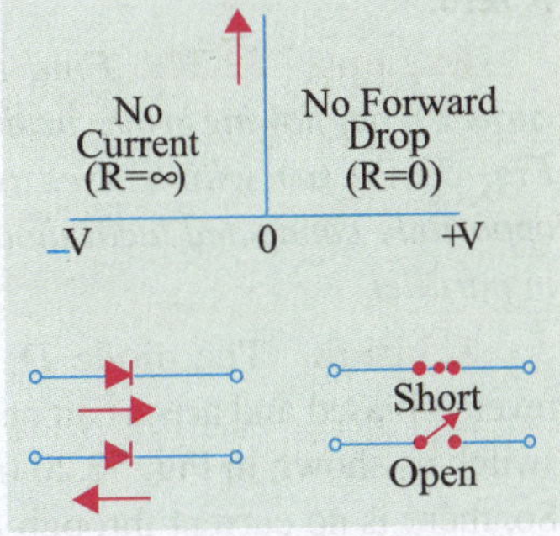

Fig. 52.17

An ideal diode may be defined as a two-terminal device which

Ideal diode

(*a*) conducts with zero resistance when forward-biased, and

(*b*) appears as an infinite resistance when reverse-biased.

In other words, such a device acts as a short-circuit in the forward direction and as an open-circuit in the reverse direction as shown in Fig. 52.17.

Also, in the forward direction, there is no voltage drop (even though current is there) since a short has zero resistance. On the other hand, there is no reverse current because reverse resistance is infinite.

It is helpful to think of an ideal diode as a bistable switch which is closed in the forward direction and open in the reverse direction. Hence, it has two stable states : ON or OFF.

Example 52.10. *Calculate the circuit current and power dissipated in the (a) ideal diode (b) 6 Ω resistor of the circuit shown in Fig. 52.18 (a).*

Solution. (*a*) The diode is an ideal one and is forward-biased. Hence, it can be replaced by a short (closed switch) as shown in Fig. 52.18 (*b*). The circuit current, as given by Ohm's law, is

$$I = 12/6 = \mathbf{2\ A}$$

Since there is no voltage drop across the diode, power consumed by it is **zero.**

As we know, there is no power when either the voltage or current is zero. In the forward direction, there is current but no voltage drop, hence power dissipated by the ideal diode is zero. In the reverse direction, there is voltage but no current. Hence, power dissipated by the diode is again zero. In fact, an ideal diode never dissipates any power.

(*b*) power consumed by 6 Ω resistor = $2^2 \times 6$ = **24 W.**

Example 52.11. *Calculate the current in the circuit of Fig. 52.19 (a).*

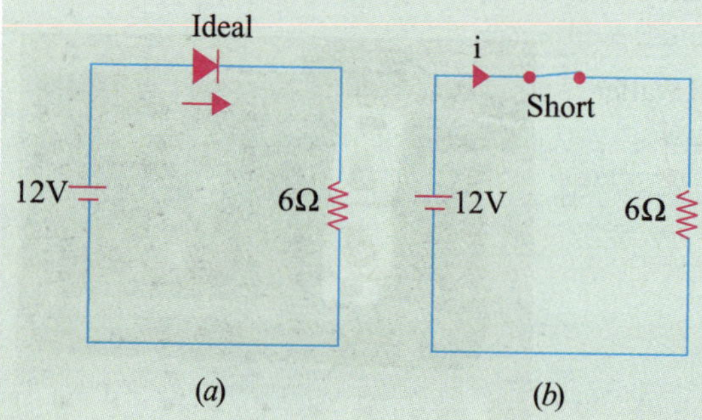

Fig. 52.18

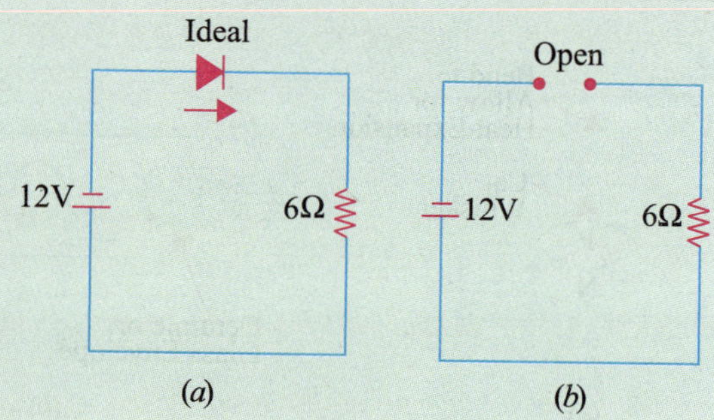

Fig. 52.19

Solution. As seen from the polarity of the battery terminals, the diode is reverse-biased. Hence, it acts like an open switch as shown in Fig. 52.18 (*b*). Obviously, current in such a circuit is zero.

Example 52.12. *Find the current, if any flowing in the circuit of Fig. 52.20 (a) which uses two oppositely-connected ideal diodes in parallel.*

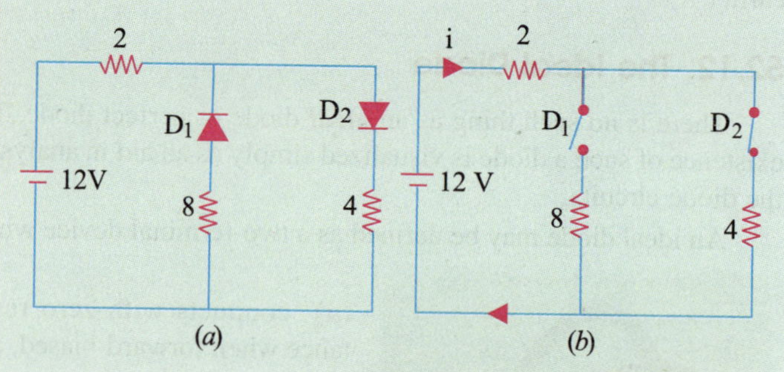

Fig. 52.20

Solution. The diode D_1 is reverse-biased and acts as an open switch as shown in Fig. 52.20 (*b*). So, there is no current through D_1 and the 8 Ω resistor.

However, D_2 is forward-biased and acts like a short-current or closed switch. The current drawn is $I = 12/(2 + 4) = \mathbf{2\ A}$.

Example. 52.13. *A sinusoidal voltage of peak value 50 V is applied to a diode as shown in Fig. 52.21. Sketch the waveform of voltage V_O treating the diode as an ideal one.*

Solution. First, consider the positive half-cycle of the input signal. The diode acts as a short in the forward direction and the moment V_{in} exceeds battery voltage of 10 V, current will start flowing through the circuit. The value of V_O will remain steady at 10 V, the balance of 40 V dropping across 5 K resistance. It is seen that value of V_O is set by the battery voltage.

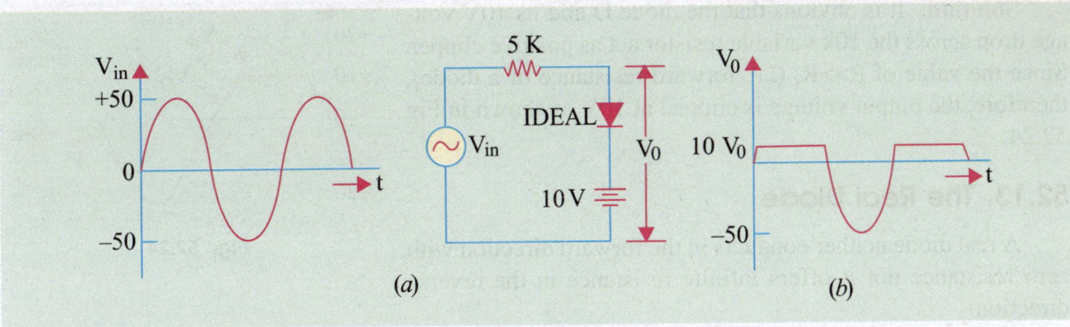

Fig. 52.21

Next, consider the negative input half-cycle. Now, the diode acts like an open switch. Consequently, there is no circuit current and thus no voltage drop across 5 K resistor. Hence, V_0 equals source voltage of peak value 50 V.

It is interesting to note that the above circuit acts as a ***positive*** clipper with a clipping level of 10 V *i.e.* equal to battery voltage. All positive parts of V_{in} above 10 V have been clipped off [Fig. 52.21 (*b*)].

Example 52.14. *Sketch the waveform of the output voltage V_O of the circuit shown in Fig. 52.22 (a)], considering diodes as ideal diode.*

Solution. It is obvious that diode D_1 and its 10 V battery act as positive clipper with positive clipping level at +10 V. Similarly, D_2 and its 10 V battery act as negative clipper with a clipping level at –10 V.

It is clear that output voltage V_0 would be clipped during both the positive and negative half-cycles as shown in Fig. 52.22 (*b*).

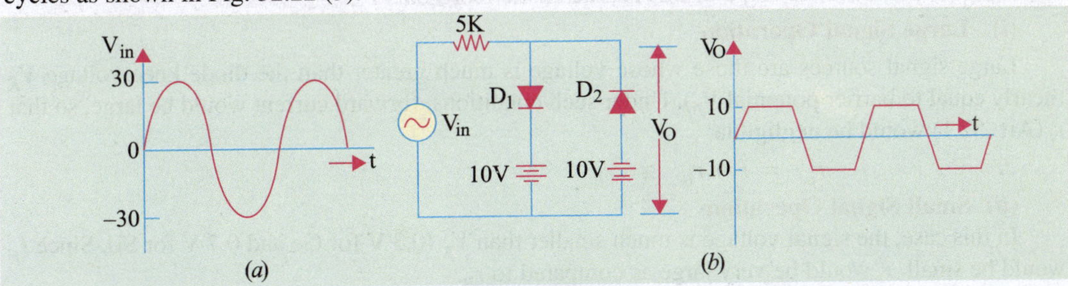

Fig. 52.22

Incidentally, the above example represents one way of obtaining an approximate square wave-form from a sine wave. In fact, clipping takes place at +10 V and –10 V regardless of the shape of the input wave *i.e.* whether it is triangular or sawtooth etc.

Example 52.15. *In Fig 52.23 draw input and output waveforms to scale. Consider diode forward resistance of 50 ohms.* **(Nagpur Univ. Winter 2003)**

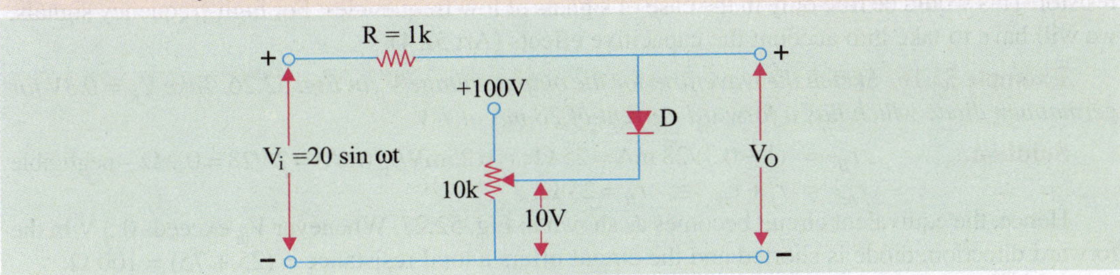

Fig. 52.23

Solution. It is obvious that the diode D and its 10V voltage drop across the 10k variable resistor act as positive clipper. Since the value of R>>R_f (*i.e.* forward resistance of a diode), therefore, the output voltage is clipped at 10V as shown in Fig 52.24.

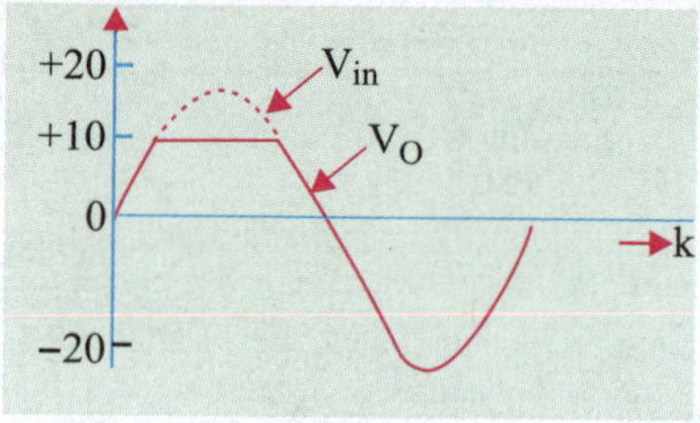

Fig. 52.24

52.13. The Real Diode

A real diode neither conducts in the forward direction with zero resistance nor it offers infinite resistance in the reverse direction.

(*a*) Forward Direction

In this case, we have to take two factors into account. One is that forward current does not start flowing until the voltage applied to the diode exceeds its threshold or knee voltage V_K (0.3 V for Ge and 0.7 V for Si). Hence, a real diode is shown as equivalent to an ideal diode in series with a small ***oppositely-connected*** battery of e.m.f. V_K as shown in Fig. 52.25 (*a*).

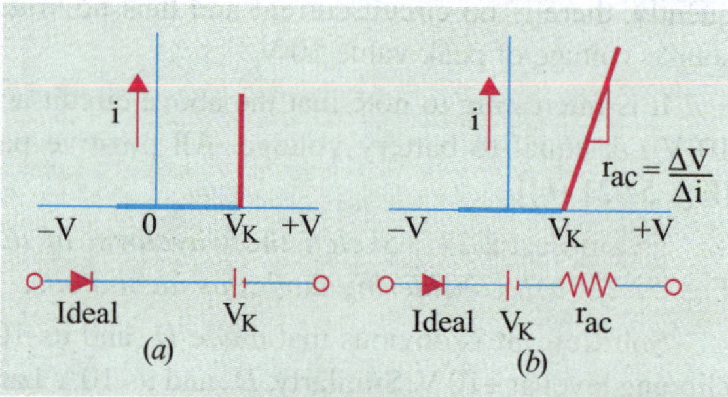

Fig. 52.25

The second factor to be considered is the forward dynamic or ac resistance (r_{ac}) offered by the circuit. So far, we considered this resistance to be zero implying that forward characteristic is a straight vertical line [Fig. 52.25 (*a*)]. If we take r_{ac} into account, the forward characteristic becomes as shown in Fig. 52.25 (*b*). Here, the reciprocal of the slope of this characteristic represents r_{ac}.

(*i*) Large Signal Operation

Large signal sources are those whose voltage is much greater than the diode knee voltage V_K (nearly equal to barrier potential V_B). Under such conditions, forward current would be large, so that r_j (Art. 52.1) would be negligible.

$$\therefore \quad r_{ac} = r_j + r_B \cong r_B$$

(*ii*) Small Signal Operation

In this case, the signal voltage is much smaller than V_K (0.3 V for Ge and 0.7 V for Si). Since I_F would be small, r_j would be very large as compared to r_B.

$$\therefore \quad r_{ac} = r_j + r_B \cong r_j$$

(*b*) Reverse Direction

An actual or real diode does not have infinite resistance in the reverse direction because it will always have some reverse saturation current prior to breakdown. For example, if with a V_R = 50 V, I_R is 10 μA, then $R_R = 5 \times 10^6\ \Omega$ = 5 M. Silicon diodes have reverse resistance of many thousands of megaohms. Hence, an actual diode in the reverse direction can be thought of as equivalent to a high resistor. This would be true only in the case of signals of low frequencies. For high-frequency signals, we will have to take into account the capacitive effects (Art 52.4).

Example 52.16. *Sketch the waveform for the output voltage V_0 in Fig. 52.26. Take V_K = 0.3V for germanium diode which has a forward current of 28 mA at 1 V.*

Solution. r_B = (1 − 0.3)/28 mA = 25 Ω ; r_j = 25mV/I_F (mA) = 25/28 = 0.9 Ω − negligible

$$\therefore \quad r_{ac} = r_j + r_B \cong r_B = 25\ \Omega.$$

Hence, the equivalent circuit becomes as shown in Fig. 52.27. Whenever V_{in} exceeds 0.3 V in the forward direction, diode is shorted and the circuit offers a total resistance = (25 + 75) = 100 Ω.

The peak voltage = 20 – 0.3 = 19.7 V. Hence, peak value of forward current is = 19.7/100 = 197 mA.

$\therefore$ peak value of output voltage $= 197 \times 10^{-3} \times 75 = 14.77$ V

Hence, the peak value of the half-wave rectified voltage is 14.77 V as shown in Fig. 52.27 (*b*). During the negative half-cycle of the applied voltage, diode acts as an open switch so that $V_0 = 0$. The waveform sketch of V_0 is shown in Fig. 52.27 (*b*).

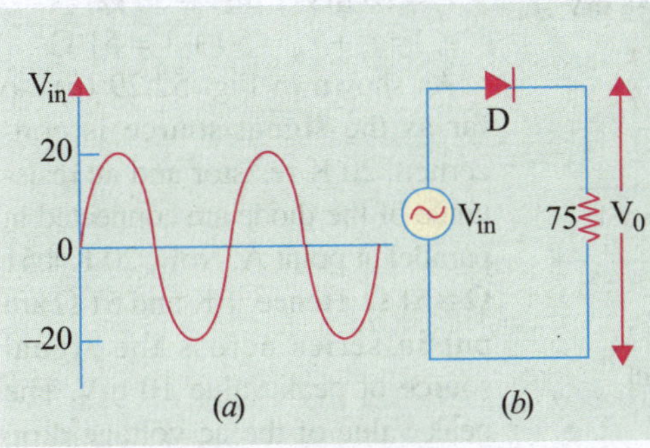

Fig. 52.26

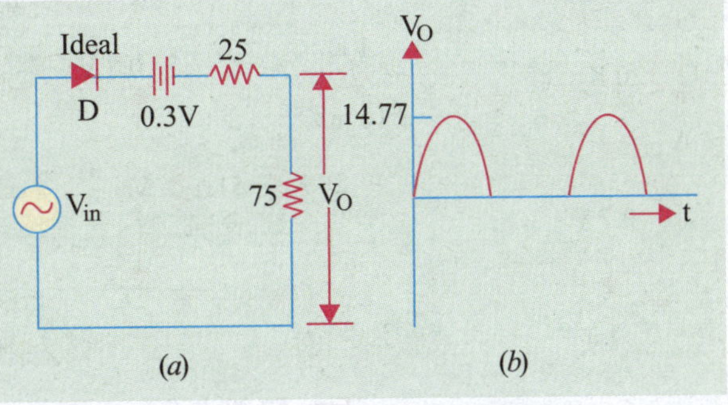

Fig. 52.27

52.14. Diode Circuits with DC and AC Voltage Sources

We will often come across diode and transistor circuits which will contain both dc and ac voltage sources. Such circuits can be easily analysed by using Superposition Theorem (Art 2.17). We will first draw the dc equivalent circuit while neglecting ac-sources and find the required current and voltage values. Next, we will draw the ac equivalent circuit while neglecting the dc sources and again find the voltage and current values. Finally, we will superimpose the two sets of values to get the final result.

While drawing the equivalent circuits, following points must be remembered :

1. Direct current cannot flow 'through' a capacitor. Hence, all capacitors look ***like an open switch to a dc source.***

2. Usually, capacitors used in most circuits are large enough to ***look like a short to an ac source*** particularly one of very high frequency. Similarly, dc batteries would also act as short circuits unless they have very high internal resistances.

Example 52.17. *The silicon diode shown in Fig. 52.28 has a bulk resistance of 1 Ω. The frequency of the 10 mV (peak) signal is so high that the reactance of the coupling, capacitor may be taken as zero. Sketch the approximate waveform of the total voltage 'V' across the diode.*

(Basic Electronics, Punjab Univ.)

Solution. We will apply Superposition theorem to find *V*. First, voltage drop due to dc source would be found and then it would be added to the drop due to the ac source.

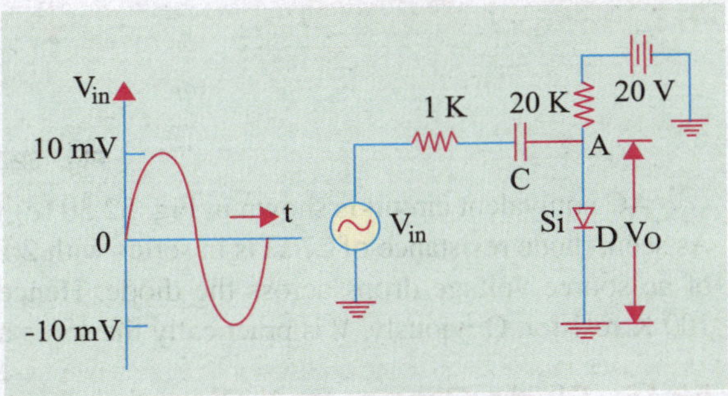

Fig. 52.28

(*i*) DC Equivalent Circuit

It is seen from Fig. 52.28 that the circuit to the left of point *A* is 'open' to the dc source of 20 V because of capacitor *C*. Hence, the equivalent dc circuit is as shown in Fig. 52.29 (*a*). As seen, diode is forward-biased by the battery. Hence, only 0.7 V dc appear across the diode.

The dc current $I = (20 - 0.7)/20 \text{ K} \cong 1 \text{ mA}$.

(ii) AC Equivalent Circuit

Here, the capacitor C and the 20-V battery would be treated as shorts thereby giving us the ac equivalent circuit of Fig. 52.29 (b).

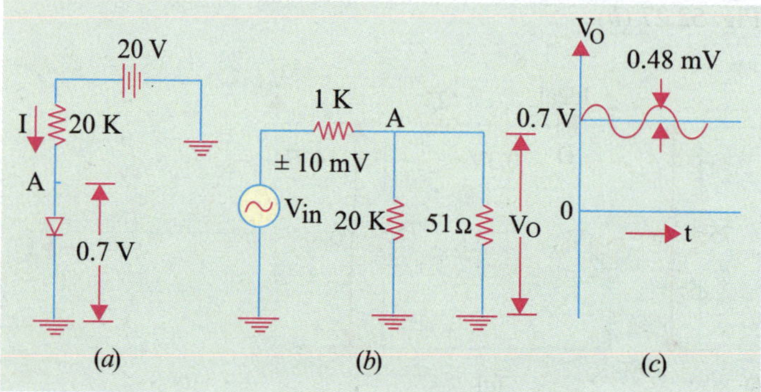

Fig. 52.29

Since, it is a silicon diode

$r_j = 50 \text{ mV}/1 \text{ mA} \cong 50\ \Omega$

$r_{ac} = r_j + r_B = 50 + 1 = 51\ \Omega$

As shown in Fig. 52.29 (b), so far as the signal source is concerned, 20 K resistor and ac resistance of the diode are connected in parallel at point A. Now, 20 K || 51 Ω ≅ 51 Ω. Hence, 1 K and 51 Ω are put in series across the signal source of peak value 10 mV. The peak value of the ac voltage drop over 51 Ω resistance is

$$= \frac{10 \times 10^{-3}}{1{,}051} \times \frac{20 \times 10^3}{20{,}051} \times 51 = \mathbf{0.48\ mV}$$

The total drop across the diode is the sum of the ac and dc drops. The combined voltage waveform is shown in Fig. 52.29 (c). It consists of a dc voltage of 0.7 V over which rides an ac voltage of peak value ± 0.48 mV.

Example 52.18. *The Ge diode shown in Fig. 52.30 has a bulk resistance of 2 Ω. The two capacitors offer negligible reactance to the a.c. signal. Sketch the waveform of the output voltage.*

Solution. As seen from Fig. 52.30 (b), no dc current can pass through 100 K resistor because it is blocked by C_2. Hence, no dc voltage appears across 100-K resistor. The battery current passes through 10 K resistor, through diode, through 20 K resistor and finally to ground.

$$I = (30 - 0.3)/(10 + 20) \times 10^3 = 0.99 \text{ mA} \cong 1 \text{ mA}$$

$$r_{ac} = r_j + r_B = (25 \text{ mV}/1 \text{ mA}) + 2 = \mathbf{27\ \Omega}$$

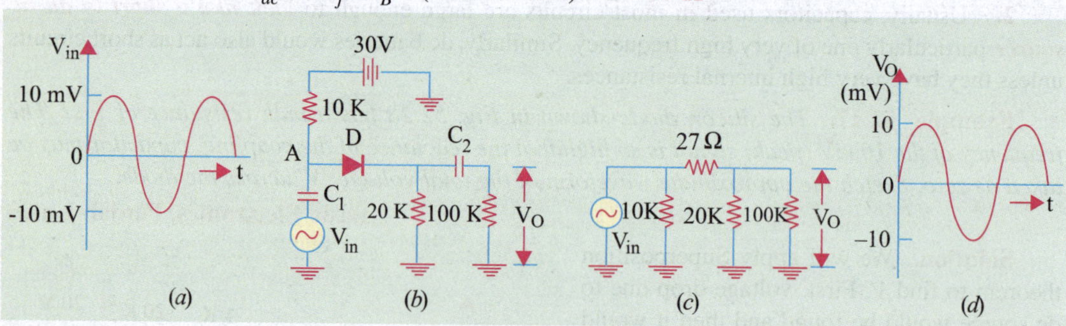

Fig. 52.30

AC equivalent circuit is shown in Fig. 52.30 (c) where capacitors and battery have been shorted. As seen, diode resistance of 27 Ω is in series with 20 K || 100 K. It means that only a very small part of ac source voltage drops across the diode. Hence, almost all a.c.source voltage appears across 100 K resistor. Obviously, V is practically the same as ac source voltage as shown in Fig. 52.30 (d).

52.15. Diode Clipper and Clamper Circuits

These are diode waveshaping circuits *i.e.* circuits meant to control the **shape** of the voltage and current waveforms to suit various purposes. Each performs the waveshaping function indicated by its name. The output of the clipping circuit appears as if a portion of the input signal were ***clipped off***. But clamper circuits simply clams (*i.e.* lift up or down) the input signal to a different dc level.

52.16. Clippers

A clipping circuit requires a minimum of two components *i.e.* a diode and a resistor. Often, dc battery is also used to fix the clipping level. The input waveform can be clipped at different levels by simply changing the battery voltage and by interchanging the position of various elements. We will use an ideal diode which acts like a closed switch when forward-biased and as an open switch when reverse-biased.

Such circuit are used in radars and digital computers etc. when it is desired to remove signal voltages above or below a specified voltage level. Another application is in radio receivers for communication circuits where noise pulses that rise well above the signal amplitude are clipped down to the desired level.

Example 52.19. *For the simple parallel clipper of Fig. 52.31, find the shape of the output voltage V_O across the diode if the input sine wave signal is as shown in Fig. 52.31 (a). What will happen when diode and resistor are inter-changed ?*

Solution. When positive half-cycle of the signal voltage is applied to the clipper *i.e.* when *A* is positive with respect to *B*, the diode *D* is reverse-biased. Hence, it acts as an open switch. Consequently, the entire input voltage appears across it.

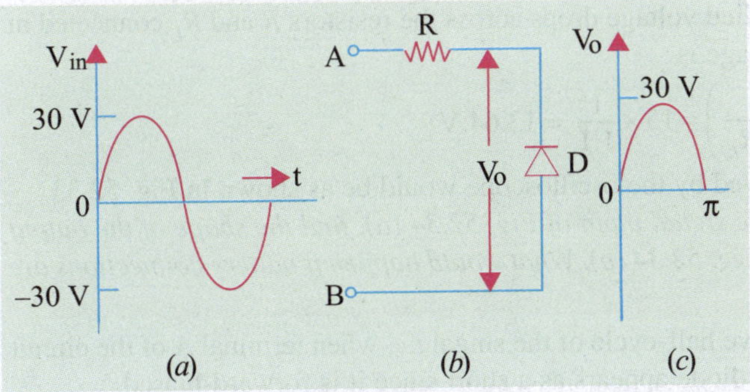

Fig. 52.31

During the negative half-cycle of the signal voltage when circuit terminal *B* becomes positive with respect to *A*, the diode is forward-biased. Hence, it acts like a closed switch (or short) across which no voltage is dropped. Hence, the waveshape of V_0 is as shown in Fig. 52.31 (*c*). It is seen that the negative portion of the signal voltage has been removed. Hence, such a circuit is called a **negative clipper**.

When Diode and Resistor are Interchanged

In this case, the circuit becomes as shown in Fig. 52.32 (*b*). Now, the output voltage V_0 is that which is dropped across *R*. During the positive half-cycle of the signal voltage, *D* acts as an open switch. Hence, all applied voltage drops across *D* and none across *R*. So, there is no output signal voltage.

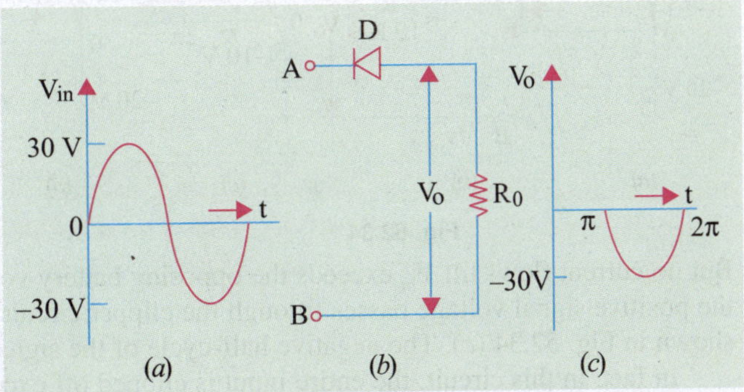

Fig. 52.32

During the negative input half-cycle, terminal is positive and so it is forward-biases which acts as a short. Hence, there is no voltage drop across *D*. Consequently, all the applied signal voltage drops across *R* and none across *D*. As a result, the negative half-cycle of the input signal is allowed to pass through the clipper circuit. Obviously, now the circuit acts as a **positive** clipper.

Example 52.20. *What would be the output waveform displayed by the oscilloscope in Fig. 52.33? The silicon diode has a barrier voltage of 0.7 V.*

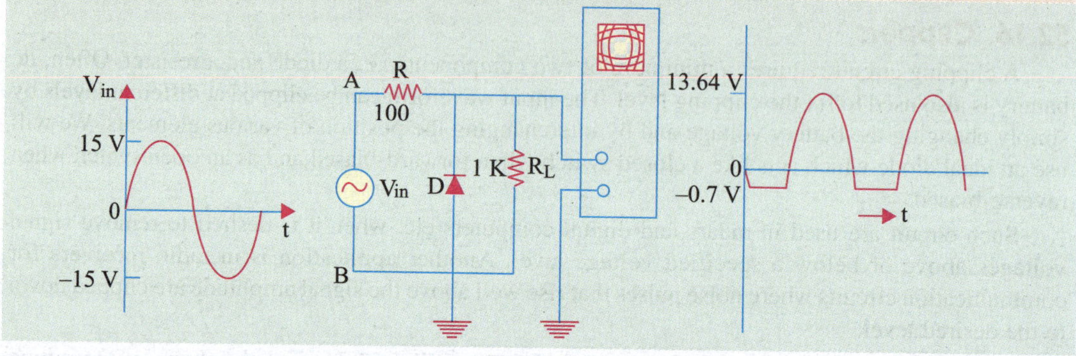

Fig. 52.33

Solution. Consider the negative input half-cycle first *i.e.* when point *B* is positive with respect to point *A*. The diode starts conducting when applied voltage exceeds 0.7 V. Since *D* and R_L are in parallel, voltage across them cannot exceed 0.7 V. Obviously, negative half-cycle beyond 0.7 V gets clipped. Hence, circuit behaves like a negative clipper.

During the positive input half-cycle when point *A* is positive, diode becomes reverse-biased and hence, becomes open-circuited. The applied voltage drops across the resistors *R* and R_L connected in series. The peak value of the output voltage is

$$= 15\left(\frac{R_L}{R + R_L}\right) = 15 \times \frac{1}{1.1} = 13.64 \text{ V}$$

Hence, the output voltage as displayed by the oscilloscope would be as shown in Fig. 52.33.

Example. 52.21. *With the sine wave signal input of Fig. 52.34 (a), find the shape of the output signal V_0 in the biased series clipper of Fig. 52.34 (b). What would happen if battery connections are reversed ?*

Solution. Let us consider the positive half-cycle of the singal *i.e.* when terminal *A* of the circuit becomes positive with respect to *B*. The diode appears as a short since it is forward-biased.

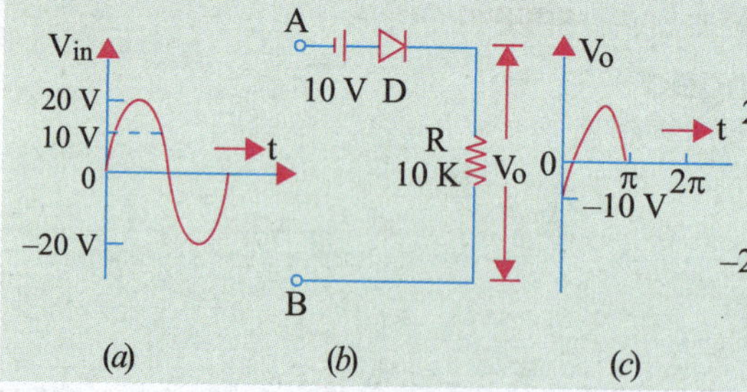

Fig. 52.34

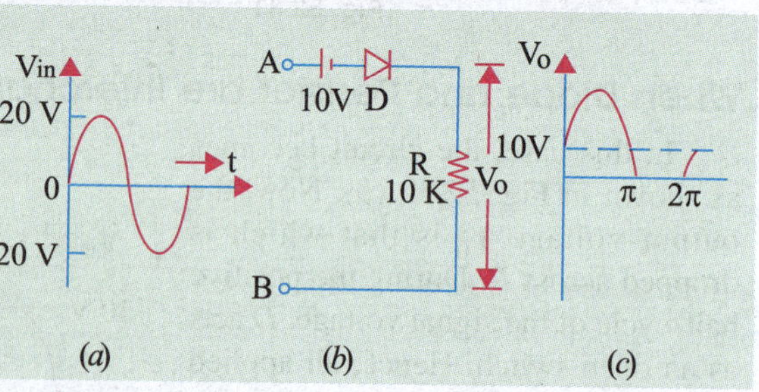

Fig. 52.35

But no current flows till V_{in} exceeds the opposing battery voltage of 10 V. Hence, only upper part of the positive signal voltage passes through the clipper circuit and appears as V_0 across *R*. Its shape is shown in Fig. 52.34 (*c*). The negative half-cycle of the signal voltage is clipped off.

In fact, in this circuit, the entire input is clipped off except positive peak portions.

Reversed Battery Connections

The battery connections have been reversed in Fig. 52.35. In this case, during the positive half-cycle of the signal, the voltage across *R* would be the sum of the signal voltage and the battery voltage *i.e.* signal voltage would be lifted up by 10 V as shown in Fig. 52.35 (*c*).

During the negative input half-cycle, the lower peak portions of the signals would be clipped off because of the 10 V battery.

Example. 52.22. *The triangular voltage of Fig. 52.36 (a) is applied to the biased parallel clipper circuit of Fig 52.36 (b). Find the wave-shape of the output voltage.*

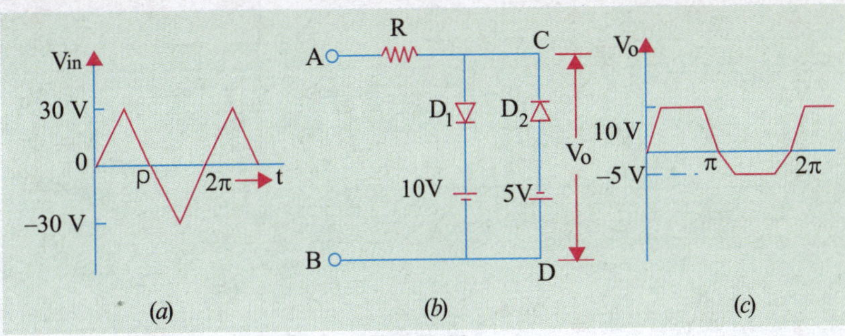

Fig. 52.36

Solution. During the positive half-cycle, D_1 would conduct but D_2 will act as an open-circuit. However, value of V_0 cannot exceed 10V because points *C* and *D* are electrically connected across the 10 V battery since D_1 is shorted. Hence, signal voltage above 10 V level would be clipped off as shown in Fig. 52.36 (*c*).

During the negative half-cycle, D_1 is open but D_2 conducts. Again, V_0 cannot exceed 5 V since it is the voltage across points *C* and *D* whose value is fixed by the battery connected in that branch. Hence, signal voltage beyond 5 V is clipped off. The wave-shape of V_0 is as shown in Fig. 52.36 (*c*).

52.17. Some Clipping Circuits

For the following circuits, a sinusoidal input signal as shown in Fig. 52.37 would be assumed.

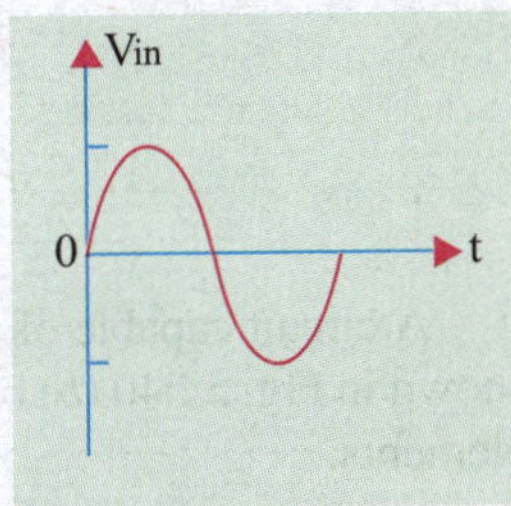

Fig. 52.37

(*a*) Biased Series Clippers

The output voltage has the waveform as shown in Fig. 52.38.

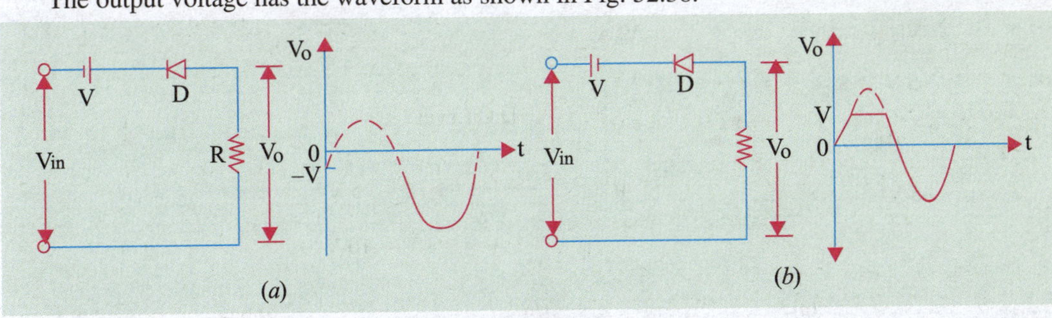

Fig. 52.38

(*b*) Biased Parallel Clippers

The waveforms of the output voltage are as shown in Fig. 52.39.

Clipping has been changed by changing the battery and diode connections.

52.18. Clampers

To put it simply, clamping is the process of introducing a dc level into an ac signal. Clampers are also sometimes known as dc restorers.

By way of illustration, consider the signal shown in Fig. 52.40 (*a*). It is a sine wave with equal positive and negative swings of ± 5 V about 0 V. Hence, its average value over one cycle is zero (it has no dc level).

In Fig. 52.40 (*b*), the signal waveform has been lifted up so as to just touch the horizontal axis. It is now said to have acquired a dc level of 5 V. This output wave-form is said to be ***positively*** clamped at 0 V. Fig. 52.40 (*c*) shows an output waveform which is negatively clamped at 0 V.

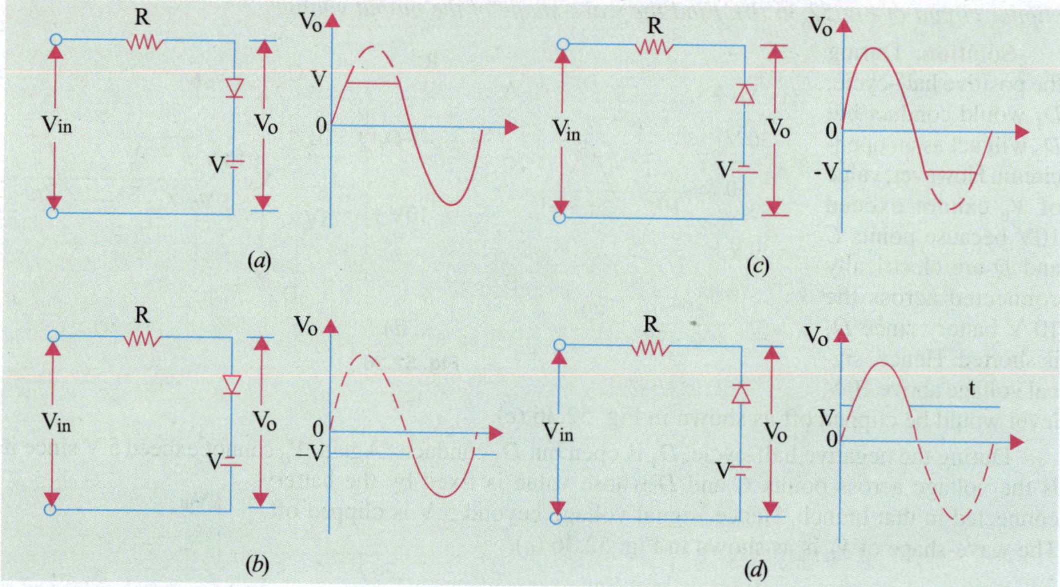

Fig. 52.39

A circuit capable of accepting the input signal shown in Fig. 52.40 (*a*) and delivering the output shown in Fig. 52.40 (*b*) or (*c*) is called a ***clamper***. Such a circuit has a minimum requirement of ***three*** elements.

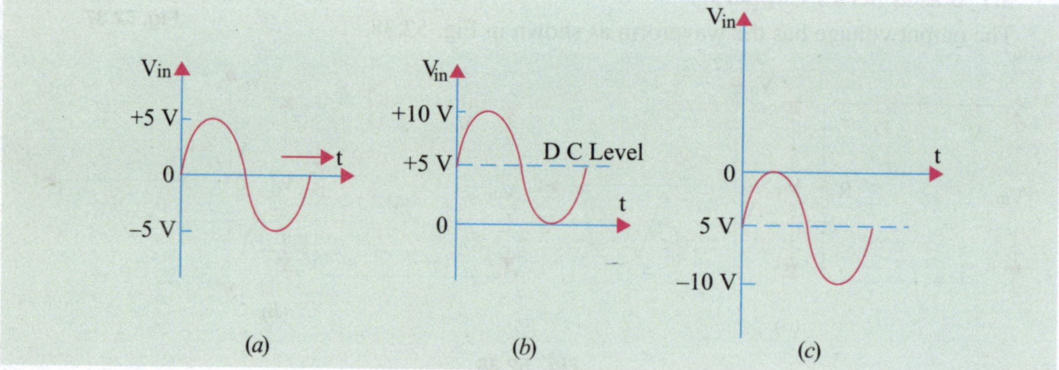

Fig. 52.40

1. a diode **2.** a capacitor and **3.** a resistor.

It will generally need a dc battery also. Following additional points regarding clamper circuits are worth keeping in mind.

1. both *R* and *C* affect the waveform.
2. values of *R* and *C* should produce a time constant ($\lambda = CR$) which is large enough to ensure that capacitor remains almost fully charged during the time-period of the signal. In other words, time constant $\lambda \gg T/2$ where *T* is the time-period of the input signal. For good clamping action, the *RC* time constant should be at least ten times the time-period of the input signal voltage.

3. it is advantageous to first consider the condition under which the diode becomes forward biased.
4. for all clamping circuits, voltage swing of the input and output waveforms is the same.
5. such circuits are often used in TV receivers as dc restorers. The incoming composite video signal is normally processed through capacitively-coupled amplifiers which eliminate the dc component thereby losing the black and white reference levels and the blanking level. These reference levels have to be restored before applying the video signal to the picture tube.

Example 52.23. *The input signal of Fig. 52.41 (a) is applied to the clamper circuit shown in Fig. 52.41 (b). Draw the waveform of the output voltage V_0. How will it change if R is made 100 kΩ ?* **(Electronic circuits, Bangalore Univ.)**

Solution. As seen, time-period of the input signal is $T = 1/1000$ second = 1 ms

$\therefore \quad 0 \rightarrow t_1 = t_1 \rightarrow t_2 = t_2 \rightarrow t_3 = \ldots\ldots\ldots = T/2 = 0.5$ ms.

$\lambda = CR = 1 \times 10^{-6} \times 10 \times 10^3 = 10$ ms

As seen, λ » *T*/2. Hence, once charged, the capacitor will have hardly any time to discharge by the time signal polarity reverses.

(a) Positive Input Half-cycle

When positive half-cycle of the input signal voltage is applied to the clamper circuit, its terminal *A* becomes positive with respect to terminal *B*. Hence, *D* acts like a short as shown in Fig. 52.41 (*c*). A steady positive voltage of 5 V remains applied to *A* for 0.5 ms. At the same time, *R* is also shorted out [Fig. 52.41 (*c*)] because it is in parallel with *D*. Hence, C will rapidly charge to 5 V. Being across a short, $V_0 = 0$ during positive half-cycle as shown in Fig. 52.41 (*d*).

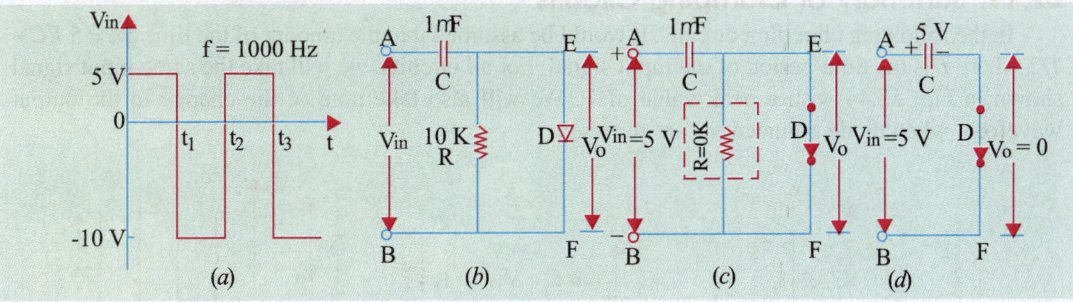

Fig. 52.41

(b) Negative Input Half-cycle

In this case, terminal *B* becomes positive and so reverse-biases *D* by 10 V. Hence, *D* acts like an open switch as shown in Fig. 52.42 (*a*). Now, *R* and *C* get connected in series so that their λ = *RC* = 10 ms. As stated earlier, capacitor will take a time of 5λ = 50 ms to get fully discharged. But the input signal will allow it just 0.5 ms during which to discharge. Obviously, *C* would hardly get discharged in this extremely short time interval of 0.5 ms. Hence, it can be assumed to be still fully charged with the original polarity during this negative half-cycle.

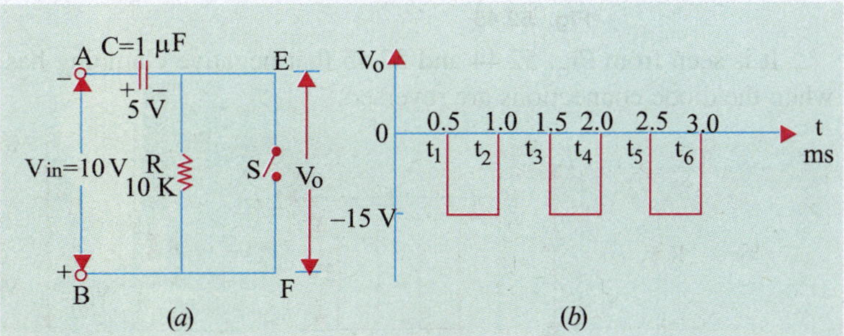

Fig. 52.42

The output voltage V_0 across the 'open' will be

$= \text{voltage from } E \rightarrow A \rightarrow B \rightarrow F$ – Fig. 52.40 (*a*)

$= 5 + 10 = 15\text{ V}$ – with *E* negative

The waveform of the output voltage is shown in Fig. 52.42 (*b*). It has same frequency as that of the input signal. However, it has been clamped down in the negative region. It is seen that voltage swing of both input and output circuits is the same *i.e.* 15 V. It is never the case in clipping circuits.

When R = 100K

Now, $\lambda = 100 \times 10^3 \times 1 \times 10^{-6}$ ms = 0.1 ms. Hence, the capacitor which is almost instantaneously charged to +5 V during the positive input half-cycle, will be almost completely discharged during the negative half-cycle because, now, 5 λ (full discharge time) equals the half time-period (0.5 s) of the signal. Hence, in this case, V_0 would be momentarily equal to – 15 V at the beginning of the negative half-cycle but will fall off to almost 0 V before the signal reverses its polarity (Fig. 52.43). As seen, υ_0 consists of voltage spikes of amplitude – 15 V.

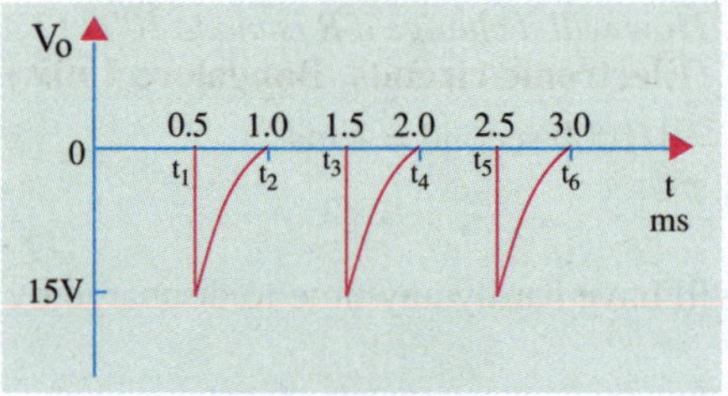

Fig. 52.43

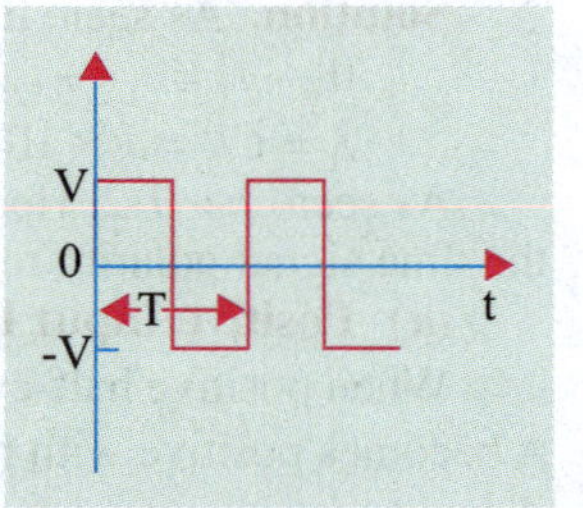

Fig. 52.44

52.19. Summary of Clamping Circuits

In the following clamping circuits, it would be assumed that the amount of the time $5\lambda = 5\,RC \gg T/2$ where *T* is the time-period of the input signal. For all circuits, we will take the same input signal shown in Fig. 52.44 with a peak value of *V*. We will also take note of the change in the output waveform when diode connections are reversed.

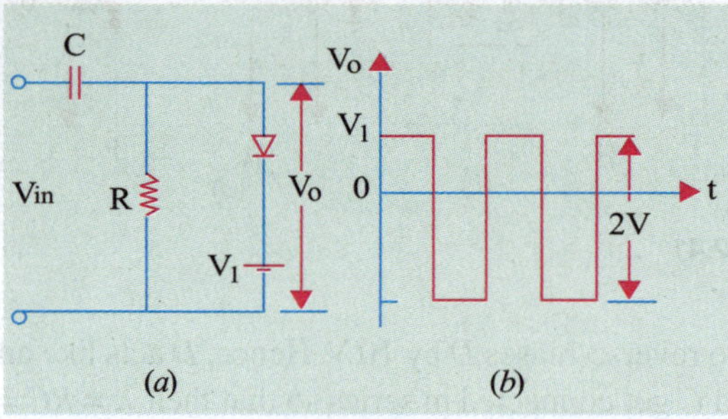

Fig. 52.45

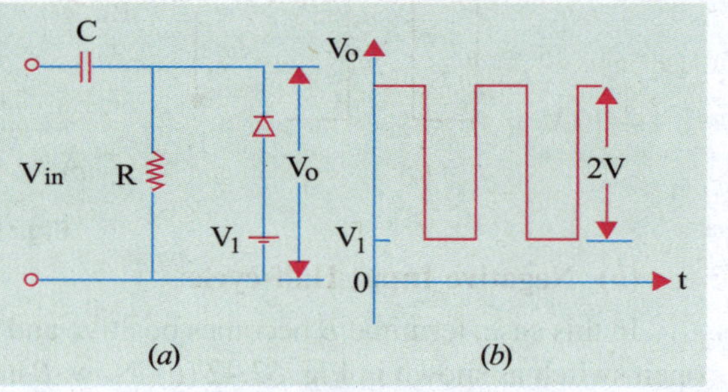

Fig. 52.46

It is seen from Fig. 52.44 and 52.45 that negative clamping has changed to positive clamping when the diode connections are reversed.

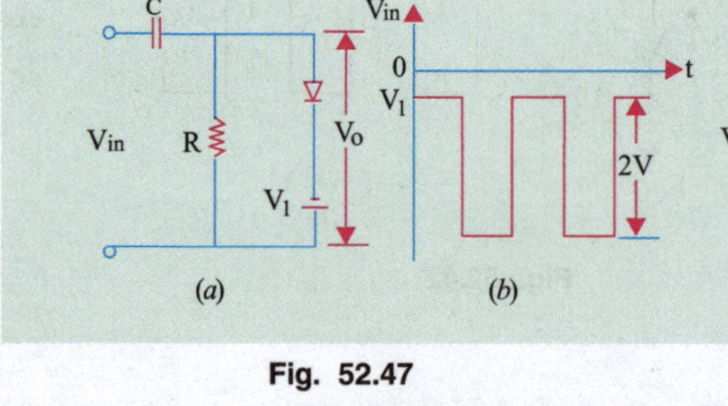

Fig. 52.47

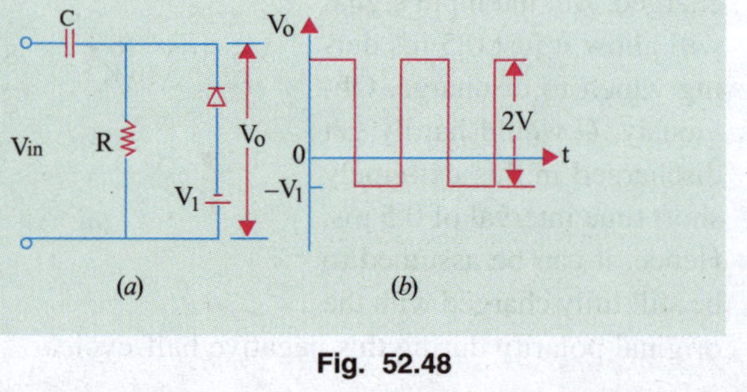

Fig. 52.48

The same thing happens in the case of clamping circuits shown in Fig. 52.47 and 52.48.

Tutorial Problems No. 52.1

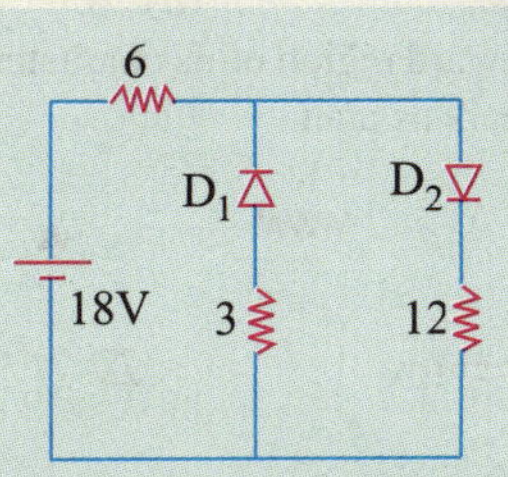

Fig. 52.49

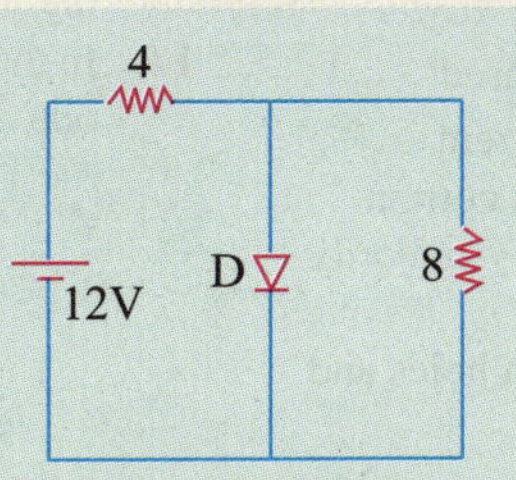

Fig. 52.50

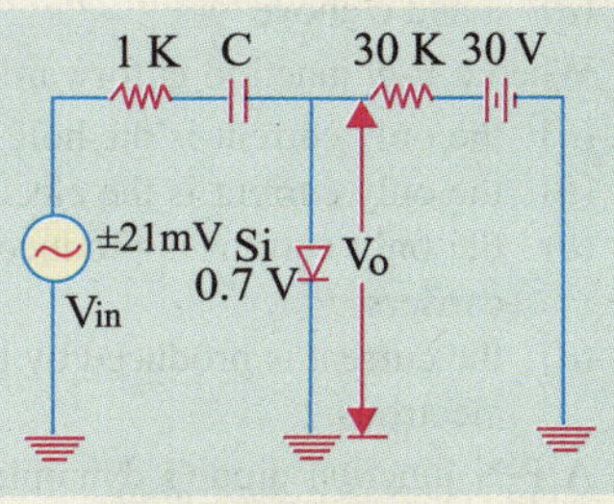

Fig. 52.51

1. Find the current supplied, if any, by the battery in the circuit of Fig. 52.49 which uses two oppositely-connected ideal diodes in parallel. **[1 A]**
2. What is the current supplied by the battery in Fig. 52.50 if D is an ideal diode. **[3 A]**
3. In Fig. 52.51, signal voltage has a peak value of ± 21 mV and its frequency is so high that reactance of the coupling capacitor can be neglected. If bulk resistance of the silicon diode is neglected, what

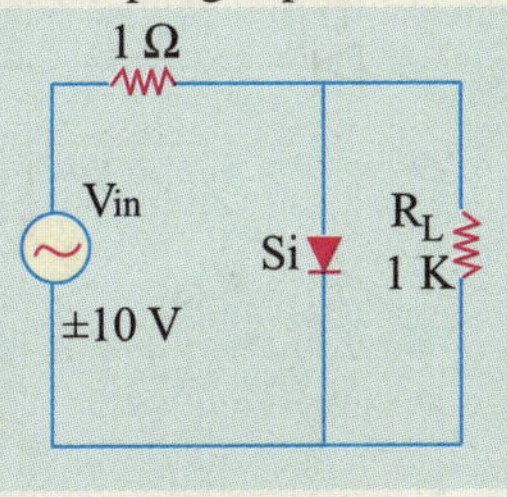

Fig. 52.52

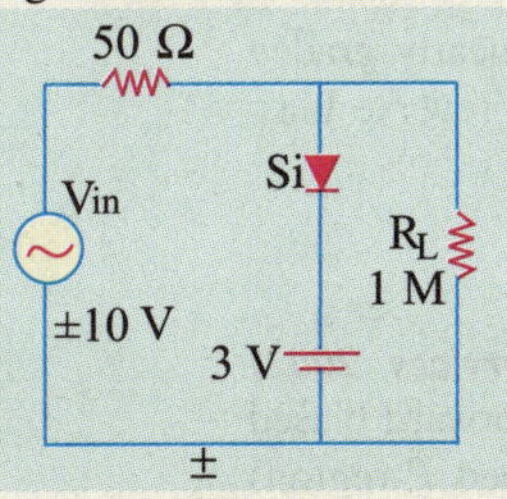

Fig. 52.53

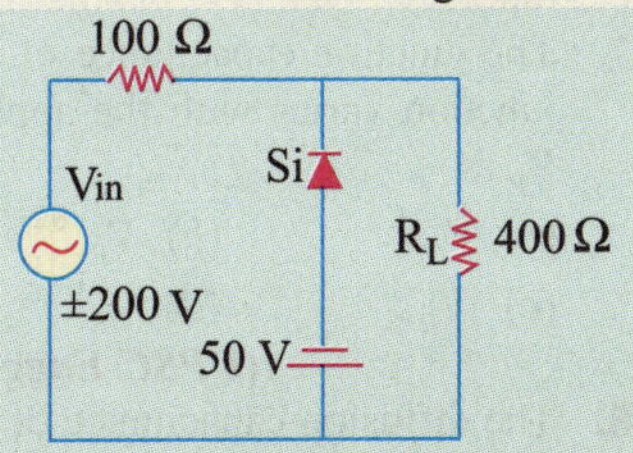

Fig. 52.54

would be the approximate waveform of the total voltage 'V_0' across the diode. **[0.7 V dc on which rides an ac voltage of peak value = 1 mV]**

4. Sketch the voltage across R_L in the clipper circuit of Fig. 52.52.
5. Sketch the voltage across R_L in the clipper circuit of Fig. 52.53.
6. What is the waveform of the voltage across R_L in Fig. 52.54 ?
7. What is the approximate dc component in the output of a positive diode clamper with a peak input of 20 V ? **[20 V]**

OBJECTIVE TESTS – 52

1. For a silicon diode, the value of the forward-bias voltage typically
   (*a*) must be greater than 0.3 V
   (*b*) must be greater than 0.7 V
   (*c*) depends on the width of the depletion region
   (*d*) depends on the concentration of majority carriers.
2. When forward biased, a diode
   (*a*) blocks current
   (*b*) conducts current
   (*c*) has a high resistance
   (*d*) drops a large voltage
3. The term bias means
   (*a*) the ratio of majority carriers to minority carriers
   (*b*) the amount of current across the P-N junction
   (*c*) a dc voltage applied across the P-N junction to control its operation
   (*d*) none of the above
4. To forward-bias a P-N junction, diode,
   (*a*) an external voltage is applied that is positive at the anode and negative at the cathode
   (*b*) an external voltage is applied that is negative at the anode and positive at the cathode.

(c) an external voltage is applied that is positive at the P-region and negative at the N-region
(d) a and c above

5. When a P-N junction is forward-biased
(a) the only current is the hole current
(b) the only current is the electron current
(c) the only current is produced by majority carriers
(d) the current is produced by both holes and electrons.

6. A P-N junction diode's dynamic conductance is directly proportional to
(a) the applied voltage
(b) the temperature
(c) the current
(d) the thermal voltage
(Hint : Conductance is reciprocal of the resistance)

7. The junction capacitance of a linearly graded junction varies with the applied reverse bias, V_R as
(a) V_R^{-1} (b) $V_R^{-1/2}$
(c) $V_R^{-1/3}$ (d) $V_R^{1/2}$
(UPSC Engg. Services 2002)

8. The diffusion capacitance of a forward biased $P^+ N$ (P^+ indicates heavily doped P-region) junction diode with a steady current I depends on
(a) width of the depletion region
(b) mean life-time of holes
(c) mean life-time of electrons
(d) junction-area
(UPSC Engg. Services 2002)

9. The width of depletion layer of a P-N junction
(a) decreases with light doping
(b) increases with heavy doping
(c) is independent of applied voltage
(d) is increased under reverse bias.

10. At room temperature of 25°C, the barrier potential for silicon is 0.7 V. Its value at 125°C is .... volt.
(a) 0.5 (b) 0.3
(c) 0.9 (d) 0.7.

11. Junction breakdown occurs
(a) under high temperature condition
(b) with forward bias
(c) under reverse bias
(d) because of manufacturing defect.

12. Avalanche breakdown is primarily dependent on the phenomenon of
(a) collision (b) doping
(c) ionization (d) recombination.

13. Reverse current in a silicon junction nearly doubles for every.......... °C rise in temperature.
(a) 10 (b) 2
(c) 6 (d) 5.

14. In the forward region of its characteristic, a diode appears as a/an

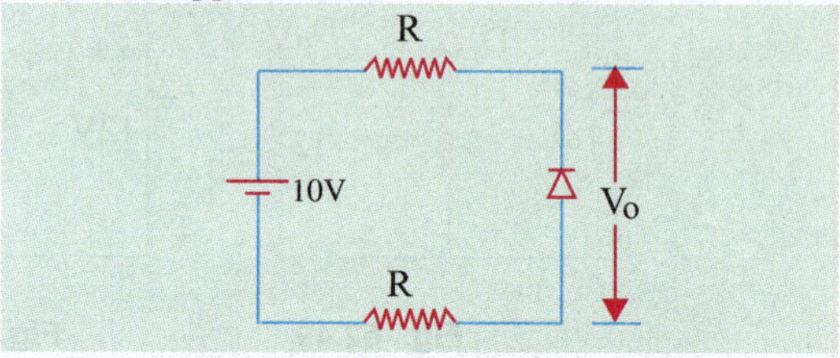

Fig. 52.55

(a) OFF switch (b) high resistance
(c) capacitor (d) ON switch.

15. The approximate value of V_O across the diode in Fig. 52.53 is

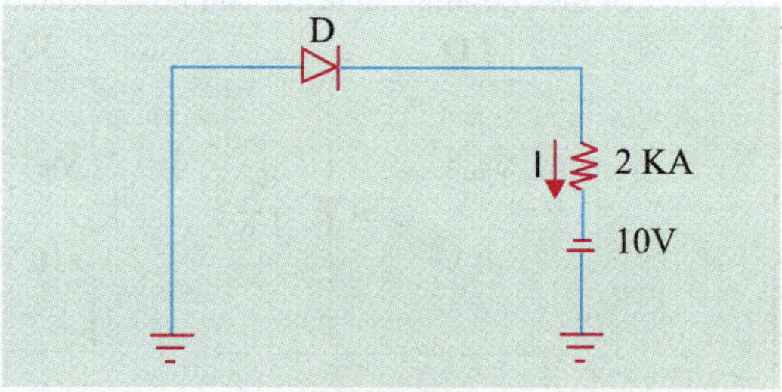

Fig. 52.56

(a) zero (b) 10 V
(c) 5 V
(d) dependent onvalue of R.

16. The diode 'D' is ideal in the network shown in Fig. 52.56. The current 'I' will be.
(a) 2 n A (b) zero
(c) 2 m A (d) 4 mA

17. The voltages at V_1 and V_2 of the arrangement shown in Fig. 52.57 will be respectively.

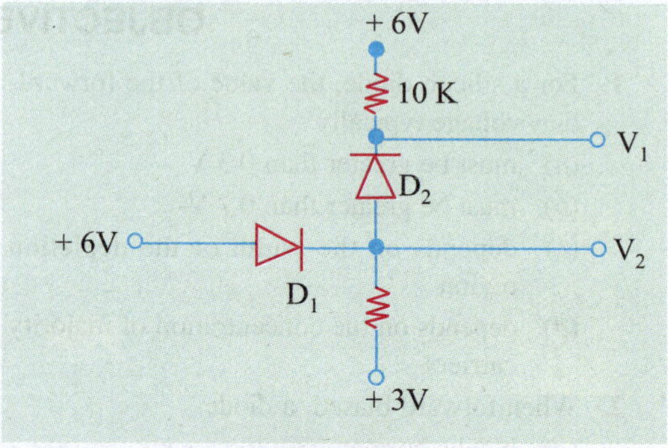

Fig. 52.57

(a) 6 V and 5.4 V (b) 5.4 V and 6 V
(c) 3 V and 5.4 V (d) 6 V and 5 V

18. Without a dc source, a clipper acts like a
 (a) rectifier (b) clamper
 (c) demodulator (d) chopper

19. The primary function of a clamper circuit is to
 (a) suppress variations in signal voltage
 (b) raise positive half-cycle of the signal
 (c) lower negative half-cycle of the signal
 (d) introduce a dc level into an ac signal

20. For an input $V_S = 5 \sin t$ (assuming ideal diode), the circuit shown in Fig. 52.58 will be behave as a
 (a) clipper, sine wave clipped at –2 V
 (b) clamper, sine wave clamped at –2 V
 (c) clamper, sine wave clamped at zero volt
 (d) clipped, sine wave clipped at 2 V

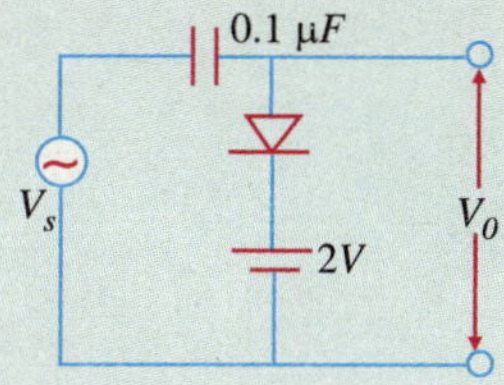

Fig. 52.58

21. A clipping circuit is shown in Fig. 52.59. Its transfer characteristic will be

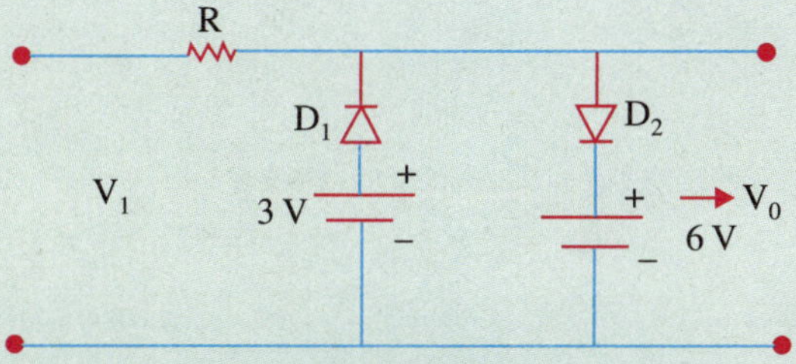

Fig. 52.59

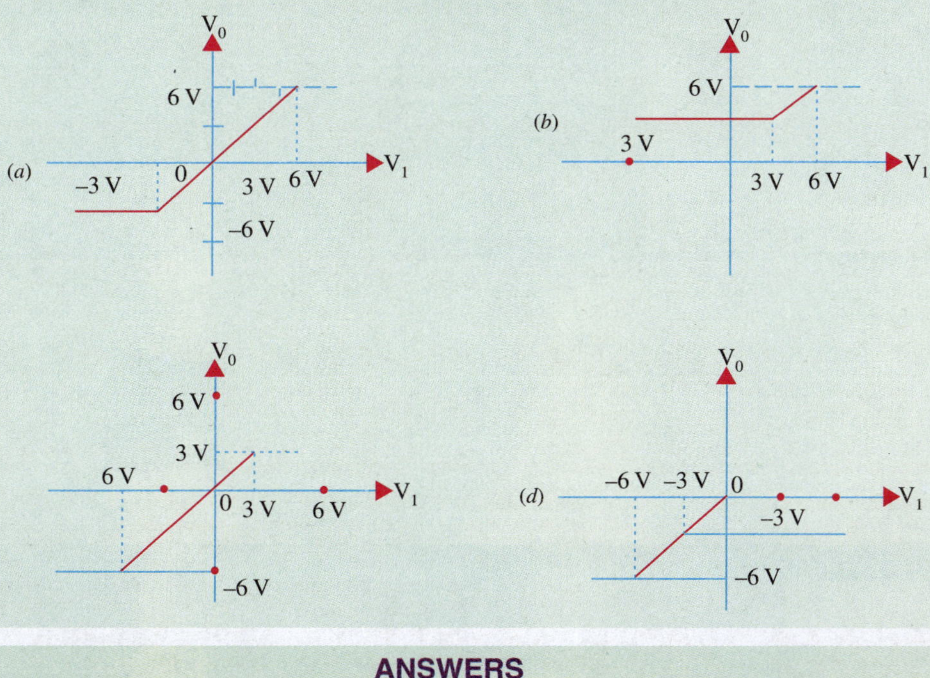

ANSWERS

1. (b)	2. (b)	3. (c)	4. (d)	5. (d)	6. (c)	7. (c)
8. (c)	9. (d)	10. (c)	11. (c)	12. (a)	13. (c)	14. (d)
15. (b)	16. (c)	17. (a)	18. (a)	19. (d)	20. (b)	21. (b)

ROUGH WORK

CHAPTER 53

OPTO-ELECTRONIC DEVICES

Learning Objectives

- Fundamentals of Light
- Light Emitting Diode (LED)
- Use of LEDs in Facsimile Machines
- Liquid Crystal Displays
- P-N Junction Photodiode
- Dust Sensor
- Photoconductive Cell
- Phototransistor
- Photodarlington
- Photo voltaic or Solar Cell
- Laser Diode
- Optical Disks
- Read-only Optical Disk Equipment
- Printers Using Laser Diodes
- Hologram Scanners
- Laser Range Finder
- Light-activated SCR(LASCR)
- Optical Isolators
- Optical Modulators
- Optical Fibre Communication Systems
- Optical Fibre Data Links

Researchers have demonstrated a new type of nanometer scale optoelectronic device that performs addition and other complex logic operations

53.1. Fundamentals of Light

According to the Quantum Theory, light consists of discrete packets of energy called photons. The energy contained in a photon depends on the ***frequency*** of the light and is given by the relation $E = hf$ where h is Plank's constant (6.625×10^{-34} Joule-second). In this equation, energy E is in Joules and frequency f is in hertz (Hz). As seen, photon energy is directly proportional to frequency: higher the frequency, greater the energy. Now, velocity of light is given by $c = f\lambda$ where c is the velocity of the light (3×10^8 m/s) and λ is the wavelength of light in metres. The wavelength of light determines its colour in the visible range and whether it is ultraviolet or infrared outside the visible range.

Now, $$E = hf = hc/\lambda \quad \text{or} \quad \lambda = hc/E \text{ metres}$$

$\therefore$ $$\lambda = (6.625 \times 10^{-34}) \times (3 \times 10^{8})/E = (19.875 \times 10^{-26})/E \quad \text{—}E \text{ in Joules}$$

If E is in electron-volt (eV), then since $1 \text{ eV} = 1.6 \times 10^{-19}$ J

$\therefore$ $$\lambda = (19.875 \times 10^{-26})/(E \times 1.6 \times 10^{-19}) = (12.42 \times 10^{-7})/E \text{ metre}$$

or $$\lambda = \frac{1.242}{E} \mu\text{m}$$

In a forward-biased *P-N* junction, electrons and holes both cross the junction. In the process, some electrons and holes recombine with the result that electrons lose energy. The amount of energy lost is equal to the difference in energy between the conduction and valence bands, this being known as the semiconductor energy band gap E_g. The value of E_g for silicon is 1.1 eV, for GaAs is 1.43 eV and for InAs is 0.36 eV. For example, the wavelength of light emitted by silicon *P-N* junction is $\lambda = 1.242/E_g = 1.242/1.1 = 1.13$ μm.

53.2. Light Emitting Diode (LED)

(*a*) Theory

As the name indicates, it is a forward-biased *P-N* junction which emits visible light when energised. As discussed earlier (Art. 51.42), charge carrier recombination takes place when electrons from the *N*-side cross the junction and recombine with the holes on the *P*-side.

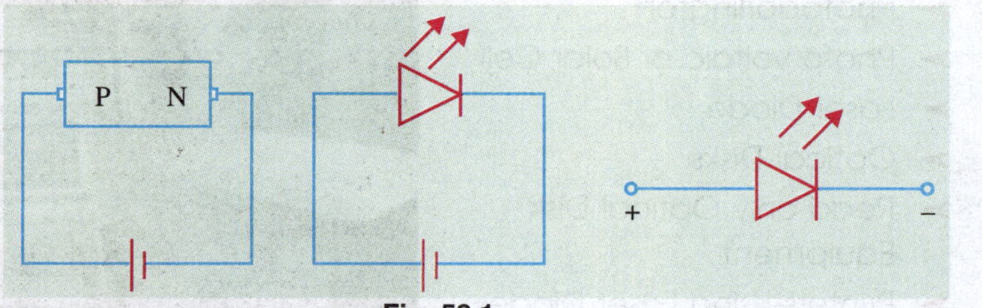

Fig. 53.1

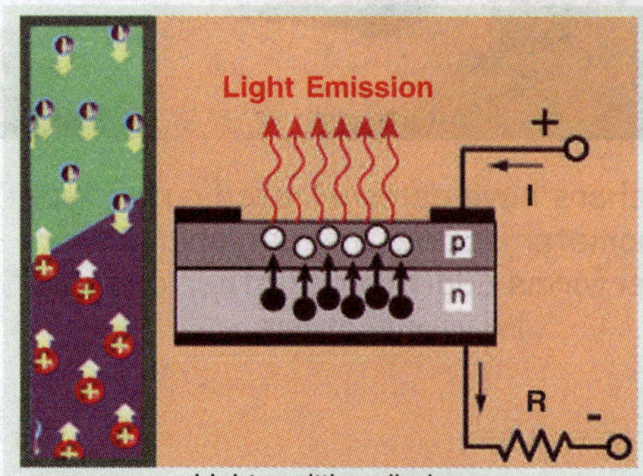

Light emitting diode

Now, electrons are in the higher conduction band on the *N*-side whereas holes are in the lower valence band on the *P*-side. During recombination, some of the energy difference is given up in the form of heat and light (*i.e.* photons). For Si and Ge junctions, greater percentage of this energy is given up in the form of heat so that the amount emitted as light is insignificant. But in the case of other semiconductor materials like gallium arsenide (GaAs), gallium phosphide (GaP) and gallium-arsenide-phosphide (GaAsP), a greater percentage of energy released during recombination is given out in the form of light. If the semiconductor material is translucent, light is emitted and the junction becomes a light source *i.e.* a light-emitting diode (LED) as shown schematically in Fig. 53.1. The colour of the emitted light depends on the type of material used as given on the next page.

1. GaAs — infrared radiation (invisible).
2. GaP — red or green light.
3. GaAsP — red or yellow (amber) light.

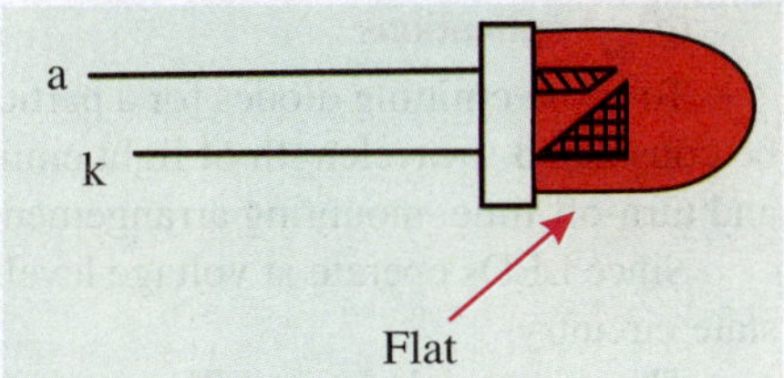

LEDs that emit ***blue*** light are also available but red is the most common. LEDs emit no light when reverse-biased. In fact, operating LEDs in reverse direction will quickly destroy them. Fig. 53.1 shows a picture of LEDs that emits different colours of light.

(*b*) Construction

Broadly speaking, the LED structures can be divided into two categories :

1. ***Surface-emitting LEDs* :** These LEDs emit light in a direction **perpendicular** to the *PN* junction plane.
2. ***Edge-emitting LEDs* :** These LEDs emit light in a direction **parallel** to the *PN* junction plane.

Fig. 53.2 shows the construction of a surface-emitting LED. As seen from this figure, an *N*-type layer is grown on a substrate and a *P*-type layer is deposited on it by diffusion. Since carrier recombination takes place in the *P*-layer, it is kept upper most. The metal anode connections are made at the outer edges of the *P*-layer so as to allow more central surface area for the light to escape. LEDs are manufactured with domed lenses in order to lessen the reabsorption problem.

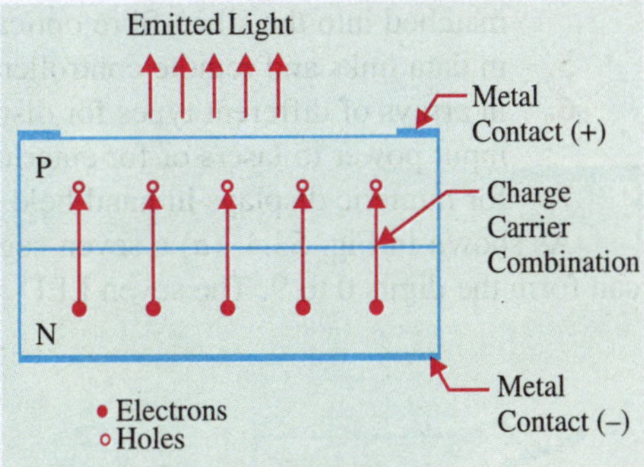

Fig. 53.2

A metal (gold) film is applied to the bottom of the substrate for reflecting as much light as possible to the surface of the device and also to provide cathode connection. LEDs are always encased in order to protect their delicate wires.

Being made of semiconductor material, it is rugged and has a life of more than 10,000 hours.

(*c*) Working

The forward voltage across an LED is considerably greater than for a silicon *PN* junction diode. Typically the maximum forward voltage for LED is between 1.2 V and 3.2 V depending on the device. Reverse breakdown voltage for an LED is of the order of 3 V to 10 V. Fig. 53.3 (*a*) shows a simple circuit to illustrate the working of an LED. The LED emits light in response to a sufficient forward current. The amount of power output translated into light is directly proportional to the forward current as shown in Fig. 53.3 (*b*). It is evident from this figure that greater the forward current, the greater the light output.

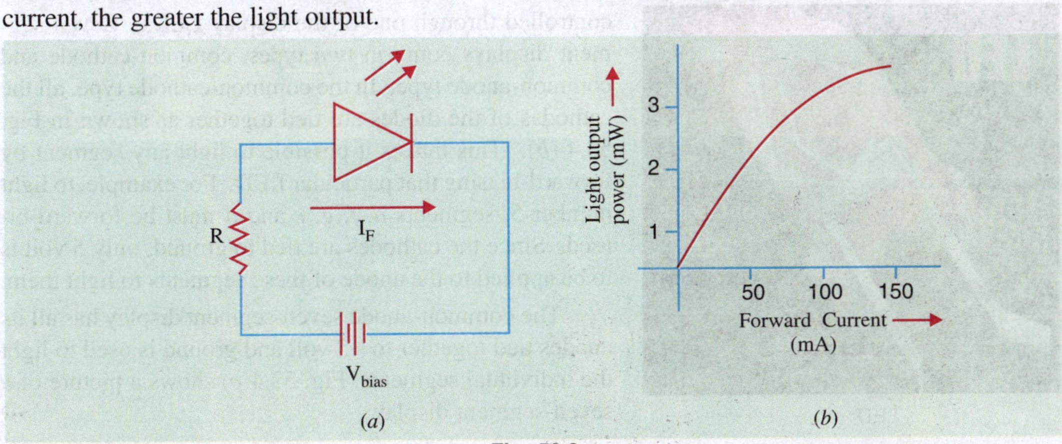

Fig. 53.3

(*d*) Applications

To chose emitting diodes for a particular application, one or more of the following points have to be considered : wavelength of light emitted, input power required, output power, efficiency, turn-on and turn-off time, mounting arrangement, light intensity and brightness etc.

Since LEDs operate at voltage levels from 1.5 V to 3.3 V, they are highly compatible with solid-state circuitry.

Their uses include the following :

1. LEDs are used in burglar-alarm systems;
2. for solid-state video displays which are rapidly replacing cathode-ray tubes (CRT);
3. in image sensing circuits used for 'picturephone';
4. in the field of optical fibre communication systems where high-radiance GaAs diodes are matched into the silica-fibre optical cable;
5. in data links and remote controllers;
6. in arrays of different types for displaying alphanumeric (letters and numbers) or supplying input power to lasers or for entering information into optical computer memories;
7. for numeric displays in hand-held or pocket calculators.

As shown in Fig. 53.4 (*a*) a seven-segment display consists of seven rectangular LEDs which can form the digits 0 to 9. The seven LED segments are labelled '*a*' to '*g*'. Each of this segments is

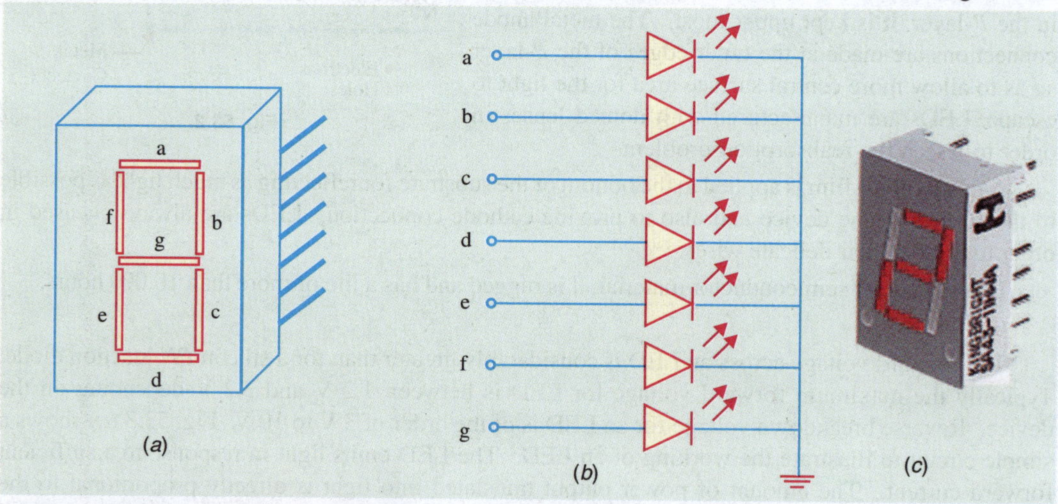

Fig. 53.4

controlled through one of the display LEDs. Seven-segment displays come in two types, common-cathode and common-anode type. In the common-cathode type, all the cathodes of the diodes are tied together as shown in Fig. 53.4 (*b*). This makes it possible to light any segment by forward-biasing that particular LED. For example, to light number 5, segments *a*, *f*, *g*, *c* and *d* must be forward-biased. Since the cathodes are tied to ground, only 5 volt is to be applied to the anode of these segments to light them.

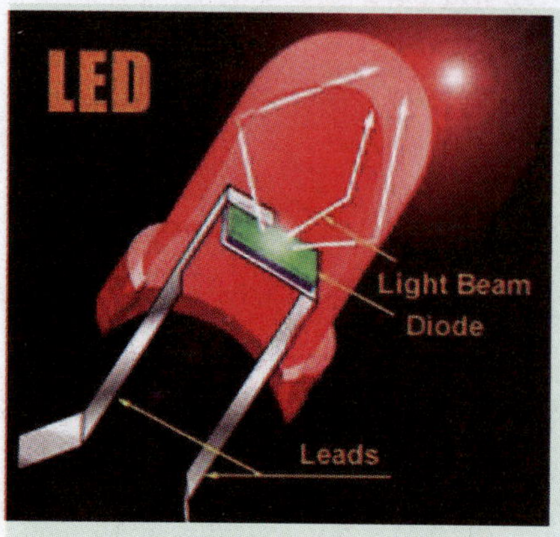

LED

The common-anode seven-segment display has all its anodes tied together to +5 volt and ground is used to light the individual segments. Fig. 53.4(*c*) shows a picture of a seven-segment display.

(e) Multicoloured LEDs

LEDs are available which gives out light in either two or three colours. There are also blinking LEDs. A two-colour LED is a three-terminal device as shown in Fig. 3.5. The longest lead is the cathode and the remaining two leads are the anodes. When leads R and C are forward-biased, the LED emits red light and when leads G and C are forward-biased, LED emits green light. The tricolour LED looks similar to the ordinary LED but emits, red, green or yellow light depending on operating conditions. It has two leads and each of these acts as both anode and cathode. When dc current flows through it in one direction, LED emits red light but when current flows in the opposite direction, LED emits green light. However, with ac current, yellow light is given out.

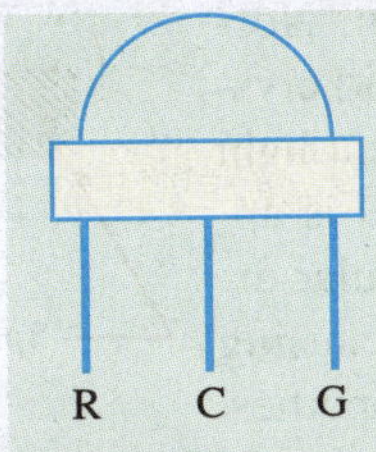

Fig. 53.5

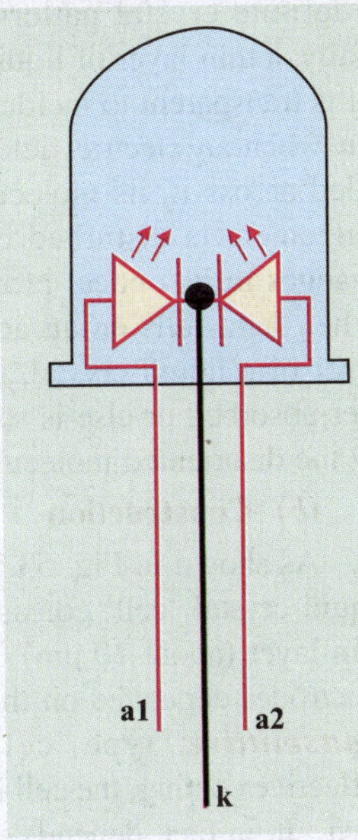

The blinking LED is a combination of an oscillator and a LED in one package. Since it has an anode and a cathode lead, it looks like an ordinary LED. The blinking frequency is usually 3 Hz when the diode forward bias is 5 V. It conducts about 20 mA of current when ON and 0.9 mA when OFF.

53.3. Use of LEDs in Facsimile Machines

Fig. 53.6 shows a simplified schematic diagram of a facsimile (or fax) machine. As seen, the light from the LED array is focussed on the document paper. The light reflected at the paper is focussed on a charge-coupled device (CCD) by a combination of mirror and a lens. This causes the optical information to be converted into electrical information. The electrical information is then sent through the data-processing unit to its destination via telephone line.

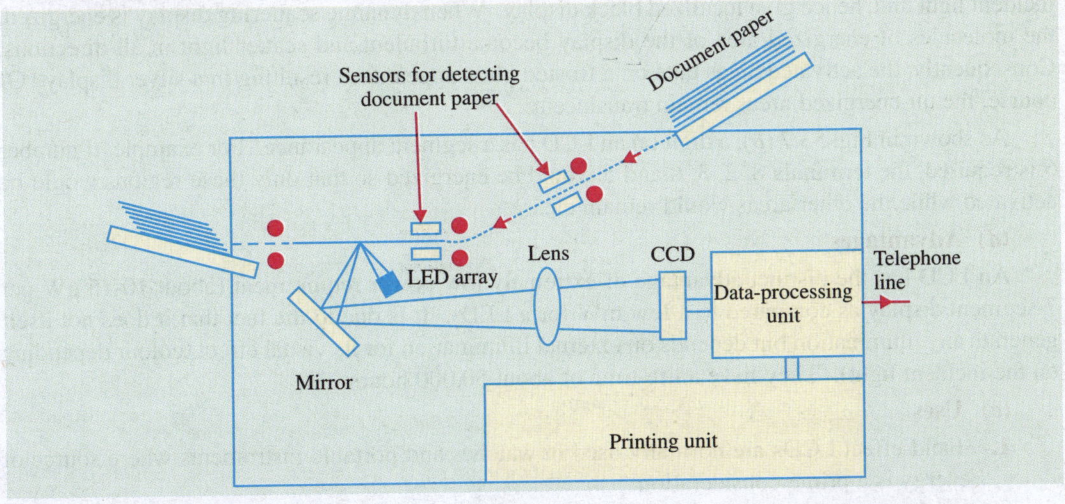

Fig. 53.6

53.4. Liquid Crystal Displays

(a) General

A liquid crystal is a material (usually, an organic compound) which flows like a liquid at room temperature but whose molecular structure has some properties normally associated with solids (examples of such compounds are : cholesteryl nonanoate and p-azoxyanisole). As is well-known, the molecules in ordinary liquids have random orientation but in a liquid crystal they are oriented in a definite crystal pattern. Normally, a thin layer of liquid crystal is transparent to incident light but when an electric field is applied across it, its molecular arrangement is disturbed causing changes in its optical properties. When light falls on an activated layer of a liquid crystal, it is either absorbed or else is scattered by the disoriented molecules.

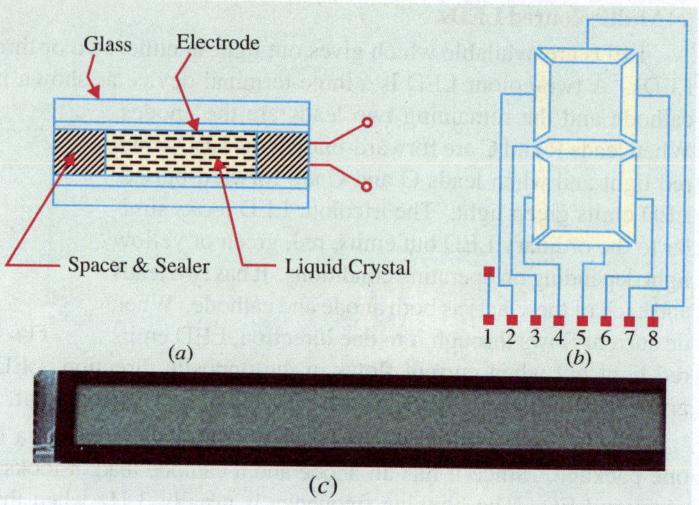

Fig. 53.7

(b) Construction

As shown in Fig. 53.7 (*a*), a liquid crystal 'cell' consists of a thin layer (about 10 μm) of a liquid crystal sandwiched between two glass sheets with transparent electrodes deposited on their inside faces. With both glass sheets transparent, the cell is known as ***transmittive*** type cell. When one glass is transparent and the other has a reflective coating, the cell is called ***reflective*** type. The LCD does not produce any illumination of its own. It, in fact, depends entirely on illumination falling on it from an external source for its visual effect.

(c) Working

The two types of display available are known as **(*i*) field-effect display** and **(*ii*) dynamic scattering display.** When field-effect display is energized, the energized areas of the LCD absorb the incident light and, hence give localized black display. When dynamic scattering display is energized, the molecules of energized area of the display become turbulent and scatter light in all directions. Consequently, the activated areas take on a frosted glass appearance resulting in a silver display. Of course, the un-energized areas remain translucent.

As shown in Fig. 53.7 (*b*), a digit on an LCD has a segment appearance. For example, if number 5 is required, the terminals 8, 2, 3, 6 and 5 would be energized so that only these regions would be activated while the other areas would remain clear.

(d) Advantages

An LCD has the distinct advantage of extremely low power requirement (about 10-15 μW per 7-segment display as compared to a few mW for a LED). It is due to the fact that it does not itself generate any illumination but depends on external illumination for its visual effect (colour depending on the incident light). They have a life-time of about 50,000 hours.

(e) Uses

1. Field-effect LCDs are normally used in watches and portable instruments where source of energy is a prime consideration.
2. Thousands of tiny LCDs are used to form the picture elements (pixels) of the screen in one type of B & W pocket TV receiver.
3. Recent desk top LCD monitors.
4. Note book computer display
5. Cellular phone display, to display data on personal digital assistant (PDAs) such as Palm V*x* etc.

The liquid crystal display (LCDs) commonly used on notebook computers and handheld PDAs are also appearing on desktop. These flat panel displays promise great clarity at increasingly high resolutions and are available in screen sizes upto 15 inches. The LCD monitor offers benefits and drawbacks. The first benefit is size. Because of the need to house the tube itself, cathode-ray tube (CRT) monitors are big and heavy. LCD monitors are only a few inches deep and they are much lighter in weight. However LCD monitors are expensive than CRTs at present. Another problem is the viewing angle. The optimal viewing angle of an LCD is from straight in front and as you move further to the side the screen becomes harder to read, much more so than with a CRT. Moreover screen resolutions generally reach only as high as 1,024 × 768, which is insufficient for some applications. Fig. 53.7(*c*) shows the picture of an LCD used in portable instrument.

53.5. P-N Junction Photodiode

It is a two-terminal junction device which is operated by first reverse-biasing the junction and then illuminating it. A reverse-biased *P-N* junction has a small amount of reverse saturation current I_s (or I_0) due to thermally-generated electron-hole pairs. In silicon, I_s is in the range of nanoamperes. The number of these minority carriers depends on the intensity of light incident on the junction. When the diode is in glass package, light can reach the junction and thus change the reverse current.

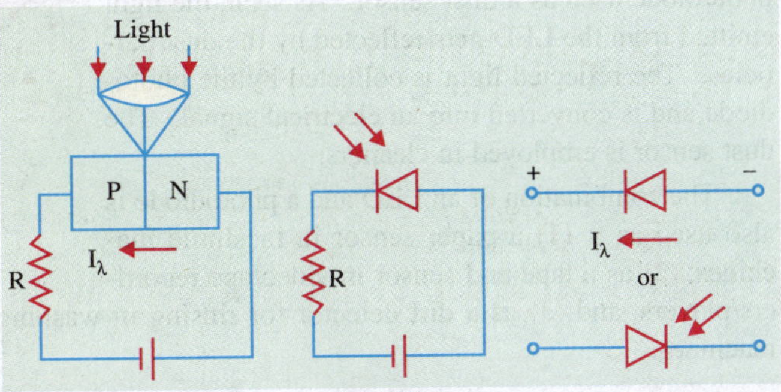

Fig. 53.8

Photodiode

The basic biasing arrangement, construction and symbols of a photodiode are shown in Fig. 53.8. As seen, a lens has been used in the cap of the unit to focus maximum light on the reverse-biased junction. The active diameter of these devices is about 2.5 mm but they are mounted in standard TO-5 packages with a window to allow maximum incident light.

The characteristics of Fig. 53.9 show that for a given reverse voltage, I_λ (or I_s) increases with increase in the level of illumination. The dark current refers to the current that flows when no light is incident. By changing the illumination level, reverse current can be changed. In this way, reverse resistance of the diode can be changed by a factor of nearly 20.

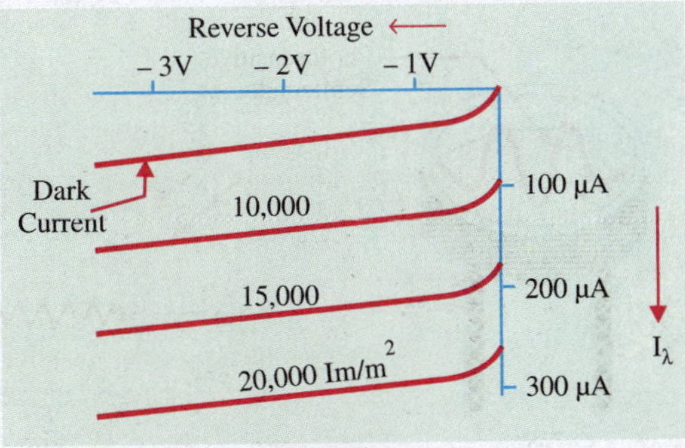

Fig. 53.9

A photodiode can turn its current ON and OFF in nanoseconds. Hence, it is one of the fastest photodetectors. It is used where it is required to switch light ON and OFF at a maximum rate. Applications of a photodiode include

1. detection, both visible and invisible ;
2. demodulation ;
3. switching ;
4. logic circuit that require stability and high speed ;
5. character recognition ;
6. optical communication equipment ;
7. encoders etc.

53.6. Dust Sensor

Fig. 53.10 shows a combination of an LED and a photodiode used as a dust sensor. As seen, the light emitted from the LED gets reflected by the dust particles. The reflected light is collected by the photodiode and is converted into an electrical signal. The dust sensor is employed in cleaners.

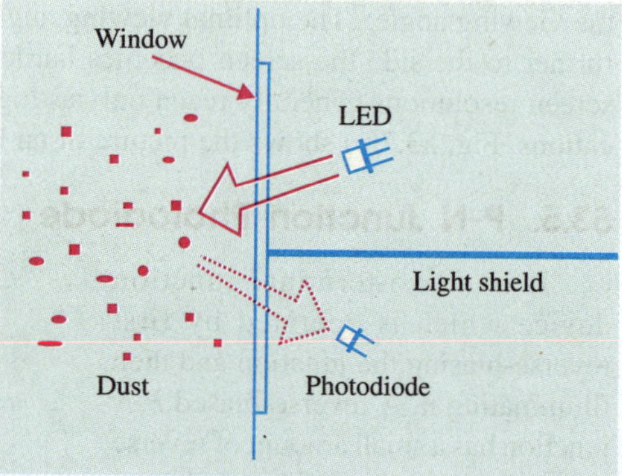

Fig. 53.10

The combination of an LED and a photodiode is also used as : **(1)** a paper sensor in facsimile machines, **(2)** as a tape-end sensor in videotape recorders/players, and **(3)** as a dirt detector for rinsing in washing machines.

53.7. Photoconductive Cell

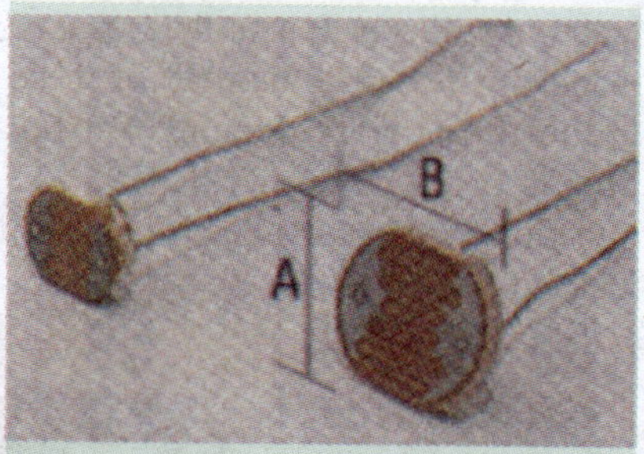

CdS photo sensitive detectors

It is a semiconductor device whose resistance varies inversely with the intensity of light that falls upon it. It is also known as ***photoresistive*** cell or ***photoresistor*** because it operates on the principle of photoresistivity.

(*a*) Theory

The resistivity (and, hence, resistance) of a semiconductor depends on the number of free charge carriers available in it. When the semiconductor is not illuminated, the number of charge carriers is small and, hence, resistivity is high. But when light in the form of photons strikes the semiconductor, each photon delivers energy to it. If the photon energy is greater than the energy band gap of the semiconductor, free mobile charge carriers are liberated and, as a result, resistivity of the semiconductor is decreased.

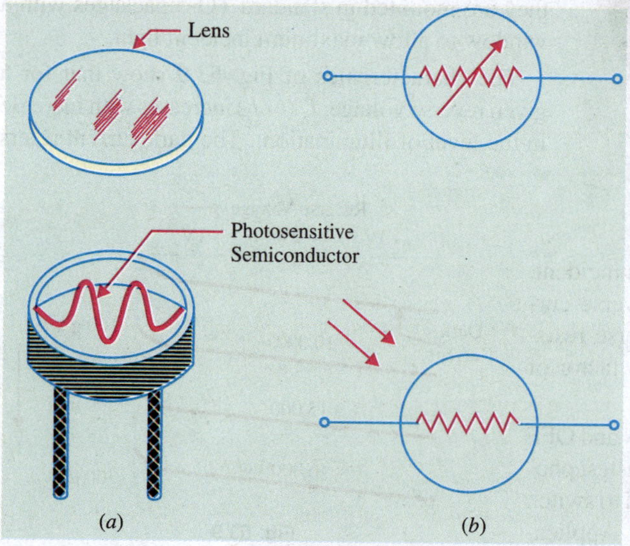

Fig. 53.11

(*b*) Construction and Working

Photoconductive cells are generally made of cadmium compounds such as cadmium sulphide (CdS) and cadmium selenide (CdSe). Spectral response of CdS cell is similar to the human eye, hence such cells are often used to simulate the human eye. That is why they find use in light metering circuits in photographic cameras.

The construction of a typical photo conductive cell and its two alternative circuit symbols are shown in Fig. 53.11 (*a*) and (*b*) respectively. As seen, a thin layer of photosensitive semiconductor material is deposited in the form of a long strip zigzagged across a disc-shaped ceramic base with protective sides. For added protection, a glass lens or plastic cover is used. The two ends of the strip are brought out to connecting pins below the base.

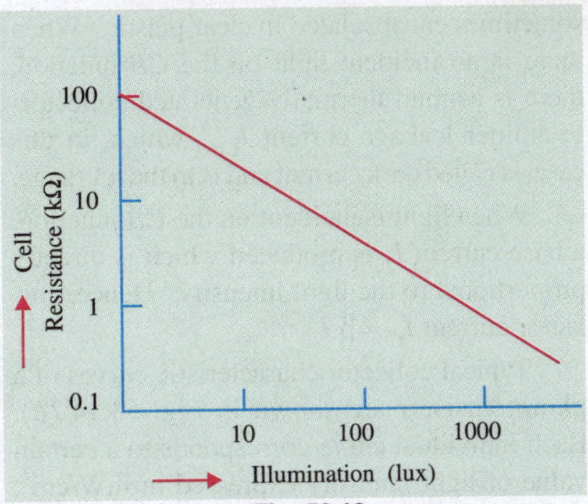

Fig. 53.12

The terminal characteristic of a photoconductive cell is shown in Fig. 53.12. It depicts how the resistance of the cell varies with light intensity. Typically, the dark resistance of the cell is 1 MΩ or larger. Under illumination, the cell resistance drops to a value between 1 and 100 kΩ depending on surface illumination.

(*c*) Applications

A photoconductive cell is an inexpensive and simple detector which is widely used in OFF/ON circuits, light-measurement and light-detecting circuits.

Example 53.1. *A relay is controlled by a photoconductive cell which has resistance of 1kΩ when illuminated and 100 kΩ when in the dark. The relay is supplied with 10 mA from a 30-V supply when cell is illuminated and is required to be de-energized when the cell is in the dark. Sketch a suitable circuit and calculate the required series resistance and value of dark current.*

(Optoelectronic Devices, Gujarat Univ. 1993)

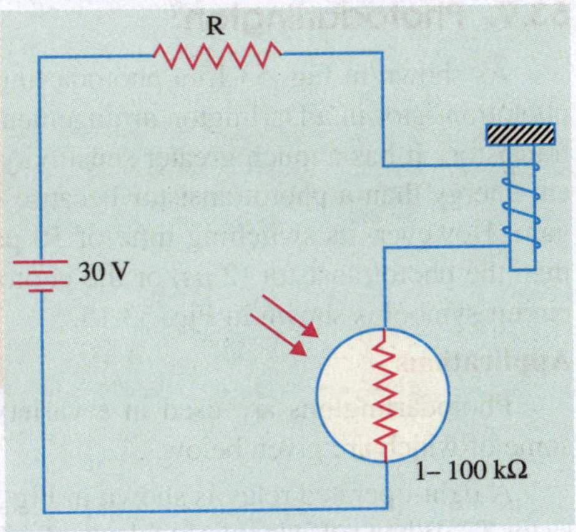

Fig. 53.13

Solution. The circuit is as shown in Fig. 53.13 where R is a current-limiting resistor.

$$I = 30/(R + r)$$

—where r is cell resistance

$$\therefore \quad R = (30/I) - r$$

When illuminated

$$R = (30/10 \times 10^{-3}) - 1 \times 10^{3} = 2 \times 10^{3} = \mathbf{2\ k\Omega}$$

Dark current is given by

$$I_d = 30/(2 + 100) \times 10^{3} = 0.3 \times 10^{-3}\ \text{A} = \mathbf{0.3\ mA}$$

53.8. Phototransistor

It is a light-sensitive transistor and is similar to an ordinary bipolar junction transistor (BJT) except that it has no connection to the base terminal. Its operation is based on the photodiode that exists at the *CB* junction. Instead of the base current, the input to the transistor is provided in the form of light as shown in the schematic symbol of Fig. 53.14 (*a*).

Phototransistor

Silicon NPNs are mostly used as photo transistors. The device is usually packed in a TO-type can with a lens on top although it is sometimes encapsulated in clear plastic. When there is no incident light on the *CB* junction, there is a small thermally-generated collector-to-emitter leakage current I_{CEO} which, in this case, is called dark current and is in the *nA* range.

When light is incident on the *CB* junction, a base current I_λ is produced which is directly proportional to the light intensity. Hence, collector current $I_C = \beta I_\lambda$

Typical collector characteristic curves of a phototransistor are shown in Fig. 53.14 (*b*). Each individual curve corresponds to a certain value of light intensity expressed in mW/cm². As seen, I_C increases with light intensity.

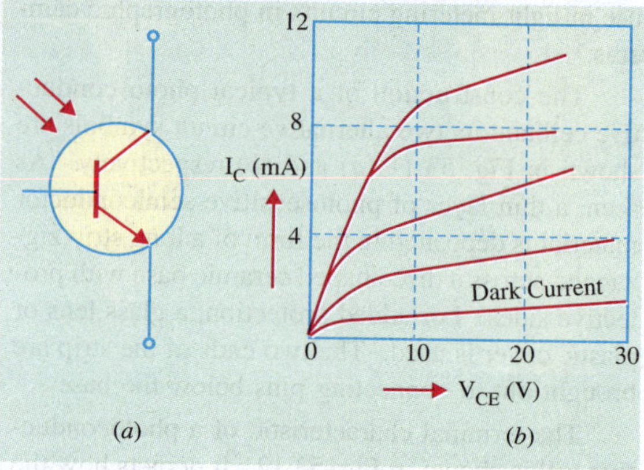

Fig. 53.14

The phototransistor has applications similar to those of a photodiode. Their main differences are in the current and response time. The photo-transistor has the advantages of greater sensitivity and current capacity than photodiodes. However, photodiodes are faster of the two, switching in less than a nanosecond.

53.9. Photodarlington

As shown in Fig. 53.15 a photodarlington consists of a phototransistor in a Darlington arrangement with a common transistor. It has a much greater sensitivity to incident radiant energy than a phototransistor because of higher current gain. However, its switching time of 50 μs is much longer than the phototransistor (2 μs) or the photodiode (1 ns). Its circuit symbol is shown in Fig. 53.15.

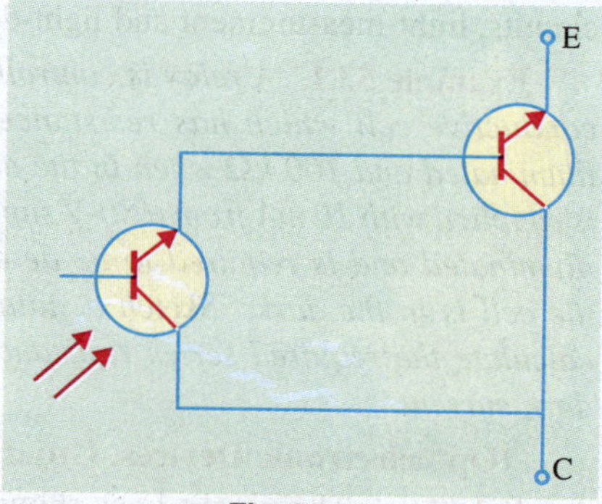

Fig. 53.15

Applications

Photodarlingtons are used in a variety of applications some of which are given below.

A light-operated relay is shown in Fig. 53.16 (*a*). The phototransistor T_1 drives the bipolar transistor T_2. When sufficient light falls on T_1, T_2 is driven into saturation so that I_C is increased manifold. This collector current while passing through the relay coil energizes the relay.

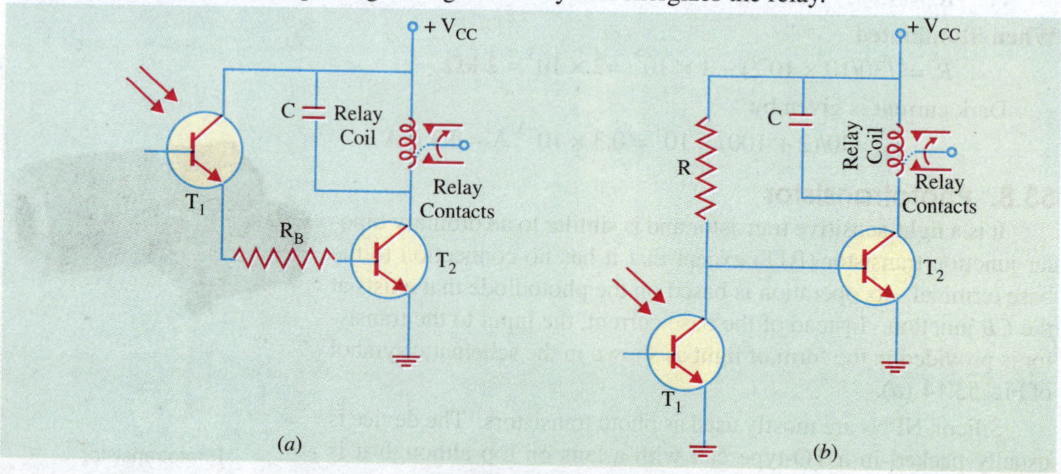

Fig. 53.16

Fig. 53.16 (*b*) shows a dark-operated relay circuit *i.e.* one in which relay is deenergized when light falls on the phototransistor. Here, T_1 and R form a potential divider across V_{CC}. With insufficient light incident on T_1, transistor T_2 is biased ON thereby keeping the relay energized. However, when there is sufficient light, T_1 turns ON and pulls the base of T_2 low thereby turning T_2 OFF and hence, deenergizing the relay.

Such relays are used in many applications such as *(i)* automatic door activators, *(ii)* process counters and *(iii)* various alarm systems for smoke or intrusion detection.

53.10. Photo voltaic or Solar Cell

Such cells operate on the principle of photovoltaic action *i.e.* conversion of light energy into electrical energy. This action occurs in all semiconductors which are constructed to absorb energy.

(a) Construction

As shown in Fig. 53.17 (*a*), a basic solar cell consists of *P*-type and *N*-type semiconductor material (usually, silicon or selenium) forming a *P-N* junction. The bottom surface of the cell (which is always away from light) covered with a continuous conductive contact to which a wire lead is attached. The upper surface has a maximum area exposed to light with a small contact either along the edge or around the perimeter. The surface layer of *P*-type material is extremely thin (0.5 mm) so that light can penetrate to the junction.

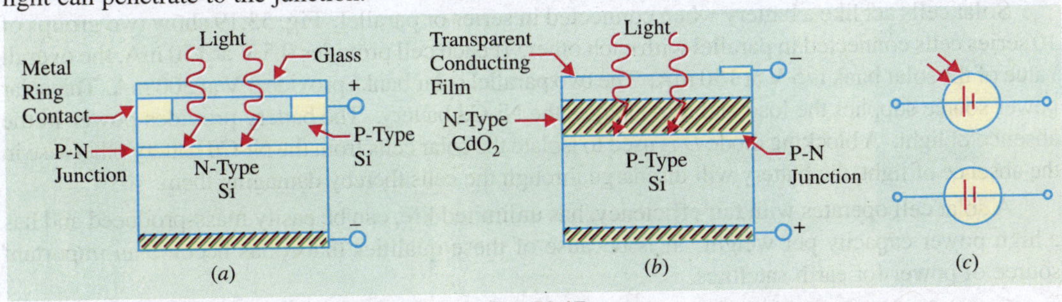

Fig. 53.17

Although silicon is commonly used for fabricating solar cells, another construction consists of *P*-type selenium covered with a layer of *N*-type cadmium oxide to form *P-N* junction as shown in Fig. 53.17 (*b*). Two alternative circuit symbols are shown in Fig. 53.17 (c). Power solar cells are also fabricated in flat strips to form efficient coverage of available surface area. Incidentally, the maximum efficiency of a solar cell in converting sunlight into electrical energy is nearly 15 per cent at the present.

(b) Theory

When the *P-N* junction of a solar cell is illuminated, electron-hole pairs are generated in the same way, as in photovoltaic cell. An electric field is established near the *P-N* junction by the positive and negative ions created due to the production of electron-hole pairs which leads to the development of potential across the junction. Since the number of electron-hole pairs far exceeds the number needed for thermal equilibrium, many of the electrons are

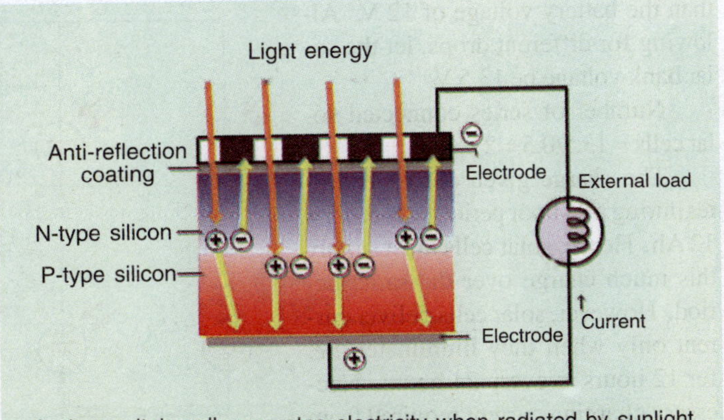

A photovoltaic cell generates electricity when radiated by sunlight

pulled across the junction by the force of the electric field. Those that cross the junction contribute to the current in the cell and through the external load. The terminal voltage of the cell is directly proportional to the intensity of the incident light. The voltage may be as high as 0.6 V depending on the external load. Usually a large number of cells are arranged in an array in order to obtained higher voltages and currents as shown in Fig. 53.18.

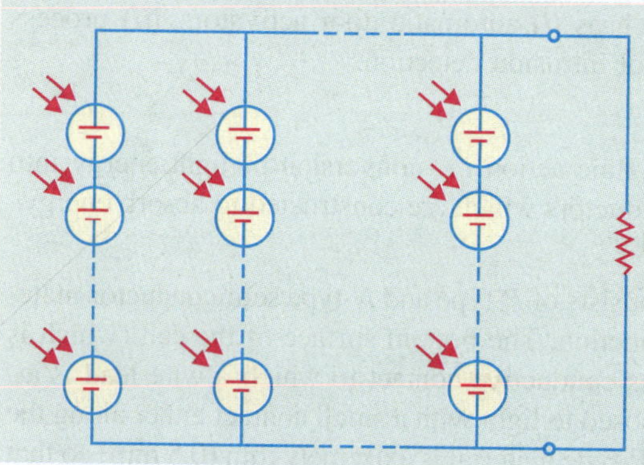

Fig. 53.18

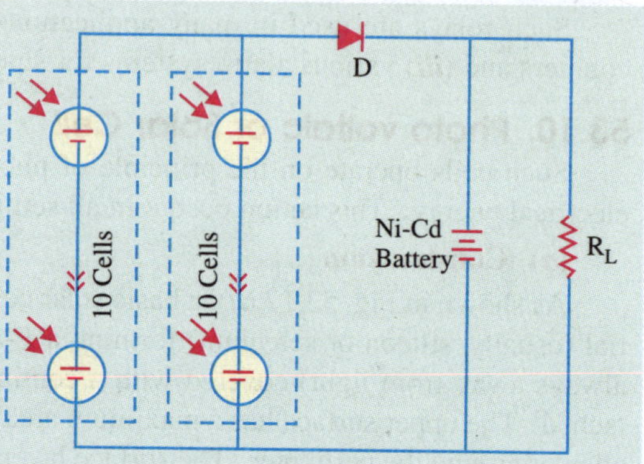

Fig. 53.19

Solar cells act like a battery when connected in series or parallel. Fig. 53.19 show two groups of 10 series cells connected in parallel with each other. If each cell provides 0.5 V at 150 mA, the overall value of the solar bank is 5 V at 150 mA. The two parallel solar banks provide 5 V at 300 mA. This solar power source supplies the load and also charges the Ni-Cd battery. The battery provides power in the absence of light. A blocking diode *D* is used to isolate the solar cells from the Ni-Cd battery otherwise in the absence of light, the battery will discharge through the cells thereby damaging them.

A solar cell operates with fair efficiency, has unlimited life, can be easily mass-produced and has a high power capacity per weight. It is because of these qualities that it has become an important source of power for earth satellites.

Example 53.2. *An earth satellite has on board 12-V battery which supplies a continuous current of 0.5 A. Solar cells are used to keep the battery charged. The solar cells are illuminated by the sun for 12 hours in every 24 hours. If during exposure, each cell gives 0.5 V at 50 mA, determine the number of cells required.* **(Optoelectronics Devices, Gujarat Univ. 1994)**

Solution. The solar cell battery-charging circuit is shown in Fig. 53.20. The cells must be connected in series to provide the necessary voltage and such groups must be connected in parallel to provide the necessary current. The charging voltage has to be greater than the battery voltage of 12 V. Allowing for different drops, let the solar bank voltage be 13.5 V.

Number of series connected solar cells = 13.5/0.5 = 27

The charge given out by batteries during a 24 hour period = 24 × 0.5 = 12 Ah. Hence, solar cells must supply this much charge over the same period. However, solar cells deliver current only when they illuminated *i.e.* for 12 hours in every 24 hours. Necessary charging current required from the solar cells = 12 Ah/12 h = 1 A.

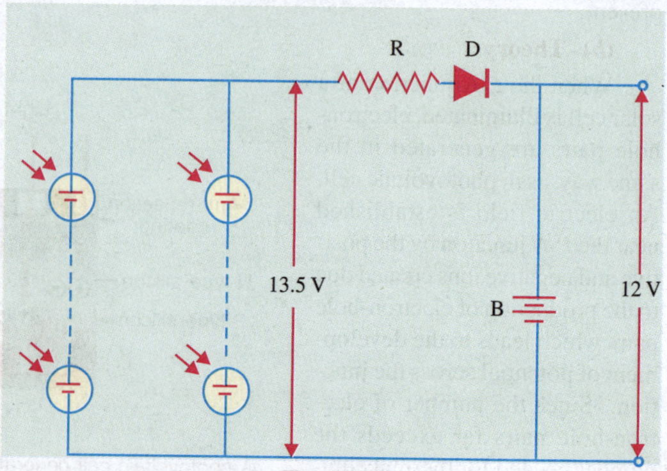

Fig. 53.20

Total number of groups of solar cells required to be connected in parallel is

$= \text{output current / cell current} = 1 / 50 \times 10^{-3} = 20$

$\therefore$ total number of solar cells required for the earth satellite $= 27 \times 20 =$ **540**

53.11. Laser Diode

Like LEDs, laser diodes are typical *PN* junction devices used under a forward-bias. The word LASER is an acronym for ***Light Amplification*** by ***Stimulated Emission of Radiation.*** The use of laser is (becoming increasing common) in medical equipment used in surgery and in consumer products like compact disk (CD) players, laser printers, hologram scanners etc.

(a) Construction

Broadly speaking, the laser diode structure can be divided into two categories :

1. ***Surface-emitting laser diodes :*** These laser diodes emit light in a direction **perpendicular** to the PN junction plane.
2. ***Edge-emitting laser diodes :*** These laser diodes emit light in a direction **parallel** to the *PN* junction plane.

Fig. 53.21 (*a*) shows the structure of an edge-emitting laser diode. This type of structure is called Fabry-Perot type laser. As seen from the figure, a *P-N* junction is formed by two layers of doped gallium arsenide (GaAs). The length of the *PN* junction bears a precise relationship with the wavelength of the light to be emitted. As seen, there is a highly reflective surface at one end of the junction and a partially reflective surface at the other end. External leads provide the anode and cathode connections.

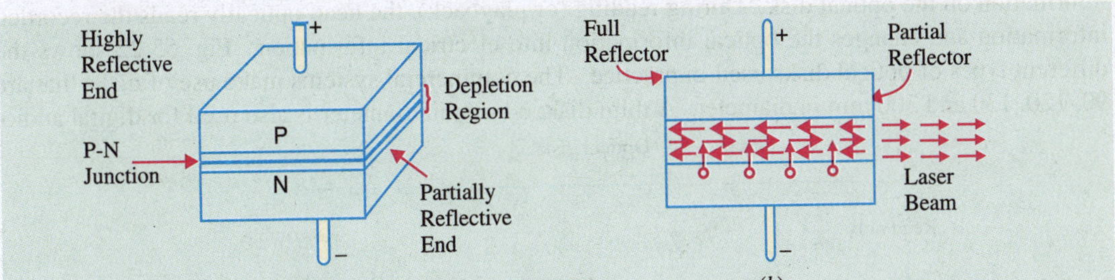

Fig. 53.21

(b) Theory

When the *P-N* junction is forward-biased by an external voltage source, electrons move across the junction and usual recombination occurs in the depletion region which results in the production of photons. As forward current is increased, more photons are produced which drift at random in the depletion region. Some of these photons strike the reflective surface perpendicularly. These reflected photons enter the depletion region, strike other atoms and release more photons. All these photons move back and forth between the two reflective surfaces. [Fig. 53.21 (*b*)] The photon activity becomes so intense that at some point, a strong beam of laser light comes out of the partially reflective surface of the diode.

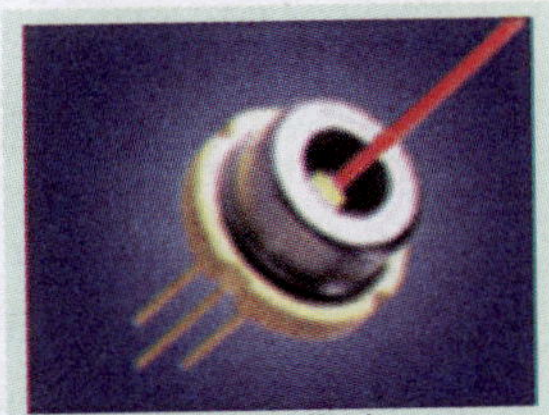

Laser diode

(c) Unique Characteristics of Laser Light

The beam of laser light produced by the diode has the following unique characteristics :

1. It is coherent *i.e.* there is no path difference between the waves comprising the beam;
2. It is monochromatic *i.e.* it consists of one wavelength and hence one colour only;

3. It is collimated *i.e.* emitted light waves travel parallel to each other.

Laser diodes have a threshold level of current above which the laser action occurs but below which the laser diode behaves like a LED emitting incoherent light. The schematic symbol of a laser diode is similar to that of LED. Incidentally, a filter or lens is necessary to view the laser beam.

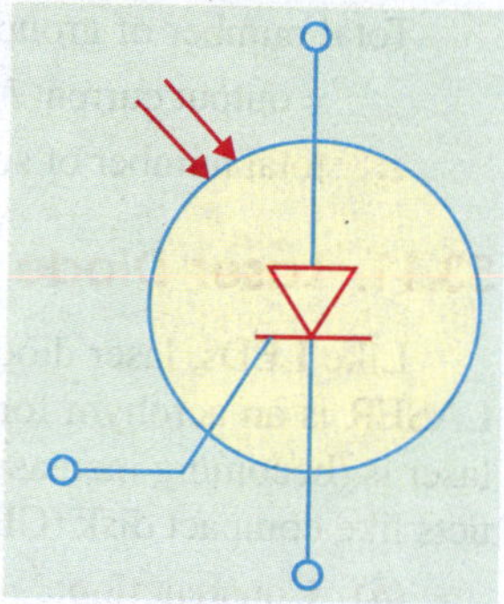

Fig. 53.22

(*d*) Applications

Laser diodes are used in variety of applications ranging from medical equipment used in surgery to consumer products like optical disk equipment, laser printers, hologram scanners etc. Laser diodes emitting visible light are used as pointers. Those emitting visible and infrared light are used to measure range (or distance). The laser diodes are also widely used in parallel processing of information and in parallel interconnections between computers. Some of these applications are discussed in the following articles.

53.12. Optical Disks

The major application field for laser diodes is in optical disk equipment. This equipment is used for reading or recording information and can be broadly divided into two groups :

1. Reading-only and **2.** Recording-and-reading type.

The optical disk equipment of either type make use of a laser diode, lenses and photodiodes. During recording, it changes electrical information into optical information and then records the information on the optical disk. During reading (or playback), the head optically reads the recorded information and changes the optical information into electrical information. Fig. 53.22 shows the different types of optical disks used in practice. The commercial systems make use of disks that are 90, 120, 130 and 300 mm in diameter. A mini disk, 64 mm in diameter is also used for digital audio.

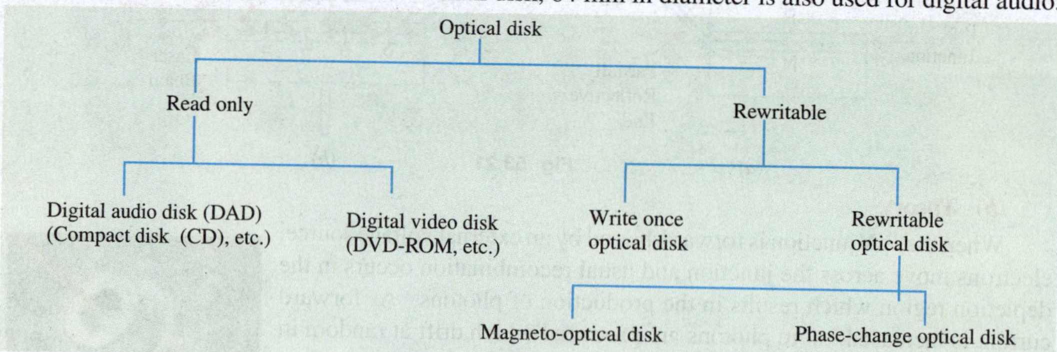

Fig. 53.23

The optical disks have several advantages over semiconductor memories. Some of these include their larger data storage capacity, shorter access time and smaller size. Therefore they are used in terminal equipment of computers as well as in audio visual equipment.

53.13. Read-only Optical Disks Equipment

Fig. 53.24 shows an optical equipment for reading data from digital audio (compact) disks. Compact disks (CDs) which are 120 mm in diameter are typical digital audio disks. Compact disks usually means digital audio compact disk, but it also includes the read-only memory (CD-ROM) for data memory and interactive compact disk (CD-I) for multimedia use.

Audio information (*i.e.* sound) is digitally recorded in stereo on the surface of a CD in the form of microscopic "pits" and "flats". As seen from Fig. 53.24, the light emitted from the laser diode passes through the lens and is focussed to a diameter of about 1 μm on the surface of a disk. As the CD rotates, the lens and beam follow the track under control of a servo motor. The laser light which is altered by the pits and flats along the recorded track is reflected back from the track through the lens and optical system to infrared photodiodes. The signal from the photodiodes is then used to reproduce the digitally recorded sound.

A CD-Rom

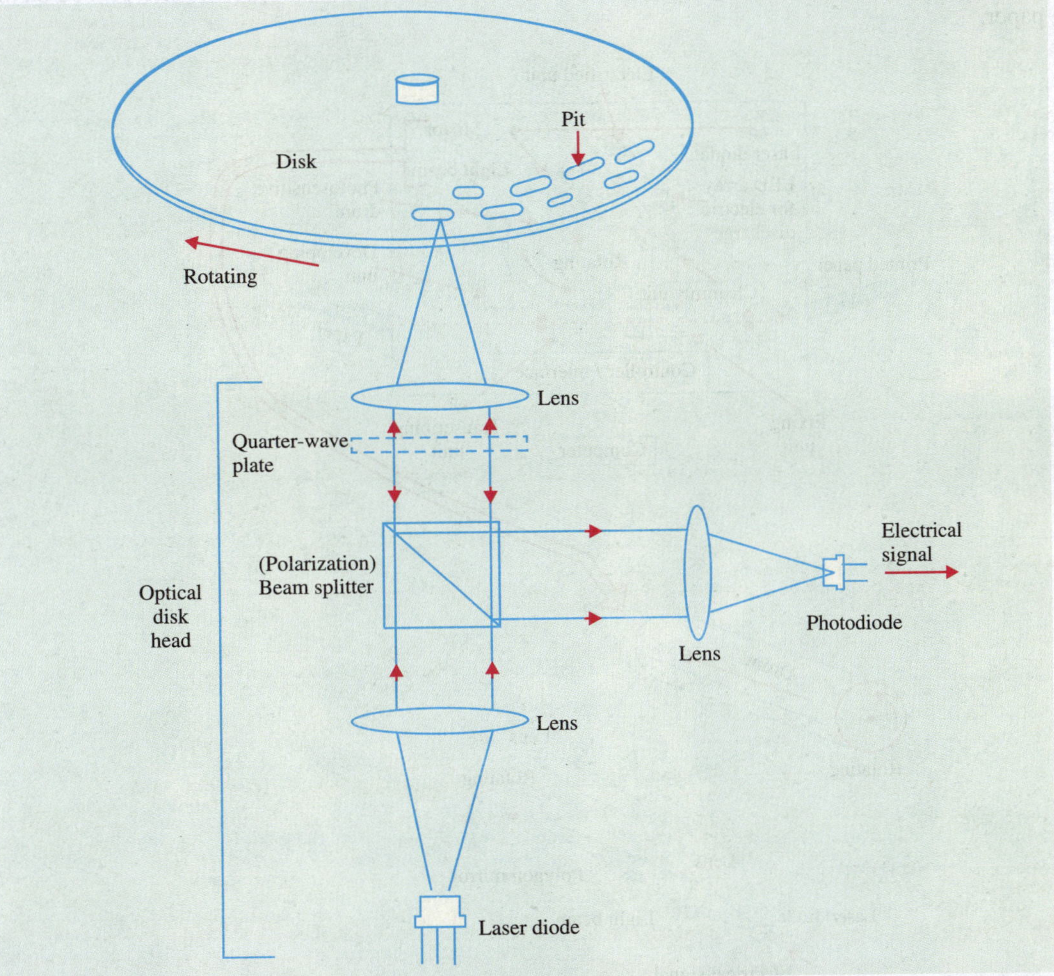

Fig. 53.24

53.14. Printers Using Laser Diodes

There are two types of optical sources usually used in printers ; **(1)** laser diodes and **(2)** LED arrays. The printers using laser diodes are called laser beam printers (or simply laser printers). These are one of the most attractive type of equipment in office automation in today's world. Words and figures can be printed rapidly and more easily by a laser printer than by other types of printers.

(Courtesy optical semiconductor devices by M.Fukuda published by John Weliy & Sons Inc.)

Fig. 53.25 shows a simplified diagram of a laser printer. As seen the laser diode is driven by modulated signals from the computer. The optical beam after passing through the lens is reflected by the rotating polygon mirror and scanned on the photosensitive drum. The drum is homogeneously charged when it passes through the charging unit consisting of an LED array. The homogeneous electrification is partially erased in accordance with the scanned optical beam. This is because of the fact that the electrical resistance at the light-irradiated part decreases and the electric charge is released. This causes the signals (*i.e.* data) from the computer to be written on to the drum. At the developing unit, an electrically charged powder (called toner) is electrostatically attached to the written parts. At the transcribing unit, the powder is transferred to the paper. Next, the transferred pattern is fixed by heating and pressing at the fixing unit. The data from the computer is thus printed on the paper.

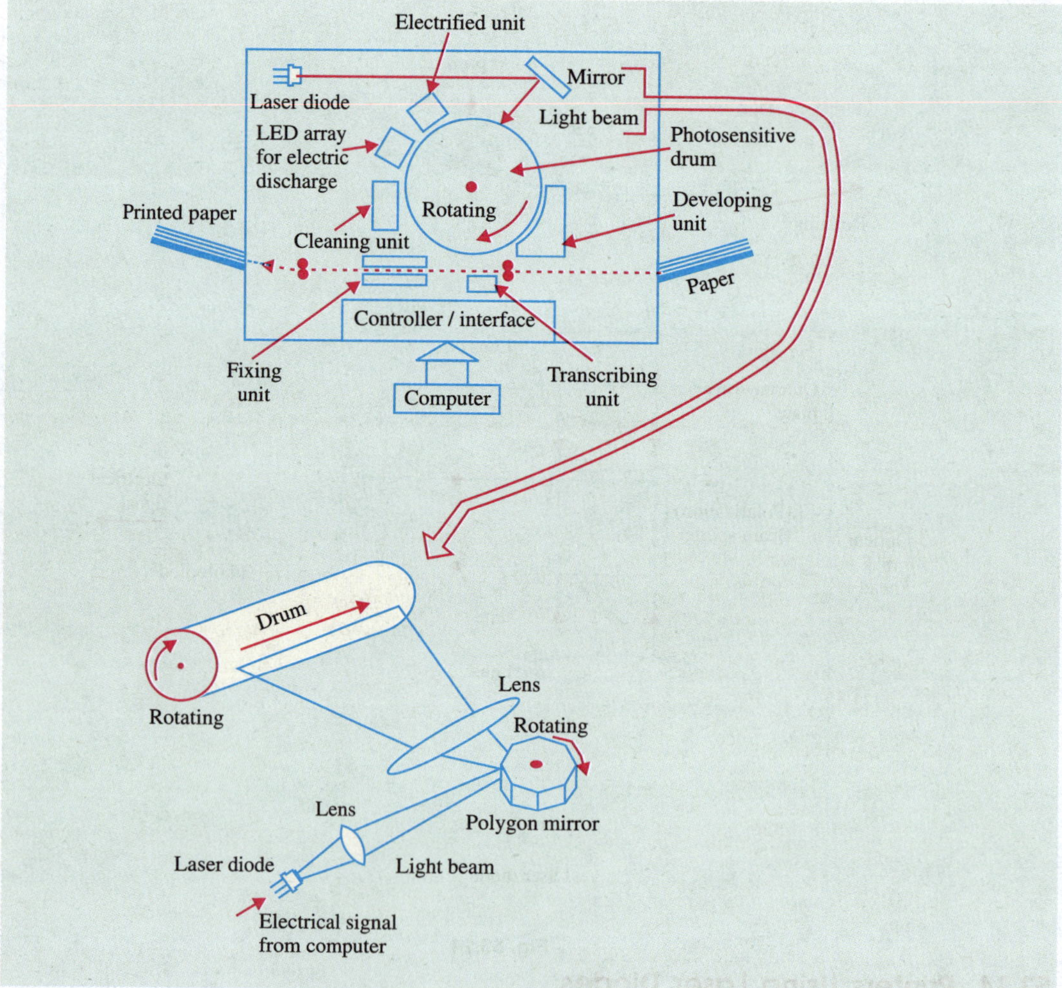

Fig. 53.25

53.15. Hologram Scanners

The hologram scanner is widely used in various equipment and is ordinarily used in bar-code readers in point-of-sale systems (such as super marked checkout counters). It is also used in laser printers for scanning the laser beam on the drum precisely.

Fig. 53.26 shows a simplified schematic of a hologram scanner. As seen, the optical beam for reading the bar-code is focussed by a lens through the hologram disk and scanned on the bar-code by rotating the hologram disk. Gratings with coaxial circles are formed on the hologram disk. This

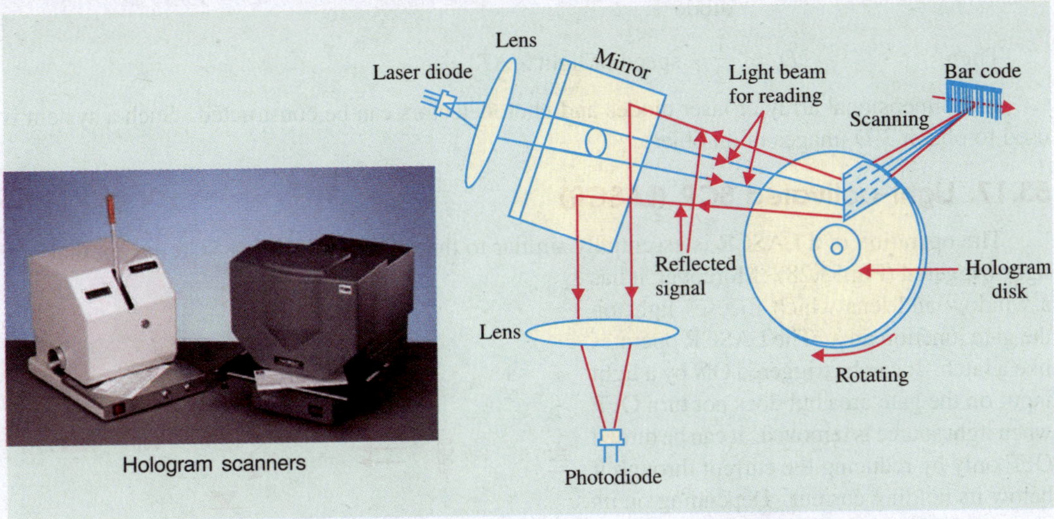

Fig. 53.26

causes the incident laser beam to bend at the grating by an amount determined by the grating pitch. The reflected light modulated according to the bar-code is reflected by the mirror and monitored by the photodiode. The monitored optical signal is then translated into an electrical signal.

53.16. Laser Range Finder

The laser diodes along with photodiodes can be used to measure the range (*i.e.* a distance) of an object. Fig. 53.27 shows a simple schematic of a laser range finder. As seen, the laser diode is modulated with high current pulses. The pulsed high-power beam is emitted in the direction of an object. The beam is reflected from the object. The reflected beam is detected with a photo detector (or photodiode). The range is calculated as the difference between the time the light was emitted from the laser diode and the time it was detected by the photodiode.

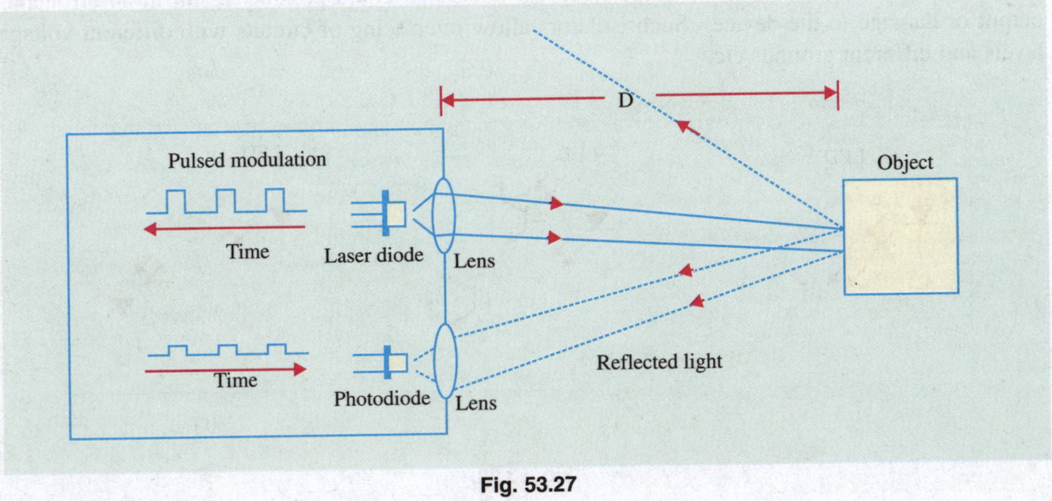

Fig. 53.27

Let D = distance between the laser range finder and the object.

ΔT = Time difference between the instance when the light was emitted from the laser diode and the instance when it was detected by the photo-diode

Then
$$D = \frac{1}{2} \times \text{speed of light} \times \Delta T$$

A 2-dimensional array of laser diodes and photodetectors can be constructed. Such a system is used to obtain 3-D images of an object.

53.17. Light-activated SCR (LASCR)

The operation of a LASCR is essentially similar to that of a conventional SCR except that it is light-triggered (Fig.53.28). Moreover, it has a window and lens which focuses light on the gate junction area. The LASCR operates like a latch. It can be triggered ON by a light input on the gate area but does not turn OFF when light source is removed. It can be turned OFF only by reducing the current through it below its holding current. Depending on its size, a LASCR is capable of handling larger amount of current that can be handled by a photodiode or a photo-transistor.

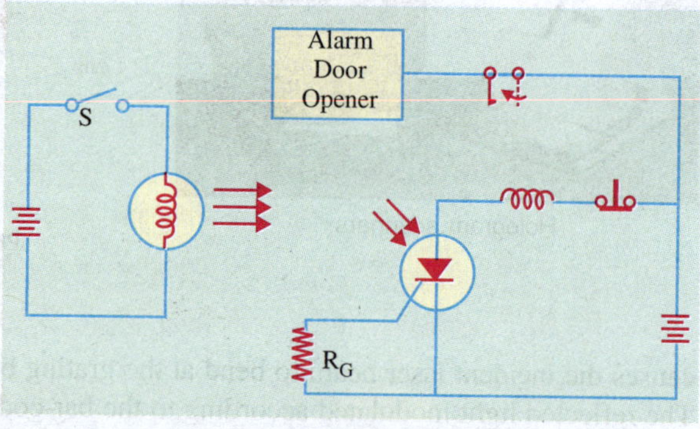

Fig. 53.28

Fig. 53.28 shows how a LASCR can be used for energizing a latching relay. The input dc source turns on the electric lamp and the resulting incident light triggers the LASCR into conduction. The anode current energizes the relay and closes the contact. It is seen that the input dc source is electrically isolated from the rest of the circuit.

53.18. Optical Isolators

Optical isolators are designed to electrically isolate one circuit from another while allowing one circuit to control the other. The usual purpose of isolation is to provide protection from high-voltage transients, surge voltages and low-level electrical noise that could possibly result in an erroneous output or damage to the device. Such isolators allow interfacing of circuits with different voltage levels and different grounds etc.

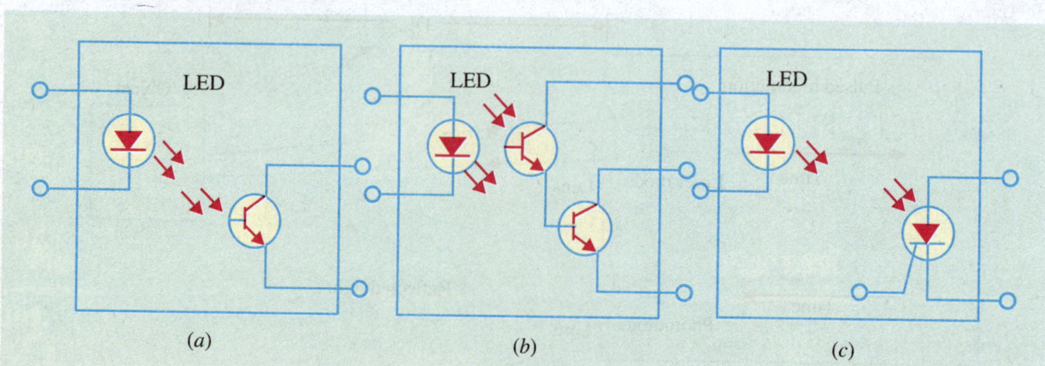

Fig. 53.29

An optical isolator (or coupler) consists of a light source such as LED and a photodetector such as a photo transistor as shown in Fig. 53. 29 (*a*) and is available in a standard *IC* package.

When LED is forward-biased, the light produced by it is transferred to the phototransistor which is turned ON thereby producing current through the external load.

Fig. 53.29 (*b*) shows a Darlington transistor coupler which is used when increased output current capability is needed beyond that provided by the phototransistor output. The LASCR output coupler of Fig. 53.29 (*c*) can be used in applications where a low-level input voltage is required to latch a high voltage relay for activating some kind of electro-mechanical device.

53.19. Optical Modulators

Light emitting *PN* junction devices such as LEDs and laser diodes are easily modulated by superimposing signals on to the injected current. This is **direct modulation.** Laser diodes in high-bit rate and long-span optical communication systems are frequently used under direct modulation.

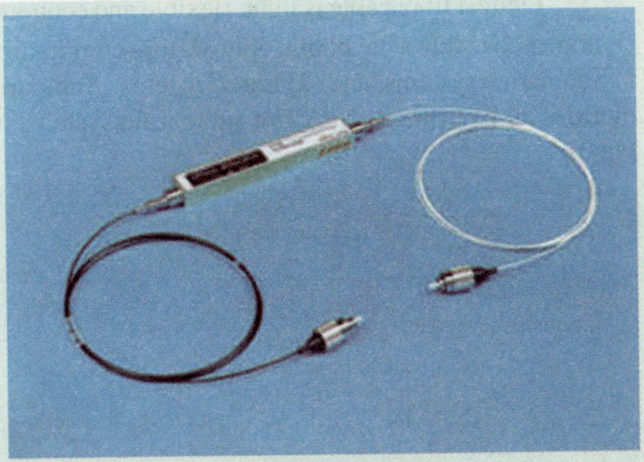

Optical modulator

However direct modulation results in ***chirping*** which limits transmission quality because of ***dispersion*** in optical fibres. An optical modulator can modulate the light output from laser diodes with little or no chirping. There are two types of optical modulators :

1. The semiconductor optical modulators
2. Optical modulators composed of dielectric materials such as ***lithium nitrate*** ($LiNO_3$)

The semiconductor optical modulators are *PN* junction diodes and can further be subdivided into two types :

1. Devices used under forward bias (as LEDs and laser diodes are used). The optical modulation in these devices is carried out by changing gain or loss within the modulators.
2. Devices used under reverse bias (*i.e.*, as photodiodes are used). Most high-performance semiconductor optical modulators are used under reverse bias. The reverse bias is needed to generate strong electric field. Optical modulation is basically performed by modulating the ***refractive index* or *optical absorption coefficient*** of the modulators. The devices which make use of refractive index phenomenon for modulation are called phase modulation type devices while those that use optical absorption coefficient phenomenon are called intensity modulation type devices.

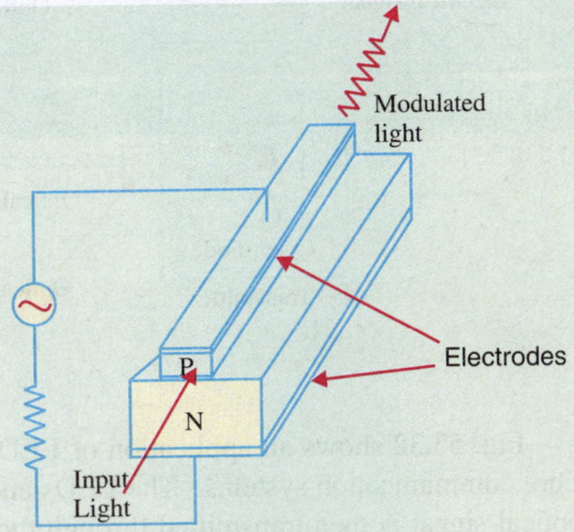

Fig. 53.30

There are several different types of optical modulators available today. But the waveguide type optical modulator is more common in use. Further there are several different waveguide type optical modulator structures possible. Fig. 53.30 shows a mesa type optical modulator structure.

It may be noted that although we have shown the structure making use of a simple *N*- and *P*-layer but in reality each layer (*N*-type or *P*-type) is made up of several different semiconductors.

53.20. Optical Fibre Communication Systems

The optical fibre communication systems (such as public communication networks and data links) are the basic infrastructure of the information hungry society. There are several advantages of the optical fibre system over metallic transmission systems as listed below :

1. Data can be transmitted at a very high-frequency over longer distances without much loss.
2. Electromagnetic induction (EMI) noise is never induced during transmission through optical fibre cables.
3. Optical fibre cable is light, flexible and economical.

Fig. 53.31 shows the public optical fibre communication system broadly divided into two groups: (1) Submarine systems, and (2) Land systems. Submarine systems have already been used to connect countries all over the world. The submarine systems help people to talk overseas without any time delay.

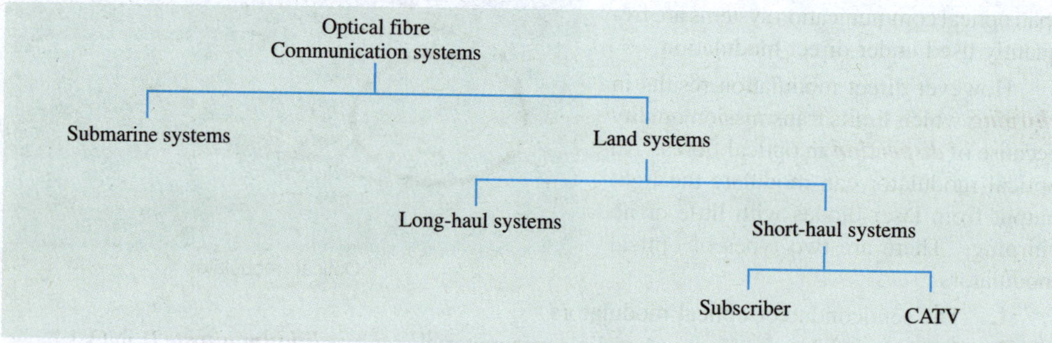

Fig. 53.31

In land systems, long-haul systems have been connected between large cities. The land systems also include systems such as subscriber systems and CATV (*i.e.* community or common antenna television, cable and telecommunication television system, or cable television system).

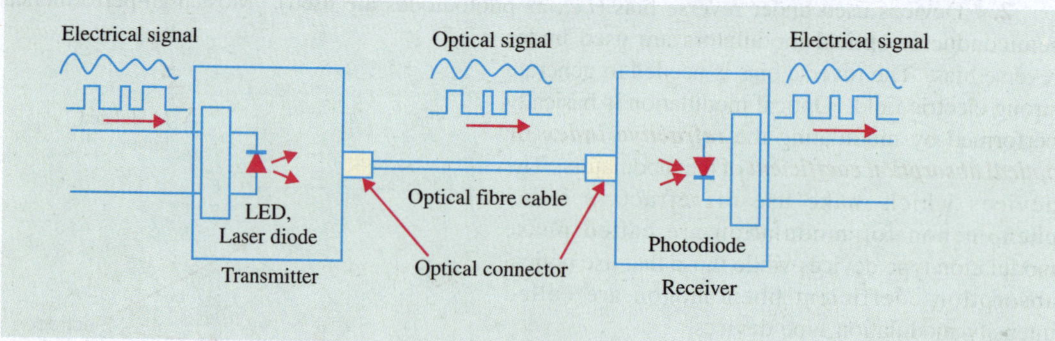

Fig. 53.32

Fig. 53.32 shows an application of LEDs, laser diodes and photodiodes in a simplified optical fibre communication systems. The LEDs and laser diodes emit light modulated with a signal. The optical signal is then transmitted through the optical fibre and is received with photodiodes on the destination side. In this type of a system LEDs or laser diodes emit the light directly through the optical fibre and therefore is referred to as direct modulation type systems. But in more recent systems, the **optical modulators** modulate the light emitted from the laser diodes and then the modulated light is transmitted through the optical fibre [refer to Fig. 53.33].

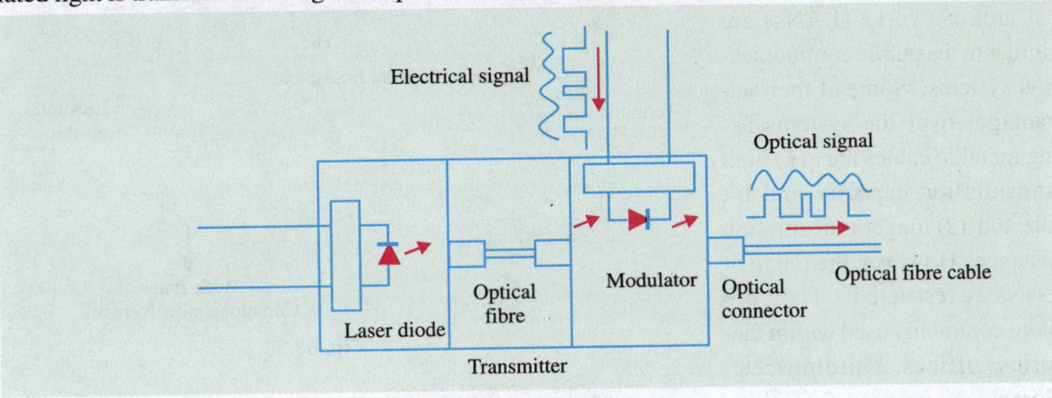

Fig. 53.33

In long-haul systems, **repeaters** (which include photodiodes and laser diodes and electronic circuits) are inserted. In the repeater, the weak optical signal being transmitted through the optical fibre is detected by the photodiode. The detected signal is reformed and amplified by the electronic circuits. The amplified signal is converted again into an optical signal by a laser diode and transmitted again through the optical fibre cable. Fig. 53.34 shows a simple schematic of a repeater.

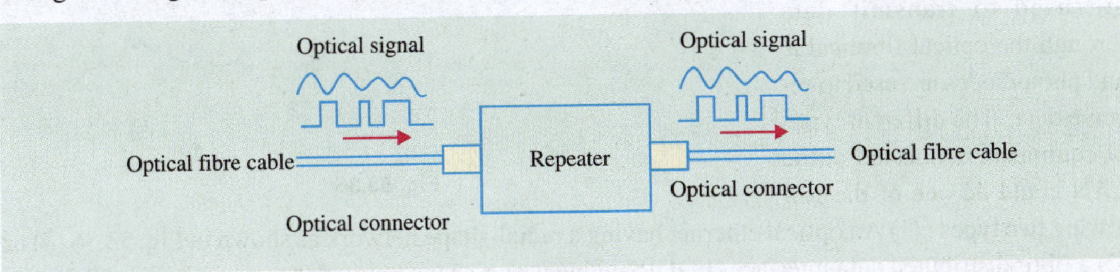

Fig. 53.34

From the modulation point of view, the optical fibre communication systems can be divided into ***digital systems*** and ***analog systems***. Most long-haul and large capacity optical fibre communication systems are digital systems. The analog systems are used for transmitting information over a short distance.

53.21. Optical Fibre Data Links

The use of optical fibre data links has wide spread in the past few decades. Its application ranges from local area networks (LANs) to the computer, digital audio and mobile fields. Several different types of LEDs and laser diodes emitting light at wavelengths ranging from visible to the infrared are used as optical sources. The transmission data rate is a function of transmission distance and varies from application to application. For computer links where the distance varies from 1 m to 100 m, the data transmission rate varies from 1 M bits/s to 100 M bits/s. For local-area-networks used in factory, office and building automation, the data transmission rate varies from 10 K bits/s to 5 M bits/s. In digital audio field, where the distance is below 1 m the data transmission rate varies from 500 bits/s to over 10 M bits/s. Similarly in mobile fields (such as ship, aircraft, train and automotive applications) where the distance could vary from 1 m to 100 m, the data transmission rate varies from 1 K bits/s to 1 M bits/s.

1. Optical fibre local area networks. The optical fibre local area networks (LANs) are similar to the public communication systems. Some of their advantages over the systems using metallic cables are : **(1)** high transmission capacity and bit rate and **(2)** longer transmission distance. However, the range of LANs is restricted. They are more commonly used within factories, offices, buildings etc. Computers, printers, facsimile machines and other office equipment are connected with each other by optical fibre cables as shown in Fig. 53.35.

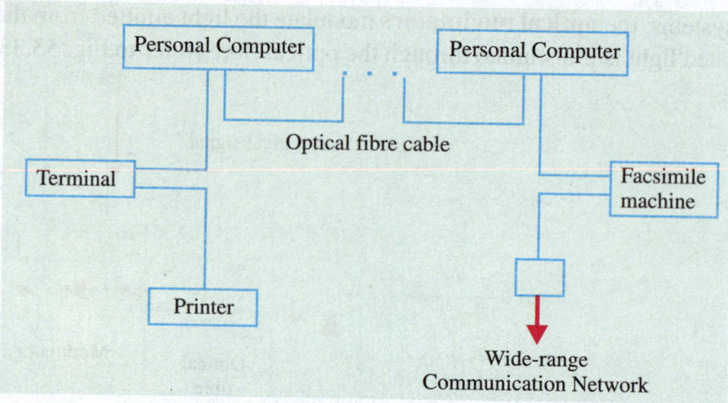

Fig. 53.35

The LEDs and laser diodes are used to transmit data through the optical fibre cable and photodiodes are used to receive data. The different types of equipment connected in the LAN could be one of the following two types : (1) An optical ethernet having a radial-shape network as shown in Fig. 53.36 (*a*) or (2) a fibre-distributed data interface (FDDI) having a ring-shape network as shown in Fig. 53.36 (*b*).

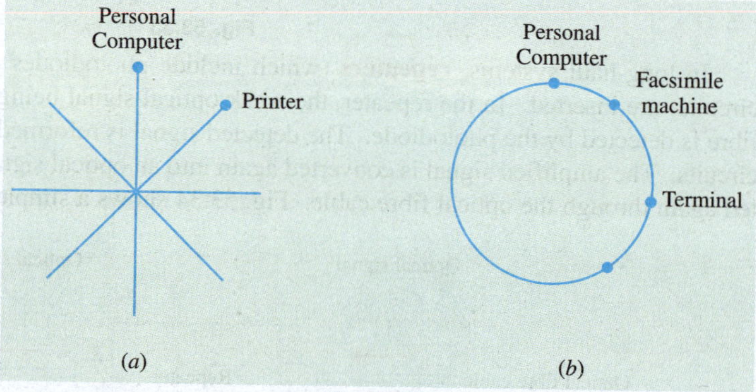

Fig. 53.36

2. Digital audio field. Fig. 53.37 shows an example of a data link in digital audio field. As seen, the optical fibre cable is used to connect compact disk (CD) player, laser disk (LD) player, digital audio tape (DAT) and tuner with the amplifier and speaker. The connection between the amplifier and everything except DAT is unidirectional. The audio digital signals from CD, LD player, DAT and tuner are converted into optical signals by LEDs or laser diodes at one end of the fibre optic cable and then transmitted through the cable to the opposite end. At the opposite end, the signals are received by photodiodes and converted into an electrical signal for amplification and finally speaker for reproduction to a sound.

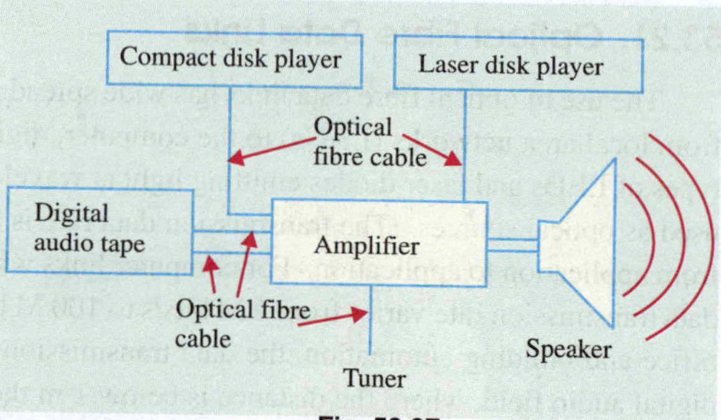

Fig. 53.37

3. Mobile fields. The optical data links are very suitable in mobile fields such as ship, aircraft, train, automotive etc. The reason is that optical data links are very compact, and light in weight than metallic data links. In addition to this, the optical data links are not subjected to noise induced by electromagnetic induction.

OBJECTIVE TESTS – 53

1. LEDs are commonly fabricated from gallium compounds like gallium arsenide and gallium phosphide because they
   (a) are cheap
   (b) are easily available
   (c) emit more heat
   (d) emit more light.
2. A LED is basically a .... P-N junction.
   (a) forward-biased
   (b) reverse-biased
   (c) lightly-doped
   (d) heavily-doped.
3. As compared to a LED display, the distinct advantage of an LCD display is that it requires
   (a) no illumination
   (b) extremely low power
   (c) no forward-bias
   (d) a solid crystal
4. Before illuminating a P-N junction photodiode, it has to be
   (a) reverse-biased
   (b) forward-biased
   (c) switched ON
   (d) switched OFF.
5. In a photoconductive cell, the resistance of the semiconductor material varies .... with the intensity of incident light.
   (a) directly
   (b) inversely
   (c) exponentially
   (d) logarithmically.
6. A photoconductive cell is known as .... cell.
   (a) phototransistor
   (b) photoresistor
   (c) photovoltaic
   (d) both (a) and (b).
7. A phototransistor excels a photodiode in the matter of
   (a) faster switching
   (b) greater sensitivity
   (c) higher current capacity
   (d) both (a) and (b)
   (e) both (b) and (c).
8. A photodarlington comprises of
   (a) a phototransistor
   (b) a transistor
   (c) a photodiode
   (d) both (a) and (b).
9. A solar cell operates on the principle of
   (a) diffusion
   (b) recombination
   (c) photo voltaic action
   (d) carrier flow.
10. Solar cells are used as source of power in earth satellites because they have
    (a) very high efficiency
    (b) unlimited life
    (c) higher power capacity per weight
    (d) both (b) and (c)
    (e) both (a) and (b).
11. The device possessing the highest sensitivity is a
    (a) photo conductive cell
    (b) photovoltaic cell
    (c) photodiode
    (d) phototransistor
12. The unique characteristics of LASER light are that it is
    (a) coherent
    (b) monochromatic
    (c) collimated
    (d) all of the above
13. The LASCR operates like a
    (a) latch (b) LED
    (c) photodiode (d) phototransistor.
14. Optical couplers are designed to .... one circuit from another.
    (a) control (b) isolate
    (c) disconnect (d) protect.
15. The main purpose of using optical isolators is to provide protection to devices from
    (a) high-voltage transients
    (b) surge voltages
    (c) low-level noise
    (d) all of the above.

16. A LED emits visible light when its ....

(*a*) P-N junction is reverse-biased

(*b*) depletion region widens

(*c*) holes and electrons recombine

(*d*) P-N junction becomes hot.

17. In LED, light is emitted because

(*a*) recombination of charge carriers takes place

(*b*) diode gets heated up

(*c*) light falling on the diode gets amplified

(*d*) light gets reflected due to lens action.

18. GaAs, LEDs emit radiation in the

(*a*) ultraviolet region

(*b*) violet-blue green range of the visible region

(*c*) visible region

(*d*) infra-red region

19. Phototransistors respond much like a conventional transistor except that, in their case, light energy is used to ....

(*a*) alter leakage current

(*b*) change base voltage

(*c*) switch it ON

(*d*) alter emitter current.

ANSWERS

1. (*d*) **2.** (*a*) **3.** *b* **4.** (*a*) **5.** (*b*) **6.** (*d*) **7.** (*e*) **8.** (*d*) **9.** (*c*) **10.** (*d*) **11.** (*d*) **12.** (*d*)
13. (*a*) **14.** (*b*) **15.** (*d*) **16.** (*c*) **17.** (*a*) **18.** (*d*) **19.** (*c*)

CHAPTER 54

SPECIAL DIODES

Learning Objectives

- Zener Diode
- Voltage Regulation
- Zener Diode as Peak Clipper
- Meter Protection
- Zener Diode as a Reference Element
- Tunneling Effect
- Tunnel Diode
- Tunnel Diode Oscillator
- Varactor Diode
- PIN Diode
- Schottky Diode
- Step Recovery Diode
- Gunn Diode
- IMPATT Diode

A major application for zener diodes is voltage regulation in dc power supplies. Zener diode maintains a nearly constant dc voltage under the proper operating conditions.

54.1. Zener Diode

It is a reverse-biased heavily-doped silicon (or germanium) *P-N* junction diode which is operated in the ***breakdown region*** where current is limited by both external resistance and power dissipation of the diode. Silicon is perferred to Ge because of its higher temperature and current capability. As seen from Art. 52.3, when a diode breaks down, both Zener and avalanche effects are present although usually one or the other predominates depending on the value of reverse voltage. At reverse voltages less than 6 V, Zener effect predominates whereas above 6 V, avalanche effect is predominant. Strictly speaking, the first one should be called Zener diode and the second one as avalanche diode but the general practice is to call both types as Zener diodes.

Zener breakdown occurs due to ***breaking of covalent bonds by the strong electric field set up in the depletion region by the reverse voltage.*** It produces an extremely large number of electrons and holes which constitute the reverse saturation current (now called Zener current, I_z) whose value is limited only by the external resistance in the circuit. ***It is independent of the applied voltage.*** Avalanche breakdown occurs at higher reverse voltages when thermally-generated electrons acquire sufficient energy to produce more carriers by collision.

(*a*) V/I Characteristic

A typical characteristic is shown by Fig. 54.1 in the negative quadrant. The forward characteristic is simply that of an ordinary forward-biased junction diode. The important points on the reverse characteristic are :

V_z = Zener breakdown voltage

$I_{z\,min}$ = minimum current to sustain breakdown

$I_{z\,max}$ = maximum Zener current limited by maximum power dissipation.

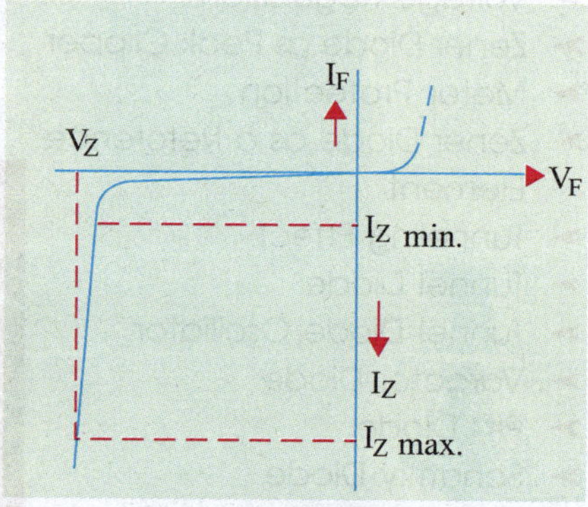

Fig. 54.1

Since its reverse characteristic is not exactly vertical, the diode possesses some resistance called ***Zener dynamic impedance****. However, we will neglect it assuming that the characteristic is truly vertical. In other words, we will assume an ideal Zener diode for which voltage ***does not change once it goes into breakdown.*** It means that V_z remains constant even when I_z increases considerably.

The schematic symbol of a Zener diode and its equivalent circuit are shown in Fig. 54.2 (*a*). The complete equivalent circuit is shown in Fig. 54.2 (*b*) and the approximate one in Fig. 54.2 (*c*) where it looks like a battery of V_z volts.

The schematic symbol of Fig. 54.2 (*a*) is similar to that of a normal diode except that the line representing the cathode is bent at both ends. With a little mental effort, the cathode symbol can be imagined to look like the letter Z for Zener.

(*b*) Zener Voltages

Zener diodes are available having Zener voltages of 2.4 V to 200 V. This voltage is temperature dependent. Their power dissipation is given by the product $V_z I_z$... maximum ratings vary from 150 mW to 50 W.

(*c*) Zener Biasing

For proper working of a Zener diode in any circuit, it is essential that it must

1. be reverse-biased;
2. have voltage across it greater than V_z;

---

* Its value is given by $Z_z = \Delta V_z / \Delta I_z$. It is negiligible as compared to large external resistance connected in the circuit.

3. be in a circuit where current is less than $I_{z\,max}$;

(d) Diode Identification

Physically, a Zener diode looks like any other diode and is recognized by its IN number such as IN 750 (10 W power) or IN 4000 (high power). Fig 54.2(*d*) shows a picture of a zener diode with $V_z = 4.7V$.

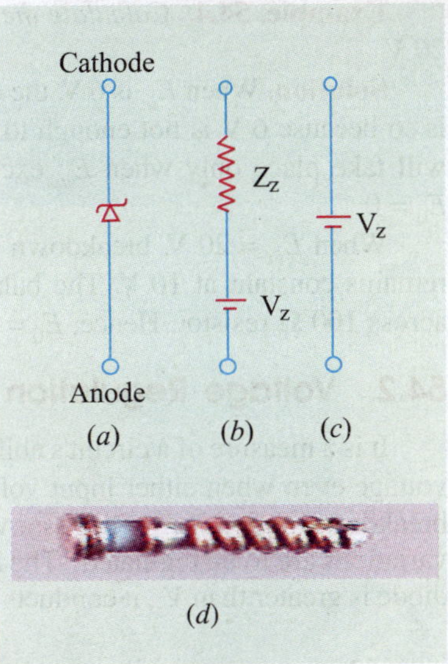

Fig. 54.2

(e) Uses

Zener diodes find numerous applications in transistor circuitry. Some of their common uses are :

1. as voltage regulators;
2. as a fixed reference voltage in a network for biasing and comparison purposes and for calibrating voltmeters;
3. as peak clippers or voltage limiters;
4. for metre protection against damage from accidental application of excessive voltage;
5. for reshaping a waveform.

Example 54.1. *Determine whether the ideal Zener diode of Fig. 54.3 is properly biased. Explain why ?*

Solution. Since positive battery terminal is connected to its cathode, the diode is reverse-biased.

Since applied voltage is less than V_z, the diode is not properly voltage-biased.

Example 54.2. *Find out if the Zener diode of Fig. 54.4 is properly-biased. If so, find diode current assuming it to be an ideal one.*

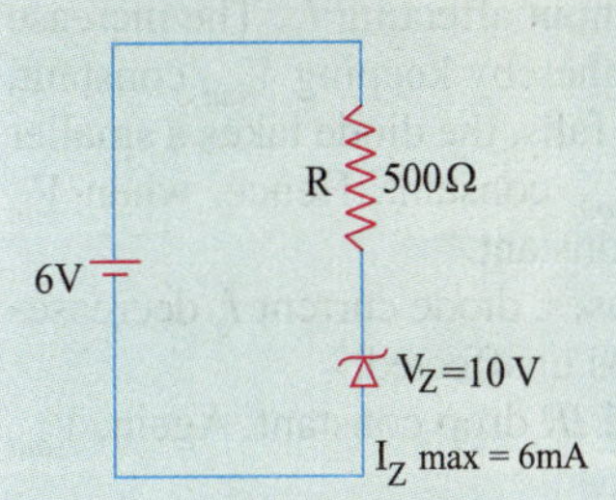

Fig. 54.3

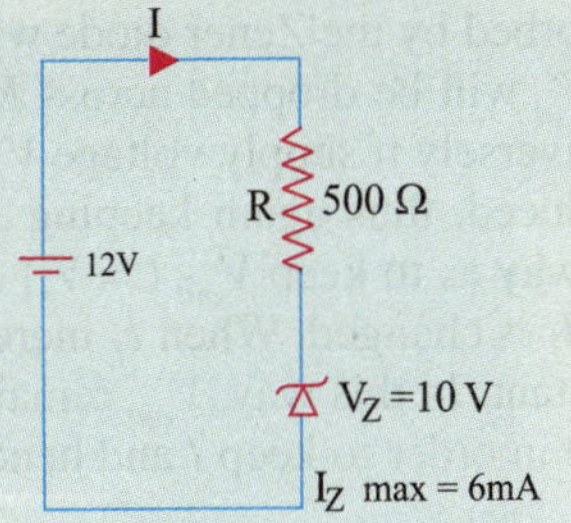

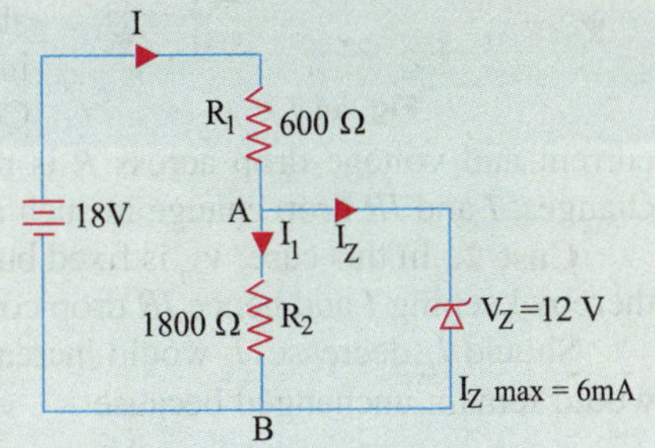

Fig. 54.4Fig. 54.5

Solution. Polarity-wise, the diode is properly-biased. Since applied voltage is greater than V_z, the diode is properly voltage-biased.

Drop across $R = 12 - 10 = 2\text{ V} \quad \therefore \quad I = 2/500 = $ **4 mA**

Since this current is less than the maximum diode current, the diode is properly-biased according to the criteria laid down in Art. 54.1 (*c*).

Example. 54.3. *Determine if the Zener diode of Fig. 54.5 is biased properly. If so, find I_z and the power dissipated by the diode.*

Solution. Since its anode is connected to the negative battery terminal, the Zener diode is correctly reverse-biased.

Now, $V_{AB} = V_Z = 12$ V.-Hence, drop across $R_1 = 18 - 12 = 6$ V

$$\therefore \quad I = 6/600 = 0.01\text{ A} = 10\text{ mA}$$

$$I_1 = 12/1800 = 6.7 \times 10^{-3}\text{ A} = 6.7\text{ mA}$$

$$I_z = I - I_1 = 10 - 6.7 = 3.3\text{ mA}$$

Since I_z is less than $I_{z\,max}$, the diode is properly-biased in every respect as per Art 54.1. Power dissipated $= V_z I_z = 12 \times 3.3 =$ **39.6 mW.**

Example. 54.4. *Calculate the value of E_0 in the given circuit of Fig. 54.6. Given E_{in} = 6 V and 20 V.* **(Electrical Engg. II, Indore Univ.)**

Solution. When E_{in} is 6 V, the diode acts like an open circuit. It is so because 6 V is not enough to cause Zener break-down which will take place only when E_{in} exceeds 10 V. Hence, in this case, $E_0 = 0$.

When E_{in} = 20 V, breakdown occurs but voltage across diode remains constant at 10 V. The balance (20 – 10) = 10 V appears across 100 Ω resistor. Hence, E_0 = drop across R = **10 V.**

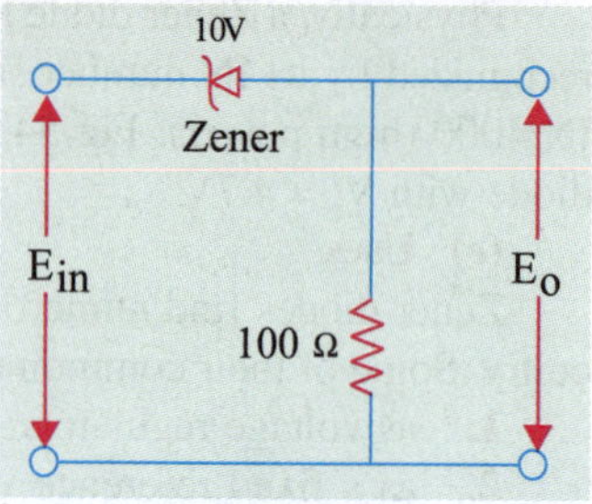

Fig. 54.6

54.2. Voltage Regulation

It is a measure of a circuit's ability to maintain a constant output voltage even when either input voltage or load current varies. A Zener diode, when working in the breakdown region, can serve as a voltage regulator. In Fig. 54.7, V_{in} is the input dc voltage whose variations are to be regulated. The Zener diode is reverse-connected across V_{in}. When p.d. across the diode is greater than V_z, it conducts and draws relatively large current through the series resistance R.

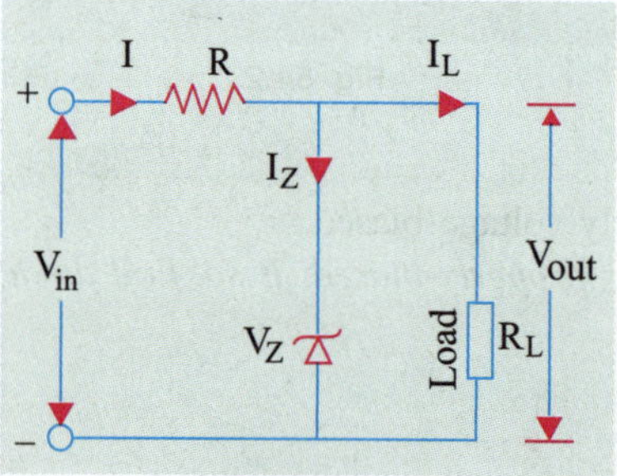

Fig. 54.7

The load resistance R_L across which a constant voltage V_{out} is required, is connected in parallel with the diode. The total current I passing through R equals the sum of diode current and load current *i.e.* $I = I_z + I_L$.

It will be seen that under all conditions $V_{out} = V_z$. Hence, $V_{in} = IR + V_{out} = IR + V_z$.

Case 1. Suppose R_L is kept fixed but supply voltage V_{in} is increased slightly. It will increase I. This increases in I will be absorbed by the Zener diode without affecting I_L. The increase in V_{in} will be dropped across R thereby keeping V_{out} constant. Conversely if supply voltage V_{in} falls, the diode takes a smaller current and voltage drop across R is reduced, thus again keeping V_{out} constant. Hence, when V_{in} changes, I and IR drop change in such a way as to keep V_{out} (= V_z) constant.

Case 2. In this case, V_{in} is fixed but I_L is changed. When I_L increases, diode current I_z decreases thereby keeping I and hence IR drop constant. In this way, V_{out} remains unaffected.

Should I_L decrease, I_z would increase in order to keep I and hence IR drop constant. Again, V_{out} would remain unchanged because

$$V_{out} = V_{in} - IR = V_{in} - (I_z + I_L)\,R$$

Incidentally, it may be noted that $R = (V_{in} - V_{out}) / (I_z + I_L)$

It may also be noted that when diode current reaches its maximum value, I_L becomes zero. In that case

$$R = \frac{V_{in} - V_{out}}{I_{Zmax}}$$

In Fig. 54.7, only one reference voltage level is available. Fig. 54.8 shows the circuits for establishing two reference levels. Here, two diodes having different Zener voltages have been connected in series.

Example 54.5. *Calculate the battery current I, I_z and I_L in the circuit of Fig. 54.9. How will these values be affected if source voltage increases to 70 V? Neglect Zener resistance.* **(Industrial Electronics, Pune Univ.)**

Solution. When V_{in} = **40 V**

Now, $V_{AB} = V_z = 10$ V

∴ drop across 3K series (or line) resistor is = 40 – 10 = 30 V

∴ I = 30/3 K = **10 mA**

$I_L = V_z/R_L = V_{AB}/R_L = 10/2\text{ K} =$ **5 mA**

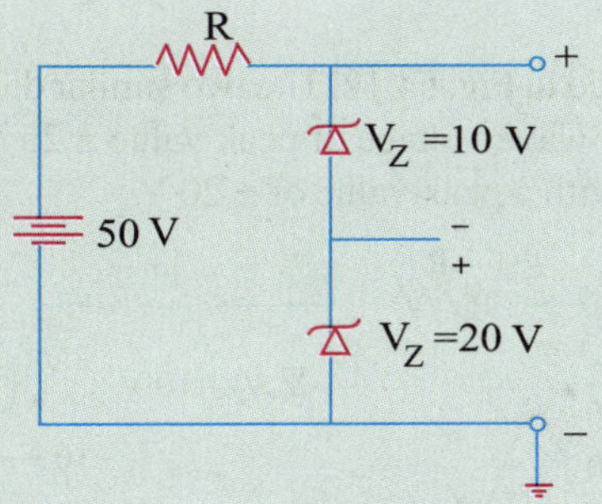

Fig. 54.8

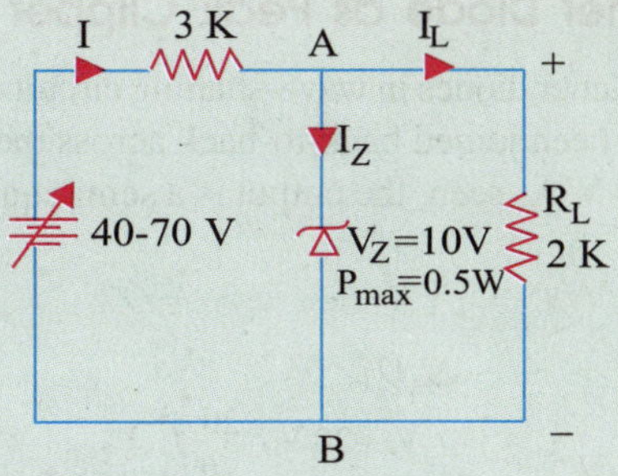

Fig. 54.9

$$\therefore \quad I_z = I - I_L = 10 - 5 = \mathbf{5\ mA}$$

Now, $\quad P_{max} = V_z . I_{z\,(max)} \quad$ or $\quad 0.5 = 10 \times I_{z\,(max)}$

or $\quad I_{z\,(max)} = 0.5/10 = 0.05\ A = 50\ mA$

Obviously, diode current of 5 mA is very much within the current range of the diode.

(b) When, $\quad V_{in} = \mathbf{70\ V}$

Drop across $\quad R = 70 - 10 = 60\ V \quad \therefore \quad I = 60/3\ K = \mathbf{20\ mA}$

$$I_L = \mathbf{5\ mA} \text{ (as before)}; \quad I_z = I - I_L = 20 - 5 = \mathbf{15\ mA}$$

Example 54.6. *Using the ideal Zener approximations, find current through the diode of Fig. 54.10 when load resistance R_L is (i) 30 K (ii) 5 K (iii) 3 K.* **(Electronics, Madurai Kamraj Univ.)**

Solution. (i) $\quad R_L = \mathbf{30\ K}$

$$V_{AB} = V_z = 30\ V; \quad \text{drop across } R = 60 - 30 = 30\ V$$

$$\therefore \quad I = 30/3\ K = 10\ mA;\ I_L = V_{AB}/R_L = 30/30\ K = 1\ mA$$

$$\therefore \quad I_z = I - I_L = 10 - 1 = 9\ mA$$

(ii) When $\quad R_L = \mathbf{5\ K}$

$$I = 10\ mA \text{ — as before};\ I_L = 30/5\ K = 6\ mA \quad \therefore\ I_z = 10 - 6 = \mathbf{4\ mA}$$

(iii) When $\quad R_L = \mathbf{3K}$

$$I = 10\ mA \text{ — as before};\ I_L = 30/3\ K = 10\ mA \quad \therefore\ I_z = I - I_L = 10 - 10 = 0$$

In this case, it is obvious that the diode is just on the ***verge of coming out of breakdown region***. If R_L is reduced further, the diode will come out of breakdown region and would no longer act as a voltage regulator.

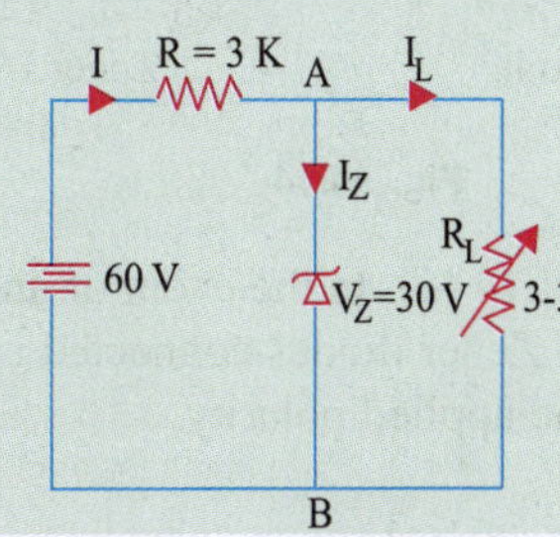

Fig. 54.10

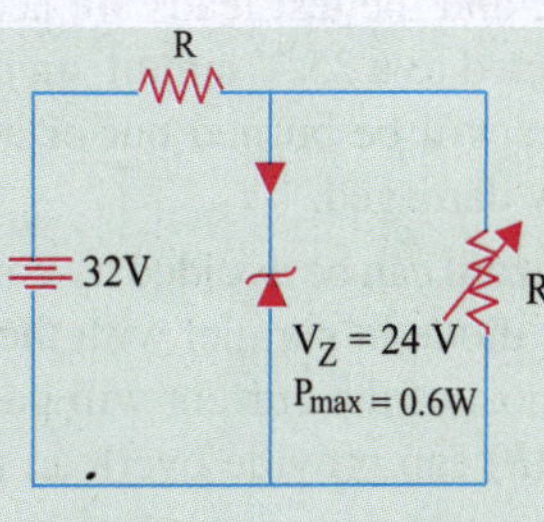

Fig. 54.11

Example 54.7. *A 24-V, 600-mW Zener diode is to be used for providing a 24-V stabilized supply to a variable load (Fig. 54.11). If input voltage is 32 V, calculate the (i) series resistance R required (ii) diode current when R_L = 1200 Ω.* **(Applied Electronics, Punjab Univ. 1991)**

Solution. (i) $\quad V_z I_z(max) = 600\ mW; \quad I_{z(max)} = 600/24 = 25\ mA$

$$\therefore \quad R = \frac{V_{in} - V_{out}}{I_{z\,(max)}} = \frac{32 - 24}{25 \times 10^{-3}} = \mathbf{320\ \Omega}$$

(ii) When $\quad R_L = 1200\,\Omega, \quad I_L = V_z/R_L = 24/1200 = 20\ mA;\ I_z = 25 - 20 = \mathbf{5\ mA}$

54.3. Zener Diode as Peak Clipper

Use of Zener diodes in wave-shaping circuits is illustrated in Fig. 54.12. The two similar diodes D_1 and D_2 have been joined back-to-back across the input sine wave voltage of peak value ± 25 V. Both have V_z = 20 V. As seen, the output is a semi-square wave with a peak value of ± 20 V.

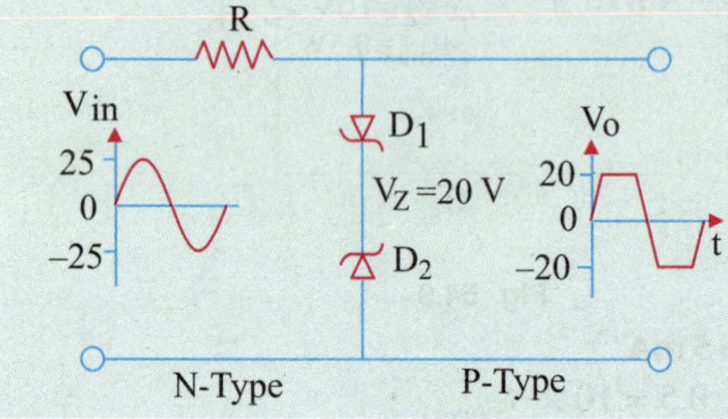

Fig. 54.12

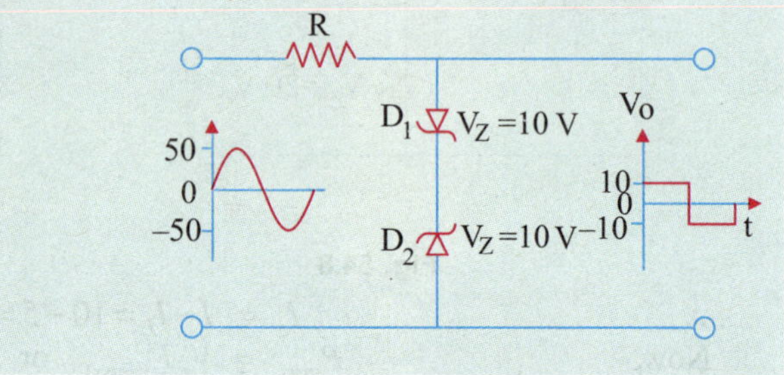

Fig. 54.13

It is well-known that a Zener diode acts like *a* 'short' (or very low resistance) in the forward direction and an 'open' in the reverse direction till it goes into breakdown at V_z. During positive input half-cycle, D_1 is shorted (being forward-biased) but D_2 acts like an open upto 20 V. Thereafter, it goes into breakdown and holds the output voltage constant till input voltage falls below 20 V in the later part of the half-cycle. At that point, D_2 comes out of the breakdown and again acts like an open across which the entire input voltage is dropped.

During the negative input half-cycle, roles of D_1 and D_2 are reversed. As a result, the output wave is clipped on both peaks as shown in Fig. 54.12.

If we increase the peak value of the input signal voltage and use Zener diodes of lesser V_z value, we can get an almost square output voltage wave from a sinusoidal input wave as shown in Fig. 54.13.

54.4. Meter Protection

Zener diodes are frequently used in volt-ohm-milliammeters (VOM) for protecting meter movement against burn-out from accidental overloads. If VOM is set to its 2.5 V range and the test leads are accidentally connected to a 25 V circuit, an unprotected meter will be burned out or at least get severely damaged.

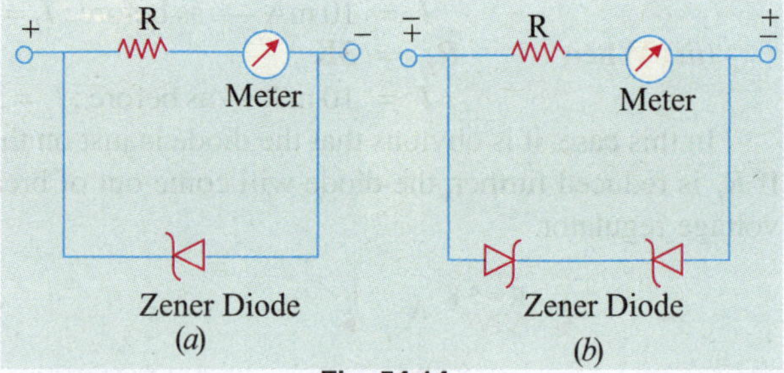

Fig. 54.14

This hazard can be avoided by connecting a Zener diode in parallel with the meter as shown in Fig. 54.14 (*a*). In the event of an accidental overload, most of the current will pass through the diode. Two Zener diodes connected as shown in Fig. 54.14 (*b*) can provide overload protection regardless of the applied polarity.

54.5. Zener Diode as a Reference Element

In many electronic circuits, it is desirable to maintain a constant voltage between two points and use it as a reference voltage for comparing other voltages against it. The difference between the two voltages is amplified and then used for performing some control function. This type of arrangement is used for power supply voltage regulator circuits, measurement circuits and servomechanism circuits. The constant-voltage characteristic in its breakdown region makes a Zener diode desirable for this application. Fig. 54.15 shows a circuit in which Zener diode is used as a reference element. The reference voltage equals the Zener breakdown voltage. The value of R is so chosen that the diode operates well within its breakdown region. The difference $(E_{in} - E_{ref})$ gives the control output.

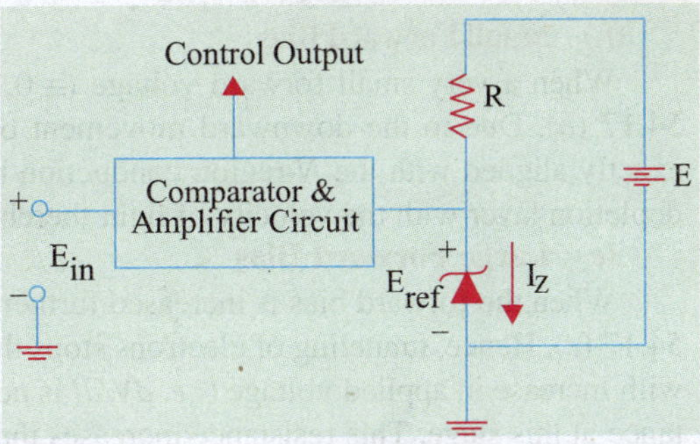

Fig. 54.15

54.6. Tunneling Effect

In a normally-doped P-N junction, the depletion layer is relatively wide and a potential barrier exists across the junction. The charge carriers on either side of the junction cannot cross over unless they possess sufficient energy to overcome this barrier (0.3 V for Ge and 0.7 V for Si). As is well-known, width of the depletion region depends directly on the doping density of the semiconductor. If a P-N junction is doped very heavily (1000 times or more)*, its depletion layer becomes extremely thin (about 0.00001 mm). It is found that under such conditions, many carriers can 'punch through' the junction with the speed of light even when they do not possess enough energy to overcome the potential barrier. Consequently, large forward current is produced even when the applied bias is much less than 0.3 V.

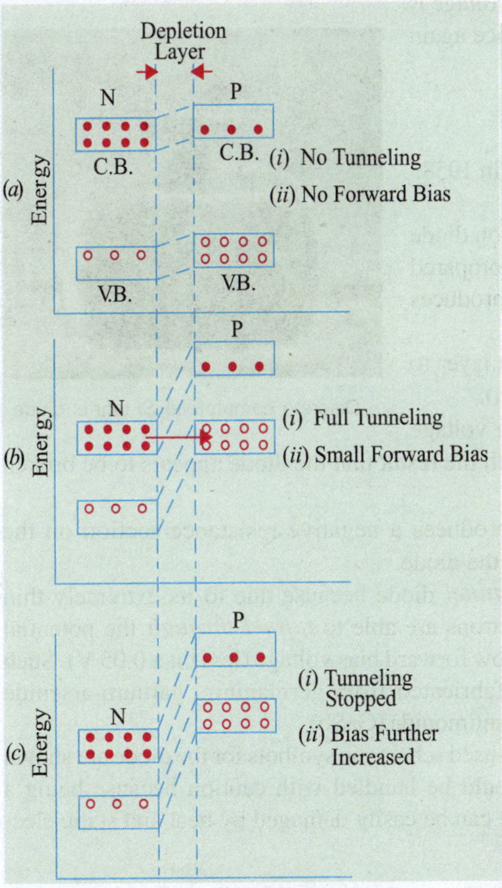

Fig. 54.17

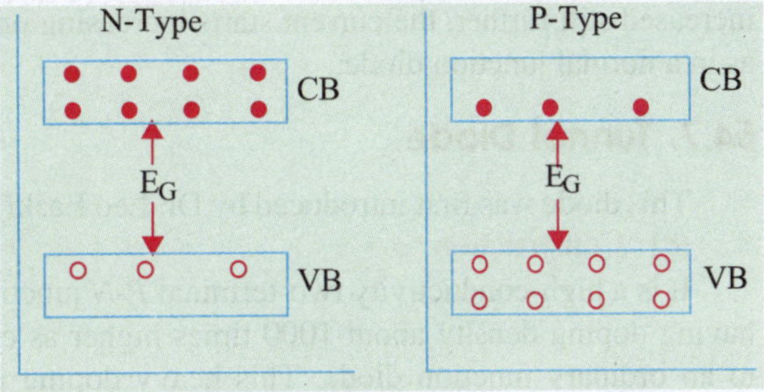

Fig. 54.16

This conduction mechanism in which charge carriers (possessing very little energy) bore through a barrier directly instead of climbing over it is called **tunneling.**

Explanation

Energy band diagrams (EBD) of *N*-type and *P*-type semiconductor materials can be used to explain this tunneling phenomenon. Fig. 54.16 shows the energy band diagram of the two types of silicon separately. As explained earlier (Art. 51.21), in the *N*-type semiconductor, there is increased concentration of electrons in the conduction band. It would be further increased under heavy doping. Similarly, in a *P*-type material, there is increased concentration of holes in the valence band for similar reasons.

(a) No Forward Bias

When the *N*-type and *P*-type materials are joined, the EBD under no-bias conditiion becomes as shown in Fig. 54.17 (*a*). The junction barrier produces only a rough alignment of the two materials and their respective valence and conduction bands. As seen, the depletion region between the two is extremely narrow due to very heavy doping on both sides of the junction. The potential hill is also increased as shown.

* Much more than even for a Zener diode.

(b) Small Forward Bias

When a very small forward voltage (≅ 0.1 V) is applied, the EBDs become as shown in Fig. 54.17 (*b*). Due to the downward movement of the *N*-region, the *P*-region valence band becomes exactly aligned with the *N*-region conduction band. At this stage, electrons tunnel through the thin depletion layer with the velocity of light thereby giving rise to a large current called peak current I_p.

(c) Large Forward Bias

When the forward bias is increased further, the two bands get out of alignment as shown in Fig. 54.17 (*c*). Hence, tunneling of electrons stops thereby decreasing the current. Since current decreases with increase in applied voltage (*i.e.* *dV/dI* is negative), the junction is said to possess negative resistance at this stage. This resistance increases throughout the negative region.

However, it is found that when applied forward voltage is increased still further, the current starts increasing once again as in a normal junction diode.

Discrete commercial *Si* tunnel diode

54.7. Tunnel Diode

This diode was first introduced by Dr. Leo Easki in 1958.

(a) Construction

It is a high-conductivity two-terminal *P-N* junction diode having doping density about 1000 times higher as compared to an ordinary junction diode. This heavy doping produces following three unusual effects :

1. Firstly, it reduces the width of the depletion layer to an extremely small value (about 0.00001 mm).
2. Secondly, it reduces the reverse breakdown voltage to a very small value (approaching zero) with the result that the diode appears to be broken down for any reverse voltage.
3. Thirdly, it produces a negative resistance section on the *V/I* characteristic of the diode.

It is called a ***tunnel*** diode because due to its extremely thin depletion layer, electrons are able to ***tunnel through*** the potential barrier at relatively low forward bias voltage (less than 0.05 V). Such diodes are usually fabricated from germanium, gallium-arsenide (GaAs) and gallium antimonide (GaSb).

The commonly-used schematic symbols for the diode are shown in Fig. 54.18. It should be handled with caution because being a low-power device, it can be easily damaged by heat and static electricity.

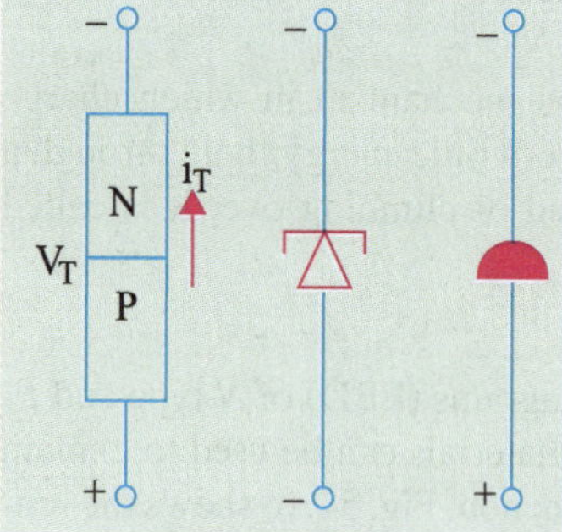

Fig. 54.18

(b) V/I Characteristic

It is shown in Fig. 54.19. As seen, forward bias produces immediate conduction *i.e.* as soon as forward bias is applied, significant current is produced. The current quickly rises to its peak value I_p when the applied forward voltage reaches a value V_p (point *A*). When forward voltage is increased further, diode current starts decreasing till it achieves its minimum value called valley current I_v corresponding to valley voltage V_v (point *B*). For voltages greater than V_v, current starts increasing again as in any ordinary junction diode.

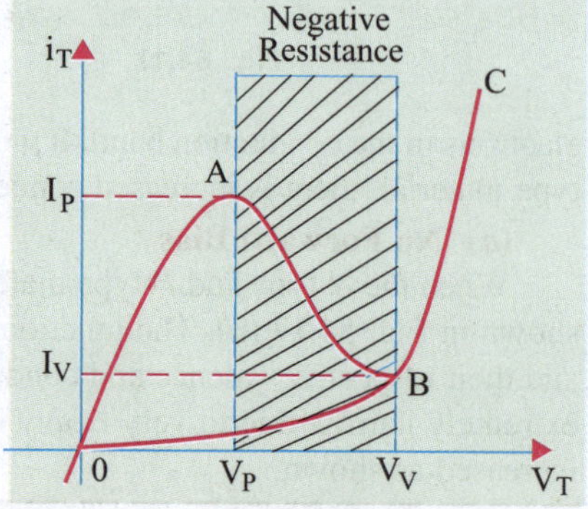

Fig. 54.19

As seen from Fig. 54.19, between the peak point *A* and valley point *B*, current decreases with increase in the applied voltage. In other words, tunnel diode possesses negative resistance (— R_N) in this region. In fact, this constitutes the most useful property of the diode. Instead of absorbing power, a negative resistance produces power. By offsetting losses in *L* and *C* components of a tank circuit, such a negative resistance permits oscillations. Hence, a tunnel diode is used as a very high frequency oscillator.

Another point worth noting is that this resistance increases as we go from point *A* to *B* because as voltage is increased, current keeps decreasing which means that diode negative resistance keeps increasing.

(c) Tunneling Theory

At zero forward bias, the energy levels of conduction electrons in *N*-region of the junction are slightly out of alignment with the energy levels of holes in the *P*-region. As the forward voltage is slightly increased, electron levels start getting aligned with the hole levels on the other side of junction thus permitting some electrons to cross over. This kind of junction crossing is called tunneling.

As voltage is increased to peak voltage (V_P), all conduction band electrons in the *N*-region are able to cross over to the valence band in the *P*-region because the two bands are exactly aligned. Hence, maximum current (called peak current I_P) flows in the circuit.

After V_P, as the applied voltage is increased, current starts decreasing because the two bands start gradually getting out of alignment. It reaches minimum value (called valley current) when the two are totally out of alignment at a forward bias of V_V (valley voltage).

For voltages greater than V_V, current starts increasing again exactly as it does in the case of an ordinary P-N junction diode.

Tunneling is much faster than normal crossing which enables a tunnel diode to switch ON and OFF much faster than an ordinary diode. That is why a tunnel diode is extensively used in special applications requiring very fast switching speeds like high-speed computer memories and high frequency oscillators etc.

(d) Diode Parameters

(i) Negative Resistance (– R_N). It is the resistance offered by the diode within the negative-resistance section of its characteristic (shown shaded in Fig. 54.19). It equals the reciprocal of the slope of the characteristic in this region.

It may also be found from the following relation $R_N = - dV/dI$.

Its value depends on the semiconductor material used (varying from – 10 Ω to – 200 Ω).

(ii) I_P/ I_V Ratio

It is almost as important a factor (particularly for computer applications) as the negative resistance of the diode.

Silicon diodes have a low I_P/I_V ratio of 3 : 1 and their negative resistance can be approximated from $R_N = - 200/I_P$. Such diodes are used mainly for switches operating in high ambient temperatures.

Germanium diodes have an I_P/I_V ratio of 6 : 1 and negative resistance formula $R_N = - 120/I_P$. GaAs diodes (used exclusively in oscillators) have an I_P/I_V ratio of about 10 : 1 and a negative resistance nearly equal to that of silicon diodes.

The minimum I_P/I_V ratio for GaSb diode is about 12 : 1 and has the lowest resistance of all given by $R_N = - 60/I_P$. Hence, such diodes have the lowest noise.

(e) Equivalent Circuit

The equivalent circuit of a tunnel diode is shown in Fig. 54.20. The capacitance *C* is the junction diffusion capacitance (1 to 10 pF) and (– R_N) is the negative resistance. The inductor L_S is mainly due to the terminal leads (0.1 to 4 nH). The resistance R_S is due to the leads, ohmic contact and semiconductor materials (1 – 5 Ω). These factors limit the frequency at which the diode may be used. They are also important in determining the switching-speed limit.

(f) Biasing the Diode

The tunnel diode has to be biased from some dc source for fixing its Q-point on its characteristic when used as an amplifier or as an oscillator and modulator.

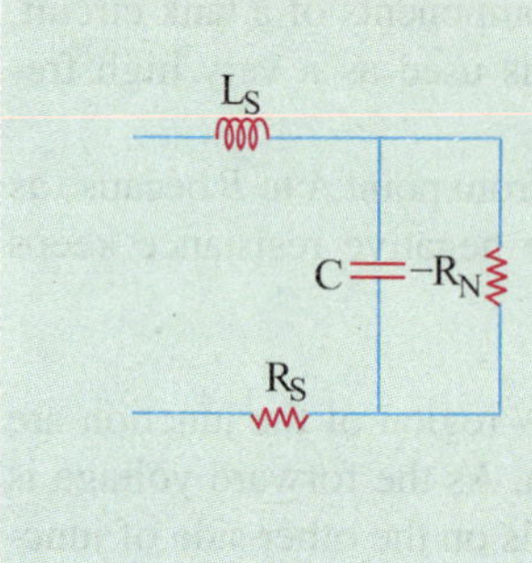

Fig. 54.20

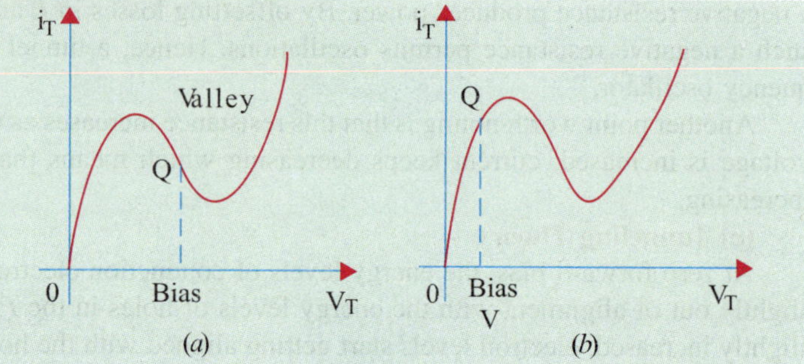

Fig. 54.21

The diode is usually biased in the negative region [Fig. 54.21 (*a*)]. In mixer and relaxation oscillator applications, it is biased in the positive-resistance region nearest zero [Fig. 54.21 (*b*)].

(g) Applications

Tunnel diode is commonly used for the following purposes :

1. as an ultra-high-speed switch-due to tunneling mechanism which essentially takes place at the speed of light. It has a switching time of the order of nanoseconds or even picoseconds;
2. as logic memory storage device – due to triple-valued feature of its curve for current.
3. as microwave oscillator at a frequency of about 10 GHz – due to its extremely small capacitance and inductance and negative resistance.
4. in relaxation oscillator circuits – due to its negative resistance. In this respect, it is very similar to the unijunction transistor.

(h) Advantages and Disadvantages

The advantages of a tunnel diode are :

1. low noise,
2. ease of operation,
3. high speed,
4. low power,
5. Insensitivity to nuclear radiations

The disadvantages are :

1. the voltage range over which it can be operated properly is 1 V or less;
2. being a two-terminal device, it provides no isolation between the input and output circuits.

54.8. Tunnel Diode Oscillator

The basic job of an oscillator is to convert dc power into ac power. Ordinarily, we do not expect an ac signal from a circuit which has no input ac source. But the circuit shown in Fig. 54.22 does exactly that as explained below.

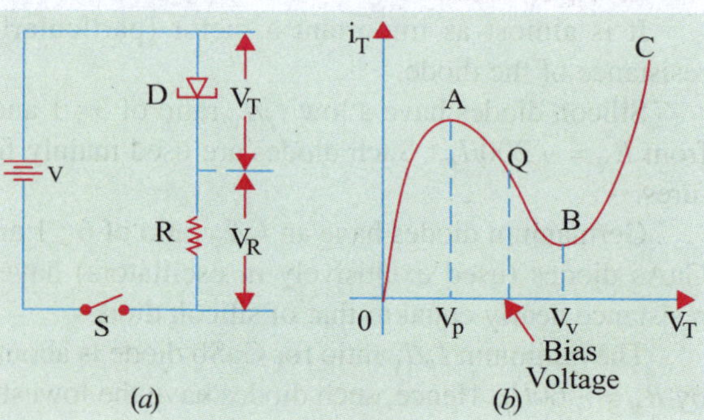

Fig. 54.22

The value of R is so selected as to bias the diode D in the negative-resistance region $A - B$. The working or quiescent point Q is almost at the centre of the curve $A - B$. When S is closed, the current immediately rises to a value determined by R and the diode resistance which are in series. The applied voltage V divides across D and R according to the ratio of their resistances.

However, as V_T exceeds V_P, diode is driven into the negative area and its resistance starts to increase (Art 54.7). Hence, V_T increases further till it becomes equal to valley voltage V_V (point B). At this point, further increase in V_T drives the diode into the positive- resistance region BC [Fig. 54.22

(*b*)]. The resulting increase in current now increases V_R but correspondingly decreases V_T, thereby bringing the diode back into the negative-resistance region. This decrease in V_T increases the circuit current till point *A* is reached when V_T equals V_P.

It describes one cycle of operation. In this way, the circuit will continue to oscillate back and forth through the negative-resistance region *i.e.* between points *A* and *B* on its characteristic. Its output across *R* is like a sine wave.

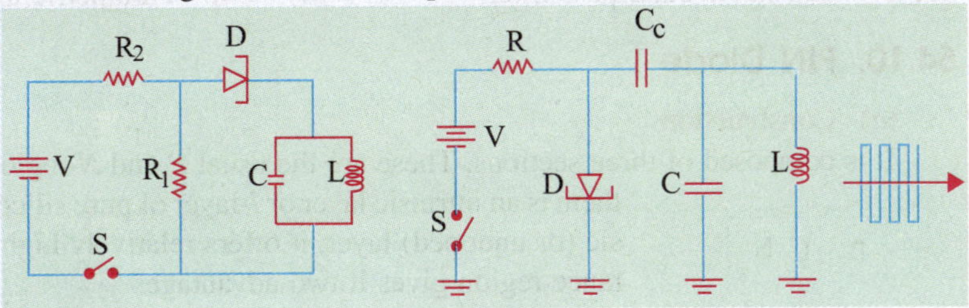

Fig. 54.23

Fig. 54.23 shows a practical circuit drawn in two slightly different ways. Here, R_2 sets the proper bias level for the diode whereas R_1 (in parallel with the *LC* tank circuit) sets proper current level for it. The capacitor C_C is the coupling capacitor. As the switch *S* is closed, the diode is set into oscillations whose frequency equals the resonant frequency of the tank.

54.9. Varactor Diode

The varactor diode is a semiconductor, voltage-dependent variable capacitor alternatively known as varicap or voltacap or voltage-variable capacitor (*VVC*) diode. Basically, it is just a reverse-biased junction diode whose mode of operation depends on its transition capacitance (C_T). As explained earlier in Art. 52.4, reverse-biased junctions behave like capacitors whose capacitance is $\propto 1/V_R^n$ where *n* varies from 1/3 to 1/2. As reverse voltage V_R is increased, depletion layer widens thereby decreasing the junction capacitance. Hence, we can change diode capacitance by simply changing V_R. Silicon diodes which are optimised for this variable capacitance effect are called varactors.

The picture, schematic symbol and a simple equivalent circuit for a varactor are shown in Fig. 54.24.

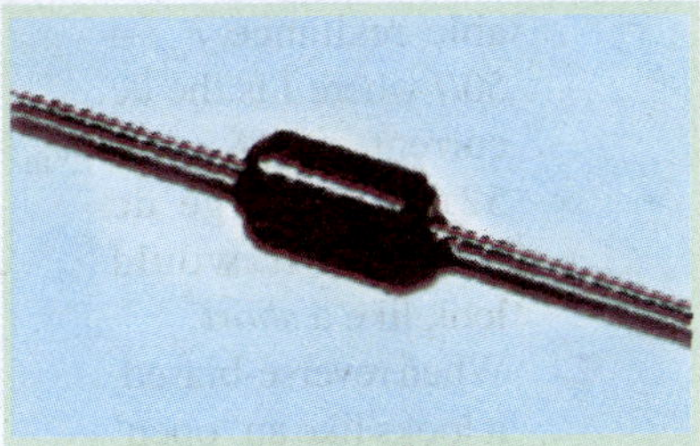
Varactor diode

Varactors may be of two types as shown in Fig. 54.25. The doping profile of the abrupt-junction diode is shown in Fig. 54.25 (*a*) and that of the hyperabrupt-junction diode in Fig. 54.25 (*b*). The abrupt-junction diode has uniform doping and a capacitive tuning ratio (*TR*) of 4 : 1. For example, if its maximum transition capacitance is 100 pF and minimum 25 pF, then its *TR* is 4 : 1 which is not enough to tune a broadcast receiver over its entire frequency range of 550 to 1050 kHz.

The hyperabrupt-junction diode has highest impurity concentration near the junction. It results in narrower depletion layer and larger capacitance. Also, changes in V_R produce larger capacitance changes. Such a diode has a tuning range of 10 : 1 enough to tune a broadcast receiver through its frequency range of nearly 3 : 1.

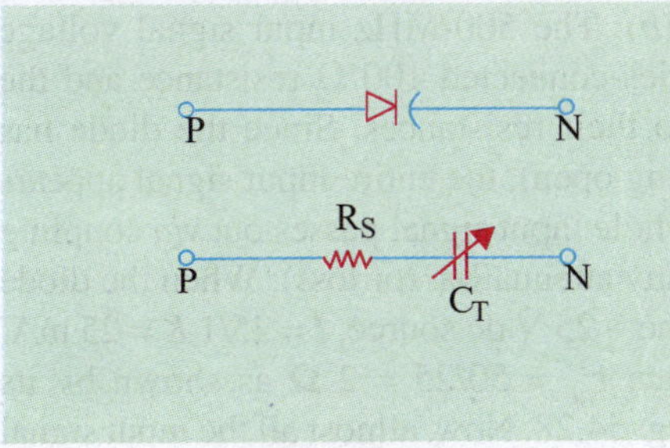

Fig. 54.24

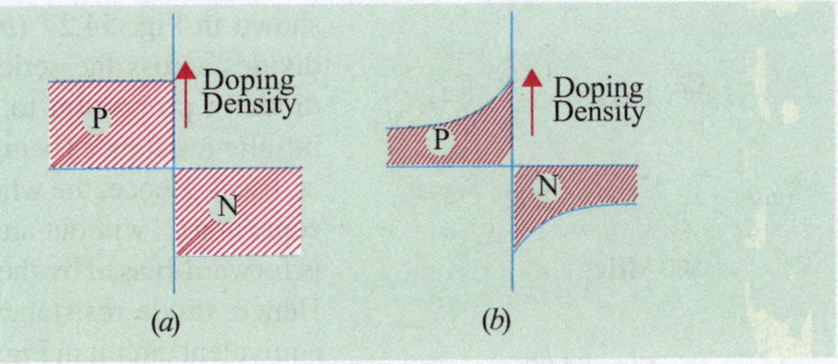

Fig. 54.25

Applications

Since the junction capacitance of a varactor is in the pF range, it is suitable for use in high-frequency circuits. Its main applications are as

1. automatic frequency control device,
2. FM modulator,
3. adjustable band-pass filter,
4. Parametric amplifier.

54.10. PIN Diode

(a) Construction

It is composed of three sections. These are the usual *P* and *N*-regions but sandwiched between them is an intrinsic layer or *I*-layer of pure silicon (Fig. 54.26). Being intrinsic (or undoped) layer, it offers relatively high resistance. This high-resistance region gives it two advantages as compared to an ordinary *P-N* diode. The advantages are :

Fig. 54.26

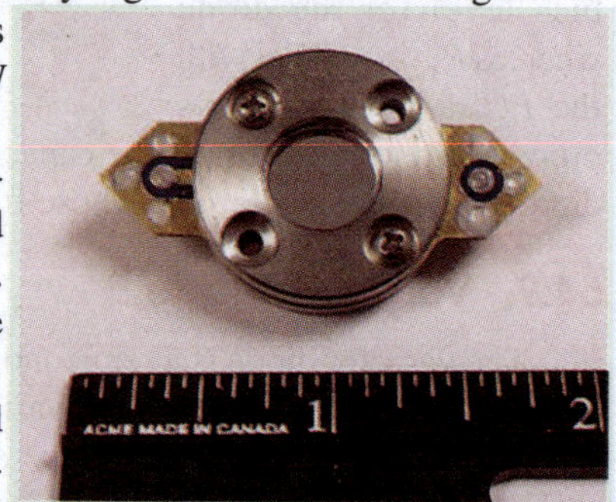

Pin diode

1. decrease in capacitance C_{pn} because capacitance is inversely proportional to the seperation of *P*-region and *N*-region. It allows the diode a faster response time. Hence, PIN diodes are used at high frequencies (more than 300 MHz);
2. possibility of greater electric field between the *P*-and *N*-junctions. It enhances the electron-hole pair generation thereby enabling PIN diode to process even very weak input signals.

(b) Diode Resistance

1. When forward-biased, it offers a variable resitance $r_{ac} \cong 50/I$ where *I* is the dc current in mA (Art. 52.2). For large dc currents, it would look like a *short*.
2. When reverse-biased, it looks like an 'open' *i.e.* it offers infinite resistance in the reverse direction.

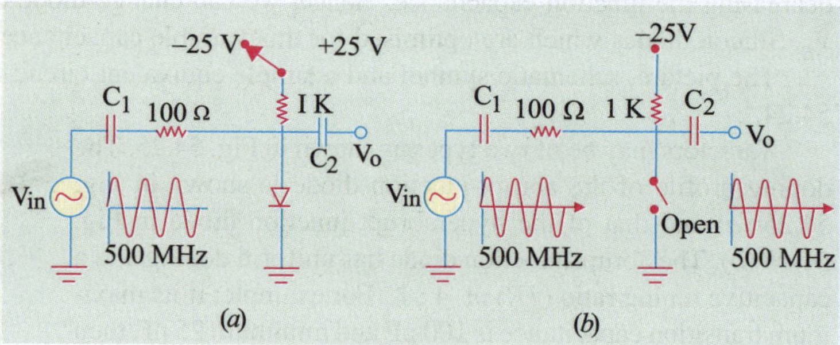

Fig. 54.27

(c) Operation

(i) High Frequency Switching. Its use in electronic high frequency switching is illustrated in Fig. 54.27.

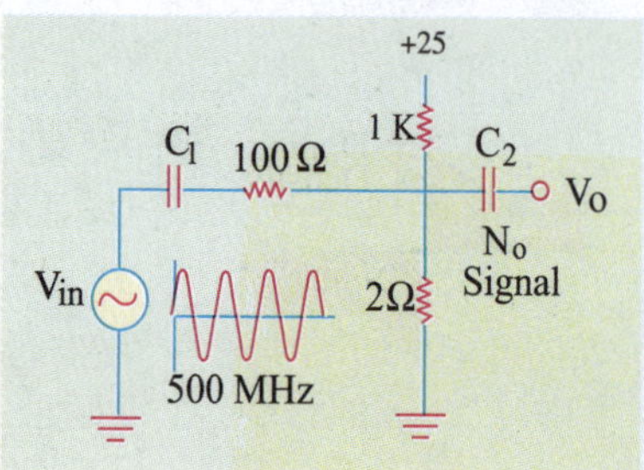

Fig. 54.28

When the diode is reverse-biased, it looks like an 'open' as shown in Fig. 54.27 (*b*). The 500-MHz input signal voltage divides across the series-connected 100 Ω resistance and the diode in proportion to their resistances. Since the diode has infinite resistance (being open), the entire input signal appears across it. Hence, the whole input signal passes out *via* coupling capacitor C_2 without any attenuation (or loss). When the diode is forward-biased by the + 25 V dc source, $I = 25/1\ K = 25$ mA. Hence, diode resistance $r_{ac} = 50/25 = 2\ \Omega$ as shown by its equivalent circuit in Fig. 54.28. Now, almost all the input signal voltage drops across 100 Ω resistance and practically none across the 2 Ω resistance. Hence, there is hardly any signal output.

In practice instead of mechanically switching the diode-biasing supply from – 25 V to + 25 V, a transistor is used to do this switching operation. In this way, we can turn a very high frequency signal (MHz range) OFF and ON with the speed of a transistor switching circuit.

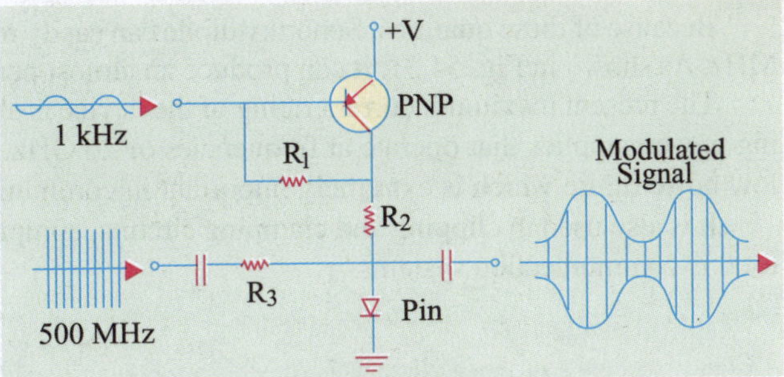

Fig. 54.29

***(ii)* Use as AM Modulator.** The way in which the 500-MHz signal is modulated at 1 kHz rate is illustrated in Fig. 54.29.

A 1 kHz signal is fed into a *PNP* transistor where it varies its dc output current at the same rate. This varying dc current is applied as biasing current to the PIN diode as shown in Fig. 54.29. It varies the diode ac resistance as seen by the 500 MHz signal. Hence, the signal is modulated at 1 kHz rate as shown.

***(d)* Applications**

(i) as a switching diode for signal frequencies upto GHz range;

(ii) as an AM modulator of very high frequency signals.

54.11. Schottky Diode

It is also called Schottky barrier diode or ***hot-carrier*** diode. It is mainly used as a rectifier at signal frequencies exceeding 300 MHz. It has more uniform junction region and is more rugged than PIN diode – its main rival.

Schottky barrier diode family expands to meet needs of power related application

***(a)* Construction**

It is a metal-semiconductor junction diode with ***no depletion layer***. It uses a metal (like gold, silver, platinum, tungsten etc.) on the side of the junction and usually an *N*-type doped silicon semiconductor on the other side. The diode and its schematic symbol are shown in Fig. 54.30.

***(b)* Operation**

When the diode is unbiased, electrons on the *N*-side have lower energy levels than electrons in the metal. Hence, they cannot surmount the junction barrier (called Schottky barrier) for going over to the metal.

When the diode is forward-biased, conduction electrons on *N*-side gain enough energy to cross the junction and enter the metal. Since these electrons plunge into the metal with very large energy, they are commonly called ***'hot-carriers'***. That is why this diode is often referred to as hot-carrier diode.

***(c)* Applications**

This diode possesses two unique features as compared to an ordinary *P-N* junction diode :

1. it is a unipolar device because it has electrons as majority carriers on both sides of the junction. An ordianry *P-N* junction diode is a bipolar device because it has both electrons and holes as majority carriers;
2. since no holes are available in metal, there is ***no depletion layer or stored charges*** to worry about. Hence, Schottky diode can switch OFF faster than a bipolar diode.

Because of these qualities, Schottky diode can easily rectify signals of frequencies exceeding 300 MHz. As shown in Fig. 54.31, it can produce an almost perfect half-wave rectified output.

The present maximum current rating of the device is about 100 A. It is commonly used in switching power supplies that operate at frequencies of 20 GHz. Another big advantage of this diode is its low noise figure which is extremely important in communication receivers and radar units etc.

It is also used in clipping and clamping circuits, computer gating, mixing and detecting networks used is communication systems.

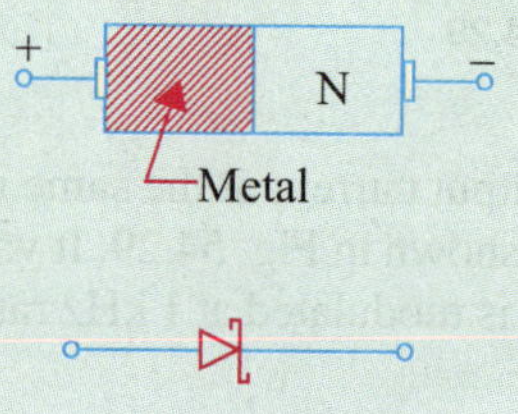

Fig. 54.30

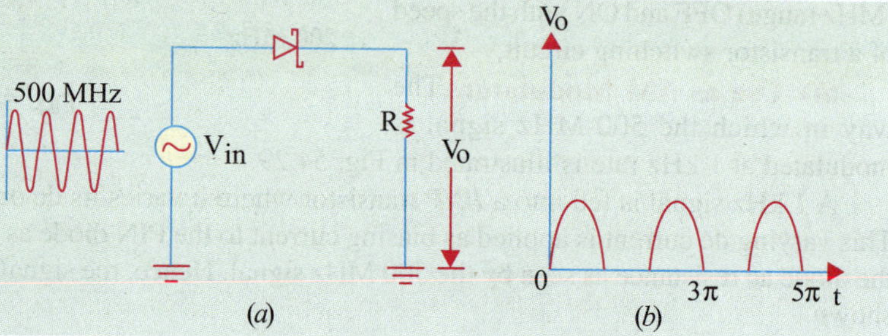

Fig. 54.31

54.12. Step Recovery Diode

It is another type of VVC diode having a ***graded doping profile*** where doping density decreases near the junction as shown in Fig. 54.32. This results in the production of strong electric fields on both sides of the junction.

(*a*) Theory

At low frequencies, an ordinary diode acts as a rectifier. It conducts in the forward direction but not in the reverse direction *i.e.* it recovers immediately from ON state to the OFF state. However, it is found that when driven forward-to-reverse by a high-frequency signal (above a few MHz), the diode does not recover immediately. Even during the negative half-cycle of the input signal when the diode is reverse-biased, it keeps conducting for a while after which the reverse current ceases abruptly in one step. This reverse conduction is due to the fact that charges stored in the depletion layer during the period of forward bias take time to drain away from the junction.

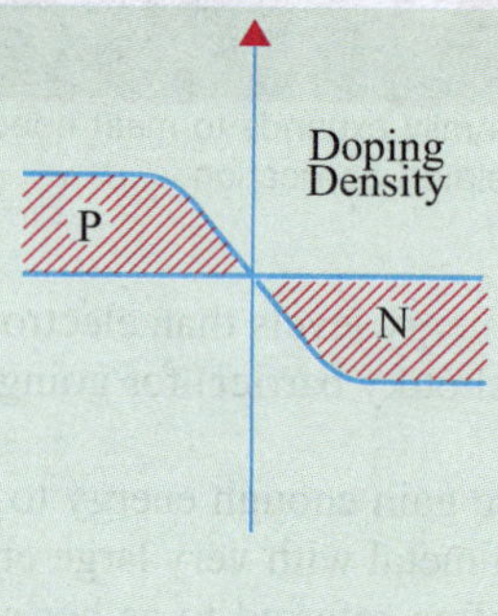

Fig. 54.32

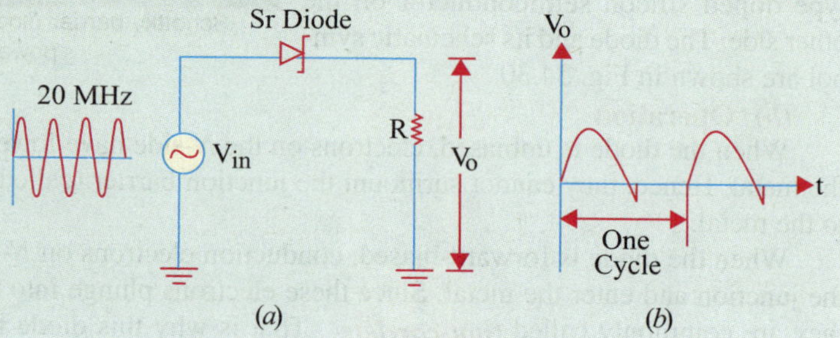

Fig. 54.33

Fig. 54.33 (*a*) shows a step-recovery diode being driven by a 20-MHz signal source. As seen from Fig. 54.33 (*b*), it conducts in the forward direction like any diode. During the reverse half-cycle, we get reverse current due to the draining of the stored charge after which current suddenly drops to zero. It looks as though diode has suddenly ***snapped open*** during the early part of the reverse cycle. That is why it is sometimes called a ***snap diode.***

The step or sudden recovery from reverse current ON to reverse current OFF gives the diode its name.

(b) Applications

Its main use is in high-frequency harmonic generator circuits as a frequency multiplier as explained below.

It is found that whenever a waveform has sudden step or transition, it contains all the harmonics of the input signal (*i.e.* multiples of its fundamental frequency). For example, the output waveform of Fig. 54.33 (*b*) contains waves of frequencies 40 MHz, 60 MHz, 80 MHz and so on.

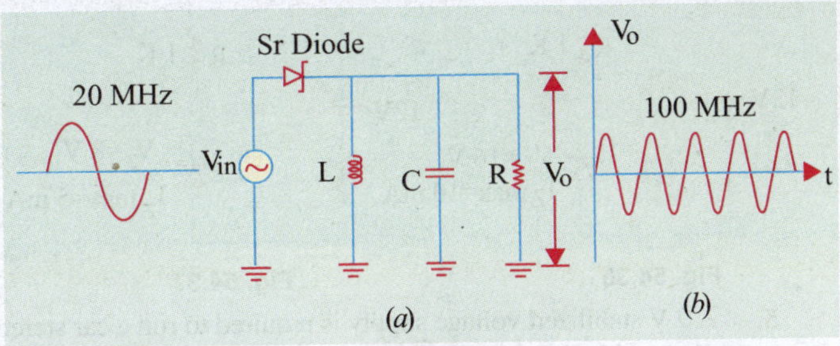

Fig. 54.34

Fig. 54.34 shows how the output of a step recovery diode can be used to drive a tuning circuit which can be made to tune out all harmonics except one *i.e.* fifth in this case (100 MHz). With an input signal of 20 MHz, the step recovery diode generates harmonics of different multiple frequencies listed above. However, the resonant *L-C* circuit is tuned to 5th harmonic of $f = 100$ MHz. Hence, all except this harmonic are filtered out of the circuit. The signal appearing across *R* is almost a pure sine wave with $f = 100$ MHz as shown separately in Fig. 54.34 (*b*).

Step-recovery diodes are also used in pulse and digital circuits for generating very fast pulses with rise time of less than 1 nanosecond.

54.13. Gunn Diode

It is a negative-resistance microwave device for oscillator applications.

As shown in Fig. 54.35, it consists of a thin slice of *N*-type gallium arsenide sandwiched between two metal conductors. The central section is *N*-gallium arsenide whereas the two outer sections are epitaxially grown from GaAs with increased doping and higher conductivity. As an oscillator, its frequencies range from 5 GHz and 100 mW output upto 35 GHz and 1 mW output. Efficiencies of 3 to 5 per cent are possible at present. Fig. 54.35(*b*) shows the picture of a Gunn diode.

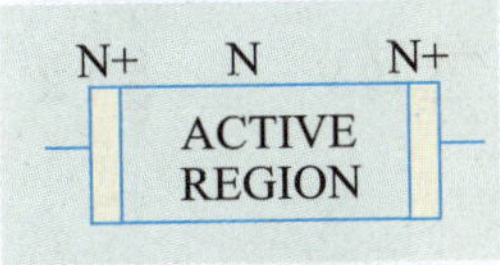

Fig. 54.35(*a*)

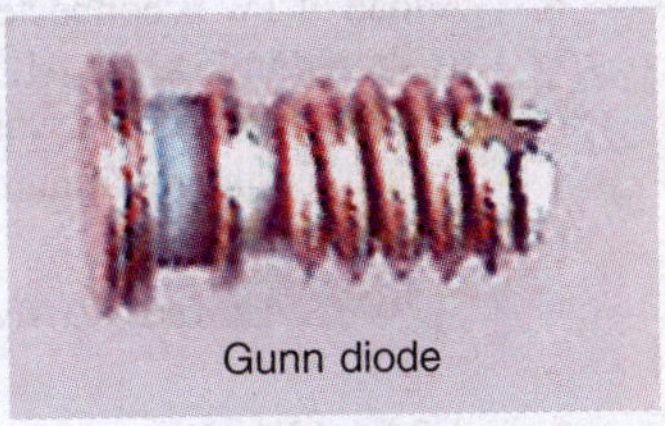
Gunn diode

Fig. 54.35(*b*)

54.14. IMPATT Diode

IMPATT stands for ***impact avalanche and transit time*** diode. As the name indicates, it is a microwave diode that utilizes the delay time required for attaining an avalanche condition plus transit time to produce a negative-resistance characteristic. It is used as a microwave oscillator within a frequency range of 10-100 GHz.

Tutorial Problems 54.1

1. Is the ideal Zener diode shown in Fig. 54.36 properly biased ? If not, explain why ? **[No]**
2. Check up if the diode in Fig. 54.37 is biased properly for normal operation. What is the current taken by the diode ? **[Yes : 4 mA]**
3. Check up if Zener diode of Fig. 54.38 is reverse-biased as well as properly voltage-biased. Calculate diode current and power dissipation. **[6 mA, 96 mW]**
4. Using ideal Zener diode approximations, find the minimum and maximum currents through the diode in Fig. 54.39. **[4 mA, 8 mA]**

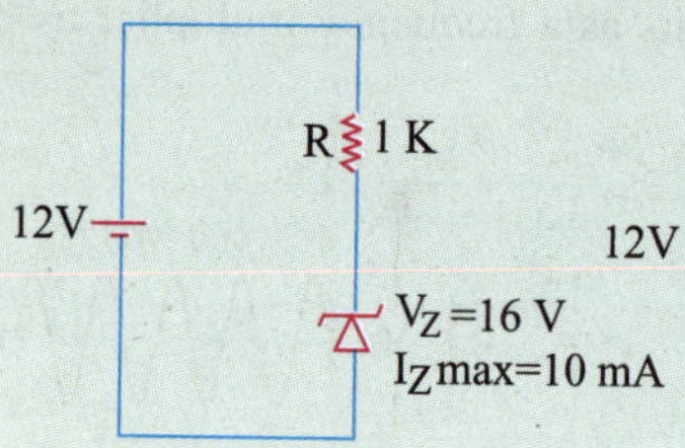

Fig. 54.36

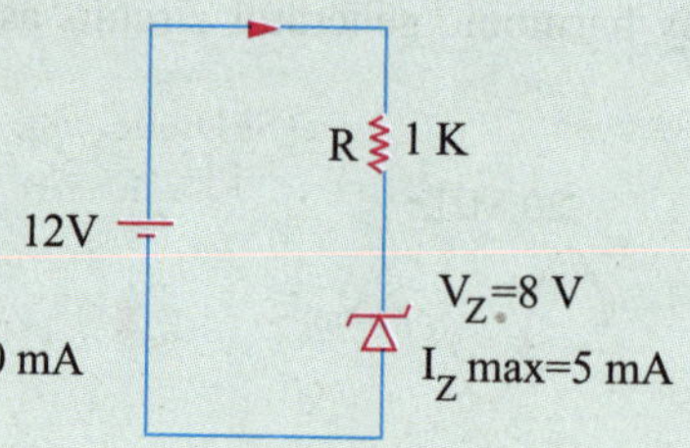

Fig. 54.37

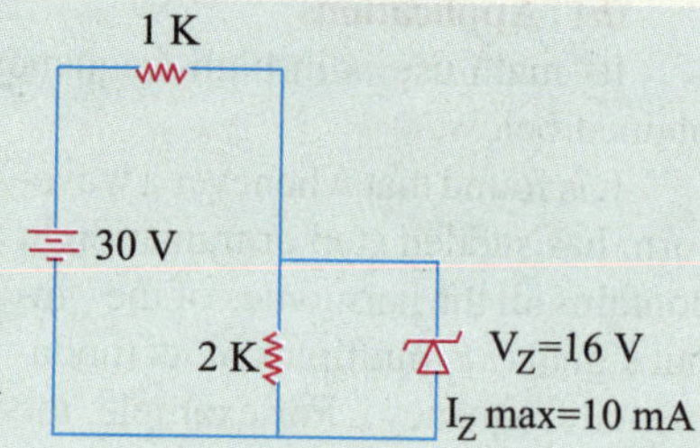

Fig. 54.38

5. A 9 V stabilized voltage supply is required to run a car stereo system from car's 12 V battery. A Zener diode with V_z = 9 V and P_{max} = 0.25 W is used as a voltage regulator as shown in Fig. 54.40.

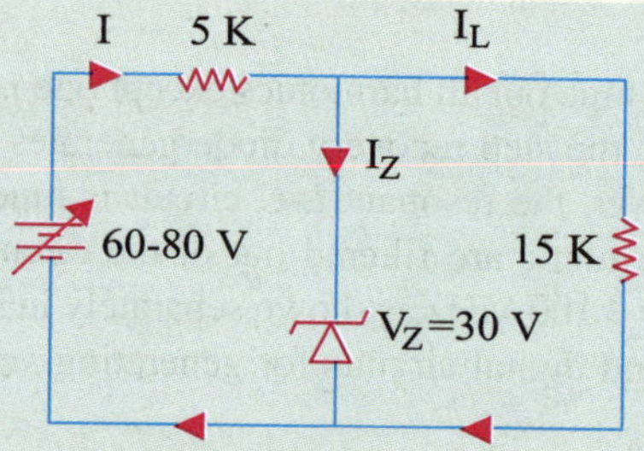

Fig. 54.39

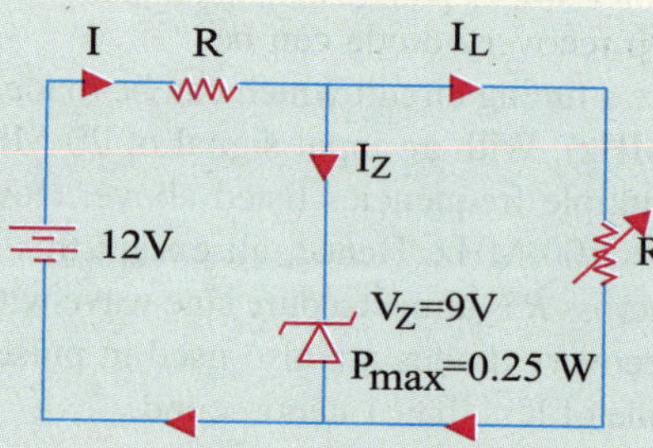

Fig. 54.40

Find the value of the series resistor R. **[108 Ω]**

6. A load of 1W is connected across a 10 V Zener regulator as shown in Fig. 54. 41. The zener current can vary between 5mA to 55mA while maintaining the voltage constant. Find the minimum and maximum voltage level at input. [62.5V & 87.5 V]

(Electronic Devices and Circuits, Nagpur Univ. Summer, 2004)

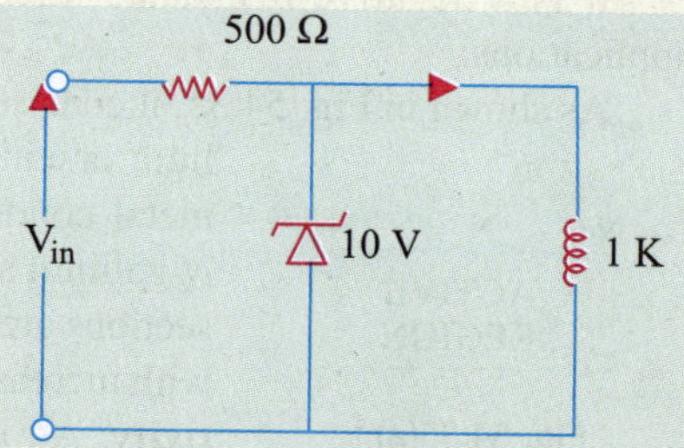

Fig. 54.41

7. A 24V, 600 mW zener diode is used for providing a 24V stabilized supply to a variable load. If the input voltage is 32V, calculate

(*i*) the value of series resistance required.

(*ii*) diode current when the load is 1200 Ω

[320 Ω, 5 mA]

(Electronics Engg., Bangalore Univ. 2003)

OBJECTIVE TESTS – 54

1. Silicon is preferred for manufacturing Zener diodes because it
   (*a*) is relatively cheap
   (*b*) needs lower doping level
   (*c*) has higher temperature and current capacity
   (*d*) has lower break-down voltage.
2. When used in a circuit, a Zener diode is always
   (*a*) forward-biased
   (*b*) connected in series
   (*c*) troubled by overheating
   (*d*) reverse-biased.
3. The main reason why electrons can tunnel through a *P-N* junction is that
   (*a*) they have high energy
   (*b*) barrier potential is very low
   (*c*) depletion layer is extremely thin
   (*d*) impurity level is low.
4. The I_P/I_V ratio of a tunnel diode is of primary importance in

(a) determining tunneling speed of electrons
(b) the design of an oscillator
(c) amplifier designing
(d) computer applications.

5. Mark the INCORRECT statement. A varactor diode
(a) has variable capacitance
(b) utilizes transition capacitance of a junction
(c) has always a uniform doping profile
(d) is often used as an automatic frequency control device.

6. The microwave device used as an oscillator within the frequency range 10-100 GHz is .... diode.
(a) Schottky
(b) IMPATT
(c) Gunn
(d) Step Recovery.

7. A PIN diode is frequently used as a
(a) peak clipper
(b) voltage regulator
(c) harmonic generator
(d) switching diode for frequencies upto GHz range.

8. Mark the WRONG statement. A Schottky diode
(a) has no depletion layer
(b) has metal-semiconductor junction
(c) has fast recovery time
(d) is a bipolar device
(e) is also called hot-carrier diode
(f) can easily rectify high-frequency signals.

9. A special purpose diode which uses metals like gold, silver or platinum on one side of the junction, *n*-type doped silicon on another side and has almost no charge storage in the junction, is a
(a) Schotty diode
(b) tunnel diode
(c) varactor diode
(d) zener diode

10. A step-recovery diode
(a) has an extremely short recovery time
(b) conducts equally well in both directions
(c) is mainly used a harmonic generator
(d) is an ideal rectifier of high-frequency signals.

11. A semiconductor device that resembles a voltage variable capacitor is called diode.
(a) tunnel
(b) PIN
(c) Schottky
(d) varactor

12. A diode that has no depletion layers and operates with hot carriers is called ...... diode.
(a) Schottky
(b) Gunn
(c) step recovery
(d) PIN

13. In switching devices, gold doping is used to
(a) improve bonding
(b) reduce storage time
(c) increase the mobility of the carrier
(d) protect the terminals against corrosion

14. When the reverse bias voltage of a varactor diode increases, its
(a) capacitance decreases
(b) lealkage current decreases
(c) negative resistance increases
(d) depletion zone decreases.

15. Which of the following are negative-resistance microwave diodes oscillator applications ?
(a) Gunn
(b) IMPATT
(c) step recovery
(d) both (a) and (b)
(e) both (b) and (c).

16. A negative-resistance microwave diode having a thin slice of a semiconductor material sandwiched between two metal conductors is called .... diode.
(a) Schottky
(b) PIN
(c) Gunn
(d) varactor.

17. Zener diodes are used primarily as
(a) rectifiers
(b) voltage regulators
(c) oscillators
(d) amplifiers.

18. The diode which is often used for voltage regulation in electronic circuits is called ...... diode

(*a*) zener (*b*) varactor
(*c*) silicon (*d*) germanium.

19. Avalanche photodiodes are preferred over PIN diodes in optical communication systems because of

(*a*) speed of operation
(*b*) higher sensitivity
(*c*) larger bandwidth
(*d*) larger power handling capacity

20. The current through the Zener diode in Fig. 54. 42.

(*a*) 33 mA
(*b*) 3.3 mA
(*c*) 2 mA
(*d*) 0 mA

(GATE ; 2004)

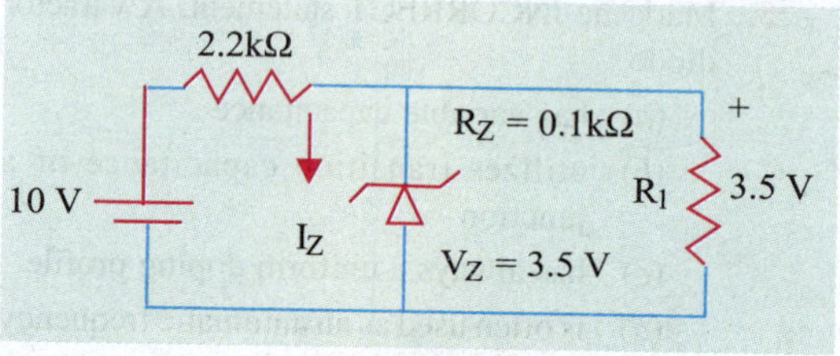

Fig. 54.42

ANSWERS

1. (*c*)	**2.** (*d*)	**3.** (*c*)	**4.** (*d*)	**5.** (*c*)	**6.** (*b*)	**7.** (*d*)	**8.** (*d*)	**9.** (*a*)	**10.** (*c*)
11. (*d*)	**12.** (*a*)	**13.** (*b*)	**14.** (*a*)	**15.** (*d*)	**16.** (*b*)	**17.** (*b*)	**18.** (*a*)	**19.** (*a*)	**20.** (*c*)

CHAPTER 55

DC POWER SUPPLIES

Learning Objectives

- Unregulated Power Supply
- Regulated Power Supply
- Rectifiers
- Single-phase Half-wave Rectifier
- Six-phase Half-wave Rectifier
- Filters
- Shunt Capacitor Filter
- Effect of increasing Filter Capacitance
- Series Inductor Filter
- The Choke Input of L-C Filter
- The R-C Filter
- The C-L-C or Pi Filter
- Bleeder Resistor
- Voltage Dividers
- Complete Power Supply
- Voltage Multipliers
- Half-wave Voltage Doubler
- Full-wave Voltage Doubler
- Voltage Tripler and Quadrupler Circuits
- Troubleshooting Power Supplies
- Controlled Rectification
- Output Voltage and Current Values in Controlled Rectifiers
- Average Values for FW Controlled Rectifier
- Silicon Controlled Rectifier
- 180° Phase Control of SCR
- SCR Controlled Load Circuit
- UJT Controlled Load Circuit
- Chopper
- Inverters
- Single Phase Inverter
- Push-pull Inverter

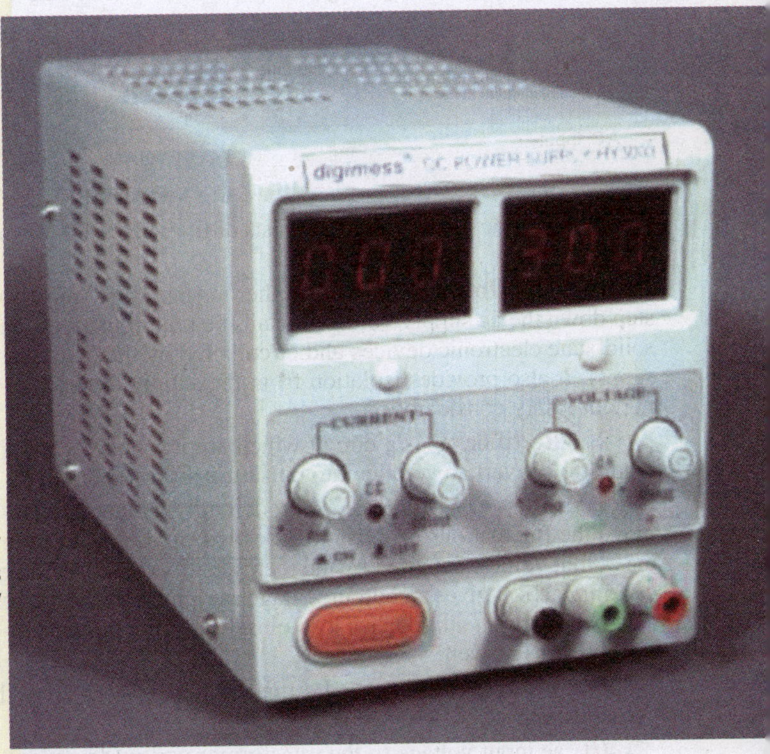

Low cost DC Power Supply

55.1. Introduction

Most of the electronic devices and circuits require a dc source for their operation. Dry cells and batteries are one form of dc source. They have the advantage of being portable and ripple-free. However, their voltages are low, they need frequent replacement and are expensive as compared to conventional dc power supplies. Since the most convenient and economical source of power is the domestic ac supply, it is advantageous to convert this alternating voltage (usually, 220 V rms) to dc voltage (usually smaller in value). This process of converting ac voltage into dc voltage is called ***rectification*** and is accomplished with the help of a

(i) rectifier *(ii)* filter and *(iii)* voltage regulator circuit.

These elements put together constitute dc power supply.

55.2. Unregulated Power Supply

An unregulated power supply is one whose dc terminal voltage is affected significantly by the amount of load. As the load draws more current, the dc terminal voltage becomes less.

55.3. Regulated Power Supply

It is that dc power supply whose terminal voltage remains almost constant regardless of the amount of current drawn from it. An unregulated supply can be converted into a regulated power supply by adding a voltage regulating circuit to it (Art 56.5).

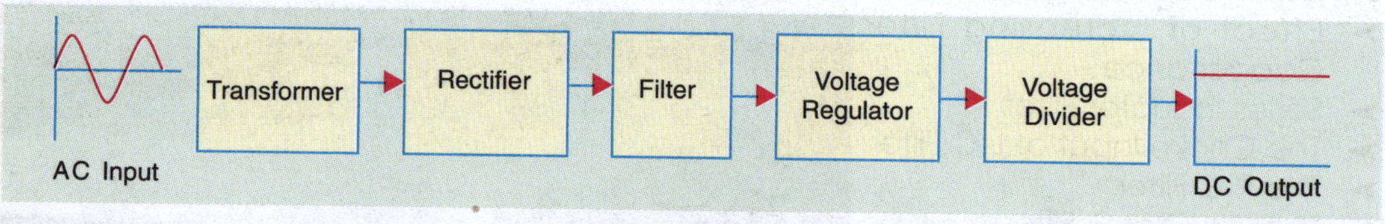

Fig. 55.1

A typical dc power supply consists of five stages as shown in Fig. 55.1.

1. Transformer. Its job is either to step up or (mostly) step down the ac supply voltage to suit the requirement of the solid-state electronic devices and circuits fed by the dc power supply. It also provides isolation from the supply line–an important safety consideration.

2. Rectifier. It is a circuit which employs one or more diodes to convert ac voltage into pulsating dc voltage.

3. Filter. The function of this circuit element is to remove the fluctuations or pulsations (called ripples) present in the output voltage supplied by the rectifier. Of course, no filter can, in practice, gives an output voltage as ripple-free as that of a dc battery but it approaches it so closely that the power supply performs as well.

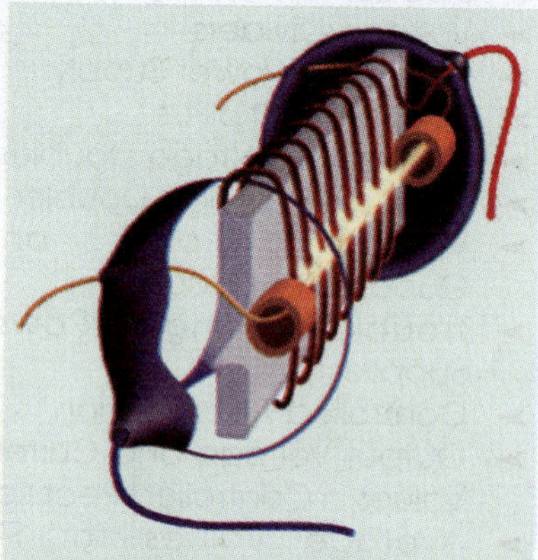

Single filament Rectifier

4. Voltage Regulator. Its main function is to keep the terminal voltage of the dc supply constant even when

(i) ac input voltage to the transformer varies (deviations from 220 V are common); or

(ii) the load varies.

Usually, Zener diodes and transistors are used for voltage regulation purposes. Again, it is impossible to get 100% constant voltage but minor variations are acceptable for most of the jobs.

5. Voltage Divider. Its function is to provide different dc-voltages needed by different electronic circuits. It consists of a number of resistors connected in series across the output terminals of the voltage regulator. Obviously, it eliminates the necessity of providing separate dc power supplies to different electronic circuits working on different dc levels.

Comments. Strictly speaking, all that is really required for conversion from ac to dc is a transformer and a rectifier (in fact, even the transformer could be eliminated if no voltage transformation is required). The filter, voltage regulator and voltage divider are mere refinements of a dc power supply though they are essential for most applications except for battery charging and running small dc motors etc.

55.4. Rectifiers

We will consider the following circuits :

1. single-phase half-wave rectifier,
2. single-phase full-wave rectifier,
3. full-wave bridge circuit,
4. three-phase half-wave rectifier,
5. three-phase full-wave rectifier,
6. six-phase half-wave rectifier,
7. three-phase bridge circuit,
8. voltage multiplier circuits.

Many semiconductor devices or systems (like car stereo systems) require a negative dc source or both a negative and a positive dc source. For the sake of simplicity, we will analyse only the positive dc power supplies. However, a positive dc supply can be converted into a negative one by simply reversing the two leads in the same way as we reverse the polarity of a dry cell.

Quite a number of integrated circuits (*ICs*) require both positive and negative source with common ground. In that case, the polarised components in the negative portion of the supply will have to be reversed. For example, its rectifier, filter capacitor and voltage/current regulation devices will have to be reversed as compared to the positive supply.

55.5. Single-phase Half-Wave Rectifier

The basic circuit of a half-wave rectifier with a resistive load (but no filter circuit) is shown in Fig. 55.2 (*a*). The alternating secondary voltage is applied to a diode connected in series with a load resistor R_L. Let the equation of the alternating secondary voltage be $V_s = V_{sm}\ wt$.

(a) Working

During the positive half-cycle of the input ac voltage, the diode *D* is forward-biased (ON) and conducts. While conducting, the diode acts as a short-circuit so that circuit current flows and hence, positive half-cycle of the input ac voltage is dropped across R_L. It constitutes the output voltage V_L as shown in Fig. 55.2 (*b*). Waveform of the load voltage is also shown in Fig. 55.2 (*b*). It consists of half-wave rectified sinusoids of peak value V_{LM}.

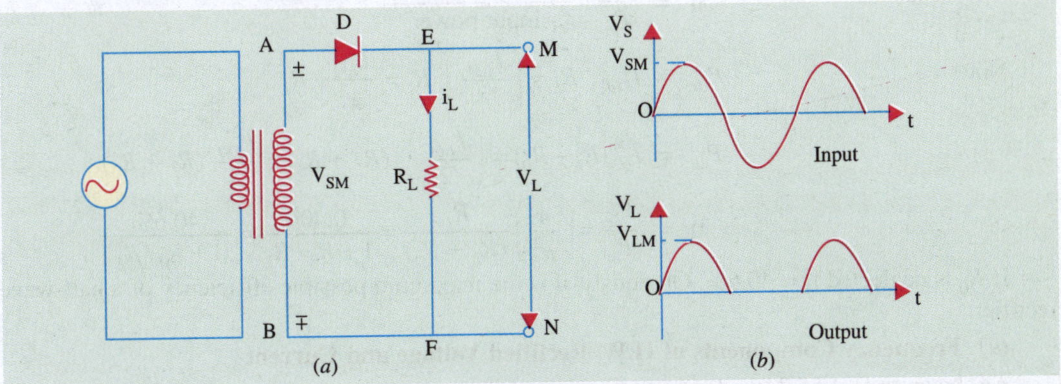

Fig. 55.2

During the negative input half-cycle, the diode is reverse-biased (OFF) and so, does not conduct *i.e.* there is no current flow. Hence, there is no voltage drop across R_L. In other words $i_L = 0$ and

$V_L = 0$. Obviously, the negative input half-cycle is suppressed *i.e.* it is not utilized for delivering power to the load. As seen, the output is not a steady dc but ***only a pulsating dc wave*** having a ripple frequency equal to that of the input voltage frequency. This wave can be observed by an oscilloscope connected across R_L. When measured by a dc meter, it will show some ***average*** positive value both for voltage and current. Since only one half-cycle of the input wave is used, it is called a half-wave rectifier. It should be noted that forward voltage drop across the diode has been neglected in the above discussion. We have, in fact, assumed an ideal diode (having zero forward resistance and infinite reverse resistance).

(*b*) Average and RMS Values

Let

$$V_{sm} = \text{maximum value of transformer secondary voltage}$$
$$V_s = \text{rms value of secondary voltage}$$
$$V_{LM} = \text{maximum value of load voltage}$$
$$= V_{sm} - \text{diode drop} - \text{secondary resistance drop}$$
$$V_L = \text{rms value of load voltage}$$
$$I_L = \text{rms value of load current}$$
$$V_{L(dc)} = \text{average value of load voltage}$$
$$I_{L(dc)} = \text{average value of load current}$$
$$I_{LM} = \text{maximum value of load current}$$
$$R_L = \text{load resistance}$$
$$R_S = \text{transformer secondary resistance}$$
$$r_d = \text{diode forward resistance}$$
$$R_0 = R_S + r_d$$
$$I_{LM} = \frac{V_{sm} - V_B}{(R_S + r_d) + R_L} = \frac{V_{sm} - V_B}{R_0 + R_L}, \quad V_{LM} = I_{LM} . R_L$$
$$V_{L(dc)} = \frac{V_{LM}}{\pi} = 0.318\, V_{LM}, \quad I_{L(dc)} = \frac{I_{LM}}{\pi} = 0.318\, I_{LM}$$
$$I_L = I_{LM}/2 = 0.5\, I_{LM} = 0.5\, V_{LM}/R_L$$

(*c*) Efficiency

The efficiency of rectification is given by the ratio of the output dc power to the total amount of input power supplied to the circuit. It is also called the conversion efficiency.

$$\eta = \frac{P_{dc}}{P_{in}} = \frac{\text{power in the load}}{\text{input power}}$$

Now,

$$P_{dc} = I_{L(dc)}{}^2 R_L = \left(\frac{I_{LM}}{\pi}\right)^2 R_L = \frac{I_{LM}^2}{\pi^2} . R_L$$
$$P_{in} = I_L{}^2 (R_L + R_0) = \left(\frac{I_{LM}}{2}\right)^2 (R_L + R_0) = \frac{I_{LM}^2}{4} (R_L + R_0)$$

$$\therefore \quad \eta = \frac{P_{dc}}{P_{in}} = \left(\frac{4}{\pi^2}\right) \frac{R_L}{(R_L + R_0)} = \frac{0.406}{1 + R_0/R_L} = \frac{40.6\%}{(1 + R_0/R_L)}$$

If R_0 is neglected $\eta = 40.6\%$. Obviously, it is the maximum possible effficiency of a half-wave rectifier.

(*d*) Frequency Components of H.W. Rectified Voltage and Current

As shown in Fig. 55.3, the load current I_L consist of a dc component $I_{L(dc)}$ and an ac component $I_{L(ac)}$. The Fourier series of the half-wave rectified current flowing through the load is found to be

$$i_L = I_{LM}\left(\frac{1}{\pi} + \frac{1}{2}\sin \omega t - \frac{2}{3\pi}\cos 2\omega t - \frac{2}{15\pi}\cos 4\,\omega t + ...\right)$$

As seen, the half-wave rectified current consists of a large number of ac components (which constitute the ripple) in addition to the dc component. The first term is I_{LM}/π which represents the dc component $I_{L(dc)}$. The second term $(I_{LM}/2)$ sin ωt has peak value of $(I_{LM}/2)$. It is called the fundamental or first harmonic component and its rms value is $I_{L1} = I_{LM}/2\sqrt{2}$.

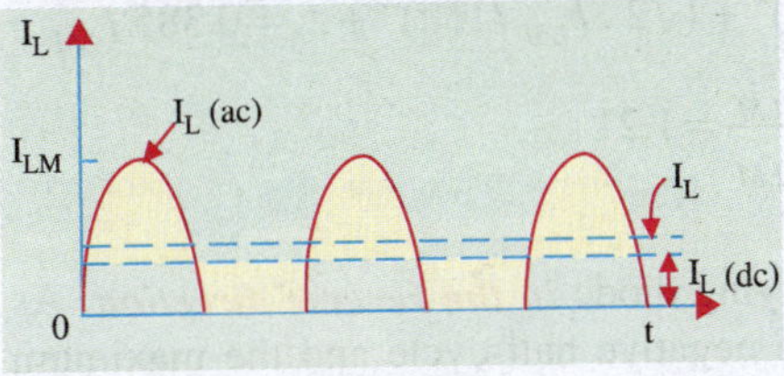

Fig. 55.3

The third term represents the second harmonic component whose frequency is double that of the supply frequency. The rms value is I_{L2} = peak value/$\sqrt{2} = 2\,I_{LM}/3\,\pi\,\sqrt{2} = \sqrt{2}\,I_{LM}/3\pi$.

The fourth term represents the third harmonic component whose frequency is four times the supply frequency. Its rms value is $2\,I_{LM}/15\,\pi \times \sqrt{2} = \sqrt{2}\,.\,I_{LM}/15\,\pi$.

The rms values of other components can be similarly calculated. However, they are found to be of continuously diminishing value.

As discussed above, the rectified output (or load) current consists of

(*i*) dc component, $I_{L(dc)} = I_{LM}/\pi$ and

(*ii*) ac components of rms values I_{L1}, I_{L2} and I_{L3} etc. Their combined rms value is given by

$$I_{L(ac)} = \sqrt{I_{L1}^2 + I_{L2}^2 + I_{L3}^2 + \ldots\ldots}$$

The rms (or effective) value of the total load current is given by

$$I_L = \sqrt{I_{L(dc)}^2 + I_{L(ac)}^2} = \sqrt{I_{L(dc)}{}^2 + (I_{L1}{}^2 + I_{L2}{}^2 + I_{L3}{}^2 + \ldots\ldots}$$

Similarly, the Fourier series of the load voltage is given by

$$V_L = V_{LM}\left(\frac{1}{\pi} + \frac{1}{2}\sin\omega t - \frac{2}{3\pi}\cos 2\,\omega t - \frac{2}{15\,\pi}\cos 4\,\omega t \ldots\ldots\right)$$

It also consists of

(*i*) a dc component, $V_{L(dc)} = V_{LM}/\pi$

(*ii*) ac components of rms values V_{L1}, V_{L2} and V_{L3} etc. which are given by

$$V_{L1} = V_{LM}/2\sqrt{2},\; V_{L2} = \sqrt{2}\,.\,V_{LM}/3\,\pi\,;\; V_{L3} = \sqrt{2}\,V_{LM}/15\,\pi \text{ etc.}$$

Again, $V_{L(ac)} = \sqrt{V_{L1}^2 + V_{L2}^2 + V_{L3}^2 + \ldots\ldots}$

The rms value of the entire load voltage is given by

$$V = \sqrt{V_{L(dc)}^2 + V_{L(ac)}^2} = \sqrt{V_{L(dc)}^2 + V_{L1}^2 + V_{L2}^2 + V_{L3}^2 + \ldots\ldots}$$

(*e*) Ripple Factor

When defined in terms of voltage, it is given by

$$\gamma = \frac{\text{rms value of ac components}}{\text{dc value of load voltage}} = \frac{V_{L(ac)}}{V_{L(dc)}} = \frac{V_{(rms)}}{V_{L(dc)}}$$

In terms of current, we have $\gamma = I_{L(ac)}/I_{L(dc)}$

As seen from above, $I_{L(ac)} = \sqrt{I_L^2 - I_{L(dc)}^2}$

$$\gamma = \frac{I_{L(ac)}}{I_{L(dc)}} = \frac{\sqrt{I_L^2 - I_{L(dc)}^2}}{I_{L(dc)}} = \sqrt{\left(\frac{I_L}{I_{L(dc)}}\right)^2 - 1}$$

Now, $I_L/I_{L(dc)}$ = form factor K_f (Art 11.18) $\quad\therefore\quad \gamma = \sqrt{K_f^2 - 1}$

In the case of a half-wave rectifies with resistive load but no filter $K_f = \pi/2 = 1.57$

$$\therefore \qquad \gamma = \sqrt{1.57^2 - 1} = 1.21$$

Alternatively, the value of γ could be found as under :

If we neglect fourth and higher harmonics in the load current, then as seen from above

$$I_{L(ac)} = \sqrt{I_{L1}^2 + I_{L2}^2 + I_{L3}^3 + ...}$$

$$= \sqrt{(I_{LM}/2\sqrt{2})^2 + (\sqrt{2}I_{LM}/3\pi)^2 + (\sqrt{2}\,.\,I_{LM}/15\pi)^2 + ...} = 0.385\,I_{LM}$$

$$\therefore \quad \gamma = \frac{I_{L(ac)}}{I_{L(dc)}} = \frac{0.385\,I_{LM}}{I_{LM}/\pi} = \frac{0.385\,I_{LM}}{0.318\,I_{LM}} = 1.21$$

(f) Peak Inverse Voltage (PIV)

It is the maximum voltage that occurs across the rectifying diode ***in the reverse direction***. As seen from Fig. 55.2, the diode is reverse-biased during the negative half-cycle and the maximum voltage applied across it equals the maximum secondary voltage *i.e.* V_{sm}.

(g) Transformer Utilization Factor (TUF)

While designing any power supply, it is necessary to determine the rating of the transformer. It can be done provided *TUF* is known. The value of *TUF* depends on the amount of power to be delivered to the load and the type of rectifier circuit to be used.

$$TUF = \frac{\text{dc power delivered to the load}}{\text{ac rating of transformer secondary}}$$

$$= \frac{P_{dc}}{P_{ac\,.\,rated}} = \frac{P_{dc}}{P_{in\,.\,rated}}$$

At first sight it might appear as if the above ratio is the same as the conversion efficiency. Actually, it is not so because the rating of the transformer secondary is different from the actual power delivered by the secondary.

$$P_{dc} = V_{L(dc)} \cdot I_{L(dc)} = \frac{V_{LM}}{\pi} \cdot \frac{V_{LM}}{\pi R_L} = \frac{V_{LM}^2}{\pi^2 R_L}$$

$$= \frac{V_{sm}^2}{\pi^2 R_L} \quad \text{—if drop over } R_0 \text{ is neglected}$$

Now, the rated voltage of transformer secondary is $V_{sm}/\sqrt{2}$ but the actual current flowing through the secondary is $I_L = I_{LM}/2$ (and not $I_{LM}/\sqrt{2}$) since it is a half-wave rectified current.

$$P_{ac.rated} = \frac{V_{sm}}{\sqrt{2}} \cdot \frac{I_{LM}}{2} = \frac{V_{sm}}{\sqrt{2}} \cdot \frac{V_{LM}}{2\,R_L} = \frac{V_{sm}^2}{2\sqrt{2}\,R_L}$$

$$TUF = \frac{V_{sm}^2/\pi^2 R_L}{V_{sm}^2/2\sqrt{2}\,R_L} = \frac{2\sqrt{2}}{\pi^2} = 0.287$$

However, due to saturation effects produced by the flow of direct current through the tranformer secondary, the value of *TUF* is further reduced to 0.2.

Obviously, dc power delivered to the load=ac transformer rating × *TUF*

If, we have a 1-kVA transformer, then the power which it would be able to deliver to a resistive load in a half-wave rectifier without over-heating would be = 0.2 × 1000 = 200 W.

Example 55.1. *In the half-wave rectifier circuit of Fig. 55.4, determine*

(i) maximum and rms values of load voltage,

(ii) peak and rms values of load current,

(iii) power absorbed by the load,

(iv) PIV of the diode,

(v) rms value of ripple voltage.

Neglect resistance of transformer secondary and that of the diode.

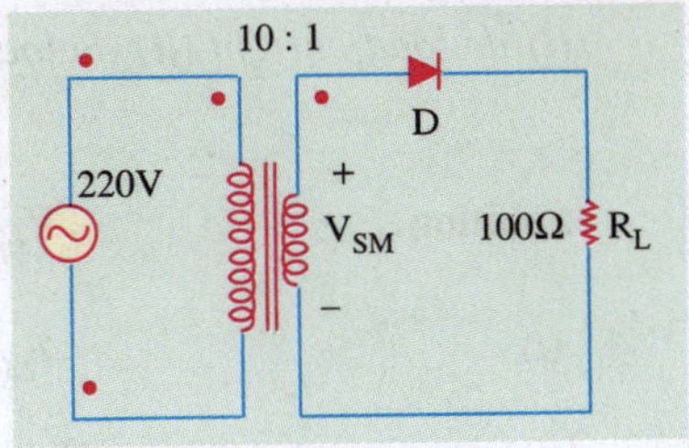

Fig. 55.4

Solution. Here, $K = N_2/N_1 = 1/10$. Peak primary voltage is

$$V_{pm} = 220\sqrt{2} = 311\text{ V.}$$

Hence, $V_{sm} = KV_{pm} = 311/10 = \mathbf{31.1V \approx 31\ V}$

(i) $V_{LM} = V_{sm} = \mathbf{31\ V}$

$V_L = V_{LM}/2 = 31/2 = \mathbf{15.5\ V}$

(ii) $I_{LM} = V_{LM}/R_L = 31/100 = \mathbf{0.31\ A}$

$I_L = I_{LM}/2 = 0.31/2 = \mathbf{0.155\ A}$

(iii) $P_L = V_L I_L = 15.5 \times 0.155 = \mathbf{2.4\ W}$

(iv) $PIV = 2V_{sm} = 2 \times 31 = \mathbf{62\ V}$

(v) $V_{L(ac)} = \sqrt{V_L^2 - V_{L(dc)}^2}$

Now, $V_L = V_{LM}/2$ and $V_{L(dc)} = V_{LM}/\pi$

$\therefore\quad V_{L(ac)} = \sqrt{(V_{LM}/2)^2 - (V_{LM}/\pi)^2} = 0.385\,V_{LM}$

$\therefore\quad V_{r(rms)} = V_{L(ac)} = 0.385 \times 31 = \mathbf{11.9\ V}$

It represents the rms value of the ripple voltage.

55.6. Equivalent Circuit of a HW Rectifier

Such a circuit is shown in Fig. 55.5. Here, the diode has been replaced by its equivalent circuit (Art. 52.13). The transformer secondary of Fig. 55.2 has been replaced by an ac sinusodial generator having a peak value of V_{sm}. Resistance R_S represents transformer secondary resistance. Obvisouly.

$$I_{LM} = \frac{V_{sm} - V_B}{(R_S + r_d) + R_L} = \frac{V_{sm} - V_B}{R_0 + R_L}$$

$$V_{LM} = I_{LM} \cdot R_L$$

$$V_{L(dc)} = V_{LM}/\pi,\ I_{L(dc)} = I_{LM}/\pi$$

$$V_L = V_{LM}/2 \text{ and } I_L = I_{LM}/2$$

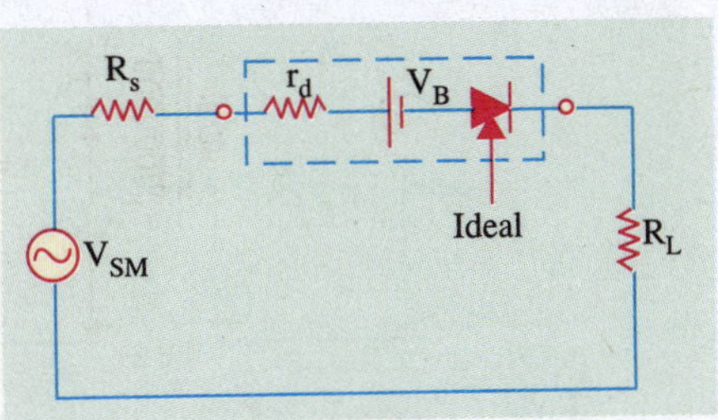

Fig. 55.5

(i) $$\eta = \left(\frac{4}{\pi^2}\right)\frac{R_L}{(R_S + r_d) + R_L}$$

$$= \left(\frac{4}{\pi^2}\right)\frac{R_L}{R_0 + R_L} = \frac{40.6\%}{1 + R_0/R_L}$$

(ii) Voltage regulation is given by

$$V_R = \frac{V_{NL} - V_{FL}}{V_{FL}} \times 100$$

(iii) Under no-load condition *i.e.* when no output current flows, the voltage has maximum value. When rectifier is fully loaded *i.e.* when output current flows, there is drop over R_0. Hence, output voltage is decreased by this much amount.

$$V_{FL} = V_{NL}\frac{R_L}{R_0 + R_L}$$

Substituting this value in the above equation, we get $V_R = R_0/R_L$

Example 55.2. *A half-wave rectifier using silicon diode has a secondary emf of 14.14 V (rms) with a resistance of 0.2 Ω. The diode has a forward resistance of 0.05 Ω and a threshold voltage of 0.7 V. If load resistance is 10 Ω, determine*

(i) dc load current (ii) dc load voltage (iii) voltage regulation and (iv) efficiency.

(Applied Electronics-I, Punjab University, 1992)

Solution. $V_{sm} = \sqrt{2} \times 14.14 = 20\text{ V}, R_0 = 0.2 + 0.05 = 0.25\ \Omega$

(i) $I_{LM} = \dfrac{V_{sm} - V_B}{R_0 + R_L} = \dfrac{20 - 0.7}{10.25} = 1.88\text{ A}; I_{L(dc)} = \dfrac{I_{LM}}{\pi} = \dfrac{1.88}{\pi} = \mathbf{0.6\ A}$

(ii) $V_{L(dc)} = I_{L(dc)} . R_L = 0.6 \times 10 = \mathbf{6\ V}$

(iii) $V_R = R_0/R_L = 0.25/10 = 0.025$ or **2.5%**

(iv) $\eta = \dfrac{40.6}{1 + 0.25/10} = \mathbf{39.6\%}$

55.7. Single-phase Full-wave Rectifier

In this case, both half-cycles of the input are utilized with the help of two diodes working alternately. For full-wave rectification, use of a transformer is essential (though it is optional for half-wave rectification).

The full-wave rectifier circuit using two diodes and a centre-tapped transformer shown in 55.6 (*a*). The centre-tap is usually taken as the ground or zero voltage reference point.

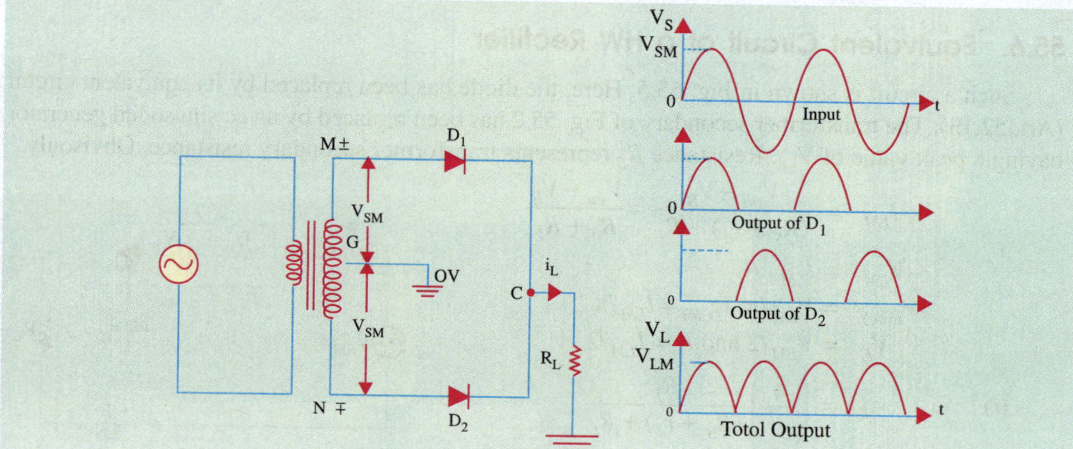

(*a*) **Fig. 55.6** (*b*)

Fig. 55.7 shows two different ways of drawing the circuit. In Fig. 55.7 (*a*), R_L becomes connected to point *G via* the earth whereas in Fig. 55.7 (*b*). It is connected directly to *G*.

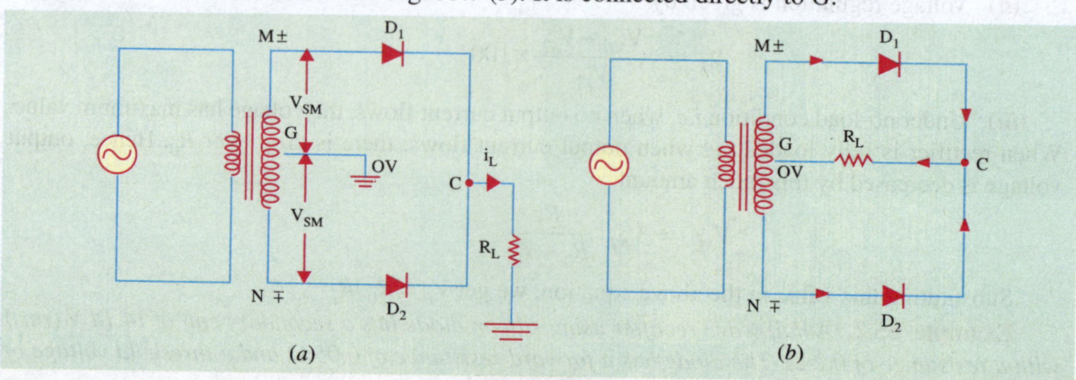

(*a*) (*b*)

Fig. 55.7

(a) Working

When input ac supply is switched on, the ends M and N of the transformer secondary become +ve and – ve alternately. During the positive half-cycle of the ac input, terminal M is +ve, G is at zero potential and N is at –ve potential. Hence, being forward-biased, diode D_1 conducts (but not D_2 which is reverse-biased) and current flows along MD_1CG. As a result, positive half-cycle of the voltage appears across R_L.

During the negative half-cycle, when terminal N becomes +ve, then D_2 conducts (but not D_1) and current flows along ND_2CG. So, we find that current keeps on flowing through R_L in the *same* direction (*i.e.* from C to G) in both half-cycles of ac input. It means that both half-cycles of the input ac supply are utilized as shown in Fig. 55.6 (*b*). Also, the frequency of the rectified output voltage is twice the supply frequency. Of course, this rectified output consists of a dc component and many ac components of diminishing amplitudes.

(b) Average and RMS Values

As proved earlier in and now shown in Fig. 55.8

$$V_L = V_{LM}/\sqrt{2} = 0.707\,V_{LM}\,;\; V_{L(dc)} = 2\,V_{LM}/\pi = 0.636\text{ V}$$

$$V_{L(ac)} = \text{rms value of ac components in the output voltage}$$

$$= \sqrt{V_L^2 - V_{L(dc)}^2}$$

Similarly, $$I_{LM} = \frac{V_{LM}}{R_L}\;;\; I_L = \frac{I_{LM}}{\sqrt{2}} = 0.707\,I_{LM}$$

$$I_{L(dc)} = \frac{2\,I_{LM}}{\pi} = 0.636\,I_{LM}\,;\, I_{L(ac)} = \sqrt{I_L^2 - I_{L(dc)}^2}$$

Incidentally, $I_{L(ac)}$ is the same thing as $I_{r(rms)}$.

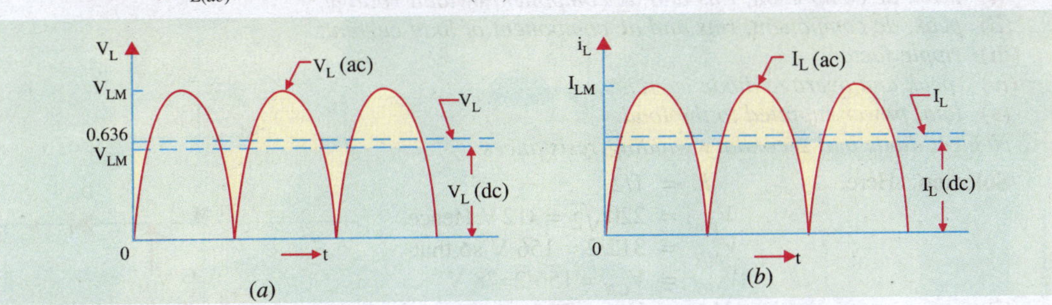

Fig. 55.8

(c) Efficiency

$$P_{in} = I_L^2\,(R_0 + R_L) = \left(\frac{I_{LM}}{\sqrt{2}}\right)^2 (R_0 + R_L) = \frac{1}{2}\,I_{LM}^2\,(R_0 + R_L)$$

$$P_{dc} = I_{L(dc)}^2\,(R_L) = \left(\frac{2\,I_{LM}}{\pi}\right)^2 (R_L) = \frac{4\,I_{LM}^2}{\pi^2}\,(R_L)$$

$$\therefore \quad \eta = \frac{P_{dc}}{P_{in}} = \left(\frac{8}{\pi^2}\right)\left(\frac{R_L}{R_0 + R_L}\right) = \frac{0.812}{(1 + R_0/R_L)} = \frac{81.2\%}{(1 + R_0/R_L)}$$

It is twice the value for the half-wave rectifier for the simple reason that a full-wave rectifier utilizes both half-cycles of the input ac supply.

(d) Frequency Components

As in the case of a HW rectifier, the output of a full-wave rectifier also consists of (*i*) a dc component and *(ii)* a number of ac components which form the ripple. The Fourier series for rectifier output voltage is

$$V_L = V_{LM}\left(\frac{2}{\pi} - \frac{4}{3\pi}\cos 2\,\omega t - \frac{4}{15\,\pi}\cos 4\,\omega t - \frac{4}{35}\cos 6\,\omega t - ....\right)$$

As seen, $V_{L(dc)} = \frac{2\,V_{LM}}{\pi}$; $V_{L1} = \frac{4\,V_{LM}}{3\pi\sqrt{2}}$, $V_{L2} = \frac{4\,V_{LM}}{15\pi\sqrt{2}}$ etc.

$$V_{L(ac)} = \sqrt{V_{L1}^2 + V_{L2}^2} = \sqrt{\left(\frac{4\,V_{LM}}{\sqrt{2}\,.\,3\,\pi}\right)^2 + \left(\frac{4\,V_{LM}}{\sqrt{2}\,.\,15\,\pi}\right)^2} = 0.305\,V_{LM}$$

Similalry, $I_{L(ac)} = I_{r(rms)} = \sqrt{I_{L1}^2 + I_{L2}^2} = 0.305\,I_{LM}$

(e) Ripple Factor

$$\gamma = \frac{V_{L(ac)}}{V_{L(dc)}} = \frac{V_{r(rms)}}{V_{L(rms)}} = \frac{0.305\,V_{LM}}{0.636\,V_{LM}} = 0.482 \text{ (approx)}$$

It is much less as compared to 1.21 for half-wave rectifier.

(f) PIV

Its value is = 2 V_{sm}

(g) TUF

Its value is found by considering the primary and secondary windings of the transformer separately. Its value is 0.693 (as compared to 0.287 for a half-wave rectifier). In such a rectifier, there is no problem due to dc saturation of flux in the core because the dc currents in the two halves of the secondary flow in opposite directions.

Example 55.3. *With reference to the full-wave rectifier of Fig. 55.9, determine*
(i) peak, dc component, rms and ac component of load voltage,
(ii) peak, dc component, rms and ac component of load current,
(iii) ripple factor,
(iv) peak and average diode currents,
(v) total power supplied to the load.
Neglect diode and secondary winding resistances.

Solution. Here,

$$K = 1/2$$
$$V_{pm} = 220\sqrt{2} = 312 \text{ V. Hence,}$$
$$V_{MN} = 312/2 = 156 \text{ V so that}$$
$$V_{MG} = V_{GN} = 156/2 = 78 \text{ V}$$

(i)

$$V_{sm} = V_{LM} = \mathbf{78\ V}$$
$$V_{L(dc)} = 0.636\,V_{LM} = 0.636 \times 78 = \mathbf{49.6\ V}$$
$$V_L = 0.707\,V_{LM} = 0.707 \times 78 = \mathbf{55\ V}$$
$$V_{L(ac)} = \sqrt{55^2 - 49.6^2} = \mathbf{23.8\ V}$$

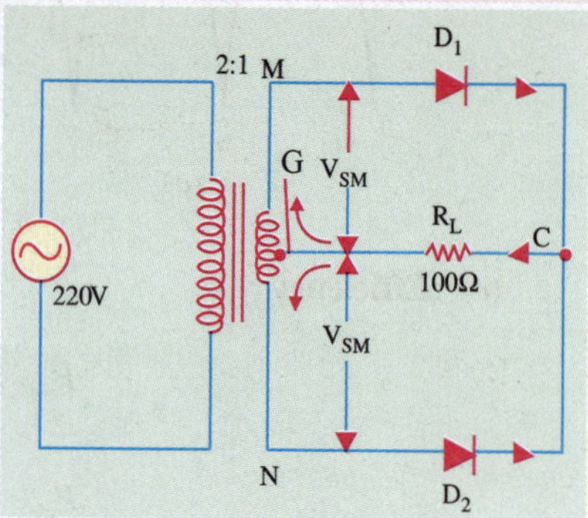

Fig. 55.9

It also represents $V_{r(rms)}$.

(ii)

$$I_{LM} = V_{LM}/R_L = 78/100 = \mathbf{0.78\ A}$$
$$I_{L(dc)} = 0.636 \times 0.78 = \mathbf{0.496\ A}$$
$$I_L = 0.707 \times 0.78 = \mathbf{0.55\ A}$$
$$I_{L(ac)} = \sqrt{0.55^2 - (.496)^2} = \mathbf{0.238\ A}$$

It also represents $I_{r(rms)}$.

(iii)

$$\gamma = \frac{I_{L(ac)}}{I_{L(dc)}} = \frac{I_{r(max)}}{I_{L(dc)}} = \frac{0.238}{0.496} = \mathbf{0.48}$$

(*iv*) peak diode current = peak load current = **0.78 A**

For finding the average current of a diode, it must be remembered that each diode carries current for one half-cycle only.

$$I_{D(av)} = I_{D(max)}/\pi = 0.318 \times 0.78 = \mathbf{0.25\ A}$$

(*v*) $$P_L = V_L I_L = 55 \times 0.55 = \mathbf{30.25\ W}$$

Example 55.4. *A 1-ϕ, full-wave rectifier supplies power to a 1 k W load. The ac voltage applied to the diode is 300-0-300 V (rms). If diode resistance is 25 W and that of the transformer secondary negligible, determine*

(*i*) *average load current,* (*ii*) *average value of load voltage,*

(*iii*) *rms value of ripple,* (*iv*) *efficiency.*

(Applied Electronics, Bombay Univ.)

Solution. It may be noted that rms value of ac voltage across each secondary half is 300 V.

(*i*) $$V_{sm} = 300\sqrt{2} = 424\ \text{V}, I_{LM} = V_{sm}/(r_d + R_L) = 424/1025 = 0.414\ \text{A}$$

$$I_{L(dc)} = 2 I_{LM}/\pi = 2 \times 0.414/\pi = \mathbf{0.263\ A}$$

(*ii*) $$V_{L(dc)} = I_{L(dc)} \cdot R_L = 0.263 \times 1000 = \mathbf{263\ V}$$

(*iii*) $$\gamma = \frac{V_{L(ac)}}{V_{L(dc)}} = \frac{V_{r(rms)}}{V_{L(dc)}}$$

$\therefore$ $$V_{r(rms)} = \gamma \cdot V_{L(dc)} = 0.482 \times 263 = \mathbf{126.8\ V}$$

(*iv*) $$\eta = \frac{81.2\%}{1 + r_d/R_L} = \frac{81.2\%}{1 + 25/1000} = \mathbf{79.2\%}$$

Example 55.5. *A full-wave rectifier is built up by using the same components as in Ex. 55.2. Determine :*

(*i*) *dc load current,* (*ii*) *dc load voltage,*

(*iii*) *voltage regulation,* (*iv*) *circuit efficiency and*

(*v*) *diode PIV and current rating.*

Solution.

(*i*) $$I_{L(dc)} = 0.636 \times 1.88 = \mathbf{1.2\ A}$$

(*ii*) $$V_{L(dc)} = I_{L(dc)} \cdot R_L = 1.2 \times 10 = \mathbf{12\ V}$$

(*iii*) $$VR = 0.25/10 = 0.025 \text{ or } \mathbf{2.5\%}$$

(*iv*) $$\eta = \frac{81.2\%}{1 + R_0/R_L} = \frac{81.2\%}{1 + 0.25/10} = \mathbf{79.2\ \%}$$

(*v*) $$PIV = 2\,V_{sm} = 2 \times 20 = \mathbf{40\ V}$$

With a safety factor of 1.5, *PIV* = 40 × 1.5 = 60 V. A dc current rating of about 2 A would be satisfactory.

Example 55.6. *Silicon diodes are used in a two-diode full-wave rectifier circuit to supply a load with 12 volts D.C. Assuming ideal diodes and that the load resistance is 12 ohms, compute* (*i*) *the transformer secondary voltage,* (*ii*) *the load ripple voltage,* (*iii*) *the efficiency of the rectifier. Derive equations used.* ***(Electronics-I, Bangalore Univ.)***

Solution.

(*i*) $V_{L(dc)} = 2\,V_{sm}/\pi$; $V_{sm} = \pi \times 12/2 = \mathbf{18.8\ V.}$

It is the maximum value of the voltage across one half of the secondary.

(*ii*) RMS value of ripple voltage = $\gamma\, V_{L(dc)} = 0.482 \times 12 = \mathbf{5.78\ V}$

(*iii*) Since diodes are ideal ones, their forward resistance is negligible. Hence,

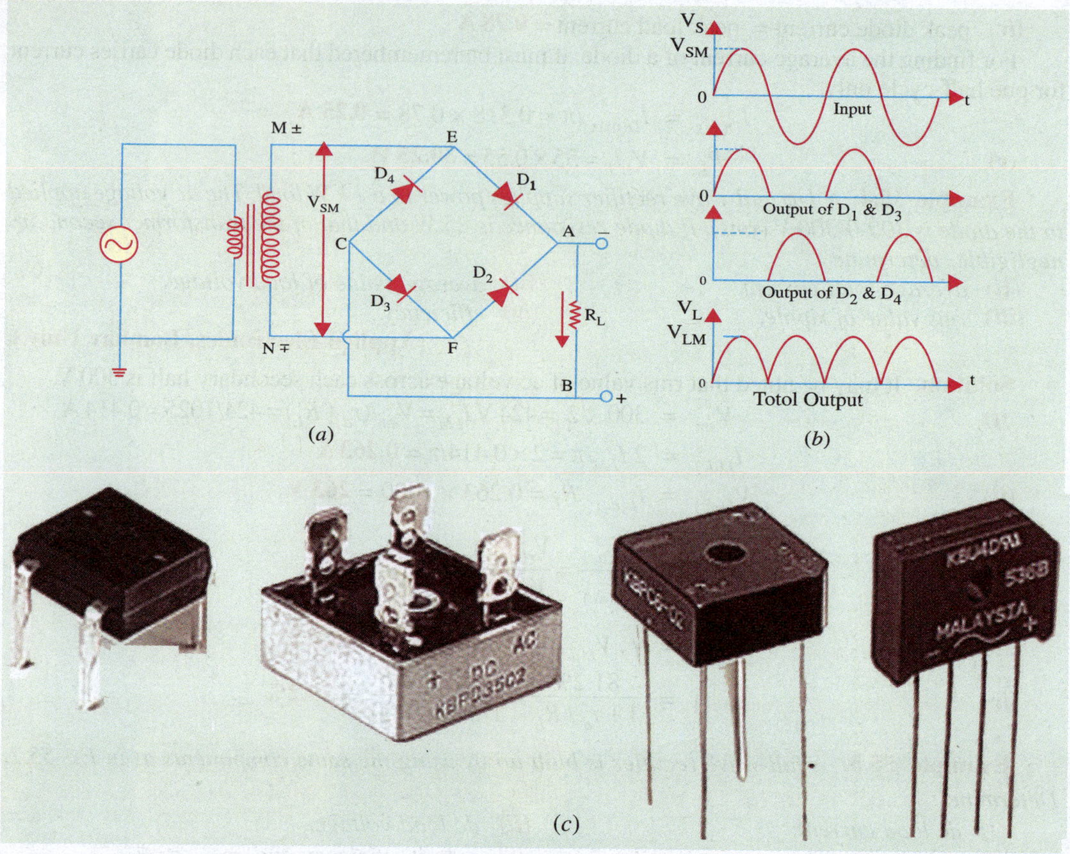

Fig. 55.10

$$\eta = \frac{81.2\%}{(1 + r_d / R_L)} = \frac{81.2\%}{1 + 0/12} = \mathbf{81.2\%}$$

55.8. Full-Wave Bridge Rectifier

It is the most frequently-used circuit for electronic dc power supplies. It requires four diodes but the transformer used is not centre-tapped and has a maximum voltage of V_{sm}. The full-wave bridge-rectifier is available in three distinct physics forms.

1. four discrete diodes, **2.** one device inside a four-terminal case,
3. as part of an array of diodes in an IC.

The circuit using four discrete diodes is shown in Fig. 55.10 (*a*) and 55.10 (*c*) shows some pictures of the bridge rectifier available as one device in a four terminal case.

(*a*) Working

During the positive input half-cycle, terminal *M* of the secondary is positive and *N* is negative as shown separately in Fig. 55.11 (*a*). Diodes D_1 and D_3 become forward-biased (ON) whereas D_2 and D_4 are reverse-biased (OFF). Hence, current flows along *MEABCFN* producing a drop across R_L.

During the negative input half-cycle, secondary terminal *N* becomes positive and *M* negative. Now, D_2 and D_4 are forward-biased. Circuit current flows along *NFABCEM* as shown in Fig. 55.11 (*b*). Hence, we find that current keeps flowing through load resistance R_L in the same direction *AB* during both half-cycles of the ac input supply. Consequently, point *A* of the bridge rectifier always

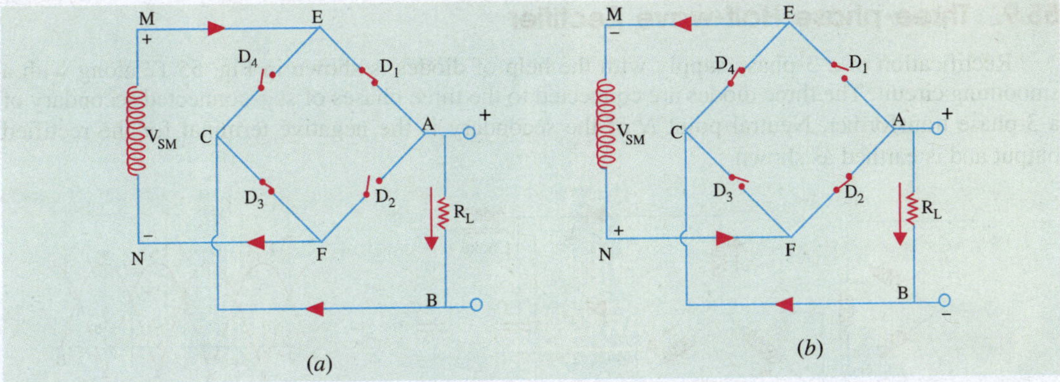

Fig. 55.11

acts as an anode and point *C* as cathode. The output voltage across R_L is as shown in Fig. 55.10 (*b*). Its frequency is twice that of the supply frequency.

(*b*) Average and RMS Values

These are the same as for the centre-tapped full-wave rectifier discussed in Art 55.7.

(*c*) Efficiency

$$\eta = \frac{81.2}{1 + 2\, r_d / R_L} \%$$

(*d*) Ripple Factor

It is the same as for a full-wave rectifier *i.e.* $\gamma = 0.482$

(*e*) PIV

The *PIV* rating of each of the four diodes is equal to V_{SM} —the entire voltage across the secondary.

When Secondary and Diode Resistances are considered

(*i*) $I_{LM} = \dfrac{V_{sm} - 2\,V_B}{(R_S + 2\,r_d) + R_L} = \dfrac{V_{sm} - 2\,V_B}{R_0 + R_L}$ (*ii*) $V_{LM} = I_{LM} \cdot R_L$

(*iii*) $\eta = \left(\dfrac{8}{\pi^2}\right)\left(\dfrac{R_L}{(R_S + 2\,r_d) + R_L}\right) = \left(\dfrac{8}{\pi^2}\right)\dfrac{R_L}{R_0 + R_L}$

(*iv*) $VR = \dfrac{R_S + 2\,r_d}{R_L} = \dfrac{R_0}{R_L}$

(*f*) Advantages

After the advent of low-cost, highly-reliable and small-sized silicon diodes, bridge circuit has become much more popular than the centre-tapped transformer *FW* rectifier. The main reason for this is that for a bridge rectifier, a much smaller transformer is required for the same output because it utilizes the transformer secondary continuously unlike the 2-diode FW rectifier which uses the two halves of the secondary alternately.

So, the advantages of the bridge rectifier are :

1. no centre-tap is required on the transformer;
2. much smaller transformers are required;
3. it is suitable for high-voltage applications ;
4. it has less *PIV* rating per diode.

The obvious disadvantage is the need for twice as many diodes as for the centre-tapped transformer version. But ready availability of low-cost silicon diodes has made it more economical despite its requirement of four diodes.

55.9. Three-phase Half-wave Rectifier

Rectification of a 3-phase supply with the help of diodes is shown in Fig. 55.12 along with a smoothing circuit. The three diodes are connected to the three phases of star-connected secondary of a 3-phase transformer. Neutral point N of the secondary is the negative terminal for the rectified output and is earthed as shown.

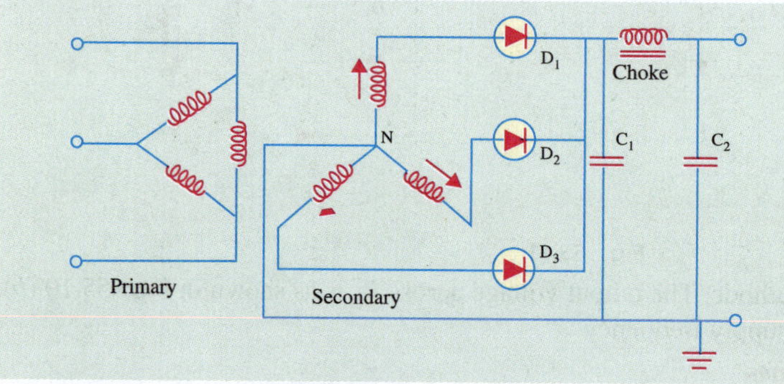

Fig. 55.12

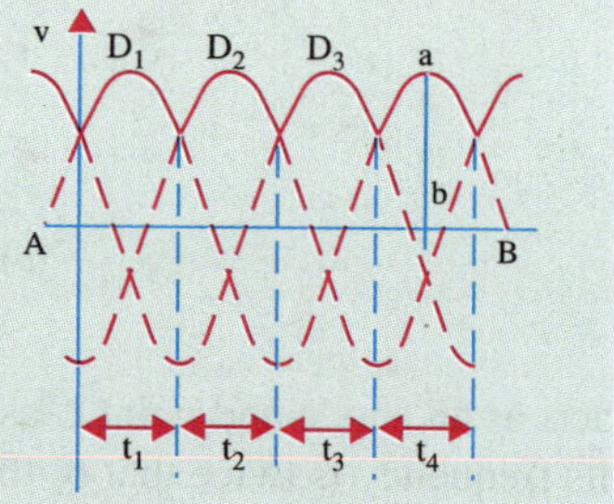

Fig. 55.13

The shape of the output is shown in Fig. 55.13. The horizontal line AB represents the potential of the negative d.c. terminal output and the sine waves 1, 2 and 3 each represent the anode potentials of the three diodes.

During one-third of the cycle *i.e.,* during time t_1, only diode D_1 will conduct. It will cease conducting at t_2 and then D_2 will conduct current upto t_2 after which D_3 will take over and will supply anode current till t_3. When one diode conducts, the other two remain inactive because then their anodes are more –ve than their cathodes. This process repeats itself during each ensuing cycle, with the conducting period of each diode being as indicated. The output is given by the vertical distance 'ab' between the upper envelope and the line AB. Obviously, the output fluctuates between the maximum and minimum values thrice in each cycle. The variations of output lie between V_{sm} and 0.5 V_{sm} (neglecting voltage drop in diodes) and has a mean value of $V_{dc} = 0.83\ V_{sm}$ or 1.17 V_S where V_S is the r.m.s. value of the secondary phase voltage. Its maximum conversion $\eta = 96.5\%$ and $\gamma = 0.17$. But it should be noted that the magnitude of these fluctuations or pulsations is lesser than for a 1-phase, full-wave rectified output since the current never touches zero. It is further smoothened up by a *C-L-C* filter circuit as shown in Fig. 55.12.

It is seen that direct current of each diode appears in the secondary phase winding and so causes transformer saturation resulting in large primary current. It can be avoided by using zig-zag secondary.

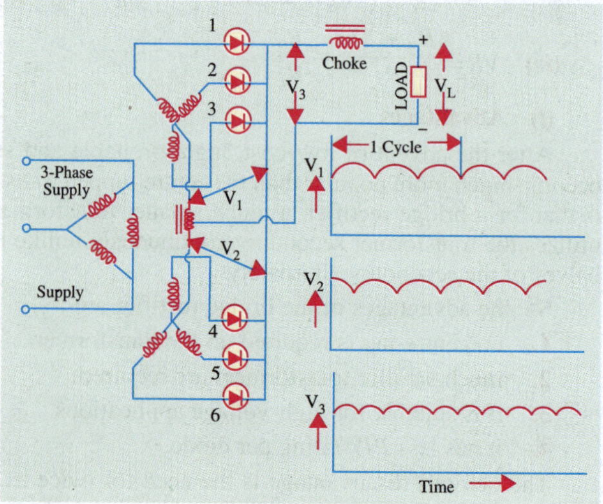

Fig. 55.14

55.10. Full-wave Rectification of 3-phase Currents

As shown in Fig. 55.14, a three-phase full-wave rectifier requires a transformer with six secondary windings connected to give two separate three-phase supplies 180° out of phase with each other. The centre taps are connected by an interphase transformer which enables the two rectifier units to operate independently of each other. Obviously, each diode conducts for one-third cycle. However, the addition of two outputs cancels the lowest frequency component of the ripple.

The output voltage has a mean value of 0.83 V_{sm} (less the diode voltage drop) and a ripple having a fundamental frequency six times the supply frequency.

Three-phase full-wave rectifier circuit is preferred for high powers because

(*i*) each secondary carries current for one-third of a cycle;

(*ii*) each primary carries current for two-thirds of a cycle;

(*iii*) Cu loss in the tranformer windings is comparatively lower.

55.11. Six-phase Half-wave Rectifier

Such a rectifier can be operated from a 3-phase supply using a transformer with three centre-tapped secondary windings with all the centre taps connected together as shown in Fig. 55.15. Obviously, each diode conducts for one-sixth of a cycle. The output voltage has a mean value of 0.955 V_{sm} (less the voltage drop in the diode). Ripple has a very small value and a fundamental frequency six times the supply frequency.

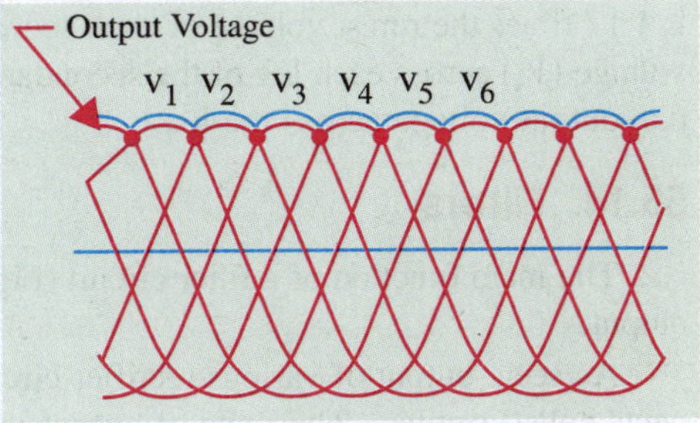

Fig. 55.15

55.12. Three-phase Bridge Circuit

This circuit is very frequently used because apart from being simple, it does not require centre-tap transformer. Only that diode supplies the load whose phase voltage is more positive than the others. For example, when D_1 is forward-biased, D_2 and D_3 are reverse-biased. The current returns to the supply *via* D_5 and D_6.

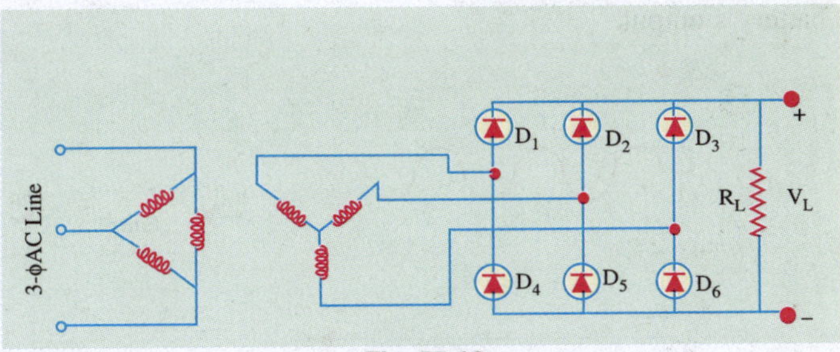

Fig. 55.16

In this circuit, I_{dc} is 0.955 times the peak current through each diode and only one-third of it flows through each diode (rather than one-sixth as in Fig. 55.15). Similarly, V_{dc} is twice of that in 3-ϕ, half-wave rectifier or $2 \times 1.17 = 2.34$ times the r.m.s. a.c. voltage across each secondary leg. Accordingly, r.m.s. voltage across each secondary leg need be only $1/2.34 = 0.428$ times the desired d.c. output voltage.

Example 55.7. *The r.m.s. value of transformer secondary voltage per leg (V_S) in a full-wave, D/Y, 6-f rectifier is 150 V. If average value of load current is 2 A, find* **(i)** V_{dc} **(ii)** *peak and average current through each diode and* **(iii)** *average power delivered to the load i.e.* P_{dc}.

Solution. (*i*) $V_{dc} = 2.34 \times V_S = 2.34 \times 150 =$ **351 V**

(*ii*) peak current/diode $= (1/0.955) \times I_{dc} = 1.05 \times 2 = 2.1$ A

average d.c. current/diode $= 2/3 =$ **0.667 A**

(In *D/Y* half-wave rectifier, its value is 1/6th rather than 1/3rd)

(*iii*) $P_{dc} = V_{dc}\, I_{dc} = 351 \times 2 = 702$ W

55.13. Calculations with Resistive Load

Let us suppose that no filter is connected across the rectifier (Fig. 55.12) but only a load of resistance R. Further, we will neglect transformer resistance and leakage and internal diode drops. In that case,

$$V_{dc} = 0.827\, V_{sm}\,;\; I_{dc} = 0.827\, V_{sm}/R$$

$$P_{dc} = V_{dc} I_{dc} = (0.827\, V_{sm})^2/R; \qquad P_{ac} = P_{in} = 0.706\, V_{sm}^2/R$$

$$\therefore \quad \text{rectifier } \eta = \frac{P_{dc}}{P_{ac}} = \frac{(0.827\, V_{sm})^2/R}{0.706\, V_{sm}^2/R} = 0.965 = 96.5\%$$

$$\gamma = \sqrt{(1.014)^2 - 1} = 0.17 \text{ or } 17\%.$$

Conclusions

Here, average d.c. current is 0.827 times the peak current as compared to 0.318 times for 1-ϕ, half-wave circuit and 0.636 times for 1-ϕ, full-wave circuit. Moreover, V_{dc} is also correspondingly high, it is 1.17 times the r.m.s. voltage of each secondary leg ($\because$ $1.414 \times 0.827 = 1.17$). Conversely, r.m.s. voltage (V_S) across each leg of the secondary need only be 1/1.17 = 0.855 times the average desired d.c. output voltage across the load.

55.14. Filters

The main function of a filter circuit (Fig. 55.17) is to minimize the ripple content in the rectifier output.

As seen, output of various rectifier circuits is pulsating. It has a dc value and some ac components called ripples. This type of output is not useful for driving sophisticated electronic circuits/ devices. In fact, these circuits require a very steady dc output that approaches the smoothness of a battery's output.

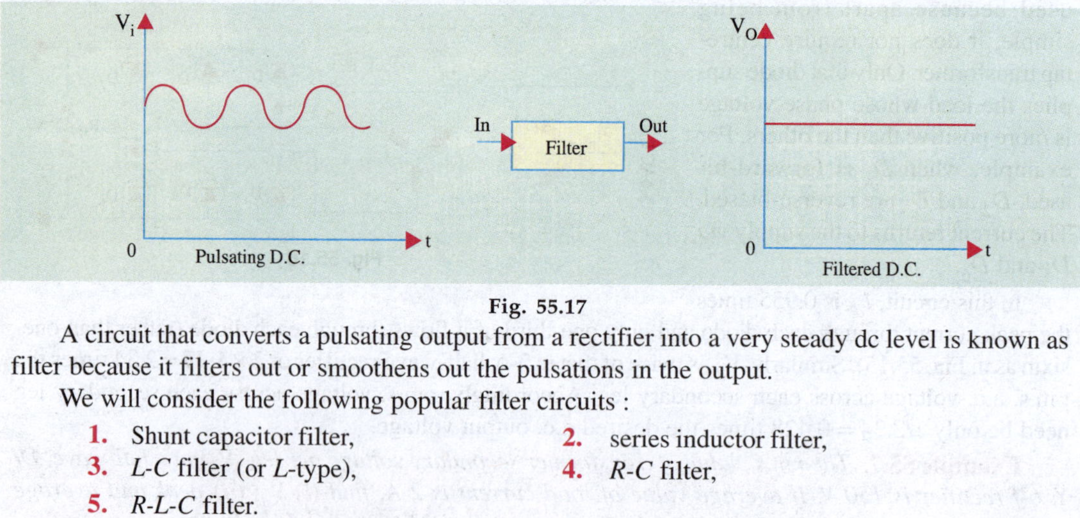

Fig. 55.17

A circuit that converts a pulsating output from a rectifier into a very steady dc level is known as filter because it filters out or smoothens out the pulsations in the output.

We will consider the following popular filter circuits :

1. Shunt capacitor filter,
2. series inductor filter,
3. *L-C* filter (or *L*-type),
4. *R-C* filter,
5. *R-L-C* filter.

55.15. Shunt Capacitor Filter

In this circuit, a suitable single capacitor C is connected across the rectifier and in parallel with the load R_L to achieve filtering action. This type of filter is known as ***capacitor input filter.***

This filter circuit depends for its operation on the property of a capacitor to charge up (*i.e.* store energy) during conducting half-cycle and to discharge (*i.e.* deliver energy) during the non-conducting half-cycle. In simple words, a capacitor opposes any change in voltage. When connected across a pulsating d.c. voltage, it tends to smoothen out or filter out the voltage pulsations (or ripples). The filtering action of the simple capacitor filter when used in a half-wave rectifier can be understood with the help of Fig. 55.18.

(a) Circuit Analysis

When positive half-cycle of the ac input is applied, the diode is forward-biased and hence is turned ON. This allows *C* to quickly charge up to peak value of input voltage V_{ip} [point *b* in Fig. 55.18 (*b*)] because charging time constant is almost zero. It is so because there is no resistance in the charging path except diode forward resistance which is negligible. Hence, capacitor follows the charging voltage as shown. After being fully charged, the capacitor holds the charge till input ac supply to the rectifier goes negative. During the negative half-cycle, the capacitor attempts to discharge. However, it cannot discharge through diode which, being now reverse-biased, is OFF. Hence, it discharges through R_L from point *b* to *c* in Fig. 55.18 (*c*) and its voltage decreases somewhat. The discharging time constant (= CR_L) is usually 100 times more than the charging time. Hence, *C* does not have sufficient time to discharge appreciably. It is seen that even during negative half-cycle of the input supply, the capacitor maintains a sufficiently large voltage across R_L.

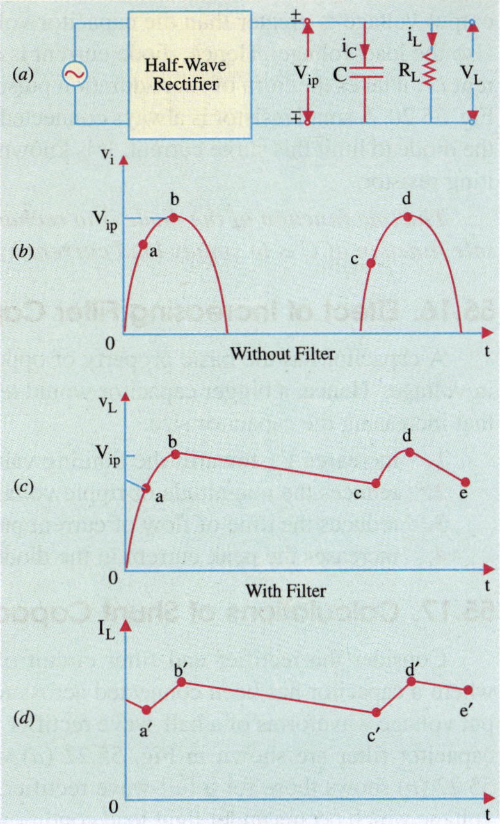

Fig. 55.18

During the next positive half-cycle, when rectifier voltage exceeds the capacitor voltage represented by point *c* in Fig. 55.18 (*c*), *C* is again charged quickly to V_{ip} as represented by point *d*. Once more, input voltage goes negative, opening the diode and forcing *C* to discharge through R_L during the interval *de*. In this way, R_L sees a nearly constant dc voltage across it at all times.

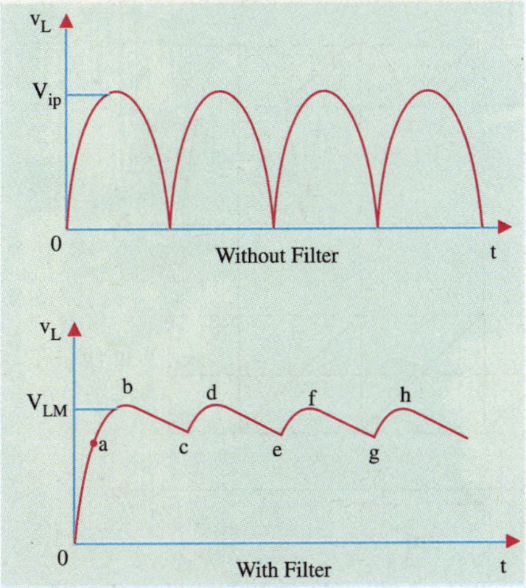

Fig. 55.19

The filtering action of this simple capacitor filter on a full-wave rectifier is shown in Fig. 55.19. It is seen that as compared to a HW rectifier.

(i) dc load voltage increases slightly towards V_{ip},

(ii) ripple voltage has been reduced by half.

The decreased ripple is because of shorter discharge time before the capacitor is reenergised by another pulse of current.

(b) Load Current

The load current has the same wave-shape as v_L because load is purely resistive. It is shown in Fig. 55.18 (*d*). During periods *a′b′* and *c′d′* etc., current is supplied by the diode and during periods *b′c′* and *d′e′* etc. by the capacitor.

(c) Diode Current

Diode current flows during short intervals of time like *ab* and *cd* etc. in Fig. 55.18 (*c*) which is reproduced in Fig. 55.20. During these intervals, diode output voltage is greater than the capacitor voltage which is also the load voltage. Hence, diode current is a surging current *i.e.* it takes the form of short-duration pulses as shown in Fig. 55.20. A small resistor is always connected in series with the diode to limit this surge current. It is known as surge limiting resistor.

The sole function of the diode is to recharge C and the sole function of C is to supply load current by discharge.

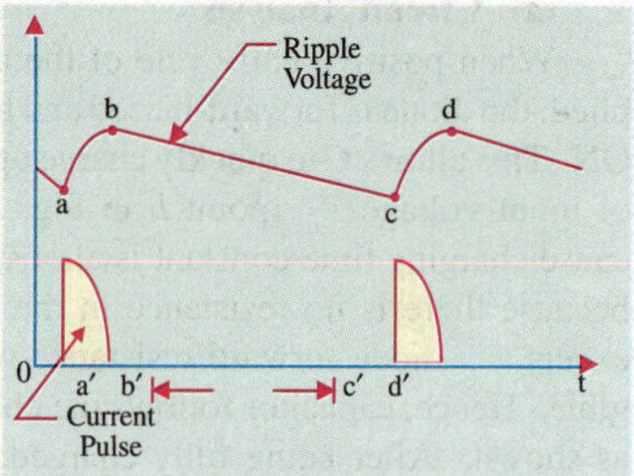

Fig. 55.20

55.16. Effect of Increasing Filter Capacitance

A capacitor has the basic property of opposing changes in voltage. Hence, a bigger capacitor would tend to reduce the ripple magnitude. It has been found that increasing the capacitor size.

1. increases V_{dc} towards the limiting value V_{ip} ;
2. reduces the magnitude of ripple voltage ;
3. reduces the time of flow of current pulse through the diode ;
4. increases the peak current in the diode.

55.17. Calculations of Shunt Capacitor Filter

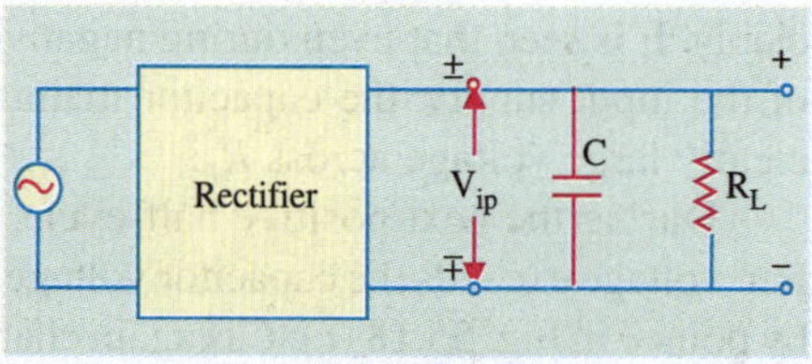

Fig. 55.21

Consider the rectifier and filter circuit of Fig. 55.21 where a capacitor has been connected across R_L. The output voltage waveforms of a half-wave rectifier with a shunt capacitor filter are shown in Fig. 55.22 (*a*) whereas Fig. 55.22 (*b*) shows those for a full-wave rectifier. The ripple voltage which occurs under light load conditions can be approximated by a triangular wave which has a peak-to-peak value of $V_{r(p-p)}$ and a time period of T_r* centred around the dc level.

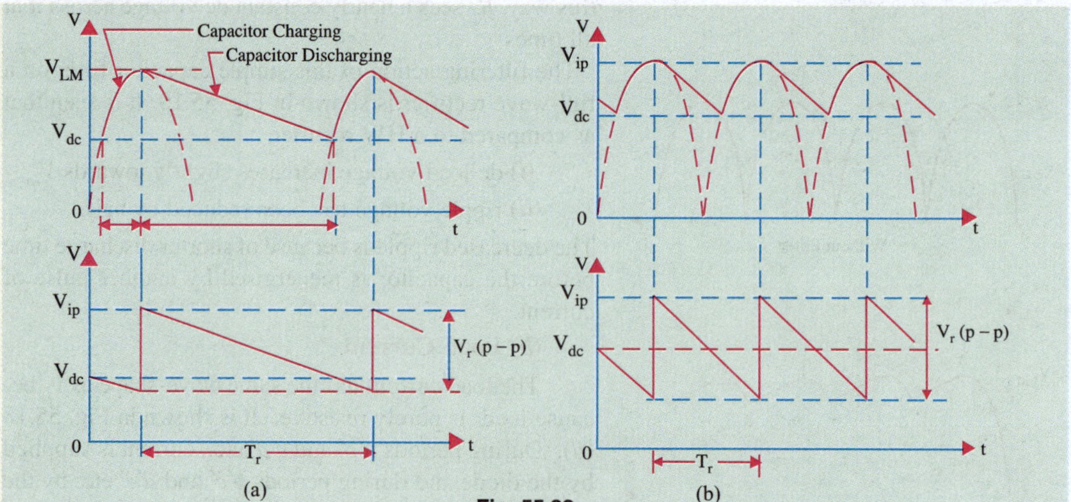

Fig. 55.22

In fact, $V_{r(p-p)}$ is the amount by which capacitor voltage falls during discharge period T_r. This discharge is actually exponential** but can be approximated by a straight line discharge if we assume

* Since charging time is negligibly small, the approximate discharging time represents the full time-period.

** given by $V_c = V_{ip}\, e^{-t}/CR_L$

the discharge rate to remain constant at the dc level I_{dc}. In that case, charge lost dQ in time T_r is $I_{dc}\,T_r$.

$$\therefore \quad V_{r(p-p)} = \frac{dQ}{C} = \frac{I_{dc}T_r}{C} = \frac{V_{dc}}{f_r C R_L} \qquad \left(\because\ I_{dc} = \frac{V_{dc}}{R_L}\right)$$

The triangular ripple has an rms value given by $V_{r(rms)} = V_{r(p-p)}/2\sqrt{3}$

$$\therefore \quad V_{r(rms)} = \frac{V_{r(p-p)}}{2\sqrt{3}} = \frac{V_{dc}}{2\sqrt{3}f_r C R_L} \quad \therefore \quad \gamma = \frac{V_{r(rms)}}{V_{dc}} = \frac{1}{2\sqrt{3}f_r C R_L}$$

Now, f_r is the frequency of the ripple voltage. For a half-wave rectifier, f_r equals the rectifier line input frequency whereas it is double the line input frequency for a full-wave rectifier. If f is the line frequency, then

$$\gamma \cong \frac{1}{2\sqrt{3}f\,CR_L} \qquad \text{— for HW rectifier}$$

$$\cong \frac{1}{4\sqrt{3}f\,CR_L} \qquad \text{— for FW rectifier}$$

It can be further proved that

$$\gamma = \frac{1}{2\sqrt{3}f\,CR_L} = \frac{I_{dc}}{4\sqrt{3}f\,C}\left(\frac{1}{V_{ip}} - \frac{1}{V_{dc}}\right) \qquad \text{— half-wave rectifier}$$

$$= \frac{1}{4\sqrt{3}f\,CR_L} = \frac{I_{dc}}{4\sqrt{3}f\,CV_{ip}} \qquad \text{— for full-wave rectifier}$$

It is seen from above that ripple increases with increase in load (*i.e.* output) current.

Incidentally, $\quad V_{dc} = V_{ip} - \dfrac{V_{r(p-p)}}{2}$

where $\quad V_{ip}$ = peak rectifier output voltage.

Substituting the value of $V_{r(p-p)} = V_{dc}/f_r CR_L$, we get

$$V_{dc} = V_{ip} - \frac{V_{dc}}{2f_r CR_L} \quad \text{or} \quad V_{dc} = \frac{V_{ip}}{1 + \dfrac{1}{2f_r CR_L}}$$

$$\therefore \quad V_{dc} = V_{ip}\left(\frac{2fCR_L}{1+2fCR_L}\right) = \frac{V_{ip}}{1 + \dfrac{I_{dc}}{2fCV_{ip}}} \qquad \text{—half-wave rectifier}$$

$$= V_{ip}\left(\frac{4fCR_L}{1+4fCR_L}\right) = \frac{V_{ip}}{1 + I_{dc}/4fCV_{ip}} \qquad \text{—full-wave rectifier}$$

Example 55.8. *A half-wave rectifier has a peak output voltage of 12.2 V at 50 Hz and feeds a resistive load of 100 Ω. Determine (i) the value of the shunt capacitor to give 1 percent ripple factor and (ii) the resulting dc voltage across the load resistor.*

Solution. (*i*) $\quad \gamma = 1/2\sqrt{3}\,fCR_L \quad \text{or} \quad 0.01 = 1/2\sqrt{3}\times 50\times C\times 100 \therefore C = 5773\,\mu F$

(*ii*) $\quad V_{dc} = \dfrac{V_{ip}}{1+1/2fCR_L} = \dfrac{12.2}{1+1/2\times 50\times 5773\times 10^{-6}\times 100} = \mathbf{12\ V}$

Example 55.9. *Find the ripple factor and dc output voltage for the filtered bridge rectifier shown in Fig. 55.23. Each silicon diode has a threshold voltage of 0.7 V.*

Solution. Peak primary voltage $= 230\times\sqrt{2}$

$= 325$ V

$$\text{Peak secondary voltage} = 325 \times 1/10 = 32.5\,\text{V}$$

Peak full-wave rectified voltage at the filter input

$$V_{ip} = 32.5 - 2\times 0.7 = 31.1\ \text{V}$$

$$\gamma = \frac{1}{4\sqrt{3}\, fCR_L} = \frac{1}{4\sqrt{3}\times 50\times 5\times 10^{-6}\times 20\times 10^{3}}$$

$= 0.028$ or **2.8 %**

$$V_{dc} = \frac{V_{ip}}{1+\dfrac{1}{4fCR_L}} = \frac{31.1}{1+\dfrac{1}{4\times 50\times 5\times 10^{-6}\times 20\times 10^{3}}} = \mathbf{29.6\ V}$$

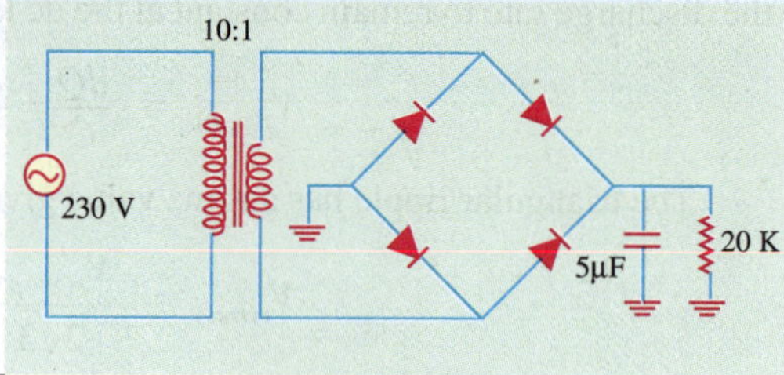

Fig. 55.23

Example 55.10. *Derive an expression for the ripple in the output of a full-wave rectifier circuit, with a simple capacitor element as the filter.*

The load current from the above circuit operating from 200-V, 50-Hz supply is 12 mA. Calculate minimum value of filter capacitor which is required to keep the ripple voltage below 2%.

(App. Electronics and Circuits ; Grad. I.E.T.E.)

Solution. As seen from Art. 55.17 $\gamma = I_{dc}/4\sqrt{3}\, fCV_{ip}$

$$\therefore \quad C = I_{dc}/4\sqrt{3}\, f\gamma V_{ip} = 12\times 10^{-3}/4\sqrt{3}\times 50\times 0.02\times 200\sqrt{2} = 6.12\,\mu\text{F}$$

55.18. Series Inductor Filter

This filter consists of a choke in series with the load resistor R_L as shown in Fig. 52.24. The operation of such a filter depends on the fundamental property of an inductor to oppose any sudden changes in the current flowing through it. Since this inductor presents high impedance to the ac components in the filter output, it reduces their amplitude with respect to the dc component thereby producing only a small ripple as shown in Fig. 55.24 (*b*).

The Fourier series for the rectifier output voltage is

$$v_i = V_{ip}\left(\frac{2}{\pi} - \frac{4}{3\pi}\cos 2\,\omega t - \frac{4}{15\,\pi}\cos 4\,\omega t - ....\right)$$

For finding the ripple factor, we will calculate the dc as well as ac drop over R_L. If we neglect choke resistance (R_C), then the entire dc component of filter output is available across R_L and its value is $V_{dc} = 2\,V_{ip}/\pi$ *.

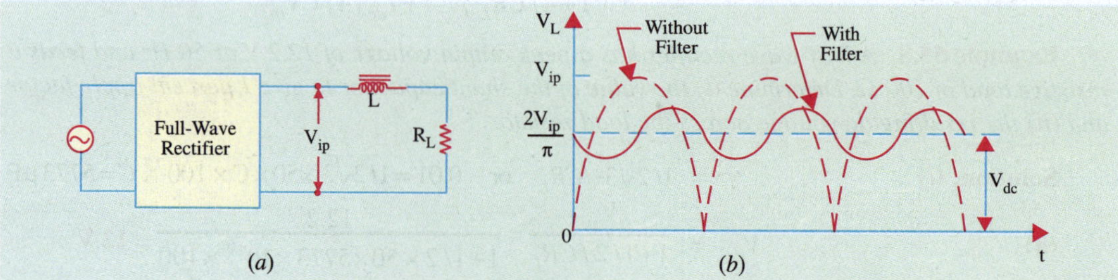

Fig. 55.24

* If R_C is taken into account, then filter output dc voltage drops partly on R_C and partly on R_L. The drop over R_L would be

$$= \frac{2V_{ip}}{\pi}\cdot\frac{R_L}{R_L + R_C}$$

We will consider only the second harmonic voltage $(4V_{ip}/3\pi)\cos 2\omega t$ of frequency 2ω and neglect higher harmonic voltages. This ac voltage partly drops over X_L and partly over R_L. Since choke and R_L are connected in series, the maximum value of drop over R_L is

$$= \frac{4V_{ip}}{3\pi}\cdot\frac{R_L}{\sqrt{R_L^2+X_L^2}}$$

The rms value of this ac voltage drop across R_L is

$$V_{ac} = \frac{4V_{ip}}{3\pi\sqrt{2}}\frac{R_L}{\sqrt{R_L^2+X_L^2}}$$

$$\therefore \quad \gamma = \frac{V_{ac}}{V_{dc}} = \frac{4V_{ip}}{3\pi\sqrt{2}}\frac{R_L}{\sqrt{R_L^2+X_L^2}}\times\frac{\pi}{2V_{ip}}$$

$$= \frac{\sqrt{2}R_L}{3\sqrt{(R_L^2+X_L^2)}} = \frac{\sqrt{2}}{3\sqrt{(1+X_L^2/R_L^2)}}$$

Since $X_L = 2\omega L$, hence

$$\gamma = \frac{\sqrt{2}}{3(1+4\omega^2L^2/R_L^2)^{1/2}}$$

If $4\omega^2L^2/R^2_L \gg 1$, then $\gamma = \dfrac{\sqrt{2}R_L}{3\times 2\omega L} = \dfrac{R_L}{3\sqrt{2}.\omega L}$

It is seen that ripple decreases as R_L decreases or load current increases (just the opposite of what happens in the case of shunt capacitor filter).

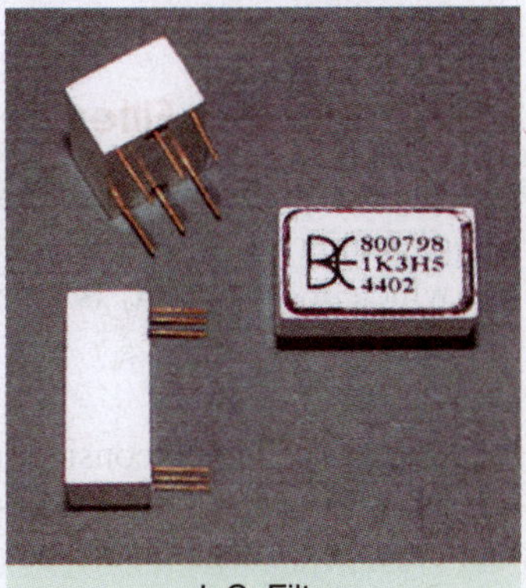

L.C. Filter

55.19. The Choke Input or L-C Filter

It is a combination of two filters considered in Art. 55.15 and 55.18 and provides a lower ripple than is possible with either L or C alone. As is known, in an inductor filter, ripple increases with R_L but decreases in a capacitor filter. The combination of L and C (*i.e.* L–section) filter makes the ripple independent of R_L. Fig. 55.25 (*a*) shows the filter and (*b*) the voltage variations.

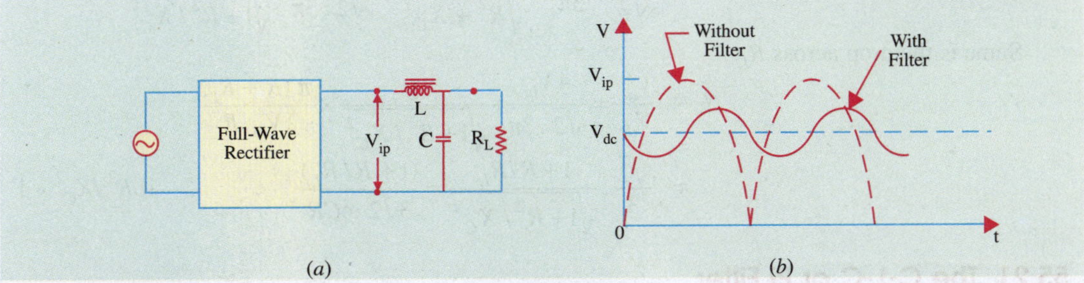

Fig. 55.25

Ripple Factor

If choke resistance R_C is neglected, then dc voltage available across $R_L = 2V_{ip}/\pi$. The ac drop over R_L is the same as across C. Since $X_C \ll R_L$, the parallel combination of R_L and X_C has impedance $\cong X_C$. The second harmonic voltage $(4V_{ip}/3\pi)\cos 2\omega t$ can be assumed to drop over the L–C series combination because R_L is effectively not there.

Maximum value of ac drop over C is

$$= \frac{4V_{ip}}{3\pi} \cdot \frac{X_C}{(X_L + X_C)}$$

$$RMS \text{ value} = \frac{4V_{ip}}{\sqrt{2}.3\pi} \cdot \frac{X_C}{(X_L + X_C)}$$

$$\therefore \quad \gamma = \frac{V_{ac}}{V_{dc}} = \frac{4V_{ip}}{\sqrt{2}.3\pi} \cdot \frac{X_C}{X_L + X_C} \times \frac{\pi}{2V_{ip}} = \frac{\sqrt{2}X_C}{3X_L} \text{—if } X_C \ll X_L$$

$$= \frac{\sqrt{2}}{3(2\omega C)(2\omega L)} = \frac{\sqrt{2}}{12\omega^2 LC} = \frac{1.19}{LC} \text{—if } C \text{ is in } \mu\text{F and } L \text{ in henrys}$$

Now, $I_{dc} = 2V_{ip}/\pi RL$; maximum value of second harmonic current $I_{2h} = 4V_{ip}/3\pi \times 2\omega L$. The critical (or minimum) value of choke inductance essential for proper working of the filter is reached when $I_{dc} = I_{2h}$ or $2V_{ip}/\pi R_L = 4V_{ip}/3\pi.2\omega L$ or $L = R_L/3\omega$. For $f = 50$ Hz, $L = R_L/940$.

Example 55.11. *A single-phase full-wave rectifier uses 300-0-300 V, 50-Hz transformer. For a load current of 60 mA, design an L-filter using 10 H coil and a suitable capacitor to ensure a ripple factor of not more than 1%.* **(Electronics-II, Bangalore Univ.)**

Solution.

$$\gamma = 1.19/LC$$

$$0.01 = 1.19/LC; \quad \therefore \quad LC = 119$$

Hence,

$$10 \times C = 119$$

$$C = 11.9\ \mu\text{F} \cong 12\ \mu\text{F}.$$

55.20. The R-C Filter

Such a filter is shown in Fig. 55.26. Suppose that it is connected to a full-wave rectifier having a filtered output voltage of V_{ip}. The dc component voltage which drops over R_L is

$$= \frac{2V_{ip}}{\pi} \cdot \frac{R_L}{R + R_L} \quad \text{... (i)}$$

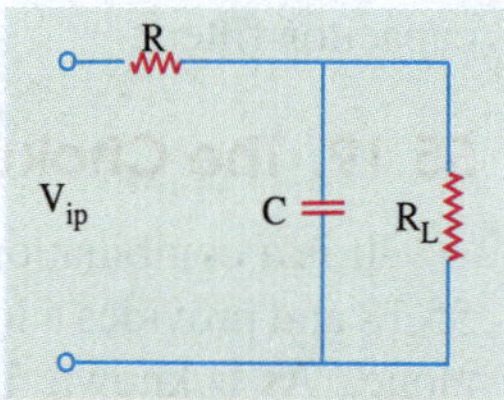

Fig. 55.26

Again, we would consider only the second harmonic voltage $(4V_{ip}/3\pi)\cos 2\omega t$. As before, it will be assumed that $X_C \ll R_L$ so that $R_L \parallel X_C \cong X_C$. In that case, ac voltage would be assumed to drop across *R-C* combination.

$$V_{ac} = \frac{1}{\sqrt{2}} \cdot \frac{4V_{ip}}{3\pi} \cdot \frac{X_C}{\sqrt{R^2 + X_C^2}} = \frac{4V_{ip}}{\sqrt{2}.3\pi} \cdot \frac{1}{\sqrt{1 + R^2/X_C^2}}$$

Same is the drop across R_L.

$$\therefore \quad \gamma = \frac{V_{ac}}{V_{dc}} = \frac{4V_{ip}}{\sqrt{2}.3\pi} \cdot \frac{1}{\sqrt{1 + R^2/X_C^2}} \times \frac{\pi(R + R_L)}{2V_{ip}.R_L}$$

$$= \frac{\sqrt{2}}{3} \cdot \frac{1 + R/R_L}{\sqrt{1 + R^2/X_C^2}} = \frac{(1 + R/R_L)}{3\sqrt{2}\,\omega CR} \quad \text{—if } R^2/X_C^2 \gg 1$$

55.21. The C-L-C or Pi Filter

As shown in Fig. 55.27, it consists of one inductor and two capacitors connected across its each end. The three components are arranged in the shape of the Greek letter π. It is also called capacitor input π-filter. The input capacitor C_1 is selected to offer very low reactance to the ripple frequency. Hence, major part of filtering is done by C_1. Most of the remaining ripple is removed by the combined action of L and C_2.

The charging and discharging action of C_1 is exactly the same as described in Art. 55.15. The output voltage waveform is also like that shown in Fig. 55.25 (*b*).

This circuit gives much better filtering than LC filter circuit. However, C_1 is still directly connected across the supply and would need high pulses of current if load current is large. Since these high peak current pulses are likely to damage the rectifier diode, this filter is used ***with low-current equipment.***

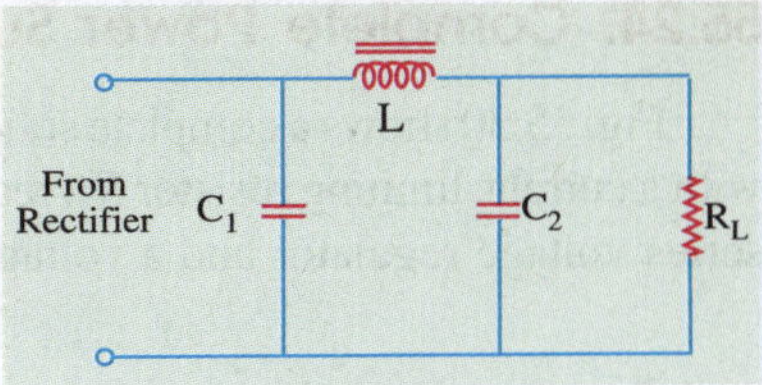

Fig. 55.27

Though this filter-gives somewhat higher output voltage, its voltage regulation is inferior to that of the LC filter.

The ripple factor of this filter is given by

$$\gamma = \sqrt{2}\,\frac{X_{C1}\,X_{C2}}{R_L X_L} = \frac{\sqrt{2}}{8\omega^3 C_1\,C_2\,LR_L} = \frac{5700}{LC_1C_2R_L} \qquad \text{—when } f = 50 \text{ Hz}$$

Here, C_1, C_2 are in μF, L in henrys and R_L in ohms.

55.22. Bleeder Resistor

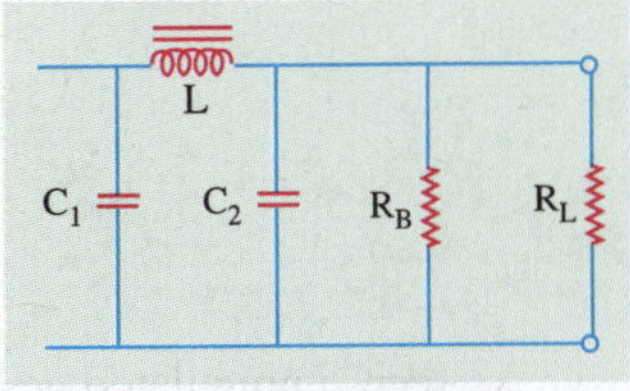

Fig. 55.28

Very often, a resistor (called bleeder resistor) is placed across the filter output (Fig. 55.28) because it provides the following advantages :

1. It improves voltage regulation of the supply.
   By acting as a pre-load on the supply, it causes an initial voltage drop. When the real load is connected, there is only a small amount of additional drop. In this way, difference between no-load and full-load voltage is reduced thereby improving the regulation.
2. It provides safety to the technicians handling the equipment.
   When power supply is switched off, it provides a path for the filter capacitor to discharge through. That is why it is called bleeder resistor. Without it, the capacitor will retain its charge for quite sometime even when the power supply is switched off. This high voltage can be dangerous for people working with the equipment.
3. By maintaining a minimum current through the choke, it improves its filtering action. Value of R_B should be such as to conduct 10 per cent of the total load current.

55.23. Voltage Dividers

Often more than one dc voltage is needed for the operation of electronic circuits. A single power supply can provide as many voltages as are needed by using a voltage divider. As shown in Fig. 55.29 a voltage divider is a single tapped resistor connected across the output terminals of the supply. The tapped resistor may consist of two or three resistors connected in series across the supply. In fact, bleeder resistor may also be used as a voltage divider.

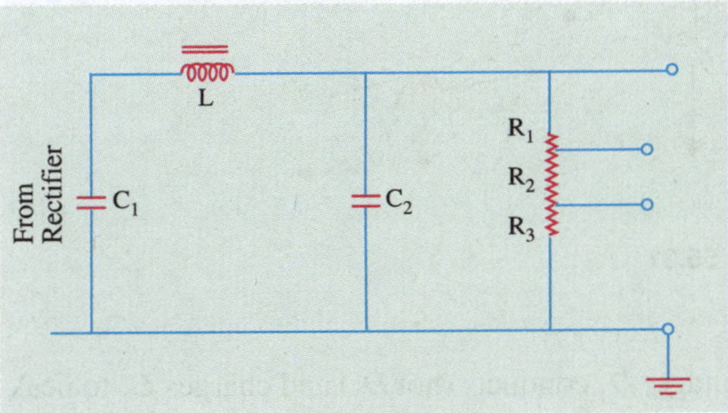

Fig. 55.29

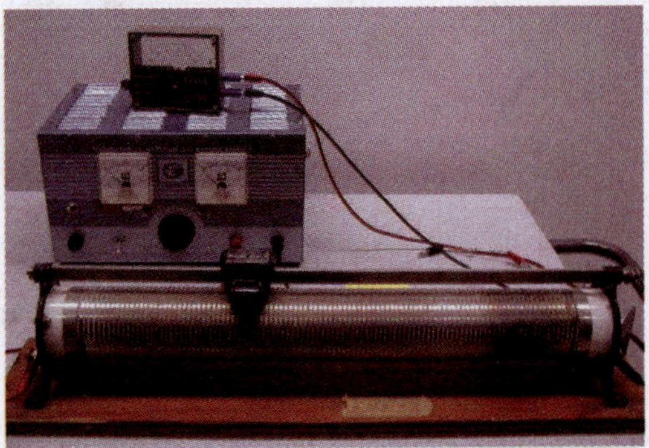

Voltage Dividers

55.24. Complete Power Supply

Fig. 55.30 shows a complete solid-state power supply. From left to right, it consists of a transformer with a current-limiting resistor R_1, rectifier diodes for full-wave rectification, a π-type filter, a transistor series voltage regulator and a voltage divider.

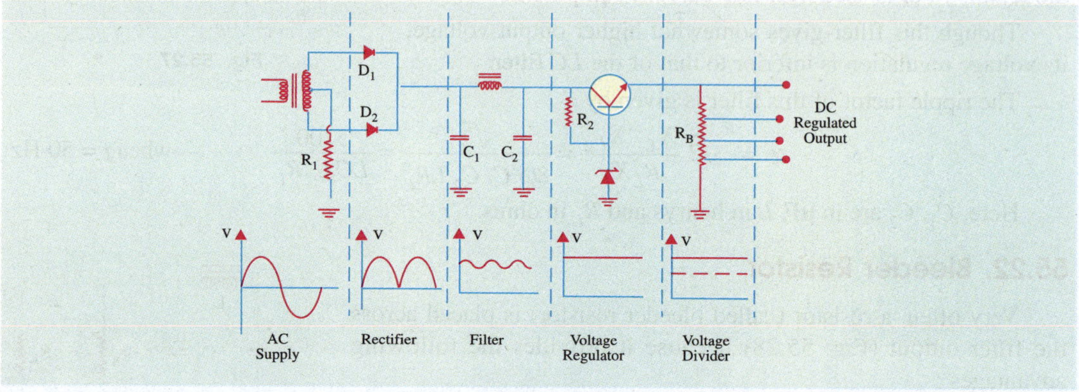

Fig. 55.30

As seen, unregulated ac voltage is fed from the transformer through a full-wave rectifier. It is then filtered by the *CLC* filter and finally regulated by a transistor regulator. The regulated dc supply becomes available across voltage divider resistance R_B. The output is practically ripple-free.

55.25. Voltage Multipliers

A voltage multiplier is a circuit which produces a greater dc output voltage than ac input voltage to the rectifiers. Multipliers are required in many circuit applications where it is necessary to have high voltages with low currents as for electron accelerating purposes in a cathode-ray tube (*CRT*).

We will consider the following circuit :

1. half-wave voltage doubler,
2. full-wave voltage doubler,
3. voltage tripler,
4. voltage quadrupler.

55.26. Half-wave Voltage Doubler

It is also known as cascade voltage doubler. The circuit is shown in Fig. 55.31.

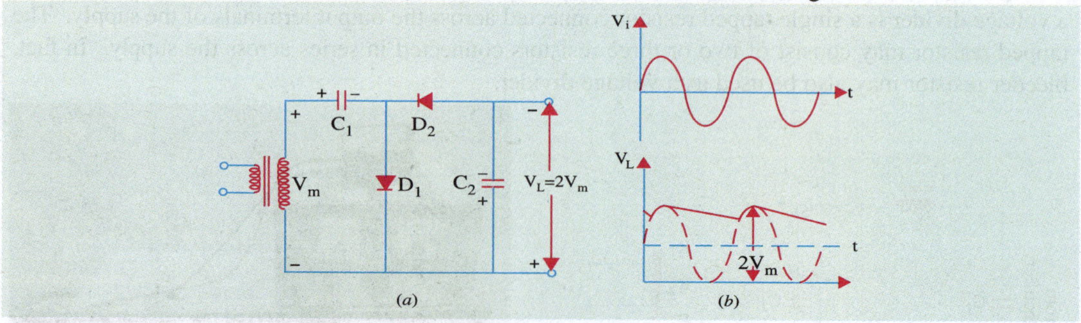

Fig. 55.31

Circuit Analysis

During the positive half-cycle of the input voltage, D_1 conducts (not D_2) and charges C_1 to peak value of secondary voltage (V_m) with the polarity as shown in Fig. 55.31 (*a*). During the negative half-cycle, D_2 conducts (not D_1) and charges C_2. The voltage across C_2 is the sum of peak supply voltage and the voltage across C_1 (Fig. 55.31). It can be proved by applying *KVL* to the outer loop. Starting from the bottom of the transformer secondary in Fig. 55.32 and going clockwise, we get

$$-V_m - V_m + V_{C2} = 0$$

or $\quad V_{C2} = 2V_m = 2 \times \text{peak input voltage}$

During the next positive half-cycle, D_2 is open and C_2 will discharge through the load if it is connected. If no load is connected across C_2, then both capacitors stay charged *i.e.* C_1 to V_m and C_2 to $2V_m$. If there is a load connected across C_2, it will discharge a little bit and, as a result of it, voltage across it will drop slightly. But it will get recharged in the next half-cycle.

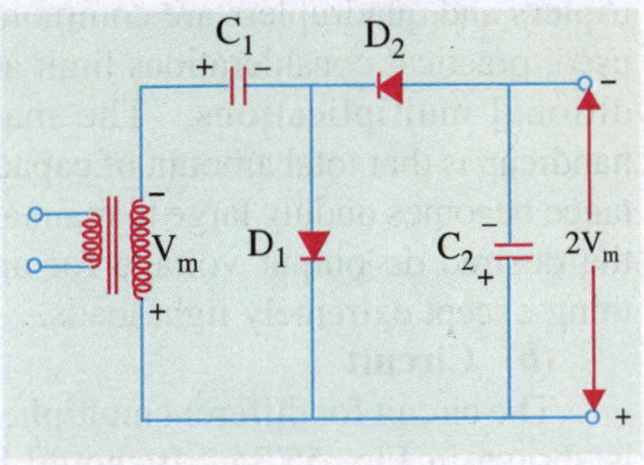

Fig. 55.32

The output waveform shown in Fig. 55.31 (*b*) is that of a half-wave rectifier filtered by a shunt capacitor. *PIV* across each diode is 2 V_m. Ripple frequency is equal to the supply frequency.

This circuit has very poor regulation and its ripple content is also high. This circuit has a common connection between the line and load (which a full-wave doubler does not have).

55.27. Full-wave Voltage Doubler

This circuit is shown in Fig. 55.33 (*a*). During the positive half-cycle of the input voltage, D_1 conducts (but not D_2) and charges capacitor C_1 to the peak voltage V_m with the polarity as shown.

During the negative half-cycle, D_2 conducts (but not D_1) charging C_2 to V_m. As far as the load is concerned, voltages across C_1 and C_2 are in series-aiding. If there is no load connected across the output, then load voltage $V_L = 2V_m$ as shown in Fig. 55.33 (*a*). For example, if 220-V, 50-Hz is the supply, then $V_{dc} = 2V_m = 2 \times 220\sqrt{2} = 622$ V. Of course, if a load is connected across the output terminals, then V_L would be less than 2 V_m.

The waveform of the output voltage is shown in Fig. 55.33 (*b*).

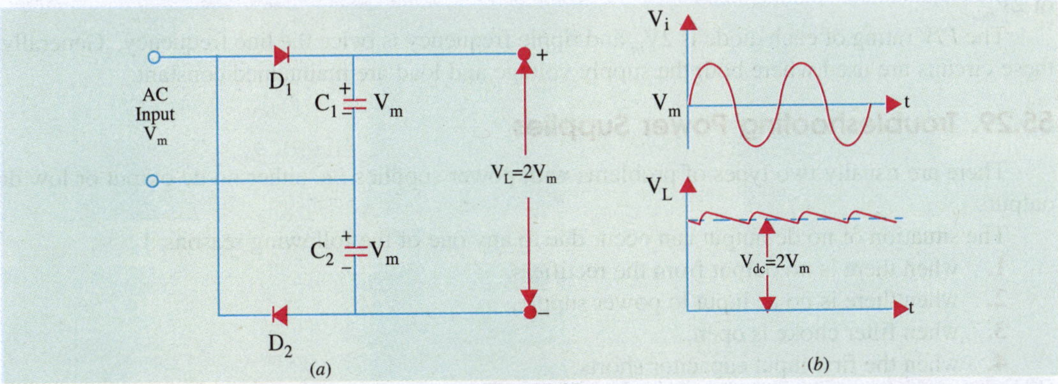

Fig. 55.33

The *PIV* rating of each diode is $2V_m$. Ripple frequency is twice the supply frequency. As seen, there is no common connection between the supply line and the load.

It is worth noting that where the expense of a line transformer is justified, it is preferable to use the superior conventional full-wave rectifier (Art. 55.7).

55.28. Voltage Tripler and Quadrupler Circuits

(*a*) General

The half-wave voltage doubler circuit (Fig. 55.31) can be extended to obtain any multiple of the peak input voltage (V_m) *i.e.* $3V_m$, $4V_m$, $5V_m$ etc. Theoretically speaking, there is no upper limit to the amount of voltage multiplication that can be obtained. Though voltage triplers and quadruplers are commonly used, practical considerations limit additional multiplications. The main handicap is that total amount of capacitance becomes unduly large to maintain the desired dc output voltage for any thing except extremely light loads.

(b) Circuit

The circuit for different multipliers is shown in Fig. 55.34. It should be obvious from the repetitive pattern of the circuit connections how additional diodes and capacitors may be connected to the doubler circuit for obtaining higher multiplications of the peak output voltage V_m.

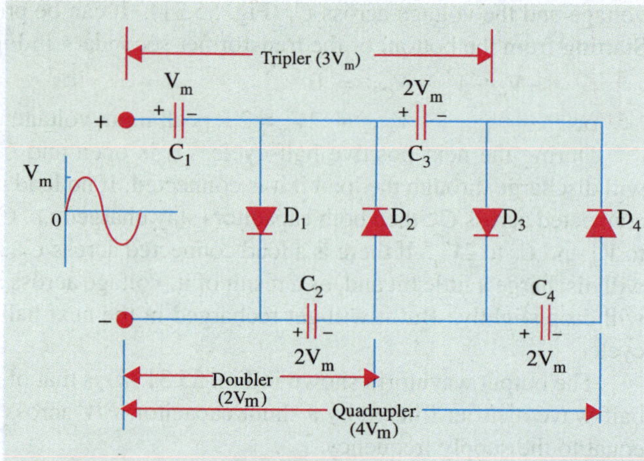

Fig. 55.34

(c) Analysis

During the first positive half-cycle, C_1 charges to V_m as diode D_1 conducts. During negative half-cycle, C_2 is charged through D_2 to $2V_m$ *i.e.* to the sum of voltage across C_1 and peak input voltage V_m (Art 55.26).

During the second positive half-cycle, D_3 conduct and voltage across C_2* charges C_3 to same voltage 2 V_m. During the negative half-cycle, diodes D_2 and D_4 conduct allowing C_3 to charge C_4 to the same peak voltage $2V_m$.

If is seen from Fig. 55.34 that voltage across C_2 is $2V_m$, across C_1 and C_3 is $3V_m$ and across C_2 and C_4 is $4V_m$.

If additional diodes and capacitors are used, each capacitor would be charged to a peak voltage of $2V_m$.

The *PIV* rating of each diode is $2V_m$ and ripple frequency is twice the line frequency. Generally, these circuits are used where both the supply voltage and load are maintained constant.

55.29. Troubleshooting Power Supplies

There are usually two types of problems with power supplies *i.e.* either no dc output or low dc output.

The situation of no dc output can occur due to any one of the following reasons :

1. when there is no output from the rectifiers,
2. when there is no ac input to power supply,
3. when filter choke is open,
4. when the first input capacitor shorts.

A low dc output can occur in the following situations :

1. decreased input ac voltage,
2. open input capacitor of the filter circuit,
3. partial short across the load.

55.30. Controlled Rectification

It is that rectification in which the output of a rectifier circuit can be varied by controlling the point in the ac cycle at which the circuit is turned ON. A thyristor or *SCR* can be used for the purpose

* C_1 cannot charge C_3 because it is shorted by D_1.

because under proper firing conditions, it can control the conduction angle of the rectifier circuit.

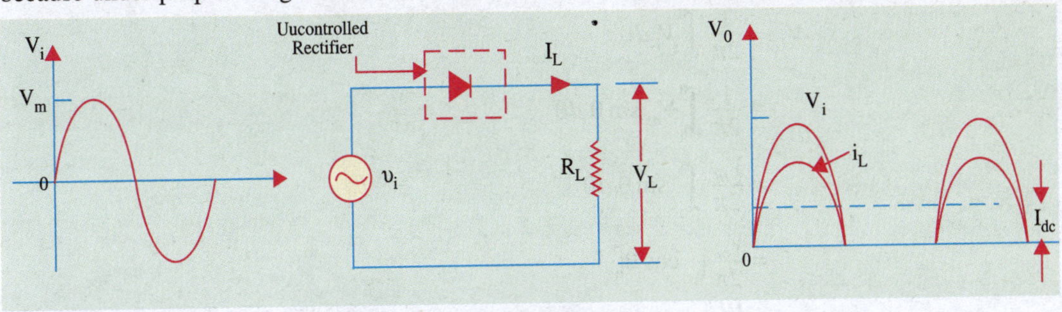

Fig. 55.35

In an ordinary diode rectifier circuit, current flows through the diode whenever instantaneous value of the ac supply voltage is greater than the voltage across the load at that instant.

For a resistive load shown in Fig. 55.35, load current flows at all times during the positive half-cycles of the supply.

In a controlled rectifier, on the other hand, load current flows only when a control signal is applied to turn on the rectifier at a specific point [like *A* in Fig. 55.36 (*b*)] in the ac cycle. Point *A* corresponds to the angle θ_1 so that conduction is delayed by this much period.

Once the HW rectifier is turned ON, it remains in conduction for the rest of the positive half-cycle *i.e.* upto 180°. Obviously, the firing point *A* is determined by the angle of delay in a plying the firing signal by the control circuit. As θ_1 increases, conduction occurs later in the cycle thereby decreasing the load current further.

The curve for voltage drop across rectifier diode is shown in Fig. 55.36 (*c*). During the positive half-cycle when the rectifier is fired into conduction, it acts like a short and voltage across it drops to zero (neglecting forward voltage of the diode). During negative half-cycle when diode is reverse-biased (and hence open) the full supply voltage appears across it as shown.

The curve for load voltage is shown in Fig. 55.36 (*d*). Since current obeys Ohm's law, it follows the load voltage.

Average Value of Load Voltage

Let the equation of the supply voltage be $V_i = V_m \sin\theta$

For a half-wave rectifier, average value is found by taking average over the whole cycle (2π) even though conduction takes

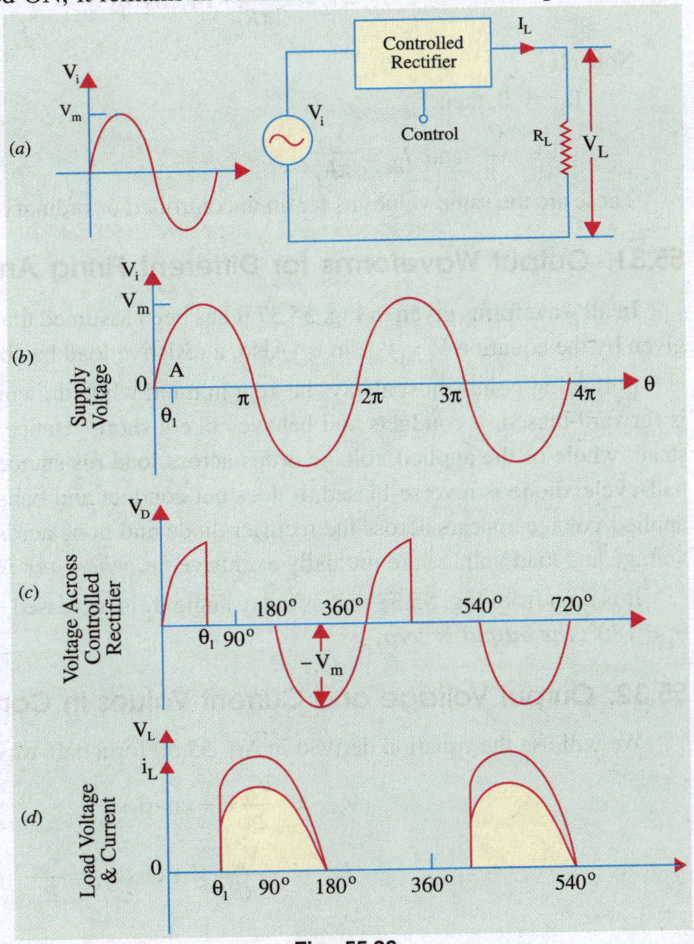

Fig. 55.36

place only from $\theta_1 - \pi$ in the entire cycle.

$$\therefore \quad V_L = \frac{1}{2\pi}\int_{\theta_1}^{\pi} V_i \, d\theta$$

$$= \frac{1}{2\pi}\int_{\theta_1}^{\pi} V_m \sin\theta \, d\theta$$

$$= \frac{V_m}{2\pi}\int_{\theta_1}^{\pi} \sin\theta \, d\theta$$

$$= \frac{V_m}{2\pi}\left|-\cos\theta\right|_{\theta_1}^{\pi}$$

$$= \frac{V_m}{2\pi}\left|(-\cos\pi) - (-\cos\theta_1)\right|$$

$$= \frac{V_m}{2\pi}(1+\cos\theta_1)$$

$$\therefore \quad V_{dc} = V_L = \frac{V_m}{2\pi}(1+\cos\theta_1)$$

For a resistive load,

$$I_{dc} = I_L = \frac{V_m}{2\pi R_L}(1+\cos\theta_1)$$

Note : If

$\theta_1 = 0$, then V_{dc}

$$= \frac{V_m}{\pi} \text{ and } I_{dc} = \frac{V_m}{\pi R_L}$$

These are the same values as for an uncontrolled or ordinary HW rectifier (Art 55.6).

55.31. Output Waveforms for Different Firing Angles

In all waveforms given in Fig. 55.37 it has been assumed that the ac supply voltage is sinusoidal given by the equation $V_i = V_m \sin\theta$. Also, a resistive load has been assumed.

Following points must always be kept in mind while drawing these diagrams. When the diode is forward-biased, it conducts and behaves like a short. Hence, drop across it is almost zero. Instead, whole of the applied voltage drops across load resistance. When during the negative input half-cycle, diode is reverse-biased, it does not conduct and behaves like an 'open'. Hence, all the applied voltage appears across the rectifier diode and none across R_L. In other words, diode anode voltage and load voltage are mutually exclusive ***i.e. when one is there, the other is not.***

It is seen from Fig. 55.38 that as delay angle θ_1 is increased, the output keeps decreasing till for $\theta_1 = 180°$, ***the output is zero.***

55.32. Output Voltage and Current Values in Controlled Rectifiers

We will use the equation derived in Art 55.30 for a half-wave rectifier *i.e.*

$$V_{dc} = \frac{V_m}{2\pi}(1+\cos\theta_1)$$

$$I_{dc} = \frac{V_m}{2\pi R_L}(1+\cos\theta_1) = \frac{V_{dc}}{R_L}$$

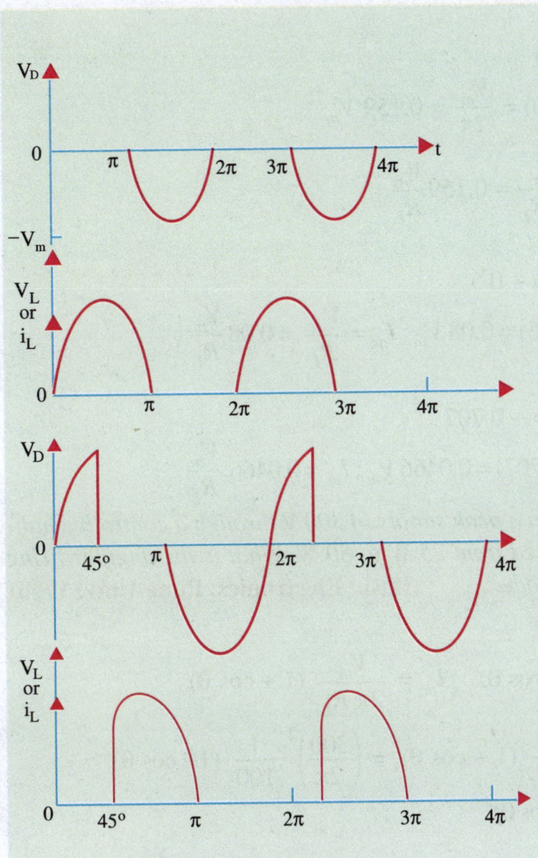

Fig. 55.37

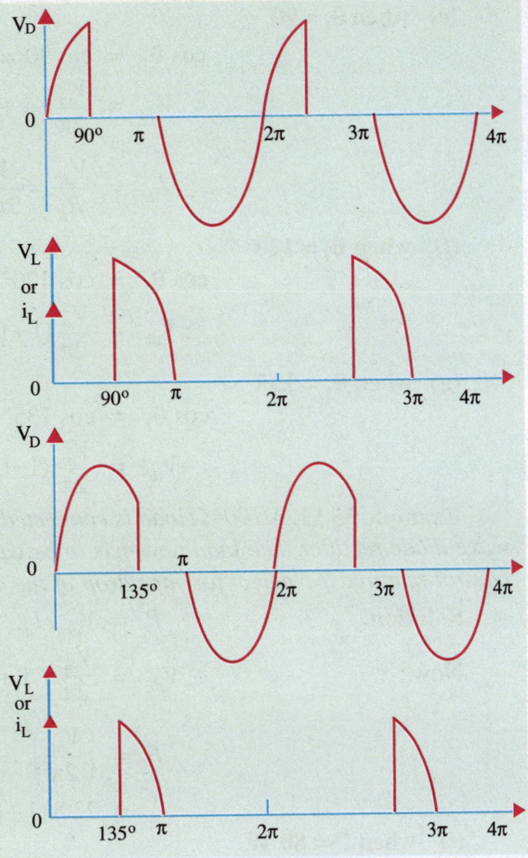

Fig. 55.38

(a) when $\theta_1 = 0°$

$$\cos\theta_1 = \cos 0° = 1$$

$$V_{dc} = \frac{V_m}{2\pi}(1+1) = \frac{V_m}{\pi} = 0.318\, V_m$$

$$I_{dc} = \frac{V_{dc}}{R_L} = \frac{V_m}{\pi\, R_L} = 0.318\,\frac{V_m}{R_L}$$

(b) when $\theta_1 = 30°$

$$\cos\theta_1 = \cos 30° = 0.866$$

$$\therefore \quad V_{dc} = \frac{V_m}{2\pi}(1+0.866) = 0.297\, V_m \; ; \quad I_{dc} = \frac{V_{dc}}{R_L} = 0.297\,\frac{V_m}{R_L}$$

(c) when $\theta_1 = 45°$

$$\cos\theta_1 = \cos 45° = 0.707$$

$$\therefore \quad V_{dc} = \frac{V_m}{2\pi}(1+0.707) = 0.272\, V_m\, ; I_{dc} = \frac{V_{dc}}{R_L} = 0.272\,\frac{V_m}{R_L}$$

(d) when $\theta_1 = 60°$

$$\cos\theta_1 = \cos 60° = 0.5$$

$$\therefore \quad V_{dc} = \frac{V_m}{2\pi}(1+0.5) = 0.239\, V_m\, ; I_{dc} = \frac{V_{dc}}{R_L} = 0.239\,\frac{V_m}{R_L}$$

(e) when $\theta_1 = 90°$

$$\cos\theta_1 = \cos 90° = 0$$

$$V_{dc} = \frac{V_m}{2\pi}(1+0) = \frac{V_m}{2\pi} = 0.159\,V_m$$

$$I_{dc} = \frac{V_{dc}}{R_L} = \frac{V_m}{2\pi R_L} = 0.159\frac{V_m}{R_L}$$

(f) when $\theta_1 = 120°$

$$\cos\theta_1 = \cos 120° = -0.5$$

$$\therefore \quad V_{dc} = \frac{V_m}{2\pi}(1-0.5) = 0.08\,V_m\,;\ I_{dc} = \frac{V_{dc}}{R_L} = 0.08\frac{V_m}{R_L}$$

(g) when $\theta_1 = 135°$

$$\cos\theta_1 = \cos 135° = -0.707$$

$$V_{dc} = \frac{V_m}{2\pi}(1-0.707) = 0.0466\,V_m\,;\ I_{dc} = 0.0466\frac{V_m}{R_L}$$

Example 55.12. *A 100–Ω load is connected to a peak supply of 300 V through a controlled half-wave diode rectifier. The load power is to be varied from 25 W to 80 W. What is the angular firing control required ? Neglect forward drop of the diode.* **(Basic Electronics, Pune Univ. 1990)**

Solution.
$$P = V_{dc}\cdot I_{dc}$$

Now,
$$V_{dc} = \frac{V_m}{2\pi}(1+\cos\theta).\quad I_{dc} = \frac{V_m}{2\pi R_L}(1+\cos\theta)$$

$$\therefore \quad P = \left(\frac{V_m}{2\pi}\right)^2\cdot\frac{1}{R_L}(1+\cos\theta)^2 = \left(\frac{300}{2\pi}\right)^2\cdot\frac{1}{100}(1+\cos\theta)^2$$

$$= 22.8\,(1+\cos\theta)^2$$

(i) when P = 80 W

$$\therefore \quad 80 = 22.8\,(1+\cos\theta)^2 \quad \text{or} \quad (1+\cos\theta) = 1.8732$$

$$\therefore \quad \cos\theta = 0.8732 \quad \text{or} \quad \theta = \mathbf{29.17°}$$

(ii) when P = 25 W

$$\therefore \quad 25 = 22.8\,(1+\cos\theta)^2 \quad \therefore\ \cos\theta = 0.0471 \quad \text{or} \quad \theta = \mathbf{87.30°}$$

Example 55.13. *In a controlled half-wave rectifier, peak supply voltage is 200 V and the value of load resistor is 1 k Ω. Calculate the power delivered to the load circuit for firing angles of (i) 0° (ii) 45° (iii) 90° and (iv) 135°.* **(Solid State Devices and Ckts, BHU)**

Solution. (i) $V_{dc} = 0.318\times200 = 63.6\text{ V}$; $I_{dc} = 63.6/1000 = 63.6\text{ mA}$

$P = V_{dc}.I_{dc} = 63.6\times63.6 = 4045\text{ mW} = \mathbf{4.045\ W}$

(ii) $V_{dc} = 0.27\times200 = 54\text{ V}$; $I_{dc} = 54/1000 = 54\text{ mA}$

$P = 54\times54 = 2916\text{ mW} = \mathbf{2.916\ W}$

(iii) $V_{dc} = 0.159\times200 = 31.8\text{ V}$; $I_{dc} = 31.8\text{ mA}$

$P = 31.8\times31.8 = 1011\text{ mW} = \mathbf{1.011\ W}$

(iv) $V_{dc} = 0.0466\times200 = 9.3\text{ V}$; $I_{dc} = 9.3\text{ mA}$

$P = 9.3\times9.3 = 86.5\text{ mW} = \mathbf{0.0865\ W}$

55.33. Average Values for FW Controlled Rectifier

In this case, the average values would be doubled because there are two half sinusoids to be averaged as shown in Fig. 55.39 for $\theta_1 = 60°$.

$$V_{dc} = 2 \times \frac{V_m}{2\pi}(1 + \cos\theta_1) = \frac{V_m}{\pi}(1 + \cos\theta_1)$$

$$\left[I_{dc} = \frac{V_{dc}}{R_L} = \frac{V_m}{\pi R_L}(1 + \cos\theta_1) \right]$$

In the present case, $\theta_1 = 60°, \cos\theta_1 = 0.5$

$$\therefore \quad V_{dc} = \frac{V_m}{\pi}(1 + 0.5) = 0.477\, V_m$$

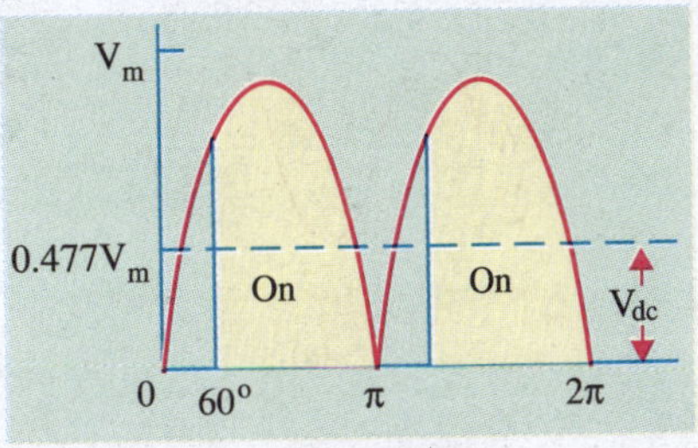

Fig. 55.39

55.34. Silicon Controlled Rectifier (*SCR*)

It is a trijunction *PNPN* device having three external connections : anode (*A*), cathode (*C*) and gate (*G*) as shown in Fig. 55.40.

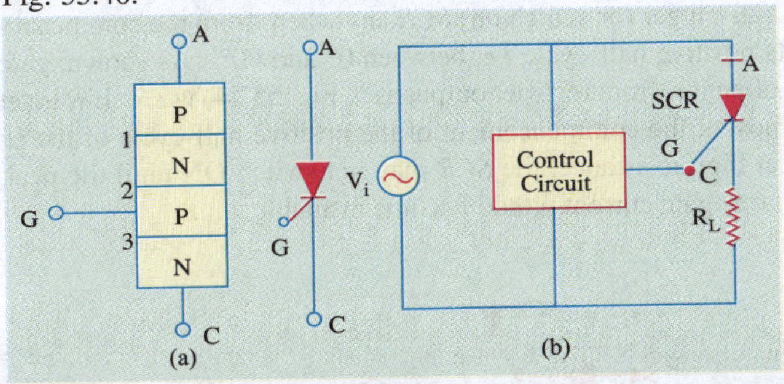

Fig. 55.40

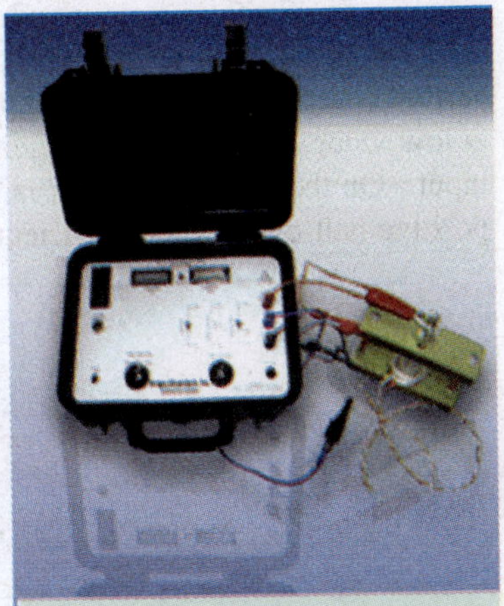
SCR and diode tester

When *A* is made positive with respect to *C*, the junctions marked 1 and 3 are forward-biased but the centremost junction marked 2 is reverse-biased. Hence, the *SCR* does not conduct. However, if a sufficiently positive voltage pulse is applied to this junction via the gate *G*, the *SCR* starts conducting like an ordinary diode. Gate has no control over *SCR* once it has been triggered into conduction. Conduction can only be terminated by removing positive voltage from its anode.

As shown in Fig. 55.40 (*b*), *SCR* needs a control circuit which triggers it into conduction by a gate pulse. Two popular ways of turning ON an *SCR* are:

1. Amplitude firing–in which gate current flows for the whole period of conduction.
2. *UJT* oscillator firing–in this method, gate current is supplied only momentarily. This method allows control over the power delivered to *SCR*'s load from zero watt to fully ON.

Some of the simple ways of controlling the conduction of an *SCR* are discussed below.

55.35. Pulse Control of *SCR*

The simplest *SCR* control circuit of this type is shown in Fig. 55.41. If *SCR* were an ordinary rectifier, it would produce half-wave rectified ac voltage across R_L. However, *SCR* will not conduct even during positive half-cycle of the input ac voltage unless *G* is given positive voltage to forward-bias its centremost junction. By applying a trigger pulse at any time during the positive input half-cycle, it can be fired into conduction.

The resultant load waveform across R_L will consist of a portion of the positive half-cycle commencing at the instant at which the *SCR* is triggered [Fig. 55.41 (*c*)].

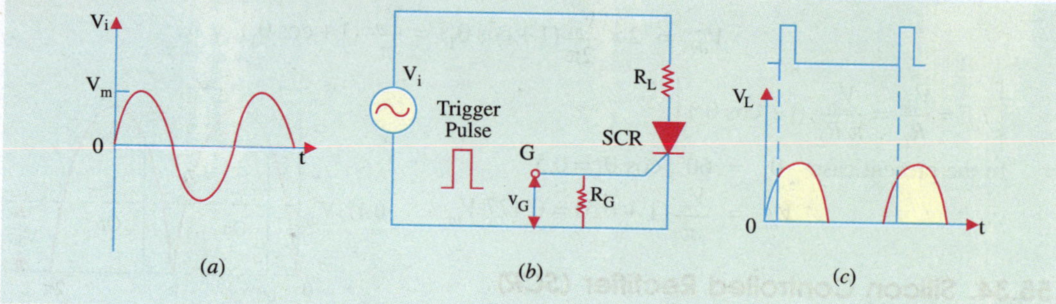

Fig. 55.41

55.36. 90° Phase Control of *SCR*

Fig. 55.42 shows a circuit which can trigger (or switch on) *SCR* any where from the commencement of the ac cycle to the peak of its positive half-cycle *i.e.* between 0° and 90°. As shown, gate current is derived from the ac supply (often it is from rectifier output as in Fig. 55.44) *via R*. If *R* is set at low value, the *SCR* will trigger almost at the commencement of the positive half-cycle of the ac input. On the other hand, if *R* is set at high resistance, the *SCR* may not switch ON until the peak positive half-cycle when sufficiently large gate current would become available.

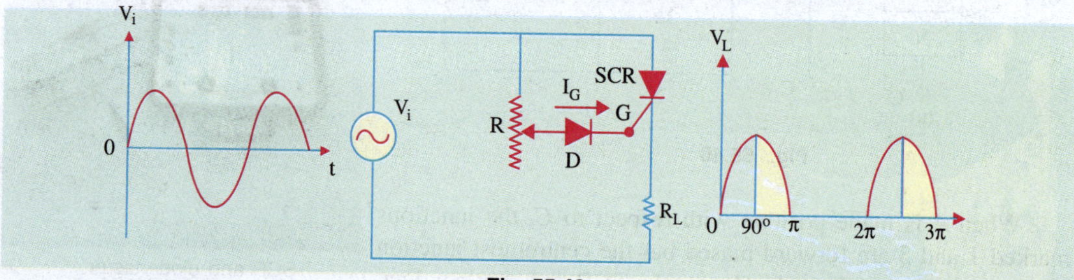

Fig. 55.42

If I_G is not large enough even at peak positive half-cycle, then *SCR* will not trigger at all because I_G has the greatest value at the peak input and then falls off as the voltage falls. The purpose of diode *D* is to protect the *SCR* gate from negative voltage during the negative input half-cycle. This method is also known as the amplitude firing of an *SCR*.

55.37. 180° Phase Control of *SCR*

The circuit shown in Fig. 55.43 can trigger the *SCR* from 0° to 180° of the input waveform.

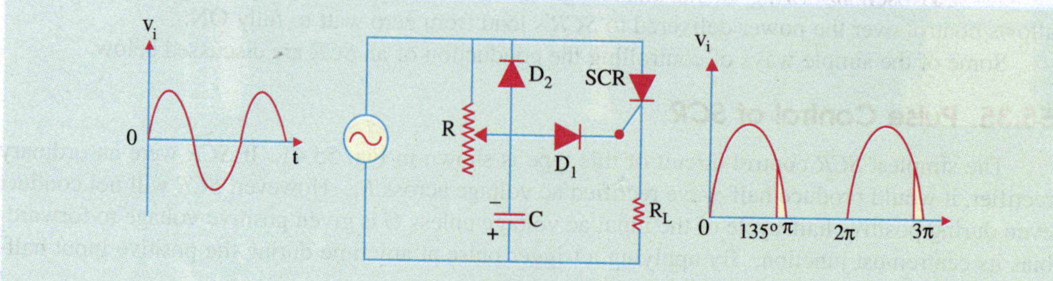

Fig. 55.43

The circuit is identical to that of Fig. 55.42 except for the addition of the diode D_2 and capacitor *C*.

We will start the analysis with the negative half-cycle. During the negative half-cycle of the input, C is charged immediately (with the polarity as shown) to the peak of the input voltage because D_2 is forward-biased. When peak of the negative half-cycle passes over, D_2 becomes reverse-biased because its anode (connected to C) becomes more negative than its cathode (connected to the supply). Hence, C starts to discharge through R. Depending on the time constant (= CR), C may almost be completely discharged at the commencement of the oncoming positive half-cycle or may retain partial charge until almost 180° of positive half-cycle has passed. So long as C remains negatively-charged, D_1 is reverse-biased and the gate cannot become positive to trigger the *SCR* into conduction. Hence, R and/or C can be adjusted to trigger the *SCR* anywhere from 0° to 180° of the input ac cycle.

55.38. *SCR* Controlled Load Circuit

We will now consider a circuit where an amplitude-fired *SCR* is used to control the power in a full-wave rectifier circuit.

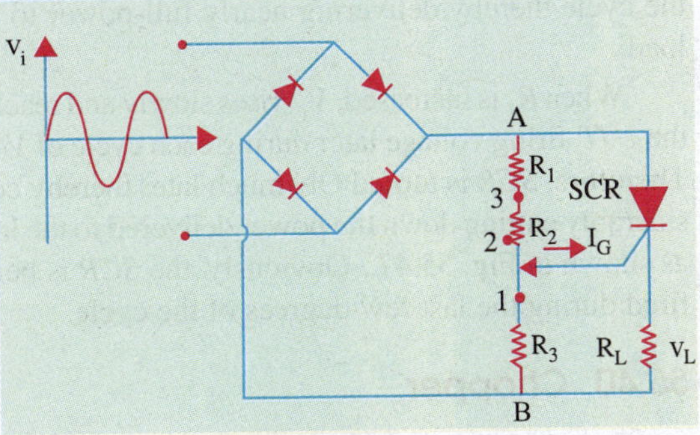

Fig. 55.44

In Fig. 55.44, R_L is the load resistance which may be a lamp load, heater or a small dc motor whose power we wish to control. The full-wave bridge rectifier furnishes the rectified output shown in Fig 55.45 (*a*). Obviously, control over both half-cycles of the ac input is possible.

The rectifier output voltage is dropped across the potentiometer *AB*. When potentiometer is set at point 1, drop across R_3 is not enough to provide sufficient gate current to switch ON the *SCR*. Hence, there is no load current and, consequently, no drop across the load. When potentiometer is set at point 2 and then 3, the *SCR* gets fired yielding load voltage waveforms shown in Fig. 55.45 (*b*) and (*c*). Hence, changing the moving contact on the potentiometer changes the conduction angle and hence the amount of power delivered to the load.

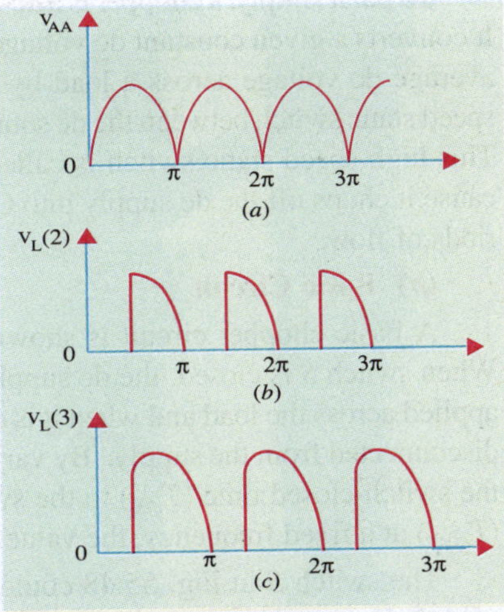

Fig. 55.45

55.39. UJT Controlled Load Circuit

This circuit employs a *UJT* oscillator to control the firing of the *SCR* as shown in Fig. 55.46. The full-wave rectified output voltage V_{AF} is available across two parallel paths *AF* and *BF*. When *SCR* is not conducting (*i.e.* it acts as an open), then whole output voltage drops across it and none across R_L. When it conducts (*i.e.* acts as a short), whole of V_{AF} drops across R_L and none across it.

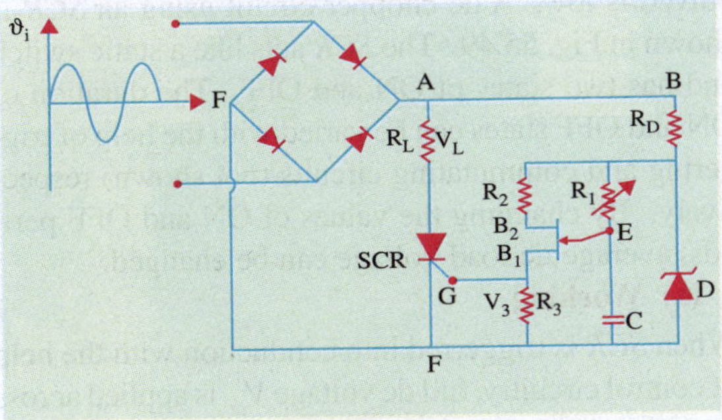

Fig. 55.46

Now, consider the other parallel circuit *BF*. R_D is the voltage dropping resistor so that a suitable voltage V_u is applied to the *UJT* oscillator circuit for the *UJT* to work without damage. The full-wave rectified voltage is clipped at a convenient level by Zener diode *D* for proper operation of *UJT*. Varying R_1 varies the time it takes for *C* to charge to the *UJT*'s firing voltage. Charging of *C* always starts when V_U is at 0 V dc or at 0° of V_{AF}.

If R_1 is very small, *C* charges very quickly (∵ time constant CR_1 is very short) and reaches the

UJT firing voltage soon after V_{AF} starts to go positive. Hence, *UJT* fires early on every cycle of V_{AF}. When *UJT* fires, it becomes almost a short between its emitter (*E*) and B_1. Thus, it discharges *C* very quickly through R_3 which is in parallel with the *SCR*'s gate. This triggers the *SCR* early in the cycle of V_{AF} and the *SCR* latches itself ON until V_{AF} falls to zero. When V_{AF} goes to zero, the *SCR* is shut OFF. The cycle repeats itself with the *SCR* being turned ON early in the cycle thereby delivering nearly full-power to the load.

When R_1 is increased, V_C rises slowly and reaches the *UJT* firing voltage later during each cycle of *VAF*. Therefore, *SCR* is turned ON much later thereby considerably cutting down the power delivered to the load as shown in Fig. 55.47. Obviously, the *SCR* is being fired during the last few degrees of the cycle.

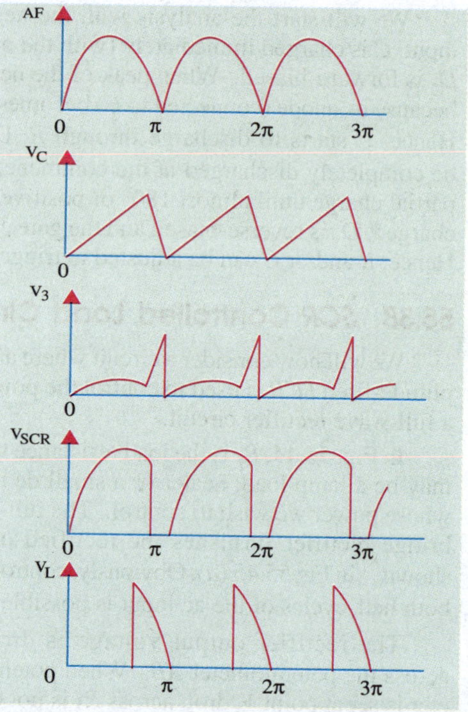

Fig. 55.47

55.40 Chopper

(*a*) Definition

To put it simply, a chopper is a ***dc-to-dc converter.*** It converts a given constant dc voltage into a variable average dc voltage across a load by placing a high-speed static switch between the dc source and the load. This high-speed static switch is called a chopper because it chops off the dc supply into ON and OFF periods of flow.

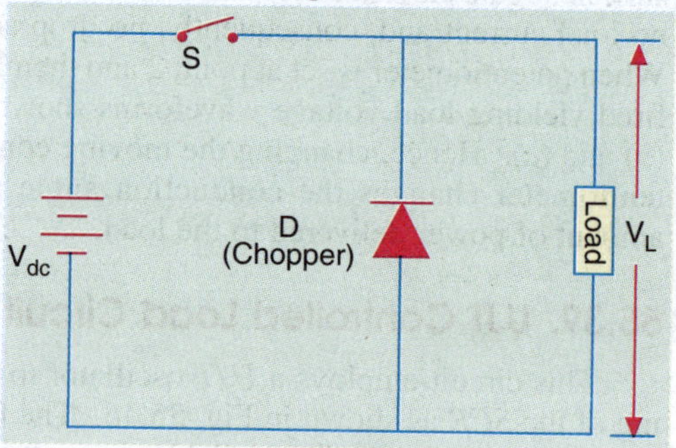

Fig. 55.48

(*b*) Basic Circuit

A basic chopper circuit is shown in Fig. 55.48. When switch *S* is closed, the dc supply voltage V_{dc} is applied across the load and when it is open, the load is disconnected from the supply. By varying the ratio of the switch-closed time (T_{ON}) to the switch-open time (T_{OFF}) at a fixed frequency, the value of the average output dc voltage can be controlled.

The switch *S* in Fig. 55.48 could be either a transistor or an *SCR* depending on the amount of power involved. An *SCR* is used in high-power applications whereas transistors are used when power involved is low. A dc chopper circuit using an *SCR* is shown in Fig. 55.49. The *SCR* acts like a static switch and has two states of ON and OFF. The duration of ON and OFF states can be varied with the help of triggering and commutating circuits (not shown) respectively. By changing the values of ON and OFF periods, average dc load voltage can be changed.

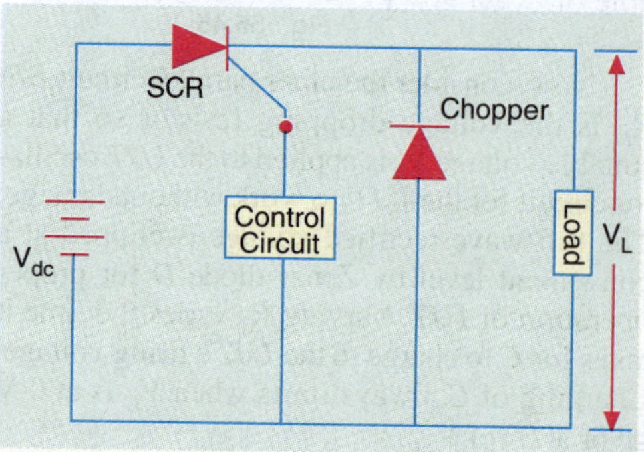

Fig. 55.49

(*c*) Working

When *SCR* is triggerred into conduction with the help of control circuitry, full dc voltage V_{dc} is applied across the load for a period of T_{ON}. When the *SCR* is switched OFF by the control circuitry, there is no voltage across the load. The output load voltage is in the form of a square wave as shown in Fig. 55.50. The waveshapes of the load current and chopper current for an inductive load are as shown. The freewheeling diode *D* provides path for the stored inductive energy to flow.

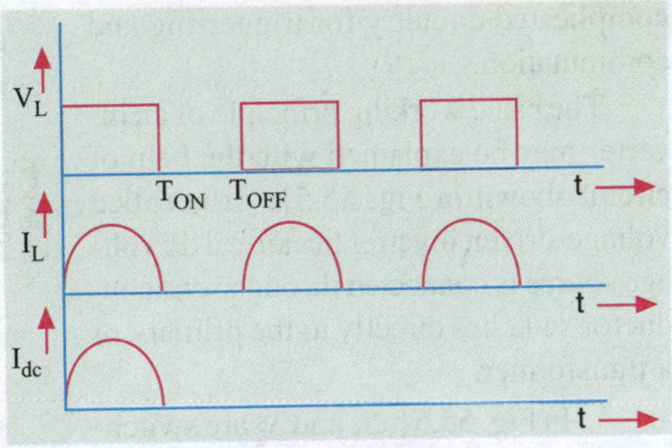

Fig. 55.50

(*d*) Calculations

If T_{ON} is the ON time and T_{OFF} is the OFF time of the chopper, the duty cycle of the chopper is given by

$$\text{duty cycle} = \frac{T_{ON}}{T} = \frac{T_{ON}}{T_{ON} + T_{OFF}}$$

The load voltage V_L is given by

$$V_L = V_{dc}\left(\frac{T_{ON}}{T}\right) = V_{dc} \times \text{duty cycle} = f\, V_{dc}\, T_{ON}$$

where $f\ (= 1/T)$ is the switching frequency of the chopper.

It is seen from the above that V_L depends on the duty cycle since V_{dc} is constant. Hence, following different methods of controlling V_L are available.

1. keeping T_{OFF} constant, varying T_{ON} ;
2. keeping T_{ON} constant, varying T_{OFF} ;
3. varying the ratio T_{ON}/T_{OFF} ;
4. any combination of the above.

Example 55.14. *A dc chopper has ON time of 30 μs and OFF time of 10 μs. Calculate (i) chopper duty cycle, (ii) chopping frequency.*

Solution. (*i*) duty cycle $= T_{ON}/T = 30/(30 + 10) = 0.75$

(*ii*) chopping frequency, $f = 1/T = 1/(30+10)\times 10^{-6}$
$= 25{,}000$ Hz = **25 kHz**

Example 55.15. *A chopper supplied by a 200 V dc has ON time of 30 ms and OFF time of 10 ms. Determine the value of the average dc output voltage.*

(Industrial Electronics, Mysore Univ. 1993)

Solution.

$$T_{ON} = 30\times 10^{-3}\ s,\quad T_{OFF} = 10\times 10^{-3}\ s,$$
$$T = 40\times 10^{-3}\ s$$

Duty cycle of the chopper $= 30 \times 10^{-3} / 40 \times 10^{-3} = 0.75$

$$V_L = V_{dc} \times \text{duty cycle} = 200 \times 0.75 = \textbf{150 V}$$

55.41. Inverters

An inverter is a device that ***changes dc power into ac power*** (just the opposite of converters). The inversion process can be achieved with the help of transistors, *SCR*s and tunnel diodes etc. For low and medium outputs, transistorised inverters are suitable but for high power outputs, *SCR* inverters are essential. For very low voltage and high current requirements, tunnel diode invertors are used.

Power Inverter

For inverter applications, transistors have definite advantages over *SCR*s regarding the switching speed, simplicity of control circuitry, high efficiency and greater reliability. It is mainly due to this fact the *SCR* inverters require complicated circuitry for triggerring and commutation.

The basic working principle of an inverter may be explained with the help of circuit shown in Fig. 55.51. It is called voltage-driven inverter because a dc voltage source is connected through semiconductor switches directly to the primary of a transformer.

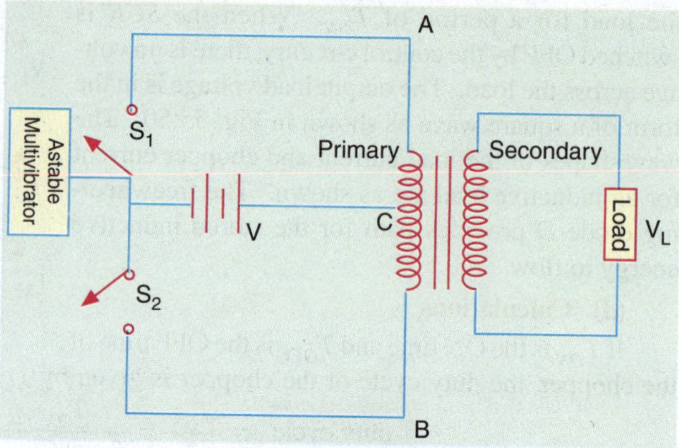

Fig. 55.51

In Fig. 55.51, S_1 and S_2 are switching devices (transistors or *SCR*s) which open and close alternately at regular intervals of time. The two switching devices are generally driven by an astable multi-vibrator operating at the desired frequency. When S_1 is closed, the entire dc source voltage V is applied across points A and C of the transformer primary. S_1 remains closed for a certain period of time after which it is cut off and S_2 closes. It also remains closed for the same period of time during which the source voltage V is impressed across points B and C of the primary. S_2 then opens out and S_1 closes. In this way, an alternating voltage is applied across the primary which induces an ac voltage in the secondary. Since dc supply voltage is connected directly across the primary, the output waveform of the secondary voltage is a square wave (Fig. 55.52) irrespective of the type of load and load power factor. However, the waveforms of both the primary and secondary currents depend on the type of load whether resistive, inductive or capacitive.

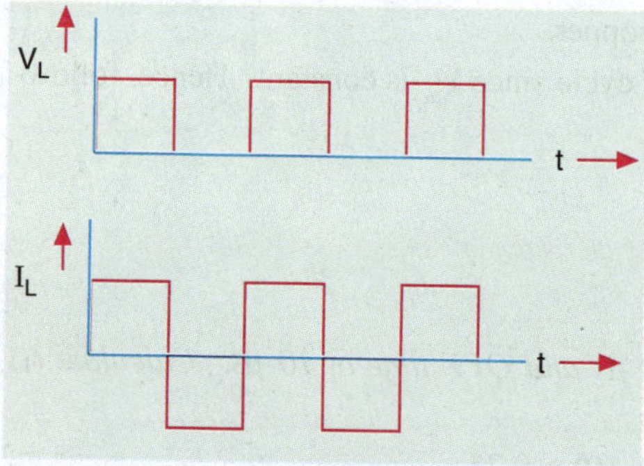

Fig. 55.52

55.42. Single-phase Inverter

Fig. 55.53 (*a*) shows a single-phase inverter with a load resistor using 4 *SCR*s working in pairs. The triggerring and commutating circuitry of the *SCR*s has not been shown in the figure. The two thyristors SCR_1 and SCR_4 are triggerred simultaneously so that load current passes through R_L from left to right. Exactly when these two *SCR*s are switched off by the commutating circuitry, thyristors SCR_2 and SCR_3 are triggerred into conduction thereby sending current through R_L from right to left. Hence, an ac voltage is developed across the load whose waveform is as shown in Fig. 55.53 (*b*).

Single-phase inverter

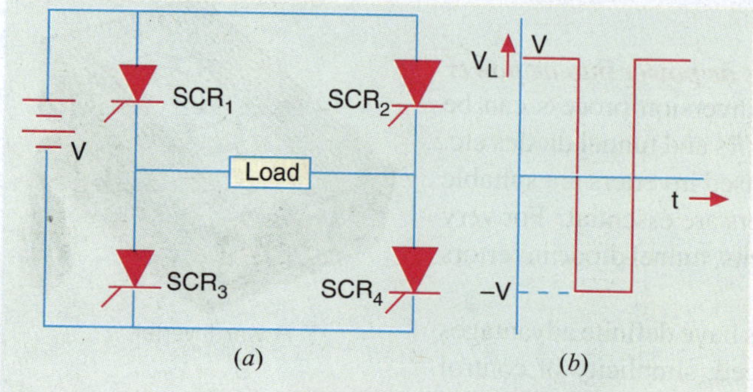

Fig. 55.53

55.43. Push pull Inverter

Fig. 55.54 shows an inverter which

employs two *SCR*s and one transformer. These two *SCR*s are triggerred into conduction alternately for the same period of time. As a result, current through the primary becomes alternating which induces an ac voltage across the secondary and hence the load. As explained earlier in Art 55.42, the secondary ac voltage has a square waveform. The capacitor *C* is connected across the anodes of the two *SCR*s and provides commutation *i.e.* switching off of the *SCR*s. The capacitor charges to double the supply voltage as a result of transformer action between the two halves of the primary winding. This large voltage is sufficient to reverse-bias the *SCR*s and drive the holding current below its rated value.

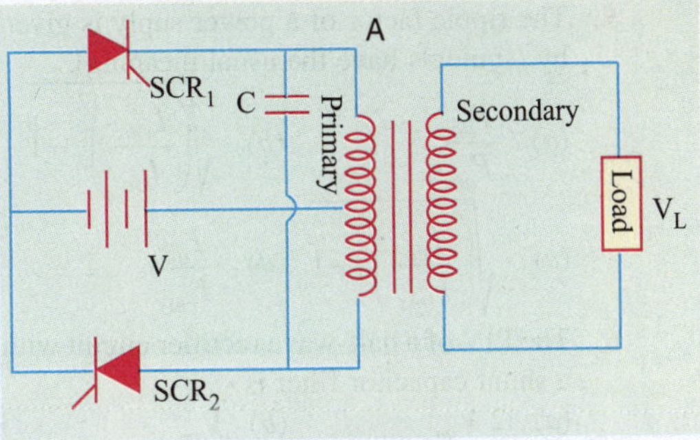

Fig. 55.54

Tutorial Problems No. 55.1

1. A 1-ϕ half-wave rectifier using a 10 : 1 transformer supplies power to a 9 Ω load. If the primary input voltage has an r.m.s. value of 200 V and forward diode resistance is 0.2 Ω and transformer secondary resistance is 0.8 Ω, determine
(*i*) $I_L(dc)$, (*ii*) r.m.s. ripple voltage and (*iii*) efficiency **[(*i*) 0.9 A(*ii*) 9.8 V(*iii*) 36.54%]**

2. A single-phase full-wave rectifier using a power transformer has secondary voltage of 100-0-100V (r.m.s.). It supplies a load of 1 KΩ. Neglecting transformer losses and forward voltage drop of the diode, determine :
(*a*) dc output voltage, (*b*) dc output current, (*c*) ripple voltage and *PIV* rating of the diodes.
[(*a*) 90 V (*b*) 90 mA (*c*) 43.4 V (rms.) ; 282 V]

3. A single-phase half-wave rectifier supplies power to a 1 KΩ load. The rectifier voltage is 200 V (r.m.s.). Neglecting diode resistance, calculate (*i*) dc load voltage, (*ii*) dc load current and (*iii*) r.m.s. ripple voltage. **[(*i*) 90 V(*ii*) 90 mA(*iii*) 1.09 V]**

4. A single-phase half-wave diode rectifier supplies power to a 2-kΩ resistive load. The input ac supply voltage has a peak value of 300 V. Neglecting forward drop of the diode, calculate
(*a*) V_{dc}, (*b*) I_{dc} , (*c*) power delivered to the load, (*d*) ripple voltage (rms value)
[(*a*) 95.4 V(*b*) 47.7 mA (*c*) 4550 mW (*d*) 115.4 V]

5. A full-wave diode rectifier supplies a load of 10 kΩ. The ac voltage applied to the diode is 300-0-300 V rms. It diode resistance is neglected, calculate: (*a*) V_{dc}, (*b*) I_{dc}, (*c*) I_{rms}, (*d*) form factor, (*e*) ripple voltage. **[(*a*) 270 V (*b*) 27 mA (*c*) 30 mA (*d*) 1.11 (*e*) 130.1 V]**

6. A dc and an ac voltmeter are used to measure the output voltage of a filter circuit. The readings of the two voltmeters are 50 V and 5 V respectively. Calculate the ripple factor of the filter. **[10%]**

7. In a controlled full-wave rectifier, peak supply voltage is 200 V and load resistance 1 kW. Calculate the power delivered to the load for firing angles of (*a*) 60° and (*b*) 120°. **[(*a*) 9.1 W (*b*) 1.01 W]**

OBJECTIVE TESTS – 55

1. The ripple factor of a power supply is a measure of
(*a*) its filter efficiency
(*b*) its voltage regulation
(*c*) diode rating
(*d*) purity of power output.

2. The basic reason why a FW rectifier has twice the efficiency of a HW rectifier is that
(*a*) it makes use of a transformer
(*b*) its ripple factor is much less
(*c*) it utilizes both half-cycle of the input
(*d*) its output frequency is double the line frequency.

3. The output of a half-wave rectifier is suitable only for
(*a*) running car radios
(*b*) running ac motors
(*c*) charging batteries
(*d*) running tape-recorders.

4. The ripple factor of a HW rectifier is
(*a*) 0.406 (*b*) 0.812
(*c*) 1.21 (*d*) 1.11

5. The ripple factor of a power suply is given by (symobls have the usual meaning).
(a) $\frac{P_{dc}}{P_{ac}}$ (b) $\sqrt{\left(\frac{I_{rms}}{I_{dc}}\right)^2 - 1}$
(c) $\sqrt{\left(\frac{I_{dc}}{I_{rms}}\right)^2 - 1}$ (d) $\frac{I_{dc}}{I_{rms}}$

6. The PIV of a half-wave rectifier circuit with a shunt capacitor filter is
(a) $2 V_{sm}$ (b) V_{sm}
(c) $V_{sm}/2$ (d) $3 V_{sm}$.

7. The primary function of a rectifier filter is to
(a) minimise ac input variations
(b) suppress odd harmonics in the rectifier output
(c) stabilise dc level of the output voltage
(d) remove ripples from the rectified output

8. In a rectifier, larger the value of shunt capacitor filter
(a) larger the *p-p* value of ripple voltage
(b) larger the peak current in the rectifying diode
(c) longer the time that current pulse flows through the diode
(d) smaller the dc voltage across the load.

9. In a LC filter, the ripple factor,
(a) increases with the load current
(b) increases with the load resistance
(c) remains constant with the load current
(d) has the lowest value.

10. The main reason why a bleeder resistor is used in a dc power supply is that it
(a) keeps the supply ON
(b) improves voltage regulation
(c) improves filtering action
(d) both (b) and (c).

11. Which stage of a dc power supply uses a Zener as the main component?
(a) rectifier (b) voltage divider
(c) regulator (d) filter.

12. Which rectifier requires four diodes ?
(a) half-wave voltage doubler
(b) full-wave voltage doubler
(c) full-wave bridge circuit
(d) voltage tripler.

13. For a half-wave controlled rectifier, the average value of output dc voltage is given by
(a) $V_{dc} = \frac{V_m}{2\pi}(1 - \cos\theta_1)$
(b) $V_{dc} = \frac{2V_m}{\pi}(1 + \cos\theta_1)$
(c) $V_{dc} = \frac{V_m}{\pi}(\cos\theta_1 - 1)$
(d) $V_{dc} = \frac{V_m}{2\pi}(\cos\theta_1 + 1)$.

14. If, by mistake, ac source in a bridge rectifier is connected across the dc terminals, it will burn out and hence short .... diodes.
(a) one (b) two
(c) three (d) four.

15. The circuit in Fig. 55.55 shows a full-wave rectifier. The input voltage is (rms) single-phase ac. The peak reverse voltage across the diodes D1 and D2 .

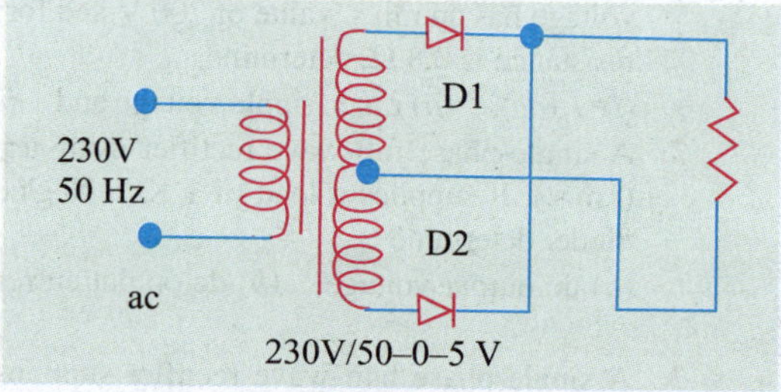

Fig. 55.55

(a) $100\sqrt{2}$ V (b) 100 V
(c) $50\sqrt{2}$ V (d) 50V

(GATE ; 2004)

16. The circuit in Fig. 55.56 shows a 3-phase half-wave rectifier. The source is a symmetrical, 3-phase four-wire system. The line-to-line voltage of the source is 100 V. The supply frequency is 400 Hz. The ripple frequency at the output is

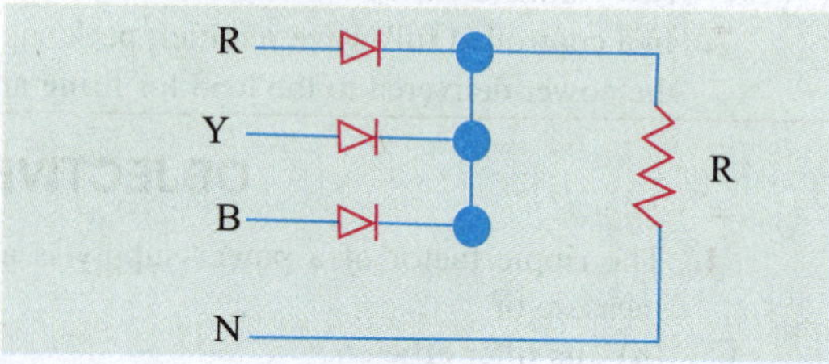

Fig. 55.56

(a) 400 Hz (b) 800 Hz
(b) 1200 Hz (d) 2400 Hz

(GATE ; 2004)

ANSWERS

1. (a) 2. (c) 3. (b) 4. (c) 5. (b) 6. (a) 7. (d) 8. (b) 9. (c) 10. (d) 11. (c) 12. (c) 13. (d) 14. (d) 15. (a) 16. (c)

CHAPTER 56

REGULATED POWER SUPPLY

Learning Objectives

- General
- Voltage Regulation
- Zener Diode Shunt Regulator
- Transistor Series Voltage Regulator
- Controlled Transistor Series Regulator
- Transistor Shunt Voltage Regulator
- Transistor Current Regulator
- Variable Feedback Regulator
- Basic OP-AMP Series Regulator
- Basic OP-AMP Shunt Regulator
- Switching Regulators
- Step-down Switching Regulator
- Step-up Switching Regulator
- Inverting Switching Regulator
- IC Voltage Regulators
- Fixed Positive Linear Voltage Regulators
- Fixed Negative Linear Voltage Regulators
- Adjustable Positive Output Linear Voltage Regulators
- Adjustable Negative Output Linear Voltage Regulators
- Use of External Pass Transistor with Linear Voltage Regulators
- Use of Linear Voltage Regulator as a Current Regulator
- Switching Voltage IC Regulators

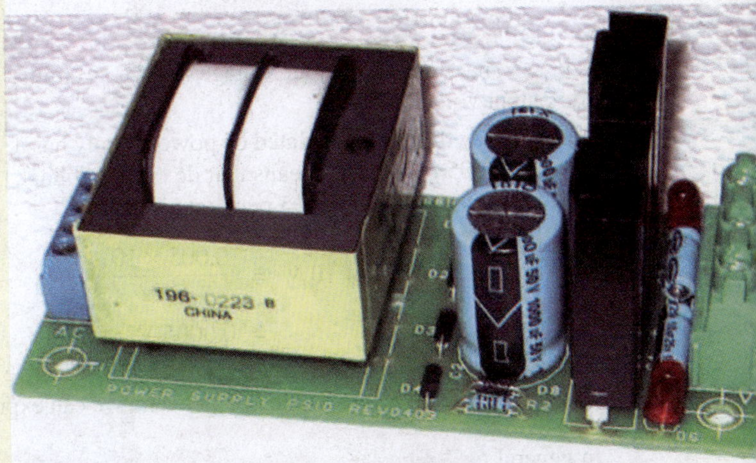

Voltage regulator provides a constant dc output voltage that is essentially independent of the input voltage, output load current and temperature.

56.1. General

The various power-supply circuits considered in Chapter 55 suffer from the drawback that their dc output voltage changes with changes in load or input voltage. Such a dc power supply is called unregulated power supply. Regulated power supply can be obtained by using a voltage regulator circuit. A regulator is an electronic control circuit which is capable of providing a nearly constant dc output voltage even when there are variations in load or input voltage. A source of regulated dc power is essential for all communication, instrumentation, computers or any other electronic system.

We will consider both ***linear*** regulators and ***switching*** regulators which are also available in integrated circuit form. In linear regulators, the transistor operates somewhere between saturation and cut-off. It is always ON and dissipates power. Hence, its efficiency (output power/input power) is 50 per cent or less. In switching regulators, the transistor operates like a switch *i.e.* it is either saturated or cut-off. Hence, its power efficiency is 90 per cent or more.

The linear regulators are of two basic type *i.e.* series regulators and shunt regulators. Likewise switching regulators can be of three basic types *(i)* step-down type, *(ii)* step-up type and *(iii)* inverting type.

56.2. Voltage Regulation

As stated above, in an unregulated power supply, output voltage changes whenever input supply voltage or load resistance changes. It is never constant. The change in voltage from no-load to full-load condition is called ***voltage regulation.*** The aim of a voltage regulator circuit is to reduce these variations to zero or, at least, to the minimum possible value.

The percentage regulation or, simply, regulation of a power supply is given by

$$\%\ \text{regulation} = \frac{V_{max} - V_{min}}{V_{min}} \times 100$$

where V_{max} = maximum dc output voltage and V_{min} = minimum dc output voltage.

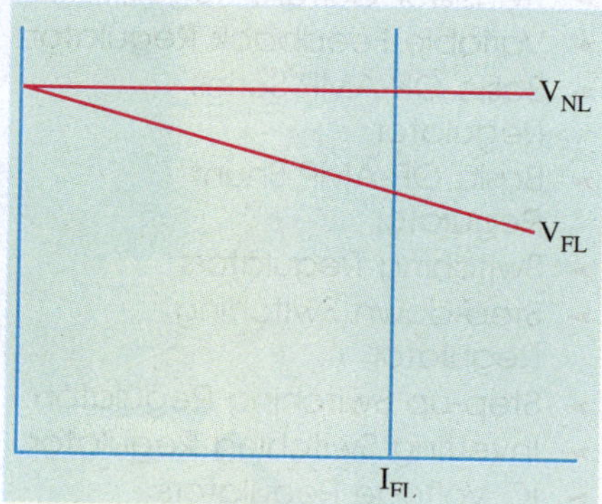

Fig. 56.1

When we say that 10 V regulated dc power supply has a regulation of 0.005 per cent, it means that dc output voltage will vary within an envelope 0.005 per cent of 10 V.

Now, $\quad 0.005\%\ \text{of}\ 10\ \text{V} = \dfrac{0.005 \times 10}{100}$

$= 0.0005\ \text{V} = 0.5\ \text{mV}$

Hence, output voltage will vary by ± 0.25 mV. So, we see that instead of expressing voltage regulation by unwieldly expression 0.005 per cent, we can express it by a simple figure of ± 0.25 mV.

In general, $\quad \%\ \text{regn.} =$

where, $\quad V_{NL}$ = no-load or open-circuit terminal voltage of the supply (Fig. 56.1).

V_{FL} = full-load terminal voltage of the supply

In an ideal or perfectly regulated dc power supply, the percentage voltage regulation is zero. This voltage regulation is also called ***load regulation.***

56.3. Zener Diode Shunt Regulator

A simple shunt voltage regulating system using a zener diode is shown in Fig. 56.2. The input voltage V_{in}, in fact, is the unregulated output of a rectifier. This simple regulator restricts output voltage variations within reasonable limits around V_z in the face of changing load current or changing input voltage. Obviously, the Zener diode will regulate so long as it is kept in reverse conduction.

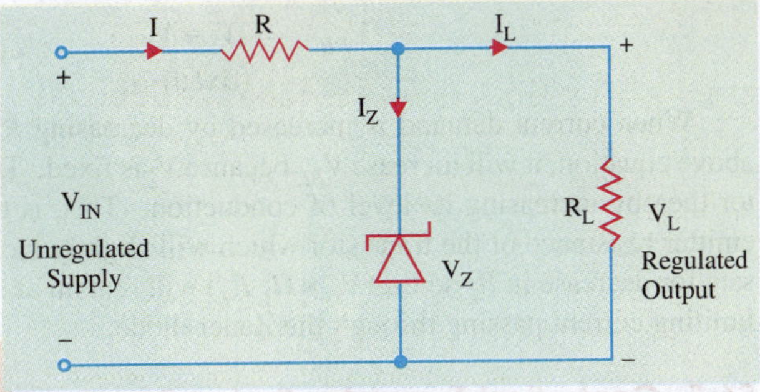

Fig. 56.2

Example 56.1. *The Zener diode of Fig. 56.2 has the following ratings :*

$$V_z = 6.8 \text{ V} \quad at \quad I_z = 50 \text{ mA}$$
$$r_z = 2\,\Omega \quad at \quad I_z = 50 \text{ mA}$$
$$I_{z(min)} = 5 \text{ mA} \qquad I_{z(max)} = 150 \text{ mA}$$

What would be the load voltage when load current I_L varies from 10 mA to 120 mA ? Also, calculate voltage regulation of the regulator. **(Power Electronics-I, Punjab Univ. 1992)**

Solution. We will call V_z = 6.8 V and I_z = 50 mA as reference values and calculate changes in voltage with respect to these values.

(i) I_L = 120 mA

Obviously, $I_z = 150 - 120 = 30$ mA

Deviation from 50 mA $= 30 - 50 = -20$ mA

Drops across diode $= I_z.r_z = -20 \times 2 = -40$ mV

$$V_L = V_z + I_z.r_z = 6.8 + (-40 \times 10^{-3}) = 6.76 \text{ V}$$

(ii) I_L = 10 mA

Now, $I_z = 150 - 10 = 140$ mA

Deviation from reference value $= 140 - 50 = 90$ mA

Diode drop $= I_z r_z = 90 \times 10^{-3} \times 2 = 0.18$ V

$$\therefore \quad V_L = V_z + I_z r_z = 6.8 + 0.18 = 6.98 \text{ V}$$

$$\% \text{ regn.} = \frac{6.98 - 6.76}{6.76} \times 100 = 3.25\,\%$$

For many applications, a change in load voltage of 3.15% is acceptable but, in some, it may be intolerable. This regulation can be reduced to 1% or less with the help of circuits discussed below.

56.4. Transistor Series Voltage Regulator

The circuit is shown in Fig. 56.3. It is also called ***emitter-follower*** regulator because the voltage at the emitter follows the base voltage. In this set-up, the transistor behaves like a variable resistor whose resistance is determined by the base current. It is called ***pass*** transistor because total current to be regulated passes through it.

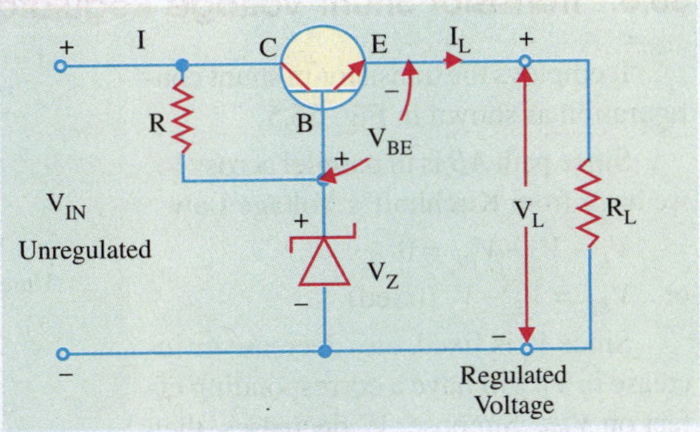

Fig. 56.3

Keeping in mind the polarities of different voltages, they are related by the equation derived from *KVL*

$$V_L + V_{BE} - V_z = 0$$

$$\therefore \quad V_{BE} = V_z - V_L \text{ (fixed)}$$

When current demand is increased by decreasing R_L, V_L tends to decrease. As seen from the above equation, it will increase V_{BE} because V_z is fixed. This will increase forward bias of the transistor thereby increasing its level of conduction. This, is turn, will lead to decrease in the collector-emitter resistance of the transistor which will slightly increase the input current in order to compensate for decrease in R_L so that $V_L = (I_L R_L)$ will remain at a constant value. Incidentally, R is used for limiting current passing through the Zener diode.

56.5. Controlled Transistor Series Regulator

The circuit employing a second transistor T_2 as a sensing element is shown in Fig. 56.4. It has the additional feature of control with the help of potentiometer $R_1 - R_2$. In the discussion to follow, it will be assumed that I is much greater than I_{B2}. Now, there is a drop of V_L on $(R_1 + R_2)$ and a drop of $(V_z + V_{BE2})$ across R_2.

$$\frac{V_L}{V_z + V_{BE2}} = \frac{R_1 + R_2}{R_2} \quad \text{or} \quad V_L = \frac{R_1 + R_2}{R_2}(V_z + V_{BE2})$$

Now, $(R_1 + R_2)$ and $(V_z + V_{BE2})$ both have constant values so that $V_L \propto 1/R_2$. If the potentiometer is adjusted so that R_2 decreases, then V_L increases and *vice versa*.

Suppose R_L is decreased. Then, I_L increases but V_L decreases. Decreases in V_L decreases I_{B2} and I_{C2}. Assuming I_3 to be relatively constant (or decreasing only slightly), I_{B1} is increased thereby decreasing the terminal (collector-emitter) resistance of T_1. This leads to decrease in V_{CE1} thereby offsetting the decrease in V_L which is, therefore, returned to its original value.

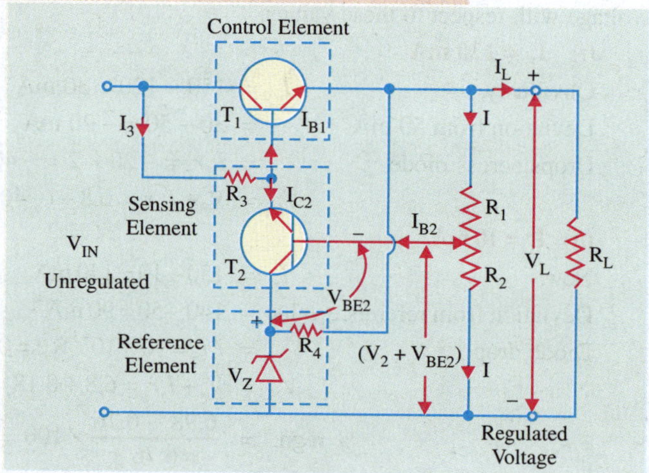

Fig. 56.4

In sequential logic, we have

$$V_L \downarrow \qquad I_{B2} \downarrow \qquad I_{C2} \downarrow \qquad I_{B1} \uparrow \qquad V_{CE1} \downarrow \qquad V_2 \uparrow$$

56.6. Transistor Shunt Voltage Regulator

It employs the transistor in shunt configuration as shown in Fig. 56.5.

Since path AB is in parallel across V_L, we have from Kirchhoff's Voltage Law

$$V_L - V_z - V_{BE} = 0$$

or $\quad V_{BE} = V_L - V_z$ (fixed)

Since V_z is fixed, any decrease or increase in V_L will have a corresponding effect on V_{BE}. Suppose, V_L decreases, then as seen from the above relation, V_{BE} also decreases. As a result, I_B decreases, hence,

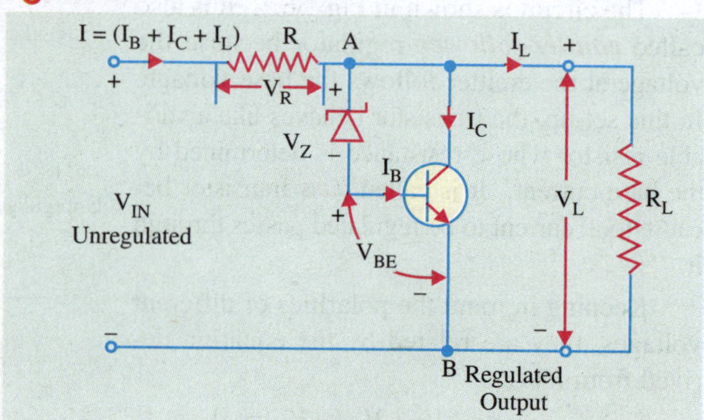

Fig. 56.5

I_C (= βI_B) decreases, thereby decreasing I and hence V_R (= IR). Consequently, V_L increases because at all times

$$V_{in} = V_R + V_L \quad \text{or} \quad V_L = V_{in} - V_R$$

In sequential logic, $V_L \downarrow \quad V_{BE} \downarrow \quad I_B \downarrow \quad I_C \downarrow \quad I_R \downarrow \quad V_R \downarrow \quad V_L \uparrow$

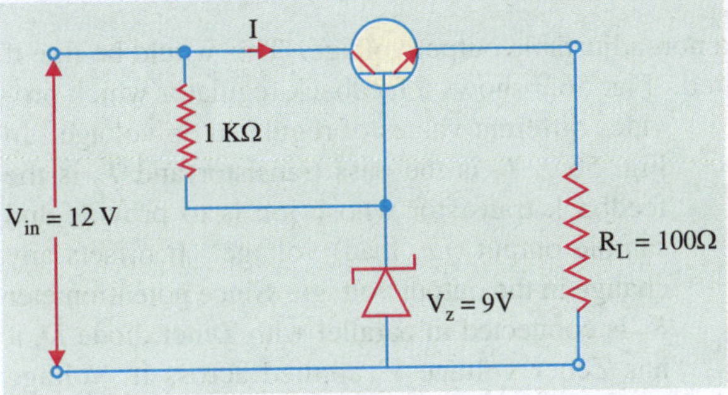

Fig. 56.6

Same line of logic applies in case V_L tries to increase.

Example 56.2. *In the NPN emitter-follower regulator circuit of Fig. 56.6, calculate (i) V_L, (ii) V_{CE}, (iii) I_E and (iv) power dissipated. Take V_{BE} = 0.7 V.*

Solution.

(*i*) $V_L = V_{out} = V_Z - V_{BE}$
$= 9 - 0.7 = \mathbf{8.3\ V}$

(*ii*) $V_{CE} = V_{in} - V_{out} = 12 - 8.3 = \mathbf{3.7\ V}$

(*iii*) $I_E = I_L = V_L/R_L$
$= 8.3/100 = \mathbf{83\ mA}$

(*iv*) Power dissipated $= V_{CE} I_E$
$= 3.7\ V \times 83\ mA = \mathbf{307\ mW}$

Example 56.3. *Compute the output voltage V_{out} for the op-amp series regulator shown in Fig. 56.7.* **(Industrial Electronics, Bombay Univ. 1992)**

Solution. We are given that V_{REF}
$= 6$ V and $R_2 = R_3 = 10$ K

$V_{out} = V_{REF}(1 + R_2/R_3)$
$= 6(1 + 10/10)$
$= 6 \times 2 = \mathbf{12\ V}$

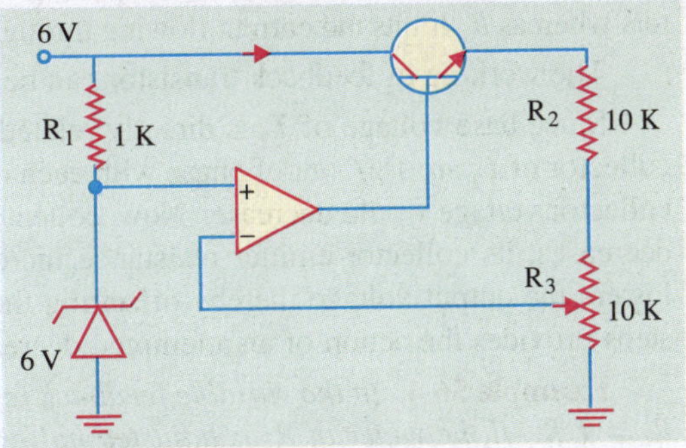

Fig. 56.7

56.7. Transistor Current Regulator

The main function of a current regulator is to maintain a fixed current through the load despite variations in the terminal voltage. Such a circuit employing a Zener diode and a *PNP* transistor is shown in Fig. 56.8. Suppose, due to drop in V_L, current I_L (= I_C) is decreased. This will decrease I_E (≅ I_C). Hence, drop across R_E *i.e.* V_{RE} will decrease. As per Kirchoff's Voltage Law

$$-V_{RE} - V_{BE} + V_z = 0, \quad \text{or} \quad V_{BE} = V_z - V_{RE} \text{ (fixed)}$$

Transistor current regulator

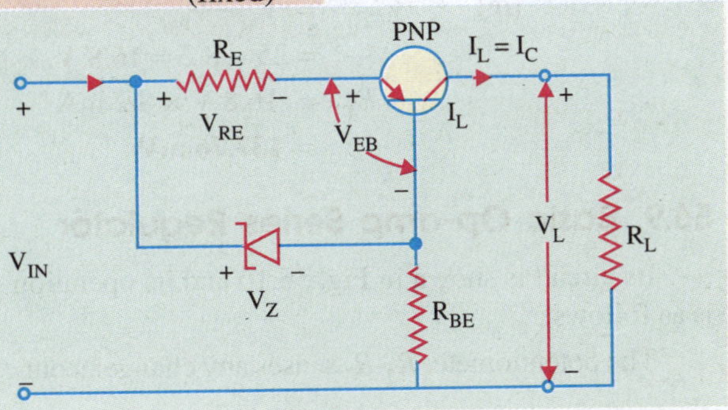

Fig. 56.8

Hence, a decrease in V_{RE} will increase V_{BE} and, hence, the conductivity of the transistor thereby keeping I_L at a fixed level.

A similar logic applies when there is increase in V_L.

56.8. Variable Feedback Regulator

The regulators considered so far provide a non-adjustable output voltage. This would be fine if only single value of regulated voltage is required. Fig. 56.9 shows a feedback regulator which provides different values of regulated dc voltage. In Fig. 56.9, T_1 is the pass transistor and T_2 is the feedback transistor whose job is to provide and sample output (*i.e.* load) voltage. It offsets any change in the output voltage. Since potentiometer R_3 is connected in parallel with Zener diode D, it has Zener voltage V_z applied across it. Voltage across the wiper varies from 0 to V_z. Capacitor C ensures that voltage across D and R_3 does not change suddenly.

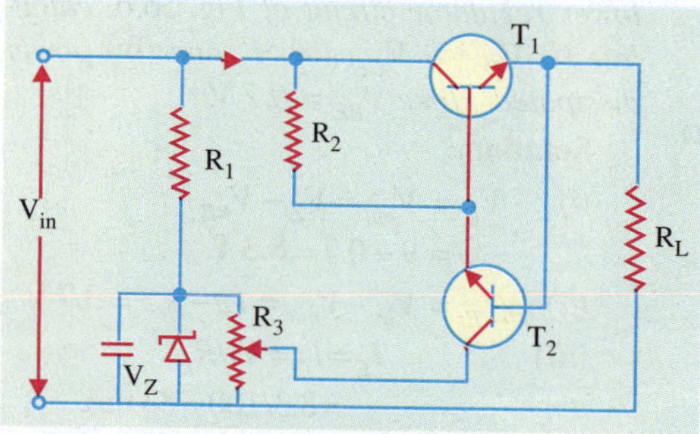

Fig. 56.9

Voltage at the base of T_2 is 0.7 V more positive than the voltage at its emitter. Its emitter voltage and hence the base voltage can be changed with the help of R_3. Since base of T_2 is tied to the output, it is responsible for providing output or load voltage. The voltage V_{CE1} across the pass transistor is given by the difference of input voltage and output voltage. The current through T_1 is equal to the load current. R_2 prevents saturation of transistors whereas R_1 limits the current flowing through D.

The working of feedback transistor can be explained as follows :

Since base voltage of T_2 is directly related to V_{out} it will change if V_{out} changes. The base and collector of T_2 are 180º out of phase with each other. If base voltage increases due to increase in V_{out} collector voltage would decrease. Now, collector of T_2 controls base of T_1. As the base voltage of T_1 decreases, its collector-emitter resistance increases which lowers the load current. This, in turn, lowers the output voltage thereby offsetting the attempted increases in V_{out}. The opposite of these steps provides the action of an attempted decrease in output voltage.

Example 56.4. *In the variable feedback regulator circuit of Fig. 56.9, V_{in} = 25 V, V_z = 15 V and R_L = 1 K. If the wiper of R_3 is adjusted half-way and assuming silicon transistor, compute* **(i)** V_{out} **(ii)** I_L **(iii)** I_{E1} **(iv)** P_1.

Solution. (*i*) V_{out} = voltage at wiper + V_{BE2} = (15/2) + 0.7 = **8.2 V**

(*ii*) $I_L = V_{out}/R_L$ = 8.2 V/1 K = **8.2 mA**

(*iii*) $I_{E1} = I_L$ = **8.2 mA**

(*iv*) $V_{CE1} = V_{in} - V_{out}$
= 25 − 8.2 = **16.8 V**

∴ P_1 = 16.8 V × 8.2 mA
= **137.76 mW**

56.9. Basic Op-amp Series Regulator

Its circuit is shown in Fig. 56.10 and its operation is as follows :

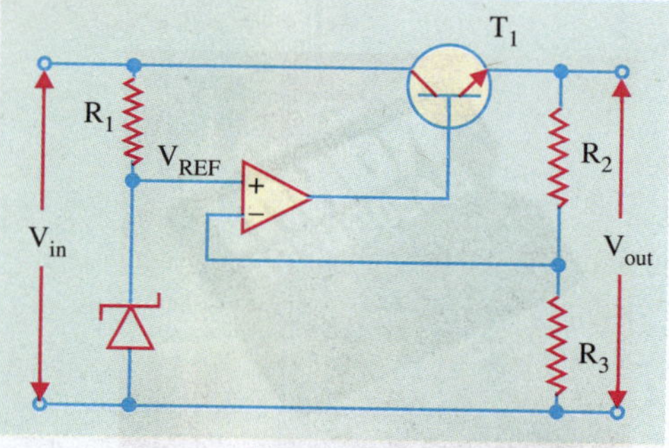

Fig. 56.10

The potentiometer R_2-R_3 senses any change in output voltage V_{out}. When V_{out} attempts to decrease because of decrease in V_{in} or because of the decrease in I_L, a proportional voltage decrease is applied to the inverting output of the op-amp by the potentiometer. Since, the other op-amp input is held by the Zener voltage at a fixed reference voltage V_{REF}, a small difference voltage (called error voltage) is developed across the two inputs of the op-amp. This difference voltage is amplified and op-amp's outout voltage increases. This increase in voltage is applied to the base of T_1 causing the emitter voltage (= V_{out}) to increase till the voltage to the inverting input again equals the reference (Zener) voltage. This action offsets attempted decrease in the output voltage thus keeping it almost constant. The opposite action occurs if the output voltage tries to increase.

Calculations

It will be seen that the op-amp of Fig. 56.10 is actually connected as a non-inverting amplifier where V_{REF} is the input at the non inverting terminal and the R_2/R_3 voltage divider forms the negative feed-back network. The closed-loop voltage gain is given by $A = 1 + (R_2/R_3)$. Neglecting base-emitter voltage of T_1, we get

$$V_{out} = V_{REF}\,(1 + R_2/R_3)$$

It is seen that V_{out} depends on Zener voltage and potential divider resistors R_2 and R_3 but is independent of input voltage V_{in}.

56.10. Basic Op-amp Shunt Regulator

Such a shunt type linear regulator is shown in Fig. 56.11. Here, the control element is a series resistor R_1 and a transistor T_1 in parallel with the load. In such a regulator, regulation is achieved by controlling the current through T_1.

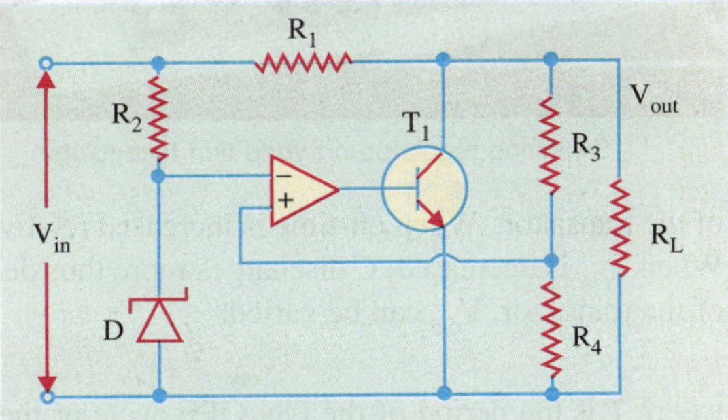

Fig. 56.11

Working

When output voltage tries to decrease due to change in either the input voltage or load current or temperature, the attempted decrease is sensed by R_3 and R_4 and applied to the non-inverting input of the op-amp. The resulting difference in voltage reduces the op-amp's output, driving T_1 less thus reducing its collector current (shunt current), and increasing its collector-to-emitter resistance. Since collector-to-emitter resistance acts as a voltage divider with R_1, this action offsets the attempted decrease in output voltage and hence, maintains it at a constant value. The opposite action occurs when output voltage tries to increase. The shunt regulator is less efficient than the series type but offers inherent short-circuit protection.

56.11. Switching Regulators

In the linear regulators considered so far, the control element *i.e.* the transistor conducts all the time, the amount of conduction varying with changes in output voltage or current. Due to continuous power loss, the efficiency of such a regulator is reduced to 50 per cent or less.

A switching regulator is different because its control element operates like a switch *i.e.* either it is saturated (closed) or cut-off (open). Hence, there is no unnecessary wastage of power which results in higher efficiency of 90% or more.

Switching regulators are of three basic types : *(i)* step-down regulator, *(ii)* step-up regulator and *(iii)* inverting regulator.

56.12. Step-down Switching Regulator

In this regulator (Fig. 56.12), V_{out} is always less than V_{in}. An unregulated positive dc voltage is applied to the collector of the *NPN* transistor. A series of pulses from an oscillator is sent to the base of transistor *T* which gets saturated (closed) on each of the positive pulses. It is so because an *NPN* transistor needs a positive voltage pulse on its base in order to turn ON. A saturated transistor acts as a closed switch, hence it allows V_{in} to send current through *L* and charge *C* to the value of output voltage during the on-time (T_{ON}) of the pulse. The diode D_1 is reverse-biased at this point and hence, does not conduct.

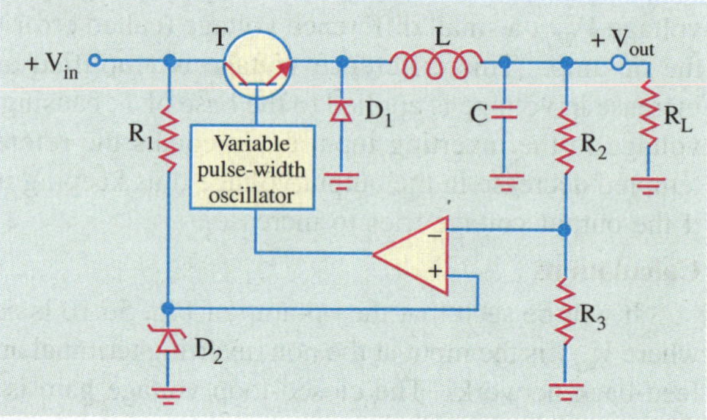

Fig. 56.12

Switching regulator in hybrid film technology

Eventually when positive pulse turns to zero, *T* is cut-off and acts like an open switch during the off period (T_{OFF}) of the pulse. The collapsing magnetic field of the coil produces self-induced voltage and keeps the current flowing by returning energy to the circuit.

The value of output voltage depends on input voltage and pulse width *i.e.* on-time of the transistor. When on-time is increased relative to off-time, *C* charges more thus increasing V_{out}. When T_{ON} is decreased, *C* discharges more thus decreasing V_{out}. By adjusting the duty cycle (T_{ON}/T) of the transistor, V_{out} can be varied.

$$\therefore \quad V_{out} = V_{in}\,(T_{ON}/T)$$

where *T* is the period of the ON-OFF cycle of the transistor and is related to frequency by $T = 1/f$. Also, $T = T_{ON} + T_{OFF}$ and the ratio (T_{ON}/T) is called the ***duty cycle.***

The regulating action of the circuit is as follows :

When V_{out} tries to decrease, on-time of the transistor is increased causing an additional charge on the capacitor *C* to offset the attempted decrease. When V_{out} tries to increase, T_{ON} of the transistor is decreased causing *C* to discharge enough to offset the attempted increase.

56.13. Step-up Switching Regulator

The circuit is shown in Fig. 56.13. When transistor *T* turns ON on the arrival of the positive pulse at its base, voltage across L increases quickly to $V_{in} - V_{CE(sat)}$ and magnetic field of *L* expands quickly. During on-time of the transistor, V_L keeps decreasing from its initial maximum value. The longer transistor is ON, the smaller V_L becomes.

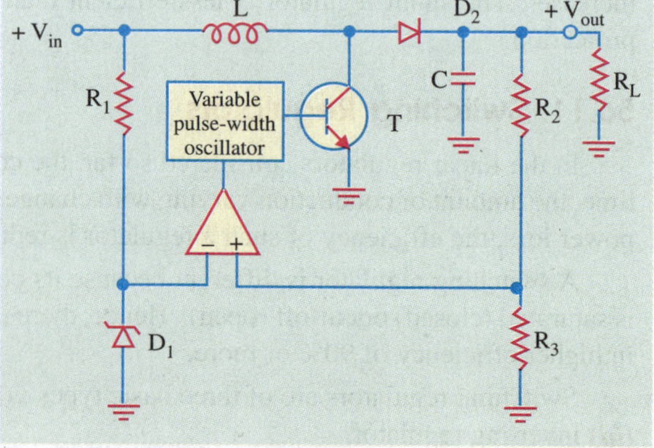

Fig. 56.13

When transistor turns OFF, magnetic field of *L* collapses and its polarity reverses so that its voltage adds to the input voltage thus producing an output voltage greater than the input voltage. During off-time of the transistor, D_2 is forward-biased and allows C to charge. The variations in V_{out} due to charging and discharging action are sufficiently smoothed by filtering action of L and C.

It may be noted that shorter the on-time of the transistor, greater the inductor voltage and hence greater the output voltage (because greater V_L adds to V_{in}). On the other hand, the longer the on-time, the smaller the inductor voltage and hence, lesser the output voltage (because smaller V_L adds to V_{in}).

The regulating action can be understood as follows :

When V_{out} tries to decrease (because of either increasing load or decreasing V_{in}), transistor on time decreases thereby offsetting attempted decrease in V_{out}. When V_{out} tries to increase, on-time increases and attempted increase in V_{out} is offset.

As seen, the output voltage is inversely related to the duty cycle.

$$\therefore \quad V_{out} = V_{in}(T/T_{ON})$$

Inverting Switching Regulator

56.14. Inverting Switching Regulator

The basic diagram of such a regulator is shown in Fig. 56.14. This regulator provides an output voltage that is opposite in polarity to the input voltage.

When transistor T turns ON by the positive pulse, the inductor voltage V_L jumps to $V_{in} - V_{CE(sat)}$ and the magnetic field of the inductor expands rapidly. When transistor is ON, the diode D_2 is reverse-biased and V_L decreases from its initial maximum value. When transistor turns OFF, the magnetic field collapses and inductor's polarity reverses. This forward-biases D_2, charges C and produces a negative output voltage. This repetitive ON-OFF action of the transistor produces a repetitive charging and discharging that is smoothed by LC filter action. As in the case of a step-up regulator, lesser the time for which transistor is ON, greater the output voltage and ***vice versa.***

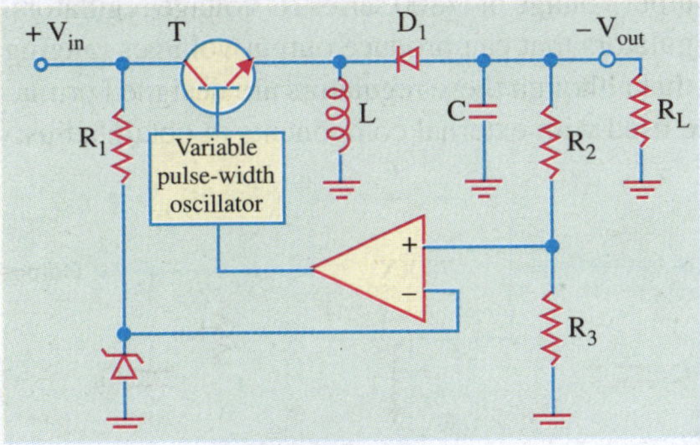

Fig. 56.14

56.15. IC Voltage Regulators

Due to low-cost fabrication technique, many commercial integrated-circuit (*IC*) regulators are available since the past two decades. These include fairly simple, fixed-voltage types of high-quality precision regulators. These *IC* regulators have much improved performance as compared to those made from discrete components. They have a number of unique build-in features such as current limiting, self-protection against overtemperature, remote control operation over a wide range of input voltages and foldback current limiting.

Now we will study the following types of *IC* voltage regulators : **(1)** fixed positive linear voltage regulators, **(2)** fixed negative linear voltage regulators, **(3)** adjustable positive linear voltage regulators, and **(4)** adjustable negative linear voltage regulators.

56.16. Fixed Positive Linear Voltage Regulators

There are many *IC* regulators available in the market that produces a fixed positive output voltage. But 7800 series of *IC* regulators is representative of three terminal devices that are available with several fixed positive output voltages making them useful in a wide range of applications. Fig. 56.15 (*a*) shows a standard configuration of a fixed positive voltage *IC* regulator of 7800 series. Notice that it has three terminals labelled as input, output and ground. The last two digits (marked xx) in the part number designate the output voltage. For example, *IC* 7805 is a +5 V regulator. Similarly *IC* 7812 is a +12 V regulator and *IC* 7815 is a +15 V regulator. The capacitor C_1 (typically 0.33 μF) is required only if the power supply filter is located more than 3 inches from the *IC* regulator. The capacitor C_2 (typically 0.01 μF) acts basically as a line filter to improve transient response.

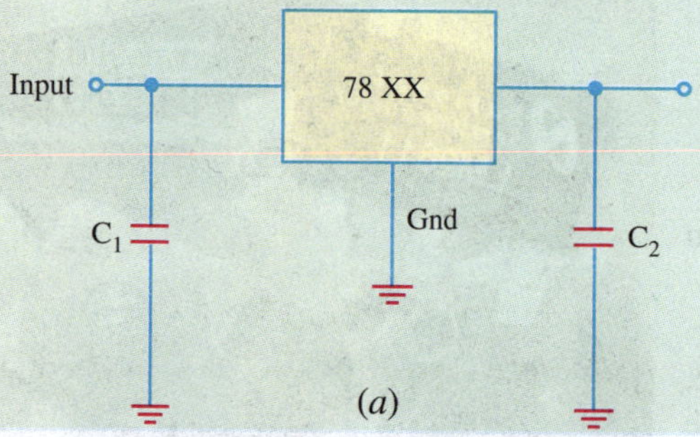

(*a*)

Part Number	Output Voltage
7805	+5.0 V
7806	+6.0 V
7808	+8.0 V
7809	+9.0 V
7812	+12.0 V
7815	+15.0 V
7818	+18.0 V
7824	+24.0 V

(*b*)

Fig. 56.15

Fig. 56.15 (*b*) shows the part number and the output voltage of 7800 series *IC* voltage regulators. As seen from this figure, the 7800 series has *IC* regulators that can produce output voltages ranging from +5.0 to +24.0 volt. It may be carefully noted that although these regulators are designed primarily to produce fixed output voltage but they can be used with external components to obtain adjustable output voltage and current.

Fig. 56.16 shows the circuit indicating the use of 78XX as an adjustable voltage regulator. The output voltage is given by the equation,

$$V_{out} = V_{fixed} + \left(\frac{V_{fixed}}{R_1} + I_Q\right) R_2$$

For example, for a 7805 *IC* regulator, $V_{fixed} = 5$ V. Let $R_1 = R_2 = 1$ kΩ and $I_Q = 5$ mA, then its output voltage is,

$$V_{out} = 5 + \left(\frac{5}{1\text{ k}\Omega} + 5\text{ mA}\right) \times 1\text{ k}\Omega$$

$$= 15\text{ V}$$

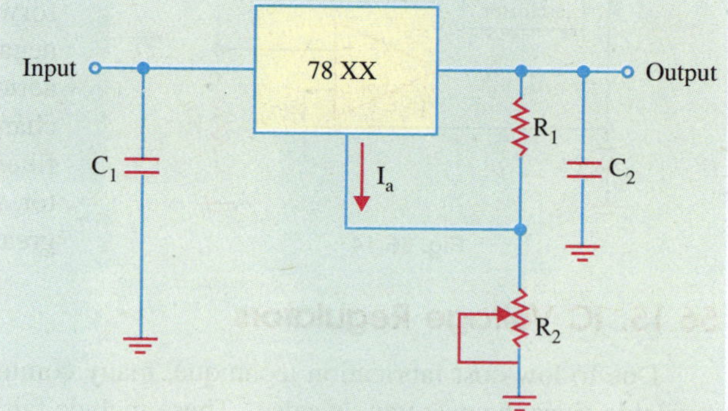

Fig. 56.16

Thus output voltage of *IC* 7805 regulator can be adjusted anywhere between 5 V to 15 V.

This example indicates that the output of *IC* 7805 regulator is adjusted to 15 V using external resistances R_1 and R_2.

IC 7805 regulator

The standard 7800 series can produce output current in excess of 1 A when used with adequate heat sink. It is available in aluminium can package TO-3 (indicated by *K* in the part number) and plastic package TO-220 (indicated by *T* in the part number), The 78L00 series can provide up to 100 mA and is available in TO-92 and metal TO-39 low profile packages. The 78M00 series can provide upto 0·5 A and is available in plastic TO-202 package. Fig. 56.17 shows the typical metal and plastic packages for the *IC* voltage regulators.

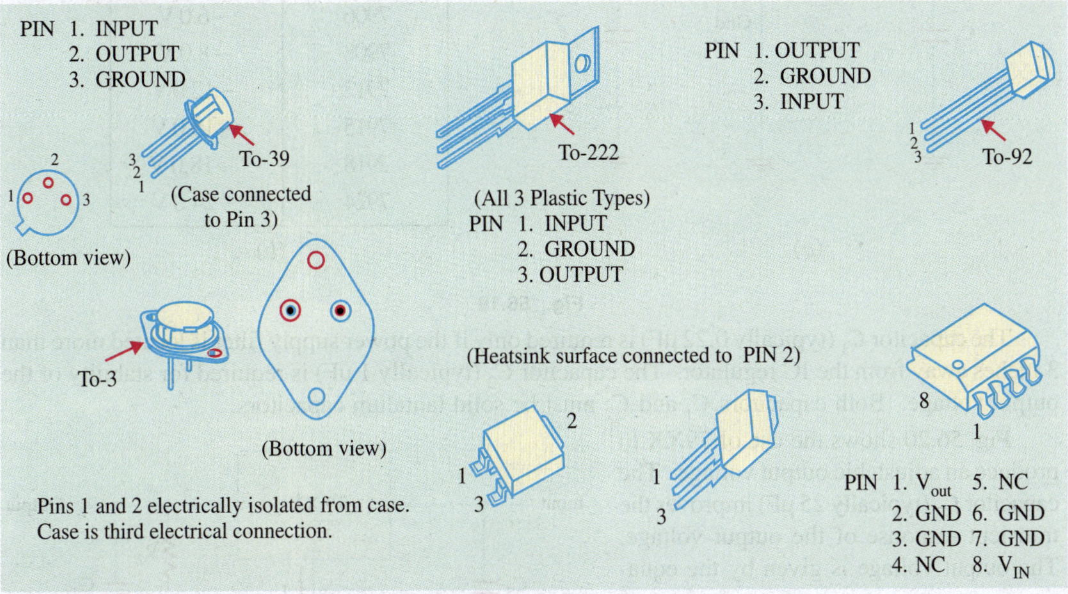

Fig. 56.17

It may be noted that the input voltage for the *IC* regulator must be at least 2 V above the output voltage. This is required in order to maintain regulation. The input voltage should not be more than 35 or 40 volts depending upon the part number. The circuit inside all the *IC* regulators have internal thermal overload protection and short-circuit current-limiting features. Thermal overload in a *IC* regulator occurs whenever the internal power dissipation becomes excessive and the temperature of the device exceeds a certain value.

In India, Bharat Electronic Limited, Bangalore manufactures the *IC* voltage regulators with output voltages of 5 and 12 V. These are available in the market with part numbers BEL 7805 and BEL 7812 respectively.

Fig 56.18 shows a picture of a transformer bridge rectifier and voltage regulator in a dc power supply.

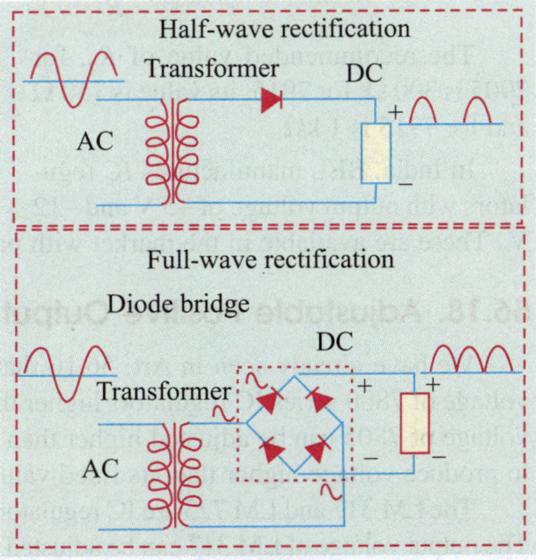

Fig. 56.18

56.17. Fixed Negative Linear Voltage Regulators

The 7900 series is typical of three-terminal *IC* regulators that provide a fixed negative output voltage. This series is a negative-voltage counterpart of the 7800 series and shares most of the same features, characteristics and package types. Fig. 56.19 (*a*) indicates the standard configuration and Fig. 56.19 (*b*), the part numbers with corresponding output voltages that are available in 7900 series.

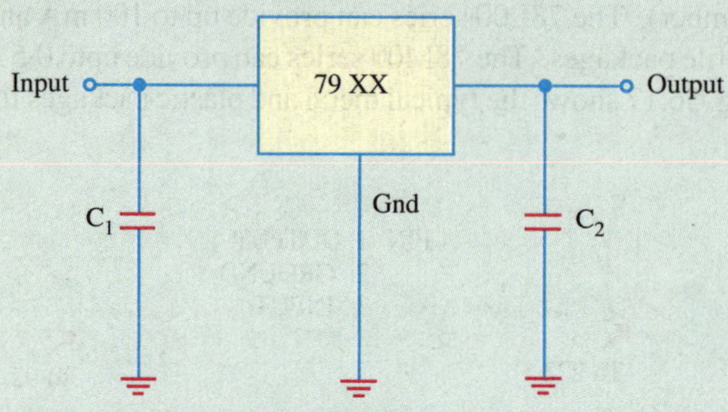

(a)

Part Number	Output Voltage
7905	−5.0 V
7905·2	−5.2 V
7906	−6.0 V
7908	−8.0 V
7912	−12.0 V
7915	−15.0 V
7918	−18.0 V
7924	−24.0 V

(b)

Fig. 56.19

The capacitor C_1 (typically 0.22 μF) is required only if the power supply filter is located more than 3 inches away from the IC regulator. The capacitor C_2 (typically 1 μF) is required for stability of the output voltage. Both capacitors C_1 and C_2 must be solid tantalum capacitors.

Fig. 56.20 shows the use of 79XX to produce an adjustable output voltage. The capacitor C_3 (typically 25 μF) improves the transient response of the output voltage. The output voltage is given by the equation,

$$V_{out} = V_{fixed}\left(\frac{R_1 + R_2}{R_2}\right)$$

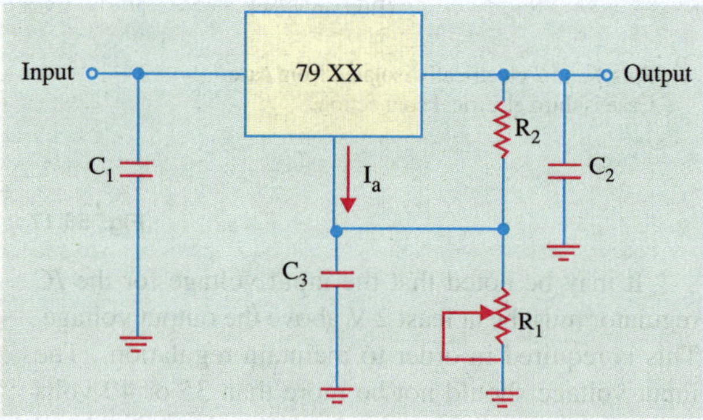

Fig. 56.20

The recommended value of R_2, for 7905 is 300 Ω, for 7915, its value is 750 Ω and for 7915 is 1 kΩ.

In India, BEL manufactures *IC* regulators with output voltage of −5 V and −12 V. These are available in the market with part numbers BEL 7905 and BEL 7915 respectively.

56.18. Adjustable Positive Output Linear Voltage Regulators

We have already seen in Art. 56.16 that by adding external resistors, we can adjust the output voltage of 7800 series IC regulators higher than their fixed (or set) voltages. Fot example, the output voltage of 7805 can be adjusted higher than 5 V. But the performance and reliability of 7800 series to produce voltage higher than its fixed value is not considered to be good.

The LM 317 and LM 723 are IC regulators whose output voltage can be adjusted over a wide range. The output voltage of LM 317 can be adjusted from 1·2 V to 37 V, it can supply output current of 100 mA and is available in TO-92 package *i.e.*, it is also a 3 terminal IC regulator. On the other hand, the output voltage of LM 723 can be adjusted from 2 V to 37 V, it can supply output current of 150 mA without external transistor. But with the addition of external transistor, the output current capability can be increased in excess of 10 A. The LM 723 is available in dual-in-line package and in a metal can package.

Fig. 56.21 (*a*) shows the LM 317 connected to the external resistors R_1 and R_2 to produce an adjustable output voltage.

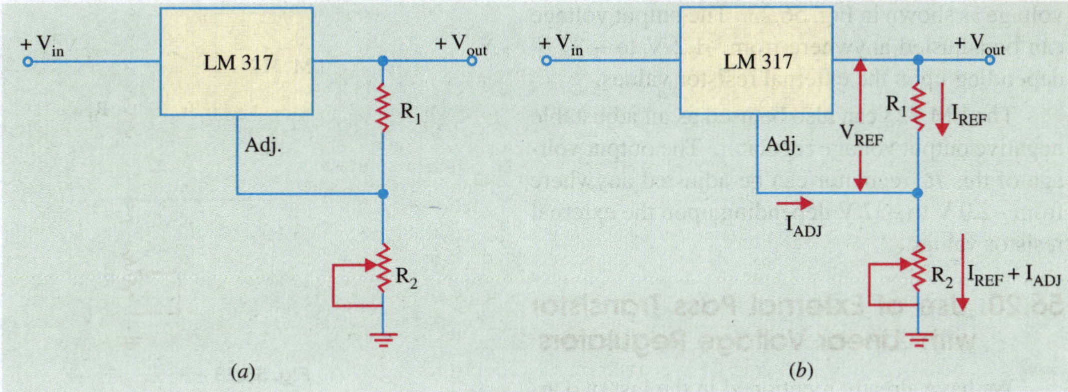

Fig. 56.21

In operation, the LM 317 develops a constant 1·25 V reference voltage (V_{REF}) between the output and adjustment terminal. This constant reference voltage produces a constant current, (I_{REF}) through R_1, regardless of the value of R_2 Fig. 56.20 (*b*). Notice that the value of current through R_2 is the sum of I_{REF} and I_{ADJ}, where I_{ADJ} is a very small current at the adjustment terminal. The value of I_{ADJ} is typically around 100 μA. It can be shown that the output voltage.

$$V_{out} = V_{REF}\left(1+\frac{R_2}{R_1}\right) + I_{ADJ} \cdot R_2$$

It is evident from the above equation that the output voltage is a function of R_1 and R_2. Usually the value of R_1 is recommended to be around 220 Ω. Once the value of R_1 is set, the output voltage is adjusted by varying R_2.

Example 56.5. *Calculate the minimum and maximum output voltages for the IC voltage regulator shown in Fig. 56.22. Assume I_{ADJ} = 100 μA, V_{in} = + 35 V.*

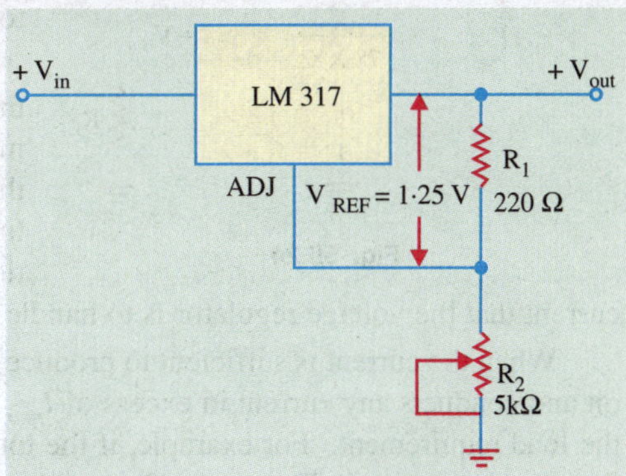

Fig. 56.22

Solution. The equation for output voltage of the IC voltage regulator is given by,

$$V_{out} = V_{REF}\left(1+\frac{R_2}{R_1}\right) + I_{ADJ} \cdot R_2$$

When R_2 is set at its minimum value (*i.e.* 0 Ω), the output voltage,

$$V_{out(min)} = 1.25\left(1+\frac{0}{220}\right) + (100 \times 10^{-6}) \times 0$$

$$= \mathbf{1.25\ V}$$

When R_2 is set at its maximum values (*i.e.* 5 kΩ), the output voltage,

$$V_{out(max)} = 1.25\left(1+\frac{5000}{220}\right) + (100 \times 10^{-6}) \times 5000$$

$$= 29.66 + 0.5 = \mathbf{30.16\ V.}$$

56.19. Adjustable Negative Output Linear Voltage Regulators

A good example of this type of regulators is LM 337. The regulator is a negative output counterpart of LM 317. The LM 337 (like LM 317) requires two external resistors for adjustment of output voltage as shown in Fig. 56.23. The output voltage can be adjusted anywhere from −1.2 V to − 37 V depending upon the external resistor values.

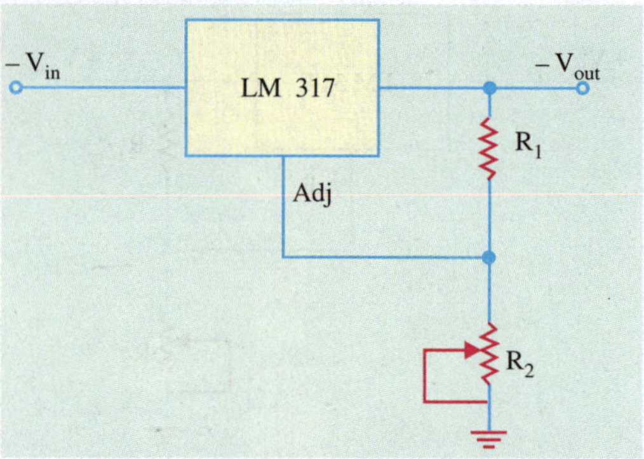

Fig. 56.23

The LM 723 can also be used as an adjustable negative output voltage regulator. The output voltage of this *IC* regulator can be adjusted anywhere from −2.0 V to −37 V depending upon the external resistor values.

56.20. Use of External Pass Transistor with Linear Voltage Regulators

We have already mentioned in the last two articles that a linear voltage regulator (7800 and 7900 series) is capable of delivering only a certain amount of output current to a load. For example, the 7800 series regulators can handle a maximum output current of at least 1.3 A and typical 2.5 A. If the load current exceeds the maximum allowable value, there will be a **thermal overload** and the regulator will **shut down.** ***A thermal overload condition means that there is excessive power dissipation inside the regulator.***

If an application requires a larger value of load current than the maximum current that the regulator can deliver, we will have to use an external pass transistor as shown in Fig. 56.24. The value of R_{ext} (current-sensing resistor) determines the value of current at which the external pass transister (T_{ext}) begins to conduct because it sets the base-to-emitter voltage of the transistor.

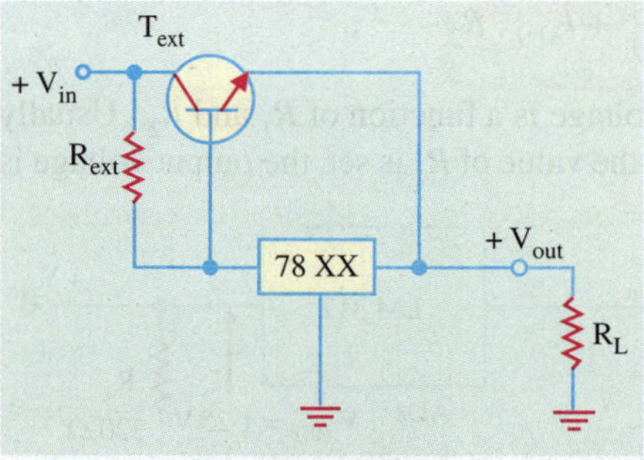

Fig. 56.24

As long as the current is less than the value set by R_{ext}, the transistor T_{ext} is off and the regulator operates normally. This is because the voltage drop across R_{ext} is less than 0.7 V (*i.e.* the base to emitter voltage required to turn T_{ext} on. The value of R_{ext} is determined by the equation $R_{ext} = 0.7\ V/I_{max}$ where I_{max} is the maximum value of current that the voltage regulator is to handle **internally**.

When the current is sufficient to produce at least a 0.7 V drop across R_{ext}, the transistor T_{ext} turns on and conducts any current in excess of I_{max}. The transistor T_{ext} will conduct current depending on the load requirement. For example, if the total load current is 5 A and I_{max} was selected to be 1 A, then the external pass Transistor (T_{ext}) will conduct 4 A of current through it.

It may be noted that the external pass transistor is a power transistor with heat sink that must be capable of handling a maximum power given by the equation,

$$P_{ext} = I_{ext}(V_{in} - V_{out})$$

One major drawback of the circuit shown in Fig. 56.24 is that the external pass transistor is not protected from excessive current, such as would result from a shorted output. This drawback can be overcome by using an additional current limiting circuit as shown in Fig. 56.25.

The operation of this circuit may be explained as follows. The current sensing resistor, R_{lim} sets the base-to-emitter voltage of T_{lim}. The base-to-emitter voltage of T_{ext} is now determined by $\left(V_{R_{ext}} - V_{R_{\lim}}\right)$ because they have opposite polarities. So for normal operation, the drop across R_{ext} must be sufficient to overcome the opposing drop across R_{lim}.

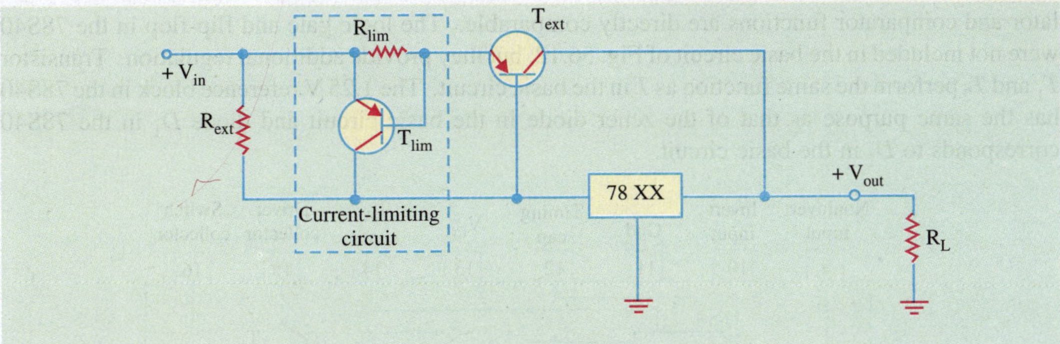

Fig. 56.25

If the current through T_{ext} exceeds a certain maximum value, ($I_{ext(max)}$) because of a shorted output or a faulty load, the voltage across R_{lim} reaches 0.7 V and turns T_{lim} on. As a result, T_{lim} now conducts current away from T_{ext} and through the regulator. This forces a thermal overload to occur and shut down the regulator. Remember, the *IC* voltage regulator circuitry is internally protected from thermal overload as part of its design. This way the external pass transistor is protected from excessive current.

56.21. Use of Linear Voltage Regulators as a Current Regulator

The 3-terminal linear voltage regulator can be used as a current source when an application requires that a constant current be supplied to a variable load. The basic circuit is shown in Fig. 56.26. Here R_1 is the current-setting resistor. The regulator provides a fixed output voltage, V_{out} between the ground terminal and the output terminal. However, it may be noted that the ground pin of the regulator is not connected to the circuit ground. The constant current supplied to the load, is given by the equation,

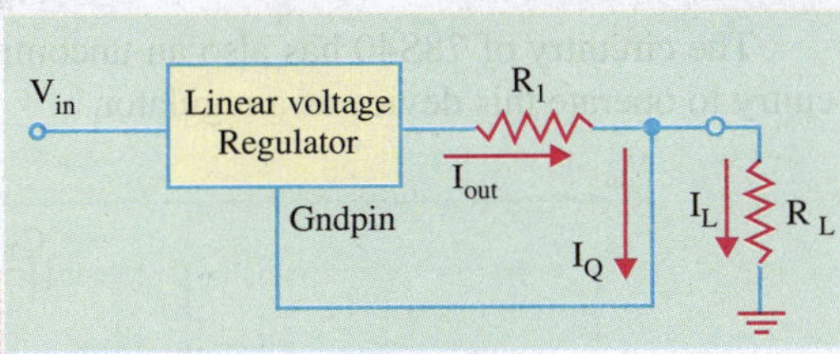

Fig. 56.26

$$I_L = \frac{V_{out}}{R_L} + I_Q$$

Usually the current, I_Q is very small as compared to the output current and hence can be neglected, therefore

$$I_L \approx \frac{V_{Out}}{R_L}$$

For example, if we use 7805 regulator to provide a constant current of 1 A to a variable load, then

$$R_1 = \frac{5}{1} = 5\ \Omega$$

Please note that input voltage must be at least 2 V greater than the output voltage. Thus for 5 V regulator, V_{in} must be greater than 7 V.

56.22. Switching Voltage IC Regulators

There are several switching voltage *IC* regulators available in the market. The choice depends upon the desired application and the cost. However, we will illustrate it with the *IC* 78S40. This device is a universal device that can be used with external components to provide step-up, step down and inverting operation.

Fig. 56.27 shows the internal circuitry of the *IC* 78540. The circuitry can be compared to the basic switching regulators discussed in Art. 56.12, 56.13, and 56.14. As seen from this diagram, the oscillator and comparator functions are directly comparable. The logic gate and flip-flop in the 78S40 were not included in the basic circuit of Fig. 56.12, but they provide additional regulation. Transistor T_1 and T_2 perform the same function as T in the basic circuit. The 1.25 V reference block in the 78S40 has the same purpose as that of the zener diode in the basic circuit and diode D_1 in the 78S40 corresponds to D_1 in the basic circuit.

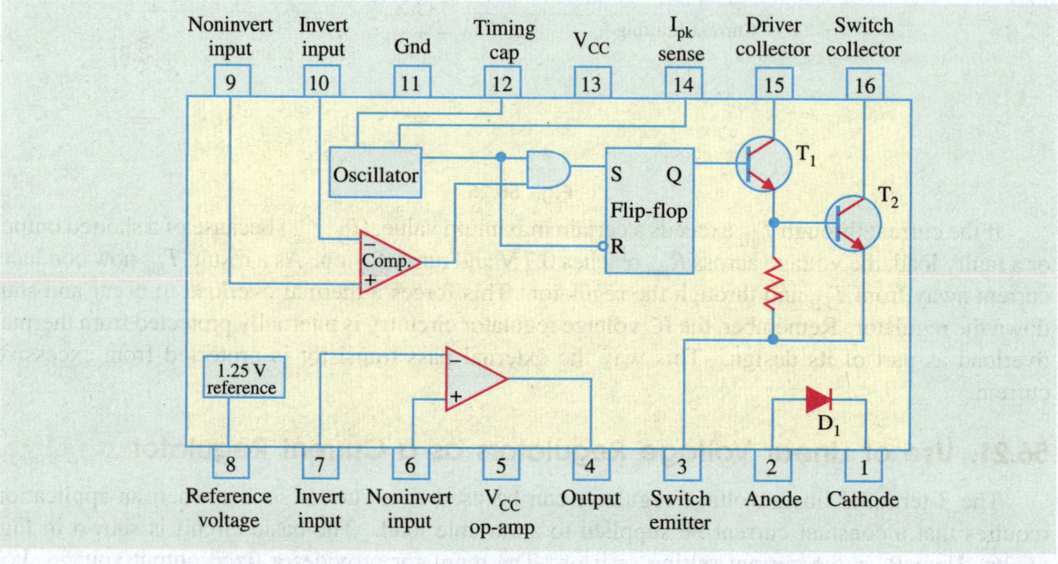

Fig. 56.27

The circuitry of 78S40 has also an uncommitted or unused OP-AMP. We require external circuitry to operate this device as a regulator.

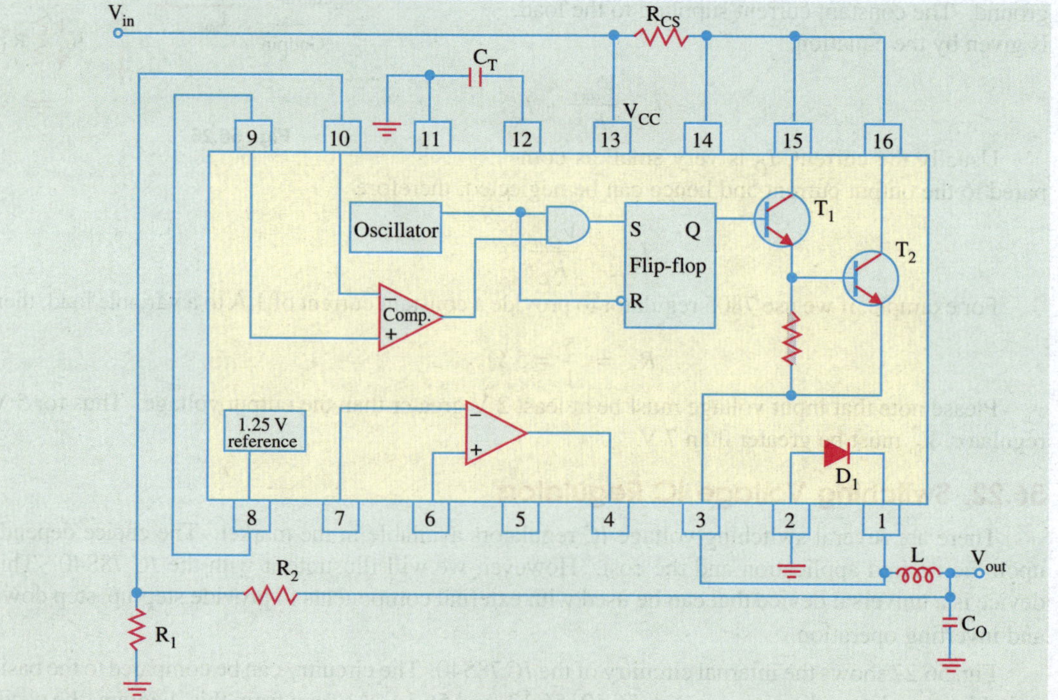

Fig. 56.28

Fig. 56.28 shows the external connections of the IC 78S40 for a step-down switching regulator configuration. Note that in this configuration the circuit produces an output voltage which is less than input voltage.

Fig. 56.28 shows the IC 78S40 connected to the external components for a step-up switching regulator configuration. In this case the output voltage is greater than the input voltage. An inverting configuration is also possible but is not shown here.

In both the circuits of Fig. 56.28 and Fig. 56.29, the capacitor C_T (called timing capacitor) controls the pulse width and frequency of the oscillator and thus establishes the on-time of transistor T_1. The voltage across the resistor R_{CS} (called current-sensing resistor) is used internally by the oscillator to vary the duty cycle based on the desired peak current. The voltage divider made up of R_1 and R_2 reduces the output voltage to a value equal to the reference voltage. If the output voltage (V_{out}) exceeds its set value, the output of the comparator switches to the low state, disabling the gate to turn T_2 off until the output decreases. This regulating action is in addition to that produced by the duty cycle variation of the oscillator.

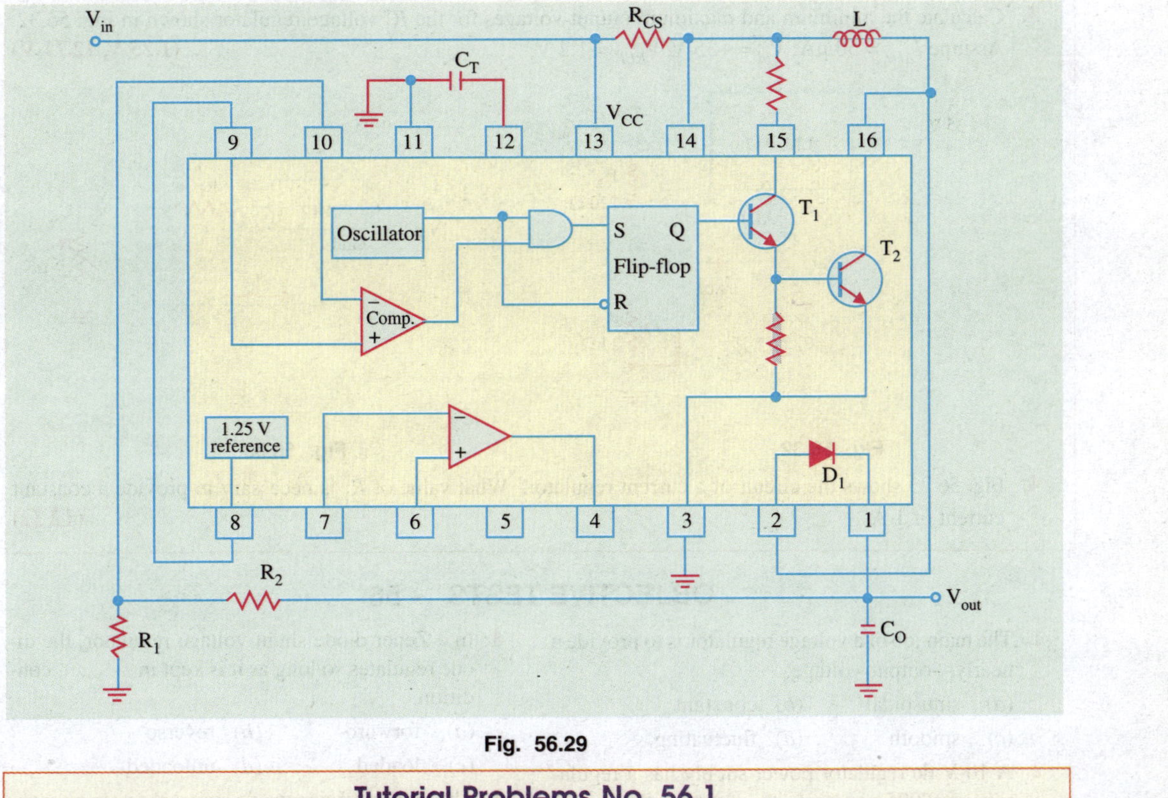

Fig. 56.29

Tutorial Problems No. 56.1

1. The Zener diode of Fig. 56.30 has the following ratings :

$V_z = 15$ V at $I_z = 17$ mA

$r_z = 14\ \Omega$ at $I_z = 17$ mA

$I_{z(min)} = 0.25$ mA $\quad I_{z(max)} = 66$ mA

What would be the load voltage when load current I_L varies from 1 mA to 60 mA. Also calculate the voltage regulation of the regulator.

(14.846 V, 15.672 V, 5.27%)

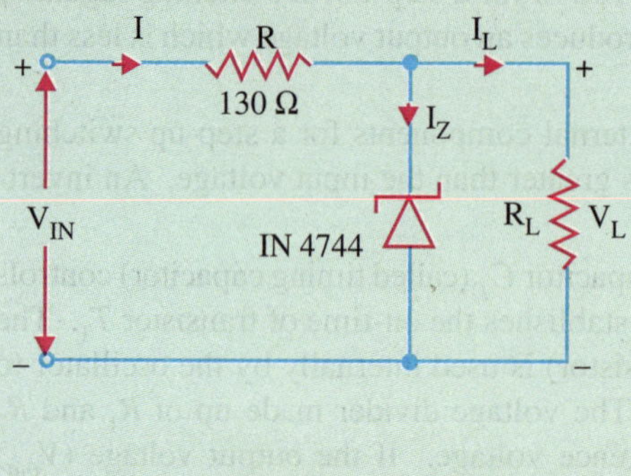

Fig. 56.30

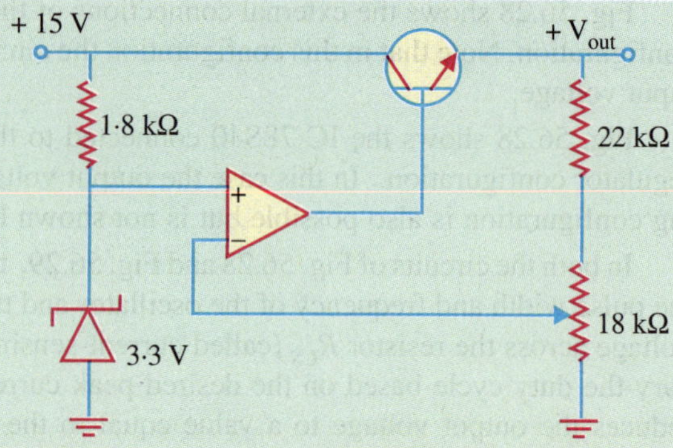

Fig. 56.31

2. Determine the output voltage, V_{out} for the op-amp series regulator shown in Fig. 56.31. **(7.3 V)**

3. Calculate the minimum and maximum output voltages for the *IC* voltage regulator shown in Fig. 56.32. Assume I_{ADJ} = 50 μA, V_{in} = +35 V, V_{REF} = 1.2 V **(1.25 V, 12.71 V)**

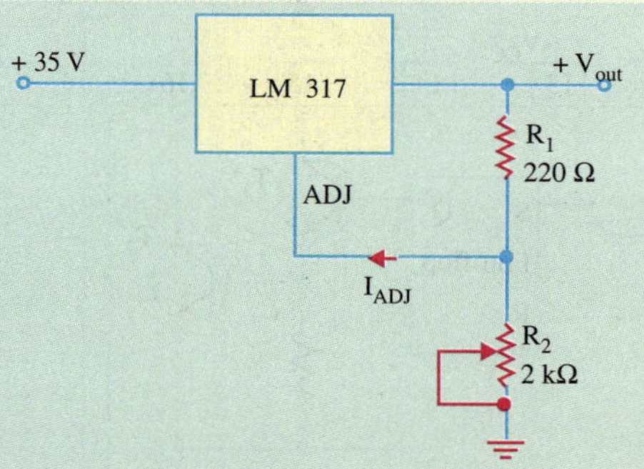

Fig. 56.32

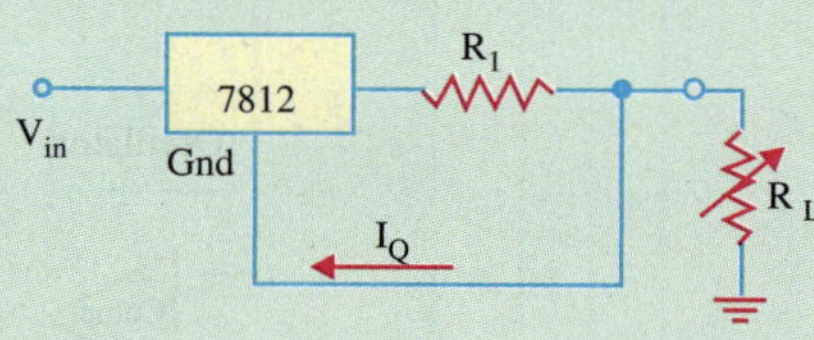

Fig. 56.33

4. Fig. 56.33 shows the circuit of a current regulator. What value of R_1 is necessary to provide a constant current of 1 A. **(12 Ω)**

OBJECTIVE TESTS – 56

1. The main job of a voltage regulator is to provide a nearly—output voltage.
   (*a*) sinusoidal (*b*) constant
   (*c*) smooth (*d*) fluctuating.
2. A 10-V dc regulator power supply has a regulation of 0.005 per cent. Its output voltage will vary within an envelope of .... millivolt.
   (*a*) ± 2.5 (*b*) ± 0.5
   (*c*) ± 5 (*d*) ± 0.05
3. An ideal voltage regulator has a voltage regulation of
   (*a*) 1 (*b*) 100
   (*c*) 50 (*d*) 0.
4. In a Zener diode shunt voltage regulator, the diode regulates so long as it is kept in .... condition.
   (*a*) forward (*b*) reverse
   (*c*) loaded (*d*) unloaded
5. The power efficiency of a switching voltage regulator is much higher than that of a linear regulator because it operates.
   (*a*) in saturation (*b*) in cut-off
   (*c*) like a switch (*d*) on high duty cycle.
6. A transistor series voltage regulator is called emitter-follower regulator because the emitter of the pass transistor follows the .... voltage.
   (*a*) output (*b*) input
   (*c*) base (*d*) collector

7. In an op-amp series voltage regulator, output voltage depends on
   (*a*) Zener voltage
   (*b*) voltage divider resistors
   (*c*) input voltage
   (*d*) both (*a*) and (*b*)
8. In a feedback series regulator circuit, the output voltage is regulated by controlling the
   (*a*) magnitude of input voltage
   (*b*) gain of the feedback transistor
   (*c*) reference voltage
   (*d*) voltage drop across the series pass transistor
9. An op-amp shunt regulator differs from the series regulator in the sense that its control element is connected in
   (*a*) series with line resistor
   (*b*) parallel with line resistor
   (*c*) parallel with load resistor
   (*d*) parallel with input voltage.
10. A switching voltage regulator can be of the following type :
   (*a*) step-down (*b*) step-up
   (*c*) inverting (*d*) all of the above
11. In an inverting type switching regulator, output voltage is .... input voltage.
   (*a*) lesser than (*b*) greater than
   (*c*) equal to (*d*) opposite to.
12. The output voltage of a step-down type switching voltage regulator depends on
   (*a*) input voltage (*b*) duty cycle
   (*c*) transistor on-time(*d*) all of the above.
13. As compared to voltage regulators made up of discrete components, IC regulators have the inherent advantage/s of
   (*a*) self protection against over-temperature
   (*b*) remote control
   (*c*) current limiting
   (*d*) all of the above
14. A 12 V monolithic regulator is adjusted to obtain a higher output voltage as shown in Fig. 56.34. The V_0 will be

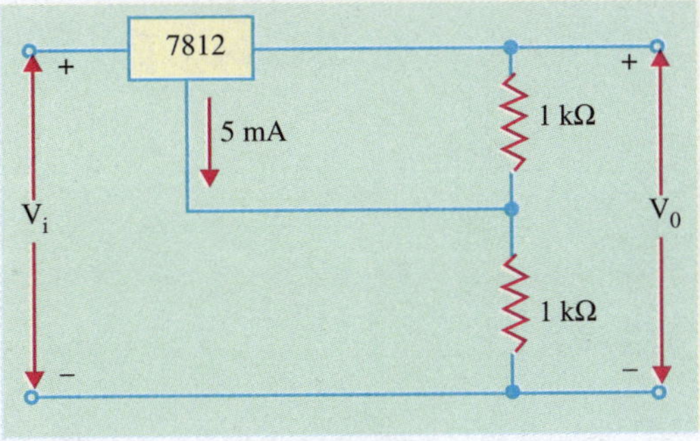

Fig. 56.34

   (*a*) 12 V (*b*) 17 V
   (*c*) 24 V (*d*) 29 V
15. A three terminal monolithic IC regulator can be used as
   (*a*) an adjustable output voltage regulator alone
   (*b*) an adjustable output voltage regulator and a current regulator
   (*c*) a current regulator and a power switch
   (*d*) a current regulator alone

ANSWERS

1. (*b*) 2. (*a*) 3. (*d*) 4. (*b*) 5. (*c*) 6. (*c*) 7. (*d*) 8. (*d*) 9. (*c*) 10. (*d*) 11. (*d*) 12. (*d*) 13. (*d*) 14. (*d*) 15. (*d*).

ROUGH WORK

CHAPTER 57

BIPOLAR JUNCTION TRANSISTOR

Learning Objectives

- Bipolar Junction Transistor
- Transistor Biasing
- Transistor Currents
- Transistor Circuit Configurations
- CB Configuration
- CE Configuration
- Relation between α and β
- CC Configuration
- Relation between Transistor Currents
- Leakage Currents in a Transistor
- Thermal Runaway
- Transistor Static Characteristics
- Common Base Test Circuit
- Common Base Static Characteristics
- Common Emitter Static Characteristics
- Common Collector Static Characteristic
- Different Ways of Drawing Transistor Circuits
- The Beta Rule
- Importance of V_{CE}
- Cut-off and Saturation Points
- BJT Operating Regions
- Active Region DC Model of BJT
- BJT Switches
- Normal DC Voltage Transistor Indications
- Transistor Fault Location
- Increase/Decrease Notation

Bipolar junction transistor is used in two broad areas-as a linear amplifier to boost or amplify an electrical signal and as an electronic switch

57.1. Bipolar Junction Transistor

The transistor was invented by a team of three scientists at Bell Laboratories, USA in 1947. Although the first transistor was not a bipolar junction device, yet it was the beginning of a technological revolution that is still continuing in the twenty first century. All of the complex electronic devices and systems developed or in use today, are an outgrowth of early developments in semiconductor transistors.

There are two basic types of transistors : **(1)** the bipolar junction transistor (BJT) which we will study in this chapter and the field-effect transistor (FET) which is covered in chapter 63. The bipolar junction transistor is used in two broad areas of electronics : **(1)** as a linear amplifier to boost an electrical signal and **(2)** as an electronic switch.

Basically, the bipolar junction transistor consists of two back-to-back *P-N* junctions manufactured in a single piece of a semiconductor crystal. These two junctions give rise to three regions called ***emitter, base*** and ***collector***. As shown in Fig. 57.1 (*a*) junction transistor is simply a sandwich of one type of semiconductor material between two layers of the other type. Fig. 57.1 (*a*) shows a layer of *N*-type material sandwiched between two layers of *P*-type material. It is described as a *PNP* transistor. Fig. 57.1 (*b*) shown an *NPN* – transistor consisting of a layer of *P*-type material sandwiched between two layers of *N*-type material.

Bipolar junction transistor

The emitter, base and collector are provided with terminals which are labelled as *E*, *B* and *C*. The two junctions are : emitter-base (*E/B*) junction and collector-base (*C/B*) junction.

The symbols employed for *PNP* and *NPN* transistors are also shown in Fig. 57.1. The arrowhead is always at the emitter (not at the collector) and in each case, its direction indicates the ***conventional*** direction of current flow. For a *PNP* transistor, arrowhead points from emitter to base meaning that emitter is positive with respect to base (and also with respect to collector)* For *NPN* transistor, it points from base to emitter meaning that base (and collector as well)* is positive with respect to the emitter.

1. Emitter

It is more heavily doped than any of the other regions because its main function is to supply majority charge carries (either electrons or holes) to the base.

2. Base

It forms the middle section of the transistor. It is very thin (10^{-6} m) as compared to either the emitter or collector and is very ***lightly-doped.***

3. Collector

Its main function (as indicated by its name) is to collect majority charge carriers coming from the emitter and passing through the base.

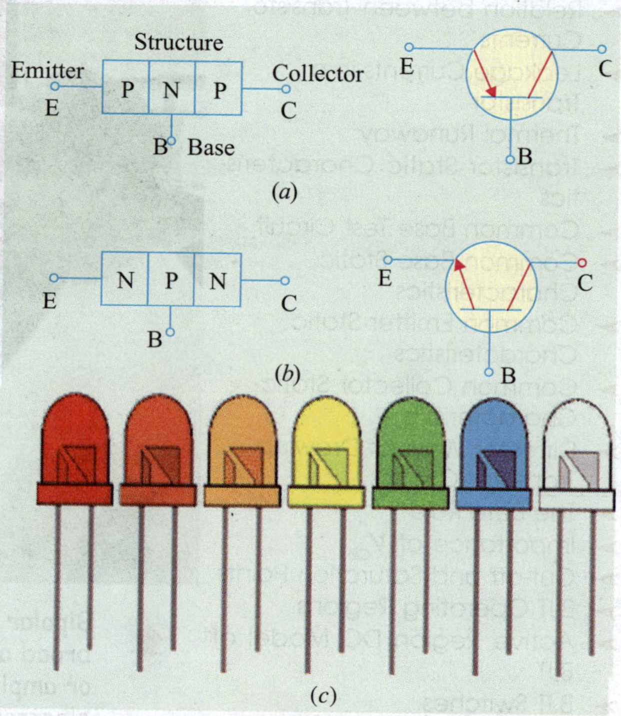

Fig. 57.1

* In a transistor, for normal operation, collector and base have the same polarity with respect to the emitter (Art. 57.3)

In most transistors, collector region is made physically larger than the emitter region because it has to dissipate much greater power. Because of this difference, there is no possibility of inverting the transistor *i.e.* making its collector the emitter and its emitter the collector. Fig 57.1 (*c*), shows the picture of C1815 (front and the back view) transistor.

57.2. Transistor Biasing

For proper working of a transistor, it is essential to apply voltages of correct polarity across its two junctions. It is worthwhile to remember that for normal operation;

1. emitter-base junction is always forward-biased and
2. collector-base junction is always reverse-biased.

This type of biasing is known as *FR* biasing.

In Fig. 57.2, two batteries respectively provide the dc emitter supply voltage V_{EE} and collector supply voltage V_{CC} for properly biasing the two junctions of the transistor. In Fig. 57.2 (*a*), **P**ositive terminal of V_{EE} is connected to **P**-type emitter in order to repel or **P**ush holes into the base.

The negative terminal of V_{CC} is connected to the collector so that it may ***attract*** or ***pull*** holes through the base. Similar considerations apply to the *NPN* transistor of Fig. 57.2 (*b*). It must be remembered that a transistor will never conduct any current if its emitter-base junction is not forward-biased.* Also refer to the picture shown in Fig. 57.2 (*c*).

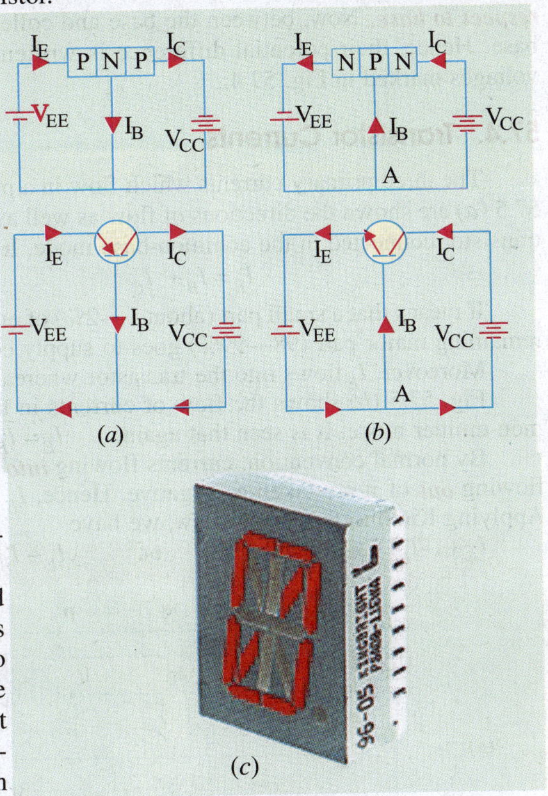

Fig. 57.2

57.3. Important Biasing Rule

For a *PNP* transistor, both collector and base are negative with respect to the emmitter (the letter *N* of Negative being the same as the middle letter of *PNP*). Of course, collector is ***more negative*** than base [Fig. 57.3 (*a*)]. Similarly, for *NPN* transistor, both collector and base are positive with respect to the emitter (the letter **P** of **P**ositive being the same as the middle letter of *NPN*). Again, collector is ***more positive*** than the base as shown in Fig. 57.3 (*b*).

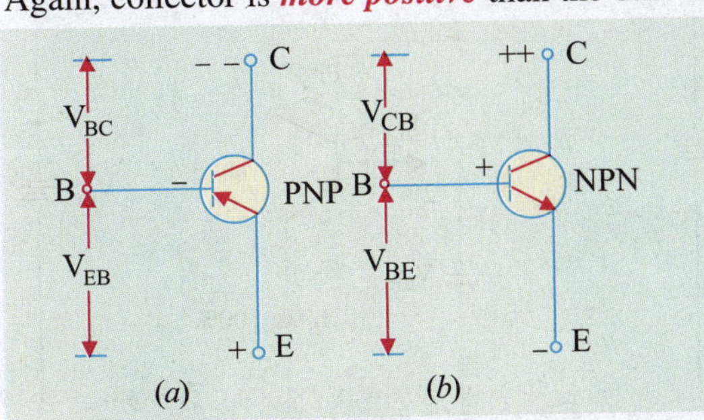

Fig. 57.3

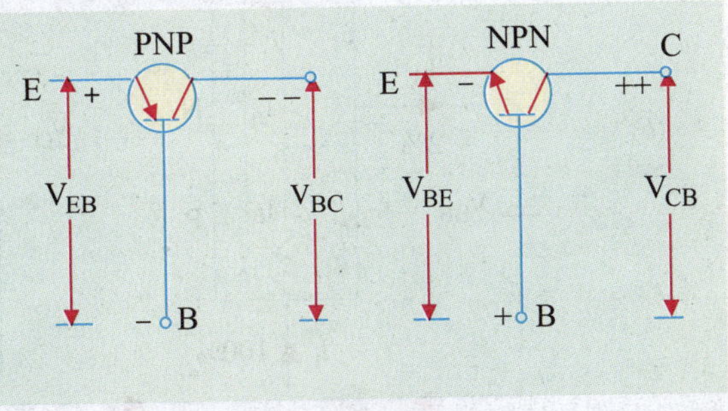

Fig. 57.4

* There would be no current due to majority charge carriers. However, there would be an extremely small current due to minority charge carriers which is called leakage current of the transistor (Art. 57.12).

It may be noted that different potentials have been designated by double subscripts. The first subscript always represents the point or terminal which is more positive (or less negative) than the point or terminal represented by the second subscript. For example, in Fig. 57.3 (*a*), the potential difference between emitter and base is written as V_{EB} (and not V_{BE}) because ***emitter is positive with respect to base.*** Now, between the base and collector themselves, collector is more negative than base. Hence, their potential difference is written as V_{BC} and not as V_{CB}. Same is the case with voltages marked in Fig. 57.4.

57.4. Transistor Currents

The three primary currents which flow in a properly-biased transistor are I_E, I_B and I_C. In Fig. 57.5 (*a*) are shown the directions of flow as well as relative magnitudes of these currents for a *PNP* transistor connected in the common-base mode. It is seen that again,

$$I_E = I_B + I_C$$

It means that a small part (about 1—2%) of emitter current goes to supply base current and the remaining major part (98—99%) goes to supply collector current.

Moreover, I_E flows into the transistor whereas both I_B and I_C flow out of it.

Fig. 57.5 (*b*) shows the flow of currents in the same transistor when connected in the common-emitter mode. It is seen that again, $I_E = I_B + I_C$

By normal convention, currents flowing ***into*** a transistor are taken as positive whereas those flowing ***out*** of it are taken as negative. Hence, I_E is positive whereas both I_B and I_C are negative. Applying Kirchhoff's Current Law, we have

$$I_E + (-I_B) + (-I_C) = 0 \quad \text{or} \quad I_E - I_B - I_C = 0 \quad \text{or} \quad I_E = I_B + I_C$$

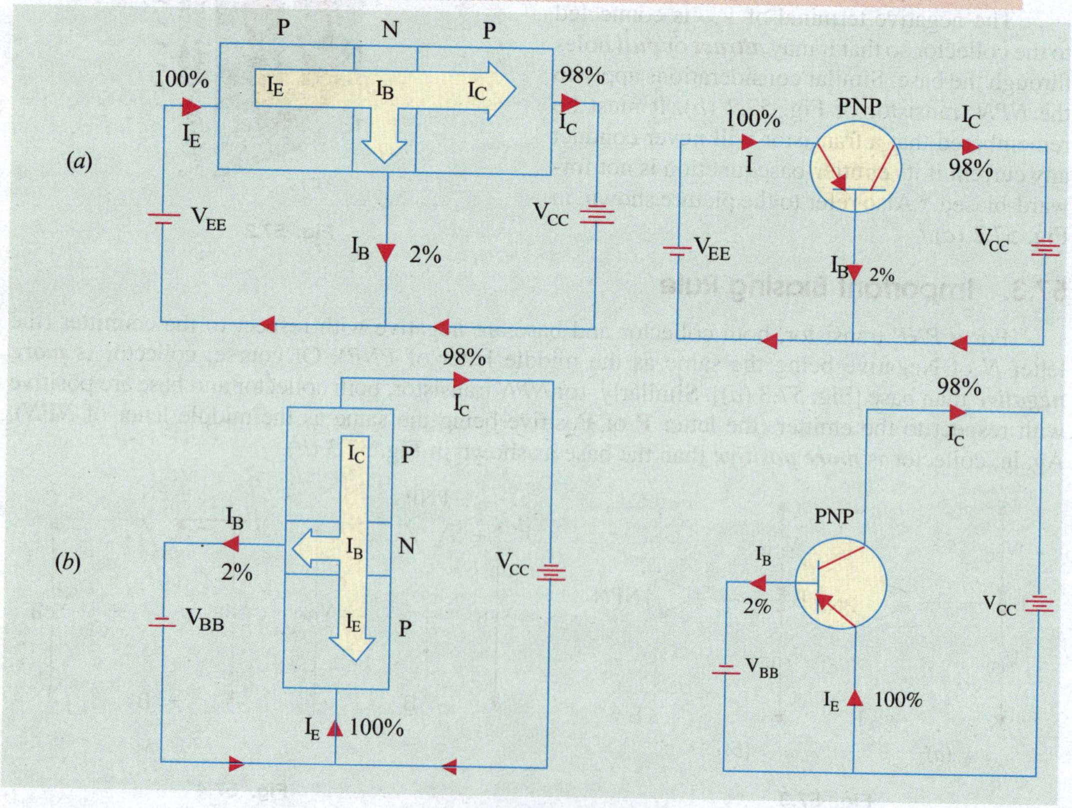

Fig. 57.5

This statement is true ***regardless of transistor type or transistor configuration.***

Note. For the time being, we have not taken into account the leakage currents which exist in a transistor (Art. 57.12).

57.5. Summing Up

The four basic guideposts about all transistor circuits are :

1. conventional current flows along the arrow whereas electrons flow against it;
2. *E/B* junction is always forward-biased;
3. *C/B* junction is always reverse-biased;
4. $I_E = I_B + I_C$.

57.6. Transistor Circuit Configurations

Basically, there are three types of circuit connections (called configurations) for operating a transistor.

1. common-base (*CB*), 2. common-emitter (*CE*), 3. common-collector (*CC*).

The term 'common' is used to denote the electrode that is common to the input and output circuits. Because the common electrode is generally grounded, these modes of operation are frequently referred to as grounded-base, grounded-emitter and grounded-collector configurations as shown in Fig. 57.6 for a *PNP* – transistor.

Since a transistor is a 3-terminal (and not a 4-terminal) device, one of its terminals has to be common to the input and output circuits.

57.7. CB Configuration

In this configuration, emitter current I_E is the input current and collector current I_C is the output current. The input signal is applied between the emitter and base whereas output is taken out from the collector and base as shown in Fig. 57.6 (*a*).

The ratio of the collector current to the emitter current is called dc alpha (α_{dc}) of a transistor.

$$\therefore \quad \alpha_{dc}{}^{*} = \frac{-I_C}{I_E}$$

The negative sign is due to the fact that current I_E flows into the transistor whereas I_C flows out of it. Hence, I_E is taken as positive and I_C as negative.

$$\therefore \quad I_C = -\alpha_{dc}.I_E$$

If we write α_{dc} simply as α**, then $\alpha = \dfrac{I_C}{I_E}$

It is also called forward current transfer ratio ($-h_{FB}$). In h_{FB}, subscript *F* stands for forward and *B* for common-base. The subscript d.c. on a signifies that this ratio is defined from dc values of I_C and I_E.

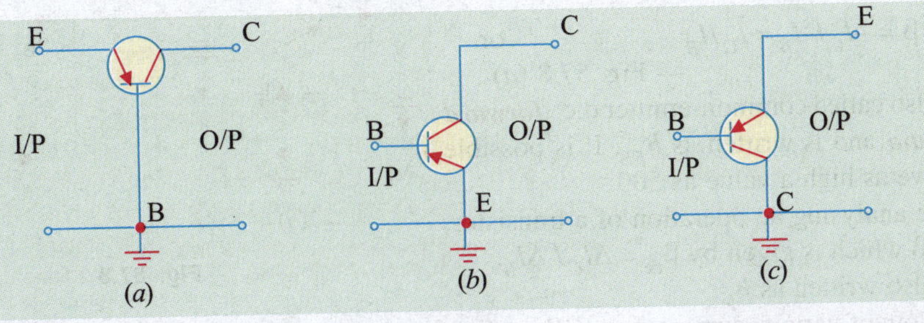

Fig. 57.6

The α of a tranistor is a measure of the quality of a transistor ; higher the value of α, better the transistor in the sense that collector current more closely equals the emitter current. Its value ranges

* More accurately, $\alpha_{dc} = \dfrac{I_C - I_{CBO}}{I_E}$ ...Art.57.12

** Negative sign has been omitted, since we are here concerned with only magnitudes of the currents involved.

from 0.95 to 0.999. Obviously, it applies only to *CB* configuration of a transistor. As seen from above and Fig. 57.7.

$$I_C = \alpha I_E. \qquad \text{Now,} \qquad I_B = I_E - \alpha I_E = (1-\alpha) I_E$$

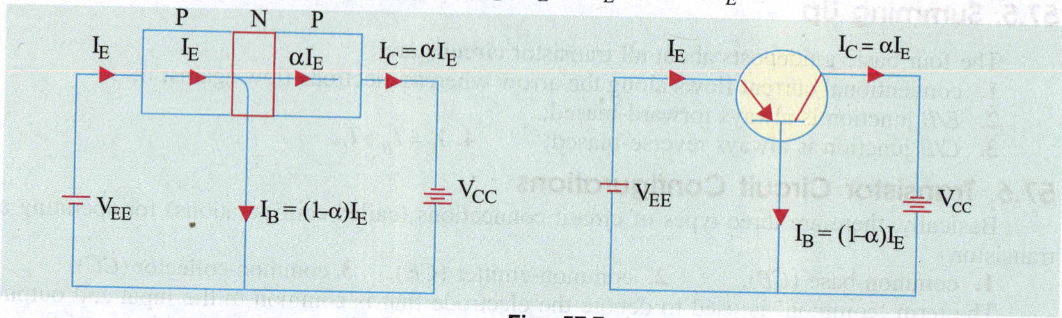

Fig. 57.7

Incidentally, there is also an a.c. α for a transistor. It refers to the ratio of ***change*** in collector current to the ***change*** in emitter current.

$$\therefore \qquad \alpha_{ac} = \frac{-\Delta I_C}{\Delta I_E}$$

It is also, known as short-circuit gain of a transistor and is written as $-h_{fb}$. It may be noted that upper case subscript '*FB*' indicates dc value whereas lower case subscript '*fb*' indicates ac value. For all practical purposes, $\alpha_{dc} = \alpha_{ac} = \alpha$.

Example 57.1. *Following current readings are obtained in a transistor connected in CB configuration : $I_E = 2$ mA and $I_B = 20$ μA. Compute the values of α and I_C.*

(Electronics-II, Punjab Univ. 1992)

Solution. $I_C = I_E - I_B = 2 \times 10^{-3} - 20 \times 10^{-6} =$ **1.98 mA**

$\alpha = I_C/I_E = 1.98/2 =$ **0.99**

57.8. CE Configuration

Here, input signal is applied between the base and emitter and output signal is taken out from the collector and emitter circuit. As seen from Fig. 57.8 (*b*), I_B is the input current and I_C is the output current.

The ratio of the d.c. collector current to dc base current is called dc beta (β_{dc}) or just β of the transistor.

$$\therefore \qquad \beta = -I_C/-I_B = I_C/I_B \qquad \text{or}$$

$$I_C = \beta I_B \qquad \text{— Fig. 57.8 } (a)$$

It is also called common-emitter d.c. ***forward transfer ratio*** and is written as h_{FE}. It is possible for β to have as high a value as 500.

While analysing ac operation of a transistor, we use ac β which is given by $\beta_{ac} = \Delta I_C / \Delta I_B$. It is also written as h_{fe}.

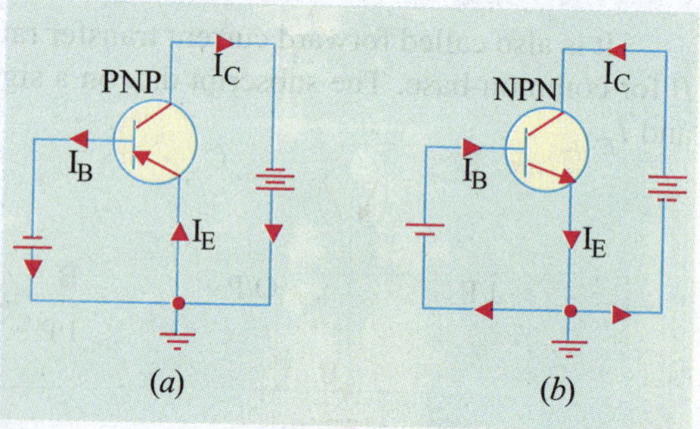

Fig. 57.8

The flow of various currents in a *CE* configuration both for *PNP* and *NPN* transistor is shown in Fig. 57.8. As seen

$$I_E = I_B + I_C = I_B + \beta I_B = (1+\beta) I_B$$

57.9. Relation Between α and β

$$\beta = \frac{I_C}{I_B} \quad \text{and} \quad \alpha = \frac{I_C}{I_E} \quad \therefore \quad \frac{\beta}{\alpha} = \frac{I_E}{I_B} \qquad \text{— only numerical value of } \alpha$$

Now, $I_B = I_E - I_C$ $\quad\therefore\quad \beta = \dfrac{I_C}{I_E - I_C} = \dfrac{I_C / I_E}{I_E / I_E - I_C / I_E}$ or $\beta = \dfrac{\alpha}{1-\alpha}$

Cross-multiplying the above equation and simplifying it, we get

$\beta(1-\alpha) = \alpha$ or $\beta = \alpha(1+\beta)$ or $\alpha = \beta/(1+\beta)$

It is seen from the about 2 equations that $1 - \alpha = 1/(1+\beta)$

57.10. CC Configuration

In this case, input signal is applied between base and collector and output signal is taken out from emitter-collector circuit [Fig. 57.6 (*c*)]. Conventionally speaking, here I_B is the input current and I_E is the output current as shown in Fig. 57.9. The current gain of the circuit is

$$\frac{I_E}{I_B} = \frac{I_E}{I_C} \cdot \frac{I_C}{I_B} = \frac{\beta}{\alpha} = \frac{\beta}{\beta/(1+\beta)} = (1+\beta)$$

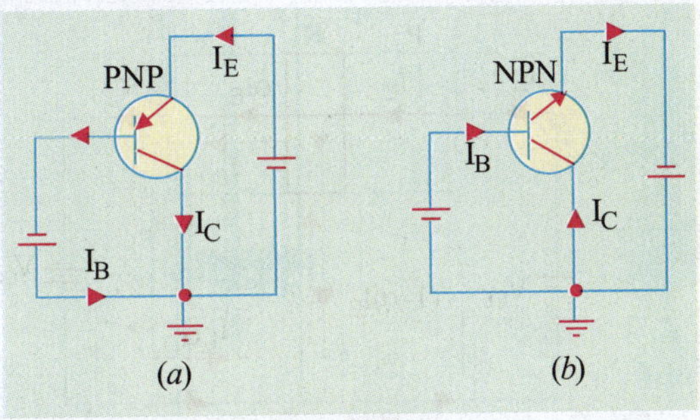

Fig. 57.9

The flow paths of various currents in a *CC* configuration are shown in Fig. 57.9. It is seen that

$I_E = I_B + I_C = I_B + \beta I_B = (1+\beta) I_B$

$\therefore$ output current $= (1+\beta) \times$ input current.

57.11. Relations Between Transistor Currents

While deriving various equations, following definitions should be kept in mind.

$$\alpha = \frac{I_C}{I_E}, \quad \beta = \frac{I_C}{I_B}, \quad \alpha = \frac{\beta}{(1+\beta)} \text{ and } \beta = \frac{\alpha}{(1-\alpha)}$$

(i) $I_C = \beta I_B = \alpha I_E = \dfrac{\beta}{1+\beta} I_E$ *(ii)* $I_B = \dfrac{I_C}{\beta} = \dfrac{I_E}{1+\beta} = (1-\alpha) I_E$

(iii) $I_E = \dfrac{I_C}{\alpha} = \dfrac{1+\beta}{\beta} I_C = (1+\beta) I_B = \dfrac{I_B}{(1-\alpha)}$

(iv) The three transistor d.c. currents always bear the following ratio*

$I_E : I_B : I_C \quad :: \quad 1 : (1-\alpha) : \alpha$

Incidentally, it may be noted that for ac currents, small letters i_e, i_b and i_c are used.

57.12. Leakage Currents in a Transistor

(a) CB Circuit

Consider the *CB* transistor circuit shown in Fig. 57.11. The emitter current (due to majority carriers) initiated by the forward-biased emitter base junction is split into two parts :

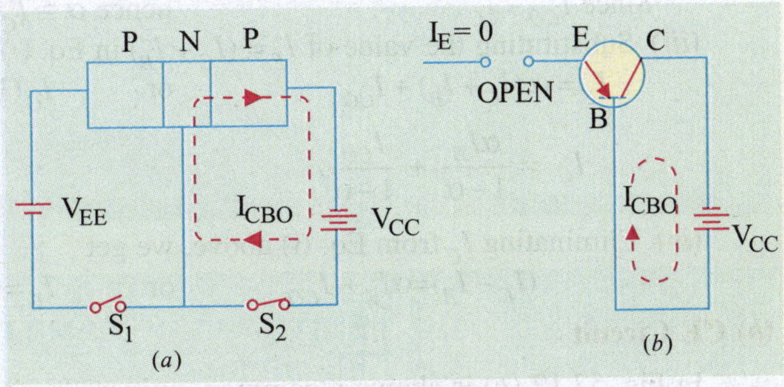

Fig. 57.10

(i) $(1-\alpha) I_E$ which becomes base current I_B in the external circuit and

(ii) αI_E which becomes collector current I_C in the external circuit.

---

* It reminds us of the power distribution relationship in an induction motor.

As mentioned earlier (Art. 57.2), though *C/B* junction is reverse-biased for majority charge carriers (*i.e.* holes in this case), it is forward-biased so far as thermally-generated minority charge carriers (*i.e.* electrons in this case) are concerned. This current flows even when emitter is disconnected from its dc supply as shown in Fig. 57.10 (*a*) where switch, S_1 is open. It flows in the ***same*** direction* as the collector current of majority carriers. It is called leakage current I_{CBO}. The subscripts *CBO* stand for 'Collector to Base with emitter Open.' Very often, it is simply written as I_{CO}.

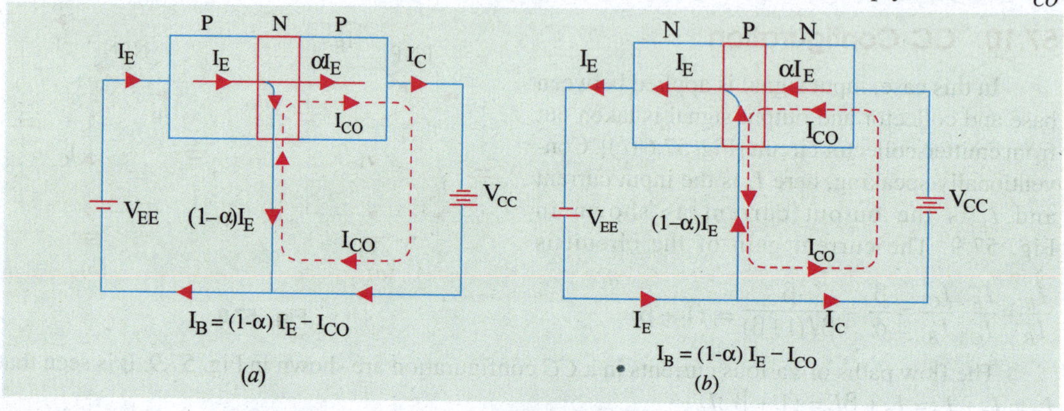

Fig. 57.11

It should be noted that

(*i*) I_{CBO} is exactly like the reverse saturation current I_S or I_0 of a reverse-biased diode discussed in Art. 52.1.

(*ii*) I_{CBO} is extremely temperature-dependent because it is made up of thermally-generated minority carriers. As mentioned earlier, I_{CBO} doubles for every 10°C rise in temperature for *Ge* and 6°C for *Si*.

If we take into account the leakage current, the current distribution in a *CB* transistor circuit becomes as shown in Fig. 57.11 both for *PNP* and *NPN* type transistors.

It is seen that total collector current is actually the sum of two components :

(*i*) current produced by normal transistor action *i.e.* component controlled by emitter current. Its value is a αI_E and is due to majority carriers.

(*ii*) temperature-dependent leakage current I_{CO} due to minority carriers.

$$\therefore \quad I_C = \alpha I_E + I_{CO} \quad \ldots(i) \qquad \therefore \quad \alpha = \frac{I_C - I_{CO}}{I_E}$$

Since $I_{CO} \ll I_C$, hence $\alpha \cong I_C / I_E$

(*iii*) Substituting the value of $I_E = (I_C + I_B)$ in Eq. (*i*) above, we get

$$I_C = \alpha\,(I_C + I_B) + I_{CO} \quad \text{or} \quad I_C(1-\alpha) = \alpha I_B + I_{CO}$$

$$\therefore \quad I_C = \frac{\alpha I_B}{1-\alpha} + \frac{I_{CO}}{1-\alpha}$$

(*iv*) Eliminating I_C from Eq. (i) above, we get

$$(I_E - I_B) = \alpha I_E + I_{CO} \quad \text{or} \quad I_B = (1-\alpha)\,I_E - I_{CO}$$

(*b*) CE Circuit

In Fig. 57.12 (*a*) is shown a common-emitter circuit of an *NPN* transistor whose base lead is

---

* Actually, electrons (which form minority charge carriers in collector) flow from negative terminal of collector battery, to collector, then to base through *C/B* junction and finally, to positive terminal of V_{CC}. However, conventional current flows in the opposite direction as shown by dotted lien in Fig. 57.10 (*a*)

open. It is found that despite $I_B = 0$, there is a leakage current from collector to emitter. It is called I_{CEO}, the subscripts *CEO* standing for 'Collector to Emitter with base Open'.

Taking this leakage current into account, the current distribution through a *CE* circuit becomes as shown in Fig. 57.12 (*c*).

$$I_C = \beta I_B + I_{CEO} = \beta I_B + (1 + \beta)I_{CO} = \beta I_B + I_{CO}/(1 - \alpha)$$

(i) $$\therefore \quad I_C = \frac{\alpha I_B}{1-\alpha} + \frac{I_{CO}}{1-\alpha}$$

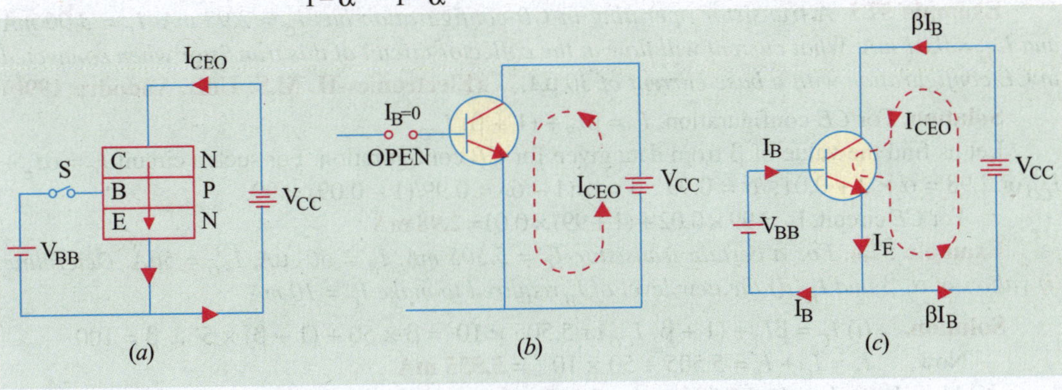

Fig. 57.12

Now, $\beta I_B = \alpha I_E$. Substituting this value above, we get,

$I_C = \alpha I_E + I_{CEO}$. Also, $I_B = I_E - I_C$

Substituting the value of I_C from above, we have

(ii) $I_B = I_E - \alpha I_E - I_{CEO} = (1 - \alpha)\, I_E - I_{CEO}$

57.13. Thermal Runaway

As seen from Art. 57.12, for a *CE* circuit

$$I_C = \beta I_B + (1 + \beta)I_{CO}$$

The leakage current is extremely temperature-dependent. It almost doubles for every 10°C rise in temperature in *Ge* and for every 6°C rise in *Si*. Any increase in I_{CO} is magnified $(1 + \beta)$ times *i.e.* 300 to 500 times. Even a slight increase in I_{CO} will affect I_C considerably. As I_C increases, collector power dissipation increases which raises the operating temperature that leads to further increase in I_C. If this succession of increases is allowed to continue, soon I_C will increase beyond safe operating value thereby damaging the transistor itself—a condition known as ***thermal runaway***. Hence, some form of stabilization is necessary to prevent this thermal runaway.

Example 57.2. *The reverse saturation current of an NPN transistor in common-base circuit is 12.5 μA. For an emitter current of 2 mA, collector current is 1.97 mA. Determine the current gain and base current.* **(Electronics-1, Gwalior Univ. 1988)**

Solution. Given : $I_{CBO} = 12.5\ \mu A$; $I_E = 2$ mA, $I_C = 1.97$ mA ; $\alpha = ?$, $I_B = ?$

$$I_C = \alpha I_E + I_{CBO} \qquad \therefore \quad \alpha = \frac{I_C - I_{CBO}}{I_E} = \frac{1.97 - 12.5\times10^{-3}}{2} = \mathbf{0.978}$$

$I_B = I_E - I_C = 2 - 1.97 =$ **0.03 mA**.

Example. 57.3. *Derive an expression for forward current gain and leakage current of common-emitter configuration in terms of current gain and leakage current of common-base configuration. If $\alpha = 0.98$, $I_{CBO} = 5\ \mu A$, calculate β and I_{CEO}.* **(Electronics-I, Mysore Univ. 1990)**

Solution. $\beta = \alpha/(1 - \alpha) = 0.98/(1 - 0.98) =$ **49**

$$I_{CEO} = (1+\beta)I_{CO} = (1+49)\times 5 = 250\ \mu\text{A} = \mathbf{0.25\ mA}.$$

Example 57.4. *For a transistor, $I_B = 100\ \mu A$, $\alpha_{dc} = 0.98$ and $I_{CO} = 5\ \mu A$. Find the values of I_C and I_E.*

Solution. As seen from Art. 57.12, $I_C = \dfrac{\alpha I_B}{1-\alpha} + \dfrac{I_{CO}}{1-\alpha} = \dfrac{0.98\times 100}{1-0.98} + \dfrac{5}{1-0.98} = \mathbf{5.15\ mA}$

$$I_E = I_C + I_B = 5.15 + 100\times 10^{-3} = \mathbf{5.25\ mA}.$$

Example 57.5. *A transistor operating in CB configuration has $I_C = 2.98$ mA, $I_E = 3.00$ mA and $I_{CO} = 0.01$ mA. What current will flow in the collector circuit of this transistor when connected in CE configuration with a base current of 30 μA.* **(Electronics-II, M.S. Univ. Vadodra 1990)**

Solution. For *CE* configuration, $I_C = \beta I_B + (1+\beta)\, I_{CO}$

Let us find the value of β from data given for *CB* configuration. For such a circuit $I_C = \alpha I_E + I_{CO}$ or $2.98 = \alpha\times 3 + 0.01$; $\alpha = 0.99$; $\beta = \alpha/(1-\alpha) = 0.99/(1-0.09) = 99$.

$\therefore$ For *CE* circuit, $I_C = 99\times 0.02 + (1+99)\times 0.01 = \mathbf{2.98\ mA}$.

Example 57.6. *For a certain transistor, $I_C = 5.505$ mA, $I_B = 50$ μA, $I_{CO} = 5$μA. Determine (i) values of α, β and I_E (ii) the new level of I_B required to make $I_C = 10$ mA.*

Solution. (*i*) $I_C = \beta I_B + (1+\beta)\, I_{CO}$ or $5.505\times 10^3 = \beta\times 50 + (1+\beta)\times 5$ $\therefore \beta = 100$

Now, $I_E = I_C + I_B = 5.505 + 50\times 10^{-3} = \mathbf{5.555\ mA}$.

Also, $I_C = \alpha I_E + I_{CO}$; $5.505 = \alpha\times 5.555 + 5\times 10^{-3}$ $\therefore$ $\alpha = 5.500/5.555 = \mathbf{0.99}$

(*ii*) As seen from Art. 57.12, $I_C = \beta I_B + (1+\beta)\, I_{CO}$

$\therefore$ $10 = 100\, I_B + 101\times 5\times 10^{-3}$; $I_B = 0.09495\text{ mA} = \mathbf{94.95\ \mu A}$.

Example. 57.7. *Discuss the operation of a PNP transistor.*

The reverse saturation current in a PNP germanium transistor type OC 71 is 8 μA. If the transistor common base current gain is 0.979, calculate the collector and emitter current for 40 μA base current. What is the collector current when base current is zero ?

(Electronics-1, Gwalior Univ. 1986)

Solution. Given : $I_{CO} = 8\text{ mA} = 0.008\text{ mA}$, $\alpha = 0.979$; $I_B = 40\ \mu\text{A} = 0.04\text{ mA}$

In a *CE* circuit: $I_C = \beta I_B + I_{CEO} = \beta I_B + I_C/(1-\alpha)$.

Now, $\beta = \alpha/(1-\alpha) = 0.979/(1-0.979) = 46.6$

$\therefore$ $I_C = 46.6\times 0.04 + (1+46.6)\times 0.008 = \mathbf{2.25\ mA}$; $I_E = I_C + I_B = 2.25 + 0.04 = \mathbf{2.29\ mA}$

57.14. Transistor Static Characteristics

There are the curves which represents relationship between different d.c. currents and voltages of a transistor. These are helpful in studying the operation of a transistor when connected in a circuit. The three important characteristics of a transistor are :

1. Input characteristic, **2.** Output characteristic, **3.** Constant-current transfer characteristic.

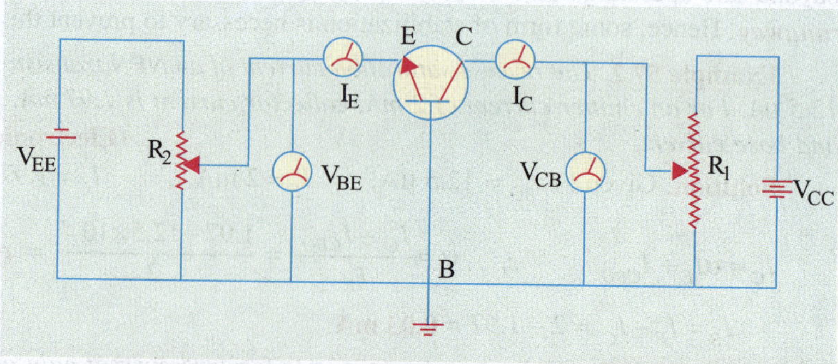

Fig. 57.13

57.15 Common Base Test Circuit

The static characteristics of an *NPN* transistor connected in common-base configuration can be determined by the use of test circuit shown in Fig. 57.13. Milliammeters are included in series with the emitter and collector circuits to measure I_E and I_C. Similarly, voltmeters are connected across *E* and *B* to measure voltage V_{BE} and across *C* and *B* to measure V_{CB}. The two potentiometer resistors R_1 and R_2 supply variable voltages from the collector and emitter dc supplies respectively.

57.16. Common Base Static Characteristics

(a) Input Characteristic

It shows how I_E varies with V_{BE} when voltage V_{CB} is held constant. The method of determining this characteristic is as follows :

First, voltage V_{CB} is adjusted to a suitable value with the help of R_1 (Fig. 57.13). Next, voltage V_{BE} is increased in a number of discrete steps and corresponding values of I_E are noted from the milliammeter connected for the purpose. When plotted, we get the input characteristic shown in Fig. 57.14, one for *Ge* and the other for *Si*. Both curves are exactly similar to the forward characteristic of a *P-N* diode which, in essence, is what the emitter-base junction is.

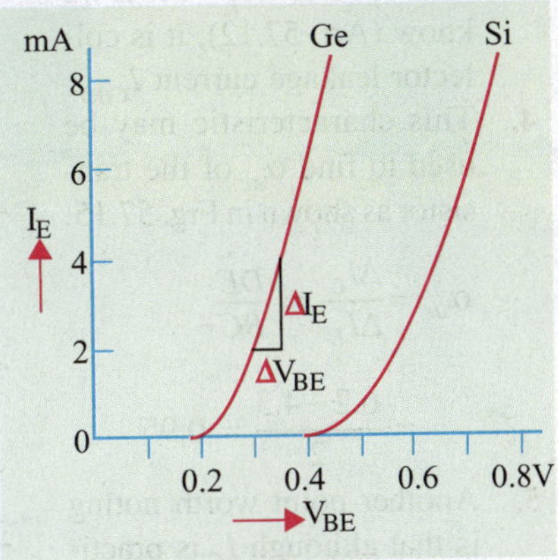

Fig. 57.14

This characteristic may be used to find the input resistance of the transistor. Its value is given by the reciprocal of its slope.

$$R_{in} = \Delta V_{BE} / \Delta I_E \qquad \text{— } V_{CB} \text{ constant.}$$

Since the characteristic is initially nonlinear, R_{in} will vary with the point of measurement. Its value over linear part of the characteristic is about 50 Ω but for low values of V_{BE}, it is considerably greater. This change in R_{in} with change in V_{BE} gives rise to distortion of signals handled by the transistor.

This characteristic is hardly affected by changes either in V_{CB} or temperature.

(b) Output Characteristic

It shows the way I_C varies with V_{CB} when I_E is held constant. The method of obtaining this characteristic is as follows:

First, movable contact, on R_2 (Fig. 57.13) is changed to get a suitable value of V_{BE} and hence that of I_E. While keeping I_E constant at this value, V_{CB} is increased from zero in a number of steps and the corresponding collector current I_C that flows is noted.

Next, V_{CB} is reduced back to zero, I_E is increased to a value a little higher than before and the whole procedure is repeated. In this way, whole family of curves is obtained, a typical family being shown in Fig. 57.15.

1. The reciprocal of the near horizontal part of the characteristic gives the output resistance R_{out} of the transistor which it would offer to an input signal. Since the characteristic is linear over most of its length (meaning that I_C is virtually independent of V_{CB}). R_{out} is very high, a typical value being 500 kΩ.

$$R_{out} = \frac{1}{\Delta I_C / \Delta V_{CB}} = \frac{\Delta V_{CB}}{\Delta I_C}$$

2. It is seen that I_C flows even when $V_{CB} = 0$. For example, it has a value = 1.8 mA corresponding to $V_{CB} = 0$ for $I_E = 2$ mA as shown in Fig. 57.15. It is due to the fact that electrons are being injected into the base under the action of forward-biased *E/B* junction and are being collected by the collector due to the action of the internal junction voltage at the *C/B* junction (Art. 57.2). For reducing I_C to zero, it is essential to neutralize this potential barrier by applying a small forward bias across *C/B* junction.

3. Another important feature of the characteristic is that a small amount of collector current flows even when emitter current $I_E = 0$. As we know (Art. 57.12), it is collector leakage current I_{CBO}.
4. This characteristic may be used to find α_{ac} of the transistor as shown in Fig. 57.15.

$$\alpha_{ac} = \frac{\Delta I_C}{\Delta I_E} = \frac{DE}{BC}$$

$$= \frac{6.2 - 4.3}{2} = 0.95$$

5. Another point worth noting is that although I_C is practically independent of V_{CB} over the working range of the transistor, yet if V_{CB} is permitted to increase beyond a certain value, I_C eventually increases rapidly due to avalanche breakdown as shown in Fig. 57.15.

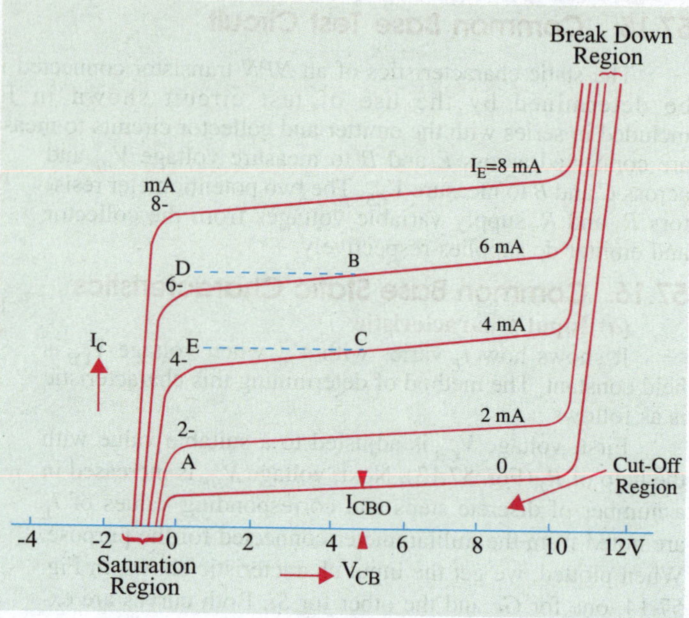

Fig. 57.15

(c) Current Transfer Characteristic

It shows how I_C varies with changes in I_E when V_{CB} is held constant. For drawing this characteristic, first V_{CB} is set to a convenient value and then I_E is increased in steps and corresponding values of I_C noted. A typical transfer characteristic is shown in Fig. 57.16 (*a*). Fig. 57.16 (*b*) shows a more detailed view of the portion near the origin.

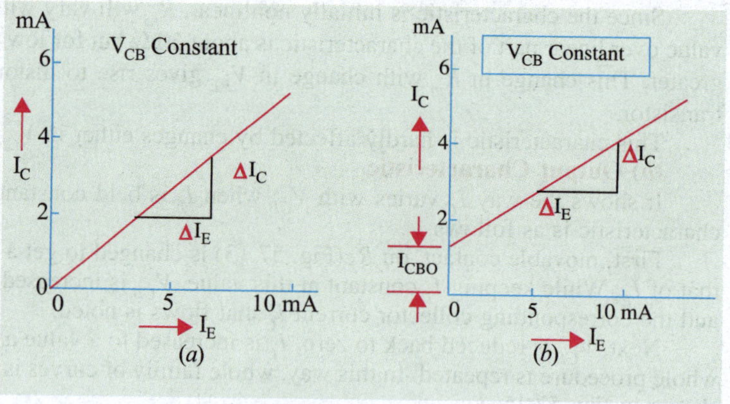

Fig. 57.16

As seen, α_{ac} may be found from the equation

$$\alpha_{ac} = \Delta I_C / \Delta I_E$$

Usually, α_{ac} is found from output characteristic than from this characteristic.

It may be noted in the end that *CB* connection is rarely employed for audio-frequency circuits because *(i)* its current gain is less than unity and *(ii)* its input and output resistances are so different.

57.17. Common Emitter Test Circuit

The static characteristics of an *NPN* transistor connected in *CE* configuration may be determined by the use of circuit diagram shown in Fig. 57.17. A milliammeter (or a microammeter in the case of a low-power transistor) is connected in series with the base to measure I_B. Similarly, a milliammeter is included in the collector circuit to measure I_C. A voltmeter with a typical range of 0 –1 V is connected across base and emitter terminals for measuring V_{BE}.

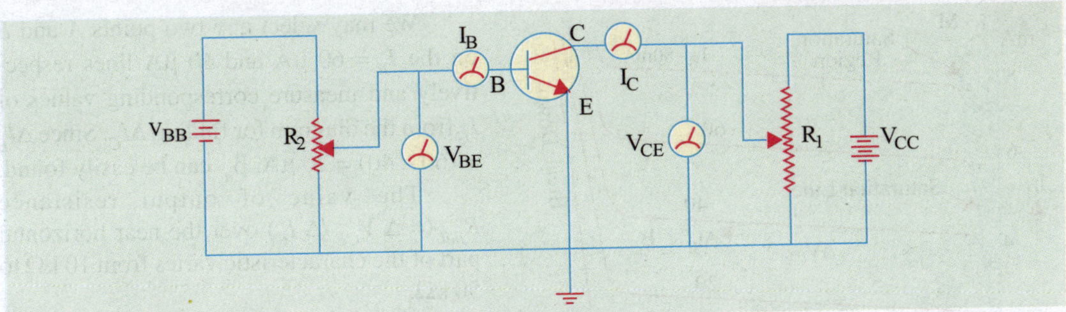

Fig. 57.17

Potentiometer R_2 connected across dc supply V_{BB} is used to vary I_B and V_{BE}. A second voltmeter with a typical range of 0–20 V is connected across collector-emitter terminals to measure the output collector-emitter voltage V_{CE}.

57.18. Common Emitter Static Characteristics

(a) Input Characteristic

It shows how I_B varies with changes in V_{BE} when V_{CE} is held constant at a particular value.

To begin with, voltage V_{CE} is maintained constant at a convenient value and then V_{BE} is increased in steps. Corresponding values of I_B are noted at each step. The procedure is then repeated for a different but constant value of V_{CE}. A typical input characteristic is shown in Fig. 57.18. Like *CB* connection, the overall shape resembles the forward characteristic of a *P-N* diode. The reciprocal of the slope gives the input resistance R_{in} of the transistor.

$$R_{in} = \frac{1}{\Delta I_B / \Delta V_{BE}} = \frac{\Delta V_{BE}}{\Delta I_B}$$

Due to initial non-linearity of the curve, R_{in} varies considerably from a value of 4 kΩ near the origin to a value of 600 Ω over the more linear part of the curve.

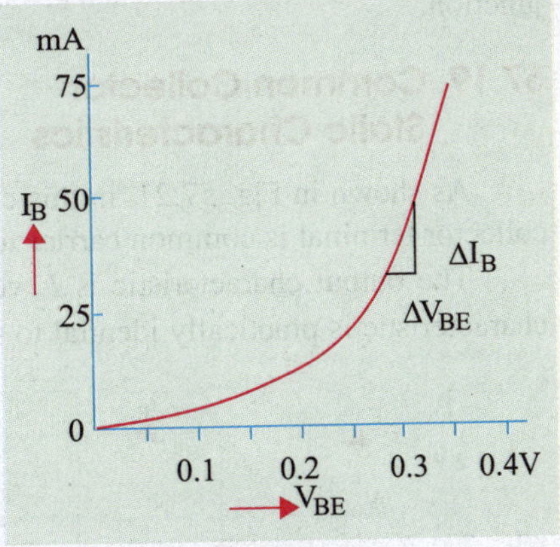

Fig. 57.18

(b) Output or Collector Characteristic

It indicates the way in which I_C varies with changes in V_{CE} when I_B is held constant.

For obtaining this characteristic, first I_B is set to a convenient value and maintained constant and then V_{CE} is increased from zero in steps, I_C being noted at each step. Next, V_{CE} is reduced to zero and I_B increased to another convenient value and the whole procedure repeated. In this way, a family of curves (Fig. 57.19) is obtained.

It is seen that as V_{CE} increases from zero, I_C rapidly increases to a near saturation level for a fixed value of I_B. As shown, a small amount of collector current flows even when $I_B = 0$. It is called I_{CEO} (Art. 57.12). Since main collector current is zero, the transistor is said to be **cut-off**.

It may be noted that if V_{CE} is allowed to increase too far, *C/B* junction completely breaks down and due to this avalanche breakdown, I_C increases rapidly and may cause damage to the transistor.

When V_{CE} has very low value (ideally zero), the transistor is said to be saturated and it operates in the saturation region of the characteristic. Here, change in I_B does not produce a corresponding change in I_C.

This characteristic can be used to find β_{ac} at a specific value of I_B and V_{CE}. It is given by $\beta_{ac} = \Delta I_C / \Delta I_B$.

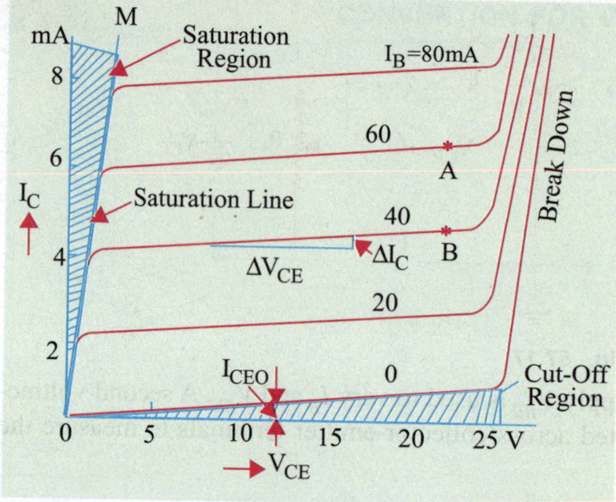

Fig. 57.19

We may select any two points A and B on the $I_B = 60$ µA and 40 µA lines respectively and measure corresponding values of I_C from the diagram for finding ΔI_C. Since $\Delta I_B = (60 - 40) = 20$ µA, β_{ac} can be easily found.

The value of output resistance R_{out} (= $\Delta V_{CE} / \Delta I_C$) over the near horizontal part of the characteristic varies from 10 kΩ to 50 kΩ.

(c) Current Transfer Characteristic

It indicates how I_C varies with changes in I_B when V_{CE} is held constant at a given value.

Such a typical characteristic is shown in Fig. 57.20 (*a*). Its slope gives

$$\beta_{ac} = \Delta I_C / \Delta I_B$$

From Fig. 57.20 (*b*), it is seen that a small collector current flows even when $I_B = 0$. It is the common-emitter leakage current $I_{CEO} = (1 + \beta)\, I_{CO}$. Like I_{CO}, it is also due to the flow of minority carriers across the reverse-biased *C/B* junction.

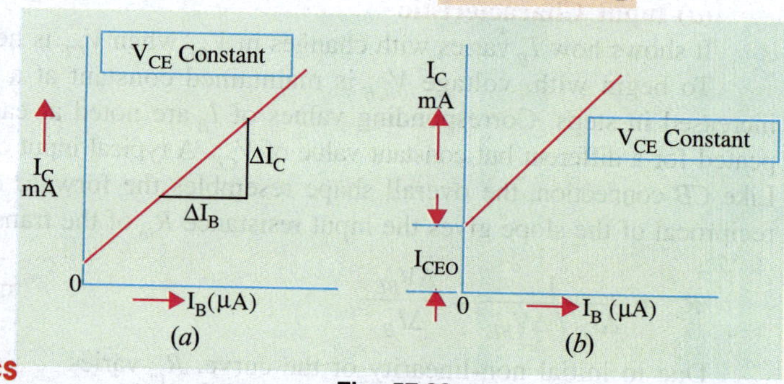

Fig. 57.20

57.19. Common Collector Static Characteristics

As shown in Fig. 57.21, in this case, collector terminal is common carrier to both the input (*CB*) and output (*CE*) carriers circuits.

The output characteristic is I_E versus V_{CE} for several fixed values of I_B. Since $I_C \cong I_E$, this characteristic is practically idential to that of the *CE* circuit and is shown in Fig. 57.22 (*a*).

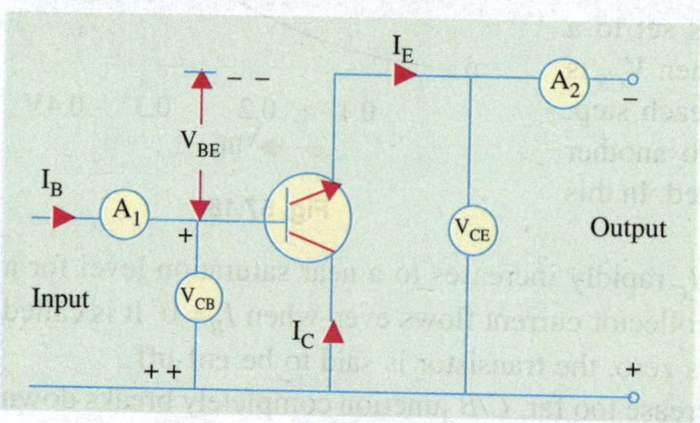

Fig. 57.21

Similarly, its current gain characteristic I_C versus I_B for different values of V_{CE} is similar to that of a *CE* circuit because $I_C \cong I_E$.

The *CC* input characteristic is a plot of V_{CB} versus I_B for different values of V_{CE} and is shown in figure 57.22 (*b*). It is quite different from those for *CB* or *CE* circuit. This difference is due to the fact that input voltage V_{CB} is largely determined by the value of *CE* voltage. Consider the input characteristic for $I_B = 100$ µA and $V_{CE} = 2$ V.

$$V_{CB} = V_{CE} - V_{BE} = 2 - 0.7 = 1.3 \text{ V} \quad \text{— for Si material}$$

Moreover, as V_{CB} is increased, V_{BE} is reduced thereby reducing I_B.

Now, consider the values $V_{CE} = 4$ V and $I_B = 100$ µA

$$V_{CB} = 4 - 0.7 = 3.3 \text{ V}$$

Again, as V_{CB} increases, I_B is decreased.

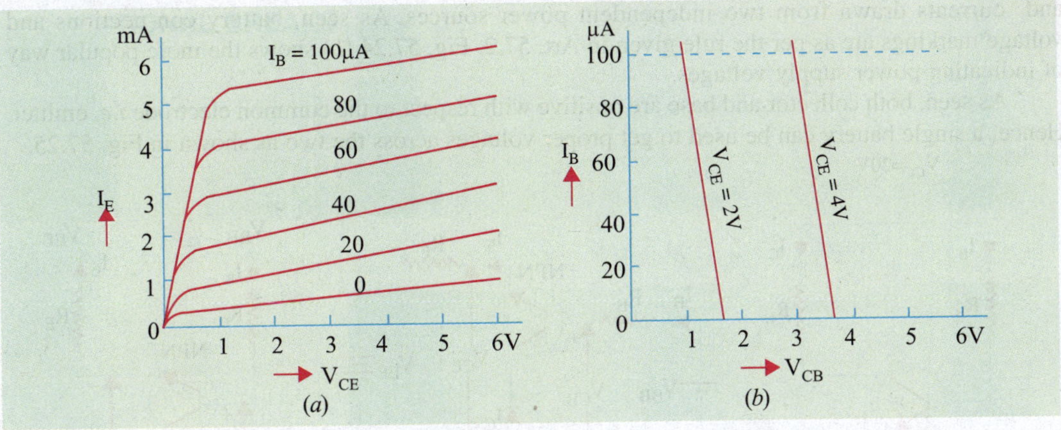

Fig. 57.22

57.20. Different Ways of Drawing Transistor Circuits

In Fig. 57.23 (*a*) is shown a *CB* transistor circuit which derives its voltage and current requirements from two independent power sources *i.e.* two different batteries. Correct battery connections can be done by remembering the transistor polarity rule (Art. 57.2) that in an PNP transistor, both collector and base have to be Negative with respect to the emitter. Of course, collector is a ***little bit more*** –ve than base which means that between themselves, collector is at a ***slightly higher –Ve*** potential with the respect to the base. Conversely, base is at a little lower potential with respect to the collector.

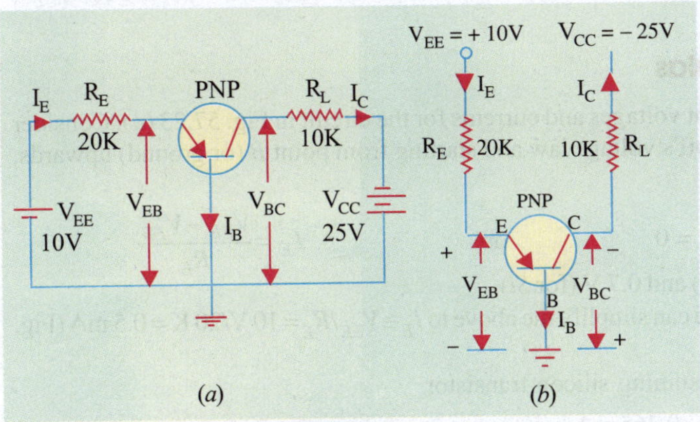

Fig. 57.23

Putting it in a slightly different way, we can say that collector is –Ve w.r.t. base and conversely, base is +Ve w.r.t. collector. That is why, potential difference between collector and base in written as V_{BC} (and not V_{CB}) because terminal at higher potential is mentioned first. Same reasoning applies to V_{EB}. Fig. 57.23 (*b*) shows another and more popular way of indicating power supply voltage. Only one terminal of the battery is shown, the other terminal is ***understood to be grounded so as to provide a complete path for the current.***

For example, +Ve terminal of V_{CC} and – Ve terminal of V_{EE} are supposed to be grounded (as is the base) even though not shown as such in the diagram.

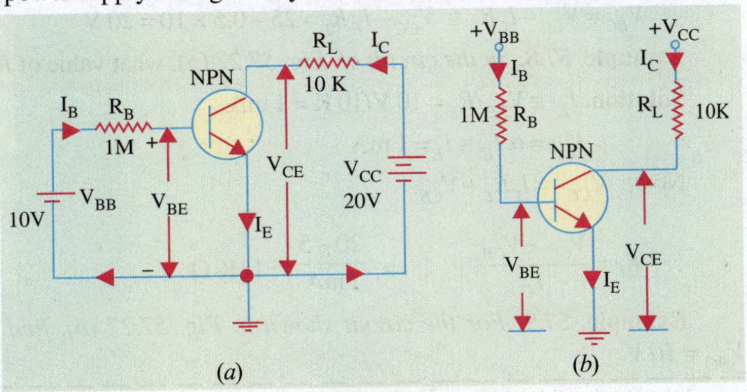

Fig. 57.24

Fig. 57.24 (*a*) shows an *NPN* transistor connected in *CE* configuration with voltages and currents drawn from two independent power sources. As seen, battery con-nections and voltage markings are as per the rule given in Art. 57.2. Fig. 57.24 (*b*) shows the more popular way of indicating power supply voltages.

As seen, both collector and base are positive with respect to the common electrode *i.e.* emitter. Hence, a single battery can be used to get proper voltages across the two as shown in Fig. 57.25.

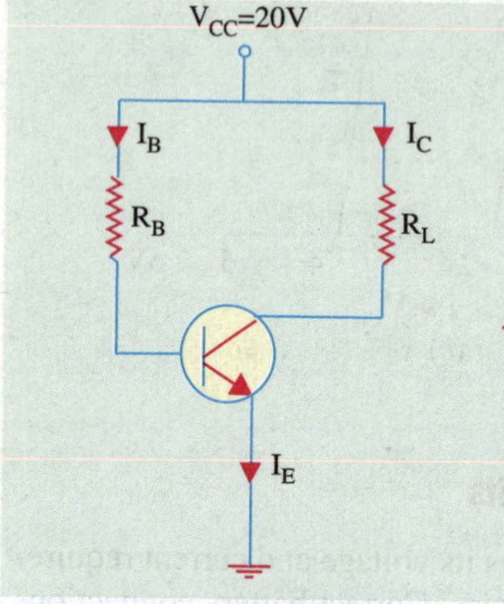

Fig. 57.25

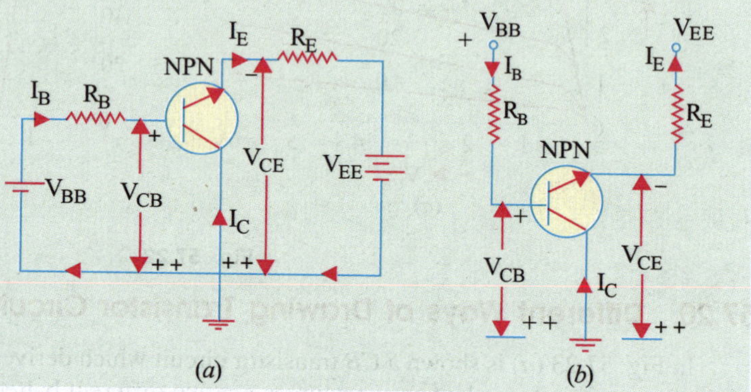

Fig. 57.26

Fig. 57.26 (*a*) shows the *CC* configuration of an *NPN* transistor and Fig. 57.26 (*b*) shows the same circuit drawn differently.

57.21. Common Base Formulas

Let us find the values of different voltages and currents for the circuit in Fig. 57.23 (*a*). Consider the circuit *MEBM*. Applying Kirchhoff's voltage law and starting from point *B* (or ground) upwards, we get

(*a*) $-V_{EB} - I_E R_E + V_{EE}{}^* = 0$ or $I_E = \dfrac{V_{EE} - V_{EB}}{R_E}$

where $V_{EB} = 0.3$ V (for *Ge*) and 0.7 V (for *Si*)

Since, generally, $V_{EE} \gg V_{EB}$, we can simplify the above to $I_E \cong V_{EE}/R_E = 10\text{ V}/20\text{ K} = 0.5$ mA (Fig. 57.23).

Taking V_{EB} into account and assuming silicon transistor

$$I_E = (10 - 0.7)\text{ V}/20\text{ K} = 0.465\text{ mA}$$

(*b*) $I_C = \alpha I_E \cong I_E = 0.5$ mA neglecting leakage current.

(*c*) From circuit *NCBN*, we get

$$V_{BC} = V_{CC} - I_C R_L \cong V_{CC} - I_E R_L = 25 - 0.5 \times 10 = 20\text{ V} \qquad (\because I_C \cong I_E)$$

Example 57.8. *In the circuit of Fig. 57.27* (*a*), what value of R_L causes $V_{CB} = 5$ V?

Solution. $I_E \cong V_{EE}/R_E = 10\text{ V}/10\text{ K} = 1$ mA

$I_C = \alpha I_E \cong I_E = 1$ mA

Now, $V_{CC} = I_C R_L + V_{CB}$

$$\therefore R_L = \frac{V_{CC} - V_{CB}}{I_C} = \frac{20 - 5}{1\text{ mA}} = \mathbf{15K\ \Omega}$$

Example. 57.9. *For the circuit shown in Fig. 57.27 (b), find the value of* R_E *which causes* $V_{BC} = 10$ *V.*

* It is taken positive because we are going from the negative to the positive terminal of the emitter battery.

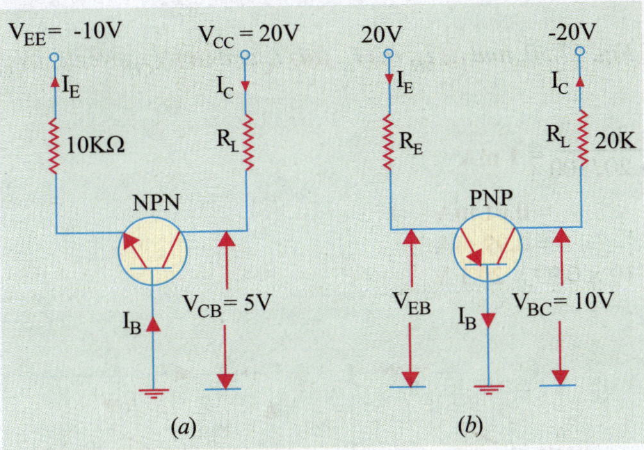

Fig. 57.27

Solution. $I_C = \dfrac{V_{CC} - V_{BC}}{R_L}$

$$= \frac{20-10}{20\text{ K}} = 0.5\text{ mA}$$

Now, $I_E = I_C/\alpha \cong I_C = 0.5$ mA

If we neglect V_{EB}, then entire V_{EE} = 20 V has to be dropped across R_E.

$\therefore \quad 0.5\,R_E = 20$

or $\quad R_E = 20/0.5\text{ mA} = 40\text{ K}\Omega$

57.22. Common Emitter Formulas

Consider the *CE* circuit of Fig. 57.28. Taking the emitter-base circuit, we have

$$I_B = \frac{V_{BB} - V_{BE}}{R_B} \cong \frac{V_{BB}}{R_B}$$

$I_C = \beta\, I_B$ — neglecting leakage current I_{CEO}

$V_{CE} = V_{CC} - I_C R_L$

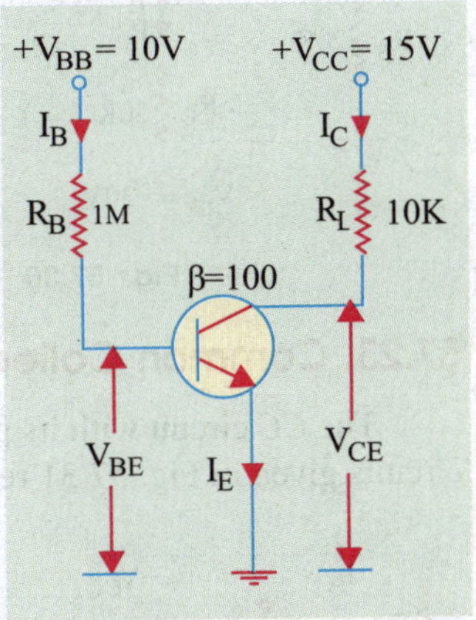

Fig. 57.28

Example 57.10. *For the circuit of Fig. 57.28, find (i) I_B (ii) I_C (iii) I_E and (iv) V_{CE}. Neglect V_{BE}.*

Sol. (*i*) $I_B \cong \dfrac{V_{BB}}{R_B} = \dfrac{10}{1M} =$ **10μA**

(*ii*) $I_C = \beta I_B = 100 \times 10\ \mu A =$ **1 mA**

(*iii*) $I_E = I_B + I_C = 1\text{mA} + 10\ \mu\text{A} =$ **1.01 mA**

(*iv*) $V_{CE} = V_{CC} - I_C R_L = 15 - 1 \times 10 =$ **5 V**

Example 57.11. *Find the exact value of emitter current I_E in the two-supply emitter bias circuit of Fig. 57.29.*

(Electronics-1, Bangalore Univ. 1989)

Solution. Let us apply Kirchhoff´s voltage law to the loop containing R_B, R_E and V_{EE}. Starting from emitter and going clock-wise, we get

$$-I_E R_E + V_{EE} - I_B R_B - V_{BE} = 0$$

or $\quad I_E R_E + I_B R_B = V_{EE} - V_{BE}$ ... (*i*)

Now $\quad \beta = I_C/I_B \cong I_E/I_B \quad \therefore \quad I_B \cong I_E/\beta$

Substituting this value in Eq. (*i*) above, we get

$$I_E R_E + \frac{I_E R_B}{\beta} = V_{EE} - V_{BE}$$

or $$I_E = \frac{V_{EE} - V_{BE}}{R_E + R_B/\beta}$$

Since, in most cases, $(R_B/\beta) \ll R_E$

$\therefore \quad I_E = (V_{EE} - V_{BE})/R_E \cong V_{EE}/R_E$

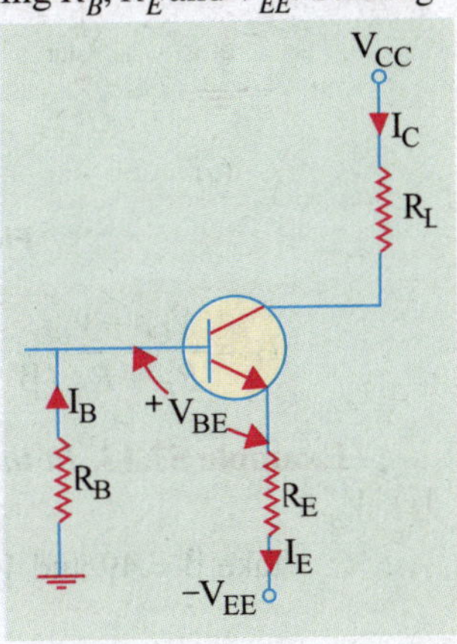

Fig. 57.29

Also, $I_B = I_E/(1+\beta) \cong I_E/\beta$

Example. 57.12. *In the circuit of Fig. 57.30, find* **(i)** I_E, **(ii)** I_B, **(iii)** I_C *and* **(iv)** V_{CE}. *Neglect* V_{BE} *and take* $\beta = 100$.

Sol. (i) $I_E = \dfrac{V_{EE}}{R_E + R_B/\beta} = \dfrac{30}{30 + 20/100} \cong$ **1 mA**

(ii) $I_B \cong I_E/\beta = 1/100$ = **0.01 mA**

(iii) $I_C = I_E - I_B = 1 - 0.01$ = **0.99 mA**

(iv) $V_{CE} = V_{CC} - I_C R_L = 30 - 10 \times 0.99 =$ **20.1 V.**

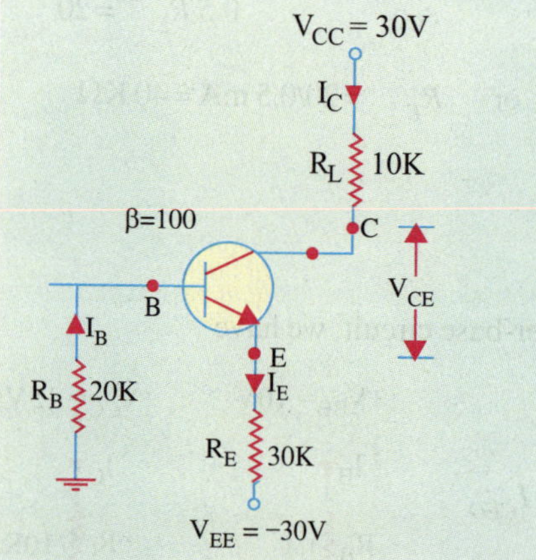

Fig. 57.30

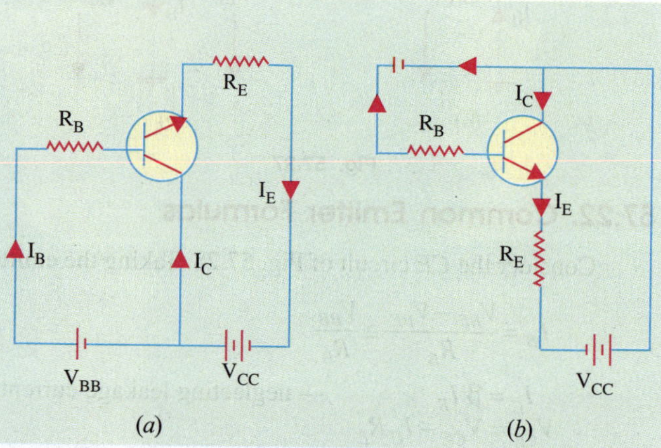

Fig. 57.31

57.23. Common Collector Formulas

The *CC* circuit with its proper d.c. biasing voltage sources is shown in Fig. 57.31 (*a*). The two circuits given in Fig. 57.31 represent the same thing.

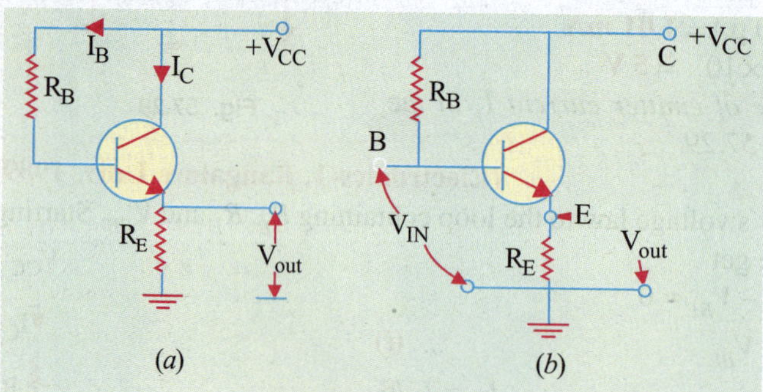

Fig. 57.32

Another way of drawing the same circuit is shown in Fig. 57.32 (*a*) where only one battery has been used. It should be noted that load resistor is not in the collector lead but in the emitter lead as shown.

Fig. 57.32 (*b*) makes the circuit connection quite clear. Input is between base and collector terminals whereas output is between emitter and collector terminals.

It is seen that

$$I_E = \frac{V_{CC} - V_{BE}}{R_E + R_B/\beta}; \quad V_{CC} = V_{CE} + I_E R_E; \quad I_E = \frac{V_{CC} - V_{BE}}{R_E + \beta R_E}; \quad I_C = \beta I_B$$

Example 57.13. *In the CC circuit of Fig. 57.33, find* **(a)** I_B, **(b)** I_E, **(c)** V_{CE}, **(d)** V_E and **(e)** V_B.

Take $\beta = 49$ and $V_{BE} = 0.7$ V.

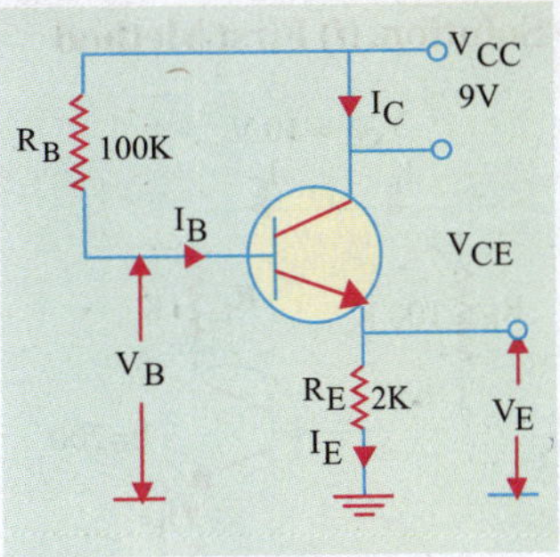

Fig. 57.33

Solution. (*a*) $I_B = \dfrac{V_{CC} - V_{BE}}{R_B + (1+\beta)R_E}$

$$= \frac{9.0 - 0.7}{100 + 50 \times 2} = \mathbf{41.5\ \mu A}$$

(*b*) $I_E = (1 + \beta)\, I_B = 50 \times 41.5 = \mathbf{2.075\ mA}$

(*c*) $V_{CE} = V_{CC} - I_E R_E = 9 - 2.075 \times 2 = \mathbf{4.85\ V}$

(*d*) $V_E = I_E R_E = 2.075 \times 2 = \mathbf{4.15\ V}$

(*e*) $V_B = V_{BE} + I_E R_E = 0.7 + 4.15 = \mathbf{4.85\ V}$

57.24. The Beta Rule

According to this rule, resistance from one part of a transistor circuit can be referred to another of its parts (as we do with the primary and secondary winding impedances of a transformer). For example, resistance R_L in the collector circuit can be referred to the base circuit and ***vice versa.*** Similarly, R_E can be referred to the base circuit and, reciprocally, R_B can be referred to the emitter circuit. Since current through R_L is I_C (= β I_B), hence β-factor comes into the picture. Similarly, current through R_E is I_E which is (1 + β) times I_B, hence (1 + β) or approximately β-factor comes into the picture again, Use of this 'β-rule' makes transistor circuit calculations quite quick and easy. It makes the calculation of I_B quite simple.

The 'β-rule' may be stated as under :

1. When referring R_L or R_C to the base circuit, **multiply** it by β. When referring R_B to the collector circuit, **divide** it by β.
2. When referring R_E to base circuit, **multiply** it by (1 + β) or just β (as a close approximation).
3. Similarly, when referring R_B to emitter circuit, **divide** it by (1 + β) or β.

Before you apply this rule to any circuit, you must remember one very important point otherwise you are likely to get wrong answers. The point is that ***only those resistances are transferred which lie in the path of the current being calculated.*** Not otherwise. The utility of this rule will be demonstrated by solving the following problems.

Example 57.14. *Calculate the value of* V_{CE} *in the collector stabilisation circuit of Fig. 57.34.*

Solution. We will use β-rule to find I_C in the following two ways.

(*i*) First Method

Here, we will transfer R_L to the base circuit.

$$I_B = \frac{V_{CC}}{R_B + \beta R_L} = \frac{20}{1000 + 100(10)} = 10\,\text{mA}$$

$$I_C = \beta I_B = 100 \times 10 = 1000\ \text{mA} = 1\ \text{A}$$

$$V_{CE} \cong V_{CC} - I_C R_L = 20 - 1 \times 10 = \mathbf{10\ V}$$

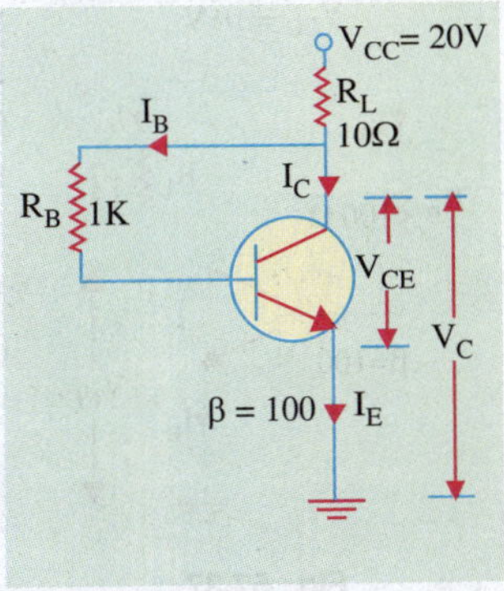

Fig. 57.34

(*ii*) Second Method

Now, we will refer R_B to collector circuit.

$$I_C \cong \frac{V_{CC}}{R_L + R_B/\beta} = \frac{20}{10 + (1000/100)} = 1\,\text{A}$$

$$V_{CE} = V_{CC} - I_C R_L = 10\,\text{V} \qquad \text{— as above}$$

It was a simple circuit because $R_E = 0$ and R_B was connected to V_{CC} through R_L and not directly (in which case, R_L would not lie in the path of I_B). Now, we will consider the case when R_E is present and R_L does not lie in the path of I_B.

Example 57.15. *Calculate the three transistor currents in the circuit of Fig. 57.35.*

Solution. ***(i)* First Method**

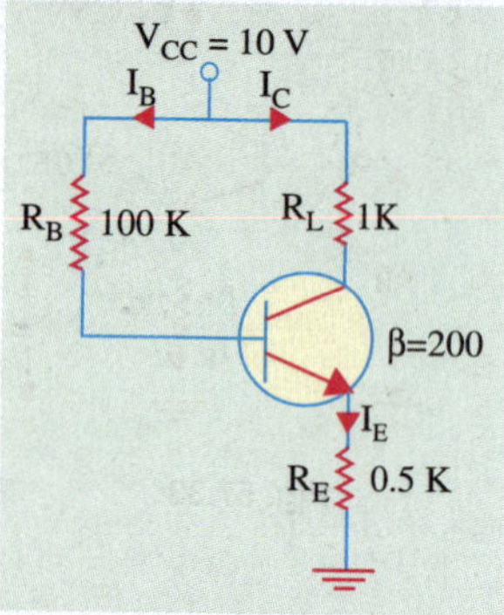

Fig. 57.35

Since R_E lies in the path of I_B

$$\therefore \quad I_B = \frac{V_{CC}}{R_B + \beta R_E} \quad \text{— neglecting } V_{BE}$$

$$= \frac{10}{100 + 200\,(0.5)} = \mathbf{0.05\ mA}$$

$$I_C = \beta I_B = 200 \times 0.05 = \mathbf{10\ mA}, \quad I_E = I_B + I_C = \mathbf{10.05\ mA}$$

***(ii)* Second Method**

Now, we will transfer R_B to emitter circuit and find I_E directly.

$$I_E = \frac{V_{CC}}{R_E + R_B/\beta} = \frac{10}{0.5 + 100/200} = 10\text{ mA (approx) — as before}$$

$$I_B = I_C/\beta \cong 10/200 = \mathbf{0.05\ mA}$$

Example 57.16. *Calculate I_E in the circuit of Fig. 57.36.* **(Electronic & Commu., Ranchi Univ. 1990)**

Solution. If we neglect V_{BE}, then as seen from the circuit of Fig. 57.36.

$$I_E = \frac{V_{EE}}{R_E + R_B/\beta} = \frac{10}{10 + 10/100} = \mathbf{0.99\ mA}$$

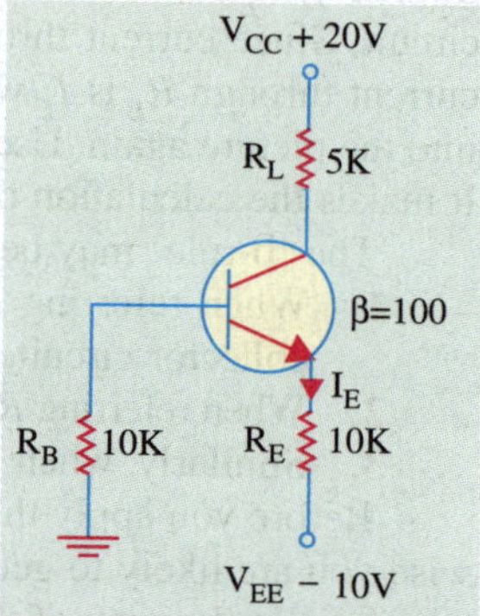

Fig. 57.36

57.25. Importance of V_{CE}

The voltage V_{CE} is very important in checking whether the transistor is

(a) defective, **(b)** working in cut-off,

(c) in saturation or well into saturation (Example 57.17 and 57.18)

When $V_{CE} = V_{CC}$, the transistor is in cut-off *i.e.* it is turned OFF. When $V_{CE} = 0$, the transistor is in saturation *i.e.* it is turned fully ON. When V_{CE} is less than zero *i.e.* negative, the transistor is said to be well into saturation. In practice, both these conditions are avoided. For amplifier operation, $V_{CE} = \frac{1}{2} V_{CC}$ *i.e.* transistor is operated at approximately $\frac{1}{2}$ ON. In this way, variations in I_B in either direction will control I_C in both directions. In other words, when I_B increases or decreases, I_C also increases or decreases. However, if I_B is OFF, I_C is also OFF. On the other hand, if collector has been turned fully ON, maximum I_C flows. Hence, no further increase in I_B can be reflected in I_C.

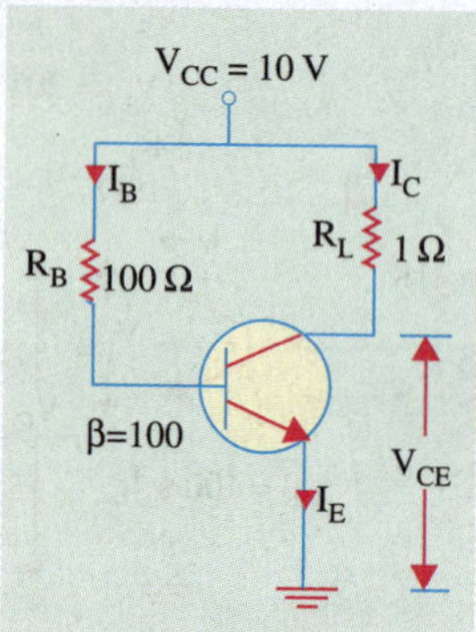

Fig. 57.37

Example 57.17. *For the CE circuit of Fig. 57.37, find the value of V_{CE}. Take β = 100 and neglect V_{BE}. Is the transistor working in cut-off or saturation ?*

Solution.

$$I_B = 10/100 = 0.1\text{ A}$$

$$I_C = \beta I_B = 100 \times 0.1 = 10\text{ A}$$

$$V_{CE} = V_{CC} - I_C R_L = 10 - 10 \times 1 = 0$$

Obviously, the transistor is operating just at saturation and not well into saturation.

Example 57.18. *Find out whether the transistor of Fig. 57.38 is working in saturation or well into saturation. Neglect V_{BE}.* **(Basic Electronics, Bombay Univ.)**

Solution.

$$I_B = 10/10 = 1\text{ A}$$

$$I_C = 100 \times 1 = 100\text{ A}$$

Obviously, I_C cannot be that large because its maximum value is given by $V_{CC}/R_L = 10/1 = 10$ A. However, let us assume that I_C takes this value temporarily. Then,

$$V_{CE} = V_{CC} - I_C R_L$$
$$= 10 - 100 \times 1 = -90 \text{ V}$$

It means that the transistor **is working well into saturation.**

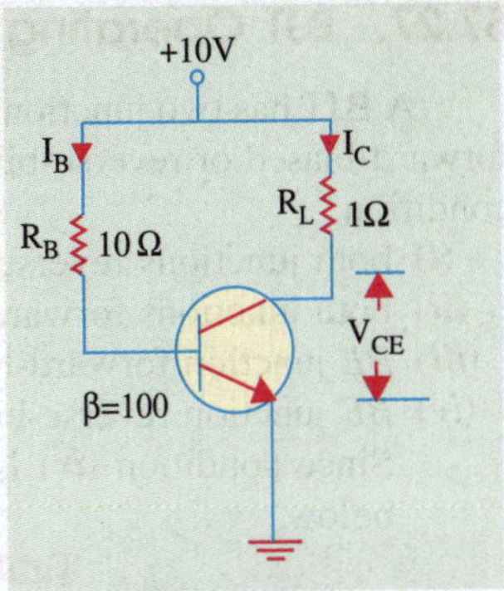

Fig. 57.38

57.26. Cut-Off And Saturation Points

Consider the circuit of Fig. 57.39 (*a*). As seen from Art 57.22,

$$V_{CE} = V_{CC} - I_C R_L.$$

Since, $I_B = 0$, $\quad \therefore \quad I_C = 0.$

Hence, $V_{CE} = V_{CC}$

Under these conditions, the transistor is said to be cut-off for the simple reason that it does not ***conduct any current.*** This value of V_{CE} is written as $V_{CE\,(cut\text{-}off)}$. Incidentally, a transistor when cut-off acts like an open switch.

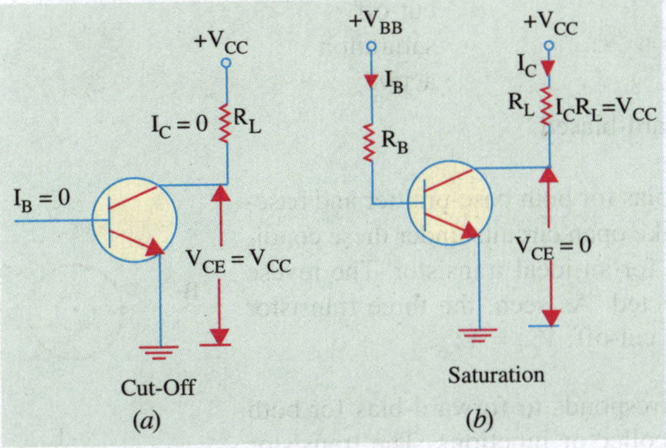

Fig. 57.39

If, in Fig. 57.39 (*b*), values of R_B and R_L are such that V_{CE} comes out to be zero, then transistor is said to be saturated. Putting $V_{CE} = 0$ in the above equation, we get

$$0 = V_{CC} - I_C R_L$$

or $\quad I_C = V_{CC} / R_L$

It should be noted that a transistor, when saturated, acts as a closed switch of negligible resistance.

It is obvious that under saturation condition,

(*i*) whole of V_{CC} drops across R_L.

(*ii*) collector current has maximum possible value called I_C (*sat*).

Normal operation of a transistor lies between the above two extreme conditions of cut-off and saturation.

Example. 57.19. *In a simple amplifier circuit (Fig. 57.40) with base resistance, $R_B = 50$ K, $R_E = 2$ K, $R_C = 3$K, $V_{CC} = 10$ V, $h_{FE} = 100$, $\beta = 100$ determine whether or not the silicon transistor is in the saturation and find I_B and I_C. Explain the saturation region in common-emitter characteristics.* **(Electronics, MS. Univ. Baroda,)**

Solution. Whether the transistor is in saturation or not will depend on the value of V_{CE}.

$$I_E = \frac{V_{BB} - V_{BE}}{R_E + R_B/\beta} \cong \frac{V_{BB}}{R_E + R_B/\beta}$$
$$= \frac{5}{2 + 50/100} = 2 \text{ mA}$$

$I_C \cong I_E = 2$ mA ; $I_B = I_C/\beta = 2/100$
$= 0.02$ mA

Now, $\quad V_{CC} = I_C R_C + V_{CE} + I_E R_E$

or $\quad V_{CE} = 10 - (2 \times 3) - (2 \times 2) = 0$

Obviously, the transistor has entered in saturation.

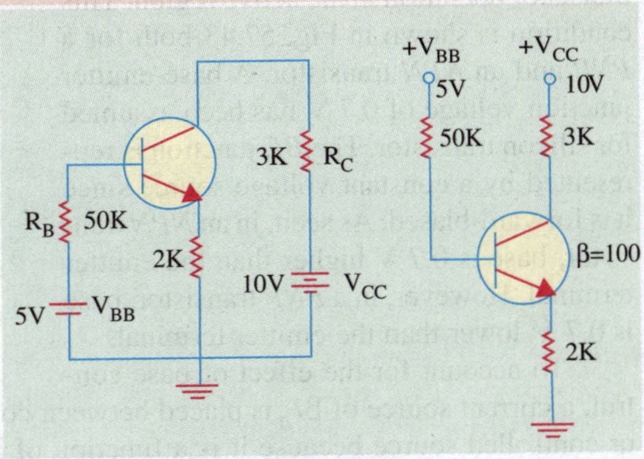

Fig. 57.40

57.27. BJT Operating Regions

A BJT has two junctions *i.e.* base-emitter and base-collector junctions either of which could be forward-biased or reverse-biased. With two junctions, there are four possible combinations of bias condition.

(*i*) both junctions reverse-biased,
(*ii*) both junctions forward-biased,
(*iii*) *BE* junction forward-biased, *BC* junction reverse-biased.
(*iv*) *BE* junction reverse-biased, *BC* junction forward-biased.

Since condition (*iv*) is generally not used, we will tabulate the remaining three conditions below.

Table No. 57.1: Transistor Operation Regions

BE Jn	BC Jn	Region
RB*	RB	cut-off
FB**	FB	saturation
FB	RB	active

* Reverse-biased, ** Forward-biased

(*a*) Cut-off

This condition corresponds to reverse-bias for both base-emitter and base-collector junctions. In fact, both diodes act like open circuits under these conditions as shown in Fig. 57.41, which is true for an ideal transistor. The revese leakage current (Art 57.12) has been neglected. As seen, the three transistor terminals are uncoupled from each other. In cut-off, $V_{CE} = V_{CC}$.

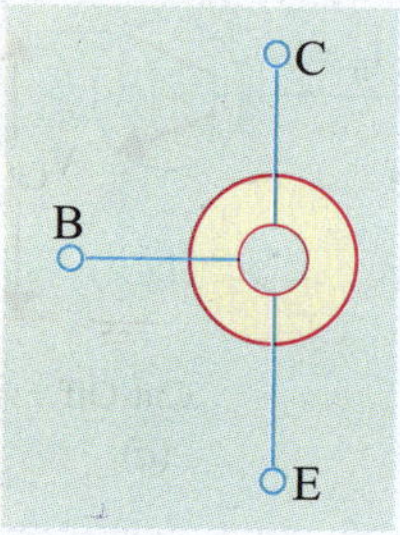

Fig. 57.41

(*b*) Saturation

This condition corresponds to forward-bias for both base-emitter and base-collector junctions. The transistor becomes saturated *i.e.* there is perfect short-circuit for both base-emitter and base-collector diodes. The ideal case is shown in Fig. 57.42, where the three transistor terminals have been connected together thereby acquiring equal potentials. In this case, $V_{CE} = 0$.

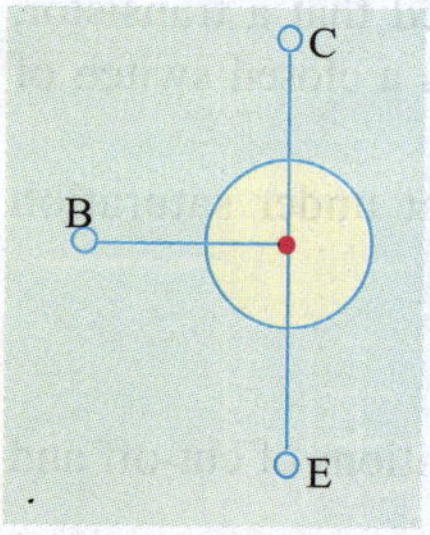

Fig. 57.42

(*c*) Active Region

This condition corresponds to forward-bias for base-emitter junction and reverse bias for base-collector junction. In this, $V_{CE} > 0$.

57.28. Active Region DC Model of a BJT

Such a model is used for predicting transistor operation in the active region. This condition is shown in Fig. 57.43 both for a *PNP* and an *NPN* transistor. A base-emitter junction voltage of 0.7 V has been assumed for silicon transistor. The *BE* junction is represented by a constant voltage source since it is forward-biased. As seen, in an *NPN* transistor, base is 0.7 V higher than the emitter terminal. However, in a *PNP* transistor, base is 0.7 V lower than the emitter terminal.

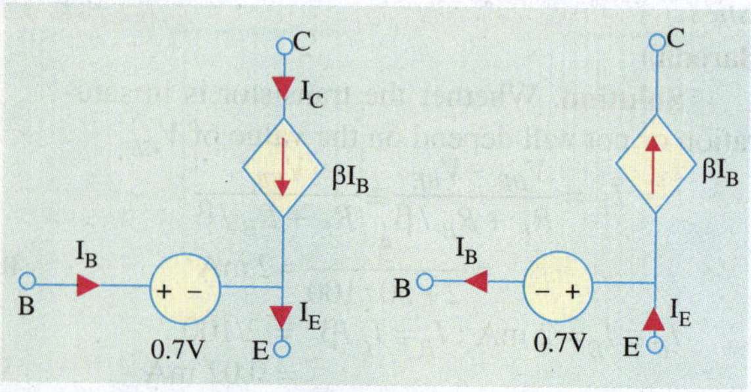

Fig. 57.43

To account for the effect of base control, a current source of βI_B is placed between collector and base terminals. It is called a dependent or controlled source because it is a function of a variable in another circuit. It may be noted that $I_E = (I_B + I_C)$ in both cases.

57.29. BJT Switches

Very often, bipolar junction transistors are used as electronic switches. With the help of such a switch, a given load can be turned ON or OFF by a small control signal. This control signal might be the one appearing at the output of a digital logic or a microprocessor. The power level of the control signal is usually very small and, hence, it is incapable of switching the load directly. However, such a control signal is certainly capable of providing enough base drive to switch a transistor ON or OFF and, hence, the transistor is made to switch the load.

When using BJT as a switch, usually two levels of control signal are employed. With one level, the transistor operates in the cut-off region (open) whereas with the other level, it operates in the saturation region and acts as a short-circuit. Fig. 57.44 (*b*) shows the condition when control signal $V_i = 0$. In this case, the *BE* junction is reverse-biased and the transistor is open and, hence acts as an open switch. However, as shown in Fig. 57.44 (*c*) if V_i equals a positive voltage of sufficient magnitude to produce saturation *i.e.* if $V_i = V_i$ the transistor acts as a closed switch.

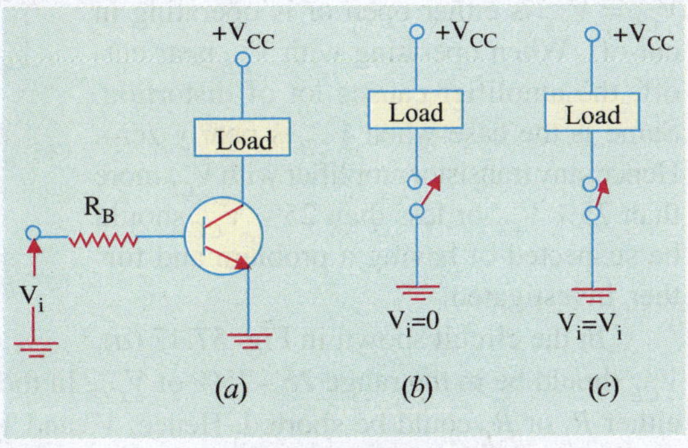

Fig. 57.44

Fig. 57.45 shows a form of series switching circuit utilizing an PNP transistor with a negative dc supply and a control signal voltage having levels of zero and $- V_i$.

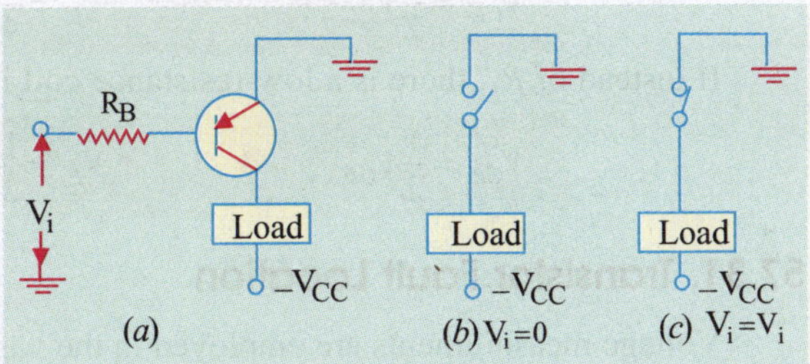

Fig. 57.45

Example 57.20. *The circuit of Fig. 57.46 is designed to produce nearly constant current through the variable collector load resistance. An ideal 6V source is used to establish the current. Determine* **(a)** *value of I_C and V_E,* **(b)** *range of R_C over which the circuit will function properly. Assume silicon transistor and α β large enough to justify the assumptions used.*

(Applied Electronics-II, Punjab Univ. 1993)

Solution. (*a*) $I_C \cong I_E = (6 - 0.7)/530 = $ **10 mA**

$V_E = 530 \times (10 \times 10^{-3}) =$ **5.3 V**.

This voltage will remain constant so long as transistor operation is confined to active region.

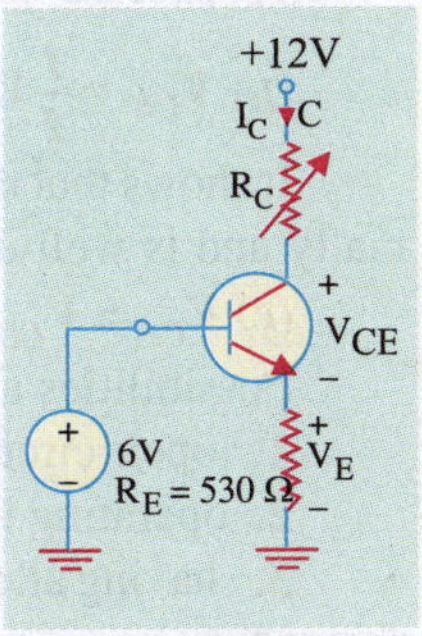

Fig. 57.46

(*b*) When $R_C = 0$

$$V_{CE} = 12 - 5.3 = 6.7 \text{ V}$$

It is certainly well within the active region. As R_C increases, its drop increases and hence, V_{CE} decreases. There will be some value of R_C at which active region operation ceases.

Now, $V_{CE} = 12 - 5.3 - I_C R_C = 6.7 - I_C R_C$

Value of $R_{C(max)}$ can be found by puting $V_{CE} = 0$

∴ $0 = 6.7 - I_C R_{C(max)}$

or $R_{C(max)} = 6.7/I_C = 6.7/0.01 =$ **670 Ω**

Hence, circuit will function as a constant current source so long as R_C is in the range $0 < R_C <$ 670 Ω. When R_C exceeds 670 Ω, the *BJT* becomes saturated.

57.30. Normal DC Voltage Transistor Indications

For a transistor to operate as an amplifier, it is desirable that $V_{CE} = \frac{1}{2} V_{CC}$. However, in actual practice, wide tolerances are allowed. Generally, V_{CE} varies between 25% to 75% of V_{CC}. Any transistor amplifier with $V_{CE} = V_{CC}$ is either open or is operating in cut-off. When operating with V_{CE} near cut-off, the amplifier causes lot of distortion. Same is the case when V_{CE} is nearly zero. Hence, any transistor amplifier with V_{CE} more than 75% V_{CC} or less than 25% V_{CC} should be suspected of having a problem and further investigated.

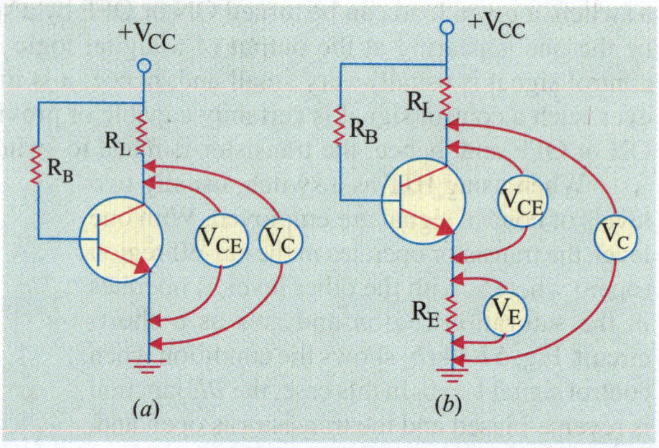

Fig. 57.47

In the circuit shown in Fig. 57.47 (*a*), V_{CE} should be in the range 25 –75% of V_{CC}. In the circuit of Fig. 57.47 (*b*), V_{CE} may be normal but either R_L or R_E could be shorted. Hence, V_C and V_E should be measured seperately. Moreover, V_{CE} could be found by subtracting V_E from V_C.

For the circuit of Fig. 57.47 (*b*), the normal voltmeter readings are

$$V_{CE} = \frac{1}{2} V_{CC} \;; \qquad V_E = \frac{1}{4} V_{CC} \;; \qquad V_C = \frac{3}{4} V_{CC}$$

If instead of R_L, there is a low-resistance coil in the circuit, then

$$V_{CE} = \frac{1}{2} V_{CC} \;; \qquad V_E = \frac{1}{2} V_{CC} \;; \qquad V_C = V_{CC}$$

57.31. Transistor Fault Location

Voltage measurements are employed in the vast majority of trouble situations because current measurements are comparatively difficult to make. Magnitude of V_{CE} is of great diagnostic value in finding and locating faults in a transistor circuit. Following possibilities are considered :

(a) $V_{CE} = 0$

Possibilities are that the transistor is

1. shorted out, **2.** operating in saturation, **3.** disconnected from V_{CC}

(b) $V_{CE} = \frac{1}{2} V_{CC}$

It shows that the circuit is operating normally and is well-designed.

(c) $V_{CE} = V_{CC}$

Possibilities are that the transistor is

1. open-circuited,

2. operating in cut-off

3. having all resistors in series with V_{CE} shorted.

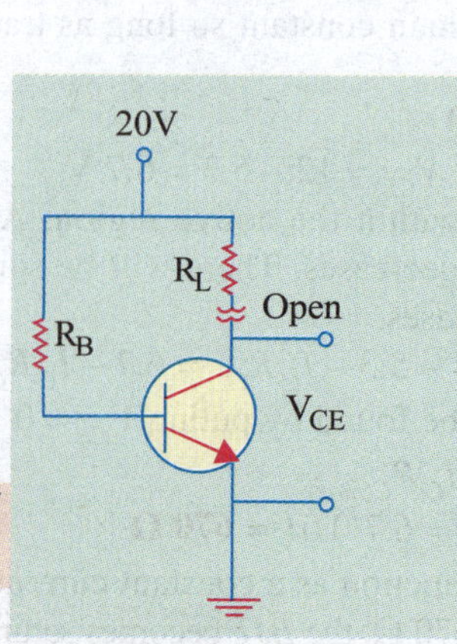

Fig. 57.48

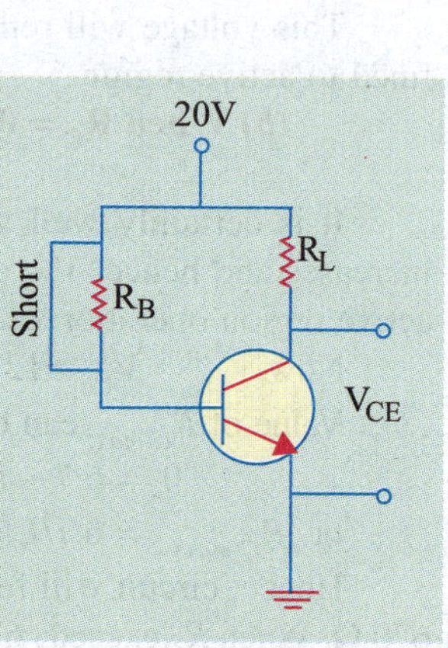

Fig. 57.49

Example 57.21. *Compute the value of V_{CE} for the CE circuit shown in Fig. 57.48.*

Solution. Since the collector is disconnected from the supply due to 'open' in the circuit, $V_{CE} = 0$. It represents fault condition No. (*a*) 3 in Art. 57.31.

Example. 57.22. *What is the value of V_{CE} in the CE circuit of Fig. 57.49.*

Solution. Since R_B is shorted out, I_B, would increase and probably burn out the *E/B* junction. But this burn out is not indicated in the question. Hence, with high base current, the transistor is operating in saturation so that $V_{CE} = 0$.

Example 57.23. *What is the value of V_{CE} in the circuit of Fig. 57.50.*

Solution. Since R_L is the only resistor in series with the transistor and is shorted out, it means that there is no voltage drop anywhere. Hence, $V_{CE} = V_{CC}$. It represents fault No. (*c*) 3 stated in Art. 7.31 above.

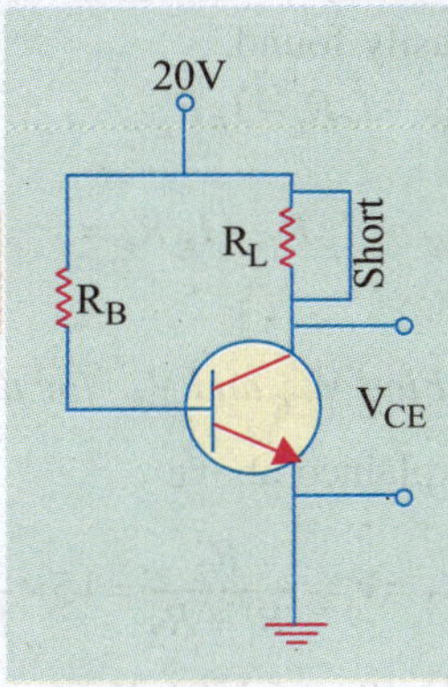

Fig. 57.50

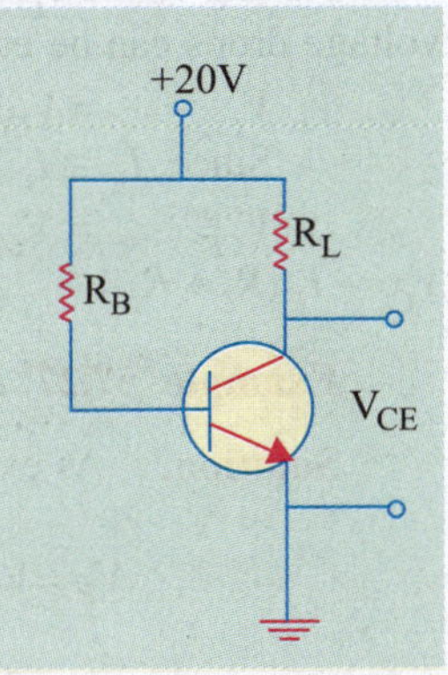

Fig. 57.51

Example 57.24. *Find the possible value of V_{CE}, V_C and V_E for the circuit shown in Fig. 57.51.*

Solution. In the circuit of Fig. 57.51, there is neither a short nor an open and the voltage polarities are correct for an *NPN* transistor. It looks like a well-designed circuit operating normally. Hence, according to Art. 57.30.

$$V_{CE} = \frac{1}{2}V_{CC} = \mathbf{10\ V};\quad V_E = \frac{1}{4}V_{CC} = \mathbf{5\ V}$$

$$V_C = \frac{3}{4}V_{CC} = \mathbf{15\ V}$$

Example 57.25. *Find the values of V_C, V_E and V_{CE} in the circuit of Fig. 57.52.*

Solution. Since $I_B = 0$, transistor is cut off.

Hence, $V_E = \mathbf{0}$

Also $V_C = \mathbf{-20\ V}$

and $V_{CE} = \mathbf{-20\ V}$

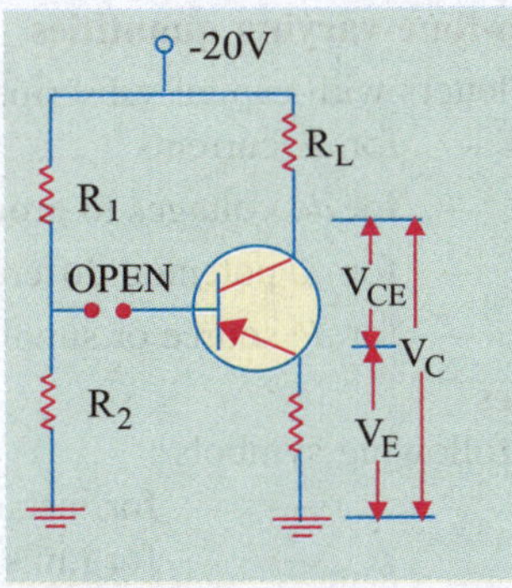

Fig. 57.52

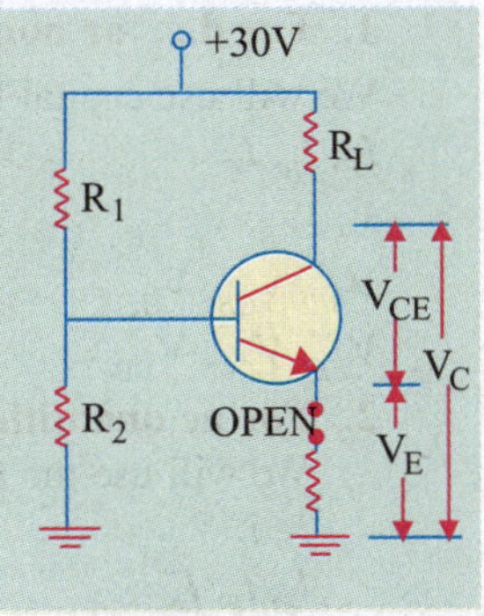

Fig. 57.53

Example 57.26. *What would be the values of V_C, V_E and V_{CE} for the circuit shown in Fig. 57.53.*

Solution. Since emitter is open, no current flows in any part of the circuit. The transistor is essentially cut off. Without IR drops, all points above the emitter are at 30 V.

$\therefore$ $V_E = \mathbf{30\ V}$;

$V_C = \mathbf{30\ V}$ and

$V_{CE} = \mathbf{0\ V}$

57.32. Solving Universal Stabilization Circuit

Such a circuit is shown in Fig. 57.54 in which ***R_E appears*** to be in parallel with R_2. But according to the β-rule (Art 57.24), R_2 is actually in parallel with $\beta\, R_E$. In a well-designed circuit, the resistance $\beta\, R_E$ is much larger than R_2. Hence, their combined resistance $= R_2 \,||\, \beta\, R_E \cong R_2$. On this assumption as well as another that I_B is practically zero, we can find voltage drop across R_2 by the Proportional Voltage Formula. Since V_{CC} is applied across $R_1 - R_2$ potential divider circuit, drop across R_2.

$$= V_{CC} \,.\, R_2/(R_1 + R_2)$$

If we neglect V_{BE}, then this drop equals V_E.

$$\therefore \qquad V_E \cong V_{CC}\frac{R_2}{R_1 + R_2}$$

and
$$I_E \cong \frac{V_E}{R_E}$$

Having found I_E, other currents and voltage drops can be easily found.

$$V_{CE} = V_{CC} - I_C R_L - I_E R_E$$

Since, $I_E \cong I_C$

$$\therefore V_{CE} = V_{CC} - I_E R_L - I_E R_E = V_{CC} - I_E (R_L + R_E)$$

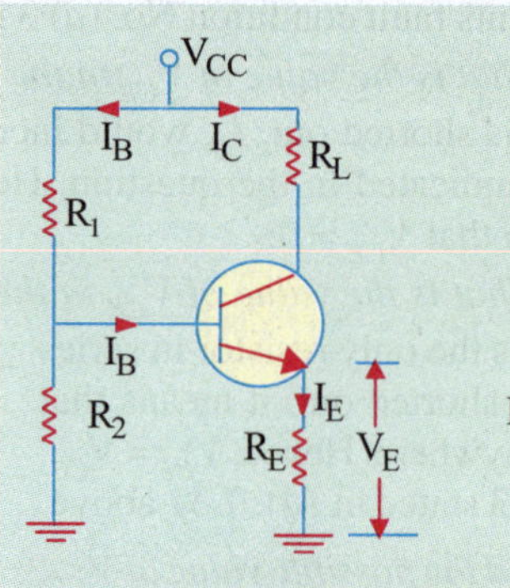

Fig. 57.54

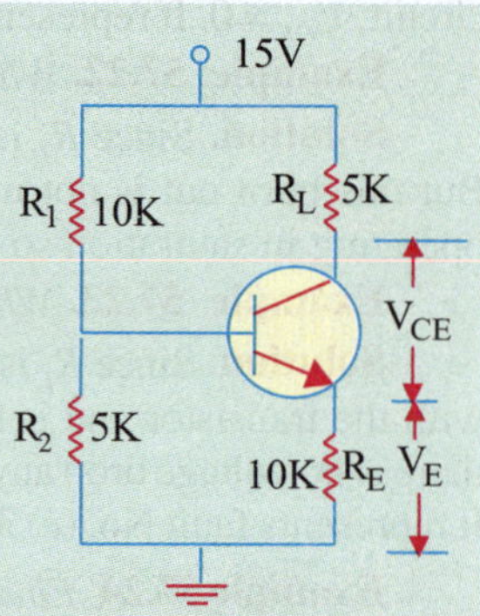

Fig. 57.55

Example 57.27. *Find V_{CE} and V_E for the circuit shown in Fig. 57.55. Neglect V_{BE}.*

Solution. As explained above

$$V_E = V_2 = V_{CC}\frac{R_2}{R_1 + R_2} = 15 \times \frac{5}{15} = \mathbf{5\ V}$$

$$I_E = V_E/R_E = 5\ \text{V}/10\ \text{K} \quad = 0.5\ \text{mA}$$

$$I_C \cong I_E = 0.5\ \text{mA},\ V_{CE} = V_{CC} - I_E(R_L + R_E) = 15 - 0.5 \times 15 = \mathbf{7.5\ V}$$

57.33. Notation for Voltages and Currents

In order to avoid confusion while dealing with dc and ac voltages and currents, following notation will be employed :

1. For d.c. or non-time-varying quantities

We will use capital letters with capital subscripts such as

I_E, I_B, I_C — for *dc* currents
V_E, V_B, V_C — for *dc* voltages to ground
V_{BE}, V_{CB}, V_{CE} — for *dc* potential differences
V_{EE}, V_{CC}, V_{BB} — for *dc* source or supply voltages

2. For ac quantities

We will use the following symbols :

i_e, i_b, i_c — for instantaneous values of ac currents
I_e, I_b, I_c — for r.m.s values of a.c. currents
v_e, v_b, v_c — for instantaneous values of a.c. voltages to ground
v_{be}, v_{eb}, v_{ce} — for a.c. voltage differences

3. Total ac and dc voltages and currents

In this case, we will use a hybrid notation. For example, i_E will be used to represent the total emitter current, *i.e.* sum of dc and ac currents in the emitter.

Fig. 57.56 illustrates the notation discussed above.

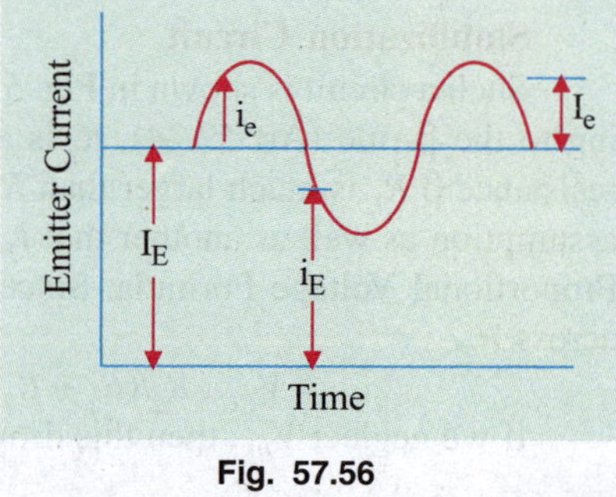

Fig. 57.56

57.34. Increase/Decrease Notation

This notation is very helpful in analysing transistor operation when ac signal is applied to it. It is simply this:
↑ means increases and ↓ means decrease.

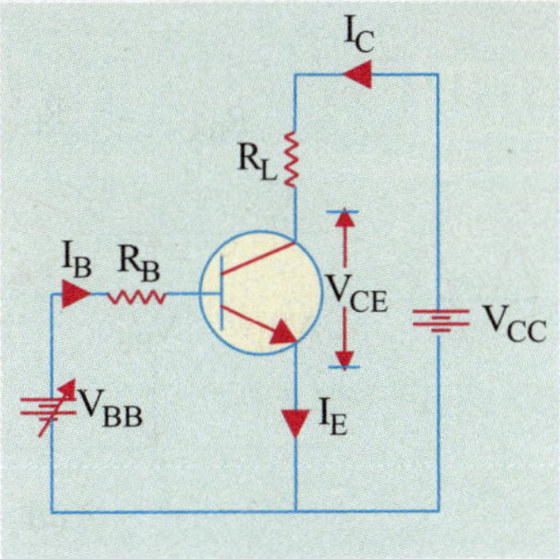

Fig. 57.57

As an illustration, consider the transistor circuit of Fig. 57.57.

If V_{BB} were increased (↑), I_B would increase (↑). This would increase I_C (↑) because it equals β I_B. The drop $I_C R_L$ would increase (↑) and, hence, V_{CE} will decrease (↓) because $V_{CE} = V_{CC} - I_C R_L$.

Using increase/decrease notation, the above sequence of changes can be written as

$$V_{BB}\uparrow, I_B\uparrow, I_C\uparrow, I_C R_L\uparrow V_{CE}\downarrow$$

At one look, we can straight away say that as input voltage is increased, output voltage is decreased.

57.35. Applying AC to a DC Biased Transistor

Suppose we want to apply an ac signal to the input emitter-base circuit of a properly-biased transistor shown in Fig. 57.58. If we apply the a.c. source directly across the *EBJ* as shown in Fig. 57.58 (*a*), it will upset the d.c. bias. It should be kept in mind that most ***ac signal sources are nearly a short to dc.*** Hence, nearly whole of I_B would pass through a.c. source rather than the base thereby spoiling the transistor bias.

In order to connect the ac source and at the same time not upset the d.c. bias, the ac source is connected via a coupling capacitor *C* as shown in Fig. 57.58 (*b*). This capacitor acts as an 'open' for dc but almost a short for ac source provided it is of sufficiently large capacitance.

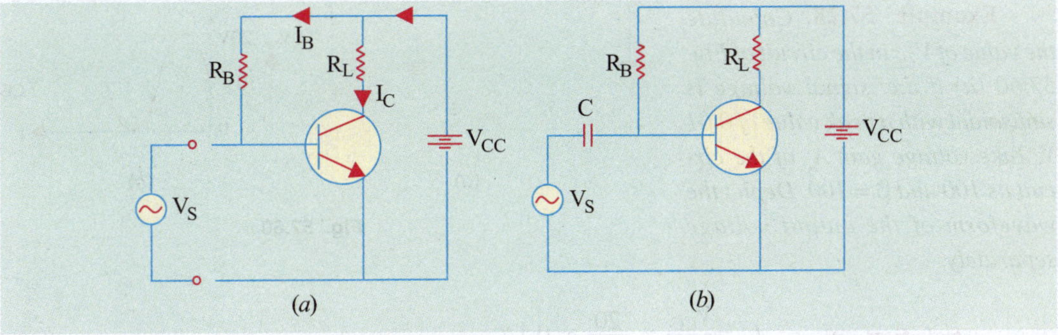

Fig. 57.58

57.36. Transistor AC/DC Analysis

In Fig. 57.59 is shown a *CE* amplifier circuit having an ac signal voltage v_{be}* applied across its *E/B* junction. This voltage will be added to the dc voltage V_{BE} as if the two were connected in series. The resultant voltage is shown in Fig. 57.59 (*b*) which shows ac voltage riding the d.c. level. The variations in the resultant output voltage V_{CE} [Fig. 57.59 (*b*)] can be expressed in terms of the increase/decrease notation. It will be assumed that V_{BE} is such as to bias V_{CE} at V_{CC} when no a.c. signal is applied.

(i) First Quarter Cycle

In the first quarter-cycle of the input signal, both V_{BE} and V_{be} increase thereby giving rise to the following sequence of changes :

$$V_{BE}\uparrow, i_B\uparrow, i_C\uparrow, i_C R_L\uparrow V_{CE}\downarrow$$

Hence, output voltage decreases as shown in Fig. 57.59 (*c*)

(ii) Second Quarter Cycle

Here, V_{be} as well as V_{BE} decrease. Hence,

$$V_{BE}\downarrow, i_B\downarrow, i_C\downarrow, i_C R_L\downarrow, V_{CE}\downarrow$$

* Normally, we will use the notation υ_i or cin or c_i while discussing amplifiers.

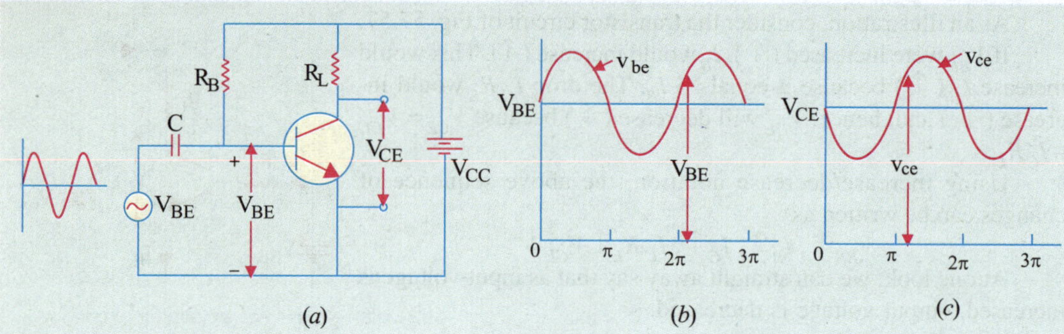

Fig. 57.59

Again, V_{CE} does the opposite of V_{BE}.

Same changes will happen in third quarter cycle as happened in the first quarter-cycle and so on. It is seen from Fig. 57.60 (*c*) that output *ac* voltage is 180° out of phase with the input voltage.

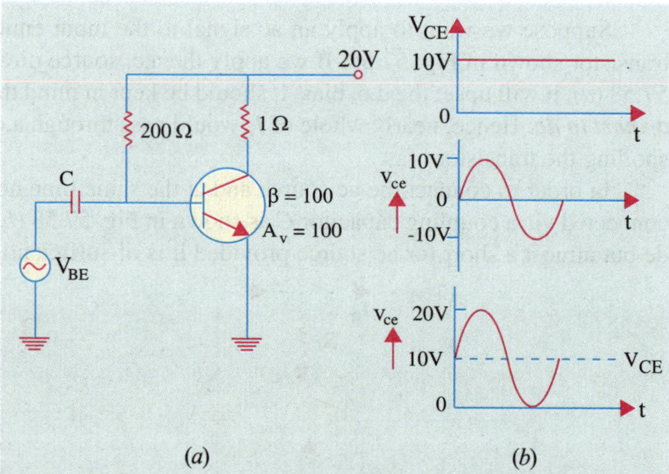

Fig. 57.60

Example 57.28. *Calculate the value of V_{CE} in the circuit of Fig. 57.60 (a) if a.c. signal voltage is sinusoidal with a peak value of 0.01 V. Take voltage gain A_v of the circuit as 100 and β = 100. Depict the waveform of the output voltage separately.*

Solution.
$$I_B = \frac{V_{CC}}{R_B} = \frac{20}{200} = 0.1\text{A}$$
$$I_C = \beta I_B = 100 \times 0.1 = 10 \text{ A}$$

$$V_{CE} = V_{CC} - I_C R_L = 20 - 10 \times 1 = 10 \text{ V};$$
$$v_{CE} = A_V v_{BE} = 100 \times 0.01 = 10 \text{ V}$$

The combined output voltage v_{CE} is the sum of v_{CE} and v_{ce} and is shown graphically in Fig. 57.60 (*b*). It is seen that 100 times amplified ac signal rides the dc voltage.

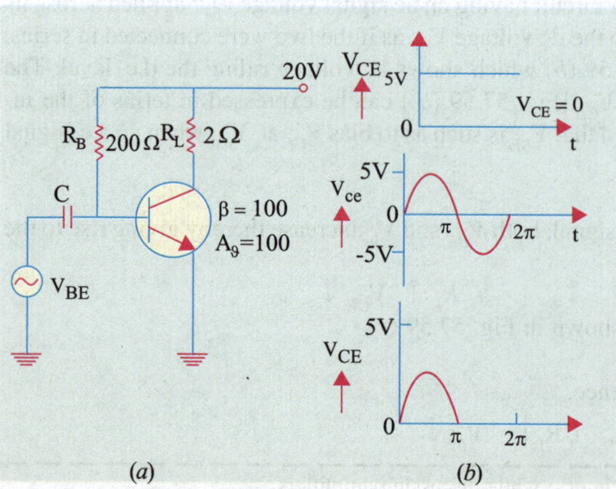

Fig. 57.61

Example 57.29. *Find v_{CE} in the circuit of Fig. 57.61 (a) and sketch its waveform. Take A_v = 100 and β = 100 and peak input signal voltage as 0.05 V.*

Solution. I_B = 20/200 = 0.1 A ; I_C = 100 × 0.1 = 10 A ; V_{CE} = 20 – (10 × 2) = 0.

Obviously, the transistor has been biased at saturation as shown in Fig. 57.61 (*b*).

The addition of v_{CE} and v_{CE}, is shown graphically in Fig. 57.61 (*b*). During the positive half-cycle of the signal, the transistor comes out of saturation and lets pass the half-cycle. However, during the negative half-cycle of input signal, transistor is further driven into saturation. Since it is already biased at V_{CC}'s most negative limit (0 volt), it cannot further go negative. Hence, the negative half-cycle of the signal is lost in saturation.

Tutorial Problems No. 57.1

1. A *CB*-connected transistor has $\alpha = 0.96$ and $I_E = 2$ mA. Find its I_C and I_B. **[1.92 mA, 80 μA]**

2. A *CB*-connected transistor has $I_B = 20$ μA and $I_E = 2$ mA. Compute the value of α and I_C. **[0.99, 1.98 mA]**

3. A *CE*-connected transistor has $\beta = 100$ and $I_B = 50$ μA. Compute the values of α, I_C and I_E. **[0.99 ; 5 mA ; 5.05 mA]**

4. The following quantities are measured in a *CE* transistor : $I_C = 5$ mA ; $I_B = 100$ μA. Determine α, β and I_E. **[0.98 ; 50 ; 5.1 mA]**

5. A transistor has $\alpha = 0.98$, $I_{CBO} = 5$ μA and $I_B = 100$ μA. Find the values of I_C and I_E. **[5.15 mA ; 5.25 mA]**

Following measurements are made in a transistor ; $I_C = 5.202$ mA, $I_B = 50$ μA, $I_{CBO} = 2.52$ Compute the values of α, β and I_E. **[0.99 ; 100 ; 5.252 mA]**

6. Following measurements were made in a certain transistor :

$I_C = 5.202$ mA; $I_B = 50$ μA ; $I_{CBO} = 2.52$ μA.

Determine (*i*) α, β and I_E (*ii*) new value of I_B required to make $I_C = 10$ mA. **[(*i*) 0.99 ; 100 ; 5.252 mA (*ii*) 98.46 μA]**

7. For the *CB* circuit of Fig. 57.62, find the value of V_{CB}. Neglect junction voltage V_{BE}. **[5 V]**

8. In the *CB* circuit of Fig. 57.63, what value of R_E causes $V_{BC} = 10$ V? Neglect V_{EB}. **[10 K]**

9. For the *CE* circuit of Fig. 57.64, calculate the values of I_B, I_C, I_E and V_{CE}. Take $\beta = 50$ and neglect V_{BE}. **[100 μA, 5 mA, 5.1 mA, 7.5 V]**

10. In the circuit of Fig. 57.65, calculate I_B, I_C, I_E and V_{CE}. Take transistor $\beta = 50$ and neglect V_{BE}. **[100 μA, 5 mA, 5.1 mA, 5V]**

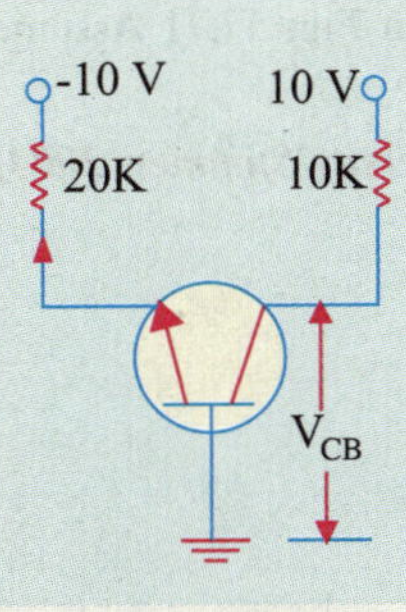

Fig. 57.62

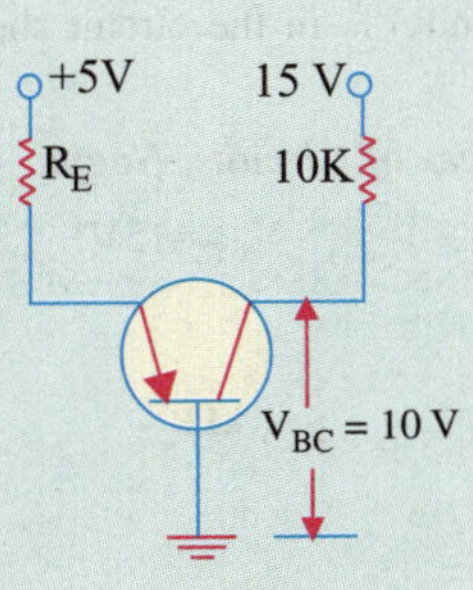

Fig. 57.63

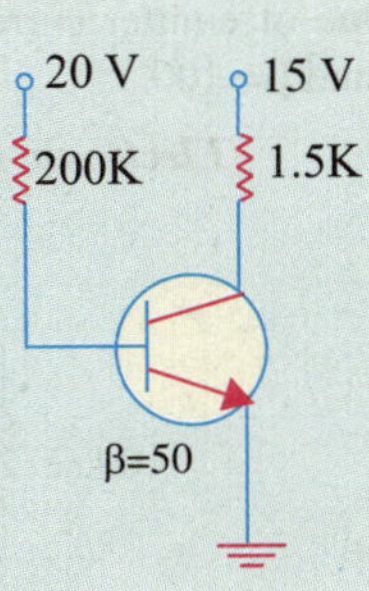

Fig. 57.64

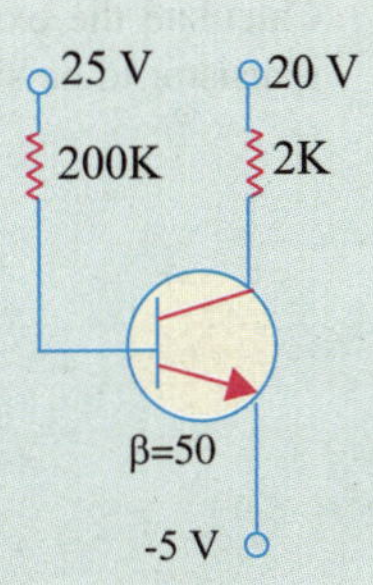

Fig. 57.65

11. In the *CC* circuit of Fig. 57.66, compute the values of I_E, I_B, I_C and V_{CE}. Neglect V_{BE}. **[0.25 mA, 2.48 μA, 0.248 mA, 5 V]**

12. In the *CC* circuit of Fig. 57.67, find I_E, I_B, I_C and V_{CE}. Neglect V_{BE}. **[0.25 mA, 2.48 μA, 0.248 mA, 5 V]**

13. In the circuit of Fig. 57.68, find the drop across R_L. The transistor $\beta = 100$. **[5 V]**

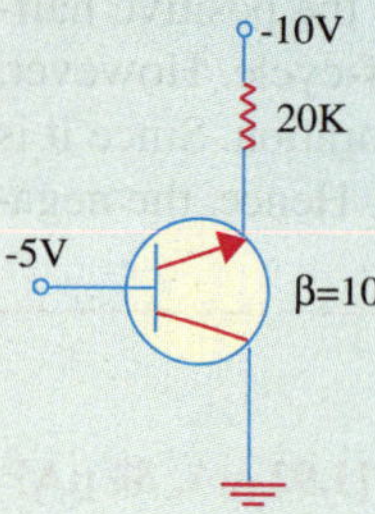

Fig. 57.66

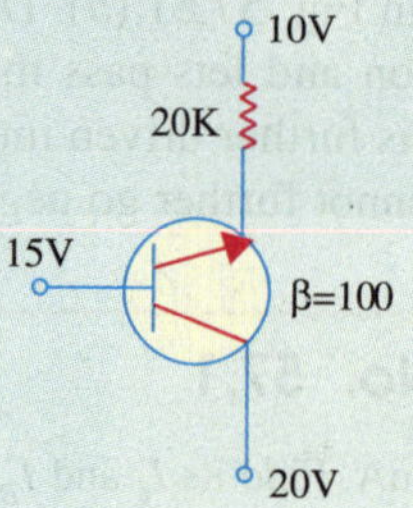

Fig. 57.67

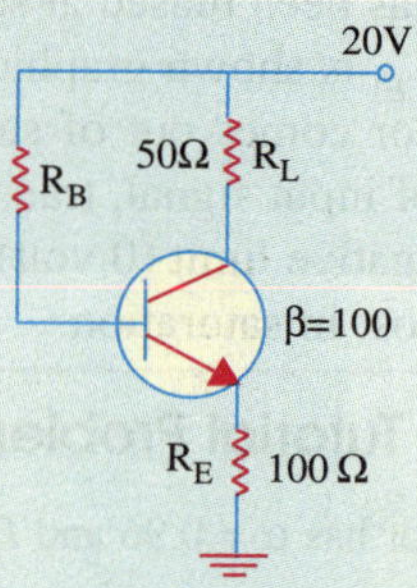

Fig. 57.68

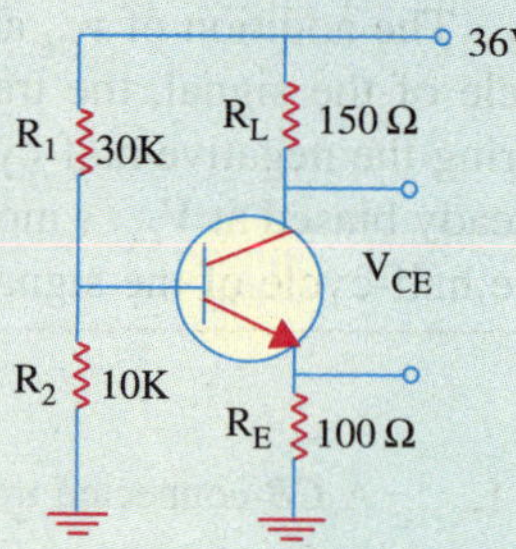

Fig. 57.69

14. Find the value of V_{CE} for the universal stabilization circuit of Fig. 57.69. **[13.5 V]**

15. (i) The reverse saturation current for the Ge transistor in Fig. 57.70 is 2μ A at room temperature (25°C) and increases by a factor of 2 for each temperature increase of 10°C. Bias voltage V_{BB} = 5V. Find the maximum allowable value for R_B if the transistor is to remain cutoff at a temperature of 75°C.

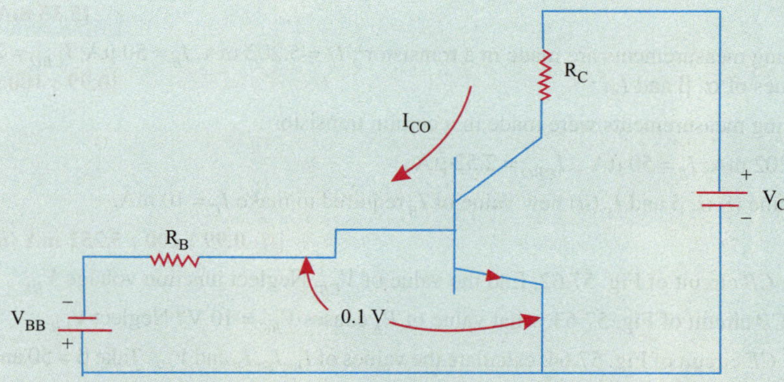

Fig. 57.70

(ii) If V_{BB} = 1.0V and R_B = 50 kΩ, how high may the temperature increase before the transistor comes out of cut-off *(Electronic Devices and Circuits Nagpur University Summer, 2004)*

16. Calculate the exact value of emitter current and V_{CE} in the circuit shown in Fig. 57.71 Assume transistor to be silicon and β= 100.

(Electronic Devices of Circuits, Nagpur University Summer, 2004)

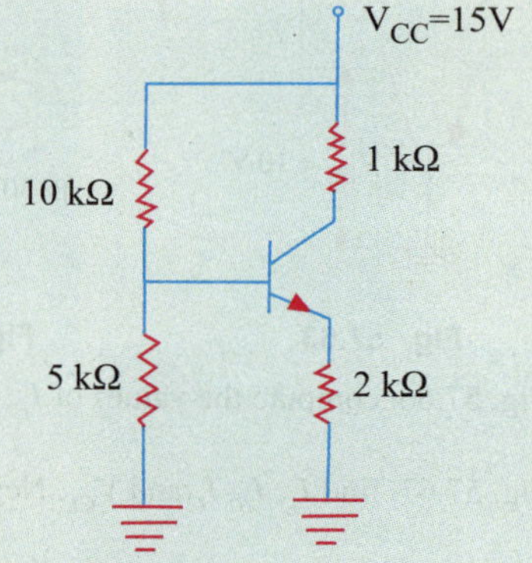

Fig. 57.71

OBJECTIVE TESTS - 57

1. The emitter of a transistor is generally doped the heaviest because it
   (a) has to dissipate maximum power
   (b) has to supply the charge carriers
   (c) is the first region of the transistor
   (d) must possess low resistance.
2. For current working of an *NPN* bipolar junction transistor, the different electrodes should have the following polarities with respect to emitter.
   (a) collector +ve, base –ve
   (b) collector –ve, base + ve
   (c) collector – ve, base –ve
   (d) collector + ve, base +ve
3. Select the CORRECT alternative.
   In a bipolar transistor
   (a) emitter region is of low/high resistivity matterial which is lightly/ heavily-doped.
   (b) collector region is of lower/higher conductivity than emitter region
   (c) base region is of high/low resistivity material which is only lightly/ heavily doped.
4. In a properly-biased *NPN* transistor, most of the electrons from the emitter
   (a) recombine with holes in the base
   (b) recombine in the emitter itself
   (c) pass through the base to the collector
   (d) are stopped by the junction barrier.
5. The following relationships between α and β are correct EXCEPT
   (a) $\beta = \frac{\alpha}{1-\alpha}$ (b) $\alpha = \frac{\beta}{1-\beta}$
   (c) $\alpha = \frac{\beta}{1+\beta}$ (d) $1-\alpha = \frac{1}{1+\beta}$
6. The value of total collector current in a *CB* circuit is
   (a) $I_C = \alpha I_E$ (b) $I_C = \alpha I_E + I_{CO}$
   (c) $I_C = \alpha I_E - I_{CO}$ (d) $I_C = \alpha I_E$.
7. In a junction transistor, the collector cut off current I_{CBO} reduces considerably by doping the
   (a) emitter with high level of impurity
   (b) emitter with low level of impurity
   (c) collector with high level of impurity
   (d) collector with low level of impurity
8. In a transistor amplifier, the reverse saturation current I_{CO}
   (a) doubles for every 10°C rise in temperature
   (b) doubles for every 1°C rise in temperature
   (c) increases linearly with the temperature
   (d) doubles for every 5°C rise in temperature
9. In the case of a bipolar transistor, α is
   (a) positive and > 1
   (b) positive and < 1
   (c) negative and > 1
   (d) negative and < 1.
10. The *EBJ* of a given transistor is forward- biased and its *CBJ* reverse-biased. If the base current is increased, then its
    (a) I_C will decrease
    (b) V_{CE} will increase
    (c) I_C will increase
    (d) V_{CC} will increase.
11. The collector characteristics of a *CE* - connected transistor may be used to find its
    (a) input resistance
    (b) base current
    (c) output resistance
    (d) voltage gain.
12. Which of the following approximations is often used in electronic circuits ?
    (a) $I_C \cong I_E$ (b) $I_B \cong I_C$
    (c) $I_B \cong I_E$ (d) $I_E \cong I_B + I_C$
13. When a transistor is fully switched ON, it is said to be
    (a) shorted (b) saturated
    (c) open (d) cut-off
14. If a change in base current does not change the collector current, the transistor amplifier is said to be
    (a) saturated (b) cut-off
    (c) critical (d) complemented.
15. When an *NPN* transistor is saturated, its V_{CE}
    (a) is zero and I_C is zero
    (b) is zero and I_C is high
    (c) equals V_{CC} and I_C is zero
    (d) equals V_{CC} and I_C is high.
16. When an *NPN* transistor is cut-off, its V_{CE}
    (a) equals V_{CC} and I_C is high
    (b) equals V_{CC} and I_C is zero
    (c) is low and I_C is high
    (d) is high and I_C is low.
17. If, in a bipolar junction transistor, I_B = 100 μA and I_C = 10 mA, in what range does the value of its beta lie ?
    (a) 0.1 to 1.0 (b) 1.01 to 10
    (c) 10.1 to 100 (d) 100.1 to 1000.
18. In a *BJT*, largest current flow occurs
    (a) in the emitter (b) in the collector
    (c) in the base
    (d) through *CB* junction.
19. In a properly-connected *BJT*, an increase in base current causes increase in
    (a) I_C only (b) I_E only
    (c) both I_C and I_E (d) leakage current.

20. When a *BJT* operates in cut-off

(*a*) $V_{CE} = 0$

(*b*) $V_{CE} = V_{cc}$

(*c*) V_{CE} has negative value

(*d*) I_C is maximum.

21. When a *BJT* is in saturation

(*a*) $I_C = 0$

(*b*) I_B controls I_C

(*c*) $V_{CE} = 0$

(*d*) V_{CE} has positive value.

22. The best approximation for V_C in the circuit shown in Fig. 54.72 will be (assume β to be high)

(*a*) 4 V (*b*) 6.8 V

(*c*) 8.7 V (*d*) 10.7 V

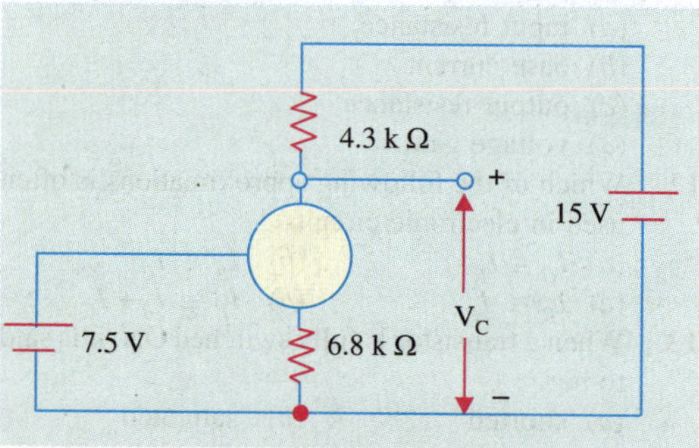

Fig. 57.72

23. Assume $V_{BE} = 0.7$ V and β = 50 for the transistor in the circuit shown in Fig. 57.73. For $V_{CE} = 2$V, the value of R_B is

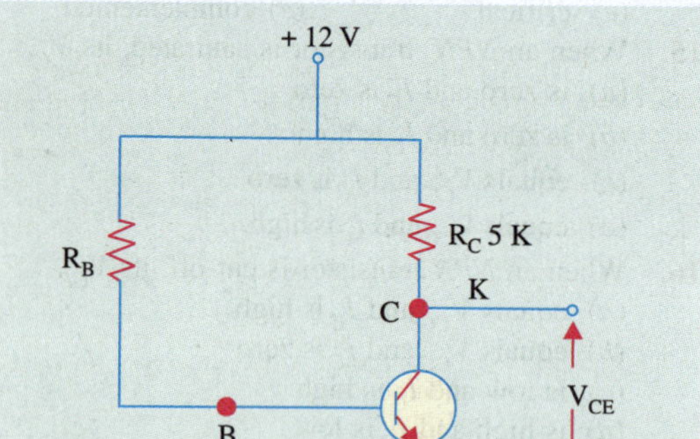

Fig. 57.73

(*a*) 200 kΩ (*b*) 242 kΩ

(*c*) 283 kΩ (*d*) 300 kΩ

24. In the circuit shown in Fig. 57.74, if $R_L = R_C = 1$ KΩ , then the value of V_0 will be

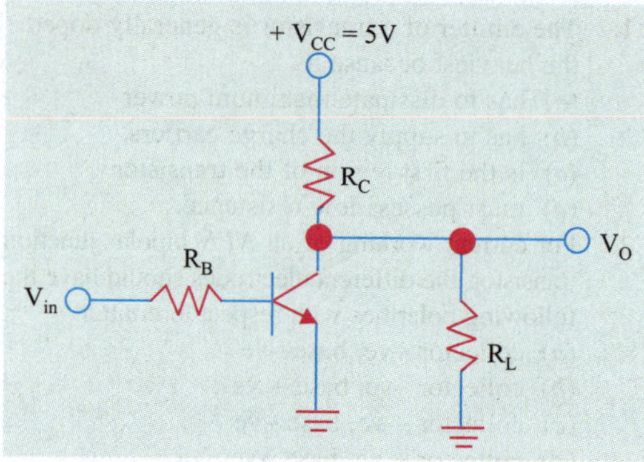

Fig. 57.74

(*a*) 4.55 V (*b*) 2.5 V

(*c*) 1 V (*d*) zero

25. A transistor is operated as a non-saturated switch to eliminate

(*a*) storage time

(*b*) turn-off time

(*c*) turn-on time

(*d*) delay time

26. Early-effect in BJT refers to

(*a*) avalanche break down

(*b*) thermal break down

(*c*) base narrowing

(*d*) zener break-down

(UPSC Engg. Services 2002)

(**Hint.** Early effect also called base-width modulation) is the variation of effective base width by the collector voltage)

27. A bipolar junction transistor (BJT) is used as power control switch by biasing it in the cut-off region (OFF state) or in the saturation region (ON state). In the ON state, for the BJT.

(*a*) both the base-emitter junction and base-collector junctions are reverse biased

(*b*) the base-emitter is reverse biased, and the base-collector junction is forward biased

(*c*) the base-emitter junction is forward biased, and the base-collector junction is reverse biased

(*d*) both the base-emitter and base-collector junctions are forward biased.

ANSWERS

1. (*b*) **2.** (*d*) **3.** (*a*) low, heavily (*b*) lower (*c*) high, lightly **4.** (*c*) **5.** (*b*) **6.** (*b*) **7.** (*d*)
8. (*a*) **9.** (*b*) **10.** (*c*) **11.** (*c*) **12.** (*a*) **13.** (*b*) **14.** (*a*) **15.** (*b*) **16.** (*b*) **17.** (*d*) **18.** (*a*)
19. (*c*) **20.** (*b*) **21.** (*c*) **22.** (*d*) **23.** (*d*) **24.** (*b*) **25.** (*a*) **26.** (*c*) **27.** (*d*)

CHAPTER 58

LOAD LINES AND DC BIAS CIRCUITS

Learning Objectives

- DC Load Line
- Q-Point and Maximum Undistorted Output
- Need for Biasing a Transistor
- Factor Affecting Bias Variations
- Stability Factor
- Beta Sensitivity
- Stability Factor for CB and CE Circuits
- Different Methods for Transistor Biasing
- Base Bias
- Base Bias with Emitter Feedback
- Base Bias with Collector Feedback
- Base Bias with Collector and Emitter Feedbacks
- Emitter Bias with two Supplies
- Voltage Divider Bias
- Load Line and Output Characteristics
- AC Load Line

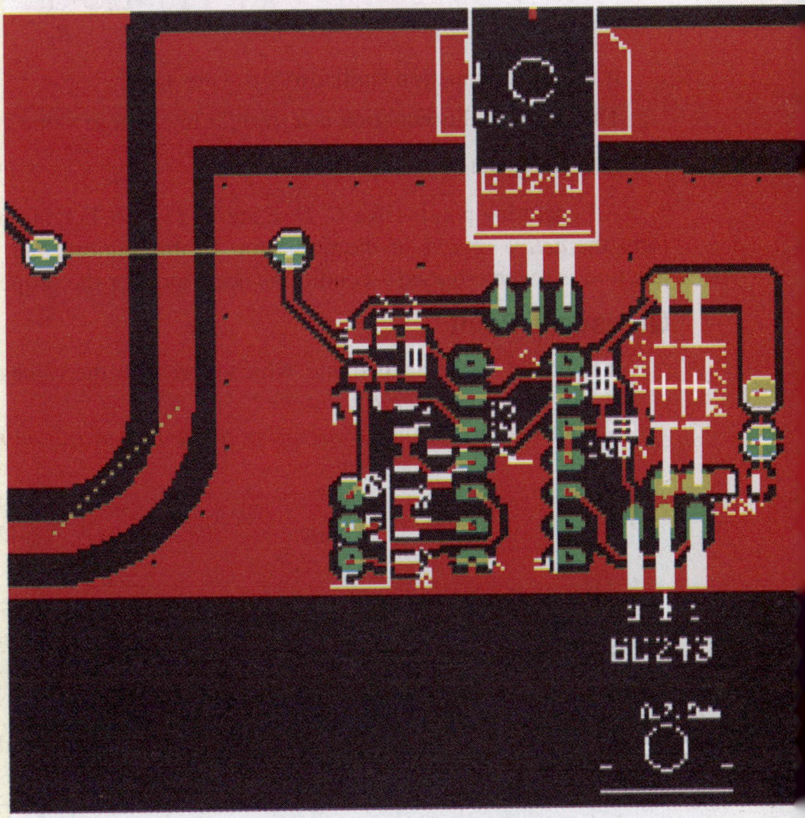

DC biasing is used to establish a steady level of transistor current and voltage called the dc operating point or quescent point (Q-Point)

58.1. D.C. Load Line

For drawing the dc load line of a transistor, one need to know only its cut-off and saturation points. It is a straight line jointing these two points. For the *CE* circuit of Fig. 57.28, the load line is drawn in Fig. 58.1. *A* is the cut-off point and *B* is the saturation point.

The voltage equation of the collector-emitter is

$$V_{CC} = I_C R_L + V_{CE} \quad \therefore \quad I_C = \frac{V_{CC}}{R_L} - \frac{V_{CE}}{R_L}$$

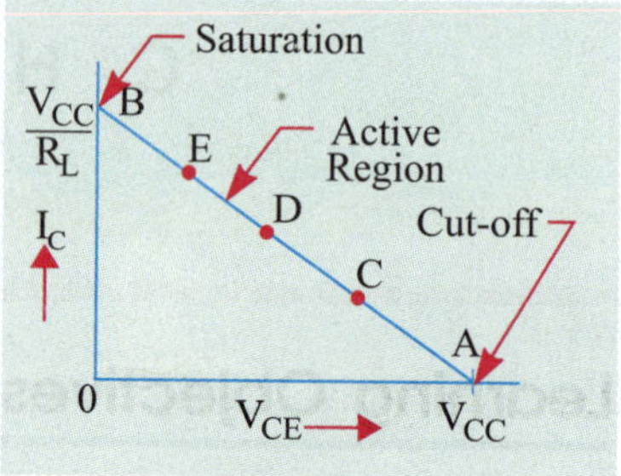

Fig. 58.1

Consider the following two particular cases :

(*i*) when $I_C = 0$, $V_{CE} = V_{CC}$

— cut-off point A

(*ii*) when $V_{CE} = 0$, $I_C = V_{CC}/R_L$

— saturation point *B*

Obviously, load line can be drawn if only V_{CC} and R_L are known. Incidentally slope of the load line $AB = -1/R_L$

Note. The above given equation can be written as

$$I_C = \frac{V_{CE}}{R_L} + \frac{V_{CC}}{R_L}$$

It is a linear equation similar to $y = -mx + c$

The graph of this equation is a straight line whose slope is $m = -1/R_L$

Active Region

All operating points (like *C, D, E* etc. in Fig. 58.1) lying between cut-off and saturation points form the ***active region*** of the transistor. In this region, *E/B* junction is forward-biased and *C/B* junction is reverse-biased—conditions necessary for the proper operation of a transistor.

Quiescent Point

It is a point on the dc load line, which represents the values of I_C and V_{CE} that exist in a transistor circuit when ***no input signal is applied.***

It is also known as the ***dc operating point or working point.*** The best position for this point is midway between cut-off and saturation points where $V_{CE} = \frac{1}{2} V_{CC}$ (like point *D* in Fig. 58.1).

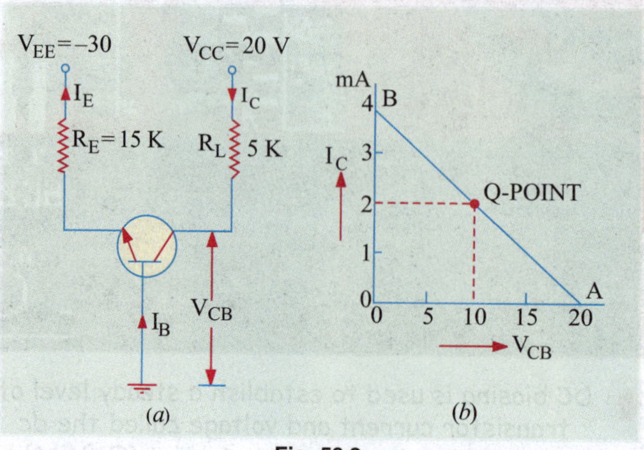

Fig. 58.2

Example 58.1. *For the circuit shown in Fig. 58.2 (a), draw the dc load line and locate its quiescent or dc working point.*

Solution. The cut-off point is easily found because it lies along *X*-axis where $V_{CB} = V_{CC} = 20$ V *i.e.* point *A* in Fig. 58.2 (*b*). At saturation point *B*, saturation value of collector current is $I_{C(sat)} = V_{CC}/R_L = 20 \text{ V}/5 \text{ K} = 4$ mA.

The line AB represents the load line for the given circuit. We will now find the actual operating point.

$I_E = V_{EE}/R_E = 30\text{ V}/15\text{ K} = 2\text{ mA}$ —neglecting V_{BE}

$I_C = \alpha I_E \cong I_E = 2\text{ mA}; \quad \therefore \quad V_{CB} = V_{CC} - I_C R_L = 20 - 2 \times 5 = 10\text{ V}$

Hence, Q-point is located at **(10 V ; 2 mA)** as shown in Fig. 58.2 (*b*)

Example 58.2. *In the CB circuit of Fig. 8.3 (a), find*

(a) dc operating point and dc load line

(b) maximum peak-to-peak unclipped signal

(c) the approximate value of ac source voltage that will cause clipping.

(Electronics and Telecom Engg. Jadavpur Univ.)

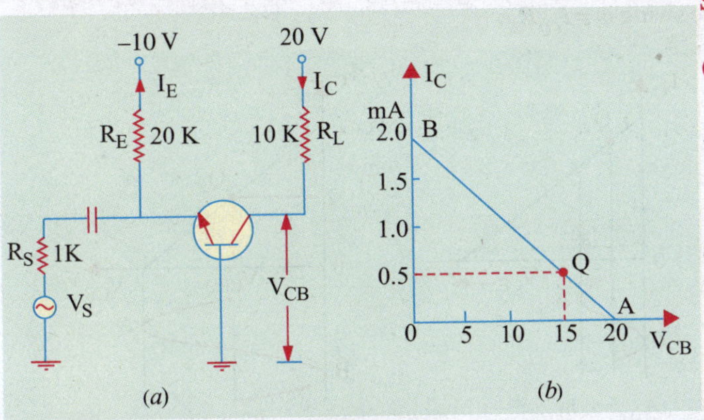

Fig. 58.3

Solution.

(*a*) $I_{C(sat)} = \dfrac{V_{CC}}{R_L} = 2\text{ mA}$ — point B

V_{CB} at cut-off $= V_{CC} = 20\text{ V}$ —point A

Hence, AB is the dc load line and is shown in Fig. 58.3 (*b*).

Now, $I_E = 10/20 = 0.5\text{ mA}$

$I_C \cong I_E = 0.5\text{ mA}$

$\therefore \quad V_{CB} = V_{CC} - I_C R_L = 20 - 0.5 \times 10 = 15\text{ V}$

The Q-point is located at **(15 V, 0.5 mA)**

(*b*) It is obvious from Fig. 58.3 (*b*) that maximum positive swing can be from 15 V to 20 V *i.e.* 5 V only. Of course, on the negative swing, the output swing can go from 15 V down to zero volt. The limiting factor being ***cut-off on positive half-cycle,*** hence maximum unclipped peak-to-peak voltage that we can get from this circuit is $2 \times 5 =$ **10 V**.

(*c*) The approximate voltage gain of the above circuit is

$$A_v = \frac{V_0}{V_S} = \frac{R_L}{R_S} = \frac{10\text{ K}}{1\text{ K}} = 10$$

It means that signal voltage will be amplified 10 times. Hence, maximum value of source voltage for obtaining unclipped or undistorted output is

$$V_s = \frac{V_0}{10} = \frac{10 V_{p-p}}{10} = \mathbf{1V_{p-p}}$$

Example 58.3. *For the CE circuit shown in Fig. 58.4 (a), draw the dc load line and mark the dc working point on it. Assume β=100 and neglect V_{BE}.*

(Applied Electronics, Punjab Univ.)

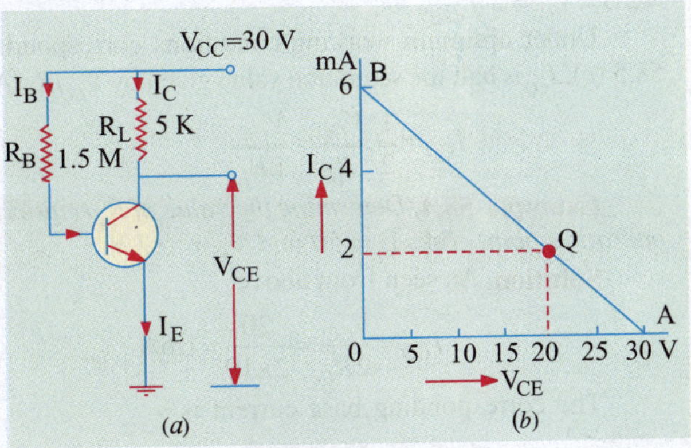

Fig. 58.4

Solution. Cut-off point A is located where, $I_C = 0$ and $V_{CE} = V_{CC} = 30$ V. Saturation point B is given where $V_{CE} = 0$ and

$I_{C(sat)} = 30\text{ V}/5\text{ K} = 6\text{ mA}.$

Line AB represents the load line in Fig. 58.4 (*b*).

Let us find the dc working point from the given values of resistances and supply voltage.

$I_B = 30\text{ V}/1.5\text{ M} = 20\ \mu\text{A}$; $\qquad I_C = \beta I_B = 100 \times 20 = 2000\ \mu\text{A} = 2\text{ mA}$;

$V_{CE} = V_{CC} - I_C R_L = 30 - 2 \times 5 = 20\text{ V}$

Hence, Q–point is **(20 V ; 2 mA)** as shown in Fig. 58.4.

58.2. Q-Point and Maximum Undistorted Output

Position of the Q-point on the dc load line determines the maximum signal that we can get from the circuit before clipping occurs. Consider the cases shown in Fig. 58.5.

In Fig. 58.5 (a), when Q is located near cut-off point, signal first starts to clip at A. It is called ***cut-off clipping*** because the positive swing of the signal drives the transistor to cut-off. In fact, as seen from Fig. 58.5 (a), maximum positive swing is $= I_{CQ} R_{ac}$.

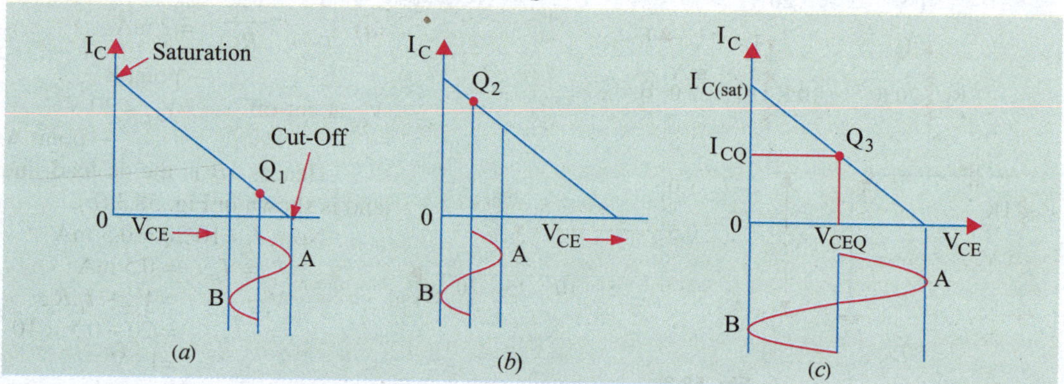

Fig. 58.5

If the Q-point Q_2 is located near saturation point, then clipping first starts at point B as shown in Fig. 58.5 (b). It is caused by saturation. The maximum negative swing $= V_{CEQ}$.

In Fig. 58.5 (c), the Q-point Q_3 is located at the centre of the load line. In this condition, we get the ***maximum possible output signal.*** The point Q_3 gives the optimum Q-point. The maximum undistorted signal $= 2V_{CEQ}$.

In general, consider the case shown in Fig. 58.6. Since $A < B$, maximum possible peak-to-peak output signal = 2 A.

If the operating point were so located that $A > B$, then maximum possible peak-to-peak output signal = 2B.

When operating point is located at the centre of the load line, then maximum undistorted peak-to-peak signal is $= 2\,A = 2\,B = V_{CC} = 2\,V_{CEQ}$.

Under optimum working conditions corresponding to Fig. 58.5 (c), I_{CQ} is half the saturation value given by V_{CC}/R_L (Art. 58.1).

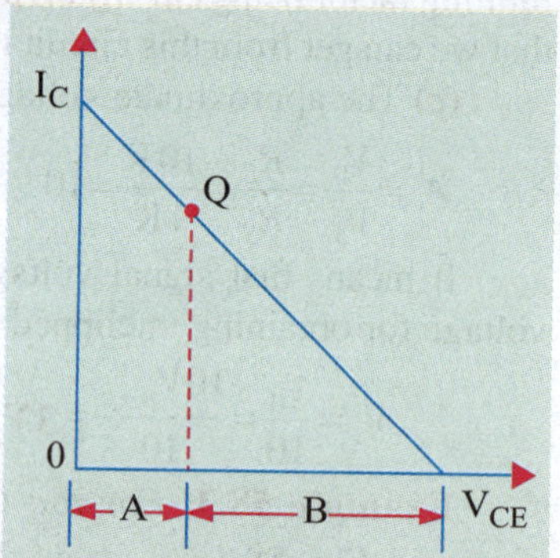

Fig. 58.6

$$\therefore \quad I_{CQ} = \frac{1}{2} \cdot \frac{V_{CC}}{R_L} = \frac{V_{CC}}{2R_L}$$

Example 58.4. *Determine the value of R_B required to adjust the circuit of Fig. 58.7 to optimum operating point. Take β = 50 and V_{BE} = 0.7 V.*

Solution. As seen from above

$$I_{CQ} = \frac{V_{CC}}{2R_L} = \frac{20}{2 \times 10} = 1\text{ mA}$$

The corresponding base current is

$$I_{BQ} = \frac{I_{CQ}}{\beta} = \frac{1}{50} = 20\mu A$$

Now, $V_{CC} = I_B R_B + V_{BE}$

$$\therefore \quad R_B = \frac{V_{CC} - V_{BE}}{I_B} = \frac{20 - 0.7}{20 \times 10^{-6}} = 965 \text{ K}\Omega$$

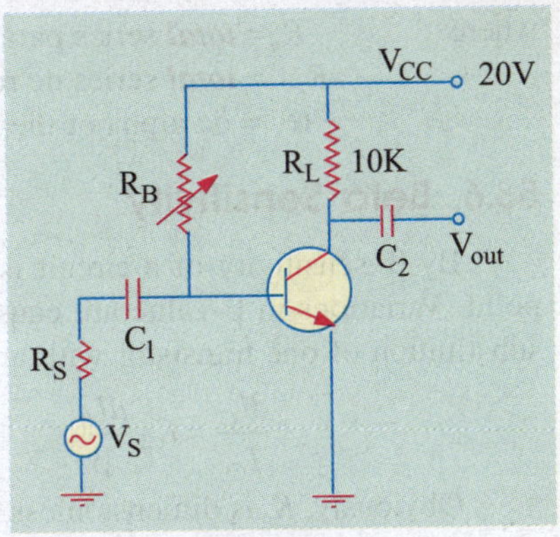

Fig. 58.7

58.3. Need For Biasing a Transistor

For normal operation of a transistor amplifier circuit, it is essential that there should be a

(*a*) forward bias on the emitter-base junction and

(*b*) reverse bias on the collector-base junction.

In addition, amount of bias required is important for establishing the *Q*-point which is dictated by the mode of operation desired.

If the transistor is not biased correctly, it would

1. work inefficiently and
2. produce distortion in the output signal.

It is desirable that once selected, the *Q*-point should remain stable *i.e.* should not shift its position due to temperature rise etc. Unfortunately, this does not happen in practice unless special efforts are made for the purpose.

58.4. Factors Affecting Bias Variations

In practice, it is found that even after careful selection, *Q*-point tends to shift its position. This bias instability is the direct result of thermal instability which itself is produced by cumulative increase in I_C that may, if unchecked, lead to thermal runaway (Art. 57.13).

The collector current for *CE* circuit is given by

$$I_C = \beta I_B + I_{CEO} = \beta I_B + (1 + \beta)\, I_{CO}$$

This equation has three variables : β, I_B and I_{CO} all of which are found to increase with temperature. In particular, increase in I_{CO} produces significant increase in collector current I_C. This leads to increased power dissipation with further increase in temperature and hence I_C. Being a cumulative process, it can lead to thermal runaway which will destroy the transistor itself.

However, if by some circuit modification, I_B is made to decrease with temperature automatically, then decrease in the term βI_B can be made to neutralize the increase in the term $(1 + \beta)\, I_{CO}$, thereby keeping I_C constant. This will achieve thermal stability resulting in bias stability.

58.5. Stability Factor

The degree of success achieved in stabilizing I_C in the face of variations in I_{CO} is expressed in terms of current stability factor *S*. It is defined as the rate of change of I_C with respect to I_{CO} when both β and $I_B\,(V_{BE})$ are held constant.

$$\therefore \quad S = \frac{dI_C}{dI_{CO}} \qquad \text{— } \beta \text{ and } I_B \text{ constant}$$

Larger the value of *S*, greater the thermal instability and *vice versa* (in view of the above, this factor should, more appropriately, be called instability factor.).

The stability factor may be alternatively expressed by using the well-known equation $I_C = \beta I_\beta + (1 + \beta)\, I_{CO}$ which, on differentiation with respect to I_C, yields.

$$1 = \beta \frac{dI_B}{dI_C} + (1+\beta)\frac{dI_{CO}}{dI_C} = \beta \frac{dI_B}{dI_C} + (1+\beta)\frac{1}{S} \qquad \therefore \qquad S = \frac{(1+\beta)}{1 - \beta(dI_B / dI_C)}$$

The stability factor of any circuit can be found by using the general formula

$$S = \frac{1 + R_B / R_E}{1 + (1 - \alpha)\,(R_B / R_E)}$$

where R_B = *total* series parallel resistance in the base
R_E = *total* series dc resistance in the emitter
α = dc alpha of the transistor

58.6. Beta Sensitivity

By β–sensitivity of a circuit is meant the influence that the β-value has on its dc operating point. Variations in β-value are caused by variations in the circuit operating conditions or by the substitution of one transistor with another. Beta sensitivity K_β is given by

$$\frac{dI_C}{I_C} = K_\beta \frac{dI_B}{\beta} \qquad \therefore \qquad K_\beta = \frac{\beta}{I_C} \cdot \frac{dI_C}{dI_B}$$

Obviously, K_β is dimensionless ratio and can have values ranging from zero to unity.

58.7. Stability Factor for CB and CE Circuits

(i) CB Circuit

Here, collector current is given by

$$I_C = \alpha I_E + I_{CO} \qquad \therefore \qquad \frac{dI_C}{dI_{CO}} = 0 + 1 \qquad \text{or} \qquad S = 1$$

(ii) CE Circuit

$$I_C = \beta I_B + (1+\beta) I_{CO} \qquad \therefore \qquad \frac{dI_C}{dI_{CO}} = (1+\beta) \qquad \text{—treating } I_B \text{ as a constant}$$

$$\therefore \qquad S = (1+\beta).$$

If β = 100, then S = 101 which means that I_C changes 101 times as much as I_{CO}.

58.8. Different Methods for Transistor Biasing

Some of the methods used for providing bias for a transistor are :

1. base bias or fixed current bias (Fig. 57.25)

It is not a very satisfactory method because bias voltages and currents do not remain constant during transistor operation.

2. base bias with emitter feedback (Fig. 58.9)

This circuit achieves good stability of dc operating point against changes in β with the help of emitter resistor which causes degeneration to take place.

3. base bias with collector feedback (Fig. 58.11)

It is also known as collector-to-base bias or collector feedback bias. It provides better bias stability.

4. base bias with collector and emitter feedbacks (Fig. 58.12)

It is a combination of (2) and (3) above.

5. emitter bias with two supplies (Fig. 58.14)

This circuit uses both a positive and a negative supply voltage. Here, base is at approximately 0 volt *i.e.* $V_B \cong 0$.

6. voltage divider bias (Fig. 58.16)

It is most widely used in linear discrete circuits because it provides good bias stability. It is also called universal bias circuit or base bias with one supply.

Each of the above circuits will now be discussed separately.

58.9. Base Bias

It has already been discussed in Art. 57.20 and is shown in Fig. 57.25. For such a circuit, $S = (1+\beta) \cong \beta$ and $K_\beta = 1$.

58.10. Base Bias with Emitter Feedback

This circuit is obtained by simply adding an emitter resistor to the base bias circuit as shown in Fig. 58.8.

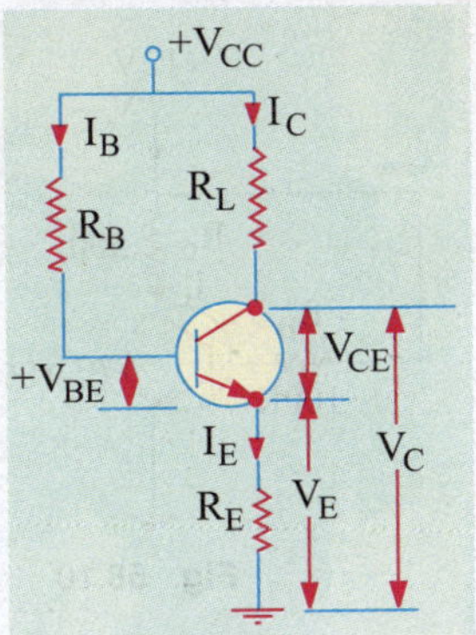

Fig. 58.8

1. At saturation, V_{CE} is essentially zero, hence V_{CC} is distributed over R_L and R_E.

$$\therefore \qquad I_{C(sat)} = \frac{V_{CC}}{R_E + R_L}$$

2. I_C can be found as follows :

Consider the supply, base, emitter and ground route. Applying Kirchhoff 's Voltage Law, we have

$$-I_B R_B - V_{BE} - I_E R_E + V_{CC} = 0$$

or $\qquad V_{CC} = I_B R_B + V_{BE} + I_E R_E \qquad ...(i)$

Now $\qquad I_B = I_C/\beta \quad$ and $\quad I_E \cong I_C$

Substituting these values in the above equation, we have

$$V_{CC} \cong \frac{I_C R_B}{\beta} + V_{BE} + I_C R_E$$

$$\therefore \qquad I_C \cong \frac{V_{CC} - V_{BE}}{R_E + R_B/\beta} \cong \frac{V_{CC}}{R_E + R_B/\beta} \qquad \text{—neglecting } V_{BE}$$

(we could have applied the β-rule given in Art. 57.24)

3. collector-to-ground voltage $\qquad V_C = V_{CC} - I_C R_L$

4. emitter-to-ground voltage $\qquad V_E = I_E R_E \cong I_C R_E$

5. $$S = \frac{1 + R_B/R_E}{1 + R_B/(1+\beta) R_E} = \frac{1 + R_B/R_E}{1 + R_B/\beta R_E}$$

6. The β-sensitivity of this circuit is $K_\beta = \dfrac{1}{1 + \beta R_E/R_B}$

Example 58.5. *For the circuit shown in Fig. 58.9, find* ***(i)*** $I_C(sat)$ *,* ***(ii)*** I_C*,* ***(iii)*** V_C*,* ***(iv)*** V_E*,* ***(v)*** V_{CE} *and* ***(vi)*** K_β.

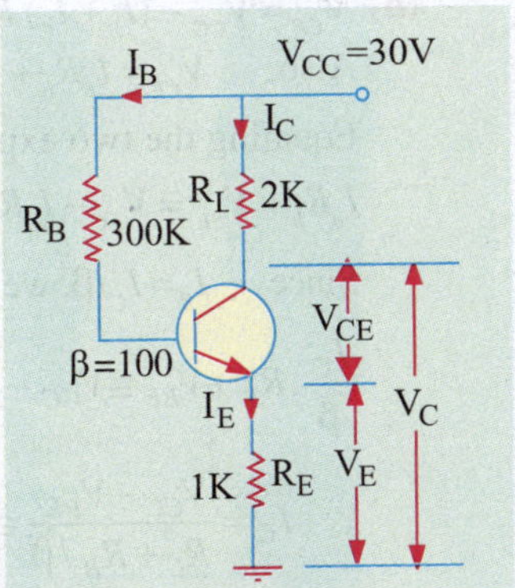

Fig. 58.9

Solution. *(i)* $\qquad I_{C(sat)} = \dfrac{V_{CC}}{R_E + R_L} = \dfrac{30}{1+2} = $ **10 mA**

(ii) $\qquad$ actual $I_C \cong \dfrac{V_{CC}}{R_E + R_B/\beta} = \dfrac{30}{1 + 300/100} = $ **7.5 mA**

(iii) $\qquad V_C = V_{CC} - I_C R_L = 30 - 2 \times 7.5 = $ **15 V**

(iv) $\qquad V_E \cong I_E R_E \cong I_C R_E = 7.5 \times 1 = $ **7.5 V**

(v) $\qquad V_{CE} = V_C - V_E = 15 - 7.5 = $ **7.5 V**

(vi) $\qquad K_\beta = \dfrac{1}{1 + 100 \times 1/300} = $ **0.75**

Example 58.6. *The base-biased transistor circuit of Fig. 58.10 is subjected to increase in junction temperature from 25°C to 75°C. If β increases from 100 to 150 with rising temperature, calculate the percentage change in Q-point values (I_C, V_{CE}) over the temperature range. Assume that V_{BE} remains constant at 0.7 V.*

Solution. At 25°C $\qquad I_B = \dfrac{V_{CC} - V_{BE}}{R_B} = \dfrac{12 - 0.7}{100 \times 10^3} = 0.113\,\text{mA}$

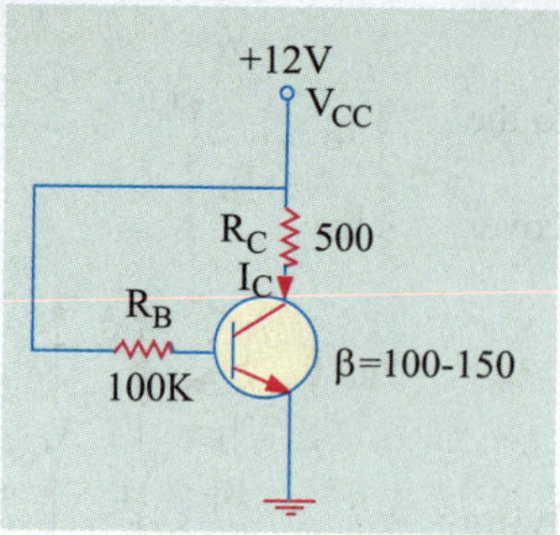

Fig. 58.10

$I_C = \beta I_B = 100 \times 0.113 = 11.3$ mA

$V_{CE} = V_{CC} - I_C R_C = 12 - (11.3 \times 10^{-3} \times 500) = 6.35$ V

At 75°C

$I_B = 0.113$ mA

$I_C = \beta I_B = 150 \times 0.113 = 16.95$ mA

$V_{CE} = 12 - (16.95 \times 10^{-3} \times 500) = 3.525$ V

$$\% \Delta I_C = \frac{16.95 - 11.3}{11.3} \times 100 = \textbf{50 (increase)}$$

$$\% \Delta V_{CE} = \frac{3.52 - 6.35}{6.35} \times 100 = \textbf{–44.57 (decrease)}$$

It is seen that Q-point is very much dependent on temperature and makes the base-bias arrangement very unstable.

58.11. Base Bias with Collector Feedback

This circuit (Fig. 58.11) is like the base bias circuit except that base resistor is returned to collector rather than to the V_{CC} supply. It derives its name from the fact that since voltage for R_B is derived from collector, there exists a negative feed back effect which tends to stabilise I_C against changes in β. To understand this action, suppose that somehow β increases. It will increase I_C as well as $I_C R_L$ but decrease V_{CE} which is applied across R_B. Consequently, I_B will be decreased which will partially compensate for the original increase in β.

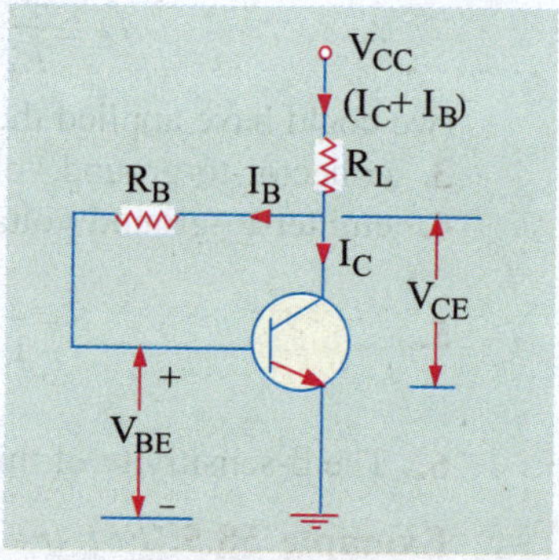

Fig. 58.11

(i) $I_{C(sat)} = V_{CC}/R_L$ —since $V_{CE} = 0$

(ii) $V_{CE} = V_{CC} - (I_B + I_C) R_L \cong V_{CC} - I_C R_L$

Also, $V_{CE} = I_B R_B + V_{BE}$

Equating the two expressions for V_{CE}, we have

$I_B R_B + V_{BE} \cong V_{CC} - I_C R_L$

Since $I_B = I_C/\beta$, we get

$$\frac{I_C}{\beta}.R_B + V_{BE} \cong V_{CC} - I_C R_L$$

$$\therefore \quad I_C = \frac{V_{CC} - V_{BE}}{R_L + R_B/\beta} \cong \frac{V_{CC}}{R_L + R_B/\beta}$$

This is also the approximate value of I_E (again, we could take the help of β-rule). The β-sensitivity factor is given by

$$K_\beta = \frac{1}{1 + \beta R_L / R_B} = 1 - \frac{I_C}{I_{C(sat)}}$$

$$S = \frac{1 + R_B/R_L}{1 + R_B/(1+\beta)R_L} \cong \frac{V_{CC}}{R_L + R_B/\beta}$$

Example 58.7. *In Fig. 58.11, $V_{CC} = 12$ V, $V_{BE} = 0.7$ V, $R_L = 1K$, $R_B = 100K$, $\beta = 100$. Find (i) I_C, (ii) V_{CE}, (iii) I_B, (iv) K_β and (v) S.*

Solution. *(i)* $I_C \cong I_E = \dfrac{12-0.7}{1+100/100} =$ **5.65 mA**

(ii) $V_{CE} \cong 12 - (5.65 \times 1) =$ **6.35 V** *(iii)* $I_B = I_C/\beta = 5.65/100 =$ **56.5 μA**

(iv) $K_\beta = \dfrac{1}{1+100\times 1/100} =$ **0.5** *(v)* $S = \dfrac{1+100/1}{100/1\times 101} =$ **50.75**

58.12. Base Bias with Collector and Emitter Feedbacks

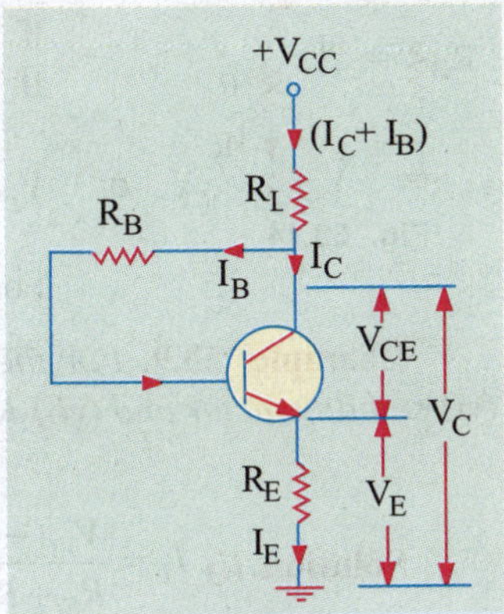

Fig. 58.12

In the circuit of Fig. 58.12, both collector and emitter feedbacks have been used in an attempt to reduce circuit sensitivity to changes in β. If β increases, emitter voltage increases but collector voltage decreases. It means that voltage across R_B is reduced causing I_B to decrease thereby partially off-setting the increase in β.

Under saturation conditions, V_{CC} is distributed over R_L and R_E. Assuming I_B to be negligible as compared to I_C, we get, $I_{C(sat)} = V_{CC}/(R_E + R_L)$.

Actual value of I_C is $= \dfrac{V_{CC} - V_{BE}}{R_E + R_L + R_B/\beta}$

—going *via* R_B because V_{CE} is unknown.

$$V_C = V_{CC} - (I_C + I_B)\, R_L \cong V_{CC} - I_C R_L$$

$$V_E = I_E R_E \cong I_C R_E; \quad V_{CE} = V_C - V_E$$

$$V_{CE} \cong V_{CC} - I_C (R_L + R_E)$$

$$S = \frac{1 + R_B/(R_E + R_L)}{1 + R_B/\beta(R_E + R_L)}$$

It can be proved that $K_\beta = \dfrac{1}{1+\beta(R_E + R_L)/R_B} = 1 - \dfrac{I_C}{I_{C(sat)}}$

Obviously, K_β will be degraded with increase in R_B.

Example 58.8. *For the circuit shown in Fig. 58.13, find (a) $I_{C(sat)}$ (b) V_{CE} and (c) K_β. Neglect V_{BE} and take β = 100.*

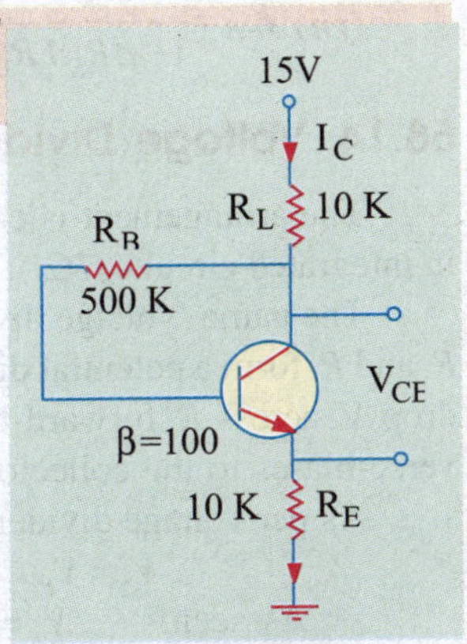

Fig. 58.13

Solution. *(a)* $I_{C(sat)} = 15/(10 + 10) =$ **0.75 mA**

(b) $I_C = \dfrac{V_{CC}}{R_E + R_L + R_B/\beta} = \dfrac{15}{10+10+500/100} = 0.6\,\text{mA}$

$V_{CE} = 15 - 0.6\,(10 + 10) =$ **3 V**

(c) $K_\beta = \dfrac{1}{1+100\,(10+10)/500} = 0.2$

or $K_\beta = 1 - I_C/I_{C(sat)} = 1 - 0.6/0.75 = 0.2$

Obviously, K_β will be degraded with increase in R_B.

58.13. Emitter Bias with Two Supplies

This circuit gives a reasonably stable *Q*-point and is widely used whenever two supplies (positive and negative) are available. Its popularity is due to the fact that I_C is essentially independent of β.

It can be shown that $V_B \cong 0$ and $V_E = -V_{BE}$

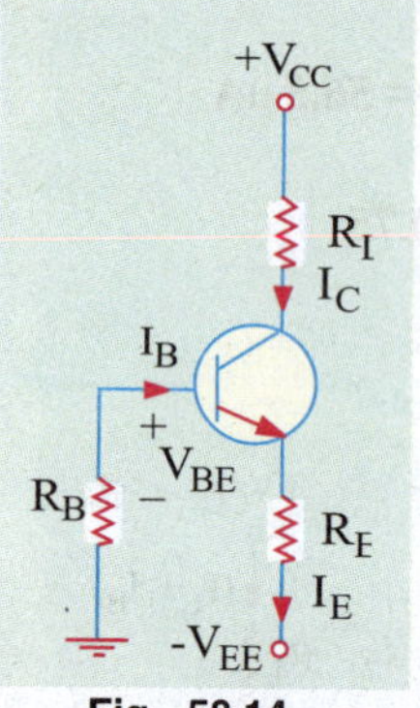

Fig. 58.14

Starting from ground and going clockwise round the base-emitter circuit, we get according to *KVL*.

$$-I_B R_B - V_{BE} - I_E R_E + V_{EE} = 0$$

or $$I_B R_B + I_E R_E = V_{EE} - V_{BE} \quad ...(i)$$

Now, $I_B = I_C/\beta \cong I_E/\beta$. Substituting this in (*i*) above we have

$$\frac{I_E R_B}{\beta} + I_E R_E = V_{EE} - V_{BE} \quad or \quad I_E = \frac{V_{EE} - V_{BE}}{R_E + R_B/\beta}$$

If $V_{EE} \gg V_{BE}$ and $R_E \gg R_B/\beta$, $I_E = V_{EE}/R_E$.

If V_E is the emitter to ground voltage, then

$$-I_B R_B - V_{BE} - V_E = 0$$

or $$V_E = -(I_B R_B + V_{BE}) = -(V_{BE} + I_C R_B/\beta) \cong -V_{BE}$$

For this circuit, $S = \dfrac{1 + R_B/R_E}{1 + R_B/\beta R_E}$ and $K_B = \dfrac{1}{1 + \beta R_E/R_B}$

Example 58.9. *For the circuit of Fig. 58.15, find* **(i)** I_E, **(ii)** I_C, **(iii)** V_C, **(iv)** V_E, **(v)** V_{CE}, **(vi)** *stability factor and* **(vii)** K_β *for a* β *of 50. Take* $V_{BE} = 0.7$ V.

(Electronics-II, Bangalore Univ. 1995)

Solution. (*i*) $I_E = \dfrac{V_{EE} - V_{BE}}{R_E + R_B/\beta} = \dfrac{10 - 0.7}{20 + 10/50} = $ **0.46 mA**

(*ii*) $I_C \cong I_E =$ **0.46 mA**

(*iii*) $V_C = V_{CC} - I_C R_L = 20 - 0.46 \times 10 =$ **15.4 V**

(*iv*) $V_E = -(V_{BE} + I_C R_B/\beta)$
$= -(0.7 + 0.46 \times 10/50) = -0.8$ V

(*v*) $V_{CE} = V_C - V_E = 15.4 - (-0.8) = 16.2$ V

(*vi*) $S = \dfrac{1 + R_B/R_E}{1 + R_B/\beta R_E} = \dfrac{1 + 10/20}{1 + 10/50 \times 20} =$ **1.485**

(*vii*) $K_\beta = \dfrac{1}{1 + \beta R_E/R_B} = \dfrac{1}{1 + 50 \times 20/10} \cong$ **0.01**

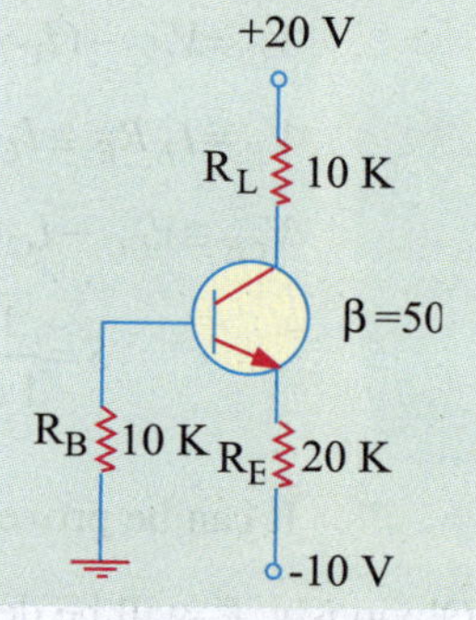

Fig. 58.15

58.14. Voltage Divider Bias

The arrangement is commonly used for transistors incorporated in integrated circuits (*ICs*).

The name 'voltage divider' is derived from the fact that resistors R_1 and R_2 form a potential divider across V_{CC} (Fig. 58.16)*. The voltage drop V_2 across R_2 forward-biases the emitter whereas V_{CC} supply reverse-biases to the collector.

As per voltage divider theorem.

$$V_2 = V_{CC} \cdot R_2/(R_1 + R_2)$$

As seen, $V_E = V_2 - V_{BE}$

$$\therefore \quad I_E = \frac{V_E}{R_E} = \frac{V_2 - V_{BE}}{R_E} \cong \frac{V_2}{R_E}$$

Also, $V_C = V_{CC} - I_C R_L$

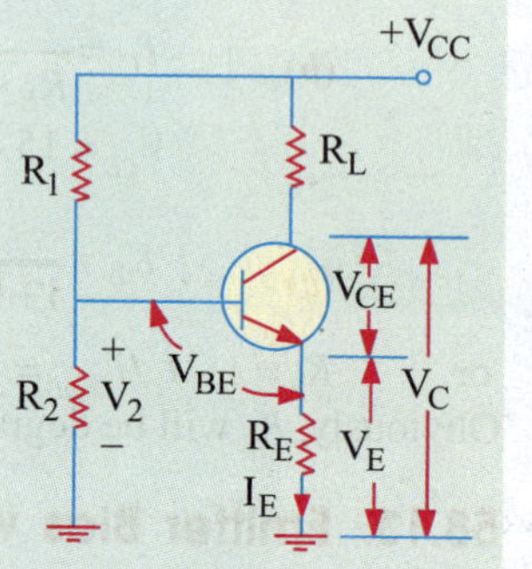

Fig. 58.16

* It is also known as Universal Bias Stabilization Circuit.

$$V_{CE} = V_C - V_E = V_{CC} - I_C R_L - I_E R_E$$
$$\cong V_{CC} - I_C(R_L + R_E) \quad \because \quad I_C \cong I_E$$

As before, $$I_{C(sat)} \cong \frac{V_{CC}}{R_L + R_E}$$

It is seen from above calculations that value of β was never used anywhere. The base voltage is set by V_{CC} and R_1 and R_2. The *dc* bias circuit is independent of transistor β. That is why it is such a very popular bias circuit.

$$K_\beta = \frac{1}{1 + \beta R_E/(R_1 \| R_2)}$$

$$S = \frac{1 + (R_1 \| R_2)/R_E}{1 + (R_1 \| R_2)/(1+\beta)R_E}$$

$$\cong \frac{1 + (R_1 \| R_2)/R_E}{1 + (R_1 \| R_2)/\beta R_E}$$

Using Thevenin's Theorem

More accurate results can be obtained by Thevenizing the voltage divider circuit as shown in Fig. 58.17. The first step is to open the base lead at point *A* and remove the transistor along with R_L and R_E thereby leaving the voltage divider circuit behind as in Fig. 58.17 (*a*) and (*b*).

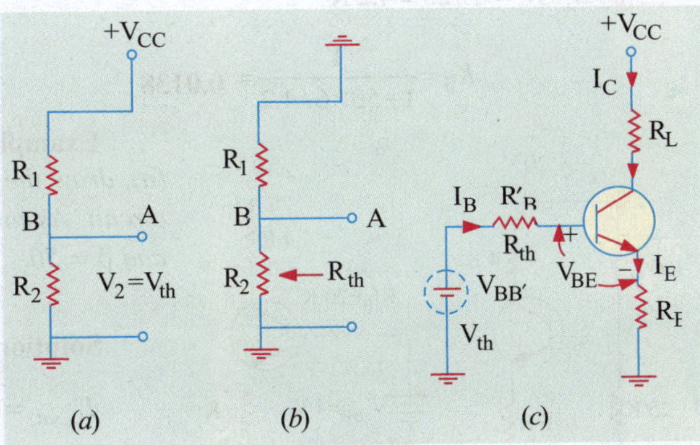

Fig. 58.17

$$V_{th} = V_2 = V_{CC} \cdot \frac{R_2}{R_1 + R_2} \quad \text{and} \quad R_{th} = R_1 \| R_2 = \frac{R_1 R_2}{(R_1 + R_2)}$$

The original circuit is reduced to that shown in Fig. 58.17 (*c*) where $V_{th} = V_{BB'}$ ∴ $R_{th} = R_B'$. Now, applying *KVL* to the base-emitter loop, we get

$$V_{BB'} - I_{B'} R_B' - V_{BE} - I_E R_E = 0$$

Substituting the value of $I_E = (1 + \beta) I_B$ in (*i*) above, we get

$$I_B = \frac{V_{BB}' - V_{BE}}{R_B' + (1+\beta)R_E}$$

However, if we substitute the value of $I_B = I_E/(1 + \beta)$, we get

$$I_E = \frac{V_{BB}' - V_{BE}}{R_E + R_B'/(1+\beta)}$$

$$V_{CE} = V_{CC} - I_C R_L - I_E R_E \cong V_{CC} - I_C(R_L + R_E)$$

The above results could also be obtained directly by applying β-rule (Art. 57.24)

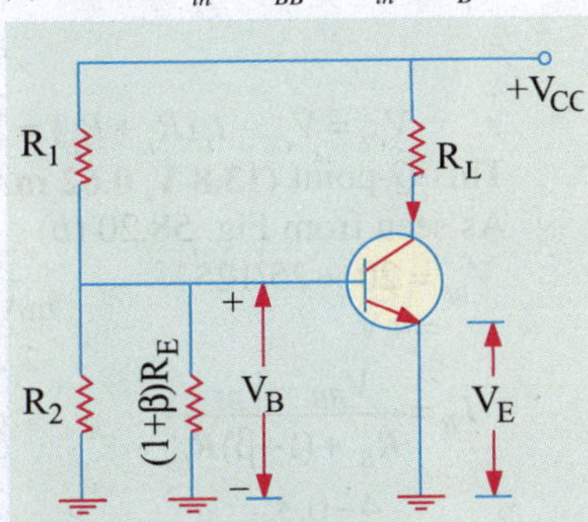

Fig. 58.18

Using β-rule

As per β-rule (Art. 57.24) when R_E is transferred to the base circuit, it becomes $(1 + \beta) R_E$ and is in parallel with R_2 as shown in Fig. 58.18. Now, V_{CC} drops over R_1 and $R_2 \| (1 + \beta) R_E$

$$\therefore\ V_B = V_{CC}\frac{R_2 \parallel (1+\beta)R_E}{R_1 + R_2 \parallel (1+\beta)R_E}$$

$$V_E = V_B - V_{BE};\ I_E = V_E/R_E \quad \text{and so on.}$$

Example. 58.10. *For the circuit of Fig. 58.19, find (a) $I_C(sat)$, (b) I_C, (c) V_{CE}, (d) K_β. Neglect V_{BE} and take $\beta = 50$.* **(Electronics, Gorakhpur Univ.)**

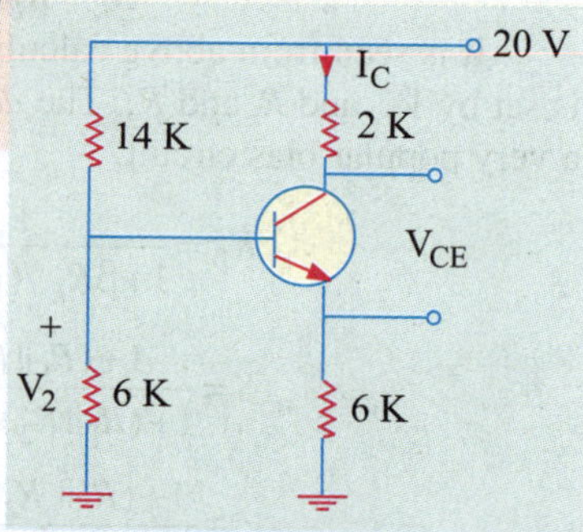

Fig. 58.19

Solution.

(*a*) $$I_{C(sat)} = \frac{V_{CC}}{R_L + R_E} = \frac{20}{2+6} = \textbf{2.5 mA}$$

(*b*) $I_C \cong I_E \cong V_2/R_E = 6/6 =$ **1 mA**

(*c*) $V_{CE} = V_{CC} - I_C(R_L + R_E) = 20 - 1\,(2+6) =$ **12 V**

(*d*) $R_1 \parallel R_2 = 84/20 = 4.2$ K

$$K_\beta = \frac{1}{1 + 50\times 6/4.2} = \textbf{0.0138}$$

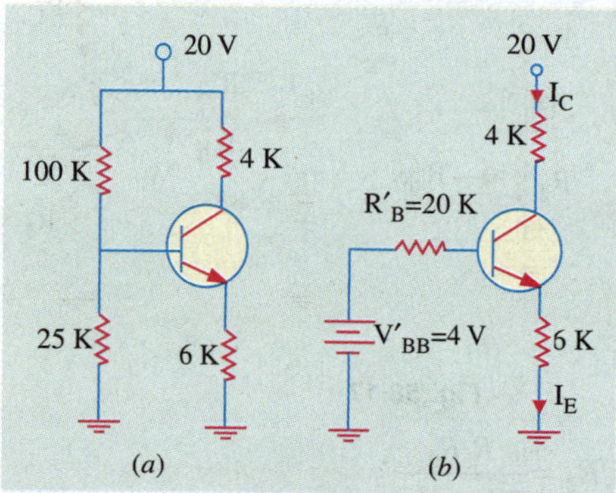

Fig. 58.20

Example 58.11. *For the circuit shown in Fig. 58.20 (a), draw the dc load line and mark the Q-point of the circuit. Assume germanium material with V_{BE} = 0.3 V and $\beta = 50$.* **(Electronics-I, M.S. Univ. 1991)**

Solution. $V_{CE(cut-off)} = V_{CC} = 20$ V

$$I_{C(sat)} = \frac{V_{CC}}{R_L + R_E} = \frac{20}{4+6} = 2 \text{ mA}$$

(*a*) Approximate Method

$$V_2 = V_{CC}\frac{R_2}{R_1 + R_2} = 20\times\frac{25}{125} = 4 \text{ V}$$

$$I_E = \frac{V_2 - V_{BE}}{R_E} = \frac{4 - 0.3}{6} = 0.62 \text{ mA}$$

$\therefore\quad V_{CE} = V_{CC} - I_C(R_L + R_E) = 20 - 0.62\times 10 =$ **13.8 V**

This *Q*-point **(13.8 V, 0.62 mA)** is shown in Fig. 58.21 (*a*).

As seen from Fig. 58.20 (*b*)

$$V_{BB} = 20 \times 25/125 = 4 \text{ V}$$

$$I_B = \frac{V_{BB}' - V_{BE}}{R_{B'} + (1+\beta)R_E} = \frac{4-0.3}{20 + 51\times 6} = 11.3\ \mu\text{A}$$

$$I_C = \beta I_B = 50 \times 11.3 = 0.567 \text{ mA}$$

$$I_E = (1+\beta)I_B = 576.3\ \mu\text{A} = 0.576 \text{ mA}$$

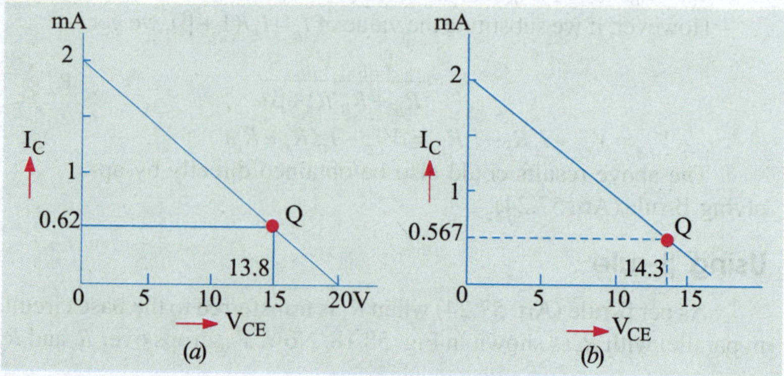

Fig. 58.21

$\therefore \quad V_{CE} = V_{CC} - I_C R_C - I_E R_E = 20 - 0.567 \times 4 - 0.576 \times 6 = \mathbf{14.276\ V}$

The new and more accurate Q-point **(14.276 V, 0.567 mA)** is shown in Fig. 58.21 (*b*).

58.15. Load Line and Output Characteristics

In order to study the effect of bias conditions on the performance of a *CE* circuit, it is necessary to superimpose the dc load line on the transistor output (V_{CE}/I_C) characteristics. Consider a silicon *NPN* transistor which is connected in *CE* configuration (Fig. 58.22) and whose output characteristics are given in Fig. 58.23. Let its $\beta = 100$.

First, let us find the cut-off and saturation points for drawing the dc load line and then mark in the Q-point.

$$I_{C(sat)} = 10/2 = 5 \text{ mA} \qquad \text{—point } B \text{ in Fig. 58.23}$$
$$V_{CE(cut\text{-}off)} = V_{CC} = 10 \text{ V} \qquad \text{—point } A \text{ in Fig. 58.23}$$

The load line is drawn in Fig. 58.23 below.

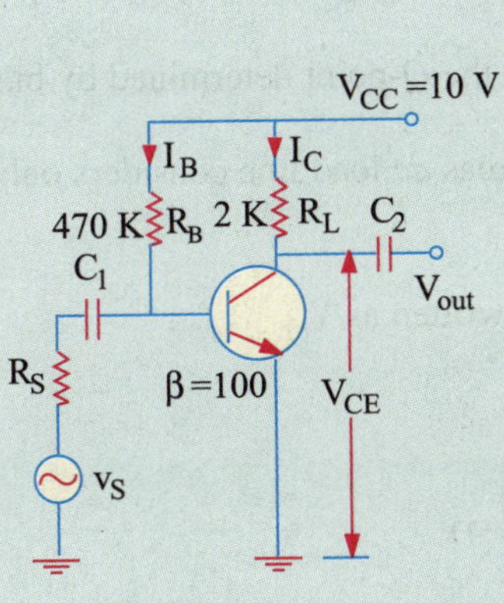

Fig. 58.22

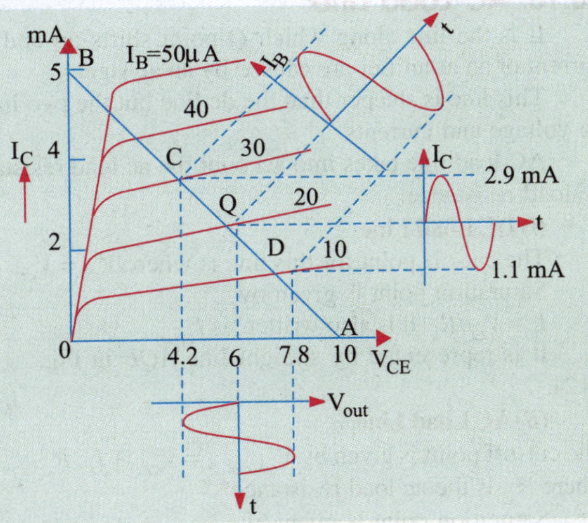

Fig. 58.23

Actual $\quad I_B = \dfrac{V_{CC} - V_{BE}}{R_B} = \dfrac{10 - 0.7}{470} = 20 \mu\text{A}$

$$I_C = \beta I_B = 100 \times 20 = 2000\ \mu\text{A} = 2 \text{ mA}$$
$$V_{CE} = V_{CC} - I_C R_L = 10 - 2 \times 2 = 6 \text{ V}$$

This locates the Q-point in Fig. 58.23.

Suposse an ac input signal voltage injects a sinusoidal base current of peak value 10 μA into the circuit of Fig. 58.22. Obviously, it will swing the operating or Q-point up and down along the load line.

When positive half-cycle of I_B is applied, the Q-point shifts to point C which lies on the (20 + 10) = 30 mA line.

Similarly, during negative half-cycle of input base current, Q-point shifts to point D which lies on the (20 – 10) = 10 μA line.

By measurement, at point C, $I_C = 2.9$ mA. Hence, $V_{CE} = 10 - 2 \times 2.9 = 4.2$ V

Similarly, at point D, I_C measures 1.1 μA.Hence, $V_{CE} = 10 - 2 \times 1.1 = 7.8$ V

It is seen that V_{CE} decreases from 6 V to 4.2 V *i.e.* by a peak value of (6 – 4.2) = 1.8 V when base current goes positive. On the other hand, V_{CE} increases from 6 V to 7.8 V *i.e.* by a peak value of (7.8 – 6) = 1.8 V when input base current signal goes negative. Since changes in V_{CE} represent changes in output voltage, it means that when input signal is applied, I_B varies according to the signal amplitude and causes I_C to vary, thereby producing voltage variations.

Incidentally, it may be noted that variations in voltage drop across R_L are exactly the same as in V_{CE}.

Steady drop across R_L when no signal is applied = 2 × 2 = 4 V. When base signal goes positive, drop across R_L = 2 × 2.9 = 5.8 V. When base signal goes negative, I_C = 1.1 mA and drop across R_L = 2 × 1.1 = 2.2 V.

Hence, voltage variation is = 5.8 − 4 = 1.8 V during positive input half-cycle and 4 − 2.2 = 1.8 V during negative input half-cycle. Obviously, rms voltage variation $\frac{1.8}{\sqrt{2}} = 1.27\,V$

Now, power dissipated in R_L by ac component of output voltage is R_L

$P_{ac} = 1.27^2/2 = 0.81$ mW and $P_{dc} = I_C^2 R = 2^2 \times 2 = 8$ mW

Total power dissipated in R_L = 8.81 mW.

58.16. AC Load Line

It is the line along which *Q*-point shifts up and down when changes in output voltage and current of an amplifier are caused by an ac signal.

This line is steeper than the dc line but the two intersect at the *Q*-point determined by biasing dc voltage and currents.

AC load line takes into account the ac load resistance whereas *dc* load line considers only the dc load resistance.

(*i*) DC Load Line

The cut-off point for this line is where $V_{CE} = V_{CC}$. It is also written as $V_{CE(cut\text{-}off)}$.

Saturation point is given by

$I_C = V_{CC}/R_L$. It is also written as $I_{C(sat)}$.

It is represented by straight line *AQB* in Fig. 58.24.

(*ii*) AC Load Line

The cut-off point is given by $V_{CE(out\text{-}off)} = V_{CEQ} + I_{CQ} R_{ac}$ where R_{ac} is the ac load resistance*.

Saturation point is given by

$$I_{C(sat)} = I_{CQ} + V_{CEQ}/R_{ac}.$$

It is represented by straight line *CQD* in Fig. 58.24.

The slope of the ac load line is given by $y = 1/R_{ac}$.

It is seen from Fig. 58.24 that maximum possible positive signal swing is = $I_{CQ} R_{ac}$. Similarly, maximum possible negative signal swing is V_{CEQ}. In other words, peak-signal handling capacity is limited to $I_{CQ} R_{ac}$ or V_{CEQ} whichever is smaller.

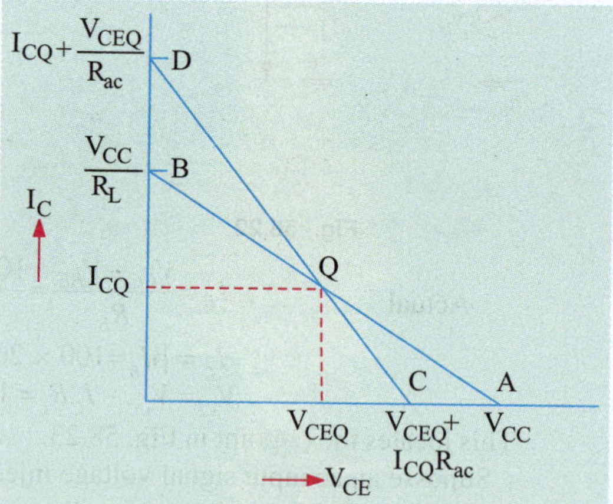

Fig. 58.24

Example 58.12. *Draw the dc and ac load lines for the CE circuit shown in Fig. 58.25 (a). What is the maximum peak-to-peak signal that can be obtained ?*

(Applied Electronics, Kerala Univ. 1991)

Solution. DC Load Line [Fig. 58.25 (*b*)]

$V_{CE(cut\text{-}off)} = V_{CC}$ = 20 V (point A) and $I_{C(sat)} = V_{CC}/(R_L + R_E)$ = 20/5 = 4mA (point B). Hence, *AQB* represents dc load line for the given circuit.

Approximate bias conditions can be quickly found by assuming that I_B is too small to affect the base bias in Fig. 58.25 (*a*).

* Written as r_L in Art. 59.4

$V_2 = 20 \times 4/(4+16) = 4$ V

If we neglect V_{BE}, $V_2 = V_E$;

$$I_E = \frac{V_E}{R_E} = \frac{V_2}{R_E} = \frac{4}{2}$$
$$= 2 \text{ mA}$$

Also, $I_C \cong I_E = 2$ mA.

Hence, $I_{CQ} = 2$ mA

The corresponding value of V_{CEQ} can be found by drawing dotted line in Fig. 58.25 (*b*) or may be calculated as under :

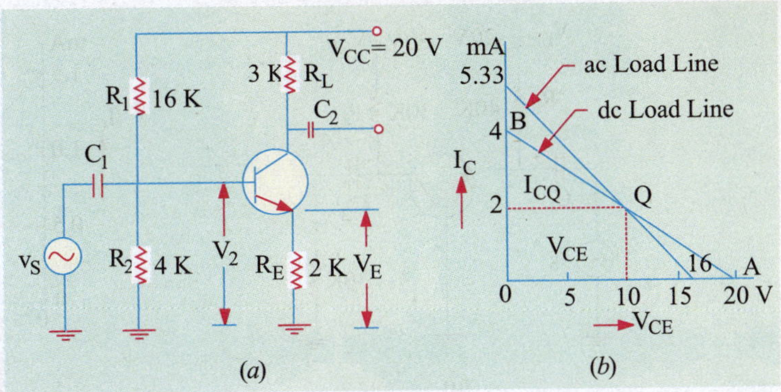

Fig. 58.25

$$V_{CEQ} = V_{cc} - I_C R_C - I_E R_E \cong V_{CEQ} \cong V_{CC} - I_c (R_C + R_E)$$
$$= 20 - 2(3+2) = 10\text{V}$$

AC Load Line

Cut-off point,

$$V_{CE(cut\text{-}off)} = V_{CEQ} + I_{CQ} R_{ac}$$

Now, for the given circuit, ac load resistance is $R_{ac} = R_C = 3$K

Cut-off point $= 10 + 2 \times 3 = 16$ V [Fig. 58.25 (*b*)].

Saturation point, $I_{c(sat)} = I_{CQ} + \dfrac{V_{CEQ}}{R_{ac}} = 2 + \dfrac{10}{3} = 5.33$ mA

Hence, line joining 16-V point and 5.33 mA point gives ac load line as shown in Fig. 58.25 (*b*). As expected, this line passes through the *Q*-point.

Now, $I_{CQ}.\ R_{ac} = 2 \times 3 = 6$ V and $V_{CEQ} = 10$ V. Taking the smaller quantity, maximum peak output signal = 6 V. Hence, peak-to-peak value = 2 × 6 = 12 V.

Example 58.13. *Find the dc and ac load lines for CE circuit shown in Fig. 58.26 (a).*

Solution. The given circuit is identical to that shown in Fig. 58.25 (*a*) except for the addition of 6-K resistor. This makes $R_{AC} = 3\text{ K} \| 6\text{ K} = 2$ K because collector feeds these two resistors in parallel. The dc loadline would remain unaffected. Change would occur only in the ac load line.

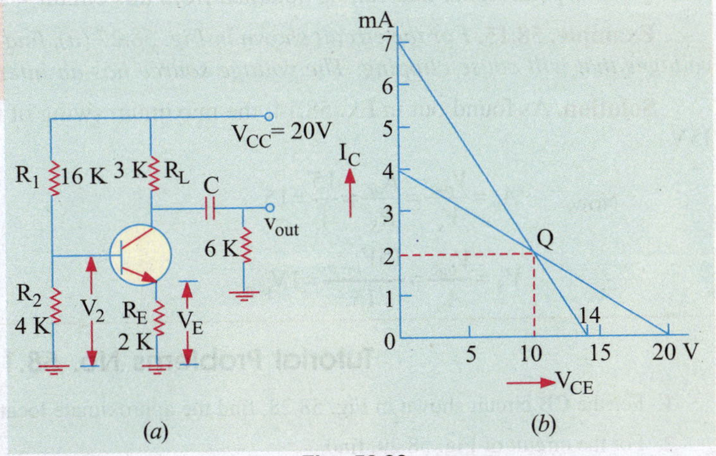

Fig. 58.26

AC Load Line

$V_{CE(cut\text{-}off)} = V_{CE} + I_{CQ} R_{ac} = 10 + 2 \times 2 =$ **14 V**

$I_{C(sat)} = I_C + V_{CE}/R_{ac} = 2 + 10/2 =$ **7 mA**

Example 58.14. *Draw the dc and ac load lines for the CB circuit shown in Fig. 58.27 (a). Which swing starts clipping first ?*

Solution. The dc load line passes through cut-off point of 30 V and saturation point of V_{CC}/R_L = 1 mA.

Now, $I_E \cong 20/40 = 0.5$ mA ; $I_C \cong I_E = 0.5$ mA; $I_{CQ} = 0.5$ mA

$V_{CB} = V_{CC} - I_C R_C = 30 - 0.5 \times 30 = 15$ V ; $V_{CBQ} = 15$ V

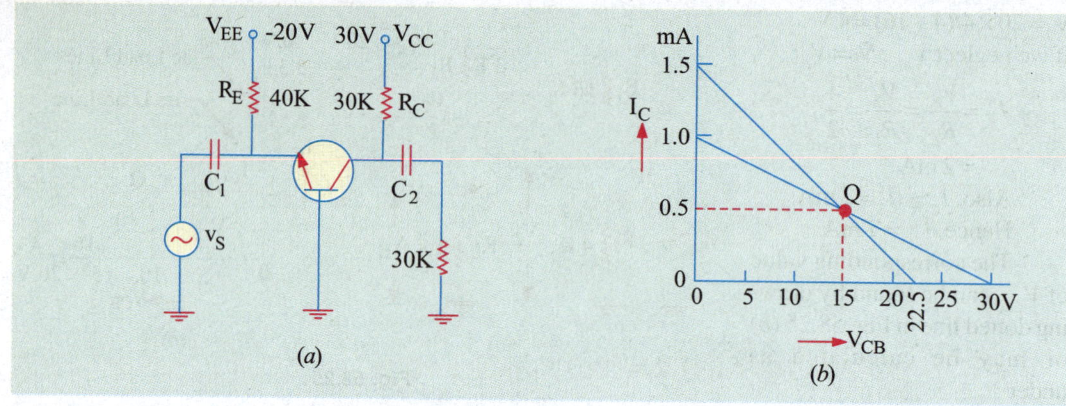

Fig. 58.27

Hence, *dc* operating point or *Q*-point is (15 V, 0.5 mA) as shown in Fig. 58.27 (*b*).

The cut-off point for ac load line is = $V_{CBQ} + I_{CQ} R_{ac}$

Since collector sees an ac load of two 30 K resistors in parallel, hence R_{ac} = 30 K || 30 K =15 K.

$\therefore \quad V_{CBQ} + I_{CQ} R_{ac}$ =15 + 0.5 × 15 = 22.5 V

Saturation current for ac line is = $I_{CQ} + V_{CBQ}/R_{ac}$ = 0.5 + 15/15 = 1.5 mA

The line joining these two points (and also passing through *Q*) gives the ac load line as shown in Fig. 58.27 (*b*).

Note. Knowing ac cut-off point and *Q*-point, we can draw the ac load line. Hence, we need not find the value of saturation current for this purpose.

As seen, positive swing starts clipping first because $I_{CQ} R_{ac}$ is less than V_{CBQ}. Obviously, maximum peak-to-peak signal that can be obtained from this circuit = 2 × 7.5 = 15 V.

Example. 58.15. *For the circuit shown in Fig. 58.27 (a), find the approximate value of source voltages that will cause clipping. The voltage source has an internal resistance of 1 K.*

Solution. As found out in Ex. 58.14, the maximum swing of the unclipped output = 2 × 7.5 = 15 V.

Now, $$A_v = \frac{V_{out}}{V_s} = \frac{R_{ac}}{R_S} = \frac{15}{1} = 15$$

$$\therefore \quad V_S = \frac{V_{out}}{A_v} = \frac{15 V_{p-p}}{15} = 1 V_{\text{p-p}}$$

Tutorial Problems No. 58.1

1. For the CB circuit shown in Fig. 58.28, find the approximate location of *Q*-point. **[10 V, 2 mA]**
2. For the circuit of Fig. 58.29, find
   (*a*) dc operating-point,
   (*b*) maximum peak-to-peak unclipped signal. **[(15 V, 0.5 mA) ; 10 V_{p-p}]**
3. What is the maximum peak-to-peak signal that can be obtained from the circuit of Fig. 58.30 ? **[20 V_{p-p}]**
4. Find the value of maximum peak-to-peak output of signal that can be obtained from the circuit of Fig. 58.31. **[10 V_{p-p}]**

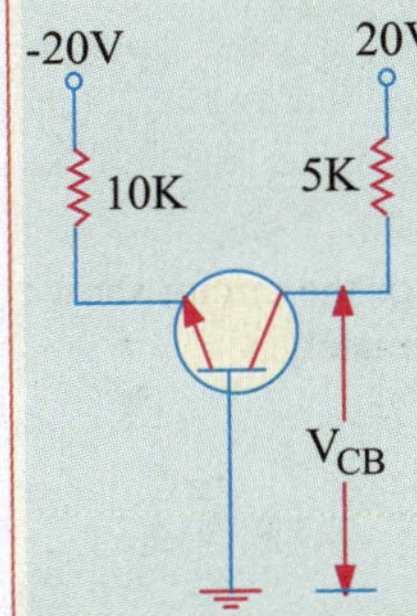

Fig. 58.28

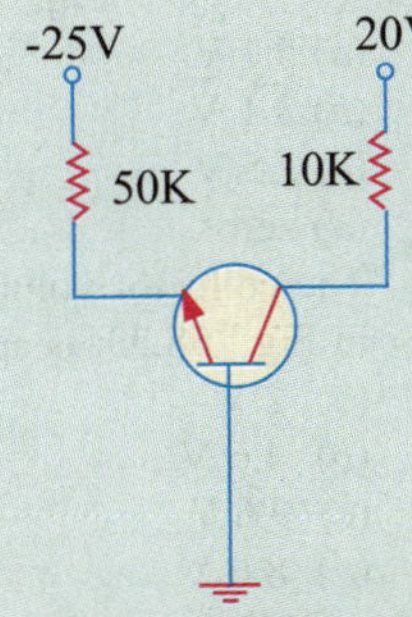

Fig. 58.29

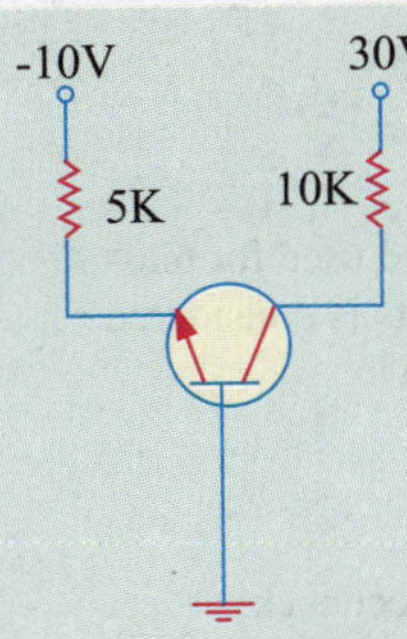

Fig. 58.30

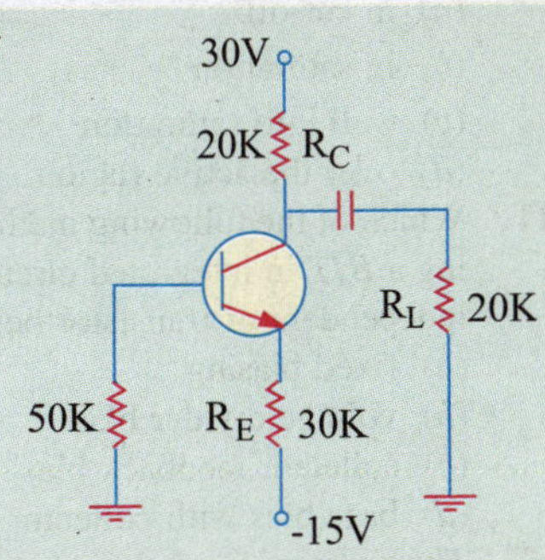

Fig. 58.31

OBJECTIVE TESTS — 58

1. The dc load line of a transistor circuit

(*a*) has a negative slope

(*b*) is a curved line

(*c*) gives graphic relation between I_C and I_B

(*d*) does not contain the *Q*-point.

2. The positive swing of the output signal in a transistor circuit starts clipping first when *Q*-point of the circuit moves

(*a*) to the centre of the load line

(*b*) two-third way up the load line

(*c*) towards the saturation point

(*d*) towards the cut-off point.

3. To avoid thermal run away in the design of analog circuit, the operating point of the *BJT* should be such that it satisfies the condition

(*a*) $V_{CE} = 1/2\ V_{CC}$

(*b*) $V_{CE} < 1/2\ V_{CC}$

(*c*) $V_{CE} > 1/2\ V_{CC}$

(*d*) $V_{CE} < 0.78\ V_{CC}$

4. In the case of a *BJT* amplifier, bias stability is achieved by

(*a*) keeping the base current constant

(*b*) changing the base current in order to keep the I_C and V_{CB} constant

(*c*) keeping the temperature constant

(*d*) keeping the temperature and the base current constant

5. For a transistor amplifier with self-biasing network, the following components are used :

$R_1 = 4\ \text{k}\ \Omega$, $R_2 = 4\ \text{k}\ \Omega$ and $R_E = 1\ \text{k}\ \Omega$

the approximate value of the stability factor '*S*' will be

(*a*) 4

(*b*) 3

(*c*) 2

(*d*) 1.5

6. A transistor circuit employing base bias with collector feedback has greater stability than the one without feedback because

(*a*) I_C decrease in magnitude

(*b*) V_{BE} is decreased

(*c*) of negative feedback effect

(*d*) I_C becomes independent of β.

7. The universal bias stabilization circuit is most popular because

(*a*) I_C does not depend on transistor characteristics

(*b*) its β-sensitivity is high

(*c*) voltage divider is heavily loaded by transistor base

(*d*) I_C equals I_E.

8. Improper biasing of a transistor circuit leads to

(*a*) excessive heat production at collector terminal

(*b*) distortion in output signal

(*c*) faulty location of load line

(*d*) heavy loading of emitter terminal.

9. The negative output swing in a transistor circuit starts clipping first when *Q*-point

(*a*) has optimum value

(*b*) is near saturation point

(*c*) is near cut-off point

(*d*) is in the active region of the load line.

10. When a *BJT* is employed as an amplifier, it operates

(*a*) in cut-off
(*b*) in saturation
(*c*) well into saturation
(*d*) over the active region.

11. Which of the following method used for biasing a *BJT* in integrated circuits is considered independent of transistor beta?
(*a*) fixed biasing
(*b*) voltage divider bias
(*c*) collector feedback bias
(*d*) base bias with collector feedback.

12. The voltage V_0 of the circuit shown in Fig. 58.32 is,

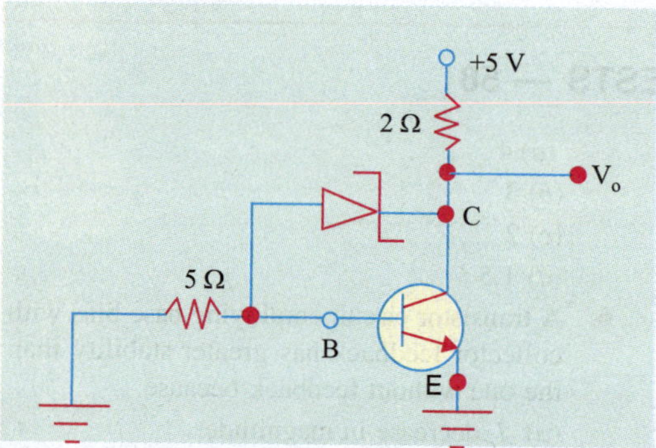

Fig. 58.32

(*a*) 5.1 V
(*b*) 3.1 V
(*c*) 2.5 V
(*d*) zero

13. The collector voltage V_C of the circuit shown in Fig. 58.33, is approximately
(*a*) 2 V
(*b*) 4.6 V
(*c*) 9.6 V
(*d*) 8.6 V

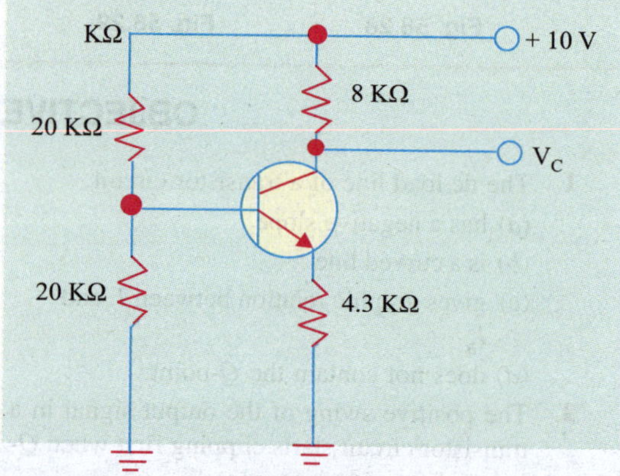

Fig. 58.33

ANSWERS

1.	(*a*)	**2.**	(*d*)	**3.**	(*c*)	**4.**	(*a*)	**5.**	(*b*)	**6.**	(*c*)
7.	(*a*)	**8.**	(*b*)	**9.**	(*b*)	**10.**	(*d*)	**11.**	(*b*)	**12.**	(*d*)
13.	(*a*)										

CHAPTER 59

TRANSISTOR EQUIVALENT CIRCUITS AND MODELS

Learning Objectives

- General
- DC Equivalent Circuit
- AC Equivalent Circuit
- Equivalent Circuit of a CB Amplifier
- Effect of Source Resistance R_s on Voltage Gain
- Equivalent Circuit of a CE Amplifier
- Equivalent Circuit of a CC Amplifier
- Small-signal Low-frequency Model or Representation
- T-Model
- Formulas for T-Equivalent of a CC Circuit
- What are h-parameters ?
- Input Impedance of a Two Port Network
- Voltage Gain of a Two Port Network
- The h-parameters of an Ideal CB Transistor
- The h-parameters of an ideal CE Transistor
- Approximate Hybrid Equivalent Circuits
- Transistor Amplifier Formulae Using h-parameters
- Typical Values of Transistor h-parameters
- Approximate Hybrid Formulas
- Common Emitter h- parameter Analysis
- Common Collector h-parameter Analysis
- Conversion of h-parameters

Generalised hybrid parameter equivalent circuit

59.1. General

We will begin by idealizing a transistor with the help of simple approximations that will retain its essential features while discarding its less important qualities. These approximations will help us to ***analyse transistor circuits easily and rapidly.***

We will discuss only the ***small signal*** equivalent circuits in this chapter. Small signal operation is that in which the ac input signal voltages and currents are in the order of ± 10 per cent of *Q*-point voltages and currents.

There are two prominent schools of thought today regarding the equivalent circuit to be substituted for the transistor. The two approaches make use of

(*a*) four *h*-parameters of the transistor and the values of circuit components,

(*b*) the beta (β) of the transistor and the values of the circuit components.

Since long, industrial and educational institutions have heavily relied on the hybrid parameters because they produce more accurate results in the analysis of amplifier circuits. In fact, hybrid-parameter equivalent circuit continues to be popular even to-day. But their use is beset with the following difficulties :

1. The values of *h*-parameters are not so readily or easily available.
2. Their values vary considerably with individual transistors ***even of the same type number.***
3. Their values are limited to a particular set of operating conditions for reasonably accurate results.

The second method which employs transistor beta and resistance values is gaining more popularity of late. It has the following advantages :

1. The required values are easily available;
2. The procedure followed is simple and easy to understand;
3. The results obtained are quite accurate for the study of amplifier circuit characteristics.

To begin with, we will consider the second method first.

59.2. DC Equivalent Circuit

(*a*) CB Circuit

In an ideal transistor, $\alpha = 1$ which means that $I_C = I_E$.

The emitter diode acts like any ***forward-biased ideal diode.*** However, due to transistor action, collector diode acts as a ***current source.*** In other words, for the purpose of drawing dc equivalent circuit, we can view an ideal transistor as nothing more than a rectifier diode in emitter and a current source in collector. In the dc equivalent circuit of Fig. 59.1 (*b*), current arrow always points in the direction of conventional current.

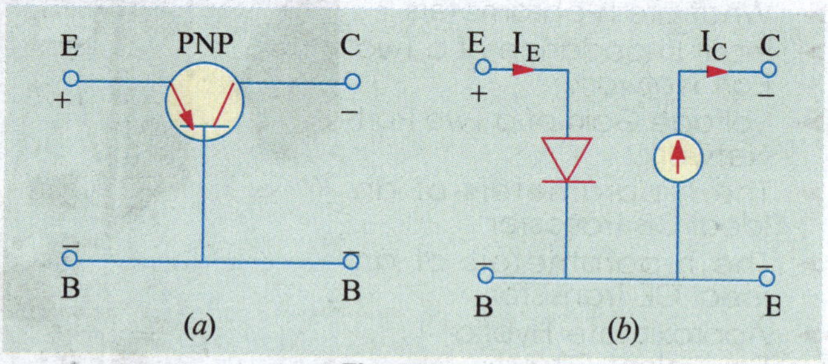

Fig. 59.1

As per the polarities of transistor terminals (Art. 57.3) shown in Fig. 59.1 (*a*), emitter current flow from *E* to *B* and collector current from *B* to *C*.

The dc equivalent circuit shown in Fig. 59.2 for an *NPN* transistor is exactly similar except that direction of current flow is opposite.

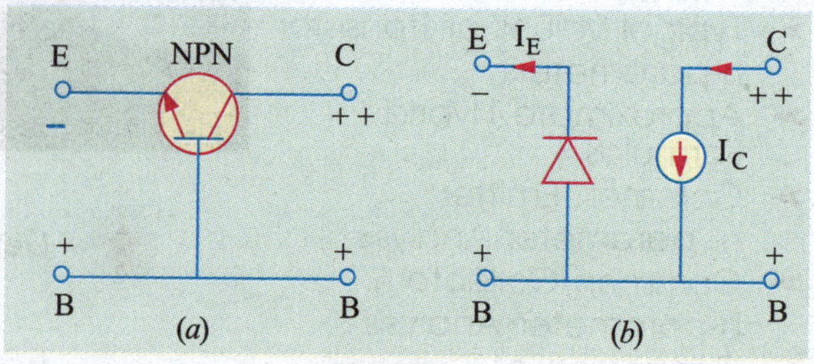

Fig. 59.2

***(b)* CE Circuit**

Fig. 59.3 shows the dc equivalent circuit of an *NPN* transistor when connected in the *CE* configuration. Direction of current flow can be easily found by remembering the transistor polarity rule given in Art 57.3. In an ideal *CE* transistor, we disregard leakage current and take a.c beta equal to dc beta.

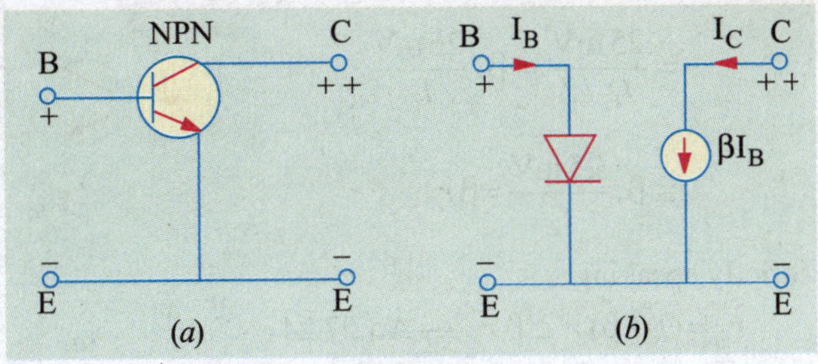

Fig. 59.3

59.3. AC Equivalent Circuit

***(a)* CB Circuit**

In the case of ***small input*** ac signals, the emitter ***diode does not rectify,*** instead it offers resistance called ac ***resistance.*** As usual, collector diode acts as a current source.

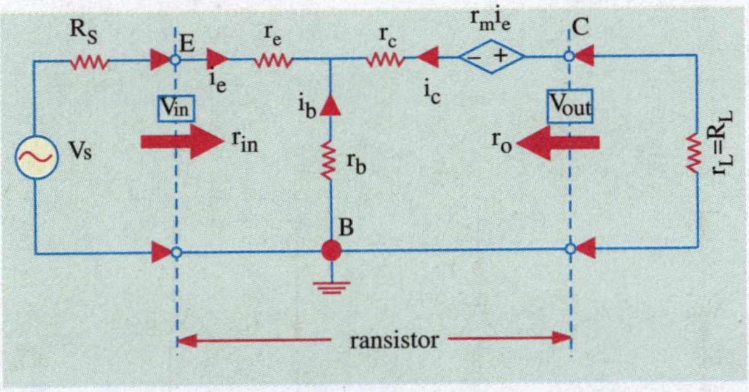

Fig. 59.4

Fig. 59.4 shows the ac equivalent circuit of a transistor connected in the *CB* configuration.* Here, ac resistance offered by the emitter diode is

r_{ac} = junction resistance (r_j) + base-spreading resistance (r_B)**

$= r_j \quad (\because \; r_B$ is negligible)

$$= \frac{25\,\text{mV}}{I_E}$$

— where I_E is dc emitter current in mA

It is written as r_e' signifying junction resistance of the emitter *i.e.* a.c resistance looking into the emitter.

$$r_e' = \frac{25\,\text{mV}}{I_E}$$

Hence, the a.c equivalent circuit of a *CB* circuit becomes as shown in Fig. 59.5. Since changes in collector current are almost equal to changes in emitter current, $\Delta i_c = \Delta i_e$.

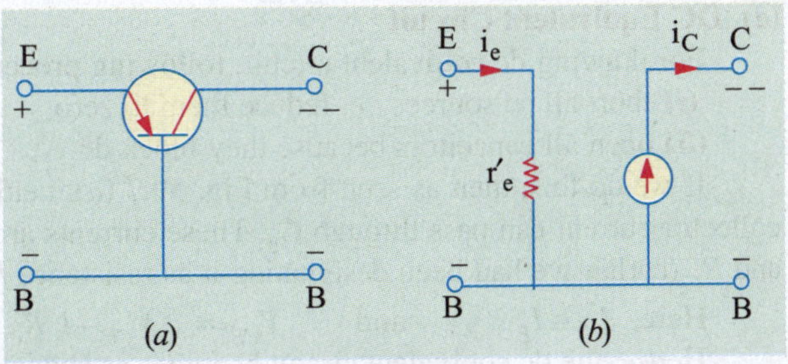

Fig. 59.5

***(b)* CE Circuit**

Fig. 59.6 (*a*) shows the equivalent circuit when an *NPN* transistor has been connected in the *CE* configuration.

The a.c resistance ***looking into the base is***

$$r_{ac} = \frac{25\,\text{mV}}{I_B} \quad \text{— d.c current is } I_B \text{ not } I_E$$

* This circuit is valid both for *PNP* as well as *NPN* transistors because difference in the direction of ac current does not matter.

** Also called bulk-resistance.

$$= \frac{25\,\text{mV}}{I_C/\beta} = \beta.\frac{25\,\text{mV}}{I_C}$$

$$\cong \beta.\frac{25\,\text{mV}}{I_E} = \beta r_e'$$

Strictly speaking,

$r_{ac} = (1+\beta)\, r' \cong \beta\, r_e'$ — Art 57.24

The a.c collector current is β times the base current *i.e.*, $i_c = \beta\, i_b$.

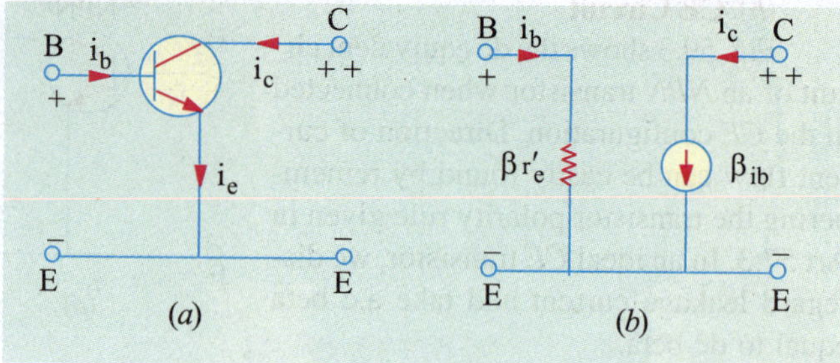

Fig. 59.6

59.4. Equivalent Circuit of a CB Amplifier

In Fig. 59.7 (*a*) is shown the circuit of a common-base amplifier. As seen, emitter is forward-biased by $-V_{EE}$ and collector is reverse-biased by $+V_{CC}$. The a.c signal source voltage υ_s drives the

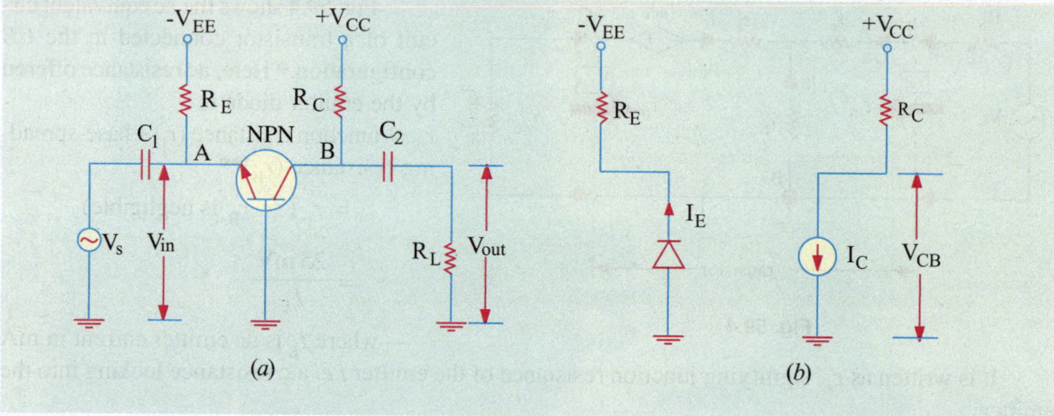

Fig. 59.7

emitter. It produces small fluctuations in transistor voltage and currents in the output circuit. It will be seen that ac output voltage is amplified because it is more than V_s.

(a) DC Equivalent Circuit

For drawing dc equivalent circuit, following procedure should be adopted :

(*i*) short all ac sources *i.e.* reduce them to zero,

(*ii*) open all capacitors because they block dc.

If we do this, then as seen from Fig. 59.7 (*a*), neither emitter current can pass through C_1 nor collector current can pass through C_2. These currents are confined to their respective resistances R_E and R_C (earlier we had been designating it as resistance R_L).

Here, $I_C \cong I_E$ and $V_{CB} = V_{CC} - I_C R_C$

Hence, the dc equivalent circuit becomes as shown in Fig. 59.7 (*b*).

(b) AC Equivalent Circuit

For drawing ac equivalent circuit, following procedure is adopted :

(*i*) all dc sources are shorted *i.e.* they are treated as ac ground,

(*ii*) all coupling capacitors like C_1 and C_2 in Fig. 59.7 (*a*) are shorted and

(*iii*) emitter diode is replaced by its a.c resistance $r_{ac} = r_e'$

$$r_e' = \frac{25\,\text{mV}}{I_E}$$

where I_E is dc emitter current in mA.

As seen by the input a.c signal, it has to feed R_E and r_e' in parallel [Fig. 59.8 (*a*)]. As looked from point A in Fig. 59.7 (*a*), R_E is grounded through V_{EE} which has been shorted and r_e' is grounded *via* the base.

Similarly, collector has to feed R_C and R_L which are connected in parallel across it *i.e* at point *B* in Fig. 59.7 (*a*). The ac signal in collector sees an output load resistance of $r_L = R_C \parallel R_L$.

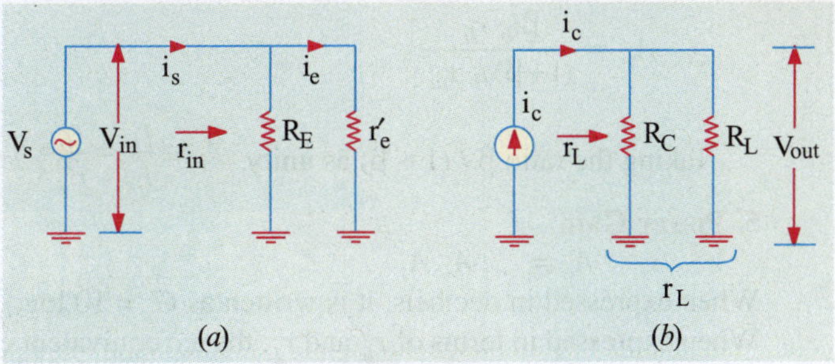

Fig. 59.8

Hence, ac equivalent circuit is as shown in Fig. 59.8 (*b*). Here, collector diode itself has been shown as a current source.

Following two points are worth noting :

(*i*) changes in collector signal current are very nearly equal to changes in emitter signal current. Hence, $\Delta i_c \cong \Delta i_e$.,

(*ii*) directions of ac currents shown in the circuit diagram are those which correspond to positive half-cycle of the a.c input voltage. That is why i_c is shown flowing upwards in Fig. 59.8 (*b*).

(c) Principal Operating Characteristics

1. Input Resistance

As seen from Fig. 59.8 (*a*), the input resistance of the circuit (or stage) is given by

$$r_{in} \cong R_E \parallel r_e' = \frac{r_e' R_E}{R_E + r_e'}$$

In practice, R_E is always much greater than r_e' so that parallel combination $R_E \parallel r_e' \cong r_e'$

$\therefore \; r_{in} \cong r_e'$ input resistance of the emitter diode

2. AC Load Resistance

The collector load as seen by output ac signal consists of a parallel combination of R_L and R_C. This has already been designated as r_L.

$$\therefore \qquad r_L = R_L \parallel R_C$$

It should be carefully noted that it is the output resistance ***as seen by collector and not the ac output resistance when looking into the collector.***

Note. In case, R_L has not been connected, then $r_L = R_C$ (Ex. 59.1)

3. Current Gain

It is given by the ratio $A_i = \dfrac{i_c}{i_e} \cong \alpha$

4. Voltage Gain

It is given by the ratio

$$A_v = \frac{V_{out}}{V_{in}}$$

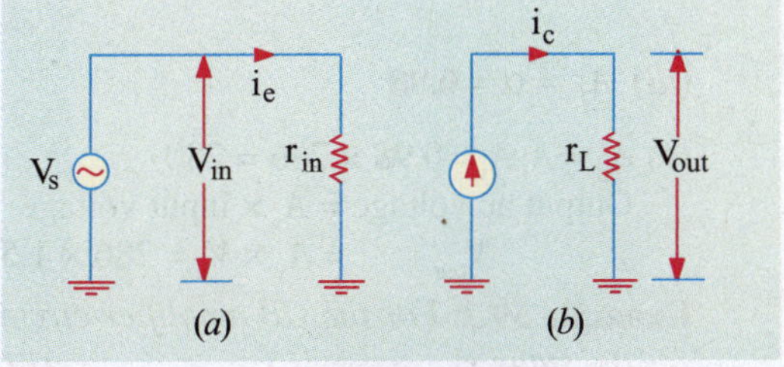

Fig. 59.9

Now, V_{in}* $= i_e\, r_{in} = (1 + \beta)\, i_b \cdot r_{in}$

$V_{out} = i_c\, r_L = \beta i_b\, r_L$

---

* Here, V_{in} equals υ_s because there is no internal resistance of the source. In case it is there, the two will not be equal (Art 59.5).

$$\therefore \quad A_v = \frac{\beta i_b \, r_L}{(1+\beta)\, i_b \, r_{in}}$$

Taking the ratio β / (1 + β) as unity, $A_V \cong \frac{r_L}{r_{in}} = \frac{r_L}{r'_e}$

5. Power Gain

$$A_p = A_v . A_i$$

When expressed in decibels, it is written as $G_p = 10 \log_{10} A_p \, dB$.

When expressed in terms of r_{in} and r_L, the a.c equivalent circuit becomes as shown in Fig. 59.9.

Example 59.1. *For the single-stage CB amplifier shown in Fig. 59.10 (a), find r_{in}, r_L, A_i, A_v and A_p. What would be the rms value of the signal voltage across the load if v_s has an rms value of 1.5 mV? Assume silicon material and transistor α = 0.98.* **(Electronics-II, Madras Univ.)**

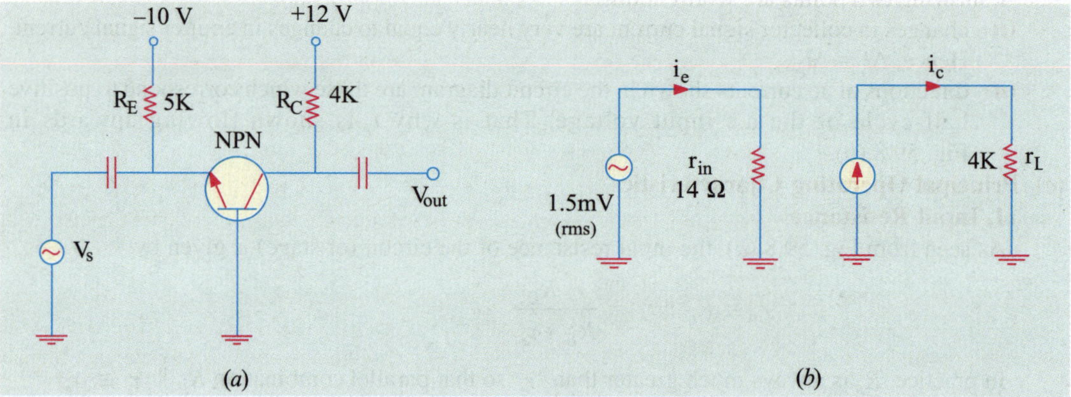

Fig. 59.10

Solution. For finding , let us first find I_E.

$$I_E = \frac{V_{EE} - V_{BE}}{R_E} = \frac{10 - 0.7}{5K} = 1.86 \text{ mA}$$

$$r'_e = \frac{25 \text{ mV}}{I_E \text{ mA}} = \frac{25}{1.86} \cong 14 \Omega$$

(*i*) $r_{in} = r'_e \parallel R_E = 14\ \Omega \parallel 5\ K \cong$ **14 Ω** (*ii*) $r_L = R_C =$ **4 K**

The ac equivalent circuit becomes as shown in Fig. 59.10 (*b*).

(*iii*) $A_i = \alpha =$ **0.98** (*iv*) $A_v = \frac{r_L}{r_{in}} = \frac{r_L}{r'_e} = \frac{4K}{14\Omega} =$ **286**

(*v*) $A_p = A_i A_v = 0.98 \times 286 = 280$ $G_p = 10 \log_{10} 280 =$ **24.5 dB**

Output ac voltage = A_v × input voltage

$$\therefore \quad V_{out} = A_v \times V_{in} = 286 \times 1.5 = 429 \text{ mV} = \textbf{0.429 V}$$

Example 59.2. *For the CB amplifier circuit shown in Fig. 59.11 (a), find*

(*i*) *stage r_{in}* (*ii*) r_L (*iii*) *a.c output voltage V_{out}* (*iv*) *voltage gain A_v.*

Take V_{BE}= 0.7 V.

Solution. $I_E = (20 - 0.7)/30\text{ K} = 0.64$ mA ; $r'_e = 25/0.64 = 39\ \Omega$

(*i*) stage $r_{in} = r'_e \parallel R_E = 39\Omega \parallel 30\text{ K} \cong 39\ \Omega$

(*ii*) $r_L = R_L \parallel R_C = 30\text{ K} \parallel 15\text{ K} =$ **10 K**

The amplifier circuit alongwith its ac equivalent circuit is shown in Fig. 59.11.

(iii) We will first find the value of V_{out} as a drop across $r_L = R_C \parallel R_L$. For that purpose, we will employ rms values.

Now, $$i_e = \frac{V_{in}}{r_{in}} = \frac{V_s}{r_{in}} = \frac{1.6}{39} = 0.04 \text{ mA}^*$$

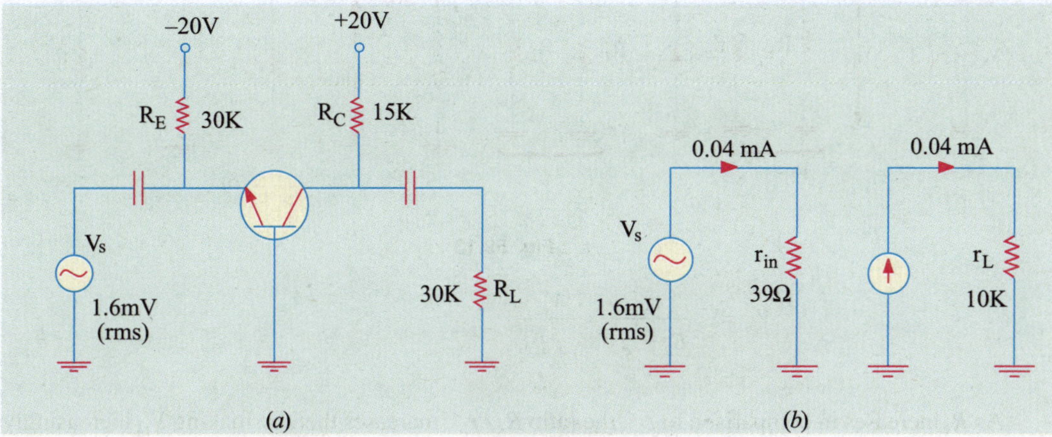

Fig. 59.11

$$i_c = i_e = 0.04 \text{ mA} ; V_{out} = i_c r_L = 0.04 \times 10 \text{ K} = 0.4 \text{ V}$$

(iv) $$A_v = \frac{r_L}{r_{in}} = \frac{r_L}{r_e'} = \frac{10 \text{ K}}{39 \Omega} = \mathbf{256}$$

The rms value of ac output voltage may be found with the help of A_v.

Now, $$A_v = \frac{V_{out}}{V_{in}} \qquad \because V_{out} = A_v . V_{in} = 256 \times 1.6 = 409.6 \text{ mV} = \mathbf{0.4 \text{ V}}$$

59.5. Effect of Source Resistance R_s on Voltage Gain

The voltage gain for the *CB* amplifier circuit has so far been calculated on the assumption that the resistance R_S of the ac signal source is negligible. The voltage gain will decrease as R_S increases because more and more of V_S will drop on R_S rather than on r_{in} and so will not appear in the output.

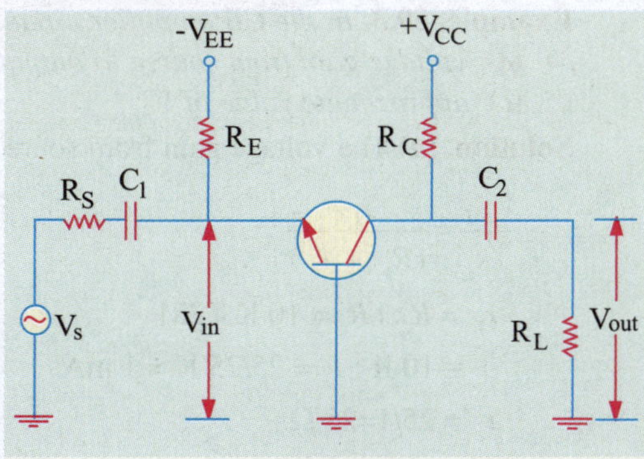

Fig. 59.12

The basic circuit is shown in Fig. 59.12 where R_S is the internal resistance of the ac signal source.

The ac equivalent circuit is shown in Fig. 59.13 (*a*). Here, R_S is in series with ($r_e' \parallel R_E$). The input ac signal voltage V_S drops on these two series resistors. Obviously, V_{in} is the drop across $r_e' \parallel R_E$ and is less than V_S. Using Proportional Voltage Formula,

* Strictly speaking, it is not i_e but i_s . a small part of the ac source current goes through R_E = 30 K and the balance (major part) goes through parallel resistance r_e'. However, if we neglect current through 30 K, then $i_e = i_s$

$$V_{in} = V_s \frac{r_e' \| R_E}{R_s + r_e' \| R_E}$$

In practice, $R_E >> r_e'$, so that $r_e' \| R_E \cong r_e'$

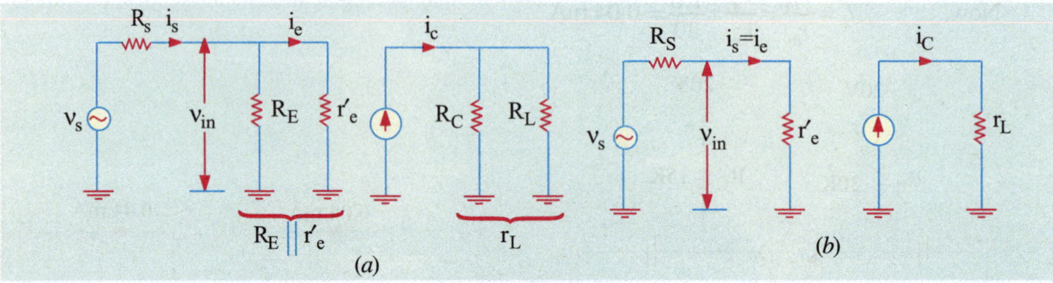

Fig. 59.13

$$\therefore \quad V_{in} = V_s \frac{r_e'}{R_s + r_e'} = \frac{V_s}{1 + \frac{R_s}{r_e'}}$$

As R_S increases in comparison to r_e', the ratio R_S / r_e' increases thereby making V_{in} increasingly less than V_s.

Moreover, as seen from the source, input resistance is

$$R_S + r_e' // R_E \cong R_S + re' \qquad \therefore \quad \frac{V_{out}}{V_s} = \frac{r_L}{(R_S + r_e')} \cong \frac{r_L}{R_S} \quad \text{— if } R_S >> r_e'$$

Making R_S much larger than r_e' is called **swamping out** the emitter diode. This makes voltage gain independent of r_e' and hence the particular transistor is used. The above gain is different from the voltage gain considered from emitter to collector which is

$$\frac{V_{out}}{V_{in}} = \frac{r_L}{r_{in}} \cong \frac{r_L}{r_e'}$$

Example. 59.3. *In the CB amplifier circuit of Fig. 59.14, find*

(a) *voltage gain from source to output* **(b)** *voltage gain from emitter to output*

(c) *approximate value of V_{in}* **(Electronics-I, Patna Univ. 1985)**

Solution. (a) The voltage gain from source to output is given by

$$\frac{V_{out}}{V_S} = \frac{r_L}{(R_S + r_e')}$$

$r_L = R_C \| R_L = 10 \text{ K} \| 2\text{M}$

$\cong 10 \text{ K}$; $I_E = 25/25 \text{ K} = 1 \text{ mA}$

$r_e' = 25/1 = 25\ \Omega$

$$\frac{V_{out}}{V_s} = \frac{10\text{K}}{(1\text{K} + 25\Omega)} \cong \mathbf{10} \qquad \ldots(i)$$

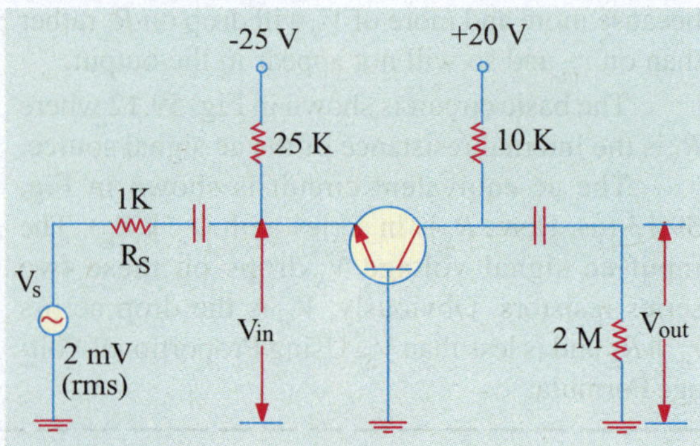

Fig. 59.14

(b) The voltage gain from emitter to output is

$$\frac{V_{out}}{V_{in}} = \frac{r_L}{r_{in}}$$

Now, $r_{in} = R_E // r_e' = 25\text{ K} \parallel 25\,\Omega \cong 25\,\Omega$ $\therefore$ $\dfrac{V_{out}}{V_{in}} = \dfrac{10\,K}{25\Omega} = \mathbf{400}$ ...(*ii*)

(c) The value of V_{in} may be found by the following two ways :

(*i*) As seen from Eq. (*ii*) above, $V_{in} = \dfrac{V_{out}}{400}$, Now, from Eq. (*i*) above, we have

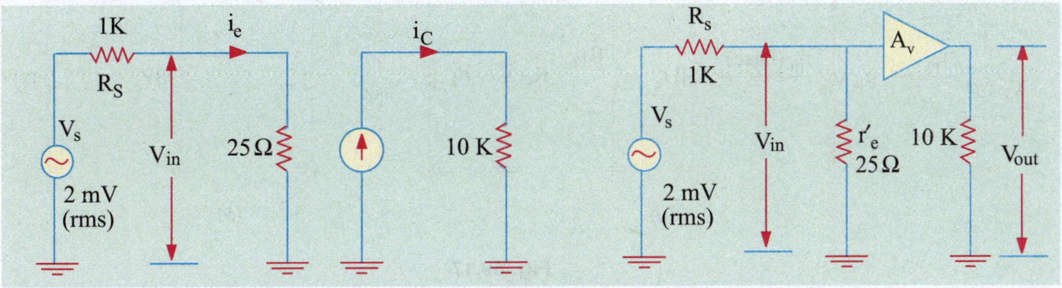

Fig. 59.15

$V_{out} = 10\,V_s = 10 \times 2 = 20\text{ mV}$ $\therefore$ $V_{in} = \dfrac{20}{400} \times 10^{-3} = \mathbf{50\mu V}$

(*ii*) As seen from the ac equivalent circuit of the amplifier (Fig. 59.15), V_s is dropped proportionally on the two series resistances R_S and r_e' (strictly speaking $r_e' \parallel R_E$). The drop across r_{in} is called V_{in}.

$$\therefore \quad V_{in} = V_s \frac{r_e'}{R_s + r_e'} = 2\text{ mV} \times \frac{25}{1000 + 25} \cong \mathbf{50\mu V}$$

59.6. Equivalent Circuit of a CE Amplifier

Consider the simple *CE* amplifier circuit of Fig. 59.16 (*a*) in which base bias has been employed.

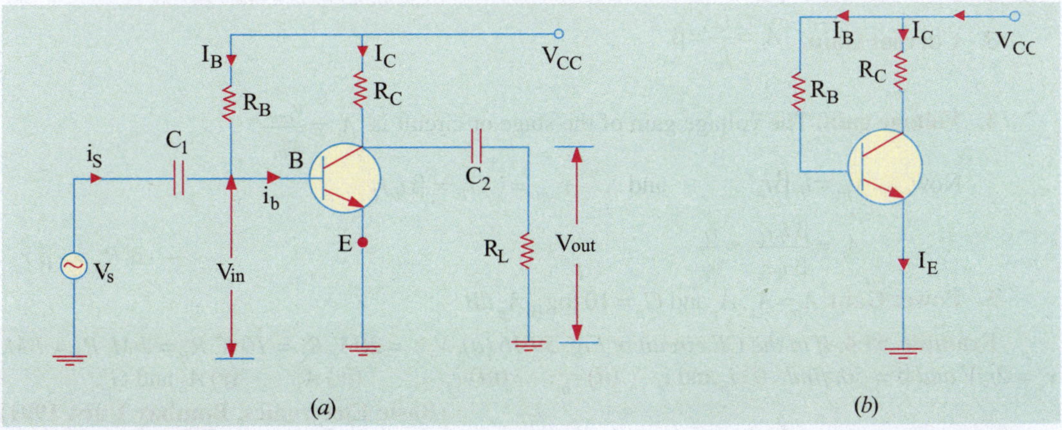

Fig. 59.16

(*a*) DC Equivalent Circuit

For drawing the dc equivalent circuit, same procedure is adopted as given in Art. 59.2 & 59.4. It is shown in Fig. 59.16 (*b*).

As seen, $I_B = \dfrac{V_{CC} - V_{BE}}{R_B} \cong \dfrac{V_{CC}}{R_B}$; $I_C = \beta \,.\, I_B$; $I_E \cong I_C \cong \beta I_B$

(*b*) AC Equivalent Circuit

Let us now analyse the ac equivalent circuit given in Fig. 59.17.

As proved in Art. 59.3, the ac resistance as seen by the input signal when looking into the base is = $\beta\, r_e'$. It may be called $r_{in(base)}$ to distinguish it from r_{in} which is resistance of the **CE stage** as shown in Fig. 59.17 (*a*).

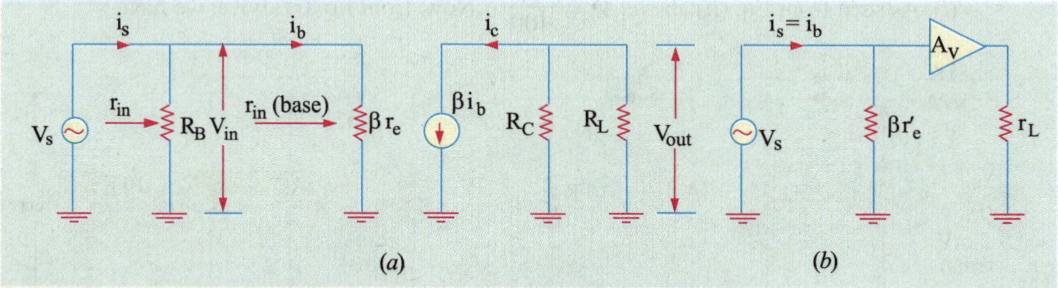

Fig. 59.17

It may be noted that in the absence of ac voltage source resistance R_S, whole of V_s acts across R_B as well as $\beta\, r_e'$ because the two are connected in parallel across it.

Another point worth noting is that major part of source current is passes through $\beta\, r_e'$ and an extremely small part (= V_s/R_B) passes through R_B. Since R_B is usually very large, current passing through it can be easily neglected. Hence, as shown in Fig. 59.17 (*b*), $i_b = i_s$.

(*c*) Principal Operating Characteristics

1. Input Resistance

As seen from Fig. 59.17, input resistance of the *stage* is

$$r_{in} = R_B \,//\, \beta\, r_e' \cong \beta\, r_e' \qquad \because R_B >> \beta\, r_e'$$

$$= \text{input resistance of the base}$$

$$r_{in(stage)} = r_{in(base)}$$

2. AC Load Resistance. $r_L = R_C \,||\, R_L$

3. Current Gain. $A_i = \dfrac{i_c}{i_b} = \beta$

4. Voltage gain. The voltage gain of the stage or circuit is $A_v = \dfrac{v_{out}}{v_{in}}$

Now, $v_{in} = i_b\, \beta r_e'$ and $v_{out} = i_c \,.\, r_L = \beta\, i_b r_L$

$$\therefore \quad A_v = \frac{\beta\, i_b r_L}{\beta\, i_b r_e'} = \frac{r_L}{r_e'} \qquad \text{— if } R_B >> \beta\, r_s'$$

5. Power Gain. $A_p = A_i \,.\, A_v$ and $G_p = 10 \log_{10} A_p\, dB$

Example. 59.4. *If in the CE circuit of Fig. 59.16 (a), $V_{CC} = 20$ V, $R_C = 10$ K, $R_B = 1$ M, $R_L = 1$ M, $v_s = 2$ mV and $\beta = 50$, find* (*i*) i_b and i_c (*ii*) r_{in} (*iii*) r_L (*iv*) A_V (*v*) A_p and G_p.

(Basic Electronics, Bombay Univ. 1991)

Solution. $I_B = 20/1\text{ M} = 20\,\mu\text{A}$; $I_C = \beta\, I_B = 50 \times 20\,\mu\text{A} = 1$ mA

$r_e' = 25\text{ mV}/1\text{ mA} = 25\,\Omega$; $r_{in(base)} = \beta r_e' = 50 \times 25 = 1250\,\Omega$

(*i*) As seen from Fig. 59.17 (*a*), ac base current is given by

$$i_b = \frac{v_s}{r_{in(base)}} = \frac{2\text{ mV}}{1250\,\Omega} = \mathbf{1.6\mu A}\ ;\ i_c = \beta i_b = 50 \times 1.6 = \mathbf{80\mu A}$$

(*ii*) stage $r_{in} = R_B \,//\, \beta r_e' = 1\text{ M} \,||\, 1250\,\Omega \cong$ **1250 Ω**

(*iii*) $r_L = R_L \,||\, R_C = 1\text{ M} \,||\, 10\text{ K} \cong$ **10 K** (*iv*) $A_v = \frac{r_L}{r_e'} = \frac{10\text{K}}{25\Omega} = \mathbf{400}$

(*v*) $A_p = A_i \cdot A_v = 50 \times 400 = 20{,}000$; $G_p = 10 \log_{10} 20{,}000 = \mathbf{43\ dB}$

Example 59.5. *In the CE amplifier circuit of Fig. 59.18 employing emitter feedback, find :*
(*i*) r_{in} (*ii*) r_L (*iii*) A_v (*iv*) A_p and (*v*) G_p
Take transistor β = 100. *How will these values change if emitter bypass capacitor is removed ?*
(Electronics—II, Madras Univ.)

Solution. It should be carefully noted that emitter bypass capacitor *C* provides ***ac ground*** to the signal *i.e.* it shorts out R_E to ground so far as ***ac signal is concerned.*** However, it plays its normal role so far as dc quantities are concerned.

Hence, the ac equivalent circuit of the amplifier becomes as shown in Fig. 59.19.

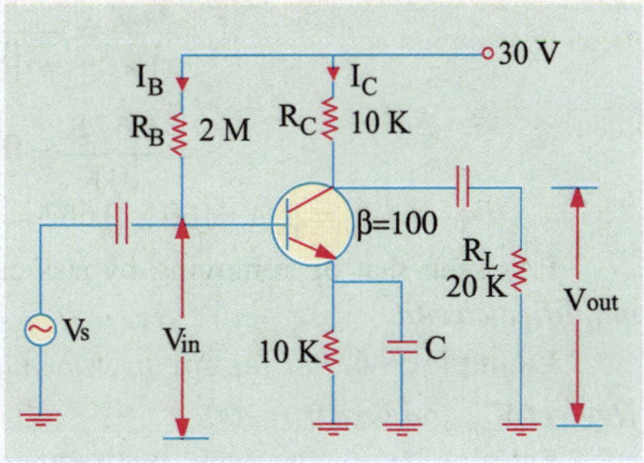

Fig. 59.18

Now, $I_B = \frac{V_{CC}}{R_B + \beta R_E} = \frac{30}{2\text{M} + 100 \times 10\text{K}}$

$= 10\ \mu\text{A}$; $I_C = \beta I_B$

$= 100 \times 10\ \mu\text{A} = 1\ \text{mA}$

$I_E \cong I_C = 1\ \text{mA}$, $r_e' = 25/1 = \mathbf{25\ \Omega}$;

$\beta r_e' = 100 \times 25 = \mathbf{2500\ \Omega}$

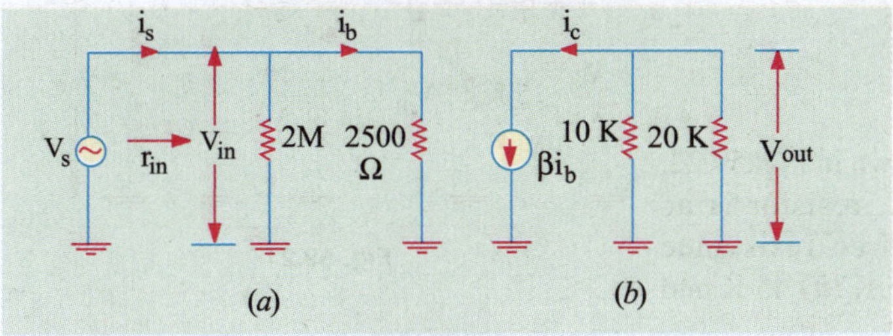

Fig. 59.19

(*i*) As seen from Fig. 59.19

$r_{in} = R_B \parallel \beta r_e'$
$= 2\text{M} \parallel 2500\ \Omega \cong \mathbf{2500\Omega}$

(*ii*) $r_L = R_C \parallel R_L = 10\ \text{K} \parallel 20\ \text{K}$
$= \mathbf{6.67\ K}$

(*iii*) $A_V = \frac{V_{out}}{V_{in}} = \frac{r_L}{r_e}$

$= \frac{6.67\ \text{K}}{25\Omega} = \mathbf{267}$

(*iv*) $A_p = A_i \cdot A_v = 100 \times 267 = \mathbf{26{,}700}$; $G_p = 10 \log_{10} 26{,}700 = \mathbf{44.3\ dB}$

When Emitter Bypass Capacitor is Removed

When bypass capacitor is removed, ac ground is removed. Now, the ac signal will have to pass through R_E also. According to β-rule of Art. 57.24, the total emitter resistance referred to base will become $(1 + \beta)(r_e' + R_E) \cong \beta (r_e' + R_E)$ because r_e' is in series with R_E as shown in Fig. 59.20 (*a*).

Now, the ac equivalent circuit becomes as shown in Fig. 59.20 (*b*).

Stage $r_{in} = R_B \parallel \beta (r_e' + R_E) \cong R_B \parallel \beta R_E$

It is so because R_E is much greater than r_e'.

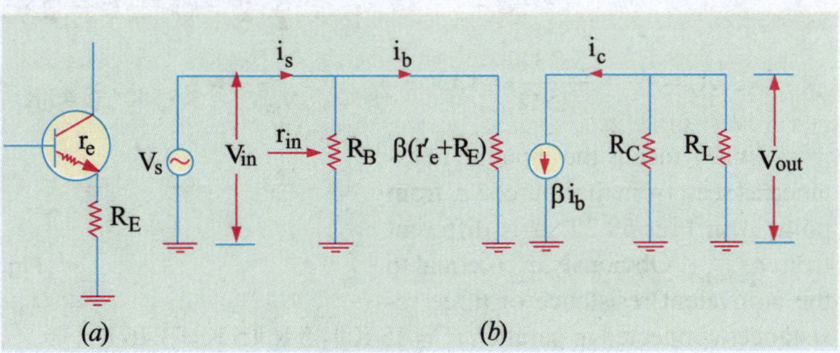

Fig. 59.20

(i) $r_e' = 25\,\Omega$ — as before

$\therefore \quad \beta(r_e' + R_E) \cong \beta R_E = 100 \times 10\text{ K} = 1\text{ M}$

$\therefore \quad r_{in} = R_B \,||\, \beta R_E = 2\text{ M} \,||\, 1\text{M}$

$= \frac{2}{3}\text{M} = \mathbf{666.7\ K}$ — much greater than before

(ii) $r_L = \mathbf{6.67\ K}$ — it remains unchanged

(iii) $V_{in} = i_b\,\beta\,(r_e' + R_E);$ $\quad V_{out} = i_c \cdot r_L = \beta\, i_b\, r_L$

$$\therefore \quad A_v = \frac{V_{out}}{V_{in}} = \frac{\beta i_b r_L}{i_b . \beta(r_e' + R_E)} = \frac{r_L}{(r_e' + R_E)} \cong \frac{r_L}{R_E}$$

$$= \frac{6.67\text{ K}}{10\text{ K}} = \mathbf{0.667} \quad \text{—reduced drastically}$$

(iv) $A_p = A_i A_v = 100 \times 0.667 = \mathbf{66.7}$ — reduced considerably

It is seen that by removing bypass capacitor, ***excessive degeneration has occurred in the amplifier circuit.***

Example 59.6. *For the circuit shown in Fig. 59.21, compute* **(i)** $r_{in\,(base)}$ **(ii)** V_{out} **(iii)** A_v **(iv)** r_{in}. *Neglect* V_{BE} *and take* $\beta = 200$.

Solution. It may be noted that voltage divider bias has been used in the circuit.

$$V_2 = \frac{15}{15 + 45} \times 30 = 7.5\text{ V}$$

$V_E = V_2 - V_{BE} \cong V_2 = 7.5\text{ V}$

$I_E = 7.5/7.5\text{ K} = 1\text{ mA}$

$r_e' = 25\text{ mV}/1\text{ mA} = 25\Omega$

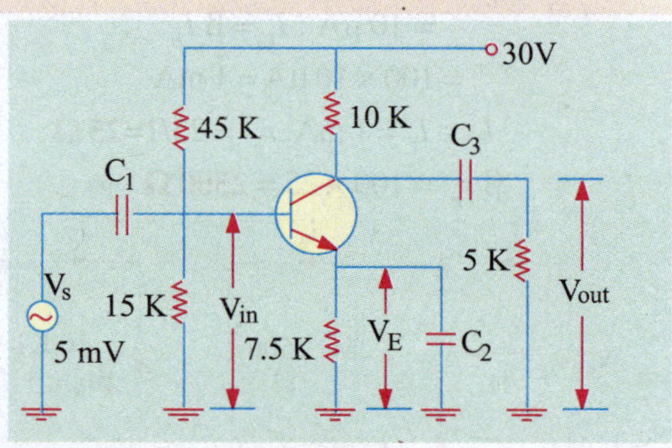

Fig. 59.21

The ac equivalent circuit is shown in Fig. 59.22. Since dc souce is shorted, 45 K, resistor is ac grounded. On the input side, three resistance become paralleled across V_s *i.e.* *(i)* 15 K *(ii)* 45 K and *(iii)* $\beta\, r_e'$ or $r_{in(base)}$.

Capacitor C_2 ac grounds the emitter resistance R_E

(i) $r_{in(base)} = \beta\, r_e' = 200 \times 25 = \mathbf{5\ K}$ *(ii)* $i_b = 5\text{ mV}/5\text{K} = 1\ \mu\text{A}$

(Obviously, i_b is not the current which leaves the source but that part of the source current is which enters the base). Now, collector load current is $i_c = \beta\, i_b = 200 \times 1 = 200\ \mu\text{A} = 0.2\text{ mA}$

$r_L = 10\text{ K} \,||\, 5\text{ K} = 10/3\text{ K} = 3.33\text{ K}, \quad V_{out} = i_c\, r_L = 0.2 \times 3.33 = \mathbf{0.666\ V}$

$$\textit{(iii)} \quad A_v = \frac{V_{out}}{V_{in}} = \frac{0.666\text{ V}}{5\text{ mV}} = \mathbf{133}$$

or $$A_v = \frac{r_L}{r_e'} = \frac{3.33\text{ K}}{25\,\Omega} = 133$$

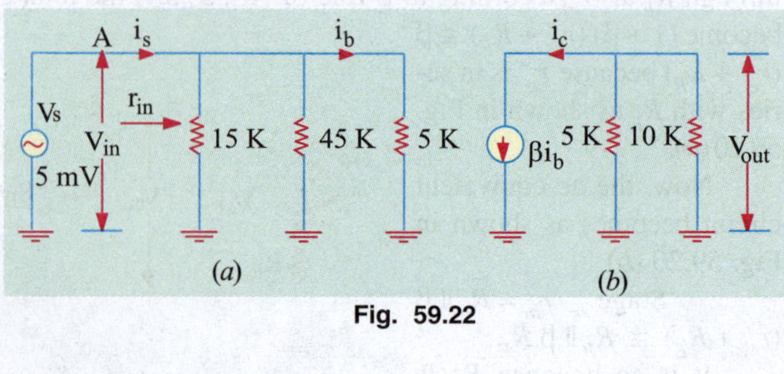

Fig. 59.22

(iv) r_{in} means the input ac resistance as seen from the source *i.e.* from point *A* in Fig. 59.22. It is different from $r_{in(base)}$. Obviously, r_{in} is equal to the equivalent resistance of three resistances connected in parallel. $r_{in} = 15\text{ K} \,||\, 45\text{ K} \,||\, 5\text{ K} = 3.46\text{ K}.$

Example 59.7. *For the CE amplifier circuit of Fig. 59.23, find out (i) r_{in} (ii) r_L (iii) A_v (iv) A_p and (v) V_{out}. Take transistor $\beta = 50$ and $R_S = 0$.* **(Applied Electronics-II, Punjab Univ. 1992)**

Solution.

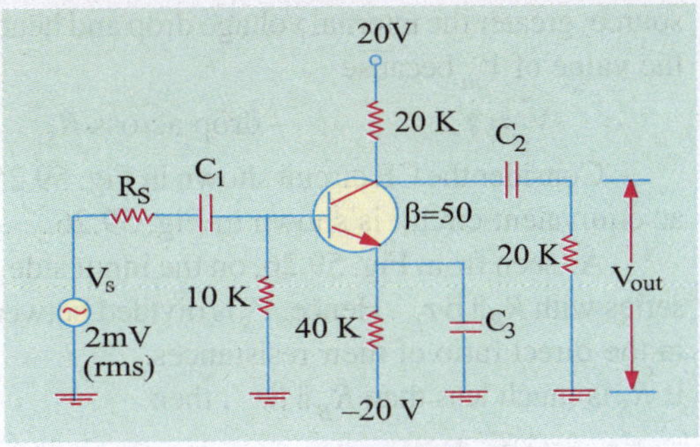

Fig. 59.23

$$I_E \cong \frac{V_{EE}}{R_E + R_B/\beta} = \frac{20}{40+10/50} = 0.5\,\text{mA}$$

$$r_e' = 25/0.5 = 50\,\Omega$$

$$\beta r_e' = 50 \times 50 = 2500\,\Omega$$

(i) $r_{in} = 10\text{ K} \parallel 2500\,\Omega =$ **$2000\,\Omega$**

(ii) $r_L = 20\text{ K} \parallel 20\text{ K} =$ **10 K**

(iii) $A_v = \dfrac{r_L}{r_e} = \dfrac{10\text{ K}}{50\,\Omega} =$ **200**

(iv) $A_p = A_v \cdot A_i$
$= 200 \times 50 = 10{,}000$

$G_p = 10 \log_{10} 10{,}000 =$ **40 dB**

(v) $V_{out} = A_v \times V_{in} = 200 \times 2 =$ **400 mV (r.m.s.)**

Since R_S is zero, whole of v_s appears across the diode base.

Example. 59.8. *For the single-stage CE amplifier circuit of Fig. 59.24, find approximate value of (i) r_{in} (ii) r_L (iii) A_v (iv) A_p and G_p. Take transistor $\beta = 100$. Use $r_e' = 50\text{ mV}/I_E$.* **(Electronics-II, Gujarat Univ. 1991)**

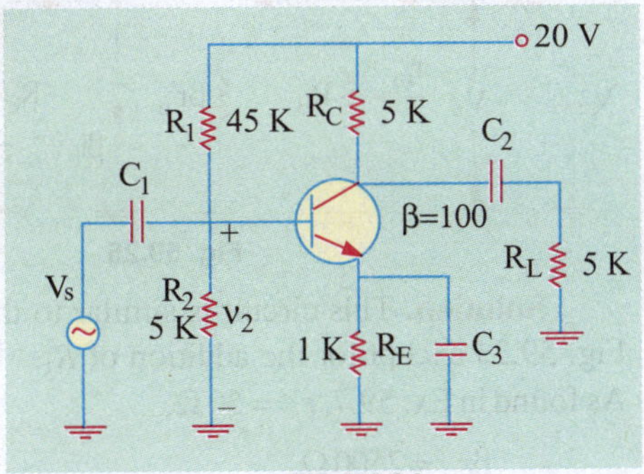

Fig. 59.24

Solution. $V_2 = 20 \times \dfrac{5}{5+45} = 2\text{ V};$

$$I_E \cong \frac{V_2}{R_E} = \frac{2}{1} = 2\text{ mA}$$

$$r_e' = 50/2 = 25\,\Omega\,;$$

$$\beta r_e' = 25 \times 100 = 2.5\text{ K}$$

(i) $r_{in} = R_1 \parallel R_2 \parallel \beta r_e' = 45 \parallel 5 \parallel 2.5 =$ **1.6 K**

(ii) $r_L = 5\text{ K} \parallel 5\text{ K} = 2.5\text{ K}$

(iii) $A_v = \dfrac{r_L}{r_{e'}} = \dfrac{2.5\text{ K}}{25\,\Omega} =$ **100**

(iv) $A_P = A_v \cdot A_i = 100 \times 100 =$ **10,000**

(v) $G_p = 10 \log_{10} 10{,}000 =$ **40 dB**

Note. R_E did not come into picture because it was ac grounded by the bypass capacitor C_3.

Example 59.9. *Find the approximate values of the quantities of Ex. 59.8 in case bypass capacitor C_3 in Fig. 59.24 is removed.*

Solution. (i) In this case,

$$r_{in} = R_1 \parallel R_2 \parallel \beta\,(r_e' + R_E) = 45\text{ K} \parallel 5\text{ K} \parallel 100\,(25 + 1\text{ K})$$

$$\cong 45\text{ K} \parallel 5\text{ K} \parallel 100\text{ K} = \textbf{4.3 K}$$

(ii) $r_L = 2.5\text{ K}$ — as before

(iii) $A_v = \dfrac{r_L}{(r_e' + R_E)} \cong \dfrac{r_L}{r_E} = \dfrac{2.5\text{ K}}{1\text{ K}} =$ **2.5**

(Please note the reduction)

(iv) $A_p = 100 \times 2.5 =$ **250**

(v) $G_p = 10 \log_{10} 250 =$ **24 dB**

59.7. Effect of Source Resistance R_s

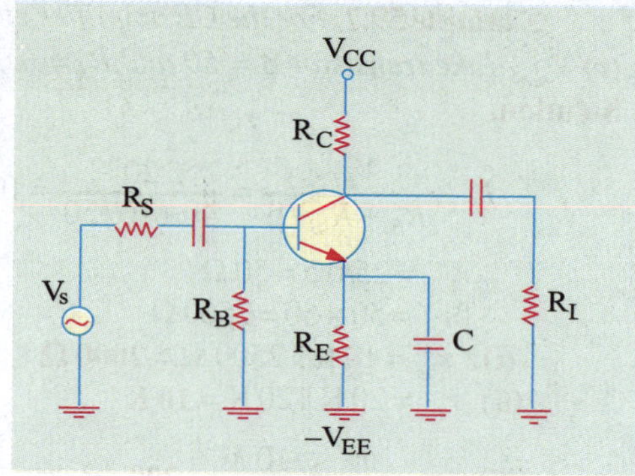

Fig. 59.25

Greater the internal resistance of the ac signal source, greater the internal voltage drop and hence lesser the value of V_{in} because

$$V_{in} = V_s \quad - \text{drop across } R_S.$$

Consider the CE circuit shown in Fig. 59.25 whose ac equivalent circuit is shown in Fig. 59.26.

As seen from Fig. 59.26, on the input side, R_S is in series with $R_B \parallel \beta r_E'$. Hence, V_S is divided between them in the direct ratio of their resistances.

If R_S is much less than $R_B \parallel \beta r_e'$, then

$$V_{in} \cong V_s$$

The voltage gain from base to output is still given by $\dfrac{V_{out}}{V_{in}} = \dfrac{r_L}{r_e'}$

If, in any question, V_s is given, we will first find out how much of it appears across the base as V_{in}. Then, for determining V_{out}, we will simply multipy this V_{in} by the voltage gain.

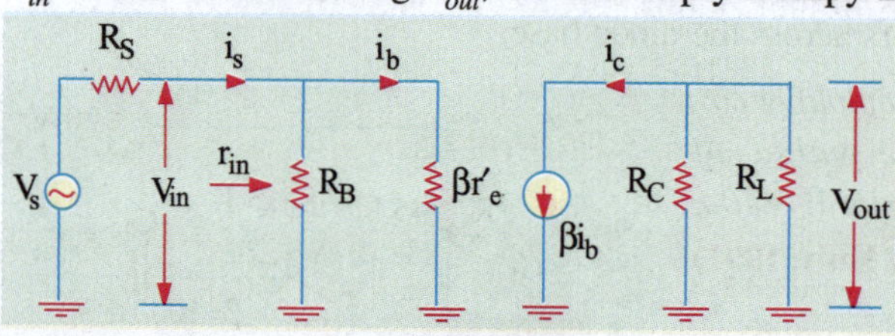

Fig. 59.26

Example 59.10. *In the CE amplifier circuit of Fig. 59.27, find the rms signal output voltage when R_S is (a) 100 Ω and (b) 3 K. Take β = 50.*

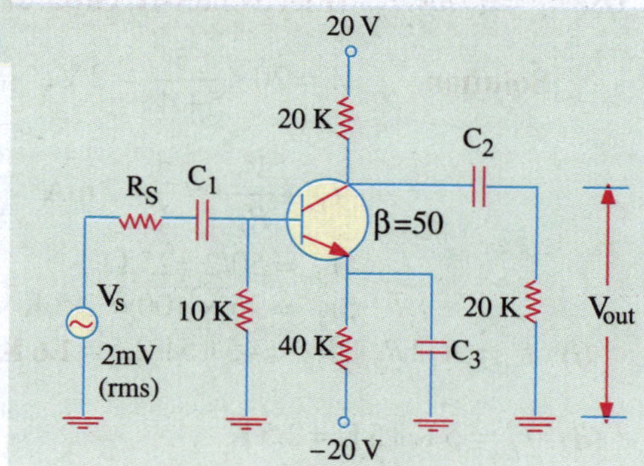

Fig. 59.27

Solution. This circuit is similar to that shown if Fig. 59.23 except for the addition of R_S.

As found in Ex. 59.7, $r_e' = 50\,\Omega$,

$$\beta r_e' = 2500\,\Omega.$$

$$\therefore \quad R_B \parallel \beta r_e' = 10\,\text{K} \parallel 22500\,\Omega = 2000\,\Omega$$

(a) When $R_S = 100\,\Omega$

In this case, 2 mV are dropped across a series combination of 100 Ω and 2000 Ω. Drop over 100 Ω is negligible as compared to that on 2000 Ω. Hence, it can be presumed that $V_s = V_{in} = 2\text{mV}$. We have already found that $A_v = 200$.

$$\therefore \quad V_{out} = A_v \times V_{in} = 200 \times 2 = \mathbf{400\ mV\ (rms)}$$

(b) When $R_S = 3$ K

In this case, drop across $R_B \parallel \beta r_e' = 2$ K is $V_{in} = V_s \dfrac{R_B \parallel \beta r_e'}{R_s + (R_B \parallel \beta r_e')} = 2 \times \dfrac{2}{3+2} = 0.8\,\text{mV}$

$$\therefore \quad V_{out} = 200 \times 0.8 = \mathbf{160\ mV}$$

Obviously, as R_S is increased, V_{out} is decreased.

59.8. Equivalent Circuit of a CC Amplifier

We will consider the two-supply emitter-bias circuit shown in Fig. 59.28 (*a*) in which the collector is placed at ac (not dc) ground. The ac input signal is coupled into the base and output signal is taken out of the emitter. This circuit is also called **emitter follower circuit** because the emitter signal ***follows the signal at the base both in magnitude and phase.***

(a) DC Equivalent Circuit

It is drawn in the usual way by opening all the capacitors and shorting all ac sources. It is shown in Fig. 59.28 (*b*).

$$I_E = \frac{V_{EE} - V_{BE}}{R_E + R_B/\beta} \cong \frac{V_{EE}}{R_E}$$

(b) AC Equivalent Circuit

It is obtained by shorting all ac sources and capacitors and is shown in Fig. 59.29.

(c) Principal Operating Characteristics

1. Input Resistance

The input resistance of the *CC* stage is given by the parallel combination of R_B and $r_{in(base)}$. Now, $r_{in(base)}$ is the input resistance looking into the base. It is found to be equal to $(1 + \beta)(r_e' + r_L) \cong \beta\, r_L$ *i.e.* β times the ac load seen by the emitter.

$\therefore\; r_{in}$ or $r_{in(stage)} = R_B \parallel r_{in(base)}$

$= R_B \parallel \beta\,(r_e' + r_L) \cong \beta\,(r_e' + r_L)$ —when R_B is very large

$\cong \beta r_L$ — when $r_e' << r_L$

$= \beta r_e'$ —if $R_L = 0$

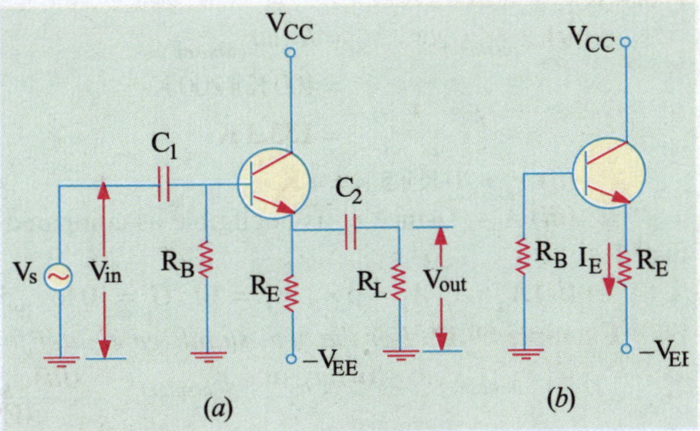

Fig. 59.28

2. AC Load Resistance : $r_L = R_E \parallel R_L$

It is the ac output resistance as seen by the emitter (and not the one when looking into the emitter).

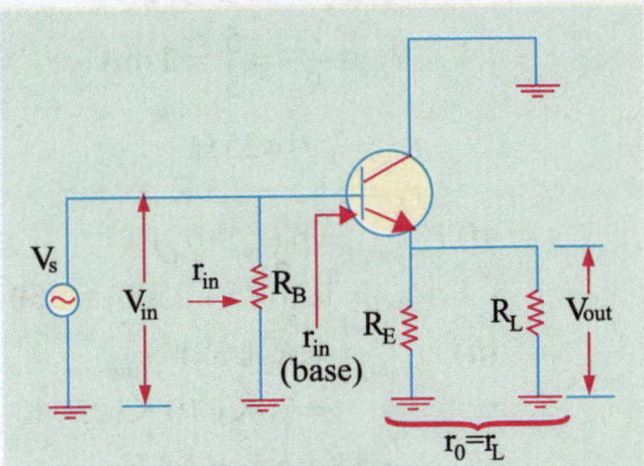

Fig. 59.29

3. Current Gain.

$$A_i = \frac{i_e}{i_b} = (1+\beta) \cong \beta$$

4. Voltage Gain.

$$V_{in} = i_b \,.\, r_{in} = i_b \,.\, \beta\,(r_e' + r_L):$$
$$V_{out} = i_e \,.\, r_L = \beta i_b \,.\, r_L$$

$$A_v = \frac{V_{out}}{V_{in}} = \frac{\beta i_b\, r_L}{i_b\,.\,\beta(r_e' + r_L)} = \frac{r_L}{r_e' + r_L} \cong \frac{r_L}{r_L} = 1$$

— if $r_e' << r_L$

It means that output signal from emitter has the **same magnitude** as the input signal at the base.

5. Power Gain

$$A_p = A_v \,.\, A_i = 1 \times (1 + \beta) = (1 + \beta) \cong \beta$$
$$G_p = 10 \log_{10} A_p \text{ dB}$$

It would be noted from above that main usefulness of the emitter follower is to **step up the impedance level** *i.e.* it transforms load impedance to a much higher value. It does not increase the signal voltage. Hence, primary application of emitter follower or *CC* stage is as an impedance matching device. It offers a higher impedance at the input terminals *i.e.* r_{in} and a low output impedance (r_L) ***– something opposite of typical basic transistor amplifier.***

Another point worth keeping in mind is that the above formulas are not exact because we have used ideal transistor approximations. However, these formulas do help in rapidly grasping the essential features of an emitter follower. Moreover, they are adequate for preliminary analysis and design.

Example. 59.11. *For the emitter follower shown in Fig. 59.30, find*
(i) r_{in} *or* $r_{in(stage)}$ *(ii)* r_L *(iii)* A_v *(iv)* A_p. *Take transistor* $\beta = 50$.
(Electronics, Delhi Univ. 1985)

Solution. $I_E = 20/20 = 1$ mA

$$r_e' = 25/1 = 25\ \Omega$$

$$r_L = R_E \,||\, R_L = 20 \,||\, 5 = 4\text{ K}$$

$$r_{in(base)} = \beta(r_e' + r_L) \cong \beta\, r_L = 50 \times 4 = 200\text{ K}$$

∵ (r_e' of 25 Ω has been neglected as compared to r_L of 4 K)

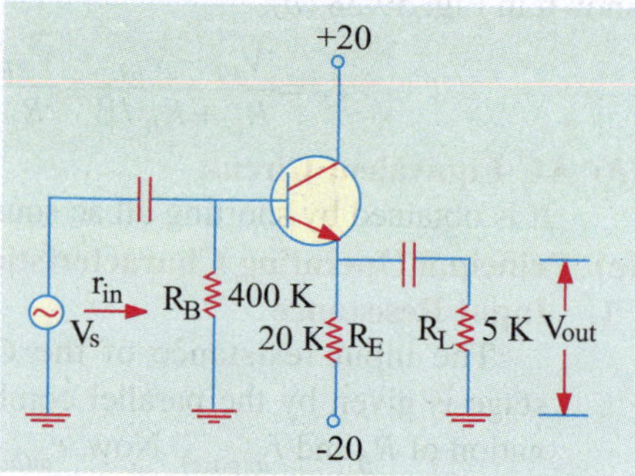

Fig. 59.30

(i) $r_{in(stage)}$ or $r_{in} = R_B \,||\, r_{in(base)}$

$$= 400\text{ K} \,||\, 200\text{ K}$$

$$= \mathbf{133.3\ K}$$

(ii) $r_L = 20\text{ K} \,||\, 5\text{ K} = \mathbf{4\ K}$

(iii) $A_v \cong 1$ since r_e' is negligible as compared to r_L. If it is not, then A_v could be even less than 1.

(iv) $A_p = A_i \,.\, A_v = \beta \times 1 = \beta = 50$; $G_p = 10 \log_{10} 50 = \mathbf{17\ dB}$

Example 59.12. *For the beta-stabilized emitter follower circuit of Fig. 59.31, find*
(i) $r_{in(base)}$ *(ii)* r_{in} or $r_{in(stage)}$ *(iii)* r_L and *(iv)* A_v. Take transistor $\beta = 100$.
(Electronics Technology, Mysore Univ. 1986)

Solution. $V_2 = V_{CC}\dfrac{R_2}{R_1 + R_2} = 20 \times \dfrac{10}{40} = 5$ V

$$I_E \cong \frac{V_2}{R_E} = \frac{5}{5} = 1\text{ mA} \quad ∵$$

$$r_e' = 25/1 = 25\ \Omega$$

$$r_L = 5 \,||\, 5 = 2.5\text{ K}$$

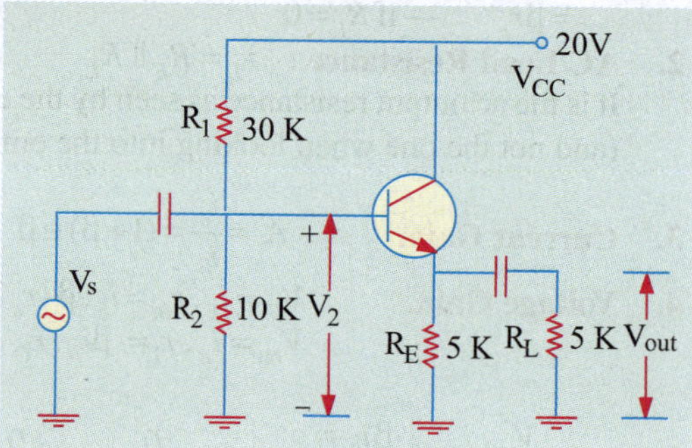

Fig. 59.31

(i) $r_{in(base)} = \beta\,(r_e' + r_L)$

$$= 100\,(25 + 2500) \cong \mathbf{250\ K}$$

(ii) $r_{in(stage)} = R_1 \,||\, R_2 \,||\, r_{in(base)}$

$$= 30\text{K} \,||\, 10\text{ K} \,||\, 250\text{ K} \cong 7\text{ K}$$

(iii) $r_L = 5\text{ K} \,||\, 5\text{ K} = \mathbf{2.5\ K}$

(iv) $A_p = A_i \,.\, A_v \cong \beta \times 1 = 100$; $G_p = 10 \log_{10} 100 = \mathbf{20dB}$

Example 59.13. *For the CE circuit of Fig. 59.32, find*
(i) r_{in} or $r_{in(stage)}$ *(ii)* A_v *(iii)* A_p. Take transistor $\beta = 200$

Solution. $I_E = 20/20 = 1$ mA

$$r_e' = 25/1 = 25\ \Omega$$

$$r_L = 20\text{ K} \,||\, 50\ \Omega = 50\ \Omega$$

$$r_{in(base)} = \beta\,(r_e' + r_L) = 200\,(25 + 50) = 15\text{ K};$$

(i) r_{in} or $r_{in(stage)} = R_B \,||\, r_{in(base)}$

$$= 100\text{ K} \,||\, 15\text{ K} = \mathbf{13\ K}$$

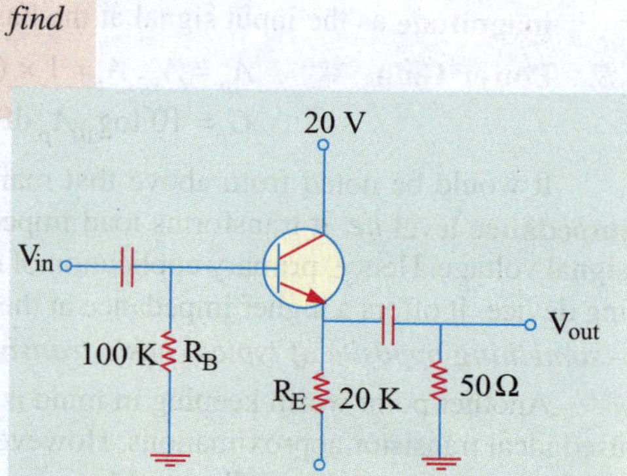

Fig. 59.32

(ii) Since r_e' is not negligible as compared to r_L, we will use the expression

$$A_v = \frac{r_L}{r_e' + r_L} = \frac{50}{25 + 50} = \mathbf{0.667}$$

(iii) $A_p = A_v \cdot A_i = 0.667 \times 200 = 133.3$

59.9. Small-signal Low-frequency Model or Representation

Although many transistor representations or models have been suggested and widely used, the equivalent *T*-model is the easiest to understand because, in this representation, component parts retain their identity in all configurations leading to rapid appreciation of a given network.

59.10. T-Model

Such models are not in common use today because they do not take into account any gain between input and output.

(a) CB Circuit

In Fig. 59.33 is shown the low-frequency *T*-equivalent circuit of a transistor connected in *CB* configuration. It utilizes *T*- or *r*-parameters. All these parameters are ac parameters and are measured under open-circuit conditions. Here, r_e represents the ac resistance of the forward-biased emitter-base junction. Its value is

$$r_e = \frac{25\text{ mV}}{I_E} \quad \text{— for Ge}$$

$$= \frac{50\text{ mV}}{I_E} \quad \text{— for Si}$$

This resistance is fairly small and depends on I_E. Also, r_c represents the ac resistance of the reverse-biased *C/B* junction. It is of the order of a few MΩ. Finally, r_b represents the resistance of the base region which is common to both junctions. Its value depends on the degree of doping. Usually, r_b is larger than r_e but much smaller than r_c.

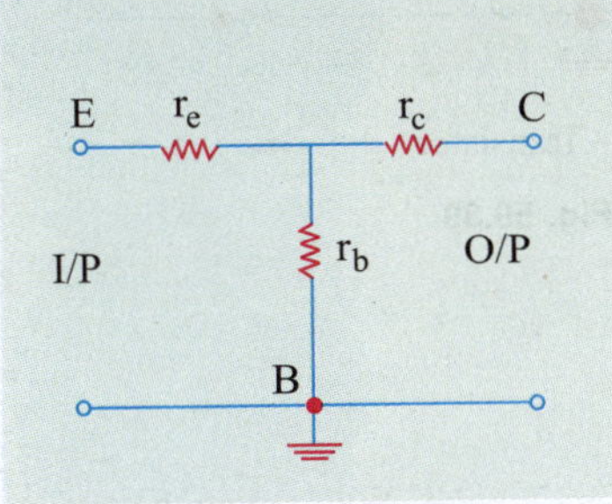

Fig. 59.33

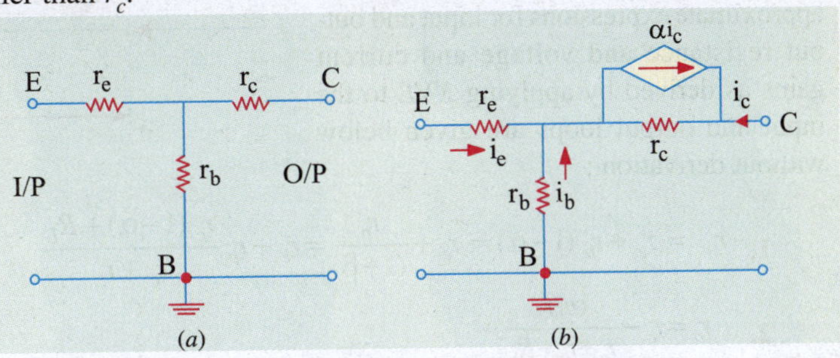

Fig. 59.34

However, circuit shown in Fig. 59.34 (*a*) is not complete because it does not illustrate the forward current transfer ratio. Since current in the output of a transistor depends on the current at the input, a dependent current-generator in parallel with r_c must be included as shown in Fig. 59.34 (*b*). As it is the usual practice, all currents are shown flowing inwards even though some of them may actually be flowing in the opposite direction.

The current generator may be replaced by a voltage generator with the help of Thevenin's theorem as shown in Fig. 59.35 (*a*). In that case, the *T*-equivalent circuit becomes as shown in Fig. 59.35 (*b*). The generator has a voltage of $\alpha i_e r_c = r_{in} i_e$ where $r_{in} = \alpha r_c$.

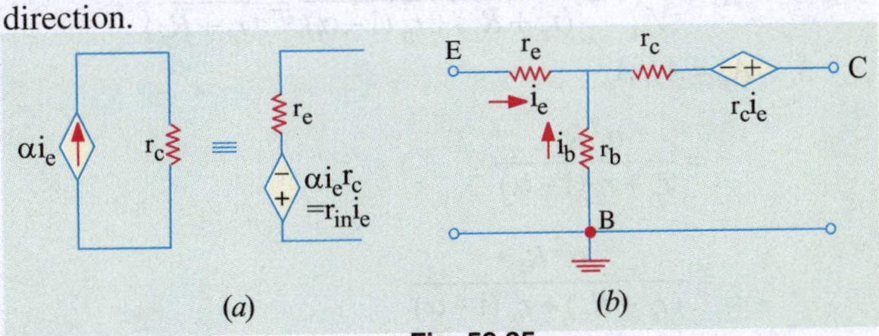

Fig. 59.35

Typical values of different parameters are :

$r_e = 25$ to $50\ \Omega$; $r_b = 100$ to $1000\ \Omega$; $r_c = 1\ \text{M}\Omega$

(b) CE Circuit

The *T*-equivalent circuit for such a configuration is shown in Fig. 59.36. Whereas circuit shown in Fig. 59.36 contains a parallel current-generator that shown in Fig. 59.37 contains a series-voltage generator.

(c) CC Circuit

The *T*-equivalent circuit for such a configuration is shown in Fig. 59.38.

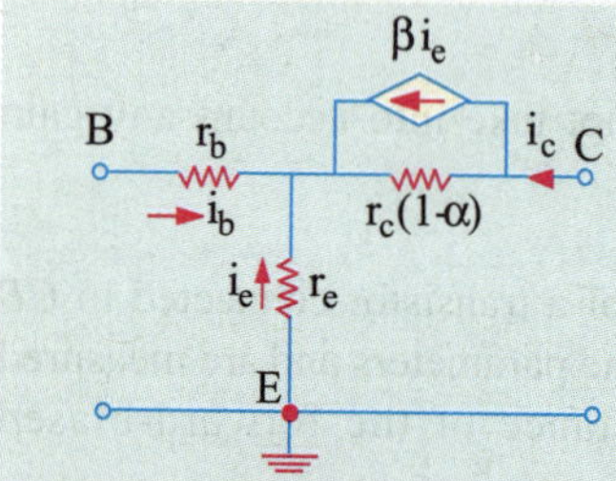

Fig. 59.36

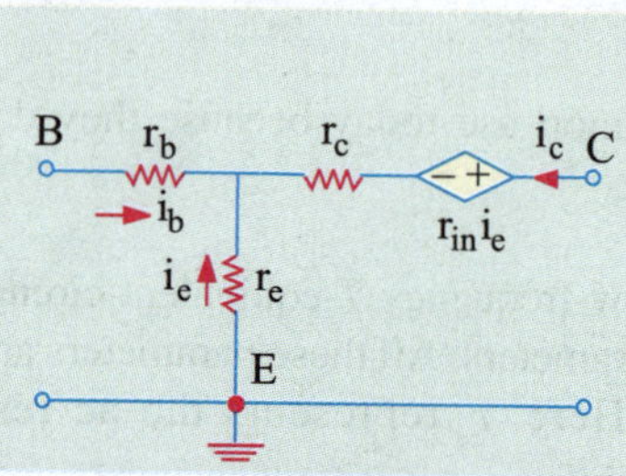

Fig. 59.37

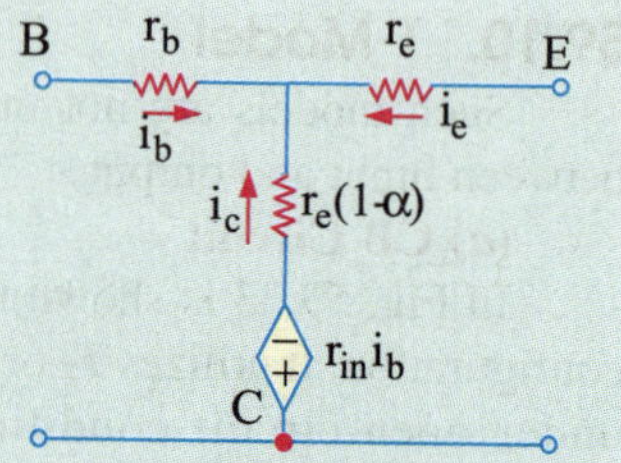

Fig. 59.38

59.11. Formulas for T-Equivalent of a CB Circuit

In Fig. 59.39 is shown a small-signal low-frequency *T*-equivalent circuit for *CB* configuration. The ac input signal source has a resistance of R_S and voltage of V_s.

Of course, dc biasing circuit has been omitted and only ac equivalent shown. The approximate expressions for input and output resistance and voltage and current gains as derived by applying *KVL* to the input and output loops are given below without derivation :

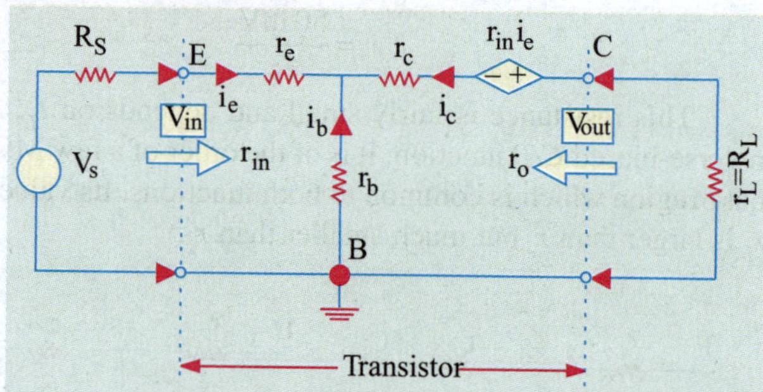

Fig. 59.39

1. $r_{in} = r_e + r_b\,(1-\alpha) = r_e + \dfrac{r_b}{(1+\beta)} = r_e + r_b\,\dfrac{r_c\,(1-\alpha)+R_L}{r_c + r_L}$

2. $r_o = r_c - \dfrac{\alpha r_b r_c}{r_e + r_b + R_S}$

3. $A_i = \alpha$

4. $A_v = \dfrac{v_{out}}{v_{in}} = \dfrac{\alpha\,R_L}{r_e + r_b\,(1-\alpha)}$

$$A_v = \frac{V_{out}}{V_{in}} = \frac{\alpha R_L}{(r_e + R_s) + r_b\,(1-\alpha)} \cong \frac{\alpha R_L}{(r_e + R_s)}$$

5. $A_P = A_i\,.\,A_v$

$$= \frac{\alpha^2 R_L}{r_e + r_b\,(1-\alpha)} \qquad \text{— no source resistance}$$

$$= \frac{\alpha^2 R_L}{(r_e + R_s) + r_b\,(1-\alpha)} \qquad \text{— with source resistance}$$

$$G_p = 10\,\log_{10} A_p$$

59.12. Formulas for T-Equivalent of a CE Circuit

The low-frequency small-signal *T*-equivalent circuit for such a configuration is shown in Fig. 59.40.

The approximate expressions for various resistances and gains as found by applying *KVL* to the input and output loops are given below :

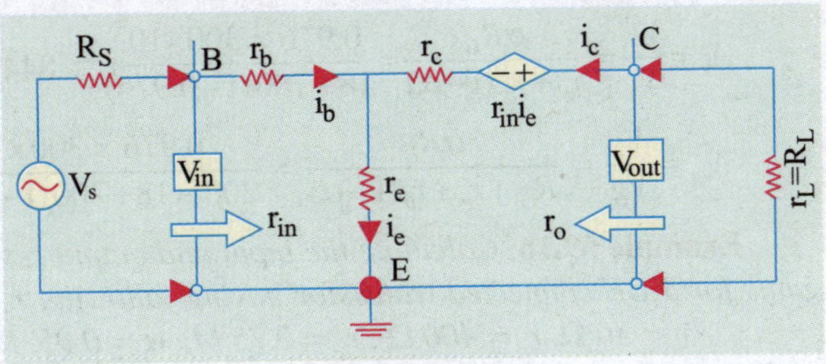

Fig. 59.40

1. $r_{in} = r_b + \dfrac{r_e}{((1-\alpha)}$
   $= r_b + (1+\beta)r_e$

2. $r_o = r_e(1-\alpha) + \dfrac{\alpha r_c r_e}{r_b + r_e + R_S}$

3. $A_i = \beta = \dfrac{\alpha}{(1-\alpha)}$

4. $A_v = \dfrac{V_{out}}{V_{in}} = \dfrac{\alpha R_L}{r_e + r_b(1-\alpha)}$ $\qquad A_{vs} = \dfrac{V_{out}}{V_s} = \dfrac{\alpha R_L}{r_e + (r_b + R_s)(1-\alpha)}$

5. $A_p = A_v . A_i = \dfrac{\alpha^2 R_L}{(1-\alpha)[r_e + r_b(1-\alpha)]} = \dfrac{\alpha^2 R_L}{(1-\alpha)[r_e + (r_b + R_S)(1-\alpha)]}$

59.13. Formulas for T-Equivalent of a CC Circuit

The low-frequency *T*-equivalent circuit for such a configuration is shown in Fig. 59.41.

The approximate expressions for various resistances and gains are given below :

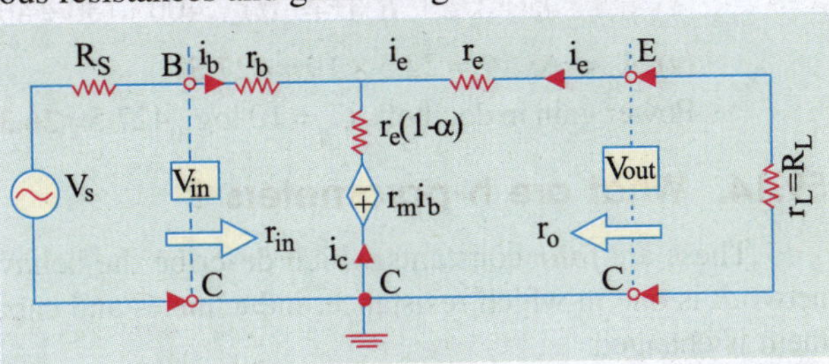

Fig. 59.41

1. $r_{in} = \dfrac{R_L}{1-\alpha} = (1+\beta)R_L$

2. $r_v = r_e + (1-\alpha)(r_b + R_S)$
   $= r_e + \dfrac{(r_b + R_S)}{(1+\beta)}$

3. $A_i = \dfrac{1}{(1-\alpha)} = (1+\beta)$

4. $A_v = 1$

5. $A_p = \dfrac{1}{(1-\alpha)} = (1+\beta) \qquad G_p = 10 \log_{10} A_p dB$

Example 59.14. *A junction transistor has $r_e = 50\,\Omega$, $r_b = 1\,K$, $r_c = 1M$ and $\alpha = 0.98$. It is used in common-base circuit with a load resistance of 10 K. Calculate the current , voltage and power gains and the input resistance.* **(Applied Electronics, Punjab Univ. 1993)**

Solution. *(i)* $A_i = \alpha = 0.98$

(ii) $A_v = \dfrac{\alpha R_L}{r_e + r_b(1-\alpha)} = \dfrac{0.98 \times 10,000}{50 + 1000(1-0.98)} = \mathbf{140}$ $\qquad$ *(iii)* $A_p = A_v . A_i = 140 \times 0.98 = 137$

(iv) $r_{in} = r_e + r_b(1-\alpha) = 50 + 1000\,(1 - 0.98) = \mathbf{70\,\Omega}$

Example 59.15. *A P-N-P junction transistor is used as a voltage amplifier in the grounded-base circuit with a load resistance being 300 K and the internal resistance of the original source being 200 Ω.*

Derive an expression for the voltage gain of the amplifier and calculate its magnitude if the transistor T-network parameters are : $r_e = 18\,\Omega$, $r_b = 700\,\Omega$, $r_c = 1$ M and $r_m = 976$ K.

(Applied Electronics and Circuits, Grad. I.E.T.E. 1987)

Solution. Here, $r_m = \alpha(r_c + r_b) \cong \alpha r_c$ $\quad \therefore \alpha = r_m / r_c = 976\,K / 1000\,K = 0.976$

$$A_v = \frac{V_2}{V_1} = \frac{\alpha R_L}{r_e + r_b(1-\alpha)} = \frac{0.976 \times 300 \times 10^3}{18 + 700(1-0.976)} = \mathbf{84413}$$

$$A_{vg} = \frac{V_2}{Vg} = \frac{\alpha R_L}{R_G + r_e + r_b(1-\alpha)} = \frac{0.976 \times 300 \times 10^3}{200 + 18 + 700(1-0.976)} = \mathbf{1247}$$

Example 59.16. *Calculate the input and output resistance, overall current, voltage and power gains for a CE connected transistor having following r-parameters :*

$r_b = 30\,\Omega$, $r_e = 400\,\Omega$, $r_c = 0.75\,M$, $\alpha = 0.95$, $R_L = 10\,K$ *and* $R_S = 400\,\Omega$

Also, calculate the power gain in decibels.

(Electronics Technology, Bangalore Univ. 1999)

Solution. *(i)* $r_{in} = r_b + \dfrac{r_e}{(1-\alpha)} = 30 + \dfrac{400}{(1-0.95)} = \mathbf{8030\,\Omega}$

(ii) $r_o = r_e(1-\alpha) + \dfrac{\alpha r_c r_e}{r_b + r_e + R_S}$

$$= 750{,}000(1-0.95) + \frac{0.95 \times 750{,}000 \times 400}{(30+400+400)} = 380{,}870 = \mathbf{0.38\,M\Omega}$$

(iii) $A_i = \beta = \dfrac{\alpha}{1-\alpha} = \dfrac{0.95}{(1-0.95)} = 19$

(iv) $A_v = \dfrac{\alpha R_L}{r_e + (r_b + R_s)(1-\alpha)} = \dfrac{0.95 \times 10 \times 10^3}{400 + (30+400) \times 0.05} = \mathbf{22.5}$

(v) $A_p = A_v \cdot A_i = 22.5 \times 19 = \mathbf{427.5}$

Power gain in decibels, $G_p = 10 \log_{10} 427.5 = \mathbf{26.3\,dB}$

59.14. What are h-parameters ?

These are ***four*** constants which describe the behaviour of a two-port linear network. A linear network is one in which resistance, inductances and capacitances remain fixed when voltage across them is changed.

Consider an unknown linear network contained in a black box as shown in Fig. 59.42. As a matter of convention, currents flowing into the box are taken positive whereas those flowing out of it are considered negative.

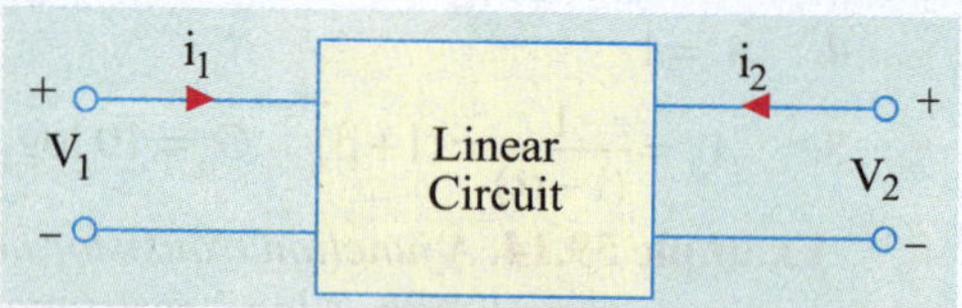

Fig. 59.42

Similarly, voltages are positive from the upper to the lower terminals and negative the other way around. The electrical behaviour of such a circuit can be described with the help of four hybrid parameters or constants designated as h_{11}, h_{12}, h_{21}, h_{22}. In this type of double-number subscripts, it is implied that ***the first variable is always divided by the other.*** The subscript 1 refers to quantities on the input side and 2 to the quantities on the ***output side.*** The letter '*h*' has come from the word ***hybrid*** which means ***mixture*** of distinctly different items. These constants are hybrid because they have different units.

Out of the four *h*-parameters, two are found by short-circuiting the output terminals 2-2 and the other two by open-circuiting the input terminals 1-1 of the circuit.

(a) Finding h_{11} and h_{21} from Short-Circuit Test

As shown in Fig. 59.43, the output terminals have been shorted so that $V_2 = 0$, because no voltage can exist on a short. The linear circuit within the box is driven by an input voltage V_1. It produces an input current i_1 whose magnitude depends on the type of circuit within the box.

$$h_{11} = \frac{V_1}{i_1} \quad \text{— output shorted}$$

$$h_{21} = \frac{i_2}{i_1} \quad \text{— output shorted}$$

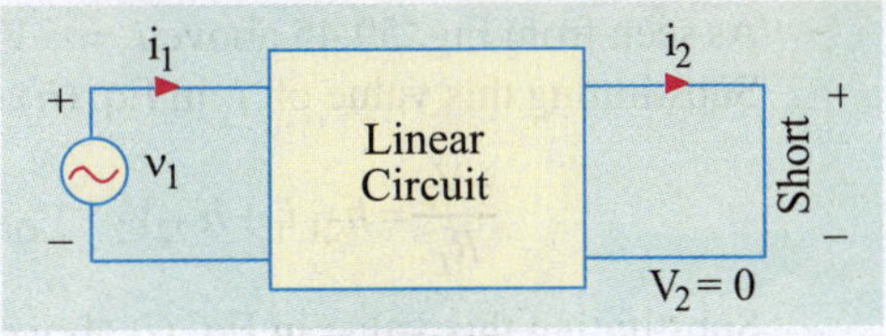

Fig. 59.43

These two constants are known as **forward** parameters.

The constant h_{11} represents input impedance with output shorted and has the unit of ohm. The constant h_{21} represents current gain of the circuit with output shorted and has no unit since it is the ratio of two similar quantities.

The voltages and currents of such a two-port network are related by the following sets of equations or *V/I* relations.

$$V_1 = h_{11}\, i_1 + h_{12}\, V_2 \quad \text{...(i)}$$

$$i_2 = h_{21}\, i_1 + h_{22}\, V_2 \quad \text{... (ii)}$$

Here, the h_s are constants for a given circuit but change if the circuit is changed. Knowlege of parameters enables us to find the voltages and currents with the help of the above two equations.

(b) Finding h_{12} and h_{22} from Open-circuit Test

As shown in Fig. 59.44, the input terminals are open so that $i_1 = 0$ but there does appear a voltage v_1 across them. The output terminals are driven by an ac voltage V_2 which sets up current i_2.

$$h_{12} = \frac{V_1}{V_2} \quad \text{— input open}$$

$$h_{22} = \frac{i_2}{V_2} \quad \text{— input open}$$

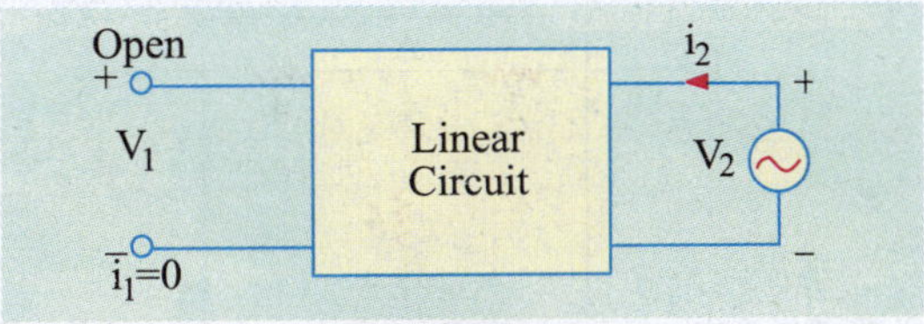

Fig. 59.44

As seen, h_{12} represents voltage gain (not forward gain which is $\frac{V_2}{V_1}$). Hence, it has no units. The constant h_{22} represents admittance (which is reverse of resistance) and has the unit of mho or Siemens, *S*. It is actually the admittance looking into the output terminals with input terminals open. Generally, these two constants are also referred to as ***reverse parameters***.

Summary of *h*-parameters

h_{11}	=	input impedance	with output shorted
h_{21}	=	forward current gain	
h_{12}	=	reverse voltage gain	with input open
h_{22}	=	output admittance	

59.15. Input Impedance of a Two Port Network

Consider the two-port linear network shown in Fig. 59.45 which has a load resistance r_L across its output terminals. The voltage source V_1 on the input side drives the circuit and sets up current i_1. As seen, $Z_{in} = V_1 / i_1$. Substituting the value of V_1 from Eq. (*i*) of Art. 59.14, we get

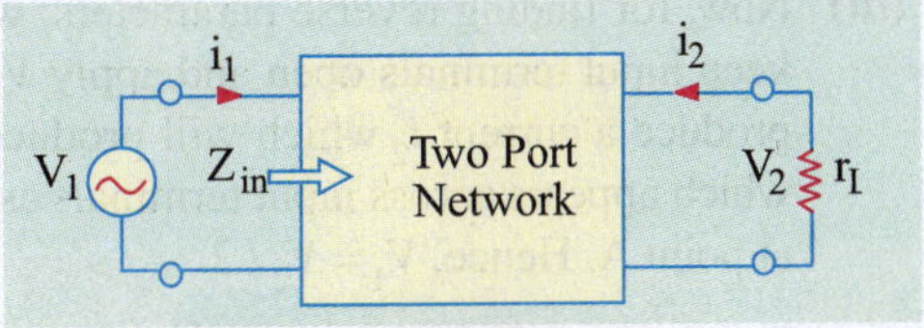

Fig. 59.45

$$Z_{in} = \frac{V_1}{i_1} = \frac{h_{11} i_1 + h_{12} V_2}{i_1}$$

$$= h_{11} + \frac{h_{12} V_2}{i_1} \quad \text{... (i)}$$

As seen from Fig. 59.45 above $i_2 = -V_2/R_L$*

Substituting this value of i_2 in Eq. (*ii*) of Art 59.14, we have

$$\frac{-V_2}{R_L} = h_{21}i_1 + h_{22}V_2 \quad \text{or} \quad \frac{V_2}{i_1} = -\frac{h_{21}}{h_{22} + 1/R_L}$$

Substituting this value in Eq. (*i*) above, we have

$$Z_{in} = h_{11} - \frac{h_{12}.h_{21}}{h_{22} + 1/R_L}$$

59.16. Voltage Gain of a Two Port Network

The voltage gain of such a circuit (Fig. 59.45) is $A_v = \frac{V_2}{V_1}$. Now $V_1 = i_1 . Z_{in}$. Hence, $A_v = V_2/i_1.Z_{in}$.

Substituting the value of V_2/i_1 as found earlier in Art. 59.15, we get

$$A_v = \frac{-h_{21}}{Z_{in}(h_{22} + 1/R_L)}$$

Example. 59.17. *Find the h-parameters of the circuit shown in Fig. 59.46 (a).*

Solution. First of all, let us find the forward parameters *hu* and h_{11}. For that purpose, a short is put across the output terminals as shown in Fig. 59.46 (*b*).

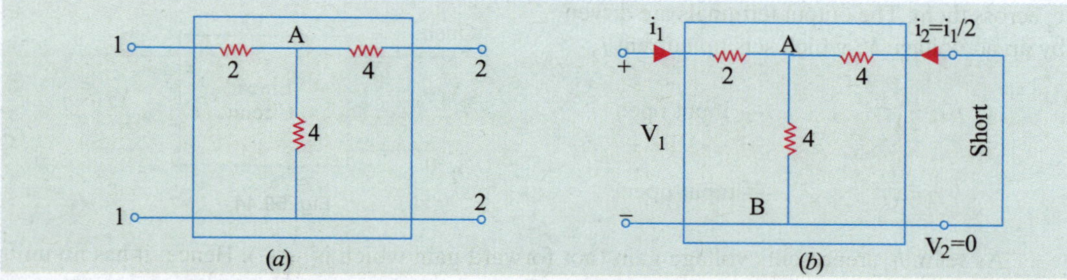

Fig. 59.46

(*i*) The input impedance of the network as viewed from input terminals is $h_{11} = 2 + 4 \parallel 4 = 4\ \Omega$

(*ii*) As seen from Fig. 59.46 (*b*), input current i_1 divides into two equal parts at point A. The output current $i_2 = -i_1/2$ (negative sign has been taken because actually it is flowing out of the box).

$$\therefore \quad h_{21} = \frac{i_2}{i_1} = \frac{-i_1/2}{i_1} = -\frac{1}{2} = -0.5$$

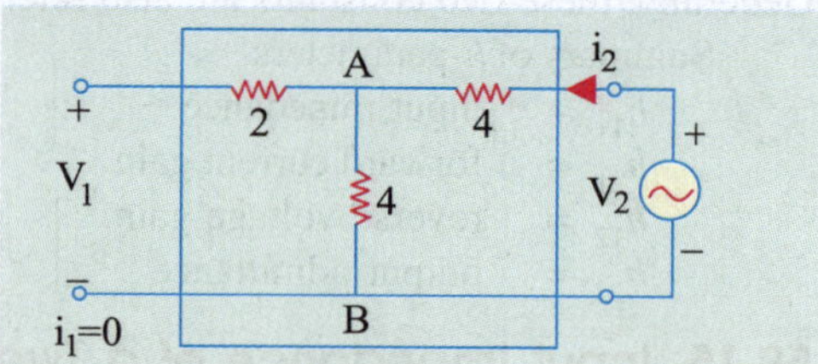

Fig. 59.47

(*iii*) Now, for finding reverse parameters, we will keep input terminals open and apply V_2 across output terminals as shown in Fig. 59.47. It will produce a current i_2 which will produce equal drops across the two 4 Ω resistors. The voltage which appears across input terminals as V_1 is the drop across the vertical 4 Ω resistor connected at point A. Hence, $V_1 = V_2/2$.

$$\therefore \quad h_{12} = \frac{V_1}{V_2} = \frac{V_2/2}{V_2} = 0.5$$

* The negative sign is used because actual load current is opposite to that shown in figure.

The input impedance of the network when viewed from output terminals with input terminals open is = 4 + 4 = 8 Ω.

$\therefore \qquad h_{22} = 1/8 = $ **0.125 Siemens (*i.e.* mho)**

Hence, for the network shown in Fig. 59.46 (*a*), the *h*-parameters are as under :

$h_{11} = 4\ \Omega \qquad h_{21} = -0.5 \qquad h_{12} = 0.5$ and $\qquad h_{22} = 0.125$ S

59.17. The h-parameter Notation for Transistors

While using *h*-parameters for transistor circuits, their numerical subscripts are replaced by the first letters for defining them.

$h_{11} = h_i$ = input impedance
$h_{21} = h_f$ = forward current gain] output shorted

$h_{12} = h_r$ = reverse voltage gain
$h_{22} = h_o$ = output admittance] input open

A second subscript is added to the above parameters to indicate the particular configuration. For example, for *CE* connection, the four parameters are written as :

$h_{ie} \qquad h_{fe} \qquad h_{re}$ and $\qquad h_{oe}$

Similarly, for *CB* connection, these are written as h_{ib}, h_{fb}, h_{rb}, h_{ob} and for *CC* connection as h_{ic}, h_{fc}, h_{rc} and h_{oc}.

59.18. The h-parameters of an Ideal Transistor

As stated earlier, every linear circuit has a set of parameters associated with it, which fully describe its behaviour. When small ac signals are involved, a transistor behaves like a linear device because its output ac signal ***varies directly as the input signal.*** Hence, for small ac signals, each transistor has its own characteristic set of *h*-parameters or constants.

The *h*-parameters depend on a number of factors such as

1. transistor type **2.** configuration **3.** operating point **4.** temperature **5.** frequency

These *h*-parameters can be found experimentally or graphically. The parameters h_i and h_r are determined from input characteristics of the *CE* transistor whereas h_f and h_0 are found from output characteristics.

59.19. The h-parameters of an Ideal CB Transistor

In Fig. 59.48 (*a*), a *CB*-connected transistor has been shown connected in a black box. Fig., 59.48 (*b*) gives its equivalent circuit. It should be noted that no external biassing resistors or any signal source has been shown connected to the transistor.

(*i*) Forward Parameters

The two forward h-parameters can be found from the circuit of Fig. 59.49 (*a*) where a short has been put across the output. The input impedance is simply r_e.

$$\therefore \qquad h_{ib} = r_e$$

The output current equals the input current *i.e.* Since it flows out of the box, it is taken as negative. The forward current gain is

$$h_{fb} = \frac{-i_e}{i_e} = -1$$

(It also called the ac α of the *CB* circuit.)

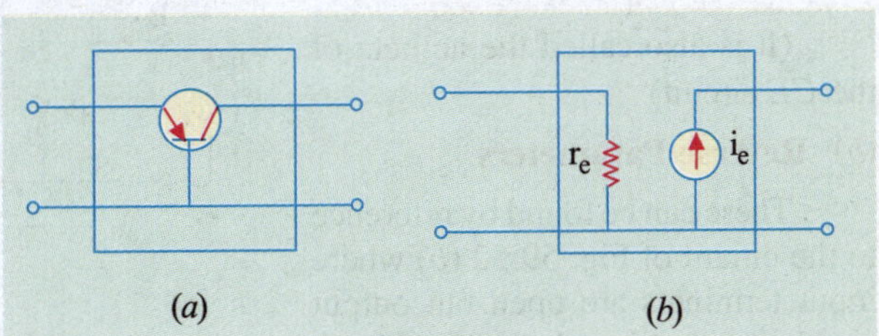

(*a*) (*b*)

Fig. 59.48

(*ii*) Reverse Parameters

The two reverse parameters can be found from the circuit diagram of Fig. 59.49 (*b*). When input terminals are open, there can be no ac emitter current. It means that ac current source (inside the box) has a value of zero and so appears as an 'open'. Because of this open, no voltage can appear across input terminals, however, large V_2 may be. Hence, $V_1 = 0$.

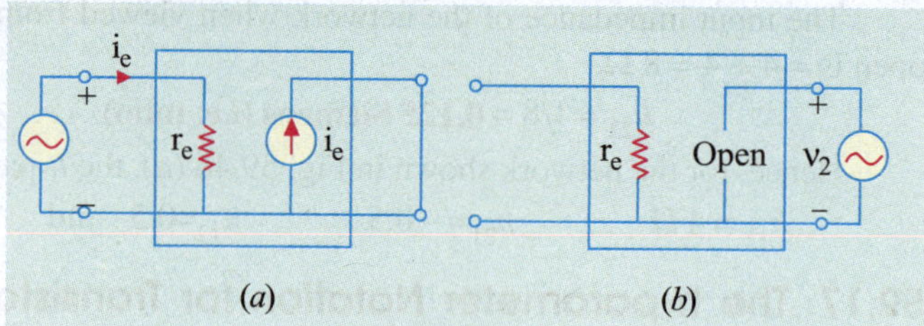

Fig. 59.49

$$\therefore \quad h_{rb} = \frac{V_1}{V_2} = \frac{0}{V_2} = 0$$

Similarly, the impedance, looking into the output terminals is infinite. Consequently, its admittance (= 1/∞) is zero.

$$\therefore \quad h_{ob} = 0$$

Summary

The four h-parameters of an ideal transistor connected in CB configuration are

$$h_{ib} = r_e \; ; \quad h_{fb} = -1, \quad h_{rb} = 0 \; ; \quad h_{ob} = 0$$

The equivalent hybrid circuit is shown in Fig. 59.50.

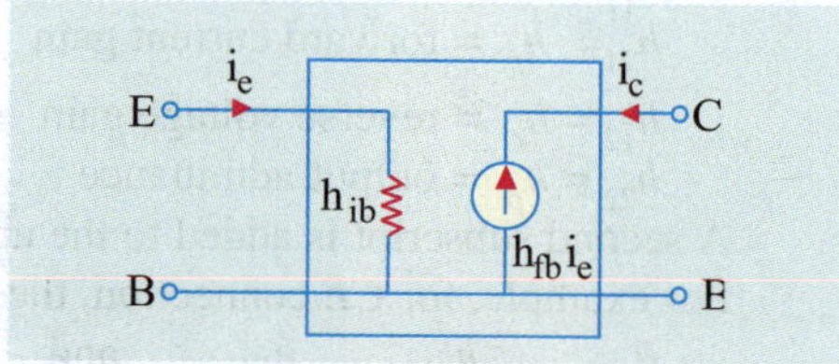

Fig. 59.50

Note. In an actual transistor, h_{rb} and h_{ob} are not zero but have some finite value though extremely small (ranging from 10^{-1} to 10^{-6}).

In reality, output impedance is not infinity but very high so that h_{ob} is extremely small. Similarly, there is some amount of feedback between the output and the input circuits (even when open) though it is very small. Hence, h_{rb} is very small.

59.20. The *h*-parameters of an ideal CE Transistor

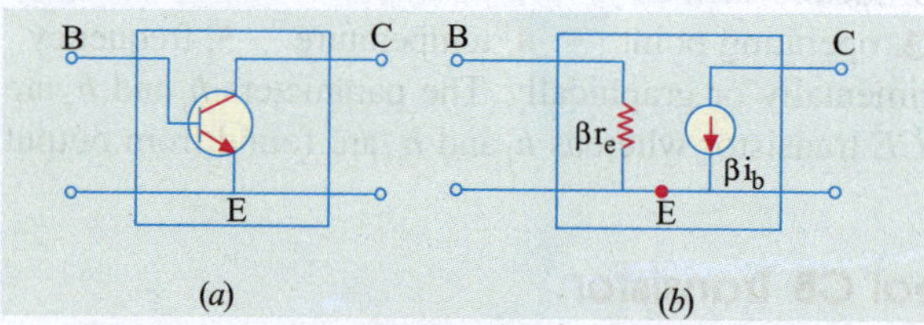

Fig. 59.51

Fig. 59.51 (*a*) shows a *CE*-connected ideal transistor contained in a black box whereas 59.51 (*b*) shows its ac equivalent circuit in terms of its β and resistance values.

(*a*) Forward Parameters

The two forward h-parameters can be found from the circuit of Fig. 59.52 (*a*) where output has been shorted. Obviously, the input impedance is simply βr_e.

$$\therefore \quad h_{ie} = \beta r_e$$

The forward current gain is given by

$$h_{fe} = \frac{i_2}{i_1} = \frac{\beta i_b}{i_b} = \beta$$

(It is also called the ac beta of the *CE* circuit)

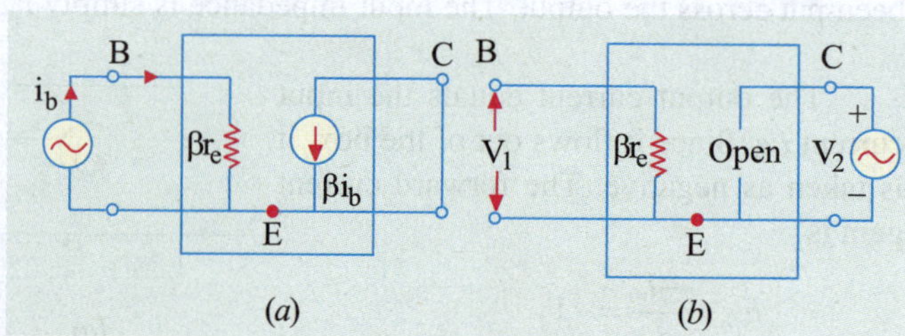

Fig. 59.52

(*b*) Reverse Parameters

These can be found by reference to the circuit of Fig. 59.52 (*b*) where input terminals are open but output terminals are driven by an ac voltage source V_2. With input terminals open, there can be no base current so that $i_b = 0$. If $i_b = 0$, then collector current source has zero value and looks like an open. Hence, there can be no V_1 due to this open.

$$\therefore \quad h_{re} = \frac{V_1}{V_2} = \frac{0}{V_2} = 0$$

Again, the impedance looking into the output terminals is infinite so that conductance is zero.

$$\therefore \quad h_{oe} = 0$$

Hence, the four h-parameters of an ideal transistor connected in CE configuration are :

$$h_{ie} = \beta r_e ; \quad h_{fe} = \beta, \quad hr_e = 0 ; \quad h_{oe} = 0.$$

The hybrid equivalent circuit of such a transistor is shown in Fig. 59.53.

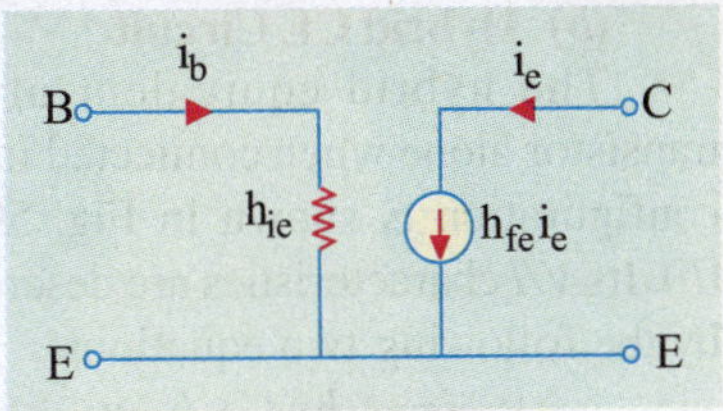

Fig. 59.53

Note. In practice, h_{re} and h_{oe} are not exactly zero but quite small for the same reasons as given in the Note to Art 59.19.

59.21. Approximate Hybrid Equivalent Circuits

So far, we did not take into account the following two factors which exist in an actual transistor (as opposed to an ideal one).

(i) because of the transistor's non-unilateral behaviour, there is a 'feedback' of the output voltage into the input voltage. This feedback is represented by a voltage-controlled generator $h_r v_2$ as shown in Fig. 59.54 and 59.55.
By definition, an ideal amplifier is one which responds only to signals applied to its input terminals. It should not do the reverse *i.e.* reproduce at the input any portion of the ac signal applied at the output. Such an ideal one-way device is called a unilateral device. A real transistor cannot be unilateral because of unaviodable interaction between its input and output circuits (after all, it consists of a single piece of a crystal). Therefore, not only its output responds to its input but, to a lesser degree, its input also responds to its output.

(ii) even when input circuits is open, there is some effective value of conductanece when looking into the transistor from its output terminals. It is represented by h_0.

We will now draw ***low-frequency small-signal*** hybrid equivalent circuits after taking into account the 'feedback' voltage generator and output admittance.

(a) Hybrid CB Circuit

In Fig. 59.54 (*a*) is shown an *NPN* transistor connected in *CB* configuration. Its ac equivalent circuit employing h-parameters is shown in Fig. 59.54 (*b*). The *V/I* relationships are given by the following two equations.

$$v_{be} = h_{ib} i_e + h_{rb} v_{cb}$$
$$i_e = h_{fb} i_e + h_{ob} v_{cb}$$

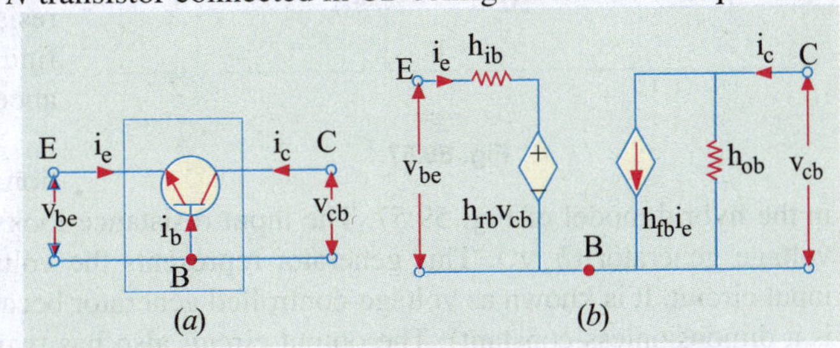

Fig. 59.54

These equations are self-evident because applied voltage across input terminals must equal the drop over h_{ib} and the generator voltage. Similarly, current i_c in the output terminals must equal the sum of two branch currents.

As per current convention stated earlier (Art. 59.14), emitter i_e is shown flowing ***inwards*** though actually this current flows ***outwards*** as shown by the arrow inside the ac current source. Similarly, ac voltage polarities have been taken by considering upper terminal positive and lower one as negative (please remember that the dc biasing rule of Art. 59.12 does not apply here).

It may be noted that no external dc biasing resistor or ac voltage sources have been connected to the equivalent circuit as yet.

Incidentally, it may be noted that the ac equivalent circuit contains a Thevenin's circuit in the input and a Norton's circuit in the output. It is all the more a reason to call it a hybrid equivalent circuit.

(*b*) Hybrid CE Circuit

The hybrid equivalent of the transistor alone when connected in *CE* configuration is shown in Fig. 59.55 (*b*). Its *V/I* characteristics are described by the following two equations.

$$v_{be} = h_{ie}\, i_b + h_{re}\, v_{ce}$$
$$i_e = h_{fc}\, i_b + h_{oe}\, v_{ce}$$

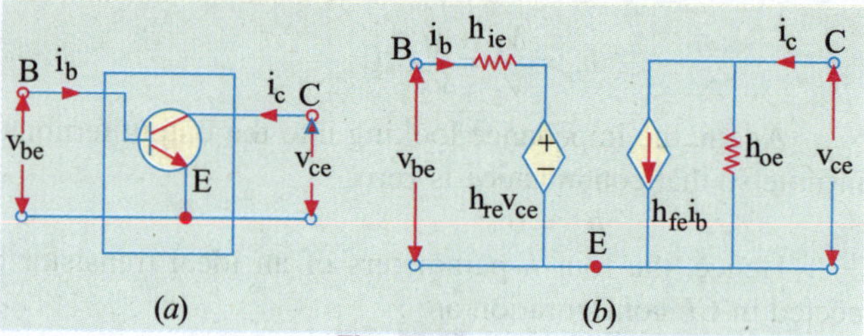

Fig. 59.55

We may connect signal input source across its input terminals and load resistance across output terminals (Art. 59.22).

(*c*) Hybrid CC Circuit

The hybrid equivalent of a transistor alone when connected in *CC* configuration is shown in Fig. 59.56 (*b*). Its *V/I* characteristics are defined by the following two equations :

$$v_{bc} = h_{ic}\, i_b + h_{rc}\, v_{ce}$$
$$i_e = h_{fe}\, i_b + h_{oc}\, v_{ce}$$

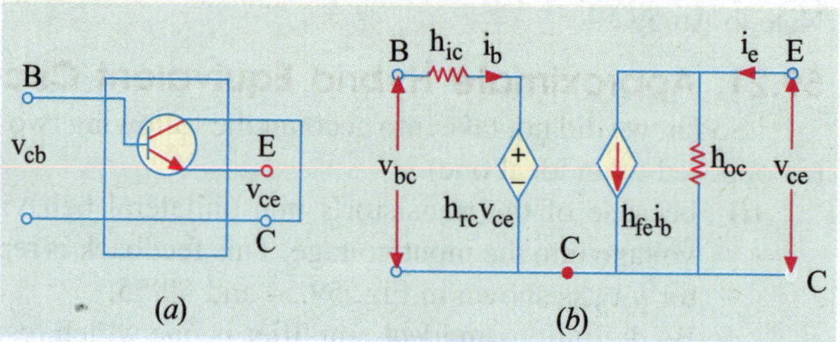

Fig. 59.56

We may connect signal input source across ioutput terminals *BC* and load resistance across output terminals *EC* to get a *CC* amplifier.

59.22. Transistor Amplifier Formulae Using *h*-parameters

As shown in Fig. 59.57, if we add a signal source across input terminals 1-1 of a transistor and a load resistor across its output terminals 2–2, we get a small-signal, low-frequency hybrid model of a transistor amplifier. It is valid for all the three configurations and holds good for all types of load whether a resistance or an impedance. We will now find expressions for its gains and impedances.

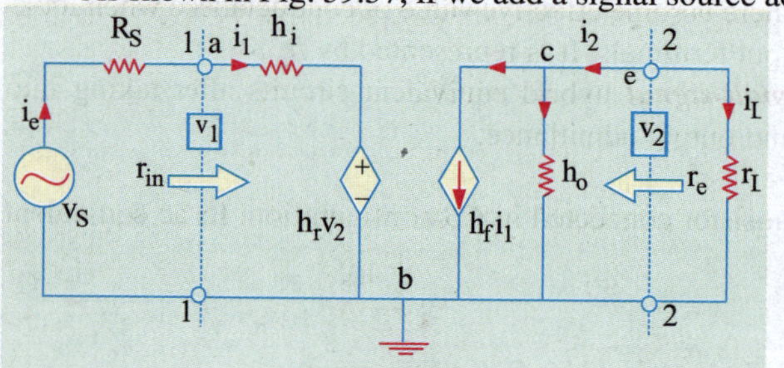

Fig. 59.57

Before undertaking the above derivations, let us consider different components in the hybrid model of Fig. 59.57. The input resistance looks like a resistance (h_i) in series with a voltage generator ($h_r\, v_2$). This generator represents the voltage feed-back from the output to the input circuit. It is known as voltage-controlled generator because its value is determined by v_2 (as h_r is a dimensionless constant). The output circuit also has two components (*i*) h_0 component which represents the conductance as seen from output terminals and (*ii*) the current-controlled generator ($h_f\, i_1$) which simulates the transistor's ability to amplify. The parameter h_f is a dimensionless constant.

The above model can be described mathematically by using the following two equations :

Input Circuit

v_1 = sum of voltage drops from *a* to *b*

$$= h_i\, i_1 + h_r\, v_2 \qquad ...(i)$$

Output Circuit

i_2 = sum of currents leaving junction *c*

$$= h_f\, i_1 + h_0\, v_2 \qquad ...(ii)$$

Now, $v_2 = -i_2 r_L$. Substituting this value in Eq. (*ii*) above, we have

$$i_2 = h_f i_1 - h_o i_2 r_L \quad \text{... (iii)}$$

Eq. (*i*) and (*iii*) can now be used to find various gains of a transistor.

(*i*) Current Gain

It is given by $A_i = i_2/i_1$

Dividing both sides of Eq. (*iii*) by i_1, we get

$$\frac{i_2}{i_1} = h_f - h_0 \frac{i_2}{i_1} r_L \quad \text{or} \quad A_i = h_f - h_0 A_i r_L$$

$$\therefore \quad A_i = \frac{h_f}{1 + h_0 r_L}$$

If $r_L = 0$ or $h_0 r_L << 1$, then $A_i = h_f$

Current Gain Taking R_S into Account

The source current is not the transistor input current because i_1 partly flows along R_S and partly along r_{in}.

To illustrate this point, consider the Norton's equivalent of the source (Fig. 59.58). The overall current gain A_{is} is given by

$$A_{is} = \frac{i_2}{i_s} = \frac{i_2}{i_1} \times \frac{i_1}{i_s} = A_i \frac{i_1}{i_s}$$

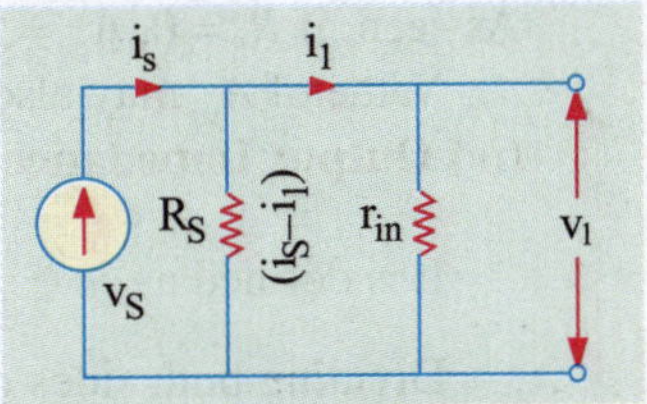

Fig. 59.58

As seen from Fig. 59.58

$$i_1 = \frac{i_s R_s}{r_{in} + R_s} \quad \text{or} \quad \frac{i_1}{i_s} = \frac{R_s}{r_{in} + R_s}$$

$$\therefore \quad A_{is} = A_i \,.\, R_S / (r_{in} + R_S)$$

(*ii*) Input Impedance

It is defined as the resistance when looking into the amplifier from its input terminals. Hence, $r_{in} = v_1/i_1$.

From Eq. (*i*) above, we have

$$r_{in} = \frac{v_1}{i_1} = \frac{h_i i_1 + h_r v_2}{i_1} = h_i + h_r \cdot \frac{v_2}{i_1}$$

Substituting the value of $v_2 = -i_2 r_L = -A_i \, i_1 r_L$, we get

$$r_{in} = h_i - h_r A_i r_L = h_i - \frac{h_f h_r r_L}{1 + h_0 r_L} = h_i - \frac{h_f h_r}{h_0 + 1/r_L}$$

$$= \frac{h_i + \Delta h r_L}{1 + h_0 r_L} \quad \text{where} \quad \Delta h = h_i h_0 - h_f h_r$$

$$\cong h_i \quad \text{– if } h_r \text{ or } r_L \text{ is very small.}$$

It is seen that r_{in} depends on r_L *i.e.* ac resistance of the load across output terminals of the transistor.

(*iii*) Voltage Gain

$A_v = v_2 / v_1$. It is also known as the internal voltage gain of the transistor. It is different from $A_{vs} = v_2 / v_s$ which is the gain from the source to the output terminals and is known as stage gain or overall gain.

As seen from above, $v_2 = -A_i i_1 r_L$ and $v_1 = i_1 r_{in}$

$$\therefore \quad A_v = \frac{v_2}{v_1} = \frac{-i_1 A_i r_L}{i_1 r_{in}} = -A_i \frac{r_L}{r_{in}}$$

$$= \frac{-h_f}{1+h_0 r_L} \times \frac{r_L(1+h_0.r_L)}{h_i + \Delta h.r_L} = \frac{-h_f.r_L}{h_i + \Delta h.r_L}$$

$$= \frac{-h_f r_L}{h_i(1+h_0 r_L) - h_f h_r r_L} \cong -h_f . \frac{r_L}{h_i}$$

Overall voltage gain is

$$A_{vs} = \frac{v_2}{v_s} = \frac{v_2}{v_1} \times \frac{v_1}{v_s} = A_v \frac{v_1}{v_s}$$

Now, v_s drops over series combination of R_S and r_{in}.

Drop across r_{in} constitutes v_1. Hence, $v_1 = v_s \times r_{in}/(R_S + R_{in})$

$$\therefore \quad \frac{v_1}{v_s} = \frac{r_{in}}{(R_S + r_{in})} \quad \text{or} \quad A_{vs} = A_v \frac{r_{in}}{(R_S + r_{in})}$$

As seen, if $R_S = 0$, $A_{vs} = A_v$

Value of A_{vs} may also be obtained by adding R_S to h_i in the expression for A_v.

(iv) Output Impedance

It is defined as $r_0 = \frac{v_2}{i_2}\Big|_{v_s=0}$ or $g_0 = \frac{i_2}{v_2}$

Dividing both sides of Eq. (ii) by v_2, we get

$$g_0 = h_f . \frac{i_1}{v_2} + h_0 \qquad \text{..... (iv)}$$

Taking $v_s = 0$ and then applying KVL to the input circuit in Fig. 59.57, we get

$$-i_1(h_i + R_S) - h_r v_2 = 0 \quad \text{or} \quad i_1/v_2 = -h_r/(h_i + R_S)$$

Substituting this value in Eq. (iv) above, we have

$$g_0 = h_0 / \frac{h_f h_r}{(h_i + R_S)} = \frac{h_o R_S + \Delta h}{(h_i + R_S)} \qquad \therefore \quad r_o = \frac{1}{g_o} = \frac{h_i + R_S}{h_o R_S + \Delta h}$$

It is seen that r_{in} depends on r_L whereas r_o depends on R_S.

If R_S is very large (*i.e.* circuit is driven by a current source) or h_r is negligible, then $r_o \cong 1/h_o$.

(v) Power Gain

$$A_p = \frac{P_2}{P_1} = \frac{v_2 i_2}{v_1 i_1} = A_v A_i = A_i^2 \frac{r_L}{r_{in}}$$

The above formulae are summarized below :

(i) $$A_i = \frac{h_f}{1+h_o r_L} \cong h_f$$

$$A_i = \frac{h_f R_S}{(1+h_o r_L)(r_{in}+R_S)} \cong \frac{h_o R_S}{(r_{in}+R_S)}$$

(ii) $$r_{in} = h_i - \frac{h_f h_r r_L}{1+h_o r_L} \cong h_i$$

(iii) $$A_v = \frac{-h_f r_2}{h_i + \Delta h.r_L} \cong \frac{-h_f r_L}{h_i}; \quad A_{vs} \cong \frac{-h_f r_L}{(h_i + R_S)}$$

(iv) $$r_o = \frac{h_i + R_S}{h_o R_S + \Delta h} = \frac{1 + h_i / R_S}{h_o + \Delta h / R_S} \cong \frac{1}{h_o}$$

59.23. Typical Values of Transistor *h*-parameters

In the table below are given typical values for each parameter for the broad range of transistors available today in each of the three configurations.

Parameter	*CB*	*CE*	*CC*
h_i	25 Ω	1 K	1 K
h_r	3×10^{-4}	2.5×10^{-4}	$\cong 1$
h_f	–0.98	50	–50
h_o	0.5×10^{-6}S	25×10^{-6}S	25×10^{-6}S

59.24. Approximate Hybrid Formulas

The approximate hybrid formulas for the three connections are listed below. These are applicable when h_o and h_r are very small and R_S is very large. The given values refer to transistor terminals. The values of $r_{in(stage)}$ or r_{in}' and $r_{o(stage)}$ will depend on biasing resistors and load resistance respectively.

Item	*CE*	*CB*	*CC*
r_{in}	h_{ie}	h_{ib}	$h_{ic} + h_{fe}R_L$
r_o	$\frac{1}{h_{oe}}$	$\frac{1}{h_{ob}}$	$\frac{h_{ic}}{h_{fe}}$
A_i	$h_{fe} = \beta$	$h_{fb} \cong 1$	$h_{fe} \cong \beta$
A_v	$\frac{-h_{fe}R_L}{h_{ie}}$	$\frac{h_{fb}}{h_{ib}}R_C$	1

59.25. Common Emitter *h*-parameter Analysis

The *h*-parameter equivalent of the *CE* circuit of Fig. 59.59 (*a*) is shown in Fig. 59.59 (*b*). In Fig. 59.59 (*a*), no emitter resistor has been connected.

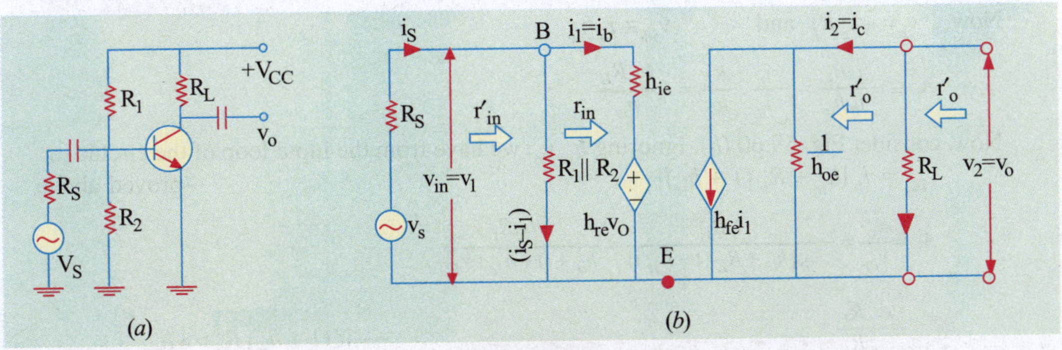

Fig. 59.59

However, Fig. 59.60 shows the *CE* circuit with an emitter resistor R_E.

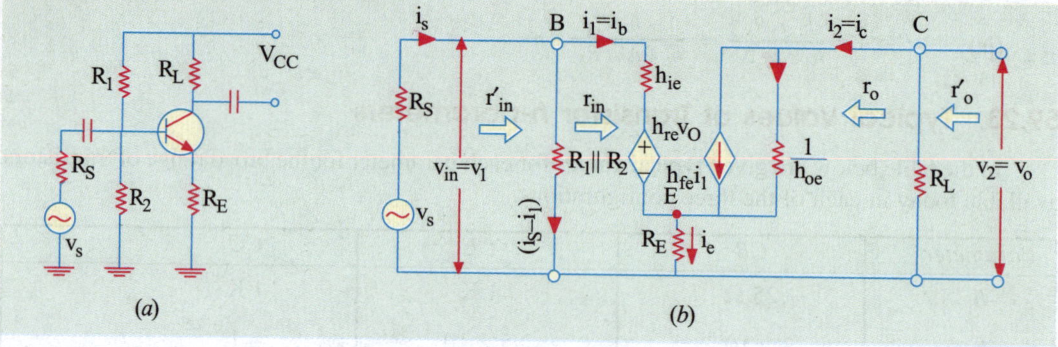

Fig. 59.60

We will now derive expressions for voltage and current gains for both these circuits.

1. Input Impedance

When looking into the base-emitter terminals of the transistor, h_{ie} is in series with h_{re} V_o. For a *CE* circuit, h_{re} is very small so that h_{re} v_0 is negligible as compared to the drop over h_{ie}. Hence, $r_{in} \cong h_{ie}$.

Now, consider the circuit of Fig. 59.60. Again ignoring $h_{re}v_o$, we have

$$v_1 = h_{ie}i_b + i_e R_E = h_{ie}\, i_b + (i_b + i_c)\, R_E$$

$$= h_{ie}i_b + i_b R_E + h_{fe}\, i_b R_E \qquad (\because\ i_c = h_{fe} i_b)$$

$$= i_b\, [h_{ie} + R_E\,(1 + h_{fe})]$$

$$\therefore \qquad r_{in} = r_{in(base)} = \frac{v_1}{i_1} = \frac{v_1}{i_b} = h_{ie} + (1 + h_{fe})\, R_E \ *$$

$$r_{in} \text{ or } r_{in(base)} = R_1 \parallel R_2 \parallel r_{in(base)}$$

2. Output Impedance

Looking back into the collector and emitter terminals of the transistor in Fig. 59.59 (*b*), $r_o \cong 1/h_{oe}$.

As seen, $r_o{}'$ or $r_{o(stage)} = r_o \parallel \mathrm{R_L} = (1/h_o) \parallel r_L$ $\qquad (\because\ r_L = R_L)$

Since $1/h_{oe}$ is typically 1 M or so and R_L is usually much smaller, $r_o{}' \cong R_L = r_L$

3. Voltage Gain

$$A_v = \frac{v_2}{v_1} = \frac{v_o}{v_{in}} \qquad \text{–Fig. 59.59 } (b)$$

Now, $\quad v_o = -\, i_c R_L$ and $\qquad v_{in} \cong i_b\, h_{ie}$

$$\therefore \qquad A_v \cong -\frac{i_c R_L}{i_b\, h_{ie}} = -\frac{i_c}{i_b} \cdot \frac{R_L}{h_{ie}} = -\frac{h_f R_L}{h_{ie}}$$

Now, consider Fig. 59.60 (*b*). Ignoring h_{re} v_o, we have from the input loop of the circuit

$$v_{in} = i_b\, [h_{ie} + R_E\,(1 + h_{fe})] \qquad \text{–proved above}$$

$$\therefore \qquad A_v = \frac{v_o}{v_{in}} = -\frac{i_c R_L}{i_b\,[h_{ie} + R_E\,(1 + h_{fe})]} = \frac{-h_{fe} R_L}{h_{ie} + (1 + h_{fe})\, R_E}$$

$$\cong -\frac{R_L}{R_E} \qquad \text{— if } (1 + h_{fe})\, R_E >> h_{ie}$$

* The above result could also be obtained by applying β-rule (Art. 57.24)

In general, $A_v = \dfrac{-h_{fe} \times \text{load resistance}}{\text{input resistance}}$

4. Current Gain

$$A_i = \frac{i_2}{i_1} = \frac{h_{fe}}{1 + h_{oe} r_L} \cong h_{fe} \qquad \text{— if } h_{oe} r_L << 1$$

$$A_{is} = \frac{h_{fe} \,.\, R_1 \parallel R_2}{r_{in} + R_1 \parallel R_2}$$

5. Power Gain

$$A_p = A_v \times A_i$$

59.26. Common Collector *h*-parameter Analysis

The *CC* transistor circuit and its *h*-parameter equivalent are shown in Fig. 59.61.

One can make quick approximations of *CC* gains and impedance if one remembers that $h_{re} = 1$ *i.e.* all of v_o is fed back to the input (Art. 59.23).

1. Input Impedance

$$v_{in} = i_b h_{ic} + h_{rc} v_o = i_b h_{ic} + v_o = i_b h_{ic} + i_e R_L$$
$$= i_b h_{ic} + h_{fe} i_b R_L = i_b (h_{ic} + h_{fe} R_L) \qquad \because i_e = h_{fe} i_b$$

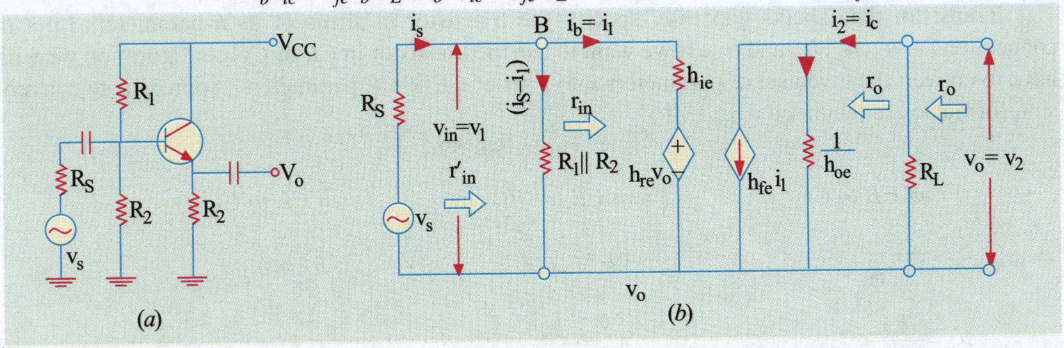

Fig. 59.61

$$\therefore \qquad r_{in} = r_{in(base)} = \frac{v_{in}}{i_b} = h_{ic} + h_{fe} R_L$$

As seen, $r_{in(stage)} = r_{in(base)} \parallel R_1 \parallel R_2 = r_{in(base)} \parallel R_B$ where $R_B = R_1 \parallel R_2$

2. Output Impedance

$$r_o = \frac{v_2}{i_2}\Big|_{v_s=0} \qquad = \frac{v_o}{i_c}\Big|_{v_s=0}$$

Now, $\qquad i_e \cong i_c = h_{fe} i_b = h_{fc} i_b$

Since $\qquad v_s = 0$, i_b is produced by $h_{rc} v_o = v_o$

Hence, considering the input circuit loop, we get

$$i_b = \frac{v_0}{h_{ic} + (R_S \parallel R_1 \parallel R_2)} = \frac{v_0}{h_{ic} + R_S \parallel R_B}$$

$$\therefore \qquad i_e \cong i_c = h_{fe} i_b = \frac{h_{fe} v_o}{h_{ic} + (R_S \parallel R_B)} \qquad \text{where } R_B = R_1 \parallel R_2$$

$$\therefore \quad r_o = \frac{V_o}{i_e} = \frac{h_{ic} + (R_S \parallel R_1 \parallel R_2)}{h_{fe}}$$

Also, r_o' or $r_{o(stage)} = r_o \parallel R_L$

3. Voltage Gain

$$A_v = \frac{v_2}{v_1} = \frac{v_o}{v_{in}}$$

Now, $v_o = i_e R_L = h_{fe} i_b R_L$ and $i_b = (v_{in} - v_o) / h_{ic}$

$$v_o = \frac{h_{fe} R_L}{h_{ic}} (v_{in} - v_o) \quad \text{or} \quad v_o \left(1 + \frac{h_{fe} R_L}{h_{ic}}\right) = \frac{h_{fe} R_L v_{in}}{h_{ic}}$$

$$\therefore \quad A_v = \frac{v_o}{v_{in}} = \frac{h_{fe} R_L / h_{ic}}{1 + h_{fe} R_L / h_{ic}} \cong 1$$

4. Current Gain

$$A_i = \frac{i_2}{i_1} = \frac{i_e}{i_b} = h_{fe}\ ;\ A_{is} = \frac{h_{fe} R_B}{r_{in} + R_B}$$

where $R_B = R_1 \parallel R_2$

59.27. Conversion of *h*-parameters

Transistor data sheets generally specify the transistor in terms of its h-parameters for *CB* connection *i.e.* h_{ib}, h_{fb}, h_{rb} and h_{ob}. If we want to use the transistor in *CE* or *CC* configuration we will have to convert the given set of parameters into a set of *CE* or *CC* parameters. Approximate conversion formulae are tabulated over leaf :

Table No. 59.2

From CB to CE	*From CE to CB*	*From CE to CC*
$h_{ie} = \frac{h_{ib}}{1 + h_{fb}}$	$h_{ib} = \frac{h_{ie}}{1 + h_{fe}}$	$h_{ic} = h_{ie}$
$h_{oe} = \frac{h_{ob}}{1 + h_{fb}}$	$h_{ob} = \frac{h_{oe}}{1 + h_{fe}}$	$h_{oc} = h_{oe}$
$h_{fe} = \frac{-h_{fb}}{1 + h_{fb}}$	$h_{fb} = \frac{-h_{fe}}{1 + h_{fe}}$	$h_{fc} = (1 + h_{fe})$
$h_{re} = \frac{h_{ib} h_{ob}}{1 + h_{fb}} - h_{rb}$	$h_{rb} = \frac{h_{ie} h_{oe}}{1 + h_{fe}} - h_{re}$	$h_{rc} = 1 - h_{re} \cong 1$

Example. 59.18. *A transistor used in CB circuit has the following set of parameters.*

h_{ib} = 36 Ω, h_{fb} = −0.98, h_{rb} = 5 × 10^{-4}, h_{ob} = 10^{-6} Siemens

With R_S = 2 K and R_C = 10 K, calculate **(i)** $r_{in(base)}$ **(ii)** r_{out} **(iii)** A_i and **(iv)** A_v.

(Applied Electronics-I; Punjab Univ. 1991)

Solution. Approximate Values

(i) $r_{in} = h_{ib} = 36\ \Omega$ ***(ii)*** $r_o = \frac{1}{h_{ob}} = \frac{1}{10^{-6}} = \mathbf{1M}$

(iii) $A_i = h_{fb} = \mathbf{-0.98}$ ***(iv)*** $A_v = -\frac{h_{fb}}{h_{ib}} \times R_C = \frac{-(-0.98)}{36\Omega} \times 10\text{K} = \mathbf{272}$

More Accurate Values

(i) $r_{in(base)} = h_{ib} - \dfrac{h_{rb} h_{fb}}{h_{ob} + 1/r_L}$ –Art 9.22

$= 36 - \dfrac{-0.98 \times 5 \times 10^{-4}}{10^{-6} + 1/10 \times 10^3}$ ($\therefore$ $r_L = R_C$ since there is no R_L)

$= 36 + 4.9 =$ **40.9 Ω**

It is the input resistance at transistor terminals.

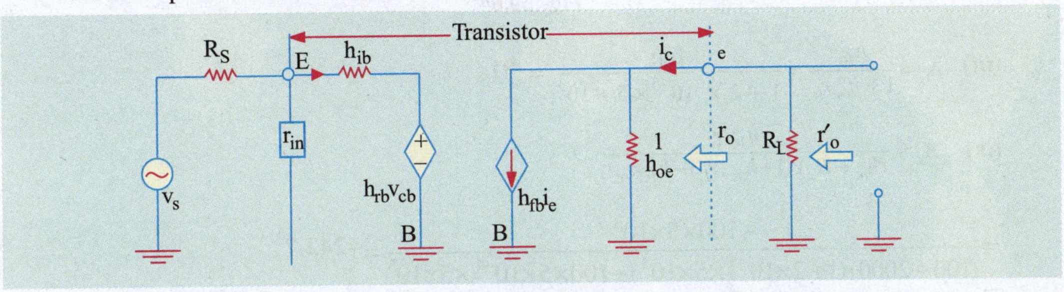

Fig. 59.62

(ii) $r_o = \dfrac{h_{ib} + R_s}{h_o(h_{ib} + R_S) - h_{fb} . h_{rb}} = \dfrac{36 + 2000}{10^{-6}(36 + 2000) - (-0.98) \times 5 \times 10^{-4}} =$ **0.8 M**

It is the output resistance at transistor terminals.

(iii) $A_i = \dfrac{h_{fb}}{1 + h_{ob} r_L} = \dfrac{-0.98}{1 + 10^{-6} \times 10^4} =$ **–0.97**

(iv) $A_v = \dfrac{-h_{fb} r_L}{h_{ib}(1 + h_{ob} r_L) - h_{fb} h_{rb} . r_L}$ ($\because r_L = R_c$)

Here, ac load $r_L = R_C = 10\text{ K} = 10^4\ \Omega$

$\therefore\ A_v = \dfrac{-(-0.98) \times 10^4}{36(1 + 10^{-6} \times 10^4) - (-0.98) \times 5 \times 10^{-4} \times 10^4} =$ **237.5**

Example 59.19. *A transistor used in CE connection (Fig. 59.63) has the following set of h-parameters : $h_{ie} = 1$ K, $h_{fe} = 100$, $h_{re} = 5 \times 10^{-4}$ and $h_{oe} = 2 \times 10^{-5}$ S. With $R_S = 2$ K and $R_C = 5$ K, determine* **(i)** r_{in} **(ii)** r_o **(iii)** A_i *and* **(iv)** A_v

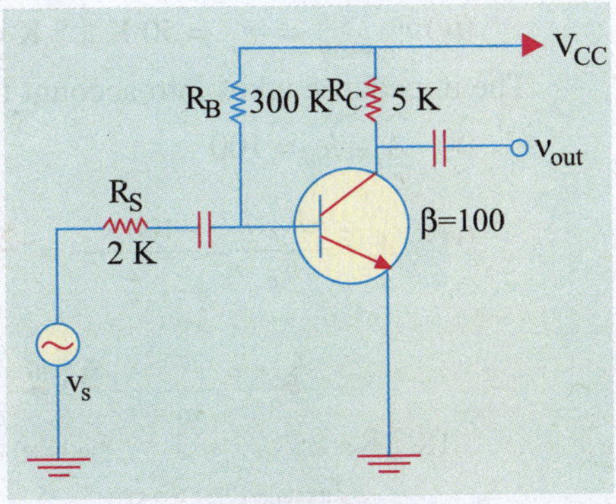

Fig. 59.63

Solution.

(i) $r_{in} = h_{ie} - \dfrac{-h_{fe} h_{re}}{h_o + 1/r_L}$

$= 1000 - \dfrac{100 \times 5 \times 10^{-4}}{2 \times 10^{-5} + 1/5 \times 10^3}$

= 773 Ω

(ii) $r_o = \dfrac{h_{ie} + R_s}{(h_{ie} + R_s) h_{oe} - h_{fe} . h_{re}} = \dfrac{1000 + 2000}{(1000 + 2000) \times 2 \times 10^{-5} - 100 \times 5 \times 10^{-4}}$

$= 300{,}000\Omega =$ **0.3 MΩ**

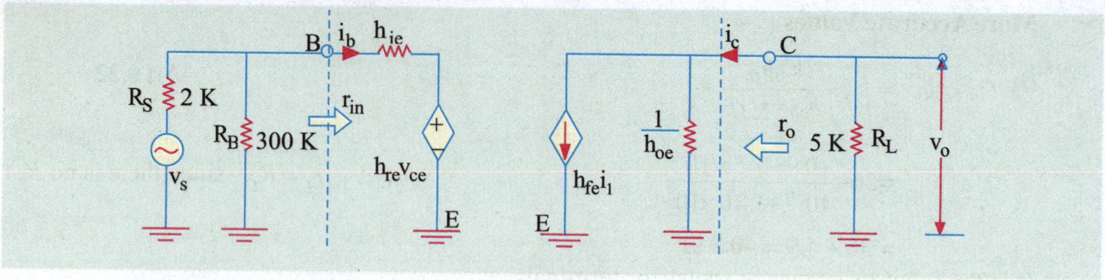

Fig. 59.64

(iii) $A_i = \dfrac{h_{fe}}{1 + h_{oe}r_L} = \dfrac{100}{1 + 2 \times 10^{-5} \times 5 \times 10^3} =$ **91**

(iv) $A_v = \dfrac{-h_{fe}r_L}{(h_{ie} + R_s)\,(1+h_{oe}.r_L) - h_{fe}h_{re}r_L}$

$$= \frac{-100\times5\times10^3}{(100+2000)(1+2\times10^{-5}\times5\times10^3)-100\times5\times10^{-4}\times5\times10^3} = \mathbf{-242.7}$$

The negative sign indicates that there is 180° phase shift between the input and output ac signals. Obviously, it is the overall (or circuit) voltage gain and not the voltage gain of the transistor alone.

Example 59.20. *In the CE circuit shown in Fig. 59.65, the transistor parameters are :*

h_{ie}=2 K, h_{fe}=100, $h_{re} = 5 \times 10^{-4}$, $h_{oe} = 2 \times 10^{-5}$ S

Calculate **(i)** $r_{in(base)}$, **(ii)** $r_{in(stage)}$, **(iii)** r_o, **(iv)** $r_{o(stage)}$, **(v)** A_i and **(vi)** A_v.

(Electronics–1, Karnataka Univ. 1988)

Solution. The hybrid equivalent circuit is shown in Fig. 59.66. We will use the approximate formulas given in Art. 59.24.

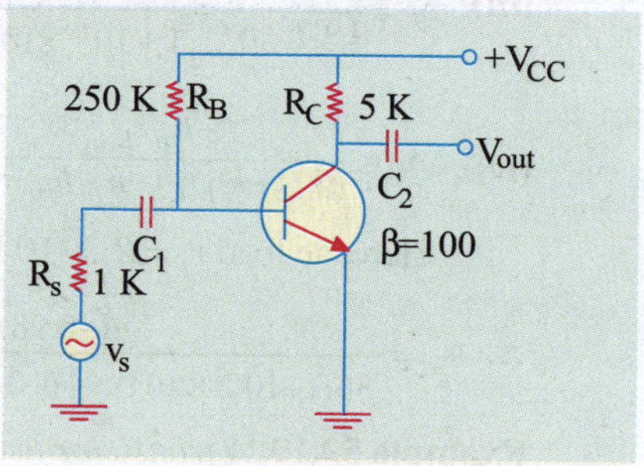

Fig. 59.65

(i) $r_{in(base)} = h_{ie}$ = **2K**

(ii) $r_{in(stage)}$ = 2 K || 250 K = **1.98 K**

(iii) $r_o = 1 / h_{oe} = 1/2 \times 10^{-5}$ = **50 K**

It is the output impedance of the transistor only.

(iv) $r_{o(stage)} = r_o' $ = 50 K || 5 K = **4.54 K**

The impedance takes into account the collector load.

(v) $A_i \cong h_{fe} = 100$

(vi) $A_v = \dfrac{-h_{fe}r_L}{h_{ie}} = \dfrac{-100 \times 5}{2} =$ **−250**

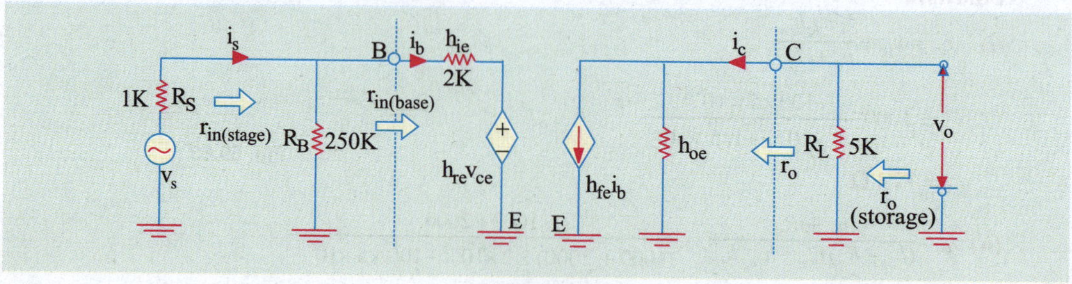

Fig. 59.66

Example 59.21. *Determine the various gains of the circuit of Fig. 59.67 if an emitter resistance of 0.5 K is included in the circuit.* **(Applied Electronics, Punjab Univ. 1991)**

Solution. The *CE* circuit with R_E included is shown in Fig. 59.67. Different performance characteristics of the circuit are as under :

(i) $r_{in(base)} \cong h_{ie} + (1+\beta)R_E$
$= 2 + 101 \times 0.5 = $ **52.5 K**

(ii) $r_{in(stage)} = R_B \,||\, r_{in} = 250 \,||\, 52.5$
$=$ **43.3K**

(iii) $r_o = 1/h_{oe} = $ **50 K**

(iv) $r_{o(stage)} = r_o' = 50\text{ K} \,||\, 5\text{ K}$
$=$ **4.54 K**

(v) $A_i \cong h_{fe} = $ **100**

(vi) $A_v = \dfrac{-h_{fe}r_L}{h_{ie}+(1+\beta)R_E} = \dfrac{-100\times 5}{2+50.5} = $ **−9.5**

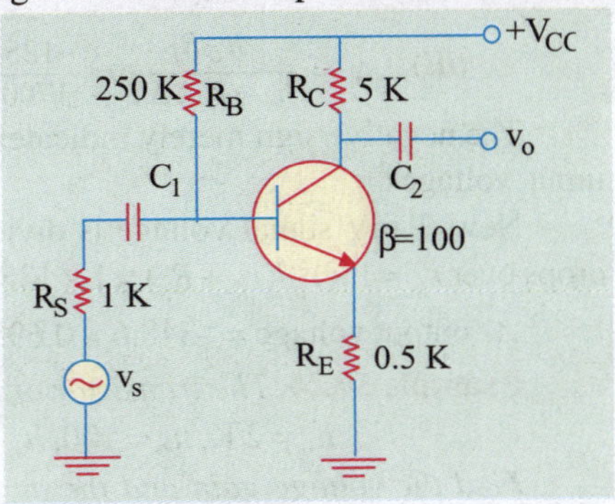

Fig. 59.67

The value is reduced from 250 to 9.5.

Example 59.22. *The transistor of Fig. 59.68 has the following set of h-parameters :*

$h_{ie} = 2\text{K}, h_{fe} = 50, h_{rs} = 4\times 10^{-4}, h_{oe} = 25\times 10^{-6}$ *Siemens*

Determine *(i)* $r_{in(base)}$ *(ii)* $r_{in(stage)}$ *(iii)* r_o *(iv)* $r_{o(stage)}$ and *(v)* A_v.

(Electronics-I, Patna Univ. 1991)

Solution. we will use the formula derived in Art. 59.25.

(i) $r_{in(base)} = h_{ie} + (1+h_{fe})R_E = 2 + (1+50)\times 5$
$=$ **257 K**

(ii) $r_{in(stage)} = r_{in(base)} \,||\, R_1 \,||\, R_2 = 257 \,||\, 80 \,||\, 40$
$=$ **24 K**

(iii) $r_o = 1/h_{oe} = 1/25\times 10^{-6} = $ **40 K**

(iv) $r_{o(stage)} = r_o \,||\, r_L$ where $r_L = 15\text{ K} \,||\, 30\text{ K} = 10\text{ K}$
$= 40\text{ K} \,||\, 10\text{ K} = $ **8 K**

(v) $A_v = \dfrac{-h_{fe}R_L}{h_{ie}+(1+h_{fe})R_E}$

$= \dfrac{-50\times 10}{257} = $ **− 1.95**

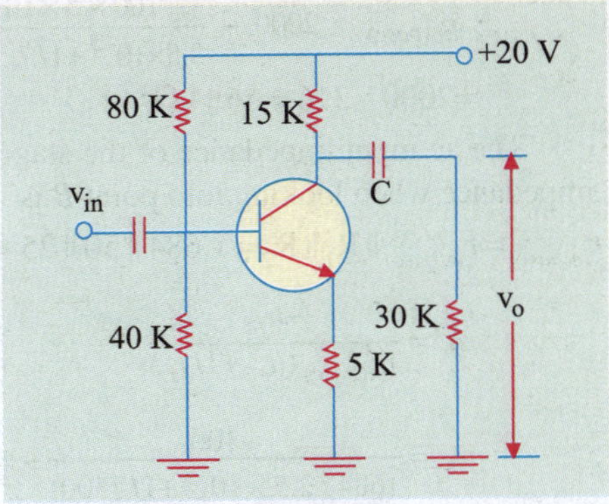

Fig. 59.68

Example 59.23. *Draw a hybrid small-signal model for a transistor in CE configuration.*

A single source with an open circuit voltage of 1 mV and internal impedance of 600 Ω is connected to the input of transistor AC 125 in CE configuration. The small-signal parameters measured at $V_{CB} = -5$ *V and* $I_E = 2$ *mA, are as follows :*

$h_{ie} = 1.7\text{ K}, h_{re} = 6.5\times 10^{-4}, h_{fe} = 125, h_{oe} = 80\ \mu\text{S}.$

Calculate the input and output impedances and signal amplification for a load of 5.6 K. Also, find the value of the signal voltage at the output. **(Electronics-I, Gwalior Univ. 1982)**

Solution. $\Delta h = 1.7\times 10^3 \times 80\times 10^{-6} - 125\times 6.5\times 10^{-4} = 0.055$

(i) $r_{in} = \dfrac{h_{ie}+\Delta h.R_L}{1+h_{oe}R_L} = \dfrac{1700 + 5.6\times 10^3 \times 55\times 10^{-3}}{1+80\times 10^{-6}\times 5.6\times 10^3} = 1386.7\Omega$ —Art. 59.22

$= 1.39\ \text{k}\Omega$

(ii) $r_o = \dfrac{h_{ie} + R_S}{h_{oe} R_S + \Delta h} = \dfrac{1700 + 600}{80 \times 10^{-6} \times 600 + 0.055} = \dfrac{2300}{0.103} =$ **22.3 K**

(iii) $A_v = \dfrac{-h_{fe} R_L}{h_{ie} + R_L . \Delta h} = \dfrac{-125 \times 5600}{1700 + 308} =$ **−348.6**

The negative sign merely indicates that there is phase reversal of 180° between the output and input voltages.

Now, 1 mV signal voltage is divided between r_{in} and R_S. The input voltage v_{in} is that which drops over $r_{in} = 1 \times r_{in}/(r_{in} + R_S) = 1 \times 1387/(1387 + 600) = 0.698$ mV

∴ output voltage = −348.6 × 0.698 = **−243.3 mV.**

Example 59.24. *The transistor of Fig. 59.69 has the following set of h-parameters :*

$$h_{ie} = 2\text{ K}, h_{fe} = 100, h_{re} = 5 \times 10^{-4}, h_{oe} = 2.5 \times 10^{-5}\text{ S}$$

Find the voltage gain and the ac impedance of the stage.

(Electronics-II, Bombay Univ. 1992)

Solution. Using somewhat exact formulas given in Art. 59.22, we have

$$r_{in(base)} = h_{ie} - \frac{h_{fe} h_{re}}{h_{oe} + 1/r_L}$$

Now, collector load

$$r_L = 10\text{ K} \parallel 30\text{ K} = 7.5\text{ K}$$

$$\therefore \quad r_{in(base)} = 2000 - \frac{100 \times 5 \times 10^{-4}}{2.5 \times 10^{-5} + 1/7.5 \times 10^3}$$

$= 2000 - 316 =$ **1684 Ω**

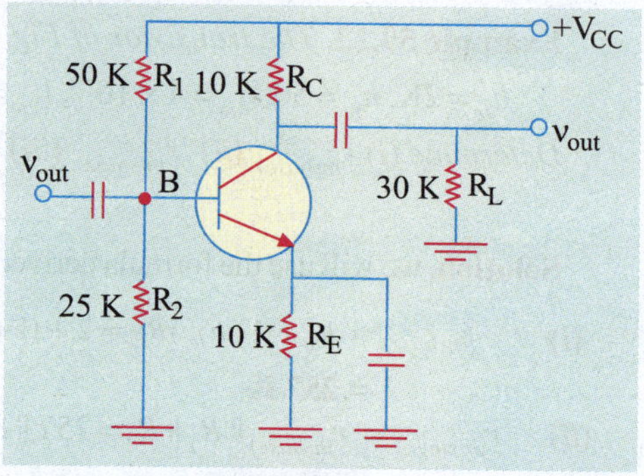

Fig. 59.69

The ac input impedance of the stage *i.e.* impedance when looking into point *B* is

$r_{in(base)} = r_{in(base)} \parallel R_1 \parallel R_2 = 1.684 \parallel 50 \parallel 25 =$ **1.53K**

$$A_v = \frac{-h_{fe}}{r_{in(base)}(h_{oe} + 1/r_L)} \qquad \text{Now,} \quad r_L = 10\text{ K} \parallel 30\text{ K} = 7.5\text{ K}$$

$$\therefore \quad A_v = \frac{-100}{1684\,(2.5 \times 10^{-5} + 1/7500)} = -\,375$$

Obviously, R_E does not come into the ac picture because it is ac grounded by the bypass capacitor.

Example 59.25. *In the CC circuit of Fig. 59.70, the transistor parameter are h_{ic} = 2K and h_{fc} = 100. Calculate the circuit input and output impedance and voltage, current and power gains.*

(Electronic Technology, Bangalore Univ. 1988)

Solution. $r_{in} \cong h_{ic} + h_{fc} R_L = 2 + 100 \times 5 =$ **502 K**

$r_{in(stage)} = R_1 \parallel R_2 \parallel r_{in}$

$= 10 \parallel 10 \parallel 502 =$ **4.95 K**

$$r_o \cong \frac{h_{ic} + (R_s \parallel R_1 \parallel R_2)}{h_{fe}}$$

$$= \frac{2 + (1 \parallel 10 \parallel 10)}{100} = 28.3\ \Omega$$

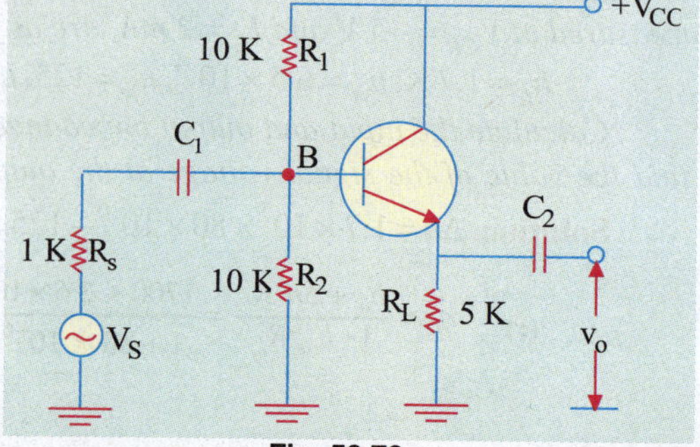

Fig. 59.70

$r_{o(stage)} = r_o \| R_L = 28.3\ \Omega \| 5K$

$= \mathbf{28.1\Omega}$

$A_v \cong 1$ and $A_i = h_{fe} = 100$

Example 59.26. *A transistor with h_{ie} = 1.5 K and h_{fe} = 75 is used in an emitter follower circuit where resistances R_1 and R_2 are used for normal biasing. Calculate* **(i)** A_i **(ii)** r_{in} **(iii)** r_o *and* **(iv)** A_v *if R_E = 860 Ω, $R_1 \| R_2$ = 20K and R_S = 1K.* **(Electronic Engg.; Indore Univ. 1992)**

Solution. Let us first convert the given *CE* values of *h*-parameters into their equivalent *CC* values with the help of Table No. 59.2. It is seen that $h_{ic} = h_{ie}$ = 1.5 K and $h_{fc} = (1 + h_{fe})$ = 76.

(*i*) $A_i = h_{fe} = 76$ (*ii*) $r_{in} = h_{ic} + h_{fc} R_L = 1.5 + 76 \times 0.860 = \mathbf{66.9\ K}$

(*iii*) $r_o = \dfrac{h_{ic} + (R_s \| R_1 \| R_2)}{h_{fe}} = \dfrac{1.5 + 1\|20}{76} = \mathbf{32.3\ \Omega}$ (*iv*) $A_v \cong 1$

Tutorial Problems No. 59.1

1. Using ideal transistor approximations for the single-stage *CB* amplifier of Fig. 59.71, find (*i*) stage r_{in} (*ii*) r_L (*iii*) A_v and (*iv*) A_p. Take transistor α = –0.99.
[(*i*) 25 Ω (*ii*) 10 K (*iii*) 400 (*iv*) 396]

2. For the single-stage *CB* amplifier circuit shown in Fig. 59.72, find (*i*) r_{in} (*ii*) r_L (*iii*) A_v and (*iv*) A_p

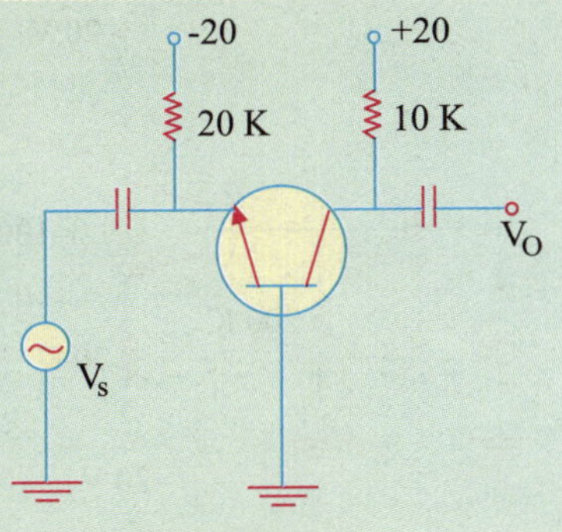

Fig. 59.71

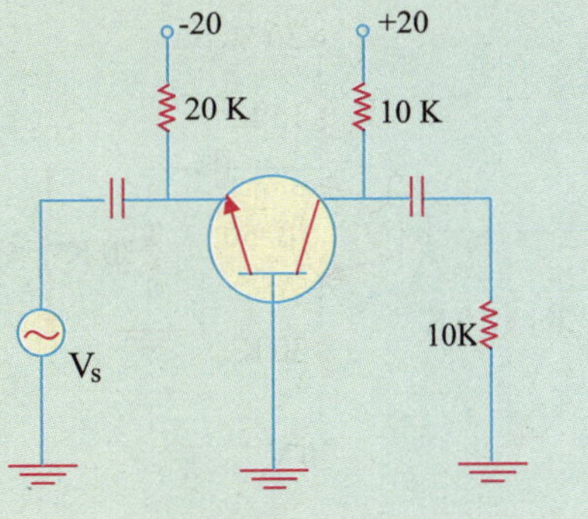

Fig. 59.72

Transistor α = – 0.98. Take G_e transistor material. **[(*i*) 25 Ω (*ii*) 5 K (*iii*) 200 (*iv*) 196]**

3. For the *CB* amplifier circuit of Fig. 59.73, find approximate values of
(*i*) stage r_{in}
(*ii*) r_L
(*iii*) $A_v = v_{out} / v_{in}$
(*iv*) $A_{v\sigma} = v_{out} / v_s$
(*v*) A_p
(*vi*) G_p
Take transistor α = –0.98
[(*i*) 25 Ω (*ii*) 5K (*iii*) 200 (*iv*) 22.2 (*v*) 196 (*vi*) 22.9 dB]

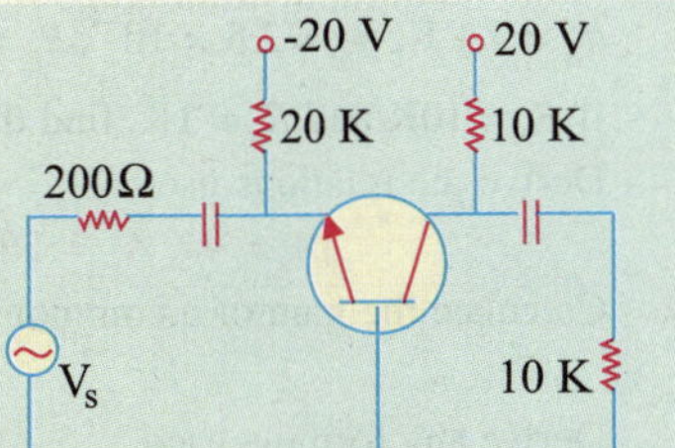

Fig. 59.73

4. Find r_{in}, r_L, A_i, A_v and G_p for the *CE* amplifier shown in Fig. 59.74.
[(*i*) 1250 Ω (*ii*) 5K (*iii*) 50 (*iv*) 200 (*v*) 40 dB]

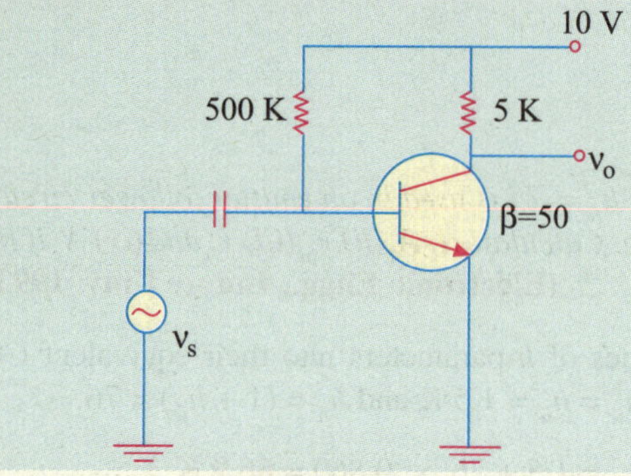

Fig. 59.74

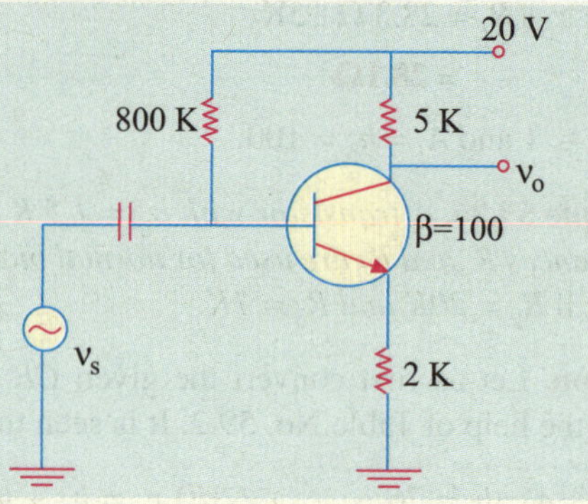

Fig. 59.75

5. For the single-stage *CE* amplifier circuit of Fig. 59.75, find
(*i*) r_{in} (*ii*) r_L (*iii*) A_v (*iv*) A_p and (*v*) G_p
Take β = 100 and use r_e = 50 mV/I_E. **[(*i*) 160 K (*ii*) 5K (*iii*) 2.5 (*iv*) 250 (*v*) 24 dB]**

6. For the *CE* amplifier of Fig. 59.76, find approximate values of
(*i*) r_{in} (*ii*) r_L (*iii*) A_v (*iv*) A_p and G_p.
Take transistor β = 50 and use r_e = 50 mV/I_E. **[(*i*) 2K (*ii*) 10K (*iii*) 200 (*iv*) 10,000 (*v*) 40 dB]**

7. For the emitter follower circuit shown in Fig. 59.77, find
(*i*) $r_{in(base)}$ (*ii*) $r_{in(stage)}$
(*iii*) stage A_v (*iv*) A_p in decibles
Take β = 100 and use r_e = 25 mV/I_E. **[(*i*) 10K (*ii*) 9.52K (*iii*) 0.75 (*iv*) 18.75 dB]**

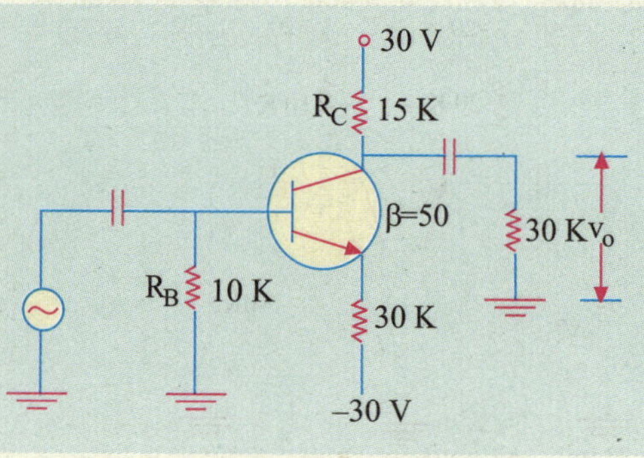

Fig. 59.76

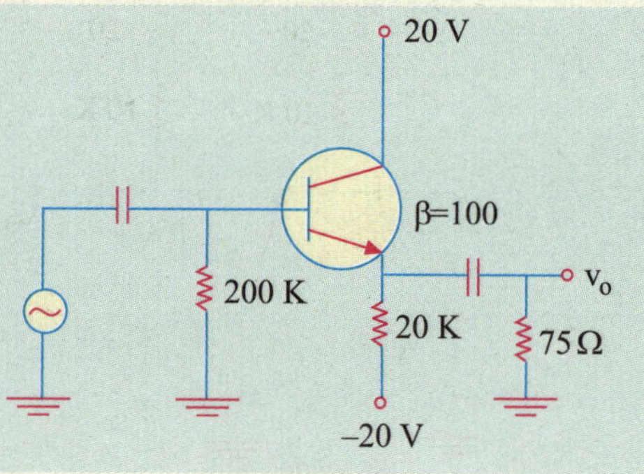

Fig. 59.77

8. An *NPN* silicon transistor is connected as a *CE* amplifier with load R_L and a source resistance R_S. The *h*-parameters are :
h_{ie} = 1.1 K, $h_{re} = 2.5 \times 10^{-4}$, h_{fe} = 50 and h_{oe} = 25 μS
If R_L = 10K and R_S = 1K, find the various gains and input and output impedances.
Derive the relations used.
[A_i = 40, A_{is} = 20, r_{in} = 1K, A_v = –400, A_{vs} = –200, r_o = 52.5 K, A_p = 16,000, G_p = 42 dB]

9. Calculate the gain of a common-emitter transistor amplifier whose hybrid parameters are:
h_{ie} = 1100 ohms, $h_{re} = 2.5 \times 10^{-4}$; h_{fe} = 50, h_{oe} = 25 μS, R_L = 5K.
Derive any formula used. ***(–200) (Electronic Engg., Indore Univ. 1980)***

10. A *CE* amplifier has h_{ie} = 1.1 K, h_{fe} = 50, $h_{re} = 2.5 \times 10^{-4}$; h_{oe} = 25 μS, $R_S = R_L$ = 500 ohm. Calculate the output impedance and voltage gain.
(r_0 = 58.2 kΩ, Av = – 15.5)
(Applied Electronics-I, Punjab Univ. Dec.)

11. A junction transistor has the following *h*-parameters $h_{ie} = 2000$ ohm, $h_{re} = 15 \times 10^{-4}$, $h_{fe} = 49$, $h_{oe} = 50\ \mu S$. Determine the current gain, voltage gain, input resistance and output resistance of the *CE* amplifier if the load resistance is 10K and source resistance is 600 ohm. Derive the expressions used.

($r_{in} = 1.5\ k\Omega$, $r_o = 109\ k\Omega$, $A_i = 32.67$, $A_v = -249.3$)
(*Applied Electronics-1, Punjab May 1986*)

12. A *CE* amplifier has $h_{ie} = 1.1$ K, $h_{fe} = 50$, $h_{re} = 2.5 \times 10^{-4}$, $h_{oe} = 25\ \mu S$, $R_S = R_L = 1K$. Calculate the output impedance and voltage gain.

($r_o = 152.5\ k\Omega$, $A_v = -23.36$)
(*Applied Electronics-1, Punjab Univ. Dec. 1985*)

13. A junction transistor has the following *h*-parameters $h_{ie} = 1k$, $h_{re} = 1$, $h_{fe} = -50$, $h_{oe} = 25mA/V$. This transistor is connected to a source of internal resistance 600 ohm and a load of 40 kΩ. Calculate the current gain, voltage gain, input resistance and output resistance of the amplifier. Derive the expressions used.

($r_{in} = 3\ k\Omega$, $r_o = 17.8\Omega$, $A_v = 0.55$, $A_v = 0.049$)
(*Applied Electronics-I, Punjab Univ. June 1985*)

14. A transistor connected in CE configuration has following *h*-parameters.

$h_{ie} = 1.1\ k\Omega$ $\qquad h_{re} = 2.5 \times 10^{-4}$
$h_{fe} = 50$ $\qquad h_{oe} = 25\ \mu$ Siemens
and $\qquad r_s = r_L = 1\ k\Omega$

Calculated current gain, input impedance and voltage gain.

($r_{in} = 1.09\ k\Omega$, $A_i = 48.8$, $A_v = -23.36$)
(*Electronics Engg. Bangalore Univ. 2001*)

OBJECTIVE TESTS – 59

1. In an ac amplifier, larger the internal resistance of the ac signal source
(*a*) greater the overall voltage gain
(*b*) greater the input impedance
(*c*) smaller the current gain
(*d*) smaller the circuit voltage gain.

2. The main use of an emitter follower is as
(*a*) power amplifier
(*b*) impedance matching device
(*c*) low-input impedance circuit
(*d*) follower of base signal.

3. An ideal amplifier is one which
(*a*) has infinite voltage gain
(*b*) responds only to signals at its input terminals
(*c*) has positive feeback
(*d*) gives uniform frequency response.

4. The smallest of the four *h*-parameters of a transistor is
(*a*) h_i (*b*) h_r (*c*) h_o (*d*) h_f

5. The voltage gain of a single-stage *CE* amplifier is increased when
(*a*) its ac load is decreased
(*b*) resistance of signal source is increased
(*c*) emitter resistance R_E is increased.
(*d*) ac load resistance is increased.

6. When emitter bypass capacitor in a *CE* amplifier is removed, its .... is considerably reduced.
(*a*) input resistance
(*b*) output load resistance
(*c*) emitter current
(*d*) voltage gain

7. Unique features of a *CC* amplifier circuit is that it
(*a*) steps up the impedance level
(*b*) does not increase signal voltage
(*c*) acts as an impedance matching device
(*d*) all of the above.

8. The input impedance h_{11} of a network with output shorted is given by the ratio

(a) v_1/i_1 (b) v_1/v_2

(c) i_2/i_1 (d) i_2/v_2

9. The h-parameters of a transistor depend on its

(a) configuration

(b) operating point

(c) temperature

(d) all of the above

10. The output admittance h_0 of an ideal transistor connected in CB configuration is .... siemens.

(a) 0 (b) 1/r

(c) $1/\beta r_e$ (d) –1.

11. A transistor has $h_{fe} = 100$, $h_{ie} = 5.2$ K Ω, and $r_{bb} = 0$. At room temperature, $V_T = 26$ mV, the collector current, $|I_C|$ will be.

(a) 10 mA (b) 5 mA

(c) 1 mA (d) 0.5 mA

ANSWERS

1. (d) 2. (b) 3. (b) 4. (c) 5. (d) 6. (d) 7. (d) 8. (a) 9. (d) 10. (a) 11. (a)

CHAPTER 60

Learning Objectives

- Classification of Amplifiers
- Common Base (CB) Amplifier
- Common Emitter (CE) Amplifier
- Common Collector (CC) Amplifier
- Comparison of Amplifier Configurations
- Class-A Amplifier
- Power Rectangle
- Power Efficiency
- Class-B Amplifier
- Maximum Values
- Class-B Push-pull Amplifier
- Crossover Distortion
- Class-C Amplifier
- Tuned Amplifier
- Distortion in Amplifiers
- Noise
- The Decibel System
- Value of 1 dB
- Cause of Amplifier Gain Variations
- Miller Effect
- Cut-off Frequencies of Cascaded Amplifiers
- The f_T of a Transistor
- Relation between f_α, f_β and f_T
- Gain-bandwidth Product

SINGLE-STAGE TRANSISTOR AMPLIFIERS

Single stage amplifier analysis is of great value in understanding the practical amplifier circuits

60.1. Classification of Amplifiers

Linear amplifiers are classified according to their mode of operation *i.e.* the way they operate according to a predetermined set of values. Various amplifier descriptions are based on the following factors :

1. **As based on its input**
   (*a*) small-signal amplifier (*b*) large-signal amplifier
2. **As based on its output**
   (*a*) voltage amplifier (*b*) power amplifier
3. **As based on its frequency response**
   (*a*) audio-frequency (AF) amplifier (*b*) intermediate-frequency (IF) amplifier
   (*c*) radio-frequency (RF) amplifier
4. **As based on its biasing conditions**
   (*a*) class-A (*b*) class-AB (*c*) class-B (*d*) class-C
5. **As based on transistor configuration**
   (*a*) common-base (CB) amplifier (*b*) common-emitter (CE) amplifier
   (*c*) common-collector (CC) amplifier

The description ***small-signal, class-A, CE, voltage amplifier*** means that input signal is small, biasing condition is class-A, transistor configuration is common-emitter and its output concerns voltage amplification.

We will first take up the basic working of a single-stage amplifier *i.e.* an amplifier having one amplifying element connected in *CB*, *CE* and *CC* configuration.

60.2. Common Base (CB) Amplifier

Both Fig. 60.1 and 60.2 show the circuit of a single-stage *CB* amplifier using *NPN* transistor. As seen, input ac signal is injected into the emitter-base circuit and output is taken from the collector-base circuit. The *E/B* junction is forward-biased by V_{EE} whereas *C/B* junction is reverse-biased by V_{CC}. The *Q*-point or dc working conditions are determined by dc batteries along with resistors R_E and R_C. In other words, values of I_E, I_B and V_{CB} are decided by V_{CC}, V_{EE}, R_E and R_C. The voltage V_{CB} is given by the equation $V_{CB} = V_{CC} - I_C R_C$.

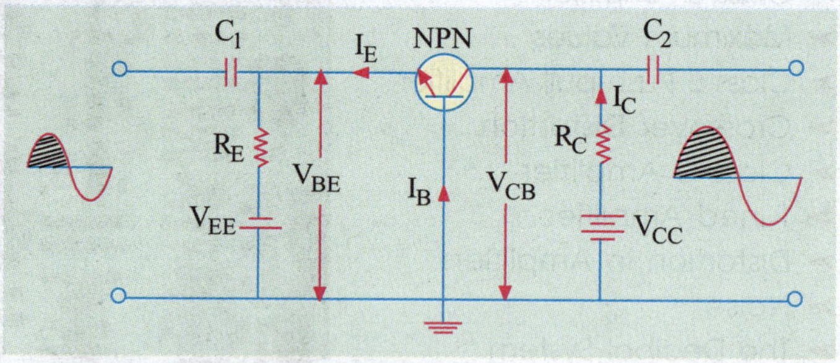

Fig. 60.1

When no signal is applied to the input circuit, the output just ***sits at the Q-point so that there is no output signal.*** Let us now see what happens when we apply an ac signal to the *E/B* junction via a coupling capacitor C_1 (which is assumed to offer no reactance to the signal).

Circuit Operation

When positive half-cycle of the signal is applied, then

1. forward bias is **decreased** because V_{BE} is already negative with respect to the ground as per biasing rule of Art. 57.3.
2. consequently, I_B is ***decreased.***
3. I_E and hence I_C are ***decreased*** (because they are both nearly β times the base current).
4. the drop $I_C R_C$ *is* ***decreased.***
5. hence, V_{CB} is ***increased*** as seen by the equation given above.

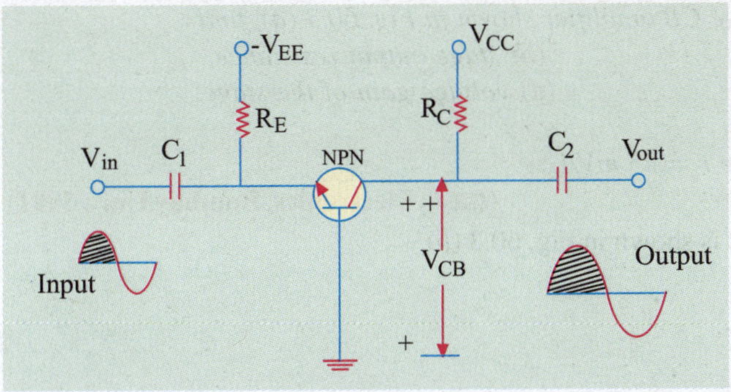

Fig. 60.2

It means that a positive output half-cycle is produced.

Since a ***positive-going*** input signal produces a ***positive-going*** output signal, there is no phase reversal between the two.

Voltage amplification in this circuit is possible by reason of relative input and output circuitry rather than current gain (α) which is always less than unity. The input circuit has low resistance whereas output circuit has very large resistance. Although changes in input and output currents are the same, the ac drop across R_L is very large. Hence, changes in V_{CB} (which is the output voltage) are much larger than changes in input ac signal. Hence, the voltage amplification.

60.3. Various Gains of a CB Amplifier

1. Input Resistance

The ac input resistance of the transistor is given by the emitter junction resistance

$$r_e = \frac{25\,\text{mV}}{I_E} \quad \text{or} \quad \frac{50\,\text{mV}}{I_E}$$

As seen from the ac equivalent circuit (Fig. 60.3)

$$r_{in} = r_e \,||\, R_E$$

2. Output Resistance

$$r_o = R_C \qquad \text{— Fig. 60.2}$$

If a load resistance R_L is connected across output terminals, then

$$r_o = R_C \,||\, R_L \qquad \text{— Fig. 60.3}$$

It is called the output resistance of the stage and is written as $r_{o(stage)}$ or r_o'

3. Current Gain $A_i = \alpha$

4. Voltage Gain $A_v = \dfrac{r_o}{r_{in}} = \dfrac{r_o}{r_e}$

5. Power Gain $A_p = A_v \,.\, A_i$

The decibel gain is given by $G_p = 10 \log_{10} A_p$ dB

60.4. Characteristics of a CB Amplifier

Common-base amplifier has

1. very low input resistance (30 – 150 Ω),
2. very high output resistance (upto 500 K),
3. a current gain α < 1,
4. large voltage gain of about 1500,
5. power gain of upto 30 dB,
6. no phase reversal between input and output voltages.

Uses

One of the important uses of a *CB* amplifier is in matching a low-impedance circuit to a high-impedance circuit.

It also has high stability of collector current with temperature changes.

Example 60.1. *For the single-stage CB amplifier shown in Fig. 60.3 (a), find*
(a) stage input resistance, *(b) stage output resistance,*
(c) current gain, *(d) voltage gain of the stage,*
(e) stage power gain in dB,
Assume $\alpha = 1$. *Neglect* V_{BE} *and use* $r_e = 25\ mV/I_E$

(Basic Electronics, Bombay Univ. 1991)

Solution. The ac equivalent circuit is shown in Fig. 60.3 (*b*).

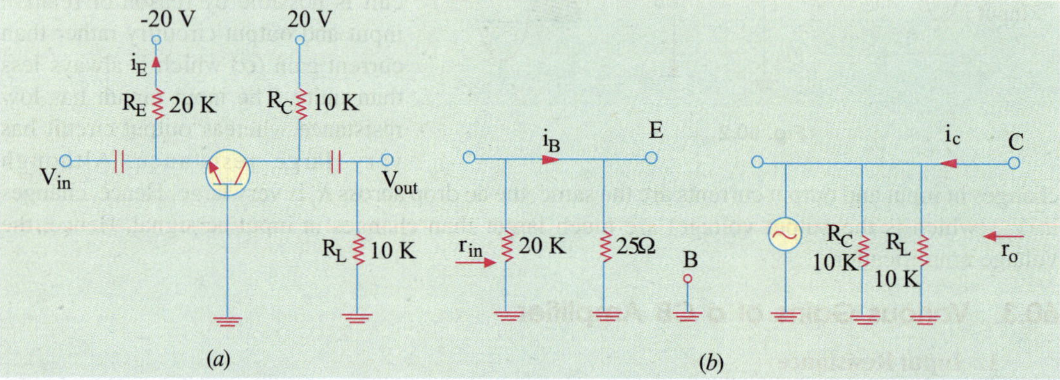

Fig. 60.3

$I_E = 20/20 = 1\text{mA}$ $\qquad r_e = 25/I_E = 25\ \Omega$

(*a*) stage $r_{in} = r_e \parallel R_E = 25\ \Omega \parallel 20\text{K} \cong$ **25 Ω**

(*b*) stage $r_o = R_C \parallel R_L$ $\qquad \therefore \qquad r_o = 10\text{K} \parallel 10\text{K} =$ **5K**

(*c*) $A_i = \alpha = 1$ $\qquad$ (*d*) $A_v = \dfrac{r_0}{r_e} = \dfrac{5000}{25} =$ **200**

(*e*) $A_p = A_v \cdot A_i = 200 \times 1 =$ **200**

$G_p = 10 \log_{10} A_p\ \text{dB} = 10 \log_{10} 200 =$ **23 dB**

60.5. Common Emitter (CE) Amplifier

Fig. 60.4 and 60.5 show the circuit of a single-stage *CE* amplifier using an *NPN* transistor. Here, base is the driven element. The input signal is injected into the base-emitter circuit whereas output signal is taken out from the collector-emitter circuit. The *E*/*B* junction is forward-biased by V_{BB} and *C*/*B* junction is reversed-biased by V_{CC} (in fact, same battery V_{CC} can provide dc power for both base and collector as in Fig. 60.5. The *Q*-point or working condition is determined by V_{CC} together with R_B and R_C. The dc equation is (Fig. 60.5).

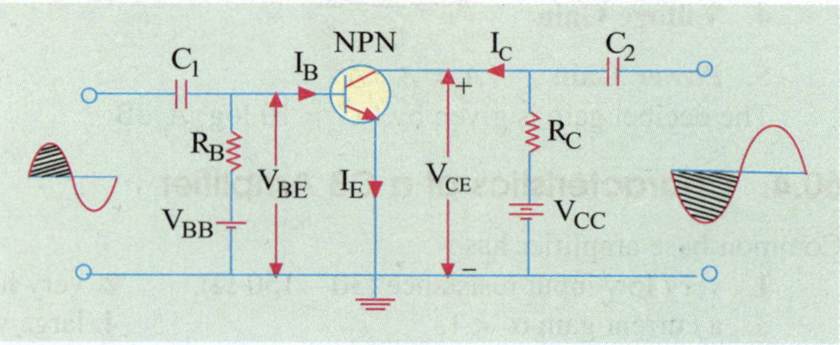

Fig. 60.4

$I_B \cong V_{cc}/R_B$ —neglecting V_{BE}

$I_C = \beta I_B$ and $V_{CE} = V_{CC} - I_C R_C$

Now, let us see what happens when an ac signal is applied at the input terminals of the circuit.

Circuit Operations

When positive half-cycle of the signal is applied (Fig. 60.4)

1. V_{BE} is **increased** because it is already positive w.r.t. the ground as per biasing rule of Art 57.3.
2. it leads to increase in forward bias of base-emitter junction
3. I_B is **increased** somewhat
4. I_C is increased by β times the **increased** in I_B.
5. drop $I_C R_C$ is **increased** considerably and consequently.
6. V_{CE} is decreased as seen from the equation given above.

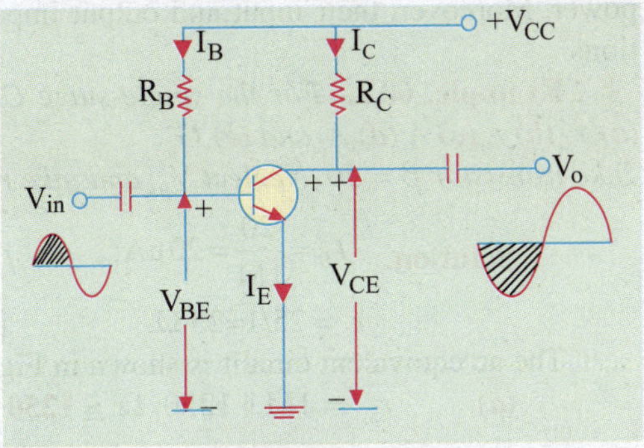

Fig. 60.5

Hence, negative half-cycle of the output is obtained. It means that a positive-going input signal becomes a negative going output signal as shown in Fig. 60.4 and 60.5.

60.6. Various Gains of a CE Amplifier

The ac equivalent of the given circuit (Fig. 60.5) is similar to the one shown in Fig. 60.6 (*b*).

1. Input Resistance

When viewed from base, ac resistance of the emitter junction is βr_e. As seen, from Fig. 60.6 (*b*), circuit input resistance is

$$r_{in}' = R_B \| \beta r_e \qquad \text{—remember } \beta\text{-rule}$$

$$\cong \beta r_e \qquad \text{—when } R_B >> \beta r_e$$

It is called input resistance of the stage *i.e.* $r_{in(stage)}$.

2. Output Resistance : $r_0 = R_C$ —Fig. 60.5

However, if a load resistor R_L is connected across the output terminals (Fig. 60.6), then

$$r_0 = R_C \| R_L = r_L \qquad \text{— Fig. 60.6 (b)}$$

It is called output resistance of the stage and is written as $r_{o(stage)}$ or r_o'.

3. **Current Gain :** $A_i = \beta$

4. **Voltage Gain :** $A_v = \beta . \dfrac{r_0'}{r_{in}'} = \beta . \dfrac{r_0'}{\beta r_e} \cong \dfrac{r_0'}{r_e}$ —if $R_B >> \beta r_e$

It is the stage voltage gain,

5. **Power Gain :** $A_p = A_v . A_i = \beta . \dfrac{r_0'}{r_e}$; $G_p = 10 \log_{10} A_p$ dB

60.7. Characteristics of a CE Amplifier

A *CE* transistor amplifier has the following characteristics :

1. it has moderately low input resistance (1 K to 2 K),
2. its output resistance is moderately large (50 K or above),
3. its current gain (β) is high (50–300),
4. it has very high voltage gain of the order of 1500 or above,
5. it produces very high power gain of the order of 10,000 times or 40 dB,
6. it produces ***phase reversal*** of input signal *i.e.* input and output signals are 180° out of phase with each other.

Uses

Most of the transistor amplifiers are of CE type because of large gains in voltage, current and power. Moreover, their input and output impedance characteristics are suitable for many applications.

Example. 60.2. *For the single-stage CE amplifier circuit shown in Fig. 60.6 (a), calculate (a) r_{in} (b) r_0 (c) A_i (d) A_v and (e) G_p*
Take transistor $\beta = 50$. Neglect V_{BE} and take $r_e = 25\ mV/I_E$

Solution. $I_B = \dfrac{20}{1M} = 20\,\mu A;$ $\qquad I_C = \beta I_B = 50 \times 20\mu A = 1\ mA$

$r_e = 25/1 = 25\ \Omega,$ $\qquad \beta r_e = 50 \times 25 = 1250\ \Omega$

The ac equivalent circuit is shown in Fig. 60.6 (*b*).

(*a*) $r_{in}' = 1M \parallel 1250\ \Omega \cong$ **1250 Ω**

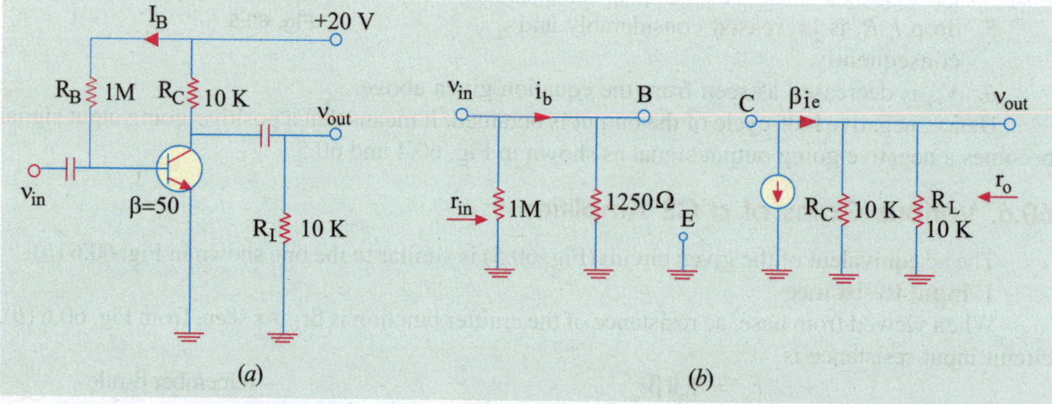

Fig. 60.6

Obviously, it is the input resistance of the stage and not that of the transistor alone.

(*b*) $r_0' = R_C \parallel R_L = 10K \parallel 10K = 5K$ $\qquad$ (*c*) $A_i = \beta = 50$

(*d*) $A_v = \dfrac{5K}{25\Omega} = 200$ $\qquad$ (*e*) $G_p = 10\log_{10}200 =$ **23** dB

60.8. Common Collector (CC) Amplifier

Fig. 60.7 and 60.8 show the circuit of a single-stage *CC* amplifier using an *NPN* transistor. The input signal is injected into the base-collector circuit and output signal is taken out from the emitter-collector circuit. The *E/B* junction is forward-biased by V_{EE} and *C/B* junction is reverse-biased by V_{CC}. The quiescent values of I_B and I_E are set by V_{CC} and V_{EE} together with R_B and R_E. As seen from Fig. 60.8.

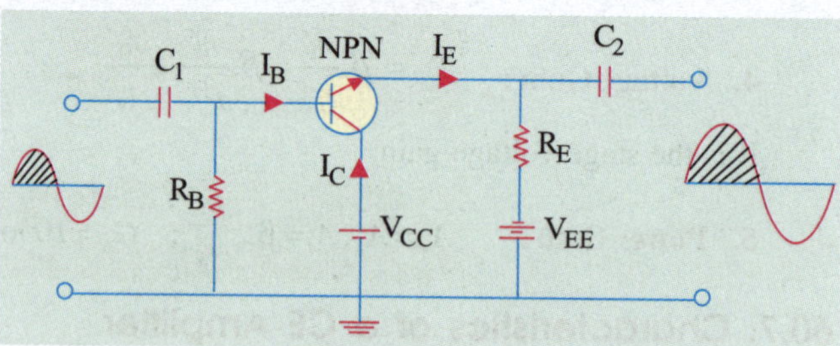

Fig. 60.7

$$I_E = \frac{V_{EE} - V_{BE}}{R_E + R_B/\beta} \cong \frac{V_{EE}}{R_E + R_B/\beta}$$

Let us now see what happens when an ac signal is applied across the input circuit.

Circuit Operation

When positive half-cycle of the signal is applied, then

1. forward bias is **increased** since V_{BE} is positive w.r.t. collector i.e. ground,
2. base current is **increased**,
3. emitter current is **increased**,
4. drop across R_E is **increased**,
5. hence, output voltage (*i.e.* drop across R_E is **increased**.

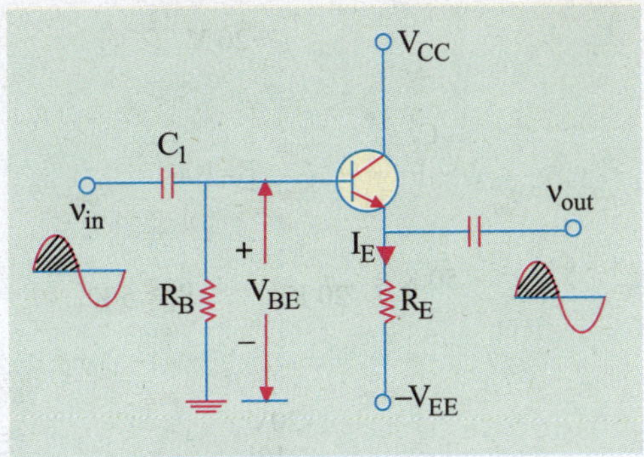

Fig. 60.8

Consequently, we get positive half-cycle of the output.

It means that a ***positive-going*** input signal results in a ***positive going*** output signal and, consequently, the input and output signals are in phase with each other as shown in Fig. 60.8.

60.9. Various Gains of a CC Amplifier

The ac equivalent circuit of the *CC* amplifier (Fig. 60.8) is given in Fig. 60.9.

1. $r'_{in} = R_B \| \beta(r_e + r_0)$
2. $r'_0 = R_E \| R_L$
3. $A_i = \beta$
4. $A_v = \dfrac{r_0'}{r_0' + r_e}$

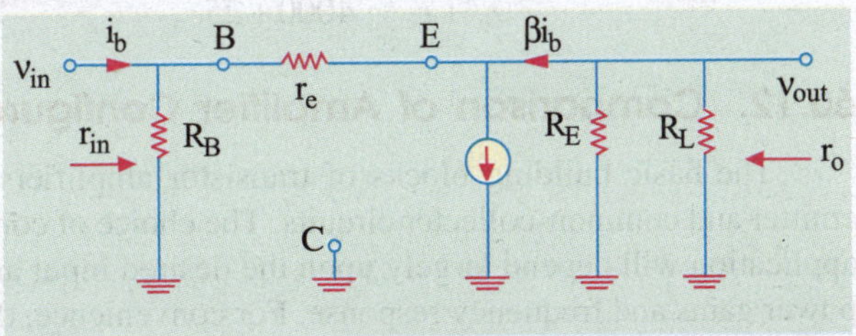

Fig. 60.9

Since usually $r_0' >> r_e'$, $A_v \cong \dfrac{r_0'}{r_0'} = 1$

5. $A_p = A_v . A_i$,
6. $G_p = 10 \log 10 A_p$ dB

60.10. Characteristics of a CC Amplifier

A *CC* amplifier has the following characteristics :

1. high input impedance (20-500 K),
2. low output impedance (50-1000 Ω),
3. high current gain of (1 + β) *i.e.* 50 – 500,
4. voltage gain of less than 1,
5. power gain of 10 to 20 dB,
6. no phase reversal of the input signal.

60.11. Uses

The CC amplifiers are used for the following purposes :

1. for impedance matching *i.e.* for connecting a circuit having high output impedance to one having low input impedance;
2. for circuit isolation;
3. as a two-way amplifier since it can pass a signal in either direction;
4. for switching circuits.

Example 60.3. *For the CC amplifier circuit of Fig. 60.10 (a), compute*
(i) r_o' *(ii)* r_{in}' *(iii)* A_v *and* *(iv)* A_p
Take transistor β = *100. Neglect* V_{BE} *and use* $r_e = 25\ mV/I_E$ **(Electronics-I, M.S. Univ. 1991)**

Solution. $I_E = \dfrac{20}{20 + 50/100} = \dfrac{20}{20.5} = 0.97 \text{ mA} \cong 1 \text{mA}$ $\quad r_e = \dfrac{25}{1} = 25\Omega$

The ac equivalent circuit is shown in Fig. 60.10 (*b*).

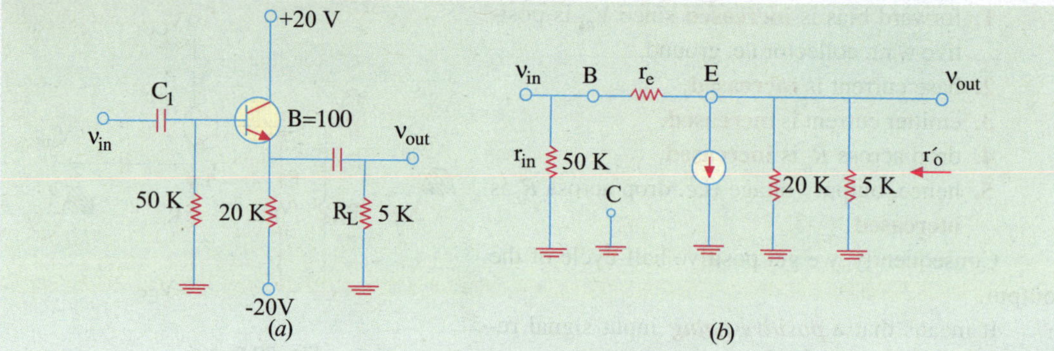

Fig. 60.10

(i) $r_o' = R_E \| R_L = 20\text{K} \| 5\text{K} = \mathbf{4K}$

(ii) $r_{in}' = R_B \| \beta(r_e + r_0) \cong R_B \| \beta r_0' = \dfrac{50 \times 400}{450} = \mathbf{44.4\ K}$

(iii) $A_v = \dfrac{r_o'}{r_o' + r_e} = \dfrac{4000}{4000 + 25} \cong 1$ (iv) $A_p = A_v A_i = 1 \times 100 = 100$

60.12. Comparison of Amplifier Configurations

The basic building blocks of transistor amplifiers are single-stage common-base, common-emitter and common-collecter circuits. The choice of configuration and type of transistor for a given application will depend largely upon the desired input and output impedances, voltage, current and power gains and frequency response. For convenience, these values are tabulated in Table No. 60.1. Here the input and output parameters are resistances and apply only to ***low-frequency conditions.***

Table No. 60.1

	Type of Circuit		
Characteristic	***Common base***	***Common emitter***	***Common collector***
Current gain	nearly unity (α)	high (β)	highest (1 + β)
Voltage gain	high	very high	nearly unity
Power gain	moderate	highest	lowest
Input impedance	lowest	moderate	highest
Output impedance	highest	moderate	lowest
Phase reversal	No	Yes	No

In each case, the amplifying action depends on low-power input circuits controlling the high-power output circuits.

In the common-base circuit, although the input and output currents are nearly equal, the low-impedance emitter circuit absorbs for less power as compared to that which is available at the high-impedance collector.

The low base current flowing into the common-emitter circuit (where the impedance is a few kilohms) gives rise to much higher collector current flowing out of the high-impedance output circuit.

The common-collector circuit with approximately equal input and output resistances requires a low input current to control much larger output current.

60.13. Amplifier Classification Based on Biasing Conditions

This classification is based on the amount of transistor bias and amplitude of the input signal. It takes into account the portion of the cycle for which the transistor conducts. The three main classifications are :

(a) Class-A Amplifier

In this case, the transistor is so biased that output current flows for the full-cycle of the input signal (360°) as shown in Fig. 60.11 (*a*). In other words, the transistor remains *FR*-biased throughout the input cycle. Hence, its conduction angle is 360°.

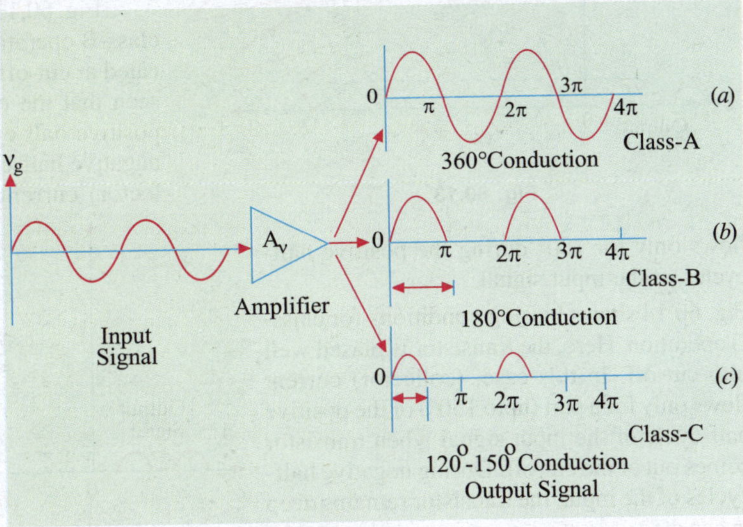

Fig. 60.11

(b) Class-B Amplifier

In this case, the transistor bias and the amplitude of input signal are such that output current flows for only half-cycle (180°) of the input signal. It means that transistor stays *FR*-biased for half the input cycle. The transistor conduction angle equals 180°.

(c) Class-C Amplifier

In this case, transistor bias and signal amplitude are such that output current flows for appreciably less than half-cycle of the input signal *i.e.* upto 120 ° or 150° angle of conduction as shown in 60.11 (*c*). In other words, transistor remains FR-biased for less than half the cycle.

(d) Class-AB Amplifier

The characteristics of such an amplifier lie in-between those of class-A and class-B. Here, biasing conditions are such that output current flows for appreciably more than half but less than the entire cycle *i.e.* current flows for more than 180° but less than 360°.

60.14. Graphic Representation

Since common-emitter is the most versatile and widely-used configuration, we will use CE output characteristic curves to differentiate between main classes of amplifiers.

Fig. 60.12 shows the biasing condition for class-A operation. It is seen that *Q*-point is located at the centre of the load line so that output (collector) current flows for the complete cycle of the input signal (conduction angle of 360°). Because of centred Q-point, the positive and negative swings of the input signal are confined to linear portions of the load line. In this linear region, equal changes in input base current produce equal changes in output (collector) current and voltage. Hence, output is an exact replica of the input. If the amplitude of the input signal is so large as to drive the *Q*-point closer to either cut-off or saturation region

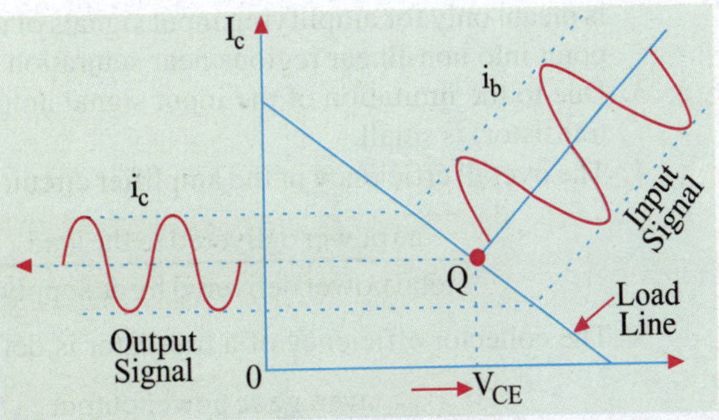

Fig. 60.12

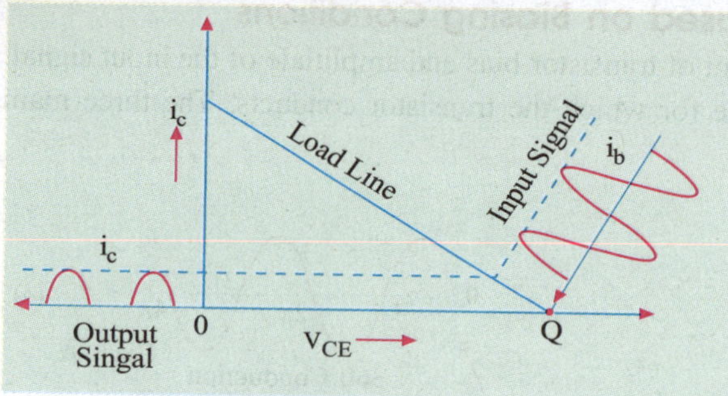

Fig. 60.13

(where base current lines are not equally spaced), the output will not be an exact replica of the input *i.e.* output signal will be distorted (Art. 60.30).

Fig. 60.13 shows biasing conditions for class-B operation. In this case, *Q*-point is located at cut-off point or very close to it. It is seen that the output current only flows for positive half-cycles of the input signal. For negative half-cycles, there is no output (collector) current. Obviously, output current flows only for 180° during the positive half-cycles of the input signal.

Fig. 60.14 shows biasing conditions for class-C operation. Here, the transistor is biased well into cut-off. In this case, (collector) current flows only for a part (upto 150°) of the positive half-cycle of the input signal when transistor comes out of the cut-off. During negative half-cycles of the input, the transistor remains deep in cut-off.

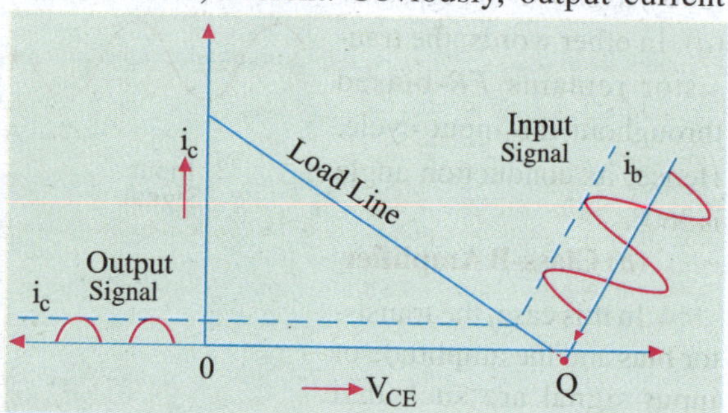

Fig. 60.14

60.15. Class-A Amplifier

It is one which has centred Q-point so that the transistor operates only over the linear region of its load line.

Another way of defining is that it is an amplifier in which the output current flows during the entire cycle of the input signal *i.e.* it has a conduction angle of 360°. In other words, the transistor remains FR-biased throughout the input cycle.

Characteristics

1. Since the transistor operates over the linear portion of the load line, the output waveform is exactly similar to the input waveform. Hence, class-A amplifiers are characterised by a high fidelity of the output. They are used where linearity or freedom from distortion is the prime requisite.
2. Since its operation is restricted only over a small central region of the load line, this amplifier is meant only for amplifying input signals of small amplitude. Large signals, will shift the Q-point into non-linear regions near saturation or cut-off and produce distortion (Art. 60.30).
3. Due to the limitation of the input signal amplitude, ac power output per active device (i.e. transistor) is small.
4. The overall efficiency of the amplifier circuit

$$= \frac{\text{ac power delivered to the load}}{\text{total power delivered by dc supply}} = \frac{\text{average ac power output}}{\text{average dc power input}}$$

5. The collector efficiency of a transistor is defined as

$$= \frac{\text{average ac power output}}{\text{average dc power input to transistor}}$$

The maximum possible collector efficiency of a class – A amplifier with resistive load is 50%

6. In case an output transformer is used, the maximum possible overall efficiency and maximum possible collector efficiency for a class-A amplifier are both 50%.

60.16. Power Distribution in a Class-A Amplifier

Fig. 60.15 (*a*) shows a *CE* connected transistor which forms the active element of a single-stage class-A amplifier. Fig. 60.15 (*b*) shows its output characteristic with a centred *Q*-point. When ac input signal is applied, *Q*-point shifts up and down from its central position. The output current will also increase or decrease from its quiescent (or no-signal) value I_{CQ}. Similarly, collector-emitter voltage V_{CE} will increase or decrease from its quiescent value V_{CEQ}. So long as signal variations are confined to linear region of the load line, average value of collector current is I_{CQ} because positive and negative input signal swings will produce equal changes in I_{CQ}.

Class-A amplifier

Hence, total average dc power drawn by the circuit from collector battery V_{CC} is

$$P_{in(dc)} = V_{CC} \cdot I_{CQ}$$

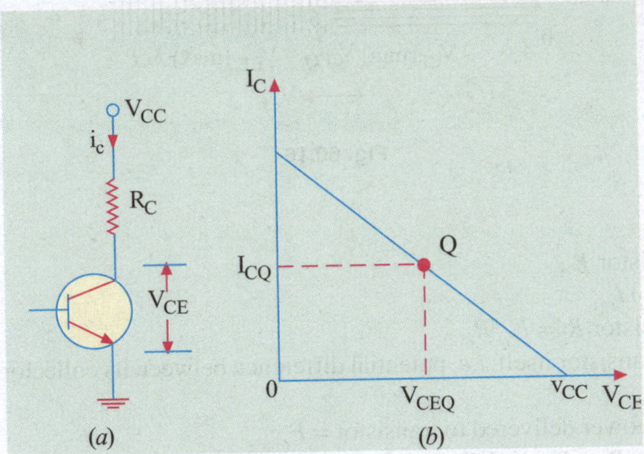

Fig. 60.15

Now, this power goes to supply the following :

(*i*) heat dissipated by the load resistor R_C connected to the collector

$$P_{RC(dc)} = I_{cQ}^2 R_c$$

(*ii*) the balance $P_{tr(dc)}$ is given to the transistor.

It is further subdivided into :

(*a*) ac power developed across the load resistor which constitutes the ac power output

$$P_{o(ac)} = I^2 R_C = \frac{V^2}{R_C} = \frac{V_m^2}{2R_C}$$

where *I* is the rms value of the ac output current through the load, *V* is rms value of ac voltage and V_m is its maximum value.

(*b*) power dissipated (in the form of heat) by the transistor itself *i.e.* its collector region. It may be called $P_{c(dc)}$.

Since, under zero-signal condition, there is no ac output power, all the power given to the transistor is wasted as heat. Hence, **a transistor dissipates maximum power under zero-signal condition**.

The power flow diagram of the transistor is as follows :

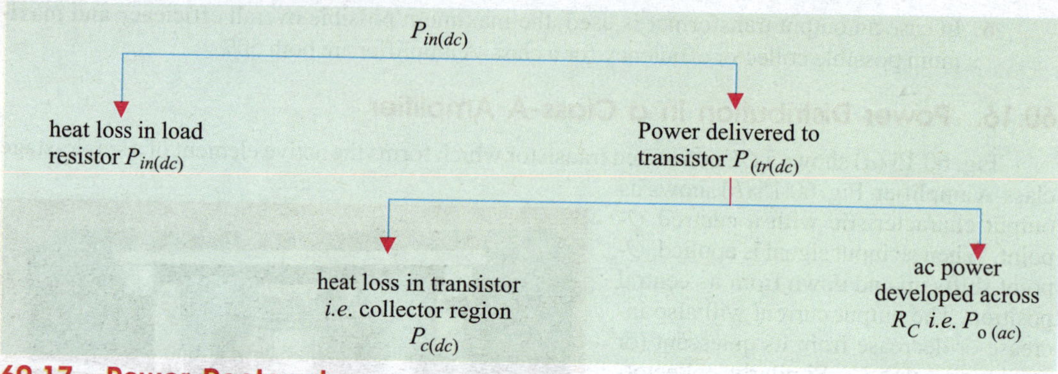

60.17. Power Rectangle

The output or collector characteristic of a *CE*-connected transistor is shown in Fig. 60.16 with centred *Q*-point for class-A operation. When input signal is applied, the *Q*-point shifts to positions Q_1 and Q_2 alternately. The output current varies around its quiescent value from maximum value of $I_{C(max)}$ to minimum value $I_{C(min)}$. Similarly, collector-emitter voltage varies from maximum value of $V_{CE(max)}$ to $V_{CE(min)}$ around its quiescent value of V_{CEQ}.

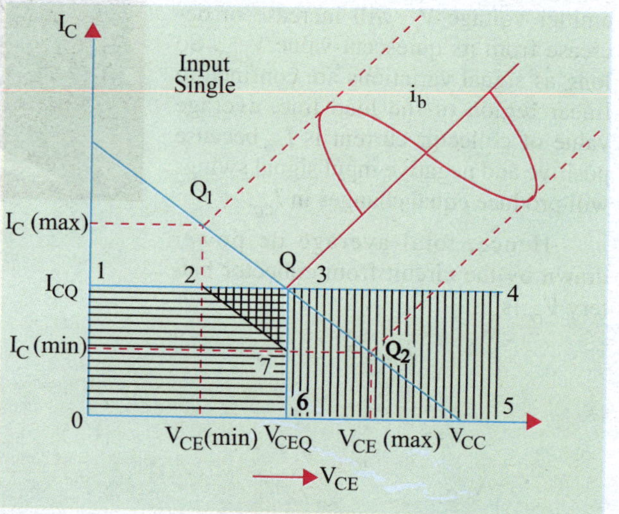

Fig. 60.16

Let us construct the various rectangles as shown in Fig. 60.16.

1. rectangle 0-1-4-5
$= V_{CC} \times I_{CQ}$
= total average power supplied to the circuit by V_{CC} battery
$= P_{in(dc)}$

Now, $V_{CC} - V_{CEQ}$
= voltage drop across load resistor R_C

2. rectangle 3-4-5-6 = $(V_{CC} — V_{CEQ})\, I_{CQ}$
= power lost as heat in load resistor $R_C = I_{CQ}^2 R_C$

Now, V_{CEQ} is the voltage drop across the transistor itself *i.e.* potential difference between its collector and emitter as shown in Fig. 60.15 (*b*).

3. rectangle 0-1-3-6 = $V_{CEQ} . I_{CQ}$ = power delivered to transistor = $P_{tr(dc)}$

4. triangle 2-3-7 = ac power across $R_C = P_{o(ac)} = I^2 R_C$

It can also be proved that

$$P_{o(ac)} = \frac{V_{CE(max)} - V_{CE(min)}}{2\sqrt{2}} \times \frac{I_{C(max)} - I_{C(min)}}{2\sqrt{2}}$$

$$= \frac{[V_{CE(max)} - V_{CE(min)}] \times [I_{c(max)} - I_{c(min)}]}{8}$$

5. area 0-1-2-7-6 = power dissipated by the collector region of the transistor
$= P_{c(dc)}$.

Obviously, when there is no input signal, there is no output signal. Hence, $P_{o(ac)} = 0$, which means that area of triangle 2-3-7 is zero. In that case, the transistor collector will have to dissipate maximum power equal to the area of the rectangle 0-1-3-6. It is called ***worst-case*** condition of the transistor and equals quiescent power.

60.18. Power Efficiency

Fig. 60.17. shows the power rectangles and triangles of a class-A amplifier in a very simple way.

$P_{in(dc)}$ = areas $A + B + C$

$P_{o(ac)}$ = area B

$P_{c(dc)}$ = area C

$P_{Rc(dc)}$ = area A

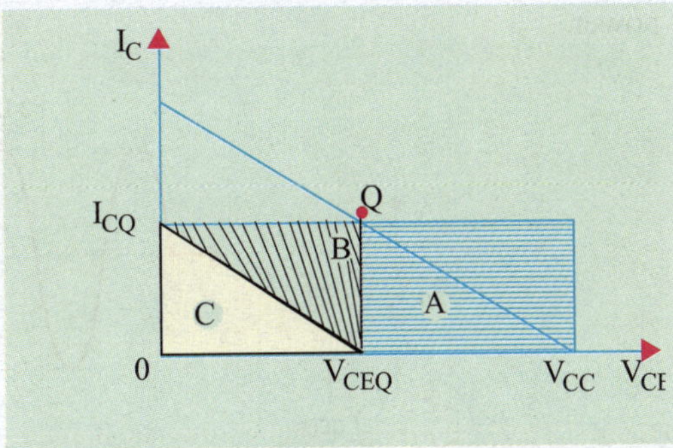

Fig. 60.17

Amplifier efficiency or circuit efficiency or overall efficiency.

$$\eta_{overall} = \frac{B}{A+B+C} = \frac{P_{o(ac)}}{P_{in(dc)}}$$

The collector efficiency is given by

$$\eta_{collector} = \frac{B}{B+C} = \frac{P_{o(ac)}}{P_{tr(dc)}}$$

Incidentally, it can be proved that $\eta_{overall} = 25\left[\frac{V_{CE(max)} - V_{CE(min)}}{V_{CC}}\right]$

Now, maximum value of $V_{CE} = V_{CC}$ and minimum value is = 0

$$\therefore \quad \text{maximum } \eta_{overall} = 25\left[\frac{V_{CC} - 0}{V_{CC}}\right]\% = 25\%$$

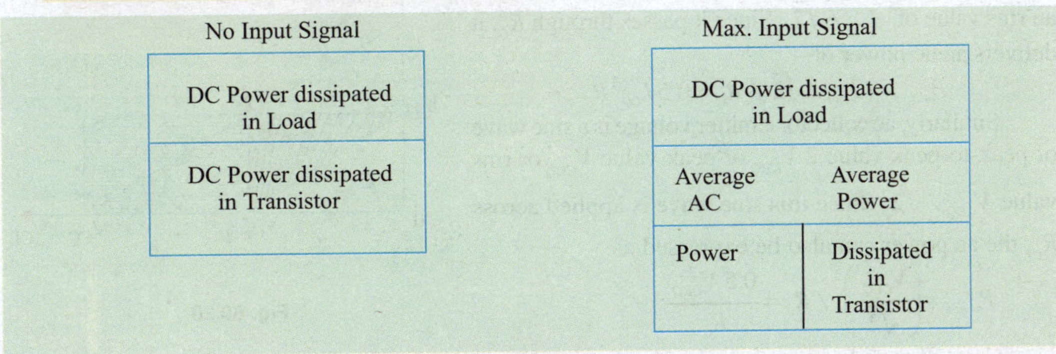

Fig. 60.18

Summary

In a direct-coupled class-A amplifier

1. maximum ac power equals half the power transistor can dissipate (or one-fourth of the total dc input power);
2. maximum ac power the transistor can dissipate is half the dc input power supplied by the battery;
3. the dc input power equals 4 times the ac output in load or twice the maximum power dissipated by the transistor.

The above facts have been illustrated with the help of power square shown in Fig. 60.18.

60.19. Maximum AC Power in Load

Fig. 60.19 (*a*) shows part of a base-driven *CE* amplifier with a collector load of R_C. Fig. 60.19 (*b*) shows its ac load line with centered *Q*-point. For the time being, we will neglect limitations of cut-off and saturation and assume (at least theoretically) that voltage and current swings down to zero are possible. Obviously, under such conditions, we will get the maximum possible ac output power.

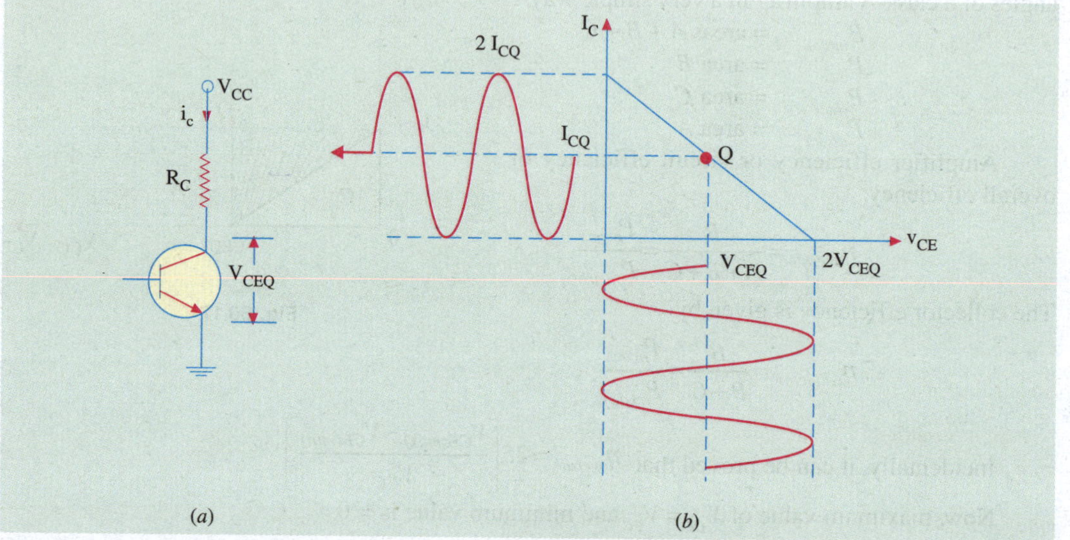

Fig. 60.19

As seen, collector current is a sine wave having peak-to-peak value of $2\,I_{CQ}$ or a maximum value of I_{CQ} or an rms value of $I_{CQ}/\sqrt{2}$. Since it passes through R_C, it delivers an ac power of

$$P_{o(ac)} = (I_{CQ}/\sqrt{2})^2 R_C = 0.5\,I_{CQ}^{\ 2} R_C$$

Similarly, ac collector-emitter voltage is a sine wave of peak-to-peak value $2\,V_{CEQ}$ or peak value V_{CEQ} or rms value $V_{CEQ}/\sqrt{2}$. Since this sine wave is applied across R_C, the ac power can also be expressed as

$$P_{o(ac)} = \left(\frac{V_{CEQ}}{\sqrt{2}}\right)^2 / R_C = \frac{0.5\,V_{CEQ}^2}{R_C}$$

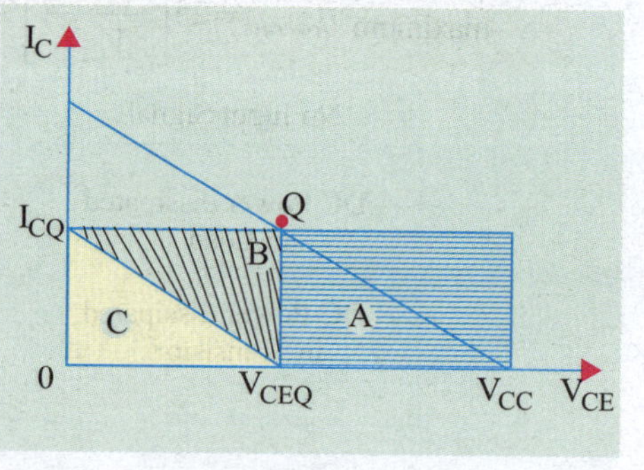

Fig. 60.20

Note. If there had been an unbypassed resistor R_E in the emitter, then ac load R_{ac} *or* $r_L = R_C + R_E$. Then, we would have used this value rather than R_C alone.

As seen from power rectangle diagram of Fig. 60.20, under maximum ac power condition

area *B* = area *C*

Also, area *A* = area *B* + area *C* = 2B

$$\therefore \quad maximum\ \eta_{overall} = \frac{B}{A+B+C}$$

$$= \frac{B}{2B+2B} = \frac{B}{4B} = 0.25 \quad \text{or} \quad 25\%$$

Example 60.4. *The 2N 1491 power transistor is used in a CE amplifier meant for class-A operation. If its zero-signal power dissipation is 10 W and ac output power is 3.5 W, find*

(a) *collector efficiency and* **(b)** *power rating of the transistor*

Solution. (*i*) $\eta_{coll} = \frac{P_{o(ac)}}{P_{tr(dc)}} = \frac{3.5}{10} = 0.35 =$ **35%**

(*ii*) Since zero-signal condition represents worst case condition for the transistor, it means that all the 10 W power is dissipated by it. Hence, transistor power rating is **10 W.**

Example 60.5. *For the class-A, CE amplifier circuit of Fig. 60.21, V_{CEQ} = 10 V and I_{CQ} = 500 mA. If collector i.e. output current varies by ± 250 mA when an input signal is applied at the base, compute*

(*i*) *total dc power taken by the circuit,*
(*ii*) *dc power dissipated by the collector load,*
(*iii*) *ac power developed across the load,*
(*iv*) *power delivered to the transistor,*
(*v*) *dc power wasted in transistor collector,*
(*vi*) *overall efficiency.*
(*vii*) *collector efficiency.*

Solution. (*i*) $P_{in(dc)} = V_{CC} . I_{CQ} = 20 \times 500 = 10^4 \text{ mW} =$ **10 W**

(*ii*) $P_{Rc(dc)} = I_{CQ}^2 R_C = 0.5^2 \times 20 = 5$ **W**

(*iii*) $P_{o(ac)} = I^2 R_C$

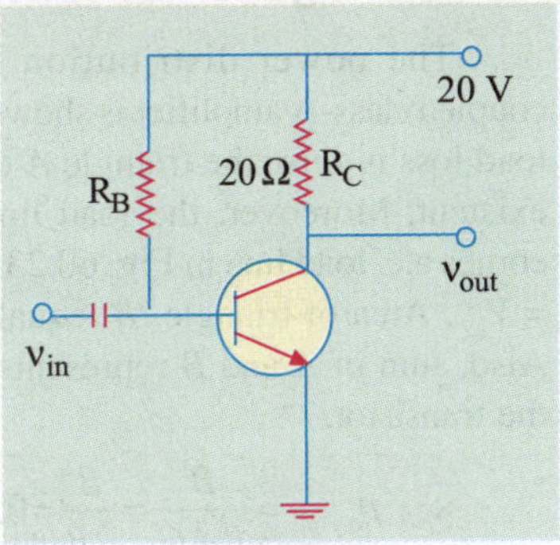

Fig. 60.21

Now, maximum value of output ac current is 250 mA = 0.25A.

Hence, rms value $I = 0.25/\sqrt{2}$ A.

$\therefore \quad P_{o(ac)} = (0.25/\sqrt{2})^2 \times 20 =$ **0.625 W**

(*iv*) $P_{tr(dc)} = 10 - 5 =$ **5 W**

(*v*) $P_{c(dc)} = 5 - 0.625 =$ **4.375 W**

(*vi*) $\eta_{overall} = \frac{0.625}{10} \times 100 =$ **6.25%**

(*vii*) $\eta_{coll} = \frac{0.625}{5} \times 100 =$ **12.5%**

It is seen that due to resistive load in the collector, efficiency of the circuit is very low.

Example 60.6. *For a class-A amplifier, $V_{CE}(max)$ = 25V, $V_{CE}(min)$ = 5V, V_{CC} = 30V. Find its overall efficiency.*

Solution. As seen from Art. 60.18

$$\% \eta_{overall} = 25 \left[\frac{V_{CE(max)} - V_{CE(min)}}{V_{CC}} \right] = 25 \left(\frac{25-5}{30} \right) = \mathbf{16.7}$$

60.20. Transformer-Coupled Class-A Amplifier

The main reason for the poor efficiency of a ***direct-coupled*** class-A amplifier is the large amount of dc power that the resistive load in collector must dissipate. This problem can be solved by using a suitable transformer for coupling the load (say, a speaker) to the amplifier stage as shown in Fig. 60.22. Since the load is not ***directly*** connected to the collector terminal, the dc collector current does not pass through it. In an ideal transformer, primary winding resistance is zero. Hence, dc power loss in the load is zero. In practice, however, there is a small dc resistance of the primary winding which does absorb some power though much less than a direct-coupled load.

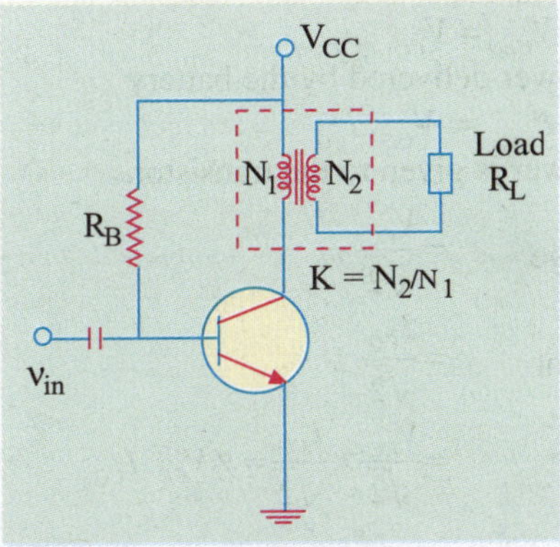

Fig. 60.22

In short, what the transformer does is to substitute ***ac load*** in place of ***ohmic*** or dc load.

The secondary load R_L when referred to primary becomes

$$R_L' = R_L / K^2 = a^2 R_L$$

where K = voltage transformation ratio=$N_2/N_1 = V_2/V_1$

a = turns ratio $N_1/N_2 = 1/K$

Since **a** is usually made much more than unity or K is much less than unity, R_L can be made to look much bigger than what it actually is

In an ideal transformer, there is no primary drop, hence $V_{CC} = V_{CEQ}$. Now, all the power supplied by V_{CC} is delivered to the transistor. Hence, the overall and collector efficiencies become equal.

$$\eta_{overall} = \frac{P_{o(ac)}}{V_{CC}\, I_{CQ}} = \frac{P_{o(ac)}}{V_{CEQ}.I_{CQ}}$$

Power Diagram

The power distribution for a transformer-coupled class-A amplifier is shown in Fig. 60.23. Since load loss is zero, the triangle A of Fig. 60.17 is non-existent. Moreover, the load line of Fig. 60.17 becomes a.c. load line in Fig. 60.23. Consequently, $V_{CEQ} = V_{CC}$. Area of triangle B equals that of triangle C. Also, sum of A and B represents power delivered to the transistor.

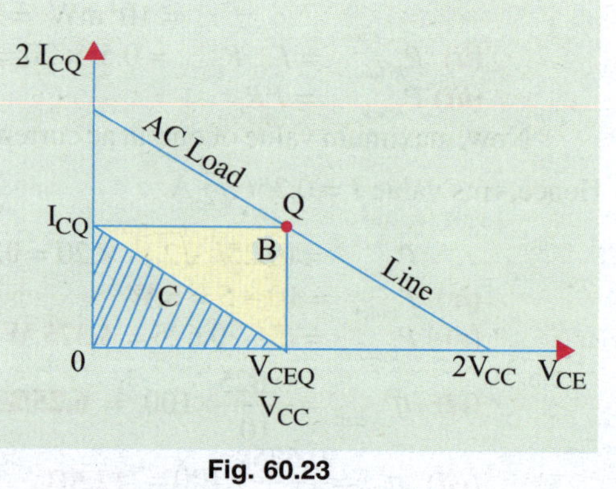

Fig. 60.23

$$\therefore \quad \eta_{overall} = \frac{B}{B+C} = \frac{B}{2B} = 0.5 \text{ or } 50\%$$

It can be proved that maximum possible overall efficiency and the maximum possible collector efficiency of a class-A amplifier using an output transformer are both 50%.

In general, overall efficiency of a transformer-coupled class-A amplifier is given by

$$\%\eta = 50\left[\frac{V_{CE(max)} - V_{CE(min)}}{2V_{CC}}\right]^2 = 50\left[\frac{V_{CE(max)} - V_{CE(min)}}{V_{CE(max)} + V_{CE(min)}}\right]^2$$

Obviously, larger the value of $V_{CE(max)}$ and smaller that of $V_{CE(min)}$ closer the efficiency to the theoretical limit of 50%.

Proof

The conditions for the development of maximum ac power are shown in Fig. 60.24. It must be kept in mind that here

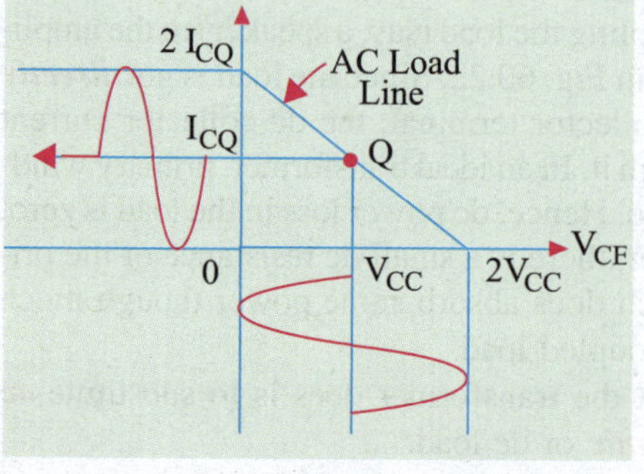

Fig. 60.24

$$V_{CEQ} = V_{CC}$$

Now, average power delivered by the battery

$$P_{in(dc)} = V_{CC} \cdot I_{CQ}$$

Whole of the power is given to the transistor.

rms output voltage $= \dfrac{V_{CC}}{\sqrt{2}}$

rms output current $= \dfrac{I_{CQ}}{\sqrt{2}}$

$\therefore$ maximum $P_{o(ac)} = \dfrac{V_{CC}}{\sqrt{2}} \times \dfrac{I_{CQ}}{\sqrt{2}} = ½ V_{CC} \cdot I_{CQ}$

= half the input ac power

Hence, maximum value of overall efficiency

$$\eta_{overall} = \frac{P_{o(ac)}}{P_{in(dc)}} = \frac{V_{CC}.I_{CC}/2}{V_{CC}.I_{CQ}} = \frac{1}{2} \text{ or } 0.5 \text{ or } 50\%$$

$$\eta_{coll} = \frac{P_{o(ac)}}{P_{tr(dc)}}$$

Example 60.7. *The optimum load resistance for a certain transistor is 200 Ω. What is the turns ratio (N_1/N_2) of a transformer required to couple an 8-Ω loud-speaker to the transistor ?*

Solution. $R_L' = a^2 R_L$ $\quad\therefore\quad 200 = a^2 \times 8 \quad$ or $\quad a = 5$

Hence, it should be a 5 : 1 *step-down* transformer.

Example 60.8. *In a transformer-coupled class-A amplifier $V_{CE(max)}$ = 27 V and $V_{CE(min)}$ = 3V. Compute its overall efficiency.*

Solution. $\eta_{overall} = 50\left(\frac{27-3}{27+3}\right)^2 = 32\%$

Example 60.9. *For the transformer-coupled optimally-biased class-A amplifier shown in Fig. 60.25, find*

(a) transformer turns ratio *(b) collector current* *(c) transistor power rating*

Solution. As shown in Fig. 60.25, if V_2 is the rms secondary voltage, then

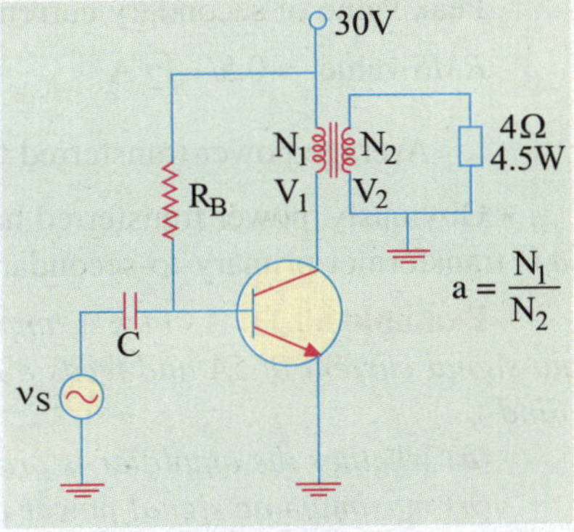

Fig. 60.25

$$P_{load} \text{ or } P_{o(ac)} = \frac{V_2^2}{R_L} = \frac{(V_{2m}/\sqrt{2})^2}{R_L} = \frac{V_{2m}^2}{2R_L}$$

where V_{2m} is the maximum value of secondary voltage.

$$\therefore \quad V_{2m} = \sqrt{2 R_L P_{load}} = \sqrt{2 \times 4 \times 4.5} = 6\text{ V}$$

Now, peak, value of primary voltage is

$$V_{1m} = a \,.\, V_{2m} = 6\,a$$

This also represents the peak value of collector voltage. For optimally-biased circuit, peak value of collector-emitter voltage equals V_{CC} = 30V.

$\therefore \quad 6a = 30 \quad$ or $a = 5 \quad \therefore \quad N_1/N_2 = 5$

(a) Hence, it is a step-down transformer with a turns ratio of **5 : 1**

(b) The overall efficiency is 50%. Hence, for a load power of 4.5 W, total dc input power supply must be 4.5/0.5 = 9 W

$\therefore \quad V_{CC} I_C = P_{in(dc)} = 9\text{W} \quad$ or $\quad 30\, I_C = 9000$ mW $\quad \therefore\ I_C = $ **300 mA**

(c) Worst case condition for the transistor is when there is no input signal. In that case, transistor has to dissipate the entire input. Therefore, transistor power rating = **9W.**

Example 60.10. *For the optimally-biased transformer -coupled class-A amplifier, maximum collector current change is 100 mA. Find the power transferred to a 4-W speaker load if it is (a) directly-coupled (b) transformer-coupled to the transistor. Find also the turn ratio of the transformer required.* **(Applied Electronics, Kerala Univ. 1985)**

Solution. (*a*) Direct-coupled Load [Fig. 60.26 (*a*)]

It should be remembered that for an optimally-biased class-A amplifier, peak change in collector-emitter voltage equals V_{CC} = 10 V.

RMS value of ac output current I = (100/$\sqrt{2}$) mA

Average power delivered to the speaker

$$= I^2 R_L = \left(\frac{100}{\sqrt{2}} \times 10^{-3}\right)^2 \times 4 = 0.02\,\text{W} = \mathbf{20\ mW}$$

(b) Transformer-coupled Load

Now, $\Delta V_{CE} = V_{CC} = 10\,\text{V}$: $\Delta I_{CQ} = 100\,\text{mA}$

If R_L' is the secondary load reflected into the primary, then

$$R_L' = \frac{\Delta V_{CE}}{\Delta I_{CQ}} = \frac{10\,\text{V}}{100\,\text{mA}} = 100\,\Omega$$

Now, $a^2 R_L = R_L'$

$\therefore\ a^2 \times 4 = 100,$

$a = 5$

Hence, it is a step-down transformer with a turns ratio of 5 : 1

Peak value of secondary voltage = 10 × 1/5 = 2 V

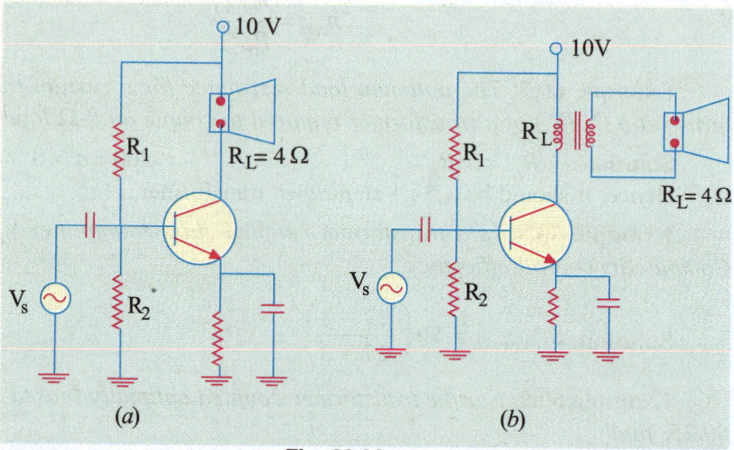

Fig. 60.26

Peak value of secondary current = 2 / 4 = 0.5 A

RMS value = $0.5/\sqrt{2}$ A

$$\text{Average power transferred to the speaker} = \left(\frac{0.5}{\sqrt{2}}\right)^2 \times 4 = 0.5\ \text{W} = \mathbf{500\ mW}$$

Obviously, power transferred has increased 500/20 = 25 times *i.e.* $a^2 = (N_1/N_2)^2$ times where a is transformer primary-to-secondary turn ratio *i.e.* $a = N_1/N_2$.

Example 60.11. *A class-A amplifier shown in Fig. 60.27 operates from $V_{CC} = 20$ V, draws a no-signal current of 5A and feeds a load of 40 Ω through a step-up transformer of $N_2/N_1 = 3.16$. Find*

(a) *whether the amplifier is properly matched for maximum power transfer,*
(b) *maximum ac signal power output,* **(c)** *maximum dc power input,*
(d) *conversion efficiency at maximum signal input.*

(Electronic Circuits, Gauhati Univ. 1990)

Solution. (*a*) Here, $a = N_1/N_2 = 1/3.16$

$R_L' = a^2 R_L = (1/3.16)^2 \times 40 = \mathbf{4\ \Omega}$

Also, $$R_L' = \frac{V_{CC}}{I_{CQ}} = \frac{20}{5} = \mathbf{4\Omega}$$

Since, the two resistance values are the same, the amplifier is properly-matched.

(*b*) As seen from Art. 60.20,

Maximum value of

$$P_{o(ac)} = \frac{1}{2} V_{CC}\, I_{CQ}$$

$$= \frac{1}{2} \times 20 \times 5 = \mathbf{50\ W}$$

(*c*) The dc power input

$= V_{CC} I_{CQ} = 20 \times 5 = \mathbf{100\ W}$

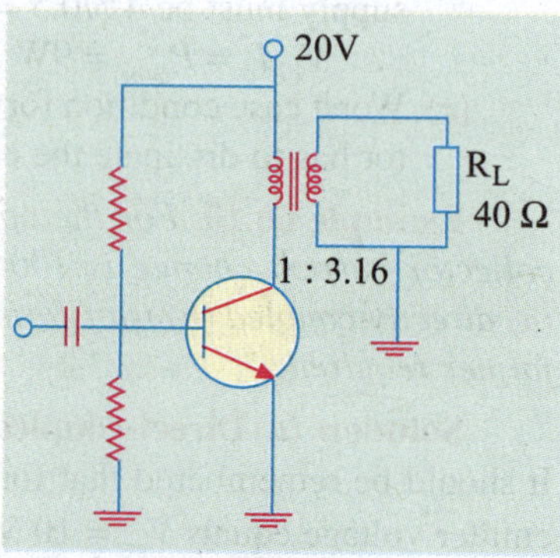

Fig. 60.27

(*d*) $\eta = \frac{50}{100} \times 100 =$ **50%**

60.21. Class-B Amplifier

Biasing condition for class-B operation has been shown both on current transfer characteristic and output characteristic for *CE* configuration in Fig. 60.28.

As seen, the transistor has been biased to cut-off. It remains FR-biased for only half-cycle of the input signal. Hence, its conduction angle is only 180°. It is obvious that with zero signal, its collector current is zero.

Characteristics

1. Since negative half-cycles are totally absent from the output, the signal distortion is high as compared to class-A amplifiers.
2. Since input voltage is large (upto V_{CC}), voltage amplification is reduced.
3. Zero-signal input represents worst condition for class-A amplifiers but best condition for class-B amplifiers.
4. In class-B amplifiers, transistor dissipates more power with increase in signal strength but opposite is the case in class-A amplifiers.
5. Average current in class-B operation is less than in class-A, hence power dissipated is less. Consequently, maximum circuit (or overall) efficiency of a class-B amplifier is 78.5% when peak signal makes $V_{CE(min)} = 0$. In general

$$\eta_{overall} = 78.5\left[1 - \frac{V_{CE(\min)}}{V_{CC}}\right]\%$$

6. The half-sinewave type of collector current contains very pronounced ***even harmonics*** (particularly the second one). These impulses can be rendered useful either by employing push-pull circuit (Art. 60.24) or by using tuned amplifiers (Art. 60.29).

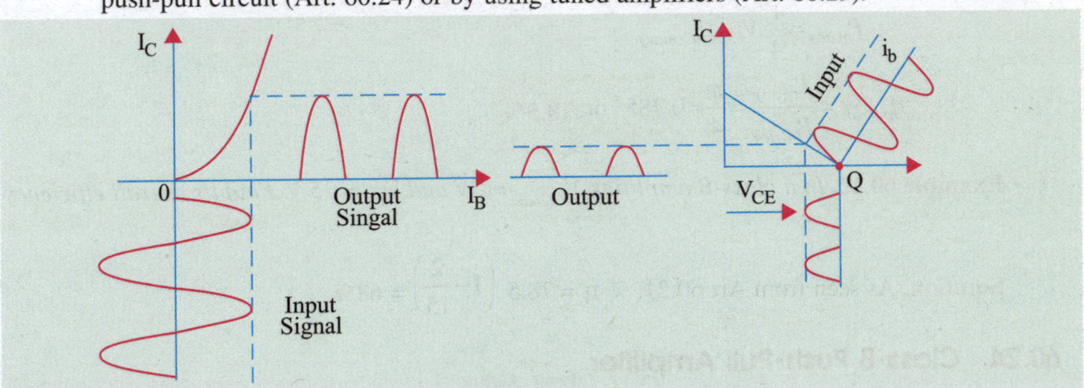

Fig. 60.28

60.22. Power Relations for Class-B Operation

(*i*) Input DC Power : $P_{in(dc)} = V_{CC} I_{dc}$

where I_{dc} is the average or dc current drawn from the supply.

If $I_{C(max)}$ is the maximum or peak value of collector or output current, then

$$I_{dc} = \frac{I_{C(\max)}}{\pi} \qquad \therefore \qquad P_{in(dc)} = V_{CC}\frac{I_{C(\max)}}{\pi}$$

(*ii*) DC Power Loss in Load : $P_{Rc(dc)} = I_{dc}^2 R_C$ where $I_{dc} = I_{C(max)}/\pi$

(iii) **AC Power Output in Load :** $P_{o(ac)} = I^2 R_C = \frac{V^2}{R_C} = \frac{V_m^2}{2R_C}$

where I = rms value of output ac current

V = rms value of output ac voltage

V_m = maximum value of output ac voltage

***(iv)* DC Power Loss in Collector Region or Transistor**

$$P_{c(d)} = P_{in(dc)} - P_{Rc(dc)} - P_{o(ac)}\ ; \quad \eta_{overall} = \frac{P_{o(ac)}}{P_{in(dc)}}$$

60.23. Maximum Values

The operation for maximum signal input is shown in Fig. 60.29. Here,

$$I_{dc} = \frac{I_{C(\max)}}{\pi} \quad \therefore \quad P_{in(dc)} = V_{CC} \cdot I_{dc} = \frac{V_{CC} \cdot I_{C(\max)}}{\pi}$$

RMS value of output or collector current

$$= \frac{I_{C(max)}}{\sqrt{2}}$$

RMS value of output voltage $= \frac{V_{CC}}{\sqrt{2}}$

Hence,ac output power during half-cycle

$$= \frac{1}{2} \times \frac{V_{CC}}{\sqrt{2}} \frac{I_{C(\max)}}{\sqrt{2}}$$

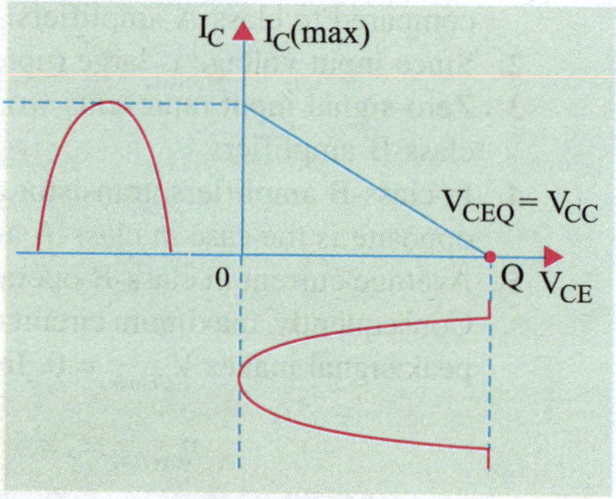

Fig. 60.29

The factor $\frac{1}{2}$ comes in because power is produced during one half-cycle only.

$$P_{o(ac)} = \frac{1}{4} . V_{CC} \cdot I_{C(max)}$$

$$\therefore \quad \eta_{overall} = \frac{P_{o(ac)}}{P_{in(dc)}} = \frac{\pi}{4} = 0.785 \quad \text{or} \quad \mathbf{78.5\%}$$

Example 60.12. *In a class-B amplifier, $V_{CE(min)}$ = 2V and V_{CC} = 15 V. Find its overall efficiency*

Solution. As seen from Art 60.21, % $\eta = 78.5 \left(1 - \frac{2}{15}\right)$ = **68%**

60.24. Class-B Push-Pull Amplifier

It employs two identical transistors operating as a ***single-stage of amplification***. As shown in Fig. 60.30, the base of the two *CE*-connected transistors have been connected to the opposite ends of the secondary of the input transformer T_1 and collectors to the opposite ends of the primary of the output transformer T_2. For getting a balanced circuit, the two emitters have been returned to the centre tap of T_1 secondary and V_{CC} connected to the centre tap on the primary of T_2. Since zero bias is required for cut-off, the two bases have been earthed. A push-pull amplifier is sometimes referred to as ***balanced amplifier***.

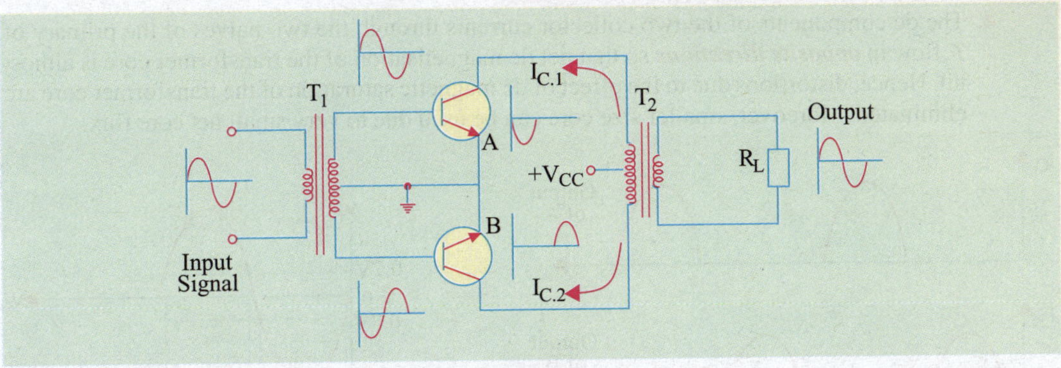

Fig. 60.30

Principle of Operation

It is seen from Fig. 60.30 that transistor *A* and *B* are driven by two input signals which are 180° out of phase with each other. These two signals are produced by T_1.

Transistor *A* takes positive half-cycles of the signal whereas *B* handles negative half-cycles. When the two outputs are combined, an almost undistorted output waveform is produced as seen from Fig. 60.31.

In Fig. 60.31, the transfer characteristic of transistor *B* has been plotted upside down with respect to that of *A* in order to get the combined output.

Detailed operation is as under :

During the positive half-cycle of the signal, *A* is turned ON because its base is driven positive. It draws collector current $I_{C\text{-}1}$ in the upward direction from V_{CC}. Meanwhile, transistor B remains OFF because its base has negative voltage. Hence, $I_{C\text{-}2} = 0$. Obviously, one positive half-cycle of the output signal appears across secondary load R_L of T_2.

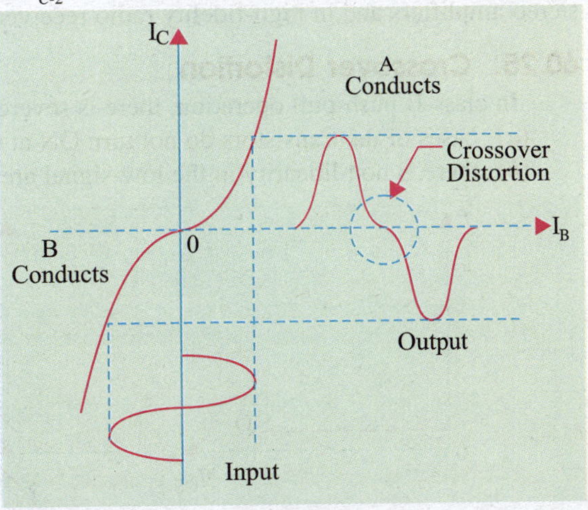

Fig. 60.31

During negative half-cycle of the input signal, *B* conducts whereas *A* remains OFF. Hence, $I_{C\text{-}2}$ is taken by B but $I_{C\text{-}1} = 0$. Now, negative output half-cycle is produced across R_L because $I_{C\text{-}2}$ is pulled down through secondary of T_2. It is obvious that in the absence of input signal, neither *A* nor *B* draws any collector current. Hence, there is no drain on V_{CC} battery.

Advantages

1. It has high efficiency, theoretical limit being 78.5%. It is primarily due to the fact that there is ***no power drawn by the circuit under zero-signal condition.***
2. Since in push-pull arrangement, 180° phase difference exists between even-order harmonics produced by each transistor, they cancel out thereby giving an almost distortion-free output. This automatic cancellation of all even-order harmonics from the output current makes class-B push-pull amplifiers highly desirable for ***communication sound equipment.***
3. This double-ended class-B amplifier provides practically four times the power supplied by a single-ended amplifier provided signal load resistance remains the same.

4. The dc components of the two collector currents through the two halves of the primary of T_2 flow in ***opposite directions*** so that net dc magnetisation of the transformer core is almost nil. Hence, distortions due to the effect of dc magnetic saturation of the transformer core are eliminated. Moreover, smaller size core can be used due to very small net core flux.

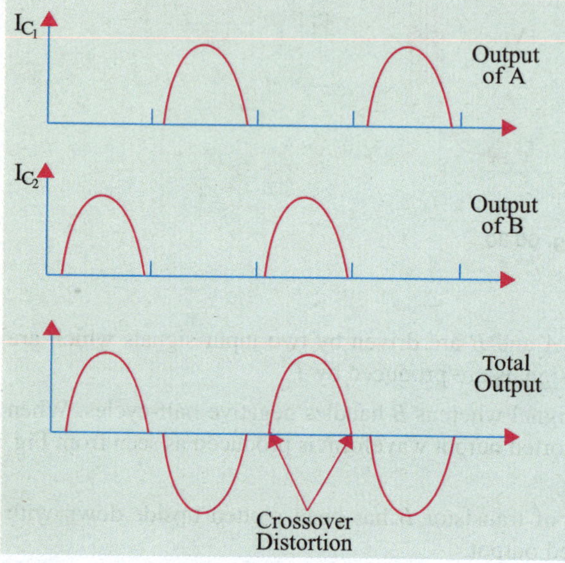

Fig. 60.32

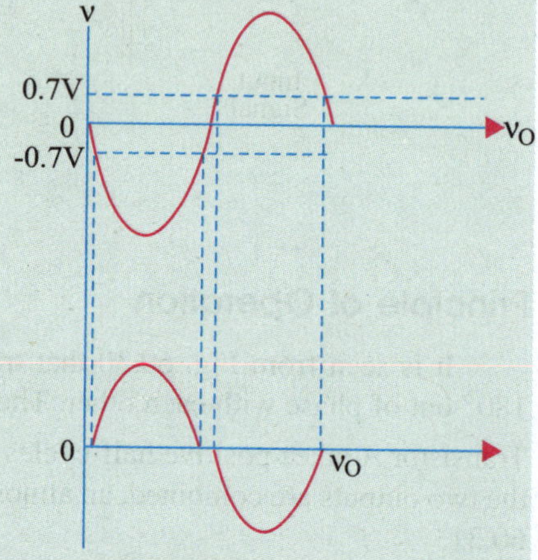

Fig. 60.33

Uses

Class-B push-pull amplifiers are extensively used for audio work in portable record players, as stereo amplifiers and in high-fidelity radio receivers.

60.25. Crossover Distortion

In class-B push-pull operation, there is severe distortion at very low signal level because

(a) bases of the transistors do not turn ON at 0 V but at 0.3 V for Ge and 0.7 V for Si

(b) there is non-linearity in the low-signal area.

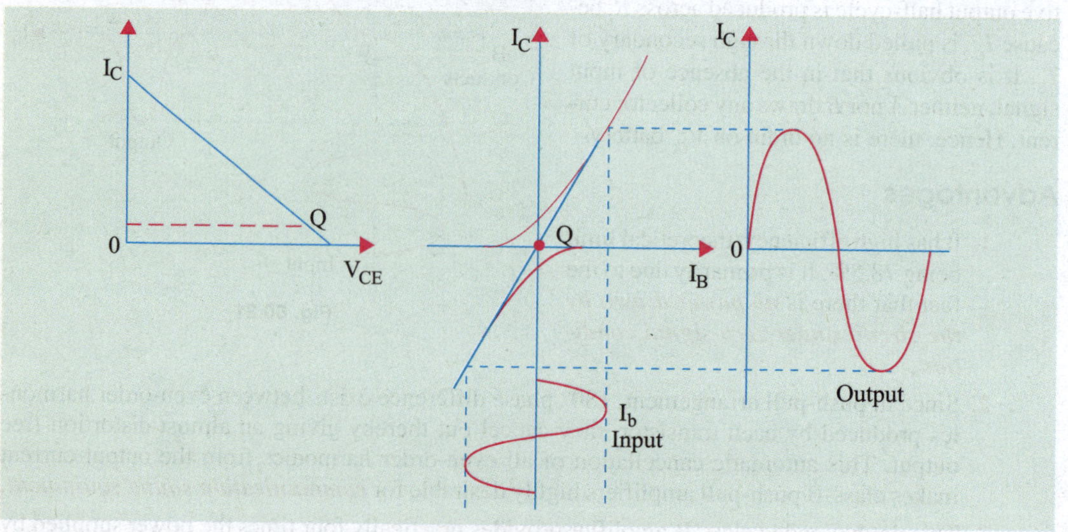

Fig. 60.34

In simple words, crossover distortion occurs as a result of one transistor cutting off before the other begins conducting. The effect is illustrated in Fig. 60.32. For silicon transistors, there is 2 × 0.7 = 1.4 V ***dead zone*** on the input signal within which neither transistor is turned ON and output is zero (Fig. 60.33).

The distortion so introduced is called ***crossover distortion*** because it occurs during the time operation ***crosses over from one transistor to the other in the push-pull amplifier.*** The same was shown earlier in Fig. 60.31 by using transfer characteristics of the two transistors.

Crossover distortion can be eliminated by applying slight forward bias to each emitter diode. It, in effect, means locating the *Q*-point of each transistor slightly above cut-off as shown in Fig. 60.34 so that each one operates for more than one half-cycle. Strictly speaking, it results in class-AB operation because each transistor may operate for about 200° (instead of 180°). However, for all practical purposes, it is still regarded as class-B push-pull operation.

60.26. Power Efficiency of Push-Pull Amplifiers

Consider the circuit shown in Fig. 60.35.

Since the two transistors are identical, $I_{C1} = I_{C2} = I_C$

$\therefore \quad I_C = I_m/\pi$

Total dc supply current I_{dc} for the two transistors is

$$I_{dc} = 2I_C = 2\frac{I_m}{\pi}$$

$$\therefore \quad P_{in(dc)} = V_{CC} \times I_{dc} = \frac{2I_m}{\pi}V_{CC}$$

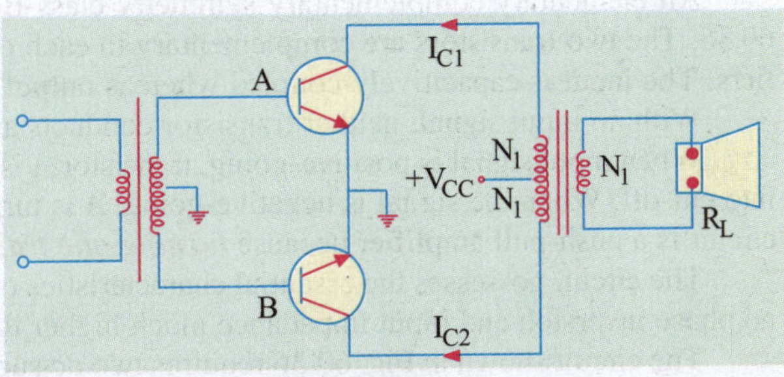

Fig. 60.35

For the sake of simplicity, we will assume transformer turns ratio as $(N_1 + N_1) : N_1$ so that for each transistor, turns ratio is $N_1 : N_1$ *i.e.* 1 because each uses half the primary of the output transformer.

If V_m is the peak voltage at either collector, then peak load voltage is also V_m. The peak current in the load is I_m. Then, ac power delivered to the load is

$$P_{o(ac)} = \frac{V_m}{\sqrt{2}} \cdot \frac{I_m}{\sqrt{2}} = \frac{V_m I_m}{2}$$

The total collector dissipation for the two transistors is

$$2.P_{c(dc)} = P_{in(dc)} - P_{o(ac)} = \frac{2I_m V_{CC}}{\pi} - \frac{V_m I_m}{2} = 2I_m\left(\frac{V_{CC}}{\pi} - \frac{V_m}{4}\right)$$

$$\eta_{overall} = \frac{P_{o(ac)}}{P_{in(dc)}} \times 100 = \frac{V_m I_m/2}{2I_m V_{cc}/\pi} \times 100 = \frac{\pi}{4} \cdot \frac{V_m}{V_{CC}} \times 100$$

Under ideal condition of maximum power in the load, $V_m = V_{CC}$

$$\therefore \quad \text{maximum } \eta_{overall} = \frac{\pi}{4} \times 100 = 78.5\%$$

Same is the value of collector efficiency.

60.27. Complementary Symmetry Push-Pull Class-B Amplifier

The push-pull amplifier discussed in Art. 60.24 suffers from two disadvantages :

(*i*) it requires a bulky and expensive output transformer

(ii) it requires two out-of- phase input signals which necessitates an input centre-tapped transformer or phase inverter. It makes the driver circuitry quite complicated.

The complementary symmetry amplifier eliminates these two disadvantages while retaining the advantages of push-pull configuration.

As we know, a standard class-B push-pull amplifier requires two power transistors of the ***same type*** with closely-matched parameters. But the chief requirement of a complementary amplifier is a pair of closely-matched but ***oppositely-doped*** power transistors. The term 'complementary' arises from the fact that one transistor is *PNP* type and the other is of *NPN* type. They have ***symmetry i.e.*** both are made with the same material and technology and have the same maximum rating.

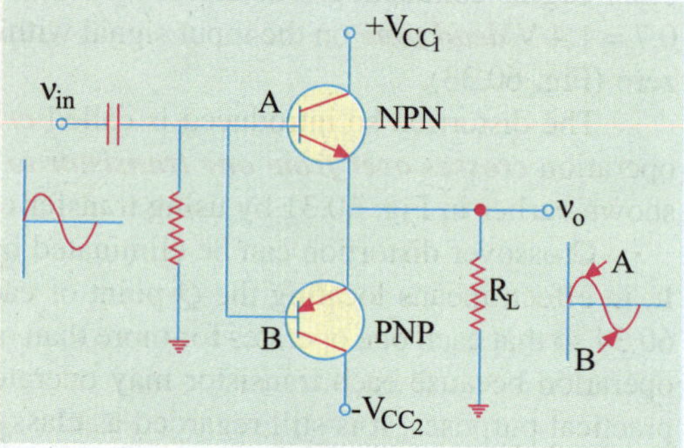

Fig. 60.36

An elementary complementary symmetry class-B push-pull amplifier is diagrammed in Fig. 60.36. The two transistors are complementary to each other and operate as emitter-follower amplifiers. The input is capacitively-coupled whereas output is direct-coupled.

With no input signal, neither transistor conducts and, therefore, current through R_L is zero.

When input signal is positive-going, transistor *A* is biased into conduction whereas *B* is driven into cut-off. When the signal is negative-going. *A* is turned OFF while *B* conducts. Obviously, this circuit is a push-pull amplifier because ***turning one transistor ON turns the other OFF.***

The circuit possesses the essential characteristics of an emitter follower *i.e.* unity voltage gain, no phase inversion and input impedance much higher than output impedance.

The circuit shown in Fig. 60.36 requires two dc supply batteries. Only one dc battery would be enough if ***'totem-pole'*** circuit configuration is used.

Due to the elimination of transformer both the high and low-frequency responses of the circuit are extended apart from reduced cost and weight.

60.28. Class-C Amplifier

In such amplifiers, the active device *i.e.* the transistor is biased much beyond cut-off. Hence,

1. output current flows only during a part of the possible half-cycle of the input signal,
2. there is no output current flow during any part of the negative half-cycle of the input signal,
3. output signal has hardly any resemblance with the input signal. It consists of short pulses only,
4. class-C amplifiers have high circuit efficiency of about 85 to 90%.

Because of this distortion, class-C amplifiers are not used for audio-frequency work. They are used for high-power output at radio frequencies (*i.e.* RF amplifiers) where harmonic distortion can be removed by simple circuits. In reality, they are used as high-frequency power switchers in radio transmitters rather than as amplifiers.

60.29. Tuned Amplifier

The gain of a transistor amplifier depends directly on the value of its load impedance. Such a high impedance can be obtained by using a high-*Q* tuned or resonant *LC* circuit as load (Fig. 60.37). The frequency response curve of the amplifier assumes the same shape as the resonance curve of the tuned circuit. Obviously, only a narrow band of frequencies around the resonant frequency f_o would be amplified well whereas other frequencies would be discriminated against.

Non-linear distortion is eliminated because of high selectivity of the load impedance. Hence, output is nearly sinusodial. With the removal of distortion, high amplifier efficiency can be achieved by operating the transistor in its nonlinear region.

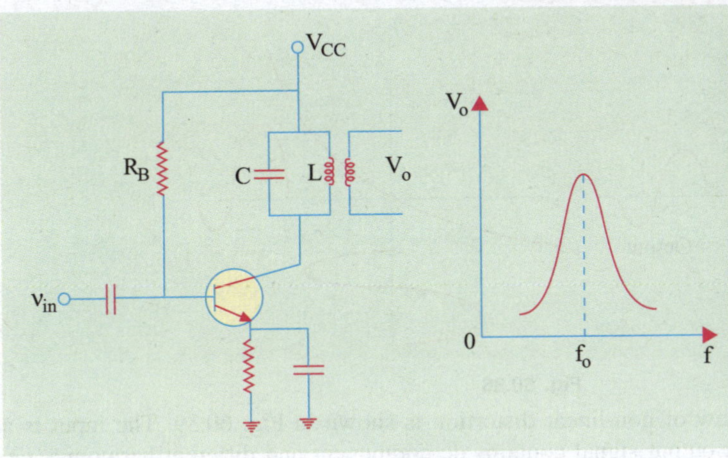

Tuned amplifier

Fig. 60.37

60.30. Distortion in Amplifiers

Amplifiers are supposed to produce an output which does not differ from the input in any respect except amplitude *i.e.* the output is expected to be larger than the input. In actual practice, it is impossible to construct such an ideal amplifier whose output is an exact duplication or replica of the input. The output is always found to differ from the input either in its waveform or frequency content. This difference between the output and input of an amplifier is called distortion.

The amplifier distortions may be divided into two broad categories depending on the region of the characteristic used by the transistor and the associated circuit and device reactances.

(*a*) non-linear distortion

This occurs when transistor operates in the non-linear region of its characteristic.

(*i*) when we visualize the signal in time domain, it is called amplitude distortion or waveform distortion;

(*ii*) when we think of the signal in the frequency domain, it is called harmonic distortion.; This distortion occurs when input signal is of one frequency, say, a pure sine wave.

(*iii*) Intermodulation (IM) distortion—when input signal has more than one frequency (like speech).

Non-linear distortion occurs in the case of large-signal inputs when the active device is driven into the non-linear regions of its characteristic.

(*b*) linear distortion

It occurs even when the active device is ***working on linear part of its characteristic with small-signal inputs***. It is primarily due to frequency-dependent reactances associated with the circuit or active device itself and occurs when input signal is composite *i.e.* has signals of different frequencies (say, fundamental and its harmonics). However, output contains no frequencies other than those at the input. It may be further subdivided into :

(*i*) ***frequency distortion***—due to unequal amplification of different frequencies present in the input signal;

(*ii*) ***phase or delay distortion***—due to unequal phase shift of various signal components.

60.31. Non-Linear Distortion

Fig. 60.38 gives time-domain view of this distortion when it is called ***amplitude distortion***. As seen, the positive half-cycle of the input signal has been amplified more than its negative half-cycle. Consequently, waveshape of the output signal differs from that of the input signal. As shown below, it is due to the appearance of new frequencies (called harmonics) at the output which are not present in the input.

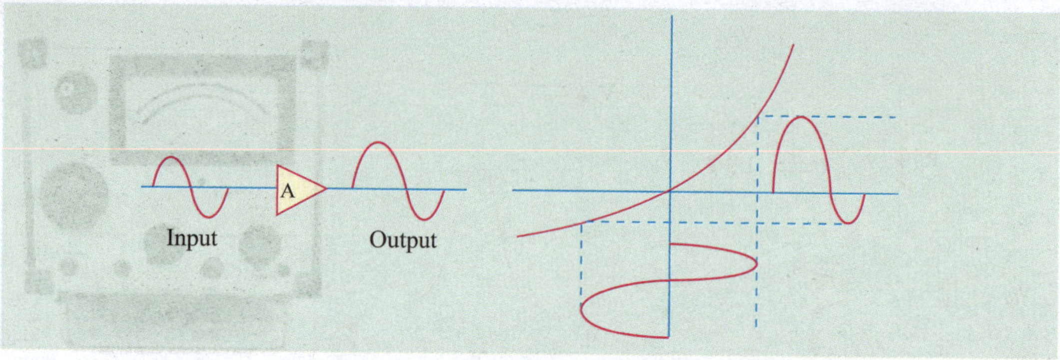

Fig. 60.38

The frequency-domain view of non-linear distortion is shown in Fig. 60.39. The input is a single-frequency (f_1) signal but output signal contains dc component and different harmonics *i.e.* frequencies which are harmonically related to each other. These harmonics are an integral multiple of the input signal frequency. Consequently, the output is distorted—the magnitude of distortion depending on the strength and number of these harmonics.

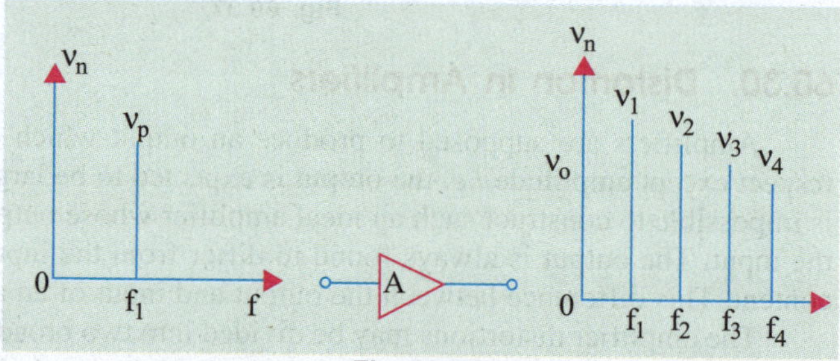

Fig. 60.39

In audio amplifiers used for amplification of speech or music, lesser the harmonic distortion, the better. For speech, harmonic distortion should not exceed 10% otherwise intelligibility will suffer. High-fidelity amplifiers have harmonic and *IM* distortion of less than 1 per cent.

60.32. Intermodulation Distortion

It occurs when input signal (like speech) consists of ***more than one frequency.*** It is also a type of non-linear distortion which generates frequency components not harmonically related to the signal frequencies. Suppose, an input signal contains two frequencies f_1 and f_2. The output signal will contain their harmonics *i.e.* f_1, $2f_1$, $3f_1$ etc. and f_2, $2f_2$, $3f_2$ etc. In adddition, there would be components $(f_1 + f_2)$ and $(f_1 - f_2)$ and also the sum and difference of the harmonics. These sum and difference frequencies are called intermodulation (*IM*) frequencies and are quite undesirable in amplifiers because they subtract from original intelligence. Since *IM* frequencies are not harmonically related to the signal, they are easily detected by human ear as noise. Hence, great care is taken to minimize them particularly in hi-fi audio amplifiers.

60.33. Frequency Distortion

It occurs even when the device is working with small-signal inputs over linear region of its characteristic. It is basically due to change in the amplifier gain with frequency *i.e.* the different input signal frequencies are amplified by different amounts.

Suppose the input signal has two frequencies of 50 Hz and 100 Hz. If they are equally amplified, their resultant output wave is as shown in Fig. 60.40.

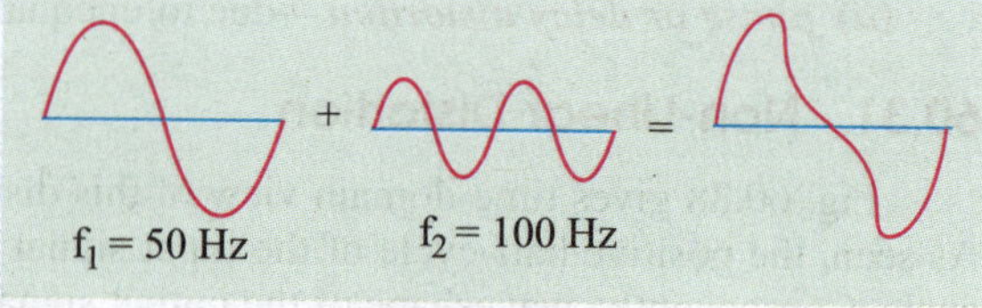

Fig. 60.40

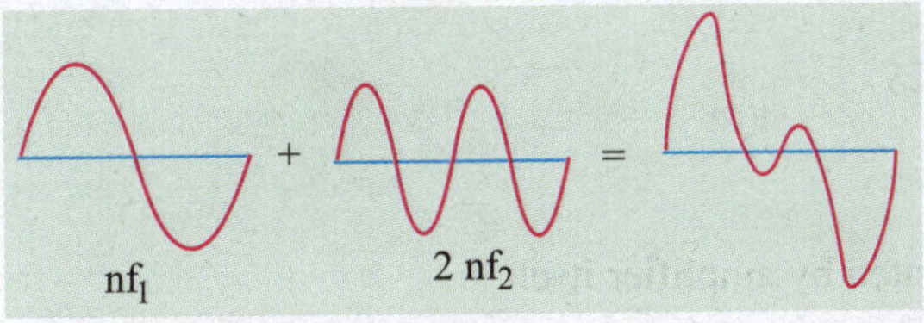

Fig. 60.41

Suppose that the amplifier amplifies the first frequency n times and the second one 2n times. Then their frequency-distorted output would be as shown in Fig. 60.41.

This distortion is due to the various frequency-dependent reactances (both capacitive and inductive) associated either with the circuit or the active device itself. In the case of audio signals the frequency distortion leads to a change in the quality of sound. Hence, in the design of untuned or wide-band amplifiers, special steps are taken to reduce variation of gain with signal frequency.

60.34. Phase or Delay Distortion

Phase distortion is said to take place when phase angles between the component waves of the output are not the same as the corresponding angles of the input. These changes in phase angles are also due to frequency-dependent capacitive and inductive reactances associated with the circuit and the active device of the amplifier.

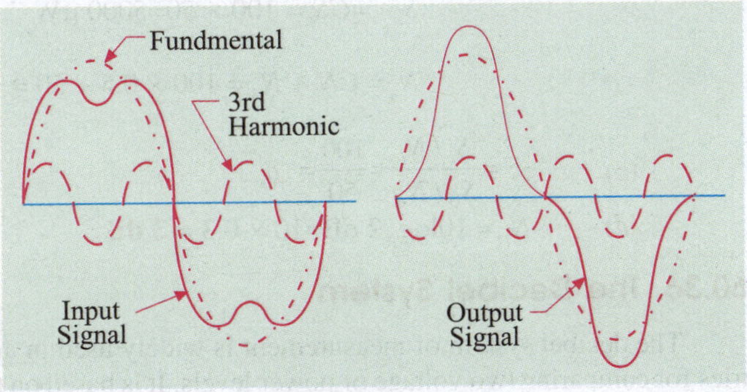

Fig. 60.42

As seen from Fig. 60.42, there has been a phase shift in the third harmonic. Hence, the resultant wave so obtained is entirely different from the input wave.

This type of distortion which is due to the non-uniform phase shift of different frequency components, is difficult to eliminate. It should be noted that if all the frequency components in a signal are shifted in phase by an integral multiple of 180°, the resultant output waveform is not changed, although the polarity of the wave may be altered but changes in polarity do not constitute distortion.

Fortunately, the human ear is unable to distinguish phase difference (though eye can) and is thus not sensitive to this distortion. Consequently, phase distortion is of no practical significance in audio amplifiers. But in amplifiers used in television sets (video amplifiers) and other systems where ear is not the final receiver, elimination of phase distortion is important.

60.35. Noise

In general , it may be defined as any kind of unwanted signal not derived from or related to the input signal. Just as distortion is the limiting factor in the amplification of large signals, noise is the limiting factor in the case of small signals.

The signal-to-noise (S/N) ratio should be high for good signal intelligibility. Since all amplifiers contribute some noise of their own to the signal being processed, their output S/N is bound to be less than their input S/N. High quality amplifiers are designed to have an output S/N as close to input S/N as possible. The amplifier performance is measured in terms of noise factor given by

Noise factor, $F = \dfrac{\text{input } S/N}{\text{output } S/N} = \dfrac{S_i/N_i}{S_0/N_0}$; obviously, ***F is greater than unity.***

When expressed in decibels, the noise factor is called noise figure (*NF*).

$\therefore \quad NF = 10 \log_{10} F \ dB$

If an amplifier could be built which generates no noise of its own, then

(i) $S_i/N_i = S_o/N_o$ *(ii)* $F = 1$ *(iii)* $NF = 0$ dB

If G is the power gain of the amplifier, then output and input signal powers are related by

$$G = \frac{S_0}{S_i} \quad i.e. \quad S_o = GS_i$$

The output noise $N_o = GN_i + N_A$

where N_i = input noise power

N_A = noise power generated by amplifier itself

Example 60.13. *The signal input to a small-signal amplifier consists of 50 μW of signal power and 0.5 μW of noise power. The amplifier generates an internal noise power of 50 μW and has a gain of 20 dB. For this amplifier, compute*

(a) input S/N, (b) output S/N, (c) noise factor, (d) noise figure.

(Electronics & Commun. Engg. Pune Univ. 1991)

Solution. (*a*) $\frac{S_i}{N_i} = \frac{50}{0.5} = \mathbf{100}$

(*b*) Now, $10 \log_{10} G = 20$ dB or $\log_{10} G = 2$ or $G = 10^2 = 100$

$S_o = GS_i = 100 \times 50 = 5000$ μW

$N_o = GN_i + N_A = 100 \times 0.5 + 50 = 100$ μW $\therefore \frac{S_0}{N_0} = \frac{5000}{100} = \mathbf{50}$

(*c*) $F = \frac{S_i / N_i}{S_0 / N_0} = \frac{100}{50} = \mathbf{2}$

(*d*) $N_F = 10 \log_{10} 2$ dB $= 10 \times 0.3 =$ **3 dB**

60.36. The Decibel System

The decibel system of measurement is widely used in audio, radio, TV and instrument industries for comparing two voltage or power levels. It is based on the established fact that an individual's response to ***seeing or hearing is nonlinear.*** It has been found that the changes in power and audio levels are related logarithmically. For example, when power is increased from 4 W to 16 W, the audio level does not increase 16/4 = 4 times but by a factor of 2 ($\because\ 4^2 = 16$). Similarly, for a power change from 4 W to 64 W, the audio level changes by a factor of 3 and not by 64/4 = 16. In logarithmic form, the relationship can be written as

increase in audio level $= \log_4 64 = 3$.

While comparing two powers, it is common practice to choose the log base of 10. Suppose, we want to compare any two powers P_1 and P_2. The simple method is to take their ratio P_2/P_1 and to state how many times P_2 is bigger or smaller than P_1 which is used as reference power. In the decibel system, we take the log of this ratio.

$\therefore$ power level $= \log_{10} R_2/P_1$ bel

Since bel is too large, we use decibel (dB) instead. Remembering that 1 bel = 10 decibel, the power level becomes

$= 10 \log_{10} P_2/P_1\, dB$

Similarly, if P_i is the input power of an amplifier and P_o its power output, then power gain of the amplifier in decibels is

$= 10 \log_{10} P_0/P_i$ dB

Also, if the output power of an amplifier changes from P_1 to P_2, the power level change is

$= 10 \log_{10} P_2/P_1$

It is obvious that dB is the unit of power change (*i.e.* increase or decrease) and not of power itself.

Another point worth remembering is that 20 dB is not twice as much power as 10 dB.

Example 60.14. *An amplifier has an input signal of 16 V peak-to-peak and an input impedance of 320 K. It gives an output voltage of 8 V peak-to-peak across a load resistor of 4 W. Calculate the dB power gain of the amplifier.* **(Electronics-I, Gujarat Univ.)**

Solution. $$P_i = \frac{V_i^2}{R_i} = \frac{(16/2\sqrt{2})^2}{320\,\text{K}} = 100\,\mu\text{W}$$

where V_i = rms value of input voltage = $\frac{V_{i \times p-p}}{2\sqrt{2}}$

If V_o is the rms value of the output voltage, then

$$P_o = \frac{V_o^2}{R_o} = \frac{(8/2\sqrt{2})^2}{4} = 2\,\text{W}$$

$\therefore$ power amplification, $\frac{P_0}{P_i} = \frac{2\,\text{W}}{100\,\mu\,\text{W}} = 20,000$

decibel power gain = $10 \log_{10} 20{,}000 = 10 \times 4.3 =$ **43 dB**

60.37. Other Expressions for Power Gain

Suppose P_o and P_i are the respective output and input powers of an amplifier, R_o and R_i are its output and input resistances and V_0 and V_i the rms values of output and input voltages, then

$$P_o = \frac{V_o^2}{R_o} \quad \text{and} \quad P_i = \frac{V_i^2}{R_i}$$

Power amplification, $A_p = P_o/P_i$. But power gain is given by

$$G_p = 10 \log_{10} P_o / P_i = 10 \log_{10} \frac{V_0^2 / R_o}{V_i^2 / R_i} = 10 \log_{10} (V_o / V_i)^2 . R_i / R_o$$

$$= 10 \left[\log_{10} (V_o / V_i)^2 + \log_{10} R_i / R_o \right] = 10 \log_{10} (V_0 / V_i)^2 + 10 \log_{10} R_i / R_0$$

$$= 20 \log_{10} V_o / V_i + 10 \log_{10} R_i / R_o$$

Also $\quad P_o = I_o^2 R_0 \quad$ and $\quad P_i = I_i^2 R_i$

$$\therefore \text{ power gain, } G_p = 10 \log_{10} P_o / P_i = 10 \log_{10} I_o^2 R_o / I_i^2 R_i = 10 \left[\log_{10} (I_o / I_i)^2 R_o / R_i \right]$$

$$= 10 [2 \log_{10} I_o / I_i + \log_{10} R_o / R_i] = (20 \log_{10} I_o / I_i + 10 \log_{10} R_o / R_i)\,\text{dB}$$

Of course, if $R_o = R_i$, then $\log_{10} R_o / R_i = \log_{10} 1 = 0$

Hence, in that case, power gain, $G_p = 20 \log_{10} I_o / I_i$ dB

60.38. Voltage and Current Levels

Though decibel was initially defined as the unit of power level, it can also be used with voltage and current levels.

Suppose, an amplifier has an input voltage of V_i and gives an output voltage of V_0. Then, its voltage amplification is $A_v = V_o/V_i$ but its decibel voltage gain is

$$G_v = 20 \log_{10} A_v \,\text{dB} = 20 \log_{10} V_o / V_i$$

It should be carefully noted that multiplying factor of 20 has been used and not of 10 as is done for finding power level. In fact, it implies that R_o has been taken equal to R_i in the power gain equation derived in Art 60.37.

Similarly, current amplification is $A_i = I_o/I_i$. However, current gain of the amplifier is

$$G_i = 20 \log_{10} I_o / I_i \,\text{dB}$$

Example 60.15. *The input and output voltages of a network are 16 V and 8V respectively. If input impedances are equal, find the voltage gain.*

Solution. $G_v = 20 \log_{10} V_o/V_i = 20 \log_{10}{}^{(8/16)}$
$= 20 [\log_{10} 8 - \log_{10} 16] = 20 (0.90 - 1.2) = 20 \times (-0.3) = -$ **6 dB.**

Alternative Method

When the voltage ratio is less than 1, *its log is negative* which is often difficult to handle. In such cases, it is best to invert the fraction and then make the result negative. The above problem could be solved thus :

$$G_v = -20 \log_{10}{}^{(16/8)} = -20 \times 0.3 = -\textbf{6 dB}$$

Example 60.16. *A microphone delivers 30 mV to the 300 Ω input of an amplifier. The ac power delivered to an 8-Ω speaker is 18 W. What is the power gain of the amplifier ?*

Solution. We may use any one of the equations given above.

$G_p = 20 \log_{10} V_o/V_i + 10 \log_{10} R_i/R_o$

Now, $P_o = \dfrac{V_o^2}{R_o}$ or $18 = \dfrac{V_o^2}{8}$ $\therefore\ V_o = \sqrt{8 \times 18} = 12\text{ V}$

$\therefore$ $G_p = 20 \log_{10} 12\text{V}/30\text{mV} + 10 \log_{10} 300/8 = 20 \log_{10} 400 + 10 \log_{10} 37.5$
$= 20 \times 2.6 + 10 \times 1.57 =$ **67.7 dB.**

Example 60.17. *The output power of an amplifier is 100 mW when the signal frequency is 5 kHz. When the frequency is increased to 25 kHz, the output power falls to 50 mW. Calculate the dB change in power.* **(Electronics, Gujarat Univ. 1991)**

Solution. The decibel change in power level
$= 10 \log_{10} 50/100 = 10 \log_{10}{}^{(1/2)} = -10 \log_{10} 2 = -10 \times 0.3 = -$**3 dB.**

Example 60.18. *The output voltage of an amplifier is 10 V at 5 kHz and 7.07 V at 25 kHz. What is the decibel change in output power level ?*

Solution. Since changes in voltage are across the same output resistance, the decibel change in power is

$= 20 \log_{10} V_2/V_1 = 20 \log_{10} 7.07/10 \text{ dB} = -20 \log_{10} 10/7.07 = -20 \log_{10} 1.414 \text{ dB}$
$= -20 \times 0.15 = -3 \text{ dB}$

It is seen from Ex. 60.17 and Ex. 60.18 that output decibel power falls by a 3 dB when

(i) absolute value of power falls to half its original value

or

(ii) output voltage falls to 0.707 or $1/\sqrt{2}$ *of its original value.*

Example 60.19. *A certain radio receiver delivers an output power of 3.6 W.*
(i) what would be the decibel gain if power output is increased to 7.2 W ?
(ii) what power output would be required to produce a power gain of 10 dB?

Solution. *(i)* $G_p = 10 \log_{10} P_2/P_1 = 10 \log_{10} 7.2/3.6 =$ **3 dB**

(ii) Here, $G_p = 10$, $P_1 = 3.6$ W, $P_2 = ?$

$\therefore$ $10 = 10 \log_{10} P_2/3.6$ $\therefore$ $\log_{10} P_2/3.6 = 1$ or $\dfrac{P_2}{3.6} = 10^1 = 10$

$\therefore$ $P_2 = 3.6 \times 10 =$ **36 W**

60.39. Characteristics of the Decibel System

1. a decibel is a measure of **ratio** and not of **an amount**. It tells us how many times one quantity is greater or lesser with respect to another or the reference quantity. It does not measure the actual (or absolute) voltage or power but only their **changes;**

2. decibel is non-linear *i.e.* 20 dB is not twice as much power or voltage as 10 dB;
3. this log-based system allows a tremendous range of power ratios to be encompassed by using only two-digit numbers. For example
   1 dB = 1.26 : 1 power ratio and 50 dB = 100,000 : 1 power ratio;
4. total dB of a cascaded amplifier can be found by simply adding the stage dBs.

60.40. Value of 1 dB

It can be proved that 1 dB represents the log of two powers which ***have a ratio of 1.26.***

$$1 \text{ dB} = 10 \log_{10} P_2/P_1$$

$$\therefore \quad \log_{10} P_2/P_1 = \frac{1}{10} = 0.1 \quad \text{or} \quad \frac{P_2}{P_1} = 10^{0.1} = 1.26$$

Hence, +1 dB represents an increase in power of 26%.

60.41. Zero Decibel Reference Level

By now, it should be clear that decibel does not measure any physical quantity like voltage or power etc., but merely ratio of two physical quantities. For determining the power levels of various powers like P_1, P_2, P_3 etc. it is essential to fix some reference power with which their ratios can be taken. If we fix P_0 as the reference power, then different power levels would be

$$10 \log_{10} P_1/P_o \quad \text{and} \quad 10 \log_{10} P_2/P_o \text{ etc.}$$

Obviously, ***P_o cannot be 0 watt i.e.*** 0 watt cannot be taken as 0 dB because, in that case, power level of P_1 would be

$$= 10 \log_{10} P_1/0 = \infty$$

In fact, any power compared with zero power is infinity. Hence, decibels would not be defined if zero watt is taken as zero reference level.

Following three zero reference levels are in common use :

1. Zero dB refers to 6 mW dissipated in a 500 Ω resistive load.
   The reference voltage value corresponding to 0 dB is
   $P = V^2/R$ or $6 \times 10^{-3} = V^2/500$ or $V = 1.73$ V
2. Zero dB refers to 1 mW dissipated in 600 Ω.
   Here, $1 \times 10^{-3} = V^2/600$; V = 0.774 V
3. Zero dB refers to a 1-mW dissipation.
   It is written as dBm indicating that it uses 1 mW as a reference.
   This reference does not depend on any particular load impedance value.
   Calculations are performed by using the relation.

$$G_p = 10 \log_{10} P_2/0.001 \text{ dB}_{in}$$

60.42. Variations in Amplifier Gain with Frequency

If the input voltage of an amplifier is kept constant but its frequency is varied, it is found that the amplifier gain

(i) remains practically constant over a sizable range of mid-frequencies,
(ii) decreases at low as well as high frequencies.

A typical frequency- versus-gain curve is shown in Fig. 60.43. While analyzing this curve, three value of frequency are important

(i) mid-frequency range, *(ii)* lower cut-off frequency, f_1,
(iii) upper cut-off frequency, f_2.

The lower and upper cut-off frequencies are defined as those frequencies.

(*a*) where voltage gain of the amplifier decreases to 0.707 times the mid-frequency gain *i.e.* to

$$\frac{1}{\sqrt{2}} A_{v(mid)} = 0.707\, A_{v(mid)}$$

or

(*b*) in terms of power, where power amplification of the amplifier decreases to half its value at mid-frequencies

i.e. $A_{p.1}$

or $A_{p.2} = \frac{1}{2} A_{p(mid)}$

(*c*) in terms of decibels, where power gain **falls by 3 dB**.

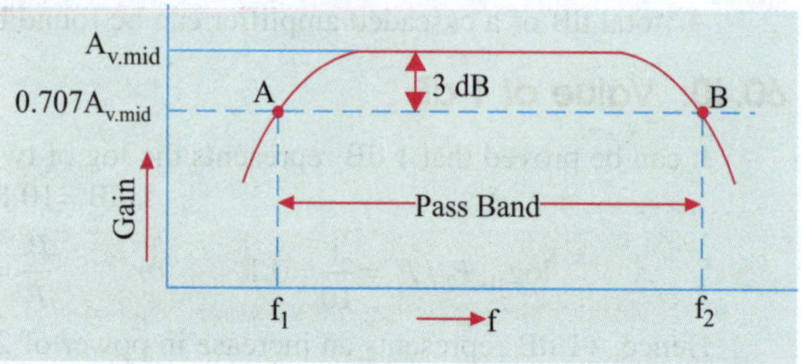

Fig. 60.43

That is why the two cut-off frequencies are referred to as

(*i*) –3 dB frequencies or
(*ii*) down 3 dB frequencies or
(*iii*) 3 dB loss frequencies.

The points A and B in Fig. 60.43 are called 3-dB points or (sometime) as ***minus 3 dB points***. The cut-off frequencies are also called roll-off frequencies because at these frequencies the amplifier gain starts rolling down from its midband value or maximum value.

The frequency span between these two cut-off frequencies is called the passband or bandwidth (BW) of the amplifier.

$\therefore \qquad \Delta f = f_2 - f_1$ = band width (BW) = passband

All frequencies lying between f_1 and f_2 are amplified almost equally.

For maximum bandwidth, the stray capacitances in the amplifier must be kept to the minimum.

60.43. Causes of Amplifier Gain Variations

The primary cause of gain variation in amplifiers is the presence of capacitances, some of which are connected in series along the signal path and some in parallel.

The different types of external capacitances present in an amplifier circuit are :

1. coupling and bypass capacitors–usually of large capacitance value, [Fig. 60.44 (a)].
2. internal or inter-element capacitances of the transistor and stray wiring capacitance [Fig. 60.44 (*b*)].

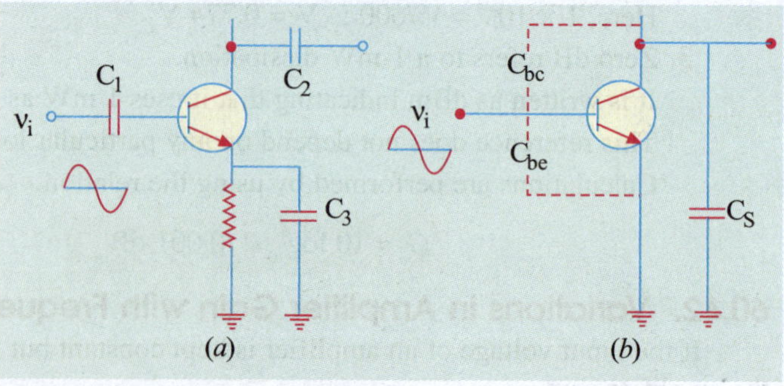

Fig. 60.44

The coupling and bypass capacitors [C_1, C_2 and C_3 in Fig. 60.44 (*a*)] are series-connected whereas interelement capacitances and stray capacitances are parallel-connected [Fig. 60.44 (*b*)] to the signal path. At mid-frequencies, C_1, C_2 and C_3 act almost as 'shorts' but C_{bc} and C_{be} act as 'open'. Hence, their effect on mid-frequencies is negligible.

However, at low frequencies, series-connected capacitors *i.e.* coupling and by-pass capacitors offer relatively large reactance thereby dropping off a large part of the input signal. Hence, amplifier gain starts decreasing as frequency is lowered.

The internal transistor capacitances and stray capacitances are *(i)* small and *(ii)* parallel-connected to the signal path. At low frequencies, they offer very high reactance and so act as effective 'open'. But with increase in frequency, their reactance keeps decreasing till at very high frequencies, they almost short or shunt the ac signal to ground both at the input and output ends. This explains why amplifier gain starts decreasing at high frequencies.

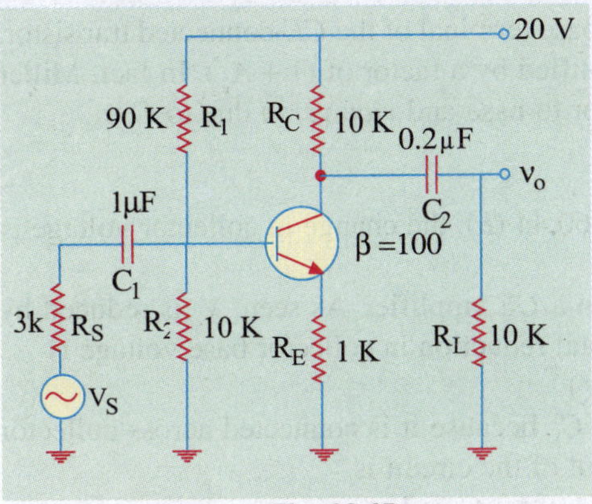

Fig. 60.45

In summary

(i) series-connected coupling and bypass capacitors cause decrease in amplifier gain at low frequencies,

(ii) parallel-connected internal transistor capacitances and stray wiring capacitances cause decrease in amplifier gain at high frequencies.

Example 60.20. *For the RC- coupled circuit of Fig. 60.45, calculate the lower cut-off frequency (i) at C_1 (ii) at C_2 and (iii) for the amplifier.*

Solution. The low-frequency cut-off for each coupling capacitor is given by

$$f_1 = \frac{1}{2\pi C R_{eq.}}$$

where R_{eq} is the resistance 'seen' by the capacitor on its right and left.

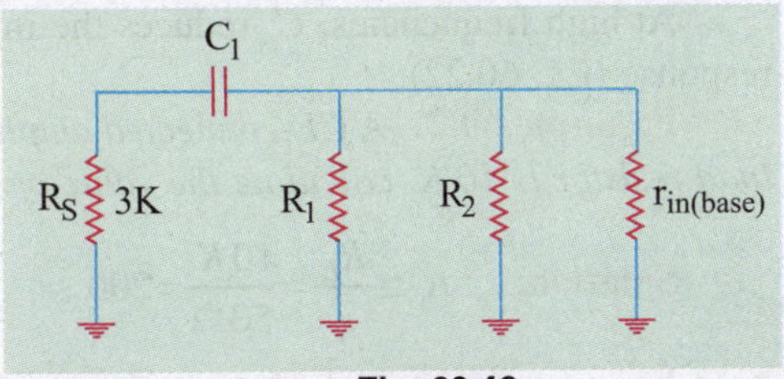

Fig. 60.46

(i) **f_1 at C_1**

For finding the resistance 'seen' by C_1, the dc and ac sources are shorted out because they have negligible resistance. The circuit becomes as shown in Fig. 60.46.

$$R_{eq} = R_S + R_1 \parallel R_2 \parallel r_{in(base)}$$

$$\text{Now, } r_{in(base)} = \beta(r_e + R_E) \cong \beta R_E = 100 \text{ K}$$

$$\therefore \quad R_{er} = 3\text{K} + 90\text{K} \parallel 10\text{K} \parallel 100\text{K} = 3 + 8.26 = 11.26 \text{ K}$$

$$\therefore \quad f_1 = \frac{1}{2\pi \times 11.26 \times 10^3 \times 1 \times 10^{-6}} = \mathbf{14.13\ Hz}$$

(ii) **f_1 at C_2**

Let us first find the resistance 'seen' by C_2. On one side, it sees $R_L = 10$ K load resistor to the ground and on the other side, it sees R_C to ground in parallel with the resistance seen looking into transistor collector.

Since *C/B* junction is reverse-biased, its resistance is very high. If we consider it as 'open', the equivalent circuit becomes as shown in Fig. 60.47.

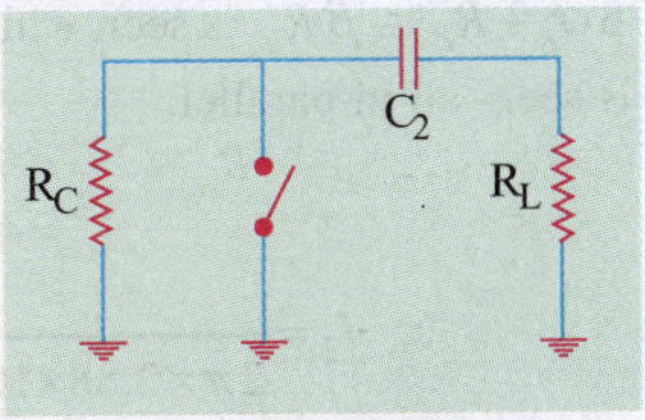

Fig. 60.47

$$\therefore \quad R_{eq} = R_L + R_C = 10 + 10 = 20 \text{ K}$$

$$\therefore \quad f_1 = \frac{1}{2\pi \times 20 \times 10^3 \times 0.2 \times 10^{-6}} \cong 40 \text{ Hz } \mathbf{Hz}$$

(iii) Since cut-off frequency for C_2 occurs at 40 Hz while cut-off for C_1 occurs way down at 14 Hz, C_1 determines the lower cut-off frequency for the amplifier *i.e.* **14 Hz.**

60.44. Miller Effect

According to this effect, when viewed from input base terminal of the *CE*-connected transistor, the capacitance C_{bc} appears as $(1 + A_v)\ C_{bc}$ *i.e.* it is amplified by a factor of $(1 + A_v)$. In fact, Miller effect takes into account the feedback from the collector to base and *vice-versa* due to C_{bc}.

Proof

When V_i is applied to the transistor's base in Fig. 60.44 (*b*), the change in collector voltage is

$$\Delta V_C = -A_v \cdot V_i$$

The negative sign is due to phase shift inherent in a *CE* amplifier. As seen, V_C is reduced by $(A_v \cdot V_i)$ when base voltage is increased by V_i. Hence, total reduction in collector base voltage is

$$\Delta\ V_{CB} = V_i + A_v . V_i = V_i (1 + A_v)$$

This also represents the change of voltage across C_{bc} because it is connected across collector and base. Using $Q = CV$, the charge supplied to the input of the circuit is

$$Q = C_{bc} \times \Delta\ V_{CB} = C_{bc} \times (1 + A_v)\ V_i = (1 + A_v)\ C_{bc} \times V_i$$

Hence, C_{bc} appears as $(1 + A_v)\ C_{bc}$ when looked from the input side of the circuit.

Incidentally, the total input capacitance to the transistor is

$$C_{in} = C_{be} \parallel (1 + A_v) C_{bc} = C_{be} + (1 + A_v)\ C_{bc}$$

At high frequencies, C_{in} reduces the input impedance of the circuit and affects the frequency response (Ex. 60.22).

Example 60.21. *A CE-connected amplifier has C_{bc} = 4 pF, C_{be} = 10 pF and r_e = 50Ω. If circuit load resistor is 10 K, calculate the value of C_{in}.*

Solution. $A_v \cong \dfrac{R_E}{r_e} = \dfrac{10\,K}{50\Omega} = 200 \quad \therefore \quad C_{in} = 10 + (1 + 200)4 =$ **814 pF**

Example 60.22. *Calculate the upper cut-off frequency of the CE amplifier shown in Fig. 60.48. Given the input wiring capacitance C_{wi} = 40 pF, C_{bc} = 8 pF, C_{be} = 10 pF and β = 100.*

(Electronic Engg-I, Osmania Univ. 1984)

Solution. $f_2 = \dfrac{1}{2\pi R_{eq} C_{in}}$

The voltage amplification of the amplifier is

$$A_v \cong \frac{R_C \parallel R_L}{R_E} = \frac{20\,K \parallel 20\,K}{400\Omega} = 25$$

The total capacitance from base to ground is

$$C_{in} = C_{wi} + C_{be} + (1 + A_v)\ C_{bc}$$
$$= 40 + 10 + (1 + 25) \times 8 = 258 \text{ pF}$$

Now, let us determine the resistance 'seen' by C_{in}. If we look to the right, a resistance of $\beta(r_e + R_E) \cong \beta R_E$ is seen while to the left $R_1 \parallel R_2 \parallel R_S$ is seen–all in parallel.

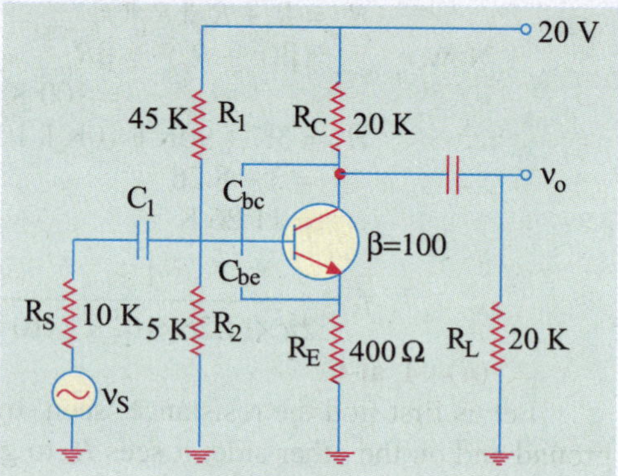

Fig. 60.48

$$\therefore \quad R_{eq} = R_1 \parallel R_2 \parallel R_S \parallel \beta R_E$$
$$= 45\,K \parallel 5K \parallel 10K \parallel 40\,K = 2.88\,K$$

$$\therefore \quad f_2 = \frac{1}{2\pi \times 2.88 \times 10^3 \times 258 \times 10^{-12}} \cong \textbf{214 kHz}$$

60.45. Cut-Off Frequencies of Cascaded Amplifiers

Cascading of stages gives higher amplification but narrow bandwidth because the product of the two remains almost constant. The approximate values of composite lower and upper cut-off frequencies of the cascaded amplifier having n identical stages are

$$f_{1.n} = 1.1\sqrt{n} \times f_1 \quad \text{per stage} = \frac{f_1 \text{ per stage}}{\sqrt{(2^{1/n} - 1)}}$$

$$f_{2.n} = \frac{f_2 \text{ per stage}}{1.1\sqrt{n}} = \sqrt{(2^{1/n} - 1)} \times f_2 \text{ per stage}$$

The reduced bandwidth of the cascaded amplifier is $= f_{2n} - f_{1.n}$

60.46. Transistor Cut-off Frequencies

Even if no external stray capacitances were present in an amplifier, there would still be an upper limit on its frequency response due to

(*i*) internal or interelement capacitances of the transistor and

(*ii*) transit time of charge carriers across the transistor junctions and through the semiconductor material.

This limitation is expressed in terms of

(*i*) alpha cut-off frequency (f_α) and (*ii*) beta cut-off frequency (f_β)

The two are defined below.

60.47. Alpha Cut-off Frequency

It is found that at high frequencies, the value of transistor α begins to fall. This decrease in α is related to the transit time effect of the charge carriers as they move from the emitter to collector.

The alpha cut-off frequency f_α is that high frequency at which the α of a *CB*-connected transistor becomes 0.707 of its low-frequency value (usually 1 kHz).

For example, if value of α at 1 kHz is 0.98, then its value at f_α would be

$$= 0.707 \times 0.98 = 0.693$$

It means that at f_α, the collector current I_C would be only 0.693 of the emitter current rather than 0.98 I_E.

It is found that f_α is

(a) inversely proportional to the square of the base width,

(b) directly proportional to the minority carrier mobility.

In this regard, NPN transistors are superior to *PNP* type because electrons have greater mobility than holes. For decreasing the base transit time, base should be as thin as possible.

It may be noted that alpha cut-off frequency of any given transistor is always greater than its beta cut-off frequency f_β. In fact, $f_\alpha \cong \beta f_\beta$.

60.48. Beta Cut-off Frequency

It is that high frequency at which the β of a *CE*-connected transistor drops to 0.707 of its low-frequency (1 kHz) value.

60.49. The f_T of a Transistor

It is another high-frequency characteristic of a transistor.

It is that high frequency of a *CE*-connected transistor where its β drops to unity *i.e.* β = 1. This frequency is much larger than f_β but less than f_α. For example, for a typical transistor, their values may be $f_\beta = 6$ MHz, $f_T = 300$ MHz and $f_\alpha = 345$ MHz.

60.50. Relation Between f_α, f_β and f_T

(*i*) For simple junction transistors $f_\alpha = 1.2\, f_T$ and (*ii*) $f_\beta = \dfrac{f_T}{\beta}$, where β refers to its low-frequency value.

60.51. Gain-Bandwidth Product

As its name indicates, it is the product of the gain and bandwidth of an amplifier. It is very useful in comparing the performance capability of various circuits.

For any amplifier, gain-bandwidth product (GBP) is constant and is equal to f_T.

If, for example, $\beta = 1$, at a frequency of 6 MHz, then f_T = **6 MHz.**

∴ gain-bandwidth product = **6 MHz.**

Hence, for a given f_T in an amplifier, increased gain may be obtained only at the expense of its bandwidth.

Example 60.23. *A transistor has f_α = 8 MHz and β = 80. When connected as an amplifier, it has stray capacitance of 100 pF at the output terminal. Calculate its upper 3 dB frequency when R_L is* **(a)** *10 K and* **(b)** *100 K.*

Solution. It should be remembered that stray capacitance would reduce the amplifier gain by 3-dB when

capacitive reactance = output resistance

i.e. $$\frac{1}{2\pi f_S C_S} = R_L \quad \text{or} \quad f_s = \frac{1}{2\pi C_S R_L}$$

First, let us find the value of f_β for comparison with f_s.

$$f_\beta = \frac{f_1}{\beta} = \frac{8\,\text{MHz}}{80} = 100\,\text{kHZ}$$

(a) **R_L = 10 K**

$$f_2 = \frac{1}{2\pi C_s R_L} = \frac{1}{2\pi \times 100 \times 10^{-12} \times 10 \times 10^3} = 159\,\text{kHz}$$

Obviously, before this frequency is reached, cut-off would have been achieved at f_β = 100 kHz. Hence, $f_2 = f_\beta$ = **100 kHz.**

(b) R_L=100 K

In this case, $f_s = \frac{10}{100} \times 159 = 15.9\,\text{kHz}$

In this case, cut-off would be achieved much earlier at 15.9 kHz before f_β is reached

∴ f_2 = **15.9 kHz.**

Tutorial Problems No. 60.1

1. An amplifier raises the power level of its 5-µW input signal by 30 dB. What is the output power ? **[5 mW]**

2. An attenuation network provides an output of 5 µW with an input of 5 mW. Calculate the decibel loss of the network. **[–30 dB]**

3. What is the decibel difference between a 100 kW and 500 kW radio transmitters? **[6.98 dB]**

4. The noise level of a certain tape recording is 30 dB below the signal level. If the signal power is 5 mW, calculate the noise power. **[5 µW]**

5. An amplifier rated at 72-W output is connected to an 8 Ω speaker

(*a*) what input power is required for full power output if power gain is 30 dB,

(*b*) what is the input voltage for rated output if amplifier voltage gain is 40 dB.

[(*a*) 72 mW (b) 240 mV]

6. The characteristics of a certain audio amplifier are such that it gives a voltage amplification of 10 at 100 Hz, 30 at 3 kHz and 60 at 10 kHz. Taking the amplification at 3 kHz as the reference level, calculate the loss or gain in decibels at the other two frequencies. **[–9.54 dB, 6.02 dB]**

7. An amplifier with full power rating of 100 W drives a speaker load of 16 Ω. The hum-level rating of the amplifier is 80 dB below its full-power rating. Calculate (*i*) hum-level in the load,

(*ii*) voltage produced by the hum across the load. **[(i) 1 μW (ii) 40 mW]**

8. Derive the voltage, current and power gain relationships and then find the voltage gain of the single-stage amplifier shown in Fig. 60.49. An *NPN* transistor that has $h_{ie} = 500\ \Omega$, h_{re} negligible, $h_{fe} = 150$, and $h_{oe} = 50\ \mu S$ is used. Neglect all coupling and stray capacitances.

Comment on the various methods available for biasing the base and discuss their merits. **[– 545]**

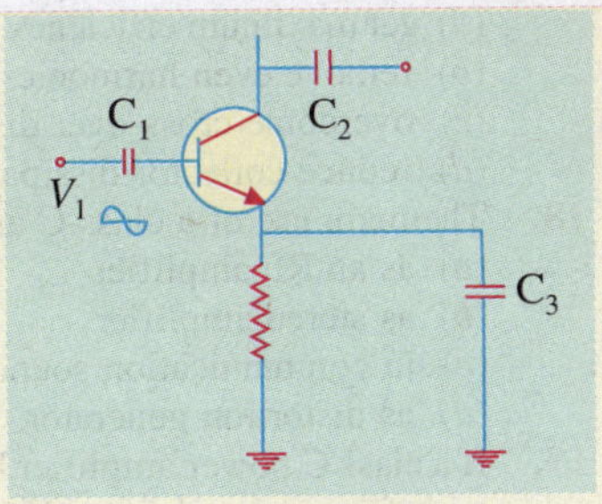

Fig. 60.49

OBJECTIVE TESTS – 60

1. A CB amplifier has very low input resistance because
(*a*) low emitter ac resistance r_e' shunts all other resistances
(*b*) it handles small input signals
(*c*) emitter bulk resistance is small
(*d*) its base is at ac ground.

2. CE amplifier is characterised by
(*a*) low voltage gain
(*b*) moderate power gain
(*c*) signal phase reversal
(*d*) very high output impedance.

3. A CC amplifier has the highest
(*a*) voltage gain (*b*) current gain
(*c*) power gain (*d*) output impedance.

4. In a CC amplifier, voltage gain
(*a*) cannot exceed unity
(*b*) depends on output impedance
(*c*) is dependent on input signal
(*d*) is always constant.

5. In a class-A amplifier, conduction extends over 360° because Q-point is
(*a*) located on load line
(*b*) located near saturation point
(*c*) centred on load line
(*d*) located at or near cut-off point.

6. The circuit efficiency of a class-A amplifier can be increased by using
(*a*) low dc power input
(*b*) direct-coupled load
(*c*) low-rating transistor
(*d*) transformer-coupled load.

7. In a class-A amplifier, worst-case condition occurs with
(*a*) zero signal input
(*b*) maximum signal input
(*c*) high load resistance
(*d*) transformer coupling.

8. The output of a class-B amplifier
(*a*) is distortion-free
(*b*) consists of positive half-cycle only
(*c*) is like the output of a full-wave rectifier
(*d*) comprises short-duration current pulses.

9. The maximum overall efficiency of a transformer-coupled class-A amplifier is – per cent.
(*a*) 78.5 (*b*) 25 (*c*) 50 (*d*) 85

10. A transistor audio amplifier is found to have an overall efficiency of 70 per cent. Most probably, it is a .... amplifier.
(*a*) class-B push-pull
(*b*) single-stage class-C
(*c*) transformer-coupled class-A
(*d*) direct-coupled class-A

11. The main purpose of using transformer coupling in a class-A amplifier is to make it more
(*a*) distortion-free (*b*) bulky
(*c*) costly (*d*) efficient.

12. A class-B push-pull amplifier has the main advantage of being free from
(*a*) any circuit imbalances
(*b*) unwanted noise
(*c*) even-order harmonic distortion
(*d*) dc magnetic saturation effects.

13. Crossover distortion occurs in .... amplifiers.
(*a*) push-pull (*b*) class-A
(*c*) class-B (*d*) class AB

14. The maximum overall efficiency of a class-B push-pull amplifier cannot exceed – per cent.
(*a*) 100 (*b*) 78.5 (*c*) 50 (*d*) 85

15. The circuit of a class B push-pull amplifier is shown in Fig. 60.50 . If the peak output voltage, V_0 is 16 V, the power drawn from the dc source would be
(*a*) 10 W (*b*) 16 W (*c*) 20 W (*d*) 32 W

16. The dissipation at the collector is zero in the quiescent state and increases with excitation in the case of a
(*a*) class A series-fed amplifier
(*b*) class A transistor coupled amplifier
(*c*) class AB amplifier
(*d*) class B amplifier

17. Class AB operation is often used in power (large signal) amplifiers in order to,

(a) get maximum efficiency
(b) remove even harmonics
(c) overcome cross-over distortion
(d) reduce collector dissipation

18. The main use of a class-C amplifier is
(a) as an RF amplifier
(b) as stereo amplifier
(c) in communication sound equipment
(d) as distortion generator.

19. If a class C power amplifier has an input signal with frequency of 200 kHz and the width of collector current pulses of 0.1μS, then the duty cycle of the amplifier will be
(a) 1% (b) 2% (c) 10% (d) 20%

20. The primary cause of linear distortion in amplifiers is
(a) change of gain with frequency
(b) unequal phase shift in component frequencies
(c) reactances associated with the circuit and active amplifying element
(d) inherent limitations of the active device.

21. An amplifier is said to suffer from distortion when its output is
(a) low
(b) different from its input
(c) noisy
(d) larger than its input.

22. While discussing amplifier performance, noise is defined as any kind of unwanted signal in the output which is
(a) unrelated to the input signal
(b) derived from the input signal
(c) not generated by the amplifier
(d) due to associated circuitry.

23. An ideal amplifier has
(a) noise figure of less than 1 dB
(b) noise factor of unity
(c) output S/N more than input S/N
(d) noise figure of more than 0 dB.

24. The decibel is a measure of
(a) power (b) voltage
(c) current (d) power level.

25. When power output of an amplifier doubles, the increase in its power level is .... decibels.
(a) 2 (b) 20 (c) 3 (d) 10

26. When output power level of a radio receiver increases by 3 dB, its absolute power changes by a factor of
(a) 2 (b) 10 (c) 1/2 (d) 3.

27. Zero watt cannot be chosen as zero decibel level because
(a) it is impossible to measure zero watt
(b) it is too small
(c) every power compared with it would be zero
(d) it would be impossible to define a decibel.

28. A minus 3 dB point on the gain versus frequency curve of an amplifier is that point where
(a) signal frequency drops to half the mid-band frequency
(b) voltage amplification becomes half of its maximum value
(c) power falls to half its maximum value
(d) upper cut-off frequency becomes twice the lower cut-off frequency

29. The bandwidth of an amplifier may be increased by
(a) decreasing the capacitance of its by pass capacitors
(b) minimizing its stray capacitances
(c) increasing input signal frequency
(d) cascading it.

30. Lower cut-off frequency of an amplifier is primarily determined by the
(a) interval capacitances of the active device used
(b) stray capacitance between its wiring and ground
(c) ac β value of its active devices
(d) capacitances of coupling and bypass capacitors.

31. The main reason for the variation of amplifier gain with frequency is
(a) the presence of capacitances, both external and internal
(b) due to interstage transformers
(c) the logarithmic increase in its output power
(d) the Miller effect.

32. The gain-bandwidth product of an amplifier is given by
(a) $f_2 - f_1$ (b) $f_\alpha - f_\beta$
(c) f_T (d) βf_α

33. A Circuit which resonates at 1MHz has *a* of 100. Bandwidth between half-power points is
(a) 10 kHz (b) 100 kHz
(c) 10 Hz (d) 100 Hz

(UPSC Engg. Services 2002)
(Hint : $BW = F_o/a$)

ANSWERS

1. (a)	2. (c)	3. (b)	4. (a)	5. (c)	6. (d)	7. (a)	8. (b)	9. (c)	10. (a)	11. (d)
12. (c)	13. (a)	14. (b)	15. (b)	16. (d)	17. (a)	18. (a)	19. (d)	20. (c)	21. (b)	22. (a)
23. (b)	24. (d)	25. (c)	26. (a)	27. (d)	28. (c)	29. (b)	30. (d)	31. (a)	32. (c)	33. (a)

C H A P T E R **61**

MULTISTAGE AND FEEDBACK AMPLIFIERS

Learning Objectives

- General
- Amplifier Coupling
- RC-coupled Two-stage Amplifier
- Impedance-coupled Two-stage Amplifier
- Advantages of Impedance Coupling
- Transformer-coupled Two-stage Amplifier
- Advantages of Transformer Coupling
- Frequency Response
- Direct-coupled Two-stage Amplifier Using Similar Transistors
- Direct-coupled Amplifier Using Complementary Symmetry of Two Transistors
- Darlington Pair
- Advantages of Darlington Pair
- Comparison Between Darlington Pair and Emitter Follower
- Special Features of a Differential Amplifier
- Common Code Input
- Differential Amplifier

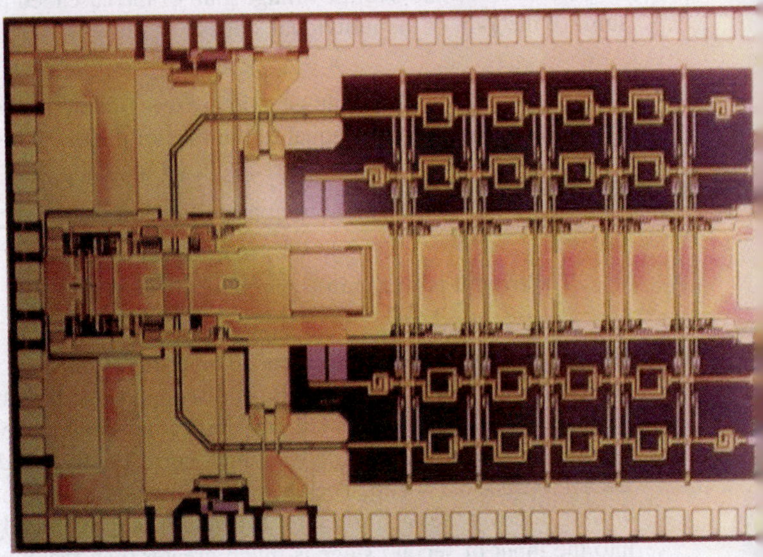

In a multistage amplifier, a number of single amplifiers are connected in cascade arrangement *i.e.* the output of first stage is connected to the input of second stage.

61.1. General

Often, the voltage amplification or power gain or frequency response obtained with a single stage of amplification is insufficient to meet the requirements of either a composite electronic circuit or a load device. Hence, two or more single stages of amplification are frequently used to achieve greater voltage or current amplification or both. In such cases, the output of one stage serves as input of the next stage as shown in Fig. 61.1. Such amplifiers may be divided into following two categories :

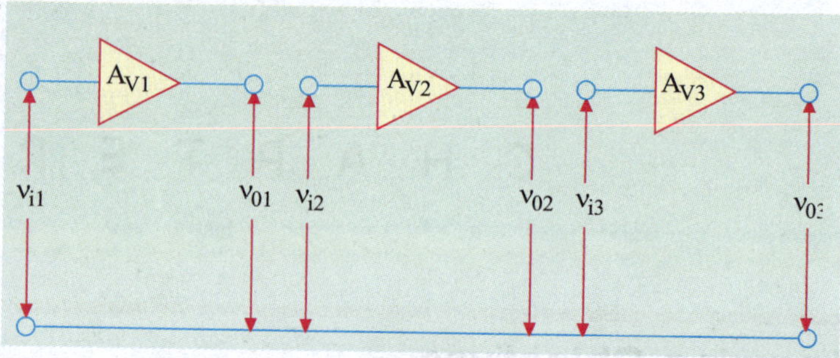

Fig. 61.1

(a) Cascaded Amplifiers

In these amplifiers, each stage as well as the type of interstage coupling used are identical.

(b) Compound Amplifiers

In these amplifiers, each stage may be different from the other (one may be *CE* and the other may be *CC* stage) and also different types of interstage couplings may be employed.

As stated above, in cascaded amplifiers, the output ac voltage of the first stage becomes the input voltage of the second stage and the ac output of the second stage becomes the input of the third stage and so on. The overall voltage gain of the cascaded amplifier is equal to the ***product*** (not the sum) of the gain of the individual stages.

$$\therefore \quad A_v = A_{v1} \times A_{v2} \times A_{v3} \times \ldots\ldots\ldots\ldots$$

However, when the voltage gain is expressed in decibels (dB), then the overall decibel gain of the multistage amplifier is equal to the ***sum*** of the dB gains of the individual stage *i.e.*

$$G = G_1 + G_2 + G_3 + \ldots\ldots\ldots$$

Similarly, the overall current amplification is given by

$$A_i = A_{i1} \times A_{i2} \times A_{i3} \times \ldots\ldots\ldots$$

The overall power gain is given by

$$A_p = A_v \, . \, A_i \quad \text{and} \quad G_p = 10 \log_{10} A_P \text{ dB}$$

Suppose in a two-stage cascaded amplifier, first stage has a voltage amplification of 2000 (dB gain of the 20 $\log_{10} 2000 = 66$ dB) and second stage has corresponding values of 1000 (60 dB). If the ac output of first stage is fed into the second stage, the overall amplification would theoretically become = $1000 \times 2000 = 2 \times 10^6$ which corresponds to a dB gain of (60 + 66) = 126 *i.e.* 20 $\log_{10} 2 \times 10^6 = 20 \times 6.3 = 126$. The above result would be true only when we ***neglect the loading effect of first stage by the second stage.*** It would be approximately true so long as the impedance looking into the input of second stage is much greater than the output impedance of the first stage. Otherwise, the overall gain would be much less.

61.2. Amplifier Coupling

All amplifiers need some ***coupling network***. Even a single-stage amplifier has to be coupled to the input and output devices. In the case of multistage systems, there is ***interstage*** coupling. The type of coupling used determines the characteristics of the cascaded amplifier. In fact, amplifiers are classified according to the coupling network used.

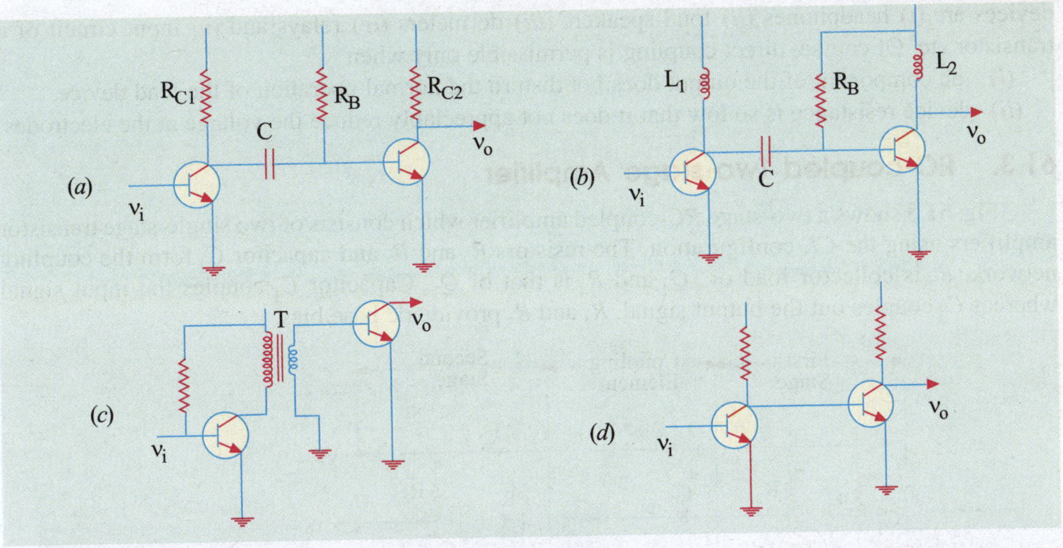

Fig. 61.2

The four basic methods of coupling are :

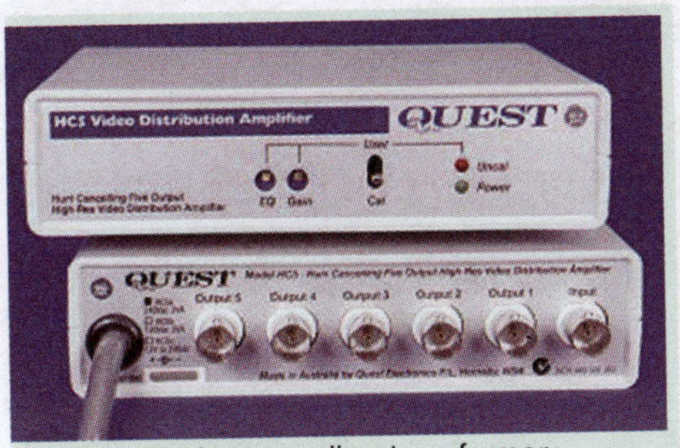

Modern coupling transformer

1. Resistance-Capacitance (RC) Coupling

It is also known as ***capacitive*** coupling and is shown in Fig. 61.2 (*a*). Amplifiers using this coupling are known as *RC*-coupled amplifiers. Here, *RC* coupling network consists of two resistors R_{C1} and R_{C2} and one capacitor *C*. The connecting link between the two stages is *C*. The function of the *RC*-coupling network is two-fold :

(*a*) to pass ac signal from one stage to the next,

(*b*) to block the passage of dc voltages from one stage to the next.

2. Impedance Coupling or Inductive Coupling

It is also known as choke-capacitance coupling and is shown in Fig. 61.2 (*b*). Amplifiers using this coupling are known as **impedance-coupled** amplifiers. Here, the coupling network consists of L_1, *C* and R_B. The impedance of the coupling coil depends on (*i*) its inductance and (*ii*) signal frequency.

3. Transformer Coupling

It is shown in Fig. 61.2 (*c*). Since secondary of the coupling transformer conveys the ac component of the signal directly to the base of the second stage, there is ***no need for a coupling capacitor.*** Moreover, the secondary winding also provides a base return path, hence there ***is no need for a base resistance.*** Amplifiers using this coupling are called ***transformer-coupled amplifiers.***

R.C. Coupled two-stage amplifier

4. Direct Coupling

It is shown in Fig. 61.2 (*d*). This coupling is used where it is desirable to connect the load directly in series with the output terminal of the active circuit element. The examples of such load devices are *(i)* headphones *(ii)* loud-speakers *(iii)* dc meters *(iv)* relays and *(v)* input circuit of a transistor etc. Of course, direct coupling is permissible only when

(i) dc component of the output does not disturb the normal operation of the load device,

(ii) device resistance is so low that it does not appreciably reduce the voltage at the electrodes.

61.3. RC-coupled Two-stage Amplifier

Fig. 61.3 shows a two-stage *RC*-coupled amplifier which consists of two single-stage transistor amplifiers using the *CE* configuration. The resistors R_2 and R_3 and capacitor C_2 form the coupling network. R_2 is collector load of Q_1 and R_4 is that of Q_2. Capacitor C_1 couples the input signal whereas C_3 couples out the output signal. R_1 and R_3 provide dc base bias.

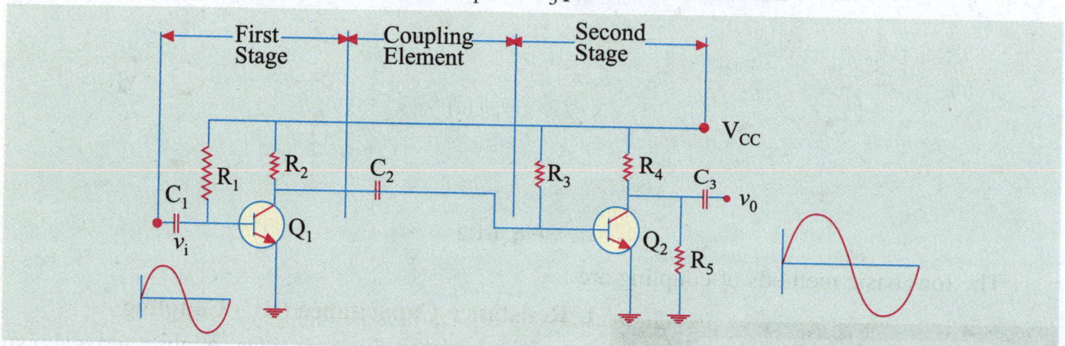

Fig. 61.3

(i) Circuit Operation

The brief circuit operation is as under :

1. the input signal V_i is amplified by Q_1. It is phase ***reversed*** (usual with *CE* connection);
2. the amplified output of Q_1 appears across R_2;
3. the output of the first stage across R_2 is coupled to the input at R_3 by coupling capacitor C_2. This capacitor is also sometimes referred to as ***blocking capacitor*** because it blocks the passage of dc voltages and currents;

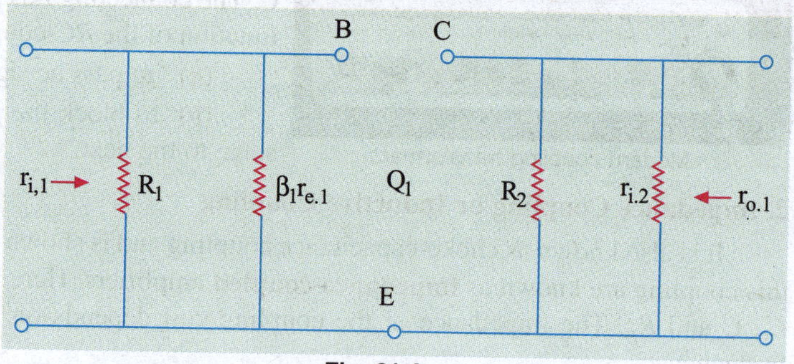

Fig. 61.4

4. the signal at the base of Q_2 is further amplified and ***its phase is again reversed;***
5. the ac output of Q_2 appears across R_4;
6. the output across R_4 is coupled by C_3 to load resistor R_5;
7. the output signal v_0 is the ***twice-amplified replica of the input signal*** v_i. It is in phase with v_i *because it has been reversed twice.*

(ii) AC Equivalent Circuit

The ac equivalent circuits for the two stages have been shown separately in Fig. 61.4 and Fig. 61.5 respectively.

In Fig. 61.4, $r_{i.1} = R_1 \parallel \beta_1.r_{e.1}$

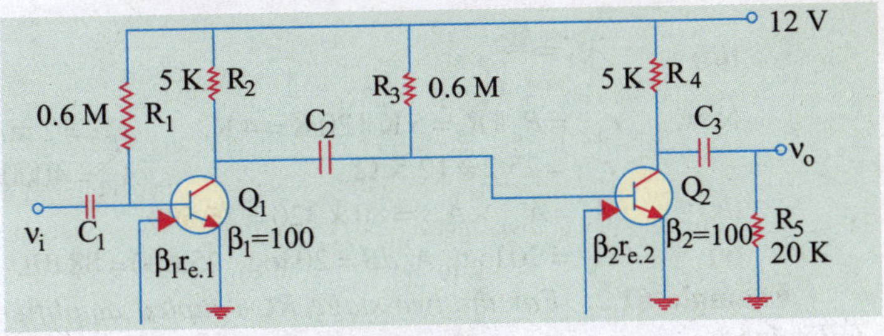

Fig. 61.5

It is the input impedance of the first stage and not $r_{in(base)}$.

The output impedance of the first stage is

$r_{0.1} = R_2 \parallel r_{i.2}$

It is so because the input of the second stage forms a part of the output of the first stage. As seen from Fig. 61.5.

$r_{i.2} = R_3 \parallel \beta_2 . r_{e2} \cong \beta_2 r_{e.2}$,

where $r_{e.1}$ and $r_{e.2}$ are ac junction resistances of the two transistors and are given by

$$r_{e.1} = \frac{25\,mV}{I_{E.1}} \text{ or } \frac{50\,mV}{I_{E.1}} \text{ and } r_{e.2} = \frac{25\,mV}{I_{E.2}} \text{ or } \frac{50\,mV}{I_{E.2}}$$

The output impedance of Q_2 is $r_{0.2} = R_4 \parallel R_5$

(iii) Voltage Gain

$r_{i.2} = \beta_2 r_{e.2}$ Now $A_{v2} = \beta_2 \dfrac{r_{0.2}}{r_{i2}}$

Also, $r_{i.2} = \beta_2 r_{e.2}$ $\therefore$ $A_{v2} = \beta_2 \dfrac{r_{0.2}}{\beta_2 . r_{e.2}} = \dfrac{r_{0.2}}{r_{e.2}}$

The voltage gain of the first stage is also given by a similar equation

$$A_{vi} = \frac{r_{0.1}}{r_{e.1}}$$

Example 61.1. *For the two-stage RC-coupled low-level audio amplifier shown in Fig. 61.6, compute the following :*

(i) r_i *(ii)* A_{v1} *(iii)* A_{v2} *and (iv)* A_v *in dB. Neglect* V_{BE} *and take* $r_e = 25\ mV/I_E$

(Electronic Circuits, Mysore Univ. 1992)

Solution. The input impedance of the cascaded amplifier is

(i) $r_i = R_1 \parallel \beta_1 . r_{e.1}$

For finding $r_{e.1}$, we need $I_{E.1}$ which approximately equals $I_{C.1}$

Now, $I_{C.1} = \beta . I_{B.1}$

Also, $I_{B1} = 12/R_1 = 12/0.6\ M = 20\ \mu A$

$\therefore$ $I_{C1} = 100 \times 20 = 2000\ \mu A = 2\ mA$

$\therefore$ $I_{E.1} = 2\ mA$

$\therefore$ $r_{e.1} = 25/2 = 12.5\ \Omega$;

$\beta_1 r_{e.1} = 100 \times 12.5 = 1250\ \Omega$

$\therefore$ $r_i = R_1 \parallel \beta_1 . r_{e.1} = 0.6\ M \parallel 1250\ \Omega \cong$ **1250 Ω**

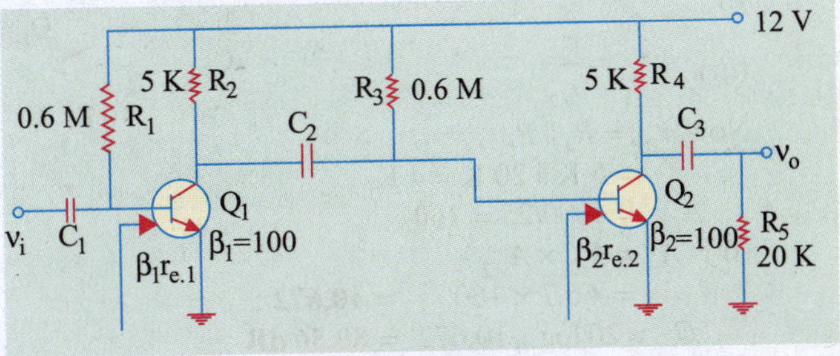

Fig. 61.6

(ii) $A_{v1} = \dfrac{r_{0.1}}{r_{e.1}}$ Now, $r_{0.1} = R_2 \parallel r_{i.2}$

$r_{i.2} = R_3 \parallel \beta_2 r_{e.2} = 0.6\ M \parallel 1250\ \Omega \cong 1250\ \Omega$

$r_{0.1} = 5\ K \parallel 1250\ \Omega = 1000\ \Omega$; $r_{e.1} = 12.5\ \Omega$ $\therefore$ $A_{v.1} = 1000/12.5 = 80$

(iii) $A_{v2} = \dfrac{r_{0.2}}{r_{e.2}}$

Now, $r_{0.2} = R_4 \| R_5 = 5\text{ K} \| 20\text{ K} = 4\text{ K}$, $\quad I_{E.2} = 2\text{ mA}$ — same as $I_{E.1}$

$r_{e.2} = 25/2 = 12.5\ \Omega \quad \therefore \quad A_{v2} = 4000/12.5 =$ **320**

(iv) $A_v = A_{v.1} \times A_{v.2} = 80 \times 320 =$ **25,600**

(v) $G_v = 20 \log_{10} A_v\, dB = 20 \log_{10} 25{,}600 =$ **88 dB.**

Example 61.2. *For the two-stage RC-coupled amplifier shown in Fig. 61.7 compute the following :*

(i) r_i, *(ii)* $A_{v.1}$, *(iii)* $A_{v.2}$, *(iv)* A_v *in decibels.*

Take $\beta_1 = \beta_2 = 100$. *Neglect* V_{BE} *and use* $r_e = 25\ mV/I_E$.

(Applied Electronics-I, Punjab Univ. 1991)

Solution. *(i)* The input impedance of the stage is $r_i = R_1 \| \beta_1 . r_{e.1}$

It should be noted that R_6 does not come into the picture because it has been ac grounded by C_4. However, it would affect the dc emitter current.

$$I_{E.1} = \frac{V_{CC}}{R_6 + R_1/\beta_1} = \frac{25}{10{,}000 + 1.5 \times 10^6/100} = 1\text{ mA}$$

$$r_{e.1} = 25/1 = 25\ \Omega; \quad \beta_1 . r_{e.1} = 100 \times 25 = 2500\ \Omega$$

$$r_i = 1.5\text{ M} \| 2500\ \Omega \cong \mathbf{2500\ \Omega}$$

(ii) $A_{v.1} = \dfrac{r_{0.1}}{r_{e.1}}$

Now, $r_{0.1} = R_2 \| r_{i.2}$ and $r_{i.2} = R_3 \| \beta_2 r_{e.2}$

Now, $I_{E.2} = \dfrac{25}{10{,}000 + 1.5 \times 10^6/100} = 1\text{ mA}$

$\therefore \quad r_{e.2} = 25/1 = 25\ \Omega\ ;\quad \beta_2 . r_{e.2} =$ **2500 Ω**

$r_{i.2} = 1.5\text{ M} \| 2500\ \Omega$

$\cong 2500\ \Omega$

$r_{0.1} = 5\text{ K} \| 2.5\text{ K}$

$= 1{,}667\ \Omega$

$\therefore \quad A_{v.1} = 1667/25 =$ **66.7**

(iii) $A_{v.2} = \dfrac{r_{0.2}}{r_{e.2}}$,

Now, $r_{0.2} = R_4 \| R_5$

$= 5\text{ K} \| 20\text{ K} = 4\text{ K}$

$\therefore \quad A_{v.2} = 4000/25 =$ **160**

(iv) $A_v = A_{v.1} \times A_{v.2}$

$= 66.7 \times 160 =$ **10,672** ;

$G_v = 20 \log_{10} 10{,}672 =$ **80.56 dB**

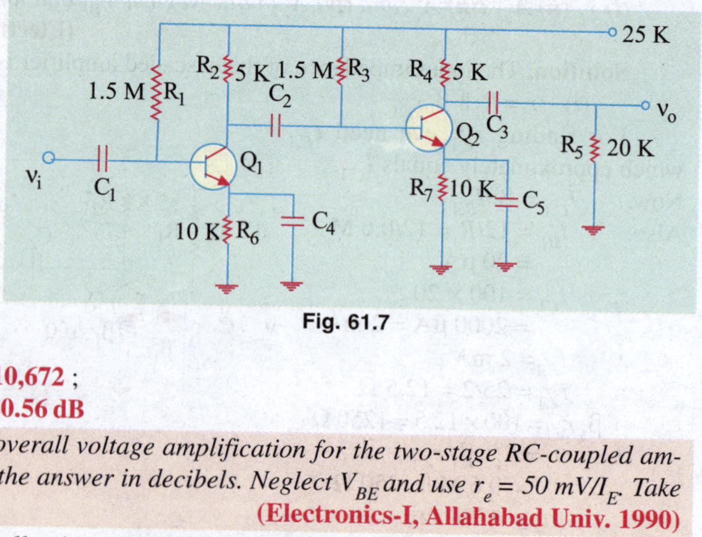

Fig. 61.7

Example 61.3. *Compute the overall voltage amplification for the two-stage RC-coupled amplifier shown in Fig. 61.8. Express the answer in decibels. Neglect* V_{BE} *and use* $r_e = 50\ mV/I_E$. *Take* $\beta_1 = \beta_2 = 100$.

(Electronics-I, Allahabad Univ. 1990)

Solution. For finding the overall gain, we will have to find each stage gain.

(i) $A_{v1} = \dfrac{r_{0.1}}{r_{e.1}}$, $\quad$ Now, $r_{0.1} = R_3 \| r_{i.2}$ and $r_{i.2} = R_5 \| R_6 \| \beta_2 . r_{e.2}$

Now, $r_{e.2} = 50/I_{E.2}$

Drop across

$$R_6 = 20 \times \frac{5}{5+45} = 2\text{V}$$

It also represents the approximate drop across R_8.

$\therefore\ I_{E2} = 2/1\text{ K} = 2\text{ mA}$

$\therefore\ r_{e.2} = 50/2 = 25\ \Omega.$

$\beta_2 . r_{e.2} = 100 \times 25 = 2.5\text{ K}$

$r_{i.2} = 45\text{ K} \parallel 5\text{ K} \parallel 2.5\text{ K} = 1.6\text{ K};$

$\therefore\ r_{0.1} = 5\text{ K} \parallel 1.6\text{ K} = 1.2\text{ K}$

$\therefore\ r_{e.1} = 50/I_{E.1} = 50/2 = 25\ \Omega$

$\therefore\ A_{v.1} = 1200/25 = 48$

$$A_{v.2} = \frac{r_{o.2}}{r_{e.2}}; \quad r_{0.2} = R_7 \parallel R_9 = 5\text{ K} \parallel 20\text{ K} = 4\text{ K}$$

$\therefore\ A_{v.2} = 4000/25 = 160 \quad \therefore\ A_v = 48 \times 160 = 7{,}680;\ G_v = 20\log_{10} 7{,}680 = $ **77.7 dB**

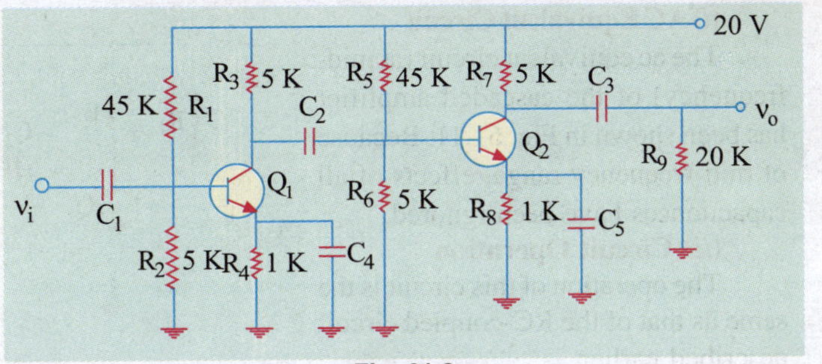

Fig. 61.8

61.4. Advantages of RC Coupling

1. It requires no expensive or bulky components and no adjustments. Hence, it is small, light and inexpensive.
2. Its overall amplification is higher than that of the other couplings.
3. It has minimum possible nonlinear distortion because it does not use any coils or transformers which might pick up undesirable signals. Hence, there are no magnetic fields to interfere with the signal.
4. As shown in Fig. 61.9, it has a very flat frequency ***versus*** gain curve *i.e.* it gives uniform voltage amplification over a wide range from a few hertz to a few megahertz because resistor values are independent of frequency changes.

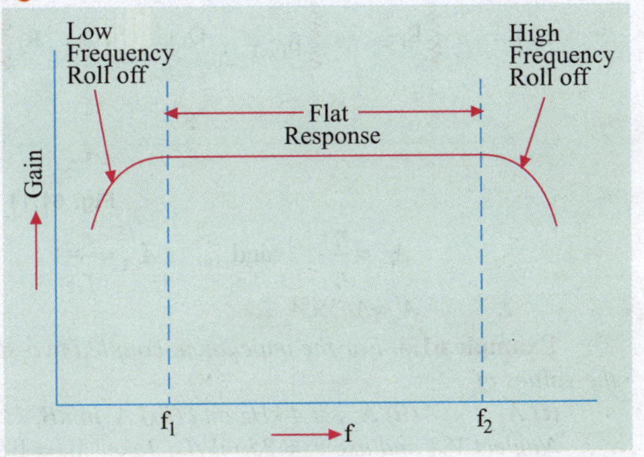

Fig. 61.9

As seen from Fig. 61.9, amplifier gain falls off at very low as well as very high frequencies. At low frequencies, the fall in gain (called ***roll-off***) is due to capacitive reactance of the coupling capacitor between the two stages. The high-frequency roll-off is due to output capacitance of the first stage, input capacitance of the second stage and the stray capacitance.

The only drawback of this coupling is that due to large drop across collector load resistors, the collectors work at relatively small voltages unless higher supply voltage is used to overcome this large drop.

61.5. Impedance-coupled Two Stage Amplifier

The circuit is shown in Fig. 61.10. The coupling network consists of L, C_2 and R_3. The only basic difference between this circuit and the one shown in Fig. 61.3 is that inductor **L *has replaced the resistor*** R_2.

(i) AC Equivalent Circuit

The ac equivalent circuit (at mid-frequency) of the cascaded amplifier has been shown in Fig. 61.11. Because of mid-frequency range, effects of all capacitances have been ignored.

(ii) Circuit Operation

The operation of this circuit is the same as that of the RC-coupled circuit described earlier.

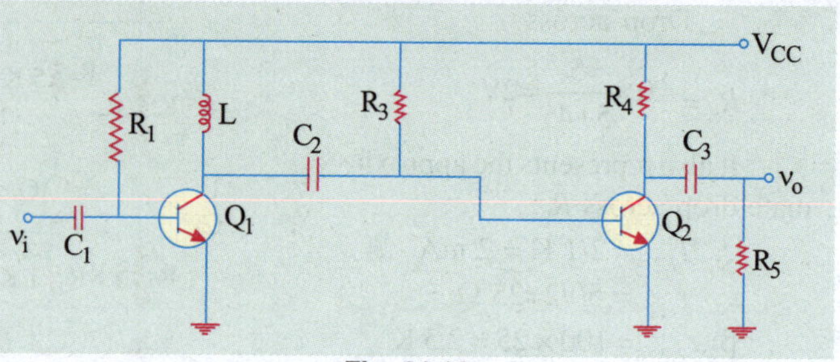

Fig. 61.10

(iii) Voltage Gain

It is given by the product of two stage gains $A_v = A_{v1} \times A_{v2}$

Now, $A_{v.1} = \dfrac{Z_{0.1}}{r_{e.1}}$ where $Z_{0.1} = X_L \| r_{i.2}$ and $r_{i.2} = R_3 \| \beta_2 . r_{e.2}$

In case, $X_L \gg r_{i.2}$, then, $Z_{0.1} \cong r_{i.2}$

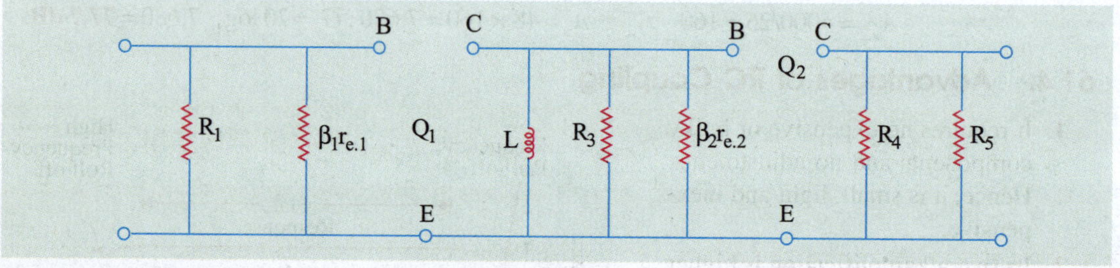

Fig. 61.11

$$\therefore \quad A_{v.1} = \frac{r_{i.2}}{r_{e.1}} \quad \text{and} \quad A_{v.2} = \frac{r_{0.2}}{r_{e.2}}$$

$$\therefore \quad A_v = A_{v.1} \times A_{v.2}$$

Example 61.4. *For the impedance-coupled two-stage amplifier shown in Fig. 61.12, compute the values of*

(i) $A_{v.2}$, (ii) $A_{v.1}$ at 4 kHz and (iii) A_v in dB.

Neglect V_{BE} and use r_e = 25 mV/I_E. Take $\beta_1 = \beta_2 = 100$.

(Industrial Electronics, Calcutta Univ. 1991)

Solution. (i) $A_{v.2} = \dfrac{r_{0.2}}{r_{e.2}}$

Now, $r_{0.2} = R_4 \| R_5 = 8\text{ K} \| 24\text{ K} = 6\text{ K}$

$I_{B2} = 12/1.2\text{M} = 10\ \mu\text{A}$, $I_{C.2} = \beta_2 . I_{B.2} = 100 \times 10 = 1000\ \mu\text{A} = 1\text{ mA}$

$\therefore \quad I_{E2} \cong 1\text{ mA}$

$\therefore \quad r_{e.2} = 25/1 = 25\ \Omega$

$$A_{v.2} = \frac{6,000}{25} = \mathbf{240}$$

(ii) $A_{v.1} = \dfrac{Z_{0.1}}{r_{e.1}} \cong \dfrac{r_{i.2}}{r_{e.1}}$

Now, $r_{e.1} = 25\ \Omega$ – equal to $r_{e.2}$

$X_L = 2\pi fL$

$= 2\pi \times 4 \times 10^3 \times 1$

$= 25,132\ \Omega$

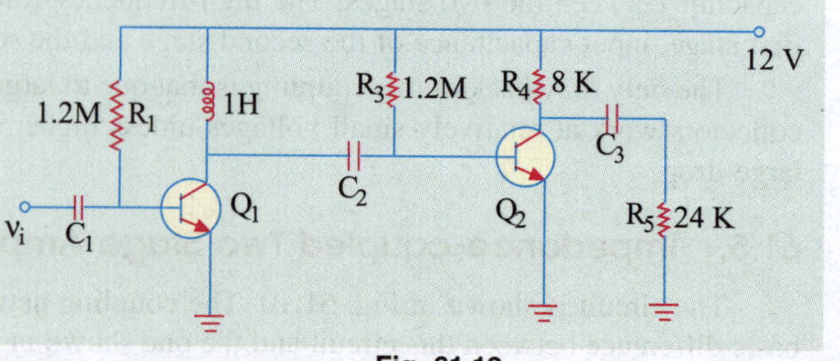

Fig. 61.12

$r_{i2} = R_3 \parallel \beta_2\, r_{e3} = 1.2\text{M} \parallel 2500\,\Omega \cong 2500\,\Omega$

Obviously, $X_L \gg r_{i.2}$ thus justifying the above approximation.

$\therefore \quad A_{v.1} = 2500/25 = \mathbf{100}$

(iii) $\quad A_v = 100 \times 240 = 24{,}000,\ G_v = 20 \log_{10} 24{,}000 = \mathbf{87.6\ dB}$

61.6. Advantages of Impedance Coupling

The biggest advantage of this coupling is that there is hardly any dc drop across L so that low collector supply voltages can be used.

However, it has many disadvantages :

1. It is larger, heavier and costlier than *RC* coupling.
2. In order to prevent the magnetic field of the coupling inductor from affecting the signal, the inductor turns are wound on a closed core and are also shielded.
3. Since inductor impedance depends on frequency, the frequency characteristics of this coupling are not as good as those of *RC* coupling. The flat part of the frequency ***versus*** gain curve is small (Fig. 61.13).

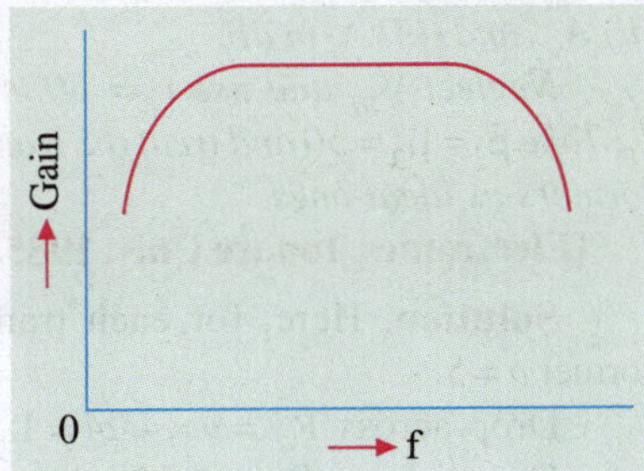

Fig. 61.13

At low frequencies, the gain is low due to large capacitance offered by the coupling capacitor just as in *RC* coupled amplifiers. The gain ***increases with frequency*** till it levels off at the middle frequencies of the audio range.

At relatively high frequencies, gain drops off again because of the increased reactance. Hence, impedance coupling is rarely used beyond audio range.

61.7. Transformer-coupled Two Stage Amplifier

The circuit for such a cascaded amplifier is shown in Fig. 61.14. T_1 is the coupling transformer whereas T_2 is the output transformer. C_1 is the input coupling capacitor whereas C_2, C_3 and C_4 are the bypass capacitors. Resistors R_1 and R_2 as well as R_4 and R_5 form voltage divider circuits whereas R_3 and R_6 are the emitter-stabilizing resistors.

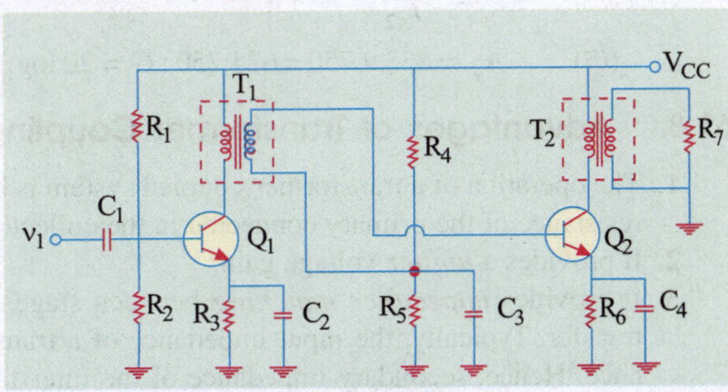

Fig. 61.14

(i) Circuit Operation

When input signal is coupled through C_1 to the base of Q_1, it appears in an amplified form in the primary of T_1. From there, it is passed on to the secondary by magnetic induction. Moreover, T_1 provides dc isolation between the input and output circuits. The secondary of T_1 applies the signal to the base of Q_2 which appears in an amplified form in the primary of T_2.

From there, it is passed on to the secondary by magnetic induction and finally appears across the matched load R_7.

(ii) Voltage Gain

$A_{v.1} = \dfrac{r_{0.1}}{r_{e.1}} \qquad \text{Now} \quad r_{0.1} = a^2 r_{i.2} \qquad \text{— where } a = N_1/N_2 \text{ for } T_1$

$r_{i.2} = R_4 \parallel R_5 \parallel \beta_2 . r_{e.2}$

Similarly, $A_{v2} = \dfrac{r_{0.2}}{r_{e.2}}$

where $r_{e.2} = a^2 R_7$

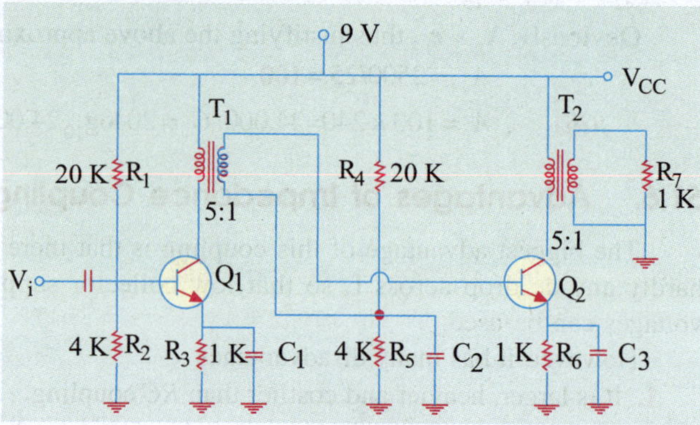

Fig. 61.15

Example 61.5. *For the transformer-coupled two-stage amplifier shown in Fig. 61.15, calculate (i) $A_{v.1}$ (ii) $A_{v.2}$ and (iii) A_v in dB.*

Neglect V_{BE} and use $r_e = 50\ mV/I_E$. Take $\beta_1 = \beta_2 = 50$ and treat the transformers as ideal ones.

(Electronics, Indore Univ. 1985)

Solution. Here, for each transformer $a = 5$

Drop across $R_2 = 9 \times 4/24 = 1.5$ V

Drop across $R_3 \cong 1.5$ V and $I_{E.1} = 1.5/1 = 1.5$ mA

$r_{e.1} = 50/1.5 = 33.3\ \Omega$

and $r_{e.2} = 33.3\ \Omega$ — same as $r_{e.1}$

$\beta_2 . r_{e.2} = 50 \times 33.3 = 1667\ \Omega$

$r_{i.2} = R_4 \parallel R_5 \parallel \beta_2 . r_{e.2}$

$= 20\ K \parallel 4\ K \parallel 1667\ \Omega = 1110\ \Omega$

$r_{0.1} = a^2 . r_{i.2} = 5^2 \times 1110 = 27{,}750\ \Omega$

(i) $A_{v.1} = \dfrac{r_{0.1}}{r_{e.1}} = \dfrac{27{,}750}{33.3} \cong$ **833**

(ii) $A_{v.2} = \dfrac{r_{0.2}}{r_{e.2}} = \dfrac{25 \times 1000}{33.3}$ **750**

(iii) $A_v = 833 \times 750 = 624{,}750$; $G_v = 20 \log_{10} 624{,}750 =$ **116 dB**

61.8. Advantages of Transformer Coupling

1. The operation of a transformer-coupled system is basically ***more efficient*** because of low dc resistance of the primary connected in the collector circuit,
2. It provides a ***higher*** voltage gain,
3. It provides ***impedance matching*** between stages which is desirable for maximum power transfer. Typically, the input impedance of a transistor stage is less than its output impedance. Hence, secondary impedance of the interstage (or coupling) transformer is typically lower than the primary impedance.

This coupling is effective when the final amplifier output is fed to a low-impedance load. For example, the impedance of a typical loud-speaker varies from 4 Ω to 16 Ω whereas output impedance of a transistor stage is several hundred ohms. Use of an output audio transformer can avoid the bad effects of such a mismatch.

Disadvantages

1. The coupling transformer is costly and bulky particularly when operated at audio frequencies because of its heavy iron core,
2. At radio frequencies, the inductance and winding capacitance present lot of problems,
3. It has poor ***frequency response*** because the transformer is frequency sensitive. Hence, the frequency range of the transformer-coupled amplifiers is limited.
4. It tends to introduce ***'hum'*** in the output,

61.9. Frequency Response

The characteristics of a coupling transformer are that *(i)* it introduces inductances in both the input and output circuits *(ii)* leakage inductance exists between the primary and secondary windings and *(iii)* both windings introduce shunting (distributed) capacitance especially at high frequencies.

A typical gain *versus* frequency curve for a transformer-coupled amplifier is shown in Fig. 61.16.

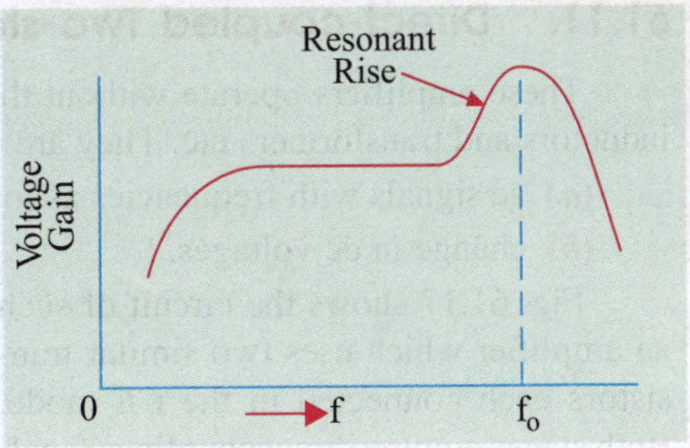

Fig. 61.16

It is seen that

1. There is decrease in gain at low frequencies and
2. Also there is decrease in gain at high frequencies except for the ***resonant rise*** in gain at resonant frequency of the tuned circuit formed by inductance and winding capacitance in the circuit.

The output voltage is equal to ac collector current multiplied by the primary reactance of the coupling transformer. Since at low frequencies, primary reactance is small, the gain is less.

At high frequencies, the distributed capacitance existing between different turns of the winding acts as a ***bypass capacitor*** and so reduces the output voltage and hence the gain. The peak or exaggerated gain occurs due to resonance or tuning effect of inductance and distributed capacitance which ***form a tuned circuit.***

Moreover, there is frequency distortion *i.e.* all frequencies are not amplified equally. In fact, the flat response part of the curve is small as compared to *RC* coupling. However, transformer-coupled amplifiers can be designed to have a flat frequency response curve and excellent fidelity over the entire audio frequency range.

61.10. Applications

Transformer coupling is often employed in the last stage of a multistage amplifier where concerted effort is made to maximise power transfer by perfect impedance matching.

Example 61.6. *In a multistage transformer-coupled amplifier, the output impedance of the first stage is 5 K and the input impedance of the second stage is 1 K. Determine the primary and secondary inductances of the tranformer for perfect impedance matching at f = 2000 Hz. If one turn gives an inductance of 10 μH, find the number of primary and secondary turns.*

(Electronic Engg.-I, Osmania Univ. 1991)

Solution. It should be clearly understood that primary has to match with the output impedance of the first stage and secondary with the input of the second stage.

$\therefore \quad X_{Lp}$ = output of 1st stage or $2\pi f L_p = 5000$

$\therefore \quad L_p = 5000/2\pi \times 2000 =$ **0.4 H**

Also, $\quad X_{Ls}$ = input of 2nd stage

$\therefore \quad 2\pi f L_s = 1000 \,;\, L_s = 1000/2\pi \times 2000 =$ **0.08 H**

Now, inductance of a coil varies as the square of its turns.

$\therefore \quad L \propto N^2 = kN^2$

When, $N = 1$, $\quad L = 10\,\mu\text{H} \quad \therefore \quad 10 \times 10^{-6} = k \times 1^2$ or $k = 10^{-5}$

For primary winding $0.4 = 10^{-5} Np^2$ or $N_p =$ **632**

For secondary winding, $\quad L_s = kN_s^2$ or $0.08 = 10^{-5} Ns^2$ or $N_s =$ **89**

As seen, it is a nearly 7 : 1 step-down transformer.

61.11. Direct-coupled Two-stage Amplifier Using Similar Transistors

These amplifiers operate without the use of frequency-sensitive components like capacitors, inductors and transformers etc. They are especially suited for amplifying

(*a*) ac signals with frequencies as low as a fraction of a hertz,

(*b*) change in dc voltages.

Fig. 61.17 shows the circuit of such an amplifier which uses two similar transistors each connected in the *CE* mode. Both stages employ direct coupling (*i*) collector of Q_1 is connected directly to the base of Q_2 and (*ii*) load resistor R_2 is connected to the collector of Q_2. The resistor R_1 establishes the forward bias of Q_1 and also indirectly that of Q_2.

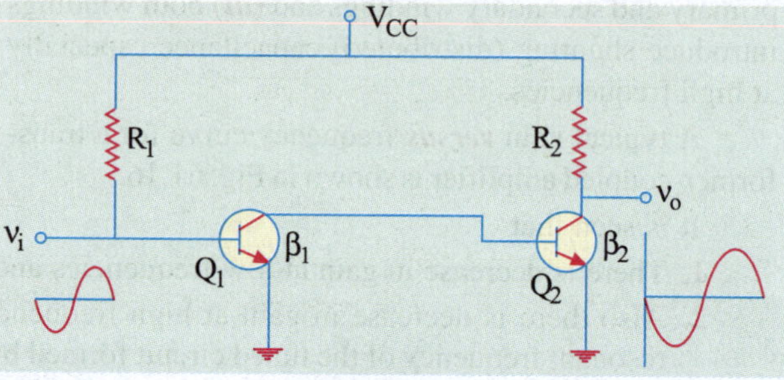

Fig. 61.17

Any signal current at the base of Q_1 is amplified β_1 times and appears at the collector of Q_1 and becomes base signal for Q_2. Hence, it is further amplified β_2 times. Obviously, signal current gain of the amplifier is

$$A_i = \beta_1 \times \beta_2 = \beta^2 \quad \text{— if transistors are identical}$$

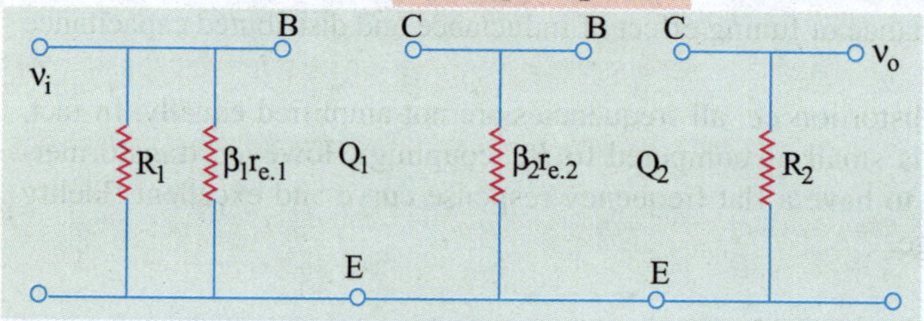

Fig. 61.18

(*i*) AC Equivalent Circuit

The ac equivalent circuit is shown in Fig. 61.18

(*ii*) Voltage Gain

The first stage *i.e.* Q_1 does not produce any voltage gain *i.e.* $A_{v \cdot 1} = 1$ as proved below :

$$A_{v.1} = \frac{r_{0.1}}{r_{e.1}} = \frac{\beta_2 \cdot r_{e.2}}{r_{e.1}} \qquad \text{... (i)}$$

As seen from Fig. 61.17, $I_{E.2} \cong \beta_2 \, . \, I_{B.2}$

Also $\quad I_{B.2} = I_{C.1} = I_{E.1} \quad \therefore \quad I_{E.2} = \beta_2 . I_{E \cdot 1}$ or $\beta_2 = \dfrac{I_{E.2}}{I_{E.1}}$

Also $\quad r_{e.1} = \dfrac{25}{I_{E.1}}$ and $r_{e.2} = \dfrac{25}{I_{E.2}}$

Substituting all these values in Eq. (*i*) above, we get

$$A_{v.1} = \frac{I_{E.2}}{I_{E.1}} \times \frac{25}{I_{E.2}} \times \frac{I_{E.1}}{25} = 1$$

It proves the statement made above.

Now, $\quad A_{v.2} = \dfrac{r_{0.2}}{r_{e.2}} \quad$ and $\quad A_v = A_{v.1} \times A_{v.2}$

(*iii*) Advantages

1. The circuit arrangement is very simple since it uses ***minimum number of components.***

2. It is quite ***inexpensive.***

3. It has the outstanding ability to ***amplify direct current*** (*i.e.* as dc amplifier) and ***low-frequency signals.***

4. It has ***no coupling or by-pass capacitors*** to cause a drop in gain at low frequencies. As seen from Fig. 61.19, the frequency-response curve is flat upto upper cut-off frequency determined by stray wiring capacitance and internal transistor capacitances.

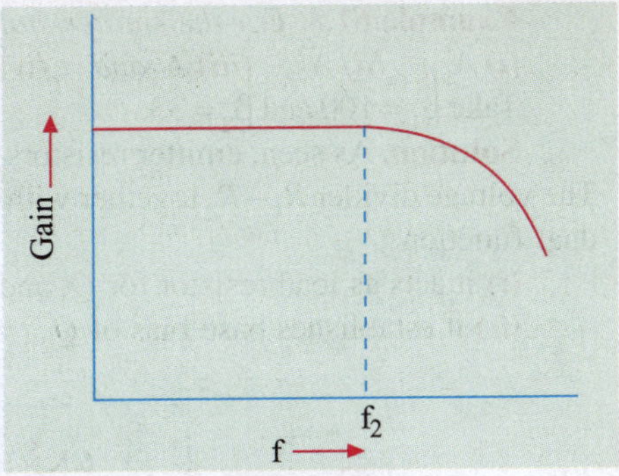

Fig. 61.19

(*iv*) Disadvantages

1. It cannot amplify high-frequency signals.

2. It has poor temperature stability.

It is due to the fact that any variation in base current (due to temperature changes) in one stage is amplified in the following stage (or stages) thereby shifting the Q-point. However, stability can be improved by using emitter-stability resistor (Fig. 61.21).

(*v*) Applications

Some of the applications of direct-coupled amplifiers are in

1. regulator circuits of electronic power supplies,

2. pulse amplifiers

3. differential amplifiers,

4. computer circuitry,

5. electronic instruments.

Example 61.7. *For the direct-coupled amplifier of Fig. 61.20, calculate*

(*a*) *current gain ,* (*b*) *voltage gain of first stage,*

(*c*) *voltage gain of second stage,* (*d*) *overall voltage gain in dB,*

(*e*) *overall power gain in dB,* (*f*) *input resistance.*

Neglect V_{BE} *and use* $r_e = 50\ mV/I_E$.

Solution. (*a*) $A_i = \beta_1\beta_2 = 100 \times 50 =$ **5000** (*b*) $A_{v.1} = 1$ — Art 61.11

(*c*) $A_{v.2} = \dfrac{r_{0.2}}{r_{e.2}}$ Now, $r_{e.2} = \dfrac{50}{I_{E.2}}$

Let us find the value of $I_{E.2}$ starting from the value of $I_{B.1}$. As seen from Fig. 61.20,

$I_{B.1} = 12/1.2M = 10\ \mu A$

$I_{C.1} = \beta_1 . I_{B1} = 100 \times 10 = 1000\ \mu A$

$I_{E.1} \cong I_{C.1} = 1000\ \mu A = 1\ mA$

$I_{B.2} = I_{C.1} = 1\ mA$

$I_{C.2} = \beta_2 . I_{B2} = 50 \times 1 = 50\ mA$

$\therefore$ $I_{E.2} = 50\ mA$

and $r_{e.2} = 50/50 = 1\ \Omega$

Also, $r_{o2} = R_2 = 200\ \Omega$

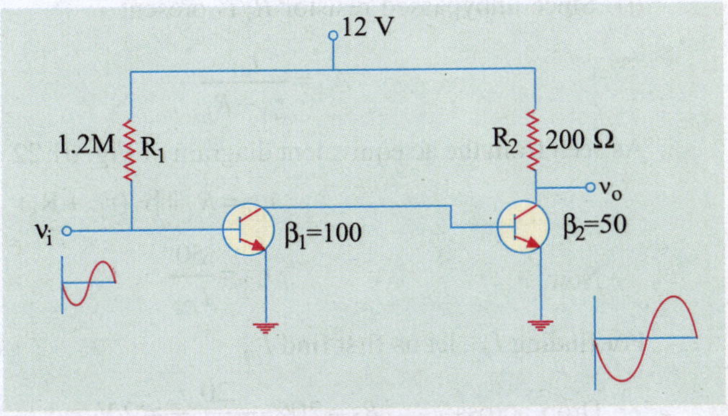

Fig. 61.20

$\therefore$ $A_{v.2} = \dfrac{200}{1} =$ **200**

(*d*) $A_v = 1 \times 200 = 200$; $G_v = 20 \log_{10} 200 =$ **46 dB**

(*e*) $A_p = A_v . A_i = 200 \times 5{,}000 = 10^6$; $Gp = 10 \log_{10} 10^6 =$ **60 dB**

(*f*) $r_i = R_1 \parallel \beta_1 . r_{e.1}$ — Fig. 61.18

$r_{e.1} = \dfrac{50}{I_{E.1}} = \dfrac{50}{1} = 50\ \Omega$ $\therefore$ $r_i = 1.2\ M \parallel (50 \times 100) \cong$ **5 K**

Example 61.8. *For the emitter-stabilized direct-coupled amplifier of Fig. 61.21, find (i) A_{v1} (ii) A_{v2} (iii) A_v and (iv) r_i. Neglect V_{BE} and use $r_e = 50\ mV/I_E$.*
Take $\beta_1 = 100$ and $\beta_2 = 33$. **(Electronics-I, Mysore Univ. 1992)**

Solution. As seen, emitter resistors R_3 and R_5 have been used to improve temperature stability. The voltage divider $R_1 - R_2$ together with R_3 determines the emitter current of Q_1. The resistor R_4 has dual function

(*i*) it acts as load resistor for Q_1 and

(*ii*) it establishes base bias of Q_2.

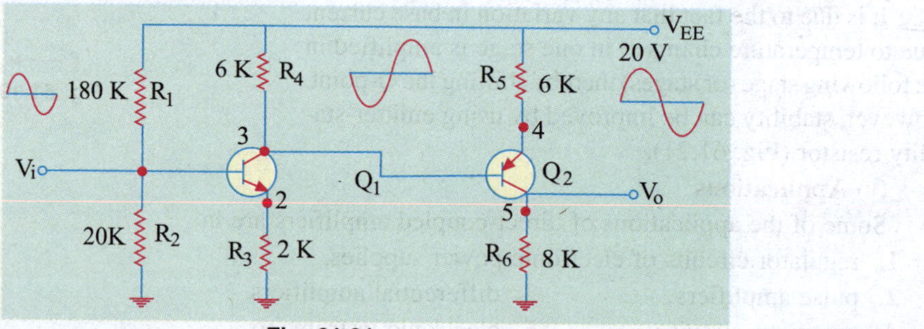

Fig. 61.21

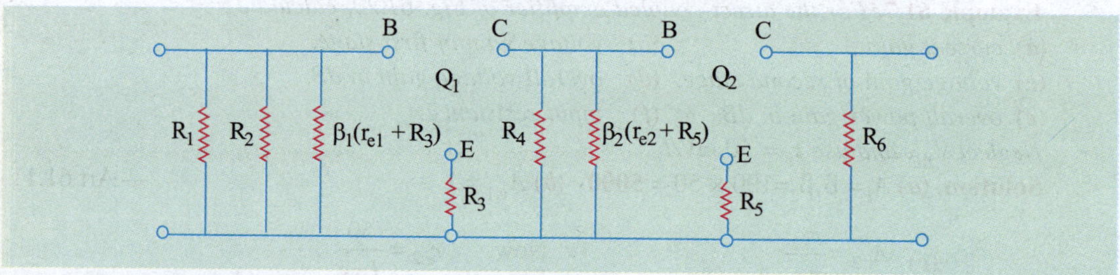

Fig. 61.22

(*i*) Since unbypassed resistor R_3 is present

$$A_{v1} = \frac{r_{01}}{r_{e1} + R_3} \qquad \text{... (i)}$$

As seen from the ac equivalent diagram of Fig. 61.22

$$r_{01} = R_4 \parallel \beta_2 (r_{e2} + R_5) \qquad \text{... (ii)}$$

Now $$r_{e2} = \frac{50}{I_{E2}}$$

For finding I_{E2}, let us first find I_{E1}

Drop across $$R_2 = 20 \times \frac{20}{180 + 20} = 2\ V$$

Same is the drop across R_3 since V_{BE} has been neglected.

$\therefore$ $$I_{E1} = \frac{2V}{2K} = 1mA$$

Now,

Voltage at point 3 = $R_3 = V_{BE} - I_{E1}.R_4$

$$= 20 - (1\ mA)\ 66k = 14\ V$$

Same is the voltage at point 4 *i.e.*, $V_4 \cong V_3 = 14\text{ V}$

$$I_{E2} = \frac{V_{CC} - R_4}{R_5} = \frac{20V - 14V}{6K} = 1\text{ mA}$$

$$r_{e2} = \frac{50}{1}\Omega = 50\Omega$$

$$\therefore \quad \beta_2(r_{e2} + R_5) = 33\,(50 + 6000) \cong 200\text{ K}\Omega$$

From equation (*ii*)

$$r_{01} = R_4 \,||\, \beta_2(r_{e2} + R_5)$$

$$= 6\text{K} \,||\, 200\text{ K} = 5.83\text{ K}$$

$$r_{e1} = \frac{50}{I_{E1}} = \frac{50}{1} = 50\Omega$$

From equation (*ii*)

$$\therefore \quad A_{v1} = \frac{5830}{(50 + 2000)} = 2.84$$

(*ii*) $A_{v2} = \dfrac{r_{02}}{(r_{e2} + R_5)} = \dfrac{8000}{(50 + 6000)} = 1.32 \quad \because\ r_o = R_6$, See fig. 61.22)

(*iii*) $A_v = A_{v1} \times A_{v2} = 2.84 \times 1.32 = \mathbf{3.75}$

(*iv*) As seen from Fig. 61.22

$$r_i = R_1 \,||\, R_2 \,||\, \beta_1(r_{e1} + R_3) = 180\text{ K} \,||\, 20\text{ K} \,||\, 200\text{ K} = 16.51\text{K} \text{ — neglecting } r_{e1}$$

$$= \mathbf{16.51\ K}$$

Example. 61.9. *In the circuit of Fig. 61.21, find dc voltage at points marked 1, 2, 3 and 4. Neglect V_{BE}.*

If a 1-V dc signal at input 1 changes by 0.01 V, what would be the voltage variation at the output?

Solution. V_1 = drop across $R_2 = 20 \times \dfrac{20}{180 + 20} V = \mathbf{2\ V}$

$V_2 \cong V_1 = \mathbf{2\ V}$;

$V_3 = V_{EE} - I_{E1} R_4 = 20 - (1\text{ mA})(6\text{ K}) = 14\text{ V}$

$V_4 \cong V_3 = 14\text{ V}$

$V_5 = I_{C2} R_6 \cong I_{E2} R_6 = (1\text{ mA})(8\text{K}) = 8\text{V}$

$\Delta V_0 = A_V . \Delta V = 3.75 \times 0.01 = 0.0375\text{ V} = 37.5\text{ mV}.$

61.12. Direct-coupled Amplifier Using Complementary Symmetry of Two Transistors

In this case, an *NPN* transistor is directly-coupled to its complementary *i.e.* a *PNP* transistor. Fig. 61.23 shows a two-stage cascaded amplifier using two complementary transistors connected in CE configuration.

The circuit differs from that shown in Fig. 61.17 in the following three ways :

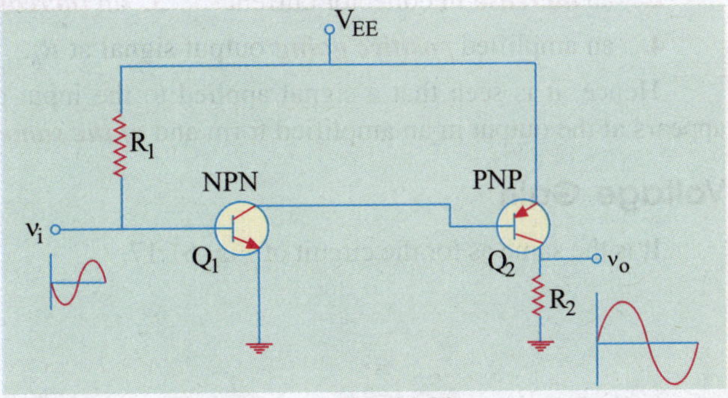

Fig. 61.23

1. It uses **complementary** transistors rather than **similar** ones,
2. Instead of V_{CC}, V_{EE} power battery has been used,
3. Output is taken directly from terminal of load resistor R_2.

More Practical Circuit

A more practical circuit of the above type is shown in Fig. 61.24. Here, bias current of Q_1 is determined by voltage divider $R_1 - R_2$ and R_3. As before, R_4 performs two functions :

1. It acts as load for Q_1 and **2.** Establishes bias voltage for Q_2

The emitter resistors R_3 and R_6, as usual, meant to improve amplifier stability.

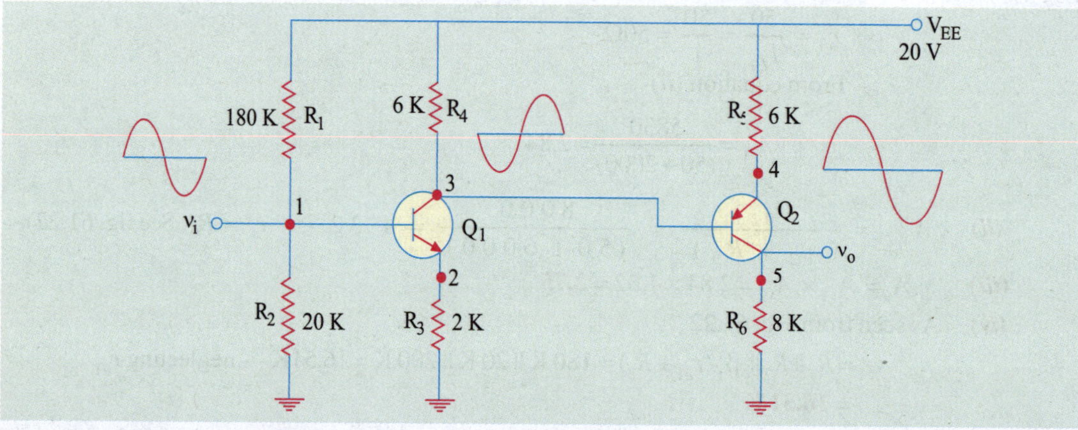

Fig. 61.24

(i) Circuit Operation

When a positive-going signal is applied to the base of Q_1, then

1. its base current ***increases,***
2. hence, its collector current increases ($I_C = \beta I_B$),
   voltage drop across R_4 ***increases,***
4. consequently, voltage at the collector of Q_1 and the base of Q_2 becomes ***less positive*** or in other words, ***more negative,***
5. hence, a negative-going signal is applied to the base of Q_2.

The negative-going signal applied to the base of Q_2 causes,

1. an ***increase*** in its forward bias (remember, it is a *PNP* transistor),
2. an ***increase*** in collector current
3. an ***increase*** in the voltage developed across R_6,
4. an amplified ***positive going*** output signal at R_6.

Hence, it is seen that a signal applied to the input of a two-stage complementary amplifier appears at the output in an amplified form and ***of the same polarity.***

Voltage Gain

It is the same as for the circuit of Fig. 61.17.

Example 61.10. *For the complementary symmetry circuit of Fig. 61.25, find* **(a)** A_v **(b)** V_1, V_2, V_3, V_4 *and* V_5. *Neglect* V_{BE} *and assume* $R_3, R_5 \gg r_e$.

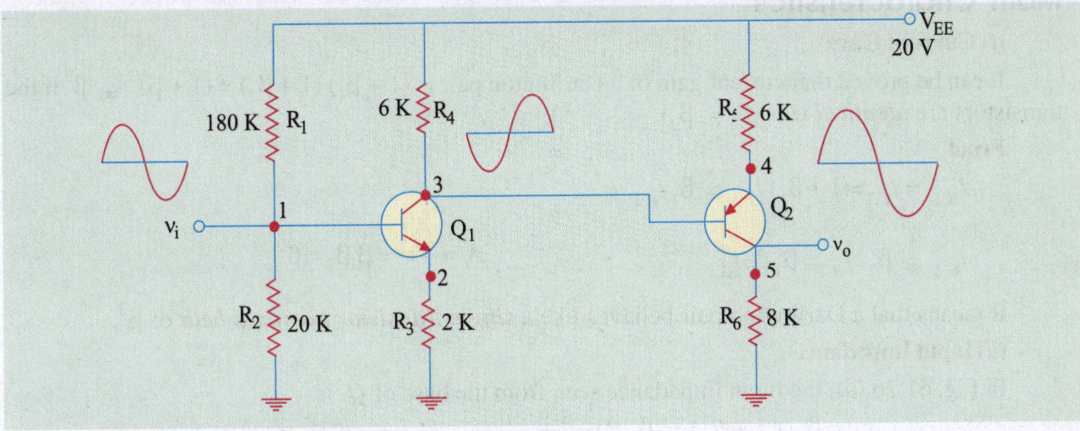

Fig. 61.25

Solution. (*a*) The overall voltage gain is

$$A_v = A_{v.1} \times A_{v.2} \cong \frac{R_4}{R_3} \times \frac{R_6}{R_5} = \frac{6}{2} \times \frac{8}{6} = 4$$

(*b*) $$V_1 = V_{EE} \frac{R_2}{R_1 + R_2} = 20 \frac{20}{180+20} = 2V \qquad \therefore\ V_2 \cong V_1 = \textbf{2V}$$

$$\therefore \quad I_{E.1} = \frac{2V}{R_3} = \frac{2V}{2K} = 1\,\text{mA}, \qquad I_{C1} \cong I_{E.1} = 1\,\text{mA}$$

$$V_3 = V_{EE} - I_{C.1} R_3 = 20 - 1 \times 6 = \textbf{14 V}, \qquad V_4 \cong V_3 = \textbf{14 V};$$

$$I_{C.2} = \frac{20-14}{6\text{K}} = 1\,\text{mA}$$

$$\therefore \quad I_{E.2} \cong I_{C.2} = 1\,\text{mA}$$

$$\therefore\ V_5 = I_{E.2} \times R_6$$
$$= 1 \times 8 = \textbf{8 V}$$

61.13. Darlington Pair

It is the name given to a ***pair of similar*** transistors so connected that emitter of one is directly joined to the base of the other as shown in Fig. 61.26 (*a*). Obviously, the emitter current of Q_1 becomes the base current of Q_2.

Darlington pairs are commercially mounted in a single package that has only three leads : base, collector and emitter as shown in Fig. 61.26 (*b*). It often forms a ***double CC stage***

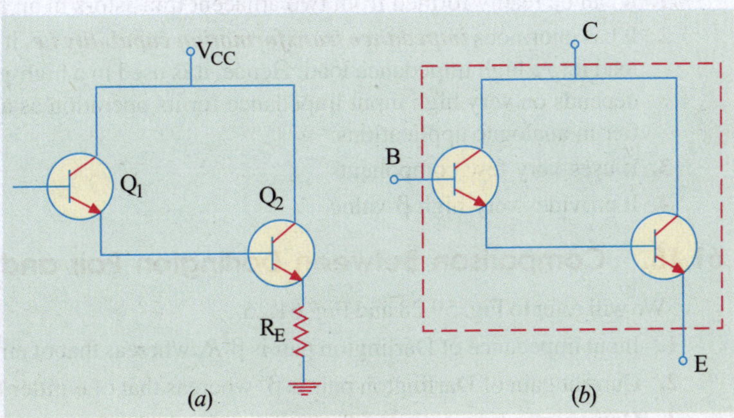

Fig. 61.26

in multistage amplifiers. It is so because a Darlington connection can be considered equivalent to two ***cascaded emitter followers.***

Main Characteristics

(i) Current Gain

It can be proved that current gain of a Darlington pair is $(1 + \beta_1)(1 + \beta_2) = (1 + \beta)^2 \cong \beta^2$ if the transistors are ***identical*** (*i.e.* $\beta_1 = \beta_2$).

Proof

$$I_{B.2} = I_{E.1} = (1 + \beta_1) I_{B.1} \cong \beta_1 I_{B.1}$$

$$I_{E.2} \cong \beta_2 . I_{B.2} = \beta_1 \beta_2 I_{B.1} \qquad \therefore \qquad A_i = \frac{I_{E.2}}{I_{B.1}} = \beta_1 \beta_2 = \beta^2$$

It means that a Darlington pair behaves like a ***single transistor having a beta*** of β^2.

(ii) Input Impedance

In Fig. 61.26 (*a*), the input impedance seen from the base of Q_2 is

$$r_{i.2} = \beta_2 (r_{e.2} + R_E) \cong \beta_2 . R_E$$

Input impedance as seen from the base of Q_1 is

$$r_{i.1} = \beta_1 (r_{e.1} + r_{i.2}) = \beta_1 r_{e.1} + \beta_1 r_{i.2} = \beta_1 r_{e.1} + \beta_1 \beta_2 R_E \cong \beta_1 \beta_2 R_E$$

or $$r_{in(base)} \text{ of } Q_1 = \beta^2 R_E$$

Note. If there is a load resistance R_L coupled to the emitter of Q_2, then

$$r_{i.1} = \beta^2 (R_E \| R_L) = \beta^2 . r_E$$

As seen, load impedance R_E has been transformed into $\beta^2 R_E$. Obviously, a Darlington pair is capable of high input impedance. In fact, whenever a load causes a severe loss in voltage gain (loading effect), it is usual to step up load impedance *via* a *FET* stage, a single *CC* stage or Darlington pair when much greater impedance transformation is required.

(iii) Voltage Gain

Assuming $r_{e.1} = r_{e.2} = r_e$ we have

$$A_v \cong \frac{R_E}{r_e + R_E} = \frac{1}{1 + \frac{r_e}{R_E}} \cong 1$$

—as in an emitter follower

61.14. Advantages of Darlington Pair

1. It can be readly formed from two adjacent transistors in an IC.
2. It has enormous ***impedance transformation capability i.e.*** it can transform a low-impedance load into a high impedance load. Hence, it is used in a high-gain operational amplifier which depends on very high input impedance for its operation as an integrator or summing amplifier in analogue applications.
3. It uses very few components.
4. It provides very high β-value.

61.15. Comparison Between Darlington Pair and Emitter Follower

We will refer to Fig. 59.28 and Fig. 61.26.

1. Input impedance of Darlington pair is $\beta^2 R_E$ whereas that of emitter follower is βR_E (Art. 59.8).
2. Current gain of Darlington pair is β^2 whereas that of emitter follower is β.
3. However, voltage gains of the two are identical.

Example. 61.11. *For the Darlington pair shown in Fig. 61.27, calculate the value of (i) β, (ii) input impedance,(iii) voltage amplification.*

Assume $\beta_1 = \beta_2 = 100$ and $R_L \gg (r_{e.1} + r_{e.2})$.

Solution. The approximate values are as under :

1. β of Darlington pair

$$= \beta_1 \times \beta_2$$

$$= 100 \times 100 = \mathbf{10{,}000}$$

2. $r_E = R_E \| R_L$

$$= 10\text{ K} \| 500\ \Omega = 476\ \Omega$$

$r_{in(base)}$ of $Q_1 \cong \beta^2 r_E$

$$= 10{,}000 \times 476 = 4.76\text{ M}$$

r_i = input impedance of the pair

$= R_B \| r_{in(base)}$ of Q_1

$$= 1\text{ M} \| 4.76\text{ M} = 0.826\text{ M} = \mathbf{826\text{ K}}$$

3. $A_v \cong 1$

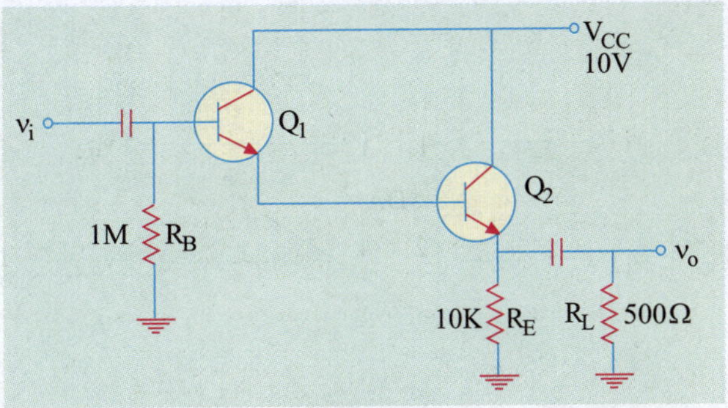

Fig. 61.27

Example 61.12. *A CE amplifier stage shown in Fig.61.28 is to drive a 100 Ω load to 10 $V_{p\text{–}p}$ level. An input signal of 1 $V_{p\text{–}p}$ is available. Find out if the stage is overloaded or not. Also, find out if this overloading has been avoided by using a Darlington pair as a buffer between Q_1 and the load as shown in Fig. 61.29. Take $\beta_1 = \beta_2 = 50$.*

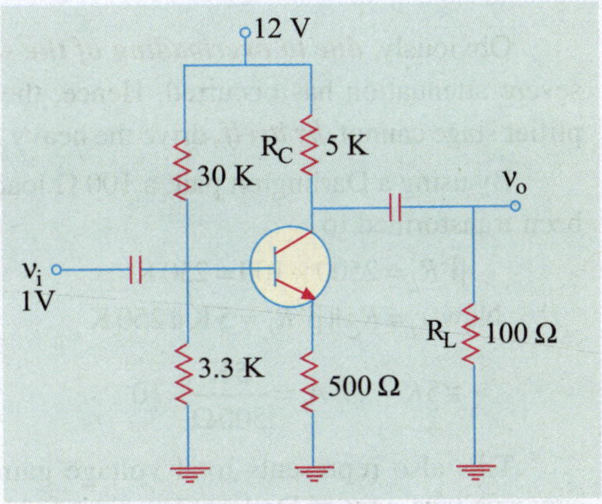

Fig. 61.28

Solution. The approximate voltage gain of Q_1 is given by

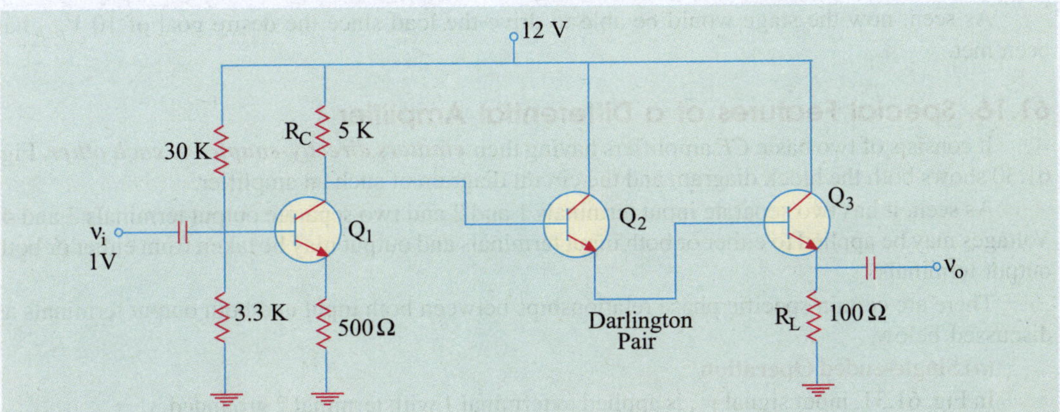

Fig. 61.29

$$A_v = \frac{r_L}{r_E} = \frac{R_C \| R_L}{R_E} = \frac{5K \| 100\,\Omega}{500\,\Omega} \cong \frac{1}{5}$$

$$\therefore \quad v_0 = 1 \times \frac{1}{5} = 0.2\,\text{mV}$$

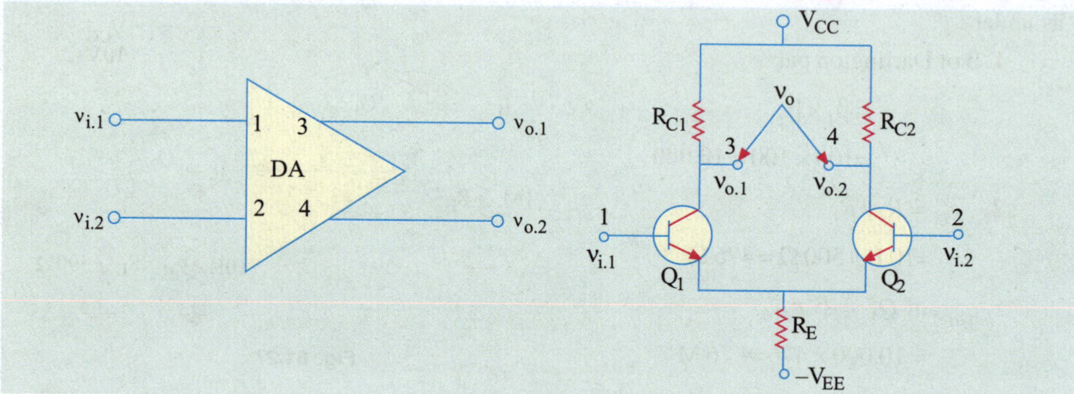

Fig. 61.30

Obviously, ***due to overloading of the stage,*** severe attenuation has occurred, Hence, the amplifier stage cannot, ***by itself,*** drive the heavy load.

By using a Darlington pair, a 100 Ω load has been transformed to

$$\beta^2 R_L = 2500 \times 100 = 250\text{ K}$$

Now, $r_L = R_C \| \beta^2 R_L = 5\text{ K} \| 250\text{ K}$

$$\cong 5K \quad \therefore A_v = \frac{5\text{K}}{500\,\Omega} = 10$$

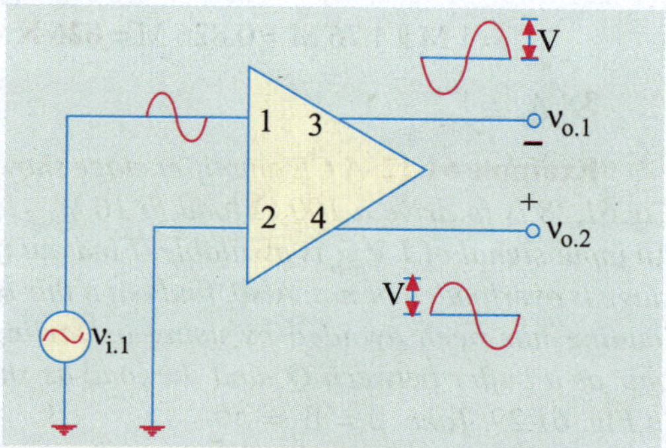

Fig. 61.31

This also represents total voltage gain because voltage gain of Darlington pair is one.

$$\therefore \quad v_0 = A_v \times v_i = 10 \times 1 = \mathbf{10\ Vp\text{–}p}$$

As seen, now the stage would be able to drive the load since the desire goal of 10 V_{P-P} has been met.

61.16. Special Features of a Differential Amplifier

It consists of two basic *CE* amplifiers having their ***emitters directly-coupled to each other.*** Fig. 61.30 shows both the block diagram and the circuit diagram of such an amplifier.

As seen, it has two separate input terminals 1 and 2 and two separate output terminals 3 and 4. Voltages may be applied to either or both input terminals and output may be taken from either or both output terminals.

There are certain specific phase relationships between both input and both output terminals as discussed below.

(*a*) Single-ended Operation

In Fig. 61.31, input signal $v_{i.1}$ is applied to terminal 1 with terminal 2 grounded.

It is seen that an ***amplified*** and ***inverted*** output signal is obtained at terminal 3 (phase inversion of a *CE* amplifier) but an ***equally-amplified*** and ***in-phase*** signal appears across output terminal 4. The differential output voltage has the polarity shown in the figure.

As shown in Fig. 61.32, when input signal v_{i2} is applied to input terminal 2, an amplified and inverted signal appears at output terminal 4 whereas ***equally-amplified*** but ***in-phase*** signal appears at terminal 3.

In summary, we can say that input at any of the two terminals causes outputs at both terminals 3 and 4. The two output are opposite in phase but of equal amplitude.

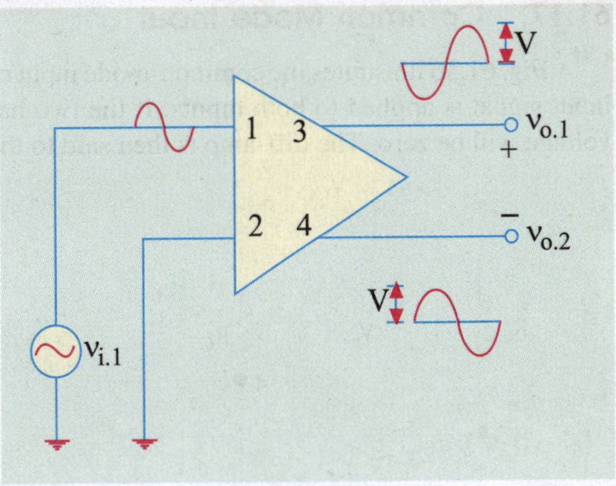

Fig. 61.32

(b) Double-ended Operation

Fig 61.33 illustrates the double ended mode of operation when two input signals of opposite phase are applied to the two input terminals.

Input signal at each input terminal causes signals to appear at both output terminals. The resultant output signals have a peak value of 2 V – twice the value for single-ended operation.

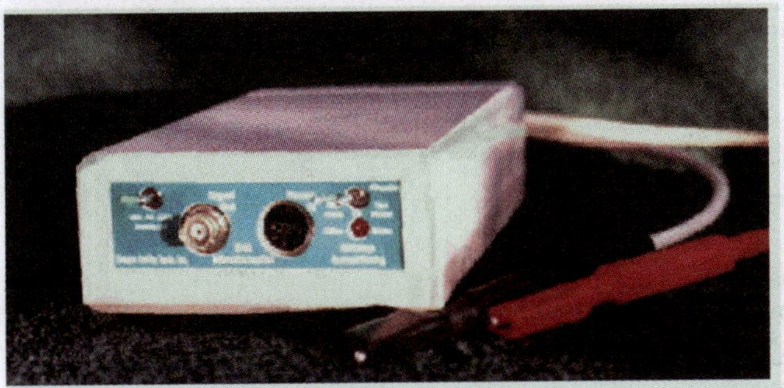

Differential amplifier

However, if two in-phase and equal signals were applied at the two input terminals, the resultant output signal at each output terminal would be zero as shown in Fig. 61.34. It means that output *between* the collectors would be zero.

If $v_{i.1}$ and $v_{i.2}$ change by *exactly* the same amount, even then output voltage between terminals 4 and 3 remains zero because of symmetry. Only when $v_{i.1}$ and $v_{i.2}$ differ from each other, we get an output voltage. When $v_{i.1}$ is more positive than $v_{i.2}$, the output terminal 4 is more positive than terminal 3.

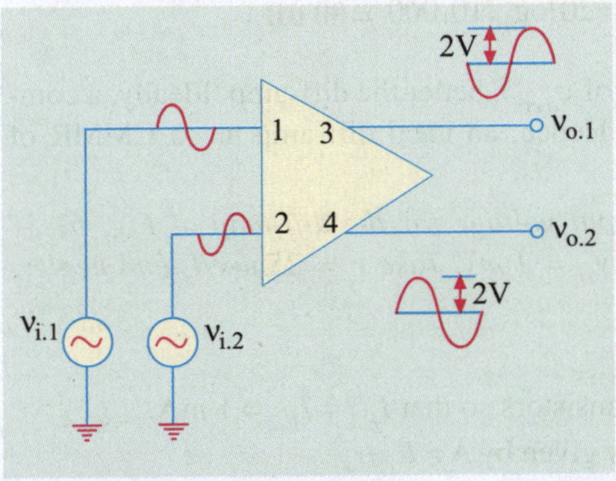

Fig. 61.33

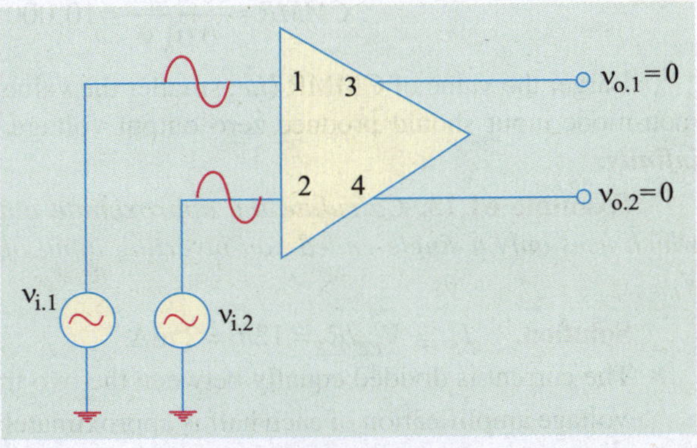

Fig. 61.34

(c) Inverting and Non-inverting Inputs

When positive $v_{i.1}$ acts alone, it produces a differential output voltage with terminal 4 positive with respect to terminal 3 as shown in Fig. 61.35. That is why the input terminal 1 is called non-inverting input terminal. How ever, when positive $v_{i.2}$ acts alone, the output voltage is inverted i.e. terminal 3 becomes positive with respect to terminal 4. That is why input terminal 2 is called ***inverting terminal.***

61.17. Common Mode Input

Fig. 61.36 illustrates the common-mode input of a differential amplifier *i.e.* when similar or same input signal is applied to both inputs. If the two halves of the diff-amp are identical, the ac output voltage will be zero. The diff-amp is then said to the perfectly balanced.

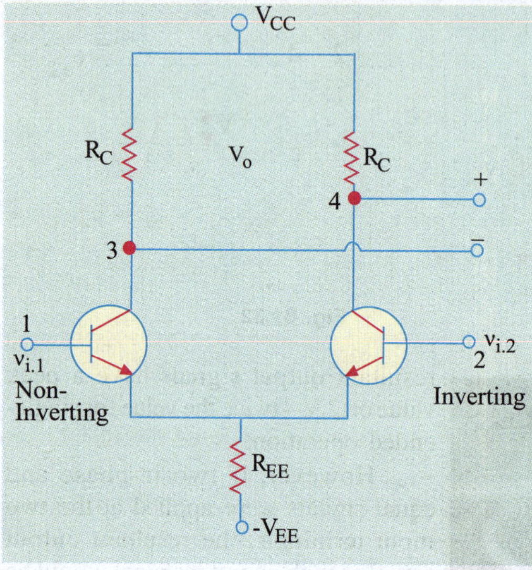

Fig. 61.35

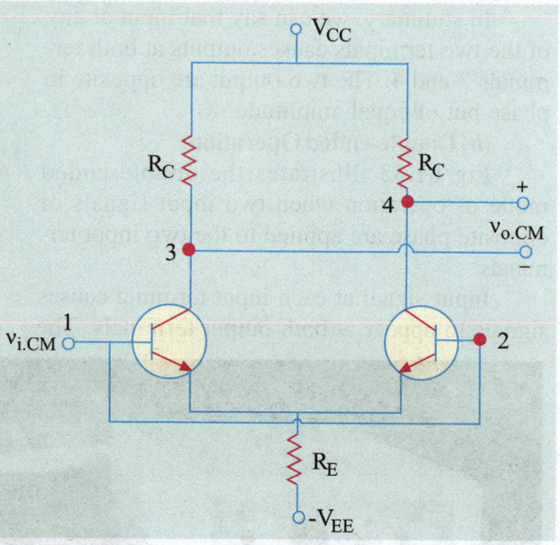

Fig. 61.36

The common-mode rejection ratio (*CMMR*) is defined as

$$CMMR = \frac{A \times v_{i.cm}}{v_{o.cm}}$$

Suppose, $v_{i.cm}$ = 1 V. In a perfectly-balanced and symmetrical diff-amp, the output should be zero. But, in practice, there is always a small output signal because of *non-symmetry*. Suppose, A = 100 and V_{ocm} = 0.01 V. Then

$$CMMR = \frac{100 \times 1\,V}{0.01\,V} = 10{,}000 = 20 \log_{10} 10{,}000 = \mathbf{80\ dB}$$

Larger the value of CMMR (*i.e.* smaller the value of $v_{o.cm}$), better the diff-amp. Ideally, a common-mode input should produce zero output voltage. Hence, an ideal diff-amp has a CMMR of ***infinity***.

Example 61.13. *Calculate the approximate output voltage for the diff-amp of Fig. 61.37 which uses only a single-ended non-inverting input of* v_{i1} *= 1 mV. Take* r_e *= 25 mV/*I_E *and neglect* V_{BE}.

Solution. $I_E \cong V_{EE}/R_E = 12/6 = 2$ mA

The current is divided equally between the two transistors so that $I_{E.1} = I_{E.2}$ = 1 mA.

Voltage amplification of each half is approximately given by $A = R_C/r_e$

Now, $r_e = 25/1 = 25\ \Omega$

$$\therefore \quad A = \frac{12{,}000}{25} = 480$$

$$\therefore \quad v_0 = A \times v_{i.1} = 480 \times 1$$

$$= \mathbf{480\ mV}$$

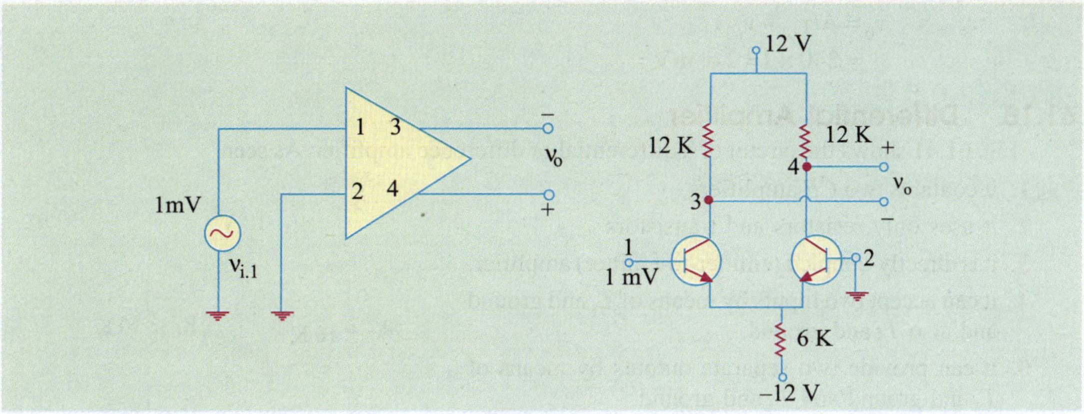

Fig. 61.37

The ac output signal is in phase with the input.

Example 61.14. *If a differential input signal of 1 mV is applied to the diff-amp shown in Fig. 61.38, calculate the output voltage. Neglect V_{BE} and take r_e = 25 mV/I_E.*

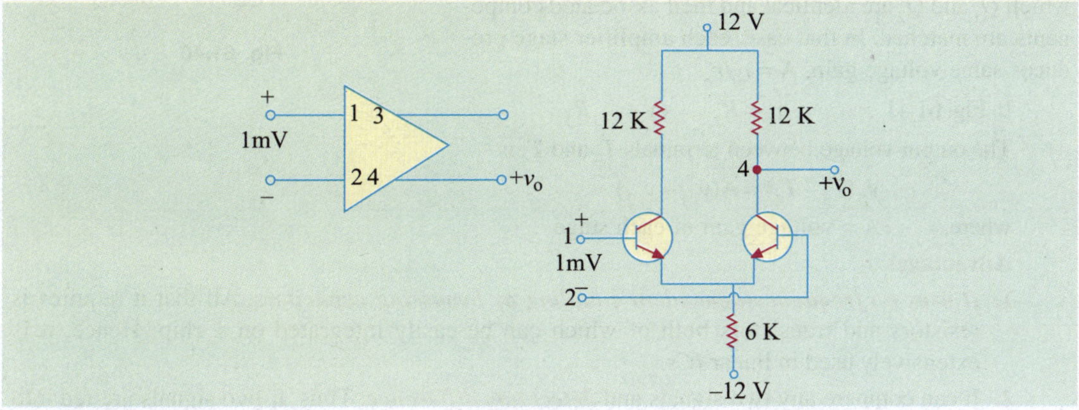

Fig. 61.38

Solution.

Here, $v_{i.1} - v_{i.2} = 1$ mV

As seen from Ex. 61.13,

A = **480**

$\therefore v_o = A(v_{i.1} - v_{i.2}) = 480 \times 1 =$ **480 mV**

Example 61.15. *A differential input signal of 1 mV is applied to the diff-amp of Fig. 61.39 when used in single-ended output mode. Calculate the approximate value of output voltage. Neglect V_{BE} and take r_e = 25 mV/I_E.*

Solution. Here, again 1 mV is the differential input signal *i.e.* $v_{i.1} - v_{i.2} = 1$ mV

Since single-ended mode is being used, A = 480/2 = **240** mV

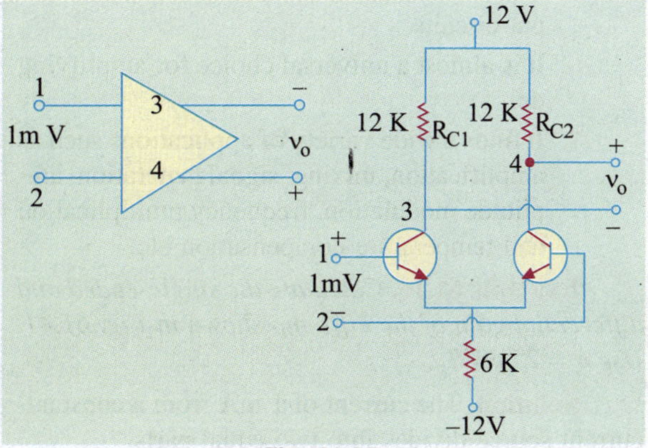

Fig. 61.39

$\therefore \quad v_0 = A(v_{i.1} - v_{i.2})$

$= 240 \times 1 = \mathbf{240\ mV}$

61.18. Differential Amplifier

Fig. 61.41 shows the circuit of a differential or difference amplifier. As seen

1. it contains two *CE* amplifiers,
2. it uses only resistors and transistors
3. it is directly-coupled (emitter-to-emitter) amplifier,
4. it can accept two inputs by means of T_1 and ground and also T_2 and ground,
5. it can provide two separate outputs by means of T_3 and ground and T_4 and ground,
6. it can provide a single output between T_3 and T_4 *i.e.* differential output.

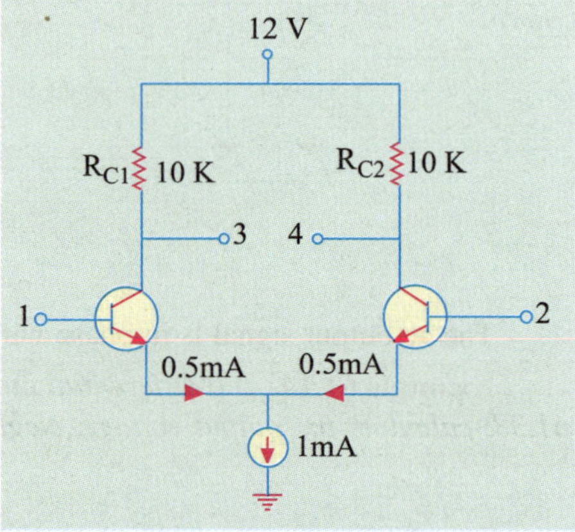

Fig. 61.40

Circuit Operation

We will consider a balanced differential amplifier in which Q_1 and Q_2 are identical and their associated components are matched. In that case, each amplifier stage produces same voltage gain. $A = r_0/r_e$

In Fig. 61.41. $\quad r_o = R_2 \quad$ or $\quad R_3$

The output voltage between terminals T_3 and T_4 is

$$v_0(T_3 - T_4) = A(v_{i.1} - v_{i.2})$$

where, $\quad A$ = voltage gain of each stage

Advantages

1. ***It uses no frequency-dependent coupling or bypassing capacitors.*** All that it requires is resistors and transistors both of which can be easily integrated on a chip. Hence, it is extensively used in linear *ICs*.
2. It can compare any two signals and ***detect any difference.*** Thus, if two signals are fed into its inputs, identical in every respect except that one signal has been slightly distorted, then only the difference between the two signals *i.e.* distortion will be amplified.
3. It gives higher gain than two cascaded stages of ordinary direct coupling.
4. It provides very ***uniform amplification*** of signal from dc upto very high frequencies.
5. It provides isolation between input and output circuits.
6. It is almost a universal choice for amplifying dc.
7. It finds a wide variety of applications such as amplification, mixing, signal generation, amplitude modulation, frequency multiplication and temperature compensation etc.

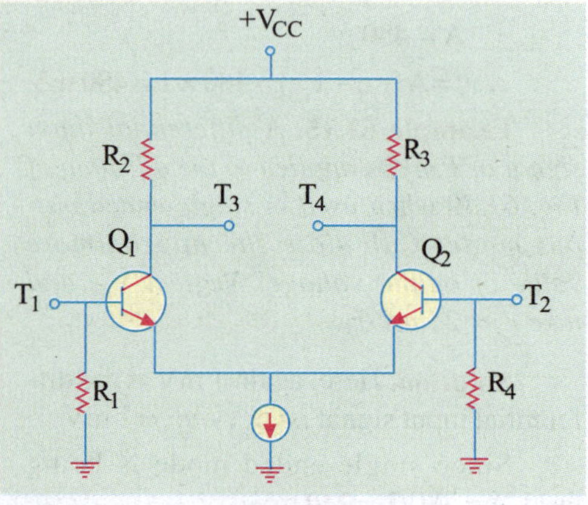

Fig. 61.41

Example 61.16. *Calculate the single-ended and differential gain of the diff-amp shown in Fig. 61.41. Use $r_e = 25\ mV/I_E$.*

Solution. The current of 1 mA from a constant-current source divides into two equal parts.

$$I_{E.1} = I_{E.2} = 1/2 = 0.5 \text{ mA}$$

$$r_{e.1} = r_{e.2} = \frac{25}{0.5} = 50\,\Omega$$

Hence, single-ended voltage gain is

$$A = \frac{r_0}{r_e} = \frac{R_C}{r_e} = \frac{10\,K}{50\,\Omega} = \mathbf{200}$$

Hence, each stage has a voltage gain of 200.

If we consider differential (double-ended) gain, its value is twice *i.e.* $2 \times 200 =$ **400.**

Example 61.17. *For the diff-amp shown in Fig. 61.41, voltage gain of each stage is 200, $v_{i.1}$ = 30 mV and $v_{i.2}$ = 20 mV. Find the voltages between*

(i) T_3 and ground, (ii) T_4 and ground,(iii) T_3 and T_4.

What are the polarities of output terminals T_3 and T_4?

(Basic Electronics, Bombay Univ. 1985)

Solution. As stated earlier in Art. 61.18.

(i) $v_0(T_3) = A_1 \times v_{i.1} = 200 \times 30 \text{ mV} = \mathbf{6\ V}$

(ii) $v_0(T_4) = A_2 \times v_{i.2} = 200 \times 20 \text{ mV} = \mathbf{4\ V}$

(iii) $v_0(T_3 - T_4) = A(v_{i.1} - v_{i.2}) = 200\,(30 - 20) \text{ mV} = \mathbf{2\ V}$

Since $v_{i.1} > v_{i.2}$; $i_{c.1} R_2 > i_{c2} R_3$, hence T_4 will be positive with respect to T_3.

Example 61.18. *Calculate the overall voltage gain of the two-stage RC coupled amplifier shown in Fig. 61.42. Neglect V_{BE} and take $\beta_1 = \beta_2 = 100$.*

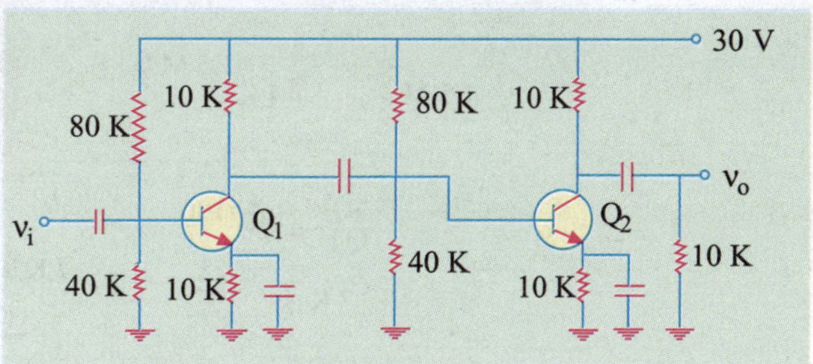

Fig. 61.42

Solution. We will first find gain of Q_2 and then multiply it with that of Q_1 to find the overall gain.

DC voltage from base to ground for Q_2 = drop across 40 K

$$= 30 \times 40/(40 + 80) = 10 \text{ V}$$

Hence, $I_{E.2} \cong 10 \text{ V}/10 \text{ K} = 1 \text{ mA}$

Assuming silicon transistor

$r_{e.2} = 50/1 \text{ mA} = 50\,\Omega$

$$\therefore \quad A_{v2} = \frac{r_{L2}}{r_{e.2}} = \frac{10\text{K} \,||\, 10\text{K}}{50} = 100$$

For finding A_{v1}, we must first calculate the ac load resistance r_L as seen by Q_1. It equals the parallel combination of 10 K, 80 K, 40 K and $\beta r_{el} = 100 \times 50 = 5$ K (because it forms part of the load on Q_1)*.

$\therefore \quad r_{L1} = 10 \text{ K} \,||\, 80 \text{ K} \,||\, 40 \text{ K} \,||\, 5 \text{ K} \cong 3 \text{ K}$

$\therefore \quad A_{v1} = r_{L1}/re_1 = 3 \text{ K}/50\,\Omega = 60$

$\therefore \quad A = A_{v1} \times A_{v2} = 100 \times 60 = \mathbf{6000}$

* (Emitter resistance does not come into the picture because it has been ac grounded by the capacitor.

Tutorial Problems No. 61.1

1. For the two-stage *RC*-coupled amplifier shown in Fig. 61.43 calculate the approximate values of

(*a*) voltage gain for the first stage,
(*b*) voltage gain for the second stage,
(*c*) voltage gain for the amplifier,
(*d*) ac input resistance of the amplifier.

Neglect V_{BE} and use $r_e = 25\text{ mV}/I_E$

[(*a*) 76.2 (*b*) 256 (*c*) – 19,507 or 85.8 dB (*d*) 1250 Ω]

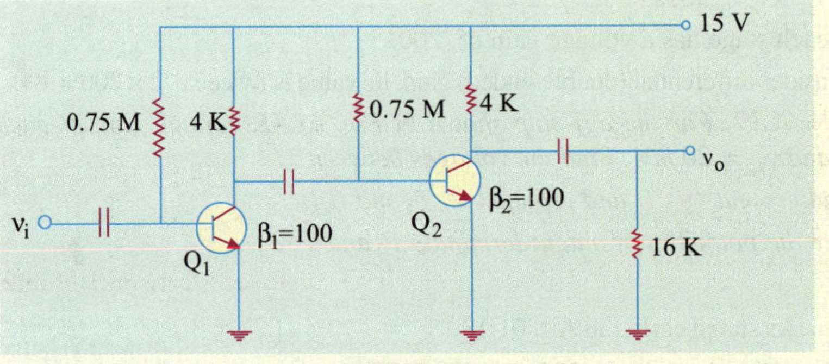

Fig. 61.43

2. For the two-stage *RC*-coupled amplifier using base and emitter bias and shown in Fig. 61.44, calculate approximate values of

(*i*) $A_{v.1}$ (*ii*) $A_{v.2}$ (*iii*) A_v (*iv*) r_i

Neglect V_{BE} and use $r_e = 25\text{ mV}/I_E$

[(*i*) 61.6 (*ii*) 128 (*iii*) 7885 or 78 dB (*iv*) 2.5 K]

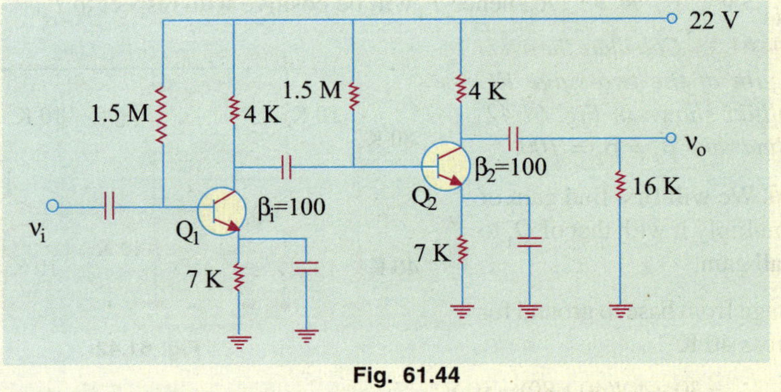

Fig. 61.44

3. Fig. 61.45 shows the circuit of a two-stage impedance-coupled amplifier. Compute the approximate values of

(*i*) voltage gain of 1st stage,
(*ii*) voltage gain of the 2nd stage ; both at *f* = 2 kHz,
(*iii*) voltage gain of the amplifier.

Neglect V_{BE} and take $r_e = 25\text{ mV}/I_E$.

[(*i*) 80 (*ii*) 4255 (*iii*) 30,040 or 90.6 dB]

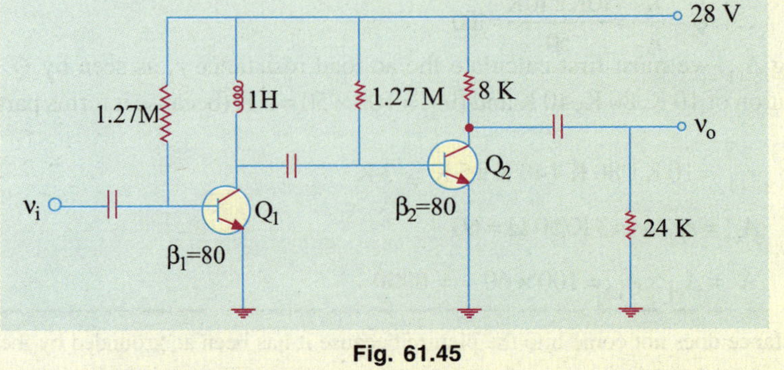

Fig. 61.45

4. For the direct-coupled amplifier shown in Fig. 61.46, calculate the approximate value of (*i*) A_i, (*ii*) $A_{v.1}$, (*iii*) $A_{v.2}$, (*iv*) A_v, (*v*) A_p and (*vi*) r_i

Neglect V_{BE} and assume $r_e = 25\ mV/I_E$. Also, $\beta_1 = 50$ and $\beta_2 = 25$.
[(*i*) 1250 (*ii*) 1 (*iii*) 240 (*iv*) 240 (*v*) 300,000 or 54.8 dB (*vi*) 1250 Ω]

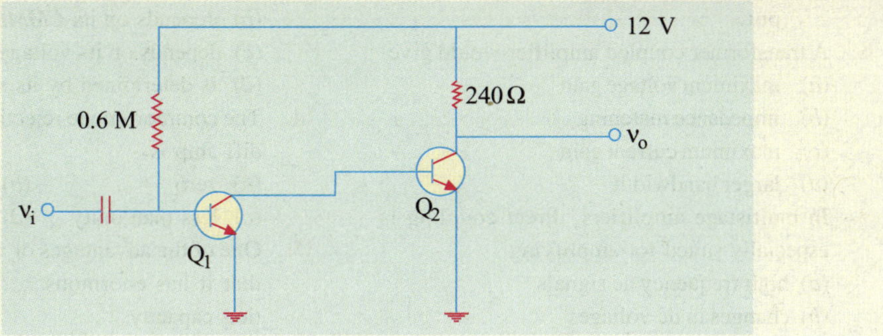

Fig. 61.46

OBJECTIVE TESTS – 61

1. The decibel gain of a cascaded amplifier equals the
   (*a*) product of individual gains
   (*b*) sum of individual gains
   (*c*) ratio of stage gains
   (*d*) product of voltage and current gains.
2. If two stages of a cascaded amplifier have decibel gains of 60 and 30, then overall gain is .... dB.
   (*a*) 90 (*b*) 1800
   (*c*) 2 (*d*) 0.5
3. Cascading two amplifiers will result in
   (*a*) reduction in overall gain and increase in overall bandwidth.
   (*b*) reduction in overall gain and reduction in overall bandwidth
   (*c*) increase in overall gain increase in overall bandwidth
   (*d*) increase in overall gain and reduction in overall bandwith
4. The overall bandwidth of two identical voltage amplifiers connected in cascade will
   (*a*) remain the same as that of a single stage
   (*b*) be worse than that of a single stage
   (*c*) be better than that of a single stage
   (*d*) be better if stage gain is low and worse if stage gain is high.
5. *RC* coupling is popular in low-level audio amplifiers because it
   (*a*) has better low frequency response
   (*b*) is inexpensive and needs no adjust- ments
   (*c*) provides an output signal in phase with the input signal
   (*d*) needs low voltage battery for collector supply.
6. Frequency response characteristic at a single-stage RC coupled amplifier is shown in Fig. 61.47. The fall in gain at both ends of the characteristic is due to
   (*a*) transistor shunt capacitances
   (*b*) bypass and coupling capacitances of the circuit
   (*c*) transistor shunt capacitances of the lower end and bypass and coupling capacitances at the higher end

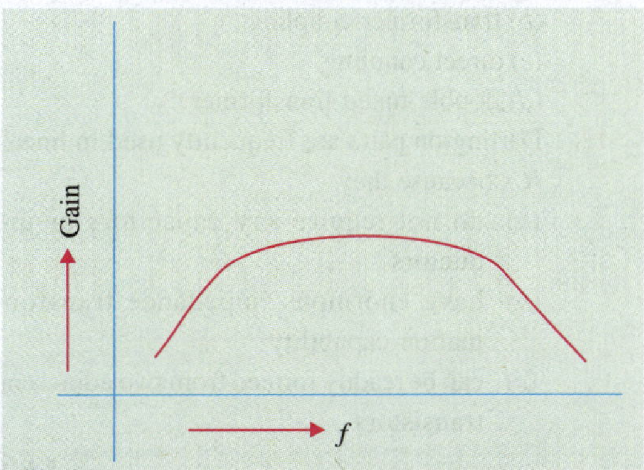

Fig. 61.47

   (*d*) transistor shunt capacitances at the higher end and bypass and coupling capacitances at the lower end.
7. The most desirable feature of transformer coupling is its
   (*a*) higher voltage gain

(b) wide frequency range
(c) ability to provide impedance matching between stages
(d) ability to eliminate hum from the output.

8. A transformer coupled amplifier would give
(a) maximum voltage gain
(b) impedance matching
(c) maximum current gain
(d) larger bandwidth

9. In multistage amplifiers, direct coupling is especially suited for amplifying
(a) high frequency ac signals
(b) changes in dc voltages
(c) high-level voltages
(d) sinusoidal signals.

10. The outstanding characteristic of a direct-coupled amplifier is its
(a) utmost economy
(b) temperature stability
(c) avoidance of frequency-sensitive components
(d) ability to amplify direct current and low-frequency signals.

11. A signal may have frequency components which lie in the range of 0.001 Hz to 10 Hz. Which one of the following types of couplings should be chosen in a multistage amplifier designed to amplify this signal ?
(a) *RC* coupling
(b) transformer coupling
(c) direct coupling
(d) double-tuned transformer

12. Darlington pairs are frequently used in linear *ICs* because they
(a) do not require any capacitors or inductors
(b) have enormous impedance transformation capability
(c) can be readily formed from two adja- cent transistors
(d) resemble emitter followers.

13. When same input signal is applied to both the inputs of an ideal diff-amp, the output
(a) is zero
(b) depends on its *CMMR*
(c) depends on its voltage gain
(d) is determined by its symmetry.

14. The common-mode rejection ratio of an ideal diff-amp is
(a) zero (b) infinity
(c) less than unity (d) greater than unity.

15. One of the advantages of a Darlington pair is that it has enormous .... transformation capacity.
(a) voltage (b) current
(c) impedance (d) power

16. A Darlington pair and an emitter follower have the same
(a) input impedance
(b) current gain
(c) voltage gain
(d) power gain.

17. Which of the following mode of operation is possible with a differential amplifier ?
(a) single-ended input
(b) differential input
(c) common-mode input
(d) all of the above.

18. The gain of a Darlington amplifier pair is determined by .... the beta values.
(a) subtracting (b) adding
(c) dividing (d) multiplying

19. The amplifier in which the emitter and collector leads of one transistor are connected to the base and collector leads of a second transistor is called an .... amplifier.
(a) push-pull (b) Darlington
(c) differential (d) complementary

ANSWERS

1. (b) 2. (a) 3. (c) 4. (c) 5. (b) 6. (d) 7. (c) 8. (b) 9. (b) 10. (d) 11. (c)
12. (c) 13. (a) 14. (b) 15. (c) 16. (c) 17. (d) 18. (d) 19. (b)

CHAPTER 62

FEEDBACK AMPLIFIER

Learning Objectives

- Feedback Amplifiers
- Principle of Feedback Amplifiers
- Advantages of Negative Feedback
- Gain Stability
- Decreased Distortion
- Feedback Over Several Stages
- Increased Bandwidth
- Forms of Negative Feedback
- Shunt-derived Series-fed Voltage Feedback
- Current-series Feedback Amplifier
- Voltage-shunt Negative Feedback Amplifier
- Current-shunt Negative Feedback Amplifier
- Noninverting Op-amp with Negative Feedback
- Effect of Negative Feedback on R_{in} and R_{out}
- R_{in} and R_{out} of Inverting Op-amp with Negative Feedback

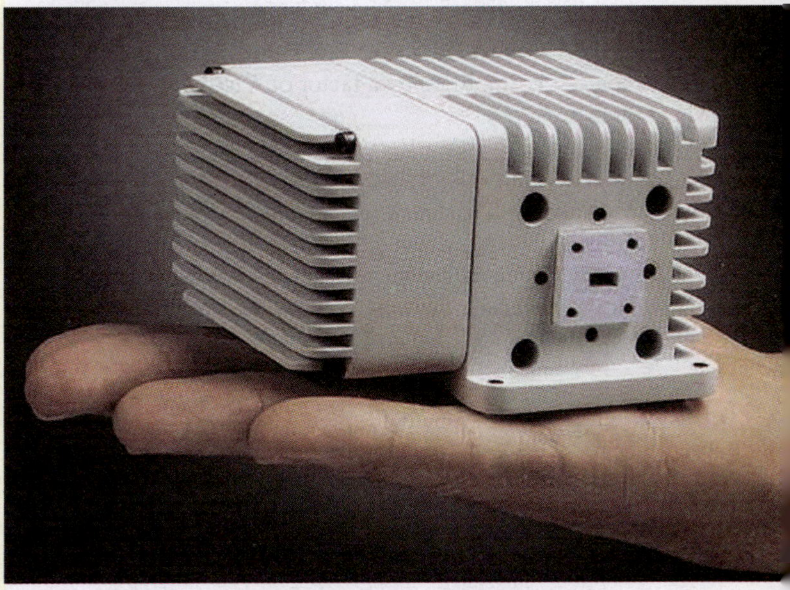

A feedback amplifier is one in which a fraction of the amplifier output is fed back to the input circuit

62.1. Feedback Amplifiers

A feedback amplifier is one in which a fraction of the amplifier output is fed back to the input circuit. This partial dependence of amplifier output on its input helps to control the output. A feedback amplifier consists of two parts : an amplifier and a feedback circuit.

***(i)* Positive feedback**

If the feedback voltage (or current) is so applied as to increase the input voltage (*i.e.* it is in phase with it), then it is called positive feedback. Other names for it are : ***regenerative* or *direct*** feedback.

Since positive feedback produces excessive distortion, it is seldom used in amplifiers. However, because it increases the power of the original signal, it is used in oscillator circuits.

***(ii)* Negative feedback**

If the feedback voltage (or current) is so applied as to reduce the amplifier input (*i.e.* it is 180° out of phase with it), then it is called negative feedback. Other names for it are : ***degenerative*** or ***inverse*** feedback.

Negative feedback is frequently used in amplifier circuits.

62.2. Principle of Feedback Amplifiers

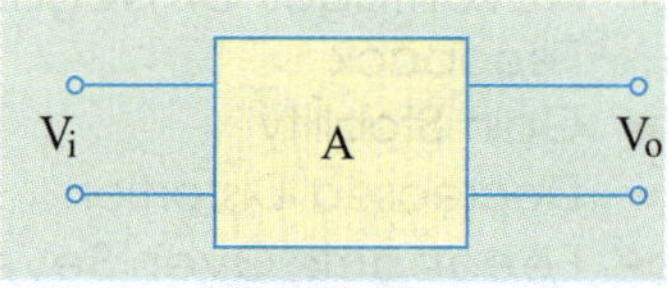

Fig. 62.1

For an ordinary amplifier *i.e.* one without feedback, the voltage gain is given by the ratio of the output voltage V_o and input voltage V_i. As shown in the block diagram of Fig. 62.1, the input voltage V_i is amplified by a factor of A to the value V_o of the output voltage.

$$\therefore \quad A = V_o / V_i$$

This gain A is often called ***open-loop*** gain.

Suppose a feedback loop is added to the amplifier (Fig. 62.2). If V_o' is the output voltage with feedback, then a fraction β* of this voltage is applied to the input voltage which, therefore, becomes $(V_i \pm \beta V_o')$ depending on whether the feedback voltage is in phase or antiphase with it. Assuming positive feedback, the input voltage will become $(V_i + \beta V_o')$. When amplified A times, it becomes $A(V_i + \beta V_o')$.

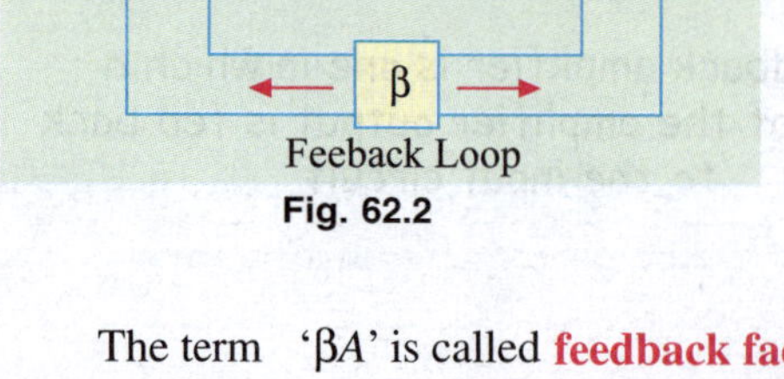

Fig. 62.2

$$\therefore \quad A(V_i + \beta V_o') = V_o'$$

$$\text{or} \quad V_o'(1 - \beta A) = AV_i$$

The amplifier gain A' with feedback is given by

$$A' = \frac{V_o'}{V_i} = \frac{A}{1 - \beta A}$$

$$\therefore \quad A' = \frac{A}{1 - \beta A} \quad \text{— positive feedback}$$

$$= \frac{A}{1 - (-\beta A)} = \frac{A}{1 + \beta A}$$

— negative feedback

The term 'βA' is called **feedback factor** whereas β is known as **feedback ratio**. The expression $(1 \pm \beta A)$ is called **loop gain**. The amplifier gain A' with feedback is also referred to as **closed-loop gain** because it is the gain obtained after the feedback loop is closed. The sacrifice factor is defined as $S = A/A'$.

---

* It may please be noted that it is not the same as the β of a transistor (Art.57.9)

(*a*) Negative Feedback

The amplifier gain with negative feedback is given by $A' = \dfrac{A}{(1+\beta A)}$

Obviously, $A' < A$ because $|1 + \beta A| > 1$.

Suppose, $A = 90$ and $\beta = 1/10 = 0.1$

Then, gain without feedback is 90 and with negative feedback is

$$A' = \frac{A}{1+\beta A} = \frac{90}{1+0.1\times 90} = 9$$

As seen, negative feedback reduces the amplifier gain. That is why it is called ***degenerative*** feedback. A lot of voltage gain is sacrificed due to negative feedback. When $|\beta A| \gg 1$, then

$$A' \cong \frac{A}{\beta A} \cong \frac{1}{\beta}$$

It means that A' depends only on β. But it is very stable because it is not affected by changes in temperature, device parameters, supply voltage and from the aging of circuit components etc. Since resistors can be selected very precisely with almost zero temperature-coefficient of resistance, it is possible to achieve highly precise and stable gain with negative feedback.

(*b*) Positive Feedback

The amplifier gain with positive feedback is given by

$$A' = \frac{A}{1-\beta A}$$ Since $|1 - \beta A| < 1, A' > A$

Suppose gain without feedback is 90 and $\beta = 1/100 = 0.01$, then gain with positive feedback is

$$A' = \frac{90}{1-(0.01\times 90)} = 900$$

Since positive feedback increases the amplifier gain. It is called ***regenerative*** feedback. If $\beta A = 1$, then mathematically, the gain becomes infinite which simply means that there is an output without any input! However, electrically speaking, this cannot happen. What actually happens is that the amplifier becomes an oscillator which supplies its own input. In fact, two important and necessary conditions for circuit oscillation are

1. the feedback must be positive, **2.** feedback factor must be unity *i.e.* $\beta A = +1$.

62.3. Advantages of Negative Feedback

The numerous advantages of negative feedback outweigh its only disadvantage of reduced gain. Among the advantages are :

1. higher fidelity *i.e.* more linear operation,
2. highly stabilized gain,
3. increased bandwidth *i.e.* improved frequency response,
4. less amplitude distortion,
5. less harmonic distortion,
6. less frequency distortion,
7. less phase distortion,
8. reduced noise,
9. input and output impedances can be modified as desired.

Example 62.1. *In the series-parallel* (*SP*) *feedback amplifier of Fig. 62.3, calculate*

(*a*) *open-loop gain of the amplifier,* (*b*) *gain of the feedback network,*

(*c*) *closed-loop gain of the amplifier,* (*d*) *sacrifice factor, S.*

(Applied Electronics-I, Punjab Univ. 1991)

Solution. (*a*) Since 1 mV goes into the amplifier and 10 V comes out

$$\therefore \quad A = \frac{10\,\text{V}}{1\,\text{mV}} = \mathbf{10{,}000}$$

(*b*) The feedback network is being driven by the output voltage of 10 V.

∴ Gain of the feedback network

$$= \frac{\text{output}}{\text{input}} = \frac{250\,\text{mV}}{10\,\text{V}} = \mathbf{0.025}$$

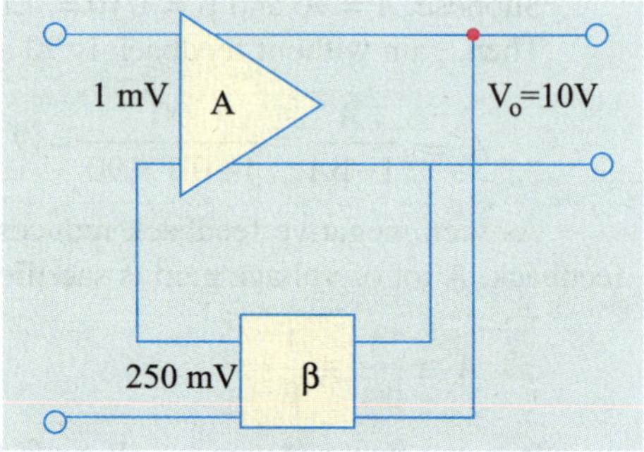

Fig. 62.3

(*c*) So far as the feedback amplifier is concerned, input is (250 + 1) = 251 mV and final output is 10 V. Hence, gain with feedback is

$$A' = 10\,\text{V}/251\,\text{mA} \cong \mathbf{40}$$

(*d*) The sacrifice factor is given by

$$S = \frac{A}{A'} = \frac{10{,}000}{40} = \mathbf{250}$$

By sacrificing so much voltage gain, we have improved many other amplifier quantities. (Art. 62.3)

Example 62.2. *Calculate the gain of a negative feedback amplifier whose gain without feedback is 1000 and β = 1/10. To what value should the input voltage be increased in order that the output voltage with feedback equals the output voltage without feedback ? Take V_i = 50 mV*

Solution. Since | β*A* | » 1, the closed-loop gain is $A' \cong \frac{1}{\beta} \cong \frac{1}{1/10} = \mathbf{10}$

The new increased input voltage is given by

$$V_i' = V_i(1+\beta A) = 50\,(1+100) = 5.05\text{V}$$

Example 62.3. *In a negative-feedback amplifier, A = 100, β = 0.04 and V_i = 50 mV. Find*
(*a*) *gain with feedback,* (*b*) *output voltage,*
(*c*) *feedback factor,* (*d*) *feedback voltage.* **(Applied Electronics, AMIEE, London)**

Solution. (*a*) $A' = \frac{A}{1+\beta A} = \frac{100}{1 + 0.04 \times 100} = \mathbf{20}$

(*b*) $V_0' = A'V_i = 20 \times 50\,\text{mV} = \mathbf{1V}$ (*c*) feedback factor $= \beta A = 0.04 \times 100 = \mathbf{4}$

(*d*) Feedback voltage $= \beta V_o' = 0.04 \times 1 = \mathbf{0.04\,V}$

Example 62.4. *An amplifier having a gain of 500 without feedback has an overall negative feedback applied which reduces the gain to 100. Calculate the fraction of output voltage feedback. If due to ageing of components, the gain without feedback falls by 20%, calculate the percentage fall in gain without feedback.* **(Applied Electronics-II, Punjab Univ. 1993)**

Solution. $A' = \frac{A}{1+\beta A}$ $\qquad \therefore \quad 1+\beta A = \frac{A}{A'}$

$$\therefore \quad \beta = \frac{1}{A'} - \frac{1}{A} = \frac{1}{100} - \frac{1}{500} = \mathbf{0.008}$$

Now, gain without feedback = 80% of 500 = 400

$$\therefore \quad \text{New} \quad A' = \frac{400}{1+0.008\times 400} \cong 95.3$$

Hence, change in the gain with feedback in the two cases = 100 – 95.3 = 4.7

$$\therefore \quad \text{Percentage fall in gain with feedback is} = \frac{4.7}{100} \times 100 = \mathbf{4.7\%}$$

Example 62.5. *An amplifier with negative feedback has a voltage gain of 100. It is found that without feedback an input signal of 50 mV is required to produce a given output whereas with feedback, the input signal must be 0.6 V for the same output. Calculate the value of voltage gain without feedback and feedback ratio.* **(Bangalore University 2001)**

Solution. $V_o' = AV_i = 100 \times 0.6 = 60$ V and $V_o = AV_i$

Since the output voltage with and without feedback are required to be the same,

$$\therefore \quad 60 = A \times 50 \text{ mV}, \qquad \therefore \quad A = \frac{60}{50 \text{ mV}} = \mathbf{1200}$$

The amplifier gain with feedback,

$$A' = \frac{A}{1+\beta A} \quad \text{or} \quad \beta = \frac{A - A'}{AA'} = \frac{1200 - 100}{1200 \times 100} = \mathbf{0.009}$$

62.4. Gain Stability

The gain of an amplifier with negative feedback is given by $A' = \dfrac{A}{1+\beta A}$

Taking logs of both sides, we have $\log_e A' = \log_e A - \log_e (1 + \beta A)$

Differentiating both sides, we get

$$\frac{dA'}{A'} = \frac{dA}{A} - \frac{\beta . dA}{1+\beta A} = dA\left(\frac{1}{A} - \frac{\beta}{1+\beta A}\right) = \frac{1}{1+\beta A}\frac{dA}{A} = \frac{(dA/A)}{1+\beta A}$$

If $\beta A \gg 1$, then the above expression becomes

$$\frac{dA'}{A'} = \frac{1}{\beta A} \cdot \frac{dA}{A}$$

Example 62.6. *An amplifier has an open-loop gain of 400 and a feedback of 0.1. If open-loop gain changes by 20% due to temperature, find the percentage change in closed-loop gain.* **(Electronics-III, Bombay 1991)**

Solution. Here, $A = 400$, $\beta = 0.1$, $dA/A = 20\% = \mathbf{0.2}$

$$\text{Now,} \quad \frac{dA'}{A'} = \frac{1}{\beta A} \cdot \frac{dA}{A} = \frac{1}{0.1 \times 400} \times 20\% = \mathbf{0.5\%}$$

It is seen that while the amplifier gain changes by 20%, the feedback gain changes by only 0.5% *i.e.* an improvement of 20/0.5 = 40 times

62.5. Decreased Distortion

Let the harmonic distortion voltage generated within the amplifier change from D to D' when negative feedback is applied to the amplifier.

Suppose $\quad D' = xD \qquad \ldots (i)$

The fraction of the output distortion voltage which is fedback to the input is

$$\beta D' = \beta x D$$

After amplification, it become $\beta x DA$ and is antiphase with original distortion voltage D. Hence, the new distortion voltage D' which appears in the output is

$$D' = D - \beta x DA \qquad \ldots (ii)$$

From (*i*) and (*ii*), we get

$$xD = D - \beta x DA \quad \text{or} \quad x = \frac{1}{1+\beta A}$$

Substituting this value of x in Eq. (*i*) above, we have $D' = \dfrac{D}{1+\beta A}$

It is obvious from the above equation that $D' < D$. In fact, negative feedback reduces the amplifier distortion by the amount of loop gain *i.e.* by a factor of $(1 + \beta A)$.

However, it should be noted that improvement in distortion is possible only when the distortion is produced by the ***amplifier itself,*** not when it is already present in the input signal.

62.6. Feedback Over Several Stages

Multistage amplifiers are used to achieve greater voltage or current amplification or both. In such a case, we have a choice of applying negative feedback to improve amplifier performance. Either we apply some feedback across each stage or we can put it in one loop across the whole amplifier.

A multistage amplifier is shown in Fig. 62.4. In Fig. 62.4 (*a*) each stage of the *n*-stage amplifier has a feedback applied to it. Let A and β_1 be the open-loop gain and feedback ratio respectively of each stage and A_1 the overall gain of the amplifier. Fig. 62.4 (*b*) shows the arrangemnent where n amplifiers have been cascaded in order to get a total gain of A^n. Let the overall feedback factor be β_2 and the overall gain A_2. The values of the two gains are given as

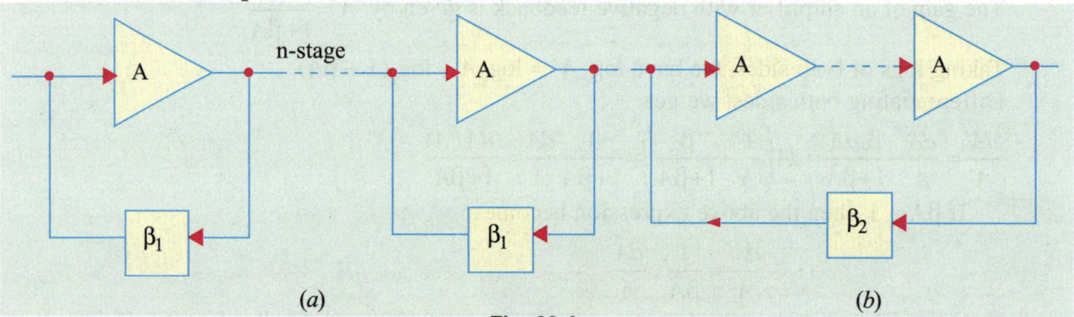

Fig. 62.4

$$A_1 = \left(\frac{A}{1+A\beta_1}\right)^n \quad \text{and} \quad A_2 = \frac{A^n}{1+A^n\beta_2} \qquad \text{... } (i)$$

taking log of both sides on the above two expressions

$$\log A_1 = n.\log\left(\frac{A}{1+A\beta_1}\right) \quad \text{and} \quad \log A_2 = n\log A^n - \log(1+A^n\beta_2)$$

$$= n.\log A - n.\log(1+A\beta_1)$$

Differentiating the above two expressions, we get

$$\frac{dA_1}{A_1} = \frac{ndA}{A} - \frac{n\beta_1 dA}{1+A\beta_1} \quad \text{and} \quad \frac{dA_2}{A_2} = \frac{ndA}{A} - \frac{n\beta_2 A^{n-1} dA}{1+A^n\beta_2}$$

$$= ndA\left(\frac{1}{A} - \frac{\beta_1}{1+A\beta_1}\right) \qquad = ndA\left(\frac{1}{A} - \frac{\beta_2 A^{n-1}}{1+A^n\beta_2}\right)$$

$$= \frac{n}{1+A\beta_1}\cdot\frac{dA}{A} \qquad = \frac{n}{1+A^n\beta_2}\cdot\frac{dA}{A}$$

For the two circuits to have the same overall gain, $A_1 = A_2$. Hence, from Eqn. (*i*) above, we get $(1 + 1A\beta_1)^n = 1 + A^n\beta_2$

$$\therefore \quad \frac{dA_2/A_2}{dA_1/A_1} = \frac{1}{(1+A\beta_1)^{n-1}}$$

If $n = 1$, then the denominator in the above equation becomes unity so that fractional gain variations are the same as expected. However, for $n > 1$ and with $(1 + A\beta_1)$ being a normally large quantity, the expression dA_2/A_2 will be less than dA_1/A_1. It means that the overall feedback would appear to be beneficial as far as stabilizing of the gain is concerned.

Example 62.7. *An amplifier with 10% negative feedback has an open-loop gain of 50. If open-loop gain increases by 10%, what is the percentage change in the closed-loop gain ?*

(Applied Electronics-I, Punjab Univ. 1991)

Solution. Let A_1' and A_2' be the closed-loop gains in the two cases and A_1 and A_2 the open-loop gains respectively.

(*i*) $A'_1 = \dfrac{A_1}{1+\beta A_1} = \dfrac{50}{1+0.1\times 50} = \mathbf{8.33}$

(*ii*) When open-loop gain changes by 10%, then $A_2 = 50 + 0.1 \times 50 = 55$

$$\therefore \quad A_2' = \frac{A_2}{1+\beta A_2} = \frac{55}{1+0.1\times 55} = 8.46$$

$\therefore$ Percentage change in closed-loop gain is

$$= \frac{A'_2 - A'_1}{A'_1} \times 100 = \frac{8.46-8.33}{8.33} \times 100 = \mathbf{1.56\%}$$

Example 62.8. *Write down formulae for* **(i)** *gain* **(ii)** *harmonic distortion of a negative feedback amplifier in terms of gain and distortion without feedback and feedback factor. If gain without feedback is 36 dB and harmonic distortion at the normal output level is 10%, what is* **(a)** *gain and* **(b)** *distortion when negative feedback is applied, the feedback factor being 16 dB.*

(Electronic Engg. II, Warangal 1991)

Solution. For first part, please refer to Art. 62.6. Distortion ratio is defined as the ratio of the amplitude of the largest harmonic to the amplitude of the fundamental.

$$A_f = A' = \frac{A}{1+\beta A}$$

Now, dB gain $= 20 \log_{10} A$ $\quad\therefore\quad$ $36 = 20 \log_{10} A, \quad \therefore A = 63$

dB feedback factor $= 20 \log_{10} \beta A$ $\quad\quad$ $16 = 20 \log_{10} \beta A$ or $\quad \beta A = 6.3$

(*a*) $A_f = A/(1+\beta A) = 63/(1+6.3) =$ **8.63** or **18.72 dB**

(*b*) $D' = 10$ per cent$/(1+6.3) \cong$ **1.4 per cent**

Example 62.9. *The overall gain of a two-stage amplifier is 150. The second stage has 10% of the output voltage as negative feedback and has 150 as forward gain. Calculate* **(a)** *gain of the first stage* **(b)** *the second harmonic distortion, if the second stage introduces 5% second harmonic without feedback. Assume that the first stage does not introduce distortion.*

(Electronics-II, Madras Univ. 1992)

Solution. (*a*) For second stage $D'_2 = \dfrac{D_2}{1+\beta A_2} = \dfrac{0.05}{1+150\times 0.1} = \mathbf{0.31\%}$

(*b*) For the second stage, gain with feedback is

$$A'_2 = \frac{A_2}{1+\beta A_2} = \frac{150}{1+150\times 0.1} = 9.38$$

Now, $\quad A_1 \times A'_2 = 150\ ; \quad A_1 = 150/9.38 = \mathbf{16}$

Example 62.10. *Determine the effective gain of a feedback amplifier having an amplification without feedback of (–200 – j300) if the feedback circuit adds to the input signal, a p.d. which is 0.5 percent of the output p.d. and lags a quarter of a cycle behind it in phase. Explain whether the feedback in this case is positive or negative.* **(Applied Electronics-II, Punjab Univ. 1992)**

Solution. $\quad A = -200 - j300 = 360 \angle -123.7°$

The feedback voltage V_β is 0.5 percent of the output voltage and lags 90° behind it.

$$\therefore \quad V_\beta = \left(\frac{0.5}{100}\angle -90°\right) V_0$$

$$\therefore \quad \beta = \frac{V_\beta}{V_0} = \frac{0.5}{100} \angle -90° = -j0.005$$

$$\therefore \quad \beta A = (-200 - j300)(-j0.005) = -1.5 + j1.0$$

In general, the stage gain with feedback is given by

$$A' = \frac{A}{1-\beta A} = \frac{360\angle -123.7°}{1-(-1.5+j1.0)} = \frac{360\angle -123.7°}{2.69\angle -21.8°} = \mathbf{134\angle -102°}$$

Since both the magnitude and the phase shift of the amplifier are reduced by feedback, the feedback must be negative.

Example 62.11. *An amplifier has a gain of 100 and 5 per cent distortion with an input signal of 1 V. When an input signal of 1 V is applied to the amplifier, calculate*
(i) output signal voltage, (ii) distortion voltage, (iii) output voltage

Solution. (*i*) Signal output voltage $V_{os} = AV_i = 100 \times 1 = $ **100 V**

(*ii*) Distortion voltage $= DV_{os} = 0.05 \times 100 = $ **5 V**

(*iii*) Amplifier output voltage $V_o = V_{os} + D = 100 + 5 = $ **105 V**

62.7. Increased Bandwidth

The bandwidth of an amplifier without feedback is equal to the separation between the 3 dB frequencies f_1 and f_2.

$$\therefore \quad BW = f_2 - f_1$$

where f_1 = lower 3 dB frequency,
and f_2 = upper 3 dB frequency.

If A is its gain, the gain-bandwidth product is $A \times BW$.

Now, when negative feedback is applied, the amplifier gain is reduced. Since the gain-bandwidth product has to remain the same in both cases, it is obvious that the bandwidth must increase to compensate for the decrease in gain. It can be proved that with negative feedback, the lower and upper 3 dB frequencies of an amplifier become.

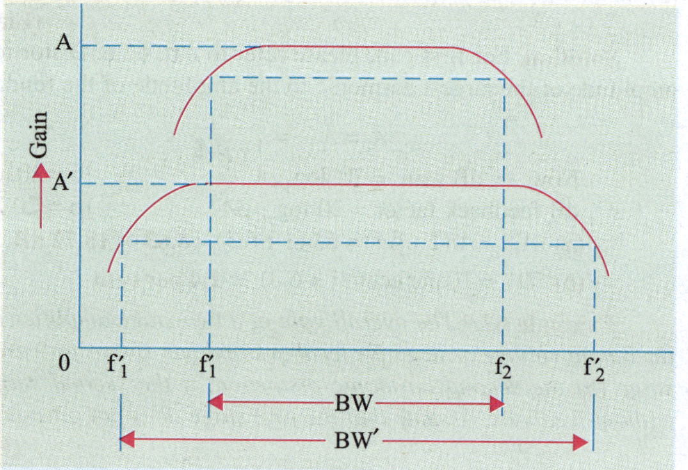

Fig. 62.5

$$(f')_1 = \frac{f_1}{(1+\beta A)} \quad \text{and} \quad (f')_2 = f_2(1+\beta A)$$

As seen from Fig. 62.5, f_1' has decreased whereas f_2' has increased thereby giving a wider separation or bandwidth. Since gain-bandwidth product is the same in both cases.

$$\therefore \quad A \times BW = A' \times BW' \text{ or } A(f_2 - f_1) = A'(f'_2 - f'_1)$$

Example 62.12. *An RC-coupled amplifier has a mid-frequency gain of 200 and a frequency response from 100 Hz to 20 kHz. A negative feedback network with β = 0.02 is incorporated into the amplifier circuit. Determine the new system performance.*

(Electronic Circuits, Mysore Univ. 1990)

Solution. $$A' = \frac{A}{1+\beta A} = \frac{200}{1+0.02\times 200} = \mathbf{40\ Hz}$$

$$f_1' = \frac{f_1}{1+\beta A} = \frac{100}{1+0.02\times 200} = \mathbf{20\ Hz}$$

$$f_2' = f_2(1 + \beta A) = 20(1 + 0.02 \times 200) = \textbf{100 kHz}$$
$$dW' = f_2' - f_1' \cong \textbf{100 kHz}$$

Incidentally, it may be proved that gain-bandwidth product remains constant in both cases.

$$dW = f_2 - f_1 \cong 20 \text{ kHz}$$
$$A \times dW = 200 \times 20 = 4000 \text{ kHz};$$
$$A' \times dW' = 40 \times 100 = 4000 \text{ kHz}$$

As expected, the two are equal.

62.8. Forms of Negative Feedback

The four basic arrangements for using negative feedback are shown in the block diagram of Fig. 62.6. As seen, both voltage and current can be fedback to the input either in series or in parallel. The output voltage provides input in Fig. 62.6 (*a*) and (*b*). However, the input to the feedback network is derived from the output current in Fig. 62.6 (*c*) and (*d*).

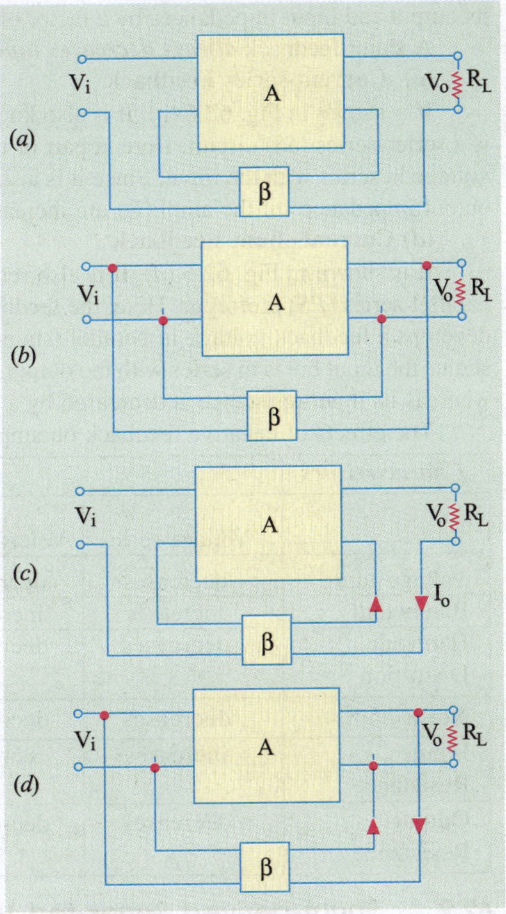

Fig. 62.6

(a) Voltage-series Feedback

It is shown in Fig. 62.6 (*a*). It is also called ***shunt-derived series-fed feedback***. The amplifier and feedback circuit are connected series-parallel. Here, a fraction of the output voltage is applied in series with the input voltage via the feedback. As seen, the input to the feedback network is in parallel with the output of the amplifier. Therefore, so far as V_o is concerned, output resistance of the amplifier is reduced by the shunting effect of the input to the feedback network. It can be proved that

$$R'_o = \frac{R_o}{(1+\beta A)}$$

Similarly, V_i sees two circuit elements in series :

(i) the input resistance of the amplifier and

(ii) output resistance of the feedback network.

Hence, input resistance of the amplifier as a whole is increased due to feedback. It can be proved that

$$R_i' = R_i(1 + \beta A)$$

In fact, ***series feedback always increases the input impedance by a factor of (1 + βA).***

Shunt Voltage

(b) Voltage-shunt Feedback

It is shown in Fig. 62.6 (*b*). It is also known as ***shunt-derived shunt-fed feedback*** *i.e.* it is parallel-parallel (*PP*) prototype. Here, a small portion of the output voltage is coupled back to the input voltage parallel (shunt).

Since the feedback network shunts both the output and input of the amplifier, it decreases both its output and input impedances by a factor of $1/(1 + \beta A)$

A shunt feedback ***always decreases input impedance.***

(c) Current-series Feedback

It is shown in Fig. 62.6 (*c*). It is also known as ***series-derived series-fed feedback.*** As seen, it is a series-series (*SS*) circuit. Here, a part of the output current is made to feedback a proportional voltage in series with the input. Since it is a series pick-up and a series feedback, both the input and output impedances of the amplifier are increased due to feedback.

(d) Current-shunt Feedback

It is shown in Fig. 62.6 (*d*). It is also referred to as ***series-derived shunt-fed feedback.*** It is a parallel-series (*PS*) prototype. Here, the feedback network picks up a part of the output current and develops a feedback voltage in parallel (shunt) with the input voltage. As seen, feedback network shunts the input but is in series with the output. Hence, output resistance of the amplifier is increased whereas its input resistance is decreased by a factor of loop gain.

The effects of negative feedback on amplifier characteristics are summarized below :

Characteristics	*Type of Feedback*			
	Voltage series	Voltage shunt	Current series	Current shunt
Voltage gain	decreases	decreases	decreases	decreases
Bandwidth	increases	increases	increases	increases
Harmonic Distortion	decreases	decreases	decreases	decreases
Noise	decreases	decreases	decreases	decreases
Input Resistance	increases	decreases	increases	decreases
Output Resistance	decreases	decreases	increases	increases

62.9. Shunt-derived Series-fed Voltage Feedback

The basic principle of such a voltage-controlled feedback is illustrated by the block diagram of Fig. 62.7. Here, the feedback voltage is derived from the voltage divider circuit formed of R_1 and R_2.

As seen, the voltage drop across R_1 forms the feedback voltage V_f

$$\therefore \quad V_f = V_o \frac{R_1}{R_1 + R_2} = \beta V_o$$

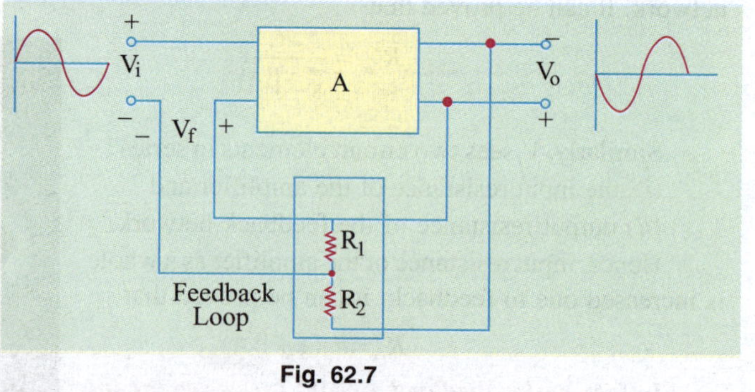

Fig. 62.7

Example 62.13. *In the voltage-controlled negative feedback amplifier of Fig. 62.8, calculate* **(a)** *voltage gain without feedback* **(b)** *feedback factor* **(c)** *voltage gain with feedback. Neglect V_{BE} and use $r_e = 25\ mV/I_E$*

Solution. (a) $A = \frac{r_L}{r_e} = \frac{R_3}{r_e}$

Now, $I_B = \dfrac{15\text{ V}}{1.5M} = 10\mu\text{A}$

$$I_E = \beta I_B = 100 \times 10 = 1\text{ mA}$$

$$r_e = 25/1 = 25\ \Omega\ ;$$

$$A = \frac{10K}{25\Omega} = \mathbf{400}$$

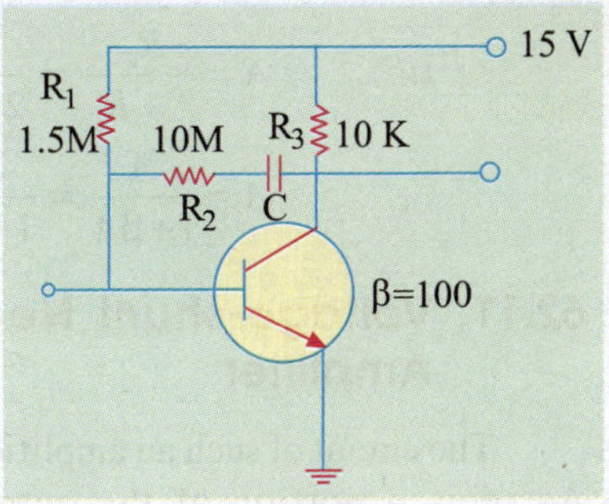

Fig. 62.8

(*b*) $\beta = \dfrac{R_1}{R_1 + R_2} = \dfrac{1.5 \times 10^6}{(1.5 + 10) \times 10^6} = 0.13$

$\therefore\ \beta A = 0.13 \times 400 = \mathbf{52}$

(*c*) $A' = \dfrac{A}{1 + \beta A} = \dfrac{400}{1 + 52} = \mathbf{7.55}$

62.10. Current-series Feedback Amplifier

Fig. 62.9 shows a series-derived series-fed feedback amplifier circuit. Since the emitter resistor is unbypassed, it effectively provides current-series feedback. When I_E passes through R_E, the feedback voltage drop $V_f = I_E R_E$ is developed which is applied in phase opposition to the input voltage V_i. This negative feedback reduces the output voltage V_0. This feedback can, however, be eliminated by either removing or bypassing the emitter resistor.

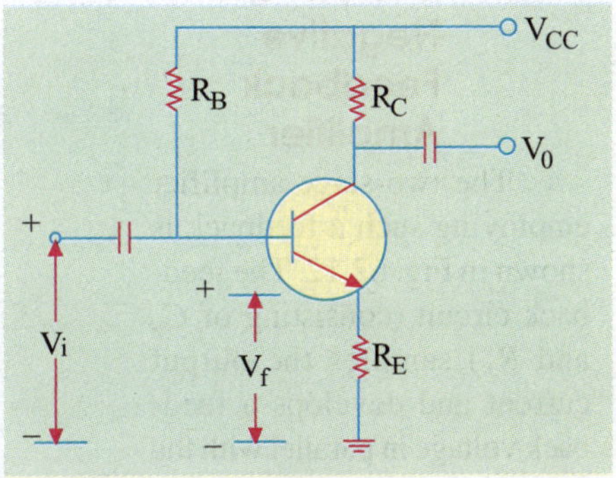

Fig. 62.9

It can be proved that

$$\beta = \frac{R_E}{R_C}\ ;\quad A' = \frac{R_C}{r_e + R_E}\ ;\quad A = \frac{R_C}{r_e}$$

Example 62.14. *For the current-series feedback amplifier of Fig. 62.10, calculate* **(*i*)** *voltage gain without feedback,* **(*ii*)** *feedback factor,* **(*iii*)** *voltage gain with feedback. Neglect V_{BE} and use r_e = 25 mV/I_E.* **(Electronics-I, Madras Univ. 1990)**

Solution. (*i*) $A = \dfrac{R_C}{r_e}$

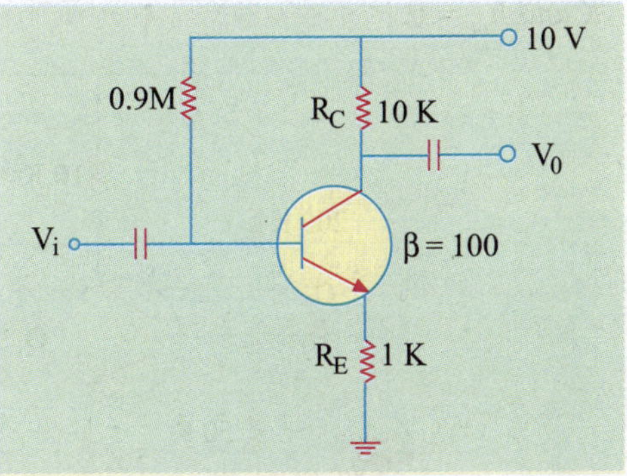

Fig. 62.10

Now, $I_E = \dfrac{V_{CC}}{R_E + R_B/\beta}$

$$= \frac{10}{1 + 900/100} = 1\text{ mA}$$

$\therefore\quad r_e = 25/I_E = 25\ \Omega$

$\therefore\quad A = \dfrac{10K}{25\Omega} = \mathbf{400}$

(*ii*) $\beta = \dfrac{R_E}{R_C} = \dfrac{1}{10} = 0.1$

$$\beta A = 0.1 \times 400 = \mathbf{40}$$

(*iii*) $A' = \frac{R_C}{r_e + R_E} = \frac{10,000}{25+1000} = \mathbf{9.756}$

or $A' = \frac{A}{1+\beta A} = \frac{400}{1+40} = \mathbf{9.756}$

62.11. Voltage-shunt Negative Feedback Amplifier

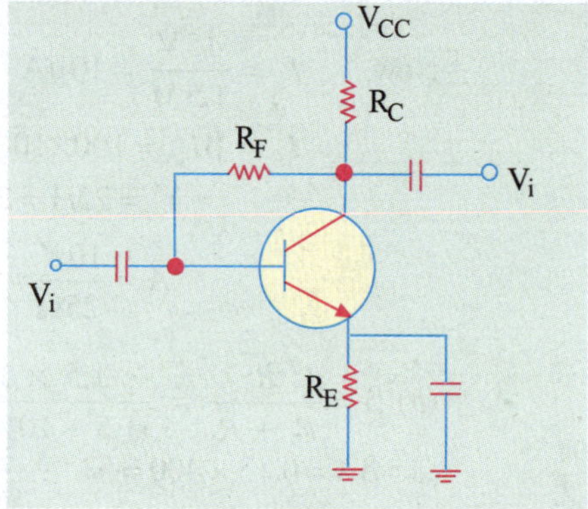

Fig. 62.11

The circuit of such an amplifier is shown in Fig. 62.11. As seen, a portion of the output voltage is coupled through R_F in parallel with the input signal at the base. This feedback stabilizes the overall gain while decreasing both the input and output resistances. It can be proved that $\beta = R_C/R_F$.

62.12. Current-shunt Negative Feedback Amplifier

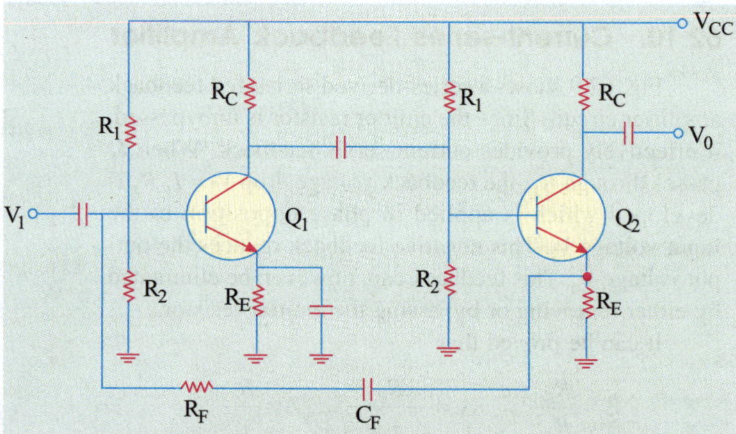

Fig. 62.12

The two-stage amplifier employing such a feedback is shown in Fig. 62.12. The feedback circuit (consisting of C_F and R_F) samples the output current and develops a feedback voltage in parallel with the input voltage. The unbypassed emitter resistor of Q_2 provides current sensing.

The polarity of the feedback voltage is such that it provides the negative feedback.

Example 62.15. *Calculate A, $r_{in(stage)}$ and $I_{o(stage)}$ of the cascaded amplifier shown in Fig. 62.13 with and without voltage series feedback. The transistor parameters are : h_{fe} = 100,. h_{ie}. = 2 K and h_{oe} = 0.* **(Applied Electronics-I, Punjab Univ. 1992)**

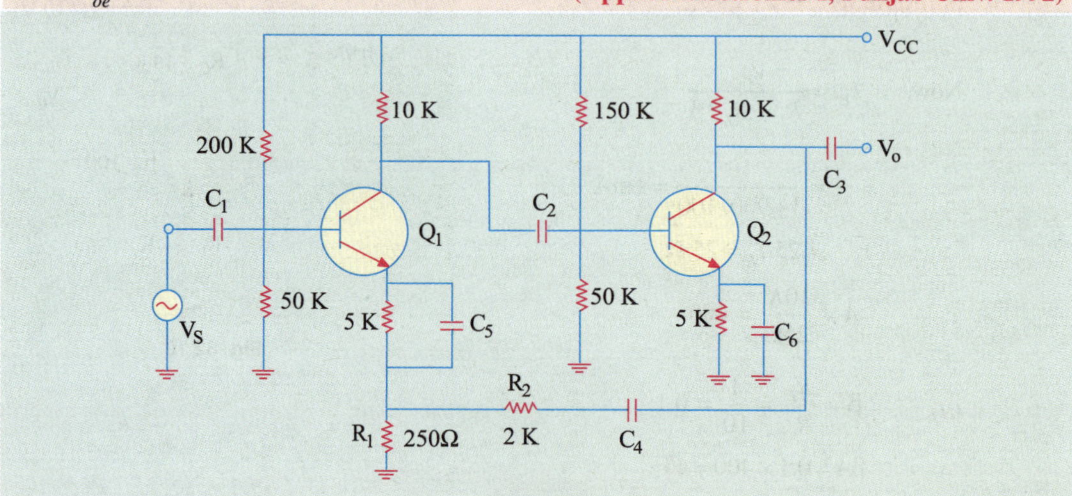

Fig. 62.13

Solution. *(i)* Without Feedback. The $r_{in(base)}$ for Q_1 is $= h_{ie} = 2K$. Same is the value for Q_2.

Also, $r_{in(stage)}$ or r_{i-1} for $Q_1 = 200\text{ K} \| 50\text{ K} \| 2\text{ K} = 1.9\text{ K}$

$r_{0.2}$ or $r_{L.2}$ for $Q_2 = 10\text{ K} \| (2.0 + 0.25)\text{ K} = 1.83\text{K}$

$r_{0.1}$ or $r_{L.1}$ for $Q_1 = 10\text{ K} \| 150\text{ K} \| 50 \| 2\text{K} = 1.6\text{ K}$

$$\therefore \quad A_{v1} = \frac{h_{fe.1}\, r_{0.1}}{h_{ie}} = \frac{h_{fe.1}\, r_{L.1}}{h_{ie}}$$

$$= \frac{100 \times 1.6}{2} = 80$$

$$A_{v2} = \frac{h_{fe.2} r_{0.2}}{h_{ie}} = \frac{h_{fe.2}\, r_{L2}}{h_{ie}} = \frac{100 \times 1.83}{2} = 91.5$$

Overall gain, $A_v = A_{v1} \cdot A_{v'2} = 0.80 \times 92 = \mathbf{73.20}$

***(ii)* With Feedback**

The feedback factor, $\beta = \dfrac{R_1}{R_1 + R_2} = \dfrac{0.25}{0.25 + 2.0} = \dfrac{1}{9}$

$$r_{02f} = \frac{r_{o2}}{1 + \beta A} = \frac{1.83}{1 + 7320/9 \times 7360} = \mathbf{2.2\ \Omega}$$

$$r_{i.1f} = r_{i1}(1 + \beta A) = 1.9 \times 814 = \cong \mathbf{1547\ K}$$

$$A_f = \frac{A}{(1 + \beta A)} = \frac{7320}{814} \cong 9.$$

Example 62.16. *In the two-stage R_C coupled amplifier (Fig. 62.14) using emitter feedback, find the overall gain. Neglect V_{BE} and take $\beta_1 = \beta_2 = 100$.*

Solution. In this amplifier circuit, voltage gain has been stabilized to some extent with the help of 500 Ω unbypassed emitter resistance. This 500 Ω resistance swamps out r_e.

$$\therefore \quad A_{v.2} = \frac{r_{L.2}}{r_e + r_E} \cong \frac{r_{L.2}}{r_E} = \frac{10\text{K} \| 10\text{K}}{500\,\Omega} = 10$$

Now, $\beta r_E = 100 \times 500 = 50\text{ K}$

$r_{i-2} = 80\text{ K} \| 40\text{ K} \| 50\text{ K}$

$r_{L.1} = R_{C.1} \| r_{i.2}$

$= 10\text{ K} \| 80\text{ K} \| 40\text{ K} \| 50\text{ K} = 6.3\text{ K}$

$$\therefore A_{v.1} = \frac{r_{L1}}{r_E} = \frac{6.3 \times 10^3}{500} = 12.6$$

$\therefore A = 10 \times 12.6 = \mathbf{126}$

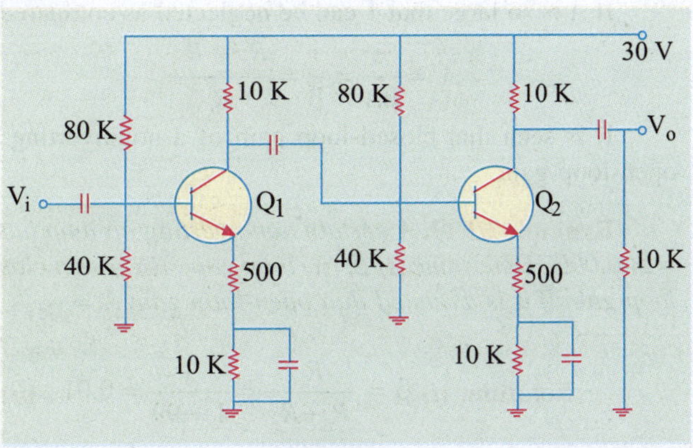

Fig. 62.14

62.13. Noninverting Op-amp With Negative Feedback

The closed-loop noninverting op-amp circuit using negative feedback is shown in Fig. 62.15. The input signal is applied to the noninverting input terminal. The output is applied back to the input terminal through the feedback network formed by R_i and R_f.

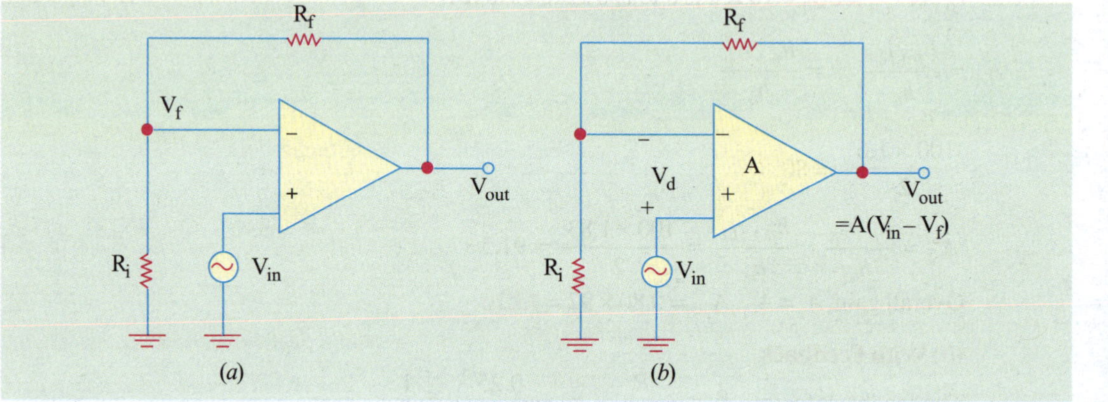

Fig. 62.15

The op-amp acts as both the difference circuit and the open-loop forward gain. The differential input to the op-amp is $(V_{in} - V_f)$. This differential voltage is amplified A times and an output voltage is produced which is given by

$$V_{out} = A_{v1} = A(V_{in} - V_f) \; ; \qquad \text{where } A \text{ is the open-loop gain of the op-amp}$$

Since, $(R_i + R_f)$ acts as voltage divider across V_{out},

$$\therefore \qquad V_f = V_{out} \frac{R_i}{R_i + R_f}$$

Now, $\beta = R_i/(R_i + R_f)$, hence $V_f = \beta V_{out}$

Substituting this value in the above equation, we get

$$V_{out} = A\,(V_{in} - \beta\, V_{out}) \text{ or } V_{out}\,(1 + \beta A) = AV_{in}$$

Hence, voltage gain A' with negative feedback is

$$A' = \frac{V_{out}}{V_{in}} = \frac{A}{1 + \beta A} = \frac{A}{1 + AR_i/(R_i + R_f)}$$

If A is so large that 1 can be neglected as compared to βA, the above equation becomes

$$A' = \frac{A}{\beta A} = \frac{1}{\beta} = \frac{R_i + R_f}{R_i}$$

It is seen that closed-loop gain of a noninverting op-amp is essentially independent of the open-loop gain.

Example 62.17. *A certain noninverting op-amp has R_i = 1K, R_f = 99 K and open-loop gain A = 500,000. Determine (i) β, (ii) loop gain, (iii) exact closed-loop gain and (iv) approximate closed-loop gain if it is assumed that open-loop gain A = ∞.*

(Power Electronics, AMIE 1991)

Solution. (*i*) $\beta = \frac{R_i}{R_i + R_f} = \frac{1}{1 + 99} = 0.01$, (*ii*) loop gain $= \beta A = 500{,}000 \times 0.01 = \mathbf{5000}$

(*iii*) $A' = \frac{A}{1 + \beta A} = \frac{500{,}000}{1 + 5000} = \mathbf{99.98}$

(iv) approx. $A' = \frac{1}{\beta} = \frac{1}{0.01} = \mathbf{100}$

It is seen that the gain changes by about 0.02%.

62.14. Effect of Negative Feedback on R_{in} and R_{out}

In the previous calculations, the input impedance of an op-amp was considered to be infinite and its output resistance as zero. We will now consider the effect of a finite input resistance and a non-zero output resistance. Since the two effects are different and their values differ by several order of magnitude, we will focus on each effect individually.

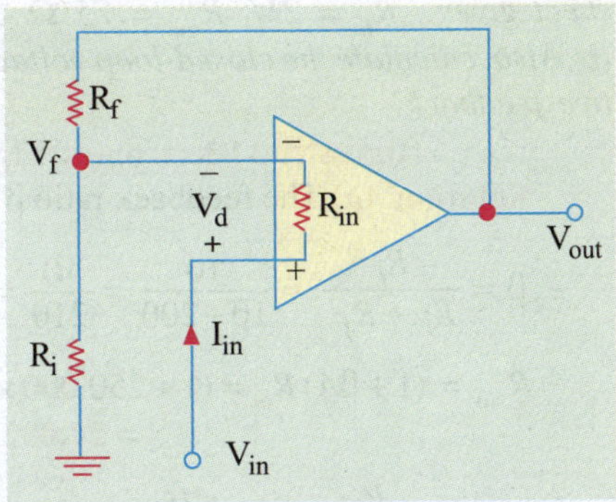

Fig. 62.16

(a) R_{in} of Noninverting Op-amp

For this analysis, it would be assumed that a small differential voltage V_d exists between the two inputs of the op-amp as shown in Fig. 62.16. It, in effect, means that neither the input resistance of the op-amp is assumed to be infinite nor its input current zero.

Now, $V_d = V_{in} - V_f$ or $V_{in} = V_d + V_f = V_d + \beta V_{out}$

Also, $V_{out} = A.\ V_d$ where A is the open-loop gain of the op-amp.

$\therefore \quad V_{in} = V_d + A\beta V_d = (1+\beta A) V_d = (1+\beta A) I_{in} R_{in} \qquad (\because V_d = I_{in} R_{in})$

where R_{in} is the open-loop impedance of the op-amp (*i.e.* without feedback)

$$\therefore \quad R'_{in} = \frac{V_{in}}{I_{in}} = (1+\beta A) R_{in}$$

where R'_{in} is the closed-loop input resistance of the non-inverting op-amp.

It will be seen that the closed-loop input resistance of the non-inverting op-amp is much greater than the input resistance without feedback.

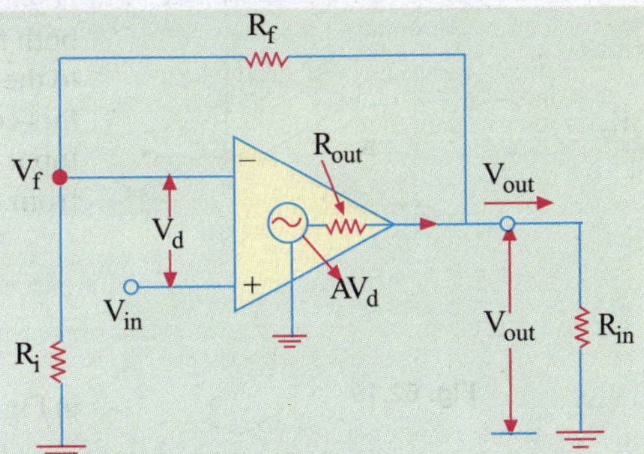

Fig. 62.17

(b) R'_{out} of Noninverting Op-amp

An expression for R'_{out} would be developed with the help Fig. 62.17. Using *KVL*, we get

$$V_{out} = AV_d - I_{out} R_{out}$$

Now, $V_d = (V_{in} - V_f)$ and neglecting $I_{out} R_{out}$ as compared to AV_d, we have

$$V_{out} = A(V_{in} - V_f) = A(V_{in} - \beta V_{out})$$

or $AV_{in} = (1+\beta A) V_{out}$

If, with negative feedback, output resistance of the noninverting op-amp is R'_{out}, then $V_{out} = I_{out}.\ R'_{out}$.

Substituting this value in the above equation, we get

$$AV_{in} = (1+\beta A) I_{out} R'_{out} \quad \text{or} \quad \frac{AV_{in}}{I_{out}} = (1+\beta A) R'_{out}$$

The term on the left is the internal output resistance R_{out} of the op-amp because without feedback, $AV_{in} = V_{out}$.

or $$R_{out} = (1+\beta A)\,R'_{out} \text{ or } R'_{out} = \frac{R_{out}}{(1+\beta A)}$$

Obviously, output resistance R'_{out} with negative feedback is much less than without feedback (i.e. R_{out}).

Example 62.18. *(a) Calculate the input and output resistance of the op-amp shown in Fig. 62.18. The data sheet gives : R_{in} = 2M, R_{out} = 75 Ω. and A = 250,000 (b) Also, calculate the closed-loop voltage gain with negative feedback.*

(Industrial Electronics, Mysore, Univ. 1992)

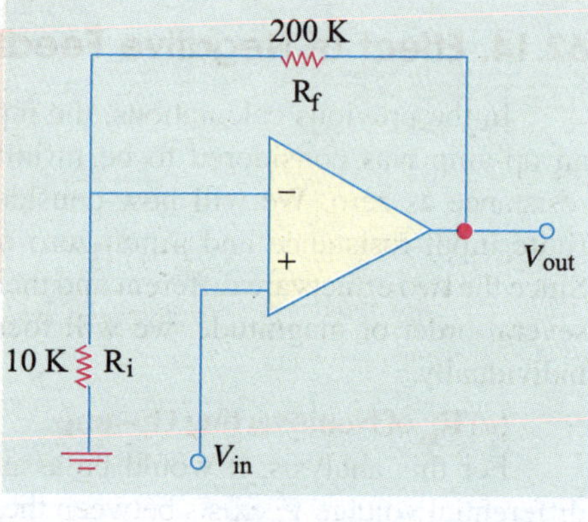

Fig. 62.18

Solution. *(a)* The feedback ratio β is given by

$$\beta = \frac{R_i}{R_i + R_f} = \frac{10}{10+200} = \frac{10}{210} = 0.048$$

$$R'_{in} = (1+\beta A)\,R_{in} = (1 + 250{,}000 \times 0.048) \times 2 = \mathbf{24{,}002\ M}$$

$$R'_{out} = \frac{R_{out}}{1+\beta A} = \frac{75}{1+12000} = \mathbf{0.006\ \Omega}$$

(b) $A' = 1/\beta = 1/0.048 = \mathbf{20.8}$

62.15. R_{in} and R_{out} of Inverting Op-amp with Negative Feedback

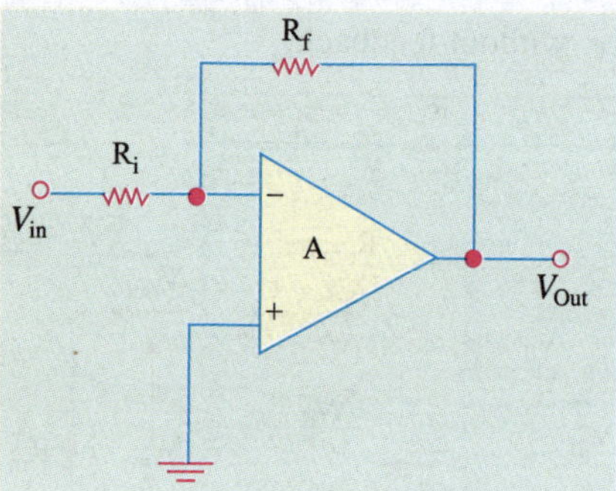

Fig. 62.19

The input resistance R_{in} of the inverting op-amp with negative feedback will be found by using Fig. 62.19. Since both the input signal and the negative feedback are applied to the inverting terminal. Miller's theorem will be applied to this configuration. According to this theorem, the effective input resistance of an amplifier with a feedback resistor from output to input is given by

$$R_{in(Miller)} = \frac{R_f}{1+A} \quad \text{and} \quad R_{out(Miller)} = R_f\left(\frac{A}{1+A}\right)$$

The Miller equivalent of the inverting op-amp is shown in Fig. 62.20 *(a)*

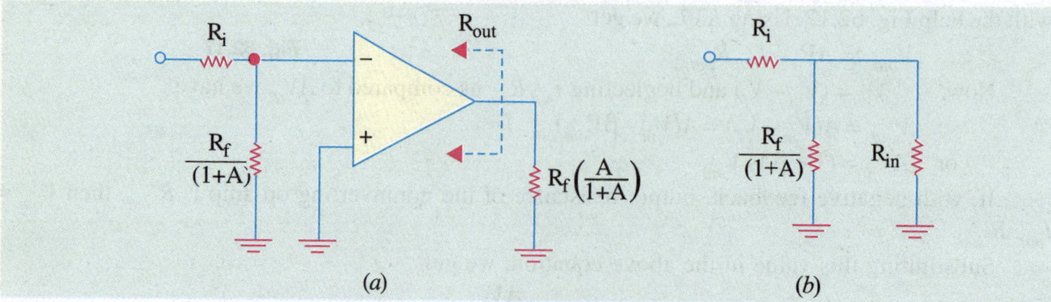

Fig. 62.20

As shown in Fig. 62.20 *(b)*, the Miller input resistance appears in parallel with the internal resistance of the op-amp (without feedback) and R_i appears in series with this

$$\therefore \quad R'_{in} = R_i + \frac{R_f}{(1+A)} \parallel R_{in}$$

Typically, the term $\dfrac{R_f}{(1+A)}$ is much less than R_{in} of an open-loop op-amp. Hence,

$$\frac{R_f}{(1+A)} \| R_{in} \cong \frac{R_f}{1+A}$$

Moreover, $A \gg 1$, hence, $\quad \therefore \quad R'_{in} \cong R_i + \dfrac{R_f}{A}$

Now, R_i appears in series with (R_f/A) and if $R_i \gg R_f/A$, we have, $R'_{in} \cong R_i$

As seen from Fig. 62.20 (*b*), Miller output resistance appears in parallel with R_{out} of the op-amp.

$$\therefore \quad R'_{out} = R_f \left(\frac{A}{1+A} \right) \| R_{out}$$

Normally, $A \gg 1$ and $R_f \gg R_{out}$ so that R'_{out} simplifies to $R'_{out} = R_{out}$

Example 62.19. *For the inverting op-amp circuit of Fig. 62.21, find (a) input and output resistances and (b) the closed-loop gain. The op-amp has the following parameters :*

$A = 100{,}000$, $\quad R_{in} = 5\ M\Omega$. and $R_{out} = 50\ \Omega$

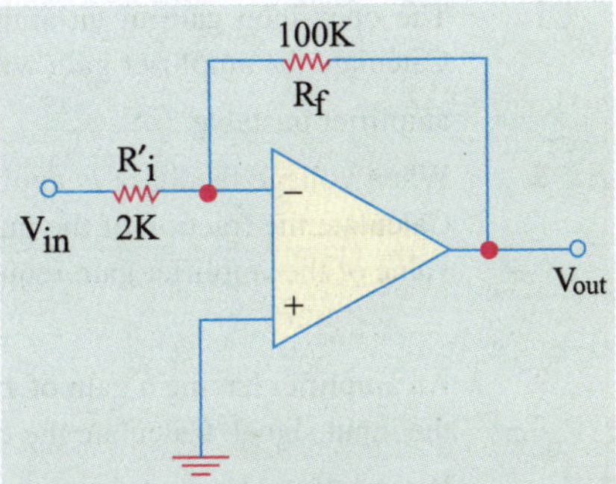

Fig. 62.21

Solution. (*a*) $R'_{in} \cong R_i \cong \mathbf{2\ k\Omega}$

$R'_{out} \cong R_{out} = \mathbf{50\ \Omega}$

(*b*) $A' = \dfrac{R_f}{R_i} = -\dfrac{100}{2} = \mathbf{50}$

The negative sign indicates the inherent sign inversion in the process.

Tutorial Problems No. 62.1

1. For the series-parallel feedback amplifier shown in Fig. 62.22. Calculate

(*i*) open-loop gain,

(*ii*) gain of feedback loop,

(*iii*) closed-loop gain,

(*iv*) sacrifice factor.

[(*i*) 10^6 (*ii*) 0.025 (*iii*) 40 (*iv*) 25.000]

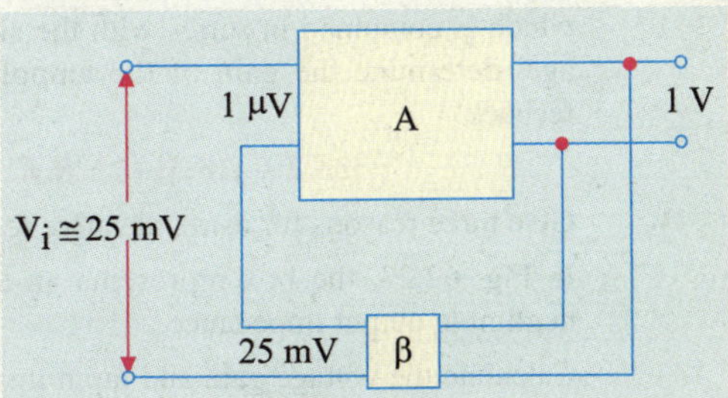

Fig. 62.22

2. A negative-feedback amplifier has the following parameters :

$A = 200$, $\beta = 0.02$ and $V_i = 5$ mV

Compute the following :

(*i*) gain with feedback, (*ii*) output voltage,

(*iii*) feedback factor, (*iv*) feedback voltage.

[(*i*) 40 (*ii*) 200 mV (*iii*) 4 (*iv*) 4 mV]

3. An amplifier has an open-loop gain of 500 and a feedback of 0.1. If open-loop gain changes by 25% due to temperature etc., find the percentage change in closed-loop gain. [0.6%]

4. An *RC*-coupled amplifier has a mid-frequency gain of 400 and lower and upper 3-dB frequencies of 100 Hz and 10 kHz. A negative feedback network with β = 0.01 is incorporated into the amplifier circuit. Calculate

(*i*) gain with feedback,
(*ii*) new bandwidth. **[(*i*) 80 (*ii*) 50 kHz (approx)]**

5. In an amplifier with constant signal input of 1 volt, the output falls from 50 to 25 V when feedback is applied. Calculate the fraction of the output which is fed back. If, due to ageing, the amplifier gain fall to 40, find the percentage reduction in stage gain
(*i*) without feedback (*ii*) with the feedback connection.
[0.02% (*i*) 20% (*ii*) 11.12%]

6. An amplifier has a gain of 1000 without feedback. Calculate the gain when 0.9 per cent of negative feedback is applied. If, due to ageing, gain without feedback falls to 800, calculate the percentage reduction in gain (*a*) without feedback and (*b*) with feedback. Comment on the significance of the results of (*a*) and (*b*) and state two other advantages of negative feedback.
[100 (*a*) 20% (*b*) 2.44%](*City & Guilds, London*)

7. The open-loop gain of an amplifier is $1000 \angle 70°$ and the feedback factor is $-0.02 \angle 20°$. Calculate the amplifier gain with negative feedback. What is the limiting value of β to make the amplifier unstable ? **[$49.9 \angle -17.1°$; $0.001 \angle -70°$] (*I.E.E. London*)**

8. When voltage feedback is applied to an amplifier of gain 100, the overall stage gain falls to 50. Calculate the fraction of the output voltage fed back. If this fraction is maintained, calculate the value of the amplifier gain required if the overall stage gain is to be 75. **[0.01 ; 300]**
(*City & Guilds, London*)

9. An amplifier having a gain of 100 has 9 per cent voltage negative feedback applied in series with the input signal. Calculate the overall stage with feedback.

If a supply voltage variation causes the gain with feedback to fall by 10 percent, determine the percentage change in gain with feedback.

[10; 52.6%] (*City & Guilds, London*)

10. If the gain of an amplifier without feedback is (800 – *j*100) and the feedback network of β = –1/(40 – *j*20) modifies the output voltage to V_{fb} which is combined in series with the signal voltage, determine the gain of the ampplifier with feeback.

[38.3 – j18.3] (*I.E.R.E., London*)

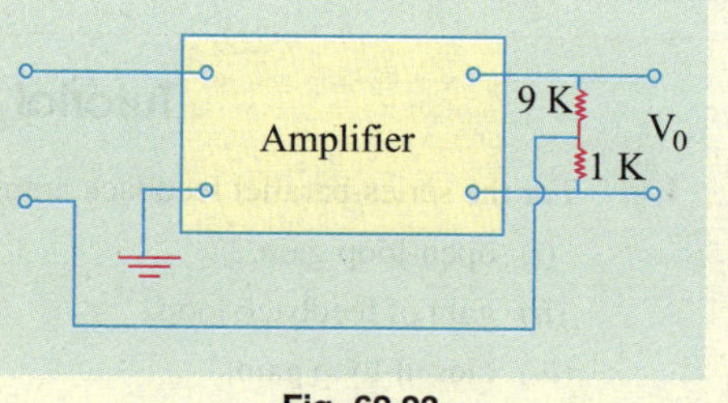

Fig. 62.23

11. Give three reasons for using negative feedback.

In Fig. 62.23, the box represents an amplifier of gain –1000, input impedance 500 kΩ and negligible output impedance.

Calculate the voltage gain and input impedance of the amplifier with feedback.

[– 9.9, 50.5 MΩ]

12. An amplifier with negative feedback has a voltage gain of 100. It is found that without feedback an input signal of 50 mV is required to produce a given output; whereas with feedback, the input signal must be 0.6V for the same output. Calculate the value of voltage gain without feedback and feedback ratio. **(A = 1200, β = 0.09)**
(Electronics Engg., Bangalore Univ. 2001)

OBJECTIVE TESTS – 62

1. The advantage of using negative feedback in an amplifier is that its gain can be made practically independent of
(*a*) temperature changes
(*b*) age of components
(*c*) frequency
(*d*) all of the above.

2. Feedback in an amplifier always helps to
   (*a*) control its output
   (*b*) increase its gain
   (*c*) decrease its input impedance
   (*d*) stabilize its gain.
3. The only drawback of using negative feedback in amplifiers is that it involves
   (*a*) gain sacrifice
   (*b*) gain stability
   (*c*) temperature sensitivity
   (*d*) frequency dependence.
4. Closed-loop gain of a feedback amplifier is the gain obtained when
   (*a*) its output terminals are closed
   (*b*) negative feedback is applied
   (*c*) feedback loop is closed
   (*d*) feedback factor exceeds unity.
5. A large sacrifice factor in a negative feedback amplifiers leads to
   (*a*) inferior performance
   (*b*) increased output impedance
   (*c*) characteristics impossible to achieve without feedback
   (*d*) precise control over output.
6. Negative feedback in an amplifier
   (*a*) lowers its lower 3 dB frequency
   (*b*) raises its upper 3 dB frequency
   (*c*) increases its bandwidth
   (*d*) all of the above.
7. Regarding negative feedback in amplifiers which statement is WRONG ?
   (*a*) it widens the separation between 3 dB frequencies
   (*b*) it increases the gain-bandwidth product
   (*c*) it improves gain stability
   (*d*) it reduces distortion.
8. Negative feedback reduces distortion in an amplifier only when it
   (*a*) comes as part of input signal
   (*b*) is part of its output
   (*c*) is generated within the amplifier
   (*d*) exceeds a certain safe level.
9. An amplifier with no feedback has a gain-bandwidth product of 4 MHz. Its closed-loop gain is 40. The new band-width is
   (*a*) 100 kHz
   (*b*) 160 MHz
   (*c*) 10 MHz
   (*d*) 20 kHz.
10. The shunt-derived series-fed feedback in an amplifier
    (*a*) increases its output impedance
    (*b*) decreases its output impedance
    (*c*) increases its input impedance
    (*d*) both (*b*) and (*c*).
11. A feedback amplifier has a closed gain of –200. It should not vary more than 50% despite 25% variation in amplifier gain *A* without feedback. The value of *A* is
    (*a*) 800
    (*b*) – 800
    (*c*) 1000
    (*d*) –1000
12. The gain of a negative feedback amplifier is 40 dB. If the attenuation of the feedback path is 50 dB, then the gain of the amplifier without feedback is
    (*a*) 78.92
    (*b*) 146.32
    (*c*) 215.51
    (*d*) 317.23
13. In a common emitter amplifier, the unbypassed emitter resistor provides
    (*a*) voltage-shunt feedback
    (*b*) current-series feedback
    (*c*) negative-voltage feedback
    (*d*) positive-current feedback

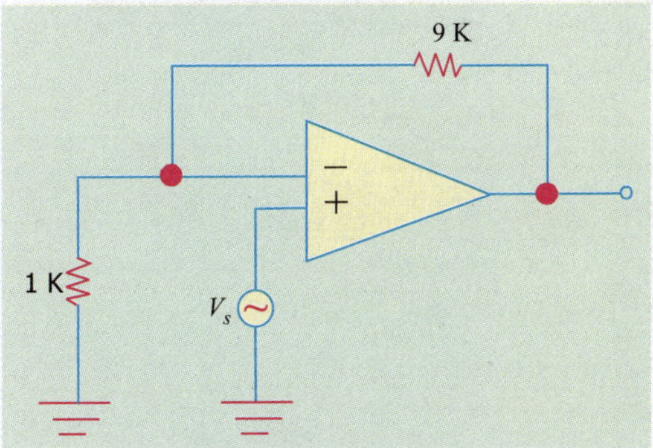

Fig. 62.24

14. The *OP-AMP* circuit shown in Fig. 62.24 has an input impedance of MΩ and an open-loop gain of 10^5. The output impedance seen by the source V_s is

(*a*) 10^{11}Ω

(*b*) 10^{10} Ω

(*c*) 10 kΩ

(*d*) 1 kΩ

15. An *OP-AMP* with an open-loop gain of 10,000, R_{in} = 2 K Ω and R_0 = 500 Ω is used in the non-inverting configuration shown in Fig. 62.25. The output resistance R'_0 is

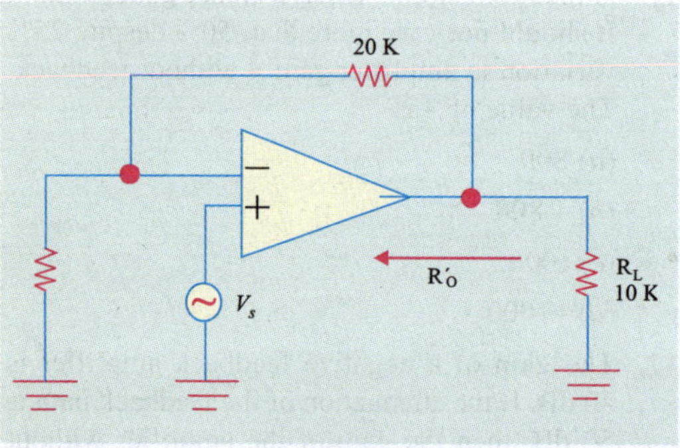

Fig. 62.25

(*a*) 250.5 Ω (*b*) 21 Ω

(*c*) 2 Ω (*d*) 0.998 Ω

16. The feedback used in the circuit shown in Fig. 62.25 can be classified as

(*a*) shunt-series feedback

(*b*) shunt-shunt feedback

(*c*) series-shunt feedback

(*d*) series-series feedback

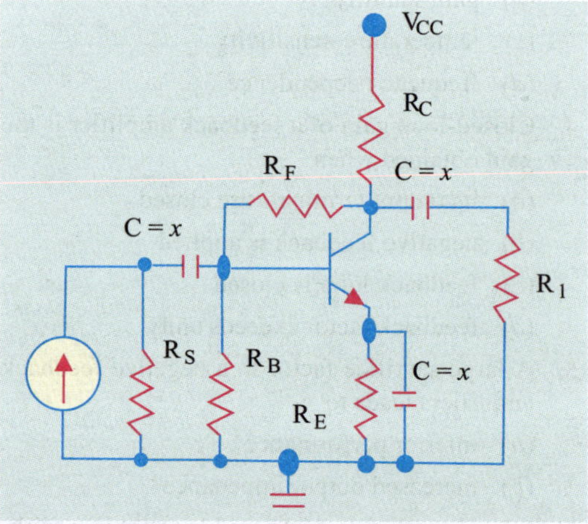

Fig. 62.25

ANSWERS

1. (*d*)	**2.** (*a*)	**3.** (*a*)	**4.** (*c*)	**5.** (*c*)	**6.** (*d*)	**7.** (*b*)	**8.** (*c*)
9. (*a*)	**10.** (*d*)	**11.** (*d*)	**12.** (*b*)	**13.** (*b*)	**14.** (*b*)	**15.** (*d*)	**16.** (*b*)

C H A P T E R 63

Learning Objectives

- ➢ What is a FET
- ➢ Junction FET (JFET)
- ➢ Static Characteristics of a JFET
- ➢ JFET Drain Characteristic with $V_{GS} = 0$
- ➢ Characteristics with External Bias
- ➢ Transfer Characteristic
- ➢ Small Signal JFET Parameters
- ➢ D.C. Baising of a JFET
- ➢ DC Load Line
- ➢ Common Source JFET Amplifier
- ➢ JFET Amplifier Gains
- ➢ Advantages of FETs
- ➢ MOSFET or IGFET—DE MOSFET
- ➢ Schematic Symbols for a DEMOSFET
- ➢ Static Characteristics of a DEMOSFET
- ➢ Enhancement-only N-Channel MOSFET
- ➢ Biasing E-only MOSFET—FET Amplifiers
- ➢ FET Applications
- ➢ MOSFET Handling

FIELD EFFECT TRANSISTORS

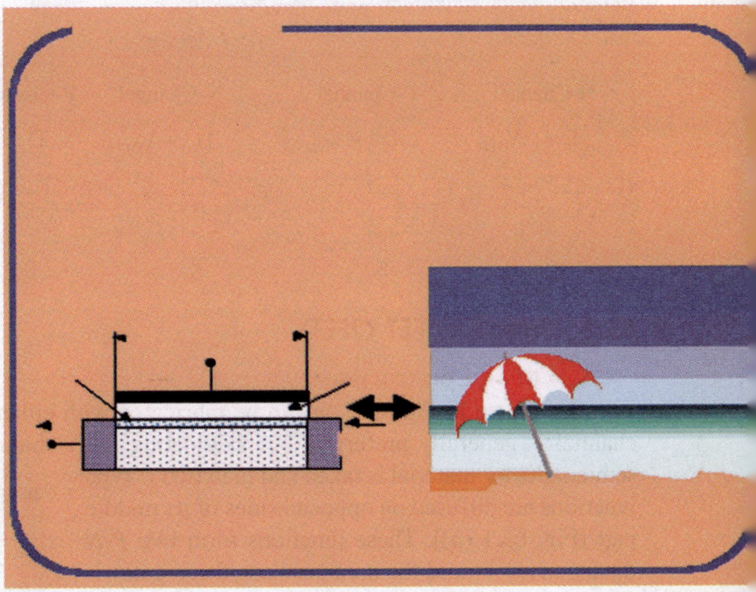

Field Effect Transistor as an Electronic Flute

63.1. What is a FET ?

The acronym 'FET' stands for **field effect transistor**. It is a three-terminal unipolar solid-state device in which current is controlled ***by an electric field*** as is done in vacuum tubes. Broadly speaking, there are two types of FETs :

(*a*) junction field effect transistor (JFET)

(*b*) metal-oxide semiconductor FET (MOSFET)

It is also called insulated-gate FET (IGFET). It may be further subdivided into :

(*i*) depletion-enhancement MOSFET *i.e.* DEMOSFET

(*ii*) enhancement-only MOSFET *i.e.* E-only MOSFET

Both of these can be either *P*-channel or *N*-channel devices.

The FET family tree is shown below :

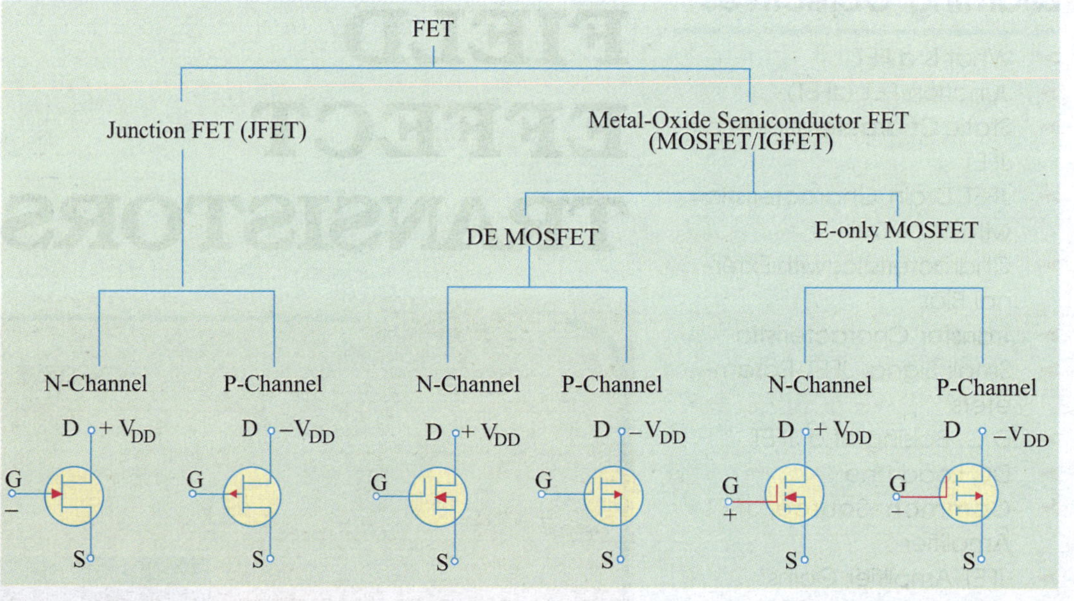

63.2. Junction FET (JFET)

(*a*) Basic Construction

As shown in Fig. 63.1, it can be fabricated with either an *N*-channel or *P*-channel though *N*-channel is generally preferred. For fabricating an *N*-channel JFET, first a narrow bar of *N*-type semiconductor material is taken and then two *P*-type junctions are diffused on opposite sides of its middle part [Fig. 63.1 (*a*)]. These junctions form two *P-N* diodes or **gates** and the area between these gates is called **channel**. The two *P*-regions are internally connected and a single lead is brought out which is called **gate terminal**. Ohmic contacts (direct electrical connections) are made at the two ends of the bar-one lead is called **source terminal** *S* and the other **drain terminal** *D*. When potential difference is established between drain and source, current flows along the length of the 'bar' through the channel located between the two *P*-regions. The current consists of **only majority carriers** which, in the present case, are electrons. *P*-channel JFET is

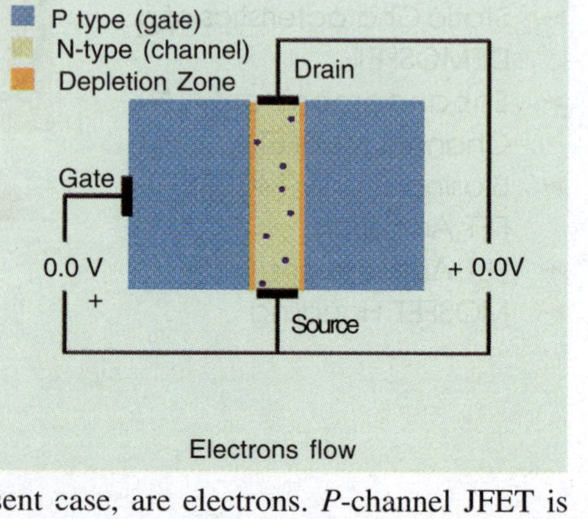

similar in construction except that it uses *P*-type bar and two *N*-type junctions. The majority carriers are holes which flow through the channel located between the two N-regions or gates.

Following FET notation is worth remembering:

1. **Source.** It is the terminal through which majority carriers enter the bar. Since carriers come from it, it is called the source.
2. **Drain.** It is the terminal through which majority carriers ***leave*** the bar *i.e.* they are drained out from this terminal. The drain-to-source voltage V_{DS} drives the drain current I_D.
3. **Gate.** These are two internally-connected heavily-doped impurity regions which form two *P-N* junctions. The gate-source voltage V_{GS} reverse-biases the gates.
4. **Channel.** It is the space between two gates through which majority carriers pass from source-to-drain when V_{DS} is applied.

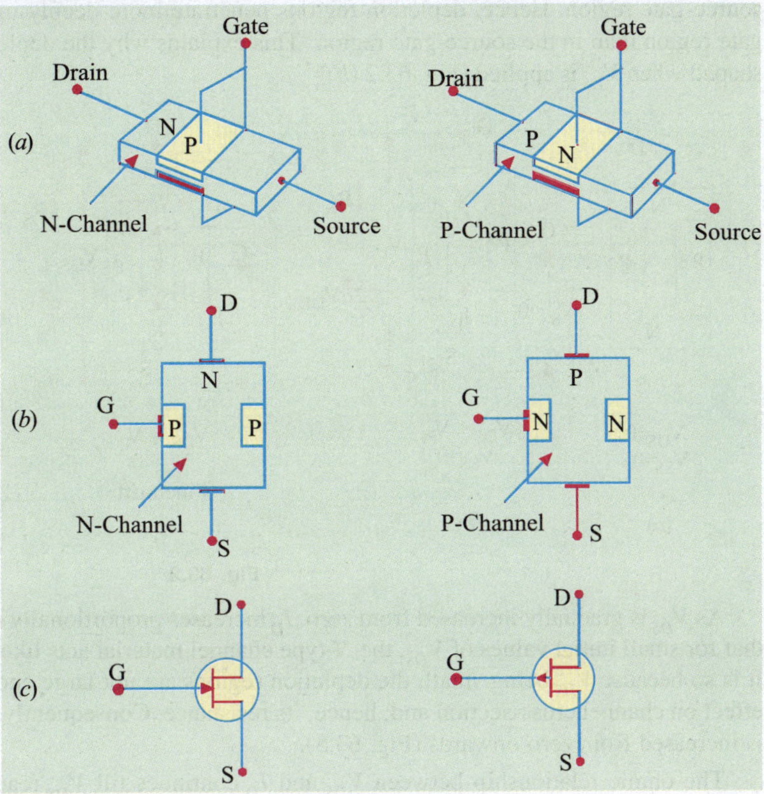

Fig. 63.1

Schematic symbols for *N*-channel and *P*-channel JFET are shown in Fig. 63.1 (*c*). It must be kept in mind that **gate arrow always points to N-type material**.

(*b*) Theory of Operation

While discussing the theory of operation of a JFET, it should be kept in mind that

1. Gates are always reversed-biased. Hence, gate current I_G is practically zero.
2. The source terminal is always connected to that end of the drain supply which provides the necessary charge carriers. In an *N*-channel JFET, source terminal *S* is connected to the negative end of the drain voltage supply (for obtaining electrons). In a *P*-channel JFET, *S* is connected to the positive end of the drain voltage supply for getting holes which flow through the channel.

Let us now consider an ***N-channel*** JFET and discuss its working when either V_{GS} or V_{DS} or both are changed.

(*i*) When $V_{GS} = 0$ and $V_{DS} = 0$

In this case, drain current $I_D = 0$, because $V_{DS} = 0$. The depletion regions around the *P-N* junctions are of equal thickness and symmetrical as shown in Fig. 63.2 (*a*)

(*ii*) When $V_{GS} = 0$ and V_{DS} is increased from zero

For this purpose, the JFET is connected to the V_{DD} supply as shown in Fig. 63.2 (*b*). The electrons (which are the majority carriers) flow from *S* to *D* whereas conventional drain current I_D flows through the channel from *D* to *S*. Now, the gate-to-channel bias at any point along the channel is = $|V_{DS}| + |V_{GS}|$ *i.e.* the numerical sum of the two voltages. In the present case, external bias

$V_{GS} = 0$. Hence gate-channel reverse bias is provided by V_{DS} alone. Since the value of V_{DS} keeps decreasing (due to progressive drop along the channel) as we go from *D* to *S*, the gate-channel bias also decreases accordingly. It has maximum value in the drain-gate region and minimum in the source-gate region. Hence, depletion regions penetrate more deeply into the channel in the drain-gate region than in the source-gate region. This explains why the depletion regions become wedge shaped when V_{DS} is applied [Fig. 63.2 (*b*)]

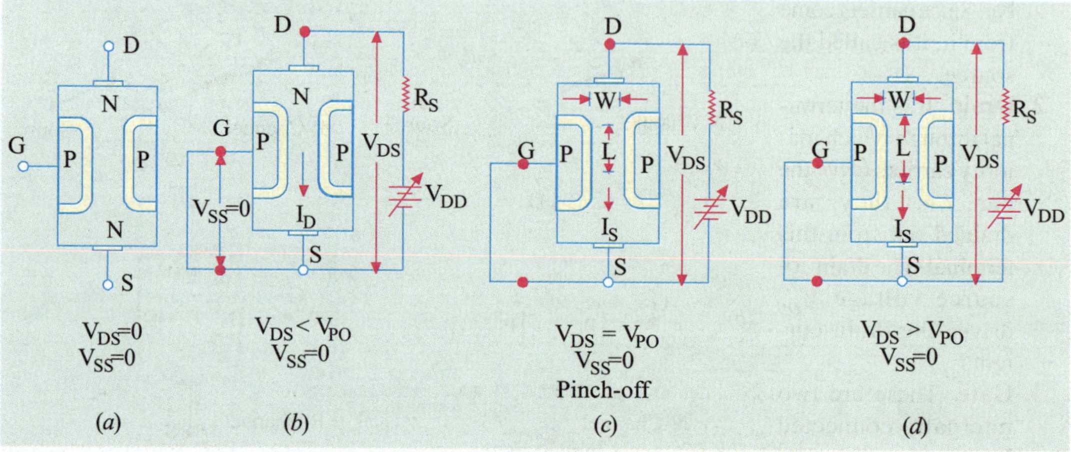

Fig. 63.2

As V_{DS} is gradually increased from zero, I_D increases proportionally as per Ohm's law. It is found that for small initial values of V_{DS}, the *N*-type channel material acts like a resistor of constant value. It is so because V_{DS} being small, the depletion regions are not large enough to have any significant effect on channel cross-section and, hence, its resistance. Consequently, I_D increases linearly as V_{DS} is increased from zero onwards (Fig. 63.5).

The ohmic relationship between V_{DS} and I_D continues till V_{DS} reaches a certain critical value called ***pinch-off voltage*** V_{PO} when drain current becomes constant at its maximum value called I_{DSS}. The *SS* in I_{DSS} indicates that the gate is shorted to source to make sure that $V_{GS} = 0$. This current is also known as ***zero-gate-voltage drain current.*** It is seen from Fig. 63.2(*c*) that under pinch-off conditions, separation between the depletion regions near the drain end reaches a minimum value *W*. It should, however, be carefully noted that ***pinch-off does not mean 'current-off'.*** In fact, I_D is maximum at pinch-off.

When V_{DS} is increased beyond V_{PO}, I_D remains constant at its maximum value I_{DSS} upto a certain point. It is due to the fact that further increase in V_{DS} (beyond V_{PO}) causes more of the channel on the source end to reach the minimum width as shown in Fig. 63.2 (*d*). It means that the channel width does not increase, instead its length *L* increases. As more of the channel reaches the minimum width, the resistance of the channel increases at the same rate at which V_{DS} increases. In other words, increase in V_{DS} is neutralized by increases in R_{DS}. Consquently, $I_D = (V_{DS} / R_{DS})$ remains unchanged even though V_{DS} is increased. Ultimately, a certain value of V_{DS} (called V_{DSO}) is reached when JFET breaks down and I_D increases to an excessive value as seen from drain characteristic of Fig. 63.5.

***(iii)* When $V_{DS} = 0$ and V_{GS} is decreased from zero**

In this case, as V_{GS} is made more and more negative, the gate reverse bias increases which increases the thickness of the depletion regions. As negative value of V_{GS} is increased, a stage comes when the two depletion regions touch each other as shown in Fig. 63.3. In this condition, the channel is said to be cut-off. This value of V_{GS} which cuts off the channel and

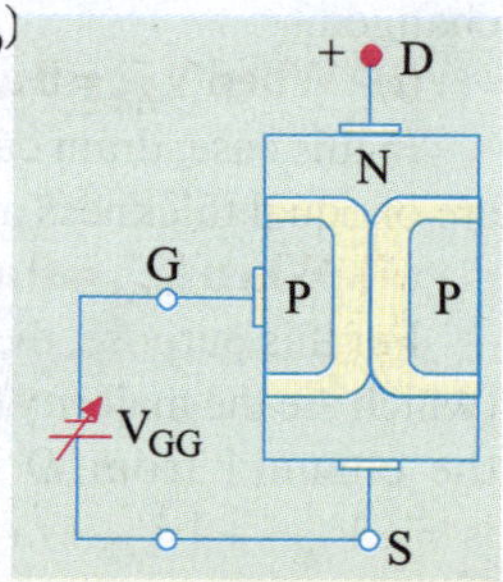

Fig. 63.3

hence the drain current is called $V_{GS(off)}$.*

It may be noted that $V_{GS(off)} = -V_{PO}$ or $|V_{PO}| = |V_{GS(off)}|$. As seen from Fig. 63.6 because V_{PO} = 4 V, $V_{GS(off)} = -4$ V. Obviously, their absolute values are equal.

(iv) When V_{GS} is negative and V_{DS} is increased

As seen from Fig. 63.6, as V_{GS} is made more and more negative, values of V_P as well as breakdown voltage are decreased.

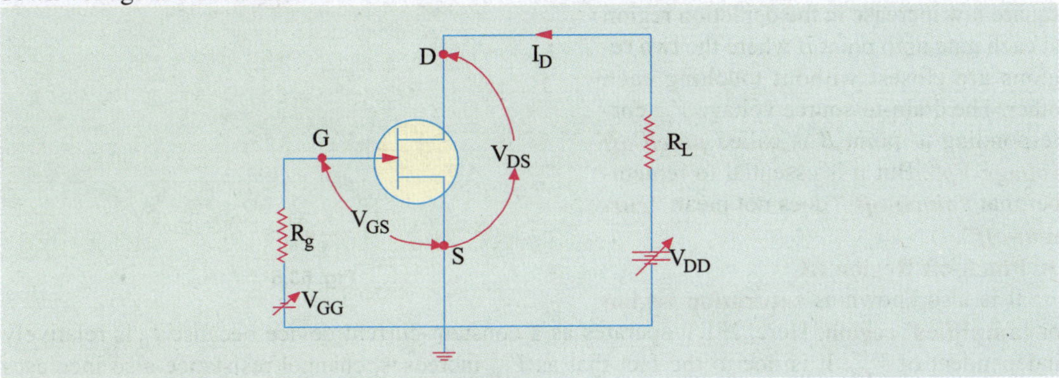

Fig. 63.4

Summary. Summarizing the above, we have that

(i) keeping V_{GS} at a fixed value (either zero or negative), as V_{DS} is increased, I_D initially increases till channel pinch-off when it becomes almost constant and finally increases excessively when JFET breaks down under high value of V_{DS}. As V_{GS} is kept fixed at progressively higher negative values, the values of V_P as well as breakdown voltage decrease.

(ii) keeping V_{DS} at a fixed value, as V_{GS} is made more and more negative, I_D decreases till it is reduced to zero for a certain value of V_{GS} called $V_{GS(off)}$.

Since gate voltage controls the drain current, JFET is called a ***voltage-controlled*** device. A *P*-channel JFET operates exactly in the same manner as an *N*-channel JFET except that current carriers are holes and polarities of both V_{DD} and V_{GG} are reversed.

Since only one type of majority carrier (either electrons or holes) is used in JFETs, they are called ***unipolar devices*** unlike bipolar junction transistors (*BJTs*) which use both electrons and holes as carriers.

63.3. Static Characteristics of a JFET

We will consider the following two characteristics:

(i) drain characteristic

It gives relation between I_D and V_{DS} for different values of V_{GS} (which is called running variable).

(ii) transfer characteristic

It gives relation between I_D and V_{GS} for different values of V_{DS}.

We will analyse these characteristics for an *N*-channel JFET connected in the common-source mode as shown in Fig. 63.4. We will first consider the drain characteristic when $V_{GS} = 0$ and then when V_{GS} has any negative value upto $V_{GS(off)}$.

63.4. JFET Drain Characteristic With $V_{GS} = 0$

Such a characteristic is shown in Fig. 63.5 and has been already discussed briefly in Art. 63.2. It can be subdivided into following four regions :

1. Ohmic Region OA

This part of the characteristic is linear indicating that for low values of V_{DS}, current varies directly with voltage following Ohm's Law. It means that JFET behaves ***like an ordinary resistor*** till

* It has negative value for an N-channel JFET but a positive value or a P-channel JFET.

point *A* (called knee) is reached.

2. Curve AB

In this region, I_D increases at reverse square-law rate upto point *B* which is called **pinch-off point**. This progressive decrease in the rate of increase of I_D is caused by the square law increase in the depletion region at each gate upto point *B* where the two regions are closest without touching each other. The drain-to-source voltage V_{DS} corresponding to point *B* is called ***pinch-off voltage Vp****. But it is essential to remember that **"*pinch-off* "** does not mean **"*current-off*"**.

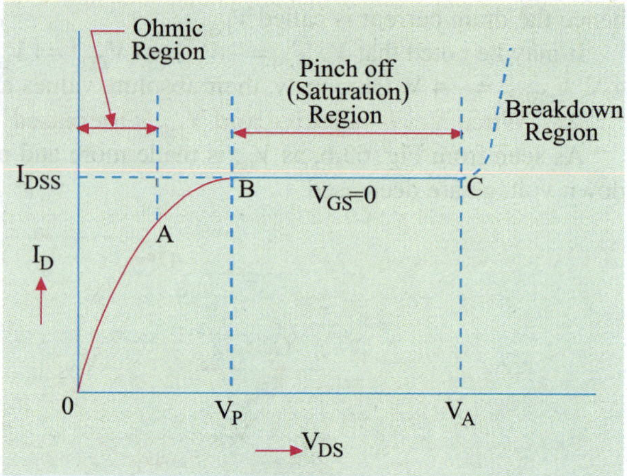

Fig. 63.5

3. Pinch-off Region BC

It is also known as **saturation region** or **'amplified' region**. Here, JFET operates as a constant-current device because I_D is relatively independent of V_{DS}. It is due to the fact that as V_{DS} increases, channel resistance also increases proportionally thereby keeping I_D practically constant at I_{DSS}. It should also be noted that the reverse bias required by the gate-channel junction is supplied entirely by the voltage drop across the channel resistance due to flow of I_{DSS} and none by external bias because $V_{GS} = 0$.

Drain current in this region is given by Shockley's equation

$$I_D = I_{DSS}\left(1-\frac{V_{GS}}{V_P}\right)^2 = I_{DSS}\left(I-\frac{V_{GS}}{V_{GS(off)}}\right)^2$$

It is the normal operating region of the JFET when used as an amplifier.

4. Breakdown Region

If V_{DS} is increased beyond its value corresponding to point *C* (called avalanche breakdown voltage), JFET enters the breakdown region where I_D increases to an excessive value. This happens because the reverse-biased gate-channel *P-N* junction undergoes avalanche breakdown when small changes in V_{DS} produce very large changes in I_D.

It is interesting to note that increasing values of V_{DS} make a JFET behave first as a resistor (ohmic region), then as a constant-current source (pinch-off region) and finally, as a constant-voltage source (breakdown region).

63.5. JFET Characteristics With External Bias

Fig. 63.6 shows a family of I_D versus V_{DS} curves for different values of V_{GS}. It is seen that as the negative gate bias voltage is increased

(i) pinch off voltage is reached at a lower value of I_D than when $V_{GS} = 0$.

(ii) value of V_{DS} for breakdown is decreased.

When an external bias of, say, −1 V is applied between the gate and source, the *P-N* junctions become reverse-biased even when $I_D = 0$. Hence,

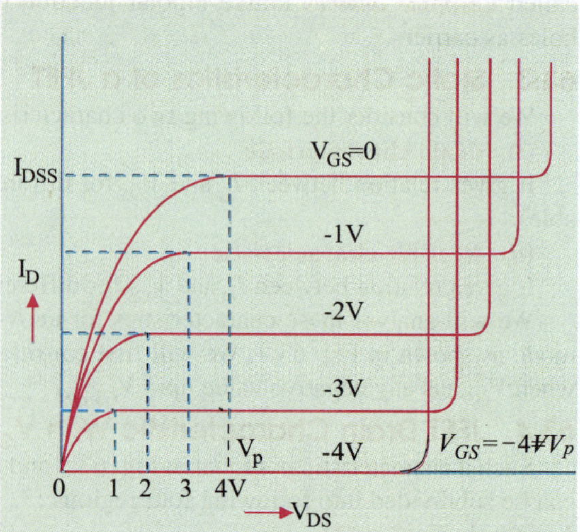

Fig. 63.6

* It is numerically equal to $V_{GS(off)}$ *i.e.* $V_p = / V_{GS(off)} /$

the depletion regions are already formed which penetrate the channel to a certain extent.

The amount of reverse bias required to be produced by I_D would, obviously, be decreased by 1V. In other words, a smaller voltage drop along the channel (*i.e.* smaller than when $V_{GS} = 0$) will increase the depletion regions to the point where they will pinch off the current. Consequently, V_P is reached at a lower I_D value than when $V_{GS} = 0$.

Now, let us see why value of V_{DS} for breakdown is decreased as the negative gate bias voltage is increased. It is simply due to the fact that V_{GS} keeps adding to the reverse bias at the junction produced by current flow.

It is seen that with $V_{GS} = 0$, I_D saturates at I_{DSS} and the characteristic shows $V_P = 4$V. When an external bias of -1 *V* is applied, gate-channel junctions still require -4 V to achieve pinch-off (remember, $V_{GS} = -V_P$). It means that a 3V drop is now required along the channel instead of the previous 4V. Obviously, this 3V drop can be achieved with a lower value of I_D. Similarly, when V_{GS} is -2V and -3V, pinch-off is achieved with 2 V and 1 V respectively along the channel.

This micrograph shows a mock field effect transistor with a layer of crystalline strontium titanate instead of silicon dioxide as the gate electrode.

These drops of 2 V and 1V are obtained with further-reduced values of I_D. As seen, when $V_{GS} = -4$ V (*i.e.* numerically equal to V_P), no channel drop is required. Hence, I_D is zero.

In general, $V_P = V_{DS(P)} - V_{GS}$ where $V_{DS(P)}$ is the pinch-off value of V_{DS} for a given value of V_{GS}.

63.6. Transfer Characteristic

It is a plot of I_D versus V_{GS} for a constant value of V_{DS} and is shown in Fig. 63.7. It is similar to the transconductance characteristics of a vacuum tube or a transistor. It is seen that when $V_{GS} = 0$, $I_D = I_{DSS}$ and when $I_D = 0$, $V_{GS} = V_P$. The transfer characteristic approximately follows the equation.

$$I_D = I_{DSS}\left(1 - \frac{V_{GS}}{V_P}\right)^2 = I_{DSS}\left(1 - \frac{V_{GS}}{V_{GS(off)}}\right)^2$$

The above equation can be written as

$$V_{GS} = V_{GS(off)}\left(1 - \sqrt{\frac{I_D}{I_{DSS}}}\right)$$

This characteristic can be obtained from the drain characteristics by reading off V_{GS} and I_{DSS} values for different values of V_{DS}.

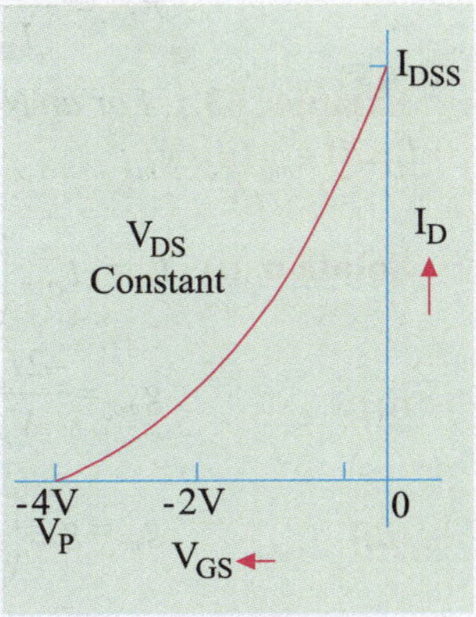

Fig. 63.7

63.7. Small Signal JFET Parameters

The various parameters of a JFET can be obtained from its two characteristics. The main parameters of a JFET when connected in common-source mode are as under :

(*i*) AC Drain Resistance, r_d

It is the ac resistance between drain and source terminals when JFET is operating ***in the pinch-off region.*** It is given by

$$r_d = \frac{\text{change in } V_{DS}}{\text{change in } I_D} \text{ at } V_{GS} \text{ constant or } r_d = \frac{\Delta V_{DS}}{\Delta I_D} \mid V_{GS}$$

An alternative name is ***dynamic drain resistance***. It is given by the slope of the drain characteristic in the pinch-off region. It is sometimes written as r_{ds} emphasizing the fact that it is the resistance from drain to source. Since r_d is usually the output resistance of a JFET, it may also be expressed as an output admittance y_{os}. Obviously, $y_{os} = 1/r_d$. It has a very high value.

(*ii*) Transconductance, g_m

It is simply the slope of transfer characteristic.

$$g_m = \frac{\text{change in } I_D}{\text{change in } V_{GS}} \text{ at } V_{DS} \text{ constant or } g_m = \frac{\Delta I_D}{\Delta V_{GS}} \,|\, V_{DS}$$

Its unit is Siemens (*S*) earlier called **mho.** It is also called ***forward transconductance*** **(g_{fs})** ***or forward transadmittance*** **y_{fs}**.

The transconductance measured at I_{DSS} is written as g_{mo}.

Mathematical Expression for g_m

The Shockley equation* is $I_D = I_{DSS}\left(1 - \frac{V_{GS}}{V_P}\right)^2$

Differentiating both sides, we have

$$\frac{dI_D}{dV_{Gs}} = 2I_{DSS}\left(1 - \frac{V_{GS}}{V_P}\right)\left(-\frac{1}{V_P}\right) \quad \text{or} \quad g_m = -\frac{2I_{DSS}}{V_P}\left(1 - \frac{V_{GS}}{V_P}\right)$$

When $V_{GS} = 0, \quad g_m = g_{mo} \quad \therefore \quad g_{mo} = -\frac{2I_{DSS}}{V_P}$

From the above two equations, we have $g_m = g_{mo}\left(1 - \frac{V_{GS}}{V_P}\right) = g_{mo}\sqrt{\frac{I_D}{I_{DSS}}}$

(*iii*) Amplification Factor, μ

It is given by $\mu = \frac{\text{change in } V_{DS}}{\text{change in } V_{GS}} \text{ at } I_D \text{ constant or } \mu = \frac{\Delta V_{DS}}{\Delta V_{GS}} \,|\, I_D$

It can be proved from above that $\mu = g_m \times r_d = g_{fs} \times r_d$

(*iv*) DC Drain Resistance, R_{DS}

It is also called the static or ohmic resistance of the channel. It is given by

$$R_{DS} = \frac{V_{DS}}{I_D}$$

Example 63.1. *For an N-channel JFET, I_{DSS} = 8.7 mA, V_P = –3 V, V_{GS} = –1 V. Find the values of (i) I_D (ii) g_{mo} (iii) g_m* **(Basic Electronics, Bombay Univ., 1985)**

Solution. (*i*) $I_D = I_{DSS}\left(1 - \frac{V_{GS}}{V_P}\right)^2 = 8.7\left(1 - \frac{-1\text{ V}}{-3\text{ V}}\right)^2$ = **3.87 A**

(*ii*) $g_{mo} = \frac{-2I_{DSS}}{V_P} = \frac{-2 \times 8.7}{-3}$ = **5.8 mS**

(*iii*) $g_m = g_{mo}\left(1 - \frac{V_{GS}}{V_P}\right) = 5.8\left(1 - \frac{-1}{-3}\right)$ = **3.87 mS**

---

* Because of the squared term in the equation, JFET and MOSFET are referred to as square-law devices.

63.8. DC Biasing of a JFET

A JFET may be biased by using either

1. a separate power source V_{GG} as shown in Fig. 63.8 (*a*),
2. some form of self-bias as shown in Fig. 63.8 (*b*),
3. source bias as in Fig. 63.8 (*c*),
4. voltage divider bias as in Fig. 63.8 (*d*).

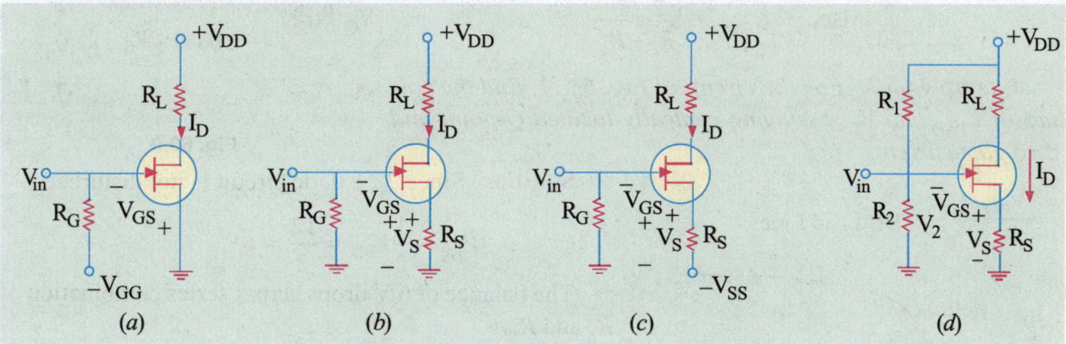

Fig. 63.8

The circuit of Fig. 63.8 (*b*) is called self-bias circuit because the V_{GS} bias is obtained from the flow of JFET's own drawn current I_D through R_S.

$\therefore \quad V_S = I_D R_S$ and $V_{GS} = -I_D R_S$

The gate is kept at this much negative potential with respect to the ground.

The addition of R_G in Fig. 63.8 (*b*), does not upset this dc bias for the simple reason that no gate current flows through it (the gate leakage current is almost zero). Hence, gate is essentially at ***dc ground.*** Without R_G, gate would be kept 'floating' which could collect charge and ultimately cut-off the JFET.

The resistance R_G additionally serves the purpose of avoiding short-circuiting of the ac input voltage, v_{in}. Moreover, in case leakage current is not totally negligible, R_G would provide it an escape route. Otherwise, the leakage current would build up static charge (voltage) at the gate which could change the bias or even destroy the JFET.

Fig. 63.8 (*c*) shows the *source* bias circuit which employs a self-bias resistor R_S to obtain V_{GS}. Here, $V_{SS} = I_D R_S + V_{GS}$ or $V_{GS} = V_{SS} - I_D R_S$.

Fig. 63.8 (*d*) shows the familiar voltage divider bias. In this case, $V_2 = V_{GS} + I_D R_S$ or $V_{GS} = V_2 - I_D R_S$

Since, $\quad V_2 = V_{DD}\dfrac{R_2}{R_1 + R_2}$ $\qquad \therefore V_{GS} = V_{DD}\dfrac{R_2}{R_1+R_2} - I_D R_S$

Example 63.2. *Find the values of V_{DS} and V_{GS} in Fig. 63.9 for $I_D = 4$ mA.*

(Applied Electronics-I, Punjab Univ. 1992)

Solution.

$$V_S = I_D R_S = 4 \times 10^{-3} \times 500 = 2.0 \text{ V}$$
$$V_D = V_{DD} - I_D R_L = 12 - 4 \times 1.5 = 6 \text{ V}$$
$$\therefore \quad V_{DS} = V_D - V_S = 6 - 2 = 4 \text{ V}$$

Since $V_G = 0$, $V_{GS} = V_G - V_S = 0 - 2.0 = -$ **2.0 V**

63.9. DC Load Line

The dc load line for a JFET can be easily drawn by remembering the following two points :

(*i*) At $\quad I_D = 0, \; V_{DS} = V_{DD}$

(*ii*) At $\quad V_{DS} = 0, \; I_D = \dfrac{V_{DD}}{R_L}$

The Q-point is generally situated at the middle point of the load line (for class-A operation) so that

$$V_{DSQ} = \frac{1}{2}V_{DD}$$

$$\text{Also,} \quad I_{DSQ} = \frac{\frac{1}{2}V_{DD}}{R_s + R_L}$$

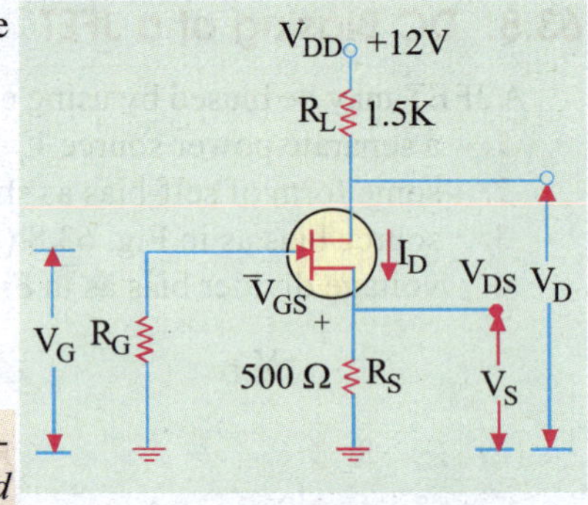

Fig. 63.9

Example 63.3. *For the circuit of Fig. 63.11, find the values of V_{DSQ} and V_{GS} assuming centrally-located Q-point and zero gate current.*

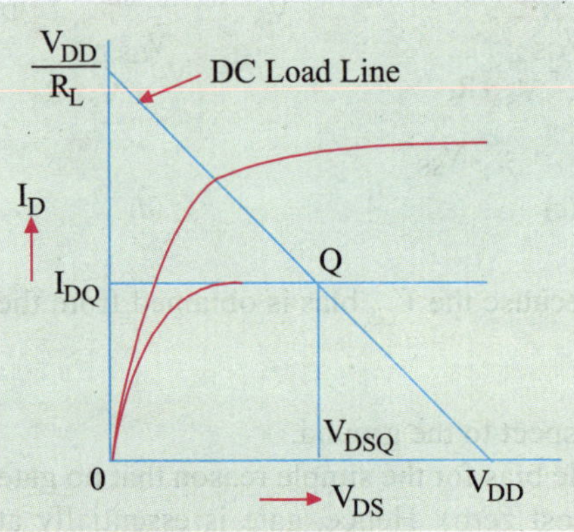

Fig. 63.10

Solution. Since $I_G = 0$, dc circuit is not disturbed.

$$V_{DS} = \frac{1}{2}V_{DD} = \frac{12}{2} = \mathbf{6V}$$

The balance of 6 V drops across series combination of R_L and R_S.

$$\therefore \quad I_D = \frac{6}{150+450} \ \mathbf{10\ mA}$$

$$\therefore \quad V_{GS} = -I_D R_S = -10 \times 150 = \mathbf{-1.5\ V}$$

It is obvious that gate is 1.5 V negative with respect to the source which is the common point. Incidentally, in common source connection of a JFET, ***gate is the most negative point in the entire circuit.***

Example 63.4. *What values of R_S and R_L are required for the circuit of Fig. 63.12 for setting up an approximate mid-point bias ? The JFET parameters are : I_{DSS} = 16 mA, $V_{GS(off)}$ = – 8 V and $V_D = \frac{1}{2}V_{DD}$.*

Solution. It should be noted that as found from Shockley's equation, for mid-point bias, $I_D \cong I_{DSS}/2$ and $V_{GS} = V_{GS(off)}/4$. Hence, for mid-point bias

$$I_D \cong 16/2 = 8 \text{ mA}$$

$$V_{GS} = \frac{1}{4}V_{GS(off)} = -\frac{8}{4} = -2 \text{ V}$$

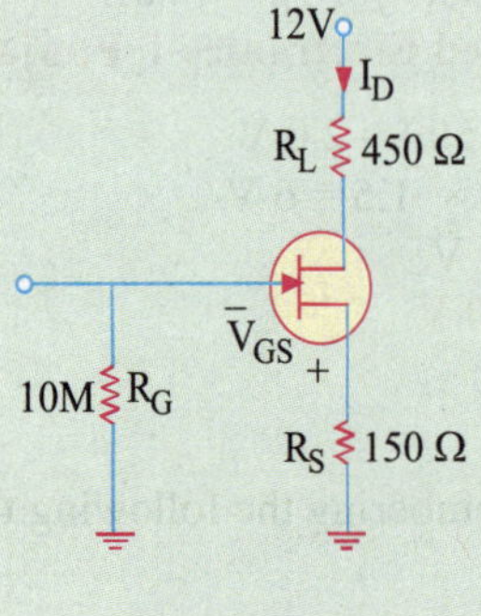

Fig. 63.11

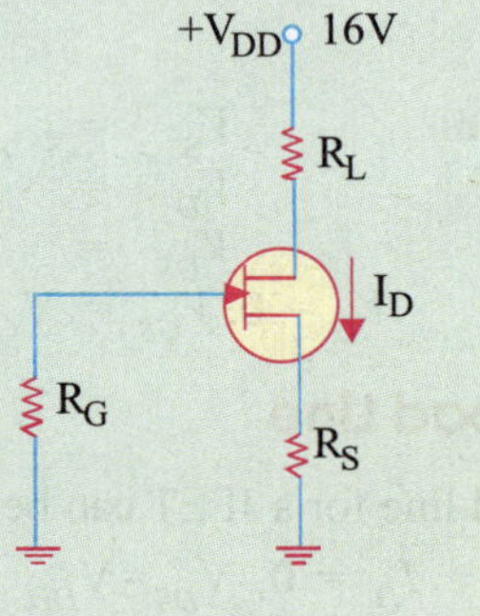

Fig. 63.12

$$R_S = \frac{V_{GS}}{I_D} = \frac{2\text{ V}}{8\,mA} = \mathbf{250\ \Omega}$$

Now, $V_D = V_{DD} - I_D R_L$, $\therefore\ R_L = \dfrac{V_{DD} - V_D}{I_D} = \dfrac{16-8}{8\text{ mA}}$

$$= \mathbf{1000\ \Omega}$$

Example 63.5. *Determine the quiescent value of V_{GS} I_D and V_{DS} for the JFET circuit of Fig. 63.13 given that I_{DSS} = 10 mA, R_S = 5 K and V_P = –5 V.*

(Electronic Devices & Circuits, Pune Univ. 1991)

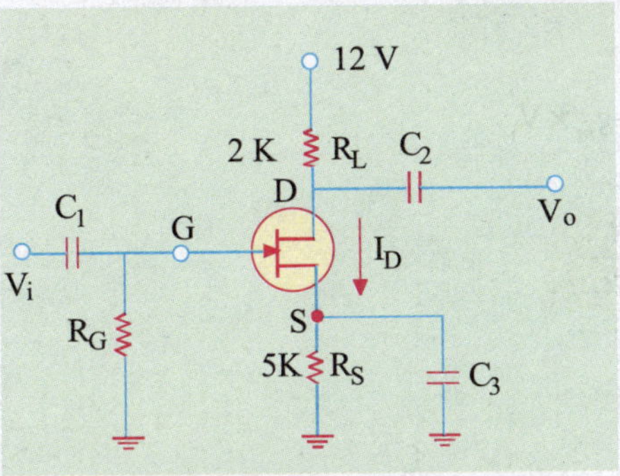

Fig. 63.13

Solution. Since $I_S \cong I_D$, $V_{GS} = -I_D R_S = -5000\,I_D$

Now, $I_D = I_{DSS}\left(1 - \dfrac{V_{GS}}{V_P}\right)^2 = 10 \times 10^{-3}\left(1 - \dfrac{V_{GS}}{-5}\right)^2$

$$= 10 \times 10^{-3}(1 + 0.2\,V_{GS})^2$$

Substituting this value in the above equation, we get

$$V_{GS} = -5000(10 \times 10^{-3})(1 + 0.2\,V_{GS})^2$$

Expanding and rearranging the above, we have

$$2V_{GS}^{\,2} + 21V_{GS} + 50 = 0$$

$$\therefore \quad V_{GS} = -3.65\text{ V} \quad \text{or} \quad -6.85\text{ V}$$

Rejecting the higher value because it is more than V_P, we have $V_{GS} = \mathbf{-3.65\ V}$

$\therefore \quad -3.65 = -5000\,I_D \quad \therefore \quad I_D = \mathbf{0.73\ mA}$

$V_D = V_{DD} - I_D R_L = 12 - 0.73 \times 2 = \mathbf{10.54\ V}$

$V_S = I_D R_S = 0.73 \times 5 = 3.65\ V \quad \therefore \quad V_{DS} = V_D - V_S = 10.54 - 3.65 = \mathbf{6.89\ V}$

63.10. Common Source JFET Amplifier

A simple circuit for such an amplifier is shown in Fig. 63.14. Here, R_G serves the purpose of providing leakage path to the gate current, R_S develops gate bias, C_3 provides ac ground to the input signal and R_L acts as drain load.

Working

When ***negative-going*** signal is applied to the input

1. gate ***bias is increased,***
2. depletion regions are ***widened,***
3. channel resistance is ***increased,***
4. ***I_D is decreased,***
5. drop across R_L is ***decreased,***
6. Consequently, a ***positive-going*** signal becomes available at the output through C_2 in Fig. 63.14.

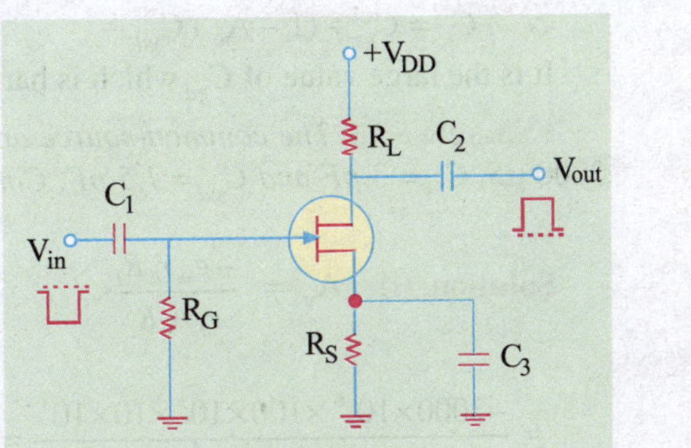

Fig. 63.14

When ***positive-going signal*** is applied at the input, then in a similar way, ***a negative-going** signal* becomes available at the output.

It is seen that there is a ***phase inversion*** between the input signal at the gate and output signal at the drain.

63.11. JFET Amplifier Gains

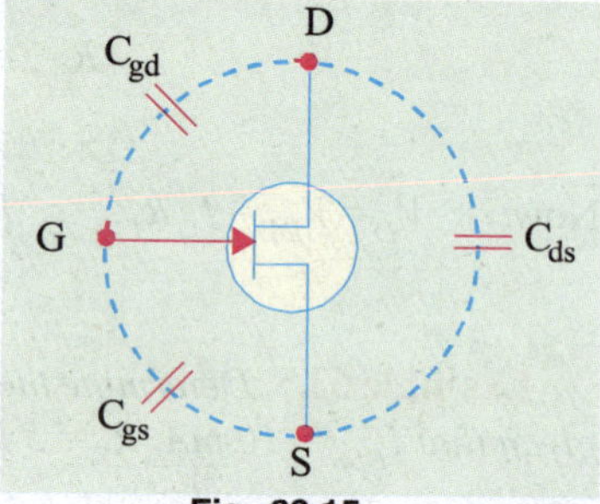

Fig. 63.15

We will now find the expressions for voltage gain, input resistance, output resistance and input capacitance of a JFET amplifier when connected in different modes. The different capacitance due to *P-N* junction and channel are shown in Fig. 63.15

(a) Common Source JFET Amplifier (Fig. 63.16)

(i) Input Resistance $r'_i = R_G \| R_{GS}$

In an ideal JFET, R_{GS} is infinite because $I_G = 0$. In an actual device, however, R_{GS} is not actually infinite but extremely high (100 M or so) as compared to R_G. Hence, $r'_i \cong R_G$.

(ii) Output Resistance $r'_o = r_d \| R_L \cong R_L$ — if $r_d \gg R_L$

(iii) Voltage Gain

$$V_o = i_d \times (r_d \parallel R_L) \qquad \text{Now,} \quad i_d = -g_m \times V_i$$

$$\therefore \quad V_o = -g_m V_i \times (r_d \| R_L)$$

$$\therefore \quad A_v = \frac{V_o}{V_i} = \frac{-g_m V_i \times (r_d \| R_L)}{V_i} = \frac{-g_m r_d R_L}{r_d + R_L} = \frac{-\mu R_L}{r_d + R_L}$$

Also, $A_v = -g_{fs} \times (r_d \| R_L)$

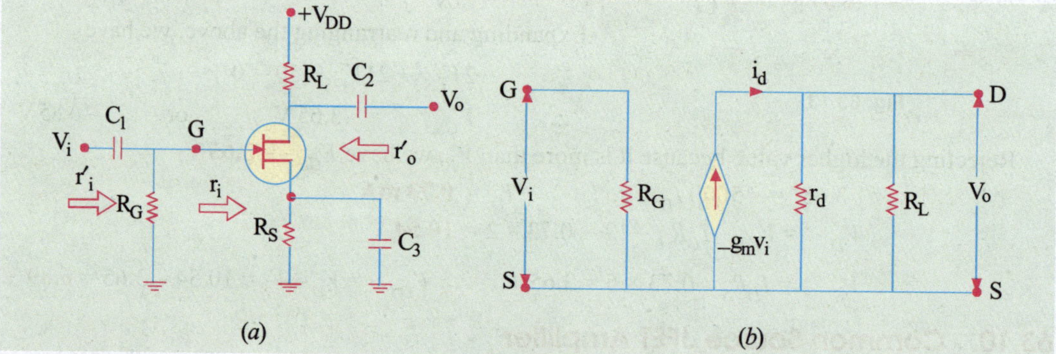

Fig. 63.16

(iv) Input Capacitance

The input capacitance is C_{gs} which is increased because of Miller effect.

$$\therefore \quad C_i = C_{gs} + (1 - A_v)\, C_{gd}.$$

It is the large value of C_{gd} which is harmful in high-frequency work.

Example 63.6. *The common-source amplifier of Fig. 63.16 (a) has $r_d = 100\ K$, $R_L = 10\ K$, $g_m = 3000\ \mu S$, $C_{gs} = 3\ pF$ and $C_{gd} = 1.5\ pF$. Compute its (i) A_v and (ii) C_i.*

Solution. (i) $$A_v = \frac{-g_m r_d R_L}{r_d + R_L}$$

$$= \frac{-3000 \times 10^{-6} \times 100 \times 10^3 \times 10 \times 10^3}{(100 \times 10^3) + (10 \times 10^3)} = \mathbf{-27.3}$$

(ii) $$C_i = C_{gs} + (1 - A_v)\, C_{gd}$$

$$= 3 + (1 + 27.3) \times 1.5 = \mathbf{45.5\ pF}.$$

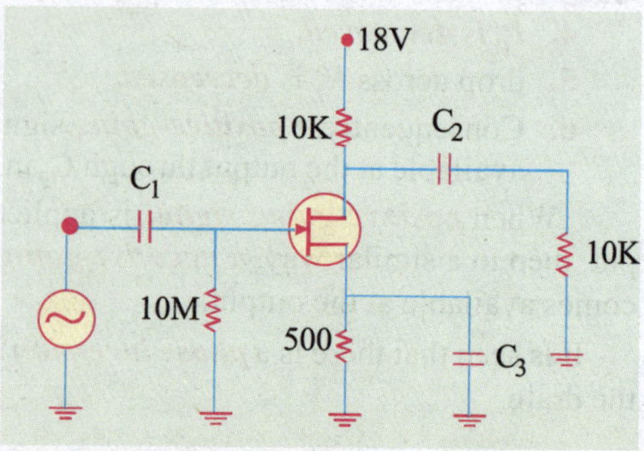

Fig. 63.17

Example 63.7. *The JFET shown in Fig. 63.17 has g_m = 3000 μS and r_{ds} = 100 K. Calculate the voltage gain of the CS amplifier circuit.* **(Basic Electronics, Bombay Univ. 1992)**

Solution. As seen from Art. 63.11 (*a*)

$$A_v = -g_m \times (r_{ds} \parallel r_L)$$

Now, $r_L = 10\text{ K} \parallel 10\text{ K} = \mathbf{5\ K}$

$\therefore \quad r_{ds} \parallel r_L = 100\text{ K} \parallel 5\text{ K} = \mathbf{4.76\ K}$

$\therefore \quad A_v = -3000 \times 10^{-6} \times 4.76 = \mathbf{-14.3 \times 10^{-3}}$

(*b*) Common Drain JFET Amplifier

In the common-drain circuit (also called source follower), the load resistance is in series with the source terminal. There is no drain resistor as shown in Fig. 63.18 (*a*). The input signal is applied to the gate through the capacitor C_1 and the output is taken out from the source *via* C_2. The common-drain equivalent circuit is shown in Fig. 63.18 (*b*). The current generator is $g_m V_{gs}$ where $v_{gs} = (V_i - V_o)$. Moreover, $R_G = R_1 \parallel R_2$.

(*i*) Voltage Gain

$$V_o = i_d \times (r_d \parallel R_L). \qquad \text{Since,} \quad i_d = g_m V_{gs} = g_m (V_i - V_o),$$

$$\therefore \quad V_o = g_m (V_i - V_o) \times (r_d \parallel R_L) = g_m (V_i - V_o) \frac{r_d R_L}{r_d + R_L}$$

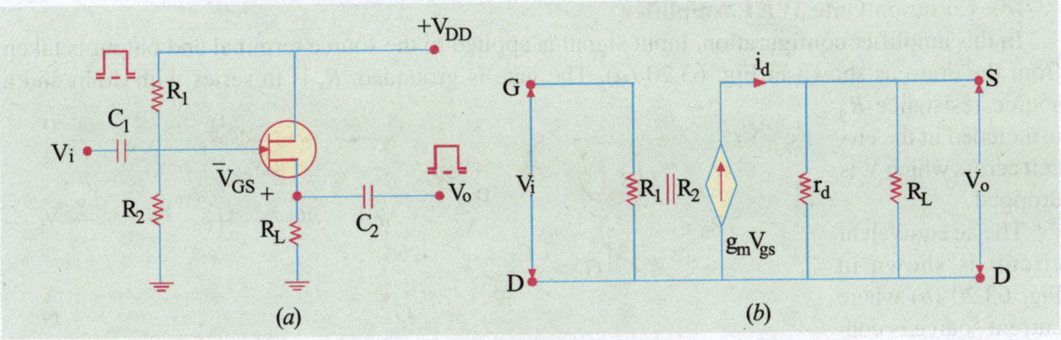

Fig. 63.18

Solving for V_o, we get

$$V_o = g_m V_i \frac{r_d R_L}{r_d + R_L + g_m r_d R_L} \qquad ...(i)$$

$$\therefore \quad A_v = \frac{V_o}{V_i} = \frac{g_m r_d R_L}{r_d + R_L + g_m r_d R_L} \cong 1 \qquad \text{– if } g_m r_d R_L \gg (r_d + R_L)$$

(*ii*) Input Resistance

$$r'_{in} = R_1 \parallel R_2 \qquad \text{– for circuit of Fig. 63.18 (a) only.}$$

(*iii*) Output Resistance

In Eq. (*i*) above, $g_m V_i$ is a current which is directly proportional to V_i and $(r_d R_L)/(r_d + R_L + g_m r_d R_L)$ = a resistance, r'_o.

$$\therefore \quad r'_o = \frac{r_d R_L}{r_d + R_L (1 + g_m r_d)} = \frac{[r_d/(1 + g_m r_d)] \times R_L}{[r_d/(1 + g_m r_d)] + R_L}$$

$$= \frac{r_d}{1 + g_m r_d} \parallel R_L \cong \frac{1}{g_m} \parallel R_L$$

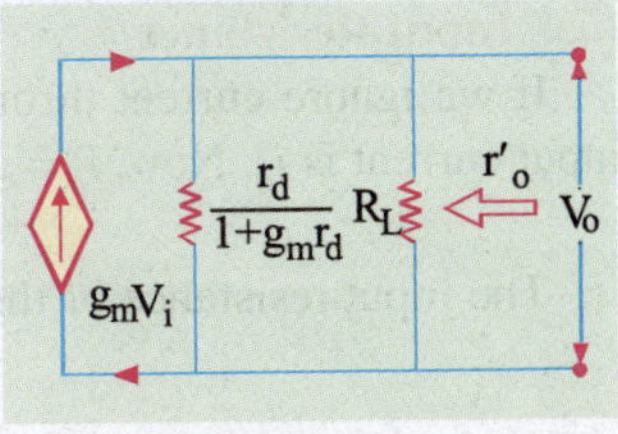

Fig. 63.19

The above result helps us to draw the modified equivalent circuit of Fig. 63.19 for the common-drain amplifier.

Example 63.8. *The common-drain circuit of Fig. 63.18 (a) uses a JFET having the following parameters :*

r_d *=100 K,* $g_m = 3000\ \mu S$ *and* R_L=10 K

Calculate **(i)** A_v *and* **(ii)** r'_o.

Solution. *(i)* $A_v = g_m \dfrac{r_d R_L}{r_d + R_L + g_m r_d R_L}$

$$= 3000 \times 10^{-6} \frac{100 \times 10^3 \times 10 \times 10^3}{(100 \times 10^3) + (10 \times 10^3) + (3000 \times 10^{-6} \times 100 \times 10^3 \times 10 \times 10^3)} = \mathbf{0.965}$$

(ii) $r'_0 = R_L \parallel \dfrac{r_d}{1 + g_m r_d}$

Now, $\dfrac{r_d}{1 + g_m r_d} = \dfrac{100 \times 10^3}{1 + 3000 \times 10^{-6} \times 100 \times 10^3} = \mathbf{332\ \Omega}$

$$r'_o = 10\text{ K} \parallel 332\ \Omega \cong \mathbf{332\ \Omega}$$

(c) Common Gate JFET Amplifier

In this amplifier configuration, input signal is applied to the source terminal and output is taken from the drain as shown in Fig. 63.20 (*a*). The gate is grounded, R_L is in series with drain and a source resistance R_S is included in the circuit across which V_i is dropped.

The ac equivalent circuit is shown in Fig. 63.20 (*b*) where current source is connected between the ***drain and the source terminals as always.*** However, since source and drain are the input and output terminals respectively, ($g_m V_{gs}$) appears between the input and the output.

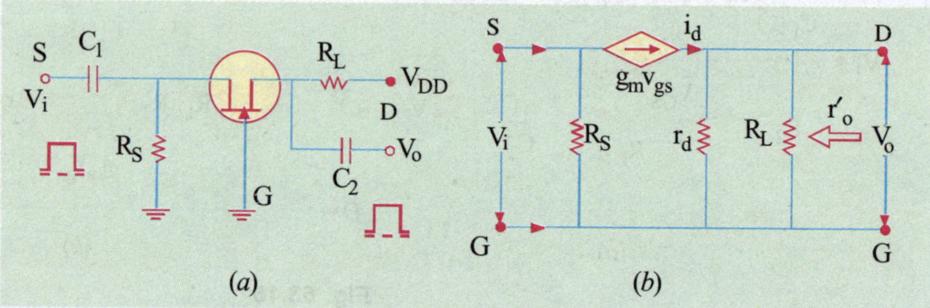

Fig. 63.20

(i) Voltage Gain

$$V_o = i_d \times (r_d \parallel R_L) = i_d \cdot \frac{r_d R_L}{r_d + R_L}$$

Now, $i_d = g_m V_{gs} = g_m V_i$ $\quad \therefore \quad V_o = g_m V_i \cdot \dfrac{r_d R_L}{r_d + R_L}$ $\quad \therefore \quad A_v = \dfrac{V_o}{V_i} = \dfrac{g_m r_d R_L}{r_d + R_L}$

It is the same as the gain for a common source amplifier except that it is a positive quantity. It means that V_o and V_i are in phase as shown in Fig. 63.20 (*a*).

(ii) Input Resistance

If we ignore current through R_S, then as seen from the ac equivalent circuit of Fig. 63.20 (*b*), input current is i_d. Now, $i_d = g_m V_{gs} = g_m V_i$.

The input resistance of the device *i.e.* JFET is $r_i = \dfrac{V_i}{i_d} = \dfrac{V_i}{g_m V_i} = \dfrac{1}{g_m}$

The input resistance of the *circuit* is $r_i' = r_i \parallel R_S = \frac{1}{g_m} \parallel R_s$

Actually speaking, r_d and R_L are also involved in r_i' but their effect is negligible.

(*iii*) Output Resistance

As seen from the equivalent circuit of Fig. 63.20 (*b*), $r_o' = r_d \parallel R_L$.

Here, only input capacitance of importance is C_{gs}. Hence, this circuit has low input capacitance as compared to common-source circuit where input capacitance is increased due to Miller effect.*

63.12. Advantages of FETs

FETs combine the many advantages of both BJTs and vacuum tubes. Some of their main advantages are :

1. high input impedance, **2.** small size, **3.** ruggedness, **4.** long life,
5. high frequency response, **6.** low noise,
7. negative temperature coefficient, hence better thermal stability,
8. high power gain,
9. a high immunity to radiations,
10. no offset voltage when used as a switch (or chopper), **11.** square law characteristics.

The only disadvantages are :

1. small gain-bandwidth product,
2. greater susceptibility to damage in handling them.

Tutorial Problems No. 63.1

1. For a particular *N*-channel JFET, $V_{GS(off)} = -4$ V. What would be the value of I_D when $V_{GS} = -6$ V? **[zero]**

2. For the *N*-channel JFET shown in Fig. 63.21, $V_P = 8$V and $I_{DSS} = 12$ mA. What would be the value of (*i*) V_{DS} at which pinch-off begins and (*ii*) value of I_D when V_{DS} is above pinch-off but below the breakdown voltage ? **[3V ; 12 mA]**

3. The data sheet of a JFET indicates that it has $I_{DSS} = 15$ mA and $V_{GS(off)} = -5$ V. Calculate the value of I_D when V_{GS} is (*i*) 0, (*ii*) –1 V and (*iii*) – 4 V. **[(*i*) 15 mA (*ii*) 9.6 mA (*iii*) 0.6 mA]**

4. The data sheet of a JEET gives the following information.
$I_{DSS} = 20$ mA, $V_{GS(off)} = -8$ V.
and $g_{mo} = 4000$ µS.
Calculate the value of I_D and g_m for $V_{GS} = -4$ V.
[5 mA ; 2000 µS]

5. For a JEET, I_{DSS} 5 mA and g_{mo} = 4000 µS.
Calculate (*i*) $V_{GS(off)}$ and (*ii*) gm at mid-point bias.
[(*i*) –5 V (*ii*) 3000 µS]

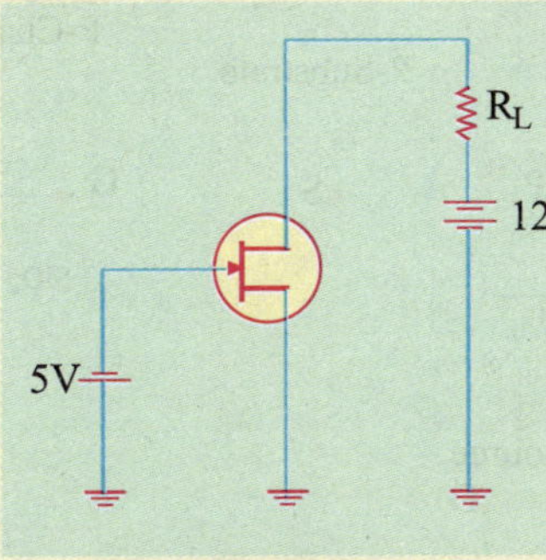

Fig. 63.21

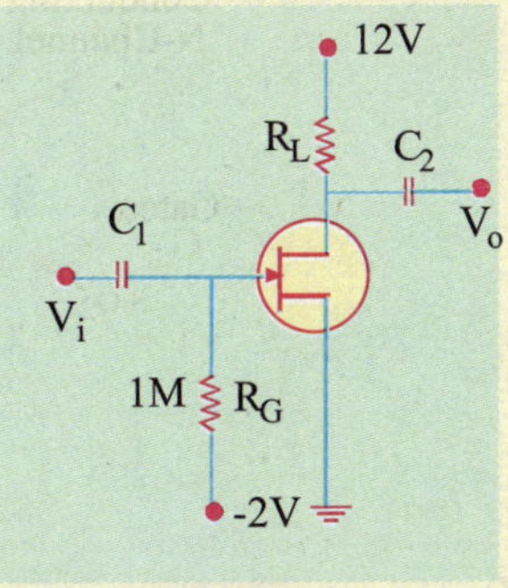

Fig. 63.22

6. At a certain point on the transfer characteristics of an N-channel JFET, following values are read : $I_{DSS} = 8.4$ mA, $V_{GS} = -0.5$ V and $V_P = -3.0$ V.
Calculate (*i*) g_{mo} and (*ii*) g_m at the point. **[(*i*) 5600 mS (*ii*) 4670 mS]**

7. For the JFET circuit of Fig. 63.22, $I_{DSS} = 9$ mA and $V_P = -3$ V.
Find the value of R_L for setting the value of V_{DS} at 7 V. **[5 K]**
(Hint : $V_{GS} = -2$ V as $I_G = 0$)

* Miller effect occurs only where output is antiphase with input as *CS* amplifier circuit.

63.13. MOSFET or IGFET

It could be further subdivided as follows :

(*i*) Depletion-enhancement MOSFET or DE MOSFET

This MOSFET is so called because it can be operated in both ***depletion*** mode and ***enhancement*** mode by changing the polarity of V_{GS}. When negative gate-to-source voltage is applied, the *N*-channel DE MOSFET operates in the depletion mode. However, with positive gate voltage, it operates in the enhancement mode. Since a channel exists between drain and source, I_D flows even when $V_{GS} = 0$. That is why DE MOSFET is known as **normally-ON MOSFET.**

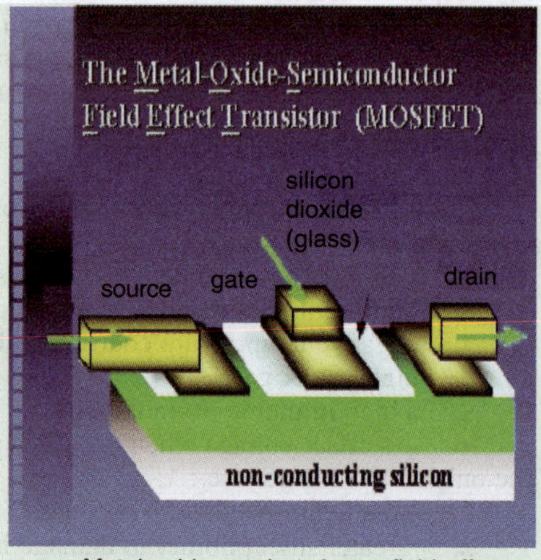

Metal-oxide semiconductor field effect transistor

(*ii*) Enhancement-only MOSFET

As its name indicates, this MOSFET operates only in the enhancement mode and has no depletion mode. It works with ***large positive*** gate voltages only. It differs in construction from the DE MOSFET in that structurally ***there exists no channel between the drain and source.*** Hence, it does not conduct when $V_{GS} = 0$. That is why it is called **normally-OFF** MOSFET.

In a DE MOSFET, I_D flows even when $V_{GS} = 0$. It operates in depletion mode with negative values of V_{GS}. As V_{GS} is made more negative, I_D decreases till it ceases when $V_{GS} = V_{GS(off)}$. It works in enhancement mode when V_{GS} is positive as shown in Fig. 63.24 (*b*).

In the case of E-only MOSFET, I_D flows only when V_{GS} exceeds $V_{GS(th)}$ as shown in Fig. 63.30 (*c*).

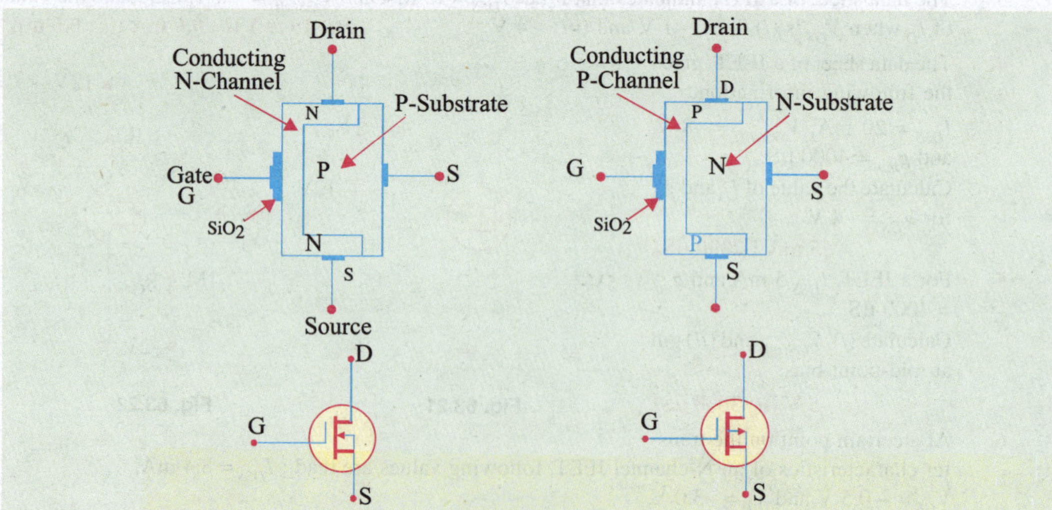

Fig. 63.23

63.14. DE MOSFET

(*a*) Construction

Like JFET, it has source, gate and drain. However, as shown in Fig. 63.23, its gate is insulated from its conducting channel by an ultra-thin metal-oxide insulating film (usually of silicon dioxide SiO_2). Because of this insulating property. MOSFET is alternatively known as insulated-gate field-effect transistor (IGFET or IGT). Here also, gate voltage controls drain current but main difference between JFET and MOSFET is that, in the latter case, we can apply ***both positive and negative voltages to the gate because it is insulated from the channel.*** Moreover, the gate, SiO_2 insulator and channel from a parallel-plate capacitor. Unlike JFET, a DE MOSFET has only one P-region or N-region called substrate. Normally, it is shorted the ***source internally***. Fig. 63.23 shows both *P*-channel and *N*-channel DE MOSFETs along with their symbols.

(*b*) Working

(*i*) Depletion Mode of N-channel DE MOSFET

When $V_{GS} = 0$, electrons can flow freely from source to drain through the conducting channel which exists between them. When gate is given negative voltage, it **depletes** the *N*-channel of its

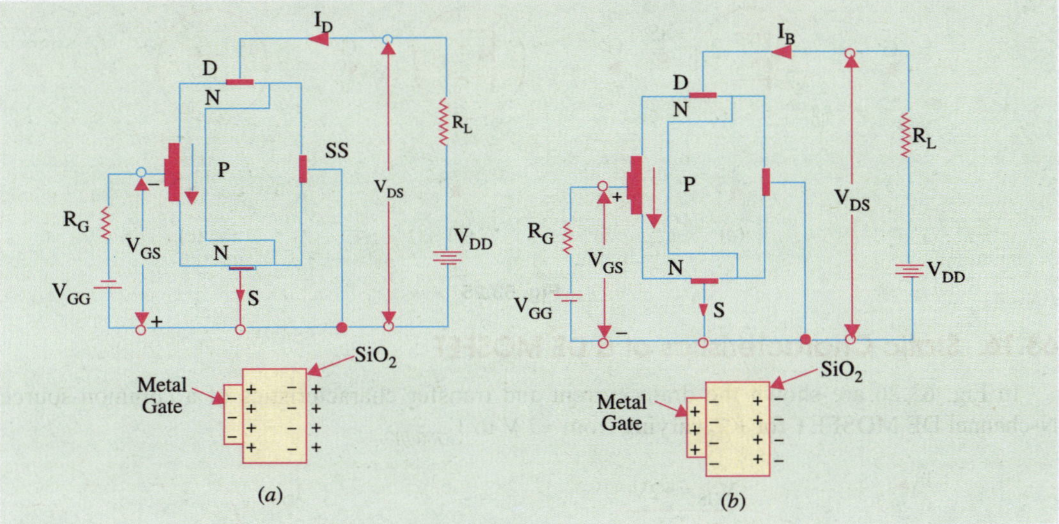

Fig. 63.24

electrons by including positive charge in it as shown in Fig. 63.24 (*a*). Greater the negative voltage on the gate, greater is the reduction in the number of electrons in the channel and, consequently, lesser its conductivity. In fact, too much negative gate voltage called $V_{GS(off)}$ can cut-off the channel. Hence, with negative gate voltage, a DE MOSFET behaves like a JFET.

For obvious reasons, negative-gate operation of a DE MOSFET is called its depletion mode operation.

(*ii*) Enhancement Mode of N-channel DE MOSFET

The circuit connections are shown in Fig. 63.24 (*b*). Again, drain current flows from source to drain even with zero gate bias. When positive voltage is applied to the gate, the input gate capacitor is able to create free electrons in the channel which increases I_D. As seen from the enlarged view of the gate capacitor in Fig. 63.24 (*b*), free electrons are induced in the channel ***by capacitor action.*** These electrons are added to those already existing there. This increased number of electrons ***increases*** or **enhances** the conductivity of the channel. As positive gate voltage is increased, the number of induced electrons is increased, so conductivity of the source-to-drain channel is increased and, consequently, increasing amount of current flows between the terminals. That is why, positive gate operation of a DE MOSFET is known as its **enhancement mode** operation.

Since gate current in both modes is negligibly small, input resistance of a MOSFET is incredibly high varying from $10^{10}\ \Omega$ to $10^{14}\ \Omega$. In fact, MOSFET input current is the leakage current of the capacitor unlike the input current for JFET which is the leakage current of a reverse-biased *P-N* junction.

63.15. Schematic Symbols for a DE MOSFET

Schematic symbol of an *N*-channel normally-ON or DE MOSFET is shown in Fig. 63.25. The gate looks like a metal plate. The arrow is on the substrate and **towards the N-channel**. When *SS* is connected to an external load, we have a 4-terminal device as shown in Fig. 63.25 (*a*) but when it is internally shorted to *S*, we get a 3-terminal device as shown in Fig. 63.25 (*b*). Fig. 63.25 (*c*) shows the symbol for a *P*-channel DE MOSFET.

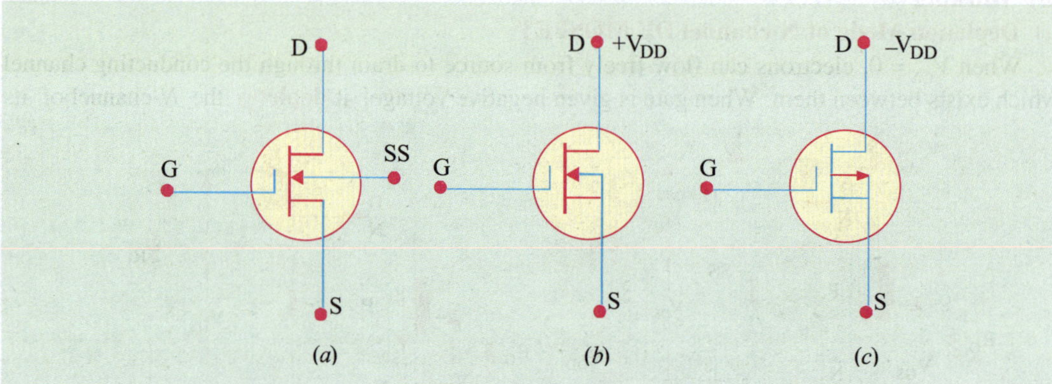

Fig. 63.25

63.16. Static Characteristics of a DE MOSFET

In Fig. 63.26 are shown the drain current and transfer characteristics of a common-source N-channel DE MOSFET for V_{GS} varying from +2 V to $V_{GS(off)}$.

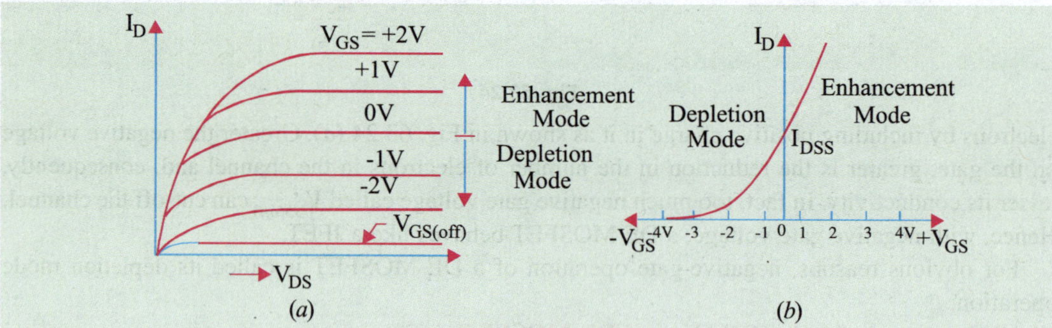

Fig. 63.26

It acts in the enhancement mode when gate is ***positive*** with respect to source and in the depletion mode when gate is ***negative.*** As usual, $V_{GS(off)}$ represents the gate-source voltage which cuts off the source-to-drain current. The transfer characteristic is shown in Fig. 63.26 (*b*). For a given V_{DS}, I_D flows even when $V_{GS} = 0$. However, keeping V_{DS} constant, as V_{GS} is made more negative, I_D keeps decreasing till it becomes zero at $V_{GS} = V_{GS(off)}$. When used in the enhancement mode, I_D increases as V_{DS} is increased positively.

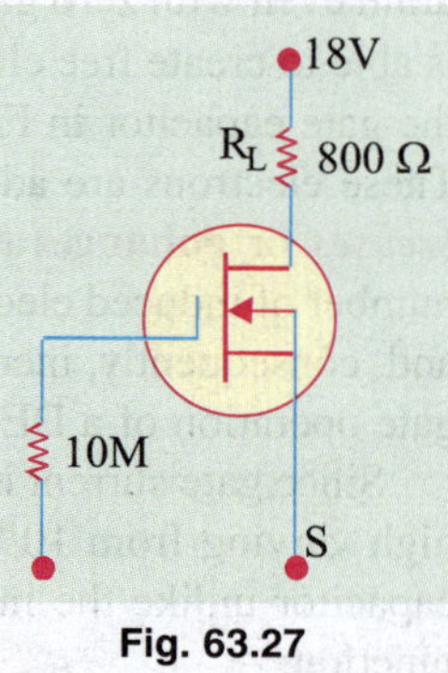

Fig. 63.27

Example 63.9. *For the N-channel zero-biased DE MOSFET circuit of Fig. 63.27, calculate V_{DS} if I_{DSS} = 10 mA and $V_{GS(off)}$ = – 6 V.*

(Electronics-I, Bangalore Univ. 1992)

Solution. Since $I_D = I_{DSS} = 10$ mA

$$\therefore \quad V_{DS} = V_{DD} - I_{DSS} R_L = 18 - 10 \times 10^{-3} \times 800 = \mathbf{10\ V}$$

63.17. Enhancement-only N-Channel MOSFET

This *N*-channel MOSFET (also called NMOS) finds wide application in digital circuitry. As shown in Fig. 63.28 in the NMOS, the *P*-type substrate extends all the way to the metal-oxide layer. Structurally, ***there exists no channel between the source and drain.*** Hence, an NMOS can never operate with a negative gate voltage because it will induce *positive* charge in the space between the drain and source which will not allow the passage of electrons between the two. Hence, it operates with ***positive gate voltage only.***

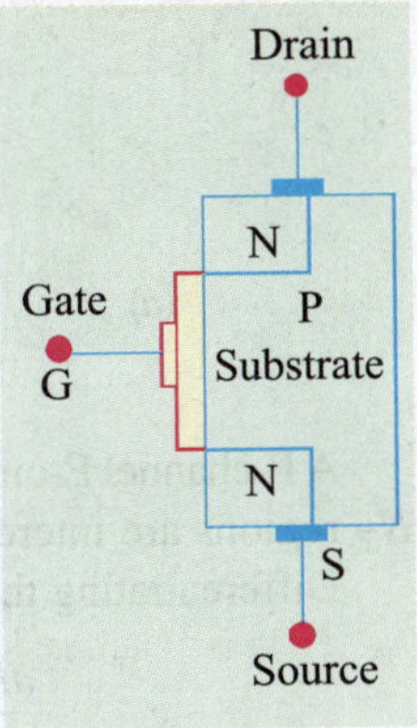

Fig. 63.28

The normal biasing polarities of this *E*-only MOSFET (both *N*-channel and *P*-channel) are shown in Fig. 63.29. With $V_{GS} = 0$, I_D is non-existent even when some positive V_{DD} is applied. It is found that for getting significant amount of drain current, we have to apply sufficiently high positive gate voltage. This voltage is found to produce a thin layer of free electrons very close to the metal-oxide film which stretches all the way from source to drain. The thin layer of *P*-substrate touching the metal-oxide film which provides channel for electrons (and hence acts like *N*-type material) is called ***N-type inversion layer or virtual*** *N*-channel [Fig. 63.29 (*a*)].

The minimum gate-source voltage which produces this *N*-type inversion layer and hence drain current is called threshold voltage $V_{GS(th)}$ as shown in Fig. 63.30 (*c*). When $V_{GS} < V_{GS(th)}$, $I_D = 0$. Drain current starts only when $V_{GS} > V_{GS(th)}$. For a given V_{DS}, as V_{GS} is increased, virtual channel deepens and I_D increases. The value of I_D is given by $I_D = K(V_{GS} - V_{GS(th)})^2$ where *K* is a constant which depends on the particular MOSFET. Its value can be determined from the data sheet by taking the specified value of I_D called $I_{D(ON)}$ at the given value of V_{GS} and then substituting the values in the above equation. Fig. 63.30 (*a*)

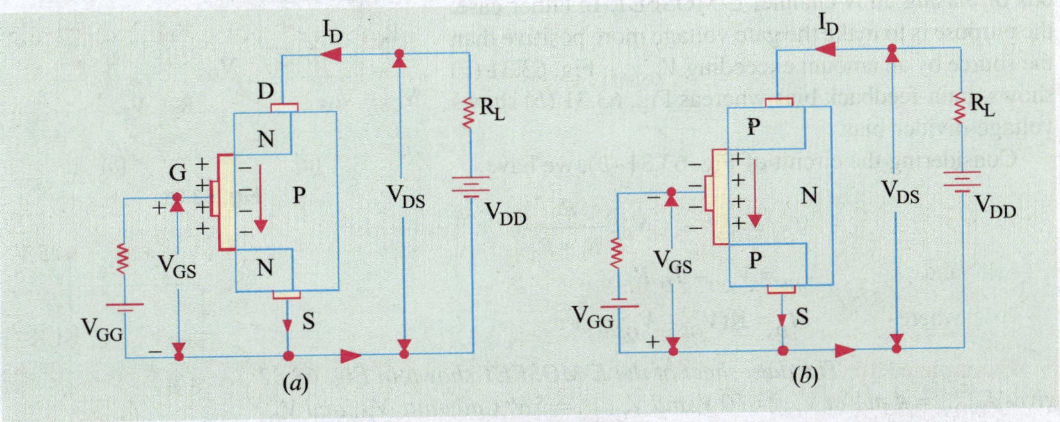

Fig. 63.29

shows the schematic symbol for an *E*-only *N*-channel MOSFET whereas Fig. 63.30 (*b*) shows its typical drain characteristics. As usual, arrow on the substrate points to the *N*-type material and the vertical line (representing-channel) is broken as a reminder of the normally-OFF condition.

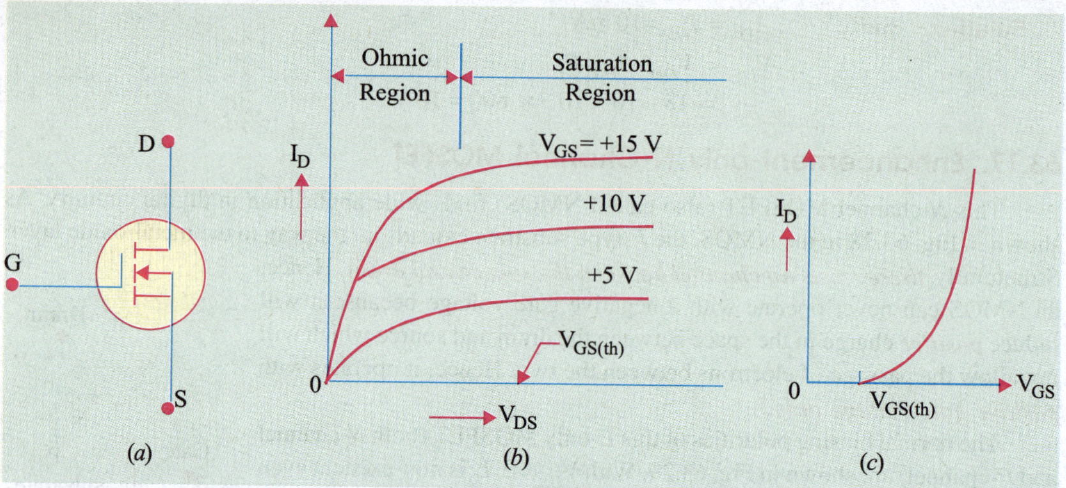

Fig. 63.30

A P-channel E-only MOSFET (PMOS) is constructed like NMOS except that all the *P*– region and *N*– regions are interchanged. It operates with ***negative*** gate voltage only.

Differentiating the above-given drain current equation with respect to V_{GS}, we get

$$\frac{dI_D}{dV_{GS}} = g_m = 2K\,(V_{GS} - V_{GS(th)})$$

Transfer Characteristics

It is shown in Fig. 63.30 (*c*). I_D flows only when V_{GS} exceeds threshold voltage $V_{GS(th)}$. This MOSFET does not have an I_{DSS} parameter as do the JFET and DE MOSFET.

63.18. Biasing E-only MOSFET

As stated earlier, enhancement-only MOSFET must have a V_{GS} greater than $V_{GS(th)}$. Fig. 63.31 shows two methods of biasing an *N*-channel E-MOSFET. In either case, the purpose is to make the gate voltage more positive than the source by an amount exceeding $V_{GS(th)}$. Fig. 63.31 (*a*) shows drain-feedback bias whereas Fig. 63.31 (*b*) shows voltage-divider bias.

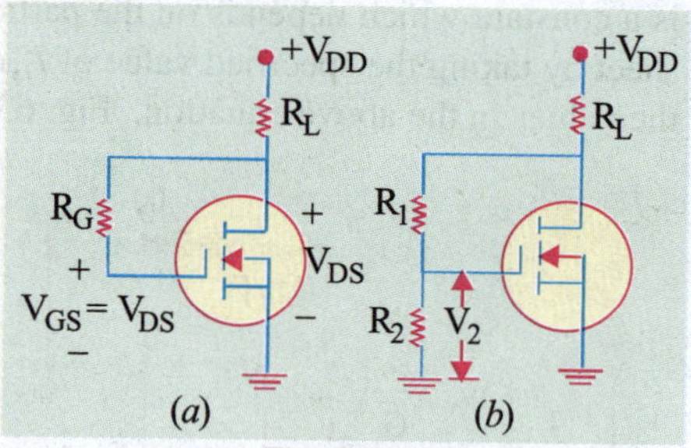

Fig. 63.31

Considering the circuit of Fig. 63.31 (*b*), we have

$$V_{GS} = V_2 = V_{DD}\frac{R_2}{R_1 + R_2}$$

and $\quad V_{DS} = V_{DD} - I_D R_L$

where $\quad I_D = K(V_{GS} - V_{GS(th)})^2$.

Example 63.10. *The data sheet of the E-MOSFET shown in Fig. 63.32 gives $I_{D(ON)}$ = 4 mA at V_{GS} = 10 V and $V_{GS(th)}$ = 5 V. Calculate V_{GS} and V_{DS} for the circuit.* **(Applied Electronics, Punjab Univ. 1991)**

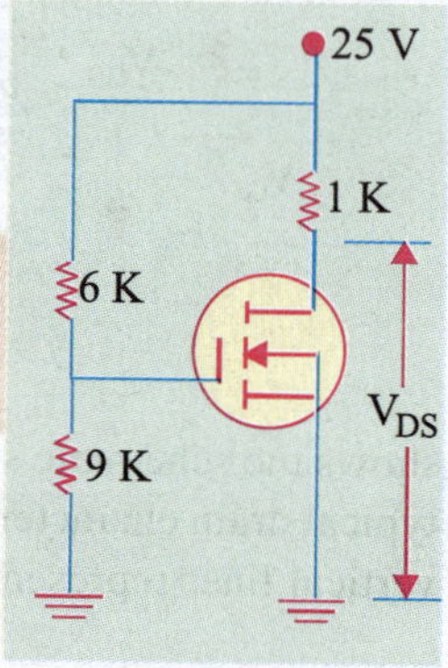

Fig. 63.32

Solution. $V_{GS} = V_2 = 25 \times (9/15) = 15$ V

Let us now find the value of *K*.

$$K = \frac{I_{D(ON)}}{(V_{GS} - V_{GS(th)})^2} = \frac{4}{(10-5)^2} = 0.16\,\text{mA/V}^2$$

Value of I_D for V_{GS}=15 V is given by

$$I_D = K(V_{GS} - V_{GS(th)})^2 = 0.16\,(15-5)^2 = 16\text{ mA}$$

$$\text{Now, } V_{DS} = V_{DD} - I_D R_L = 25 - 16 \times 1 = \mathbf{9\ V}$$

Example 63.11. *An N-channel E-MOSFET has the following parameters:*

$I_{D(ON)}$ = 4 mA at V_{GS} = 10 V and $V_{GS(th)}$ = 5 V

Calculate its drain current for V_{GS} = 8V.

Solution. $$K = \frac{I_{D(ON)}}{(V_{GS} - V_{GS(th)})^2} = \frac{4\text{mA}}{(10\text{ V} - 5\text{V})^2} = 0.16\text{mA/V}^2$$

Now, using this value of K, I_D can be found :

$$I_D = K\,(V_{GS} - V_{GS(th)})^2 = 0.16\,(8-5)^2 = \mathbf{1.44\ mA}$$

63.19. FET Amplifiers

We will consider the DE MOSFET and E-MOSFET separately.

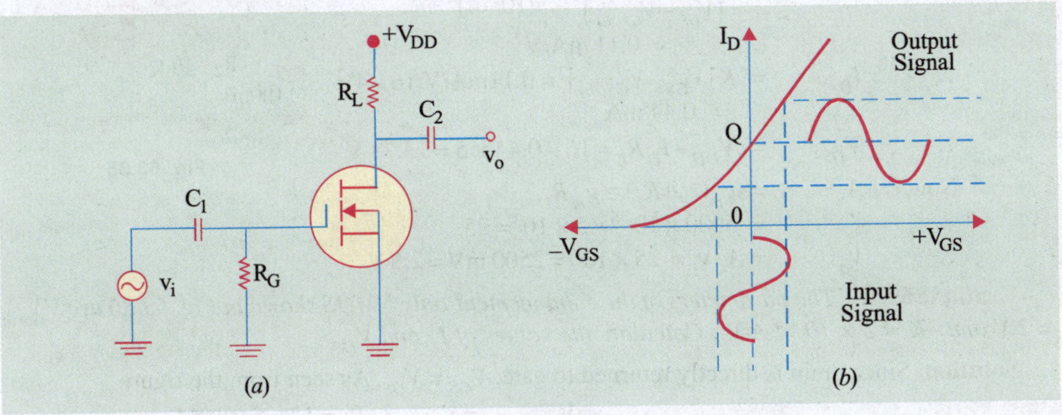

Fig. 63.33

(i) DE MOSFET Amplifier

Fig. 63.33 (*a*) shows a zero-biased *N*-channel DE MOSFET with an ac source capacitively-coupled to its gate. Since gate is at 0 volt dc and source is at ground, V_{GS} = 0. Fig. 63.33 (*b*) shows the transfer characteristic.

The input ac signal voltage V_i causes V_{GS} to swing above and below its zero value thus producing a swing in I_D as shown in Fig. 63.33 (*b*). The negative swing in V_{GS} produces depletion mode and I_D

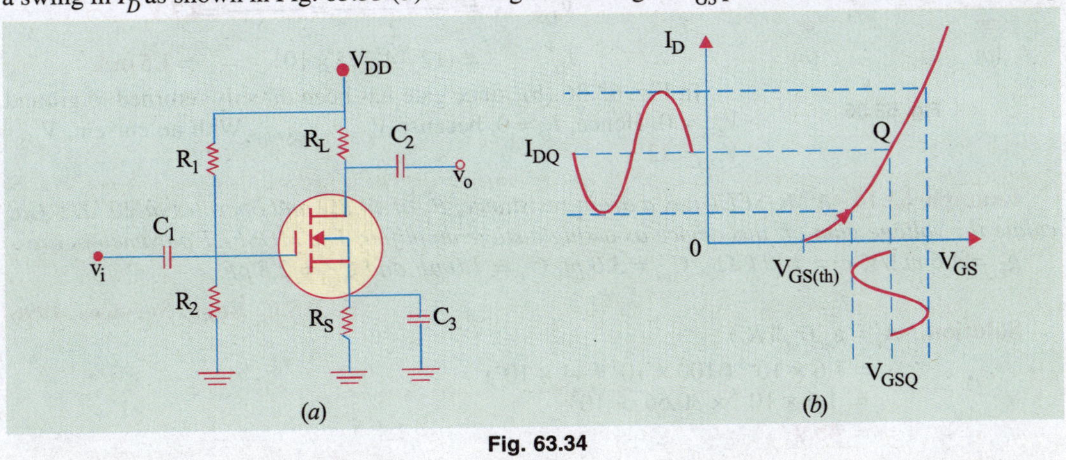

Fig. 63.34

is decreased. A positive swing in V_{GS} produces enhancement mode so that I_D is increased. This leads to a large swing in drop across R_L and can be taken out *via* C_2 as V_o.

(*ii*) E-MOSFET Amplifier

In Fig. 63.34 (*a*) is shown a voltage-divider biased *N*-channel E-MOSFET having an ac signal source coupled to its gate. The gate is biased with a positive voltage such that V_{GS} exceeds $V_{GS(th)}$.

As shown in Fig. 63.34 (*b*), the signal voltage produces a swing in V_{GS} below and above its *Q*-point value. This, in turn, causes a swing in I_D and hence in $I_D R_L$ which gives rise to V_o.

Example 63.12. *The N-channel E-MOSFET used in the common-source amplifier of Fig. 63.35 has the following parameters :*

$I_{D(ON)} = 4$ mA at $V_{CS} = 10$ V, $V_{GS(th)} = 4$ V and $g_m' = 5000$ μS.

Calculate V_{GS}, I_D, V_{DS} *and* v_o. **(Elect. and Electronic Engg., Annamalai Univ. 1992)**

Solution. $V_{GS} = V_{DD}\dfrac{R_2}{R_1+R_2} = 16\times\dfrac{30}{80} = \mathbf{6\ V}$

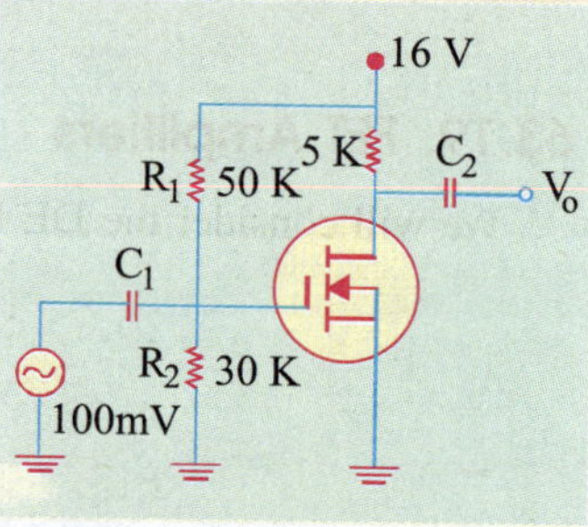

Fig. 63.35

Now, $K = \dfrac{I_{D(NO)}}{\left(V_{GS}-V_{GS(th)}\right)^2} = \dfrac{4}{(10-4)^2}$

$= 0.11\ \text{mA/V}^2$

$\therefore\quad I_D = K(V_{GS} - V_{GS(th)})^2 = 0.11\text{mA/V}^2(6-4)^2$

$= 0.44$ mA

$\therefore\quad V_{DS} = V_{DD} - I_D R_L = 16 - 0.44\times 5 = \mathbf{13.78\ V}$

Now, $A_v = g_m(r_d \| R_L) \cong g_m R_L$

$= 5000\times 10^{-6}\times 5\times 10^3 = 25$

$\therefore\quad V_o = A_v v_i = 25\times 100 = 2500\ \text{mV} = \mathbf{2.5\ V}$

Example 63.13. *The parameters of the enhancement-only NMOS shown in Fig. 63.36 are* $V_{GS(th)} = 2$ *V and* $K = 2\times 10^{-4}$ *A/V²*. *Calculate the values of* I_D *and* V_{DS}.

Solution. Since drain is directly returned to gate, $V_{GS} = V_{DS}$. As seen from the figure

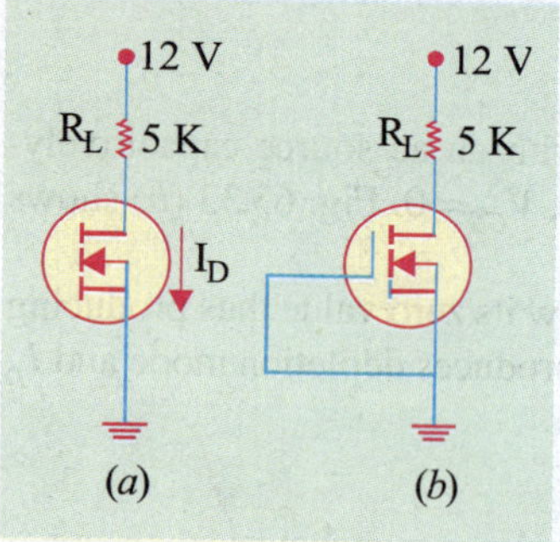

Fig. 63.36

$V_{DS} = V_{DD} - I_D R_L = 12 - 5\times 10^3 I_D$

$\therefore\quad I_D = (12 - V_{DS})/5\times 10^3$ A

Now, $I_D = K(V_{GS} - V_{GS(th)})^2 = K(V_{DS} - 2)^2$

or $\dfrac{12-V_{DS}}{5\times 10^3} = 2\times 10^{-4}(V_{DS}-2)^2$ or $V_{DS}^2 - 3V_{DS} - 8 = 0$

$\therefore\quad V_{DS} = \dfrac{3\pm\sqrt{9+32}}{2} = \dfrac{3\pm\sqrt{41}}{2} = \mathbf{4.7\ V}$

$\therefore\quad I_D = (12-4.7)/5\times 10^3 = \mathbf{1.5\ mA}$

In Fig. 63.36 (*b*), since gate has been directly returned to ground, $V_{GS} = 0$. Hence, $I_D = 0$, because $V_{GS} \ll V_{GS(th)}$. With no current, $V_{DS} = V_{DD} = \mathbf{12\ V}$.

Example 63.14. *A MOSFET has a drain resistance,* R_L *of 44 kΩ and operates at 20 kHz. Calculate the voltage gain of this device as a single stage amplifier. The MOSFET parameters are :* $g_m = 1.6$ *mA/V,* $r_d = 100$ *k Ω ,* $C_{gs} = 3.0$ *pf,* $C_{ds} = 1.0$ *pF and* $C_{gd} = 2.8$ *pF.* **(U.P.S.C. Engg. Services, 1996)**

Solution. $A_v = g_m(r_d \| R_L)$

$= 1.6\times 10^{-3}(100\times 10^3 \| 44\times 10^3)$

$= 1.6\times 10^{-3}\times 30.56\times 10^3$

$= \mathbf{48.9}$

63.20. FET Applications

FETs can be used in almost every application in which bipolar transistors can be used. However, they have certain applications which are exclusive to them :

1. As input amplifiers in oscilloscopes, electronic voltmeters and other measuring and testing equipment because their high r_{in} ***reduces loading effect to the minimum.***
2. In logic circuits where it is kept OFF when there is zero input while it is turned ON with very little power input.
3. For mixer operation of *FM* and *TV* receivers.
4. As voltage-variable resistor (*VVR*) in operational amplifiers and tone controls etc.
5. Large-scale integration (*LSI*) and computer memories because of very small size.

63.21. MOSFET Handling

MOSFETs require very careful handling particularly when ***out of circuit.*** In circuit, a MOSFET is as rugged as any other solid-state device of similar construction and size.

It is essential not to permit any stray or static voltage on the gate otherwise the ultra-thin SiO_2 layer between the channel and the gate will get ruptured. Since gate-channel junction looks like a capacitor with extremely high resistance, it requires only a few electrons to produce a high voltage across it. Even ***picking up a*** MOSFET ***by its leads can destroy it.*** Generally, grounding rings are used to short all leads of a MOSFET for avoiding any voltage build up between them. These grounding or shorting rings are removed only after MOSFET is securely wired into the circuit. Sometimes, conducting foam is applied between the leads instead of using shorting rings. As shown in Fig. 63.37, some MOSFETs have back-to-back Zener diodes internally formed to protect them against stray voltages.

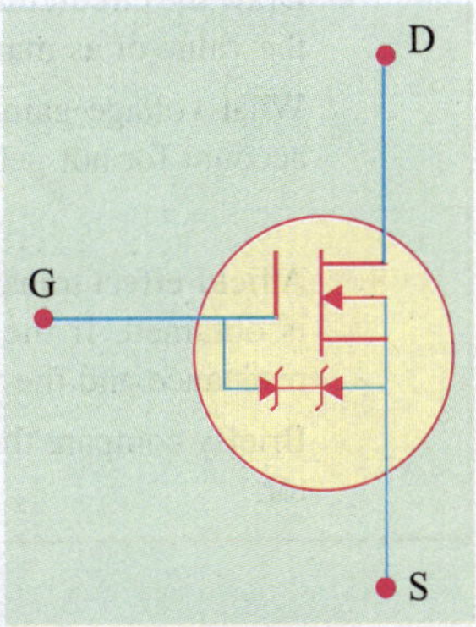

Fig. 63.37

Tutorial Problems No. 63.2

1. For a certain DE MOSFET, I_{DSS} = 10 mA and $V_{GS(off)}$ = – 8V. Calculate I_D when V_{GS} is (*i*) – 4 V and (*ii*) + 4 V. **[(*i*) 2.5 mA (*ii*) 22.5 mA]**
2. The data sheet of a certain zero-biased DE MOSFET gives I_{DSS} = 15 mA and $V_{GS(off)}$ = 5 V. What is the value of drain current ? **[15 mA]**
3. A certain *E*-only *N*-channel MOSFET has the following parameters :
   $I_{D(on)}$ =4 mA at V_{GS} = 8 V and $V_{GS(th)}$ = 2 V
   Calculate I_D for V_{GS} = 6 V **[1.78 mA]**

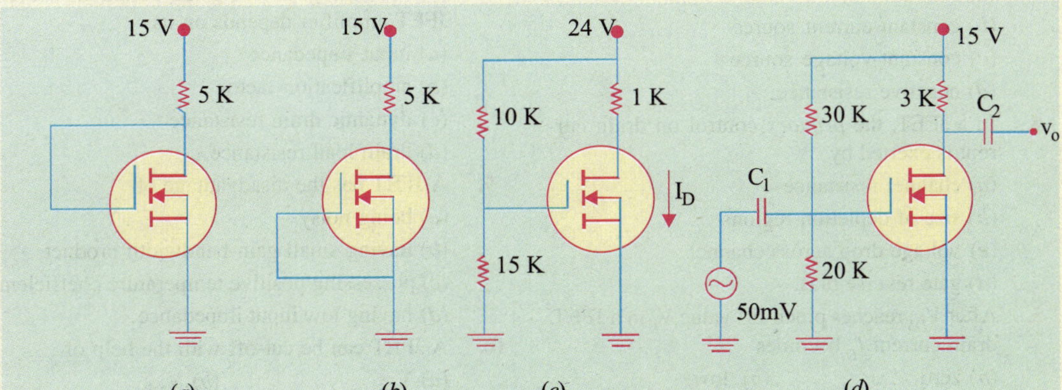

Fig. 63.38

4. The parameters of the *E*-only, *N*-channel MOSFET shown in Fig. 63.38 (*a*) are : $V_{GS(th)} = 2$ V and $K = 0.3$ mA/V^2. Determine the values of I_D and V_{DS}. If the gate is connected as shown in Fig. 63.38 (*b*), what will be the new values of I_D and V_{DS} ? **[(*i*) 2.07 mA (*ii*) 4.63 V ; 0 A, 15 V]**

5. The data sheet for the *E*-only *N*-channel MOSFET of Fig. 63.38 (*c*) gives $I_{D(on)} = 3$ mA at $V_{GS} = 10$ V and $V_{GS(th)} = 5$ V. Calculate the values of V_{GS} and V_{DS}. **[14.4 V ; 13.4 V]**

6. The amplifier circuit of Fig. 63.38 (*d*) used an *E*-only *N*-channel MOSFET having the following parameters : $I_{D(on)} = 5$ mA at $V_{GS} = 10$ V, $V_{GS(th)} = 4$ V, $g_m = 5500$ μS
   Calculate V_{GS}, I_D, V_{DS} and v_o. **[6 V ; 0.556 mA ; 13.3 V ; 0.825 V]**

7. A field-effect transistor has a small-signal equivalent circuit with input resistance = 1000 MΩ , forward transfer conductance = 4 m S and output conductance = 100 μS when at the operating point, $V_{DS} = +4$ V, $I_D = 2$ mA, $V_{GS} = -2$ V.
   Draw the circuit that you would use for a single-stage voltage amplifier. Describe the use and specify the value of as many components as possible if a 30 V supply was available.
   What voltage gain would you expect when the output was unloaded ? Give reasons which might account for not getting this gain exactly.
   [Drain load = 12 kΩ, Source Bias Resistor = 1 k Ω, – 21.8]

8. A field-effect transistor is used as a voltage amplifier and with a load resistor of 40 k Ω, a gain of 40 is obtained. If the load resistance is halved, the voltage gain drops to 30. Calculate the output resistance and the mutual conductance of the transistor.
   Briefly compare the advantages and limitations of the field-effect transistor with the bipolar transistor. **[20 kΩ, 3 mS]**

OBJECTIVE TESTS – 63

1. A FET consists of a
   (*a*) source (*b*) drain
   (*c*) gate (*d*) all of the above.
2. FETs have similar properties to
   (*a*) *PNP* transistors
   (*b*) *NPN* transistors
   (*c*) thermonic valves
   (*d*) unijunction transistors.
3. For small values of drain-to-source voltage, JFET behaves like a
   (*a*) resistor
   (*b*) constant-current source
   (*c*) constant-voltage source
   (*d*) negative resistance
4. In a JFET, the *primary* control on drain current is exerted by
   (*a*) channel resistance
   (*b*) size of depletion regions
   (*c*) voltage drop across channel
   (*d*) gate reverse bias.
5. After V_{DS} reaches pinch-off value V_P in a JFET, drain current I_D becomes
   (*a*) zero (*b*) low
   (*c*) saturated (*d*) reversed.
6. In a JFET, as external bias applied to the gate is increased
   (*a*) channel resistance is decreased
   (*b*) drain current is increased
   (*c*) pinch-off voltage is reached at lower values of I_D.
   (*d*) size of depletion regions is reduced.
7. In a JFET, drain current is maximum when V_{GS} is
   (*a*) zero (*b*) negative
   (*c*) positive (*d*) equal to Vp
8. The voltage gain of a given common- source JFET amplifier depends on its
   (*a*) input impedance
   (*b*) amplification factor
   (*c*) dynamic drain resistance
   (*d*) drain load resistance.
9. A JFET has the disadvantage of
   (*a*) being noisy
   (*b*) having small gain-bandwidth product
   (*c*) possessing positive temperature coefficient
   (*d*) having low input impedance.
10. A JFET can be cut-off with the help of
   (*a*) V_{GS} (*b*) V_{DS}
   (c) V_{DG} (*d*) V_{DD}.

11. The drain source voltage at which drain current becomes nearly constant is called
(*a*) barrier voltage
(*b*) breakdown voltage
(*c*) pick-off voltage
(*d*) pinch-off voltage

12. The transconductance 'g_m' of a JFET is equal to
(*a*) $-\frac{2I_{DSC}}{V_P}$
(*b*) $\frac{2}{|V_P|}\sqrt{I_{DSS}\, I_D}$
(*c*) $-\frac{2I_{DSS}}{V_P}\left(1-\frac{V_{GS}}{V_P}\right)$
(*d*) $\frac{I_{DSS}}{V_P}\left(1-\frac{V_{GS}}{V_P}\right)$

13. An FET source follower circuit has g_m of 2 millimho and of r_d 50 k Ω. If the source resistance R_S is 1k Ω, the output resistance of the amplifier will be
(*a*) 330 Ω (*b*) 450 Ω
(*c*) 500 Ω (*d*) 1 k Ω

14. A DE MOSFET differs from a JFET in the sense that it has no
(*a*) channel (*b*) gate
(*c*) *P-N* junctions (*d*) substrate.

15. For the operation of enhancement only *N*-channel MOSFET, value of gate voltage has to be
(*a*) high positive (*b*) high negative
(*c*) low positive (*d*) zero.

16. The extremely high input impedance of a MOSFET is primarily due to the
(*a*) absence of its channel
(*b*) negative gate-source voltage
(*c*) depletion of current carriers
(*d*) extremely small leakage current of its gate capacitor

17. The main factor which makes a MOSFET likely to break down during normal handling is its
(*a*) very low gate capacitance
(*b*) high leakage current
(*c*) high input resistance
(*d*) both (a) and (c).

18. The main factor which differentiates a DE MOSFET from an *E*-only MOSFET is the absence of
(*a*) insulated gate (*b*) electrons
(*c*) channel (*d*) *P-N* junctions.

19. The polarity of V_{GS} for *E*-only MOSFET is
(*a*) positive
(*b*) negative
(*c*) zero
(*d*) depends on P-or N-channel.

20. A tranasconductance amplifier has
(*a*) high input impedance and low output impedance
(*b*) low input impedance and high output impedance
(*c*) high input and output impedances
(*d*) low input and output impedances

21. The threshold voltage of an N-channel enhancement mode MOSFET is 0.5 V, when the device is biased at a gate voltage of 3V, pinch-off would occur at a drain voltage of
(*a*) 1.5 V (*b*) 2.5 V
(*c*) 3.5 V (*d*) 4.5 V

22. The zero gate bias channel resistance of a junction field-effect transistor is 750 and the pinch-off voltage is 3V. For a gate bias of 1.5 V and very low drain voltage, the device would behave as a resistance of
(*a*) 320 Ω (*b*) 816 Ω
(*c*) 1000 Ω (*d*) 1270 Ω

23. A MOSFET rated for 15 A, carries a periodic current as shown in Fig. 63.39. the ON state resistance of the MOSFET is 0.15 Ω. The average ON state loss in the MOSFET is

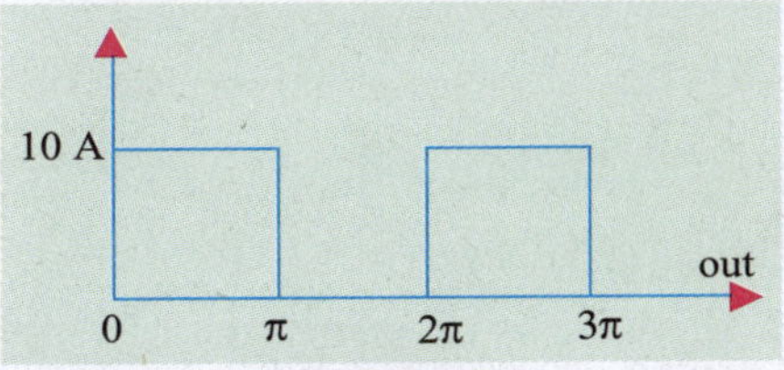

Fig. 63.39

(*a*) 33.8 W (*b*) 15.0 W
(*c*) 7.5 W (*d*) 3.8 W

ANSWERS

1. (*d*) 2. (*b*) 3. (*a*) 4. (*d*) 5. (*c*) 6. (*c*) 7. (*a*) 8. (*d*) 9. (*b*) 10. (*a*) 11. (*d*)
12. (*c*) 13. (*c*) 14. (*c*) 15. (*a*) 16. (*d*). 17. (*d*) 18. (*c*) 19. (*d*) 20. (*a*) 21. (*c*) 22. (*b*)
23. (*c*)

ROUGH WORK

CHAPTER 64

BREAKDOWN DEVICES

Learning Objectives

- What are Breakdown Devices ?
- Unijunction Transistor
- UJT Relaxation Oscillator
- Programmable UJT(PUT)
- Silicon Controlled Rectifier
- Comparison between Transistors and Thyristors
- Transient Effects in an SCR
- Phase Control
- Theft Alarm
- Emergency Lighting System
- Light Activated SCR (LASCR)
- The Shockley Diode
- Triac
- Diac
- Silicon Controlled Switch (SCS)

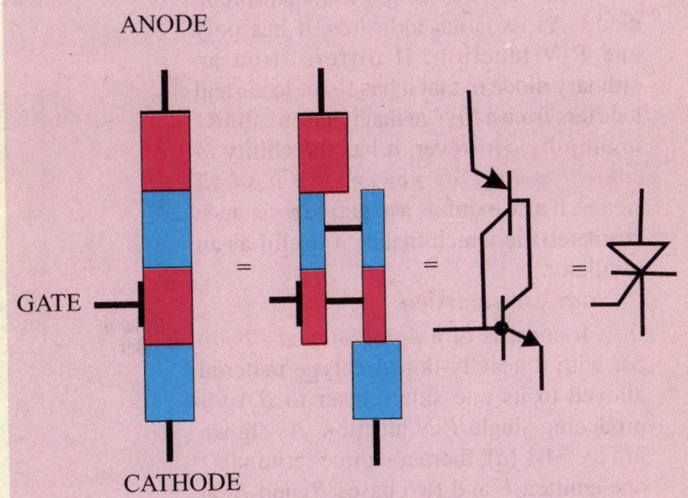

The Silicon Controlled Rectifier, usually referred to as an SCR, is one of the family of semiconductors that includes transistors and diodes

64.1. What are Breakdown Devices ?

These are solid-state devices whose working depends on the ***phenomenon of avalanche breakdown.*** They are sometimes referred to by the generic name of ***thyristor*** which is a semiconductor switch whose bistable action depends on *P-N-P-N* regenerative feedback. We will discuss the following devices :

1. Unijunction Transistor (UJT).
2. Silicon Controlled Rectifier (SCR).
3. Light Activated SCR (LASCR).
4. Triac (short for 'triode ac').
5. Diac (short for 'diode ac').
6. Silicon Controlled Switch (SCS).

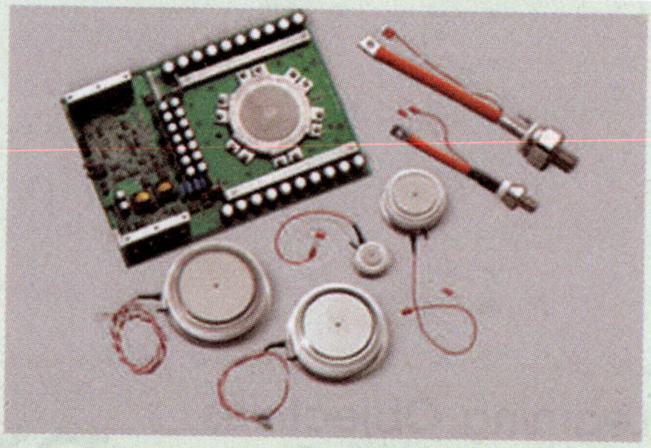

Cat-Transistor

These devices have two or more junctions and can be switched ON or OFF at ***an extremely fast rate.*** They are also referred to as **latching** devices. A latch is a kind of switch which initially once closed, remains closed until someone opens it.

64.2. Unijunction Transistor

Basically, it is a three-terminal silicon diode. As its name indicates, it has only one *P-N* junction. It differs from an ordinary diode in that it has **three** leads and it differs from a FET in that it has **no ability to amplify**. However, it has the ability ***to control a large ac power with a small signal.*** It also exhibits a negative resistance characteristic which makes it useful as an oscillator.

(*a*) Construction

It consists of a ***lightly-doped silicon*** bar with a heavily-doped *P*-type material alloyed to its one side (closer to B_2) for producing single *P-N* junction. As shown in Fig. 64.1 (*a*), there are three terminals : one emitter, *E* and two bases B_2 and B_1 at the top and bottom of the silicon bar. The emitter leg is drawn at an angle to the vertical and arrow points in the direction of ***conventional*** current when UJT is in the conducting state.

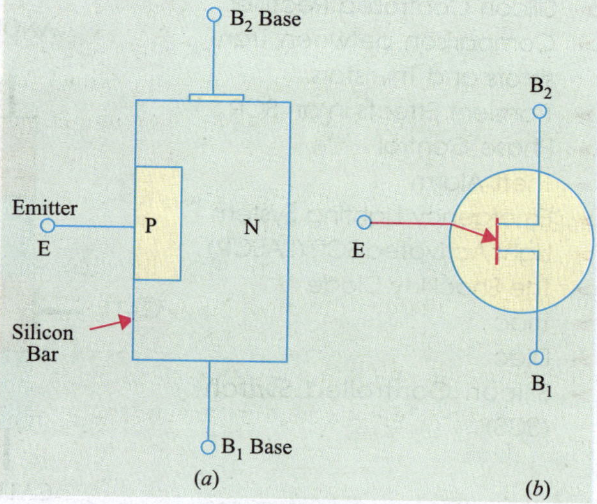

Fig. 64.1

(*b*) Interbase Resistance (R_{BB})

It is the resistance between B_2 and B_1 *i.e.* it is the total resistance of the silicon bar from one end to the other with emitter terminal open [Fig. 64.2 (*a*)].

From the equivalent circuit of Fig. 64.2 (*b*), it is seen that $R_{BB} = R_{B2} + R_{B1}$

It should also be noted that point *A* is such that $R_{B1} > R_{B2}$. Usually, R_{B1} = 60% of R_{BB}. The resistance R_{B1} has been shown as a variable resistor because its value varies inversely as I_E.

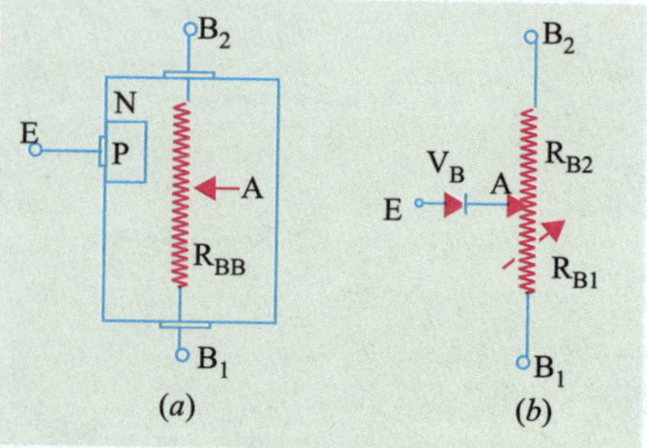

Fig. 64.2

(c) Intrinsic Stand-off Ratio

As seen from Fig. 64.3 (*a*), when a battery of 30 V is applied across $B_2 B_1$, there is a progressive fall of voltage over R_{BB} provided *E* is open. It is obvious from Fig. 64.3 (*b*) that emitter acts as a voltage-divider tap on fixed resistance R_{BB}.

With emitter open, $I_1 = I_2$, the interbase current is given by Ohm's Law.

$$I_1 = I_2 = V_{BB}/R_{BB}$$

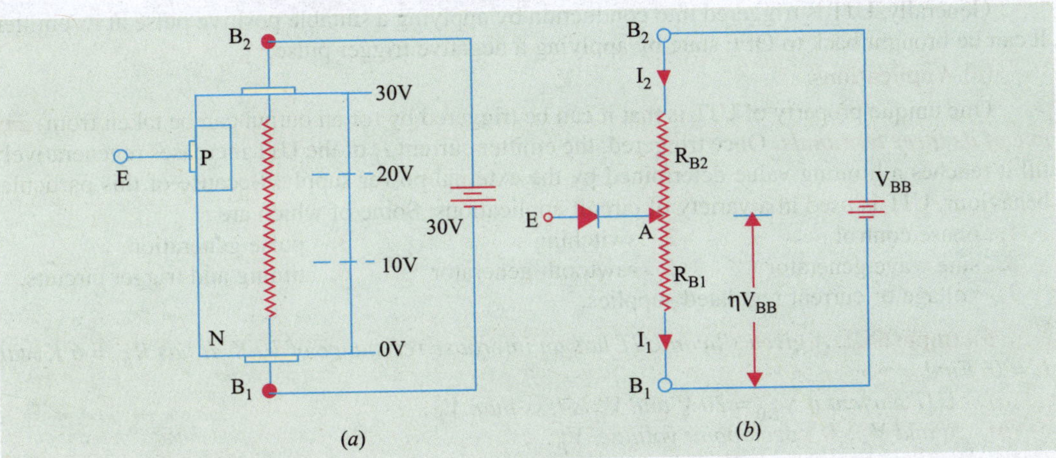

Fig. 64.3

For example, if V_{BB} = 30 V and R_{BB} = 15 K, $I_1 = I_2$ = 2 mA.

It may be noted that part of V_{BB} is dropped over R_{B2} and part on R_{B1}. Let us call the voltage drop across R_{B1} as V_A. Using simple voltage divider relationship,

$$V_A = V_{BB}\frac{R_{B1}}{R_{B1}+R_{B2}}$$

The voltage division factor is given a special symbol (η) and the name of **'intrinsic stand-off ratio'**.

$$\eta = \frac{R_{B1}}{R_{B1}+R_{B2}} \quad \therefore \quad V_A = \eta V_{BB}$$

The intrinsic stand-off ratio is the property of the UJT and is always less than unity (0.5 to 0.85). If V_{BB} = 30 V and η = 0.6, then potential of point *A* with respect to point B_1 = 0.6 × 30 = 18 V. The remaining 12 V drop across R_{B2}.

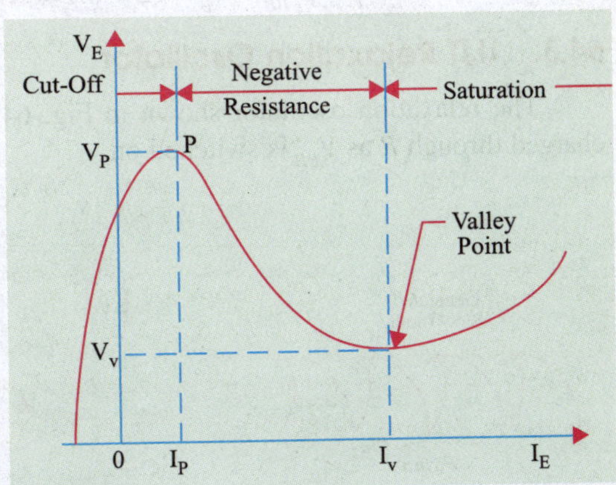

Fig. 64.4

(d) Operation

When V_{BB} is switched on, V_A is developed and ***reverse-biases the junction.*** If V_B is the barrier voltage of the *P-N* junction, then total reverse bias voltage is

$$= V_A + V_B = \eta V_{BB} + V_B$$

Value of V_B for Si is 0.7 V.

It is obvious that emitter junction will not become forward-biased unless its applied voltage V_E exceeds $(\eta V_{BB} + V_B)$. This value of V_E is called ***peak-point voltage*** V_P (Fig. 64.4). When $V_E = V_P$, emitter (peak current), I_P starts to flow through R_{B1} to ground (*i.e.* B_1). The UJT is then said to have been ***fired*** or turned ON. Due to the flow of $I_E (= I_P)$ through R_{B1}, number of charge carriers in R_{B1} is increased which ***reduces its resistance.*** As η depends on R_{B1}, ***its value is also decreased.***

Hence, we find that as V_E and hence I_E increases (beyond I_P), R_{B1} decreases, η decreases and V_A decreases. This decrease in V_A causes more emitter current to flow which causes a further reduction in R_{B1}, η and V_A. Obviously, the process is regenerative. V_A as well as V_E quickly drop as I_E increases. Since, V_E decreases when I_E increases, the UJT ***possesses negative resistance.*** Beyond the valley point, UJT is in saturation and V_E increases very little with an increasing I_E.

It is seen that only terminals *E* and B_1 are the active terminals whereas B_2 is the bias terminal *i.e.* it is meant only for applying external voltage across the UJT.

Generally, UJT is triggered into conduction by applying a suitable positive pulse at its emitter. It can be brought back to OFF state by applying a negative trigger pulse.

(e) Applications

One unique property of UJT is that it can be triggered by (or an output can be taken from) ***any one of its three terminals.*** Once triggered, the emitter current I_E of the UJT increases regeneratively till it reaches a limiting value determined by the external power supply. Because of this particular behaviour, UJT is used in a variety of circuit applications. Some of which are :

1. phase control
2. switching
3. pulse generation,
4. sine wave generator
5. sawtooth generator
6. timing and trigger circuits,
7. voltage or current regulated supplies.

Example 64.1. *A given silicon UJT has an interbase resistance of 10 K. It has R_{B1} = 6 K with I_E = 0. Find*

(a) *UJT current if V_{BB} = 20 V and V_E is less than V_P,*

(b) *η and V_A,* **(c)** *peak point voltage, V_P.*

(Applied Electronics-I, Punjab Univ. 1990)

Solution. (a) Since $V_E < V_P$, $I_E = 0$, because *P-N* junction is reverse-biased.

$$\therefore \quad I_1 = I_2 = \frac{V_{BB}}{R_{BB}} = \frac{20}{10K} = \mathbf{2mA}$$

(b) $$\eta = \frac{R_{B1}}{R_{BB}} = \frac{6}{10} = \mathbf{0.6};\ V_A = \eta V_{BB} = 0.6 \times 20 = \mathbf{12V}$$

(c) $$V_P = \eta V_{BB} + V_B = 12 + 0.7 = \mathbf{12.7\ V}$$

64.3. UJT Relaxation Oscillator

The relaxation oscillator shown in Fig. 64.5 consists of a UJT and a capacitor *C* which is charged through *R* as V_{BB} is switched on.

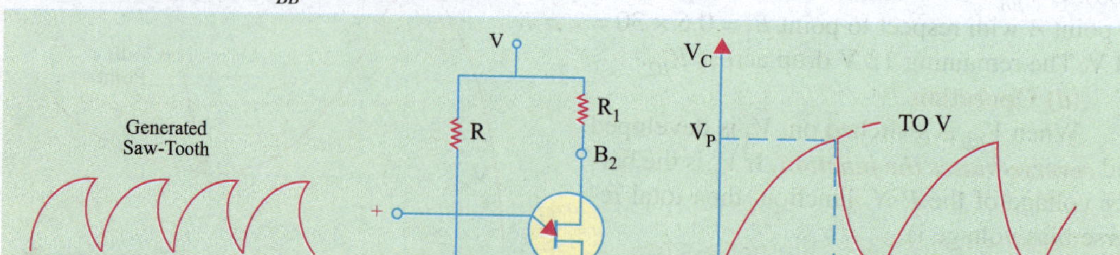

Fig. 64.5

When the capacitor voltage V_C reaches in time t_s the value of V_P, the UJT fires and rapidly discharges *C* via B_1 till the voltage falls below the minimum value V_V. The device then cuts off and

C starts to charge again. This cycle is repeated continuously thus generating a sawtooth waveform across *C*.

The inclusion of external resistances R_2 and R_1 in series with B_2 and B_1 (Fig. 64.6) provides ***spike waveforms***. When the UJT fires, the sudden surge of current through B_1 causes a drop across R_1 which produces positive going spikes. Also, at the time of firing, fall of V_{BB1} causes I_2 to increase rapidly which generates negative going spikes across R_2 as shown in Fig. 64.6.

By switching over to different capacitors, frequency of the output waveform can be changed as desired.

Condition for Turn-ON and Turn-OFF

For satisfactory working of the above oscillator, following two conditions for the turn-on and turn-off of the UJT must be met. To ensure turn-on, *R* must not limit I_E at peak point to a value less than I_P. It means that

$$V_{BB} - V_P > I_P R \text{ or } R < \frac{V_{BB} - V_P}{I_P}$$

To ensure turn-off of the UJT at valley point, *R* must be large enough to permit I_E (at valley point) to decrease below the specified value of I_V. In other words, drop across *R* at valley point must be less than $I_V R$. Hence, condition for turn-off is

$$V_{BB} - V_V < I_V R \text{ or } R > \frac{V_{BB} - V_V}{I_v}$$

Hence, for reliable turn-on and turn-off of the UJT, *R* must be in the range

$$\frac{V_{BB} - V_P}{I_P} > R > \frac{V_{BB} - V_V}{I_V}$$

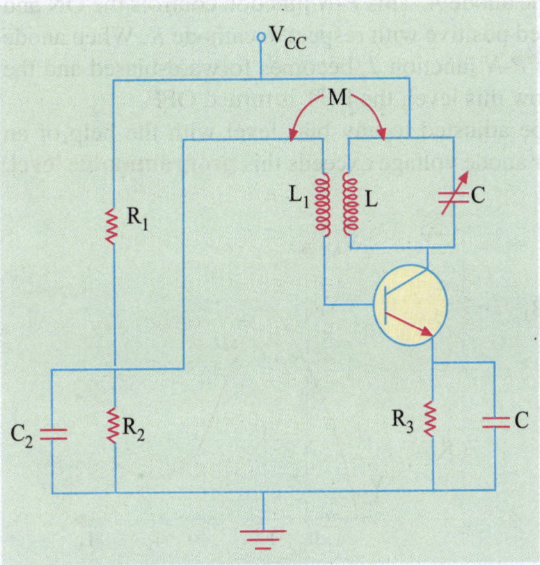

Fig. 64.6

It should be noted that charging time constant of the capacitor for voltage *V* is $T = CR$ whereas discharging time constant is $T_d = CR_{B1}$.

The time required to charge upto V_P (called ramp rise time) is $t_s = T \log_e (V - V_V) / (V - V_P)$.

Similarly, time required by the capacitor to discharge from V_P to V_V is $t_d \cong T_d \log_e V_P / V_V$. The frequency of oscillation is given by $f = 1/(t_s + t_d)$.

Example 64.2. *The windshield wiper motor of an automobile is controlled by a UJT with* $\eta = 0.6$*. The capacitor has a value of 50 μF and the charging resistor is a series combination of 50 K resistor and a 500 K potentiometer. Determine the minimum and maximum number of blade strokes per minute possible with this arrangement.*

Solution. The least value of time constant is $= 50{,}000 \times 50 \times 10^{-6} = 2.5$ second.

Maximum value of time constant when whole of potentiometer resistance is used is

$$= (50 + 500) \times 10^3 \times 50 \times 10^{-6} = 27.5 \text{ second}$$

Maximum blade strokes per minute= 60/2.5 = 24.

Minimum blade strokes per minute = 60/27.5 = **2.2.**

Example 64.3. *The oscillator circuit shown in Fig. 64.5 uses a UJT with* $R_{BB} = 10$ *K,* $\eta = 0.6$*,* $V_B = 0.7$ *V,* $V = 50$ *V,* $R_1 = 90$ *K,* $R = 100$ *K and* $C = 0.05$ *μF. When UJT is in conduction,* $R_{B1} = 10\ \Omega$ *and* $V_V = V_B$*. Find (i) ramp rise time,* t_s *(ii) approximate discharge time,* t_d *and (iii) frequency of oscillation.*

Solution. $V_{BB} = V.R_{BB}/(R_{BB} + R_1) = 50 \times 10/100 = 5$ V

$V_P = \eta\, V_{BB} + V_B = 0.6 \times 5 + 0.7 = 3.7$ V

$T = CR = 0.05 \times 10^{-6} \times 100 \times 10^3 = 5$ ms

(*i*) The capacitor charges from $Vc = V_B = 0.7$ V to V_P towards V in the time t_s given by $t_s = \text{T} \log_e (V - V_V)/(V - V_P) = 5 \log_e (50 - 0.7)/(50 - 3.7) =$ **0.315 ms.**

(*ii*) The discharge time, $t_d = CR_{B1} = 0.05 \times 10 = 0.5$ µs. The time taken by C to discharge from V_P to V_V (= 0.7 *V*) towards 0 volt is

$$T_d \cong t_d \log_e V_P/V_V = 0.5 \log_e 3.7 / 0.7 = \mathbf{0.832\ \mu s}$$

(*iii*) $f = 1/(t_s + t_d) = 1/(0.315 + 0.0016)$ = **0.0008 kHz.**

64.4. Programmable UJT (PUT)

Like a SCR, it is also a four-layer or *PNPN* device with a gate *G* as shown in Fig. 64.7. However, its gate is connected to the *N*-region adjacent to the anode *A*. This *P-N* junction controls the ON and OFF states of the *PUT*. The gate *G* is always biased positive with respect to cathode *K*. When anode voltage exceeds gate voltage by about 0.7 *V*, the *P-N* junction J_1 becomes forward-biased and the *PUT* turns ON. When the anode voltage falls below this level, the PUT is turned OFF.

As shown in Fig. 64.8 (*a*), gate bias can be adjusted to any bias level with the help of an external voltage divider circuit $R_2 - R_3$. Whenever anode voltage exceeds this **programmable level,** the *PUT* turns ON.

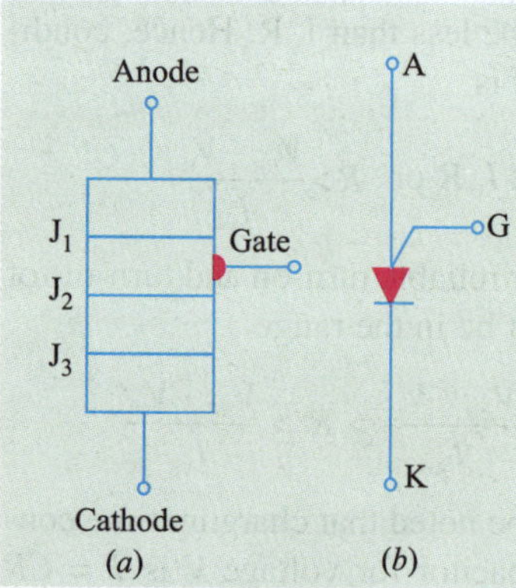

Fig. 64.7

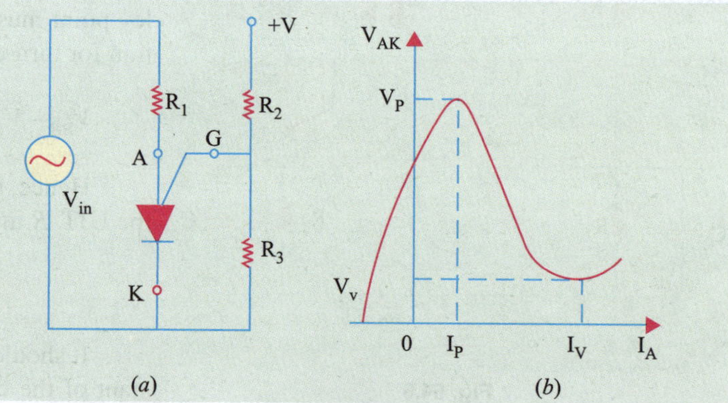

Fig. 64.8

Fig. 64.8 (*b*) shows the plot of anode-to- cathode voltage V_{AK} versus anode current I_A. It is similar to the *V/I* characteristic of a *UJT*. Hence, *PUT* replaces *UJT* in many applications, one such application is as relaxation oscillator shown in Fig. 64.9 (*a*).

Since, $R_2 = R_3$, $V_G = 12/2 = 6$ V. When dc voltage is applied, the *PUT* is off but *C* starts charging towards +12 V through R_1 [Fig. 64.9 (*b*)]. When V_C exceeds (V_G + 0.7 V), the *PUT* turns ON and, at the same time, *C* starts discharging rapidly through the low ON-resistance of the *PUT* and R_4. Consequently, a voltage spike is developed across R_4 during the discharge. As soon as *C* discharges, the *PUT* turns OFF and the charging cycle starts all over again as described above.

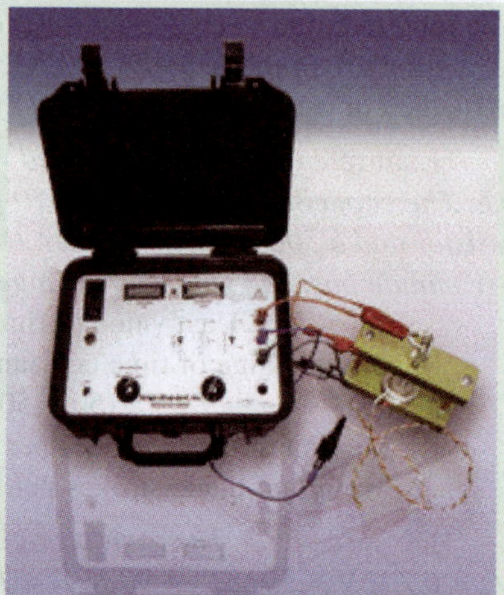

Silicon controlled rectifier and power diode

64.5. Silicon Controlled Rectifier

It is one of the prominent members of the thyristor family. It is a four-layer or PNPN device. Basically, ***it is a rectifier with a control element.*** In fact, it consists of ***three diodes*** connected back-to-back with a gate connection. It is widely used as a switching device in power control applications. It can control loads by switching current OFF and ON up to many thousand times a second. It can switch ON for variable lengths of time, thereby delivering selected amount of power to the load. Hence, it possesses the advantages of a rheostat and a switch with none of their disadvantages.

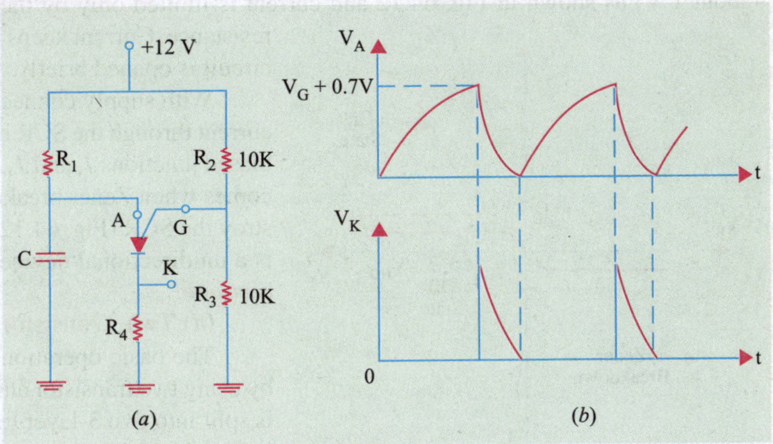

Fig. 64.9

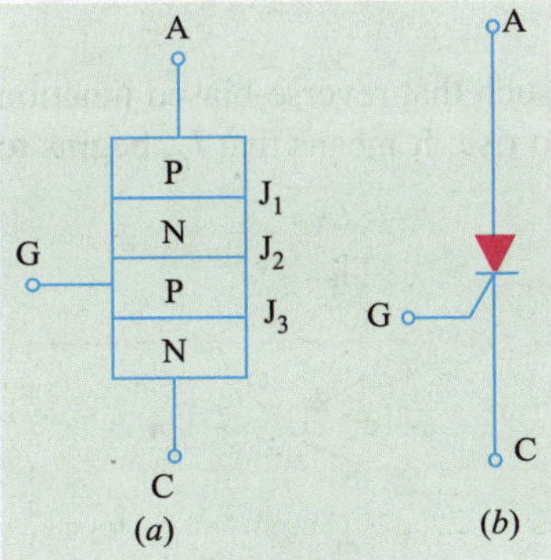

Fig. 64.10

(a) Construction

As shown in Fig. 64.10 (*a*), it is a three terminal four-layer transistor, the layers being alternately of *P*-type and *N*-type silicon. The three junctions are marked J_1, J_2 and J_3 whereas the three terminals are : anode (*A*), cathode (*C*) and gate (*G*) which is connected to the ***inner P***-type layer. The function of the gate is to control the firing of SCR. The schematic symbol is shown in Fig. 64.10 (*b*).

Since, they conduct large currents, junction areas of SCRs are very large. Commonly used stud-mounted units have their anode connected directly to the stud for good heat dissipation whereas larger units are of 'pillow' type in which many units are stacked in series and held in a pressurized clamp.

(b) Biasing

With the polarity of *V* as shown in Fig. 64.11 (*a*), the junctions J_1 and J_3 become forward-biased whereas J_2 is reverse-biased. Hence, no current (except leakage current) can flow through the SCR.

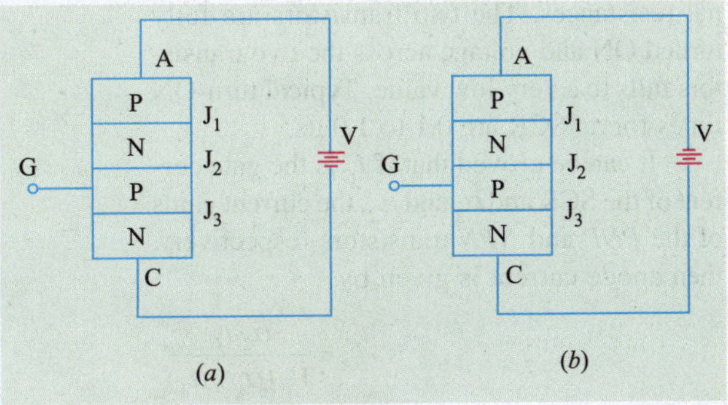

Fig. 64.11

In Fig. 64.11 (*b*), polarity of V has been reversed. It is seen that, now, junctions J_1 and J_3 become reverse-biased and only J_2 is forward-biased. Again, there is no flow of current through the SCR.

(c) Operation

In Fig. 64.11 (*a*), current flow is blocked due to reverse-biased junction J_2. However, when anode voltage is increased, a certain critical value called forward breakover voltage V_{BO} is reached when J_2 breaks down and SCR switches suddenly to a highly conducting state. Under this condition,

SCR offers very little forward resistance (0.01 Ω – 1.0 Ω) so that voltage across it drops to a low value (about 1 V) as shown in Fig. 64.12 and current is limited only by the power supply and the load resistance. Current keeps flowing indefinitely until the circuit is opened briefly.

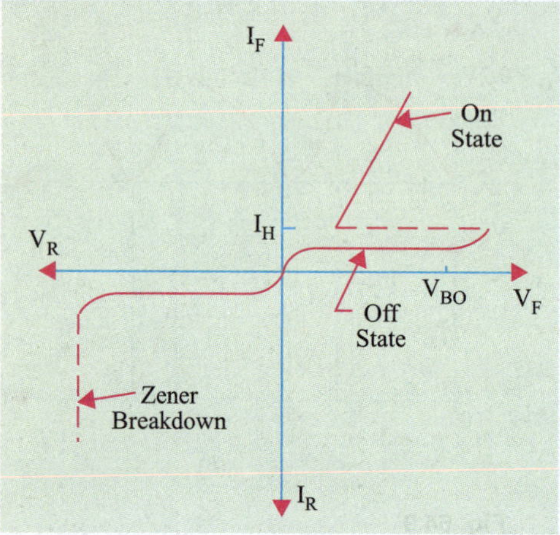

Fig. 64.12

With supply connection as in Fig. 64.11 (*b*), the current through the SCR is blocked by the two reverse-biased junctions J_1 and J_3. When *V* is increased, a stage comes when Zener breakdown occurs which may destroy the SCR (Fig. 64.12). Hence, it is seen that SCR is a unidirectional device unlike triac ***which is bi-directional.***

(*d*) Two Transistor Analogy

The basic operation of a SCR can be described by using two transistor analogy. For this purpose, SCR is split into two 3-layer transistor structures as shown in Fig. 64.13 (*a*). As seen, transistor Q_1 is a *PNP* transistor whereas Q_2 is an *NPN* device interconnected together. It will also be noted from Fig. 64.13 (*b*) that

(*i*) collector current of Q_1 is also the base current of Q_2 and

(*ii*) base current of Q_1 is also the collector current of Q_2.

Suppose that the supply voltage across terminals *A* and *C* is such that reverse-biased junction J_2 starts breaking down. Then, current through the device begins to rise. It means that I_{E1} begins to increase.

Then,

1. I_{C1} increases (remember $I_C = \alpha I_E$);
2. since $I_{C1} = I_{B2}$, I_{B2} also increases;
3. hence, I_{C2} increases (remember $I_C = \beta I_B$);
4. now, $I_{C2} = I_{B1}$, hence I_{B1} increases;
5. consequently, both I_{C1} and I_{E1} increase.

As seen, a regenerative action takes place whereby an initial increase in current produces further increase in the same current. Soon, maximum current is reached limited by external resistances. The two transistors are fully turned ON and voltage across the two transistors falls to a very low value. Typical turn-ON times for an SCR are 0.1 to 1.0 μs.

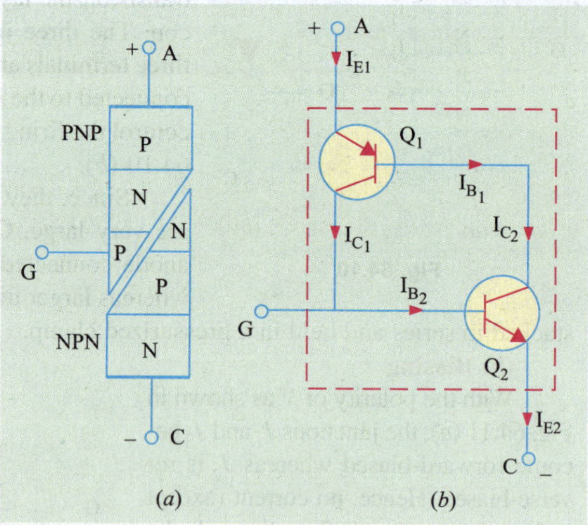

Fig. 64.13

It can be proved that if I_G is the gate current of the SCR and α_1 and α_2, the current gains of the *PNP* and *NPN* transistors respectively, then anode current is given by

$$I_A = \frac{\alpha_2 I_G}{1-(\alpha_1+\alpha_2)}$$

(*e*) Firing and Triggering

Usually, SCR is operated with an anode voltage ***slightly less*** than the forward breakover voltage V_{BO} and is triggered into conduction by a low-power gate pulse. Once switched ON, gate has no further control on the device current. Gate signals can be (*a*) dc firing signals [Fig. 64.14 (*a*)] or (*b*) pulse signals [Fig. 64.14 (*b*)].

In Fig. 64.14 (*a*) with *S* open, SCR does not conduct and the lamp is out. When *S* is closed momentarily, a positive voltage is applied to the gate which forward-biases the centre *P-N* junction. As a result, SCR is pulled into conduction and the lamp lights up. SCR will remain in the conducting state until the supply voltage is removed or reversed. Fig. 64.14 (*b*) shows triggering by timed pulses obtained from a pulse source.

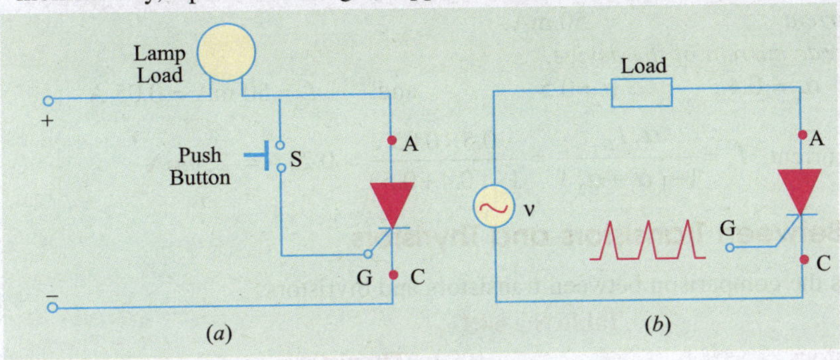

Fig. 64.14

We have discussed above the most common method of SCR triggering *i.e.* gate triggering. However, other available triggering methods are as under :

1. Thermal Triggering

In this case, the temperature of the forward-biased junction is increased till the reverse-biased junction breaks down.

2. Radiation Triggering

Here, triggering is achieved with the help of charge carriers which are produced by the bombardment of the SCR with external high-energy particles like neutrons or protons.

3. Voltage Triggering

In this case, the voltage applied across the anode and cathode of the SCR is increased which decreases the width of the depletion layer at the reverse-biased junction leading to its collapse.

4. *dv/dt* Triggering

In this case, dv / dt is made more than the value of the critical rate of rise of the voltage.

(*f*) Turning OFF

As stated earlier, once 'fired', SCR remains ON ***even when triggering pulse is removed.*** This ability of the SCR to remain ON even when gate current is removed is referred to as **latching**. In fact, SCR belongs to a class of devices known as ***latching devices.***

By now, it is clear that an SCR cannot be turned OFF by simply removing the gate pulse. Number of techniques are employed to turn an SCR off. These are :

1. anode current interruption.
2. reversing polarity of anode-cathode voltage as is done each half-cycle by v in Fig. 64.14 (*b*);
3. reducing current through SCR below the holding current I_H (Fig. 64.12). It is also called ***low-current dropout.***

(*g*) Applications

Main application of an SCR is as a ***power control device.*** It has been shown above that when SCR is OFF, its current is negligible and when it is ON, its voltage is negligible. Consequently, it never dissipates any appreciable amount of power even when controlling substantial amounts of load power. For example, one SCR requires only 150 mA to control a load current of 2500 A. Other common areas of its application include.

1. relay controls,
2. regulated power supplies,
3. static switches,
4. motor controls,
5. inverters,
6. battery chargers,
7. heater controls,
8. phase control.

SCRs have been designed to control powers upto 10 MW with individual ratings as high as 2000 A at 1.8 kV. Its frequency range of application has been extended to about 50 kHz.

Example 64.4. *The two-transistor analogy of an SCR has the following data :*

gain of PNP transistor = 0.4 ; *gain of NPN transistor = 0.5 ;*

gate current = 50 mA.

Calculate the anode current of the device.

Solution. Here, $\alpha_1 = 0.4$; $\alpha_2 = 0.5$ and $I_G = 50$ mA = 0.05 A

$$\text{anode current, } I_A = \frac{\alpha_2 I_G}{1-(\alpha_1+\alpha_2)} = \frac{0.5\times 0.05}{1-(0.4+0.5)} = 0.25\text{ A} = \mathbf{250mA}$$

64.6. Comparison Between Transistors and Thyristors

Table No. 64.1 gives the comparison between transistors and thyristors.

Table No. 64.1

Sl.No.	Transistors	Thyristors
1.	3-layers, 2-junction devices	4-layer, 2- or more junction devices
2.	fast response	very fast response
3.	high efficiency	very high efficiency
4.	highly reliable	very highly reliable
5.	small voltage drop	very small voltage drop
6.	long life	very long life
7.	small to medium power ratings	very small to very large power ratings
8.	require a continuous flow of current to remain in conducting state	require only a small pulse for triggering and thereafter remaining in conducting state.
9.	low power consumption	very low power consumption
10.	low control capability	high control capability
11.	small turn-ON and turn-OFF time	very small turn-ON and turn-OFF time

Example 64.5. *A 250 Ω resistor is connected in series with the gate of an SCR as shown in Fig. 64.15. The gate current required for firing the SCR is 8 mA. Calculate the value of the input voltage V_{in} required for causing the SCR to break down.*

(Basic Electronics, Osmania Univ. 1993)

Solution. The value of V_{in} should be such as to *(i)* overcome the barrier voltage of 0.7 *V* and *(ii)* cause 8 mA current to flow through 250 Ω resistor.

$$V_{in} = V_{GC} + I_G R = 0.7 + 8 \times 10^{-3} \times 250 = \mathbf{2.7\ V}$$

64.7. Transient Effects in an SCR

We will consider the following two effects :

***(i) di/dt* Effect**

This effect is produced due to a high initial rate-of-rise of the anode current when an SCR is just switched ON and results in the formation of a ***local hot spot near the gate connection*** as explained below :

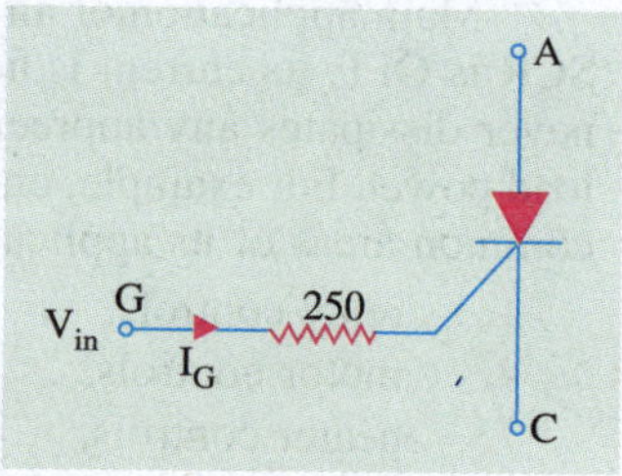

Fig. 64.15

When a trigerring pulse is applied to the gate of an SCR, the holes are injected into the P-region where they crowd together and form an initial conduction zone over a small part of the junction J_2 before spreading the conduct throughout the whole area of junction. If the anode current is allowed to rise very rapidly (as would be the case for resistive or capacitive loads), this high current will be forced to flow through this small conduction zone until the conduction has spread through the entire junction. This may result in local hot-spots in the junction which are likely to damage the SCR permanently.

The maximum allowable anode current *di/dt* can be increased (and, hence, turn-on time of an SCR decreased) by using specially-designed gate-connection geometries which result in a more rapid distribution of charge throughout the gate region.

***(ii) dv/dt* Effect**

It is found that sometimes an SCR unwantedly turns ON by itself during sudden changes of the applied anode potential at a time when there is no gate current applied and the SCR is supposed to be blocking. This false triggering is due to the capacitance possessed by the large-area junction J_2 (Fig. 64.10). When rate-of-rise of the applied anode voltage *dv/dt* is very high, the capacitive charging current may become high enough to initiate switch-on even in the absence of external gate current. False triggerings due to the *dv/dt* are prevented by using a 'snubber circuit'.

64.8. Phase Control

In the phase control circuit of Fig. 64.16, gate triggering current is derived from the supply itself. The variable resistance *R* limits the gate current during positive half-cycles of the supply. If *R* is adjusted to a low value, SCR will trigger almost immediately at the commencement of the positive half-cycle of the input.

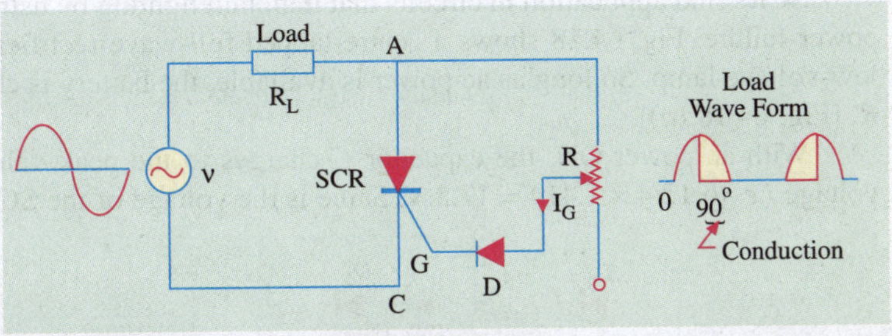

Fig. 64.16

If, on the other hand, *R* is set to a high resistance, SCR may not switch ON until the peak of the positive half-cycle. By adjusting *R* between these two extremes, SCR can be made to switch ON somewhere between the commencement and peak of the positive half-cycle *i.e.* between 0° and 90°.

It is obvious that if I_G is not enough to trigger the SCR at 90°, then the device will not trigger at all because I_G has maximum value then. This operation is sometimes referred to as ***half-wave variable-resistance phase control.*** It is an effective method of controlling the load power.

The purpose of diode *D* is to protect the gate from negative voltage which would otherwise be applied to it during the negative half-cycle of the input.

It is seen from Fig. 64.16 that at the instant of SCR switch-ON, gate current flows through R_L, *R* and *D*. Hence, at that ***instant***

$$v = I_G R_L + I_G R + V_D + V_G \qquad \therefore \quad R = \frac{V - V_D - V_G - I_G R_L}{I_G}$$

Example 64.6. *The circuit of Fig. 64.16 is connected to an ac supply v = 50 sin θ and R_L = 50 Ω. Gate current is 100 μA and V_G = 0.5 V. Determine the range of adjustment of R for the SCR to be triggered between 30° and 90°. Take V_D = 0.7 V.*

Solution. *(i)* θ = 30°

Now, v = 50 sin θ = 50 sin 30° = 25 V

$$\therefore \quad R = \frac{25 - 0.7 - 0.5 - (100 \times 10^{-6} \times 50)}{100 \times 10^{-6}} = \mathbf{238K}$$

(ii) θ = 90°

v = 50 sin 90° = 50 V

$$R = \frac{50 - 0.7 - 0.5 - 0.005}{100 \times 10^{-6}} = \mathbf{488K}$$

64.9. Theft Alarm

The circuit shown in Fig. 64.17 can be used to protect a car tape deck or a radio receiver from theft. The switch S is located at some concealed point in the car and is kept *closed*. Since gate G is grounded through the tape deck, the SCR is OFF and the horn is silent. If the tape deck is removed, G is no longer grounded. Instead, it gets connected to the car battery through R. Consequently, gate current is set up which fires the SCR. As a result, the horn starts blowing and continues to do so until S is opened.

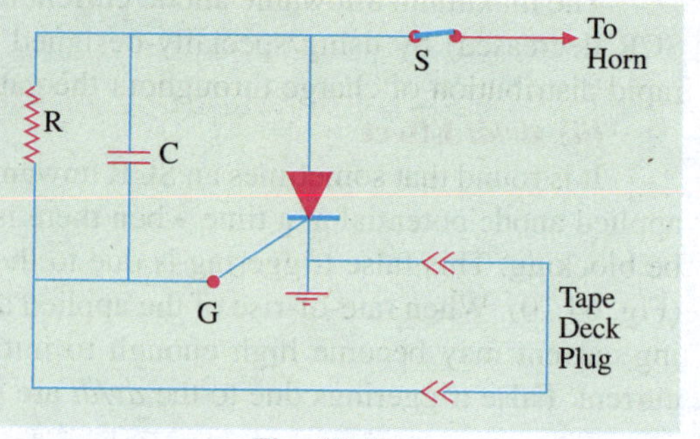

Fig. 64.17

64.10. Emergency Lighting System

SCRs find application in circuits that maintain lighting by using a backup battery in case of ac power failure. Fig. 64.18 shows a centre-tapped full-wave rectifier used for providing power to a low-voltage lamp. So long as ac power is available, the battery is charged *via* diode D_3 and resistor R_1 [Fig. 64.18 (*a*)].

With ac power ON, the capacitor C charges to the peak value of the full-wave rectified ac voltage *i.e.* to 12.4 × 1.414 = 17.5 V. Same is the voltage of the SCR cathode K.

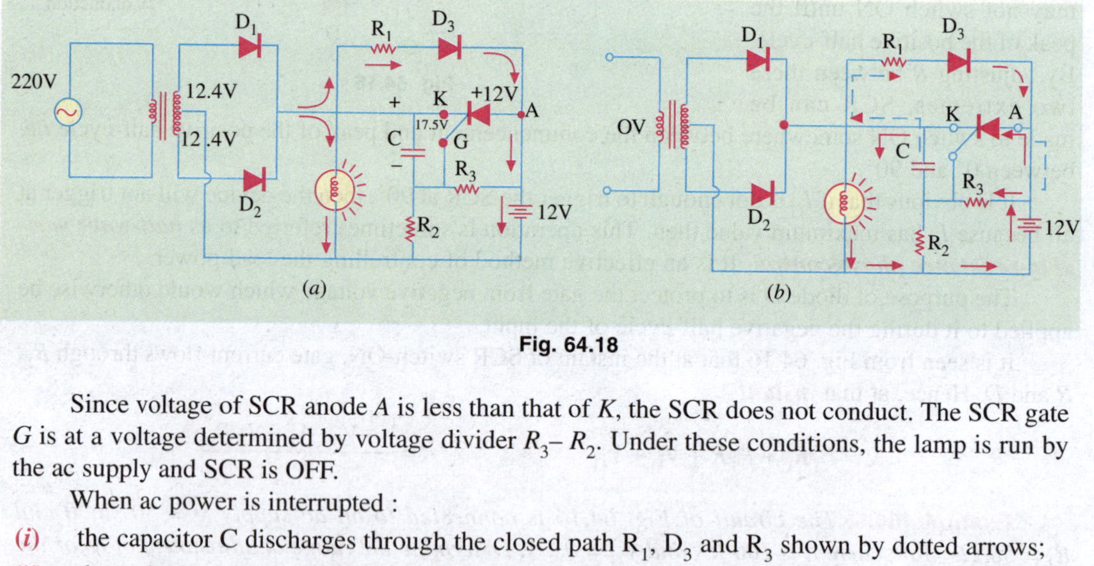

Fig. 64.18

Since voltage of SCR anode A is less than that of K, the SCR does not conduct. The SCR gate G is at a voltage determined by voltage divider $R_3 - R_2$. Under these conditions, the lamp is run by the ac supply and SCR is OFF.

When ac power is interrupted :

(*i*) the capacitor C discharges through the closed path R_1, D_3 and R_3 shown by dotted arrows;

(*ii*) the cathode voltage decreases thereby making it less positive than anode;

(*iii*) this triggers SCR into conduction which allows the battery current to pass through the lamp thus maintaining illumination.

When ac supply is restored, C recharges and the SCR turns OFF. The battery starts recharging again.

64.11. Light Activated SCR (LASCR)

It is just an ordinary SCR except that *it can also be light-triggered.* Most LASCRs also have a gate terminal for being triggered by an electrical pulse just as a convential SCR. Fig. 64.19 shows the two LASCR symbols used commonly.

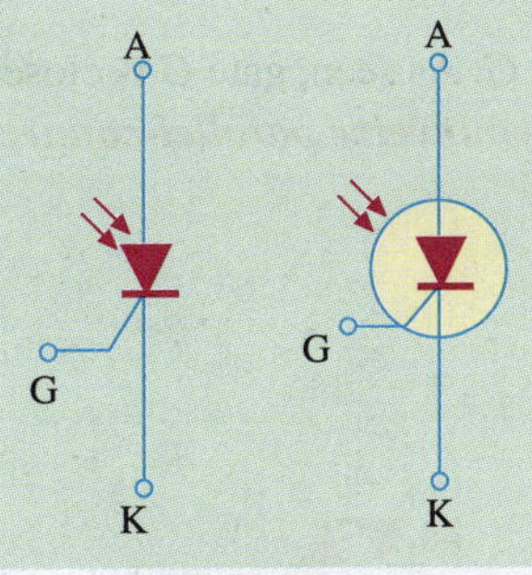

Fig. 64.19

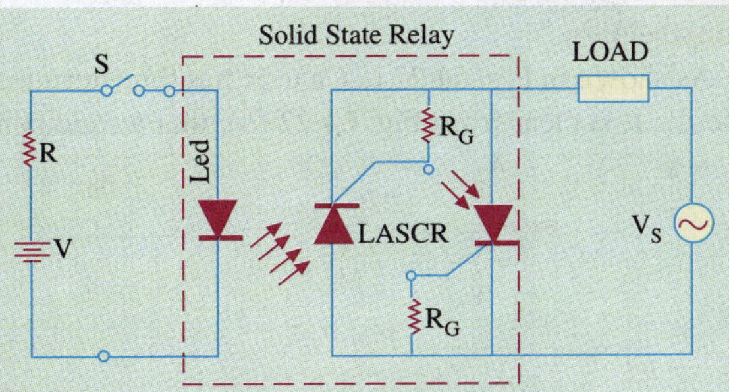

Fig. 64.20

LASCRs are manufactured mostly in relatively low-current ranges and are used for triggering larger SCRs and triacs. They are used in optical light controls, relays, motor control and a variety of computer applications. Some LASCRs have clear windows in their cases so that light sources from other devices can be coupled to them. Many have the light source device encapsulated in the same package so that a relay is formed. Since the relay action does not require direct electrical connection, such relays are often used to couple signals into very high voltage equipment and other dangerous locations. Fig. 64.20 shows the connection of such a solid-state relay. Two LASCRs are connected in reverse parallel in order to obtain conduction in both half-cycles of the applied ac voltage V_S. A single LED is used to trigger both LASCRs. Bias resistors are used to reduce the light sensitivity of the gates and prevent sporadic triggering during off-periods. Usually, all the three active devices and the two bias resistors R_G are encapsulated in the same package.

64.12. The Shockley Diode*

It is a ***two-terminal four-layer*** or *PNPN* device as shown in Fig. 64.21 along with its schematic symbol. It is essentially a low-current ***SCR without a gate***. For switching the diode ON, its anode-to-cathode voltage **(V_{AK})** must be increased to forward switching voltage **(V_S)** which is the equivalent of SCR forward breakover voltage. Like an SCR, it also has a holding current. The *PNPN* structure can be represented by an equivalent circuit consisting of a *PNP* transistor and an *NPN* transistor. One application of the diode is as a relaxation oscillator.

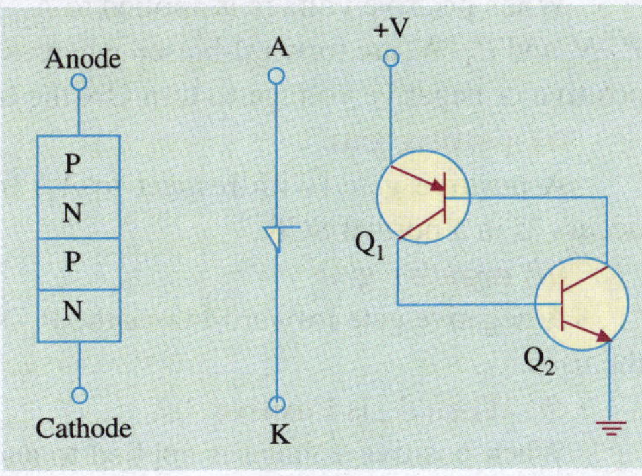

Fig. 64.21

64.13. Triac

It is a ***5-layer bi-directional device*** which can be triggered into conduction by both ***positive*** and ***negative*** voltages at its anodes and with both ***positive and negative*** triggering pulses at its gate. It behaves like two SCRs ***connected in parallel, upside down with respect to each other***. That is, the anode of one is tied to the cathode of the other and their gates are directly tied together. Hence, anode and gate voltages applied in either direction will fire a triac because they would fire at least one of the two SCRs which are in opposite directions.

Since a triac responds to both positive and negative voltages at the anode, the concept of cathode used for an SCR is dropped. Instead, the two electrodes are called anodes A_1 and A_2.

---

* After the name of its inventor William Shockley.

1. Construction

As shown in Fig. 64.22 (*a*), a triac has three terminals A_1, A_2 and G. As seen, gate G is closer to anode A_1. It is clear from Fig. 64.22 (*b*), that a triac is nothing but ***two inverse parallel-connected SCRs*** with a ***common gate terminal.*** As seen, it has six doped regions. Fig. 64.23 shows the schematic symbol which consists of two inverse-connected SCR symbols.

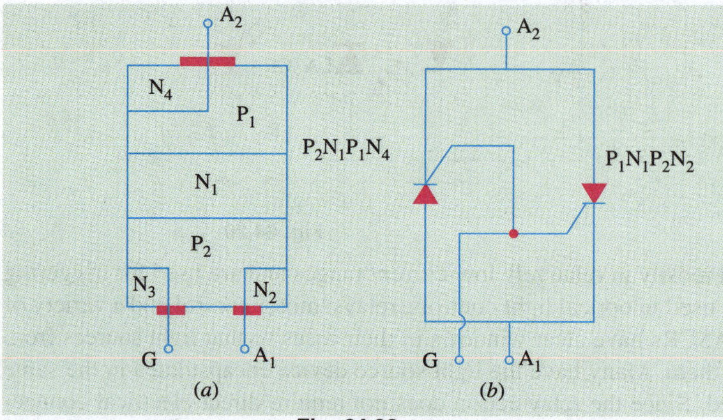

Fig. 64.22

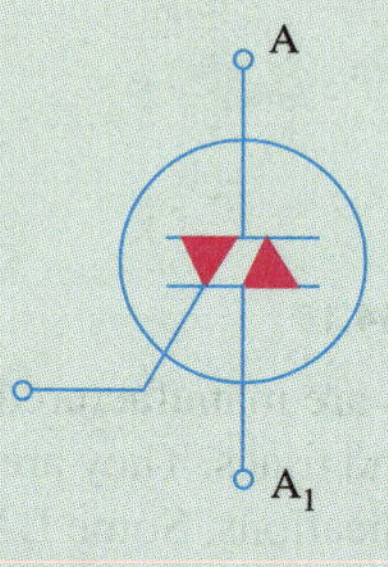

Fig. 64.23

2. Operation

(*a*) When A_2 is Positive

When positive voltage is applied to A_2, path of current flow is P_1-N_1-P_2-N_2. The two junctions P_1-N_1 and P_2 - N_2 are forward-biased whereas N_1-P_2 junction is blocked. The gate can be given either positive or negative voltage to turn ON the triac as explained below.

(*i*) positive gate

A positive gate (with respect to A_1) forward-biases the P_2-N_2 junction and the breakdown occurs as in a normal SCR.

(*ii*) negative gate

A negative gate forward-biases the P_2-N_3 junction and current carriers injected into P_2 turn on the triac.

(*b*) When A_1 is Positive

When positive voltage is applied to anode A_1, path of current flow is P_2-N_1-P_1-N_4. The two junctions P_2-N_1 and P_1-N_4 are forward-biased whereas junction N_1-P_1 is blocked. Conduction can be achieved by giving either positive or negative voltage to G as explained below.

(*i*) positive gate

A positive gate (with respect to A_1) injects current carriers by forward-biasing P_2-N_2 junction and thus initiates conduction.

(*ii*) negative gate

A negative gate injects current carriers by forward-biasing P_2-N_3 junction thereby triggering conduction.

It is seen that there are four triac-triggering modes, two each for the two anodes.

Low-current dropout is the only way to open a triac.

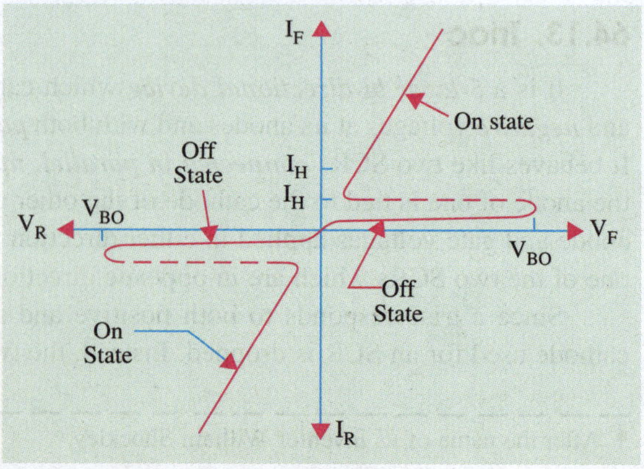

Fig. 64.24

3. V/I Characteristics

Typical characteristics of a triac are shown in Fig. 64.24.

As seen, triac exhibits same forward blocking and forward conducting characteristics as an SCR but *for either polarity of voltage applied to the main terminal.* Obviously, a triac has latch current in either direction.

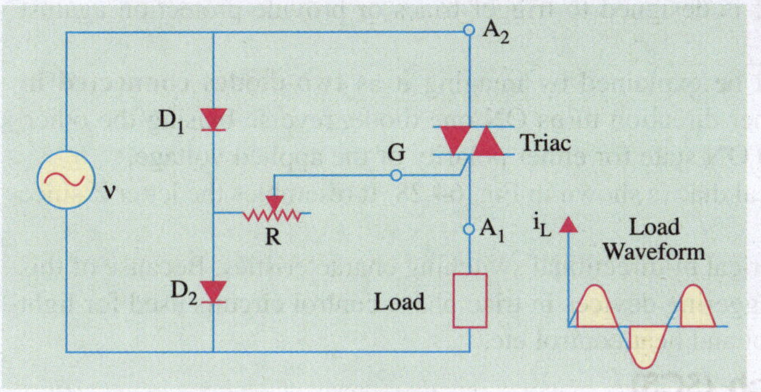

Fig. 64.25

4. Applications

One fundamental application of triac is shown in Fig. 64.25. Here, it is used to control ac power to a load by switching ON and OFF during positive and negative half-cycles of the input ac power.

During positive half-cycle of the input, diode D_1 is forward biased, D_2 is reverse-biased and gate is positive with respect to A_1. By adjusting R, the point at which conduction commences can be varied.

Diac-triac combination for ac load power control is shown in Fig. 64.26. Firing control of diac is achieved by adjusting R.

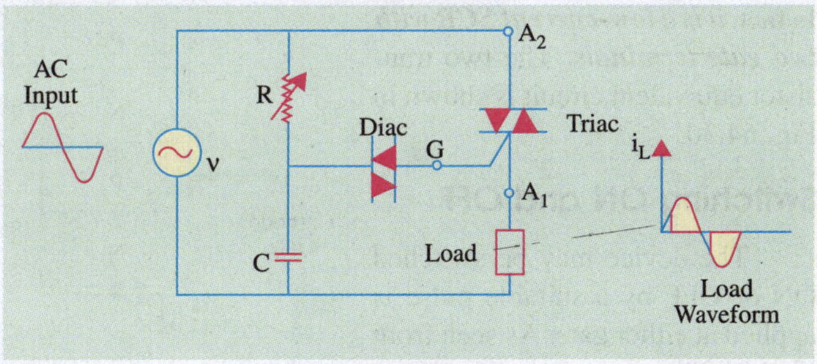

Fig. 64.26

Other applications of a triac include.

1. as static switch to turn ac power OFF and ON;
2. for minimizing radio interference;
3. for light control;
4. for motor speed control etc.

The only disadvantage of triac is that it takes comparatively longer time to recover to OFF state. Hence, its use is limited to ac supply frequencies of upto 400 Hz.

64.14. Diac

To put it simply, a diac is nothing else but a ***triac without its gate terminal*** as shown in Fig.

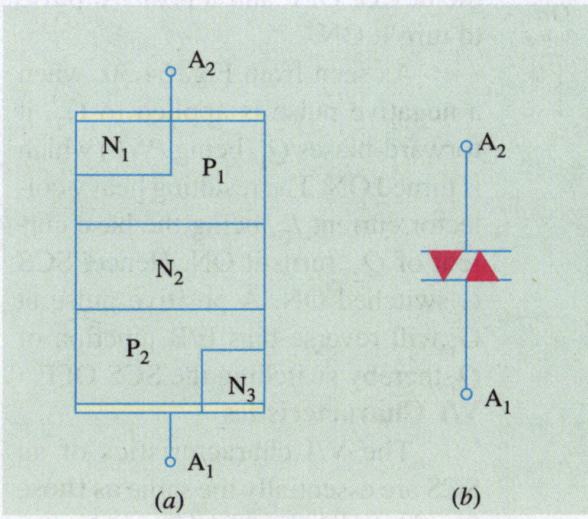

Fig. 64.27

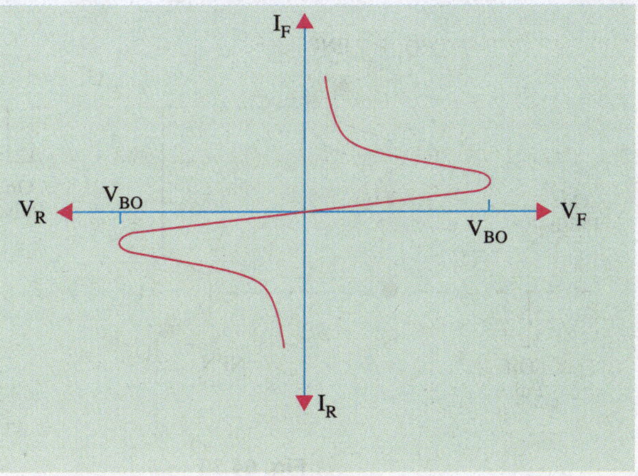

Fig. 64.28

64.27 (*a*). Its equivalent circuit is a pair of inverted four layer diodes. Its schematic symbol is shown in Fig. 64.27 (*b*). As seen, ***it can break down in either direction.***

When anode A_1 is positive, the current path is P_2-N_2-P_1-N_1. Similarly, when A_2 is positive, the current flow path is P_1-N_2-P_2-N_3. Diac is designed to trigger triacs or provide protection against over-voltages.

The operation of a diac can best be explained by imaging it as **two-diodes connected in series.** Voltage applied across it in either direction turns ON one diode, reverse-biasing the other. Hence, it can be switched from OFF to ON state for either polarity of the applied voltage.

The characteristic curve of a typical diac is shown in Fig. 64.28. It resembles the letter Z since diac breaks down in either direction.

As stated above, diac has symmetrical bi-directional switching characteristics. Because of this feature, diacs are frequently used as triggering devices in triac phase control circuits used for light dimming, universal motor speed control and heat control etc.

64.15. Silicon Controlled Switch (SCS)

It is a four-layer, four-terminal *PNPN* device having anode *A*, cathode *C*, anode gate G_1 and cathode gate G_2 as shown in Fig. 64.29. In fact, ***it is a low-current SCR with two gate terminals.*** The two transistor equivalent circuit is shown in Fig. 64.30.

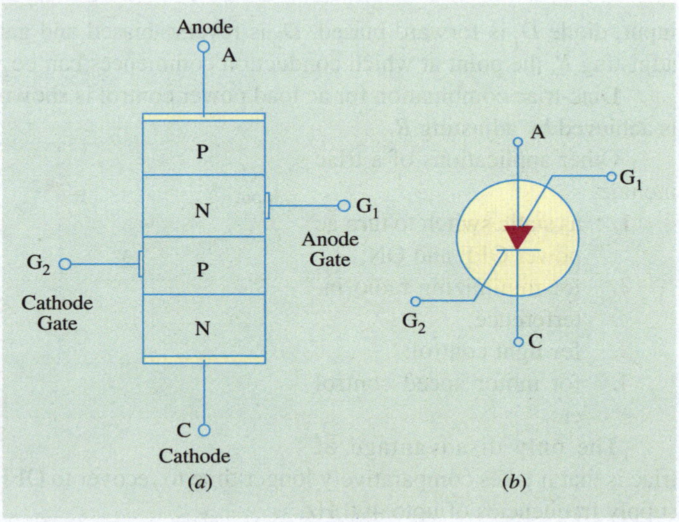

Fig. 64.29

Switching ON and OFF

The device may be switched ON or OFF by a suitable pulse is applied at either gate. As seen from Fig. 64.30, a negative pulse is required at anode gate G_1 to turn the device ON whereas positive pulse is needed to turn it OFF as explained below.

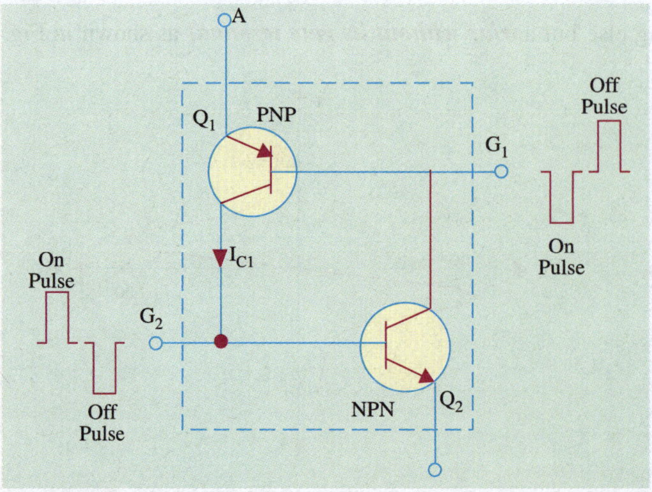

Fig. 64.30

Similarly, at cathode gate G_2, a negative pulse is required to switch the device OFF and a positive pulse to turn it ON.

As seen from Fig. 64.30, when a negative pulse is applied to G_1, it forward-biases Q_1 (being *PNP*) which is turned ON. The resulting heavy collector current I_C, being the base current of Q_2, turns it ON. Hence, SCS is switched ON. A positive pulse at G_1 will reverse bias E/B junction of Q_1 thereby switching the SCS OFF.

V/I Characteristics

The V/I characteristics of an SCS are essentially the same **as those for the SCR** (Fig. 64.13).

As compared to an SCR, an SCS has ***much reduced turn*-OFF *time*.** Moreover, it has higher control and triggering sensitivity and a more predictable firing situation.

Applications

The more common areas of SCS applications are as under :

1. in counters, registers and timing circuits of computers,
2. pulse generators,
3. voltage sensors,
4. oscillators etc.

OBJECTIVE TESTS – 64

1. A unijunction transistor has
   (*a*) anode, cathode and a gate
   (*b*) two bases and one emitter
   (*c*) two anodes and one gate
   (*d*) anode, cathode and two gates.
2. Which semiconductor device acts like a diode and two resistors ?
   (*a*) SCR (*b*) triac
   (*c*) diac (*d*) UJT.
3. A UJT has R_{BB} = 10 K and R_{B2} = 4 K. Its intrinsic stand-off ratio is
   (*a*) 0.6
   (*b*) 0.4
   (*c*) 2.5
   (*d*) 5/3.
4. An SCR conducts appreciable current when its .... with respect to cathode.
   (*a*) anode and gate are both negative
   (*b*) anode and gate are both positive
   (*c*) anode is negative and gate is positive
   (*d*) gate is negative and anode is positive.
5. After firing an SCR, the gating pulse is removed. The current in the SCR will
   (*a*) remains the same
   (*b*) immediately fall to zero
   (*c*) rise up
   (*d*) rise a little and then fall to zero.
6. An SCR may be turned OFF by
   (*a*) interrupting its anode current
   (*b*) reversing polarity of its anode- cathode voltage
   (*c*) low-current dropout
   (*d*) all of the above.
7. A triac behaves like two
   (*a*) inverse parallel-connected SCRs with common gate
   (*b*) diodes in series
   (*c*) four-layer diodes in parallel
   (*d*) resistors and one diode.
8. A triac can be triggered into conduction by
   (*a*) only positive voltage at either anode
   (*b*) positive or negative voltage at either anode
   (*c*) positive or negative voltage at gate
   (*d*) both (*b*) and (*c*).
9. A diac is equivalent to a
   (*a*) pair of SCRs
   (*b*) pair of four-layer SCRs
   (*c*) diode and two resistors
   (*d*) triac with two gates.
10. An SCS has
    (*a*) four layers and three terminals
    (*b*) three layers and four terminals
    (*c*) two anodes and two gates
    (*d*) one anode, one cathode and two gates.
11. An SCS may be switched ON by a
    (*a*) positive pulse at its anode
    (*b*) negative pulse at its cathode
    (*c*) positive pulse at its cathode gate G_2
    (*d*) positive pulse at its anode gate G_1.
12. The *dv / dt* effect in an SCR can result in
    (*a*) high rate-of-rise of anode voltage
    (*b*) increased junction capacitance
    (*c*) false triggering
    (*d*) low capacitive charging current.
13. The *di / dt* effect in an SCR leads to the formation of
    (*a*) local hot spots
    (*b*) conduction zone
    (*c*) charge spreading zone
    (*d*) none of the above.
14. SCR turns OFF from conducting state to blocking state on
    (*a*) reducing gate current
    (*b*) reversing gate voltage
    (*c*) reducing anode current below holding current value
    (*d*) applying ac to the gate
15. When a thyristor is negatively biased,
    (*a*) all the three junctions are negatively biased
    (*b*) outer junctions are positively biased and the inner junction is negatively biased.

(*c*) outer junctions are negatively biased and the inner junction is positively biased

(*d*) the junction near the anode is negatively biased and the one near the cathode is positively biased.

16. A LASCR is just like a conventional SCR except that it

(*a*) cannot carry large current

(*b*) can also be light-triggered

(*c*) has no gate terminal

(*d*) cannot be pulse-triggered.

17. The minimum value of current required to maintain conduction in an SCR is called its .... current.

(*a*) commutation

(*b*) holding

(*c*) gate trigger

(*d*) breakover

18. Diacs are primarily used as

(*a*) pulse generators

(*b*) triggering devices

(*c*) surge protection devices

(*d*) power thyristors.

ANSWERS

1. (*b*) **2.** (*d*) **3.** (*a*) **4.** (*b*) **5.** (*a*) **6.** (*d*) **7.** (*a*) **8.** (*d*) **9.** (*b*) **10.** (*d*) **11.** (*c*) **12.** (*c*) **13.** (*a*) **14.** (*c*) **15.** (*c*) **16.** (*b*) **17.** (*b*) **18.** (*b*)

C H A P T E R **65**

SINUSOIDAL AND NON-SINUSOIDAL OSCILLATORS

Learning Objectives

- What is an Oscillator?
- Classification of Oscillators
- Damped and Undamped Oscillations
- Oscillatory Circuit
- Essentials of a Feedback LC Oscillator
- Tuned Base Oscillator
- Tuned Collector Oscillator
- Hartley Oscillator
- FET Hartley Oscillator
- Colpitts Oscillator
- Clapp Oscillator
- FET Colpitts Oscillator
- Crystal Controlled Oscillator
- Transistor Pierce Cystal Oscillator
- FET Pierce Oscillator
- Phase Shift Principle
- RC Phase Shift Oscillator
- Wien Bridge Oscillator
- Pulse Definitions
- Basic Requirements of a Sawtooth Generator
- UJT Sawtooth Generator
- Multivibrators (MV)
- Astable Multivibrator
- Bistable Multivibrator (BMV)
- Schmitt Trigger
- Transistor Blocking Oscillator

An oscillator is an electronic device used for the purpose of generating a signal. Oscillators are found in computers, wireless receivers and transmitters, and audiofrequency equipment particularly music synthesizers

65.1. What is an Oscillator ?

Oscillator

An electronic oscillator may be defined in any one of the following four ways :

1. It is a circuit which converts dc energy into ac energy at a very high frequency;
2. It is an electronic source of alternating current or voltage having sine, square or sawtooth or pulse shapes;
3. It is a circuit which generates an ac output signal without requiring any externally applied input signal;
4. It is an unstable amplifier.

These definitions exclude electromechanical alternators producing 50 Hz ac power or other devices which convert mechanical or heat energy into electric energy.

65.2. Comparison Between an Amplifier and an Oscillator

As discussed in Chapter-60, an amplifier produces an output signal whose waveform is similar to the input signal but whose power level is generally high. This additional power is supplied by

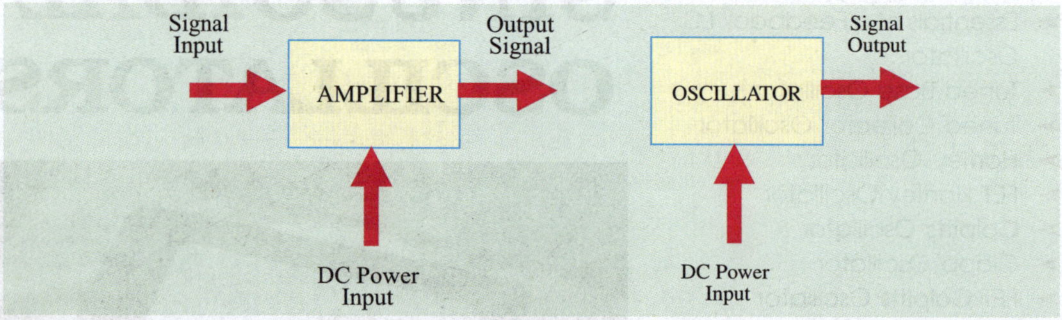

Fig. 65.1

the external dc source. Hence, an amplifier is essentially an energy convertor *i.e.* it takes energy from the dc power source and converts it into ac energy at signal frequency. The process of energy conversion is controlled by the input signal. If there is no input signal, there is no energy conversion and hence there is no output signal.

An oscillator differs from an amplifier in one basic aspect : the oscillator ***does not require an external signal*** either to start or maintain energy conversion process (Fig. 65.1). It keeps producing an output signal so long as the dc power source is connected.

Moreover, the frequency of the output signal is determined by the passive components used in the oscillator and can be varied at will.

65.3. Classification of Oscillators

Electronic oscillators may be broadly divided into following two groups :

(i) Sinusoidal (or harmonic) oscillators—which produce an output having sine waveform;

(ii) Non-sinusoidal (or relaxation) oscillators—they produce an output which has square, rectangular or sawtooth waveform or is of pulse shape.

Sinusoidal oscillators may be further subdivided into :

(a) Tuned-circuits or LC feedback oscillators such as Hartley, Colpitts and Clapp etc.;

(b) *RC* phase-shift oscillators such as Wien-bridge oscillator;

(c) Negative-resistance oscillators such as tunnel diode oscillator;

(d) Crystal oscillators such as Pierce oscillator;

(*e*) Heterodyne or beat-frequency oscillator (BFO).

The active devices (bipolars, FETs or unijunction transistors) in the above mentioned circuits may be biased class-A, B or C. Class-A operation is used in high-quality audio frequency oscillators. However, radio frequency oscillators are usually operated as class-C.

65.4. Damped and Undamped Oscillations

Sinusoidal oscillations produced by oscillators may be (*i*) damped or (*ii*) undamped.

(*i*) Damped Oscillations

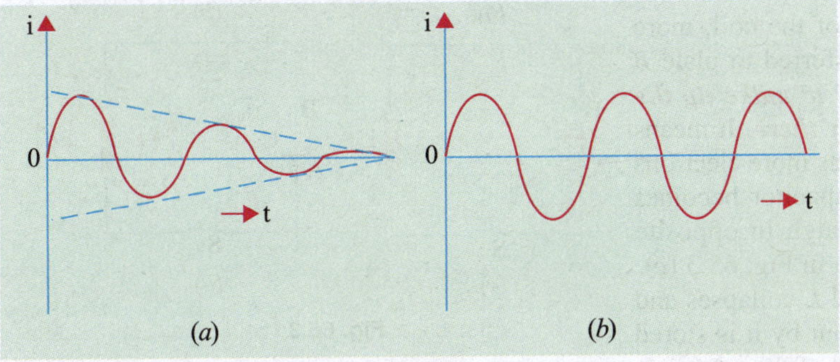

Fig. 65.2

Oscillations whose amplitude keeps decreasing (or decaying) with time are called **damped** or **decaying** oscillations. The waveform of such oscillations is shown in Fig. 65.2 (*a*). These are produced by those oscillator circuits in which I^2R losses take place continuously during each oscillation without any arrangement for compensating the same. Ultimately, the amplitude of the oscillations decays to zero when there is not enough energy to supply circuit losses. However, the frequency or time-period remains constant because ***it is determined by the circuit parameters.***

Sinusoidal oscillators serve a variety of functions in telecommunications and in electronics. The most important application in telecommunication is the use of sine waves as carrier signal in both radio and cable transmissions.

Sine wave signals are also used in frequency response testing of various types of systems and equipment including analogue communication channels, amplifiers and filters and closed-loop control systems.

(*ii*) Undamped Oscillations

Oscillations whose amplitude remains constant *i.e.* does not change with time are called undamped oscillations. These are produced by those oscillator circuits which have no losses or if they have, there is provision for compensating them. The constant-amplitude and constant-frequency sinusoidal waves shown in Fig. 65.2 (*b*) are called ***carrier waves*** and are used in communication transmitters for transmitting low-frequency audio information to far off places.

65.5. The Oscillatory Circuit

It is also called *LC* circuit or tank circuit. The oscillatory circuit (Fig. 65.3) consists of two reactive elements *i.e.* an inductor and a capacitor. Both are capable of storing energy. The capacitor stores energy in its electric field whenever there is potential difference across its plates. Similarly, a coil or an inductor stores energy in its magnetic field whenever current flows through it. Both *L* and *C* are supposed to be loss-free (*i.e.* their *Q*-factors are infinite).

As shown in Fig. 65.3 (*a*), suppose the capacitor has been fully-charged from a dc source. Since *S* is open, it cannot discharge through *L*. Now, let us see what happens when *S* is closed.

1. When *S* is closed [Fig. 65.3 (*b*)] ***electrons*** move from plate *A* to plate *B* through coil *L* as shown by the arrow (or conventional current flows from *B* to *A*). This electron flow reduces the strength of the electric field and hence the amount of energy stored in it.
2. As electronic current starts flowing, the self-induced emf in the coil opposes the current flow. Hence, rate of discharge of electrons is somewhat slowed down.
3. Due to the flow of current, magnetic field is set up which stores the energy given out by

the electric field [Fig. 65.3 (*b*)].

4. As plate *A* loses its electrons by discharge, the electron current has a tendency to die down and will actually reduce to zero when all excess electrons on *A* are driven over to plate *B* so that both plates are reduced to the same potential. At that time, there is no electric field but the magnetic field has maximum value.
5. However, due to self-induction (or electrical inertia) of the coil, more electrons are transferred to plate *B* than are necessary ***to make up the electron deficiency there.*** It means that now plate *B* has more electrons than *A*. Hence, capacitor becomes charged again though in opposite direction as shown in Fig. 65.3 (*c*).
6. The magnetic field *L* collapses and the energy given out by it is stored in the electric field of the capacitor.

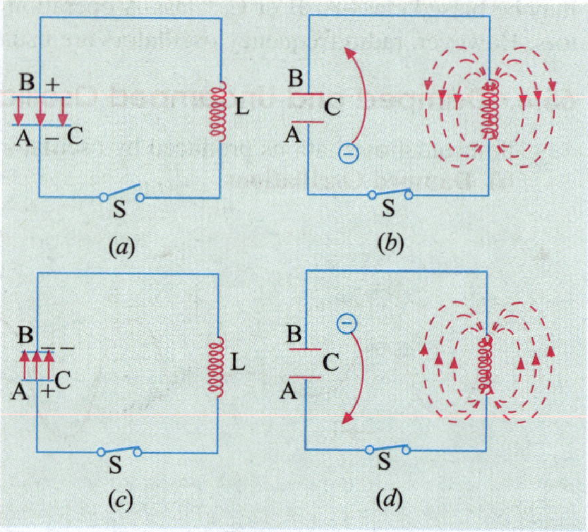

Fig. 65.3

7. After this, the capacitor starts discharging in the opposite direction so that, now, the electrons move from plate *B* to plate *A* [Fig. 65.3 (*d*)]. The electric field starts collapsing whereas magnetic field starts building up again though in the opposite direction. Fig. 65.3 (*d*) shows the condition when the capacitor becomes fully discharged once again.
8. However, these discharging electrons overshoot and again an excess amount of electrons flow to plate *A*, thereby charging the capacitor once more.
9. This sequence of charging and discharging continues. The to and fro motion of electrons between the two plates of the capacitor constitutes an oscillatory current.

It may be also noted that during this process, the electric energy of the capacitor is converted into magnetic energy of the coil and ***vice versa.***

These oscillations of the capacitor discharge are damped because energy is dissipated away gradually so that their amplitude becomes zero after sometime. There are two reasons for the loss of the energy :

(*a*) Some energy is lost in the form of heat produced in the resistance of the coil and connecting wires ;

(*b*) and some energy is lost ***in the form of electromagnetic (EM) waves*** that are radiated out from the circuit through which an oscillatory current is passing.

Both these losses subtract energy from the circuit with the result that circuit current decreases gradually till it becomes zero. The waveform of the oscillatory discharge is similar to that shown in Fig. 65.2 (*a*).

65.6. Frequency of Oscillatory Current

The frequency of time-period of the oscillatory current depends on two factors :

(*a*) Capacitance of the Capacitor

Larger the capacitor, greater the time required for the reversal of the discharge current *i.e.* lower its frequency.

(*b*) Self-inductance of the Coil

Larger the self-inductance, greater the internal effect and hence longer the time required by the current to stop flowing during discharge of the capacitor.

The frequency of this oscillatory discharge current is given by

$$f = \frac{1}{2\pi\sqrt{LC}} = \frac{159}{\sqrt{LC}} \text{ kHz}$$

where L = self-inductance in μH and C = capacitance in μF

It may, however, be pointed out here that damped oscillations so produced are not good for radio transmission purpose because of their ***limited range and excessive distortion***. For good radio transmission, we need undamped oscillations which can be produced if some additional energy is supplied in correct phase and correct direction to the LC circuit for making up the I^2R losses continually occuring in the circuit.

65.7. Frequency Stability of an Oscillator

The ability of an oscillator to maintain a constant frequency of oscillation is called its frequency stability. Following factors affect the frequency stability :

1. Operating Point of the Active Device

The Q-point of the active device (*i.e.* transistor) is so chosen as to confine the circuit operation on the linear portion of its characteristic. Operation on non-linear portion varies the parameters of the transistor which, in turn, affects the frequency stability of the oscillator.

2. Inter-element Capacitances

Any changes in the inter-element capacitances of a transistor particularly the collector- to-emitter capacitance cause changes in the oscillator output frequency, thus affecting its frequency stability. The effect of changes in inter-element capacitances, can be neutralized by adding a swamping capacitor across the offending elements—the added capacitance being made part of the tank circuit.

3. Power Supply

Changes in the dc operating voltages applied to the active device shift the oscillator frequency. This problem can be avoided by using regulated power supply.

4. Temperature Variations

Variations in temperature cause changes in transistor parameters and also change the values of resistors, capacitors and inductors used in the circuit. Since such changes take place slowly, they cause a slow change (called drift) in the oscillator output frequency.

5. Output Load

A change in the output load may cause a change in the Q-factor of the LC tuned circuit thereby affecting the oscillator output frequency.

6. Mechanical Vibrations

Since such vibrations change the values of circuit elements, they result in changes of oscillator frequency. This instability factor can be eliminated by isolating the oscillator from the source of mechanical vibrations.

65.8. Essentials of a Feedback LC Oscillator

The essential components of a feedback LC oscillator shown in Fig. 65.4 are :

1. A resonator which consists of an LC circuit. It is also known as frequency-determining network (*FDN*) or tank circuit.
2. An amplifier whose function is to amplify the oscillations produced by the resonator.

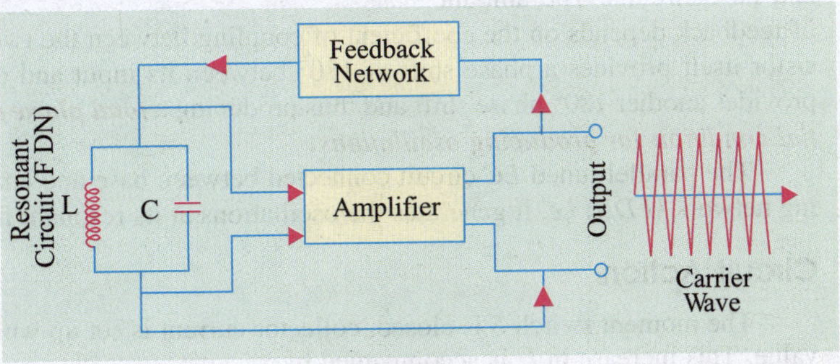

Fig. 65.4

3. A positive feedback network (*PFN*) whose function is to transfer part of the output energy to the resonant *LC* circuit in ***proper phase***. The amount of energy fed back is sufficient to meet I^2R losses in the *LC* circuit.

The essential condition for maintaining oscillations and for finding the value of frequency is

$$\beta A = 1 + j\,0 \quad \text{or} \quad \beta A \angle \phi = 1 \angle 0$$

It means that

(i) The feedback factor or loop gain $|\beta A| = 1$,

(ii) The net phase shift around the loop is 0° (or an integral multiple of 360°). In other words, ***feedback should be positive***.

The above conditions form Barkhausen criterion for maintaining a steady level of oscillation at a specific frequency.

Majority of the oscillators used in radio receivers and transmitters use tuned circuits with positive feedback. Variations in oscillator circuits are due to the different way by which the feedback is applied. Some of the basic circuits are :

1. *Armstrong or Tickler or Tuned-base Oscillator* — it employs inductive feedback from collector to the tuned *LC* circuit in the base of a transistor.

2. *Tuned Collector Oscillator*—it also employs inductive coupling but the *LC* tuned circuit is in the collector circuit.

3. *Hartley Oscillator*—Here feedback is supplied inductively.

4. *Colpitts Oscillator*—Here feedback is supplied capacitively.

5. *Clapp Oscillator*—It is a slight modification of the Colpitts oscillator.

65.9. Tuned Base Oscillator

Such an oscillator using a transistor in *CE* configuration is shown in Fig. 65.5. Resistors R_1, R_2 and R_3 determine the dc bias of the circuit. The parallel R_3—C_2 network in the emitter circuit is a stabilizing circuit to prevent signal degeneration. As usual, C_1 is the dc blocking capacitor. The mutually-coupled coils L_1 and L forming primary and secondary coils of an *RF* transformer provide the required feedback between the collector and base circuits. The amount of feedback depends on the coefficient of coupling between the two coils. The *CE* connected transistor itself provides a phase shift of 180° between its input and output circuits. The transformer provides another 180° phase shift and thus producing a ***total phase shift of 360° which is an essential condition for producing oscillations.***

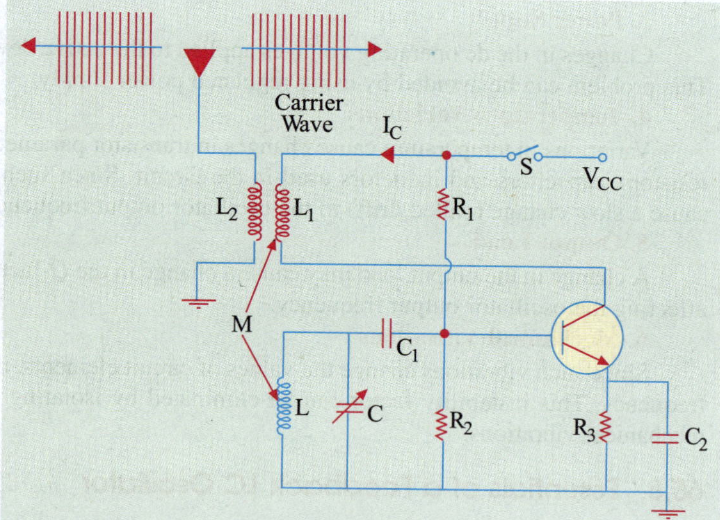

Fig. 65.5

The parallel-tuned *LC* circuit connected between base and emitter is the frequency determining network (*FDN*) *i.e.* it generates the oscillations at its resonant frequency.

Circuit Action

The moment switch *S* is closed, collector current is set up which tends to rise to its quiescent value. This increase in I_C is accompanied by :

1. An expanding magnetic field through L_1 which links with L and
2. An induced *e.m.f.* called feedback voltage in L.

Two immediate reactions of this feedback voltage are:

(i) Increase in emitter-base voltage (and base current) and
(ii) A further increase in collector current I_C.

It is followed by a succession of cycles of

1. An increase in feedback voltage,
2. An increase in emitter-base voltage and
3. An increase in I_C until saturation is reached.

Meanwhile, C gets charged. As soon as I_C ceases to increase, magnetic field of L_1 ceases to expand and thus no longer induces feedback voltage in L. Having been charged to maximum value, C starts to discharge through L. However, decrease in voltage across C causes the following sequence of reactions :

1. A decrease in emitter-base bias and hence in I_B, **2.** A decrease in I_C;
3. A collapsing magnetic field in L_1;
4. An induced feedback voltage in L though, this time, in ***opposite*** direction;
5. Further decrease in emitter-base bias and so on till I_C reaches its cut-off value.

During this time, the capacitor having lost its original charge, again becomes fully charged though with ***opposite*** polarity. Transistor being in cut-off, the capacitor will again begin to discharge through L. Since polarity of capacitor charge is opposite to that when transistor was in saturation, the sequence of reactions now will be

1. An increase in emitter-base bias, **2.** An increase in I_C,
3. An expanding magnetic field in L_1, **4.** An induced feedback voltage in L,
5. A further increase in emitter-base bias and
6. So on till I_C increases to its saturation value.

This cycle of operation keeps repeating so long as enough energy is supplied to meet losses in the LC circuit.

The output can be taken out by means of a third winding L_2 magnetically coupled to L_1. It has approximately the same waveform as collector current.

The frequency of oscillation is equal to the resonant frequency of the LC circuit.

65.10. Tuned Collector Oscillator

Such an oscillator using a transistor in CE configuration is shown in Fig. 65.6.

***(i)* Frequency Determining Network (FDN)**

It is made up of a variable capacitor C and a coil L which forms primary winding of a step-down transformer. The combination of L and C forms an oscillatory tank circuit to set the frequency of oscillation.

Resistors R_1, R_2 and R_3 are used to dc bias the transistor. Capacitors C_1 and C_2 act to by-pass R_3 and R_2 respectively so that they have no effect on the ac operation of the circuit. Moreover, C_2 provides ac ground for transformer secondary L_1.

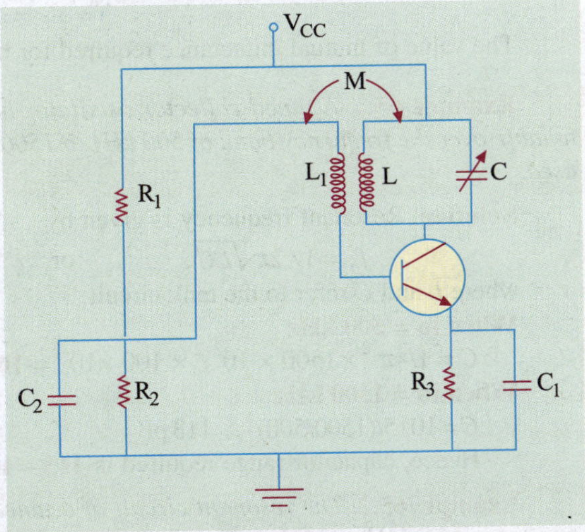

Fig. 65.6

***(ii)* Positive Feedback**

Feedback between the collector-emitter circuit and base-emitter circuit is provided by the transformer secondary winding L_1 which is mutually-coupled to L. As far as ac signals are concerned, L_1 is connected to emitter *via* low-reactance capacitors C_2 and C_1.

Since transistor is connected in CE configuration, it provides a phase shift of 180° between its input and output circuits. Another phase shift of 180° is provided by the transformer thus producing a total phase shift of 360° between the output and input voltages resulting in ***positive feedback*** between the two.

(iii) Amplifying Action

The transistor amplifier provides sufficient gain for oscillator action to take place.

(iv) Working

When the supply is first switched on, a transient current is developed in the tuned *LC* circuit as the collector current rises to its quiescent value. This transient current initiates natural oscillations in the tank circuit. These natural oscillations induce a small emf into L_1 by mutual induction which causes corresponding variations in base current. These variations in I_B are amplified β times and appear in the collector circuit. Part of this amplified energy is used to meet losses taking place in the oscillatory circuit and ***the balance is radiated out in the form of electromagnetic waves.***

The frequency of oscillatory current is almost equal to the resonant frequency of the tuned circuit.

$$\therefore \quad f_0 = \frac{1}{2\pi\sqrt{LC}}$$

65.11. Tuned Drain Oscillator (FET)

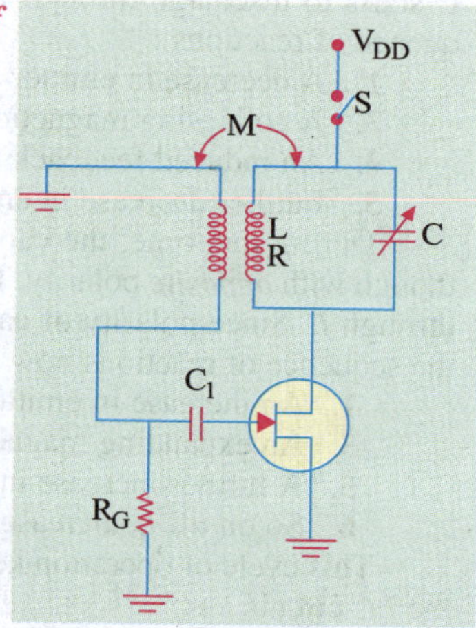

Fig. 65.7

The basic circuit is illustrated in Fig. 65.7. It is similar to the tuned collector oscillator of Fig. 65.6. Because of its high input impedance and high voltage amplification, a FET can be used to construct very simple and efficient oscillator circuit. This frequency of oscillation is given by

$$f_0 = \frac{1}{2\pi\sqrt{LC}} = \sqrt{\left(1+\frac{R}{r_d}\right)}$$

where r_d = ac drain resistance

The value of mutual inductance required for maintaining oscillations is $M = \dfrac{r_d RC + L}{\mu}$.

Example 65.1. *A tuned-collector oscillator has a fixed inductance of 100 μH and has to be tunable over the frequency band of 500 kHz to 1500 kHz. Find the range of variable capacitor to be used.* **(Principles of Telecom. Engg. Pune Univ. 1985)**

Solution. Resonant frequency is given by

$$f_0 = 1/2\pi\sqrt{LC} \quad \text{or} \quad C = 1/4\pi^2 f_o^2 L$$

where *L* and *C* refer to the tank circuit.

When fo = 500 kHz

$C = 1/4\pi^2 \times (500 \times 10^3)^2 \times 100 \times 10^{-6}$ = **1015 pF**

When fo = 1500 kHz

$C = 1015/(1500/500)^2 \cong 113$ pF

Hence, capacitor range required is **113 – 1015 pF**

Example 65.2. *The resonant circuit of a tuned-collector transistor oscillator has a resonant frequency of 5 MHz. If value of capacitance is increased by 50%, calculate the new resonant frequency.*

Solution. Using the equation for resonant frequency, we have

$$5 \times 10^6 = 1/2\pi\sqrt{LC} \quad \text{— 1st case}$$

$$f_0 = 1/2\pi\sqrt{L \times 1.5C} \quad \text{— 2nd case}$$

$$\therefore \quad \frac{f_o}{5 \times 10^6} = \frac{1}{\sqrt{1.5}} \quad \text{or} \quad f_o = \mathbf{4.08\ MHz}$$

65.12. Hartley Oscillator

In Fig. 65.8 (*a*) is shown a transistor Hartley oscillator using *CE* configuration. Its general principle of operation is similar to the tuned-collector oscillator discussed in Art. 65.10.

It uses a single tapped-coil having two parts marked L_1 and L_2 instead of two separate coils. So far as ac signals are concerned, one side of L_2 is connected to base *via* C_1 and the other to emitter *via* ground and C_3. Similarly, one end of L_1 is connected to collector *via* C_2 and the other to common emitter terminal *via* C_3. In other words, L_1 is in the output circuit *i.e.* collector-emitter circuit whereas L_2 is in the base-emitter circuit *i.e.* input circuit. These two parts are inductively-coupled and form an auto-transformer or a split-tank inductor. Feedback between the output and input circuits is accomplished through autotransformer action which also introduces a phase reversal of 180°. This phase reversal

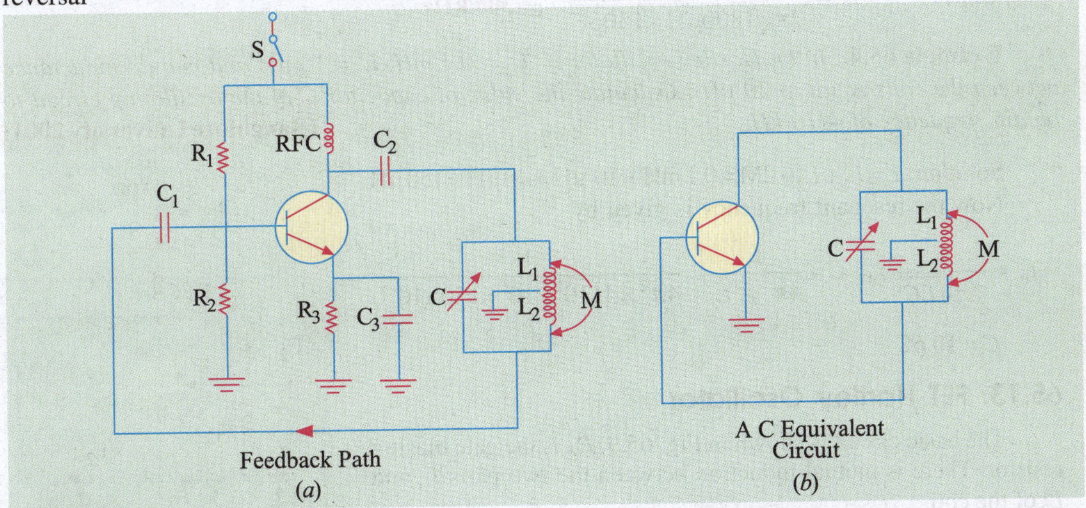

Fig. 65.8

Hartley oscillator

between two voltages occurs because they are taken from ***opposite ends*** of an inductor (L_1 - L_2 combination) with respect to the tap which is tied to common transistor terminal *i.e.* emitter which is ac grounded *via* C_3. Since transistor itself introduces a phase shift of 180°, the total phase shift becomes 360° thereby making the ***feedback positive* or *regenerative*** which is essential for oscillations (Art 65.8). As seen, positive feedback is obtained from the tank circuit and is coupled to the base *via* C_1. The feedback factor is given by the ratio of turns in L_2 and L_1 *i.e.* by N_2/N_1 and its value ranges from 0.1 to 0.5. Fig. 65.8 (*b*) shows the equivalent circuit of Hartley oscillator.

Resistors R_1 and R_2 form a voltage divider for providing the base bias and R_3 is an emitter swamping resistor to add stability to the circuit. Capacitor C_3 provides ac ground thereby preventing any signal degeneration while still providing temperature stabilisation. Radio-frequency choke (RFC) provides dc load for the collector and also keeps ac currents out of the dc supply V_{CC}.

When V_{CC} is first switched on through S, an initial bias is established by R_1 - R_2 and oscillations are produced because of positive feedback from the LC tank circuit (L_1 and L_2 constitute L). The frequency of oscillation is given by

$$f_0 = \frac{1}{2\pi\sqrt{LC}} \quad \text{where} \quad L = L_1 + L_2 + 2M$$

The output from the tank may be taken out by means of another coil coupled either to L_1 or L_2.

Example 65.3. *Calculate the oscillation frequency for the transistor Hartley oscillator circuit (refer to Fig. 65.8). Given the circuit values: L_{RFC} = 0.5 mH, L_1 = 750 μH, L_2 = 750 μH, M = 150 μH and C = 150 pF.*

Solution. $f_0 = \frac{1}{2\pi\sqrt{LC}}$ where $L = L_1 + L_2 + 2M$

$\therefore$ $L = 750\,\mu H + 750\,\mu H + 2 \times 150\,\mu H = 1800\,\mu H$

and $f_0 = \frac{1}{2\pi\sqrt{1800\mu H \times 150 pF}} \cong$ **306 kHz**

Example 65.4. *In an Hartley oscillator if L_1 = 0.1 mH, L_2 = 10 μH and mutual inductance between the coils equal to 20 μH. Calculate the value of capacitor C of the oscillating circuit to obtain frequency of 4110 kHz.* **(Bangalore University 2001)**

Solution. $L = L_1 + L_2 + 2M = 0.1\,mH + 10\,\mu H + 40\,\mu H = 150\,\mu H$

Now the resonant frequency is given by

$$f_0 = \frac{1}{2\pi\sqrt{LC}} \text{ or } C = \frac{1}{4\pi^2 f_0^2 L} = \frac{1}{4\pi^2 \times 4110^2 \times 10^6 \times 150 \times 10^{-6}}$$

C = **10 pF**

65.13. FET Hartley Oscillator

The basic circuit is shown in Fig. 65.9. R_G is the gate biasing resistor. There is mutual induction between the two parts L_1 and L_2 of the coil.

$$f_0 = \frac{1}{2\pi\sqrt{LC}}$$

where $L = L_1 + L_2 + 2M$

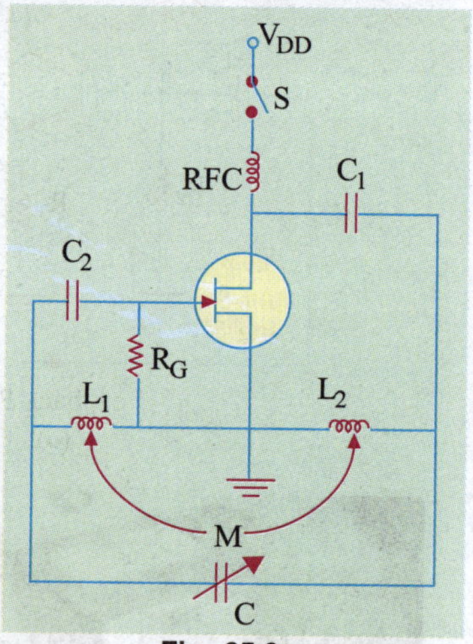

Fig. 65.9

Example 65.5. *Calculate the oscillator frequency for a FET Hartley oscillator (refer to Fig. 65.9), for the following circuit values: C = 250 pF, L_1 = 1.5 mH, L_2 = 1.5 mH, and M = 0.5 mH.*

Solution. $f_0 = \frac{1}{2\pi\sqrt{LC}}$ where $L = L_1 + L_2 + 2M$

$\therefore$ $L = 1.5\text{ mH} + 1.5\text{ mH} + 2 \times 0.5\text{ mH} = 4\text{ mH}$

and $f_0 = \frac{1}{2\pi\sqrt{4mH \times 250\,pF}}$ = **159.1 kHz**

65.14. Colpitts Oscillator

This oscillator is essentially the same as Hartley oscillator except for one difference. Colpitts oscillator uses **tapped capacitance** whereas Hartley oscillator uses **tapped inductance*. Fig. 65.10** (*a*)

* As an aid to memory, remember that Hartley begins with letter H for Henry *i.e.* coil and Colpitts begins with *C* for Capacitor.

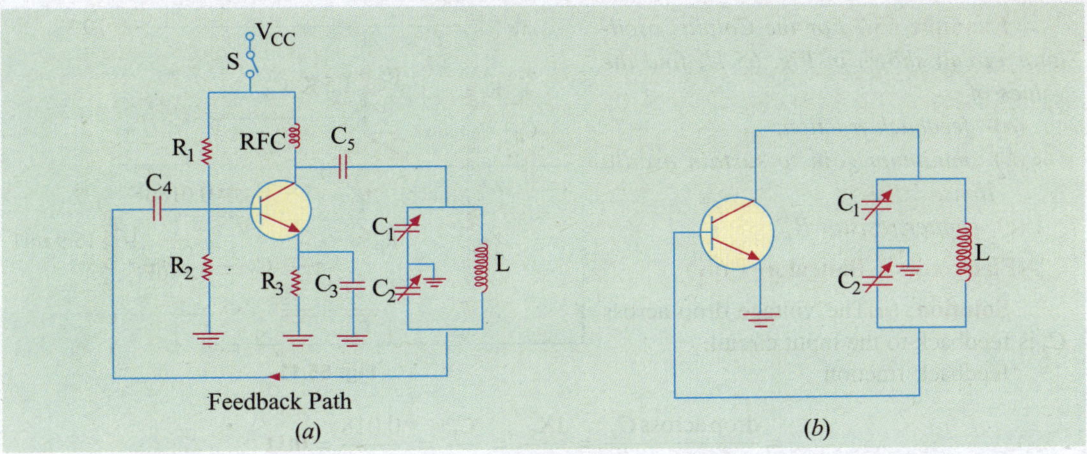

Fig. 65.10

shows the complete circuit with its power source and dc biasing circuit whereas Fig. 65.10 (*b*) shows its ac equivalent circuit. The two series capacitors C_1 and C_2 form the voltage divider used for providing the feedback voltage (the voltage drop across C_2 constitutes the feedback voltage). The feedback factor is C_1/C_2. The minimum value of amplifier gain for maintaining oscillations is

$$A_{v(min)} = \frac{1}{C_1/C_2} = \frac{C_2}{C_1}$$

The tank circuit consists of two ganged capacitors C_1 and C_2 and a single fixed coil. The frequency of oscillation (which does not depend on mutual inductance) is given by

$$f_0 = \frac{1}{2\pi\sqrt{LC}} \qquad \text{where} \qquad C = \frac{C_1 C_2}{C_1 + C_2}$$

Transistor itself produces a phase shift of 180°. Another phase shift of 180° is provided by the capacitive feedback thus giving a total phase shift of 360° between the emitter-base and collector-base circuits.

Resistors R_1 and R_2 form a voltage divider across V_{CC} for providing base bias, R_3 is for emitter stabilisation and *RFC* provides the necessary dc load resistance R_C for amplifier action. It also prevents ac signal from entering supply dc V_{CC}. Capacitor C_5 is a bypass capacitor whereas C_4 conveys feedback from the collector-to-base circuit.

When *S* is closed, a sudden surge of collector current **shock-excites** the tank circuit into oscillations which are sustained by the feedback and the amplifying action of the transistor.

Colpitts oscillator is widely used in commercial signal generators upto 1 MHz. Frequency of oscillation is varied by gang-tuning the two capacitors C_1 and C_2.

Example 65.6. *Determine the circuit oscillation frequency for a transistor Colpitts oscillator shown in Fig. 65.10(a). Given,*

$L = 100\ \mu H$, $L_{RFC} = 0.5\ mH$, $C_1 = 0.005\ \mu F$, $C_2 = 0.01\ \mu F$. $\quad C_6 = 10\ \mu F$

Solution. For a transistor Colpitts oscillator, the oscillation frequency,

$$f_0 = \frac{1}{2\pi\sqrt{LC}} \qquad \text{where} \qquad C = \frac{C_1 C_2}{C_1 + C_2}$$

$$\therefore \quad C = \frac{0.005\,\mu F \times 0.01\,\mu F}{0.005\,\mu F + 0.01\,\mu F} = 3.3\,nF$$

and $$f_0 = \frac{1}{2\pi \times \sqrt{100\,\mu H \times 3.3\,nF}} = \textbf{277 kHz}$$

Example 65.7. *For the Colpitts oscillator circuit shown in Fig. 65.11, find the values of*

(a) feedback fraction,

(b) minimum gain to sustain oscillations,

(c) emitter resistor R_E.

(Electronics-I, Bangalore Univ)

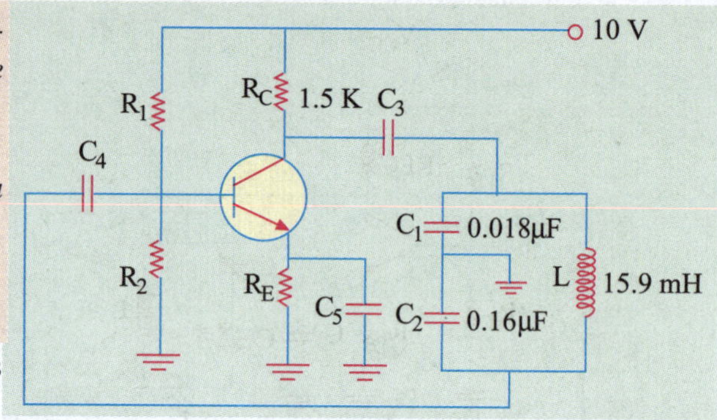

Fig. 65.11

Solution. (*a*) The voltage drop across C_2 is feedback to the input circuit.

feedback fraction

$$= \frac{\text{drop across } C_2}{\text{drop across } C_1} = \frac{IX_2}{IX_1} = \frac{C_1}{C_2} = \frac{0.018}{0.16} = \mathbf{0.11}$$

(*b*) $A_{v(min)} = \dfrac{1}{\text{feedback fraction}} = \dfrac{1}{0.11} = \mathbf{9}$

(*c*) Now, $A_v \cong \dfrac{R_C}{R_E}$ $\quad\therefore\quad$ $R_E = \dfrac{R_C}{A_v} = \dfrac{1500}{9} \cong \mathbf{167\ \Omega}$

It should be noted that the above calculations do not take into account the losses in the coil and the loading effect of the amplifier input impedance.

65.15. Clapp Oscillator

It is a variation of Colpitts oscillator and is shown in Fig. 65.12 (*a*). It differs from Colpitts oscillator in respect of capacitor C_3 only which is joined in series with the tank inductor. Fig. 65.12 (*b*) shows the ac equivalent circuit.

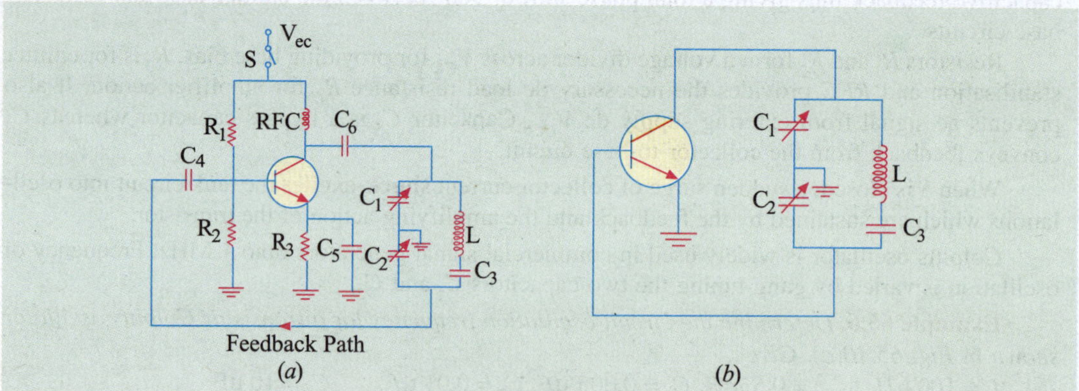

Fig. 65.12

Addition of C_3 (*i*) improves frequency stability and (*ii*) eliminates the effect of transistor's parameters on the operation of the circuit.

The operation of this circuit is the same as that of the Colpitts oscillator.

The frequency of oscillation is given by

$$f_0 = \frac{1}{2\pi\sqrt{LC}} \quad \text{where} \quad \frac{1}{C} = \frac{1}{C_1} + \frac{1}{C_2} + \frac{1}{C_3}$$

65.16. FET Colpitts Oscillator

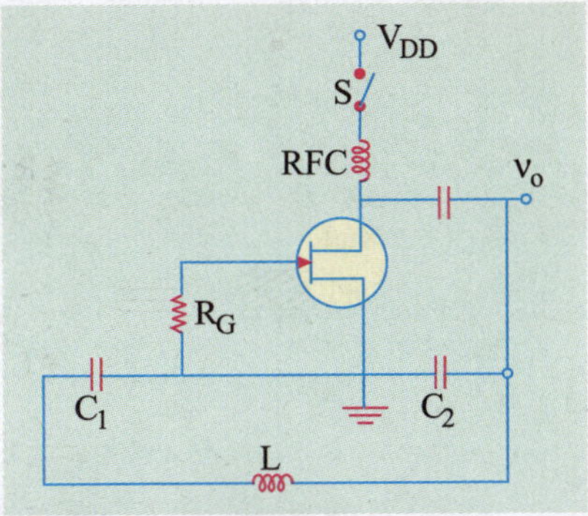

Fig. 65.13

The circuit is shown in Fig. 65.13. It is similar to the transistor circuit of Fig. 65.10. Here, R_G is the gate biasing resistor. The radio-frequency coil choke (R_{FC}) performs two functions :

(i) It keeps ac current out of the dc drain supply and

(ii) It provides drain load.

As seen, C_1 is in the input circuit whereas C_2 is in the output circuit.

The frequency of oscillation is given by

$$f_0 = \frac{1}{2\pi\sqrt{LC}} \text{ ; where, } C = \frac{C_1 C_2}{C_1 + C_2}$$

Example 65.8. *Determine the circuit oscillation frequency for the FET Colpitts oscillator (refer to Fig. 65.13), given $C_1 = 750$ pF, $C_2 = 2500$ pF and $L = 40$ μH.*

Solution. For a Colpitts oscillator, the oscillation frequency,

$$f_0 = \frac{1}{2\pi\sqrt{LC}} \qquad \text{where} \quad C = \frac{C_1 C_2}{C_1 + C_2}$$

$$\therefore \quad C = \frac{750\,\text{pF} \times 2500\,\text{pF}}{750\,\text{pF} + 2500\,\text{pF}} = \mathbf{576.9\ pF}$$

$$\text{and} \quad f_0 = \frac{1}{2\pi \times \sqrt{40\mu H \times 576.9\,pF}} = \mathbf{1.048\ MHz}$$

65.17. Crystals

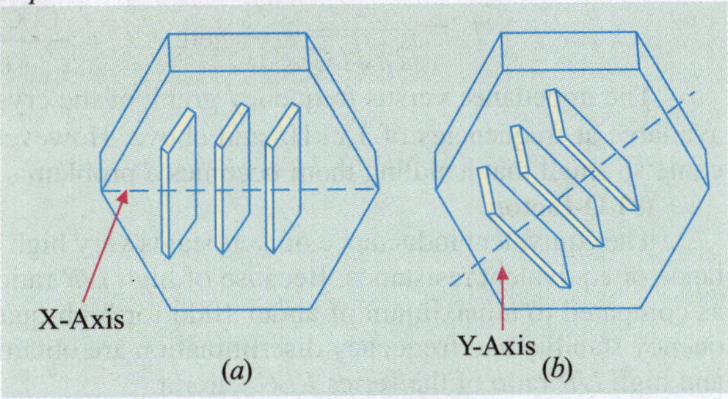

Fig. 65.14

For an exceptionally high degree of frequency stability, use of crystal oscillators is essential. The crystal generally used is a finely-ground wafer of translucent quartz (or tourmaline) stone held between two metal plates and housed in a package about the size of a postal stamp. The crystal wafers are cut from the crude quartz in two different ways. The method of 'cutting' determines the crystal's natural resonant frequency and its temperature coefficient. When the wafer is cut so that its flat surface are perpendicular to its **electrical axis,** it is called an **X-cut** crystal (Fig. 65.14). But if the wafer is so cut that its flat surfaces are perpendicular to its **mechanical axis,** it is called **Y-cut** crystal.

(a) Piezoelectric Effect

The quartz crystal described above has peculiar properties. When mechanical stress is applied across its two opposite faces, **a potential difference is developed across them.** It is called ***piezoelectric effect.*** Conversely, when a potential difference is applied across its two opposite faces, it causes the crystal to ***either expand or contract.*** If an alternating voltage is applied, the crystal wafer is set into vibrations. The frequency of vibration is equal to the resonant frequency of the crystal as determined by its structural characteristics. Where the frequency of the applied ac voltage equals the natural resonant frequency of the crystal, the amplitude of vibration will be maximum. As a general rule, thinner the crystal, ***higher its frequency of vibration.***

(b) Equivalent Electrical Circuit

The electrical equivalent circuit of the crystal is shown in Fig. 65.15 (*b*). It consists of a series RLC_1 circuit in parallel with a capacitor C_2.

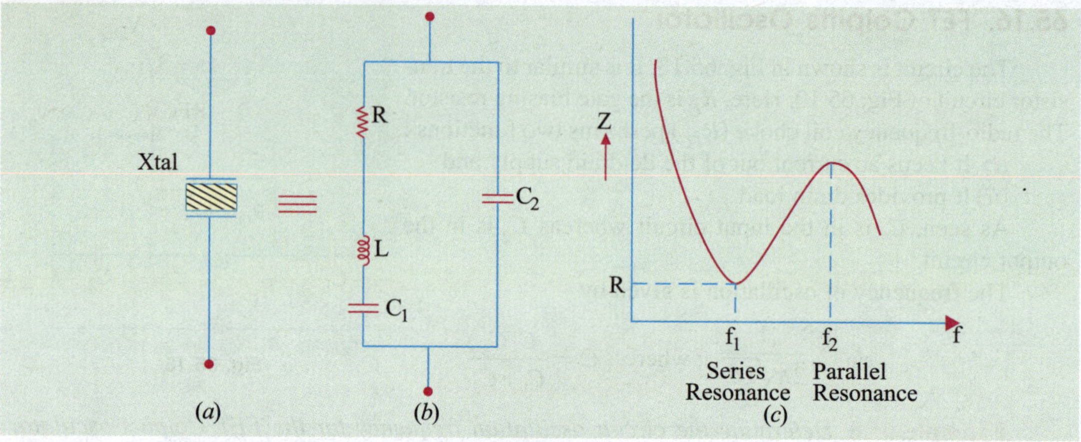

Fig. 65.15

The circuit has two resonant frequencies :

(i) one is the lower series resonance frequency f_1 which occurs when $X_L = X_{C1}$. In that case, $Z = R$ as shown in Fig. 65.15 (*c*).

$$f_1 = \frac{1}{2\pi\sqrt{LC_1}}$$

(ii) the other is the parallel resonance frequency f_2 which occurs when reactance of the series leg equals the reactances of C_2. At this frequency, the crystal offers very high impedance to the external circuit.

$$f_2 = \frac{1}{2\pi\sqrt{LC}}\,; \quad \text{where} \quad C = \frac{C_1 C_2}{C_1 + C_2}$$

The impedance versus frequency graph of the crystal is shown in Fig. 65.15 (*c*). Crystals are available at frequencies of 15 kHz and above. However, at frequencies above 100 MHz, they become so small that handling them becomes a problem.

(*c*) Q-factor

The equivalent inductance of a crystal is very high as compared to either its equivalent capacitance or equivalent resistance. Because of high *L/R* ratio, the *Q*-factor of a crystal circuit is 20,000 as compared to a maximum of about 1000 for high-quality *LC* circuits. Consequently, greater frequency stability and frequency discrimination are obtained because of extremely high *Q* (upto 10^6) and high *L/R* ratio of the series RLC_1 circuit.

(*d*) Temperature Coefficient

Temperature variations affect the resonant frequency of a crystal. The number of cycles change per million cycles for a 1°C change in temperature is called the **temperature coefficient** (TC) of the crystal. It is usually expressed in parts per million (ppm) per °C. For example, a TC of 10 ppm per °C means that frequency variation is 0.001 per cent* per °C change in temperature. It can also be expressed as 10 Hz/MHz/°C. When kept in temperature-controlled ovens, crystal oscillators have frequency stability of about ±1 ppm.

Usually, X-cut crystals have negative TC whereas *Y*-cut crystals have positive TC.

Example 65.9. *A 600 kHz X-cut crystal when calibrated at 50°C has a TC of 20 ppm/°C. What will be its resonant frequency when its temperature is raised to 60°C ?*

Solution. A TC of 20 ppm/°C is the same thing as 20 Hz/MHz/°C. Since temperature rise is 10°C and original frequency is 600 kHz = 0.6 MHz, the total change as calculated by ratio propor-

* Since 10 ppm means a change of 10 in 10^6, hence percentage change is

$$= \frac{10}{10^6} \times 100 = 10^{-3} = 0.001$$

tion is = 20 × 10 × 0.6 = 120 Hz.

Since TC is negative, the new resonance frequency of the crystal is

= 600,000 – 120 = 599,880 = **599.88 kHz.**

Example 65.10. *A certain X-cut quartz crystal resonates at 450 kHz. It has an equivalent inductance of 4.2 H and an equivalent capacitance of 0.0297 pF. If its equivalent resistance is 60 Ω, calculate its Q-factor.*

Solution. $$Q = \frac{\omega L}{R} = \frac{2\pi fL}{R}$$

$$= \frac{2\pi \times 450 \times 10^3 \times 4.2}{600} \cong \mathbf{19{,}792}$$

Example 65.11. *The parameters of a crystal oscillator equivalent circuit are : L = 0.8H; C_1 = 0.08 pF, R = 5 kΩ and C_2 = 1.0 pF. Determine the resonant frequencies f_1 and f_2*

(UPSC Engg. Services 1999)

Solution. $$f_1 = \frac{1}{2\pi\sqrt{LC_1}} = \frac{1}{2\pi\sqrt{0.8 \times 0.08 \times 10^{-12}}} = \frac{1}{2\pi \times 0.253 \times 10^{-6}}$$

$$= \frac{1}{1.5895 \times 10^{-6}} = 0.629 \times 10^6 \text{ Hz} = \mathbf{629 \text{ kHz}}$$

$$f_2 = \frac{1}{2\pi\sqrt{LC}} \text{ where } C = \frac{C_1C_2}{C_1 + C_2} = \frac{0.08 \times 1}{0.08 + 1} = 0.074 \, pF$$

$$\therefore \quad f_2 = \frac{1}{2\pi\sqrt{0.8 \times .074 \times 10^{-12}}} = \frac{1}{2\pi \times 0.243 \times 10^{-6}} \text{(Hz)}$$

$$\cong 0.655 \times 10^6 \text{ Hz} = \mathbf{655 \text{ kHz.}}$$

65.18. Crystal Controlled Oscillator

Fig. 65.16, shows the use of a crystal to stabilise the frequency of a tuned-collector oscillator which has a crystal (usually quartz) in the feedback circuit.

The *LC* tank circuit has a frequency of oscillation

$$f_0 = 1/2\pi\sqrt{LC}$$

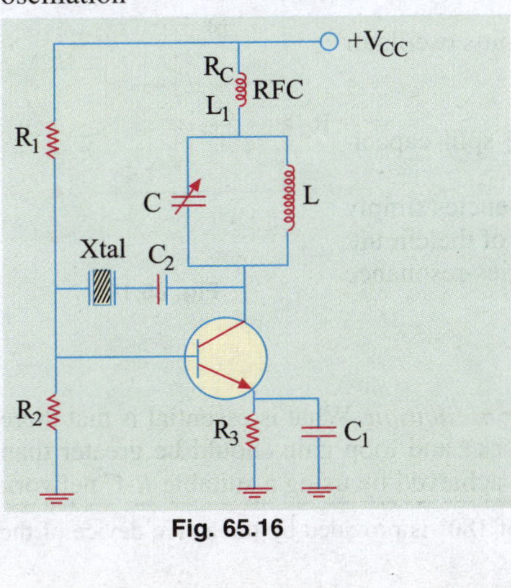

Fig. 65.16

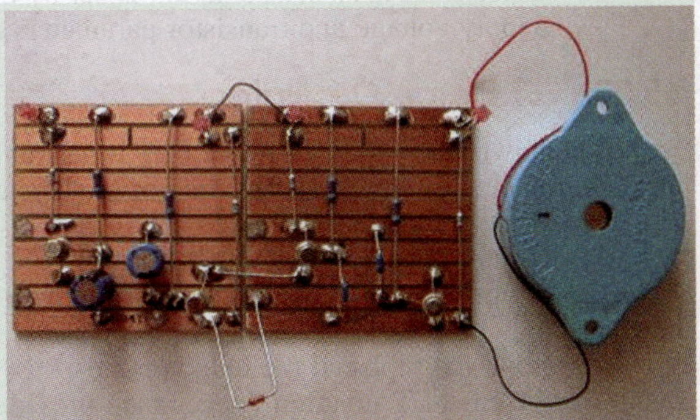
Crystal Controlled oscillator

The circuit is adjusted to have a frequency neara bout the desired operating frequency but the exact frequency is set by the crystal and stabilized by the crystal. For example, if natural frequency of vibration of the crystal is 27 MHz, the *LC* circuit is made to resonate at this frequency.

As usual, resistors R_1, R_2 and R_3 provide a voltage-divider stabilised dc bias circuit. Capacitor C_1 by-passes R_3 in order to maintain large gain. *RFC* coil L_1 prevents ac signals from entering dc line whereas R_C is the required dc load of the collector. The coupling capacitor C_2 has negligible impedance at the operating frequency but prevents any dc link between collector and base. Due to extreme stability of crystal oscillations, such oscillators are widely used in communication transmitters and receivers where frequency stability is of prime importance.

65.19. Transistor Pierce Crystal Oscillator

A typical circuit originally suggested by Pierce is shown in Fig. 65.17. Here, the crystal is excited in the series-resonance mode because it is connected as a series element in the feedback path from collector to the base. Since, in series resonance, crystal impedance is the smallest, the amount of positive feedback is the largest. The crystal not only provides the feedback but also the necessary phase shift.

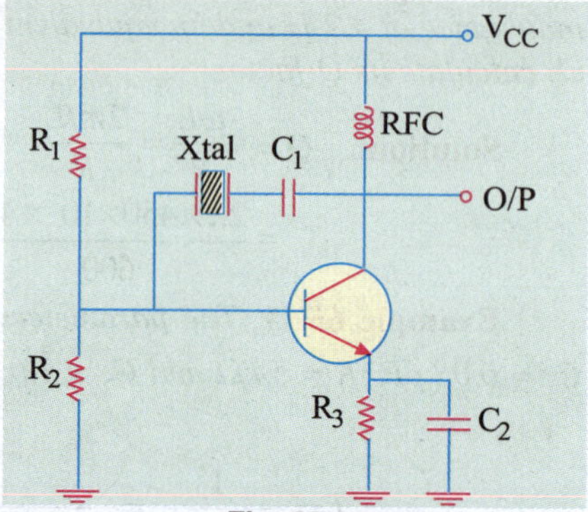

Fig. 65.17

As usual, R_1, R_2 and R_3 provide a voltage-divider stabilized dc bias circuit. C_2 bypasses R_3 to avoid degeneration. The *RFC* coil provides dc collector load and also prevents any ac signal from entering the dc supply. The coupling capacitor C_1 has negligible reactance at circuit operating frequency but ***blocks any dc flow between collector and base.*** The oscillation frequency equals the series-resonance frequency of the crystal and is given by

$$f_0 = \frac{1}{2\pi\sqrt{LC_1}} \qquad \text{— Fig 65.15 }(b)$$

Advantages

1. It is a very simple circuit because no tuned circuit other than the crystal itself is required.
2. Different oscillation frequencies can be obtained by simply replacing one crystal with another. It makes it easy for a radio transmitter to work at different frequencies.
3. Since frequency of oscillation is set by the crystal, it remains unaffected by changes in supply voltage and transistor parameters etc.

65.20. FET Pierce Oscillator

It is shown in Fig. 65.18. Use of FET is desirable because its high input impedance results in light loading of the crystal which

1. results in good stability and
2. does not lower the Q-value.

The circuit shown in Fig. 65.18 is essentially a Colpitts oscillator in which

(i) crystal has replaced the inductor and

(ii) inherent FET junction capacitance provide the split capacitance as shown in the figure.

The circuit can be made to operate at different frequencies simply by plugging in different crystals between points *A* and *B* of the circuit.

The circuit oscillation frequency is given by the series-resonance frequency of the crystal (Art. 65.17)

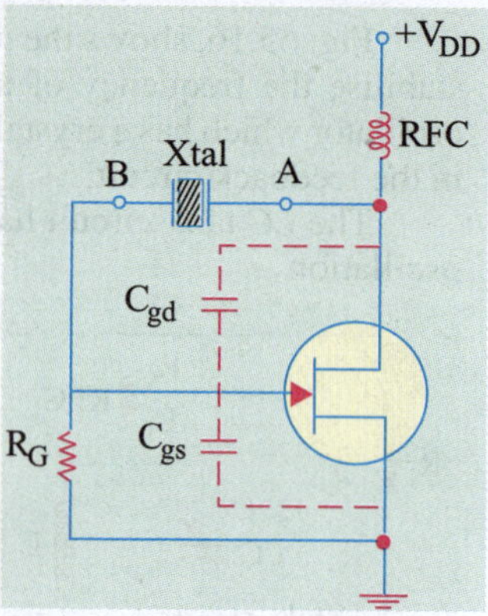

Fig. 65.18

65.21. Phase Shift Principle

Tuned circuits are ***not an essential requirement for oscillation.*** What is essential is that there should be a 180° phase shift around the feedback network* and loop gain should be greater than unity. The 180° phase shift in the feedback signal can be achieved by using a suitable *R-C* network

* Total phase shift required is 360°. However, the balance of 180° is provided by the active device of the amplifier itself.

consisting of three or four *R-C* sections. The sine wave oscillators which use *R-C* feedback network are called **phase-shift oscillators.**

65.22. RC Phase Shift Oscillator

Fig. 65.19, shows a transistor phase-shift oscillator which uses a three-section *R-C* feedback network for producing a total phase shift of 180° (*i.e.* 60° per section) in the signal fed back to the base. Since *CE* amplifier produces a phase reversal of the input signal, total phase shift becomes 360° or 0° which is essential for **regeneration and hence for sustained oscillations.**

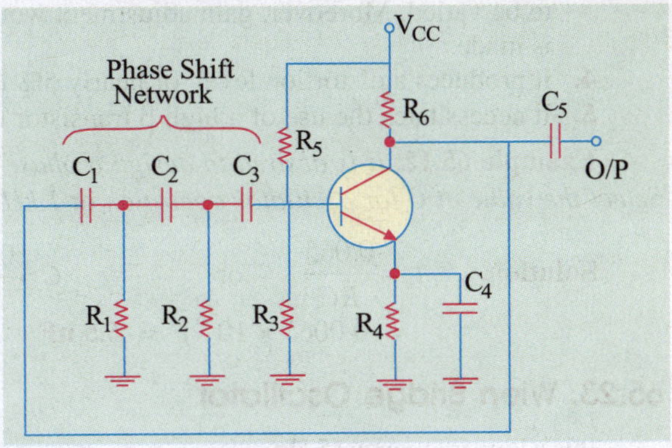

Fig. 65.19

Values of *R* and *C* are so selected that each *RC* section produces a phase advance of 60°. Addition of a fourth section improves oscillator stability. It is found that phase shift of 180° occurs ***only at one frequency*** which becomes the oscillator frequency.

(*a*) Circuit Action

The circuit is set into oscillations by any random or chance variation caused in the base current by

(*i*) noise inherent in a transistor or

(*ii*) minor variation in the voltage of the dc source.

This variation in the base current

1. is amplified in the collector circuit,
2. is then fed back to the *RC* network R_1C_1, R_2C_2 and R_3C_3,
3. is reversed in phase by the *RC* network,
4. is next applied to the base in phase with initial change in base current,
5. and hence is used to sustain cycles of variations in collector current between saturation and cut-off values.

Obviously, the circuit will stop oscillating the moment phase shift differs from 180°.

As is the case with such transistor circuits (*i*) voltage divider R_5 - R_3 provides dc emitter-base bias, (*ii*) R_6 controls collector voltage and (*iii*) R_4, C_4 provide temperature stability and prevent ac signal degeneration. The oscillator output voltage is capacitively coupled to the load by C_5.

(*b*) Frequency of Oscillation

The frequency of oscillation for the three-section *RC* oscillator when the three *R* and *C* components are equal is roughly given by

$$f_0 = \frac{1}{2\pi\sqrt{6}.RC}\text{Hz} = \frac{0.065}{RC}\text{Hz}$$

Moreover, it is found that value of β is 1/29. It means that amplifier gain must be more than 29 for oscillator operation.

(*c*) Advantages and Disadvantages

1. Since they do not require any bulky and expensive high-value inductors, such oscillators are well-suited for frequencies below 10 kHz.
2. Since only one frequency can fulfil Barkhausen phase-shift requirement, positive feedback occurs only for one frequency. Hence, pure sine wave output is possible.

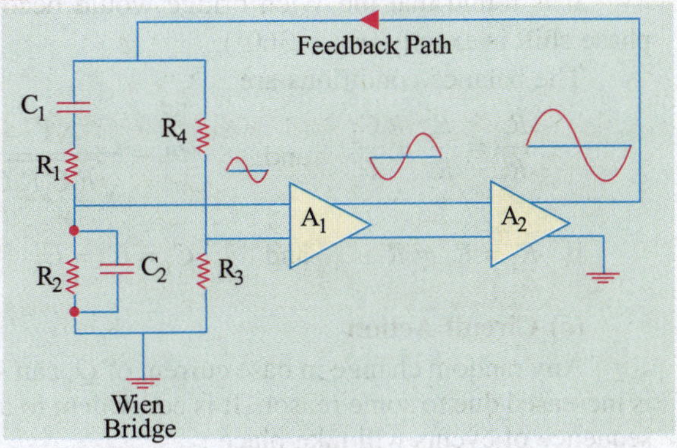

Fig. 65.20

3. It is not suited to variable frequency usage because a large number of capacitors will have to be varied. Moreover, gain adjustment would be necessary every time frequency change is made.
4. It produces a distortion level of nearly 5% in the output signal.
5. It necessitates the use of a high β transistor to overcome losses in the network.

Example 65.12. *It is desired to design a phase-shift oscillator using a BJT and R = 10 kΩ.. Select the value of C for oscillator operation at 1 kHz.*

Solution. $f_0 = \dfrac{0.065}{RC}$ or $C = \dfrac{0.065}{Rf_0} = \dfrac{0.065}{10K \times 1KHz}$
$= 0.0065 \times 10^{-6}$ F = **6.5 nF**

65.23. Wien Bridge Oscillator

It is a low-frequency (5 Hz— 500 kHz), low-distortion, tunable, high-purity sine wave generator, often used in laboratory work. As shown in the block diagrams of Fig. 65.20 and Fig. 65.21, this oscillator uses two *CE*-connected *RC*-coupled transistor amplifiers and one *RC*-bridge (called Wien bridge) network to provide feedback. Here, Q_1 serves as amplifier-oscillator and Q_2 provides phase reversal and additional amplification. The bridge circuit is used to control the phase of the feedback signal at Q_1.

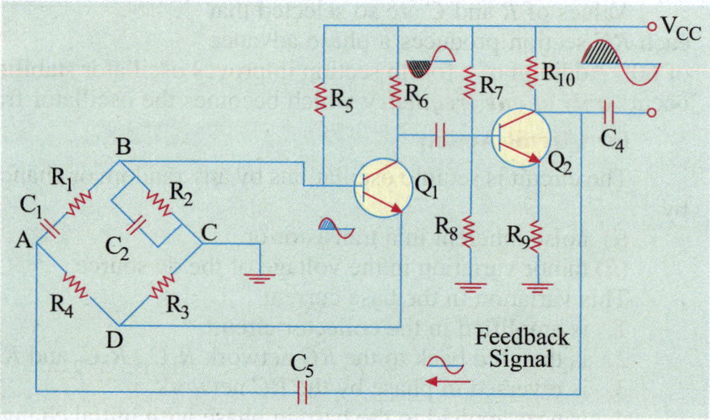

Fig. 65.21

***(a)* Phase Shift Principle**

Any input signal at the base of Q_1 appears in the amplified but phase-reversed form across collector resistor R_6 (Fig. 65.21). It is further inverted by Q_2 in order to provide a total phase reversal of 360° for positive feedback. Obviously, the signal at R_{10} is an amplified replica of the input signal at Q_1 and is of the same phase since it has been inverted twice. We could feed this signal back to the base of Q_1 directly to provide regeneration needed for oscillator operation. But because Q_1 will amplify signals over a wide range of frequencies, direct coupling would result in poor frequency stability. By adding the Wien bridge, oscillator becomes sensitive to a signal of ***only one particular frequency.*** Hence, we get an oscillator of good frequency stability.

***(b)* Bridge Circuit Principle**

It is found that the Wien bridge would become balanced at the signal frequency for which phase shift is exactly 0° (or 360°),

The balance conditions are

$$\frac{R_4}{R_3} = \frac{R_1}{R_2} + \frac{C_2}{C_1} \quad \text{and} \quad \omega_0 = \frac{1}{\sqrt{R_1C_1R_2C_2}} \quad \text{or} \quad f_0 = \frac{1}{2\pi\sqrt{R_1C_1R_2C_2}}$$

$$\text{If} \quad R_1 = R_2 = R \quad \text{and} \quad C_1 = C_2 = C, \quad \text{then} \quad f_0 = \frac{1}{2\pi RC} \quad \text{and} \quad \frac{R_4}{R_3} = 2$$

***(c)* Circuit Action**

Any random change in base current of Q_1 can start oscillations. Suppose, the base current of Q_1 is increased due to some reason. It is equivalent to applying a positive going signal to Q_1. Following sequence of events will take place :

1. An amplified but phase-reversed signal will appear at the collector of Q_1;
2. A still further amplified and twice phase-reversed signal will appear at the collector of Q_2. Having been inverted twice, this output signal will be in phase with the input signal at Q_1;
3. A part of the output signal at Q_2 is fed back to the input points of the bridge circuit (point A-C). A part of this feedback signal is applied to emitter resistor R_3 where it produces degenerative effect. Similarly, a part of the feedback signal is applied across base-bias resistor R_2 where it produces regenerative effect.

At the rated frequency f_0, effect of ***regeneration is made slightly more than that of degeneration*** in order to maintain continuous oscillations.

By replacing R_3 with a thermistor, amplitude stability of the oscillator output voltage can be increased.

(d) Advantages

Such a circuit has

1. highly stabilized amplitude and voltage amplification,
2. exceedingly good sine wave output,
3. good frequency stability.

Example 65.13K. *Calculate the resonant frequency of a Wien Bridge oscillator (shown in Fig. 65.21) when R = 10 k Ω and C = 2400 pF.*

Solution. $$f_0 = \frac{1}{2\pi RC} = \frac{1}{2\pi \times 10\,K \times 2400\,pF} = \textbf{6.63 kHz}$$

Example 65.14. *Design RC elements of a Wien Bridge oscillator (shown in Fig. 65.21), for operation at 2.5 kHz.*

Solution. Using equal values of R_1 and R_2, let us select R = 100 kΩ (you can chose any value in kilo ohms). Then from the relation,

$$f_0 = \frac{1}{2\pi RC} \quad \text{we get} \quad C = \frac{1}{2\pi f_0 R} = \frac{1}{2\pi \times 2.5\text{ kHz} \times 100\text{ K}}$$

or $C \cong$ **636 pF**

65.24. Non-sinusoidal Waveforms

Any waveform whose shape is different from that of a standard sine wave is called ***non-sinusoidal waveform***. Examples are : square, rectangular, sawtooth, triangular waveforms and pulses as shown in Fig. 65.22.

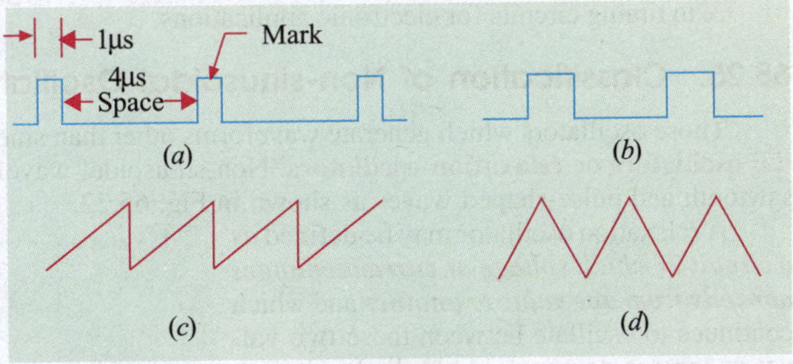

Fig. 65.22

(a) Pulses

Fig. 65.22 (*a*) shows a pulse train *i.e.* a stream of pulses at regular intervals. A pulses may, in general, be defined as a voltage or current that changes rapidly from one level of amplitude to another ***i.e. it is an abrupt discontinuity in voltage or current***. These pulses are extensively used in digital electronics.

1. Mark-to-Space Ratio (MSR)

$$MSR = \frac{\text{pulse width}}{\text{time between pulses}} = \frac{1\,\mu s}{4\,\mu s} = \textbf{0.25} \qquad \text{— Fig. 65.22 } (a)$$

Hence, mark-to-space ratio of the pulse shown in Fig. 65.22 (*a*) is 1 : 4.

This name has come from early morse-code transmission systems where a pulse was used to cause a pen to mark the paper.

2. Pulse Repetition Time (PRT)

It may be defined as ***the time between the beginning of one pulse and that of the other.***

As seen from Fig. 65.22 (*a*), $PRT = 5\ \mu s$

3. Pulse Repetition Frequency (PRF)

It is given by ***the number of pulses per second.***

$$PRF = \frac{1}{PRT} = \frac{1}{5\mu s} = \frac{10^6}{5} = 200,000\text{Hz} = 200\,\text{kHz}$$

Pulse circuits find applications in almost all electronic-based industries. Various types of pulse code modulations are employed in communication systems whereas radars utilize pulses to track targets. Digital computers require circuits that can be switched very rapidly between two states by using appropriate pulses.

(*b*) Square Wave

It is shown in Fig. 65.22 (*b*) and is, in fact, a pulse waveform with a mark-to-space ratio of 1:1 Such square waves or pulses are used

1. for audio frequency note generation,
2. for digital electronic switching as in computers,
3. in radars,
4. as synchronizing pulses in TV,
5. for switching of high-power electronic circuits such as thyristor circuits.

(*c*) Sawtooth Wave

It is shown in Fig. 65.22 (*c*). Such waves are used

1. in the scanning circuits of cathode-ray tubes (CRT),
2. in timing circuits where the time for the wave to proceed from one level to another is measured, such as that produced in an integrating circuit.

(*d*) Triangular Wave

It is shown in Fig. 65.22 (*d*). Such waves are often used

1. in scanning circuits where a uniform left-to-right scan is required as in computer displays,
2. for audio frequency note generation,
3. in timing circuits for electronic applications.

65.25. Classification of Non-sinusoidal Oscillators

Those oscillators which generate waveforms other than sine waveform are called **non-sinusoidal oscillators** or **relaxation oscillators**. Non-sinusoidal waveforms include: square, rectangular, sawtooth and pulse-shaped waves as shown in Fig. 65.22.

A relaxation oscillator may be defined ***as a circuit in which voltage or current changes abruptly from one value to another*** and which continues to oscillate between these two values as long as dc power is supplied to it.

We will consider the following three types of such oscillators :

1. Sawtooth generators **2.** Blocking oscillators **3.** Multivibrators (MV)

65.26. Pulse Definitions

Due to capacitive effects in a transistor (or to circuit elements external to it), its **output does not directly follow its input**. For example, if we apply a square input pulse to its *E/B* junction, some amount of time lapses before I_C starts to rise. Similarly, when the input becomes zero,

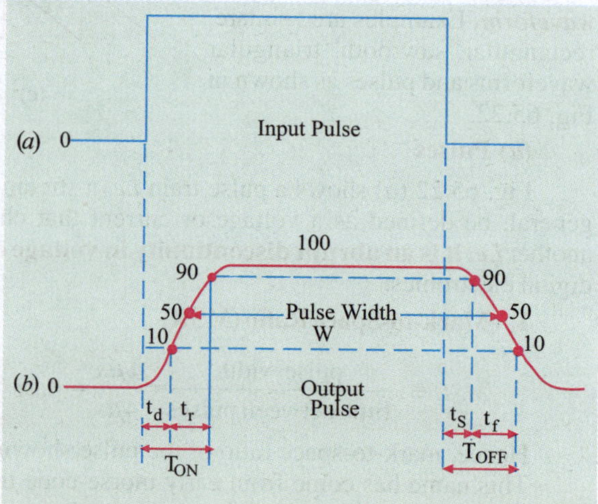

Fig. 65.23

there is some time lapse before I_C starts to decrease. There is always some time delay between the application of input and change in the output. In order to measure how quickly the output changes *i.e.* in order to define the switching (*i.e.* OFF/ON) characteristics of a transistor, we will define the following few terms. It will be assumed that perfect square wave, as shown in Fig. 65.23 (*a*), has been applied at the input. The output wave and the various time delays are shown in Fig. 65.23 (*b*).

1. Time delay, t_d

It is the time interval between the beginning of the input pulse and the time the output voltage (or current) reaches 10 per cent of its maximum value.

It depends on **(*i*)** depletion region capacitances, **(*ii*)** turn-on base current and **(*iii*)** value of transistor β.

2. Rise time, t_r

It is the time taken by the output voltage (or current) to rise from 10% to 90% of its maximum value.

It primarily depends on diffusion capacitance C_D of the transistor (Art. 61.4).

3. Turn-on time T_{ON}

It is equal to the sum of the delay time and rise time *i.e.* $T_{ON} = t_d + t_r$

4. Storage time, T_s

It is the time interval between the end of the input pulse (trailing edge) and the time when output voltage (or current) falls to 90% of its initial maximum value.

It depends on the degree of saturation. Deeper the transistor is driven into saturation, more the stored charge that has to be removed and hence longer the storage time. That is why non-saturated switching is often preferred.

5. Fall time, t_f

It is the time interval during which the output voltage (or current) falls from 90% of its maximum value to 10%.

In simple words, it is the time interval between 90% and 10% levels of the output pulse.

6. Turn-off time, T_{OFF}

It is equal to the sum of storage time and fall time *i.e.* $T_{OFF} = t_s + t_f$

For a fast switching transistor, T_{ON} and T_{OFF} must be of the order of nanoseconds.

7. Pulse Width, W

It is the time duration of the output pulse measured between two 50% levels of the rising and falling waveform.

65.27. Basic Requirements of a Sawtooth Generator

The essential requirements of a sawtooth generator are :

1. a dc power source, **2.** a switching device (neon tube, thyratron, thyristor, UJT etc.), **3.** a capacitor, **4.** a resistor.

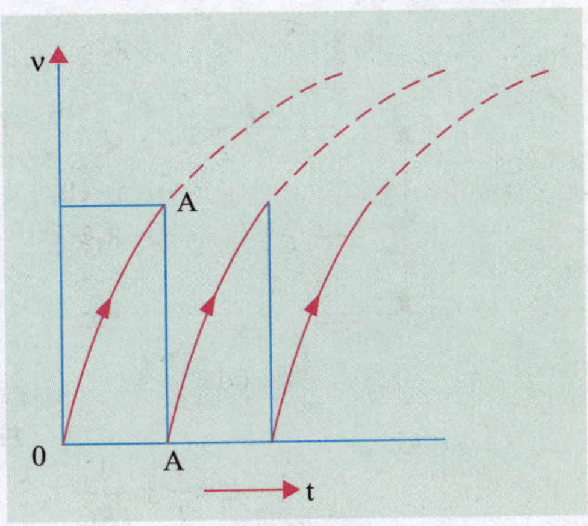

Fig. 65.24

Circuit Action

The *V/I* characteristics of the *RC* circuit play an important role in the operation of such a generator. By restricting the time interval equal to the time constant λ = *CR*, only the rising portion of the characteristic (*OA* in Fig. 65.24) which is almost a straight line, is utilized. For periods of time greater than λ , the rising portion of the characteristic is no longer a straight line and hence cannot be utilized.

The frequency of the wave is given by the reciprocal of time which elapses between the two waves. In Fig. 65.24, $f = 1/\lambda$.

65.28. UJT Sawtooth Generator

The circuit is shown in Fig. 65.25. It consists of a power source, a unijunction transistor and an *R-C* network.

Circuit Action

When *S* is initially closed, following chain of events takes place :

1. a small current is set up through R_2 and R_1 *via* B_2 and B_1 and an initial reverse bias is established across the E/B_1 junction;
2. at the same time, *C* begins to get charged through R_E and voltage across it increases exponentially with time towards the target voltage *V*;
3. when capacitor voltage equals the emitter firing (or peak point) voltage V_P, E/B_1 junction becomes forward-biased and the emitter goes into the negative region of its characteristic;
4. being forward-biased, E/B_1 junction offers very low resistance. Hence, *C* starts discharging through B_1 and R_1 at a rate determined mainly by E/B_1 junction resistance and R_1;
5. as capacitor voltage approaches zero, the E/B_1 junction again becomes reverse-biased and so stops conducting;
6. we revert to the initial state where *C* begins to charge and the whole cycle of circuit actions is repeated.

The emitter voltage waveform is shown in Fig. 65.25 (*b*). As seen, V_E rises exponentially towards the target voltage *V* but drops to a very low value after it reaches the value V_P due to sudden conduction through E/B_1 junction. Since R_E is large (10 *K* or so) charging rate is comparatively slow but discharge is much quicker since R_1 is very small (50 Ω or less). ***This slow charge and fast discharge produces a sawtooth wave.***

The time required for v_E to rise to V_P is given by

$$T = kRC \qquad \text{where} \quad k = \log_C{}^{1/(1-\eta)} \text{ and}$$

$$\eta = \text{intrinsic stand-off ratio} = \frac{R_{B1}}{R_{B1} + R_{B2}} \qquad \text{...Art 64.2}$$

Here, R_{B1} and R_{B2} are the internal inter-base resistances.

The frequency of oscillation of the *UJT* or of the output sawtooth wave is

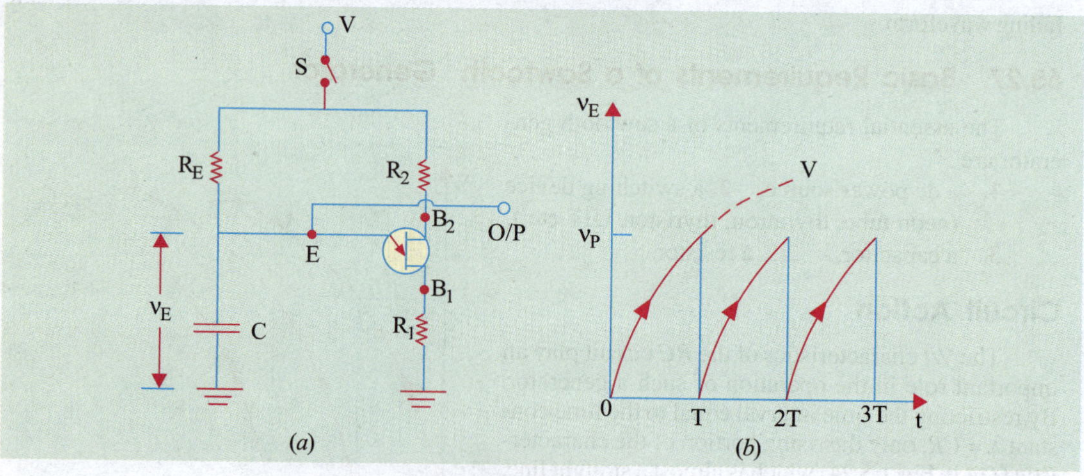

Fig. 65.25

$$f_0 = \frac{1}{T} = \frac{1}{kRC}$$

Its amplitude is determined primarily by applied voltage V and V_P.

Applications

Sawtooth voltage waves are commonly used as

1. sweep voltages at the picture tubes of TV receivers,
2. as sweep voltages of the viewing screens of oscilloscopes and radar equipment.

Example 65.15. *For the UJT oscillator circuit shown in Fig. 65.25, $R_E = 10\ K$, $\eta = 0.75$. If the required oscillator frequency is 1 kHz, find the value of C.*

(Components and Devices, Pune Univ 1984)

Solution. $k = \log_e 1/(1-0.75) = \log_e 4 = 2.3 \log_{10} 4 \cong 1.4$

Now, $f_0 = \dfrac{1}{kRC}$ $\quad\therefore\quad$ $1 \times 10^3 = \dfrac{1}{1.4 \times 10 \times 10^3 \times C}$ or $C = 0.07\ \mu F$

65.29. Multivibrators (MV)

These devices are very useful as pulse generating, storing and counting circuits. They are basically **two-stage amplifiers with positive feedback from the output of one amplifier to the input of the other**. This feedback (Fig. 65.26) is supplied in such a manner that one transistor is driven to saturation and the other to cut-off. It is followed by new set of conditions in which the saturated transistor is driven to cut-off and the cut-off transistor is driven to saturation.

There are three basic types of *MVs* distinguished by the type of coupling network employed.

1. astable multivibrator (*AVM*),
2. monostable multivibrator (*MMV*),
3. bistable multivibrator (*BMV*).

The first one is the *non-driven type* whereas the other two are the *driven type* (also called triggered oscillators).

1. Astable Multivibrator (AMV)

It is also called ***free-running relaxation oscillator***. It has no stable state but only two quasi-stable (half-stable) states between which it keeps oscillating continuously of its own accord without any external excitation.

In this circuit, neither of the two transistors reaches a stable state. When one is ON, the other is OFF and they continuously switch back and forth at a rate depending on the *RC* time constant in the circuit. Hence, it oscillates and produces pulses of certain mark-to-space ratio. Moreover, two outputs (180° out of phase with each other) are available.

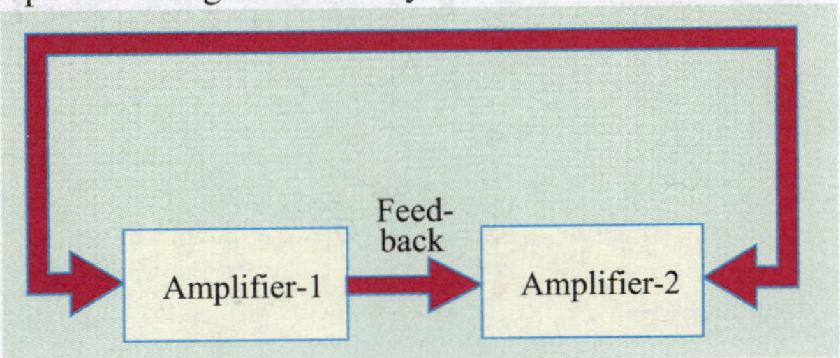

Fig. 65.26

It has two energy-storing elements ***i.e. two capacitors***.

2 . Monostable Multivibrator (MMV)

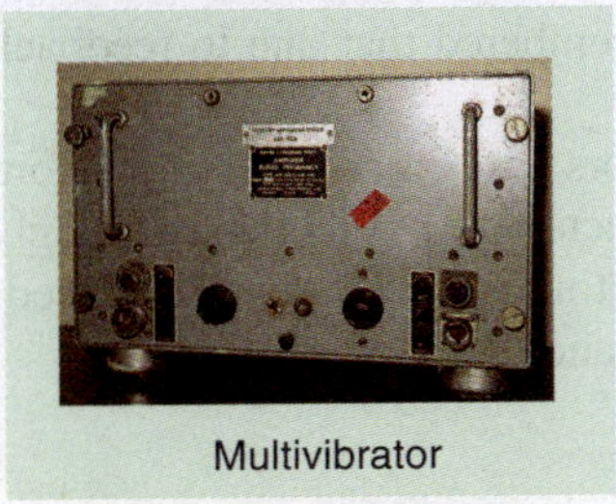

Multivibrator

It is also called a **single-shot** or **single swing** or a **one-shot** multivibrator. Other names are : ***delay multivibrator and univibrator***. It has

(i) one absolutely stable (stand-by) state and *(ii)* one quasi-stable state.

It can be switched to the quasi-stable state by an external trigger pulse but it returns to the stable condition after a time delay determined by the value of circuit components. It supplies a single output pulse of a desired duration for every input trigger pulse.

It has one energy-storing element *i.e.* **one-capacitor.**

3. Bistable Multivibrator (BMV)

It is also called Eccles-Jordan or flip-flop multivibrator. It has **two absolutely stable states.** It can remain in either of these two states unless an external trigger pulse switches it from one state to the other. Obviously, ***it does not oscillate.*** It has **no energy storage element.**

Detailed discrete circuits for the above MVs are discussed below after listing their uses.

65.30. Uses of Multivibrators

Some of their uses are :

1. as frequency dividers,
2. as sawtooth generators,
3. as square wave and pulse generators,
4. as a standard frequency source when synchronized by an external crystal oscillator,
5. for many specialised uses in radar and TV circuits,
6. as memory elements in computers.

65.31. Astable Multivibrator

Fig. 65.27 shows the circuit of a symmetrical collector-coupled AMV using two similar transistors. It, in fact, consists of two *CE* amplifier stages, each providing a feedback to the other. The feedback ratio is unity and positive because of 180° phase shift in each stage. Hence, the circuit oscillates. Because of the very strong feedback signal, the transistors are driven either to saturation or to cut-off (they do not work on the linear region of their characteristics).

The transistor Q_1 is forward-biased by V_{CC} and R_1 whereas Q_2 is forward-biased by V_{CC} and R_2. The collector-emitter voltages of Q_1 and Q_2 are determined respectively by R_{L1} and R_{L2} together with V_{CC}. The output of Q_1 is coupled to the input of Q_2 by C_2 whereas output of Q_2 is coupled to Q_1 by C_1.

Note that it is not essential to draw the coupling leads at 45° to the vertical as shown but it is usually done because it helps to identify the circuit immediately as MV.

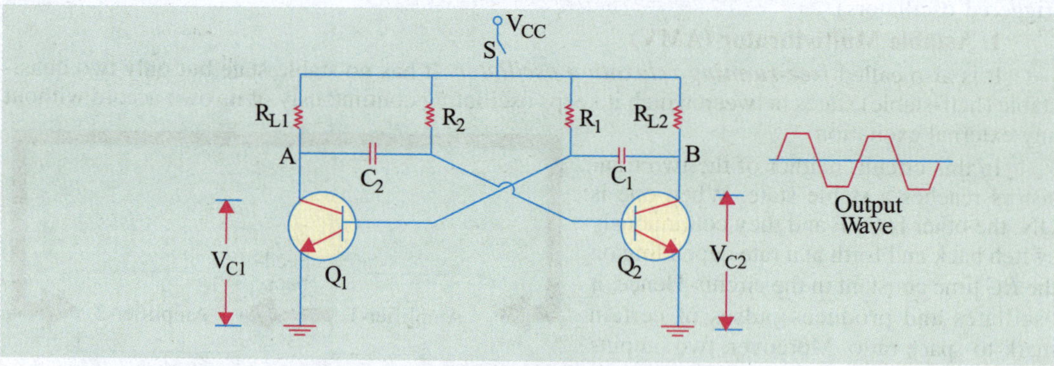

Fig. 65.27

The output can be taken either from point *A* or *B* though these would be phase-reversed with respect to each other as shown in Fig. 65.27.

Circuit Operation

The circuit operation would be easy to understand if it is remembered that due to feedback (*i*) when Q_1 is ON, Q_2 is OFF and (*ii*) when Q_2 is ON, Q_1 is OFF.

When the power is switched on by closing *S*, one of the transistors will start conducting before the other does (or slightly faster than the other). It is so because characteristics of no two seemingly similar transistors can be exactly alike. Suppose that Q_1 starts conducting before Q_2 does. The feedback system is such that Q_1 will be very rapidly driven to saturation and Q_2 to cut-off.

The following sequence of events will occur :

1. Since Q_1 is in saturation, whole of V_{CC} drops across R_{L1}. Hence, $V_{C1} = 0$ and point *A* is at zero or ground potential.
2. Since Q_2 is in cut-off *i.e.* it conducts no current, there is no drop across R_{L2}. Hence, point *B* is at V_{CC}.
3. Since *A* is at 0 V, C_2 starts to charge through R_2 towards V_{CC}.
4. When voltage across C_2 rises sufficiently (*i.e.* more than 0.7 V), it biases Q_2 in the forward direction so that it starts conducting and is soon driven to saturation.
5. V_{C2} decreases and becomes almost zero when Q_2 gets saturated. The potential of point *B* decreases from V_{CC} to almost 0 V. This potential decrease (negative swing) is applied to the base of Q_1 through C_1. Consequently, Q_1 is pulled out of saturation and is soon driven to cut-off.
6. Since, now, point *B* is at 0 V, C_1 starts charging through R_1 towards the target voltage V_{CC}.
7. When voltage of C_1 increases sufficiently, Q_1 becomes forward-biased and starts conducting. In this way, the whole cycle is repeated.

It is seen that the circuit alternates between a state in which Q_1 is ON and Q_2 is OFF and a state in which Q_1 is OFF and Q_2 is ON. The time in each state depends on *RC* values. Since each transistor is driven alternately into saturation and cut-off the voltage wavefrom at either collector (points *A* and *B* in Fig. 65.27) is essentially a square waveform with a peak amplitude equal to V_{CC} (Fig. 65.28).

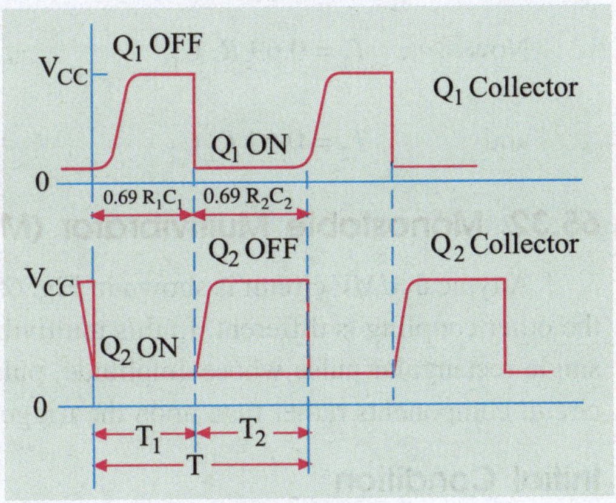

Fig. 65.28

Switching Times

It can be proved that off-time for Q_1 is $T_1 = 0.69\ R_1C_1$ and that for Q_2 is $T_2 = 0.69\ R_2C_2$.

Hence, total time-period of the wave is

$T = T_1 + T_2 = 0.69\ (R_1C_1 + R_2C_2)$

If $R_1 = R_2 = R$ and $C_1 = C_2 = C$ *i.e.* the two stages are symmetrical, then $T = 1.38\ RC$

Frequency of Oscillation

It is given by the reciprocal of time period,

$$\therefore \quad f = \frac{1}{T} = \frac{1}{1.38\,RC} = \frac{0.7}{RC}$$

Minimum Values of β

To ensure oscillations, the transistors must saturate for which minimum values of β are as under :

$$\beta_1 = \frac{R_1}{R_{L1}} \quad \text{and} \quad \beta_2 = \frac{R_2}{R_{L2}}$$

If $R_1 = R_2 = R$ and $R_{L1} = R_{L2} = R_L$. then $\beta_{min} = \dfrac{R}{R_L}$

Example 65.16. *Determine the period and frequency of oscillation for an astable multivibrator with component values: $R_1 = 2\ K$, $R_2 = 20\ K$, $C_1 = 0.01\ \mu F$ and $C_2 = 0.05\ \mu F$.*

Solution. $T_1 = 0.69 \times 2\ K \times 0.01\ \mu F = 13.8\ \mu s$

and $T_2 = 0.69 \times 20\ K \times 0.05\ \mu F = 690\ \mu s$

$\therefore \quad T = T_1 + T_2 = 13.8\ \mu s + 690\ \mu s = 703.8\ \mu s$

$$\therefore \quad f_0 = \frac{1}{703.8\,\mu s} = \textbf{1.42 kHz}$$

Example. 65.17. *In the AMV circuit of Fig. 65.27, $R_1 = R_2 = 10$ K, $C_1 = C_2 = 0.01$ μF and $R_{L1} = R_{L2} = 1$ K. Find*

(a) frequency of circuit oscillation, (b) minimum value of transistor β.

(Digital Electronics, Bombay Univ. 1985)

Solution. *(a)* $T_1 = T_2 = 0.69 \times 10 \times 10^3 \times 0.01 \times 10^{-6} = 69\ \mu s$

$\therefore \quad T = T_1 + T_2 = 2 \times 69 = 138\ \mu s$

$\therefore \quad f_0 = \dfrac{1}{138 \times 10^{-6}} = $ **7.25 kHz**

(b) $\beta_{min} = \dfrac{R}{R_L} = \dfrac{10 \times 10^3}{1 \times 10^3} = $ **10**

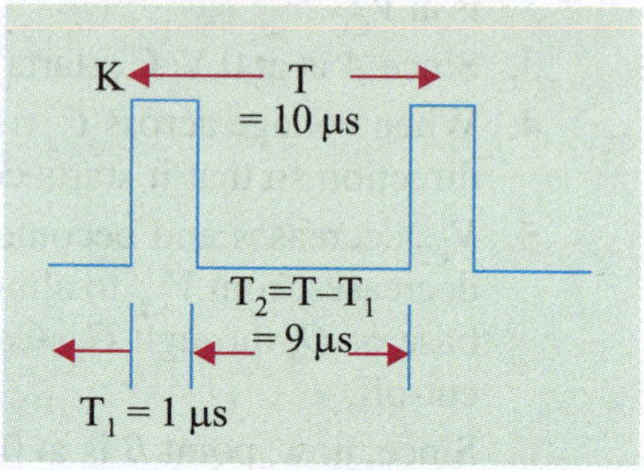

Fig. 65.29

Example 65.18. *Determine the value of capacitors to be used in an astable multivibrator to provide a train of pulses 1μs side at a repetition rate of 100 kHz. Given $R_1 = R_2 = 10$ K.*

Solution. Fig. 65.29 shows the waveform to be generated by the astable multivibrator. Note that the desired pulse width is actually the time interval $T_1 = 1\ \mu s$ and the time period *T* which equals 10 μs (*i.e.* 1/100 kHz) is the desired repetition time.

Now $\quad T_1 = 0.69\, R_1 C_1, \quad \therefore \quad C_1 = \dfrac{1\ \mu s}{0.69 \times 10\,K} = $ **145pF**

and $\quad T_2 = 0.69\, R_2 C_2, \quad \therefore \quad C_2 = \dfrac{9\ \mu s}{0.69 \times 10\,K} = $ **1304 pF**

65.32. Monostable Multivibrator (MMV)

A typical *MMV* circuit is shown in Fig. 65.30. Here, Q_1 is coupled to Q_2 base as in an *AMV* but the other coupling is different. In this multivibrator, a single narrow input trigger pulse produces a single rectangular pulse whose amplitude, pulse width and wave shape depend upon the values of circuit components rather than upon the trigger pulse.

Initial Condition

In the absence of a triggering pulse at C_2 and with *S* closed,

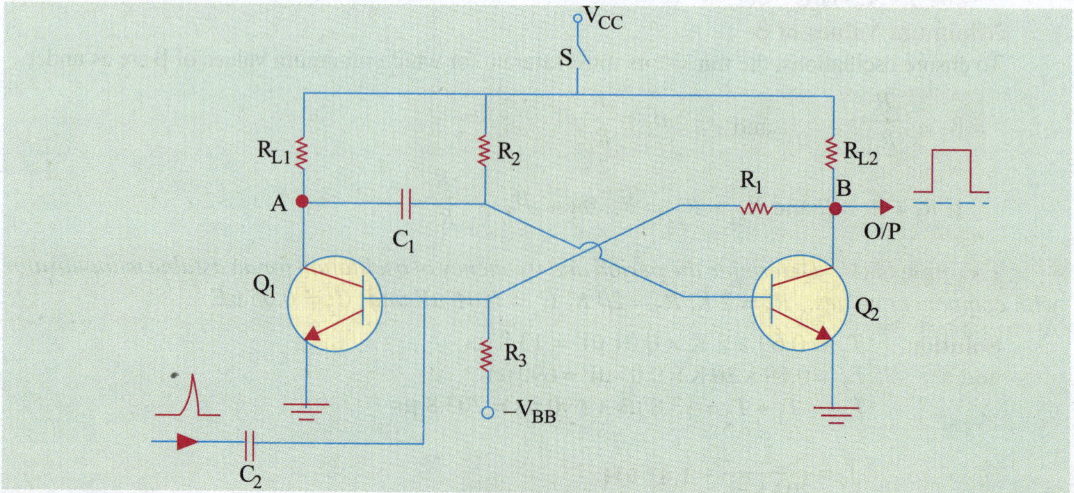

Fig. 65.30

1. V_{CC} provides reverse bias for *C/B* junctions of Q_1 and Q_2 but forward-bias for *E/B* junction of Q_2 only. Hence, Q_2 conducts at saturation.
2. V_{BB} and R_3 reverse bias Q_1 and keep it cut off.
3. C_1 charges to nearly V_{CC} through R_{L1} to ground by the low-resistance path provided by saturated Q_2.

As seen, the initial stable state is represented by
(*i*) Q_2 conducting at saturation and (*ii*) Q_1 cut-off.

When Trigger Pulse is Applied

When a trigger pulse is applied to Q_1 through C_2, *MMV* will switch to its opposite unstable state where Q_2 is cut-off and Q_1 conducts at saturation. The chain of circuit actions is as under :

1. If positive trigger pulse is of sufficient amplitude, it will override the reverse bias of the *E/B* junction of Q_1 and give it a forward bias. Hence, Q_1 will start conducting.
2. As Q_1 conducts, its collector voltage falls due to voltage drop across R_{L1}. It means that potential of point *A* falls (negative-going signal). This negative-going voltage is fed to Q_2 *via* C_1 where it decreases its forward bias.
3. As collector current of Q_2 starts decreasing, potential of point *B* increases (positive-going signal) due to lesser drop over R_{L2}. Soon, Q_2 comes out of conduction.
4. The positive-going signal at *B* is fed *via* R_1 to the base of Q_1 where it increases its forward bias further. As Q_1 conducts more, potential of point *A* approaches 0 V.
5. This action is cumulative and ends with Q_1 conducting at saturation and Q_2 cut-off.

Return to Initial Stable State

1. As point A is at almost 0 V, C_1 starts to discharge through saturated Q_1 to ground.
2. As C_1 discharges, the negative potential at the base of Q_2 is decreased. As C_1 discharges further, Q_2 is pulled out of cut-off.
3. As Q_2 conducts further, a negative-going signal from point *B* *via* R_1 drives Q_1 into cut-off.

Hence, the circuit reverts to its original state with Q_2 conducting at saturation and Q_1 cut-off. It remains in this state till another trigger pulse comes along when the entire cycle repeats itself.

As shown in Fig. 65.30, the output is taken from the collector of Q_2 though it can also be taken from point A of Q_1. The width of this pulse is determined by the time constant of C_1R_2. Since this MV produces one output pulse for every input trigger pulse it receives, it is called ***mono or one-shot multivibrator.***

The width or duration of the pulse is given by $T = 0.69\ C_1R_2$
It is also known as the one-shot period.

Uses

1. The falling part of the output pulse from *MMV* is often used to trigger another pulse generator circuit thus producing a pulse delayed by a time *T* with respect to the input pulse.
2. *MMV* is used for regenerating or rejuvenating old and worn out pulses. Various pulses used in computers and telecommunication systems become somewhat distorted during use. An *MMV* can be used to generate new, clean and sharp pulses from these distorted and used ones.

Example 65.19. *A 20 kHz, 75% duty cycle square (t_p)wave is used to trigger continuously, a monostable multivibrator with a triggered pulse duration of 5μs . What will be the duty cycle of the waveform at output (B) of the monostable multivibrator (refer to Fig. 65.30).*

Solution. Time period of the square wave

$$T = \frac{1}{f} = \frac{1}{20\,\text{kHz}} = 50\,\mu\text{s}$$

Since the duty cycle of the square wave is 75%, therefore the time interval during which the input waveform is at a higher voltage level is, 0.75 × 50 μs = 37.5 μs. Fig. 65.31 (*a*) shows a sketch of the input waveform which is used to trigger the monostable multivibrator.

Now the monostable multivibrator is triggered once each time a new pulse arives. The monostable multivibrator remains triggered only for a duration, t_p = 5 μs. A sketch of the waveform at the output (B) of the monostable multivibrator is as shown in Fig. 65.31(*b*).

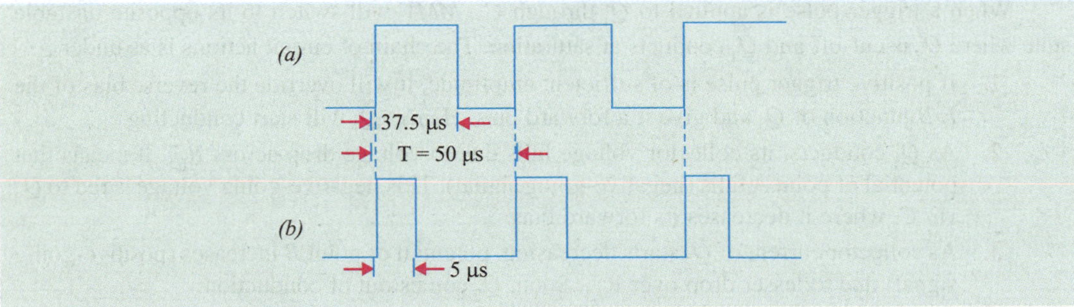

Fig. 65.31

Example 65.20. *A monostable multivibrator is required to convert a 100 kHz, 30% duty cycle square wave to a 100 kHz, 50% duty cycle square wave. Find the values of R_2 and C_2.*

Solution. Fig. 65.32 (*a*) shows a sketch of the input waveform. In order to convert it to a square waveform with 50% duty cycle, we want that the monostable multivibrator must remain triggered for 50% of the time period.

Now time period of 100 kHz square wave,

$$T = \frac{1}{f} = \frac{1}{100\,\text{kHz}} = 10\,\mu\text{s}$$

∴ The duration for which the monostable multivibrator must remain triggered = 50% × 10 μs= 5 μs.

Fig. 65.32 (*b*) shows a sketch of the required waveform at the output of the monostable multivibrator.

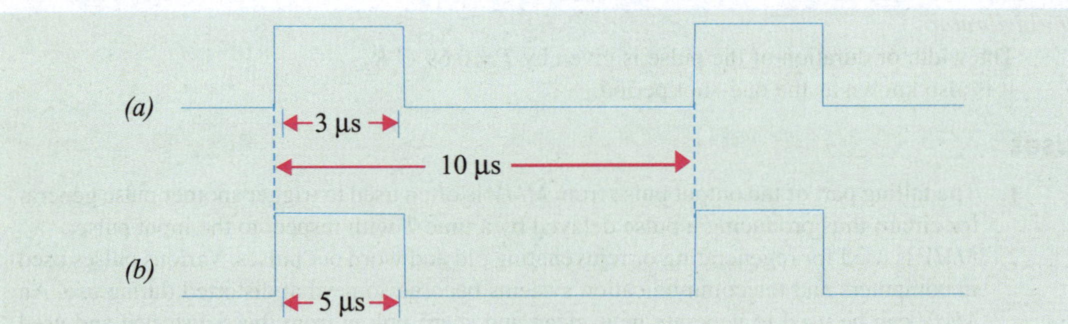

Fig. 65.32

Now width or duration of the pulse,

$$T = 0.69\, C_1 R_2$$

Let us select *C_1 = 0.001μF, then from the above equation.

* You can chose any value of C which could result in the value of R_2 in kilohms. The reason for this is that we want the current in the circuit to be limited to few milliamperes.

$$R_2 = \frac{T}{0.69C_1} = \frac{5\mu s}{0.69 \times 0.001\mu F} = 7.2\text{ K}$$

65.33. Bistable Multivibrator (BMV)

The basic circuit is shown in Fig. 65.33. As stated earlier, it has **two absolutely stable** states. It can stay ***in one of its two states indefinitely*** (as long as power is supplied) changing to the other state only when it receives a trigger pulse from outside. When it receives another triggering pulse, only then it goes back to its original state. Since one trigger pulse causes the *MV* to 'flip' from one state to another and the next pulse causes it to ***'flop'*** back to its original state, the *BMV* is also popularly known as **'flip-flop'** circuit.

The *BMV* circuit shown in Fig. 65.33 differs from the *AMV* circuit of Fig. 65.27 in the following respects :

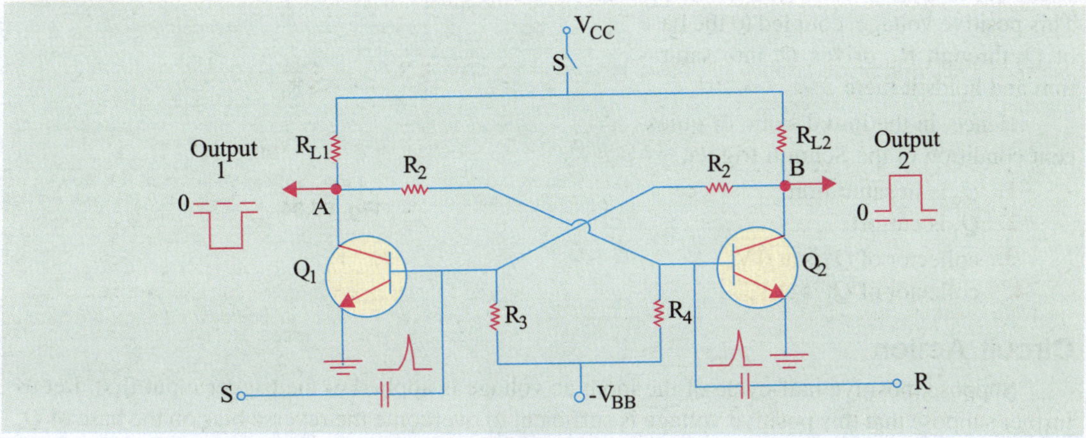

Fig. 65.33

1. the base resistors are not joined to V_{CC} but to a common source–V_{BB},
2. the feedback is coupled through two resistors (not capacitors).

Circuit Action

If Q_1 is conducting, then the fact that point *A* is at nearly 0 V makes the base of Q_2 negative (by the potential divider $R_2 - R_4$) and holds Q_2 off.

Similarly, with Q_2 OFF, the potential divider from V_{CC} to $-V_{BB}$ (R_{L2}. R_1, R_3) is designed to keep base of Q_1 at about 0.7 V ensuring that Q_1 conducts. It is seen that Q_1 holds Q_2 OFF and Q_2 holds Q_1 ON.

Suppose, now, a positive pulse is applied momentarily to *R*, it will cause Q_2 to conduct. As collector of Q_2 falls to zero, it cuts Q_1 OFF and, consequently, the *BMV* switches over to its other state.

Similarly, a positive trigger pulse applied to *S* will switch the *BMV* back to its original state.

Uses

1. in timing circuits as a frequency divider,
2. in counting circuits,
3. in computer memory circuits.

65.34. Schmitt Trigger

The Schmitt trigger (after the name of its invertor) is a **binary circuit** and **closely resembles an MV**. It has two bistable states and the magnitude of the input voltage determines which of the two is possible. It is also called emitter-coupled binary oscillator because positive feedback occurs by coupling through emitter resistor R_E.

The Quiescent Condition

As shown in Fig. 65.34, it consists of two similar transistors Q_1 and Q_2 coupled through R_E. Resistors R_1, R_3 and R_4 form a voltage divider across V_{CC} and $-V_{BB}$ which places a small positive voltage (forward bias) on the base of Q_2. Hence, when power is first switched ON, Q_2 starts conducting. The flow of its current through R_E places a small reverse bias on the base of Q_1, thereby cutting it OFF. Consequently, collector of Q_1 rises to V_{CC}. This positive voltage, coupled to the base of Q_2 through R_3, drives Q_2 into saturation and holds it there.

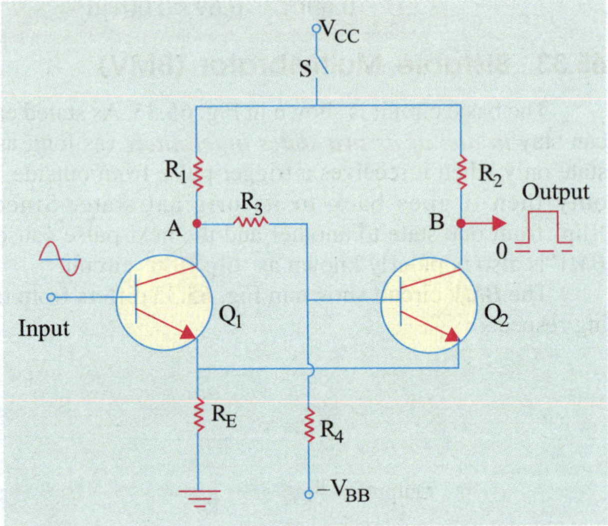

Fig. 65.34

Hence, in the initial static or quiescent condition of the Schmitt trigger,

1. Q_2 is in saturation,
2. Q_1 is cut-off,
3. collector of Q_2 is at 0 V,
4. collector of Q_1 is at V_{CC}.

Circuit Action

Suppose, positive half-cycle of the input ac voltage is applied to the trigger input first. Let us further suppose that this positive voltage is sufficient to overcome the reverse bias on the base of Q_1 placed there by the voltage drop across R_E. Then, the chain of events that follows is as under :

1. Q_1 comes out of cut-off and starts to conduct;
2. as it does so, its collector voltage drops (swings negative);
3. this negative-swinging voltage coupled to the base of Q_2 *via* R_3 reduces its forward bias and hence its emitter current;
4. with reduced emitter current, voltage drop across R_E is reduced;
5. consequently, reverse bias of Q_1 is further lowered and it conducts more heavily;
6. as a result, collector voltage of Q_1 falls further, thereby driving Q_2 still closer to cut-off.

This process is cumulative and ends up with

(*a*) Q_1 conducting at saturation with its collector voltage almost zero;

(*b*) Q_2 becoming cut-off with its collector voltage nearly V_{CC}.

Negative Half-cycle of the Input Voltage

Now, when the negative half-cycle of the input voltage is applied

1. Q_1 becomes reverse-biased. Consequently, its collector current falls and collector voltage rises (*i.e.* potential of point A increases towards V_{CC}) ;

2. this positive-swinging voltage is coupled to the base of Q_2 through R_3 and, as a result, Q_2 is driven to saturation;

3. this re-establishes the original conditions of

(*a*) Q_1 cut off with collector voltage at V_{CC} and

(*b*) Q_2 at saturation with collector voltage at 0 V.

It completes one cycle. This cycle is repeated as the input voltage rises and falls again. Hence, each cycle of the Schmitt trigger produces a positive-going pulse at its output which is taken out from the collector of Q_2 *i.e.* from point *B* in Fig. 65.34.

Output Pulse Width

It depends on the time during which Q_2 is conducting. It, in turn, depends on the input voltage, within the limits imposed by emitter resistor R_E.

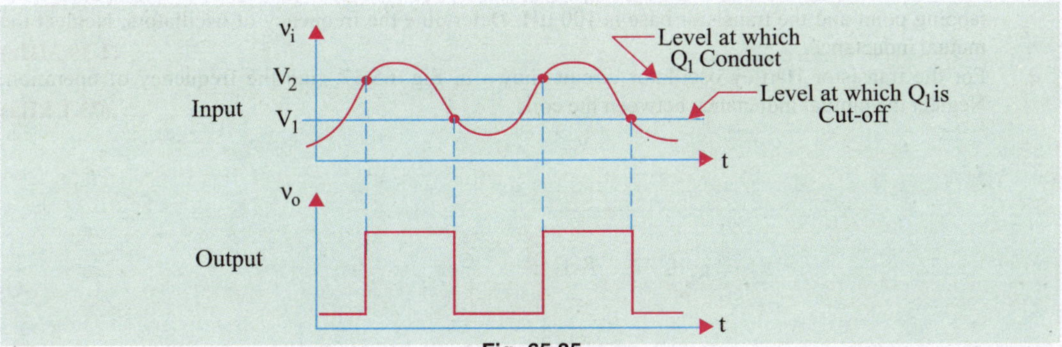

Fig. 65.35

Uses

1. It is frequently used for wave-shaping purposes. As shown in Fig. 65.35, it can convert inputs with any waveshape into output pulses having rectangular or square waveshapes. That is why Schmitt trigger is often called a **'squaring' circuit** or a **'squarer'** circuit.
2. It can reshape worn-out pulses by giving them sharp leading and trailing edges.
3. Since a change of state occurs whenever the input crosses a trigger point, the Schmitt trigger is often used as a **level detector** *i.e.* as a pulse height discriminator.

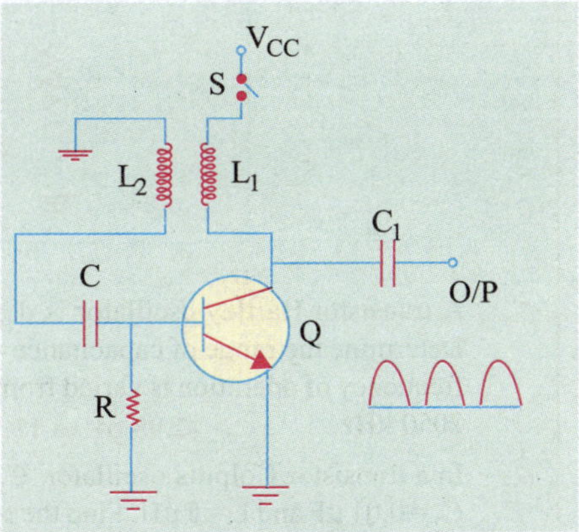

Fig. 65.36

65.35. Transistor Blocking Oscillator

The basic circuit is shown in Fig. 65.36. When *S* is closed, base current rises rapidly due to forward-bias placed by V_{CC} on the *E/B* junction of *Q*. It causes a corresponding increase in its collector current. This rising flow of I_C through L_1 produces an induced *e.m.f.* in L_2. Coupling between L_1 and L_2 is such that lower end of L_2 becomes positive and the grounded end negative.

The positive voltage from L_2 is applied to the base of *Q* through *C*. It further increases the forward-bias of the *E/B* junction which leads to further increase in I_C. Since this process is cumulative, *Q* is quickly driven to saturation. At that point, there is no further increase in I_C and hence, no induced *e.m.f.* in L_2 to be applied to *Q*.

Now, *C* which had been charged earlier, places a negative charge on the base of *Q* which reverse-biases its *E/B* junction and ultimately drives it to cut-off.

The transistor remains at cut-off as *C* now starts to discharge through *R*. When sufficient amount of charge leaks off *C* so that reverse bias of *E/B* junction is removed and forward bias is re-established, *Q* comes out of cut-off and its collector current starts rising once again. Then, the entire cycle of operation is repeated. As shown, the output consists of sharp and narrow pulses.

Tutorial Problems No. 65.1

1. A tuned collector oscillator circuit is tuned to operate at 22 kHz by a variable capacitor set to 2 nF. Find the value of tuned circuit inductance. **(0.026 H)**
2. A tuned collector oscillator operates at 2.2 MHz frequency. At what frequency will it work if its tuned circuit capacitance is reduced by 50% ? **(3.11 MHz)**
3. In a transistorized Hartley oscillator, the tank circuit has the capacitance of 100 pF. The value of inductance between the collector and tapping point is 30 μH and the value of inductance between the tapping point and the transistor base is 100 μH. Determine the frequency of oscillators. Neglect the mutual inductance. **(1.39 MHz)**
4. For the transistor Hartley oscillator circuit shown in Fig. 65.37, find the frequency of operation. Neglect the mutual inductance between the coils. **(73.1 kHz)**

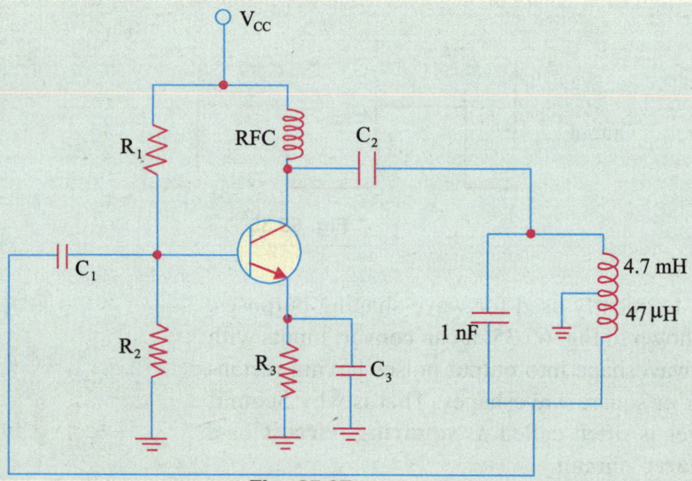

Fig. 65.37

5. A transistor Hartley oscillator is designed with $L_1 = 2$ mH, $L_2 = 20$ μH and a variable capacitance. Determine the range of capacitance values if the frequency of operation is varied from 950 kHz to 2050 kHz. **(2.98 pF to 13.9 pF)**
6. In a transistor Colpitts oscillator, $C_1 = 0.001$μF, $C_2 = 0.01$ μF and L = 5 μH. Find the required gain for oscillation and the frequency of oscillations. **(10 and 2.37 MHz)**
7. Determine the frequency of oscillations for the transistor Colpitts oscillator circuit shown in Fig. 65.38. **(24.4 kHz)**

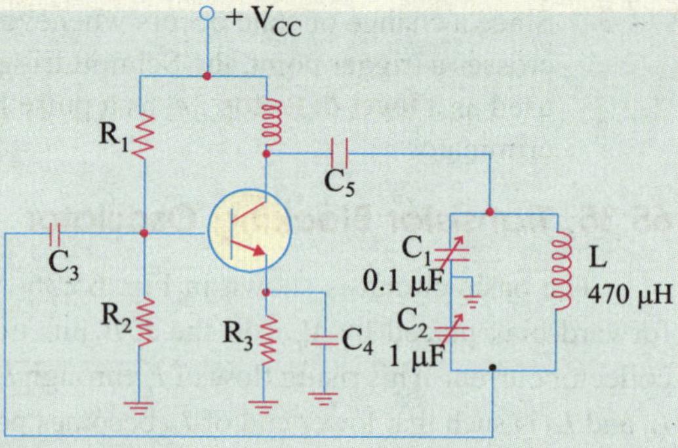

Fig. 65.38

8. A Colpitts oscillator is designed with $C_1 = 100$ pF, $C_2 = 7500$ pF. The inductance is variable. Determine the range of inductance values if the frequency of oscillation is to vary between 950 kHz and 2050 kHz. **(61 μH to 284 μH)**
9. The frequency of oscillation of a Colpitts oscillator is given by,

$$f_0 = \frac{1}{2\pi\sqrt{L\left(\frac{C_1 C_2}{C_1 + C_2}\right)}}$$

where L, C_1 and C_2 are the frequency-determining components. Such a circuit operates at 450 kHz with $C_1 = C_2$. What will be the oscillation frequency if the value of C_2 is doubled. **(389.7 kHz)**

10. Calculate the frequency of oscillations for the Clapp oscillator shown in Fig. 65.39. **(243.59 kHz)**

11. A crystal has the following parameters : $L = 0.33$ H, $C_1 = 0.065$ pF, $C_2 = 1$ pF and $R = 5.5$ KΩ. Find the series resonant frequency and Q-factor of the crystal. **(1.12 MHz and 422 approx)**

12. A Wien Bridge oscillator is used for operation at $f_0 = 10$ kHz. The value of R is 100 K, find the value of capacitor, C. Assume $R_1 = R_2 = R$ and $C_1 = C_2 = C$. **(159 pF)**

13. *A RC* phase shift oscillator has R = 1kΩ and $C = 0.01$ μF. Calculate the frequency of oscillations **(6.5 kHz)**

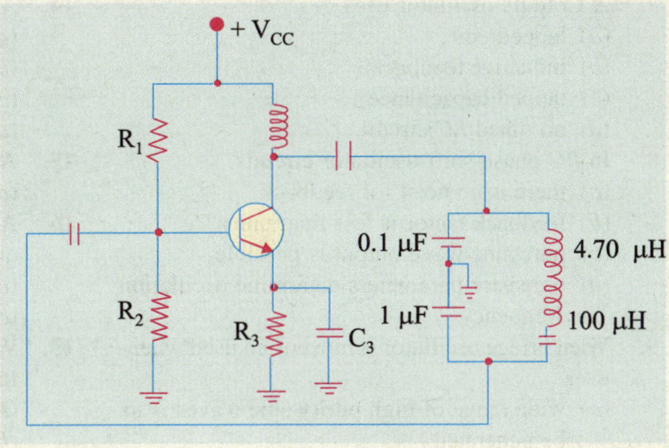

Fig. 65.39

(Electronics Engg, Annamalai Univ. 2002)

14. Calculate the frequency and the duty cycle of the output of stable multivibrator and draw the waveform obtained across the capacitor C_T as shown in Fig. 65.40. ***(Electronics Engg. Bangalore Univ. 2001)***

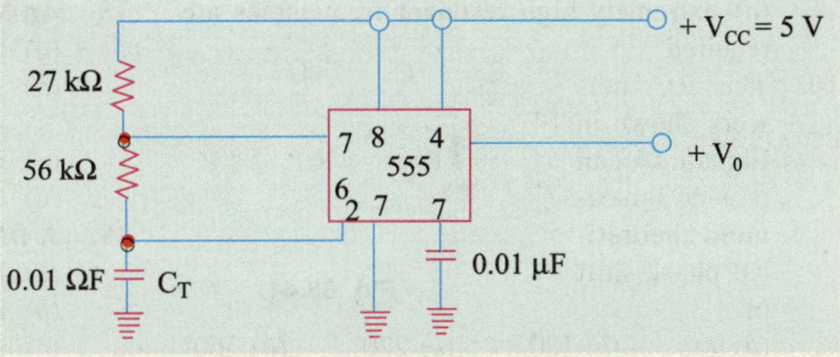

Fig. 65.40

15. In an Hartey oscillator if $L_1 = 0.1$ mH, $L_2 = 10$ μH and mutual inductance between the coils equal to 20 μH. Calculate the value of capacitor C of the oscillatory circuit to obtain frequency of 4110 kHz and also find the condition for sustained oscillations. **(10 pF)** ***(Electronics Engg: Bangalore Univ. 2001)***

16. Calculate the frequency of oscillation of a colpit oscillator with $C_1 = C_2 = 400$ pF and $L = 2$mH. **(251.6 kHz) *(Electronics Engg., Bangalore Univ. 2002)***

17. In an R-C phase shift oscillator R = 5000 Ω and C = 0.1 pF. Calculate the frequency of oscillations. **(130 MHz) (Electronics Engg., Bangalore Univ. 2002**

OBJECTIVE TESTS – 65

1. An electronic oscillator is
   (*a*) just like an alternator
   (*b*) nothing but an amplifier
   (*c*) an amplifier with feedback
   (*d*) a converter of ac to dc energy.
2. The frequency of oscillation of an elementary *LC* oscillatory circuit depends on
   (*a*) coil resistance
   (*b*) coil inductance
   (*c*) capacitance
   (*d*) both (*b*) and (*c*).
3. For sustaining oscillations in an oscillator
   (*a*) feedback factor should be unity
   (*b*) phase shift should be 0°
   (*c*) feedback should be negative
   (*d*) both (*a*) and (*b*).
4. If Barkhausen criterion is not fulfilled by an oscillator circuit, it will
   (*a*) stop oscillating
   (*b*) produce damped waves continuously
   (*c*) become an amplifier
   (*d*) produce high-frequency whistles.
5. In a transistor Hartley oscillator
   (*a*) inductive feedback is used
   (*b*) untapped coil is used
   (*c*) entire coil is in the output circuit
   (*d*) no capacitor is used
6. A Hartley oscillator is used for generating
   (*a*) very low frequency oscillation
   (*b*) radio-frequency oscillation
   (*c*) microwave oscillation
   (*d*) audio-frequency oscillation

7. A Colpitts oscillator uses
   (*a*) tapped coil
   (*b*) inductive feedback
   (*c*) tapped capacitance
   (*d*) no tuned *LC* circuit
8. In *RC* phase-shift oscillator circuits.
   (*a*) there is no need for feedback
   (*b*) feedback factor is less than unity
   (*c*) pure sine wave output is possible
   (*d*) transistor parameters determine oscillation frequency.
9. Wien bridge oscillator is most often used whenever
   (*a*) wide range of high purity sine waves is to be generated
   (*b*) high feedback ratio is needed
   (*c*) square output waves are required
   (*d*) extremely high resonant frequencies are required.
10. The RC network shown in Fig 65.40 can provide a maximum theoretical phase shift of

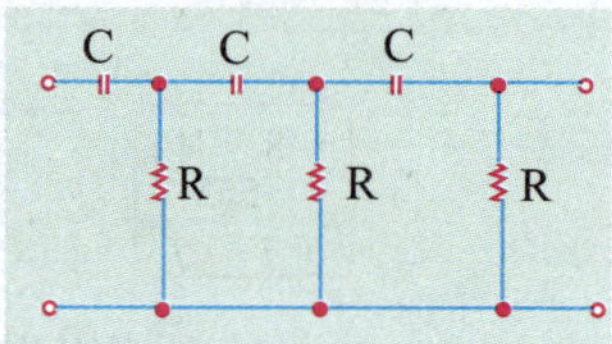

Fig. 65.41

   (*a*) 90° (*b*) 180° (*c*) 270° (*d*) 360°
   (UPSC Engineering Services 2002)
11. A *CE* amplifier can be converted into oscillator by
   (*a*) providing adequate positive feedback
   (*b*) phase shifting the output by 180° and feeding this phase-shifted output to the input.
   (*c*) using only a series tuned circuit as a load on the amplifier
   (*d*) using a negative resistance device as a load on the amplifier
   (*e*) all of the above
   (UPSC Engineering Service, 2002)
12. The primary advantage of a crystal oscillator is that
   (*a*) it can oscillate at any frequency
   (*b*) it gives a high output voltage
   (*c*) its frequency of oscillation remains almost constant
   (*d*) it operates on a very low dc supply voltage.
13. Non-sinusoidal waveforms
   (*a*) are departures from sine waveform
   (*b*) have low mark-to-space ratio
   (*c*) are much easier to generate
   (*d*) are unfit for digital operation.
14. A relaxation oscillator is one which
   (*a*) has two stable states
   (*b*) relaxes indefinitely
   (*c*) produces non-sinusoidal output
   (*d*) oscillates continuously.
15. A square pulse has a mark-to-space ratio of
   (*a*) 1 : 1 (*b*) 1 : 2 (*c*) 2 : 1 (*d*) 1 : 4.
16. Apart from a dc power source, the essential requirements of a sawtooth generator are
   (*a*) a resistor (*b*) a capacitor
   (*c*) a switching device (*d*) all of the above.
17. Which of the following statement is WRONG ? In a multivibrator
   (*a*) output is available continuously
   (*b*) feedback between two stages is 100%
   (*c*) positive feedback is employed
   (*d*) when one transistor is ON, the other is OFF.
18. An MMV circuit
   (*a*) has no stable state
   (*b*) gives two output pulses for one input trig ger pulse
   (*c*) returns to its stand-by states automatically
   (d) has no energy-storage element.
19. A BMV circuit
   (*a*) has two unstable states
   (*b*) has one energy-storage element
   (*c*) switches between its two states automatically
   (*d*) is not an oscillator.
20. The digital circuit using two inverters shown in Fig. 65.42 will act as
   (*a*) a bistable multi-vibrator
   (*b*) an astable multi-vibrator
   (*c*) a monostable multi-vibrator
   (*d*) an oscillator

Fig. 65.42

21. In the Schmitt trigger circuit shown in Fig. 65.43 if $V_{CE(sat)} = 0.1\ V$, the output logic low level (V_{OL}) is
   (*a*) 1.25 V
   (*b*) 1.35 V
   (*c*) 2.50 V
   (*d*) 5.00 V

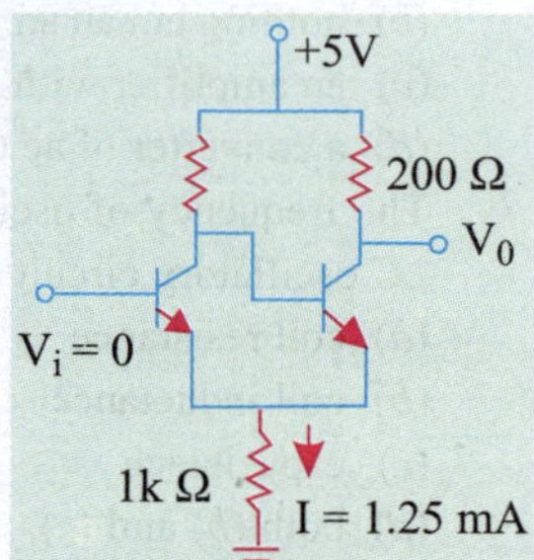

Fig. 65.43

ANSWERS

1. (*c*)	2. (*d*)	3. (*d*)	4. (*a*)	5. (*a*)	6. (*b*)	7. (*c*)	8. (*c*)	9. (*a*)	10. (*b*)	11. (*b*)
12. (*c*)	13. (*a*)	14. (*c*)	15. (*a*)	16. (*d*)	17. (*a*)	18. (*c*)	19. (*a*)	20. (*a*)	21. (*b*)	

CHAPTER 66

MODULATION AND DEMODULATION

Learning Objectives

- What is a Carrier Wave?
- Radio Frequency Spectrum
- Sound
- Need for Modulation
- Radio Broadcasting
- Modulation
- Methods of Modulation
- Amplitude Modulation
- Percent Modulation
- Upper and Lower Sidebands
- Mathematical Analysis of a Modulated Carrier Wave
- Power Relation in an AM Wave
- Forms of Amplitude Modulation
- Generation of SSB
- Methods of Amplitude Modulation
- Modulating Amplifier Circuit
- Frequency Modulation
- Modulation Index
- Deviation Ratio
- Percent Modulation
- FM Sidebands
- Modulation Index and Number of Sidebands
- Demodulation or Detection
- Essentials of AM Detection
- Transistor Detectors for AM Signals
- Quadrature Detector
- Frequency Conversion
- Standard Superhet AM Receiver
- FM Receiver
- Comparison between AM and FM
- The Four Fields of FM

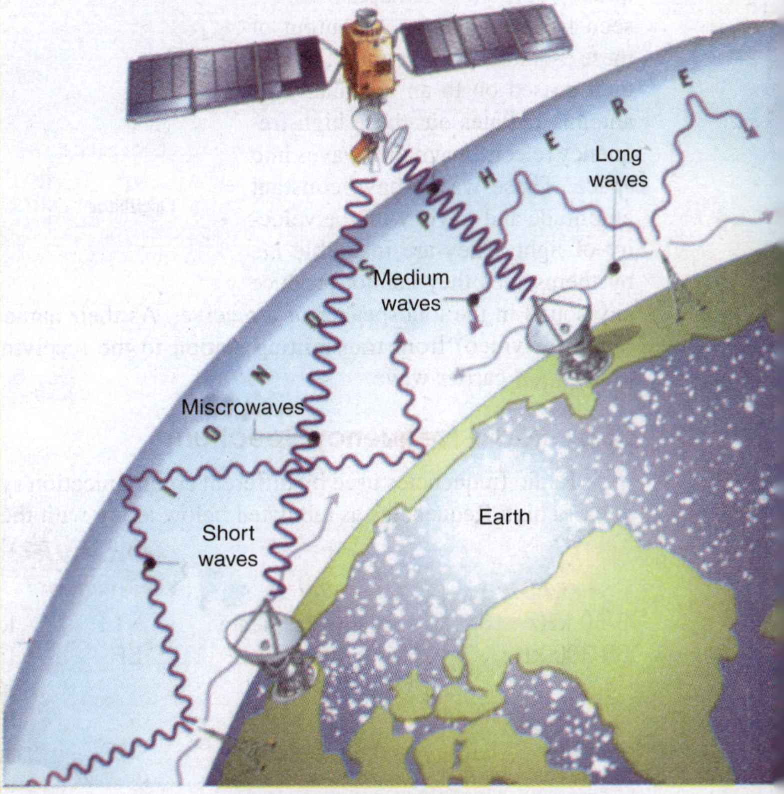

The radio waves travel through the air, bounce off a layer of the atmosphere called the ionosphere, or are relayed by satellite

66.1. Introduction

For successful transmission and reception of intelligence (code, voice, music etc.) by the use of radio waves, two processes are essential :

(*i*) modulation and (*ii*) demodulation.

Speech and music etc. are sent thousands of kilometres away by a radio transmitter. The scene in front of a television camera is also sent many kilometres away to viewers. Similarly, a Moon probe or Venus probe checking its environments, sends the information it gathers millions of kilometres through space to receivers on earth. In all these cases, the carrier is the high-frequency radio wave. The intelligence i.e. sight, sound or other data collected by the probe is impressed on the radio wave and is carried along with it to the destination.

Modulation is the process of **combining the low-frequency signal with a very high-frequency radio wave called carrier wave (CW)**. The resultant wave is called **modulated carrier wave**. This job is done at the transmitting station.

Demodulation is the process of separating or **recovering the signal from the modulated carrier wave**. It is just the opposite of modulation and is performed at the receiving end.

66.2. What is a Carrier Wave?

It is a high-frequency undamped radio wave produced by radio-frequency oscillators (Chapter 65). As seen from Fig. 66.1, the output of these oscillators is first amplified and then passed on to an antenna. This antenna radiates out these high-frequency (electromagnetic) waves into space. These waves have constant amplitude and travel with the velocity of light. They are inaudible *i.e.* by themselves they cannot produce any sound in the loudspeaker of a receiver. As their name shows, their job is to **carry the signal** (audio or video) from transmitting station to the receiving station. The resultant wave is called **modulated** carrier wave.

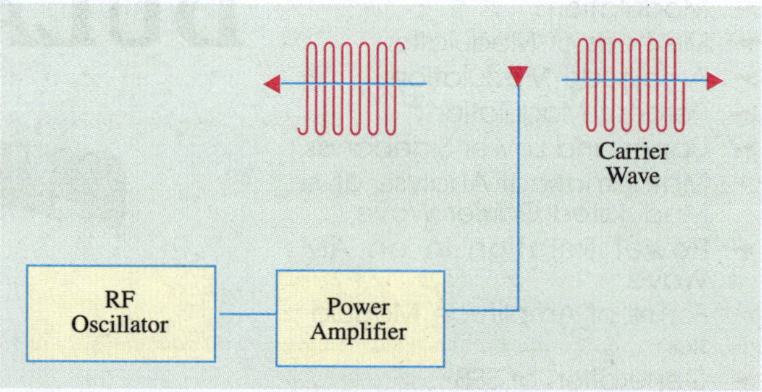

Fig. 66.1

66.3. Radio Frequency Spectrum

Radio frequencies used by different communication systems extend from very low frequencies to extra high frequencies as tabulated below along with their acronym abbreviations.

Table No. 66.1

Frequency	*Designation*	*Abbreviation*	*Uses*
3-30 kHz	very low frequency	VLF	long distance telegraphy broadcasting
30-300 kHz	low frequency	LF	long distance point-to-point service, navigational aids, sound broadcasting and line carrier systems.
300 kHz-3MHz	medium frequency	MF	sound broadcasting, ship-shore services and line carrier systems
3-30 MHz	high frequency	HF	medium and long-distance point-to-point services, sound broadcasting, linear carrier systems.

30-300 MHz	very high frequency	VHF	short-distance communication, TV and
300 MHz-3GHz	ultra high frequency	UHF	sound broadcasting, radar
3-30 GHz	super high frequency	SHF	outer-space radio communication,
30-300 GHz	extra high frequency	EHF	point-to-point microwave communication systems and radar.

Generally, none of the frequencies above 300 GHz is classified as radio waves.

66.4. Sound

It is a sort of disturbance which requires some physical medium for its propagation. Human voice consists of a series of compressions and rarefactions which travel through air with a velocity of about 345 m/s. The frequency range of human voice is from 20-4000 Hz which lies within the audible range of 20 to 20,000 Hz. Variations in human voice can be converted into corresponding variations in electric current with the help of a microphone as shown in Fig.66.2.

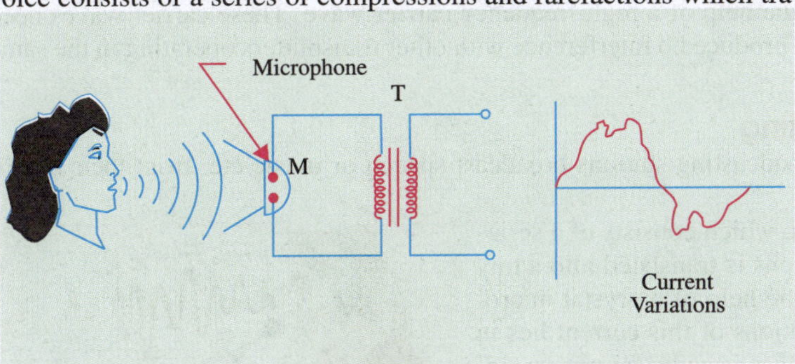

Fig. 66.2

When a sound wave strikes the microphone, it produces AF sound current. The positive half-cycles of sound current are produced by the compressions and negative half cycles by rarefactions.

As seen, human voice does not produce pure sinusoidal current because it does not consist of one frequency alone. It is quite complex and can be analysed to consist of a fundamental (or lowest frequency) and its integral multiple frequencies called **overtones** or **harmonics**.

We are interested in two main characteristics of sound :

(i) **Intensity**—It is the energy content of the wave. It depends on its amplitude. In fact, intensity of a wave is directly proportional to the square of its amplitude *i.e.* $I \propto a^2$. Sensation of loudness felt by a listener depends directly on the intensity of the wave falling on his ears.

(ii) **Frequency**—It produces the sensation called pitch. Audible sounds have a frequency range from 20 Hz to 20,000 Hz.

Sound wave strikes the microphone and it produces AF sound

Though every sound has complex frequency structure, we will consider only single-frequency sound whose current wave is a pure sine wave as shown in Fig. 66.3. It will be used as the modulating signal when discussing the process of modulation.

66.5. Need for Modulation

Sometimes, beginners question the necessity of modulation *i.e.* using a carrier wave to carry the low-frequency signal from one place to another. Why not transmit the signals directly and save lot of botheration? Unfortunately, there are three main hurdles in the process of such direct transmission of audio-frequency signals :

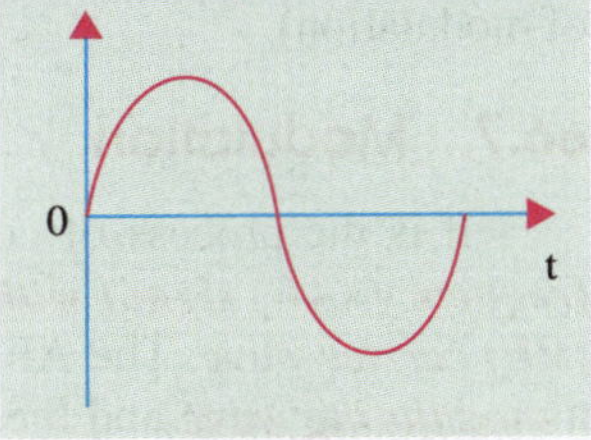

Fig. 66.3

1. They have relatively short range,
2. If everybody started transmitting these low-frequency signals directly, mutual interference will render all of them ineffective
3. Size of antennas required for their efficient radiation would be large i.e. about 75 km as explained below.

For efficient radiation of a signal, the minimum length of an antenna is one quarter wavelength ($\lambda/4$). The antenna length L is connected with the frequency of the signal wave by the relation $L = 75 \times 10^6/f$ metres. For transmitting an audio signal of $f = 1000$ Hz, $L = 75 \times 10^6/10^3 = 75{,}000$ m = 75 km ! In view of this immense size of antenna length, it is impractical to radiate audio-frequency signals directly into space.

Hence, the solution lies in modulation which enables a low-frequency signal to travel very large distances through space with the help of a high-frequency carrier wave. These carrier waves need reasonably-sized antennas and produce no interference with other transmitters operating in the same area.

66.6. Radio Broadcasting

Let us see how radio broadcasting stations broadcast speech or music etc. from their broadcasting studios.

First, the speech or music which consists of a series of compressions and rarefactions is translated into a tiny varying electric current with the help of a crystal microphone. The frequency of variations of this current lies in the audio-range, hence it is known as audio frequency signal. The audio-frequency signal cannot be radiated out from the antenna directly because transmission at audio-frequencies is not practical. For this purpose, oscillations of very high frequency or radio-frequency are produced with the help of any one of the oscillators discussed in Chapter 65. The electromagnetic waves so produced are of constant amplitude but of extremely high frequency. These waves, which are neither seen nor heard, travel through space with the velocity of light *i.e.* 3×10^8 m/s (approx). The audio-frequency signal which is to be broadcast, is then superimposed on the *RF* waves, which are known as carrier waves (because they carry *A.F.* signal through space to distant places). In a way, the carrier waves can be likened to a horse and the audio-frequency signal to a rider. The process by which *AF* signal or information is impressed on the carrier wave is known as modulation. The horse and rider travel through space. At the receiving end, they strike the receiving aerial and enter the receiver which separates the horse from the rider. The horse *i.e.* carrier wave is returned and the rider *i.e.* audio-frequency signal is converted back into sound. This process by which the *R.F.* waves and *A.F.* waves are separated is known as **detection** or **demodulation** (because it is the reverse of modulation).

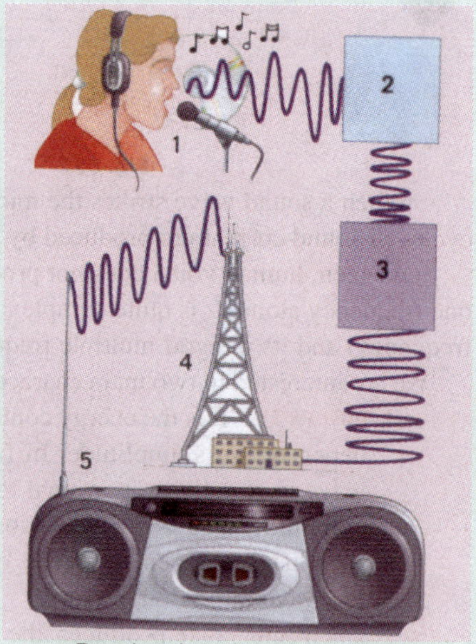

Broadcasting through waves

66.7. Modulation

It is the process of combining ***an audio-frequency (AF) signal with a radio frequency (RF) carrier wave.*** The AF signal is also called a ***modulating wave*** and the resultant wave produced is called ***modulated wave.***

During modulation, some characteristic of

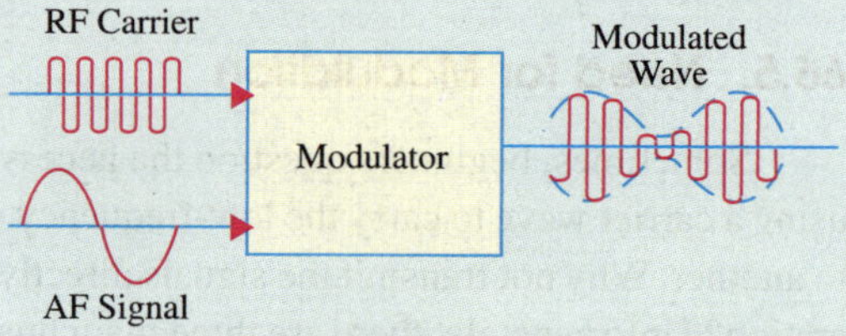

Fig. 66.4

the carrier wave is varied in time with the modulating signal and is accomplished by combining the two.

66.8. Methods of Modulation

The mathematical expression for a sinusoidal carrier wave is

$$e = E_C \sin(\omega_c t + \phi) = E_C \sin(2\pi f_c t + \phi)$$

Obviously, the waveform can be varied by any of its following three factors or parameters :

1. E_C — the amplitude, **2.** f_c — the frequency, **3.** ϕ — the phase.

Accordingly, there are three types of sine-wave modulations known as :

1. Amplitude Modulation (AM)

Here, the information or *AF* signal changes the amplitude of the carrier wave without changing its frequency or phase.

2. Frequency Modulation (FM)

In this case, the information signal changes the frequency of the carrier wave without changing its amplitude or phase.

3. Phase Modulation (PM)

Here, the information signal changes the phase of the carrier wave without changing its other two parameters.

66.9. Amplitude Modulation

In this case, the **amplitude** of the carrier wave is varied in proportion to the **instantaneous amplitude** of the information signal or AF signal. Obviously, the amplitude (and hence the intensity) of the carrier wave is changed ***but not its frequency.*** Greater the amplitude of the AF signal, greater the fluctuations in the amplitude of the carrier wave.

The process of amplitude modulation is shown graphically in Fig. 66.5. For the sake of simplicity, the *AF* signal has been assumed sinusoidal [Fig. 66.5 (*a*)]. The carrier wave by which it is desired to transmit the *AF* signal is shown in Fig. 66.5 (*b*). The resultant wave called modulated wave is shown in Fig. 66.5 (*c*).

The function of the modulator is to mix these two waves in such a way that (*a*) is transmitted along with (*b*). All stations broadcasting on the standard broadcast band (550-1550 kHz) use *AM* modulation.

If you observe the envelope of the modulated carrier wave, you will realize that it is an exact replica of the *AF* signal wave.

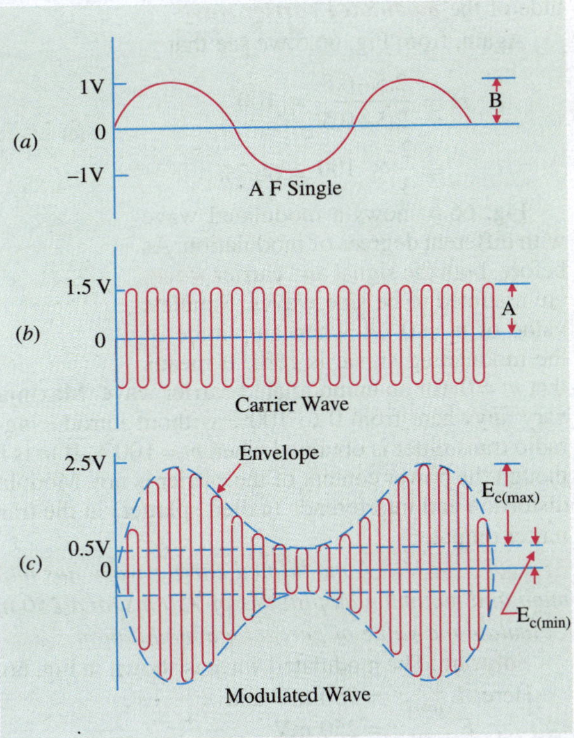

Fig. 66.5

In summary

(i) **fluctuations** in the amplitude of the carrier wave depend on the **signal amplitude**,

(ii) **rate** at which these fluctuations take place depends on the **frequency** of the audio signal.

66.10. Percent Modulation

It indicates the degree to which the AF signal modulates the carrier wave

$$m = \frac{\text{maximum value of signal wave}}{\text{maximum value of carrier wave}} \times 100 = \frac{\text{signal amplitude}}{\text{carrier amplitude}} \times 100$$

$$= \frac{B}{A} \times 100 \quad \text{—Fig. 66.5}$$

The ratio B/A expressed as a fraction is called **modulation index (MI)**

$$m = M.I. \times 100$$

From Fig. 66.5, it is seen that $B = 1$ V and $A = 1.5$ V

$$\therefore \quad m = \frac{1}{1.5} \times 100 = 66.7\%$$

Modulation may also be defined in terms of the values referred to the modulated carrier wave.

$$m = \frac{E_{c(max)} - E_{c(min)}}{E_{c(max)} + E_{c(min)}} \times 100$$

where $E_{c(max)}$ and $E_{c(min)}$ are the maximum and minimum values of the amplitude of the ***modulated carrier wave***.

Again, from Fig. 66.5 we see that

$$m = \frac{2.5 - 0.5}{2.5 + 0.5} \times 100$$

$$= \frac{2}{3} \times 100 = 66.7\%$$

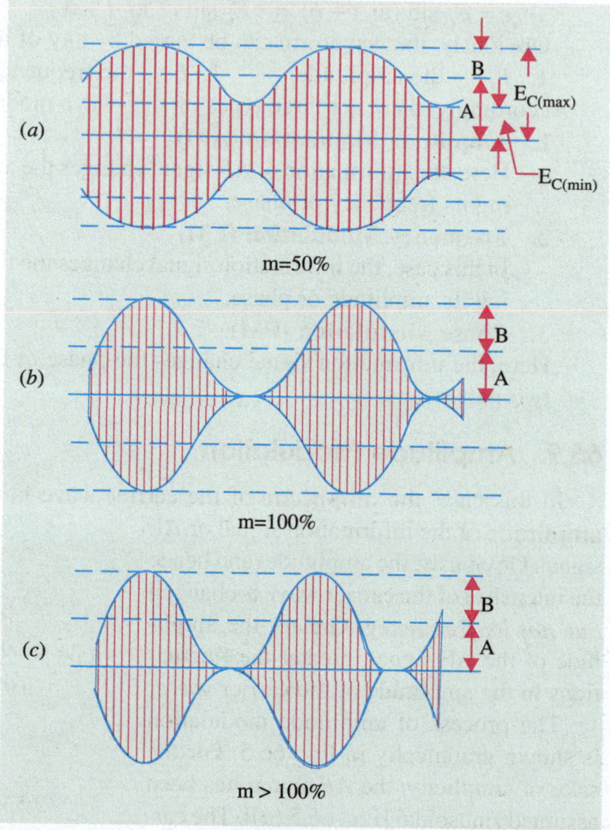

Fig. 66.6

Fig. 66.6 shows a modulated wave with different degrees of modulation. As before, both the signal and carrier **waves** are assumed to be sine waves. Smallest value of $m = 0$ *i.e.* when amplitude of the modulating signal is zero. It means that $m = 0$ for an unmodulated carrier wave. Maximum value of $m = 1$ when $B = A$. Value of m can vary anywhere from 0 to 100% without introducing distortion. Maximum undistorted power of a radio transmitter is obtained when $m = 100\%$. If m is less than 100 per cent, power output is reduced though the power content of the carrier is not. Modulation in excess of 100 per cent produces severe distortion and interference (called splatter) in the transmitter output.

Example 66.1 *A modulated carrier wave has maximum and minimum amplitudes of 750 mV and 250 mV. Calculate the value of percentage modulation.*

Solution. The modulated wave is shown in Fig. 66.7.

Here, $E_{c(max)} = 750$ mV and

$E_{c(min)} = 250$ mV

$$\therefore \quad m = \frac{E_{c(max)} - E_{c(min)}}{E_{c(max)} + E_{c(min)}} \times 100$$

$$= \frac{750 - 250}{750 + 250} \times 100 = \mathbf{50\%}$$

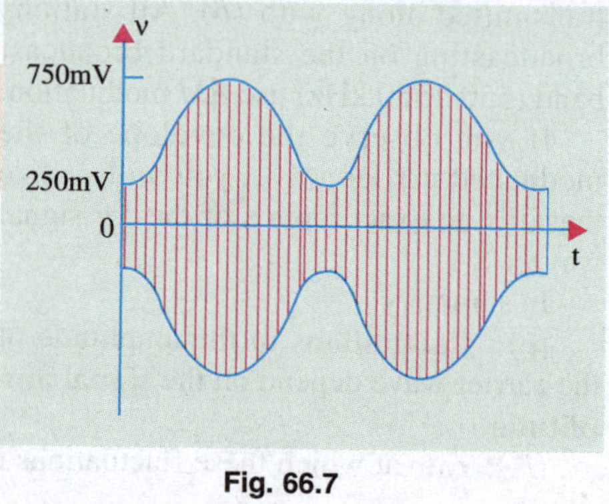

Fig. 66.7

66.11. Upper and Lower Side Frequencies

An unmodulated carrier wave consists of only one single-frequency component of frequency f_c. When it is combined with a modulating signal of frequency fm, heterodyning action takes place.

As a result, two additional frequencies called **side frequencies** are produced. The *AM* wave is found to consist of three frequency components :

1. The original carrier frequency component, f_c .
2. A higher frequency component $(f_c + f_m)$. It is called the sum component.
3. A lower frequency component $(f_c - f_m)$. It is called the difference component.

The two new frequencies are called the **upper-side frequency** (*USF*) and **lower-side frequency** (*LSF*) respectively and are symmetrically located around the carrier frequency. The modulating frequency remains unchanged but does not appear in the amplifier output because the amplifier's load presents practically zero impedance to this low frequency.

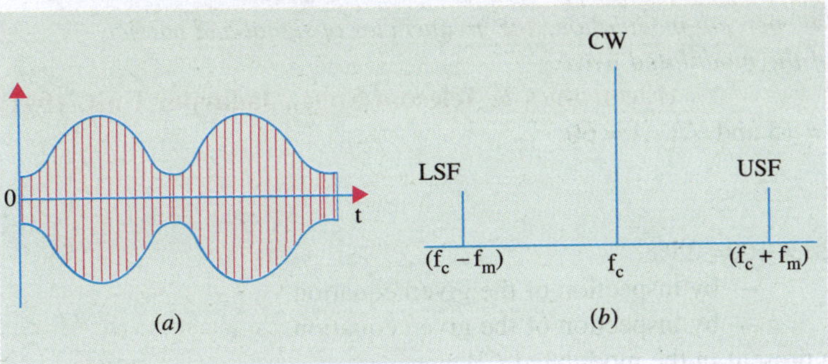

Fig. 66.8

These are shown in time domain in Fig. 66.8 (*a*) and in frequency domain in Fig. 66.8 (*b*). The amplitude of the side frequencies depends on the value of *m*. The amplitude of each side frequency = *mA*/2 where *A* is the amplitude of unmodulated carrier wave.

Example 66.2. *A 10-MHz sinusoidal carrier wave of amplitude 10 mV is modulated by a 5 kHz sinusoidal audio signal wave of amplitude 6 mV. Find the frequency components of the resultant modulated wave and their amplitudes.*

(Electronics & Comm. Engg.; Madras Univ. 1991)

Solution. Here, $f_c = 10$ MHz and $f_m = 5$ kHz = 0.005 MHz. The modulated carrier contains the following frequencies :

1. original carrier wave of frequency
   f_c = **10 MHz**
2. USF of frequency
   = 10 + 0.005 = **10.005 MHz**
3. LSF of frequency
   = 10 – 0.005 = **9.995 MHz**

The frequency spectrum is shown in Fig. 66.9

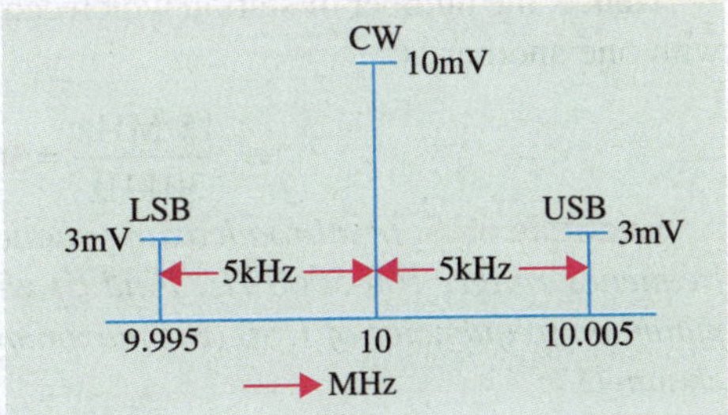

Fig. 66.9

$$\text{Here, } m = \frac{B}{A} = \frac{6}{10} = 0.6 = 0.6$$

Amplitude of *LSF* = *USF* = *mA*/2 = 0.6 × 10/2 = **3 mV** as shown.

66.12. Upper and Lower Sidebands

In Art 66.11, it was assumed that the modulating signal was composed of one frequency component only. However, in a broadcasting station, the modulating signal is the human voice (or music) which contains waves with a frequency range of 20-4000 Hz. Each of these waves has its own *LSF* and *USF*. When combined together, they give rise to an upper-side **band** (*USB*) and a lower-side **band** (LSB) as shown in Fig. 66.10. The *USB*, in fact, contains all sum components of the signal and carrier frequency whereas *LSB* contains their difference components.

The channel width (or bandwidth) is given by the difference between extreme frequencies i.e. between maximum frequency of *USB* and minimum frequency of *LSB*. As seen,

Channel width = 2 × maximum frequency of modulating signal = $2 \times f_{m\,(max)}$

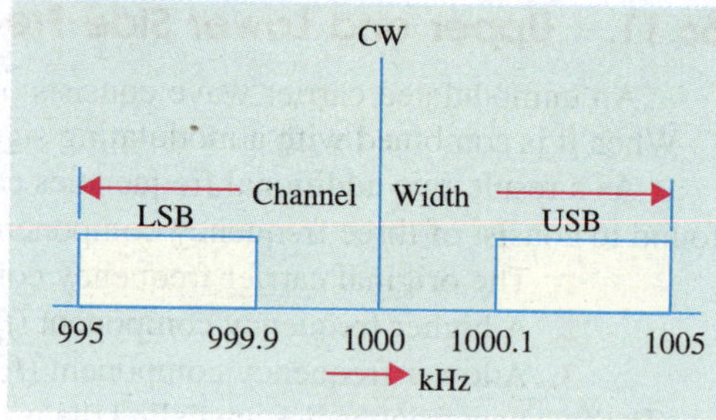

Fig. 66.10

Example 66.3. *An audio signal given by 15 sin 2π (2000 t) amplitude-modulates a sinusoidal carrier wave 60 sin 2π (100,000) t.*

Determine :

(a) modulation index, (b) percent modulation, (c) frequencies of signal and carrier, (d) frequency spectrum of the modulated wave.

(Electronics & Telecom Engg., Jadavpur Univ. 1991)

Solution. Here, $B = 15$ and $A = 60$

(a) $M.I. = \dfrac{B}{A} = \dfrac{15}{60} = \mathbf{0.25}$

(b) $m = M.I. \times 100 = 0.25 \times 100 = 25\%$

(c) $f_m = \mathbf{2000\ Hz}$ — by inspection of the given equation

$f_c = \mathbf{100{,}000\ Hz}$ — by inspection of the given equation

(d) The three frequencies present in the modulated *CW* are

(i) 100,000 Hz = **100 kHz**

(ii) 100,000 + 2000 = 102,000 Hz = **102 kHz**

(iii) 100,000 – 2000 = 98,000 Hz = **98 kHz**

Example 66.4. *A bandwidth of 15 MHz is available for AM transmission. If the maximum audio signal frequency used for modulating the carrier is not to exceed 15 kHz, how many stations can broadcast within this band simultaneously without interfering with each other?*

(Electronics & Telecom Engg.; Pune Univ. 1991)

Solution. *BW* required by each station

$= 2 f_{m(max)} = 2 \times 15 = 30$ kHz

Hence, the number of station which can broadcast within this frequency band without interfering with one another is

$$= \frac{15\ \text{MHz}}{30\ \text{kHz}} = \mathbf{500}$$

Example 66.5. *In a broadcasting studio, a 1000 kHz carrier is modulated by an audio signal of frequency range, 100-5000 Hz. Find (i) width or frequency range of sidebands (ii) maximum and minimum frequencies of USB (iii) maximum and minimum frequencies of LSB and (iv) width of the channel.*

(Electronics & Comm. Engg. IERE, London)

Solution. (i) Width of sideband = 5000 – 100 = **4900 Hz**

(ii) Max. frequency of USB

= 1000 + 5 = **1005 kHz**

Min. frequency of *USB*

= 1000 + 0.1 = **1000.1 kHz**

(iii) Max. frequency of *LSB*

= 1000 – 0.1 = **999.9 kHz**

Min. frequency of *LSB*

= 1000 – 5 = **995 kHz**

(iv) Width of channel = 1005 – 995 = **10 kHz**

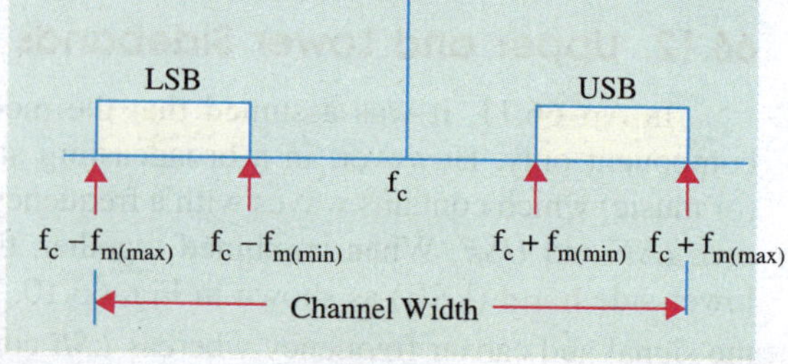

Fig. 66.11

66.13. Mathematical Analysis of a Modulated Carrier Wave

The equation of an unmodulated carrier wave is

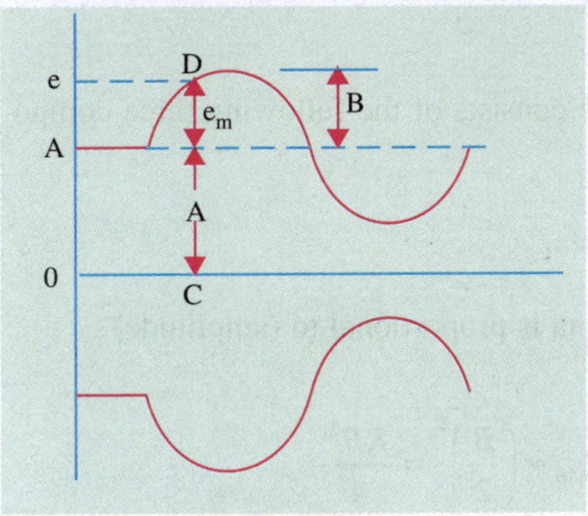

Fig. 66.12

$$e_c = E_c \sin 2\pi f_c t = A \sin 2\pi f_c t = A \sin \omega t$$

where A is the constant amplitude of the carrier wave and $\omega = 2\pi f_c$.

Let the equation of the single-frequency sinusoidal modulating signal be

$$e_m = E_m \sin 2\pi f_m t = B \sin 2\pi f_m t$$
$$= B \sin Pt \qquad \text{— where } p = 2\pi f_m$$

As seen from Fig. 66.12, the amplitude of the modulated carrier wave at any instant is

$$= A + e_m \qquad (\because A \text{ is constant})$$
$$= (A + B \sin Pt)$$

Hence, its ***instantaneous*** value is given by

$$e = (A + B \sin pt)\sin \omega t = A \sin \omega t + B \sin \omega t . \sin pt$$

$$= A \sin \omega t + \frac{B}{2} . 2 \sin \omega t . \sin pt \qquad = A \sin \omega t + \frac{B}{2}\left[\cos(\omega - p)t - \cos(\omega + p)t)\right]^{*}$$

$$= A \sin \omega t + \frac{B}{2} \cos(\omega - p)\, t - \cos(\omega + p) t$$

$$= A \sin 2\pi f_c t + \frac{B}{2} \cos 2\pi (f_c - f_m)t - \frac{B}{2} \cos 2\pi (f_c + f_m)t$$

As seen from Art. 66.10, $m = B/A$ or $B = mA$

$$\therefore \quad e = A \sin 2\pi f_c t + \frac{mA}{2} \cos 2\pi (f_c - f_m)t - \frac{mA}{2} \cos \pi 2 (f_c + f_m)t$$

It is seen that the modulated wave contains three components :

(i) $A \sin 2\pi f_c t$ — the original carrier wave

(ii) $\frac{mA}{2} \cos 2\pi (f_c + f_m)t$ — upper side frequency

(iii) $\frac{mA}{2} \cos 2\pi (f_c - f_m)t$ — lower side frequency

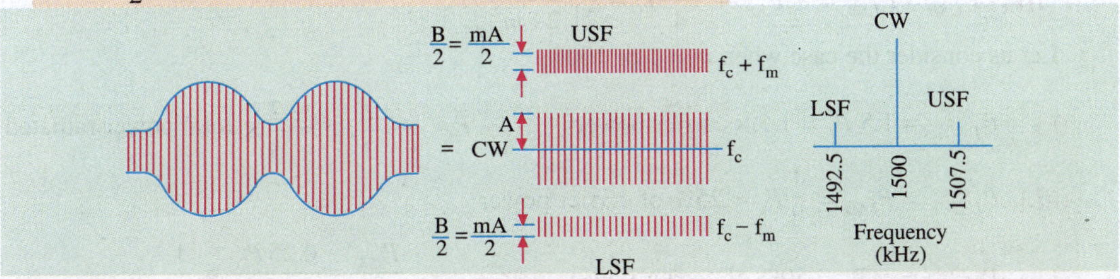

Fig. 66.13

These three frequencies are not a mathematical fiction but they actually exist. In fact, with the help of a narrow band filter, we can separate side frequencies from the carrier wave.

Example 66.6. *The tuned circuit of the oscillator in an AM transmitter uses a 40 mH coil and a 1 nF capacitor. If the carrier wave produced by the oscillator is modulated by audio frequencies upto 10 kHz, calculate the frequency band occupied by the side bands and channel width.*

(Electronics & Telecom. Engg., Calcutta Univ. 1990)

Solution. The resonant frequency is given by

$$f_c = \frac{1}{2\pi\sqrt{LC}} = \frac{1}{2\pi\sqrt{40 \times 10^{-6} \times 1 \times 10^{-9}}} \cong 796 \text{ kHz}$$

---

* $2 \sin A . \sin B = \cos (A - B) - \cos (A + B)$

The frequency range occupied by the sidebands is from **796 kHz to 806 kHz**. The channel width

$= 2 \times 10 = \mathbf{20\ kHz}$

66.14. Power Relations in an AM Wave

As discussed in Art. 66.13, a modulated carrier wave consists of the following three components :

1. original carrier wave of amplitude A
2. *USF* wave of amplitude $(mA/2)$
3. *LSF* wave of amplitude $(mA/2)$

Now, power radiated out by a wave through an antenna is proportional to (amplitude)^2.

Carrier power, $P_C \propto A^2 = KA^2$

USB power, $P_{USB} \propto \left(\frac{B}{2}\right)^2 = \frac{KB^2}{4}$; *LSB* power, $P_{LSB} \propto \left(\frac{B}{2}\right)^2 = \frac{KB^2}{4}$

Total sideband power $P_{SB} = 2 \times \frac{KB^2}{4} = \frac{KB^2}{2}$

Total power radiated out from the antenna is $P_T = P_c + P_{SB} = KA^2 + \frac{KB^2}{2}$

Substituting $B = mA$, we get $P_T = KA^2 + \frac{K}{2}(mA)^2 = KA^2\left(1+\frac{m^2}{2}\right)$

Now $P_C = KA^2$

(*i*) $\therefore \quad P_T = P_C\left(1+\frac{m^2}{2}\right) = P_C\left(\frac{2+m^2}{2}\right)$ (*ii*) $P_C = P_T\left(\frac{2}{2+m^2}\right)$

(*iii*) $P_{SB} = P_T - P_C = P_C\left(1+\frac{m^2}{2}\right) - P_C = \frac{m^2}{2}P_C = \left(\frac{m^2}{2+m^2}\right)P_T$

(*iv*) $P_{USB} = P_{LSB} = \frac{1}{2}P_{SB} = \frac{m^2}{4}P_C = \frac{1}{2}\left(\frac{m^2}{2+m^2}\right)P_T$

Let us consider the case when $m = 1$ or 100%

(*i*) $P_T = 1.5\, P_C = 1.5 \times$ carrier power (*ii*) $P_C = \frac{2}{3}P_T = \frac{2}{3} \times$ total power radiated

(*iii*) $P_{USB} = P_{LSB} = \frac{1}{4}P_C = 25\%$ of carrier power

(*iv*) $P_{SB} = \frac{1}{2}P_C = 50\%$ of carrier power (*v*) $\frac{P_{USB}}{P_T} = \frac{0.25\,P_C}{1.5\,P_C} = \frac{1}{6}$

It means that **single side-band contains 1/6th of the total power radiated out by the transmitter.** That is why single-side band (*SSB*) transmission is more power efficient.

Example 66.7. *The total power content of an AM wave is 1500 W. For a 100 percent modulation, determine*

(*i*) *power transmitted by carrier,* (*ii*) *power transmitted by each side band.*

(Electronics & Comm. Engg., Kerala Univ. 1991)

Solution. $P_T = 1500$ W, $P_C = ?$ $P_{USB} = ?$ $P_{LSB} = ?$

(*i*) $P_C = P_T\left(\frac{2}{2+m^2}\right) = P_T\left(\frac{2}{2+1^2}\right) = \frac{2}{3}P_T = \frac{2}{3} \times 1500 = \mathbf{1000\ W}$

(*ii*) $P_{USB} = P_{LSB} = \frac{m^2}{4} P_C = \frac{1^2}{4} \times 1000 =$ **250 W**

As seen, $P_{USB} = \frac{1}{6} P_T$

Example 66.8. *The total power content of an AM wave is 2.64 kW at a modulation factor of 80%. Determine the power content of*
(*i*) *carrier,* (*ii*) *each sideband.* **(Electronics & Comm., Roorkee Univ. 1991)**

Solution. (*i*) $P_C = P_T\left(\frac{2}{2+m^2}\right) = 2.64 \times 10^3 \left(\frac{2}{2+0.8^2}\right) =$ **2000 W**

(*ii*) $P_{USB} = P_{LSB} = \frac{m^2}{4} P_c = \frac{0.8^2}{4} \times 2000 =$ **320 W**

Example 66.9. *A transmitter used for radio telephone has an unmodulated carrier power of 10 kW and can be modulated to a maximum of 80 percent by a single-frequency signal before overloading. Find the value to which carrier power can be increased if a 50 percent modulation limit is imposed.* **(Electronics & Comm. Engg. Andhra Univ 1988)**

Solution. $P_T = P_C\left(\frac{2+m^2}{2}\right) = 10\left(\frac{2+0.8^2}{2}\right) =$ **13.2 kW**

Now, when $m = 0.5$, P_T is still 13.2 kW. Hence, new value of carrier power is given by

$$13.2 = P_C\left(\frac{2+0.5^2}{2}\right); \qquad P_C = \mathbf{11.73\ kW}$$

It is seen that P_C can be increased from 10 kW to 11.73 kW with a total power limit of 13.2 kW and $m = 0.5$

Example 66.10. *A certain transmitter radiates 10 kW of power with the carrier unmodulated and 11.8 kW with the carrier sinusoidally modulated.*
(*a*) *Find the modulation factor,*
(*b*) *If another wave modulated to 40% is also transmitted, calculate the radiated power.*

Solution. (*a*) $11.8 = 10(1 + m^2/2)$; $m = 0.6$ or 60%

(*b*) $P_T = 10\left(1+\frac{0.6^2}{2}+\frac{0.4^2}{2}\right) =$ **12.6 kW**

66.15. Forms of Amplitude Modulation

As shown in Art. 66.11, **one carrier** and **two sidebands** are produced in AM generation. It is found that it is not necessary to transmit all these signals to enable the receiver to reconstruct the original signal. Accordingly, we may attenuate or altogether remove the carrier or any one of the sidebands without affecting the communication process. The advantages would be

1. less transmitted power and **2.** less bandwidth required

The different suppressed component systems are :

(*a*) DSB-SC

It stands for ***double-sideband suppressed carrier*** system [Fig. 66.14 (*a*)]. Here, carrier component is suppressed thereby saving enormous amount of power. As seen from Art. 66.14, carrier signal contains 66.7 per cent of the total transmitted power for $m = 1$, Hence, power saving amounts to 66.7% at 100% modulation.

(*b*) SSB-TC

As shown in Fig. 66.14 (*b*), in this case, one sideband is suppressed but the other sideband and carrier are transmitted. It is called ***single sideband transmitted carrier*** system. For $m = 1$, power saved is 1/6 of the total transmitted power.

(c) SSB-SC

This is the **most dramatic suppression of all** because it suppresses one sideband and the carrier and transmits only the remaining sideband as shown in Fig. 66.14 (*c*). In the standard or double-sideband full-carrier (*DSB.FC*) *AM*, carrier conveys **no information but contains maximum power**. Since the two sidebands are exact images of each other, they carry the same audio information. Hence, **all information is available in one sideband only**. Obviously carrier is ***superfluous*** and one side band is ***redundant.*** Hence, one sideband and the carrier can be discarded with no loss of information. The result is *SSB* signal. The advantage of *SSB-SC* system are as follows :

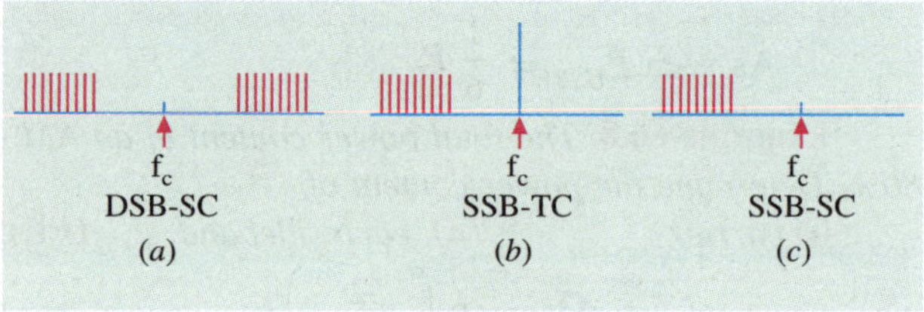

Fig. 66.14

1. Total saving of 83.3% in transmitted power (66.7% due to suppression of carrier wave and 16.6% due to suppression of one sideband). Hence, power is conserved in an SSB transmitter.
2. Bandwidth required is reduced by half *i.e.* 50%. Hence, twice as many channels can be multiplexed in a given frequency range.
3. The size of power supply required is very small. This fact assumes vital importance particularly in a spacecraft.
4. Since the SSB signal has narrower bandwidth, a narrower passband is permissible within the receiver, thereby limiting the noise pick up.

However, the main reason for wide spread use of *DSB-FC* (rather than *SSB-SC*) transmission in broadcasting is the relative simplicity of its modulating equipment.

66.16. Generation of SSB

Three main systems are employed for the generation of *SSB*.

1. Filter method, **2.** Phase cancellation method, **3.** The 'third' method.

All these methods use some form of balanced modulator to suppress the carrier.

Example 66.11. *In an AM wave, calculate the power saving when the carrier and one sideband are suppressed corresponding to*

(i) $m = 1$ *(ii)* $m = 0.5$ **(Electronics and Comm. Engg., Osmania Univ 1986)**

Solution. *(i)* When $m = 1$

$$P_T = P_C\left(1+\frac{m^2}{2}\right) = 1.5\,P_C\,; \qquad P_{LSB} = P_{USB} = \frac{m^2}{4}\,P_C = 0.25\,P_C$$

$$\therefore \quad \text{saving} = P_T - P_{USB} = 1.5\,P_c - 0.25\,P_c = 1.25\,P_c$$

$$\text{saving} = \frac{1.25}{1.5P_C} \times 100 = 83.3\%$$

***(ii)* When *m* = 0.5**

$$P_T = P_C\left(1+\frac{m^2}{2}\right) = P_C\left(\frac{2+0.5^2}{2}\right) = 1.125P_C$$

$$P_{USB} = P_{LSB} = \frac{m^2}{4}\,P_C = \frac{0.5^2}{4}\,P_C = 0.0625\,P_C$$

$$\therefore \quad \text{saving} = \frac{1.125\,P_c - 0.0625\,P_C}{1.125\,P_C} \times 100 = 94.4\%$$

66.17. Methods of Amplitude Modulation

There are two methods of achieving amplitude modulation :

(*i*) Amplifier modulation,

(*ii*) Oscillator modulation.

Block diagram of Fig. 66.15 illustrates the basic idea of amplifier modulation.

Here, carrier and *AF* signal are fed to an amplifier and the result is an *AM* output . The modulation process takes place in the active device used in the amplifier.

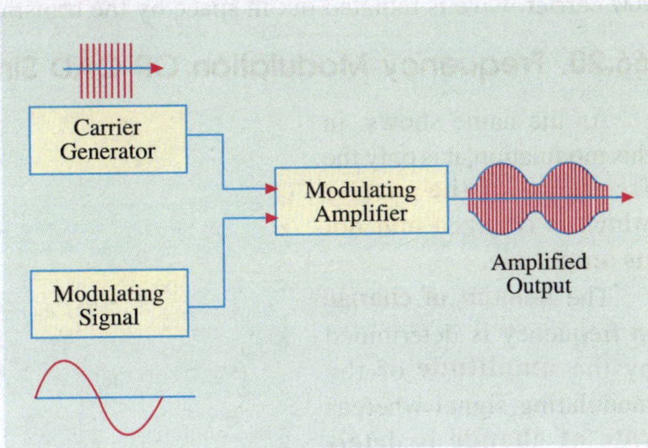

Fig. 66.15

66.18. Block Diagram of an AM Transmitter

Fig. 66.16 shows the block diagram of a typical transmitter. The carrier wave is supplied by a crystal-controlled oscillator at the carrier frequency. It is followed by a tuned buffer amplifier and an *RF* output amplifier.

The source of *AF* signal is a microphone. The audio signal is amplified by a low level audio amplifier and, finally, by a power amplifier. It is then combined with the carrier to produce a modulated carrier wave which is ultimately radiated out in the free space by the transmitter antenna as shown.

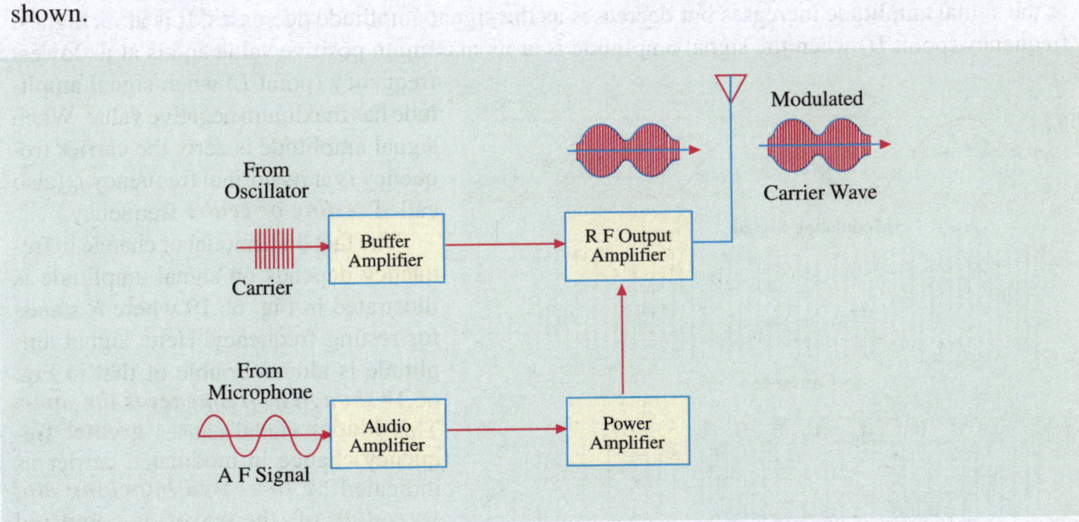

Fig. 66.16

66.19. Modulating Amplifier Circuit

It is shown in Fig. 66.17. The carrier signal from a crystal oscillator is coupled to the base of Q through transformer T_1. The low-frequency modulating signal is also coupled to the base of Q by capacitor C_1. The voltage divider $R_1 - R_2$, as usual, provides proper forward bias for the transistor whereas C_2 and C_3 are bypass capacitors.

The modulating signal applied across R_2 causes variations in the base bias in step with its own variations. The amplitude of the transistor *RF* output changes according to the changes in *AF* modulating signal. The $L_1\,C_4$ circuit is kept tuned to the carrier frequency. The amplitude-modulated carrier or *RF* current in L_1 induces similar current in L_2 which is connected to an antenna. Finally, the *AM* carrier wave is radiated out in space by the transmitting antenna.

66.20. Frequency Modulation OR CAD Simulation Diagram.

As the name shows, in this modulation, it is only the **frequency of the carrier which is changed and not its amplitude.**

The **amount of change** in frequency is determined by the **amplitude** of the modulating signal whereas **rate of change** is determined by the **frequency** of the modulating signal. As shown in Fig. 16.18, in an *FM* carrier, information (or intelligence) is carried as variations in its frequency. As seen, frequency of the modulated carrier increases as the signal amplitude increases but decreases as the signal amplitude decreases. It is at its highest frequency (point *H*) when the signal amplitude is at its maximum positive value and is at its lowest frequency (point *L*) when signal amplitude has maximum negative value. When signal amplitude is zero, the carrier frequency is at its normal frequency f_0 (also called ***resting or centre*** frequency.)

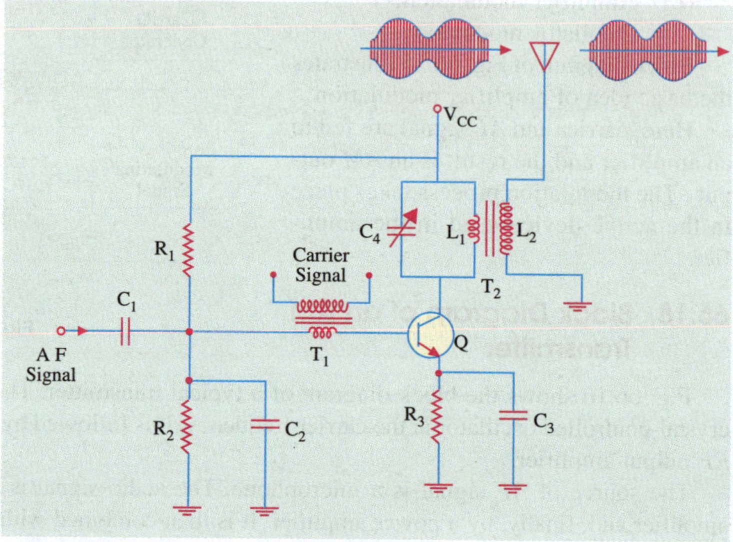

Fig. 66.17

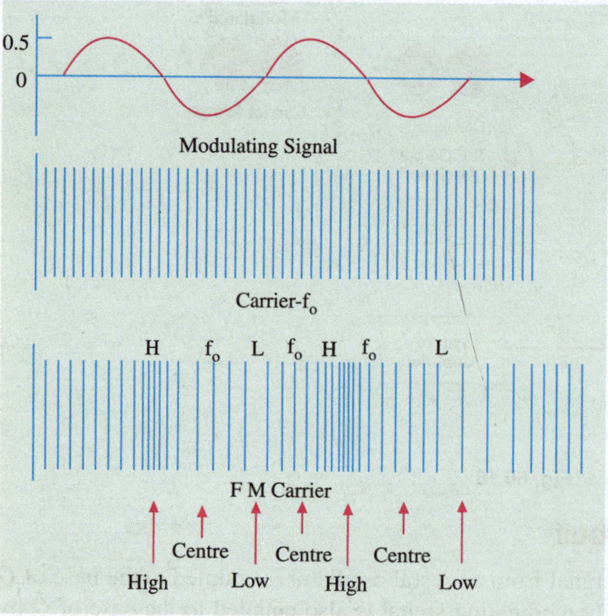

Fig. 66.18

The fact that amount of change in frequency depends on signal amplitude is illustrated in Fig. 66.19 where *R* stands for resting frequency. Here, signal amplitude is almost double of that in Fig. 66.18 ***though its frequency is the same.*** This **louder** signal causes **greater** frequency change in modulated carrier as indicated by ***increased bunching and spreading*** of the waves as compared with relatively weaker signal of Fig. 66.18.

The **rate** at which frequency shift takes place depends on the signal frequency as shown in Fig. 66.20. For example, if the modulating signal is 1 kHz, then the modulated carrier will swing between its maximum frequency and lowest frequency 1000 times per second. If f_m = 2 kHz, the rate of frequency swing would be twice as fast :

In short, we have established two important points about the nature of frequency modulation:

(i) The **amount** of frequency deviation (or shift or variation) depends on the **amplitude** (loudness) of the audio signal. ***Louder the sound, greater the frequency deviation and vice-versa.***

However, for the purposes of *FM* broadcasts, it has been internationally agreed to restrict maximum deviation to **75 kHz** on each side of the centre frequency for sounds of maximum loudness. Sounds of lesser loudness are permitted ***proportionately less frequency deviation.***

(ii) The **rate** of frequency deviation depends on the **signal frequency**.

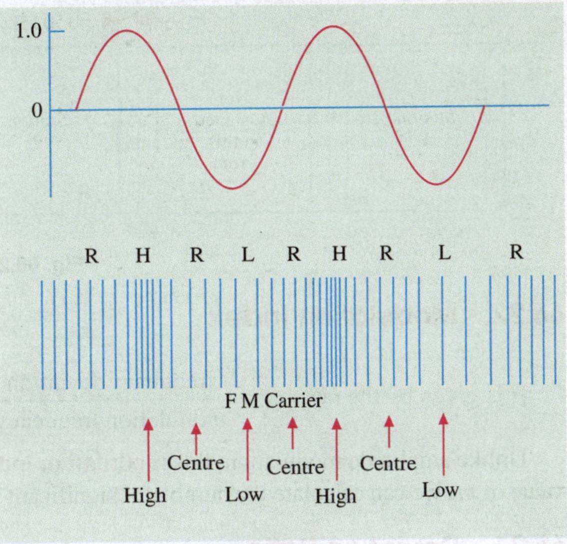

Fig. 66.19

66.21. Frequency Deviation and Carrier Swing

The frequency of an *FM* transmitter without signal input is called the **resting frequency** or **centre frequency** (f_0) and is the allotted frequency of the transmitter. In simple words, it is the carrier frequency on which a station is allowed to broadcast. When the signal is applied, the carrier frequency deviates up and down from its resting value f_0.

This change or shift either above or below the resting frequency is called frequency deviation (Δ_f)

The total variation in frequency from the lowest to the highest is called carrier swing (CS). Obviously,

carrier swing = 2 × frequency deviation of $CS = 2 \times \Delta_f$

A maximum frequency deviation of 75 kHz is allowed for commercial *FM* broadcast stations in the 88 to 168 MHz *VHF* band. Hence, *FM* channel width is 2 × 75 = 150 kHz. Allowing a 25 kHz guard band on either side, the channel width becomes = 2(75 + 25) = 200 kHz (Fig. 66.21). This guard band is meant to prevent interference between adjacent channels. However, a maximum frequency deviation of 25 kHz is allowed in the sound portion of the TV broadcast.

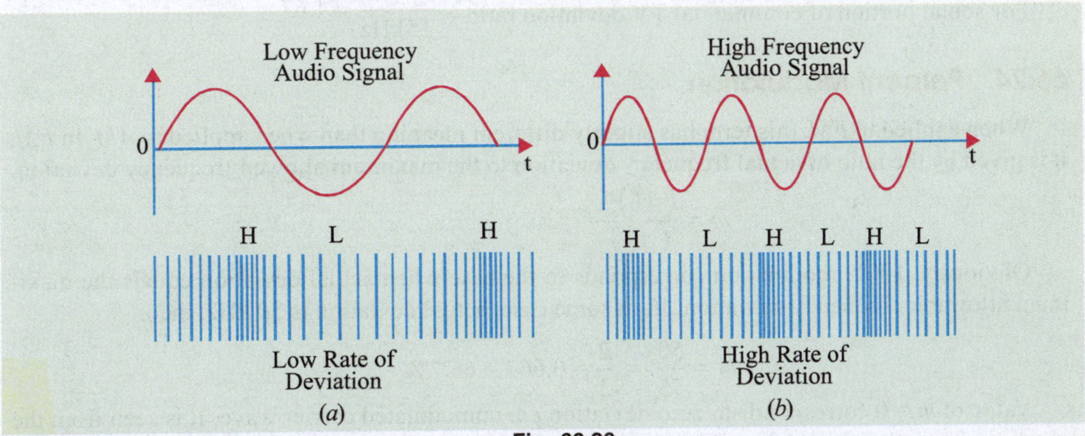

Fig. 66.20

In *FM*, the highest audio frequency transmitted is 15 kHz.

Consider an FM carrier of resting frequency 100 MHz. Since $(\Delta f)_{max}$ = 75 kHz, the carrier frequency can swing from the lowest value of 99.925 MHz to the highest value of 100.075 MHz. Of course, deviations lesser than 75 kHz corresponding to relatively softer sounds are always permissible.

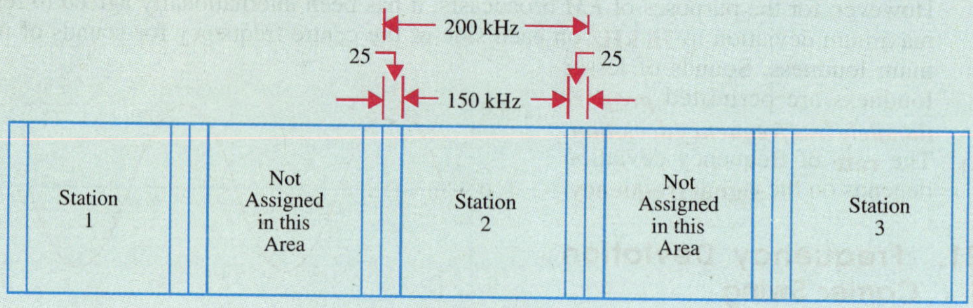

Fig. 66.21

66.22. Modulation Index

It is given by the ratio $m_f = \dfrac{\text{frequency deviation}}{\text{modulation frequency}} = \dfrac{\Delta f}{f_m}$

Unlike amplitude modulation, this **modulation index can be greater than unity**. By knowing the value of m_f, we can calculate the number of significant sidebands and the bandwidth of the *FM* signal.

66.23. Deviation Ratio

It is the worst-case modulation index in which maximum permitted frequency deviation and maximum permitted audio frequency are used.

$$\therefore \quad \text{deviation ratio} = \frac{(\Delta f)}{f_{m(max)}}$$

Now, for *FM* broadcast stations, $(\Delta f)_{max}$ = 75 kHz and maximum permitted frequency of modulating audio signal is 15 kHz.

$$\therefore \quad \text{deviation ratio} = \frac{75\,\text{kHz}}{15\,\text{kHz}} = 5$$

For sound portion of commercial TV deviation ratio $= \dfrac{25\,\text{kHz}}{15\,\text{kHz}} = 1.67$

66.24. Percent Modulation

When applied to *FM*, this term has slightly different meaning than when applied to *AM*. In *FM*, it is given by the ratio of actual frequency deviation to the maximum allowed frequency deviation.

$$m = \frac{(\Delta f)_{actual}}{(\Delta f)_{max}}$$

Obviously, 100% modulation corresponds to the case when actual deviation equals the maximum allowable frequency deviation. If, in some case, actual deviation is 50 kHz, then

$$m = \frac{50}{75} = \frac{2}{3} = 0.667 = \mathbf{66.7\%}$$

Value of $m = 0$ corresponds to zero deviation *i.e.* unmodulated carrier wave. It is seen from the above equation that $m \propto (\Delta f)_{actual}$. It means that when **frequency deviation (*i.e.* signal loudness) is doubled, modulation is doubled**.

Example 66.12. *What is the modulation index of an FM carrier having a carrier swing of 100 kHz and a modulating signal of 5 kHz ?*

Solution. $CS = 2 \times \Delta f$

$$\therefore \quad \Delta f = \frac{CS}{2} = \frac{100}{2} = \mathbf{50\ kHz} \quad \therefore \quad m_f = \frac{\Delta f}{f_m} = \frac{50}{5} = \mathbf{10}$$

Example. 66.13. *An FM transmission has a frequency deviation of 18.75 kHz. Calculate percent modulation if it is broadcast*

(i) in the 88-108 MHz band *(ii) as a portion of a TV broadcast.*

(Elect. and Comm. Engg., Madurai Kamaraj Univ. 1990)

Solution. *(i)* For this transmission band,

$$(\Delta f)_{max} = 75\ \text{kHz} \qquad \therefore \qquad m = \frac{18.75}{75} \times 100 = \mathbf{25\%}$$

(ii) In this case, $(\Delta f)_{max} = 25$ kHz $\qquad \therefore \qquad m = \frac{18.75}{25} \times 100 = \mathbf{75\%}$

Example 66.14. *An FM signal has a resting frequency of 105 MHz and highest frequency of 105.03 MHz when modulated by a signal of frequency 5 kHz. Determine*

(i) frequency deviation, *(ii) carrier swing,* *(iii) modulation index,*
(iv) percent modulation, *(v) lowest frequency reached by the FM wave.*

(Electronics and Comm. Engg., Osmania Univ. 1992)

Solution. *(i)* $\Delta f = 105.03 - 105 = 0.03$ MHz = **30 kHz**

(ii) $CS = 2 \times \Delta f = 2 \times 30 =$ **60 kHz** *(iii)* $m_f = \frac{30}{5} = \mathbf{6}$

(iv) $m = \frac{30}{75} \times 100 = \mathbf{40\%}$ *(v)* ***lowest frequency*** $= 105 - 0.03 =$ **104.97 MHz**

66.25. FM Sidebands

In *FM*, when a carrier is modulated, a number of sidebands are formed.* Though theoretically their number is infinite, their strength becomes negligible after a few sidebands. They lie on both sides of the centre frequency spaced f_m apart as shown in Fig.66.22. Sidebands at equal distances from f_0 have equal amplitudes. If f_0 is the centre frequency and f_m the frequency of the modulating signal, then *FM* carrier contains the following frequencies :

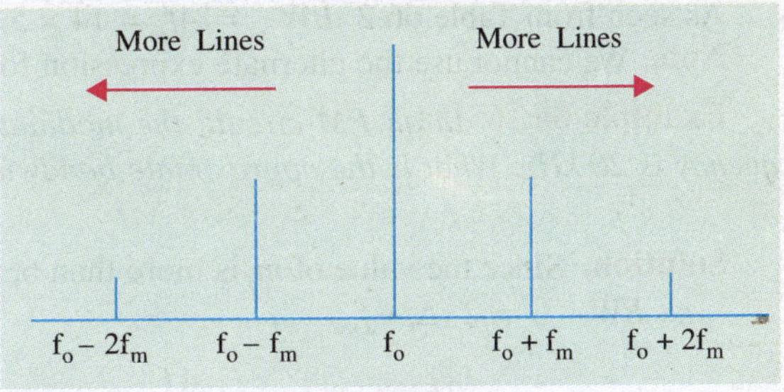

Fig. 66.22

(i) f_0 *(ii)* $f_0 \pm f_m$ *(iii)* $f_0 \pm 2f_m$ *(iv)* $f_0 \pm 3f_m$ and so on

The bandwidth occupied by the spectrum is $BW = 2nf_m$ where n is the highest order of the significant sideband.

Another approximate expression for spectrum bandwidth is $BW = 2(1 + m_f)f_m$

Now, $m_f = \frac{\Delta f}{f_m}$, hence $\qquad BW = 2(\Delta_f + f_m)$

This expression is based on the assumption that sidebands having amplitudes less than 5% of the unmodulated carrier wave are negligible or when m_f is at least 6.

66.26. Modulation Index and Number of Sidebands

It is found that the number of sidebands

1. depends ***directly*** on the amplitude of the modulating signal,
2. depends ***inversely*** on the frequency of the modulating signal.

---

* An AM signal has only two side frequencies for each modulating frequency (Art. 66.11)

Since frequency deviation is directly related to the amplitude of the modulating signal, the above two factors can be combined in one factor called **modulation index**.

Hence, number of sidebands depends on $m_f = \Delta f / f_m$

Obviously, the number of pairs of sidebands

(i) **increases** as frequency deviation (or amplitude of modulating signal) **increases**.

(ii) **increases** as the modulating signal frequency **decreases**.

Table No. 66.2

m_f	0.5	1	2	3	4	5	6	7
n	2	3	4	6	7	8	9	11
$BW = 2nf_m$	$4f_m$	$6f_m$	$8f_m$	$12f_m$	$14f_m$	$16f_m$	$18f_m$	$22f_m$
m_f	8	9	10	11	12	13	14	15
n	12	13	14	15	16	17	18	19
$BW = 2nf_m$	$24f_m$	$26f_m$	$28f_m$	$30f_m$	$32f_m$	$34f_m$	$36f_m$	$38f_m$

Table 66.2 summarizes the relation between the number (n) of sidebands on either side of the carrier and the corresponding bandwidth of the spectrum.

Example 66.15. *A 5 kHz audio signal is used to frequency-modulate a 100 MHz carrier causing a frequency deviation of 20 kHz. Determine*

(i) modulation index *(ii) bandwidth of the FM signal.*

Solution. *(i)* $m_f = \dfrac{\Delta f}{f_m} = \dfrac{20}{5} = \mathbf{4}$

As seen from Table 66.2 $BW = 14f_m = 14 \times 5 =$ **70 kHz**

Note. We cannot use the alternate expression for BW given in Art. 66.25 above because $m_f < 6$

Example 66.16. *In an FM circuit, the modulation index is 10 and the higest modulation frequency is 20 kHz. What is the approximate bandwidth of the resultant FM signal ?*

(Applied Electronics, Bombay Univ. 1990)

Solution. Since the value of m_f is more than 6, we will use the expression

$$BW = 2\,(\Delta_f + f_m)$$

Now, $m_f = \dfrac{\Delta f}{f_m}$ or $10 = \dfrac{\Delta f}{20}$ $\quad\therefore\quad \Delta_f = 200$ kHz

$\therefore\ BW = 2\,(200 + 20) =$ **440 kHz**

66.27. Mathematical Expression for FM Wave

The unmodulated carrier is given by $e_c = A \sin 2\pi f_0 t$

The modulating signal is given by $e_m = B \sin 2\pi f_m t$

The modulated carrier frequency f swings around the resting frequency f_0 thus

$$f = f_0 + \Delta f.\ \sin 2\pi f_m t$$

Hence, equation for the frequency-modulated wave becomes

$$e = A \sin 2\pi f t = A \sin [2\pi(f_0 + \Delta f.\ \sin 2\pi f_m t)t]$$

$$= A \sin \left(2\pi f_0 t + \frac{\Delta f}{f_m} + \cos 2\pi f_m t\right) = A \sin (2\pi f_0 t + m_f \cos 2\pi f_m t)$$

66.28. Demodulation or Detection

When the *RF* modulated waves, radiated out from the transmitter antenna, after travelling through space, strike the receiving aerials, they induce very weak *RF* currents and voltages in them. If these high-frequency currents are passed through headphones or loudspeakers, they produce no effect on them because all such sound-producing devices are unable to respond to such high frequencies due to large inertia of their vibrating discs etc. Neither will such *RF* currents produce any effect on human ear because their frequencies are much beyond the audible frequencies (20 to 20,000 Hz approximately). Hence, it is necessary to demodulate them first in order that the sound-producing devices may be actuated by audio-frequency current similar to that used for modulating the carrier wave at the broadcasting station.

This process of ***recovering AF signal from the modulated carrier wave is known as demodulation or detection.***

The demodulation of an *AM* wave involves two operations :

(i) rectification of the modulated wave and

(ii) elimination of the *RF* component of the modulated wave.

However, demodulation of an *FM* wave involves three operations *(i)* conversion of frequency changes produced by modulating signal into corresponding amplitude changes, *(ii)* rectification of the modulating signal and *(iii)* elimination of *RF* component of the modulated wave.

66.29. Essentials of AM Detection

For recovering the *AF* waveform from modulated wave (a mixture of *AF* wave and *RF* carrier), it is essential to find some way of reducing (or better, eliminating) one half of the modulated wave. The result of this elimination (or rectification) would be that the average value of the wave would not be zero because, now, the impulse would be all in one direction as shown in Fig. 66.23. If this new wave is now passed through a headphone shunted by a suitable capacitor, then *AF* wave will pass through the headphone whereas the *RF* wave will be by-passed by the capacitor (because the high inductance of magnet coils of the headphones will offer tremendous impendance to *RF* currents). Hence, two will become separated.

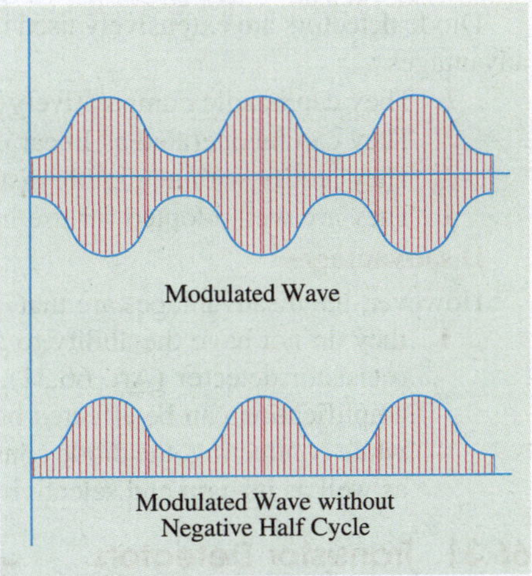

Fig. 66.23

66.30. Diode Detector for AM Signals

Diode detection is also known as **envelope-detection** or **linear detection**. In appearance, it looks like an ordinary half-wave rectifier circuit with capacitor input as shown in Fig. 66.24. It is called **envelope detection** because it recovers the *AF* signal envelope from the composite signal. Similarly, diode detector is called **linear detector** because its output ***is proportional to the voltage of the input signal*.***

Circuit Action

Of the various *RF* voltages induced in the receiver aerial, only those having the ***same*** frequency as the resonant frequency of *LC* circuit are tuned in due to electromagnetic induction between coils L_1 and *L*. By varying *C*, the resonant frequency of the *LC* circuit can be varied and hence *RF* signal of any desired frequency can be tuned in. This input signal is rectified by the diode and passed on to the low-pass filter RC_1.

* A detector circuit in which the rectified output is proportional to the square of the input signal voltage is called square-law or non-linear detector.

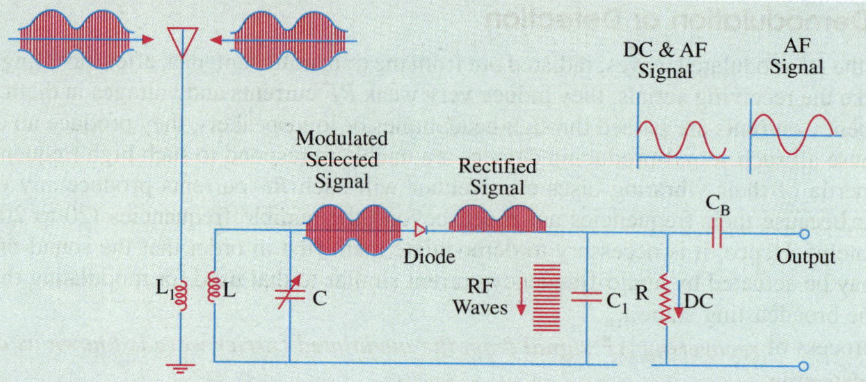

Fig. 66.24

The *RF* carrier wave is filtered out by low-reactance capacitor C_1 which is called *RF* filter capacitor or *RF* by-pass capacitor. The dc component of the remaining signal is shunted out through *R* because it cannot pass through blocking capacitor C_B. But the low, frequency *AF* signal can easily get through C_B and becomes available across the output. When passed through a suitable device, say, a headphone, the original sound can be heard.

Advantages

Diode detectors are extensively used in AM broadcast receivers because they have the following advantages :

1. They can handle comparatively large input signals;
2. They can be operated as linear or power detectors;
3. They rectify with negligible distortion and, hence, have good linearity;
4. They are well-adopted for use in simple automatic-gain control circuits.

Disadvantages

However, the disadvantages are that

1. they do not have the ability to amplify the rectified signal by themselves as is done by a transistor detector (Art. 66.31). However, it is not a very serious drawback since signal amplification can be affected both before and after rectification;
2. while conducting, the diode consumes some power which reduces the *Q* of its tuned circuit as well as its gain and selectivity.

66.31. Transistor Detectors for AM Signals

Transistors can be used as detector amplifiers *i.e.* both for rectification and amplification. As shown in Fig. 66.25, the *RF* signal is applied at the base-emitter junction where rectification takes place. The amplification of the recovered signal takes place in the emitter-collector circuit.

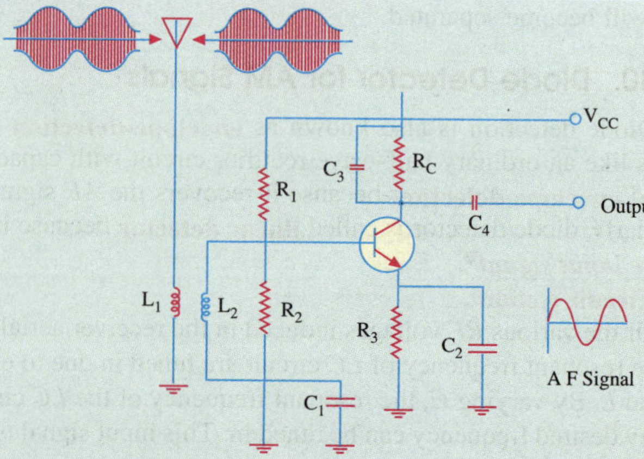

Fig. 66.25

It should be noted that when used as a detector, the transistor has to be biased at cut-off or operated as class-B. As is usual, C_2 ac grounds R_3 whereas voltage divider $R_1 — R_2$ establishes proper bias which holds the transistor just at cut-off.

Obviously, only positive half-cycles of the modulated *RF* input signal will drive the transistor into conduction, whereas negative half-cycles would be cut-off. Thus, rectification of the input signal takes place in the base-emitter circuit. The small variations in base current produced by the positive half-cycle of the input signal produce large (β time) variations in the collector current. Hence, an amplified version of the rectified signal appears across $R_C - C_3$ combination. C_3 eliminates *RF* component so that only *AF* signal voltage drops across R_C. The capacitor C_4 permits *AF* signal to pass through but blocks its dc component.

66.32. FM Detection

As discussed earlier, an *FM* carrier signal contains information (or intelligence we wish to convey) in the form of frequency variations above and below the centre frequency of the carrier. For recovering the information, we must first convert the *FM* signal in such a way that it appears as a modulated *RF* voltage across the diode. A simple method of converting frequency variations into voltage variations is to make use of the principle that reactance (of coil or capacitor) varies with frequency. When an *FM* signal is applied to an inductor, the current flowing through it varies in amplitude according to the changes in frequency of the applied signal. Now, changes in frequency of the *FM* signal depend on the amplitude of the modulating *AF* signal. Hence, the current in the inductor varies as per the amplitude of the original modulating signal. In this way, frequency changes in *FM* signal are converted into amplitude changes in current. These changes in current when passed through a resistor produce corresponding changes in voltage.

Hence, we find that, ultimately, frequency variations in *FM* signal are converted into voltage changes. Also, there exists a linear relation between the two – something essential for distortion-less demodulation.

FM demodulation may be carried out with the help of (*i*) ratio detector and (*ii*) quadrature detector.

66.33. Quadrature Detector

This detector depends on the frequency/phase relationship of a tuned circuit. It uses only one tuned circuit and is becoming increasingly popular in the integrated *FM* strips.

Theory

Let us first consider the general principle. A sinusoidal current is given by the equation

$$i = I_m \sin \theta = I_m \sin \omega t$$

Suppose, it flows through a circuit shown in Fig. 66.27 (*a*). The voltage V_L across the inductor (assumed pure) leads the current *I* by 90°

$$\therefore \quad V_L = V_L \cos \omega t$$

The voltage V_z across the parallel tuned circuit will be in phase with *I* at resonance. However, at frequencies slightly different (±1%) from the resonant frequency, the phase angle *f* will be given by

$$\tan \phi = \frac{yQR}{R} = y\,Q \quad \text{assuming } f_0/f \cong 1. \qquad \text{— Fig.66.26}$$

Here $\dfrac{\omega}{\omega_0} - \dfrac{\omega_0}{\omega} = \dfrac{f}{f_0} - \dfrac{f_0}{f} \cong \dfrac{2\Delta_f}{f_0}$

where $\Delta f = f - f_0$

f_0 = resonant frequency

f = slightly off-resonance frequency

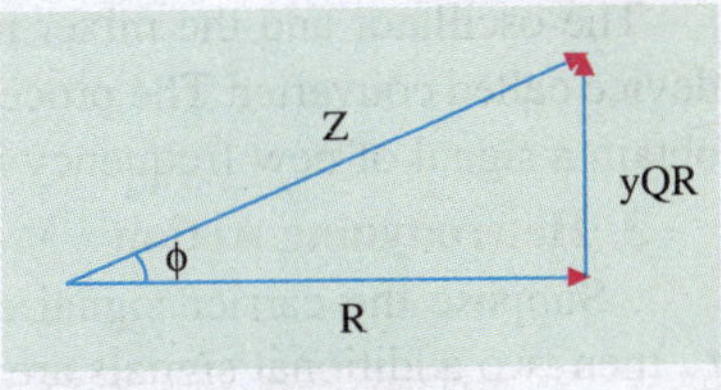

Fig. 66.26

Hence, equation for V_z is given by $V_z = V_z \sin(\omega t - \phi)$

When the two voltages V_L and V_z are applied as inputs to a multiplier, the output voltage V_0 is found to be proportional to their product as shown in Fig. 66.27(*b*).

$$V_0 \propto V_L . V_z \propto \cos \omega t . \sin (\omega t - \phi)$$
$$\propto \sin (2\omega t - \phi) - \sin \phi$$

A low-pass filter is used to reject the double frequency component sin $(2\omega t - \phi)$ and select only the low-frequency sin ϕ component.

$\therefore \quad V_0 \propto \sin\phi \propto \tan\phi$ — since ϕ is very small

$\propto yQ \propto y \propto 2.\dfrac{\Delta f}{f_0}$ — where V_m is the modulating AF voltage.

Now, $\Delta_f = f - f_0 \propto k\, V_m$

$\therefore \quad V_0 \propto V_m$

It shows that output voltage V_0 is proportional to the original modulating signal voltage.

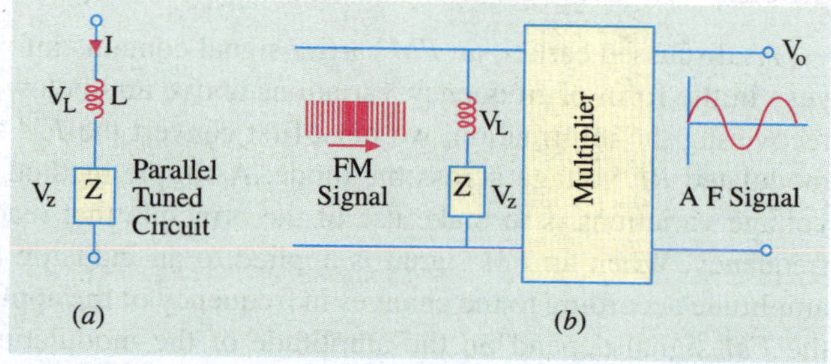

Fig. 66.27

66.34. Frequency Conversion

Let us first consider its necessity for radio broadcasting purposes and the basic principle of heterodyning action.

1. Need

It is very difficult to design amplifiers which give uniformly high gain over a wide range of radio frequencies used in commercial broadcast stations. However, it is possible to design amplifiers which can provide high-gain uniform amplification over a narrow band of comparatively lower frequencies called intermediate frequencies (*IF*). Hence, it is necessary to convert the modulated *RF* carrier into modulated *IF* carrier by using a frequency converter.

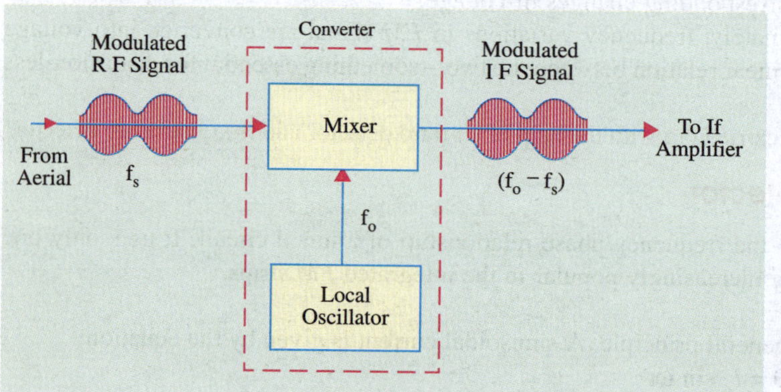

Fig. 66.28

This *IF* signal is then amplified by narrow-band *IF* amplifiers and passed on to the *AM* detectors.

2. Basic Principle

The frequency conversion can be achieved by utilizing the heterodyne principle. For this purpose, the modulated *RF* signal is mixed (in a mixer) with an unmodulated *RF* signal produced by local oscillator as shown in Fig. 66.28.

The oscillator and the mixer may be either two separate devices or may be combined into one device called converter. The process of combining two ac signals of different frequencies in order to obtain a signal of new frequency is called heterodyning action.

3. Heterodyning Action

Suppose the carrier signal of frequency f_s is heterodyned with another signal of frequency f_0, then two additional signals are produced whose frequencies are :

(i) $f_0 + f_s$ —the sum component   *(ii)* $f_0 - f_s$ — the difference component

Usually, the sum frequency is removed by bandpass filtering. The difference frequency (also called ***beat frequency***) is retained and forms the *IF* frequency in *AM* receivers.

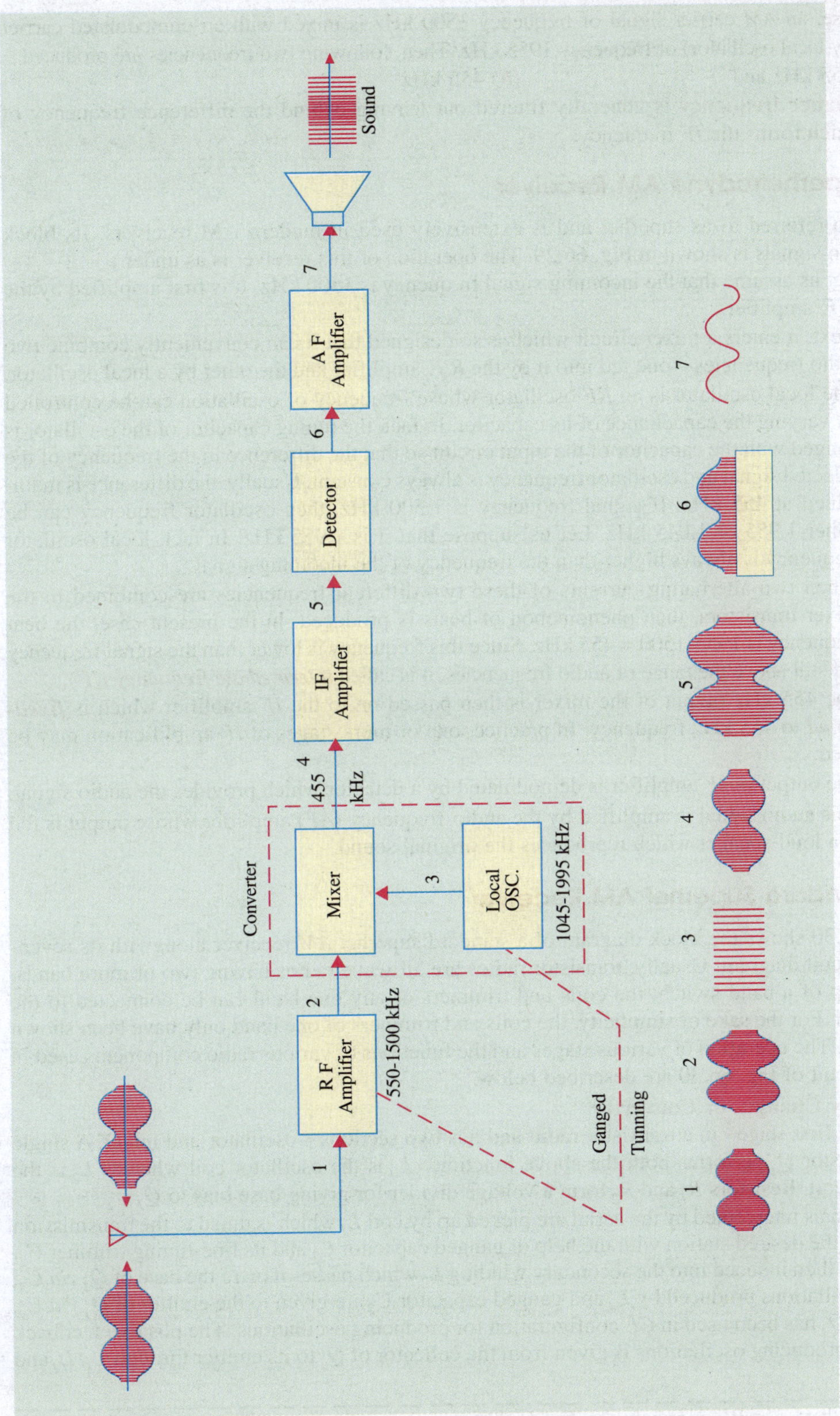
R F Amplifier
550-1500 kHz
Ganged Tuning
Converter
Mixer
Local OSC.
1045-1995 kHz
455 kHz
IF Amplifier
Detector
A F Amplifier
Sound
1
2
3
4
5
6
7

Fig. 66.29

Suppose, an *AM* carrier signal of frequency 1500 kHz is mixed with an unmodulated carrier (produced by local oscillator) of frequency 1955 kHz. Then, following two frequencies are produced

(*i*) 3455 kHz and (*ii*) 455 kHz

The **higher frequency** is generally filtered out leaving behind the difference frequency of 455 kHz which forms the *IF* frequency.

66.35. Superhetrodyne AM Receiver

It is also referred to as superhet and is extensively used in modern AM receivers. Its block diagram with signals is shown in Fig. 66.29. The operation of this receiver is as under :

(*i*) Let us assume that the incoming signal frequency is 1500 kHz. It is first amplified by the *R.F.* amplifier.

(*ii*) Next, it enters a mixer circuit which is so designed that it can conveniently combine two radio frequencies—one fed into it by the *R.F.* amplifier and the other by a local oscillator.

(*iii*) The local oscillator is an *RF* oscillator whose frequency of oscillation can be controlled by varying the capacitance of its capacitor. In fact, the tuning capacitor of the oscillator is ganged with the capacitor of the input circuit so that the difference in the frequency of the selected signal and oscillator frequency is always constant. Usually, the difference is maintained at 455 kHz. If signal frequency is 1,500 kHz, then oscillator frequency can be either 1,955 or 1,045 kHz. Let us suppose that it is 1,955 kHz. In fact, local oscillator frequency is always higher than the frequency of the incoming signal.

(*iv*) When two alternating currents of these two different frequencies are combined in the mixer transistors, then phenomenon of beats is produced. In the present case, the beat frequency is 1955-1500 = 455 kHz. Since this frequency is lower than the signal frequency but still above the range of audio frequencies, it is called ***intermediate frequency (IF).***

(*v*) The 455 kHz output of the mixer is then passed on to the *IF* amplifier which is ***fixed-tuned*** to 455 kHz frequency. In practice, one or more stages of *IF* amplification may be used.

(*vi*) The output of *IF* amplifier is demodulated by a detector which provides the audio signal.

(*vii*) This audio signal is amplified by the audio-frequency (*AF*) amplifier whose output is fed to a loud-speaker which reproduces the original sound.

66.36. Standard Superhet AM Receiver

Fig. 66.30 shows the block diagram of a standard superhet *AM* receiver alongwith its seven-transistor circuit diagram. Usually, transistor radios are all-wave radios having two or more bands. With the help of a band switch, the coils and trimmers of any one band can be connected to the circuit at will. For the sake of simplicity, the coils and trimmers of one band only have been shown in Fig. 66.30. The operation of various stages and the functions of various radio components used in the radio circuit of Fig. 66.30 are described below.

1. Frequency Changer or Converter

It is the first stage* in a transistor radio and has two sections—oscillator and mixer. A single silicon transistor Q_1 performs both the above functions. L_4 is the oscillator coil whereas L_2 is the signal input coil. Resistors R_1 and R_2 form a voltage divider for giving base bias to Q_1

The signals intercepted by the aerial are picked up by coil L_1 which is tuned to the transmission frequency of the desired station with the help of ganged capacitor C_1 and its fine-tuning trimmer C_2. This signal is then induced into the secondary winding L_2 which passes it on to the base of Q_1 *via* C_3. The local oscillations produced by L_4 and ganged capacitor C_7 are given to the emitter of Q_1 *via* C_4.

As seen, Q_1 has been used in *CE* configuration for producing oscillations. The positive feedback required for producing oscillations is given from the collector of Q_1 to its emitter through L_3, L_4 and C_4.

* Sometimes, there is an RF stage before it (Fig. 66.29)

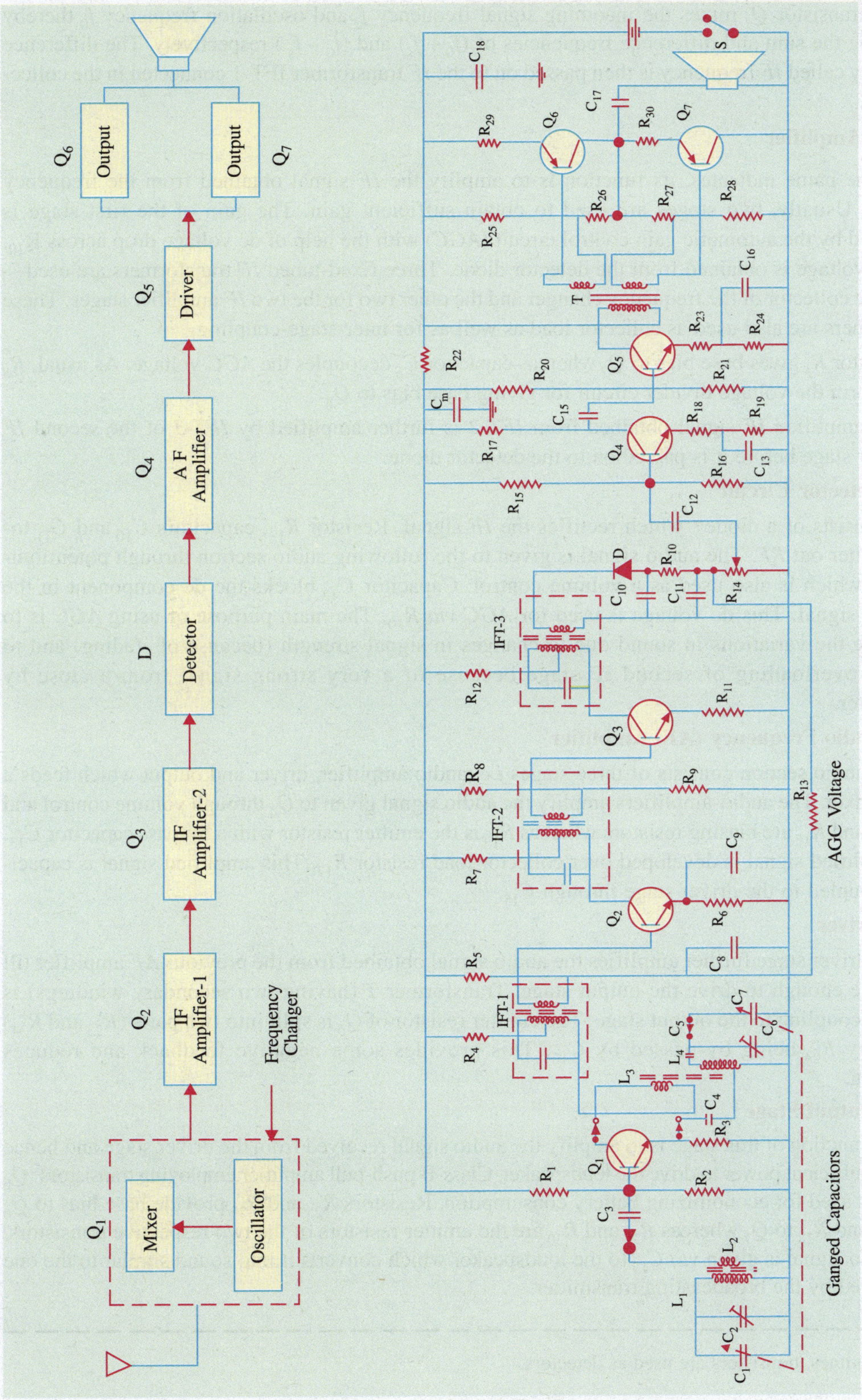
Q1
Mixer
Oscillator
Frequency Changer
Q2
IF Amplifier-1
Q3
IF Amplifier-2
D
Detector
Q4
A F Amplifier
Q5
Driver
Q6
Output
Q7
Output
IFT-1
IFT-2
IFT-3
T
S
AGC Voltage
Ganged Capacitors

Fig. 66.30

The transistor Q_1 mixes the incoming signal frequency f_s and oscillation frequency f_0 thereby producing the sum and difference frequencies of $(f_0 + f_s)$ and $(f_0 - f_s)$ respectively. The difference frequency called *IF* frequency is then passed on to the *IF* transformer IFT-1 connected in the collector of Q_1.

2. IF Amplifier

As the name indicates, its function is to amplify the *IF* signal obtained from the frequency changer. Usually, two stages are used to obtain sufficient gain. The gain of the first stage is controlled by the automatic gain control circuit (*AGC*) with the help of dc voltage drop across R_{10}. This dc voltage is obtained from the detector diode. Three fixed-tuned *IF* transformers are used—one at the collector of the frequency changer and the other two for the two *IF* amplifier stages. These transformers are also used as collector load as well as for inter-stage-coupling.

Resistor R_5 gives base bias to Q_2 whereas capacitor C_8 decouples the *AGC* voltage. As usual, R_8 and R_9 form the voltage divider circuit for giving base bias to Q_3.

The amplified *IF* signal obtained from *IFT-2* is further amplified by *IFT-3* of the second *IF* amplifier stage before it is passed on to the detector diode.

3. Detector Circuit

It consists of a diode* which rectifies the *IF* signal. Resistor R_{13}, capacitors C_{10} and C_{11} together filter out *RF*. The audio signal is given to the following audio section through potentiometer R_{14} which is also used as a volume control. Capacitor C_{12} blocks the dc component in the rectified signal. This dc voltage is used for *AGC via* R_{10}. The main purpose of using *AGC* is to minimize the variations in sound due to changes in signal strength (because of fading) and to prevent overloading of second IF stage because of a very strong signal from a close-by transmitter.

4. Audio Frequency (AF) Amplifier

The audio section consists of three stages *i.e.* audio amplifier, driver and output which feeds a loud speaker. The audio-amplifiers amplify the audio signal given to Q_4 through volume control and C_{12}, R_{15} and R_{16} are biasing resistors whereas R_{19} is the emitter resistor with a by-pass capacitor C_{13}. The amplified signal is developed over collector load resistor R_{17}. This amplified signal is capacitively-coupled to the driver stage through C_{15}.

5. Driver

The driver stage further amplifies the audio signal obtained from the previous *AF* amplifier till it is large enough to drive the output stage. Transformer *T* (having two secondary windings) is used for coupling to the output stage. The emitter resistor of Q_5 is split into two parts (R_{23} and R_{24}) with only R_{24} being by-passed by C_{16}. This provides some negative feedback and reduces distortion.

6. Output Stage

The function of this stage is to amplify the audio signal received from the driver stage and hence provide sufficient power to drive the loudspeaker. Class-B push-pull amplifier employing transistors Q_6 and Q_7 is used for economizing battery consumption. Resistors R_{25} and R_{26} provide base bias to Q_6 and R_{27} and R_{28} to Q_7 whereas R_{29} and R_{30} are the emitter resistors of the two respective transistors. The audio signal is given *via* C_{17} to the loudspeaker which converts it into sound similar to the one transmitted by the broadcasting transmitter.

---

* Sometimes, transistors are used as detectors.

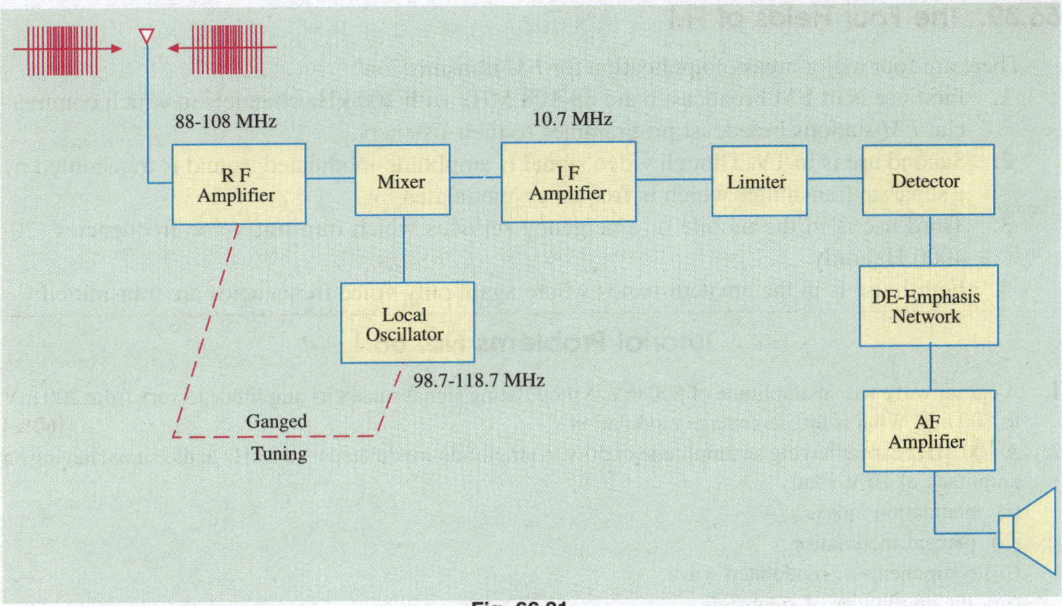

Fig. 66.31

66.37. FM Receiver

The block diagram for such a receiver is shown in Fig. 66.31. These receivers operate in the *VHF* band of 88-108 MHz and have an *IF* of 10.7 MHz with a bandpass of 2×75 kHz = 150 kHz.

Like *AM* receivers, such receivers also have a frequency converter and *IF* amplifiers.

The function of the **limiter** is to remove all ***amplitude variations*** (caused by noise) from *IF* signal which might have crept into the *FM* signal. This removal of amplitude variations is necessary for distortionless demodulation. **Limiter is a sort of clipping circuit**.

The de-emphasis network ***reduces the amplitude of high frequencies in the audio signal*** which was earlier increased by the pre-emphasis network at the transmitting station. It serves to re-establish the **tonal balance** of the speech or music etc. lost in the pre-emphasis. Without it, the sound signal would have a heavy treble effect. It is, in fact, a low-pass filter having a time constant of 75 µs.

66.38. Comparison Between AM and FM

Frequency modulation ***(FM)*** has the following advantages as compared to amplitude modulation ***(AM)*** :

1. All transmitted power in ***FM*** is useful whereas in ***AM*** most of it is in carrier which ***serves no useful purpose.***
2. It has high signal-to-noise (S/N) ratio. It is due to two reasons : firstly, there happens to be less noise at ***VHF*** band and secondly, ***FM*** receivers are fitted with amplitude limiters which remove amplitude variations caused by noise.
3. Due to 'guard-band' there is hardly any adjacent-channel interference.
4. Since only transmitter ***frequency*** is modulated in ***FM***, only fraction of a watt of audio power is required to produce 100% modulation as compared to high power required in *AM*.

However, FM has the following disadvantages :

1. It requires **much wider channel** — almost 7 to 15 times as large as needed by AM.
2. It requires complex and expensive transmitting and receiving equipment.
3. Since FM reception is limited to only line of sight, area of reception for FM is much smaller than for *AM*.

66.39. The Four Fields of FM

There are four major areas of application for *FM* transmission :

1. First use is in **FM** broadcast band 88-108 MHz with 200 kHz channels in which commercial *FM* stations broadcast programmes to their listeners.
2. Second use is in TV. Though video signal is amplitude-modulated, sound is transmitted by a separate transmitter which is frequency-modulated.
3. Third use is in the mobile or emergency services which transmit voice frequencies (20-4000 Hz) only.
4. Fourth use is in the amateur bands where again only voice frequencies are transmitted.

Tutorial Problems No. 66.1

1. A carrier wave has an amplitude of 500 mV. A modulating signal causes its amplitude to vary from 200 mV to 800 mV. What is the percentage modulation? **[60%]**
2. A 100.MHz carrier having an amplitude of 50 V is amplitude-modulated by a 5 kHz audio signal having an amplitude of 20 V. Find
   (*i*) modulation index,
   (*ii*) percent modulation,
   (*iii*) components of modulated wave,
   (*iv*) the amplitudes of sidebands.
   [(*i*) 0.4 (*ii*) 40 (*iii*) 100 MHz, 100.005 MHz, 99.995 MHz (*iv*) 10 V]
3. An audio signal given by 30 sin $(2\pi \times 2500t)$ is used for modulating a carrier wave given by the equation 60 sin $(2\pi \times 200{,}000t)$. Find
   (*i*) percent modulation,
   (*ii*) frequencies of the signal and the carrier,
   (*iii*) frequency spectrum of the modulated wave.
   [(*i*) 50 (*ii*) 2500 Hz , 200,000 Hz (*iii*) 200; 202.5; 197.5 kHz)]
4. How many AM broadcast stations can be accommodated in a 5 MHz bandwidth if each station transmits a signal modulated by an audio signal having a maximum frequency of 5 kHz. **[500]**
5. A 100 kHz bandwidth is to accomodate 5 AM broadcasts simultaneously. What is the maximum modulating frequency permissible for each station? **[10 kHz]**
6. Total power content of an AM signal is 3000 W. For 100 percent modulation, calculate
   (*i*) carrier power (*ii*) power in each sideband **[(*i*) 2 kW (*ii*) 0.5 kW]**
7. The total power content of an AM wave is 1320 W. What is the percent modulation if each sideband contains 160 W ? **[0.8]**
8. An SSB-SC signal contains 2 kW. Find power contained in
   (*i*) carrier (*ii*) sideband.
   [(*i*) 0 (*ii*) 2 kW]
9. As compared to DSB-FC, Calculate the power saving in SSB-SC when percent modulation is (*i*) 1 (*ii*) 0.8 (*iii*) 0.5
   [(*i*) 83.3% (*ii*) 87.87% (*iii*) 94.4%]
10. What is the modulation index of an FM carrier having a carrier swing of 120 kHz and a modulating signal of 10 kHz. **[6]**
11. An FM signal has a resting frequency of 100 MHz and the highest frequency of 100.05 MHz when modulated by an audio signal of 5 kHz. Determine
    (*i*) frequency deviation , (*ii*) carrier swing,
    (*iii*) modulation index, (*iv*) percent modulation.
    [(*i*) 50 kHz (*ii*) 100 kHz (*iii*) 10 (*iv*) 66.7%]
12. A 10 kHz audio signal is used to frequency-modulate a 100 MHz carrier causing a frequency deviation of 50 kHz. Determine
    (*i*) modulation index, (*ii*) bandwidth of FM signal. **[(*i*) 5 (*ii*) 160 kHz)]**

13. A carrier of 1MHz with 400 Watt of its power is amplitude modulated with a sinusoidal signal of 2500 Hz. The depth of modulation is 75%. Calculate the side band frequencies, the band width, the power in the side bands and the total power in the modulated wave.
[(*a*) 1002.5 kHz, 997.5 kHz (*b*) 5000 kHz]
[(*Electronics Engg; Bangalore Univ. 2004*)]

14. The total power content of an AM wave is 2.64 kW at a modulation factor of 80%. Determine the power content of (*i*) carrier (*ii*) each side band.
(*i*) 2000W (*ii*) 320 W] **[(*Electronics Engg. Bangalore Univ. 2002*)]**

15. A 100MHz carrier wave is frequency modulated by a 10 kHz sinusoidal modulating signal. If the maximum frequency deviation is 50 kHz. Find the modulation index.
[5] **[(*Electronics Engg. Bangalore Univ., 2002*)]**

16. A 500 W, 1MHz carrier is amplitude modulated with a sinusoidal signal of 1 kHz. The depth of modulation is 60%. Calculate the band width, power in the side bands and the total power transmitted.
[(*i*) 2kHz (*ii*) 38.14 (*iii*) 576.27 W] **[(*Electronics Engg., Bangalore Univ., 2003*)]**

17. A carrier signal has a peak amplitude of 100V. Modulation index is 40%. Power is developed across a load of 100 Ω. Determine the value of transmitted power. [108 W]
(*Electronics Engg. Bangalore Univ., 2003*)

OBJECTIVE TESTS – 66

1. The main purpose of modulation is to
(*a*) combine two waves of different frequencies
(*b*) achieve wave-shaping of the carrier wave
(*c*) transmit low-frequency information over long distances efficiently
(*d*) produce sidebands.

2. Demodulation
(*a*) is performed at the transmitting station
(*b*) removes side-bands
(*c*) rectifies modulated signal
(*d*) is opposite of modulation.

3. In amplitude modulation
(*a*) carrier frequency is changed
(*b*) carrier amplitude is changed
(c) three sidebands are produced
(d) fidelity is improved.

4. 100% modulation is produced in AM when carrier
(*a*) frequency equals signal frequency
(*b*) frequency exceeds signal frequency
(*c*) amplitude equals signal amplitude
(*d*) amplitude exceeds signal amplitude.

5. For a given carrier wave, maximum undistorted power is transmitted when value of modulation is
(*a*) 1 (*b*) 0.8
(*c*) 0.5 (*d*) 0.

6. In an AM wave with 100 percent modulation, each sideband carries — of the total transmitted power.
(*a*) one-half (*b*) one-sixth
(*c*) one-third (*d*) two-third

7. Given a carrier frequency of 100 kHz and a modulating frequency of 5 kHz, the bandwidth of AM transmission is____ kHz.
(*a*) 5 (*b*) 200
(*c*) 10 (*d*) 20

8. When modulation of an AM wave is decreased
(*a*) percentage carrier power is decreased
(*b*) percentage carrier power is increased
(*c*) total transmitted power is increased
(*d*) percentage sideband power is unaffected.

9. In AM transmission, power content of the carrier is maximum when m equals
(*a*) 0 (*b*) 1
(*c*) 0.8 (*d*) 0.5.

10. In AM transmission with m = 1, suppression of carrier cuts power dissipation by a factor of
(*a*) 6 (*b*) 2
(*c*) 3 (*d*) 4

11. As compared to DSB-FC 100% modulated transmission, power saving in SSB-SC system is ________ per cent.
(*a*) 94.4 (*b*) 50
(*c*) 100 (*d*) 83.3

12. In an AM transmission with 100% modulation, 66.7% of power is saved when ___________ is/are suppressed.
(*a*) carrier (*b*) carrier and USB
(*c*) carrier and LSB (*d*) USB and LSB

13. In an AM system, full information can be conveyed by transmitting only
(*a*) the carrier (*b*) the upper sideband
(*c*) the lower sideband (*d*) any one sideband.

14. A 10 MHz carrier of peak value 10 V is amplifide modulated by a 10 kHz signal of amplitude 6 V. The amplitude of each side band frequency is
(*a*) 3 V (*b*) 4 V
(*c*) 5 V (*d*) 6 V

15. The block diagram of an AM receiver is shown in Fig. 66.32. The missing block and its location are respectively

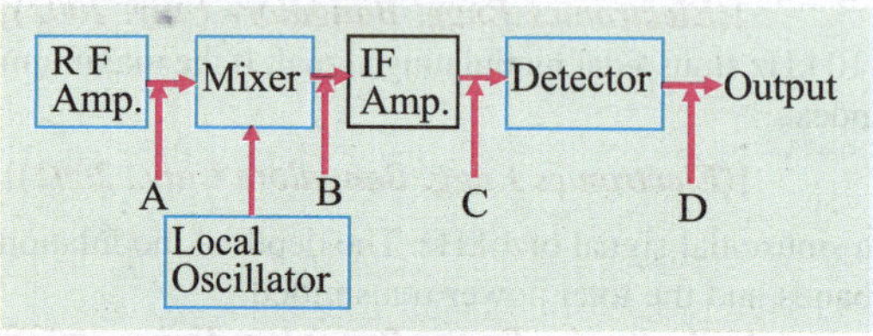

Fig. 66.32

(*a*) Band - pass filter at A
(*b*) High-pass filter at B
(*c*) Band - pass filter at C
(*d*) Low-pass filter (audio amplifier) at D.

16. In FM transmission, amplitude of the modulating signal determines
(*a*) rate of frequency variations
(*b*) amount of frequency shift
(*c*) total balance of transmission
(*d*) distance of broadcast.

17. In FM, when frequency deviation is doubled
(*a*) modulation is doubled
(*b*) modulation is halved
(*c*) carrier swing is halved
(*d*) modulation index is decreased.

18. In an FM broadcast in VHF band, channel width is ______
(*a*) 75 (*b*) 25
(*c*) 88 (*d*) 200

19. Modern FM receivers use de-emphasis circuit for
(*a*) reducing high frequency noise
(*b*) reducing the amplitude of high frequencies in the audio signal
(*c*) increasing the amplitude of higher modulating frequencies
(*d*) making demodulation easy

20. One of the serious disadvantages of FM transmission is its
(*a*) high static noise
(*b*) limited line-of-sight range
(*c*) expensive equipment
(*d*) adjacent channel interference.

21. In India, the maximum frequency deviation is commercial FM broadcasting signals is about,
(*a*) 5 kHz (*b*) 15 kHz
(*c*) 7.5 kHz (*d*) 200 kHz

22. The correct sequence of subsystems in an FM receiver is
(*a*) mixer, RF amplifier, limiter, IF amplifier, discriminator, audio amplifier
(*b*) RF amplifier, mixer, IF amplifier, limiter, discriminator, audio amplifier
(*c*) RF amplifier, mixer, limiter, discriminator, IF amplifier, audio amplifier
(*d*) mixer, IF amplifier, limiter, audio amplifier, discriminator.

ANSWERS

1. (*c*)	**2.** (*d*)	**3.** (*b*)	**4.** (*c*)	**5.** (*a*)	**6.** (*b*)	**7.** (*c*)	**8.** (*b*)	**9.** (*a*)	**10.** (*c*)	**11.** (*d*)
12. (*a*)	**13.** (*d*)	**14.** (*a*)	**15.** (*d*)	**16.** (*b*)	**17.** (*a*)	**18.** (*d*)	**19.** (*b*)	**20.** (*b*)	**21.** (*d*)	**22.** (*b*)

CHAPTER 67

INTEGRATED CIRCUITS

Learning Objectives

- What is an Integrated Circuit ?
- Advantages of ICs
- Drawbacks of ICs
- Scale of Integration
- Classification of ICs by Structure
- Comparison between Different ICs
- Classification of ICs by Function
- Linear Integrated Circuits (LICs)
- Manufacturer's Designation of LICs
- Digital Integrated Circuits
- IC Terminology
- Semiconductors Used in Fabrication of ICs and Devices
- How ICs are Made?
- Material Preparation
- Crystal Growing and Wafer Preparation
- Wafer Fabrication
- Oxidation
- Etching
- Diffusion
- Ion Implantation
- Photomask Generation
- Photolithography
- Epitaxy
- Metallization and Interconnections
- Testing, Bonding and Packaging
- Semiconductor Devices and Integrated Circuit Formation
- Popular Applications of ICs

Jack Kilby would justly be considered one of the greatest electrical engineers of all time for one invention; the monolithic integrated circuit, or microchip. He went on to develop the first industrial, commercial and military applications for this integrated circuits- including the first pocket calculator (pocketronic) and computer that used them

67.1. Introduction

Electronic circuitry has undergone tremendous changes since the invention of a triode by Lee De Forest in 1907. In those days, the active components (like triode) and passive components (like resistors, inductors and capacitors etc.) of the circuits were **separate and distinct units** connected by soldered leads.

With the invention of the transistor in 1948 by W.H. Brattain and I. Bardeen, the electronic circuits became considerably reduced in size. It was due to the fact that a transistor was not only cheaper, more reliable and less power consuming but was also much smaller in size than an electron tube. To take advantage of small transistor size, the passive components too were greatly reduced in size thereby making ***the entire circuit very small.*** Development of printed circuit boards (*PCBs*) further reduced the size of electronic equipment by eliminating bulky wiring and tie points.

In the early 1960s, a new field of **microelectronics** was born primarily to meet the requirements of the Military which wanted to reduce the size of its electronic equipment to approximately one-tenth of its then existing volume. This drive for extreme reduction in the size of electronic circuits has led to the development of microelectronic circuits called **integrated circuits (ICs)** which are so small that their actual construction is done by technicians using high powered microscopes.

67.2. What is an Integrated Circuit ?

To put it very briefly, an integrated circuit (*IC*) is **just a packaged electronic circuit.**

A more detailed definition is as under :

An *IC* is a complete electronic circuit in which both the active and passive components are fabricated on a tiny single chip of silicon.

Active components are those which have the ability to produce gain. Examples are : transistors and *FETs*.

Passive components or devices are those which do not have this ability. Examples are : resistors, capacitors and inductors.

ICs are produced by the same processes as are used for manufacturing individual transistors and diodes etc. In such circuits, different components are isolated from each other by isolation diffusion within the crystal chip and are interconnected by an aluminium layer that serves as wires.

A discrete circuit, on the other hand, is one ***that is built by connecting separate components.*** In this case, each component is produced separately and then all are assembled together to make the electronic circuit.

J.S. Kilby of Texas Instruments was the first person to develop (in 1959) an integrated circuit — a single monolithic silicon chip in which active and passive components were fabricated by successive deposition, etching and diffusions. He was soon followed by Robert Noyce of Fairchild who successfully fabricated a complete *IC* including the interconnections on a single silicon chip. Since then the evolution of this technology is fast-paced.

67.3. Advantages of ICs

As compared to **standard printed circuits** which use discrete components, *ICs* have the following advantages :

1. Extremely small physical size

Often the size is thousands of times smaller than a discrete circuit. The various components and their interconnections are distinguishable only under a powerful microscope.

2. Very small weight

Since many circuit functions can be packed into a small space, complex electronic equipment can be employed in many applications where weight and space are critical, such as in aircraft or space-vehicles.

3. Reduced cost

It is a major advantage of *ICs*. The reduction in cost per unit is due to the fact that many identical circuits can be built simultaneously on a single wafer—this process is called **batch fabrication.** Although the processing steps for the wafer are complex and expensive, the large number of resulting integrated circuits make the ultimate cost of each *IC* fairly low.

4. Extremely high reliability

It is perhaps the **most important** advantage of an *IC* and is due to many factors. Most significant factor is the absence of soldered connections. Another is the need for fewer interconnections—the major cause of circuit failures. Small temperature rise due to low power consumptions of *ICs* also improves their reliability. In fact, an *IC* logic gate has been found to be 100,000 times more reliable than a vacuum tube logic gate and 100 times more reliable than a transistor logic gate.

Obviously, higher reliability means that *ICs* will work for longer periods without giving any trouble—something most desirable from both military and consumer application point of view.

5. Increased response time and speed

Since various components of an *IC* are located close to each other ***in*** or ***on*** a silicon wafer, the time delay of signals is reduced. Moreover, because of the short distances, the chance of stray electrical pickup (called parasitic capacitance) is practically nil. Hence it makes them very suitable for small signal operation and high frequency operation. As a result, the response time or the operating speed of the system is improved.

6. Low power consumption

Because of their small size, *ICs* are more suitable for low power operation than bulky discrete circuits.

7. Easy replacement

ICs are hardly ever repaired because in case of failure, it is more economical to replace them than to repair them.

8. Higher yield

The ***yield*** is the percentage of usable devices. Because of the batch fabrication, the yield is very high. Faulty devices usually occur because of some defect in the silicon wafer or in the fabrication steps. Defects in silicon wafer can occur because of lattice imperfection and strains introduced in crystal growth, cutting and handling of the wafers. Usually such defects are extremely small, but their presence can ruin devices built on or around. Reducing the size of each device greatly increases the chance for a given device to be free of such defects. The same is true for fabrication defects such as the presence of a dust particle on the photolithographic mask.

67.4. Drawbacks of ICs

The integrated circuits suffer from the following drawbacks :

1. coils or inductors cannot be fabricated, **2.** *ICs* function at fairly low voltages,

3. they handle only limited amount of power,

4. they are quite delicate and cannot withstand rough handling or excessive heat.

However, the advantages of *ICs* far outweigh their disadvantages or drawbacks.

67.5. Scale of Integration

Level of integration in *ICs* has been increasing ever since they were developed some three and a half decades back. The number of electronic circuits or components that can be fitted into a standard size *IC* has been dramatically increasing with each passing year. In fact, whole **electronic systems** rather than just **a circuit** are incorporated in one package.

An approximate method of classifying the amount of circuit or component density is as follows:

1. SSI—***small scale integration***

In this case, the number of **circuits** contained in one *IC* package is less than 12 (or number of **components** is less than 50).

2. MSI—***medium scale integration***

Here, number of circuits per package is between 13 and 99 (or number of components is between 50 and 5000).

3. LSI—***large scale integration***

In this case, circuit density is between 100 and 9,999 (or component density is between 5000 and 100,000).

4. VLSI—***very large scale integration***

Here the number of circuits per package is between 10,000 to 99,999 (or number of components is between 100,000 – 1,000,000).

5. ULSI—***ultra large scale integration***

In this case, the circuit density is between 100,000 to 999,999 (or component density is between 1,000,000 – 10,000,000).

6. GSI—***Giga scale integration***

Here the number of circuits per package is 1,000,000 or more (or number of components are over 100,000,000).

In summary,

SSI	<	12	circuits per chip
MSI		12 – 99	
LSI		100 – 9,999	
VLSI		10,000 – 99,999	
ULSI		100,000 – 999,999	
GSI	>	1,000,000	

67.6. Classification of ICs by Structure

Structurally speaking, *ICs* can be classified into the following three types :

1. Monolithic Integrated Circuits

The word 'monolithic' means 'single stone' or more appropriately 'a single-solid structure'. In this *IC*, all circuit components (both active and passive) are fabricated insepa-rably within a single continuous piece of silicon crystalline material called **wafer (or substrate).** All components are **atomically** part of the same chip. Transistors, diodes and other passive components are fabricated at appropriate spots in the substrate using epitaxial diffusion technique.

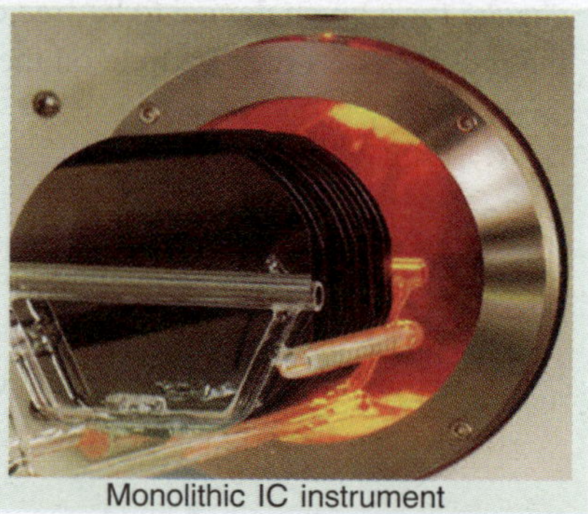
Monolithic IC instrument

Component interconnections are provided on the surface of the structure and external connecting wires are taken out to the terminals. It is a complete circuit requiring no 'add ons'.

Despite some of its distinct disadvantages, monolithic *ICs* are in wide use because for mass production, monolithic process has been found to be the most economical.

2. Thick and Thin-Film ICs

The essential difference between thick-film and thin-film *ICs* *is not their relative thickness but* the **method** of depositing the film. Both have similar appearance, properties and general characteristics though they both differ in many respects from monolithic *ICs*. These *ICs* are not formed **within** a silicon wafer but **on** the surface of an insulating substrate such as glass or a ceramic material. Moreover, **only passive components** (resistors, capacitors) are formed through thick or thin-film techniques on the insulating surface. The active elements (transistors, diodes) are added externally as **discrete elements** to complete a functional circuit. These discrete active components are frequently produced by using the monolithic process.

As stated above, the primary difference between the thick and thin film techniques is the process used for forming passive components and the metallic conduction pattern.

(*a*) Thin-film ICs

Such circuits are constructed by depositing films (typically 0.1 to 0.5 μm) of conducting material through a mask on the surface of a substrate made of glass or ceramic. Resistors and conductors are formed by varying the width and thickness of the film and by using materials of different resistivity. Capacitors are produced by sandwiching an insulating oxide film between two conducting films. Small inductors can be made by depositing a spiral formation of film. The active components like transistors and diodes etc. are externally added and inter-connected by wire bonds.

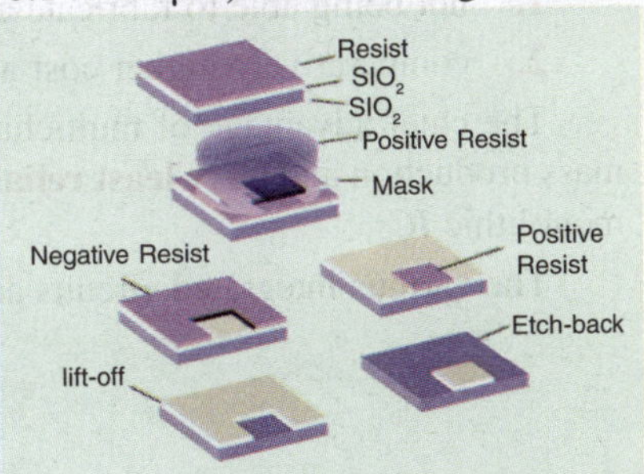

Thin film ICs

Following two methods are used to produce thin films :

(*i*) vacuum evaporation

In this method, the vaporised material is deposited through a set of masks on the glass or ceramic substrate contained in vacuum.

(*ii*) cathode sputtering

In this method, atoms from a cathode made of the desired film material are deposited on the substrate which is located between the cathode and the anode.

(*b*) Thick-film ICs

Such type of integrated circuits are sometimes referred to as **printed** thin-film circuits. They are so called because silk-screen printing techniques are employed to create the desired circuit pattern on the surface of the substrate. The screens are made of fine stainless steel wire mesh and the 'inks' are pastes (of pulverised glass and aluminium) which have conductive, resistive or dielectric properties. After printing, the circuits are high-temperature fired in a furnace to fuse the films to the insulating substrate. As with thin-film *ICs*, active elements are added externally as discrete components.

3. Hybrid or Multichip ICs

As the name implies, such circuits are formed either by inter-connecting a number of individual chips or by a combination of film and monolithic *IC* techniques. In such *ICs*, active components are first formed within a silicon wafer (using monolithic technique) which is subsequently covered with an insulating layer such as SiO_2. Film techniques are then employed to form passive components on the SiO_2 surface. Connections are made from the film to the monolithic structure through 'windows' cut in the SiO_2 layer.

Hybrid or Multichip ICs

67.7. Comparison Between Different ICs

Each type of *IC* has its own advantages and disadvantages.

Monolithic circuits have ***the advantage of lowest cost and highest reliability.*** However they have the following disadvantages :

1. isolation between components is poorer,
2. range of values of passive components used in the circuits is comparatively small,
3. inductors cannot be fabricated,
4. they afford no flexibility in circuit design because for making any changes in the circuit, **a new set of masks is required.**

The film circuits have the advantage of forming passive components with broader range of values and reduced tolerances as compared to monolithic circuits. Isolation between their components is also better since they are deposited on a substrate that is an insulator. Use of external discrete active components allows greater flexibility in circuit design. These circuits also give better high frequency performance than monolithic circuits.

However, they suffer from the disadvantages of

1. not being able to fabricate active components,
2. comparatively higher cost and 3. larger physical size.

The chief advantage of multichip *ICs* is their greater flexibility but they are **too expensive** for mass production and **have least reliability.** Hence, such circuits are generally used as prototypes for monolithic *ICs*.

The various integrated circuits are arranged in the form of an '*IC*' tree of Fig. 67.1.

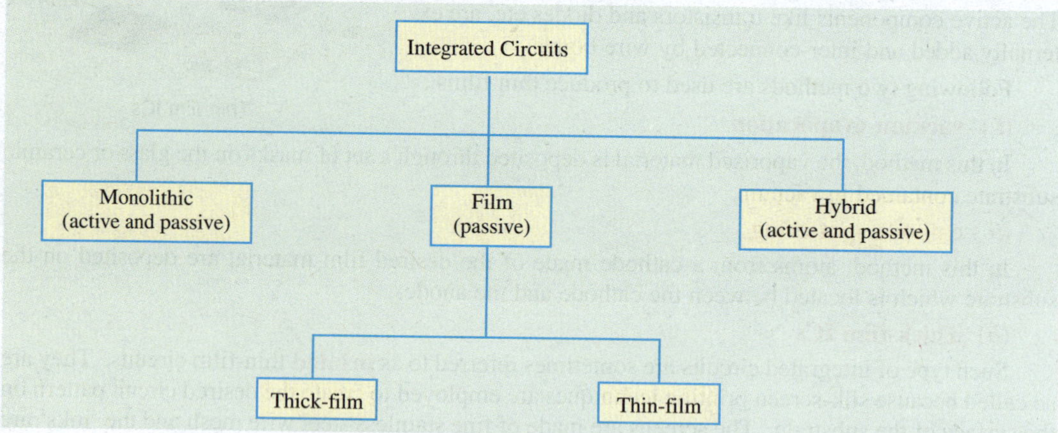

Fig. 67.1

67.8. Classification of ICs By Function

The earlier classification of ICs was based on ***their method of construction.*** However, the integrated circuits can also be classified according to their general function. The two most important categories are :

1. linear and 2. digital

The same fact has been shown in Fig. 67.2.

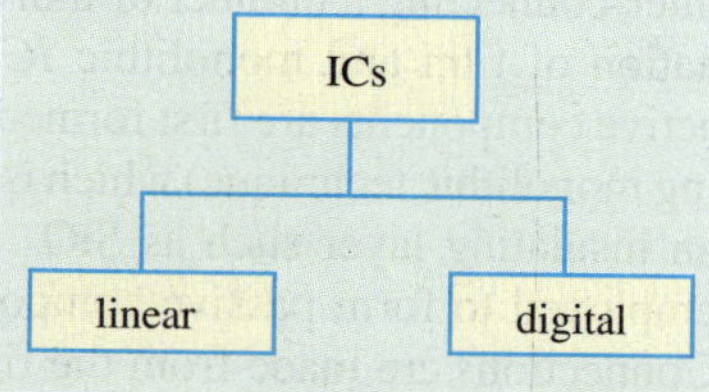

Fig. 67.2

Examples of linear *ICs* are :

1. *BEL CA*-3020—used as multipurpose wide-band power amplifier.

2. *BEL CA*-3065—it is a monolithic *IC* which combines a multistage *IF* amplifier, limiter, an *FM* detector, an electronic attenuator, a Zener diode regulated power supply and an audio amplifier. In fact, this *IC* provides a high performance multistage sub-system of a TV receiver. It is available in 14-pin dual-in-line package.

Both are manufactured by Bharat Electronics Ltd., Bangalore.

3. SSD 710—is a linear *IC* used as Differential Comparator

It is manufactured by Solid State Devices, Syed Abdullah Road, Bombay.

An example of a digital *IC* is

1. *BEL* 7400—is a *TTL* IC gate which provides designer with one of the gating logic necessary to design medium-speed digital control and data processing systems. It is available in 14-pin dual-in-line plastic package.

67.9. Linear Integrated Circuits (LICs)

LICs are also referred to as analog *ICs* because their inputs and outputs can take on a continuous range of values and the outputs are generally proportional to the inputs. As compared to digital *ICs*, *LICs* are used much less. But *LICs* are quickly displacing their discrete circuit counterparts in many applications as their cost becomes competitive. They also possess much higher reliability because so many external connections (major source of circuit failure) are eliminated. *LICs* find wide use in military and industrial applications as well as in consumer products. They are frequently used in

Linear ICs

1. operational amplifiers,
2. small-signal amplifiers,
3. power amplifiers,
4. *RF* and *IF* amplifiers,
5. microwave amplifiers,
6. multipliers,
7. voltage comparators,
8. voltage regulators etc.

Operational amplifier is by far the most versatile form for an *LIC* and is discussed separately.

67.10. Manufacturer's Designation of LICs

Each manufacturer assigns a specific code and type number to the *LICs* produced by him. For example, an internally-compensated op-amp 741 produced by Fairchild is designated as μA 741. Here, μA is the identifying code used by Fairchild. Many other manufacturers produce *LICs* similar to 741 but use their own code while retaining the same type number. For example, Fiarchild's original μA 741 is manufactured by other manufacturers with the following codes :

1. National Semiconductor —— *LM* 741
2. Motorola —— *MC* 1741
3. RCA —— *CA* 3741
4. Texas Instruments —— *SN* 5274

It is seen that the last three digits in each manufacturer's designation are the same *i.e.* 741. All these op-amps have the same specifications. Hence, in practice, the manufacturer's code number is often ignored and all such *LICs* are referred to as 741.

Many *LICs* are available in different classes such as A, B, C, E, S and SC. For example, main classes of 741 are as under :

741	——	Military grade op-amp
741 C	——	Commercial grade op-amp
741 A	——	Improved version of 741
741 E	——	Improved version of 741 C
741 S	——	Military grade op-amp with higher slew rate
741 SE	——	Commercial grade op-amp with higher slew rate

67.11. Digital Integrated Circuits

About 80 per cent of the *IC* market has been captured by digital *ICs* which are mostly utilized by the computer industry. Digital *ICs* lend themselves easily to monolithic integration because a computer ***uses a large number of identical circuits.*** Moreover, such circuits employ relatively few capacitors and values of resistances, voltages and currents are low.

Digital *ICs* contain circuits whose input and output voltages are limited to **two** possible levels—low or high. It is so because ***digital signals are usually binary.*** Sometimes, digital circuits are referred to as switching circuits. Digital *ICs* include circuits such as

1. logic gates **2.** flip-flops **3.** counters **4.** clock-chips
5. calculator chips **6.** memory chips **7.** microprocessors (μP) etc.

67.12. IC Terminology

Some of the common terms used in fabricating integrated circuits are defined below :

1. **Bonding** – attaching the die on the ceramic substrate and then connecting the leads to the package.
2. **Chip** – an extremely small part of a silicon wafer on which *IC* is fabricated. A photograph of the wafer containing hundreds of chips (or dice) and a drawing of a chip are shown in Fig. 67.3. The identical chips, each of which may vary in area from 10 to over 100 mm^2, may contain up to several million devices.

Dish silicon wafer in DHD

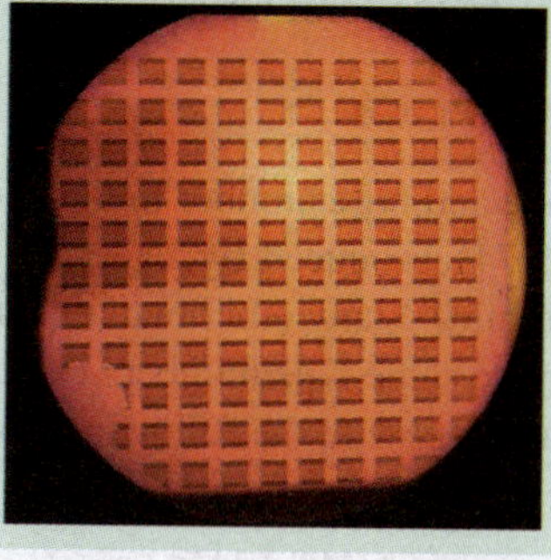

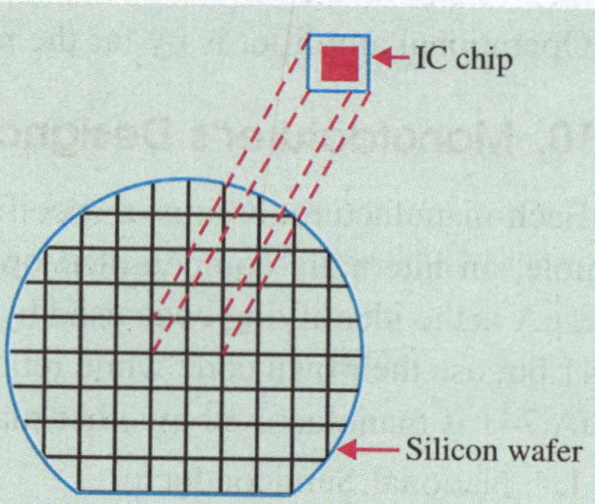

Fig. 67.3

3. **Circuit probing** – testing the electrical performance of each *IC* chip with the help of a microscope and multi-point probe.
4. **Die** – same as a chip.
5. **Diffusion** – a process that consists of the introduction of impurities into selected regions of a wafer to form junctions.
6. **Encapsulation** – putting a cap over the *IC* and sealing it in an inert atmosphere.

7. **Epitaxy** – a process of the controlled growth of a crystalline doped layer of silicon on a single crystal substrate.
8. **Etching** – a process of selective removal of regions of a semiconductor, metal or silicon dioxide.
9. **Mask** – a glass plate with desired pattern for diffusion or metallization. Usually a single mask is not sufficient to fabricate an *IC*.
10. **Metallization** – a process for providing ohmic contacts and interconnections by evaporating aluminium over the chip.
11. **Photolithography** – a process to transfer geometrical pattern from the mask to the surface of the wafer.
12. **Photoresist** – a light-sensitive material that hardens when exposed to ultraviolet light.
13. **Wafer** – a thin disk of semiconductor in which number of *ICs* are fabricated simultaneously.

67.13. Semiconductors Used in Fabrication of ICs and Devices

The fabrication of *ICs* has been based on the use of silicon (Si) as the premier semiconductor. Two other semiconductors used for *IC* fabrication are germanium and gallium arsenide (GaAs). But these semiconductors present special problems for device fabrication as discussed below.

Gallium arsenide has very attractive electrical properties but its crystals have a high density of defects which limit the performance of devices made from it. Moreover gallium arsenide is more difficult to grow in single crystal form. Both silicon and germanium do not suffer from these problems. On the plus side, gallium arsenide has an electron velocity that is larger than silicon. Because of this gallium arsenide devices are faster than silicon devices. Also, gallium arsenide has a lower saturation electric field than silicon. Because of this, the gallium arsenide devices have lower power-delay product. Devices made from substrates of gallium arsenide have lower parasitic capacitances. This property contributes to their speed advantage over the silicon devices. Another advantage of gallium arsenide results from direct band gap which makes it possible to provide certain functions not possible in silicon such as ***coherent and incoherent light emission.***

A major advantage of silicon, in addition to its abundant availability in the form of sand, is that it is possible to form a superior stable oxide (SiO_2). This oxide has superb insulating properties and provides an essential and excellent ingredient in the fabrication and protection of devices or *ICs*. On the other hand, germanium oxide is unsuited for device applications. The intrinsic resistivity of germanium is 47 Ω-cm, while that of silicon is 230,000 Ω-cm. The low resistivity of germanium would have precluded the fabrication of rectifying devices with high breakdown voltages. Thus high-voltage rectifying devices and certain infrared sensing devices are practical with silicon. Finally, there is an economic consideration and that is electronic grade germanium is now more costly than silicon. Thus, at present time, silicon remains the major semiconductor in the fabrication of *ICs*.

Some other semiconductors used in semiconductor industry are gallium phosphide (GaP), gallium nitride (GaN), zinc sulphide (ZnS), indium antimonide (InSb), compound of cadmium and selenium (CdSe). GaAs, GaP and GaN are used in high-speed devices and devices requiring emission and absorption of light such as lasers and light emitting diodes (LEDs). ZnS is used as fluorescent material such as those in television screens. In Sb and CdSe are used as light detectors.

67.14. How ICs are Made ?

The *ICs* are manufactured in four distinct stages (refer to Fig. 67.4). These are (1) material preparation, (2) crystal growing and wafer preparation, (3) wafer fabrication and (4) testing, bonding and packaging. All these stages are discussed one by one in the following pages.

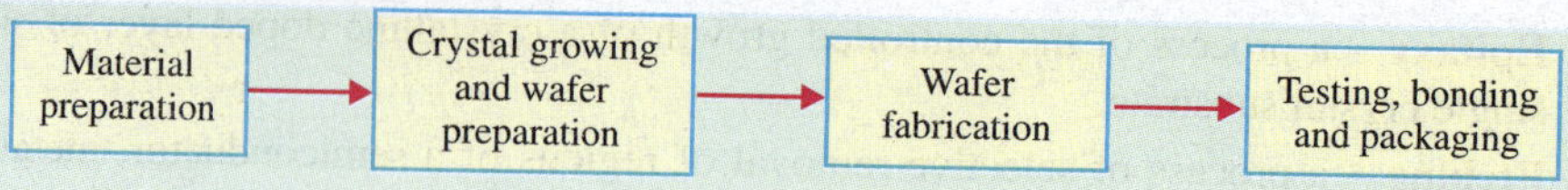

Fig. 67.4

67.15. Material Preparation

Silicon, as an element is not found in nature. However, it is found abundantly in nature in the form of silicon dioxide, which constitutes about 20% of earth's crust. Silicon is commonly found as quartz or sand. A number of processes are required to convert sand into pure silicon with a polycrystalline structure. Fig. 67.5 shows the different processes involved in the preparation of polycrystalline silicon from sand. As seen from this figure, the sand is allowed to react with a gas produced from the burning of carbon (coal, coke and wood chips). This produces silicon with 98% purity. Next silicon is further purified in a reactor to produce electronic-grade polycrystalline silicon.

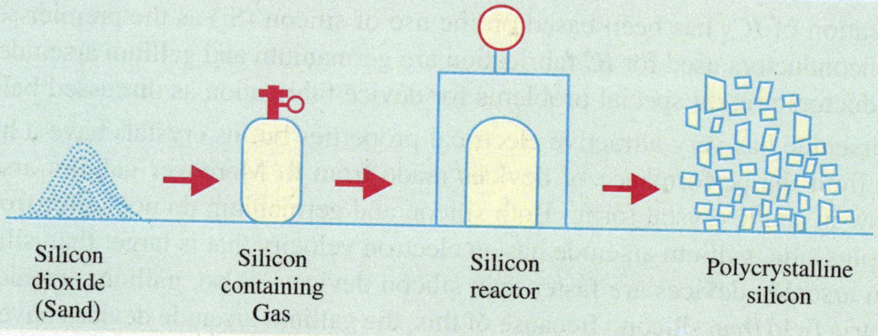

Fig. 67.5

67.16. Crystal Growing and Wafer Preparation

The polycrystalline silicon is composed of many small crystals having random orientation and containing many defects. For silicon to be used in the fabrication of *ICs*, it must be nearly perfect and crystalline in nature. We, therefore, now need to produce single crystals of silicon. This is done by a process called crystal growth. There are two methods to carryout the crystal growth : *(i)* the ***Czochralski*** and *(ii)* the ***flat zone*** process. The Czochralski process prepares virtually all the silicon used for *IC* fabrication. The flat zone process is used to prepare crystals for fabricating high-power, high voltage semiconductor devices.

The Czochralski process. The equipment used for single crystal growth (called puller) is as shown in Fig. 67.6. The puller has three main components : *(i)* a furnace which includes quartz crucible, a rotation mechanism (clockwise as shown), and a radio frequency (*RF*) heating element, ***(ii)*** a crystal pulling mechanism which includes a seed holder and a rotation mechanism (counter-clockwise), and ***(iii)*** an ambient control which includes an argon gas source, a flow control and an exhaust system. In addition the puller has a computer system to control process parameters such as temperature, crystal diameter, pull rate and rotation speed.

To grow crystals, the polycrystalline silicon is placed in the crucible. The furnace is heated to a temperature of 1690 K which is slightly greater than the melting point (1685 K) of silicon. A precisely controlled amount of dopant (boron or phosphorus) is added to the melt to make the silicon as *P*-type or *N*-type. A suitable oriented seed crystal (*i.e.*, a small highly perfect crystal) is suspended over the crucible in a seed holder. The seed is inserted into the melt and a small portion of it is allowed to melt. The seed is rotated and pulled up very slowly, while at the same time, the crucible is rotated in the opposite direction. The molten silicon attaches itself to the seed and it becomes identical to the seed in structure and orientation. As the seed is pulled up, the material that is attached to the seed solidifies (*i.e.*, freezes). Its crystal structure becomes the same as that of the seed and a larger crystal is formed. Thus using this method, cylindrical single crystal bars (called ingots) of silicon are produced.

The desired diameter of the silicon ingot is obtained by controlling both the temperature and the pulling speed. In the final step, when the bulk of the melt has been grown, the crystal diameter is decreased until there is a point contact with the melt. The resulting ingot is cooled and is removed to be made into thin discs called ***wafers***. The ingots have diameters as large as 200 mm with the latest ones approaching 300 mm. The ingot length is of the order of 1000 mm.

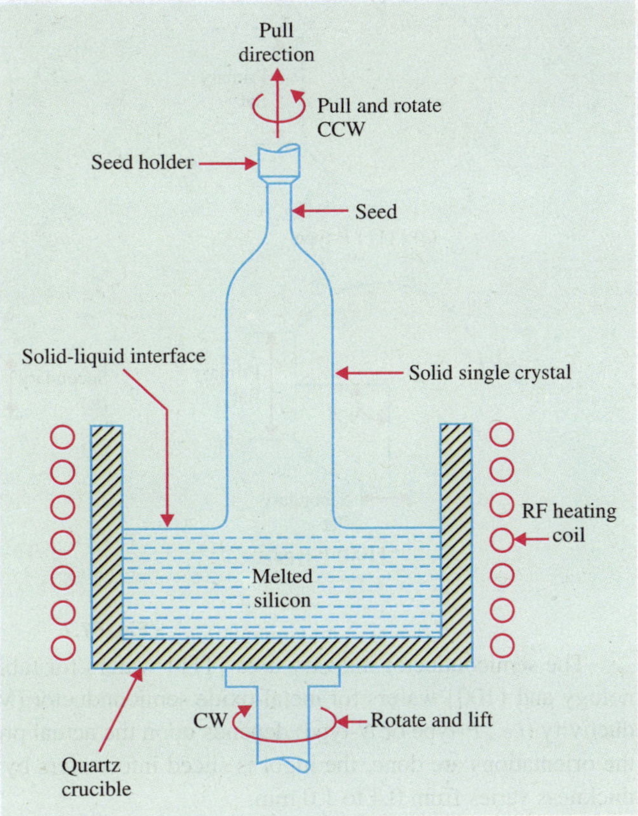

Fig. 67.6

Wafer preparation : In this stage the ingot surface is grounded throughout to an exact diameter and the top and bottom portions are cut off. After that one or more flat regions are ground along the length of the ingot. These flat regions mark the specific crystal orientation of the ingot and conductivity type (*i.e.*, *P*-type or *N*-type) of silicon material. Refer to Fig. 67.7.

Notice that the ingot is marked with two flats regions : *(i)* the larger flat (called primary flat) and *(ii)* the smaller flat or secondary flat regions. The primary flat allows a mechanical locator in automatic processing equipment to position the wafer and to orient the devices relative to the crystal in a specific manner. The secondary flat regions are used to identify the orientation and conductivity type of the crystal. The conductivity of the wafer could be either *P*-type or *N*-type and the crystal orientation, {100} or {111} (refer to Art 51.17 to know more about crystal orientation). If there is no secondary flat as shown in Fig. 67.7 (*a*), or in words the primary and the secondary flat are superimposed, the wafer is identified as *P*-type with {111}-crystal orientation. If the secondary flat is 45° with respect to the primary flat as shown in Fig. 67.7 (*b*), it is {111} *N*-type wafer. However, if the secondary flat is 90° with respect to the primary flat as shown in Fig. 67.7 (*c*), it is {100} *P*-type wafer and if the secondary flat is at 180° with respect to the primary flat as shown in Fig. 67.7 (*d*), it is {100} *N*-type wafer.

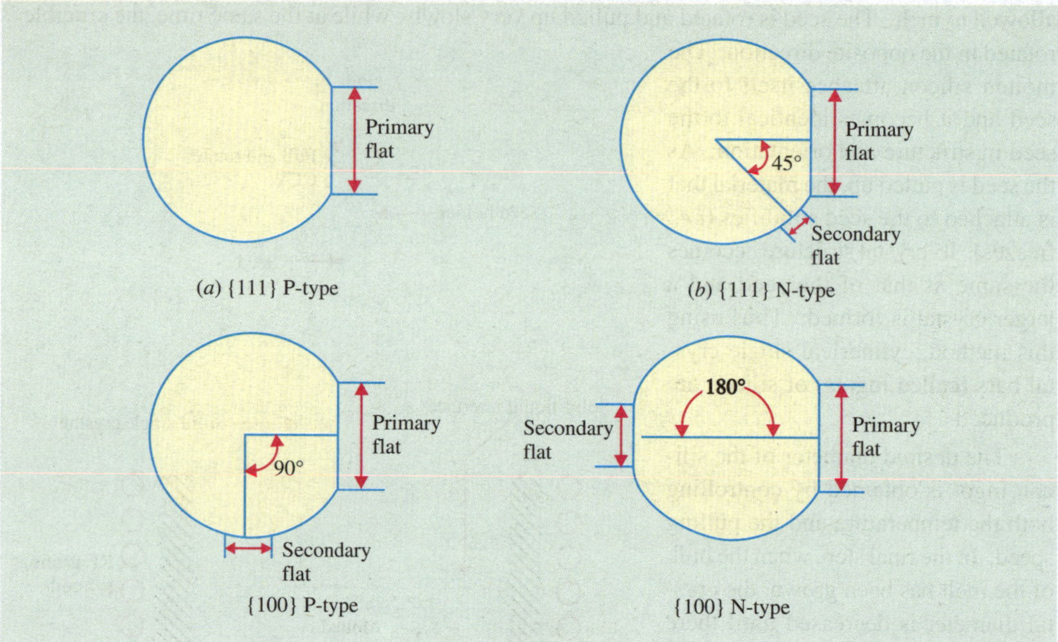

Fig. 67.7

The semiconductor industry uses {111} wafers for fabricating *ICs* with bipolar transistor technology and {100} wafers for metal-oxide semiconductor (MOS) circuits. The choice of wafer conductivity (*i.e.*, *P*-type or *N*-type) depends upon the actual process used for fabricating the *ICs*. Once the orientations are done, the ingot is sliced into wafers by a high-speed ***diamond saw***. The wafer thickness varies from 0.4 to 1.0 mm.

After slicing, both sides of the wafer are lapped to produce typical flatness uniformity with in 2 μm. The lapped orientation usually leaves the surface and edges of the wafer, damaged and contaminated. These can be removed by a process called chemical etching. The next step is to polish the wafer surface to a mirror-like finish. Finally the wafers are cleaned, rinsed and dried for use in fabrication of *ICs*. It is interesting to note that the final wafer thickness is about one-third less than after the slicing.

67.17. Wafer Fabrication

Following is the category of the processes that are used in the fabrication of *ICs* :

(*i*) oxidation (*ii*) etching (*iii*) diffusion
(*iv*) ion implantation (*v*) photolithography (*vi*) epitaxy
(*vii*) metallization and interconnections.

We will study each process separately and apply some of these later to the formation of a diode bipolar transistor and metal oxide semiconductor field effect transistor. The basic fabrication process is called ***planar process***, *i.e.*, a process in which the introduction of impurities and metallic interconnections is carried out from the top of the wafer. A major advantage of the planar process is that each fabrication step is applied to all identical circuits and each of the many wafers at the same time.

It is important to initially emphasize that the fabrication requires an extremely clean environment in addition to the precise control of temperature and humidity.

67.18. Oxidation

The process of oxidation consists of growing a thin film of silicon dioxide (SiO_2) on the surface of a silicon wafer. Silicon dioxide has several uses :

1. to serve as a mask against implant or diffusion of dopant into silicon,
2. to provide surface passivation,
3. to isolate one device from another,
4. to act as a component in MOS structures.

Fig. 67.8 shows oxide layer grown on the surface of silicon substrate. The commonly used silicon dopants, such as boron, phosphorus, arsenic, and antimony, have very low diffusion coefficients (*i.e.*, they diffuse with great difficulty) in silicon dioxide. Because of this reason, silicon dioxide is used as a shield against infiltration of these dopants. On the other hand, these dopants diffuse easily if the surface is silicon.

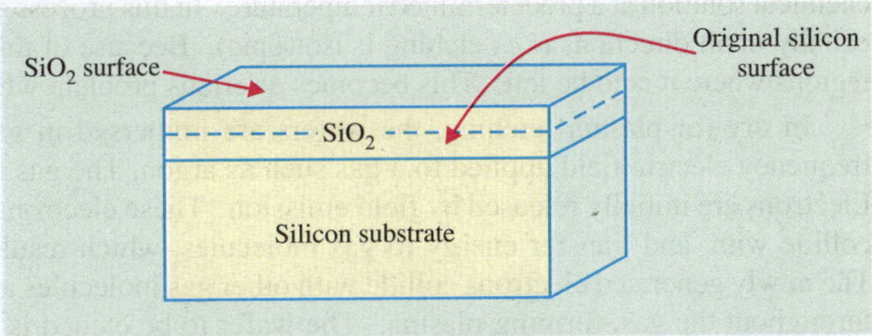

Fig. 67.8

Several techniques have been developed for forming oxide layers. Some of these are thermal oxidation, vapour phase technique [chemical vapour deposition (CVD)], and plasma oxidation. However, thermal oxidation is the more commonly used technique in *IC* processing.

Fig. 67.9 shows a thermal oxidation system. As seen, the oxidation is accomplished by placing the silicon wafers vertically into a quartz boat in a quartz tube. The quartz boat is slowly passed through a resistance heated furnace, in the presence of oxygen, operating at a temperature of about 1000ºC. The oxidizing agent may be *dry* by using dry oxygen or be using a mixture of water vapour and oxygen. A computer controls the whole operation in the thermal oxidation system. The operation include regulating the gas flow sequence, automatic insertion and removal of wafers and the furnace temperature.

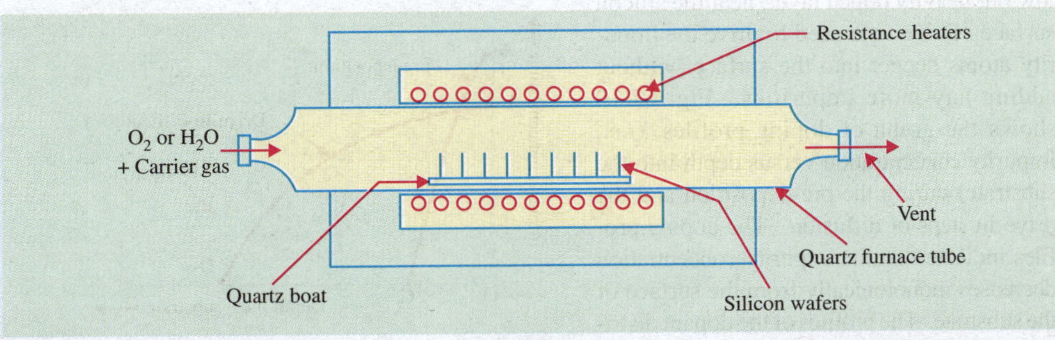

Fig. 67.9

It will be interesting to note that the oxide growth in the dry process is much ***slower*** but it produces an oxidized layer that has ***excellent*** electrical properties. For relatively thin oxides such as gate oxide in a MOSFET (typically 10 nm), dry oxidation is used. However, for thicker oxides (≥ 500 nm) such as field oxides in MOS integrated circuits and for bipolar devices, wet oxidation is used to provide both isolation and passivation.

Once the silicon dioxide layer has been formed on the surface of the wafer, it is selectively removed (etched) from those surfaces where impurities are to be introduced and kept as a shield, for the underlying silicon surface, where no dopants are to be allowed.

Oxide layers are relatively free from defects and provide stable and reliable electrical properties.

67.19. Etching

Etching is the process of selective removal of regions of a semiconductor, metal or silicon dioxide. There are two types of etching : ***wet*** and ***dry***. In wet etching, the wafers are immersed in a chemical solution at a predetermined temperature. In this process, the material to be etched is removed equally in all directions (*i.e.*, etching is isotropic). Because of this, some material is etched from the regions where it is to be left. This becomes a serious problem when dealing with small dimensions.

In dry (or plasma) etching, the wafers are immersed in gaseous plasma created by a radio-frequency electric field applied to a gas such as argon. The gas breaks down and becomes ionized. Electrons are initially released by field emission. These electrons gain kinetic energy from the field, collide with, and transfer energy to gas molecules, which result in generating ions and electrons. The newly generated electrons collide with other gas molecules and the avalanche process continues throughout the gas, forming plasma. The wafer to be etched is placed on an electrode and is subjected to the bombardment of its surface by gas ions. As a result, the transfer of momentum from the ions to the atoms removes atoms at or near the surface to be etched.

Dry etching also called reactive ion etching (RIE) is directional (or anisotropic). In other words, the material is removed only from those regions where it is required. Most modern processes use only dry etching to produce fine line patterns needed for VLSI integrated circuits.

67.20. Diffusion

This process consists of the introduction of impurities into selected regions of a wafer to form junctions. Diffusion occurs in two steps : the ***pre-deposition*** and the ***drive-in*** diffusion. In the pre-deposition step, a high concentration of dopant atoms is introduced at the silicon surface by a vapour that contains the dopant at a temperature of 1000ºC. More recently, a more accurate method of pre-deposition known as ***ion implantation*** is used.

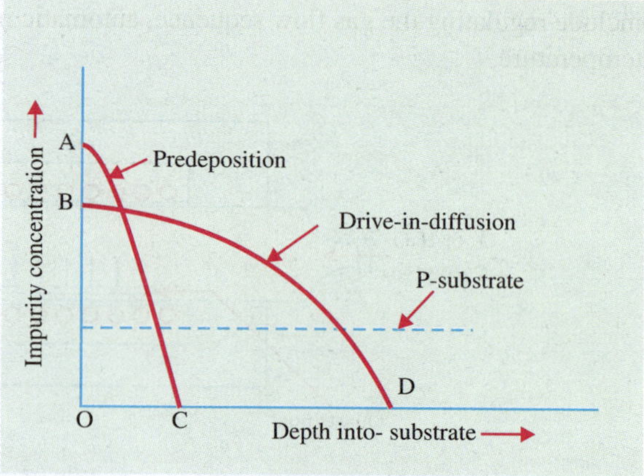

Fig. 67.10

Pre-deposition tends to produce, a shallow but heavily doped layer, near the silicon surface. Drive-in is used to drive the impurity atoms deeper into the surface, without adding any more impurities. Fig. 67.10 shows the graph of doping profiles, (*i.e.*, impurity concentration versus depth into the substrate) during the pre-deposition and the drive-in steps of diffusion. The doping profiles indicate that the impurity concentration decreases monotonically from the surface of the substrate. The profiles of the dopant distribution is determined mainly by the temperature and diffusion time. Note that the impurity concentration drops from *OA* (during pre-deposition step) to *OB* at the surface of the substrate after the drive-in step. However the depth into the substrate of impurity has increased from *OC* to *OD* as required.

Common dopants are boron for *P*-type layers and phosphorus, antimony and arsenic for *N*-type layers. However, diffusion is rarely performed using pure elements themselves. Rather, compounds of elements are used and impurities may be introduced from either solid, liquid, or gaseous substances.

Fig. 67.11 shows the equipment used in diffusion. The wafers are placed in a quartz boat within a quartz furnace tube. The furnace is heated by resistance heaters surrounding it. To introduce an impurity, phosphorus for example, phosphorus oxychloride ($POCl_3$) is placed in a container either inside the quartz tube in a region of relatively low temperature or in a container outside the furnace at a temperature that maintains its liquid form. For a *P*-type dopant, boron is used. The proper vapour pressure is maintained by a control of the temperature. Nitrogen and oxygen gas are made to pass over the container. These gases react with the silicon, forming a layer on the surface of the wafer that contains silicon, oxygen and phosphorus. At the high temperature of the furnace, phosphorus easily diffuses into the silicon.

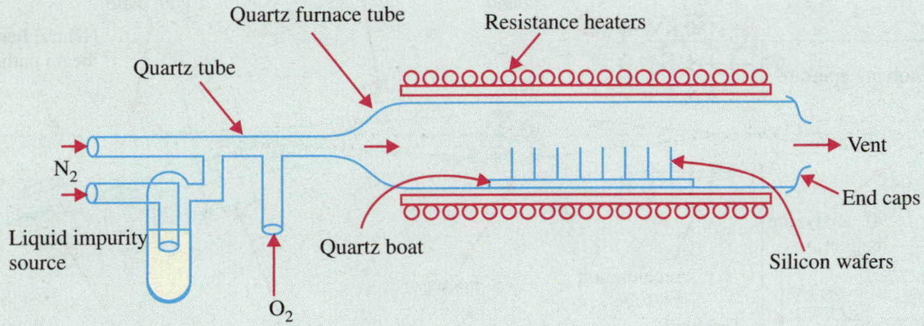

Fig. 67.11

In order that the dopant may be diffused deeper into silicon, the drive-in step follows. This is done at a higher temperature of about 1100ºC inside a furnace similar to that used for pre-deposition, except that no dopant is introduced into the furnace. The higher temperature, causes the dopant atoms to move into silicon more quickly. Diffusion depth is controlled by the time and temperature of the drive-in process. By precise control of the time and temperature, accurate junction depths of fraction of a micron can be obtained. Fig. 67.12 shows the diffusion of dopant into silicon.

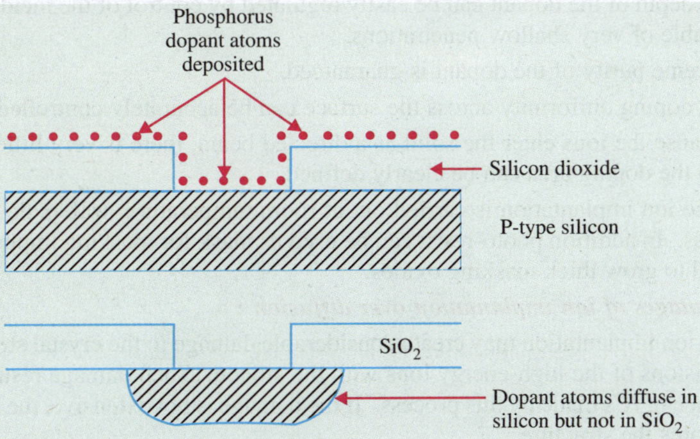

Fig. 67.12

67.21. Ion Implantation

This is a process of introducing dopants into selected areas of the surface of the wafer by bombarding the surface with high-energy ions of the particular dopant.

Fig. 67.13 shows a typical ion implantation equipment. To generate ions, such as those of phosphorus, an arc discharge is made to occur in a gas, such as phosphine (PH_3), that contains the dopant.

The ions are then accelerated in an electric field so that they acquire energy of about 20 KeV and are passed through a strong magnetic field. The ions are further accelerated so that their energy reaches several hundred KeV or MeV, whereupon they are focused on and strike the surface of the silicon wafer.

As is the case with diffusion, the ion beam is made to penetrate only into selected regions of the wafer by a process of masking (discussed later). On entering the wafer, the ions collide with silicon atoms and lose their energy. The depth of penetration of ions in ion implantation is about 0.1 to 1 μm. The higher the energy of ions and the smaller their mass, the greater is the depth of penetration.

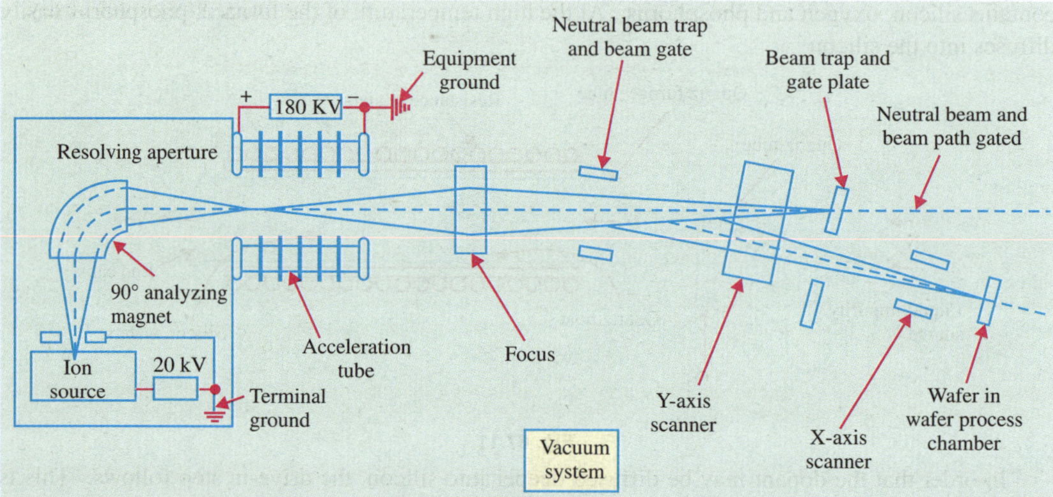

Fig. 67.13

Advantages of ion implantation over diffusion :

1. Doping levels can be precisely controlled since the incident ion beam can be accurately measured as an electric current.
2. The depth of the dopant can be easily regulated by control of the incident ion velocity. It is capable of very shallow penetrations.
3. Extreme purity of the dopant is guaranteed.
4. The doping uniformity across the surface can be accurately controlled.
5. Because the ions enter the solid as a directed beam, there is very little spread of the beam, thus the doping area can be clearly defined.
6. Since ion implantation is carried out at room temperature, wafers do not face temperature stress. In addition photo-resist can be used as mask for masking impurities, thus there is no need to grow thick masking oxides.

Disadvantages of ion implantation over diffusion :

1. The ion implantation may create considerable damage to the crystal structure because of the collisions of the high-energy ions with the silicon. Such damage results in inferior performance of *ICs* made by this process. If the damage is not extensive, the process of ***annealing*** restores the structure.
2. High initial investment and operational cost of the equipment (> US $ 1 million).
3. Uses very toxic gases for some of the dopants such as phosphorus and arsenic.

67.22. Photomask Generation

The whole process of *IC* fabrication consists of identifying selected regions of each circuit of the wafer surface into which identical dopant or metallic interconnections are made, while protecting other regions of the wafer surface. To carry out one of the many fabrication processes, a separate mask is required for each operation whose function is to expose the selected regions and protect the others. There may be hundreds of identical dies (or *ICs*) on a wafer with each circuit containing hundreds of thousands or millions of devices. Identical steps are carried out simultaneously for each process. For each process, separate mask is needed.

The mask production starts with a drawing using a computer-assisted graphics system with all the information about the drawing stored in digital form. Commands from the computer are prepared that drive a pattern generator, which uses an electron beam to write the particular pattern, for one or several dies, on a glass plate covered with thin chromium film. When the glass plate is prepared, it is called ***reticle***. A mask usually refers to a glass plate that contains a pattern for the whole wafer. The reticle pattern is projected onto the wafer and a ***wafer stepper*** is used that reduces the reticle circuit onto the photo-resist covered wafer that steps across over the surface until the entire array of circuits is built up.

It may be noted that the use of a single mask for all the circuits on a wafer is not feasible for printing very small (< 1 μm) features because of alignment problems. However, the use of single mask is still in use for fabricating simple digital and analog circuits such as light emitting diodes (LEDs).

67.23. Photolithography

It is a process in which the geometrical pattern on the glass plate (called reticle) is transferred to the surface of the wafer. This is done to open identical windows so that the diffusion (or the ion implantation) process may take place in all identical regions of the same *IC* and for all *ICs* on the wafer. As an illustration we assume that the first reticle is used over an oxidized surface as shown in Fig. 67.14 (*a*).

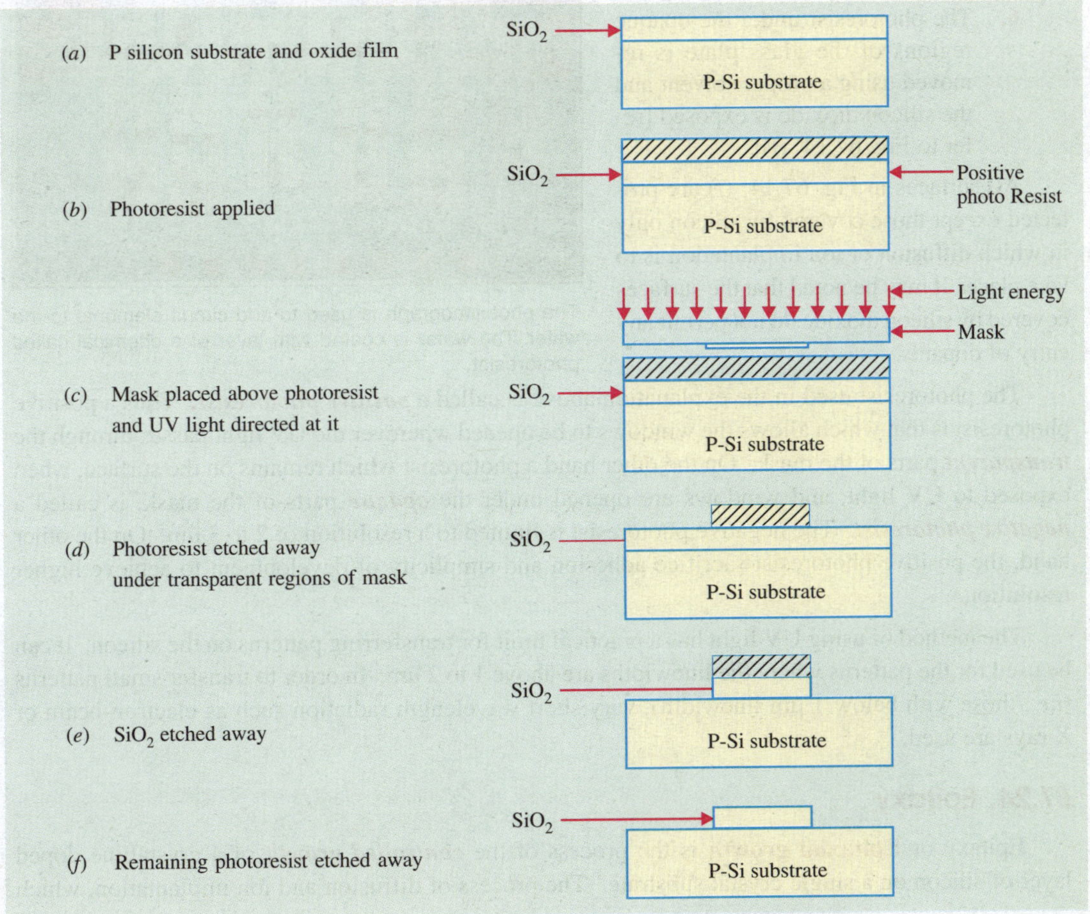

Fig. 67.14

To transfer the patterns, the wafer is coated with light-sensitive material called ***photoresist***. By applying small amount (about 1 ml) of the photoresist to the wafer surface and spinning the wafer very rapidly, a uniform film of about 1 μm thick is formed over the oxidized surface of the wafer [refer to Fig. 67.14 (*b*)]. After this the following steps are taken to open a window on the surface of the wafer :

1. The wafer is baked at 100ºC to solidify the photoresist on the wafer.
2. The glass plate is placed on the wafer and aligned by the computer control.
3. The glass plate is exposed to ultraviolet (UV) light with the transparent parts of the glass plate passing the light on to the wafer. The photoresist under the opaque regions of the glass plate is unaffected [refer to Fig. 67.14 (*c*)].
4. The exposed photoresist is chemically removed by dissolving it in an organic solvent and exposing the silicon dioxide underneath [refer to Fig. 67.14 (*d*)].
5. The exposed silicon dioxide is then etched away using hydrofluoric acid. The hydrofluoric acid dissolves silicon dioxide and not silicon. The regions under the opaque part of the glass plate are still covered by the silicon dioxide and the photoresist [refer to Fig. 67.14 (*e*)].
6. The photoresist under the opaque regions of the glass plate is removed using a proper solvent and the silicon dioxide is exposed [refer to Fig. 67.14 (*f*)].

All surfaces in Fig. 67.14 (*f*) are protected except those covered by silicon only in which diffusion or ion implantation is to take place. It may be noted that the surfaces covered by silicon dioxide do not permit any entry of dopants.

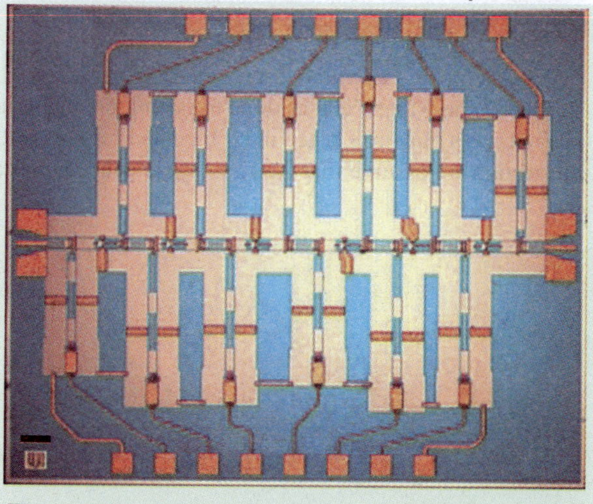

The photolithograph is used to add circuit elements to the wafer. The wafer is coated with layer of a chemical called photoresist.

The photoresist used in the explanation above is called a ***positive photoresist***. Thus a positive photoresist is that which allows the windows to be opened wherever the UV light passes through the ***transparent*** parts of the mask. On the other hand a photoresist which remains on the surface, when exposed to UV light, and windows are opened under the ***opaque*** parts of the mask, is called a ***negative photoresist***. The negative photoresist is limited to a resolution of 2 to 3 μm. On the other hand, the positive photoresist sacrifice adhesion and simplicity of development to achieve higher resolution.

The method of using UV light has a practical limit for transferring patterns on the silicon. It can be used for the patterns where the linewidths are above 1 to 2 μm. In order to transfer small patterns (*i.e.*, those with below 1 μm linewidth), very short wavelength radiation such as electron-beam or *X*-rays are used.

67.24. Epitaxy

Epitaxy or **Epitaxial growth** is the process of the ***controlled growth*** of a crystalline doped layer of silicon on a single crystal substrate. The process of diffusion and ion implantation, which were earlier described, produce a layer at the surface that is of higher doping density than that which existed before the dopant was added. It is not possible by these methods to produce, at the surface a layer of lower concentration than exists there. This can however, be accomplished by the method of epitaxy. In the processes of diffusion and ion implantation, a dopant is driven into a substrate of doped silicon. In epitaxy, a layer of doped silicon is deposited on top of the surface of the substrate. Normally this single crystal layer has different type of doping from that of the substrate.

Epitaxy is used to deposit N or N^+ (*i.e.*, heavily doped *N*-type) silicon, which is impossible to accomplish by diffusion. It is also used in isolation between bipolar transistors where N^- (*i.e.*, lightly doped *N*-type) is deposited on *P*-type layer. It may also be used to improve the surface quality of an *N*-type substrate by depositing *N*-type material over it. A variety of methods are used to grow the epitaxial layer. These methods include vapour-phase epitaxy (VPE), liquid-phase epitaxy (LPE), and molecular beam epitaxy (MPE). The system for growing an epitaxial layer using vapour-phase epitaxy (VPE) is shown in Fig. 67.15.

In this system, silicon wafers are placed in a long boat-shaped crucible made of graphite. The boat is placed in a long cylindrical quartz tube, which has inlets and outlets for the gases. The tube is heated by induction using the heating coils wound around the tube.

All the chemicals that are introduced and that take part in the reactions are in the form of gases, hence the process is known as Chemical Vapour Deposition (CVD). The epitaxial layer is grown from the vapour phase onto the silicon, which is in the solid state. The thickness of the layer varies from 3 to 30 μm and the thickness of the layer and its doping content are controlled to an accuracy of less than 2 percent. The reactions are carried out at a temperature of approximately 1200ºC. The high temperature is necessary so that the dopant atoms can acquire a sufficient amount of energy to allow them to move into the crystal to form covalent bonds and become extension of the single crystal. Because the layer is grown on the substrate, epitaxy is a growth technique where the crystal is formed without reaching the melting point of silicon.

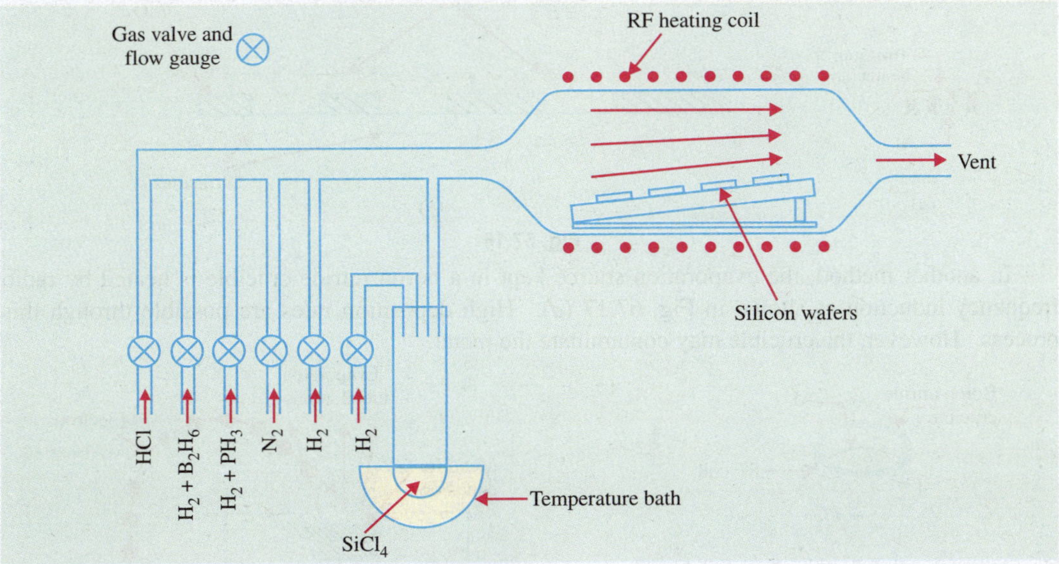

Fig. 67.15

67.25. Metallization and Interconnections

After all the fabrication steps of an *IC* are completed, it becomes necessary to provide metallic interconnections for the *IC* and for external connections to the *IC*. The requirement that must be met by the interconnections is that they have low resistance to minimize both the voltage drops on the lines as well as the capacitances between the lines so as to reduce delay times. The connections must also make ***ohmic contacts*** to semiconductors in the devices such as *P* and *N* regions of a *PN* junction diode. An ohmic contact is one that exhibits a very low resistance, allowing currents to pass easily in both directions through the contact.

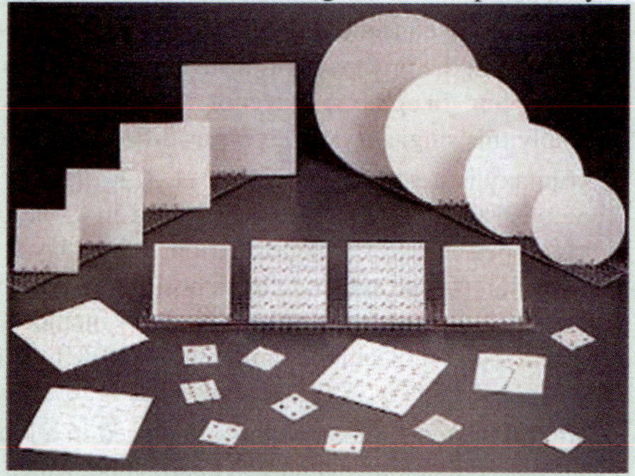

Several cycles of photolithography etching and doping are performed producing multiple layers of circuit elements on the wafer.

The high conductivity of aluminium makes it the metal of obvious choice, particularly in silicon-based devices. It also has the following advantages :

1. easy to evaporate
2. can be easily etched
3. not expensive,
4. adheres well to silicon dioxide.

There are variety of processes for depositing aluminium on silicon substrates but the following three are important from the subject point of view : **(1)** resistance heating, **(2)** electron beam heating and **(3)** sputtering.

Resistance heating : In this process, the source of the heated element and the silicon substrate are located in an evacuated chamber. The source is a small piece of aluminium attached to a coil of tungsten, which serves as a heater [refer to Fig. 67.16 (*a*)]. The heated element with a high melting point remains solid while the aluminium is vaporized. The aluminium atoms travel to the substrate where they condense, depositing an aluminium layer on the surface of the silicon. A photolithographic masking and etching method is used to remove the metal from the regions where it is not wanted. A typical interconnection between two diffused layers is as shown in Fig. 67.16 (*b*).

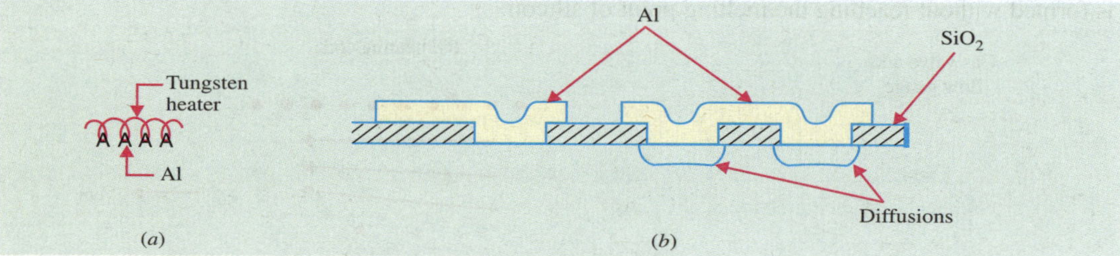

Fig. 67.16

In another method, the evaporation source kept in a boron nitride crucible is heated by radio frequency induction as shown in Fig. 67.17 (*a*). High deposition rates are possible through this process. However, the crucible may contaminate the metal.

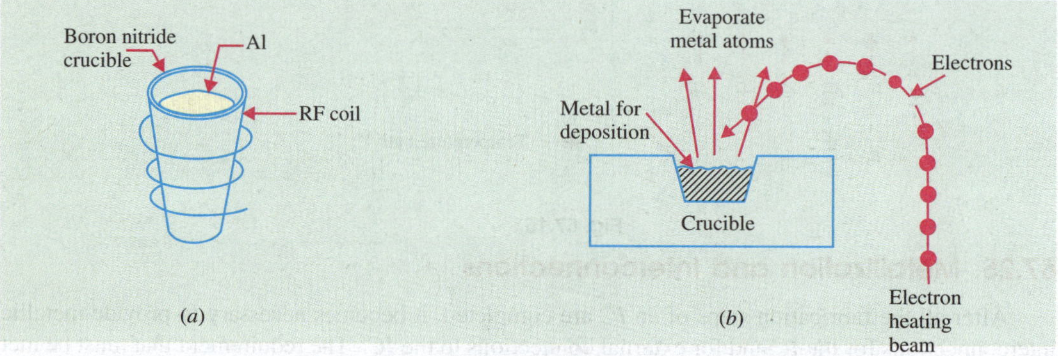

Fig. 67.17

Electron-beam heating **:** In this process, aluminium in a crucible is placed into a vacuum chamber together with the substrate. The aluminium is subjected to a high intensity electron beam formed by an electron gun, which vaporizes the aluminium as shown in Fig. 67.17 (*b*). This causes the aluminium to travels to the wafer. By the use of mask and photolithography, the aluminium is deposited on the identified regions on the wafer surface.

Sputtering **:** In this process, the material to be deposited is placed in a container maintained at low pressure in the vicinity of the substrate. The material to be deposited is labelled the cathode or the target, while the anode is the substrate. A DC or a radio-frequency high voltage is applied between anode and cathode. This high voltage ionizes the inert gas in the chamber. The ions are accelerated to the cathode (in this operation the anode is usually grounded) where, by impact with the aluminium target, atoms of aluminium are vaporized. A gas of aluminium atoms is generated and deposited on the surface of the wafer.

Following the deposition of aluminium, the silicon wafers are placed in a furnace to solidify the connections so that low resistance metallic contacts are made.

The interconnections between the elements of an *IC* are made by aluminium lines having a thickness of about 0.5 μm. These are laid on top of the silicon dioxide layer, which covers the surface of the wafer. By using photolithography, openings are made in silicon dioxide so that the aluminium layer is connected to the silicon or to the ohmic contact on the silicon. In very complicated integrated circuits, it is necessary to have two or three vertically stacked layers of interconnections separated by silicon dioxide layers. The interconnecting lines terminate at aluminium pads (called bonding pads) from which connections to the outside are made.

67.26. Testing, Bonding and Packaging

The individual *IC* chip must be connected properly to outside leads and packaged in a way that is convenient for use in a larger circuit or a system. Since the devices are handled individually once they are separated from the wafer, bonding and packaging are expensive processes.

Testing – After the wafer of monolithic circuits has been processed and the final metallization pattern defined, it is placed in a holder under a microscope and is aligned for testing by a machine called ***multiple-point probe.*** The probe contacts the various pads on an individual circuit, and a series of tests are made to verify the electrical properties of the device in a very short time. After all the circuits are tested, the wafer is removed from the testing machine, sawed between the circuits, and broken apart. Then each die that passed the test is picked up and placed in the package.

Bonding – It consists of two steps. In the first step the back of the die is mechanically attached to an appropriate mount medium such as ceramic substrate, multi-layer-ceramic package or metal lead frame. The two common die bonding methods are ***hard solders*** and ***polymers.*** In the second step, the bond pads on the circuit side of the die are connected by wires to the package as shown in Fig. 67.18. The three common schemes of interconnection to the chip bond pads are ***(i) wire bonding, (ii) tape-automated bonding*** (*TAB*), and ***(iii) flip-chip solder bonding.*** Wire bonding is further split into two different processes : ***thermosonic*** and ***thermocompression.*** The further detail of these processes is beyond the scope of this book.

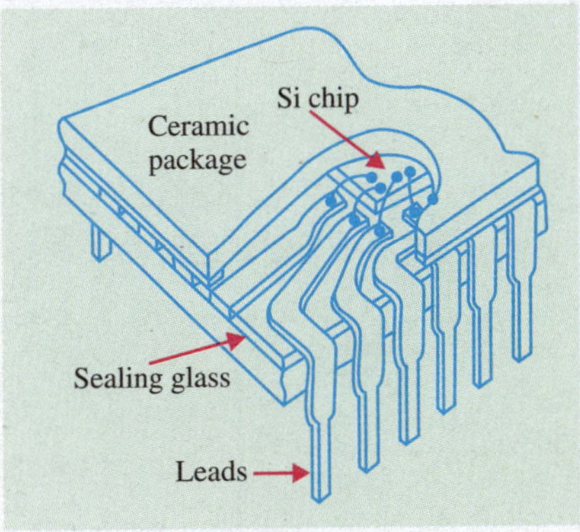

Fig. 67.18

Packaging – The final step in *IC* fabrication is packaging the device in a suitable medium that can protect it from environment of its intended application. In most cases this means the surface of the device must be isolated from the moisture and contaminants and the bonds and other elements must be protected from corrosion and mechanical shocks. The modern *ICs* are mounted in packages with many output leads. In one version, chip is mounted on a stamped metal lead frame and wire bonding is done between the chip and the leads. The package is formed by applying a ceramic or plastic case and trimming away the unwanted parts of the lead frame.

Fig. 67.19 shows various types of the packaging of *ICs* used these days. As seen from this figure, the packages can be broadly classified into the following two categories :

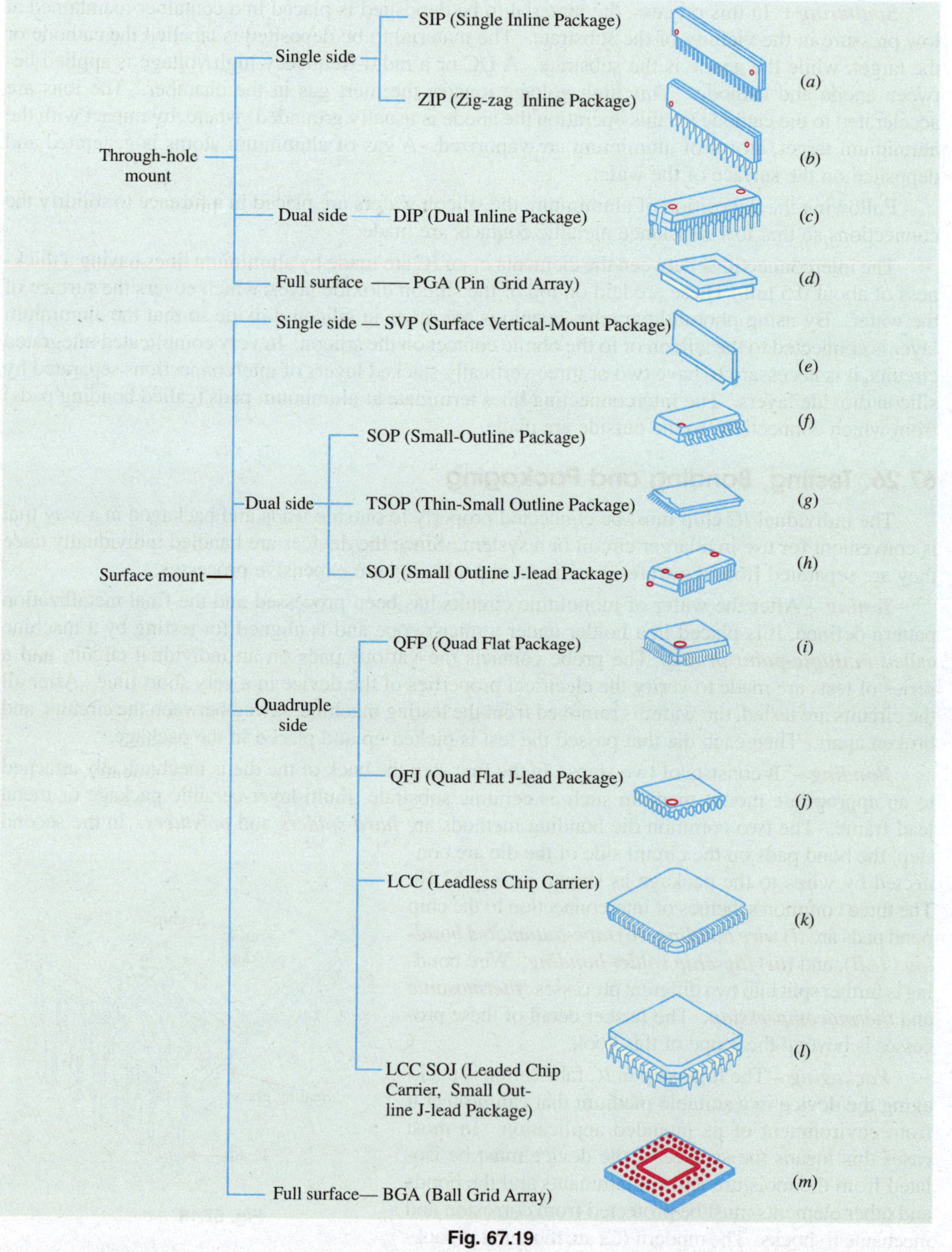

Fig. 67.19

1. ***through-hole mount*** that involve inserting the package pins through holes on the printed circuit board (PCB) before soldering [Fig. 67.19 (*a*) through (*d*)].
2. ***surface mount type*** where the leads do not pass through holes in the PCB. Instead, surface-mounted package leads are aligned to electrical contacts on the PCB, and are connected simultaneously by solder reflow [Fig. 67.19 (*e*) through (*m*)]. As refer to the picture shown in Fig 67.20.

Most packages can be made using ceramic or plastic. The *ICs* are hermetically sealed for protection from the environment. The pins can be on one side (single inline or zigzag pattern of leads), two sides (dual inline package or DIP), or four sides of the package (quad package). Most advanced packages have leads distributed over a large portion of the surface of the package as in through-hole pin grid arrays (PGAs) or surface-mounted ball grid arrays (BGAs).

67.27. Semiconductor Devices and Integrated Circuit Formation

There are literally thousands of different semiconductor device structures. They have been developed to achieve specific performances, either as discrete components or in *ICs*. Accordingly there are many different structures. However, there are basic structures required for each of the major device and circuit types.

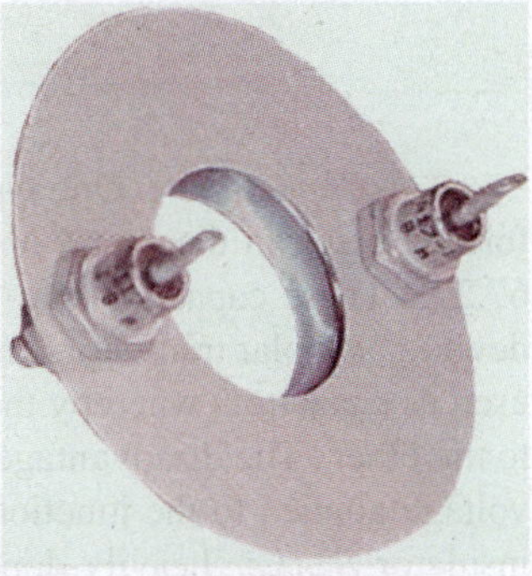

Fig. 67.20

1. ***Resistors*** **:** Most of the resistors in *ICs* are formed by the same processes that are used to form devices (*i.e.*, a sequence of oxidation, masking and doping operation). Fig. 67.21 (*a*) shows a resistor made of a *P*-type region diffused into an *N*-type epitaxial layer. Notice the metallic contacts made at the two ends of the epitaxial layer. The section of the resistor, as dictated by the diffusion, is very nearly rectangular in shape, as shown in Fig. 67.21 (*b*). The value of a resistor (in ohms) is given by :

$$R = \rho \cdot l/a \quad \text{or} \quad \rho . l/w.d$$

where ρ is the resistivity of the layer in ohm-cm, *l* is the length of the resistive region, and *a* is the cross-sectional area of the resistive region. The value of *a* is equals to *w.d.* where *w* is the width and *d* is the depth of the resistive region.

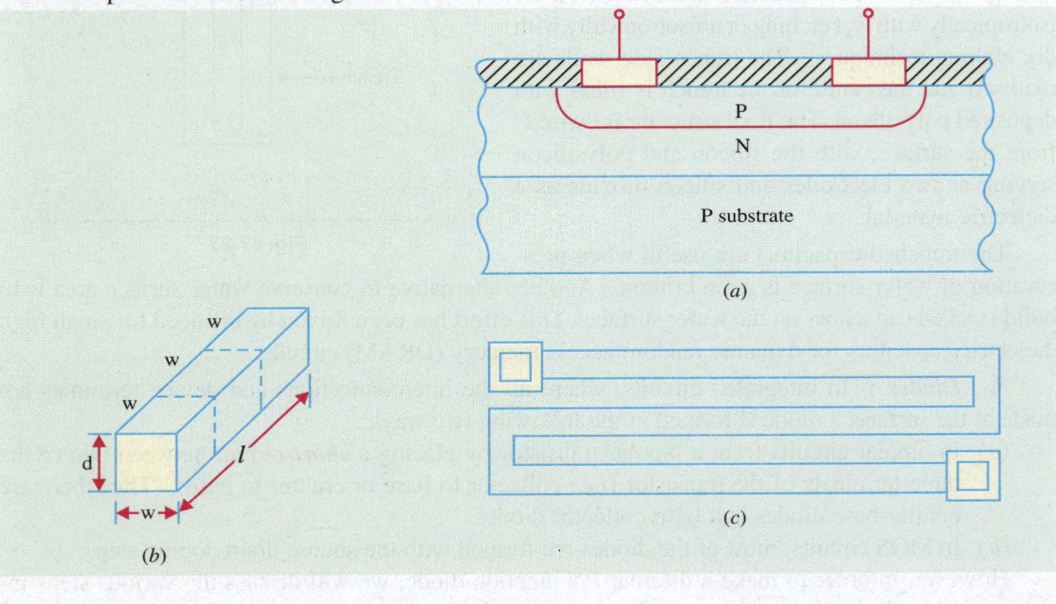

Fig. 67.21

Resistance in monolithic circuits are defined by the term called ***sheet resistance.*** The sheet resistance is the resistance of a square region having $w = l$ and has units of ohms per square. Assuming a sheet resistance of 100 to 200 ohms per square, practical resistors may have values ranging from 100 to several kilohms. Higher resistance values are obtained by using a meander pattern as shown in Fig. 67.21 (*c*). The major problem with the resistors of high values is that they tend to occupy a large area on the chip. For example, a resistor of 50 kΩ uses up an area of the wafer that may be occupied by hundreds of the transistors. Ion-implantation instead of diffusion can be used to make resistors with precise values of resistance.

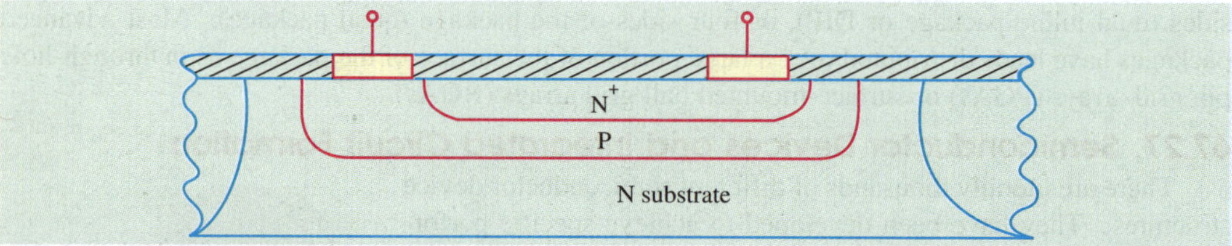

Fig. 67.22

2. *Capacitors :* One type of capacitor in monolithic circuits is made by using the capacitance formed between the *P* and *N* regions of a reverse-biased diode. Such a capacitor is shown in Fig. 67.22. These capacitors are formed by using the same diffusion processes that are used to form devices. Bipolar transistors are made of three regions in which either of the two *PN* junctions may be used as capacitors whereby the breakdown voltage of the capacitor may vary considerably from one to the other. The disadvantage of the junction capacitors is the dependence of the capacitance on the voltage applied to the junction. Capacitors that are voltage-independent can be formed from metal insulator N^+ (*i.e.*, heavily-doped) semiconductor layers as used in MOS structures.

In dense circuits, an oxide/nitride/oxide dielectric sandwich is used. The combination film has a lower dielectric constant, allowing a capacitor area smaller than a conventional silicon dioxide capacitor.

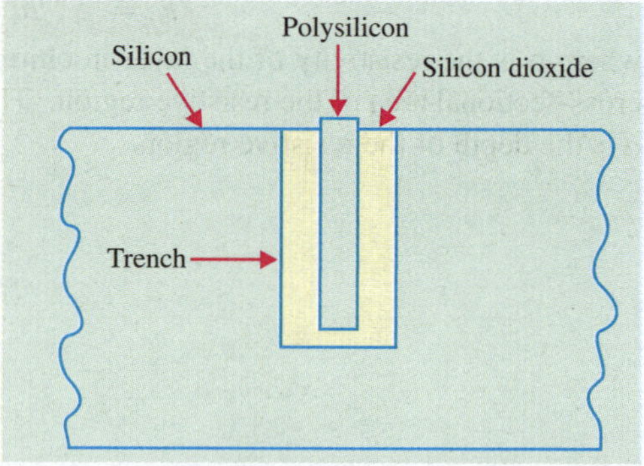

Fig. 67.23

Capacitors can also be fabricated by creating a trench etched vertically into the wafer surface (refer to Fig. 67.23). The trenches are etched either isotropically with wet etching or anisotropically with dry etching techniques. The trench side walls are oxidised and the centre of the trench is filled with deposited polysilicon. The final structure is "wired" from the surface, with the silicon and polysilicon serving as two electrodes and silicon dioxide as a dielectric material.

The trenched capacitors are useful when preservation of wafer surface is main criteria. Another alternative to conserve wafer surface area is to build stacked capacitors on the wafer surface. This effort has been driven by the need for small high dielectric capacitors for dynamic random access memory (DRAM) circuits.

3. *Diodes :* In integrated circuits, where all the interconnections and device terminals are made at the surface, a diode is formed in the following two ways :

(a) In bipolar circuits from a bipolar transistor by placing a ***short-circuit*** between two of the three terminals of the transistor (*i.e.*, collector to base or emitter to base). Thus there are emitter-base diodes and base-collector diodes.

(b) In MOS circuits, most of the diodes are formed with the source-drain doping step.

However, in order to make a discrete *PN* junction diode, we will discuss the various steps involved in the fabrication.

The starting material is a heavily doped *N*-type (N^+) substrate about 150 μm thick. A layer of *N*-type silicon (1 to 5 μm) is grown on the substrate by epitaxy as shown in Fig. 67.24 (*a*). Then a layer of silicon dioxide (SiO_2) is deposited by oxidation as shown in Fig. 67.24 (*b*). Next, the surface is coated with positive photoresist as shown in Fig. 67.24 (*c*). Then a mask is placed on the surface of silicon, aligned and exposed to ultraviolet light (UV) as shown in Fig. 67.24 (*d*). Next, mask is removed, resist is removed and silicon dioxide layer under the exposed resist is etched as shown in Fig. 67.24 (*e*). Then boron is diffused to form P-type region a shown in Fig. 67.24 (*f*). Note that boron diffuses easily in silicon but not in silicon dioxide. Next, a thin aluminium film is deposited over the surface as shown in Fig. 67.24 (*g*). Then the metallized area is covered with resist another mask is used to identify areas where metal is to be preserved. Wafer surface is etched to remove unwanted metal. Resist is then dissolved and we get a structure as shown in Fig. 67.24 (*h*). Finally, contact metal is deposited on the back surface and ohmic contacts are made by heat treatment.

4. *Bipolar Transistor* : Let us consider the fabrication of an NPN bipolar junction transistor (BJT) on a silicon wafer. While our discussion will be focussed on the BJT, it is understood that the surface of the whole wafer is being processed.

The starting material is a lightly doped *P*-type wafer. We will consider a small area on this wafer where we will form a BJT. The base on which the transistor is made is known as the substrate. The function of the substrate is to act as a mechanical support for the device. The reason for the use of a *P*-type substrate for the NPN transistor will be clarified when the term isolation is discussed. The substrate has a resistivity of 3-10 Ω cm with a thickness between 500 and 700 μm for wafers having diameters over 100 mm.

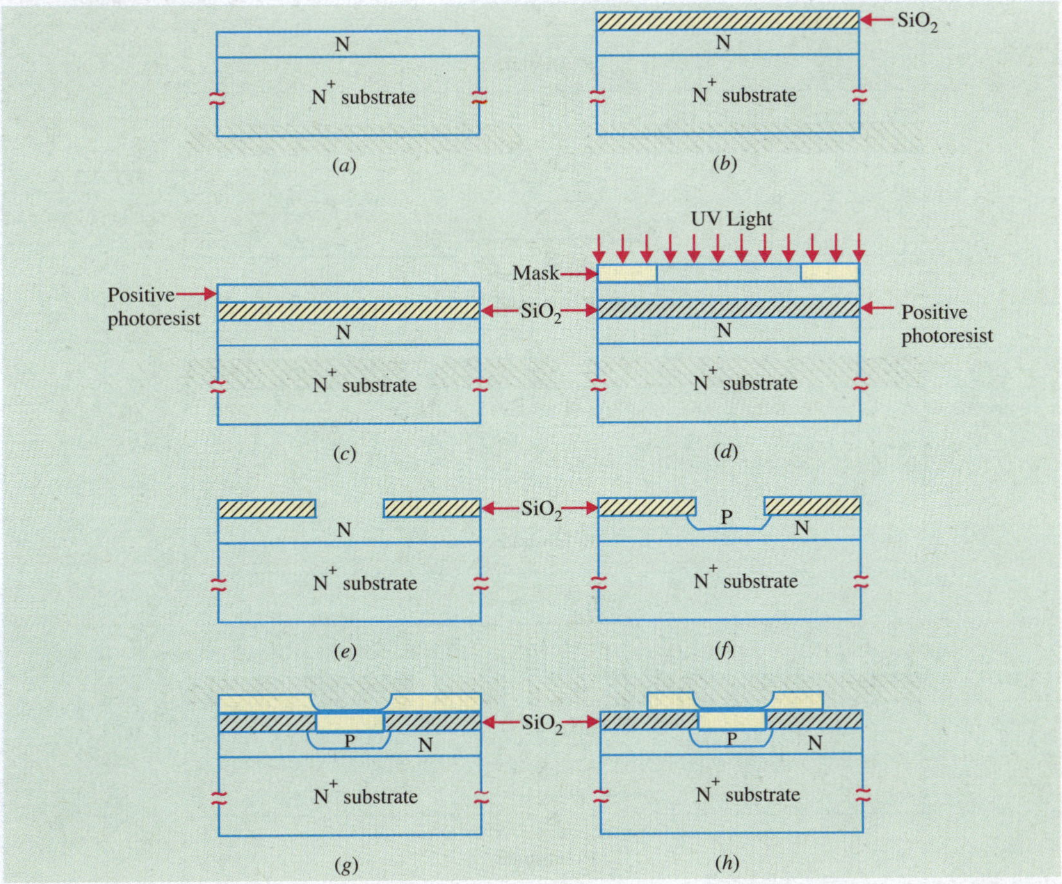

Fig. 67.24

Fig. 67.25 shows the fabrication steps up to, and including the metal contacts to the three regions (*i.e.*, emitter, base and collector).

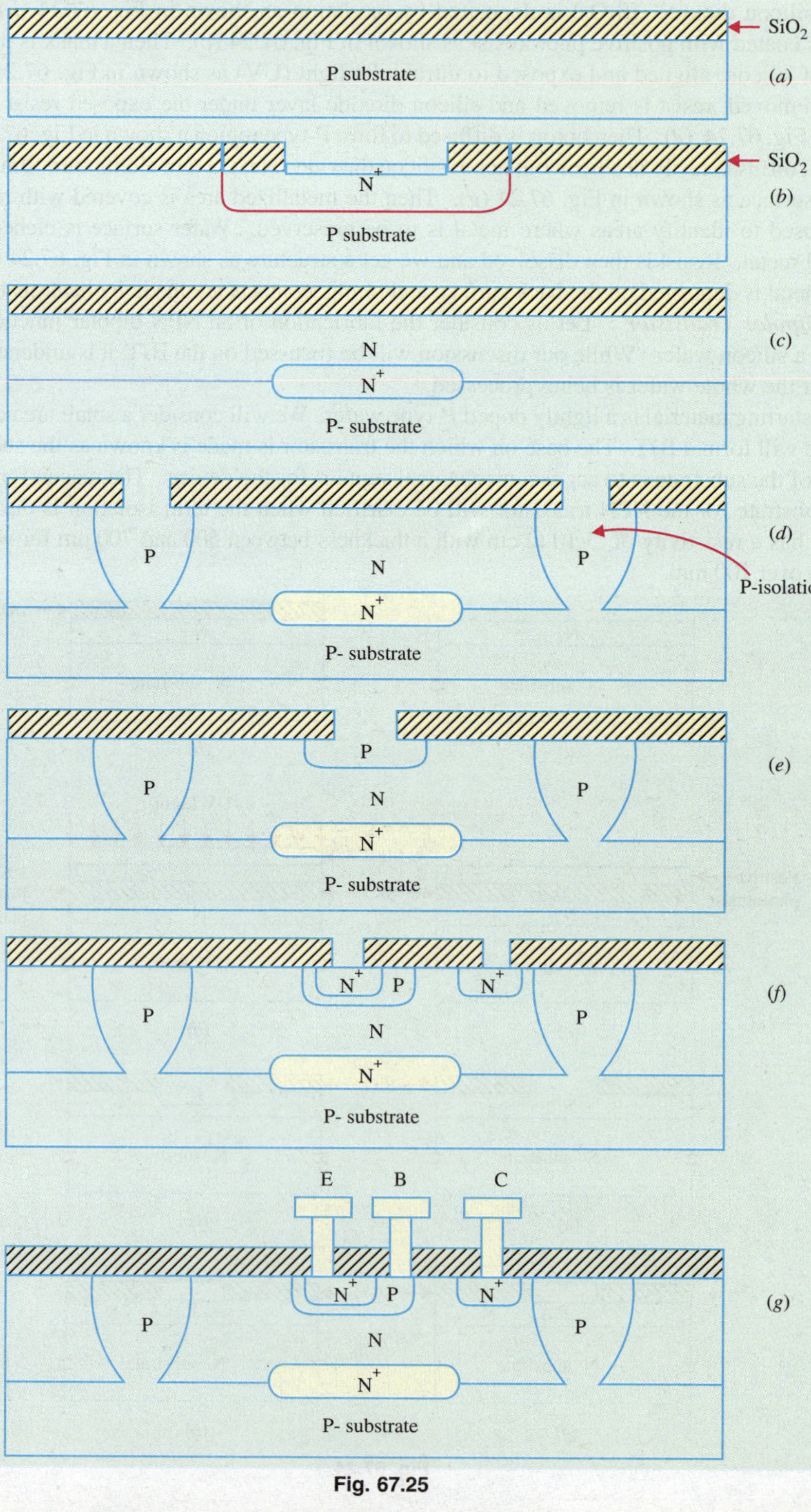

Fig. 67.25

First a layer of silicon dioxide (SiO_2), about 0.5 μm thick, is deposited on the surface of the substrate by thermal oxidation as shown in Fig. 67.25 (*a*). Using the ***first mask*** and the photolithographic process, windows are opened in the oxide for the *buried layer*. The N^+ (*i.e.*, heavily doped *N*-type) buried layer is diffused to a depth of about 3 μm as shown in Fig. 67.25 (*b*). This layer serves to collect the carriers that have crossed the base on their way to the collector terminal. It serves as a sub-collector and is used to reduce the collector ohmic-resistance. After the buried layer is diffused, the wafer is stripped of all oxide to permit the next deposition. It is to be noted that during the subsequent high-temperature processes, the buried layer tends to diffuse out.

The next operation is the deposition of a phosphorus-doped *N*-type ***epitaxial layer*** on the whole wafer. This layer has a resistivity of 0.1 to 1 Ω cm. The thickness of this layer is 0.5 to 5 μm for high-speed digital circuit applications and 10-20 μm for linear analog circuits. A layer of silicon dioxide (SiO_2) about 0.5 to 1 μm thick is grown thermally on the surface of the epitaxial layer as shown in Fig. 67.25 (*c*).

Since the collector of NPN transistor is *N*-type, and so are the collectors of the adjacent transistors, there is an obvious need to isolate the collectors from each other. The ***second mask*** is used to etch windows for this isolation regions, which are formed by the subsequent diffusion of boron extending from the surface down to the substrate as shown Fig. 67.25 (*d*). An *N*-type epitaxial layer separates the isolation regions, thus serving as the tub in which each transistor is formed. The diffusion of the isolation region is followed by oxidation of the wafer surface.

The ***third mask*** is used to open a window for the *P*-type base of the transistor, *P*-type diffusion or ion implantation is driven to form the base to a depth of about 2-3 μm as shown in Fig. 67.25 (*e*). This is followed by the deposition of an oxide layer. The ***fourth mask*** is used to open windows in the oxide for the N^+ emitter and the collector contacts. The phosphorus or arsenic diffusion is driven to a depth of about 2 μm as shown in Fig. 67.25 (*f*). The need for N^+ collector is to form a good ohmic contact. The ohmic contact permits easy current flow in both directions. To form a good ohmic contact to an *N* material, an N^+ region is needed between the metal on the top and the *N* region. Following the N^+ diffusion, an oxidation layer is formed over the entire wafer surface.

The ***fifth mask*** is used to open windows for the formation of metallic contacts to the bipolar transistor terminals. Then an aluminium film 0.5 and 1 μm thick is deposited, by evaporation or sputtering, on the top surface of the *IC* wafer. Assuming that the complete circuits are to be formed on the surface of the wafer, the ***sixth mask*** is used to define the interconnection pattern in the circuits. These interconnections are etched into the metal that has been deposited on the surface, as shown in Fig. 67.25 (*g*).

In order to protect the surface of the wafer from moisture and chemical contamination, a silicon nitride (Si_3N_4), called a ***passivation layer,*** is deposited on the surface. Contacts to the integrated circuits are made on pads that are located on the periphery of the *IC* chip. Since the *IC* chip will be bonded to an *IC* package, connections are to be made from the package leads to the bonding pads on the *IC* chip.

The ***seventh*** mask is used to define the bonding holes over the aluminium pads for external connections. Following the seven masks, the circuits are tested by a computer-controlled system and all faulty chips are identified and marked. The wafer is then sawed into chips, which are bonded on *IC* packages. Gold wires about 25 μm in diameter are used to connect the package leads to the bonding pads on the chip.

Fig. 67.26 shows an additional detail on the buried layer. It indicates the path taken by the carriers on their way from the emitter to the base and to the collector. This path is considerably longer than the path in the discrete BJT shown in Fig. 67.26 (*a*). Because of this the collector series resistance, labelled the ***parasitic resistance,*** is quite large and is of the order of hundreds of ohms. To reduce this resistance, we have placed the low resistivity buried layer in Fig. 67.26 (*b*) into the path of the carriers which acts as a subcollector. The use of the buried layer reduces the collector resistance by as much as a factor of 20. The result of this is to improve gain-bandwidth product of the transistor by the same factor.

5. ***MOSFET* :** Let us consider the fabrication of Enhancement mode *N*-channel metal oxide semiconductor field-effect transistor. The starting material is a lightly doped *P*-type silicon substrate having a resistivity of about 5 Ω cm. The doping density is determined by the drain-substrate breakdown voltage of 20-30 V. A thin (about 20 nm) layer of silicon dioxide is formed over the substrate to provide stress relief to the wafer. Next a thin (about 20 nm) layer of silicon nitride (Si_3N_4) is deposited by the chemical vapour deposition (CVD) process on top of the silicon dioxide. The silicon dioxide permits selective oxidation so that a thick oxide (about 500 nm) can be formed in the field region. These operations result in Fig. 67.27 (*a*).

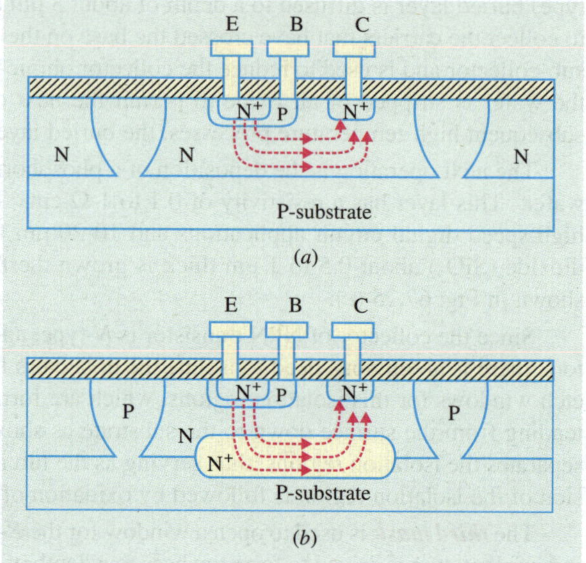

Fig. 67.26

The ***first mask*** defines the field-effect transistor areas, so that the silicon dioxide and the silicon nitride are chemically etched out except where the transistor is formed.

The next step is to diffuse (or ion-implant) boron in the field-regions to form P^+ (*i.e.* heavily doped *P*-type) islands in the substrate. The function of P^+ islands is to help increase the threshold voltage, V_T and prevent the formation of ***"parasitic" transistors*** (or electrical cross talk) between adjacent devices on the wafer. This is followed by the formation of a field oxide layer (about 500 nm thick) over the P^+ implanted region as shown in Fig. 67.27 (*b*). The field oxide layer also helps to increase V_T.

The remaining silicon nitride and silicon dioxide are etched away and an ultra thin (about 5 to 10 nm) layer of SiO_2 is grown over the transistor area (not above the field oxide). This forms the gate-oxide of the transistor. Refer to Fig. 67.27 (*c*). Next, a layer of heavily doped (typically N^+) polysilicon is deposited over the entire wafer surface. This is shown in Fig. 67.27 (*d*).

The ***second*** mask defines the gate region, whereupon the polysilicon is etched away except over the gate as shown in Fig. 67.27 (*e*). The heavily doped polysilicon region above the gate behaves electrically like a metal electrode.

By ion-implantation and using the polysilicon gate and the field oxide, as the mask, the source and drain N^+ regions are formed as shown in Fig. 67.27 (*e*). The N^+ layers literally diffuse a sufficient distance to ensure proper alignment so that the channel length is well defined. The dopants do not penetrate the field oxide. A thin layer of silicon dioxide is then grown, by the CVD process over the wafer.

The ***third mask*** is used to open windows for the metal contacts to the transistor regions as shown in Fig. 67.27 (*f*). The thin layer of silicon dioxide is etched away and aluminium is deposited over the surface of the wafer by evaporation or sputtering. The ***fourth mask*** defines the interconnection pattern that is etched in the aluminium, as shown in Fig. 67.27 (*g*). This is followed by the deposition of a protective passivation layer of phosphosilicate glass (called *P*-glass) over the entire surface of the wafer.

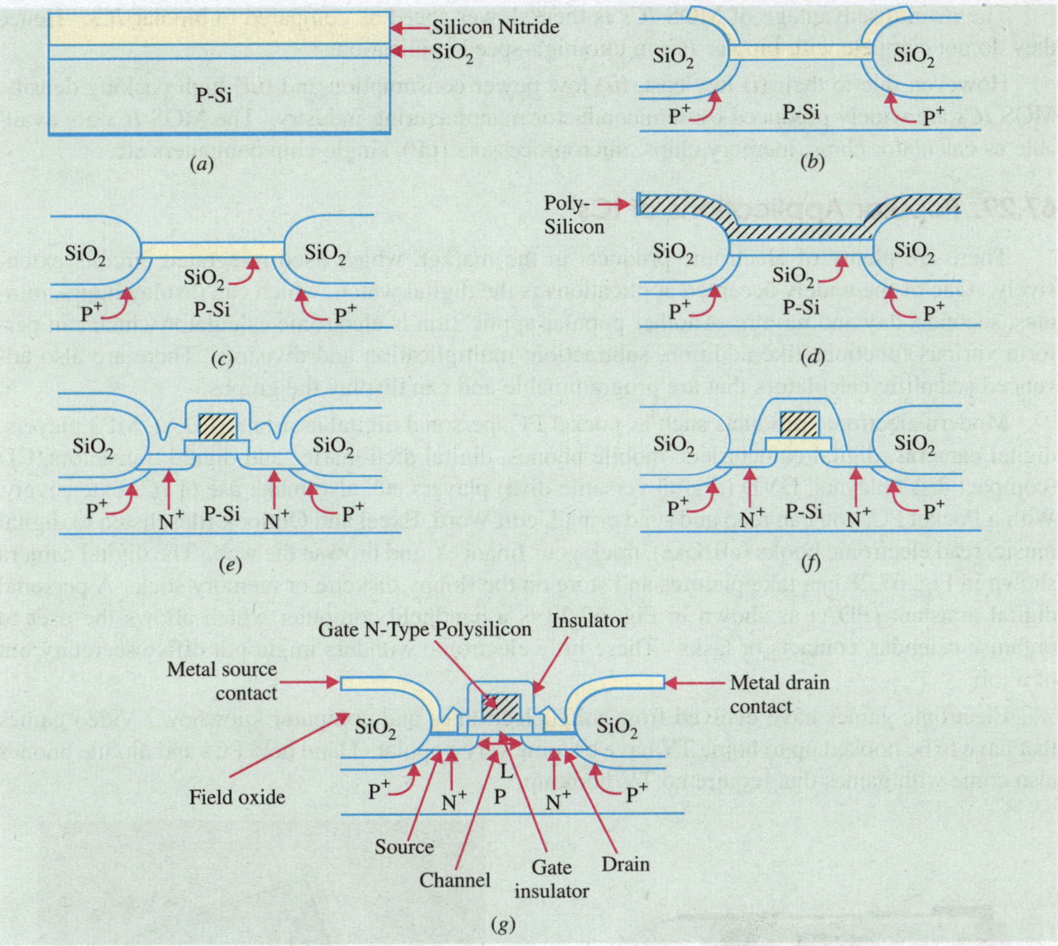

Fig. 67.27

Just as with the BJT, the ***fifth mask*** is used to open windows, so that bonding wires can be connected to the pads on the *IC* chip.

67.28. Comparison of ICs based on MOS and Bipolar Transistor Technology

1. The *ICs* based on MOS are less costly to fabricate as compared to those using bipolar transistor. The reason for this is that MOS devices are self-isolating. Bipolar transistors require tubs to isolate devices from each other on one integrated circuit. Isolation in MOS devices is provided by heavy doping and a thick oxide in the regions between adjacent devices.

2. MOS circuits consume less DC power as compared to bipolar transistor circuits.

3. The MOS transistor has lower value of transconductance (g_m) as compared to the bipolar transistor. This feature makes the bipolar transistor *ICs* superior to MOS circuits for analog circuit applications.

4. For the same channel length and the base width, the limiting cut-off frequency of the MOS transistor is better than a bipolar transistor. Therefore the MOS transistor has a higher bandwidth than the bipolar transistor.

5. The packing density of MOS *ICs* is at least 10 times more than that for bipolar *ICs*. Also, a MOS resistor occupies less than 1% of the area of a conventional diffused resistor. This high packing density makes MOS *ICs* especially suited for LSI, VLSI and ULSI circuits.

The main disadvantage of MOS *ICs* is their slower speed as compared to bipolar *ICs*. Hence they do not compete with bipolar *ICs* in ultrahigh-speed applications.

However, due to their *(i)* low cost, *(ii)* low power consumption and *(iii)* high packing density, MOS *ICs* are widely produced by semiconductor manufacturing industry. The MOS *ICs* are available as calculator chips, memory chips, microprocessors (μP), single-chip computers etc.

67.29. Popular Applications of ICs

There are plenty of electronic products in the market, which used integrated circuits extensively. One of the widely accepted applications is the digital watch, which can display hours, minutes, seconds, day and month. Another popular application is electronic calculator which can perform various functions like addition, subtraction, multiplication and division. There are also advanced scientific calculators that are programmable and can display the graphs.

Modern electronic products such as pocket PC, personal digital assistant (PDA), MP3 players, digital cameras, digital camcorders, mobile phones, digital dictionaries and digital translators, CD (compact disk) players, DVD (digital versatile disk) players etc. also make use of *ICs* extensively. With a Pocket PC, you can read and send e-mail, edit Word, Excel and Outlook files, listen to digital music, read electronic books (eBooks), track your finances, and browse the web. The digital camera shown in Fig. 67.28 can take pictures and store on the floppy diskette or memory stick. A personal digital assistant (PDA) as shown in Fig. 67.29 is a handheld computer which allows the user to organise calendar, contacts or tasks. These little electronic wonders might put office secretary out of a job.

Electronic games have evolved from microelectronics and computer knowhow. Video games that have to be hooked up to home TV have become very popular. Hand held PCs and mobile phones also come with games that require no TV hook-up.

Fig. 67.28
(courtesy Sony Corporation, Japan)

Fig. 67.29
(courtesy : 3Com, USA)

OBJECTIVE TESTS – 67

1. First integrated circuit chip was developed by
   (*a*) C.V. Raman
   (*b*) W.H. Brattain
   (*c*) J.S. Kilby
   (*d*) Robert Noyce.

2. An integrated electronic circuit is
   (*a*) a complicated circuit
   (*b*) an integrating device
   (*c*) much costlier than a single transistor
   (*d*) fabricated on a tiny silicon chip.

3. Most important advantage of an IC is its
   (*a*) easy replacement in case of circuit failure
   (*b*) extremely high reliability
   (*c*) reduced cost
   (*d*) low power consumption.
4. In monolithic ICs, all components are fabricated by .... process.
   (*a*) evaporation
   (*b*) sputtering
   (*c*) diffusion
   (*d*) oxidisation.
5. Monolithic ICs are fabricated within a
   (*a*) soft stone
   (*b*) single stone
   (*c*) silicon layer
   (*d*) ceramic base.
6. As compared to monolithic ICs, film ICs have the advantage of
   (*a*) better high-frequency response
   (*b*) much reduced cost
   (*c*) smaller size
   (*d*) less flexibility in circuit design.
7. Monolithic technique is ideally suited for fabricating .... ICs.
   (*a*) thin-film
   (*b*) digital
   (*c*) linear
   (*d*) thick-film
8. In the context of IC fabrication, metallisation means
   (*a*) connecting metallic wires
   (*b*) forming interconnecting conduction pattern and bonding pads
   (*c*) depositing SiO_2 layer
   (*d*) covering with a metallic cap.
9. Monolithic transistors are formed in the epitaxial *N*-layer
   (*a*) in one operation
   (*b*) by evaporation process
   (*c*) by successive impurity diffusions
   (*d*) by oxidisation.
10. The major component of a MOS IC is a/an
   (*a*) FET (*b*) MOSFET
   (*c*) BJT (*d*) SCR.
11. Processing of MOS ICs is less expensive than bipolar ICs primarily because they
   (*a*) use cheaper components
   (*b*) need no component isolation
   (*c*) require much less diffusion steps
   (*d*) have very high packing density.
12. In an integrated circuit, the SiO_2 layer provides
   (*a*) electrical connection to the external circuit
   (*b*) physical strength
   (*c*) isolation
   (*d*) conducting path.
13. The foundation on which an IC is built is called an
   (*a*) insulator (*b*) base
   (*c*) wafer (*d*) plate
14. Monolithic IC construction uses
   (*a*) discrete components
   (*b*) high-value resistors
   (*c*) connecting wires
   (*d*) extensive number of components

ANSWERS

1. (*c*) 2. (*d*) 3. (*b*) 4. (*c*) 5. (*c*) 6. (*a*) 7. (*b*) 8. (*b*) 9. (*c*) 10. (*b*) 11. (*c*) 12. (*c*) 13. (*c*) 14. (*d*)

ROUGH WORK

CHAPTER 68

OP-AMP AND ITS APPLICATIONS

Learning Objectives

- What is an OP-AMP?
- OP-AMP Symbol
- Polarity Conventions
- Ideal Operational Amplifier
- Virtual Ground and Summing Point
- Why V_i is Reduced to Almost Zero?
- OP-AMP Applications
- Linear Amplifier
- Unity Follower
- Adder or Summer
- Subtractor
- Integrator
- Differentiator
- Comparator
- Audio Amplifier
- OP-Amp Based Oscillator Circuits
- OP-Amp Based Wien Bridge Oscillator
- OP-Amp Based Crystal Oscillator
- A Triangular-wave Oscillator
- A Voltage- controlled Sawtooth Oscillator (VCO)
- A Square-wave Relaxation Oscillator
- High-impedance Voltmeter
- Active Filters
- Low-pass Filter
- High-pass Filter
- Band-pass Filter
- Notch Filter

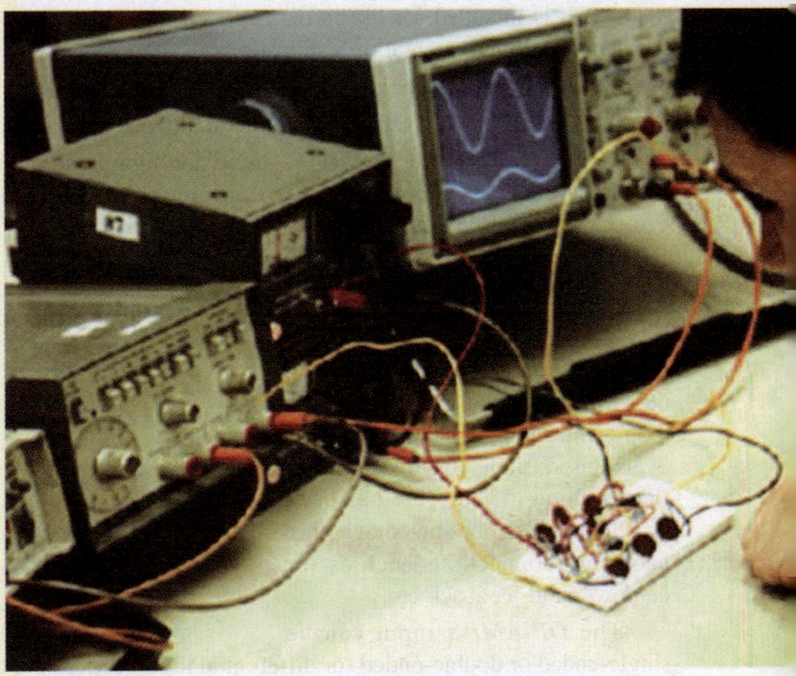

The operational amplifier was designed to perform mathematical operations. Although now superseded by the digital computer, op-amps are a common feature of modern analog electronics

68.1 What is an OP-AMP ?

It is a very high-gain, high-r_{in} directly-coupled negative-feedback amplifier which can amplify signals having frequency ranging from **0 Hz to a little beyond 1 MHz.** They are made with different internal configurations in linear *ICs*. An *OP-AMP* is so named because it was originally designed to perform mathematical operations like summation, subtraction, multiplication, differentiation and integration etc. in analog computers. Present day usage is much wider in scope but the popular name *OP-AMP* continues.

Typical uses of *OP-AMP* are : scale changing, analog computer operations, in instrumentation and control systems and a great variety of phase-shift and oscillator circuits. The *OP-AMP* is available in three different packages *(i)* standard dual-in-line package (*DIL*) *(ii)* *TO*-5 case and *(iii)* the flat-pack.

Although an *OP-AMP* is a complete amplifier, it is so designed that external components (resistors, capacitors etc.) can be connected to its terminals to change its external characteristics. Hence, it is relatively easy to tailor this amplifier to fit a particular application and it is, in fact, due to this versatility that *OP-AMPs* have become so popular in industry.

An *OP-AMP IC* may contain two dozen transistors, a dozen resistors and one or two capacitors.

Example of OP-AMPs

1. μA 709—is a high-gain operational amplifier constructed on a single silicon chip using planar epitaxial process.

   It is intended for use in dc servo systems, high-impedance analog computers and in low-level instrumentation applications.

   It is manufactured by Semiconductors Limited, Pune.
2. [*LM* 108 – *LM* 208]— Manufactured by Semiconductors Ltd. Bombay,
3. *CA* 741 *CT* and *CA* 741 *T*—these are high-gain operational amplifiers which are intended for use as *(i)* comparator, *(ii)* integrator, *(iii)* differentiator, *(iv)* summer, *(v)* dc amplifier, *(vi)* multivibrator and *(vii)* bandpass filter.

   Manufactured by Bharat Electronics Ltd (*BEL*), Bangalore.

68.2. OP-AMP Symbol

Standard triangular symbol for an *OP-AMP* is shown in Fig. 68.1 (*a*) though the one shown in Fig. 68.1 (*b*) is also used often. In Fig. 68.1 (*b*), **the common ground line has been omitted.** It also does not show other necessary connections such as for dc power and feedback etc.

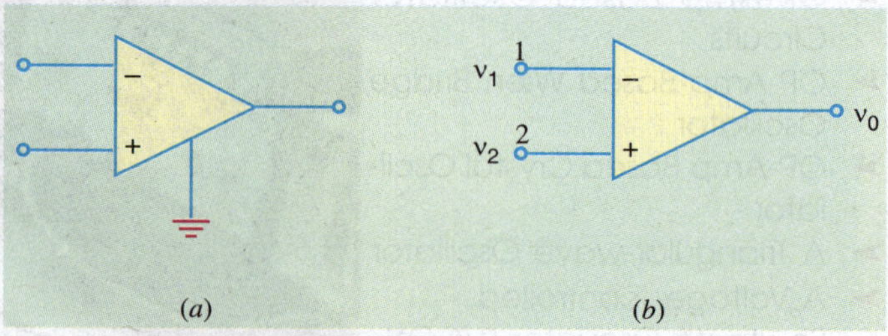

Fig. 68.1

The *OP-AMP's* input can be single-ended or double-ended (or differential input) depending on whether input voltage is applied to one input terminal only or to both. Similarly, amplifier's output can also be either single-ended or double-ended. The most common configuration is ***two input terminals and a single output.***

All OP-AMPs have a minimum of five terminals :

1. inverting input terminal,
2. non-inverting input terminal,
3. output terminal,
4. positive bias supply terminal,
5. negative bias supply terminal.

68.3. Polarity Conventions

In Fig. 68.1 (*b*), the input terminals have been marked with minus (–) and plus (+) signs. These are meant to **indicate the inverting and non-inverting terminals only** [Fig. 68.2]. It simply means that a signal applied at negative input terminal will appear amplified but **phase-inverted at the output terminal** as shown in Fig. 68.2 (*b*). Similarly, signal applied at the positive input terminal will appear amplified and **inphase** at the output. Obviously, these plus and minus polarities **indicate phase reversal only**. It does not mean that voltage v_1 and v_2 in Fig. 68.2 (*a*) are negative and positive respectively. Additionally, it also does not imply that a ***positive input voltage has to be connected to the plus-marked non-inverting terminal 2 and negative input voltage to the negative-marked inverting terminal 1.*** In fact, the amplifier can be used 'either way up' so to speak. It may also be noted that all input and output voltages are referred to a common reference usually the ground shown in Fig. 68.1 (*a*).

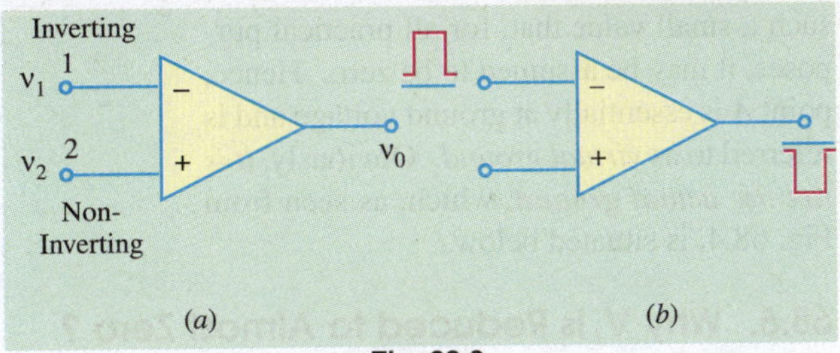

Fig. 68.2

68.4. Ideal Operational Amplifier

When an *OP-AMP* is operated without connecting any resistor or capacitor from its output to any one of its inputs (*i.e.*, without feedback), it is said to be in the **open-loop condition**. The word 'open loop' means that ***feedback path or loop is open.*** The specifications of *OP-AMP* under such condition are called open-loop specifications.

An ideal *OP-AMP* (Fig. 68.3) has the following characteristics :

1. its open-loop gain A_v is ***infinite*** *i.e.*, $A_v = -\infty$
2. its input resistance R_i (measured between inverting and non-inverting terminals) is ***infinite*** *i.e.*, $R_i = \infty$ ohm
3. its output resistance R_0 (seen looking back into output terminals) is ***zero*** *i.e.*, $R_0 = 0\ \Omega$
4. it has ***infinite bandwith*** *i.e.*, it has flat frequency response from dc to infinity.

Though these characteristics cannot be achieved in practice, yet an ideal *OP-AMP* serves as a convenient reference against which real *OP-AMPs* may be evaluated.

Following additional points are worth noting :

1. infinte input resistance means that input current $i = 0$ as indicated in Fig. 68.3. It means that an ideal *OP-AMP* is a voltage-controlled device.
2. $R_0 = 0$ means that v_0 is not dependent on the load resistance connected across the output.
3. though for an ideal *OP-AMP* $A_v = \infty$, for an actual one, it is extremely high *i.e.*, about 10^6. However, it does not mean that 1 V signal will be amplified to 10^6 V at the output. Actually, the maximum value of v_0 is limited by the basis supply voltage, typically ± 15 V. With $A_v = 10^6$ and $v_0 = 15$, V_2 the maximum value of input voltage is limited to $15/10^6 = 15\ \mu$V. Though 1 μV in the *OP-AMP*, can certainly become 1 V.

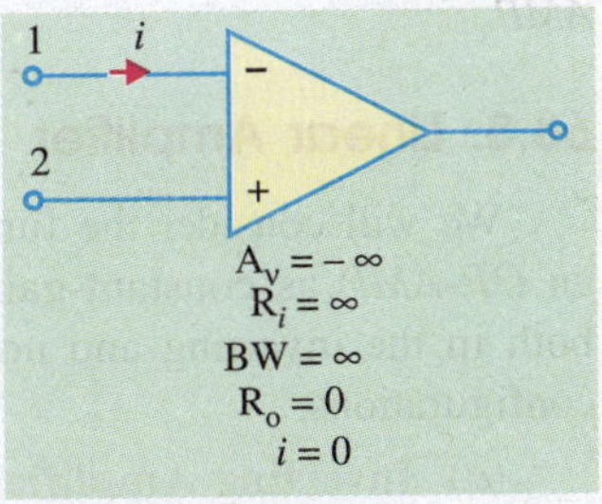

Fig. 68.3

68.5. Virtual Ground and Summing Point

In Fig. 68.4 is shown an *OP-AMP* which *employs* **negative feedback** with the help of resistor R_f which feeds a portion of the output to the input.

Since input and feedback currents are algebraically added at point *A*, it is called the **summing point**.

The concept of **virtual** ground arises from the fact that input voltage v_i at the inverting terminal of the *OP-AMP* is forced to such a small value that, for all practical purposes, it may be assumed to be zero. Hence, point *A* is essentially at ground voltage and is referred to as *virtual ground*. Obviously, ***it is not the actual ground,*** which, as seen from Fig. 68.4, is situated below.

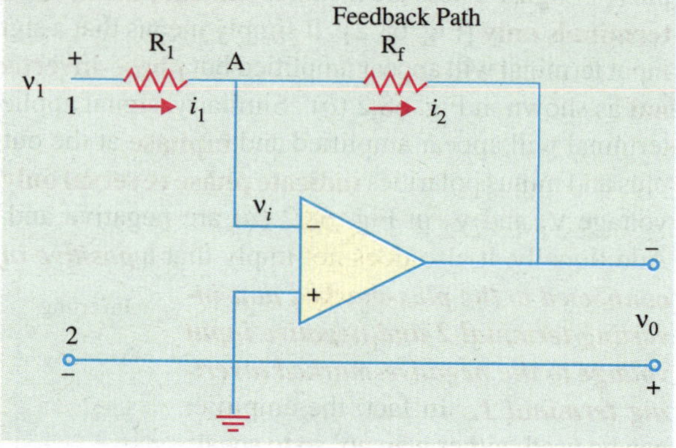

Fig. 68.4

68.6. Why V_i is Reduced to Almost Zero ?

When v_1 is applied, point *A* attains some positive potential and at the same time v_0 is brought into existence. Due to negative feedback, some fraction of the output voltage is fed back to point *A* antiphase with the voltage already existing there (due to v_1).

The algebraic sum of the two voltages is almost zero so that $v_i \cong 0$. Obviously, v_i will become exactly zero when **negative feedback voltage at A is exactly equal to the positive voltage produced by v_1 at A.**

Another point worth considering is that there exists a virtual short between the two terminals of the *OP-AMP* because $v_i = 0$. It is virtual because no current flows (remember $i = 0$) despite the existence of this short.

68.7. OP-AMP Applications

We will consider the following applications :

1. as scalar or linear (*i.e.*, small-signal) constant-gain amplifier both inverting and non-inverting,
2. as unity follower,
3. Adder or Summer
4. Subtractor,
5. Integrator
6. Differentiator
7. Comparator.

Now, we will discuss the above circuits one by one assuming an ideal *OP-AMP*.

68.8. Linear Amplifier

We will consider the functioning of an *OP-AMP* as constant-gain amplifier both in the inverting and non-inverting configurations.

(*a*) Inverting Amplifier or Negative Scale.

As shown in Fig. 68.5, noninverting

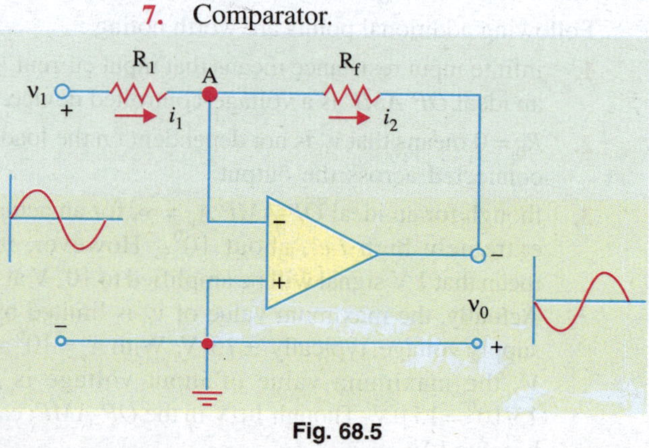

Fig. 68.5

terminal has been grounded, whereas R_1 connects the input signal v_1 to the inverting input. A feedback resistor R_f has been connected from the output to the inverting input.

Gain

Since point A is at ground potential*, $i_1 = \frac{v_{in}}{R_1} = \frac{v_1}{R_1}$

$$i_2 = \frac{-v_0}{R_f} \quad \text{Please note –ve sign}$$

Using KCL (Art. 2.2) for point A,

$$i_1 + (-i_2) = 0 \quad \text{or} \quad \frac{v_1}{R_1} + \frac{v_0}{R_f} = 0 \quad \text{or} \quad \frac{v_0}{R_f} = -\frac{v_1}{R_1} \quad \text{or} \quad \frac{v_0}{v_1} = -\frac{R_f}{R_1}$$

$$\therefore \quad A_v = -\frac{R_f}{R_1} \quad \text{or} \quad A_v = -K \quad \text{Also, } v_0 = -Kv_{in}$$

It is seen from above, that closed-loop gain of the inverting amplifier depends on the ratio of the two external resistors R_1 and R_f and is independent of the amplifier parameters.

It is also seen that the *OP-AMP* works as a negative scaler. It scales the input *i.e.*, it multiplies the input by a minus constant factor K.

(*b*) Non-inverting Amplifier or Positive Scaler

This circuit is used when there is need for an output which is equal to the input multiplied by a positive constant. Such a positive scaler circuit which uses negative feedback but provides an output that equals the input multiplied by a positive constant is shown in Fig. 68.6.

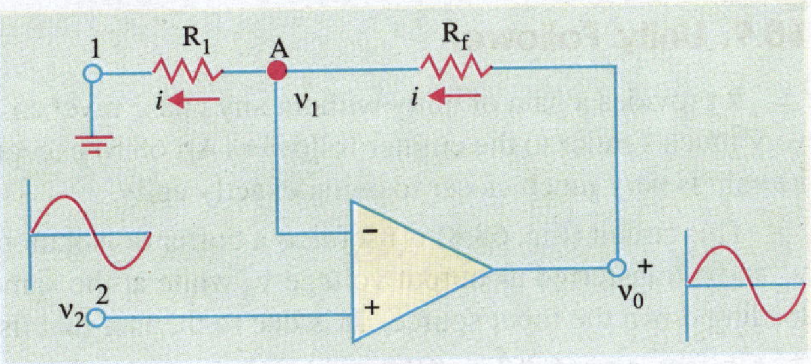

Fig. 68.6

Since input voltage v_2 is applied to the non-inverting terminal, the circuits is also called **non-inverting amplifier.**

Here, polarity of v_0 is the same as that v_2 *i.e.*, both are positive.

Gain

Because of virtual short between the two *OP-AMP* terminals, voltage across R_1 is the input voltage v_2. Also, v_0 is applied across the series combination of R_1 and R_f.

$$\therefore \quad v_{in} = v_2 = iR_1, \; v_0 = i(R_1 + R_f)$$

$$\therefore \quad A_v = \frac{v_0}{v_{in}} = \frac{i(R_1 + R_f)}{iR_1} \quad \text{or} \quad A_v = \frac{R_1 + R_f}{R_1} = \left(1 + \frac{R_f}{R_1}\right)$$

Alternative Derivation

As shown in Fig. 68.7, let the currents through the two resistors be i_1 and i_2.

The voltage across R_1 is v_2 and that across R_f is $(v_0 - v_2)$.

$$\therefore \quad i_1 = \frac{v_2}{R_1} \quad \text{and} \quad i_2 = \frac{v_0 - v_2}{R_f}$$

Applying KCL to junction A, we have

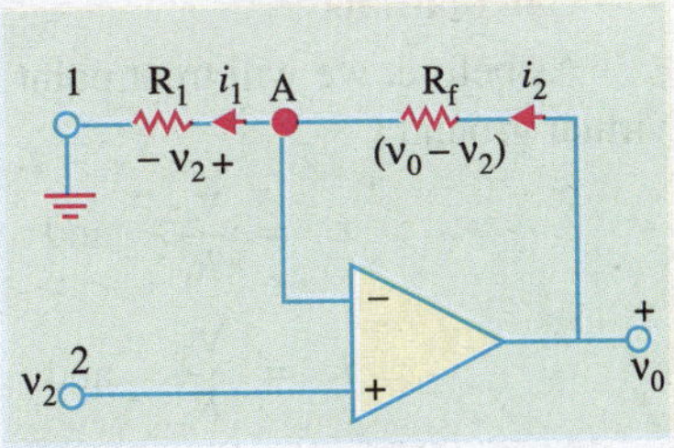

Fig. 68.7

* If not, then $i_1 = \frac{v_1 - v_i}{R_1}$ and $i_2 = \frac{v_0 - v_1}{R_f}$

$$(-i_1)+i_2 = 0 \quad \text{or} \quad -\frac{v_2}{R_1}+\frac{(v_0-v_2)}{R_f}=0$$

$$\therefore \quad \frac{v_0}{R_f} = v_2\left(\frac{1}{R_1}+\frac{1}{R_f}\right)=v_2\frac{R_1+R_f}{R_1 R_f}$$

$$\therefore \quad \frac{v_0}{v_2} = \frac{R_1+R_f}{R_1} \quad \text{or} \quad A_v = 1+\frac{R_f}{R_1} \quad \text{—as before}$$

Example 68.1. *For the inverting amplifier of Fig. 68.5, R_1 = 1 K and R_f = 1 M. Assuming an ideal OP-AMP amplifier, determine the following circuit values :*

(a) voltage gain, *(b) input resistance,* *(c) output resistance*

Solution. It should be noted that we will be calculating values of the circuit and not *for* the *OP-AMP proper.*

(a) $$A_v = -\frac{R_f}{R_1} = -\frac{1000\,K}{1\,K} = \mathbf{-1000}$$

(b) Because of virtual ground at *A*, $R_{in} = R_1 = $ **1 K**

(c) Output resistance of the circuit equal the output resistance of the *OP-AMP i.e.*, **zero ohm**.

68.9. Unity Follower

It provides a gain of unity without any phase reversal. It is very much similar to the emitter follower (Art 68.8) except that its gain is very much closer to being exactly unity.

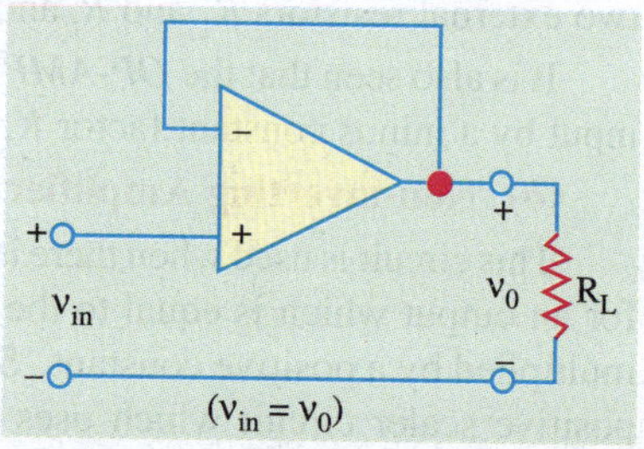

Fig. 68.8

This circuit (Fig. 68.8) is useful as a buffer or isolation amplifier because it allows, input voltage v_{in} to be transferred as output voltage v_0 while at the same time preventing load resistance R_L from loading down the input source. It is due to the fact that its $R_i = \infty$ and $R_0 = 0$.

In fact, circuit of Fig. 68.8 can be obtained from that of Fig. 68.6 by putting

$$R_1 = R_f = 0$$

68.10. Adder or Summer

The adder circuit provides an output voltage proportional to or equal to the algebraic sum of two or more input voltages each multiplied by a constant gain factor. It is basically similar to a scaler (Fig. 68.5) except that it has more than one input. Fig. 68.9 shows a three-input inverting adder circuit. As seen, *the* ***output voltage is phase-inverted.***

Calculations

As before, we will treat point *A* as virtual ground

$$i_1 = \frac{v_1}{R_1} \quad \text{and} \quad i_2 = \frac{v_2}{R_2}$$

$$i_3 = \frac{v_3}{R_3} \quad \text{and} \quad i = -\frac{v_0}{R_f}$$

Applying *KCI* to point *A*, we have

$$i_1 + i_2 + i_3 + (-i) = 0$$

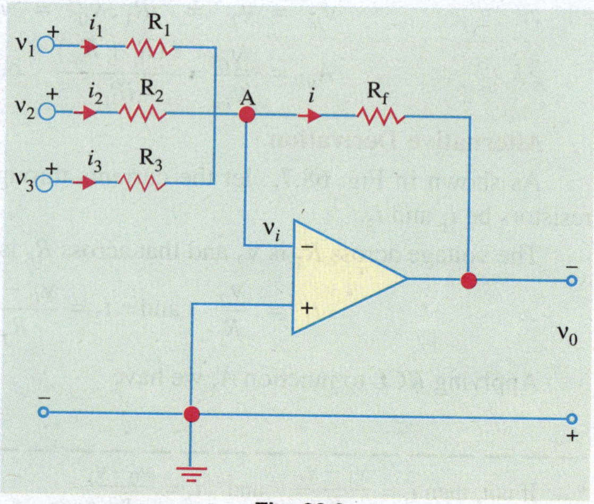

Fig. 68.9

or $$\frac{v_1}{R_1}+\frac{v_2}{R_2}+\frac{v_3}{R_3}-\left(\frac{-v_0}{R_f}\right)=0$$

$\therefore$ $$v_0=-\left(\frac{R_f}{R_1}v_1+\frac{R_f}{R_2}v_2+\frac{R_f}{R_3}v_3\right)$$

or $$v_0=-(K_1 v_1+K_2 v_2+K_3 v_3)$$

The overall negative sign is unavoidable because we are using the inverting input terminal.

If $R_1 = R_2 = R_3 = R$, then

$$v_0 = -\frac{R_f}{R}(v_1+v_2+v_3)=-K(v_1+v_2+v_3)$$

Hence, output voltage is proportional to **(*not equal to*)** the **algebraic sum** of the three input voltages.

If $R_f = R$, then output exactly equals the sum of inputs. However, if $R_f = R/3$

then $$v_0 = -\frac{R/3}{R}(v_1+v_2+v_3) = -\frac{1}{3}(v_1+v_2+v_3)$$

Obviously, the output is equal ***to the average of the three inputs.***

68.11. Subtractor

The function of a subtractor is to provide an output proportional to or equal to the difference of two input signals. As shown in Fig. 68.10 we have to apply the inputs at the inverting as well as non-inverting terminals.

Calculations

According to Superposition Theorem (Art. 2.17) $v_0 = v_0' + v_0''$

where v_0' is the output produced by v_1 and v_0'' is that produced by v_2.

Now, $$v_0' = -\frac{R_f}{R_1}\cdot v_1 \qquad \text{...Art 68.8 (a)}$$

$$v_0'' = \left(1+\frac{R_f}{R_1}\right)v_2 \qquad \text{...Art 68.8 (b)}$$

$\therefore$ $$v_0 = \left(1+\frac{R_f}{R_1}\right)v_2-\frac{R_f}{R_1}\cdot v_1$$

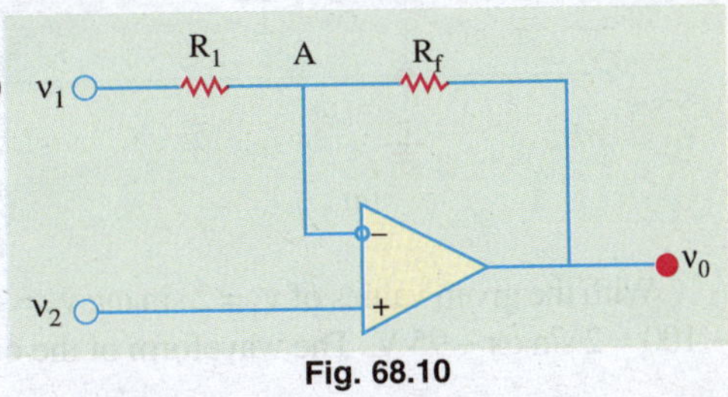

Fig. 68.10

Since $R_f \gg R_1$ and $R_f/R_1 \gg 1$, hence

$$v_0 \cong \frac{R_f}{R_1}(v_2-v_1)=K(v_2-v_1)$$

Further, If $R_f = R_1$, then

$$v_0 = (v_2 - v_1) = \text{difference of the two input voltages}$$

Obviously, if $R_f \neq R_1$, then a scale factor is introduced.

Example 68.2. *Find the output voltages of an OP-AMP inverting adder for the following sets of input voltages and resistors. In all cases, R_f = 1 M.*

$$v_1 = -3\ V, v_2 = +3V, v_3 = +2V; R_1 = 250\ K, R_2 = 500\ K, R_3 = 1\ M$$

[Electronic Engg. Nagpur Univ. 1991]

Solution. $$v_0 = -(K_1 v_1+K_2 v_2+K_3 v_3)$$

$$K_1 = \frac{R_f}{R_1}=\frac{1000\ K}{250\ K}=4, K_2=\frac{1000}{500}=2, K_3=\frac{1M}{1M}=1$$

$$\therefore \qquad v_0 = -[(4\times -3)+(2\times 3)+(1\times 2)] = +4\text{V}$$

Example 68.3. *In the subtractor circuit of Fig. 68.10, $R_1 = 5$ K, $R_f = 10$ K, $v_1 = 4$ V and $v_2 = 5$ V. Find the value of output voltage.*

Solution.
$$v_0 = \left(1+\frac{R_f}{R_1}\right)v_2 - \frac{R_f}{R_1}v_1 = \left(1+\frac{10}{5}\right)5 - \frac{10}{5}\times 4 = +7\text{V}$$

Example 68.4. *Design an OP-AMP circuit that will produce an output equal to $-(4\,v_1 + v_2 + 0{\cdot}1\,v_3)$. Write an expression for the output and sketch its output waveform when $v_1 = 2 \sin \omega t$, $v_2 = +5$ V dc and $v_3 = -100$ V dc.* **[Banglore University 2001]**

Solution.
$$v_0 = -\left[\frac{R_f}{R_1}v_1 + \frac{R_f}{R_2}v_2 + \frac{R_f}{R_3}v_3\right] \qquad ...(1)$$

and also
$$v_0 = -(4\,v_1 + v_2 + 0{\cdot}1\,v_3) \qquad ...(2)$$

Comparing equations **(1)** and **(2)**, we find,

$$\frac{R_f}{R_1} = 4, \quad \frac{R_f}{R_2} = 1, \quad \frac{R_f}{R_3} = 0{\cdot}1$$

Therefore if we assume $R_f = 100$ K, then $R_1 = 25$ K, $R_2 = 100$ K and $R_3 = 1000$ K. With there values of R_1, R_2 and R_3, the *OP-AMP* circuit is as shown in Fig. 68.11(*a*)

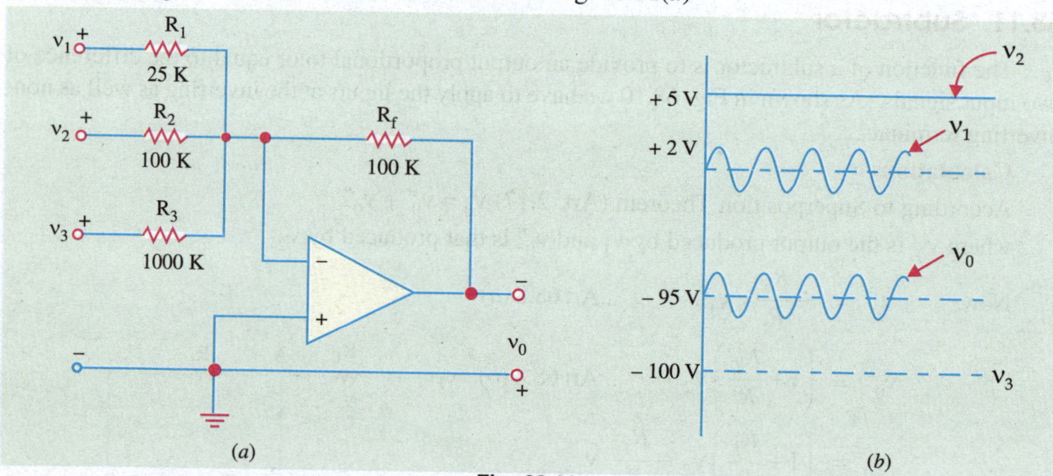

Fig. 68.11

With the given values of $v_1 = 2 \sin \omega t$, $v_2 = +5$V, $v_3 = -100$ V dc, the output voltage, $v_0 = 2 \sin \omega t + 5 - 100 = 2 \sin \omega t - 95$ V. The waveform of the output voltage is sketched as shown in Fig. 68.11 (*b*).

68.12. Integrator

The function of an integrator is to provide an output voltage which is proportional to the integral of the input voltage.

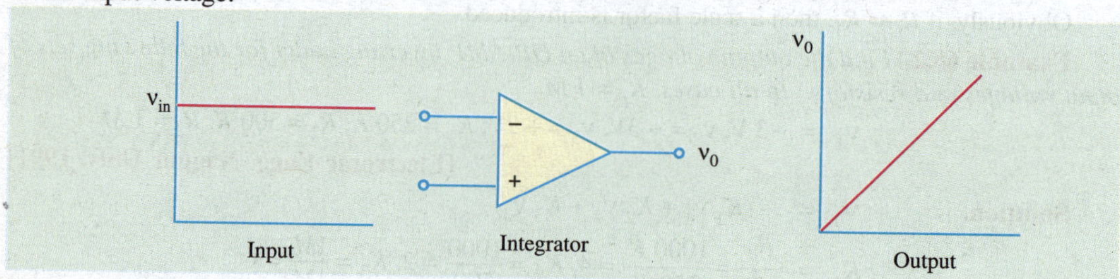

Fig. 68.12

A simple example of integration is shown in Fig. 68.12 where input is dc level and its integral is ***a linearly-increasing ramp output***. The actual integration circuit is shown in Fig. 68.13. This circuit is similar to the scaler circuit of Fig. 68.5 except that **the feedback component is a capacitor C instead of a resistor R_f.**

Calculations

As before, point *A* will be treated as virtual ground.

$$i_1 = \frac{v_1}{R}\ ;\ i_2 = -\frac{v_0}{X_C} = -\frac{v_0}{1/j\omega C} = -\frac{v_0}{1/sC} = -s\,C\,v_0$$

where $s = j\,\omega$ in the Laplace notation.

Now $i_1 = i_2$ ...Art. 68.8 (*a*)

$$\therefore \quad \frac{v_1}{R} = -s\,Cv_0$$

$$\therefore \quad \frac{v_0}{v_{in}} = \frac{v_0}{v_1} = -\frac{1}{s\,CR} \quad \text{...}(i)$$

$$\therefore \quad A_v = -\frac{1}{sCR}$$

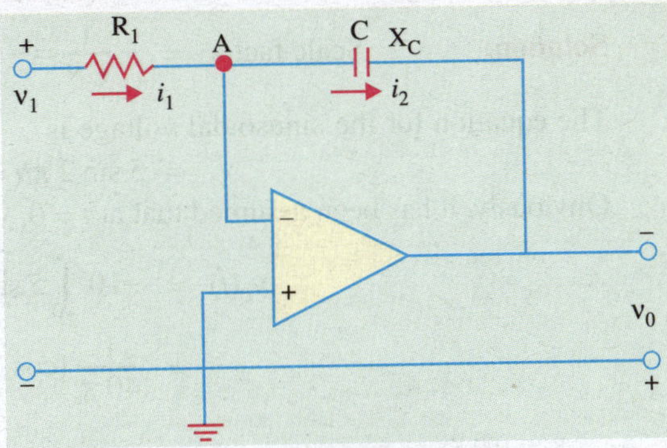

Fig. 68.13

Now, the expression of Eq. (*i*) can be written in time domain as

$$v_0(t) = \frac{1}{40\,\pi}(\cos 2000\,\pi t - 1)$$

It is seen from above that output (right-hand side expression) is an integral of the input, with an inversion and a scale factor of 1/*CR*.

This ability to integrate a given signal enables an analog computer solve differential equations and to set up a wide variety of electrical circuit analogs of physical system operation. For example, let

$R = 1$ M and $C = 1\ \mu$F. Then

$$\text{scale factor} = -\frac{1}{CR} = -\frac{1}{10^6 \times 10^{-6}} = -1$$

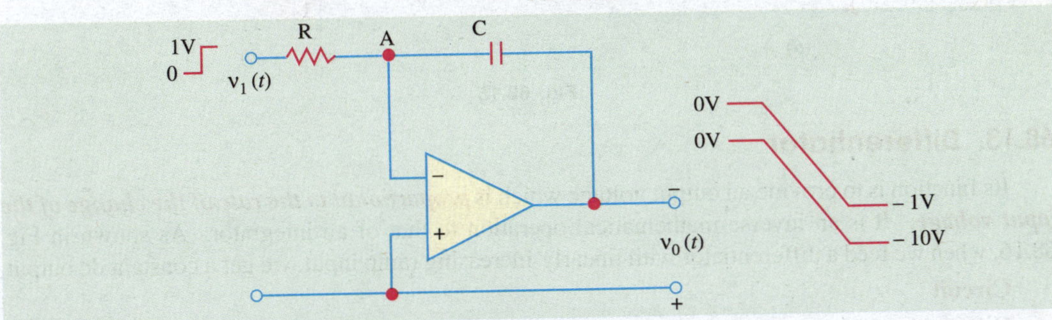

Fig. 68.14

As shown in Fig. 68.14 the input is a step voltage, whereas output is a ramp (or linearly-changing voltages) with a scale multiplier of −1. However, when *R* = 100 K, then

$$\text{scale factor} = -\frac{1}{10^5 \times 10^{-6}} = -10$$

$$\therefore \quad v_0(t) = -10\int v_1(t)\,.\,dt$$

It is also shown in Fig. 68.14. Of course, we can integrate more than one input as shown below in Fig. 68.15. With multiple inputs, the output is given by

$$v_0(t) = -\left[K_1 \int v_1(t)\,dt + K_2 \int v_2(t)\,dt + k_3 \int v_3(t)\,dt\right]$$

where $$K_1 = \frac{1}{CR_1}, K_2 = \frac{1}{CR_2} \text{ and } K_3 = \frac{1}{CR_3}$$

Fig. 68.15 (*a*) shows a summing integrator as used in an analog computer. It shows all the three resistors and the capacitor. The analog computer representation of Fig. 68.15 (*b*) indicates only the scale factor for each input.

Example 68.5. *A 5-mV, 1-kHz sinusoidal signal is applied to the input of an OP-AMP integrator of Fig. 64.37 for which R = 100 K and C = 1 μF. Find the output voltage.*

[Electronic & Comm. Engg. Kurukshetra Univ. 1990]

Solution. Scale factor $= -\frac{1}{CR} = -\frac{1}{10^5 \times 10^{-6}} = \mathbf{-10}$

The equation for the sinusoidal voltage is

$$v_1 = 5 \sin 2\pi ft = 5 \sin 2000\,\pi t$$

Obviously, it has been assumed that at $t = 0$, $v_1 = 0$

$$\therefore \quad v_0(t) = -10 \int_0^t 5 \sin 2000\,\pi t = -50 \left| \frac{-\cos 2000\,\pi t}{2000\pi} \right|_0^t$$

$$= \frac{1}{40\,\pi} (\cos 2000\,\pi t - 1)$$

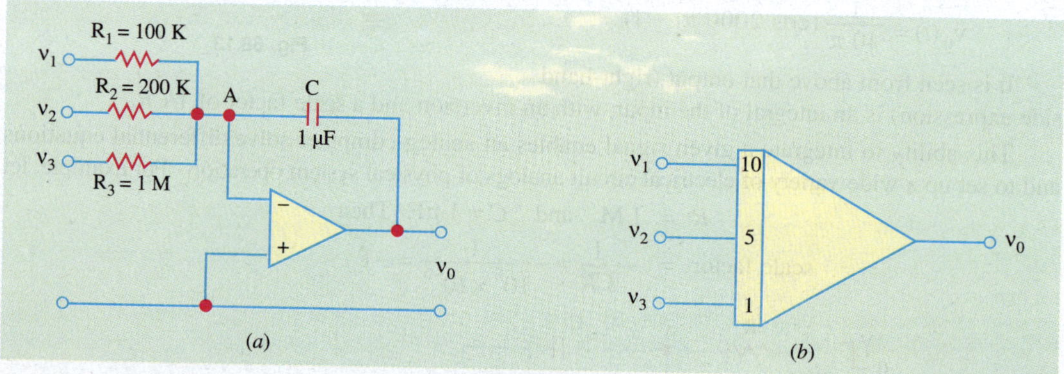

Fig. 68.15

68.13. Differentiator

Its function is to provide an output voltage which is ***proportional to the rate of the change of the input voltage.*** It is an inverse mathematical operation to that of an integrator. As shown in Fig. 68.16, when we feed a differentiator with linearly-increasing ramp input, we get a constant dc output.

Circuit

Differentiator circuit can be obtained by interchanging the resistor and capacitor of the integrator circuit of Fig. 68.13.

Let i = rate of change of charge

$= \frac{dq}{dt}$

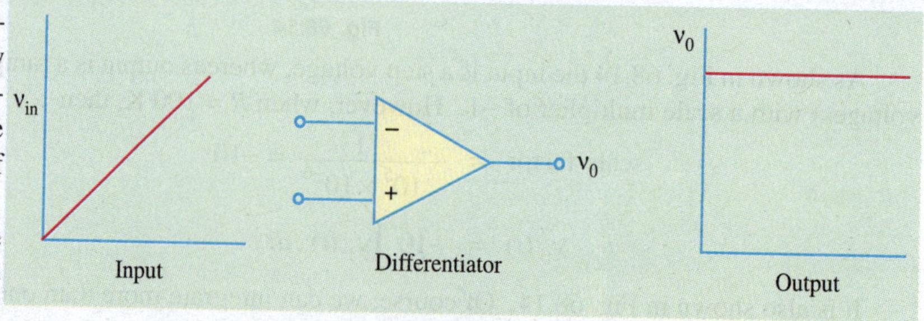

Fig. 68.16

Now, $q = Cv_c$

$$\therefore \quad i = \frac{d}{dt}(Cv_c) = C\frac{dv_c}{dt}$$

Taking point *A* as virtual ground

$$v_0 = -iR = -\left(C.\frac{dv_c}{dt}\right)R = -CR.\frac{dv_c}{dt}$$

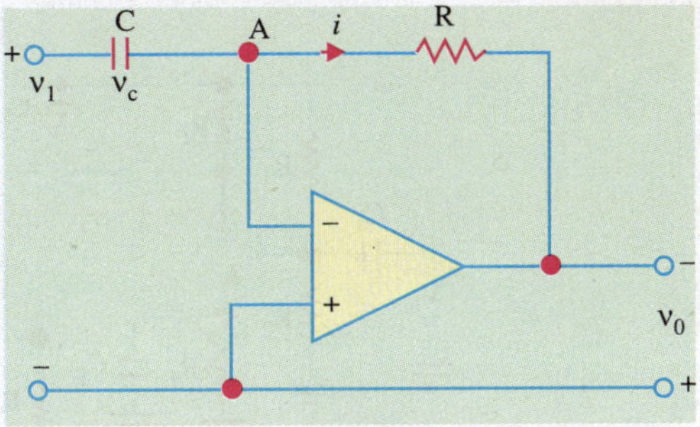

Fig. 68.17

As seen, output voltage is proportional to the derivate of the input voltage, the constant of proportionality (*i.e.,* scale factor) being (*–RC*).

Example 68.6. *The input to the differentiator circuit of Fig. 68.17 is a sinusoidal voltage of peak value of 5 mV and frequency 1 kHz. Find out the output if R = 1000 K and C = 1 μF.*

Solution. The equation of the input voltage is

$$v_1 = 5 \sin 2\pi \times 1000\,t = 5 \sin 2000\,\pi t \text{ mV}$$

$$\text{scale factor} = -CR = -10^{-6} \times 10^5 = 0\cdot1$$

$$v_0 = -0\cdot1\frac{d}{dt}(5 \sin 2000\,\pi t) = (0\cdot5 \times 2000\,\pi) \cos 2000\,\pi t = -1000\,\pi \cos 2000\,\pi t \text{ mV}$$

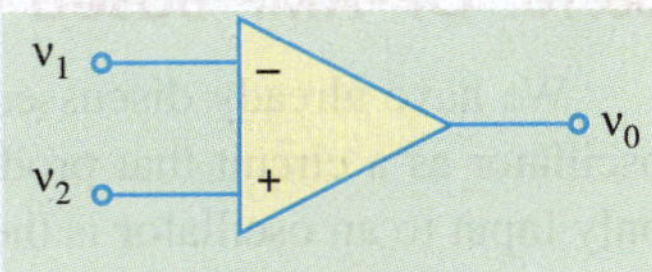

Fig. 68.18

As seen, output is a cosinusoidal voltage of frequency 1 kHz and peak value 1000 π mV.

68.14. Comparator

It is a circuit which compares two signals or voltage levels. The circuit is shown in Fig. 68.18 and (like that of the unity follower) is the simplest because it needs no additional external components.

If v_1 and v_2 are equal, then v_0 should idealy be zero. Even if v_1 differs from v_2 by a very small amount, v_0 is large because of amplifier's high gain. Hence, circuit of Fig. 68.18 can detect very small changes which is another way of saying that it compares two signals,

68.15. Audio Amplifier

As a matter of fact, in most communication receivers, the final output stage is the audio amplifier. The ideal audio amplifier will have the following characteristics :

1. High gain
2. Minimum distortion in the audio frequency range (*i.e.,* 20 Hz to 20 kHz range).
3. High input resistance (or impedance).
4. Low output resistance (or impedance) to provide optimum coupling to the speaker.

The use of *OP-AMP* is an audio amplifier will fullfill the requirements listed above very nicely. An *OP-AMP* audio amplifier is shown in Fig. 68.19.

Note that the *OP-AMP* as supplied only for + *V* volt power supply, the –*V* terminal is grounded. Because of this, the output will be between the limits of (+ *V* – 1) volts and +1 volt approximately. Also notice the use of a coupling capacitor C_2 between the *OP-AMP* and speaker. This capacitor is necessary to reference the speaker signal around ground. The capacitor C_s, is included in the V_{CC} line to prevent any transient current caused by the operation of *OP-AMP* from being coupled back to Q_1 through the power supply. The high gain requirements is accomplished by the combination of two amplifier stages. The high R_m/Low R_{out} of the audio amplifier is accomplished by the *OP-AMP* itself, as the low distortion characteristic.

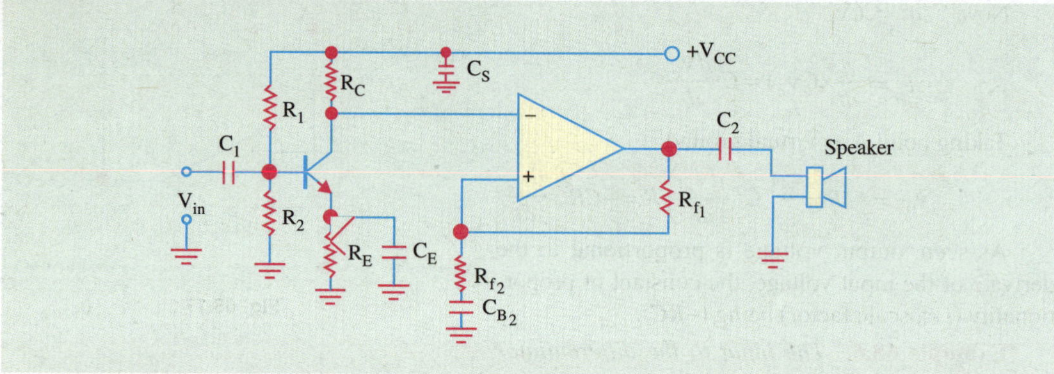

Fig. 68.19

68.16. OP-AMP Based Oscillator Circuits

We have already discussed about sinusoidal oscillators in Chapter 65. There we defined the oscillator as a circuit that produces an output waveform without any external signal source. The only input to an oscillator is the dc power supply. As such, the oscillator can be viewed as being a signal generator. We have also discussed in the same chapter, about the different types of oscillator circuits (like Wien Bridge oscillator, Colpitts Oscillator and Crystal Oscillator) using bipolar junction and field-effect transistor. Now we still study these oscillator circuits using *OP-AMP*.

68.17. OP-AMP Based Wien Bridge Oscillator

Fig. 68.20 shows the basic version of a Wien Bridge oscillator. The circuits uses an *OP-AMP* and *RC* bridge circuit. Note the basic bridge connection carefully. Resistors R_1, R_2 and capacitors $C_1$and C_2 form the frequency adjustment elements while resistors R_3 and R_4 form part of the feedback path. The *OP-AMP* output is connected as the bridge input as points '*A*' and '*C*'. The bridge circuit output at points '*B*' and '*D*' is the input to the *OP-AMP* shows an alternative way of connecting the *RC* bridge circuit to the *OP-AMP*. In a typical Wien Bridge oscillator, $R_1 = R_2 = R$, and $C_1 = C_2 = C$. This means that the two *RC* circuits will have the same cut-off frequency. The cut-off frequency is given by ($1/2\ \pi RC$).

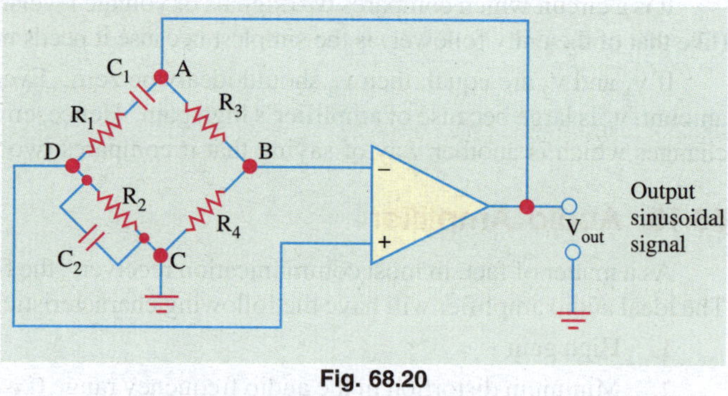

Fig. 68.20

Example 68.7. *Fig. 68.21 shows the circuit of a Wien-Bridge Oscillator using OP-AMP as an amplifier. Notice the components R_1, C_1, R_2, C_2, R_3 and R_4 are connected in the bridge configuration in the same way as shown in Fig. 68.22.*

Calculate the frequency of the Wien Bridge oscillator.

Solution. The frequency of oscillations,

$$f_0 = \frac{1}{2\pi\ RC} = \frac{1}{2\pi \times 51\ K \times 0{\cdot}001\ \mu\text{F}} = \mathbf{3{\cdot}12\ kHz}$$

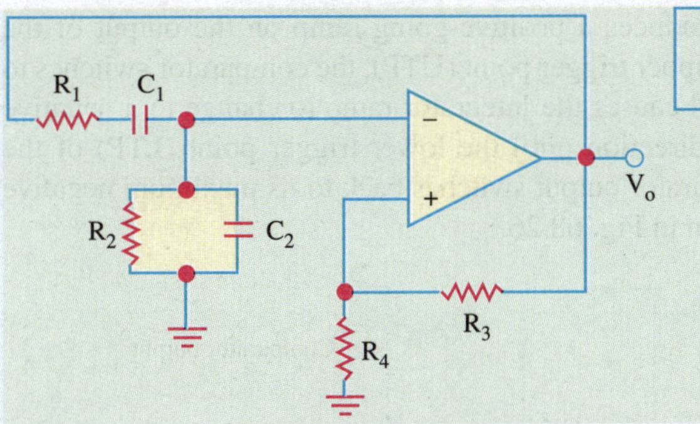

Fig. 68.21

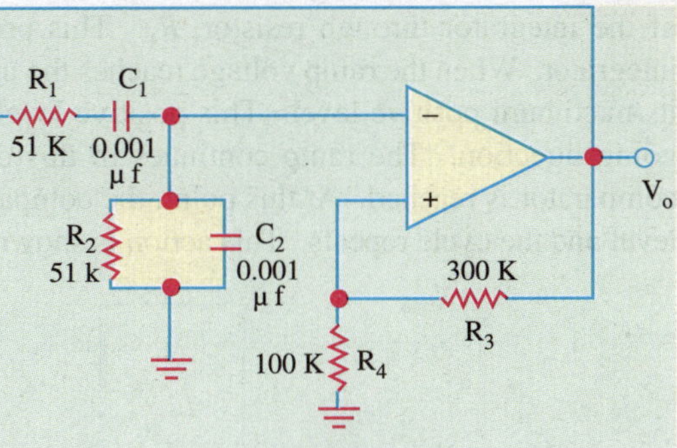

Fig. 68.22

68.18. OP-AMP Based Colpitts Oscillator

An *OP-AMP* based Colpitts oscillator is as shown in Fig. 68.23. Here the *OP-AMP* provides the basic amplification needed while a *LC* feedback network of Colpitts configuration sets the oscillator frequency.

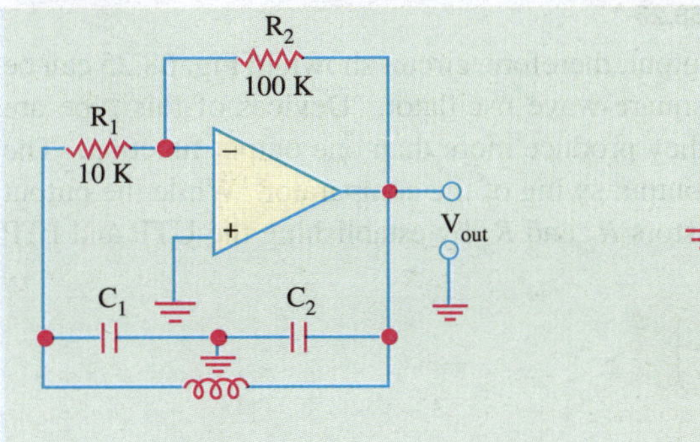

Fig. 68.23

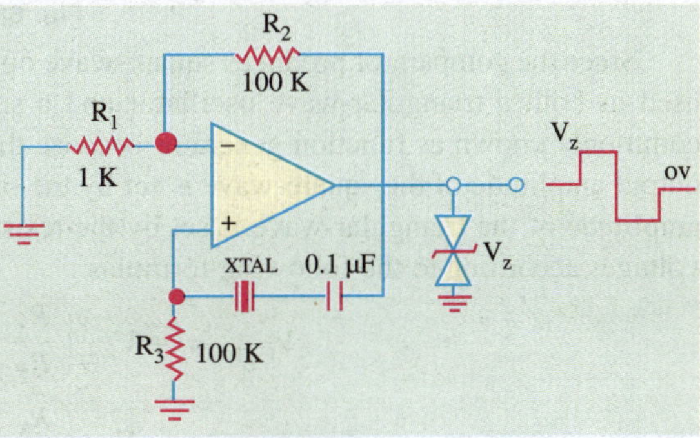

Fig. 68.24

68.19. OP-AMP Based Crystal Oscillator

An *OP-AMP* can be used in a crystal oscillator as shown in Fig. 68.24. The crystal is connected in the series resonant path and operates at the crystal series resonant frequency. The present circuit has high gain so that an output square-wave results are shown in the figure. A pair of Zener diodes is shown at the output to provide output amplitude at exactly the Zener voltage (V_Z).

68.20. A Triangular-Wave Oscillator

Fig. 68.25 shows an *OP-AMP* circuit to generate a triangular-wave from a square-wave. The circuit makes use of two *OP-AMP* : one of them is used as a comparator and the other as an integrator. The operation of the circuit is as given below :

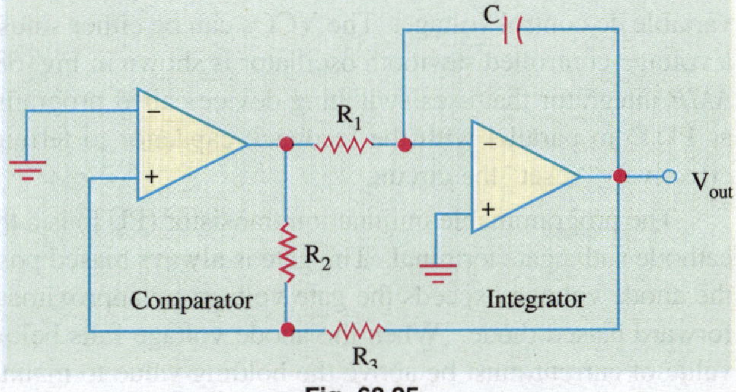

Fig. 68.25

To begin with, let us assume that the output voltage of the comparator is at its maximum negative level. This output is connected to the inverting input of the integrator through resistor, R_1. This produces a positive-going ramp on the output of the integrator. When the ramp voltage reaches the upper trigger point (UTP), the comparator switches to its maximum positive level. This positive level causes the integrator ramp to change to a negative going direction. The ramp continues in this direction until the lower trigger point (LTP) of the comparator is reached. At this point, the comparator output switches back to its maximum negative level and the cycle repeats. This action is shown in Fig. 68.26.

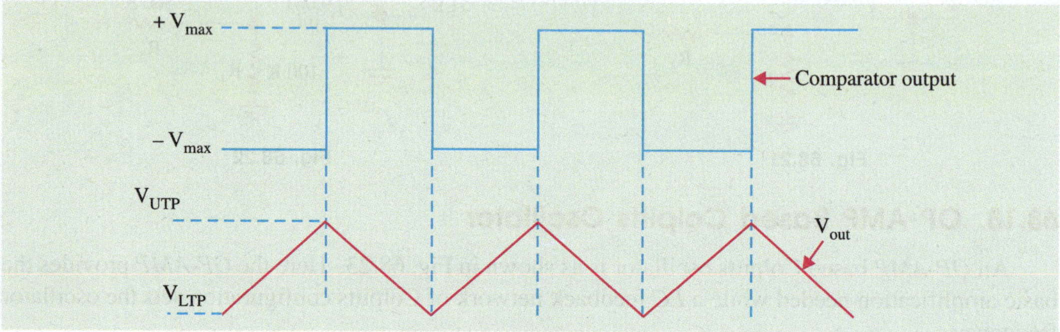

Fig. 68.26

Since the comparator produces square-wave output, therefore, circuit shown in Fig. 68.25 can be used as both a triangular-wave oscillator and a square-wave oscillator. Devices of this type are commonly known as function generator because they produce more than one output function. The output amplitude of the square-wave is set by the output swing of the comparator. While the output amplitude of the triangular-wave is set by the resistors R_2 and R_3 by establishing the UTP and LTP voltages according to the following formulas :

$$V_{\text{UTP}} = +V_{max}\left(\frac{R_3}{R_2}\right)$$

$$V_{\text{LTP}} = -V_{max}\left(\frac{R_3}{R_2}\right)$$

It may be noted that the comparator output levels, $+ V_{max}$ are equal. The frequency of both waveforms depend on the $R_1\,C$ time constant as well as the amplitude-setting resistors R_2 and R_3. By varying R_1, the frequency of oscillation can be adjusted without changing the output amplitude.

$$f = \frac{1}{4\,R_1\,C}\left(\frac{R_2}{R_3}\right)$$

68.21. A Voltage-Controlled Sawtooth Oscillator (VCO)

The voltage-controlled oscillator (VCO) is an oscillator whose frequency can be changed by a variable dc control voltage. The VCOs can be either sinusoidal or nonsinusoidal. One way to build a voltage-controlled sawtooth oscillator is shown in Fig. 68.27 (*a*). This circuit makes use of an *OP-AMP* integrator that uses switching device called programmable unijunction transistor (abbreviated as PUT) in parallel with the feedback capacitor to terminate each ramp at a prescribed level and effectively "reset" the circuit.

The programmable unijunction transistor (PUT) is a three terminal device *i.e.*, it has an anode, a cathode and a gate terminal. The gate is always biased positively with respect to the cathode. When the anode voltage exceeds the gate voltage by approximately 0·7 V, the PUT turns on and acts as a forward biased diode. When the anode voltage falls below this level, the PUT turns off. Also, the value of current must be above the holding value to maintain conduction.

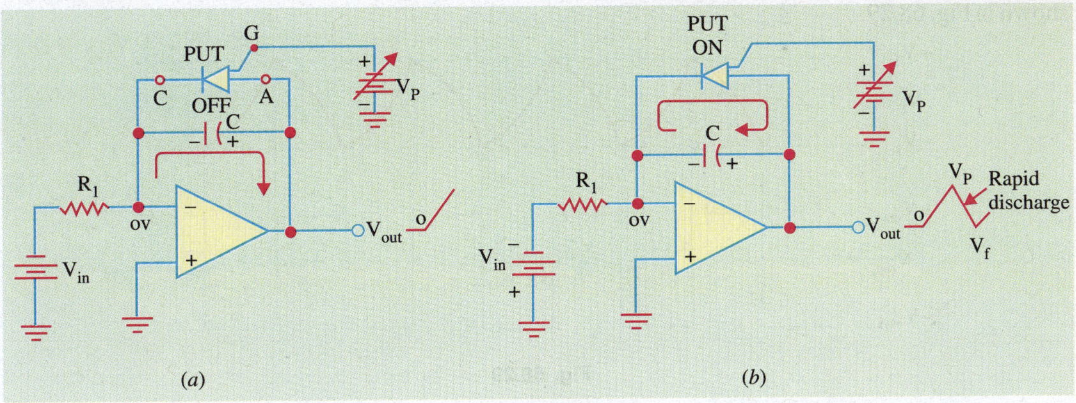

Fig. 68.27

The operation of the circuit may be explained as below. The negative dc input voltage, $-V_{IN}$, produces a positive-going ramp on the output. During the time that ramp is increasing the circuit acts as a regular integrator. When the ramp voltage (*i.e.*, voltage at PUT anode) exceeds the gate voltage by 0·7 V, the PUT turns on. This forces the capacitor to discharge rapidly as shown in Fig. 68.27 (*b*). However, the capacitor does not discharge completely to zero because of the PUT's forward voltage, V_F. Discharging of the capacitor continues until the current through PUT drops below the holding value. At this point, the PUT turn off and capacitor begins to charge again, thus generating a new output ramp. The cycle repeats and the resulting output is a repetitive sawtooth waveform as shown in the figure. It may be noted that the sawtooth amplitude and period can be adjusted by varying the PUT gate voltage.

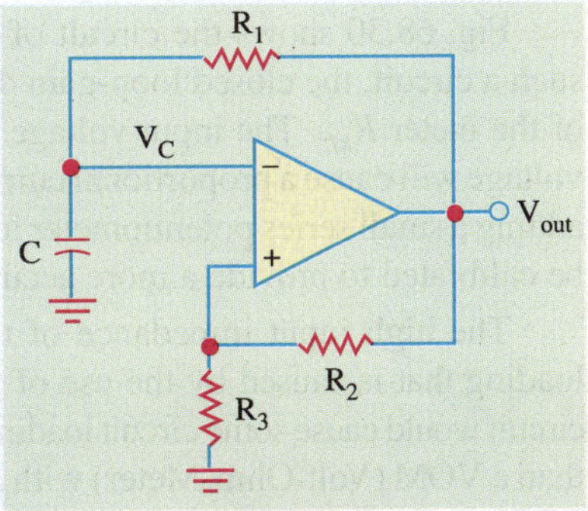

Fig. 68.28

The frequency of smooth curve is given by the relation :

$$f = \frac{|V_{IN}|}{R_1 C}\left(\frac{1}{V_P - V_F}\right)$$

It is evident from the above equation that the frequency depends upon the time constant "$R_1 C$" of the integrator and the peak voltage set by the PUT. The time period of the sawtooth wave is the reciprocal of the frequency, *i.e.*,

$$T = \frac{V_P - V_F}{|V_{IN}|/R_1 C}$$

68.22. A Square-wave Relaxation Oscillator

Fig. 68.28 shows the circuit of a basic relaxation oscillator. Its operation depends upon charging and discharging of a capacitor. Notice that the *OP-AMP*'s inverting input is the capacitor voltage and the noninverting input is a portion of the output fed-back through resistors R_2 and R_3.

When the circuit is first turned on, the capacitor is uncharged. Because of this, the inverting input is at 0 V. This makes the output a positive maximum, and capacitor begins to charge towards V_{out} through R_1. When the capacitor voltage reaches a value equal to the feedback voltage on the noninverting input, the switches to the maximum negative stage. At this point, the capacitor begins to discharge from $+V_f$ toward $-V_f$. When the capacitor voltage reaches $-V_f$ the *OP-AMP* switches back to the maximum positive state. This action continues to repeat and a square wave input is obtained as shown in Fig. 68.29.

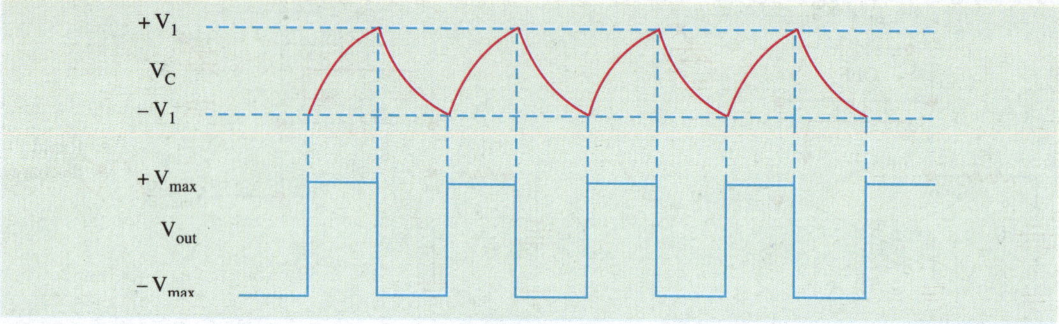

Fig. 68.29

68.23. High-impedance Voltmeter

Fig. 68.30 shows the circuit of a high impedance voltmeter. In such a circuit, the closed loop-gain depends on the internal resistance of the meter R_M. The input voltage will be amplified and the output voltage will cause a proportional current to flow through the meter. By adding a small series potentiometer in the feedback loop, the meter can be calibrated to provide a more accurate reading.

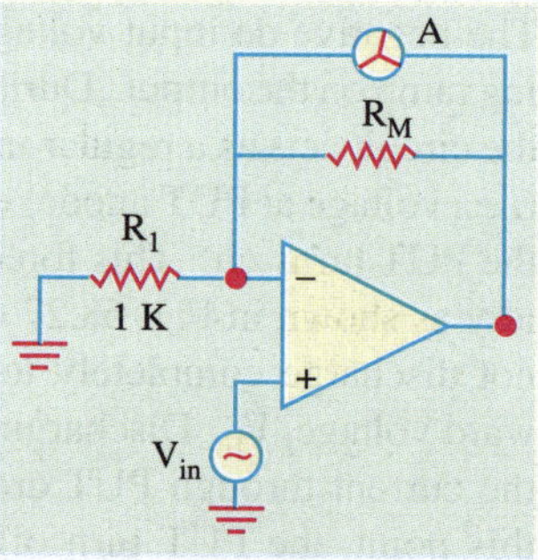

Fig. 68.30

The high input impedance of the *OP-AMP* reduces the circuit loading that is caused by the use of the meter. Although this type of circuit would cause some circuit loading, it would be much more accurate than a VOM (Volt-Ohm-Meter) with an input impedance of 20 KΩ.

68.24. Active Filters

We have already discussed in Chapter 61, about tuned amplifiers. Such amplifiers are designed to amplify only those frequencies that are within certain range. As long as the input signal is within the specified range, it will be amplified. If it goes outside of this frequency range, amplification will be drastically reduced. The tuned amplifier circuits using *OP-AMP* are generally referred to as active filters. Such circuits do not require the use of inductors. The frequency response of the circuit is determined by resistor and capacitor values.

A filter circuit can be constructed using passive components like resistors and capacitors. But an active filter, in addition to the passive components makes use of an *OP-AMP* as an amplifier. The amplifier in the active filter circuit may provide voltage amplification and signal isolation or buffering.

There are four major types of filters namely, low-pass filter, high-pass filter, band-pass filter and band-stop or notch filter. All these four types of filters are discussed one by one in the following pages.

68.25. Low-Pass Filter

A filter that provides a constant output from dc up to a cut-off frequency (f_{OH}) and then passes no signal above that frequency is called an ideal low-pass filter. The ideal response of a low-pass filter is as shown in Fig. 68.31 (*a*). Notice that the response shows that the filter has a constant output (indicated by a horizontal line AB) from dc or zero frequency up to a cutt-off frequency (f_{OH}). And beyond f_{OH}, the output is zero as indicated by the vertical line 'BC' in the figure.

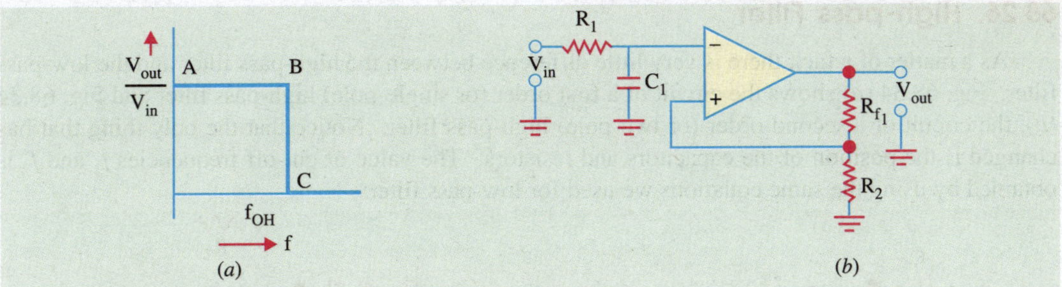

Fig. 68.31

Fig. 68.31 (*b*) shows the circuit of a low-pass active filer using a single resistor and capacitor. Such a circuit is also referred to as first-order (or single-pole) low-pass filter. It is called first-order because it makes use of a single resistor and a capacitor. The response of such a first-order low pass filter is as shown in Fig. 68.32. Notice that the response below the cut-off frequency (f_{OH}) shows a constant gain (indicated by a horizontal line 'AB')

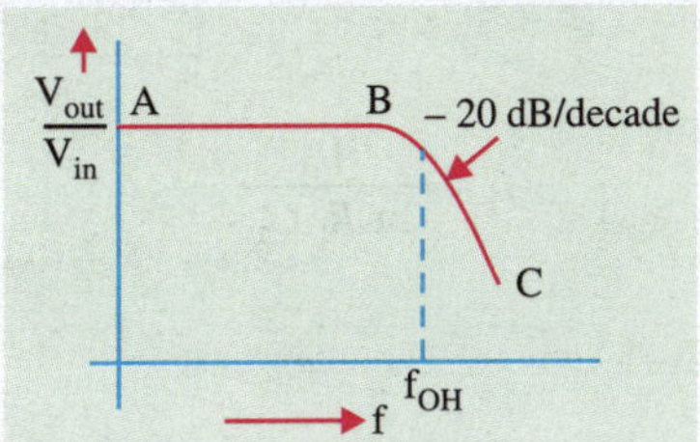

Fig. 68.32

However, beyond the cut-off frequency, the gain does not reduce immediately to zero as expected in Fig. 68.31 (*a*) but reduces with a slope of 20 dB/decade (means that the output voltage reduces by a factor of 100 when the fequency increases by a factor of 10). The voltage gain for a low-pass filter below the cut-off frequency (f_{OH}) is given by the relation.

$$A_v = 1 + \frac{R_3}{R_1}$$

And the cut-off frequency is determined by the relation

$$f_{OH} = \frac{1}{2\pi R_1 C_1}$$

It is possible to connect two sections of the filter together as shown in Fig. 68.33 (*a*). Such a circuit is called second-order (or two-pole) low pass filter. Fig. 68.33 configuration of the second-order low-pass filter.

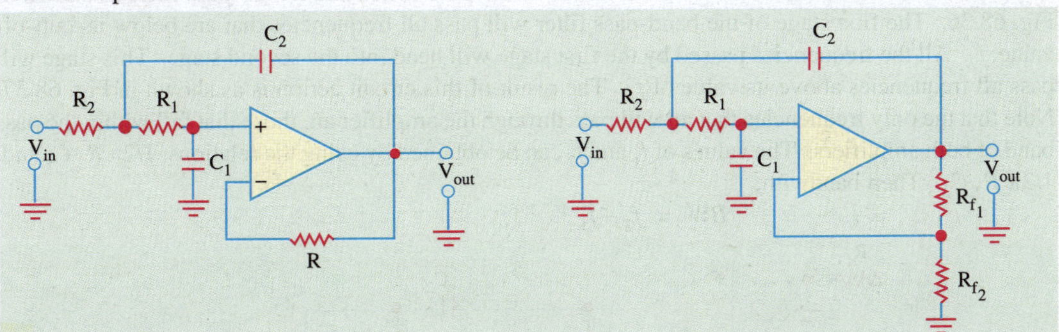

Fig. 68.33

Each circuit shown in Fig. 68.33 has two RC circuits, $R_1 - C_1$ and $R_2 - C_2$. As the operating frequency increases beyond f_2, each circuit will be dropping the closed-loop gain by 20 dB, giving a total roll-off rate of 40 dB/decade when operated above f_2. The cut-off frequency for each of the circuit is given by,

$$f_2 = \frac{1}{2\pi\sqrt{R_1 R_2 C_1 C_2}}$$

68.26. High-pass Filter

As a matter of a fact, there is very little difference between the high-pass filter and the low-pass filter. Fig. 68.34 (*a*) shows the circuit of a first order (or single-pole) high-pass filter and Fig. 68.34 (*b*), the circuit of a second-order (or two-pole) high-pass filter. Notice that the only thing that has changed is the position of the capacitors and resistors. The value of cut-off frequencies f_1 and f_2 is obtained by using the same equations we used for low-pass filter.

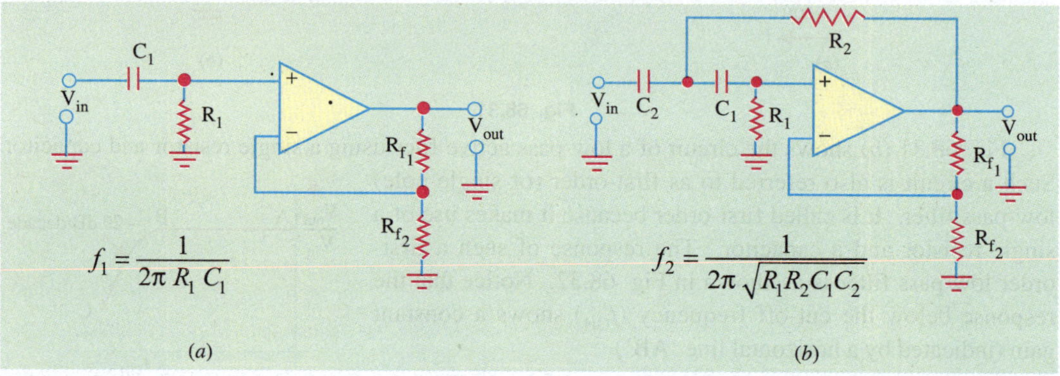

Fig. 68.34

Fig. 68.35 shows the gain versus frequency response of a high-pass filter. Notice that the solid line indicates the ideal response while the dashed line, corresponds to the actual response of the filter circuit. The ideal curve indicates that the filter has a zero output for the frequencies below f_{OL} (indicated by the line 'AB'). And beyond f_{OL}, it has a constant output. The actual response curve may correspond to the roll-off gain by 20 dB/decade for first order to 40 dB/decade for second-order low-pass filters.

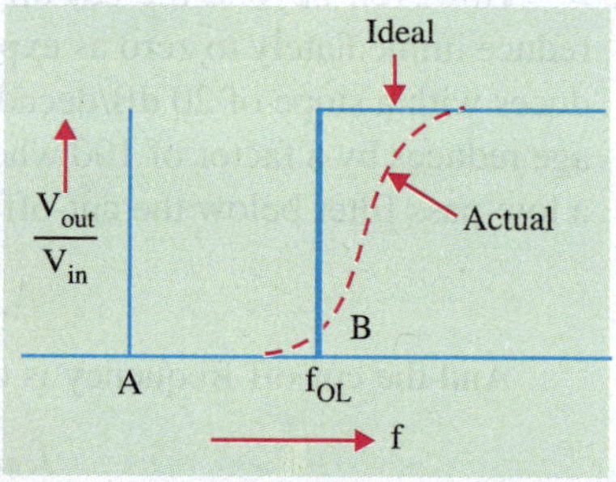

Fig. 68.35

68.27. Band-pass Filters

A band-pass filter is the one that is designed to pass all frequencies within its bandwith. A simple way to construct a band-pass filter is to cascade a low-pass filters and a high-pass filter as shown in Fig. 68.36. The first stage of the band-pass filter will pass all frequencies that are below its cutt-off value, f_2. All the frequencies passed by the first stage will head into the second stage. This stage will pass all frequencies above its value of f_1. The result of this circuit action is as shown in Fig. 68.37. Note that the only frequencies that all will pass through the amplifier are those that fall within the pass band of both amplifiers. The values of f_1 and f_2 can be obtained by using the relations, $1/2\pi R_1 C_1$ and, $1/2\pi R_2 C_2$. Then bandwith,

$$BW = f_2 - f_1$$

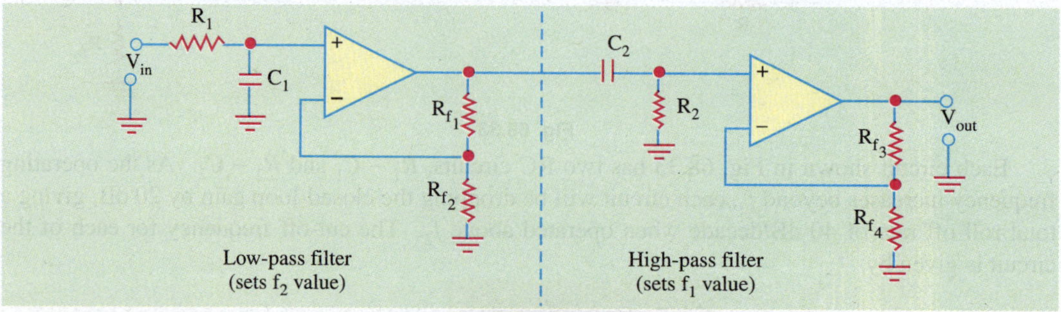

Fig. 68.36

And the centre frequency,

$$f_0 = f_1 \cdot f_2$$

The Quality-factor (or Q-factor) of the band-pass filter circuit.

$$Q = \frac{f_0}{BW}$$

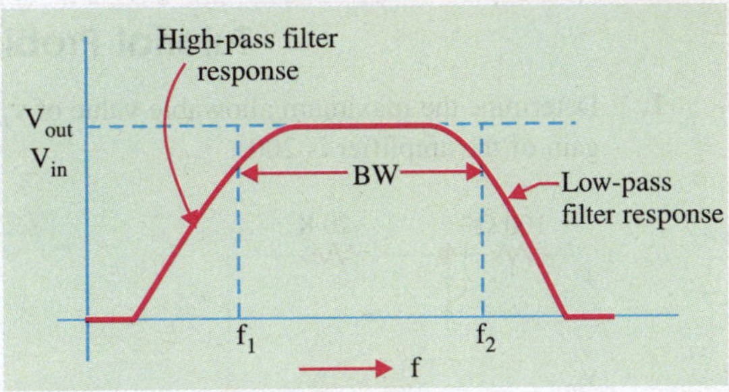

Fig. 68.37

68.28. Notch Filter

The notch filter is designed to block all frequencies that fall within its bandwith Fig. 68.38 (*a*) shows a block diagram and 68.38 (*b*), the gain versus frequency response curve of a multistage notch filter.

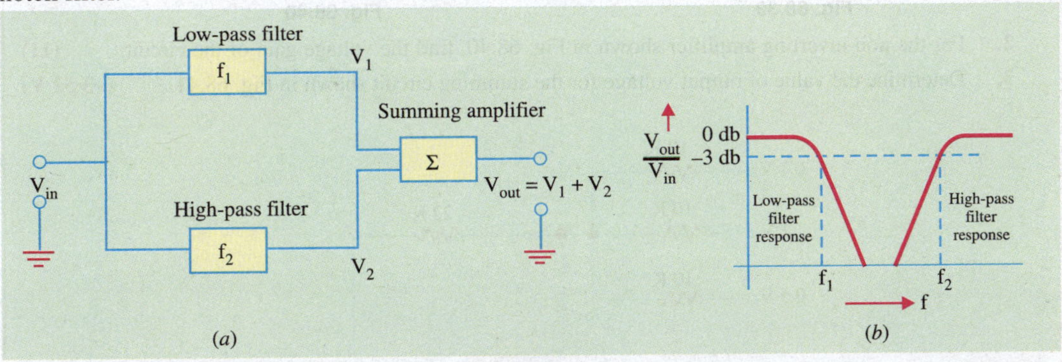

Fig. 68.38

The block diagram shows that the circuit is made up of a high-pass filter, a low-pass filter and a summing amplifier. The summing amplifier produces an output that is equal to a sum of the filter output voltages. The circuit is designed in such a way so that the cut-off frequency, f_1 (which is set by a low-pass filter) is lower in value than the cut-off frequency, f_2 (which is set by high-pass filter). The gap between the values of f_1 and f_2 is the bandwidth of the filter.

When the circuit input frequency is lower than f_1, the input signal will pass through low-pass filter to the summing amplifier. Since the input frequency is below the cut-off frequency of the high-pass filter, v_2 will be zero. Thus the output from the summing amplifier will equal the output from the low-pass filter. When the circuit input frequency is higher than f_2, the input signal will pass through the high-pass filter to the summing amplifier. Since the input frequency is above the cut-off frequency of the low-pass filter, v_1 will zero. Now the summing amplifier output will equal the output from the high-pass filter.

It is evident from the above discussion that frequencies below f_1 and those above f_2, have been passed by the notch filter. But when the circuit frequency between f_1 and f_2, neither of the filters will produce an output. Thus v_1 and v_2 will be both zero and the output from the summing amplifier will also be zero.

The frequency analysis of the notch filter is identical to the band-pass filter. Fisrt, determine the cut-off frequencies of the low-pass and the high-pass filters. Then using these calculated values, determine the bandwidth, center frequency and Q values of the circuit.

Tutorial Problems No. 68

1. Determine the maximum allowable value of v_{in} for the circuit shown in Fig. 68.39. Assume that the gain of the amplifier is 200. **(40 mV)**

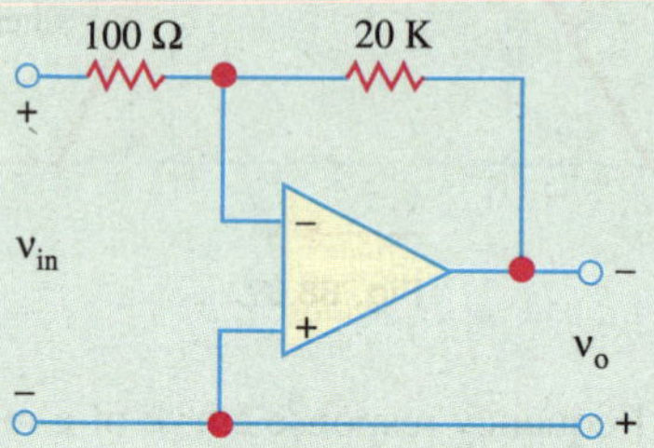

Fig. 68.39

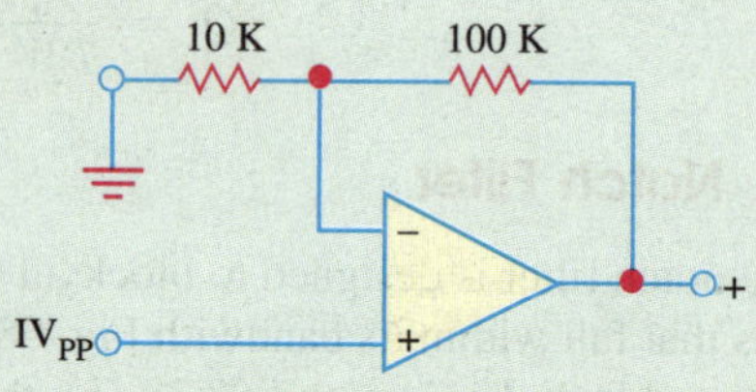

Fig. 68.40

2. For the non-inverting amplifier shown in Fig. 68.40, find the voltage gain of the circuit. **(11)**
3. Determine the value of output voltage for the summing circuit shown in Fig. 68.41. **(–3·52 V)**

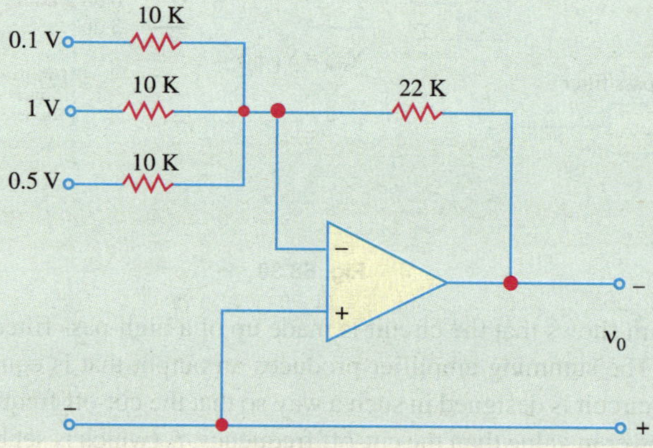

Fig. 68.41

OBJECTIVE TESTS – 68

1. An OP-AMP can be classified as .... amplifier.
   (*a*) linear
   (*b*) low-r_{in}
   (*c*) positive feedback
   (*d*) *RC*-coupled.
2. An ideal OP-AMP has
   (*a*) infinite A_v
   (*b*) infinite R_i
   (*c*) zero R_0
   (*d*) all the above.
3. OP-AMP have become very popular in industry mainly because
   (*a*) they are dirt cheap
   (*b*) their external characteristics can be changed to suit any application
   (*c*) of their extremely small size
   (*d*) they are available in different packages.
4. Since input resistance of an ideal OP-AMP is infinte
   (*a*) its output resistance is zero
   (*b*) its output voltage becomes independent of load resistance
   (*c*) its input current is zero
   (*d*) it becomes a current-controlled device.
5. The gain of an actual OP-AMP is around
   (*a*) 1,000,000 (*b*) 1000
   (*c*) 100 (*d*) 10,000
6. When an input voltage of 1 V is applied to an OP-AMP having $A_v = 10^6$ and bias supply of + 15 V, the output voltage available is
   (*a*) 15×10^6 V (*b*) 10^6 V
   (*c*) 15 μV (*d*) 15 V.

7. An inverting amplifier has $R_f = 2$ M and $R_1 = 2$ K, Its scale factor is
(a) 1000 (b) −1000
(c) 10^{-3} (d) -10^{-3}

8. In an inverting amplifier, the two input terminals of an ideal OP-AMP are at the same potential because
(a) the two input terminals are directly shorted internally
(b) the input impedance of the OP-AMP is infinity
(c) common-mode rejection ratio is infinity
(d) the open-loop gain of the OP-AMP is infinity.

9. The open-loop gain of an operational amplifier is 10^5. An input signal of 1 mV is applied to the inverting input with the non-inverting input connected to the ground. The supply voltages are ±10 V. The output of the amplifier will be
(a) + 100 V
(b) − 100 V
(c) + 10 V (approximately)
(d) − 10 V (approximately)

10. The output voltage of the circuit shown in Fig. 68.42 is

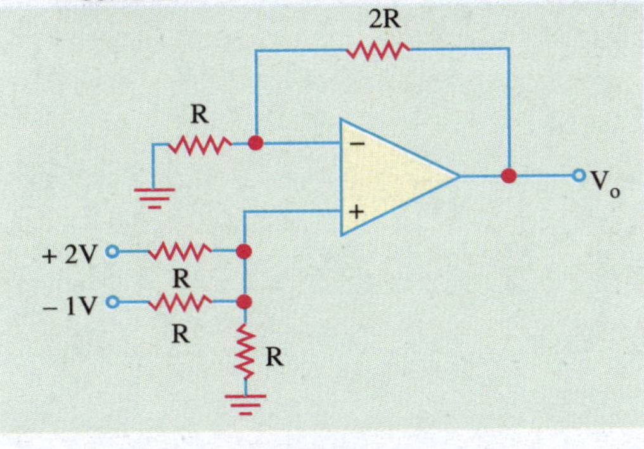

Fig. 68.42

(a) 1·0 V (b) 1·5 V
(c) 2·0 V (d) 2·5 V

11. In the circuit shown in Fig. 68.43, the value of output, v_0 is
(a) + 3 V (b) − 3 V
(c) − 7 V (d) + 7 V

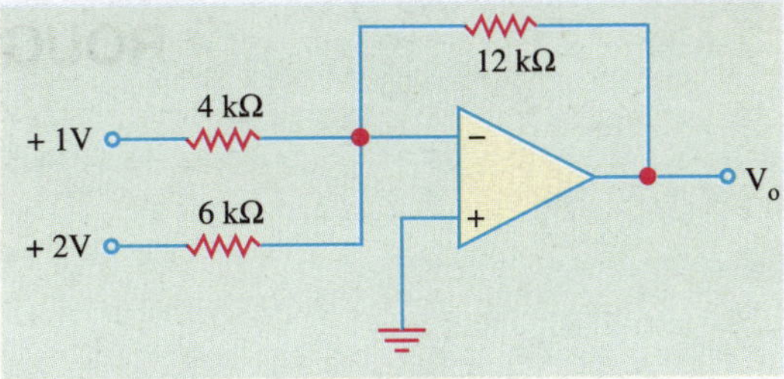

Fig. 68.43

12. When in a negative scaler, both R_1 and R_f are reduced to zero, the circuit functions as
(a) integrator (b) subtractor
(c) comparator (d) unity follower.

13. The two input terminals of an OP-AMP are known as
(a) positive and negative
(b) differential and non-differential
(c) inverting and non-inverting
(d) high and low.

14. The purpose of comparator is to
(a) amplify an input voltage
(b) detect the occurrence of a changing input voltage
(c) maintain a constant output when the dc input voltage changes.
(d) produce a change in output voltage when an input voltage equals the reference voltage.

15. The OP-AMP comparator circuit uses
(a) positive feedback
(b) negative beedback
(c) regenerative feedback
(d) no feedback

16. The feedback path in an OP-AMP integrator consists of
(a) a resistor
(b) a capacitor
(c) a resistor and a capacitor in series
(d) a resistor and a capacitor in parallel.

17. The feedback path in an OP-AMP differentiator consists of
(a) a resistor
(b) a capacitor
(c) a resistor and a capacitor in series
(d) a resistor and a capacitor in parallel.

ANSWERS

1. (a) 2. (d) 3. (b) 4. (c) 5. (a) 6. (b) 7. (b) 8. (b) 9. (b) 10. (c) 11. (c) 12. (d)
13. (c) 14. (d) 15. (d) 16. (b) 17. (a)

ROUGH WORK

C H A P T E R 69

NUMBER SYSTEMS AND CODES

Learning Objectives

- Number Systems
- The Decimal Number System
- Binary Number System
- Binary to Decimal Conversion
- Binary Fractions
- Double-Dadd Method
- Decimal to Binary Conversion
- Shifting the Place Point
- Binary Operations
- Binary Addition
- Binary Subtraction
- Complement of a Number
- 1's Complemental Subtraction
- 2's Complemental Subtraction
- Binary Multiplication
- Binary Division
- Shifting a Number to Left or Right
- Octal Number System
- Octal to Decimal Conversion
- Decimal to Octal Conversion
- Binary to Octal Conversion
- Octal to Binary Conversion
- Binary to Hexadecimal Conversion
- Decimal to Hexadecimal Conversion
- Hexadecimal to Decimal Conversion
- Digital Coding
- Binary Coded Decimal (BCD) Code
- Octal Coding
- Hexadecimal Coding
- Excess-3 Code
- Gray Code
- Excess-3 Gray Code
- ASCII Code

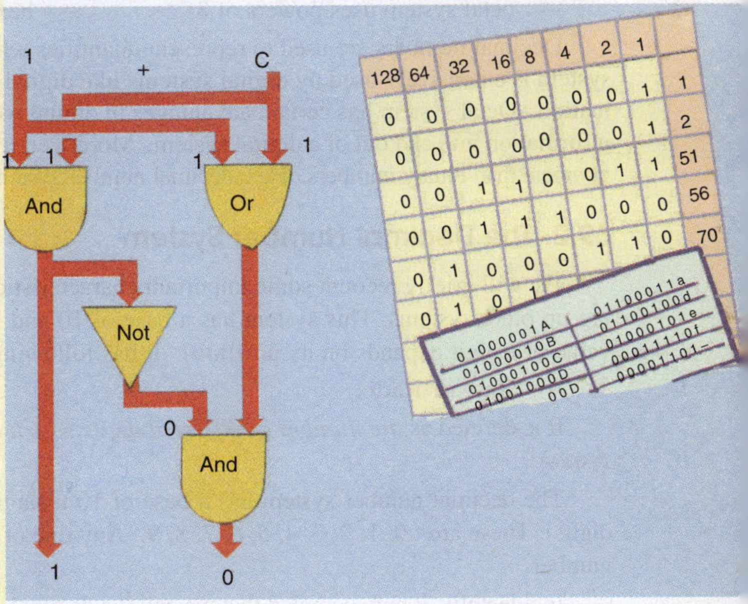

Logic gates use switches that control the flow of an electrical current. '1' is 'true' and '0' is 'false'. The columns in the binary system wave the values 1, 2, 4, 8, 16, 32 and so no.

69.1. Number Systems

The number systems are used quite frequently in the field of digital electronics and computers. However the type of number system used in computers could be different at different stages of the usage. For example, when a user key-in some data into the computer, s(he), will do it using decimal number system *i.e.* the system we all have used for several years for doing arithmetic problems. But when the information goes inside the computer, it needs to be converted to a form suitable for processing data by the digital circuitry. Similarly when the data has to be displayed on the monitor for the user, it has to be again in the decimal number system. Hence the conversion from one number system to another one is an important topic to be understood.

There are four systems of arithmetic which are often used in digital circuits. These systems are:

1. **Decimal**—it has a base (or radix) of 10 *i.e.* it uses 10 different symbols to represent numbers.
2. **Binary**—it has a base of 2 *i.e.* it uses only two different symbols.
3. **Octal**—it has a base of 8 *i.e.* it uses eight different symbols.
4. **Hexadecimal**—it has a base of 16 *i.e.* it uses sixteen different symbols.

All these systems use the same type of **positional notation** except that

— decimal system uses powers of 10 — binary system uses power of 2

— octal system uses powers of 8 — hexadecimal system uses powers of 16.

Decimal numbers are used to represent quantities which are outside the digital system. Binary system is extensively used by digital systems like digital computers which operate on binary information. Octal system has certain advantages in digital work because it requires less circuitry to get information into and out of a digital system. Moreover, it is easier to read, record and print out octal numbers than binary numbers. Hexadecimal number system is particularly suited for microcomputers.

69.2. The Decimal Number System

We will briefly recount some important characteristics of this more-familiar system before taking up other systems. This system has a base of 10 and is a **position-value system** (meaning that value of a digit depends on its **position**). It has following characteristics :

(*i*) Base or Radix

It is defined as ***the number of different digits which can occur in each position in the number system.***

The decimal number system has a base of 10 meaning that it contains ten unique symbols (or digits). These are : 0, 1, 2, 3, 4, 5, 6, 7, 8, 9. Any one of these may be used in each position of the number.

Incidentally, it may be noted that we call it a decimal (10's) system although ***it does not have a distinct symbol of 10.*** As is well-known, it expresses 10 and any number above 10 as a combination of its ten unique symbols.

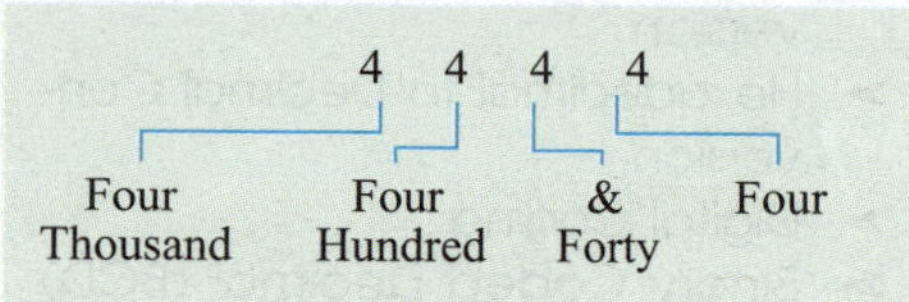

Fig. 69.1

(*ii*) Position Value

The absolute value of each digit is fixed but its **position value** (or place value or weight) is determined by its ***position*** in the overall number. For example, position value of 3 in 3000 is not the same as in 300. Also, position value of each 4 in the number 4444 is different as shown in Fig. 69.1.

Similarly, the number 2573 can be broken down as follows :

$$2573 = 2 \times 10^3 + 5 \times 10^2 + 7 \times 10^1 + 3 \times 10^0$$

It will be noted that in this number, 3 is the **least significant digit** (*LSD*) whereas 2 is the **most significant digit** (*MSD*).

Again, the number 2573.469 can be written as

$$2573.469 = 2 \times 10^3 + 5 \times 10^2 + 7 \times 10^1 + 3 \times 10^0 + 4 \times 10^{-1} + 6 \times 10^{-2} + 9 \times 10^{-3}$$

It is seen that position values are found by raising the base of the number system (*i.e.* 10 in this case) to the power of the position. Also, powers ***are numbered to the left of the decimal point starting with 0 and to the right of the decimal point starting with –1.***

69.3. Binary Number System

Like decimal number (or denary) system, it has a radix and it also uses the same type of position value system.

(*i*) Radix

Its base or radix is **two** because it uses only two digits 0 and 1 (the word 'binary digit' is contracted to **bit**). All binary numbers consist of a string of 0s and 1s. Examples are 10, 101 and 1011 which are read as one-zero, one-zero-one and one-zero-one-one to avoid confusion with decimal numbers. Another way to avoid confusion is to add a subscript of 10 for decimal numbers and of 2 for binary numbers as illustrated below.

10_{10}, 101_{10}, 5742_{10} —decimal number and 10_2, 101_2, 110001_2 — binary numbers.

Binary numbers represent all values within computers

It is seen that the subscript itself is in decimal. It may be noted that binary numbers need ***more places for counting because their base is small***

(*ii*) Position Value

Like the decimal system, binary system is also positionally-weighted. However, in this case, the position value of each bit corresponds to some power of 2. In each binary number, the value increases in powers of 2 starting with 0 to the left of the binary point and decreases to the right of the binary point starting with power of –1. The position value (or weight) of each bit alongwith a 7-bit binary number 1101.011 is shown in Fig. 69.2.

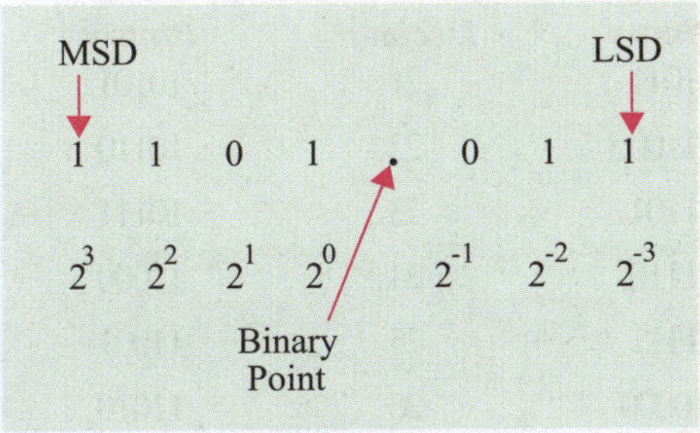

Fig. 69.2

As seen, the fourth bit to the left of binary point **carries the maximum weight** (*i.e.* it has the highest value) and is called most significant digit (*MSD*). Similarly, the third bit to the right of the binary point is called least significant digit (*LSD*). The decimal equivalent of the binary number may be found as under

$$1101.011_2 = (1 \times 2^3) + (1 \times 2^2) + (0 \times 2^1) + (1 \times 2^0) + (0 \times 2^{-1}) + (1 \times 2^{-2}) + (1 \times 2^{-3})$$

$$= 8 + 4 + 0 + 1 + 0 + \frac{1}{4} + \frac{1}{8} = 13.375_{10}$$

As stated earlier, position values of different bits are given by **ascending powers of 2** to the **left** of binary point and by **descending power** of 2 to the **right** of binary point. The different digit positions of a given binary number have the following decimal weight (Fig. 69.3)

Position value ←	2^5	2^4	2^3	2^2	2^1	2^0	.	2^{-1}	2^{-2}	2^{-3}	2^{-4} →
							↑ Binary Point ↓				
Position value ←	32	16	8	4	2	1	.	$\frac{1}{2}$	$\frac{1}{4}$	$\frac{1}{8}$	$\frac{1}{16}$ →

Fig. 69.3

Binary numbers are used extensively by all digital systems primarily due to the nature of electronics itself. The bit 1 may be represented by a saturated (fully-conducting) transistor, a light turned ON, a relay energised or a magnet magnetised in a particular direction. The bit 0, on the other hand, can be represented as a cut-off transistor, a light turned OFF, a relay de-energised or a magnet magnetised in the opposite direction. In such cases, there are only two values which a device can assume.

69.4. Binary to Decimal Conversion

Following procedure should be adopted for converting a given binary integer (whole number) into its equivalent decimal number :

Step 1. Write the binary number *i.e.* all its bits in a row.

Step 2. Directly under the bits, write 1, 2, 4, 8, 16, ...starting from *right to left*.

Step 3. Cross out the decimal weights which lie under 0 bits.

Step 4. Add the remaining weights to get the decimal equivalent.

Example 69.1. *Convert 11001_2 to its equivalent decimal number.*

Solution. The four steps involved in the conversion are as under

Step 1.	1	1	0	0	1
Step 2.	16	8	4	2	1
Step 3.	16	8	~~4~~	~~2~~	1
Step 4.	16 + 8 + 1 = 25			∴	$11001_2 = 25_{10}$

It is seen that the number contains 1 sixteen, 1 eight, 0 four's, 0 two's and 1 one. Certain decimal and binary equivalent numbers are tabulated below in Table No. 69.1

Table No. 69.1

Decimal	*Binary*	*Decimal*	*Binary*	*Decimal*	*Binary*
1	1	11	1011	21	10101
2	10	12	1100	22	10110
3	11	13	1101	23	10111
4	100	14	1110	24	11000
5	101	15	1111	25	11001
6	110	16	10000	26	11010
7	111	17	10001	27	11011
8	1000	18	10010	28	11100
9	1001	19	10011	29	11101
10	1010	20	10100	30	11110

69.5. Binary Fractions

Here, procedure is the same as for binary integers except that the following weights are used for different bit positions .

●	2^{-1}	2^{-2}	2^{-2}	2^{-4}	→
↑	$\frac{1}{2}$	$\frac{1}{4}$	$\frac{1}{8}$	$\frac{1}{16}$	→

Binary Point

Example 69.2. *Convert the binary fraction 0.101 into its decimal equivalent.*

Solution. The following four steps will be used for this purpose.

Step 1.	0	●	1	0	1
Step 2.			$\frac{1}{2}$	$\frac{1}{4}$	$\frac{1}{8}$
Step 3.			$\frac{1}{2}$	$\cancel{\frac{1}{4}}$	$\frac{1}{8}$

Step 4. $\frac{1}{2}+\frac{1}{8} = 0.625$

$\therefore \quad 0.101_2 = 0.625_{10}$

Example 69.3. *Find the decimal equivalent of the 6-bit binary number 101.101_2.*

Solution.

1	0	1	●	1	0	1	
4	2	1		$\frac{1}{2}$	$\frac{1}{4}$	$\frac{1}{8}$	
4	$\cancel{2}$	1		$\frac{1}{2}$	$\cancel{\frac{1}{4}}$	$\frac{1}{8}$	$= 5 + \frac{1}{2} + \frac{1}{8} = 5.625$

$\therefore \quad 101.101_2 = 5.625_{10}$

69.6. Double-Dadd Method

This method of converting binary integers into decimal equivalents is much simpler and quicker than the method given in Art. 69.4 especially in the case of large numbers. Following three steps are involved :

1. Double the first bit to the extreme left and add this doubled value to the next bit on the right.
2. Double the sum obtained and add the doubled value to the next bit.
3. Continue step 2 until the last bit has been added to the previously-doubled sum.

The conversion of 11001_2 is shown in Fig. 69.4. It is seen that $11001_2 = 25_{10}$

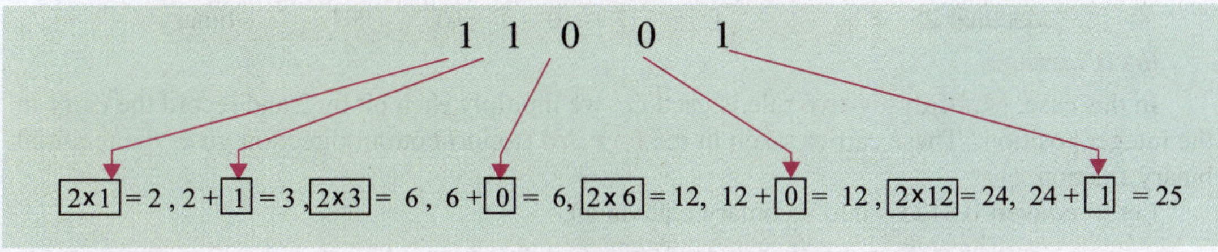

Fig. 69.4

Using double-dadd method, let us convert 111010_2 into its binary equivalent.

1. $2 \times 1 = 2,$ add next bit 1 so that $2 + 1 = 3$
2. $2 \times 3 = 6,$ add next bit 1 so that $6 + 1 = 7$
3. $2 \times 7 = 14,$ add next bit 0 so that $14 + 0 = 14$
4. $2 \times 14 = 28,$ add next bit 1 so that $28 + 1 = 29$

5. $2\times29 = 58$, add next bit 0 so that $58+0 = 58$

$\therefore$ $111010_2 = 58_{10}$

69.7. Decimal to Binary Conversion

(*a*) Integers

Such conversion can be achieved by using the so-called **double-dabble method.** It is also known as **divide-by-two** method. In this method, we progressively divide the given decimal number by 2 and write down the remainders after each division. These remainders taken in the *reverse* order (*i.e.* from bottom-to-top) form the required binary number. As an example, let us convert 25_{10} into its binary equivalent.

$25 \div 2 = 12$ + remainder of 1 — TOP
$12 \div 2 = 6$ + remainder of 0
$6 \div 2 = 3$ + remainder of 0
$3 \div 2 = 1$ + remainder of 1
$1 \div 2 = 0$ + remainder of 1
$\therefore\ 25_{10} = 11001_2$ — BOTTOM

The above process may be simplified as under :

Successive Divisions	Remainders
2) 25	
2) 12	1
2) 6	0
2) 3	0
2) 1	1
2) 0	1

(read upward ↑)

Reading the remainders from bottom to top, we get $25_{10} = 11001_2$

It may also be put in the following form :

$25 \div 2 = 12 + 1$
$12 \div 2 = 6 + 0$
$6 \div 2 = 3 + 0$
$3 \div 2 = 1 + 1$
$1 \div 2 = 0 + 1$

$\therefore$ decimal 25 = 1 1 0 0 1 binary

(*b*) Fractions

In this case, **Multiply-by-two** rule is used *i.e.* we multiply each bit by 2 and record the carry in the integer position. These carries taken in the **forward** (top-to-bottom)direction gives the required binary fraction.

Let us convert 0.8125_{10} into its binary equivalent.

$0.8125 \times 2 = 1.625 = 0.625$ with a carry of 1
$0.625 \times 2 = 1.25 = 0.25$ with a carry of 1
$0.25 \times 2 = 0.5 = 0.5$ with a carry* of 0
$0.5 \times 2 = 1.0 = 0.0$ with a carry of 1 ↓
$\therefore\ 0.8125_{10} = 0.1101_2$

* It is so because there is 0 at the interger position *i.e.* to the left of the decimal point.

Please note that ***we have to add the binary point from our side.*** Let us now convert 0.77_{10} into its binary equivalent.

$$\begin{aligned} 0.77\times 2 &= 1.54 &&= 0.54 \text{ with a carry of } 1\\ 0.54\times 2 &= 1.08 &&= 0.08 \text{ with a carry of } 1\\ 0.08\times 2 &= 0.16 &&= 0.16 \text{ with a carry of } 0\\ 0.16\times 2 &= 0.32 &&= 0.32 \text{ with a carry of } 0\\ 0.32\times 2 &= 0.64 &&= 0.64 \text{ with a carry of } 0\\ 0.64\times 2 &= 1.28 &&= 0.28 \text{ with a carry of } 1 \end{aligned}$$

We may stop here but the answer would be approximate. $\therefore\ 0.77_{10} \cong 0.110001_2$

Example 69.4. *Convert 25.625_{10} into its binary equivalent.*

Solution. We will do the conversion in two steps *(i)* first for the integer and *(ii)* then for the fraction.

(a) Integer

$$\begin{aligned} 25 \div 2 &= 12 + 1\\ 12 \div 2 &= 6 + 0\\ 6 \div 2 &= 3 + 0\\ 3 \div 2 &= 1 + 1\\ 1 \div 2 &= 0 + 1\\ \therefore\quad 25_{10} &= 11001_2 \end{aligned}$$

(b) fraction

$$\begin{aligned} 0.625 \times 2 &= 1.25 = 0.25 + 1\\ 0.25 \times 2 &= 0.5 = 0.5 + 0\\ 0.5 \times 2 &= 1.0 = 0.0 + 1\\ \therefore\quad 0.625_{10} &= 0.101_2 \end{aligned}$$

Considering the complete number, we have $25.625_{10} = 11001.101_2$

Obviously, binary system needs more bits to express the same number than decimal system.

69.8. Shifting the Place Point

In a decimal number if the decimal point is moved one place to the right, **the number is multiplied by 10**. For example, when decimal point in 7.86 is shifted one place to the right, it becomes 78.6 *i.e.* it increases the value of the number 10 times. Moving the decimal point one place to the left reduces its value to one-tenth.

In binary numbers, shifting the binary point by one place multiplies or divides the number by 2. For example, 111.0_2 is equal to 7_{10} but 1110.0_2 is 14_{10}. As seen, 7 is doubled to 14 by moving the binary point one place to the right.

Similarly, 11.1_2 is $(2 + 1 + \frac{1}{2}) = 3.5_{10}$. Hence, 111.0_2 is halved to 3.5_{10} by moving its binary point one place to the left.

69.9. Binary Operations

We will now consider the following four binary operations :

1. addition **2.** subtraction **3.** multiplication **4.** division

Addition is the most important of these four operations. In fact, by using **'complements'**, subtraction can be reduced to addition. Most digital computers subtract by complements. It leads to reduction in hardware because only adding type of circuits are required. Similarly, multiplication is nothing but repeated addition and, finally, division is nothing but repeated subtraction.

69.10. Binary Addition

Addition is simply the manipulation of numbers for combining physical quantities. For example, in the decimal number system, 2 + 3 = 5 means the combination of • • with • • • to give a total of • • • • •. Addition of binary numbers is similar to the decimal addition.

Following points will help in understanding the rules of binary addition.

1. When 'nothing' is combined with 'nothing', we get nothing.
   Binary representation of the above statement is : $0 + 0 = 0$
2. When nothing is combined with •, we get •.
   In binary language $0 + 1 = 1$
3. Combining • with nothing, gives •.
   The binary equivalent is $1 + 0 = 1$
4. When we combine • with •, we get • •.
   The binary representation of the above is $1 + 1 = 10$

It should be noted that the above sum is not 'ten' but 'one-zero' *i.e.* it represents • • and not • • • • • • • • • •. In other words, it is 10_2 which represents decimal 2. ***It is not decimal ten.***

The last rule is often written as $1 + 1 = 0$ with a carry of 1

The above rules for binary addition can be summarized as under :

$0 + 0 = 0 \qquad 0 + 1 = 1$

$1 + 0 = 1 \qquad 1 + 1 = 0$ with a carry of 1 or $=10_2$

It is worth noting that 'carry-overs' are performed in the same manner as in decimal arithmetic.

The rules of binary addition could also be expressed in the form of a table as shown below

↓ →	0	1
0	0	1
1	1	10_2

or

↓ →	0	1
0	0	1
1	1	0

with a carry 1

As an illustration, let us add 101 and 110.

```
   101      — first column      1+0 = 1
 + 110      — second column     0+1 = 1
  1011      — third column      1+1 = 10
```

(*i.e.* 0 with carry 1)

Similarly,

```
   1 carry
   111      — 1st column      1 + 0 = 1
 + 110      — 2nd column      1 + 1 = 0   with carry 1
  1101      — 3rd column      1 + 1 + carry of 1
                                    = 10 + 1 = 11₂
```

Let us consider one more example :

```
                    ←             ←              ←
  1 carry       1 1 carry      1 1 carry      1 1 carry
  1011          1011           1011           1011
+ 1001        + 1001         + 1001         + 1001
     0            00            100          10100
```

Hence, we find from the above examples that the only two possible combinations with a carry are :

(*a*) $1 + 1 =$ sum of 0 with a carry of 1.
It is binary 10 *i.e.* 10_2 which equals *decimal* 2.

(*b*) $1 + 1 +$ carry of $1 =$ a sum of 1 with a carry of 1.
It equals binary 11 *i.e.* 11_2 or decimal 3.

Example 69.5. *Add 110011_2 to 101101_2* **(Digital Electronics, Bombay Univ. April 1987)**

Solution.

$$\begin{array}{r} 110011 \\ 101101 \\ \hline 1100000 \end{array}$$

1. **first column :** 1 + 1 = 0 with a carry of 1. Hence, we put down zero there and carry 1 to the second column.
2. **second column :** 1 + 0 = 1. But combined with carry 1 from first column, it gives 1 + 1 = 0 and a carry of 1. Hence, we put down 0 there and carry 1 further to the third column.
3. **third column :** 0 + 1 = 1. Again, when it is combined with carry of 1 from the second column, we get 1 + 1 = 0 with a carry of 1. Hence, again, we put down 0 and carry 1 to the fourth column.
4. **fourth column :** Here, 0 + 1 = 1. When combined with carry 1 from third column, we get 1 + 1 = 0 with a carry of 1. Hence, we put down 0 there and carry 1 to the fifth column.
5. **fifth column :** Here, 1 + 0 = 1 When combined with the carry 1 from fourth column, we get 1 + 1 = 0 with a carry of 1. Hence, we put down 0 there and carry 1 to the sixth column.
6. **sixth column :** Here, it is a case of 1 + 1 + carry of 1 = 11_2 as stated earlier in (*b*) above.

69.11. Binary Subtraction

It is also performed in a manner similar to that used in decimal subtraction. Because binary system has only two digits, binary subtraction requires more borrowing operations than decimal subtraction. The four rules for binary subtraction are as under:

1. 0 – 0 = 0, **2.** 1 – 0 = 1, **3.** 1 – 1 = 0,

4. 0 – 1 = 1 with a borrow of 1 from the next column of the minuend

or 10 – 1 = 1

The last result represents • • – • – • which makes sense.

While using Rule 4, it should be borne in mind that borrow reduces the remaining minuend by 1. It means that a borrow will cause a 1 in the next column to the left in the minuend to become 0. If the next column also happens to contain 0, it is changed to a 1 and the succeeding 0s in the minuend are changed to 1s until a 1 is found which is then changed to a 0.

Example 1

Let us subtract 0101_2 from 1110_2. The various steps are explained below :

0 1 borrow	0 1	0 1	0 1
1 1 ~~1~~ 0	1 1 ~~1~~ 0	1 1 ~~1~~ 0	1 1 ~~1~~ 0
–0 1 0 1	–0 1 0 1	–0 1 0 1	–0 1 0 1
1	0 1	0 0 1	1 0 0 1

Explanation

1. In the first column, since we cannot subtract 1 from 0, we borrow 1 from the next column to the left. Hence, we put down 1 in the answer and change the 1 of the next left column to a 0.
2. We apply Rule 1 to next column *i.e.* 0 – 0 = 0
3. We apply Rule 3 to the 3rd column *i.e.* 1 – 1 = 0
4. Finally, we apply Rule 2 to the last *i.e.* fourth column it. 1 – 0 = 1

   As a check, it may be noted that talking in terms of decimal numbers, we have subtracted 5 from 14. Obviously, the answer has to be 9 (1001_2).

Example 2.

Let us now try subtracting 0001_2 from 1000_2.

Step 1	Step 2	Step 3	Step 4
1	11	111	111
10Ø0	1ØØ1	1̸ØØ1	1̸ØØ1
−0001	−0001	−0001	−0001
1	11	111	0111

Since there happened to be a 0 in the second column, it was changed to 1. Again, there was a 0 in the third column, so it was also changed to a 1. Finally, we met a 1 in the forth column which was changed to a 0 and the final answer was written down as shown above.

Example 69.6. *Subtract 0111_2 from 1001_2* **(Digital Computations, Punjab Univ. 1991)**

Solution.

1 0 0 1	1st column	:	1 − 1 = 0
− 0 1 1 1	2nd column	:	0 − 1 = 1 with a borrow of 1
0 0 1 0	3rd column	:	1 (after borrow) − 1 = 0
	4th column	:	0 (after borrow) − 0 = 0

Example 69.7. *Subtract 01011_2 from 10110_2* **(Computer Technology, Pune Univ. 1988)**

Solution. Step-wise the solution is as under :

Step 1	Step 2	Step 3	Step 4	Step 5
0 1 borrow	001	001	01001	01001
1011̸0	101̸1̸0	101̸1̸0	1̸01̸1̸0	1̸01̸1̸0
− 01011	−01011	−01011	−01011	−01011
1	11	011	1011	01011

69.12. Complement of a Number

In digital work, two types of complements of a binary number are used for ***complemental subtraction*** :

(*a*) 1's complement

The 1's complement of a binary number is obtained by changing its each 0 into a 1 and each 1 into a 0. It is also called ***radix-minus-one complement.*** For example, 1's complement of 100_2 is 011_2 and of 1110_2 is 0001_2.

(*b*) 2's complement

The 2's complement of a binary number is obtained by adding 1 to its 1's complement.

2's complement = 1's complement + 1

It is also known **as true** complement. Suppose we are asked to find 2's complement of 1011_2. Its 1's complement is 0100_2. Next, add 1 to get 0101_2. Hence, 2's complement of 1011_2 is 0101_2.

The complement method of subtraction reduces **subtraction to an addition process.**

This method is popular in digital computers because

1. only adder ciruits are needed thus simplifying the circuitry,

2. it is easy with digital circuits to get the complements.

69.13. 1's Complemental Subtraction

In this method, instead of subtracting a number, we add its 1's complement to the minuend. The last carry (whether 0 or 1) is then added to get the final answer. The rules for subtraction by 1's complement are as under :

1. compute the 1's complement of the subtrahend by changing all its 1s to 0s and all its 0s to 1s.
2. add this complement to the minuend
3. perform the end-around carry of the last 1 or 0
4. if there is no end-around carry (*i.e.* 0 carry), then the answer must be recomplemented and a negative sign attached to it.
5. if the end-around carry is 1, no recomplementing is necessary.

Suppose we want to subtract 101_2 from 111_2. The procedure is as under :

```
  1 1 1
 +0 1 0      ← 1's complement of subtrahend 101
 -------
 1 0 0 1
      1      ← end-round carry
 -------
  0 1 0
```

As seen, we have removed from the addition sum the 1 carry in the last position and added it onto the remainder. It is called ***end-around carry.***

Let us now subtract 1101_2 from 1010_2.

```
  1 0 1 0
 +0 0 1 0      ← 1's complement of 1101
  1 1 0 0
  NO CARRY
```

As seen, there is no end-around carry in this case. Hence, as per Rule 4 given above, answer must be recomplemented to get 0011 and a negative sign attached to it. Therefore, the final answer becomes—0011.

Finally, consider the complemental subtraction of 1110_2 from 0110_2.

```
   0 1 1 0
 + 0 0 0 1      ← 1's complement of 1110₂
 ---------
     1 1 1
   NOCARRY
```

As seen, there is no carry. However, we may add an extra 0 from our side to make it a 0 carry as shown below.

```
   0 1 1 0
 + 0 0 0 1      ← 1's complement of subtrahend
 ---------
 0 0 1 1 1
         0      ← end-around carry
 ---------
   0 1 1 1
```

After recomplementing, it becomes 1000. When negative sign is attached, the final answer becomes -1000_2.

Example 69.8. *Using 1's complemental method, subtract 01101_2 from 11011_2.*

Solution.

```
    1 1 0 1 1
  + 1 0 0 1 0      ← 1's complement of subtrahend
  -----------
  1 0 1 1 0 1
            1      ← end-around carry
  -----------
    0 1 1 1 0
```

Since end-around carry is 1, we take the final answer as it is (Rule 5).

Example 69.9. *Use 1's complement to subtract 11011_2 from 01101_2.*

(Computer Science, Allahabad Univ. 1988)

Solution.

$$\begin{array}{rl} 01101 & \\ +00100 & \leftarrow \text{1's complement of } 11011_2 \\ \hline 10001 & \rightarrow -01110 \\ \text{NO CARRY} & \end{array}$$

Since there is no final carry, we recomplement the answer and attach a minus sign to get the final answer—01110_2.

69.14. 2's Complemental Subtraction

In this case, the procedure is as under :

1. find the 2's complement of the subtrahend,
2. add this complement to the minuend,
3. drop the final carry,
4. if the carry is 1, the answer is positive and needs no recomplementing,
5. if there is no carry, recomplement the answer and attach minus sign.

Example 69.10. *Using 2's complement, subtract 1010_2 from 1101_2.*

Solution. The 1's complement of 1010 is 0101. The 2's complement is 0101 + 1 = 0110. We will add it to 1101.

$$\begin{array}{rl} 1101 & \\ +\ 0110 & \leftarrow \text{2's complement of } 1010_2 \\ \hline 10011 & \\ \text{DROP} & \end{array}$$

The final answer is 0011_2.

Example 69.11. *Use 2's complement to subtract 1101_2 from 1010_2.*

(Digital Computations, Punjab Univ. 1992)

Solution. The 1's complement of 1101 is 0010. The 2's complement is 0011.

$$\begin{array}{rl} 1010 & \\ +\ 0011 & \leftarrow \text{2's complement of } 1101_2. \\ \hline 1101 & \\ \text{NO CARRY} & \end{array}$$

In this case, there is no carry. Hence, we have to recomplement the answer. For this purpose, we first subtract 1 from it to get 1100. Next, we recomplement it to get 0011. After attaching the minus sign, the final answer becomes $-\ 0011_2$.

Talking in terms of decimal numbers, we have subtracted 13 from 10. Obviously the answer is –3.

69.15. Binary Multiplication

The procedure for this multiplication is the same as for decimal multiplication though it is comparatively much easier. The four simple rules are as under:

1. $0 \times 0 = 0$,
2. $0 \times 1 = 0$,
3. $1 \times 0 = 0$,
4. $1 \times 1 = 1$.

The rules of binary multiplication could be summarized in the form of a table as shown.

x	0	1
0	0	0
1	0	1

As in the decimal system, the procedure is

1. copy the multiplicand when multiplier digit is 1 but not when it is 0
2. shift as in decimal multiplication
3. add the resulting binary numbers according to the rules of binary addition.

Example 69.12. *Multiply 111_2 by 101_2 using binary multiplication method.*

(Electronics-1, Indore Univ. 1991)

Solution.

```
    111
   ×101
   ----
    111
   000      — shift left, no add
  111       — shift left and add
 ------
 100011
```

Example 69.13. *Multiply 1101_2 by 1100_2.* **(Digital Electronics, Bombay Univ. 1990)**

Solution.

```
     1101
    ×1100
    -----
     0000
    0000
   1101
  1101
 --------
 10011100
```

Example 69.14. *Multiply 1111 by 0111_2.*

Solution. This example has been included for the specific purpose of explaining how to handle the addition if multiplication results in ***columns with more than two 1s***.

1. Result of the first column is 1.
2. In the second column, addition of $1 + 1 = 10_2$. Hence, we put down 0 there and carry 1 to the third column.
3. In the third column, $1 + 1 + 1 + 1 = 100_2$ (decimal 4). We keep one 0 there, put the second 0 in fourth column and pass on 1 to the fifth column.
4. In the fourth column, $1 + 1 + 1 + 0 = 11$, (decimal 3). Hence, one 1 is kept there and the other 1 is passed on to the fifth column.
5. In the fifth column, $1 + 1 + 1 + 1 = 100_2$ (decimal 4). Again, one 0 is retained there, second 0 is passed on to the sixth column and 1 to the seventh column.
6. In the sixth column, $1 + 0 = 1$.
7. The seventh column already has 1 given by the addition of the fifth column.

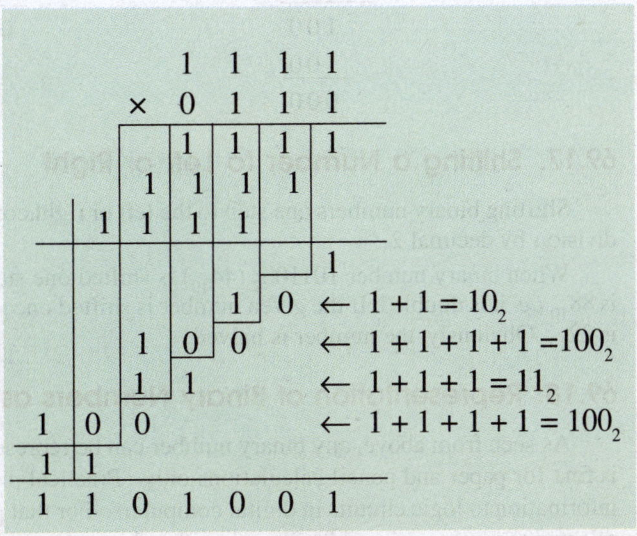

Fig. 69.5

69.16. Binary Division

It is similar to the division in the decimal system. As in that system, here also division by 0 is meaningless. Rules are :

1. $0 \div 1 = 0$ or $\frac{0}{1} = 0$, **2.** $1 \div 1 = 1$ or $\frac{1}{1} = 1$.

Example 69.15. *Carry out the binary division 11001 ÷ 101*

Solution.

```
          101
      ________
 101 ) 11001
       101
       ----
        0010
          101
          101
        -----
          000
```

After we bring down the next 0 bit, the number 10 so formed is not divisible by 101. Hence, we put a 0 in the quotient. Therefore, the answer is 101_2 (decimal 5).

Incidentally, it may be noted that the dividend 25_{10} and divisor is 5_{10} so that the result of division, as expected, is 5_{10}.

Example 69.16. *Divide 11011_2 by 100_2* **(Digital Computations, Punjab Univ. 1990)**

Solution.

```
          110.11
       ________
 100 ) 11011
       100
       ---
        101
        100
       ----
         0011
          110
          100
          ---
          100
          100
          ---
          000
```

```
     6.75
    _____
 4 ) 27
     24
     --
     30
     28
     --
     20
     20
     --
     00
```

69.17. Shifting a Number to Left or Right

Shifting binary numbers one step to the left or right corresponds respectively to multiplication or division by decimal 2.

When binary number 101100_2 (44_{10}) is shifted one step **to the left**, it becomes 1011000_2 which is 88_{10} *i.e.* it is doubled. If the given number is shifted one step **to the right**, it becomes 10110_2 which is 22_{10}. Obviously, the number is halved.

69.18. Representation of Binary Numbers as Electrical Signals

As seen from above, any binary number can be represented as a string of 0s and 1s. However, it is fine for paper and pencil calculations only. Practical problem is how to apply the desired binary information to logic circuits in digital computers. For that purpose, two types of electrical signals are selected to represent 1 and 0. Since speed and accuracy are of primary importance in digital circuits, the two electrical signals chosen to represent 1 and 0 must meet very rigid requirements.

1. they must be suitable for use in high-speed circuitry,
2. the signals should be very easy to tell apart,
3. they must be hard to confuse with each other.

The second and third statements may look alike but they, in fact, are not so. It is found that all transistor circuits distort, to some extent, the electrical signals that pass through them. Sometime, these distorted signals can look confusingly alike. Hence, this effect of distortion or degradation has to be kept in mind while selecting the two signals.

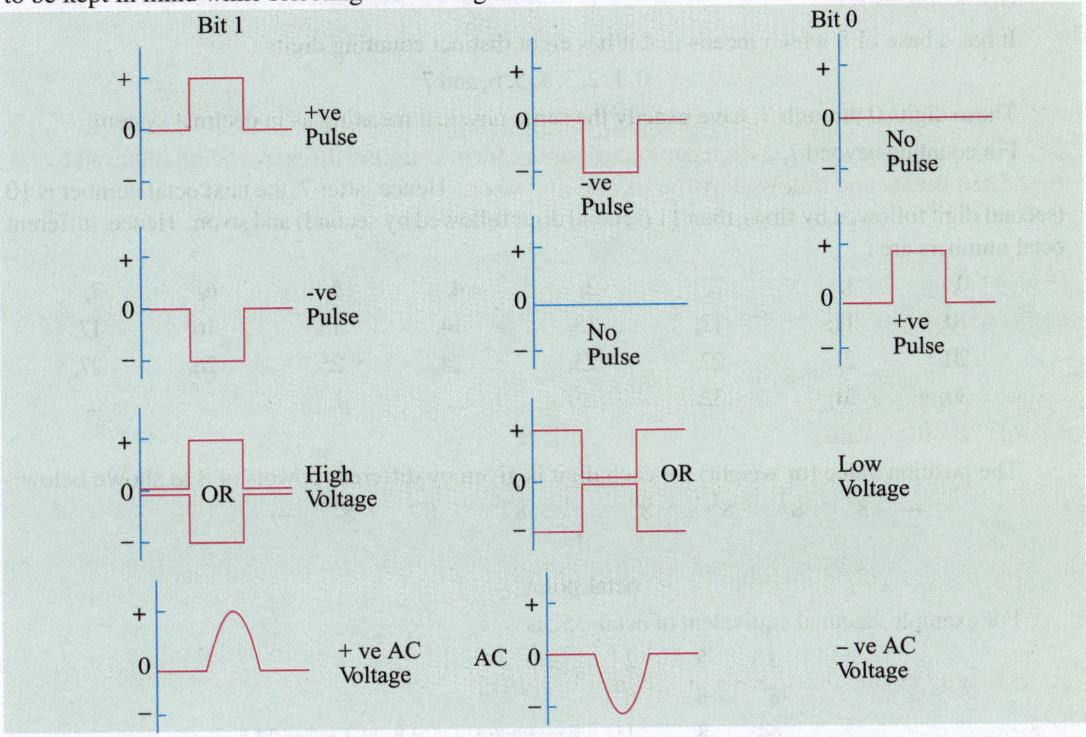

Fig. 69.6

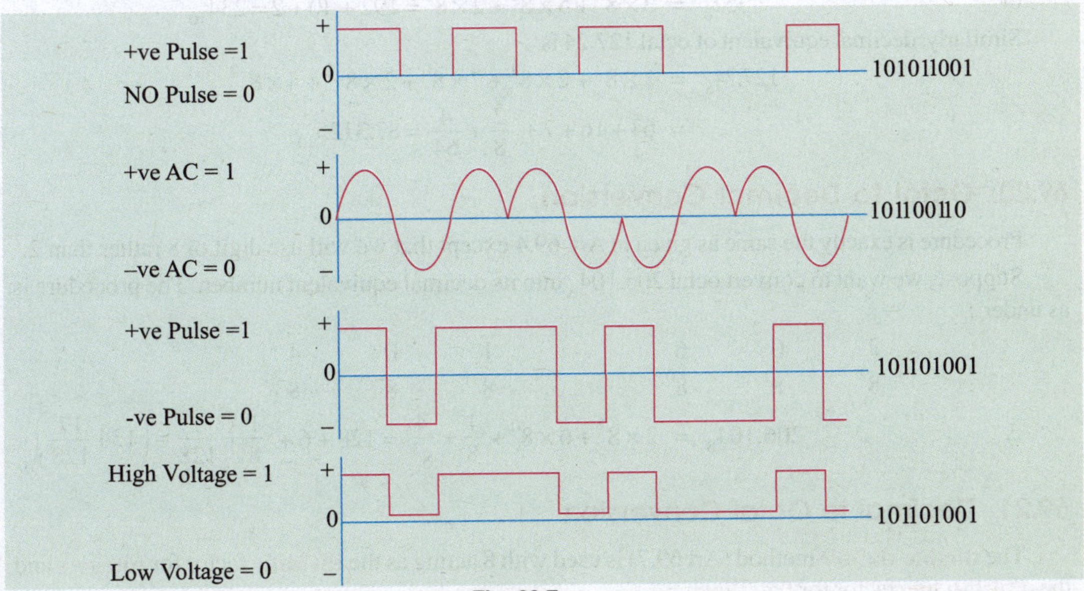

Fig. 69.7

In Fig. 69.6 are shown several signal pairs that meet the above requirements. It will be noted that it is impossible to distort a positive pulse (representing 1) to look like the no pulse or negative pulse (representing 0).

Fig. 69.7 shows how signal pairs can be used to represent different binary numbers.

69.19. Octal Number System

(i) **Radix or Base**

It has a base of 8 which means that it has eight distinct counting digits :

0, 1, 2, 3, 4, 5, 6, and 7

These digits 0 through 7, have exactly the same physical meaning as in decimal system.

For counting beyond 7, **2-digit combinations are formed taking the second digit followed by the first, then the second followed by the second and so on.** Hence, after 7, the next octal number is 10 (second digit followed by first), then 11 (second digit followed by second) and so on. Hence, different octal numbers are :

0,	1,	2,	3,	4,	5,	6,	7,
10,	11,	12,	13,	14,	15,	16,	17,
20,	21,	22,	23,	24,	25,	26,	27,
30,	31,	32,	...	...	...	...	...

(ii) **Position Value**

The position value (or weight) for each digit is given by different powers of 8 as shown below:

$$\leftarrow \quad 8^3 \quad 8^2 \quad 8^1 \quad 8^0 \quad \bullet \quad 8^{-1} \quad 8^{-2} \quad 8^{-3} \quad \rightarrow$$

↑

octal point

For example, decimal equivalent of octal 352 is

$$\begin{array}{ccc} 3 & 5 & 2 \\ 8^2 & 8^1 & 8^0 \\ 64 & 8 & 1 \end{array} = 3 \times 64 + 5 \times 8 + 2 \times 1 = 234_{10}$$

or

$$352_8 = 3 \times 8^2 + 5 \times 8^1 + 2 \times 8^0 = 192 + 40 + 2 = 234_{10}$$

Similarly, decimal equivalent of octal 127.24 is

$$127.24_8 = 1 \times 8^2 + 2 \times 8^1 + 7 \times 8^0 + 2 \times 8^{-1} + 4 \times 8^{-2}$$

$$= 64 + 16 + 7 + \frac{2}{8} + \frac{4}{64} = 87.3125_{10}$$

69.20. Octal to Decimal Conversion

Procedure is exactly the same as given in Art. 69.4 except that we will use digit of 8 rather than 2.

Suppose, we want to convert octal 206.104_8 into its decimal equivalent number. The procedure is as under :

$$\begin{array}{ccccccc} 2 & 0 & 6 & . & 1 & 0 & 4 \\ 8^2 & 8^1 & 8^0 & & 8^{-1} & 8^{-2} & 8^{-3} \end{array}$$

$$\therefore \quad 206.104_8 = 2 \times 8^2 + 6 \times 8^0 + \frac{1}{8} + \frac{4}{8^3} = 128 + 6 + \frac{1}{8} + \frac{1}{128} = \left(134 \frac{17}{128}\right)_{10}$$

69.21. Decimal to Octal Conversion

The **double-dabble** method (Art 69.7) is used with 8 acting as the **dividing** factor for **integers** and the **multiplying** factor for **fractions.**

Let us see how we can convert 175_{10} into its octal equivalent.

$175 \div 8 = 21$	with 7 remainder	↑
$21 \div 8 = 2$	with 5 remainder	
$2 \div 8 = 0$	with 2 remainder	

Taking the remainders in the **reverse order**, we get 257_8. $\therefore \quad 175_{10} = 257_8$

Let us now take decimal fraction 0.15. Its octal equivalent can be found as under:

$0.15 \times 8 = 1.20 = 0.20$	with a carry of 1	↓
$0.20 \times 8 = 1.60 = 0.60$	with a carry of 1	
$0.60 \times 8 = 4.80 = 0.80$	with a carry of 4	

$\therefore \quad 0.15_{10} \cong 0.114_8$

As seen, here carries have been taken in the **forward direction** *i.e.* from top to bottom.

Using positional notation, the first few octal numbers and their decimal equivalents are shown in Table 69.2.

Table No. 69.2

Octal	*Decimal*	*Octal*	*Decimal*	*Octal*	*Decimal*
0	0	12	10	24	20
1	1	13	11	25	21
2	2	14	12	26	22
3	3	15	13	27	23
4	4	16	14	30	24
5	5	17	15	31	25
6	6	20	16	32	26
7	7	21	17	33	27
10	8	22	18	34	28
11	9	23	19	35	29

69.22. Binary to Octal Conversion

The simplest procedure is to use **binary-triplet** method. In this method, the given binary number is arranged into groups of 3 bits starting from the octal point and then each group is converted to its equivalent octal number. Of course, where necessary, extra 0s can be added **in front** (*i.e.* left end) of the binary number to complete groups of three.

Suppose, we want to convert 101011_2 into its octal equivalent. Converting the bits into groups of three, we have $\therefore$

101 011

Now, 101_2 is 5 octal and 011 is 3 octal.

$\therefore$ 101 011
↓ ↓
5 3 $\qquad \therefore \quad 101\ 011_2 = 53_8$

Now, take 111110111_2. We will first split it into groups of **three bits** (space is left between the groups for easy reading). Then, each group is given its octal number as shown below.

111 110 111
↓ ↓ ↓
7 6 7 $\qquad \therefore \quad 111\ 110\ 111_2 = 767_8$

Finally, take the example of a mixed binary number 10101.11_2. Here, we will have to add one 0 in front of the integral part as well as to the fractional part

$$\begin{array}{ccc} 010 & 101 \cdot & 110 \\ \downarrow & \downarrow & \downarrow \\ 2 & 5 & 6 \end{array} \qquad \therefore \quad 10101.11_2 = 25.6_8$$

The equivalence between binary triplets and octal numbers is given in Table 69.3.

Table No. 69.3

Binary	*Octal*	*Binary*	*Octal*
000	0	1010	12
001	1	1011	13
010	2	1100	14
011	3	1101	15
100	4	1110	16
101	5	1111	17
110	6	10000	20
111	7	10001	21
1000	10	10010	22
1001	11	10011	23

69.23. Octal to Binary Conversion

The procedure for this conversion is just the opposite of that given in Art 69.22. Here, each digit of the given octal number is converted into its equivalent binary triplet. For example, to change 75_8 into its binary equivalent, proceed as under:

$$\begin{array}{cc} 7 & 5 \\ \downarrow & \downarrow \\ 111 & 101 \end{array} \qquad \therefore \quad 75_8 = 111\ 101_2$$

Similarly, 74.562_8 can be converted into binary equivalent as under:

$$\begin{array}{ccccc} 7 & 4 \cdot & 5 & 6 & 2 \\ \downarrow & \downarrow & \downarrow & \downarrow & \downarrow \\ 111 & 100 & 101 & 110 & 010 \end{array} \qquad \therefore \quad 74.562_8 = 111\ 100.101\ 110\ 010_2$$

Incidentally, it may be noted that number of digits in octal numbers is **one-third of that in equivalent binary numbers**. In the present case, it is five versus fifteen.

69.24. Usefulness of Octal Number System

We have already discussed the octal number system and conversion from the binary and decimal numbers to octal and vice versa. The ease with which conversions can be made between octal and binary makes the octal system attractive as a "shorthand" means of expressing large binary numbers, In computer work, binary number with up to 64 bits are not uncommon. These binary numbers, as we shall see, do not always represent a numerical quantity but are often some type of code that conveys non numerical information. In computers, binary numbers might represent :

1. actual numerical data
2. numbers corresponding to a location called (address) in memory,
3. an instruction code

4. a code representing alphabetic and other non numerical characters,
5. group of bits representing the status of devices internal or external to the computer

When dealing with a large quantity of binary numbers of many bits, it is convenient and more efficient for us to write the numbers in octal rather than binary. However keep in mind that the digital circuits and systems work strictly in binary. We use octal numbers only as a convenience for the operators of the system.

69.25. Hexadecimal Number System

The characteristics of this system are as under :

1. it has a base of 16. Hence, it uses sixteen distinct counting digits 0 through 9 and A through F as detailed below :
   0, 1, 2, 3, 4, 5, 6, 7, 8, 9, A, B, C, D, E, F
2. place value (or weight) for each digit is in **ascending powers** of 16 for integers and **descending powers** of 16 for fractions.

The main use of this system is in connection with **byte-organised machines.** It is used for specifying addresses of different binary numbers stored in computer memory.

69.26. How to Count Beyond F in Hex Number System ?

As usual, we resort to **2-digit combinations.** After reaching **F**, we take the second digit followed by the first digit, then second followed by second, then second followed by third and so on. The first few 'hex' numbers and their decimal equivalents are given in Table 69.4.

Table No. 69.4

Hexadecimal	*Decimal*	*Hexadecimal*	*Decimal*	*Hexadecimal*	*Decimal*
0	0	B	11	16	22
1	1	C	12	17	23
2	2	D	13	18	24
3	3	E	14	19	25
4	4	F	15	1A	26
5	5	10	16	1B	27
6	6	11	17	1C	28
7	7	12	18	1D	29
8	8	13	19	1E	30
9	9	14	20	1F	31
A	10	15	21	20	32
				21	33

69.27. Binary to Hexadecimal Conversion

The simple method is to split the given binary number into **4-bit groups** (supplying 0s from our own side if necessary) and then give each group its 'hex' value as found from Table 69.5.

Table No. 69.5

Binary	*Hex.*	*Binary*	*Hex.*	*Binary*	*Hex.*
0000	0	0110	6	1100	C
0001	1	0111	7	1101	D
0010	2	1000	8	1110	E
0011	3	1001	9	1111	F
0100	4	1010	A	10000	10
0101	5	1011	B	10001	11

Let us see how we would convert 10001 100_2 into its hexadecimal equivalent number. We will first split the given binary number into 4-bit groups and then give each group its proper value from Table 69.5.

$$\begin{array}{cc} 1000 & 1100 \\ \downarrow & \downarrow \\ 8 & C \end{array} \qquad \therefore \qquad 1000\,1100_2 = 8C_{16}$$

Let us now consider 1011010111_2. Following the above procedure, we have

$$\begin{array}{ccc} 0010 & 1101 & 0111 \\ \downarrow & \downarrow & \downarrow \\ 2 & D & 7 \end{array} \qquad \therefore \qquad 1011010111_2 = 2D7_{16}$$

It is seen that two 0s have been added to complete the 4-bit groups.

69.28. Hexadecimal to Binary Conversion

Here, the procedure is just the reverse of that given in Art. 69.27. Each hexadecimal digit is converted into its equivalent 4-bit binary.

Suppose, we want to convert $23A_{16}$ into its binary equivalent. It can be done as given below:

$$\begin{array}{ccc} 2 & 3 & A \\ \downarrow & \downarrow & \downarrow \\ 0010 & 0011 & 1010 \end{array} \qquad \therefore \qquad 23A_{16} = 0010\ 0011\ 1010_2$$

Incidentally, hex numbers contain **one-fourth** the number of bits contained in the equivalent binary number. Optionally, we could drop off the two 0s in front of the binary equivalent.

69.29. Decimal to Hexadecimal Conversion

Two methods are available for such a conversion. One is to go from decimal to binary and then to hexadecimal. The other method called **hex dabble** method is similar to the double-dabble (or divide-by-two) method of Art 69.7 except that we use 16 (instead of 2) for successive divisions. As an example let us convert decimal 1983 into hexadecimal by consulting Table No. 69.4 for remainders.

Hence, $1983_{10} = 7BF_{16}$

$$\begin{aligned} 1983 \div 16 &= 123 + 15 \rightarrow F \\ 123 \div 16 &= 7 + 11 \rightarrow B \\ 7 \div 16 &= 0 + 7 \rightarrow 7 \end{aligned} \quad \uparrow$$

69.30. Hexadecimal to Decimal Conversion

Two methods are available for such a conversion. One is to convert from hexadecimal to binary and then to decimal. The other direct method is as follows :

Instead of using powers of 2, use power of 16 for the weights. Then, sum up the products of hexadecimal digits and their weights to get the decimal equivalent. As an example, let us convert F6D9 to decimal.

$$\begin{aligned} \text{F6D9} &= \text{F}(16^3) + 6(16^2) + \text{D}(16^1) + 9(16^0) = 15 \times 16^3 + 6 \times 16^2 + 13 \times 16^1 + 9 \times 16^0 \\ &= 61{,}440 + 1536 + 208 + 9 = 63{,}193_{10} \end{aligned}$$

Example 69.17. *Find the binary, octal and hexadecimal equivalents of the following decimal numbers (i) 32 (ii) 256 (iii) 51.* **(Digital Computations, Punjab Univ. May 1990)**

Solution. (i) Decimal number 32

As seen from Art :

32	16	8	4	2	1
1	Ø	Ø	Ø	Ø	Ø

$\therefore \quad 32_{10} = 100000_2$

For octal conversion :

$$\begin{array}{cc} 100 & 000 \\ \downarrow & \downarrow \\ 4 & 0 \end{array} \qquad \therefore \qquad 32_{10} = 40_8$$

For hexadecimal conversion 0010 0000 → 2 0 $\therefore\ 32_{10} = 20_{16}$

As will be seen, the binary number has been divided into two 4-bit groups for which purpose two 0s have been added to the left.

(*ii*) In the same way, it can be found that $256_{10} = 100000000_2 = 400_8 = 100_{16}$

(*iii*) Also $51_{10} = 110011_2 = 63_8 = 33_{16}$

Example 69.18. *Convert the following numbers to decimal*

(*i*) $(11010)_2$ (*ii*) $(AB60)_{16}$ (*iii*) $(777)_8$

(Digital Computations, Punjab Univ. 1992)

Solution. (*i*) For binary to decimal conversion, we will follow the procedure given in Art. 69.4.

$$\begin{array}{ccccc} 1 & 1 & 0 & 1 & 0 \\ 16 & 8 & \not{4} & 2 & \not{1} \end{array}$$

$\therefore\ 11010_2 = 16 + 8 + 2 = 26_{10}$

(*ii*) Following the procedure given in Art. 69.30, we have,

$$AB60_{16} = A(16^3) + B(16^2) + 6(16^1) + 0(16^0) = 10 \times 16^3 + 11 \times 16^2 + 6 \times 16^1 + 0 \times 16^0$$
$$= 43{,}872_{10}$$

(*iii*) As per the procedure given in Art. 69.20, $777_8 = 7(8^2) + 7(8^1) + 7(8^0) = 511_{10}$.

Example 69.19. *A computer is transmitting the following groups of bytes (each consisting of 8-bits) to some output device. Give the equivalent octal and hexadecimal listings.*

1000 1100 0011 1010
0010 1110 1001 0101
0101 1111 1011 0110
0111 1011 0101 1011

Solution.

Binary		*Octal*			*Hexadecimal*	
1000	1100	10	001	100	1000	1100
		2	1	4	8	C
0010	1110	00	101	110	0010	1110
		0	5	6	2	E
0101	1111	01	011	111	0101	1111
		1	3	7	5	F
0111	1011	01	111	011	0111	1011
		1	7	3	7	B
0011	1010	00	111	010	0011	1010
		0	7	2	3	A
1001	0101	10	010	101	1001	0101
		2	2	5	9	5
1011	0110	10	110	110	1011	0110
		2	6	6	B	6
0101	1011	01	011	011	0101	1011
		1	3	3	5	B

Hence, the groups of given memory bytes when expressed in different number systems become as under :

Binary	*Octal*	*Hexadecimal*
10001100	214	8C
00101110	056	2E
01011111	137	5F
01111011	173	7B
00111010	072	3A
10010101	225	95
10110110	266	B6
01011011	133	5B

It is clear from above that so far as the computer operator (or programmer) is concerned, it is much easier to handle this data when expressed in octal or hexadecimal system than in the binary system. For example, it is much easier and less error-prone to write the hexadecimal 8C than the binary 10001100 or 6AF than 011010101111. Of course, when the need arises, the operator can easily convert from octal or hexadecimal to binary.

69.31. Digital Coding

In digital logic circuits, each number or piece of information is defined by an equivalent combination of binary digits. A complete group of these combinations which represents numbers, letters or symbols is called a ***digital code.***

Codes have been used for security reasons so that others may not be able to read the message even if it is intercepted. In modern digital equipment, codes are used to represent and process numerical information. The choice of a code depends on the function or purpose it has to serve. Some codes are suitable where arithmetic operations are performed whereas others have high efficiency *i.e.* they give more information using fewer bits.

In certain applications, use of one code or the other simplifies and reduces the circuitry required to process the information. By limiting the switching circuitry, reliability of the digital system is increased. Of continuing importance are other codes which allow for error detection or correction. These codes enable the computers to determine whether the information that was coded and transmitted is received correctly and, if there is an error, to correct it. Since coding itself is a detailed subject, only few of the more familiar codes will be discussed.

69.32. Binary Coded Decimal (BCD) Code

It is a binary code in which *each* decimal digit is represented by a group of four bits. Since the left-right-to weighting of the 4-bit positions is 8-4-2-1, it is also called an 8421 code. It is a weighted numerical code. As said above, here each decimal digit from 0 through 9 requires a 4-bit binary-coded number. For example, the decimal number 35 in *BCD* code is 00110101. The coding of ten decimal digits is given in Table No. 69.6. Lest you think that *BCD* code is the same thing as binary numbers, consider the following.

Table No. 69.6

Decimal	*BCD*
0	0000
1	0001
2	0010
3	0011
4	0100
5	0101
6	0110
7	0111
8	1000
9	1001

In the binary system, ten is represented by 1010 but in *BCD* code, it is 0001 0000. Seventeen in binary is 10001 but in *BCD* code, it is 0001 0111. See the difference ! Actually, the confusion is due to the fact that the first nine numbers in *BCD* and binary are exactly similar (Table No. 69.6). After that, they become quite different (Table No. 69.7).

It should be realized that with four-bits, sixteen num bers (2^4) can be represented although in the *BCD* code only ten of these are used. The following six combinations are *invalid* in the *BCD* code : 1010, 1011, 1100, 1101, 1110 and 1111.

Table No. 69.7

Decimal	BCD			
26			0010	0110
59			0101	1001
673		0110	0111	0011
2498	0010	0100	1001	1000

The main advantage of *BCD* code is that it can be read and recognised easily although special adders are needed for arithmetic operations.

Any decimal number can be expressed in *BCD* code by replacing each decimal digit by the appropriate 4-bit combination. Conversely, a *BCD* number can be easily converted into a decimal number by dividing the coded number into groups of four bits (starting with *LSB*) and then writing down the decimal digit represented by each four-bit group.

Example 69.20. *Write the decimal number 369 in BCD code.*

Solution. For writing this 3-digit number in *BCD*, the value of each digit must be replaced by its 4-bit equivalent from the *BCD* code. From Table No. 69.6, we get

3 = 0011,   6 = 0110,   9 = 1001

$\therefore \quad 369_{10} = 001101101001_{BCD}$

Example 69.21. *Typically digital thermometers use BCD to drive their digital displays. How many BCD bits are required to drive a 3-digit thermometer display? What 12 bits are sent to display for a temperature of 157 degrees.*

Solution. There are 12 *BCD* bits required to drive a 3-digit thermometer display because each *BCD* digit is represented by a group of four bits.

In order to display a temperature of 157 degrees, we know that we have to send 12-bits. These bits can be determined by replacing each decimal digit by its equivalent four bit binary. Thus,

$$\begin{array}{ccc} 1 & 5 & 7 \\ \downarrow & \downarrow & \downarrow \\ 0001 & 0101 & 0111 \end{array}$$

$$\therefore \quad 157_{10} = 000101010111_{BCD}.$$

Example 69.22. *Find the equivalent decimal value for the BCD code number 0001010001110101.*

(Applied Electronics, A.M.I.E.E., London)

Solution. Starting from the *LSB*, the given number can be divided into groups of four bits as 0001 0100 0111 0101. As seen from Table No. 69.6,

0001 = 1,   0100 = 4,
0111 = 7   and 0101 = 5

Hence, $\quad 0001010001110101_{BCD} = 1475_{10}$

Example 69.23. (*a*) *Convert the hexadecimal number F8E6 to the corresponding decimal number.*

(*b*) *Convert the decimal number 2479 to the corresponding hexadecimal number.*

(*c*) *Encode the following decimal numbers into 8421 BCD numbers (i) 59 (ii) 39 and (iii) 584.*

(*d*) *Decode the following 8421 BCD numbers (i) 0101 (ii) 0111.*

(Digital Computations, Punjab Univ. 1990)

Solution. (*a*)
$$F8E6 = F(16^3) + 8(16^2) + E(16^1) + 6(16^0)$$
$$= 15 \times 16^3 + 8 \times 16^2 + 14 \times 16^1 + 6 \times 16^0 = 63{,}718_{10}$$

(*b*) We would use the hex-dabble method explained in Art 69.29

$$2479 \div 16 = 154 + 15 \rightarrow F$$
$$154 \div 16 = 9 + 10 \rightarrow A$$
$$9 \div 16 = 0 + 9 \rightarrow 9$$

$\therefore \quad 2479_{10} = 9AF_{16}$

(*c*) As seen from Table No. 69.6

5	9	3	9	5	8	4
↓	↓	↓	↓	↓	↓	↓
0101	1001	0011	1001	0101	1000	0100

(*d*) Again, consulting Table No. 69.6, we have

0101	0111
↓	↓
5	7

69.33. Octal Coding

It involves grouping the bits in *three's*. For example, $(1756)_8 = (001\ 111\ 101\ 110)_2 = (001111101110)_2$. Similarly, the 24-bit number stored in the computer memory such as 101 010 011 100 010 111 000 110 can be read in the octal as

101	010	011	100	010	111	000	110
↓	↓	↓	↓	↓	↓	↓	↓
5	2	3	4	2	7	0	6

Apart from ease of recognition and conversion to binary, one important feature of the octal code is that its numbers are straight binary numbers which can be manipulated mathematically. For example, octal 25 expressed in octal code is 010 101 which can be read as binary 010101

$$(25)_8 = 2 \times 8^1 + 5 \times 8^0 = (21)_{10}$$
$$(010101)_2 = 0 \times 2^5 + 1 \times 2^4 + 0 \times 2^3 + 1 \times 2^2 + 0 \times 2^1 + 1 \times 2^0 = (21)_{10}$$

You might recall that in *BCD* code, the resulting number is always a 4-bit group and a special adder is needed to convert it into decimal. In octal coding, 3-bit grouping is used but the resulting binary number can be considered a single number in natural binary form.

69.34. Hexadecimal Coding

The advantage of this coding is that four bits are expressed by a single character. However, the disadvantage is that new symbols have to be used to represent the values from 1010 to 1111 binary. As seen from Table No. 69.5, the binary number 1010 0101 is hex number A5. Similarly, hexadecimal number *C*7 is $(11000111)_2$. To prove that the resulting binary number is the same as the hexadecimal value, consider the following example:

$$(3D)_{16} = 3 \times 16^1 + D \times 16^0 = 3 \times 16 + 13 \times 1 = (61)_{10}$$
$$(3D)_{16} = (00111101)_2 = 1 \times 2^5 + 1 \times 2^4 + 1 \times 2^3 + 1 \times 2^2 + 0 \times 2^1 + 1 \times 2^0 = (61)_{10}$$

Table No. 69.8

Decimal	*XS–3*
3	0110
26	0101 1001
629	1001 0101 1100
3274	0110 0101 1010 0111

69.35. Excess-3 Code

It is an unweighted code and is a modified form of *BCD*. It is widely used to represent numerical data in digital equipment. It is abbreviated as XS-3. As its name implies, each coded number in XS-3 is *three larger* than in *BCD* code. For example, six is written as 1001. As compared to *BCD*, the XS-3 has poorer recognition but it is more desirable for arithmetic operations. A few numbers using Excess-

3 code are given in Table 69.8.

69.36. Gray Code

It is an unweighted code for numbers 0 through 9 and is largely used in mechanical switching systems. As seen from Table No. 69.9, only a single bit changes between each successive word. Because of this, the amount of switching is minimized and the reliability of the switching system is improved.

Table No. 69.9

Decimal	Gray	XS-3 Gray
0	0000	0010
1	0001	0110
2	0011	0111
3	0010	0101
4	0110	0100
5	0111	1100
6	0101	1101
7	0100	1111
8	1100	1110
9	1101	1010

69.37. Excess-3 Gray Code

It is shown in Table No. 69.9 and is the original gray code shifted by three binary combinations. It exhibits the same properties as the Gray Code.

Example 69.24. *Express the number 43_{10} in XS-3 code.*

Solution. Let us first represent each decimal digit by its 4-bit XS-3 code.

$4 = 0111,\qquad 3 = 0110\qquad \therefore\quad 43_{10} = 01110110_{XS-3}$

Example 69.25. *The number 0110 1001 is expressed in XS-3 code. What is its decimal value ?*

Solution. Starting from least significant bit (*LSB*), the given number is first separated into groups of four and then each group is replaced by its equivalent value *i.e.* actual value decreased by 3.

$0110 = 6 - 3 = 3;\ 1001 = 9 - 3 = 6$

$\therefore\quad 01101001_{XS-3} = (36)_{10}$

69.38. Other Codes

Some of the other codes which are presently popular are given below:

(a) **4-bit codes**

The different 4-bit weighted *BCD* codes for decimal numbers 0 through 9 in use are : 5421, 2*421, 7421, $74\overline{21}$, etc. and are tabulated below:

Table No. 69.10

Decimal	5421	2*421	7421	$74\overline{21}$
0	0000	0000	0000	0000
1	0001	0001	0001	0111
2	0010	0010	0010	0110
3	0011	0011	0011	0101
4	0100	0100	0100	0100
5	1000	1011	0101	1010
6	1001	1100	0110	1001
7	1010	1101	1000	1000
8	1011	1110	1001	1111
9	1100	1111	1010	1110

(b) **5-bit Codes**

(i) **2-out-of-5 codes** is an unweighted *BCD* code and allows easy error detection. It has been used in communications and telephone operation.

(ii) **51111 Code** is a weighted *BCD* code and is much easier to operate with electronic circuitry.

(iii) **Shift-counter (Johnson) Code** is an unweighted *BCD* code and because of its pattern is easily operated on with electronic circuitry.

(*c*) **7-bit Biquinary Code**— it uses a group of seven bits to represent decimal numbers and has code features which provide easy error detection and ease of operation.

(*d*) **Ring-counter Code**— it is also called 10-bit code because it uses a group of 10 bits to represent a decimal number. Though it requires as many as 10 positions, the ease of error detection with the code and of operating electronic circuits to implement the code make it quite attractive.

(*e*) **Alphanumeric Code**— In addition to numerical data, a computer must be able to handle non-numerical information used in input/output (I/O) processing. In other words, a computer should recognize codes that represent letters of the alphabet, punctuation marks, and other special characters as well as numbers. These codes are called alphanumeric codes. A complete alphanumeric code would include the 26 lowercase letters, 26 uppercase letters, 10 numeric digits, 7 punctuation marks, and anywhere from 20 to 40 other characters, such as +, –, /, #, $, “, and so on. We can say that an alphanumeric code represents all of the various characters and functions that are found on a computer keyboard. The most widely used alphanumeric code is the American standard code for Information Interchange (ASCII). Another similarly I/O-oriented code is EBCDIC (Extended Binary Coded Decimal Interchange Code).

69.39. ASCII Code

The ASCII code (Pronounced “askee) is a seven-bit code, and so it has 2^7 (=128) possible code groups. This is more than enough to represent all of the standard keyboard characters as well as control functions such as the (RETURN) and (LINEFEED) functions. Table No. 69.11 shows a partial listing of the ASCII code. In addition to the binary code group for each character, the table gives the octal and hexadecimal equivalents. The complete list of the ASCII code is given in the Appendix.

Table No. 69.11

Character	*7-Bit ASCII*	*Octal*	*Hex*	*Character*	*7-Bit ASCII*	*Octal*	*Hex*
A	100 0001	101	41	Y	101 1001	131	59
B	100 0010	102	42	Z	101 1010	132	5A
C	100 0011	103	43	0	011 0000	060	30
D	100 0100	104	44	1	011 0001	061	31
E	100 0101	105	45	2	011 0010	062	32
F	100 0110	106	46	3	011 0011	063	33
G	100 0111	107	47	4	011 0100	064	34
H	100 1000	110	48	5	011 0101	065	35
I	100 1001	111	49	6	011 0110	066	36
J	100 1010	112	4A	7	011 0111	067	37
K	100 1011	113	4B	8	011 1000	070	38
L	100 1100	114	4C	9	011 1001	071	39
M	100 1101	115	4D	blank	010 0000	040	20
N	100 1110	116	4E	.	010 1110	056	2E
O	100 1111	117	4F	(	010 1000	050	28
P	101 0000	120	50	+	010 1011	053	2B
Q	101 0001	121	51	$	010 0100	044	24
R	101 0010	122	52	*	010 1010	052	2A

S	101 0011	123	53	)	010 1001	051	29
T	101 0100	124	54	—	010 1101	055	2D
U	101 0101	125	55	/	010 1111	057	2F
V	101 0110	126	56	,	010 1100	054	2C
W	101 0111	127	57	=	011 1101	075	3D
X	101 1000	130	58	<RETURN	000 1101	015	0D
				<LINEFEED>	000 1010	012	0A

Example. 69.26. *The following is a message encoded in ASCII code. What is the message ?*

1001000 1000101 1001100 1001100 1001111

Solution. Convert each seven-bit code to its equivalent hexadecimal number. The resulting values are:

48 45 4C 4C 4F

Now locate these hexadecimal values in Table No. 69.11 and determine the character represented by each. The results are:

"H E L L O".

Tutorial Problems No. 69.1

1. Find the decimal equivalents of the following binary numbers:
(*a*) 101 (*b*) 1001 (*c*) 10.011 **[(*a*) 5_{10} (*b*) 9_{10} (*c*) 3.0375_{10}]**

2. What are the decimal equivalents of the following binary numbers ?
(*a*) 1111 (*b*) 10100 (*c*) 11011001 (*d*) 10011001
[(*a*) 15_{10} (*b*) 20_{10} (*c*) 109_{10} (*d*) 153_{10}]

3. Express the following binary numbers into their equivalent decimal numbers :
(*a*) 11.01 (*b*) 101.11 (*c*) 110.01 **[(*a*) 3.25_{10} (*b*) 5.75_{10} (*c*) 6.25_{10}]**

4. Convert the following decimal numbers into their binary equivalents:
(*a*) 25 (*b*) 125 (*c*) 0.85
[(*a*) $(11001)_2$ (*b*) $(1111101)_2$ (*c*) $(0.110110)_2$]

5. What are the binary equivalents of the following decimal numbers ?
(*a*) 27 (*b*) 92 (*c*) 64
[(*a*) $(11011)_2$ (*b*) $(1011100)_2$ (*c*) $(1000000)_2$]

6. Perform the following binary additions:
(*a*) 1011 + 1001 (*b*) 1011 + 110 (*c*) 101101 + 1101101 (*d*) 1011.01 + 1001.11
(*e*) 0.0011 + 0.1110 (*f*) 1111 + 111 + 1111
[(*a*) $(10100)_2$ (*b*) $(10001)_2$ (*c*) $(10011010)_2$ (*d*) $(10101)_2$ (*e*) $(1.0001)_2$ (*f*) $(1101001)_2$]

7. Add the following binary numbers 1011 and 1001
[10100](*Digital Computations, Punjab Univ. Dec.*)

8. Perform the following binary subtractions :
(*a*) 1101 – 1011 (*b*) 111 – 101 (*c*) 1000 – 11 (*d*) 101011 – 10010
[(*a*) $(0010)_2$ (*b*) $(010)_2$ (*c*) $(40101)_2$ (*d*) $(011001)_2$]

9. Carry out the following subtractions using binary number system:
(*a*) $64_{10} - 32_{10}$ (*b*) $128_{10} - 64_{10}$ (*c*) $93.5_{10} - 42.75_{10}$ (*d*) $7_{10} - 11_{10}$
[(*a*) $(100000)_2$ (*b*) $(1000000)_2$ (*c*) $(110010.11)_2$ (*d*) $(-100)_2$]

10. Subtract the following binary numbers: 100011 – 111010
[–10111](*Digital Computations, Punjab Univ., Dec. 1984*)

11. Find the 1's complements of the following binary numbers:
(*a*) 01101 (*b*) 1101 (*c*) 1001 (*d*) 1010
[(*a*) $(10010)_2$ (*b*) $(0010)_2$ (*c*) $(0110)_2$ (*d*) $(0101)_2$]

12. What are 2's complements of the following binary numbers ?
(*a*) 1011 (*b*) 11011 (*c*) 11011.01 (*d*) 10011.11
[(*a*) $(0101)_2$ (*b*) $(00101)_2$ (*c*) $(00100.11)_2$ (*d*) $(01100.01)_2$]

13. Use the 1's and 2's complements for the following binary subtractions :
(*i*) 1111 – 1011 (*ii*) 110011 – 100101 (*ii*) 100011 – 111010
[(*i*) 0100 (*ii*) 1110 (*iii*) –10111] (*Digital Computations, Punjab Univ. Dec. 1985*)

14. Multiply the following binary numbers:
(*a*) 1100 × 101 (*b*) 10101 × 101 (*c*) 10111 × 101 (*d*) 1110 × 111
[(*a*) $(111100)_2$ (*b*) $(1101001)_2$ (*c*) $(1110011)_2$ (*d*) $1100010)_2$]

15. Perform the following binary divisions :
(*a*) 11011 ÷ 100 (*b*) 1110011 ÷ 101 (*c*) 1100010 ÷ 111
[(*a*) $(110.11)_2$ (*b*) $(10111)_2$ (*c*) $(1110)_2$]

16. Convert the following binary numbers into their octal equivalents :
(*a*) 1001 (*b*) 11011 (*c*) 10 101 111 (*d*) 1101.0110111
(*e*) 11 111 011 110 101
[(*a*) 11_8 (*b*) 33_8 (*c*) 25_8 (*d*) $(15.334)_8$ (*e*) $(37365)_8$]

17. Convert the undergiven octal numbers into their binary equivalents:
(*a*) 13 (*b*) 11 (*c*) 713 (*d*) 3674
[(*a*) $(011)_2$ (*b*) $(1001)_2$ (*c*) $(111\ 001\ 011)_2$ (*d*) $(11110111100)_2$]

18. Convert the following numbers :
(*a*) 357_8 to decimal (*b*) 6421_8 to decimal (*c*) 1359_{10} to octal (*d*) 7777_{10} to octal
[(*a*) 239_{10} (*b*) 3345_{10} (*c*) 2517_8 (*d*) $(17141)_8$]

19. Convert the following real numbers to the binary numbers:
(*i*) 12.0 (*ii*) 25.0 (*iii*) 0.125
[(*i*) 1100 (*ii*) 11001 (*iii*) 0.001] (*Digital Computations, Punjab Univ. 1984*)

20. Convert the following binary numbers into their equivalent hexadecimal numbers:
(*a*) 1101 0111 (*b*) 1010 0110 (*c*) 1001 1110 (*d*) $11009E_{16}$ 1111
[(*a*) $D7_{16}$ (*b*) $A6_{16}$ (*c*) $9E_{16}$ (*d*) CF_{16}]

21. Convert the following decimal numbers to binary numbers by converting them to octal, then to binary.
(*i*) 850 (*ii*) 7563
[(*a*) 1101010010 (*ii*) 1110110001011] (*Digital Computations, Punjab Univ. Dec. 1985*)

22. Convert the following numbers to decimal
(*i*) $(11010)_2$ (*ii*) $(777)_8$
[(*i*) 26 (*ii*) 511] (*Digital Computations, Punjab Univ. June 1985*)

23. What are the binary equivalents of the following hexadecimal numbers ?
(*a*) 9F (*b*) A2 (*c*) ED (*d*) 27
[(*a*) $(10011111)_2$ (*b*) $(10100010)_2$ (*c*) $(11101101)_2$ (*d*) $(00110111)_2$]

24. Convert the decimal number 4397 and the octal number 2735 to hexadecimal numbers.
[(*i*) $(112D_{16})$ (*ii*) $(50D)_{16}$] (*Power & Digital Electronics, Punjab Univ. Dec. 1984*)

25. Convert the following decimal numbers into *BCD* code
(*a*) 18 (*ii*) 92 (*iii*) 321 and (*iv*) 4721
[(*i*) 0001 1000 (*ii*) 1001 0010 (*iii*) 0011 0010 0001 (*iv*) 0100011100100001]

26. Find the decimal values for the BCD-coded numbers
(*i*) 0110 1000 (*ii*) 0111 0100 1001 and (*iii*) 1000 0100 0111 0110]
[(*i*) 68 (*ii*) 749 (*iii*) 8476]

27. Convert the following hexadecimal numbers into binary numbers:
(*i*) E 5 (*ii*) B4D (*iii*) 7AF4
(*Digital Computations, Punjab Univ. May 1994*)
[(*i*) 11100101 (*ii*) 101101001101 (*iii*) 1001 1010 1111 0100]

28. Encode the following decimal numbers into 8421 numbers.
(*i*) 45 (*ii*) 732 (*iii*) 94,685
(*Digital Computations, Punjab Univ. May 1984*)
[(*i*) 0100 0101 (*ii*) 111 0011 0010 (*iii*) 001 0100 0110 1000 0101]

29. Decode the following 8421 *BCD* numbers :
(*i*) 0011 1000 0111 (*ii*) 1001 0110 0111 1000 0111 0011
[(*i*) 387 (ii) 967,873] (*Digital Computations, Punjab Univ. Dec. 1985*)

30. Decode the following 8421 *BCD* numbers :
(*i*) 0011 (*ii*) 0111 (*iii*) 0101 (*iv*) 0110
(*v*) 1000 **[(*i*) 3 (*ii*) 7 (*iii*) 5 (*iv*) 6 (*v*) 8] (*Digital Computations, Punjab Univ. June 1985*)**

31. Convert the following binary numbers into octal code :
(*i*) 100110111 and (*ii*) 101010001 **[(*i*) $(467)_8$ (*ii*) $(521)_8$]**

32. Convert the following decimal numbers into XS-3 code:
(*i*) 35 (*ii*) 159 (*iii*) 340
[(*i*) 01101000 (*ii*) 010010001100 (*iii*) 011001110011]

33. What is the decimal value of the following numbers expressed in Excess-3 code ?
(*i*) 10010110 and (*ii*) 100101011100 **[(*i*) 63 (*ii*) 629]**

OBJECTIVE TESTS – 69

1. The digital systems usually operate on ....system.
(*a*) binary (*b*) decimal
(*c*) octal (*d*) hexadecimal.

2. The binary system uses powers of ....for positional values.
(a) 2 (*b*) 10
(*c*) 8 (*d*) 16

3. After counting 0, 1, 10, 11, the next binary number is
(*a*) 12 (*b*) 100
(*c*) 101 (*d*) 110.

4. The number 1000_2 is equivalent to decimal number
(*a*) one thousand
(*b*) eight (*c*) four
(*d*) sixteen.

5. In binary numbers, shifting the binary point one place to the right.
(*a*) multiplies by 2 (*b*) divides by 2
(*c*) decreases by 10 (*d*) increases by 10.

6. The binary addition 1 + 1 + 1 gives
(a) 111 (*b*) 10
(*c*) 110 (*d*) 11

7. The cumulative addition of the four binary bits (1 + 1 + 1 + 1) gives
(*a*) 1111 (*b*) 111
(*c*) 100 (*d*) 1001

8. The result of binary subtraction (100 – 011) is
(*a*) –111 (*b*) 111
(*c*) 011 (*d*) 001.

9. The 2's complement of 1000_2 is
(*a*) 0111 (*b*) 0101
(*c*) 1000 (*d*) 0001

10. The main reason why digital computers use complemental subtraction is that it
(*a*) simplifies their circuitry
(*b*) is a very simple process
(*c*) can handle negative numbers easily
(*d*) avoids direct subtraction.

11. The result of binary multiplication $111_2 \times 10_2$ is
(*a*) 1101 (*b*) 0110
(*c*) 1001 (*d*) 1110

12. The binary division $11000_2 \div 100_2$ gives
(a) 110 (*b*) 1100
(*c*) 11 (*d*) 101

13. The number 12_8 is equivalent to decimal
(*a*) 12 (*b*) 20
(*c*) 10 (*d*) 4

14. The number 100101_2 is equivalent to octal
(*a*) 54 (*b*) 45
(*c*) 37 (*d*) 25.

15. The number 17_8 is equivalent to binary
(*a*) 111 (*b*) 1110
(*c*) 10000 (*d*) 1111.

16. Which of the following is NOT an octal number ?
(*a*) 19 (*b*) 77
(*c*) 15 (*d*) 101

17. Hexadecimal number system is used as a shorthand language for representing ....numbers.
(*a*) decimal (*b*) binary
(*c*) octal (*d*) large

18. The binary equivalent of A_{16} is
(*a*) 1010 (*b*) 1011
(*c*) 1000 (*d*) 1110.

19. *BCD* code is
(*a*) non-weighted
(*b*) the same thing as binary numbers
(c) a binary code
(*d*) an alphanumeric code.

20. Which of the following 4-bit combinations is/are invalid in the *BCD* code ?
(*a*) 1010 (*b*) 0010
(*c*) 0101 (*d*) 1000.

21. Octal coding involves grouping the bits in
(a) 5's (b) 7's
(c) 4's (d) 3's.

22. In Excess-3 code each coded number is ....than in BCD code.
(a) four larger (b) three smaller
(c) three larger (d) much larger.

23. Which numbering system uses numbers and letters as symbols ?
(a) decimal (b) binary
(c) octal (d) hexadecimal

24. To convert a whole decimal number into a hexadecimal equivalent, one should divide the decimal value by.......
(a) 2 (b) 8
(c) 10 (d) 16.

ANSWERS

1. (a)	2. (a)	3. (b)	4. (b)	5. (a)	6. (d)	7. (c)	8. (d)	9. (c)	10. (a)
11. (d)	12. (a)	13. (c)	14. (b)	15. (d)	16. (a)	17. (b)	18. (a)	19. (c)	20. (a)
21. (d)	22. (c)	23. (d)	24. (d)						

C H A P T E R 70

LOGIC GATES

Learning Objectives

- Definition of a Logic Gate
- Positive and Negative Logic
- The OR Gate
- Equivalent Relay Circuit of an OR Gate
- Diode OR Gate
- Transistor OR Gate
- OR Gate Symbolizes Logic Addition
- Three Input OR Gate
- Exclusive OR Gate
- The AND Gate
- Equivalent Relay Circuit of an AND Gate
- Diode AND Gate
- Transistor AND Circuit
- AND Gate Symbolizes Logic Multiplication
- The NOT Gate
- The NOT Operation
- Bubbled Gates
- The NOR Gate
- The NAND Gate
- The XNOR Gate
- Logic Gate at Glance
- Digital Signals Applied to Logic Gates
- Sequential Logic Circuits
- Adders and Subtractors
- Half Adder
- Full Adder
- Parallel Binary Adder
- Half Subtractor
- Full Subtractor

A logic gate is a circuit that has one or more input signals but only one output signal. All logic gates can be analysed by constructing a truth table

70.1. Definition of a Logic Gate

A logic gate is an electronic circuit **which makes logic decisions**. It has one output and one or more inputs. The output signal appears only for certain combinations of input signals. Logic gates are the basic building blocks from which most of the digital systems are built up. They implement the hardware logic function based on the logical algebra developed by George Boole which is called Boolean algebra in his honour. A unique characteristic of the Boolean algebra is that variables used in it **can assume only one of the two values** *i.e.* either 0 or 1. Hence, every variable is either a 0 or a 1.

These gates are available today in the form of various IC families. The most popular families are: transistor-transistor logic (*TTL*), emitter-coupled logic (*ECL*), metal-oxide-semiconductor (*MOS*) and complementary metal-oxide-semiconductor (*CMOS*).

In this chapter, we will consider the *OR*, *AND*, *NOT*, *NOR*, *NAND*, exclusive *OR* (*XOR*) and exclusive *NOR* (*XNOR*) gates along with their truth tables.

70.2. Positive and Negative Logic

In computing systems, the number symbols 0 and 1 represent two possible states of a circuit or device. It makes no difference if these two states are referred to as *ON* and *OFF*, *CLOSED* and *OPEN*, *HIGH* and *LOW*. *PLUS* and *MINUS* or *TRUE* and *FALSE* depending on the circumstances. Main point is that **they must be symbolized by two opposite conditions**.

In positive logic, a 1 represents

1. an *ON* circuit **2.** a *CLOSED* switch **3.** a *HIGH* voltage
4. a *PLUS* sign **5.** a *TRUE* statement

Consequently, a 0 represents

1. an *OFF* circuit **2.** an *OPEN* switch **3.** a *LOW* voltage **4.** a *MINUS* sign **5.** a *FALSE* statement.

In negative logic, just opposite conditions prevail.

Suppose, a digital system has two voltage levels of 0V and 5V. If we say that symbol 1 stands for 5V and symbol 0 for 0V, then we have positive logic system. If, on other hand, we decide that a 1 should represent 0 V and 0 should represent 5V, then we will get negative logic system.

Main point is that in ***positive logic*, the more positive** of the two voltage levels represents the 1 while in negative logic, **the more negative** voltage represents the 1. Moreover, it is not essential that a 0 has to be represented by 0V although in some cases the two may coincide. Suppose, in a circuit, the two voltage levels are 2V and 10V. Then for positive logic, the 1 represents 10V and the 0 represents 2V (*i.e.* lesser of the two voltages). If the voltage levels are – 2V and – 8V, then, in positive logic, the 1 represents – 2V and the 0 represents – 8V (*i.e.* lesser of the two voltages).

Unless stated otherwise, we will be using only ***positive logic*** in this chapter.

70.3. The OR Gate

The electronic symbol for a two-input *OR* gate is shown in Fig. 70.1 (*a*) and its equivalent switching circuit in Fig. 70.1 (*b*). The two inputs have been marked as *A* and *B* and the output as *X*. It is worth reminding the reader that as per Boolean algebra, the three variables *A*, *B* and *X* can have only one of the two values *i.e.* either 0 or 1.

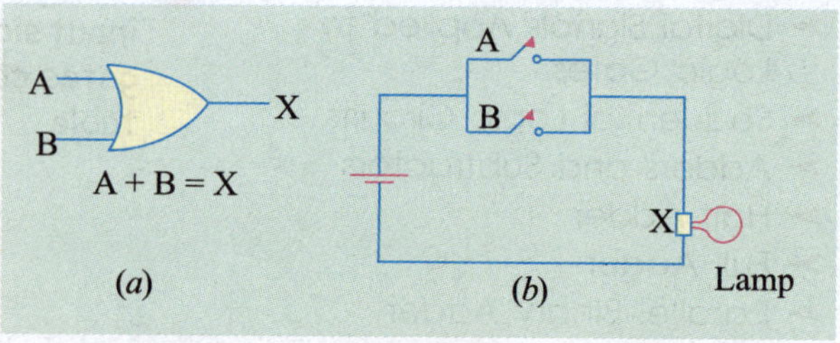

Fig. 70.1

Logic Operation

The *OR* gate has an **output of 1 when either A or B or both are 1**.

In other words, it is an **any-or-all** gate because an output occurs when any or all the inputs are present.

As seen from Fig. 70.1 (*b*), the lamp will light up (logic 1) when either switch *A* or *B* or both are closed.

Obviously, the output would be 0 **if and only if both its inputs are 0**. In terms of the switching conditions, it means that lamp would be *OFF* (logic 0) only when both switches *A* and *B* are *OFF*.

The *OR* gate represents the Boolean equation $A + B = X$

The meaning of this equation is that *X* is true when either *A* is true or *B* is true or both are true. Alternatively, it means that output *X* is 1 when either *A* or *B* or both are 1.

Table 70.1

A	B	X
0	0	0
0	1	1
1	0	1
1	1	1

Fig. 70.2

The above logic operation of the *OR* gate can be summarised with the help of the truth table given in Fig. 70.2. A truth table may be defined as a table which ***gives the output state for all possible input combinations***. The *OR* Table 70.1 gives outputs for all possible *AB* inputs of 00, 01, 10 and 11.

We may interpret the truth table as follows:

When both inputs are 0 (switches are *OPEN*), output *X* is 0 (lamp is *OFF*). When *A* is in logic state 0 (switch *A* is *OPEN*) but *B* is in logic state 1 (switch *B* is *CLOSED*), the output *X* is logic state 1 (lamp is *ON*). Lamp would be also *ON* when *A* is *CLOSED* and *B* is *OPEN*. Of course, lamp would be *ON* when both switched are *CLOSED*. It is so because an *OR* gate is equivalent to a **parallel *circuit in its logic function***.

Another point worth remembering is that the above *OR* gate is called **inclusive** *OR* gate because **it includes the case when both inputs are true**.

70.4. Equivalent Relay Circuit of an OR Gate

In Fig. 70.3, the relay contacts have been wired ***in parallel***. When +5V is applied to *A*, relay K_1 is energised and pulls *M* down thereby closing the contact. Hence, supply voltage of +5V appears at the output *X*.

Similarly, when +5V are applied to input *B*, K_2 is energised and pulls *N* down thereby bringing *X* in contact with *S*. Of course, when both *A* and *B* are at +5V, *X* is at +5V. Incidentally, when inputs at *A* and *B* are 0, *X* is also 0.

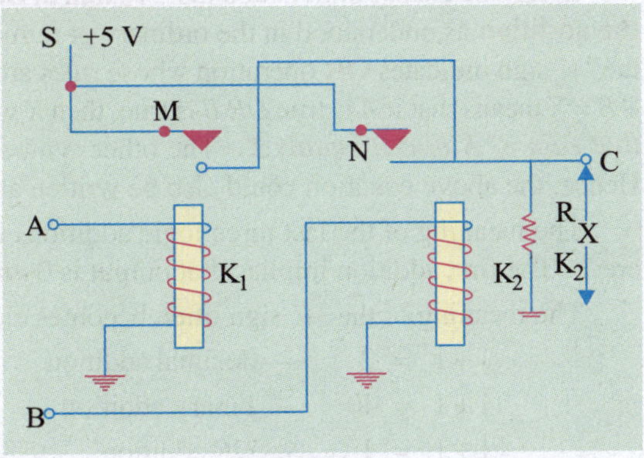

Fig. 70.3

70.5. Diode OR Gate

Fig. 70.4. shows the diode *OR* gate consisting of two ideal diodes D_1 and D_2 connected in parallel across the output *X*.

1. When *A* is at +5V, D_1 is forward-biased and hence conducts. The circuit current flows *via R* dropping 5V across it. In this way, point *X* achieves potential of +5 V.
2. When + 5V is applied to *B*, D_2 conducts causing point *X* to go to +5 V.
3. When both *A* and *B* are +5V, the drop across *R* is 5V because voltages of *A* and *B* ***are in parallel.***

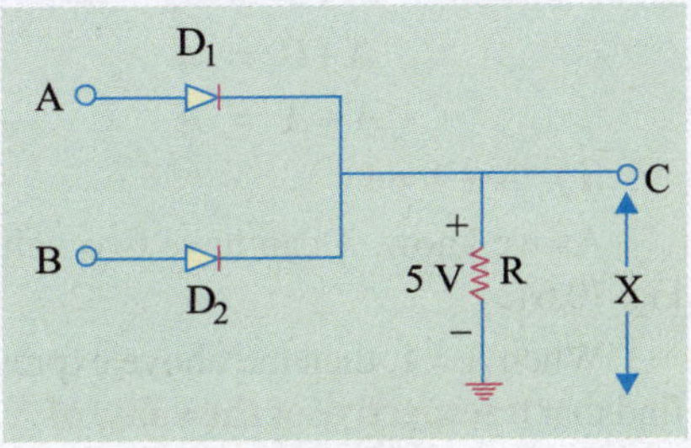

Fig. 70.4

Again, point X is driven to +5 V.

4. Obviously, when there is no voltage either at A or B, output X remains 0.

70.6. Transistor OR Gate

Fig. 70.5 illustrates a possible transistor *OR* gate consisting of three interconnected transistors Q_1, Q_2, and Q_3 supplied from a common supply V_{cc} = +5 V.

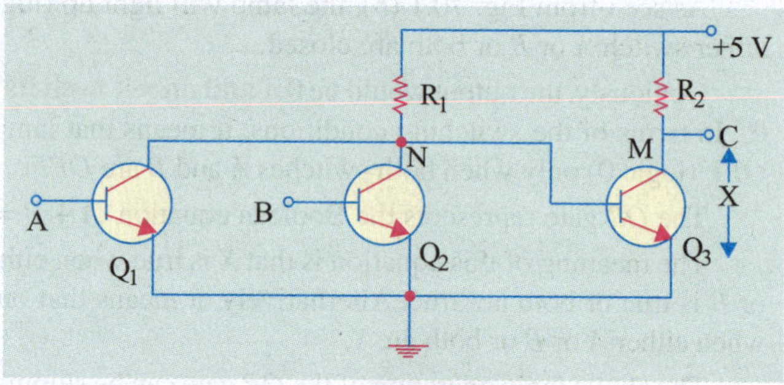

Fig. 70.5

1. When +5 V is applied to A, Q_1 is forward-biased and so it conducts. Assuming that Q_1 is saturated, entire V_{cc} = 5 V drops across R_1 thus causing N to go to ground. This, in turn, cuts off Q_3 thereby causing X to go to V_{cc} *i.e.* + 5V.
2. When +5 V is applied to B, Q_2 conducts thereby driving N to ground *i.e.* 0V. With no forward bias on its base, Q_3 is cut-off thus driving X again to V_{cc} *i.e.* +5 V.
3. If both inputs A and B are grounded, Q_1 and Q_2 are cut-off driving N to +5 V. As a result, Q_3 becomes forward-biased and conducts fully. In that case, entire V_{cc} drops across R_2 driving M and hence X to ground.

70.7. OR Gate Symbolizes Logic Addition

According to Boolean algebra, *OR* gate performs **logical addition**. Its truth table can be written as given below:

$$0 + 0 = 0$$
$$0 + 1 = 1$$
$$1 + 0 = 1$$
$$1 + 1 = 1$$

It must be clearly understood that '+' sign in Boolean algebra **does not stand for the addition** as understood in the ordinary or numerical algebra. In symbolic logic, the '+' sign indicates **OR** operation whose rules are given above. In logic algebra, $A + B = X$ means that if A is true *OR* B is true, then X will be true. ***It does not mean here that sum of A and B equals X.*** The other symbols used for '+' sign are U and V. Hence, the above equation could also be written as $AUB = X$ or $AVB = X$.

The meaning of the last three logic additions is that output is 1 when either input A or B or both are 1. The first addition implies that output is 0 ***only when both inputs are 0.***

The meaning of the '+' sign often becomes clear from the context as shown below:

$1 + 1 = 2$ — decimal addition

$1 + 1 = 10$ — binary addition

$1 + 1 = 1$ — *OR* addition

We can put the above *OR* laws in more general terms

$$A + 1 = 1$$
$$A + 0 = A$$
$$A + A = A \quad \text{— not } 2A$$

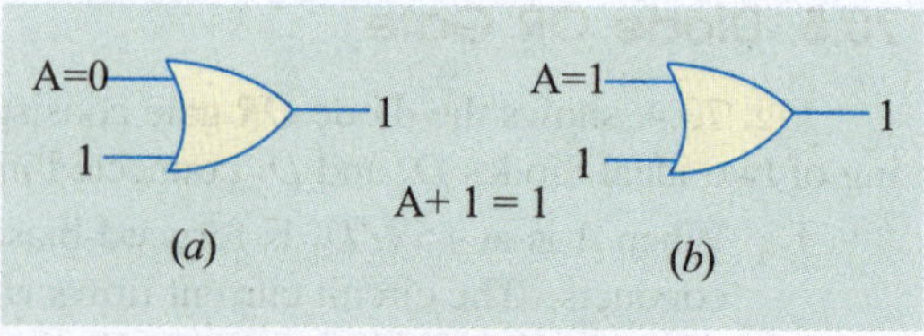

Fig. 70.6

(*i*) $A + 1 = 1$

As we know, A can have two values: 0 or 1. When A is 0, then we have 0 + 1 = 1 as shown in Fig. 70.6 (*a*).

When A = 1, then the above expression becomes : 1 + 1 = 1 as shown in Fig. 70.6 (*b*), Hence, we find that **irrespective of the value of A.**

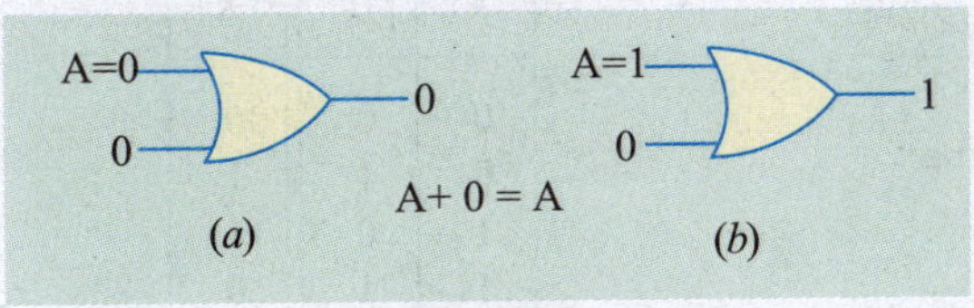

Fig. 70.7

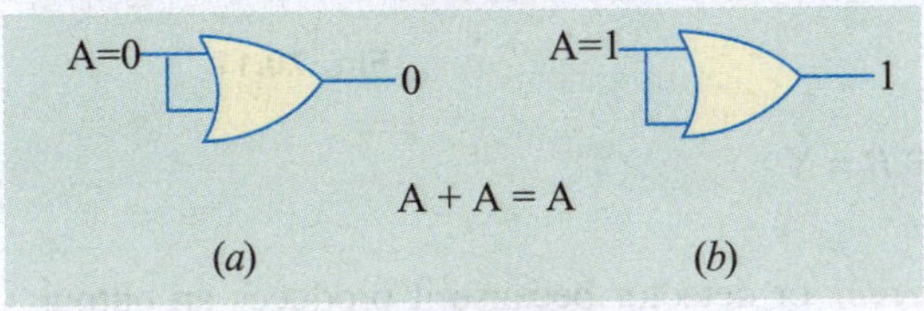

Fig. 70.8

$$A + 1 = 1$$

(ii) **A + 0 = A**

If $A = 0$, then $0 + 0 = 0$ *i.e.* output is 0 which is correct and is shown in Fig. 70. 7 (*a*). The output is what the value of *A* is.

As shown in Fig. 70.7 (*b*), when $A = 1$, output is 1 because $1 + 0 = 1$. Again, output is what the value of A is.

(iii) **A + A = A**

With *A* set to 0, the output is 0 because $0 + 0 = 0$ as shown in Fig. 70.8 (*a*).

With *A* set to 1, the output is 1 because $1 + 1 = 1$ as shown in Fig. 70.8 (*b*). Obviously, the output in both cases is *A*.

70.8. Three Input OR Gate

The electronic symbol for a 3-input (fan-in of 3) ***inclusive OR*** gate is shown in Fig. 70.9. As is usual in logic algebra, the inputs *A*, *B*, *C* as well as the output *X* can have only one of the two values *i.e.* 0 or 1.

Truth Table

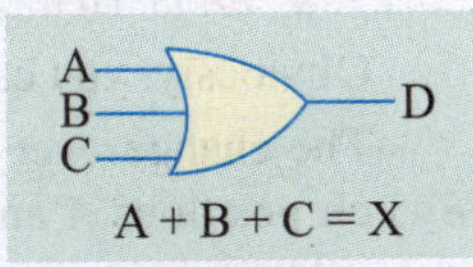

Fig. 70.9

It is shown in Table 70.2. Following points are worth noting:

1. The number of rows in the table is $2^3 = 8$ *i.e.* there are eight ways of combining the three inputs. In general, the number of horizontal rows is 2^n where *n* is the number of inputs.
2. In first column *A*, logic values alternate between 0 and 1 every four rows twice.
3. The second input column *B* alternates between 0 and 1 every two rows four times.
4. The third input column *C* alternates between 0 and 1 every other row eight times.

The truth table symbolizes the Boolean equation $A + B + C = X$ which means that output *X* is 1 when ***either A or B or C is 1 or all are 1.*** Alternatively, *X* is true when either *A* or *B* or *C* is true or all are true.

Table No. 70.2

A	B	C	X
0	0	0	0
0	0	1	1
0	1	0	1
0	1	1	1
1	0	0	1
1	0	1	1
1	1	0	1
1	1	1	1

70.9. Exclusive OR Gate

Its electronic symbol is shown in Fig. 70.10 (*a*) and its equivalent switching circuit in Fig. 70.10 (*b*).

In this gate, output is 1 if its either input **but not both**, is 1. In other words, it has an output 1 **when its inputs are different**. The output is 0 only when inputs **are the same**.

To put it a bit differently, this logic gate has output 0 **when inputs are either all 0 or all 1.**

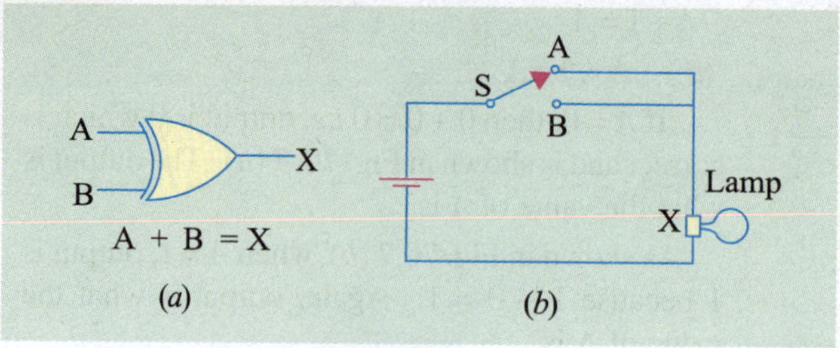

Fig. 70.10

A	B	X
0	0	0
0	1	1
1	0	1
1	1	0

Fig. 70.11

This gate works on the Boolean equation $A \oplus B = X$

The circle around plus (+) sign is worth noting.

The circuit is also called an ***inequality comparator*** or detector because it produces an output only **when the two inputs are different**.

Explanation

The ***inclusive*** *OR* gate exemplifies the everyday usage of the word *OR* which stands for one or the other or both. Take the following statement:

To qualify for a competition, you might have to subscribe to a magzine OR belong to a club. Obviously, there is no bar on your doing both. But now take **exclusive** statement:

You can be rich OR you can be poor.

Obviously, you cannot be both at the same time.

The change-over switching ***circuit*** of Fig. 70.10 (*b*) simulates the ***exclusive OR (XOR)*** gate. Switch positions *A* and *B* will individually light up the lamp but a combination of *A* and *B* is not possible.

The truth table for a 2-input *XOR* gate is given in Table No.70.3. It is instructive to compare it with that for an *inclusive OR* gate (Table 70.1).

70.10. The AND Gate

The electronic (or logic) symbol for a 2-input *AND* gate is shown in Fig. 70.12 (*a*) and its equivalent switching circuit in Fig. 70.12 (*b*). It is worth reminding the readers once again that the three variables *A*, *B*, *C* can have a **value of either 0 or 1**.

Logic Operation

1. The *AND* gate gives an output only **when all its inputs are present**.
2. The *AND* gate has a 1 output when both *A* **and** *B* are 1. Hence, this gate is an **all-or-nothing** gate whose output occurs only when all its inputs are present.
3. In True/False terminology, the output of an *AND* gate will be **true** only if **all its inputs are true**. Its output would be false if ***any of its inputs is false***.

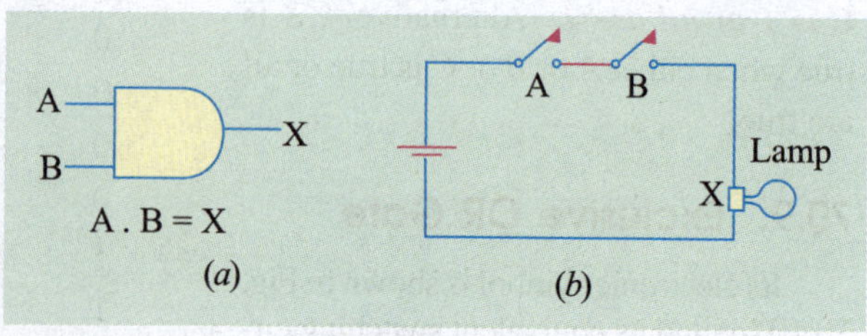

Fig. 70.12

The *AND* gate works on the Boolean algebra

$$A \times B = X \quad \text{or} \quad A \,.\, B = X \quad \text{or} \quad AB = X$$

It is a **logical** multiplication and is different from the **arithmetic** multiplication. Often the sign '×' is replaced by a dot which itself is generally omitted as shown above. The logical meaning of the above equation is that

1. output X is 1 only when both A *and* B are 1.
2. output X is true only when both A *and* B are true.

Table 70.4

A	B	X
0	0	0
0	1	0
1	0	0
1	1	1

Fig. 70.13

Table 70.5

A	B	C	X
0	0	0	0
0	0	1	0
0	1	0	0
0	1	1	0
1	0	0	0
1	0	1	0
1	1	0	0
1	1	1	1

ABC = X

Fig. 70.14

As seen from Fig. 70.12 (*b*), the lamp would be *ON* when both switches *A* and *B* are closed. Even when one switch is open, the lamp would be *OFF*. Obviously, an *AND* gate is equivalent to a **series switching circuit.**

Truth Table Fig. 70.13 shows truth table for a 2-input *AND* gate and Fig. 70.14 gives the same for a 3-input *AND* gate.

As seen, X is at logic 1 only when *all* inputs are at logic 1, not otherwise. The procedure for writing down the first three columns is the same as explained in Art. 70.8 earlier.

70.11. Equivalent Relay Circuit of an AND Gate

The *AND* gate can be physically realized with the help of relay circuit shown in Fig. 70.15. Here, the two relay contacts have been **wired in series.**

When +5 V is applied to both input circuits, relays K_1 and K_2 are energized thereby pulling M and N downwards which brings X in contact with the supply point S. Hence, X goes to +5 V.

It is obvious that energizing only ***one relay*** will not make X go to + 5 V.

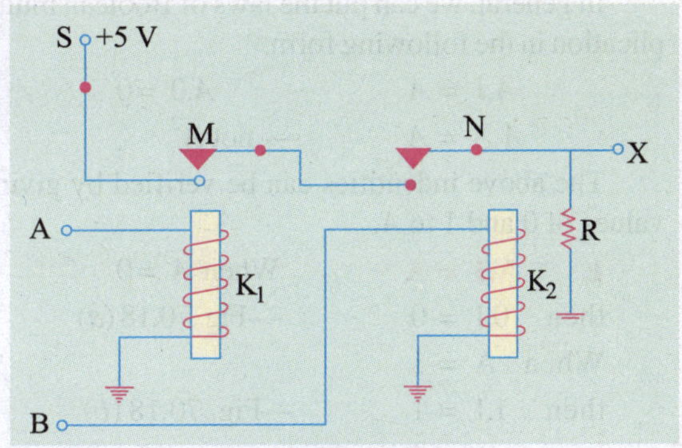

Fig. 70.15

70.12. Diode AND Gate

It is shown in Fig. 70.16. Its logical operation is as under :

1. When A is at 0 V, diode D_1 conducts and the supply voltage of +5 V drops across R. Consequently, point N and hence point X are driven to 0 V. Therefore, the output C is 0.
2. Similarly, when B is at 0 V, D_2 conducts thereby driving N and hence X to ground.
3. Obviously, when both A and B are at 0 V, both diodes conduct and, again, the output X is 0.
4. There is no supply current and hence no drop across R ***only when both A and B are at +5 V.*** Only in that case, the output X goes to supply voltage of +5 V.

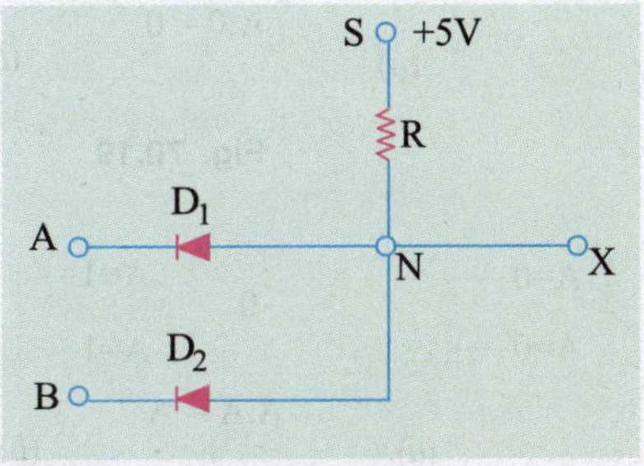

Fig. 70.16

70.13. Transistor AND Circuit

It is shown in Fig. 70.17. When both A and B are at +5 V, the two transistors Q_1 and Q_2 conduct. The current so produced drops the supply voltage of +5 V across R_1 thereby driving point N and hence base of Q_3 to ground or 0V. This cuts off Q_3 so that X goes to supply voltage of +5 V.

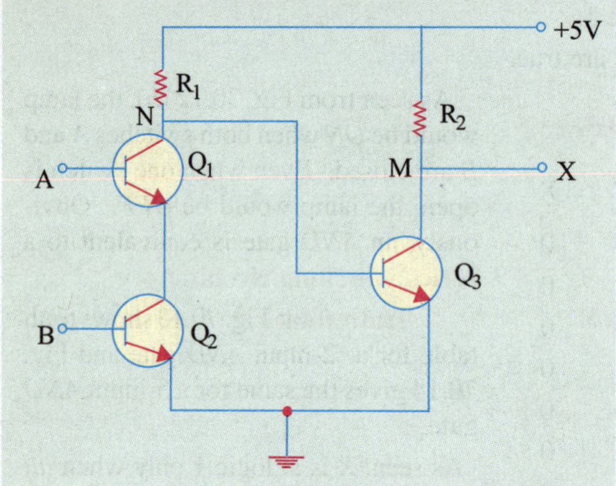

Fig. 70.17

Obivously, there is an output at X only when there is an input at A *and* B.

If either A or B is at 0 V, then Q_1 or Q_2 will be cut off and no drop will take place across R_1. Hence, point N will go to supply voltage of +5V. Consequently, Q_3 will conduct and whole of supply voltage will be dropped across R_2. As a result, point M and hence output X will go to 0 V.

70.14. AND Gate Symbolizes Logic Multiplication

According to Boolean algebra, the *AND* gate performs logical multiplication on its inputs as given below:

$$0.0 = 0$$
$$0.1 = 0$$
$$1.0 = 0$$
$$1.1 = 1$$

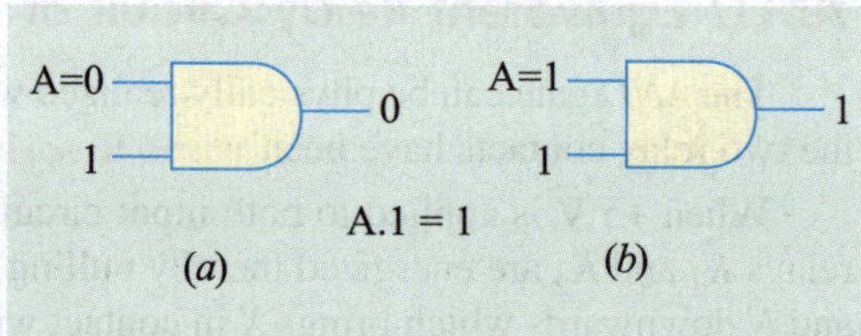

Fig. 70.18

In general, we can put the laws of Boolean multiplication in the following form:

$A.1 = A$ $\quad\quad A.0 = 0$

$A.A = A$ $\quad\quad$ — not A^2

The above indentities can be verified by giving values of 0 and 1 to A.

1. $A.1 = A$ When $A = 0$

then $0.1 = 0$ —Fig. 70.18 (*a*)

When $A = 1$

then $1.1 = 1$ —Fig. 70.18 (*b*)

It is seen that in each case, output has the same value as that of A.

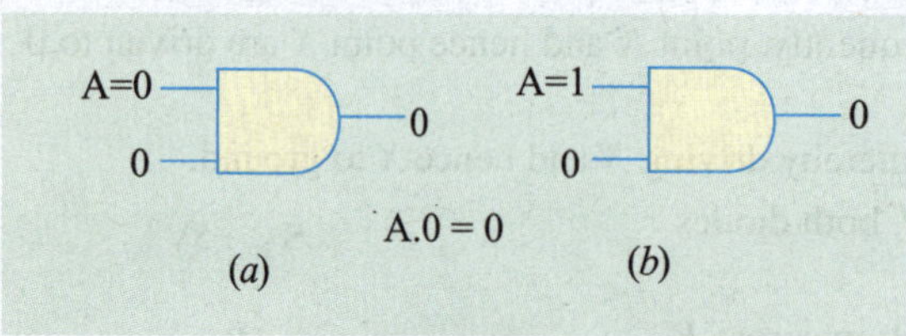

Fig. 70.19

2. $A.0 = 0$

When $A = 0$

then $0.0 = 0$ —Fig. 70.19 (*a*)

When $A = 1$

then $1.0 = 0$ —Fig. 70.19 (*b*)

It is seen that output is always 0 ***whatever the value of A.***

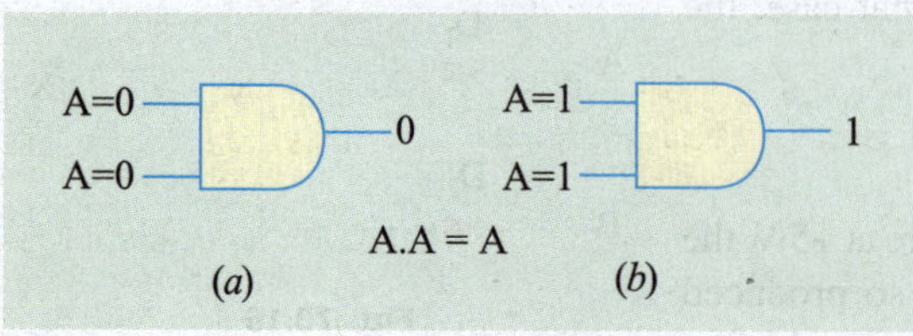

Fig. 70.20

3. $A.A = A$

When $A = 0$, then $0.0 = 0$ — Fig 70.20 (*a*)

When $A = 1$, then $1.1 = 1$ — Fig. 70.20 (*b*)

It is seen that output always ***takes on the value of A.***

70.15. The NOT Gate

It is so called because *its output is NOT the* **same as its input.** It is also called an ***inverter*** because it inverts the input signal. It has **one** input and **one** output as shown in Fig. 70.21 (*a*). All it does is to invert (or complement) the input as seen from its truth table of Fig. 70.21 (*b*).

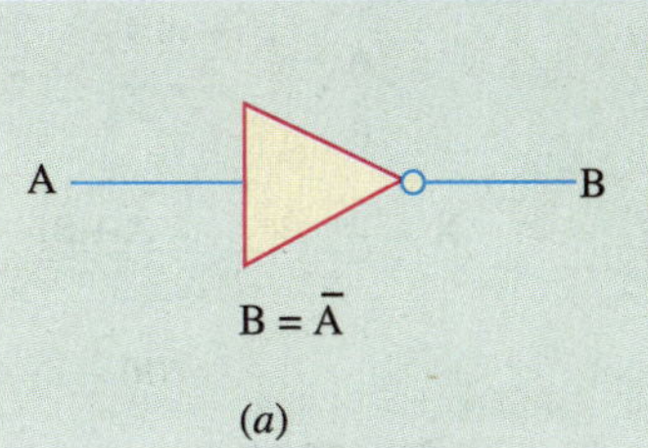

(*a*)

A	B
0	1
1	0

(*b*)

Fig. 70.21

The schematic symbol for inversion is a small circle as shown in Fig. 70.21 (*a*). The logical symbol for inversion or negation or complementation is a bar over the function to indicate the opposite state.

Sometimes, a prime is also used as *A′*. For example, $\overline{A}$ means not-A. Similarly, $\overline{(A + B)}$ means the complement of $(A + B)$.

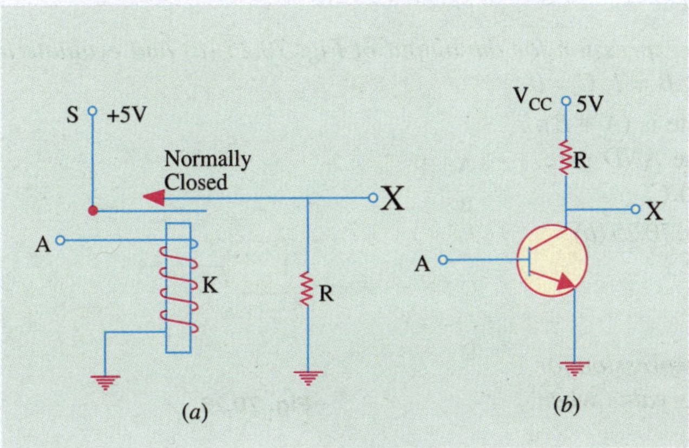

Fig. 70.22

70.16. Equivalent Circuits for a NOT Gate

The relay circuit of Fig. 70.22 (*a*) is a physical realization of the complementation operation of the Boolean algebra. When +5 V is applied to input *A*, *K* is energised and opens the ***normally-closed*** contact thereby driving output *X* to 0 V. Of course, when *A* is at 0V, *X* has the supply voltage of +5 V applied to it because the relay contact is normally closed.

In the equivalent transistor circuit of Fig. 70.22 (*b*) when +5V is applied to *A*, the transistor will be fully turned *ON*, drawing maximum ***collector*** current. Hence, whole of V_{CC} = 5 V will drop across *R* thereby sending *X* to 0 V. With 0 V applied at *A*, the transistor will be cut *OFF* and the output *X*, therefore, will go to V_{CC} *i.e.* + 5 V. Obviously, in each case, **output is the opposite of input**.

70.17. The NOT Operation

It is a ***complementation*** operation and its symbol is an **overbar**. It can be defined as under:

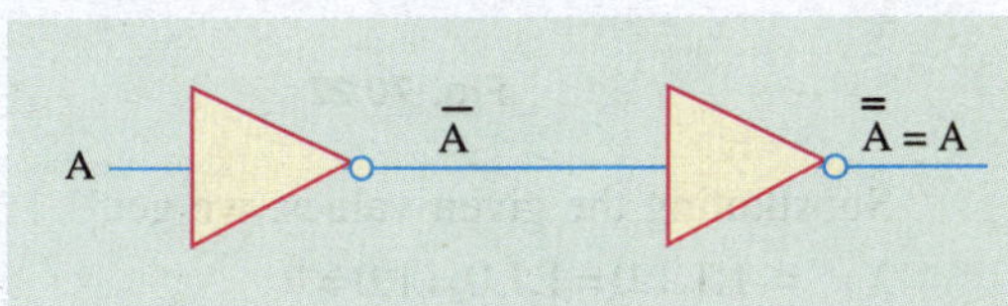

Fig. 70.23

As stated earlier, $\overline{0}$ means taking the negation or complement of 0 which is 1.

$$\overline{0} = 1$$

$$\overline{1} = 0$$

It should also be noted that complement of a value can be taken repeatedly. For example,

$\overline{\overline{1}} = \overline{0} = 1$ or $\overline{\overline{0}} = \overline{1} = 0$

As seen double complementation gives the original value as shown in Fig. 70.23.

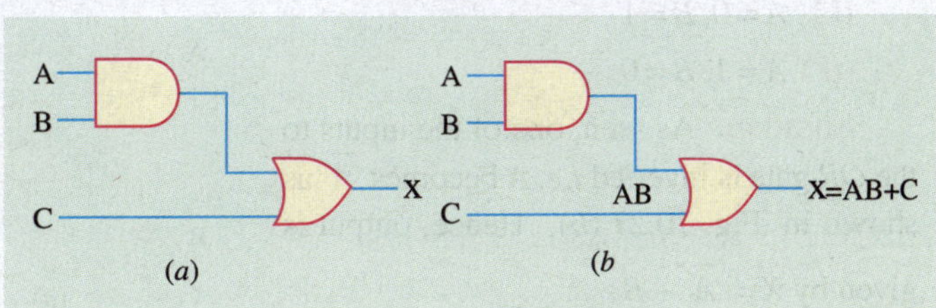

Fig. 70.24

Example 70.1. *Find the Boolean equation for the output X of Fig. 70.24 (a). Evaluate X when (i) A = 0, B = 1, C = 1 (ii) A = 1, B = 1, C = 1.* **[Computer Engg., Pune Univ. 1991]**

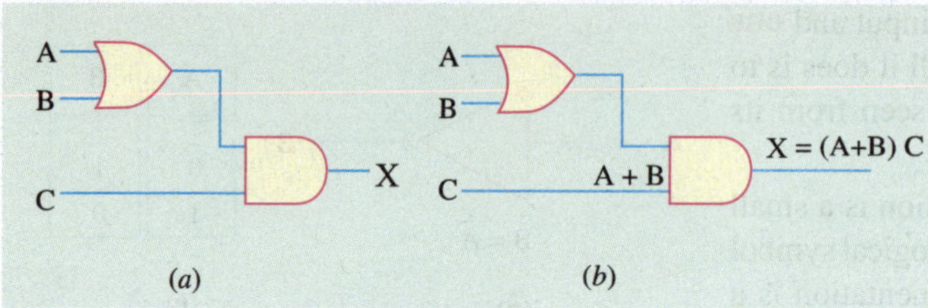

Fig. 70.25

Solution. The output of the *AND* gate is *AB*. It then becomes one of the inputs for the 2 input *OR* gate. When *AB* is ORed with *C*, we get $(AB + C)$.

$\therefore \quad X = AB + C$ —Fig. 70.24 (*b*)

(*i*) $X = 0.1 + 1 = 0 + 1 = 1$ —Art 70.7

(*ii*) $X = 1.1 + 1 = 1 + 1 = 1$

Example 70.2. *Find the Boolean expression for the output of Fig. 70.25 (a) and evaluate it when (i) A = 0, B = 1, C = 1, (ii) A = 1, B = 1, C = 0.*

Solution. The output of the *OR* gate is (*A* + *B*). Afterwards, it becomes the input of the *AND* gate. When *AND*ed with *C*, it becomes (*A* + *B*).*C*.

$\therefore \quad X = (A + B).C$ —Fig. 70.25 (*b*)

(*i*) $X = (0 + 1).1 = 1.1 = 1$

(*ii*) $X = (1 + 1).0 = 1.0 = 0$

Example 70.3. *Find the Boolean expression for the output of Fig. 70.26 and compute its value when A = B = C = 1 and D = 0.* **(Digital Computations, Punjab Univ. 1990)**

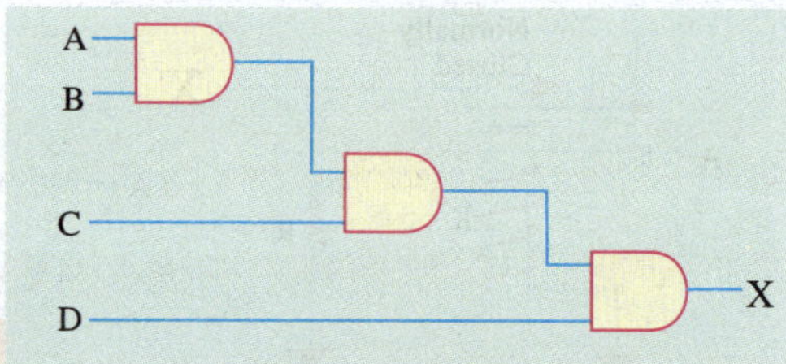

Fig. 70.26

Solution. The circuit is made up of three *AND* gates. Obviously, it is equivalent to a single 4-input *AND* gate *i.e.* an *AND* gate with a fan-in of four.

Output of the first gate is *AB*, that of the second is *ABC* and that of the third is *ABCD*. Hence, final output is $X = ABCD$.

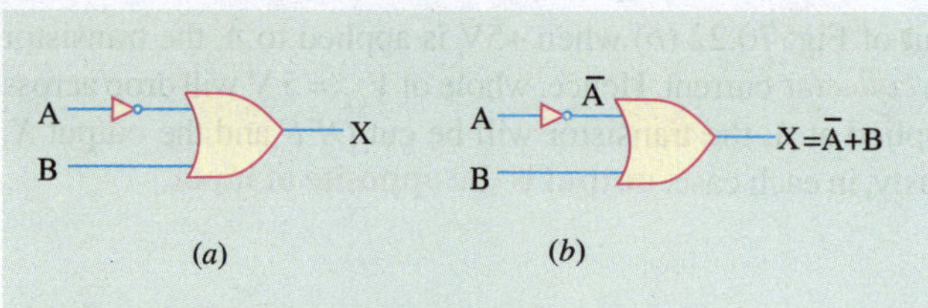

Fig. 70.27

Substituting the given values, we get

$X = 1.1.1.0 = 1.1.0 = 1.0 = 0$

Example 70.4. *Find the Boolean expression for the output X of Fig. 70.27 (a) and compute its value when*

(*i*) $A = 0, B = 1$

(*ii*) $A = 1, B = 0$

Solution. As seen, one of the inputs to the *OR* gate is inverted *i.e.* *A* becomes $\overline{A}$ as shown in Fig. 70.27 (*b*). Hence, output is given by $X = \overline{A} + B$

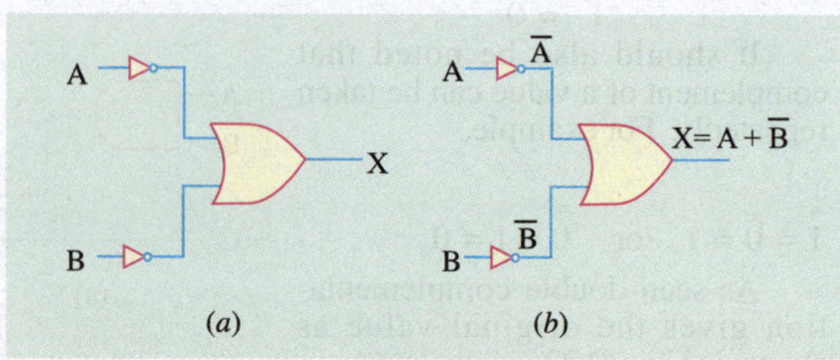

Fig. 70.28

(*i*) $X = \overline{0} + 1 = 1 + 1 = \mathbf{1}$ (*ii*) $X = \overline{1} + 0 = 0 + 0 = \mathbf{0}$

Example 70.5. *What is the Boolean expression for the output X of Fig. 70.28 (a) ? Compute the value of X when*

(*i*) $A = 0, B = 0$ (*ii*) $A = 1, B = 1$

Solution. As seen, in this case, both inputs to the OR gate have been inverted. Hence as shown in Fig. 70. 28 (*b*), the inputs become $\overline{A}$ and $\overline{B}$. Therefore, Boolean expression for the output becomes $X = \overline{A} + \overline{B}$.

(*i*) $X = \overline{0} + \overline{1} = 0 + 1 = \mathbf{1}$

(*ii*) $X = \overline{1} + \overline{0} = 0 + 1 = \mathbf{1}$

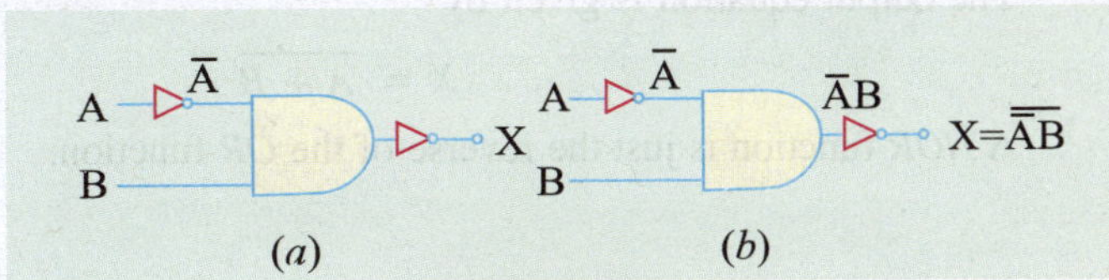

Fig. 70.29

Example 70.6. *Write down the Boolean equation for the output X of Fig. 70.29 (a). Compute its value when*

(*i*) $A = 0, B = 0$

(*ii*) $A = 0, B = 1$

(*iii*) $A = 1, B = 0$

(*iv*) $A = 1, B = 1$

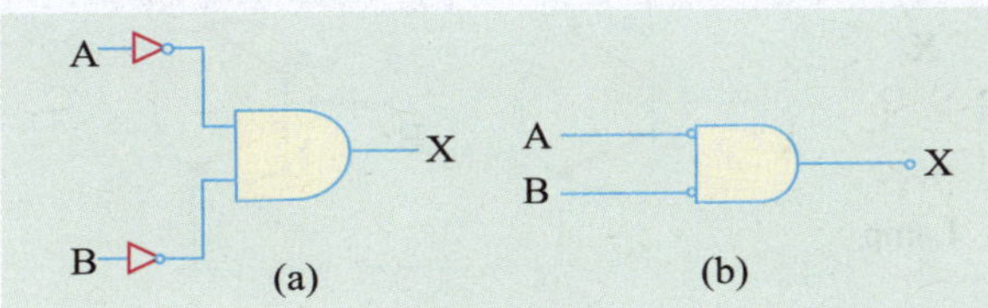

Fig. 70.30

Solution. As seen, inputs to the *AND* gate are $\overline{A}$ and B [Fig. 70.29 (*a*)]. The output of the *AND* gate is $\overline{A}.B$. However, this output is inverted by the second inverter connected in the output. Hence, final output equation is

$X = \overline{\overline{A}B}$

(*i*) $X = \overline{\overline{0}.0} = \overline{1.0} = \overline{0} = \mathbf{1}$

(*ii*) $X = \overline{\overline{0}.1} = \overline{1.1} = \overline{1} = \mathbf{0}$

(*iii*) $X = \overline{\overline{1}.0} = \overline{0.0} = \overline{0} = \mathbf{1}$

(*iv*) $X = \overline{\overline{1}.1} = \overline{0.1} = \overline{0} = \mathbf{1}$

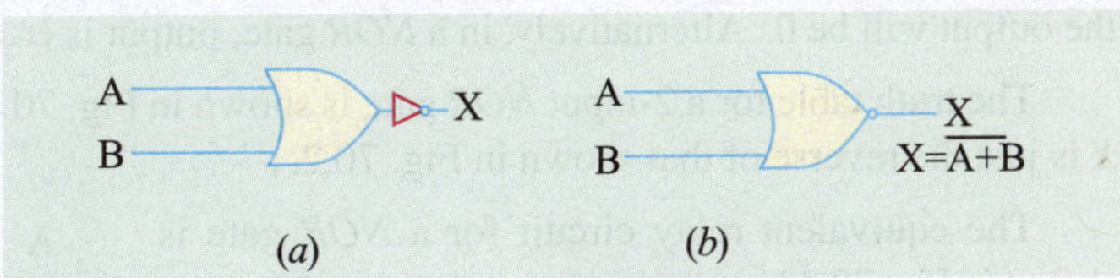

Fig. 70.31

While evaluating expressions of the above type, you must remember the following two points :

1. take the *NOT i.e.* inversion of the individual term first.
2. When a *NOT* or ***inversion*** is applied to more than one term (like 1.0), you should work out the *OR* (or *AND*) operation first and then take the *NOT* of the result so obtained.

70.18. Bubbled Gates

A bubbled gate is one whose inputs are *NOT*ed or inverted *i.e.* it is a negated gate. Fig. 70.30 (*a*) shows an *AND* gate whose both input are inverted.

In practice, instead of this logic symbol, the one shown in Fig. 70.30 (*b*) is widely used. As seen, the inverter triangles have been eliminated and the two small circles or bubbles have been moved to the inputs of the gate. Such a gate is called a ***bubbled AND*** gate, the bubbles acting as a reminder of the inversion or complementation that takes place before *AND*ing the inputs.

It would be shown later that a bubbled *AND* gate is equivalent to a *NOR* gate.

Table 70.6

A	B	X
0	0	1
0	1	0
1	0	0
1	1	0

Fig. 70.32

Similarly, a bubbled *OR* gate is equivalent to a *NAND* gate.

70.19. The NOR Gate

In fact, it is a *NOT-OR* gate. It can be made out of an *OR* gate by connecting an inverter in its output as shown in Fig. 70.31 (*a*).

The output equation is given by

$$X = \overline{A + B}$$

A *NOR* function is just the reverse of the *OR* function.

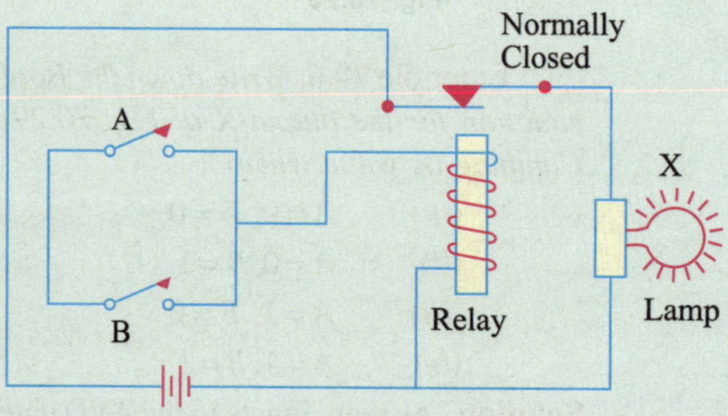

Fig. 70.33

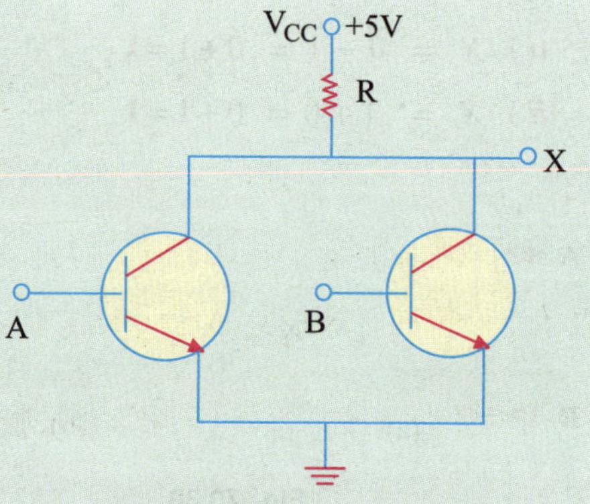

Fig. 70.34

Logic Operation

A *NOR* gate will have an output of 1 **only when all its inputs are 0**. Obviously, if any input is 1, the output will be 0. Alternatively, in a *NOR* gate, output is **true** only **when all inputs are false**.

The truth table for a 2-input *NOR* gate is shown in Fig. 70.32. It will be observed that the output *X* is just the reverse of that shown in Fig. 70.2.

The equivalent relay circuit for a *NOR* gate is shown in Fig. 70.33.

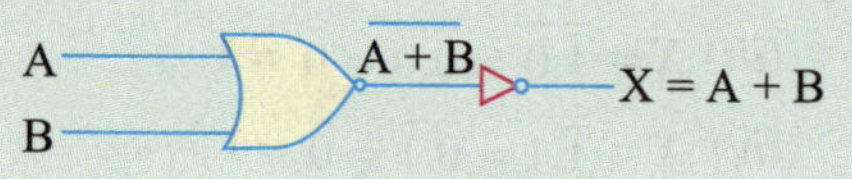

Fig. 70.35

It is seen that the lamp glows under 00 input condition only but not under 01, 10, 11 input conditions.

The transistor equivalent of the *NOR* gate is shown in Fig. 70.34. As seen, output *X* is 1 only when both transistors are cut-off *i.e.* when $A = 0$ and $B = 0$. For any other input condition like 01, 10 and 11, one or both transistors saturate forcing point *X* to go to ground.

70.20. NOR Gate is a Universal Gate

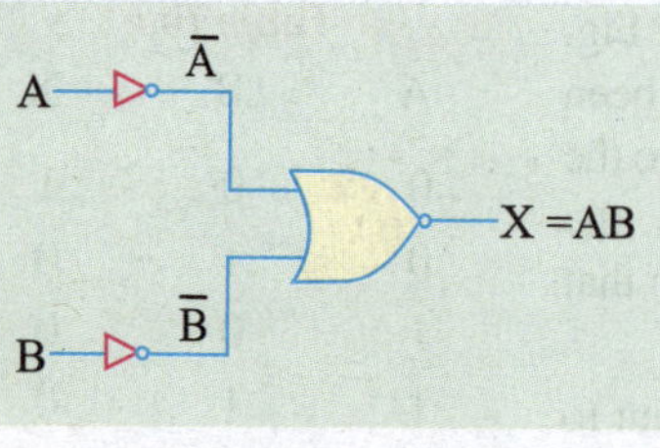

Fig. 70.36

It is interesting to note that a *NOR* gate can be used to realize the basic logic functions : *OR*, *AND* and *NOT*. That is why it is often referred to as a ***universal*** gate. Examples are given below:

1. As OR Gate

As shown in Fig. 70.35, the output from *NOR* gate is $\overline{A+B}$. By using another inverter in the output, the final output is inverted and is given by $X = A + B$ which is the logic function of a normal *OR* gate.

2. As AND Gate

Here, two inverters have been used, one for each input (Fig. 70.36). The inputs have, thus, been **inverted before they are applied** to the *NOR* gate.

The output is $\overline{\overline{A}+\overline{B}}$ which can be proved (with the help of De Morgan's theorem) to be equal to *AB*.

Incidentally, we could have used a bubbled *NOR* gate for the above purpose.

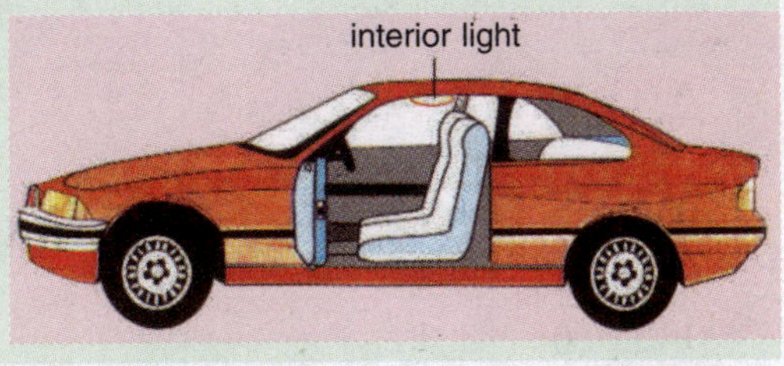

3. As NOT Gate

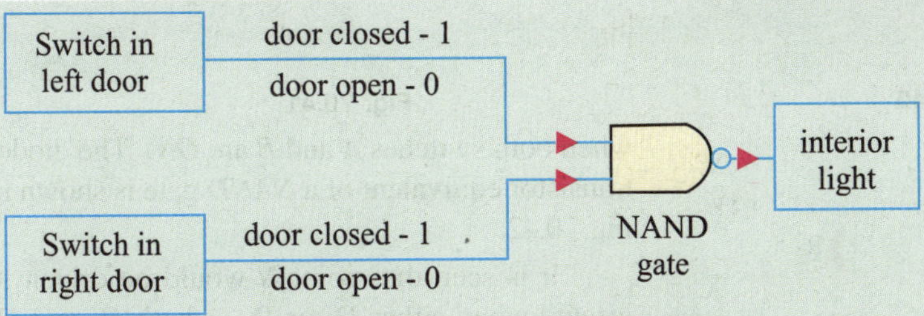

The NAND gate is used to design an interior lighting system of a car such that the light is switched off only when both doors are shut

The two inputs have been tied together as shown in Fig. 70.37 (*a*). The output is $\overline{A+A}$ which can be proved to be equal to $\overline{A}$ with the help of De Morgan's theorem. Instead of the first symbol, the second symbol shown in Fig. 70.37 (*b*) is widely used where only single input has been used.

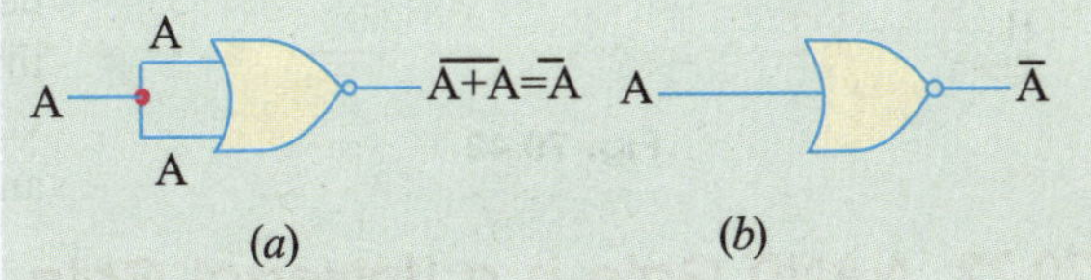

Fig. 70.37

Here is a different way of making *OR* and *AND* gates. Fig. 70.38 (*a*) shows how we can use *NOR* gates to produce an *OR* gate. Similarly, Fig. 70.38 (*b*) shows the formation of an *AND* gate from three *NOR* gates. Knowledge of De Morgan's theorem is needed to understand their logic operation $X = \overline{AB}$.

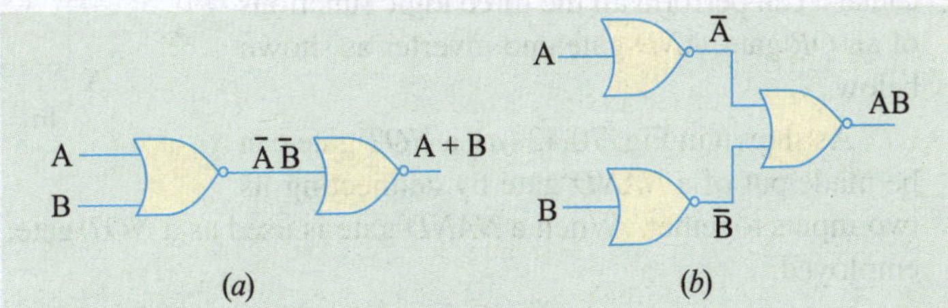

Fig. 70.38

70.21. The NAND Gate

It is, in fact, a *NOT-AND* gate. It can be obtained by connecting a NOT gate in the output of an AND gate as shown in Fig. 70.39. Its output is given by the Boolean equation.

This gate gives an output of 1 if its ***both inputs are* not 1**. In other words, it gives an output 1 if **either A or B or both are 0**.

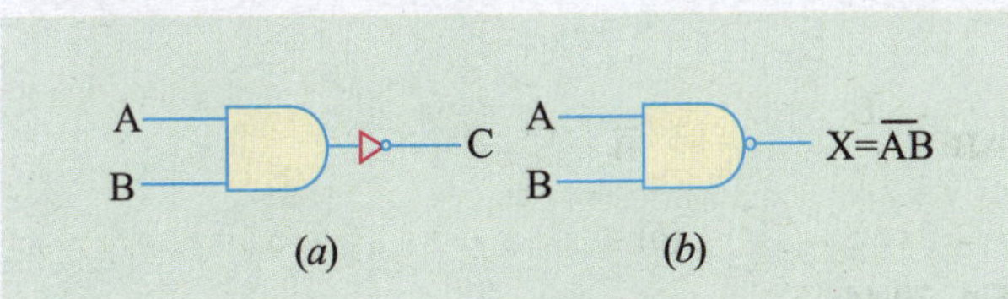

Fig. 70.39

The truth table for a 2-input *NAND* gate is given in Fig. 70.40. It is just the opposite of the truth for *AND* gate. It is so because *NAND* gate performs reverse function of an *AND* gate.

The equivalent relay circuit of a *NAND* gate is shown in Fig. 70.41.

It is seen that lamp glows under input conditions of 00, 01, 10 but not under 11 input condition when both switches A and B are ON. The diode-transistor equivalent of a $NAND$ gate is shown in Fig. 70.42.

A	B	X
0	0	1
0	1	1
1	0	1
1	1	0

Fig. 70.40

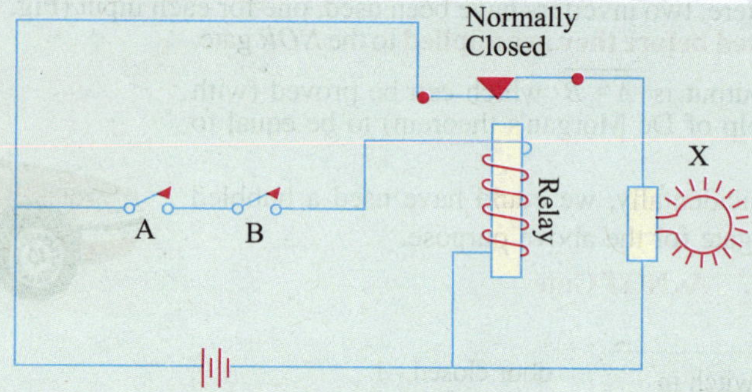

Fig. 70.41

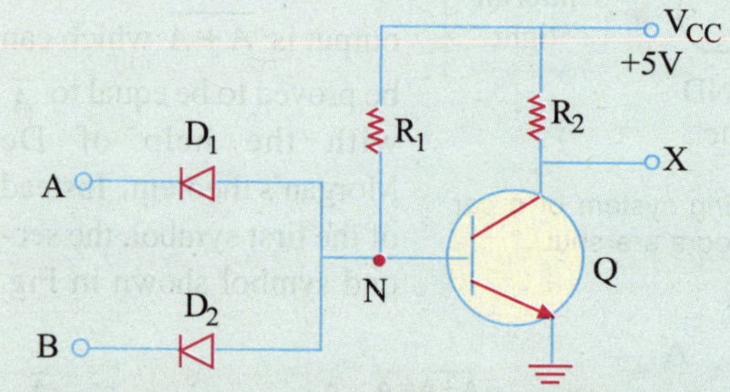

Fig. 70.42

It is seen that point N would be driven to ground when either D_1 or D_2 or both D_1 and D_2 conduct. It represents input conditions of 10, 01 and 00*. Under such conditions, Q is cut off and hence X goes to V_{CC} meaning logic 1 state. Only time X is 0 is when A = 1 and B = 1 (*i.e.* input voltages at A and B are at +5V) so that N is +5 V and Q is saturated.

70.22. NAND Gate is a Universal Gate

NAND gate is also called universal gate because it can perform all the three logic functions of an *OR* gate, *AND* gate and inverter as shown below.

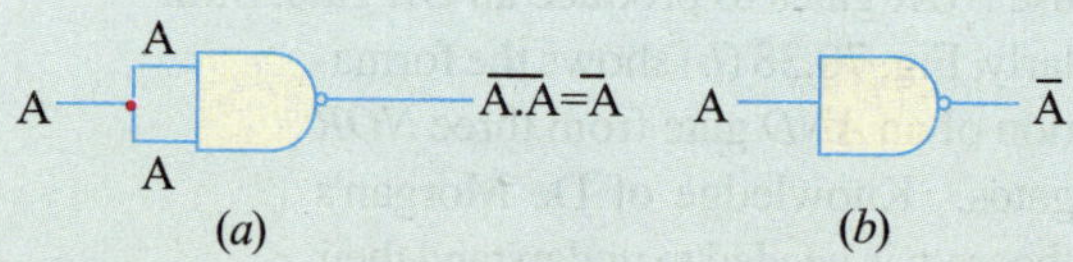

Fig. 70.43

As shown in Fig. 70.43 (*a*), a *NOT* gate can be made out of a *NAND* gate by connecting its two inputs together. When a *NAND* gate is used as a *NOT* gate, the logic symbol of Fig. 70.43 (*b*) is employed.

The use of two *NAND* gates to produce an *AND* gate is shown in Fig. 70.44 (*a*).

Similarly, Fig. 70.44 (*b*) shows how *OR* gate can be made out of three *NAND* gates. The *OR* function may not be clear from the figure because we need De Morgan's theorem to prove that $\overline{\overline{A}\,\overline{B}} = A + B$.

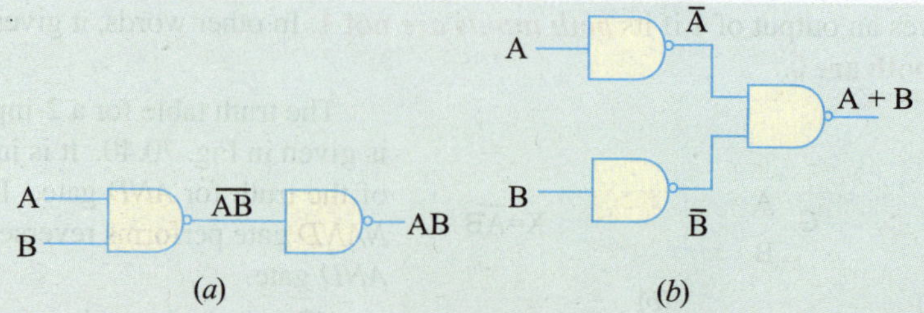

Fig. 70.44

* In this case, $V_A = V_B = 0$ V

70.23. The XNOR Gate

It is known as a NOT-*XOR* gate *i.e.* $\overline{XOR}$ gate. Its logic symbol and truth table are shown in Fig. 70.45.

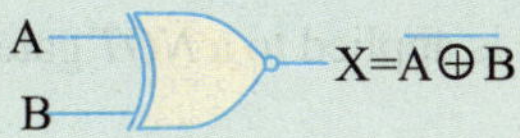

A	B	X
0	0	1
0	1	0
1	0	0
1	1	1

Fig. 70.45

Its logic function and truth table are ***just the reverse of those for XOR gate*** (Art 70.9).

This gate has an output 1 if **its both inputs are either 0 or 1**. In other words, for getting an output, its both inputs should be ***at the* same** *logic level* of either 0 or 1. Obviously, it produces **no** output if its two inputs are at the **opposite** logic level.

70.24. Logic Gates at a Glance

In Fig. 70.46 is shown the summary of all the 2-output logic gates considered so far along with their truth tables.

Following points should prove helpful when writing these truth tables:

1. In first column *A*, logic values alternate between 0 and 1 every two rows
2. In second column *B*, logic values alternate every other row
3. Column *X* is filled up as per the logic function it performs

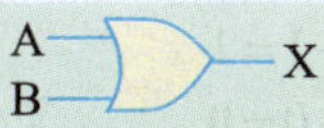

$A + B = X$

A	B	X
0	0	0
0	1	1
1	0	1
1	1	1

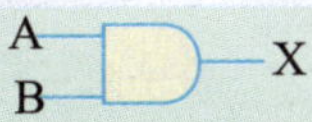

$AB = X$

A	B	X
0	0	0
0	1	0
1	0	0
1	1	1

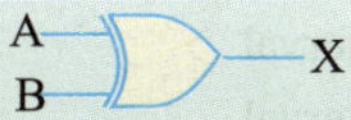

$A \oplus B = X$

A	B	X
0	0	0
0	1	1
1	0	1
1	1	0

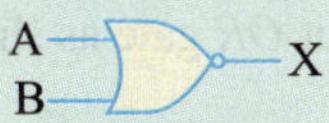

$\overline{A + B} = X$

A	B	X
0	0	1
0	1	0
1	0	0
1	1	0

$\overline{AB} = X$

A	B	X
0	0	1
0	1	1
1	0	1
1	1	0

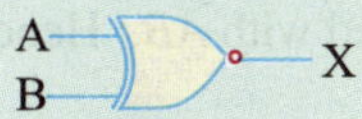

$\overline{A \oplus B} = X$

A	B	X
0	0	1
0	1	0
1	0	0
1	1	1

Fig. 70.46

4. Truth tables for *NOR*, *NAND* and *XNOR* (or $\overline{XOR}$) gates are ***just the opposite*** of those for *OR*, *AND* and *XOR* gates.

Example 70.7. *An electrical signal is expressed as 101011. Explain its meaning. If this signal is applied to a NOT gate, what would be the output signal ?*

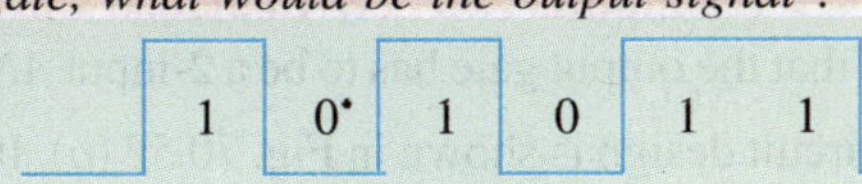

Fig. 70.47

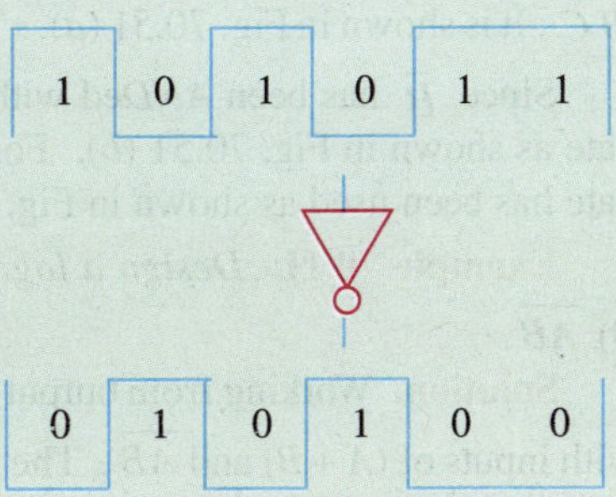

Fig. 70.48

Solution. The signal represents binary number 101011_2. It is electrically represented as a train of pulses. Taking positive logic, 1 will represent high voltage and 0 will represent low (or zero) voltage as shown in Fig. 70.47.

When such a signal is applied to a *NOT* gate, it would be inverted or complemented as shown in Fig. 70.48.

The *NOT* output will represent the binary number 010100_2.

Example 70.8. *Two electrical signals represented by A = 101101 and B = 110101 are applied to a 2-input AND gate. Sketch the output signal and the binary number it represents.*

Solution. The pulse trains corresponding to *A* and *B* are shown in Fig. 70.49.

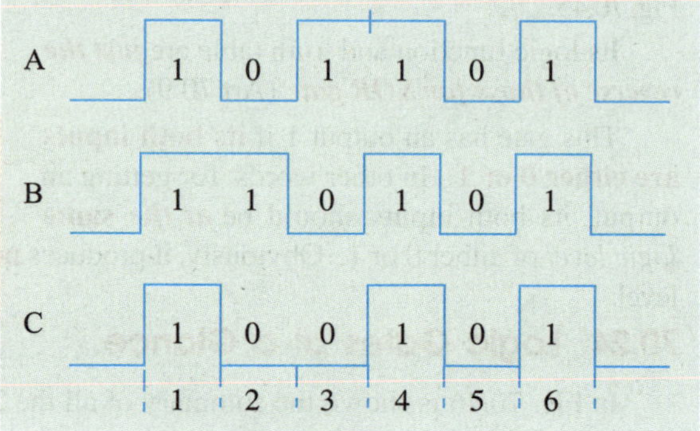

Fig. 70.49

Remember that in an *AND* gate, *C* is 1 only when both *A* and *B* are 1. It is an ***all-or-nothing gate***. The output can be found in different time intervals as under :

1. 1st interval : $1 + 1 = 1$
2. 2nd interval : $0 + 1 = 0$
3. 3rd interval : $1 + 0 = 0$
4. 4th interval : $1 + 1 = 1$
5. 5th interval : $0 + 0 = 0$
6. 6th interval : $1 + 1 = 1$

Hence, ouput of the *AND* gate is 100101_2. It is sketched in Fig. 70.49.

Example 70.9. *Convert the Boolean expression (AB + C) into a logic circuit using different logic gates.* **(Computer Engg. Pune Univ. 1992)**

Solution. In such cases, it is best to start with the ***output and work towards the input***. As seen, *C* has been *OR*ed with *AB*. Hence, the output gate must be a 2-input *OR* gate as shown in Fig. 70.50 (*a*).

Now, term *AB* is an *AND* function. Hence, we need an *AND* gate with inputs *A* and *B*. The complete logic circuit is shown in Fig. 70.50 (*b*).

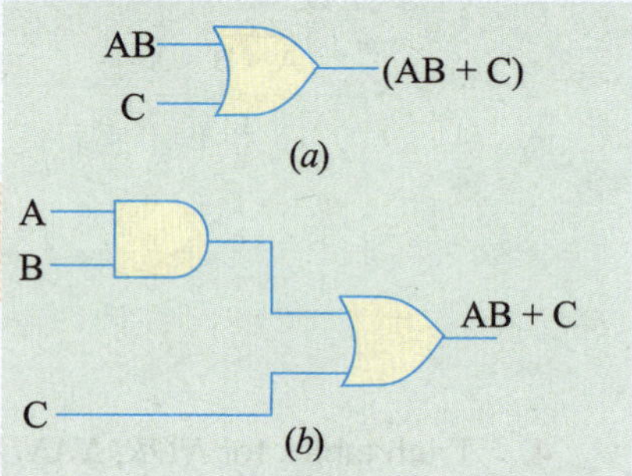

Fig. 70.50

Example 70.10. *Design logic hardware based on the Boolean expression* $(A + \overline{B}C)$.

Solution. We will work from ***output to input***. It is seen that the last gate is a 2-input *OR* gate with inputs of *A* and $\overline{B}C$. It is shown in Fig. 70.51 (*a*).

Since $\overline{B}$ has been *AND*ed with *C*, it requires an *AND* gate as shown in Fig. 70.51 (*b*). For inversion of *B*, a *NOT* gate has been used as shown in Fig. 70.51 (*c*).

Example 70.11. *Design a logic circuit whose output is given by the Boolean expression* $(A + B).\overline{AB}$. **(Computer Science, Allahabad Univ. 1992)**

Solution. Working from output to input, we find that the output gate has to be a 2-input *AND* gate with inputs of $(A + B)$ and $\overline{AB}$. The first step of the circuit design is shown in Fig. 70.52 (*a*). It is also seen that the input to the entire circuit consists of *A* and *B* only.

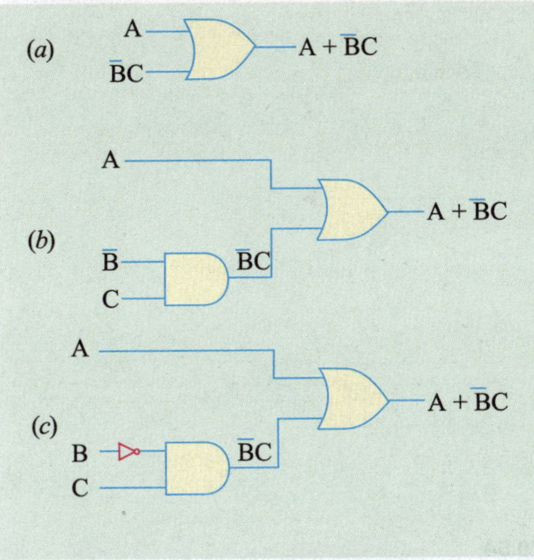

Fi.g 70.51

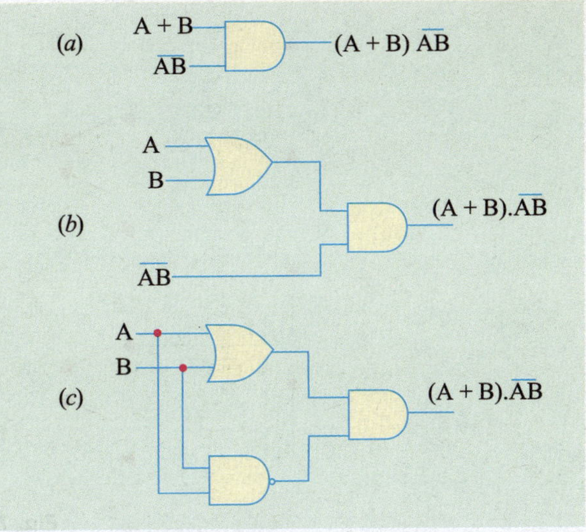

Fig. 70.52

The input of $(A + B)$ has been obtained with the help of an *OR* gate as shown in Fig. 70.52 (*b*).

Finally, a *NAND* gate is connected **in parallel** with the *OR* gate for getting its inputs of *A* and *B* and thereafter for supplying an output of $\overline{AB}$. The complete circuit is shown in Fig. 70.52 (*c*).

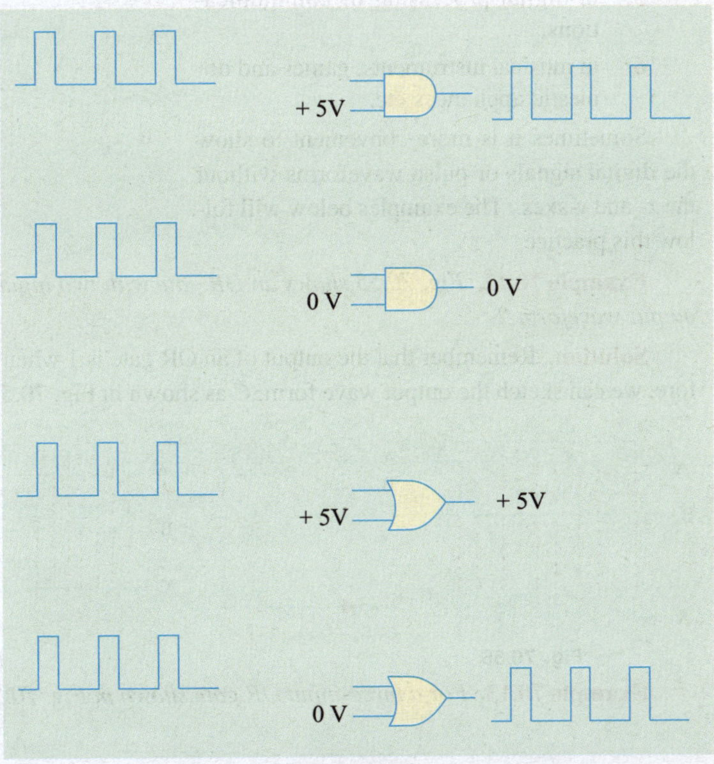

Fig. 70.53

70.25. Digital Signals Applied to Logic Gates

A binary digital signal applied to a logic gate is nothing else but the application of a time sequence of 1's and 0's. The response of *AND* and *OR* gates to various periodic digital signals is shown in Fig. 70.53.

Fig. 70.54 shows how two waveforms can be *OR*ed by using two input gates.

70.26. Applications of Logic Gates

Range of application of the logic gates is very wide but the main headings would include.

1. to build more complex devices like binary counters etc.,
2. for decision making in automatic control of machines and various industrial processes,
3. in calculators and computers,
4. in digital measuring techniques,

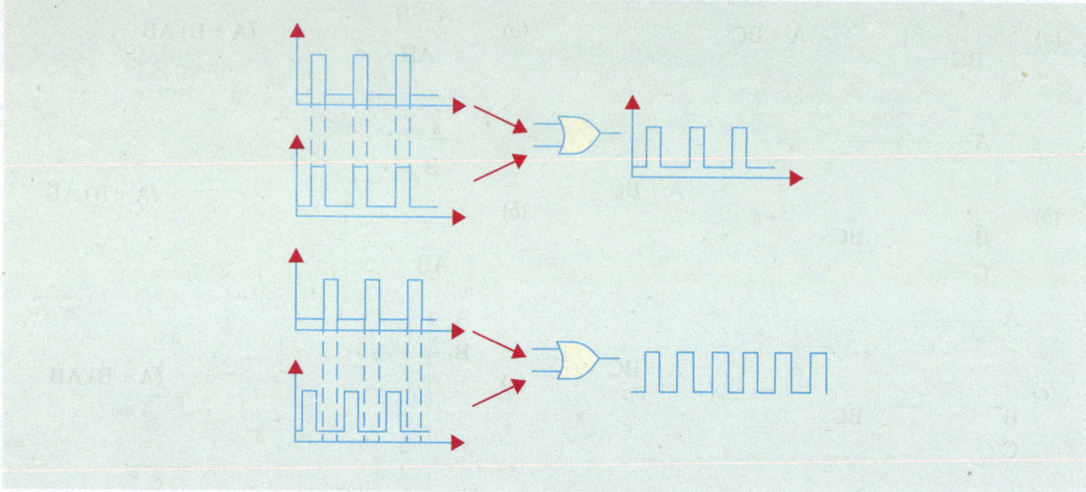

Fig. 70.54

5. in digital processing of communications,
6. in musical instruments, games and domestic appliances etc.

Sometimes it is more convenient to show the digital signals or pulse waveforms without the *x*- and *y*-axes. The examples below will follow this practice.

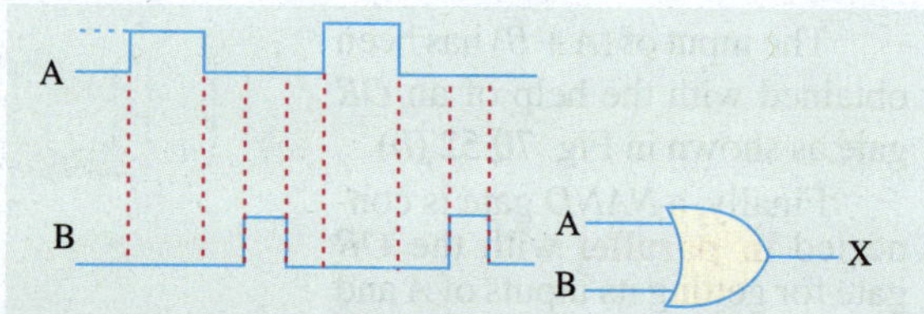

Fig. 70.55

Example 70.12. *Fig. 70.55 shows an OR gate with two input waveforms. What is the resulting output waveform ?*

Solution. Remember that the output of an OR gate is 1 when either or both inputs are 1. Therefore, we can sketch the output wave form, *C* as shown in Fig. 70.56.

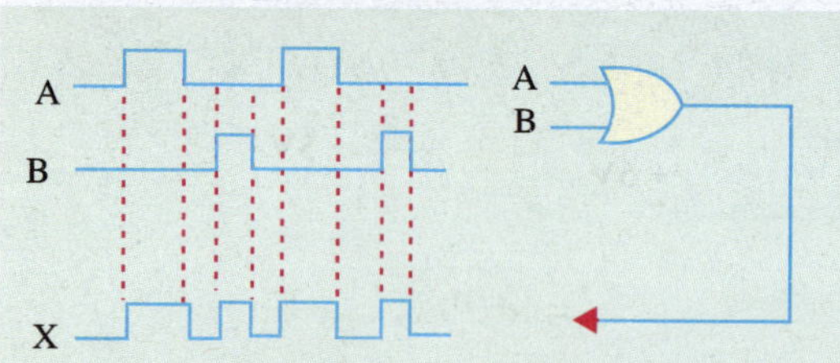

Fig. 70.56

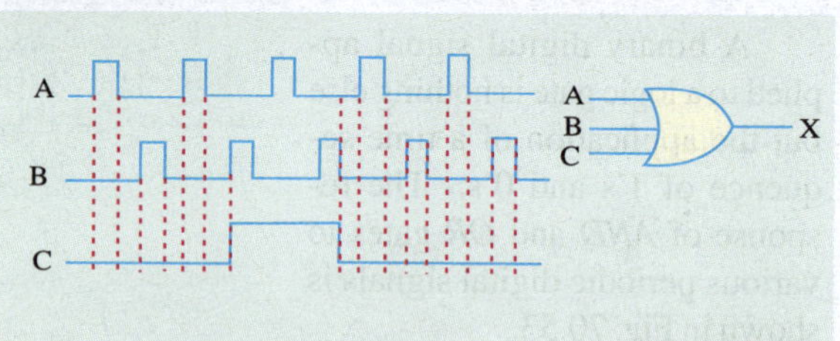

Fig. 70.57

Example 70.13. *For a three-input OR gate shown in Fig. 70.57 determine the output waveform.*

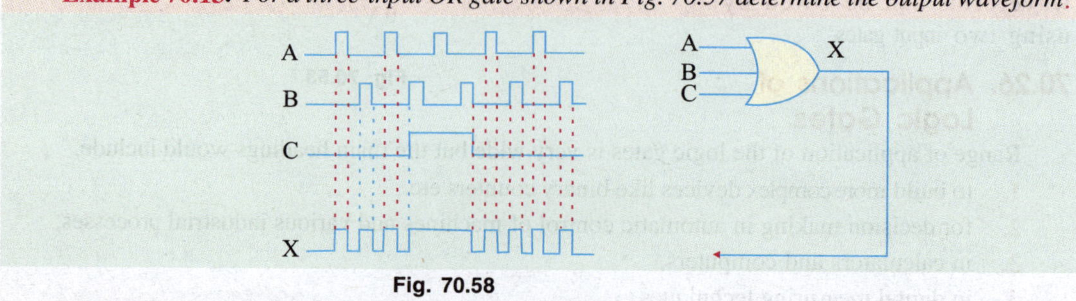

Fig. 70.58

Solution. Remember the output of a three-input *OR* gate is 1 when one or more of the inputs are

1. Therefore, we can sketch the output waveform, *X* as shown in Fig. 70.58.

Example 70.14. *Fig. 70.59 shows a 2-input AND gate with waveforms A and B. Sketch the resulting output waveform.*

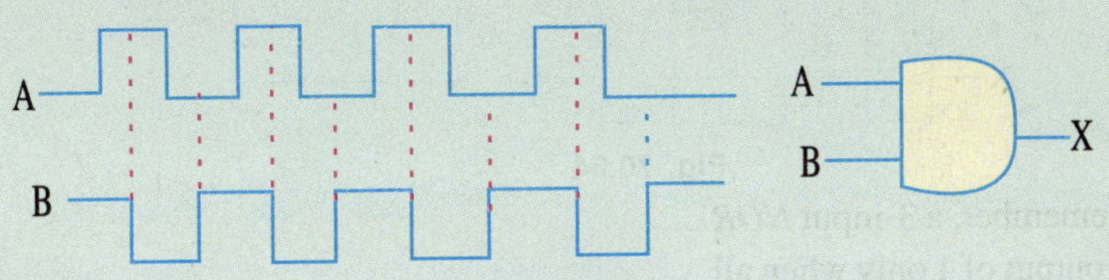

Fig. 70.59

Solution. Remember the *AND* gate produces an output 1 only when all its inputs are present. Thus the output of *AND* gate is 1 when both *A* and *B* are 1. Its output is 0 when any of its inputs is 0. Using this concept, we can sketch the output waveform as shown in Fig. 70.60.

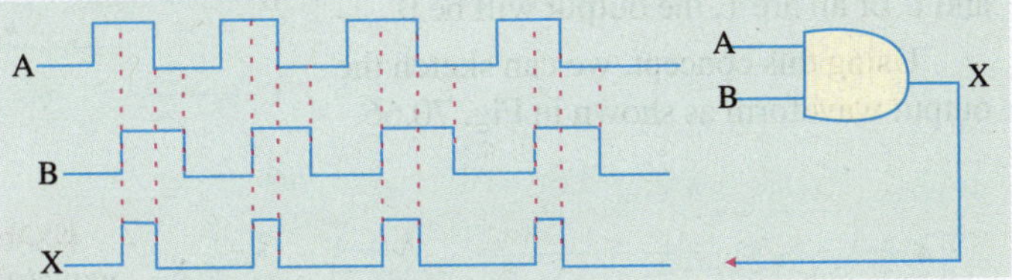

Fig. 70.60

Example 70.15. *For a 3-input AND gate with waveforms A, B and C at its inputs as shown in Fig. 70.61, determine the resulting output waveform.*

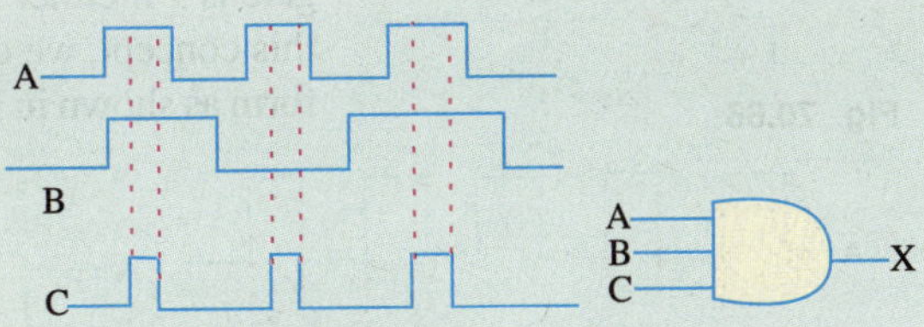

Fig. 70.61

Solution. A 3-input AND gate produces an output 1 only when all the three inputs *A*, *B* and *C* are 1. Its output is 0 when any one of three inputs is 0. Using this concept, the resulting output waveform is as shown in Fig. 70.62.

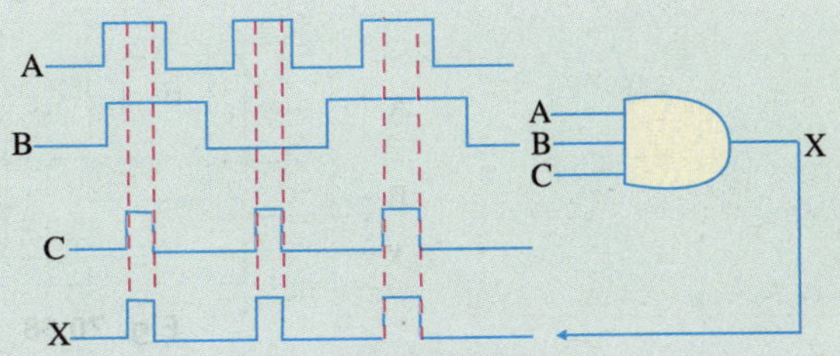

Fig. 70.62

Example 70.16. *Fig. 70.63 shows the two waveforms applied to the NOR gate inputs. Sketch the resulting output waveform.*

Solution. Remember, a *NOR* gate will have an output of 1 only when all its inputs are 0. Obviously, if any input is 1, the output will be 0. Using this concept, we can sketch the output waveform as shown in Fig. 70.64.

Example 70.17. *Sketch the output waveform for a 3-input NOR gate shown in Fig. 70.65. Showing the proper time relationship to the inputs.*

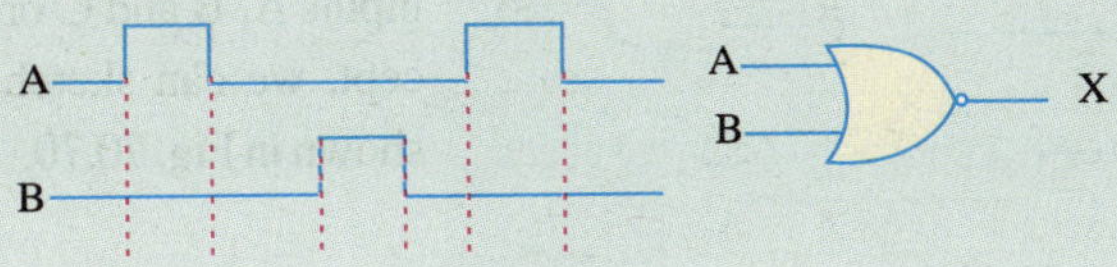

Fig. 70.63

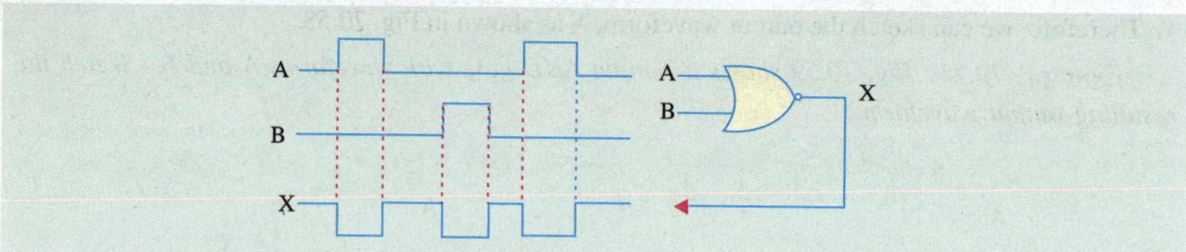

Fig. 70.64

Solution. Remember, a 3-input *NOR* gate will have an output of 1 only when all the three inputs *A*, *B* and *C* are 0. Obviously if any one of the three inputs *A*, *B* and *C* or all are 1, the output will be 0.

Using this concept, we can sketch the output waveform as shown in Fig. 70.66.

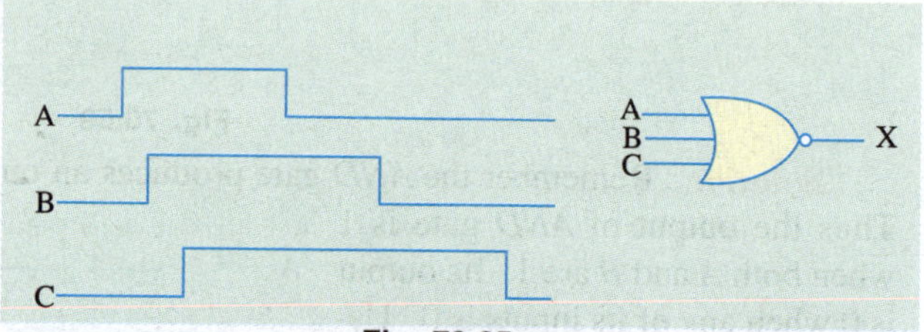

Fig. 70.65

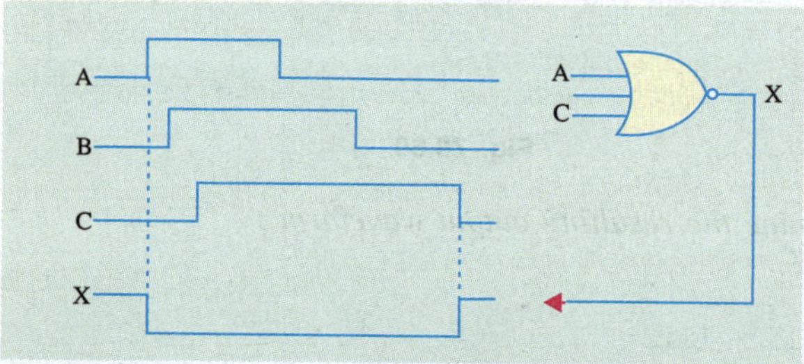

Fig. 70.66

Example 70.18. *Fig. 70.67 shows the two waveforms A and B applied to the NAND gate inputs. Determine the resulting output waveform.*

Solution. Remember, the output of *NAND* gate is 1 if either *A* or *B* or both are 0. Using this concept, we can sketch the output waveform as shown in Fig. 70.68.

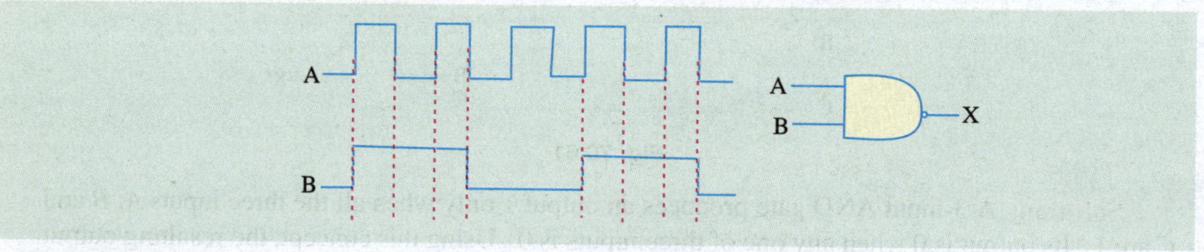

Fig. 70.67

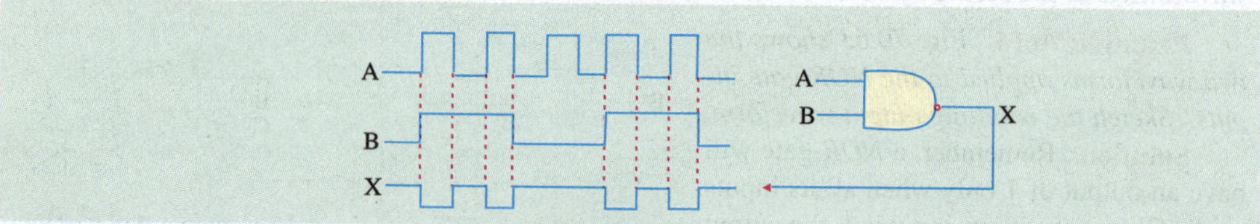

Fig. 70.68

Example 70.19. *Sketch the output waveform for a 3-input NAND gate shown in Fig. 70.69*

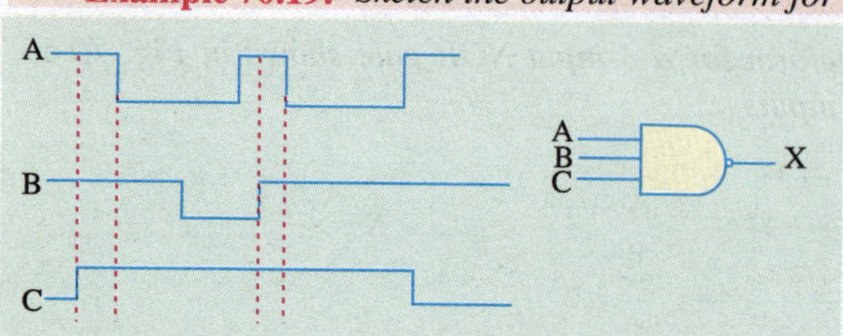

Fig. 70.69

Solution. Remember, the output of a 3-input NAND gate is 1 only if either any one of the inputs A, B and C or all are 0. Using this concept, we can sketch the output waveform as shown in Fig. 70.70.

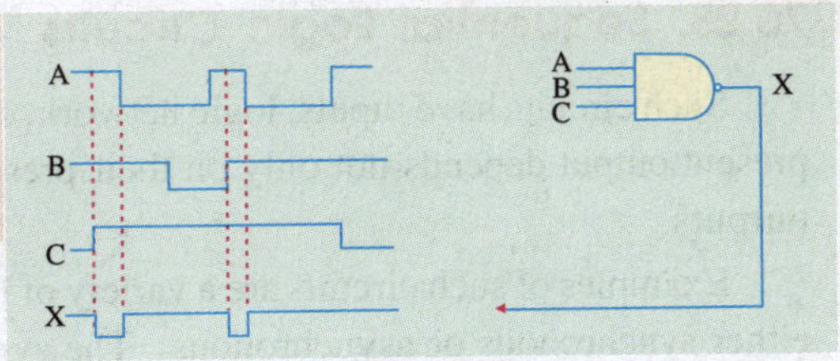

Fig. 70.70

Example 70.20. *The waveforms A and B are applied as an input to the XOR and XNOR gate as shown in Fig. 70.71. Determine the output waveforms of these logic gates.*

Solution. In exclusive–*OR* (*XOR*) gate, output is 1 if its either input but not both, is 1. In other words, it has an output 1 when its inputs are different. The output is 0 only when inputs are the same. Using this concept, we can sketch the *XOR* output waveform as shown in Fig. 70.72.

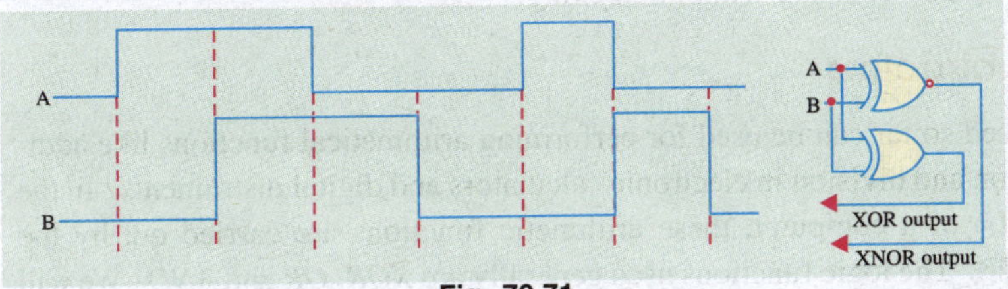

Fig. 70.71

In *XNOR* gate, output is 1 if its both inputs are either 0 or 1. It produces 0 output if its two inputs are at the opposite logic level. Using this concept, we can sketch the *XNOR* output waveform as shown in Fig. 70.72

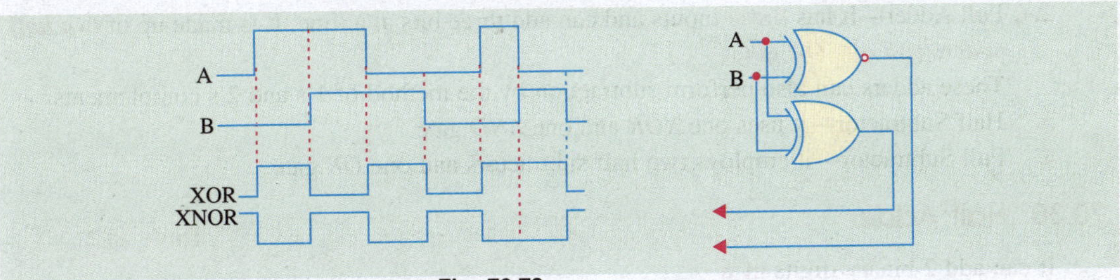

Fig. 70.72

70.27. Combinational Logic Circuit

It is a circuit built from various logic gate combinations. The circuit possesses a set of inputs, a ***memoryless*** logic network to operate on the inputs and a set of outputs as shown in Fig. 70.73 (*a*). The output from a combinational logic circuit depends solely on the ***present*** input values and not on the previous ones. Moreover, output combinational networks are used to make logical decisions and control the operation of different circuits in digital electronic systems. For a given set of input conditions, the output of such a circuit is the same. Consequently, a truth table can fully describe the operation of such a circuit. Examples of such a circuit are : decoders, adders, multiplexer and demultiplexers etc.

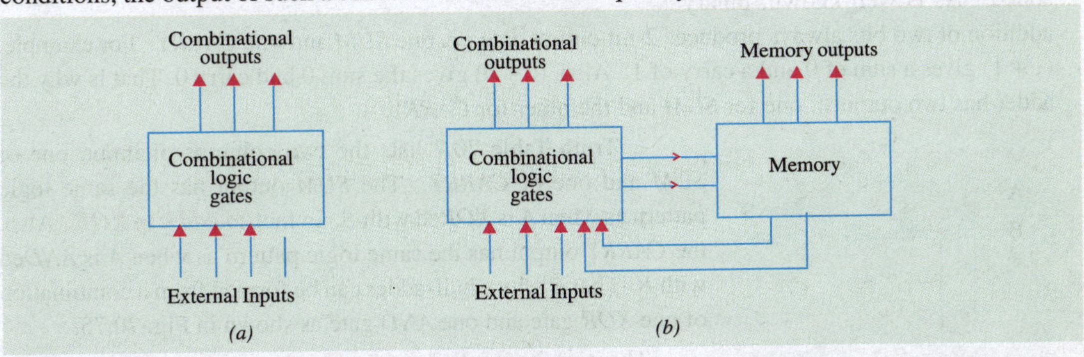

Fig. 70.73

70.28. Sequential Logic Circuits

Such circuits have inputs, logic network, outputs and a memory as shown in Fig. 70.73 (*b*). Their present output depends not only on their present inputs but also on the previous logic states of the outputs.

Examples of such circuits are a variety of latches and flip-flops. Sequential logic circuits may be either synchronous or asynchronous. The synchronous sequential circuits are built to operate at a clocked rate whereas asynchronous ones are without clocking.

70.29. Adders and Subtractors

The logical gates discussed so far can be used for performing arithmetical functions like addition, subtraction, multiplication and division in electronic calculators and digital instruments. In the central processing unit (*CPU*) of a computer, these arithmetic functions are carried out by the arithmetic and logic unit (*ALU*). The logic functions used generally are *XOR*, *OR* and *AND*. We will consider the following:

1. Half Adder—it is 1-bit adder and carries out binary addition with the help of *XOR* and *AND* gates. It has two inputs and two outputs.
2. Full Adder—It has three inputs and can add three bits at a time. It is made up of ***two half adders and one OR gate.***

   These adders can also perform subtraction by the method of 1's and 2's complements.
3. Half Subtractor—it uses one *XOR* and one *AND* gate.
4. Full Subtractor—it employs two half subtractors and one *OR* gate.

70.30. Half Adder

It can add 2 binary digits at a time and produce a 2-bit data *i.e.* sum and carry according to the binary addition rules (Art. 69.10).

Block Diagram

It is shown in Fig. 70.74. As can be seen, it has two inputs for applying the two binary digits to be added. As is well known, binary addition of two bits always produces 2-bit output data *i.e.* one *SUM* and one *CARRY*. For example, (1 + 1) gives a sum of 0 and a carry of 1. Also, (0 + 0) gives the sum 0 and carry 0. That is why the adder has two outputs : one for *SUM* and the other for *CARRY*.

Table 70.7

Input		Output	
A	*B*	*S*	*C*
0	0	0	0
0	1	1	0
1	0	1	0
1	1	0	1

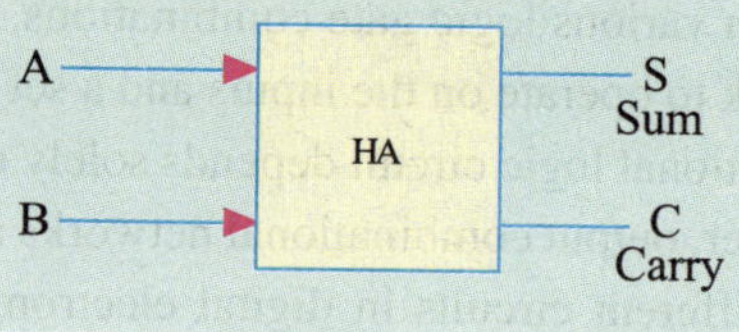

Fig. 70.74

Truth Table 70.7 lists the two columns of input, one of *SUM* and one of *CARRY*. The *SUM* output has the same logic pattern as when *A* is *XOR*ed with *B*. In fact, ***to add it to XOR.*** Also, the *CARRY* output has the same logic pattern as when *A* is *AND*ed with *B*. That is why a half-adder can be formed from a combination of one *XOR* gate and one *AND* gate as shown in Fig. 70.75.

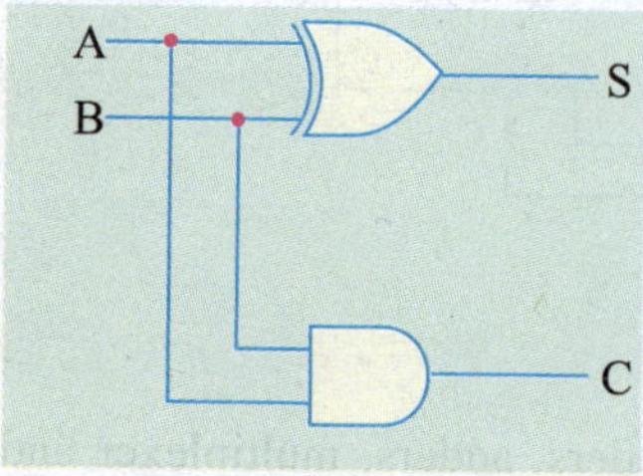

Fig. 70.75

The circuit is called ***half-adder*** because it cannot accept a carry-in from previous additions. For that purpose, we need a 3-*input* adder called full-adder.

Incidentally, the logical equations for the *SUM* and *CARRY* are $S = A \oplus B$ and $C = A.B$

Table 70.8

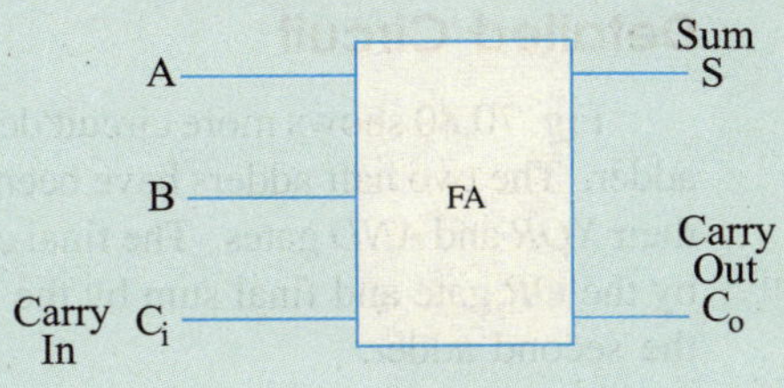

Input			Output	
A	B	C_i	S	C_o
0	0	0	0	0
0	0	1	1	0
0	1	0	1	0
0	1	1	0	1
1	0	0	1	0
1	0	1	0	1
1	1	0	0	1
1	1	1	1	1

Fig. 70.76

70.31. Full Adder

As shown in the block diagram of Fig. 70.76, it has ***three inputs and two outputs.*** It can add 3 digits (or bits) **at a time**. The bits *A* and *B* which are to be added come from the two registers and the third input comes from the carry generated by the previous addition. It produces two outputs; *SUM* and *CARRY-OUT*.

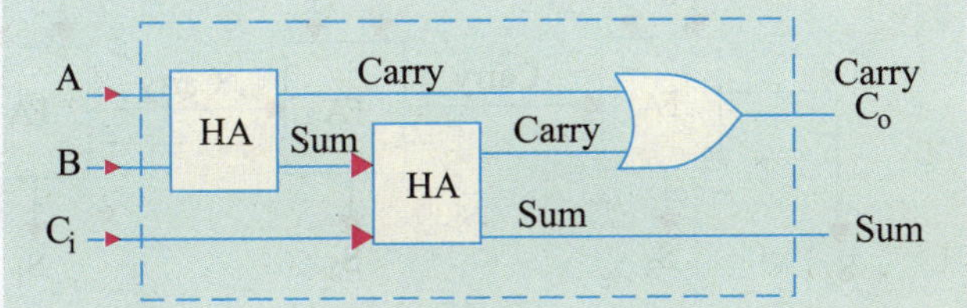

Fig. 70.77

The truth table 70.8 gives all possible input/output relationships for the full adder. *A* and *B* are the inputs from the respective digits of the registers to be added and C_i is the input for any carry generated by the previous stage. The *SUM* output gives binary addition of *A*, *B* and C_i. The other output generates the carry C_0 to be added to the next stage.

The full-adder can be constructed from two half-adders and one *OR* gate (Fig. 70.77).

Working. Let us illustrate, with the help of two examples, how this full adder adds three bits.

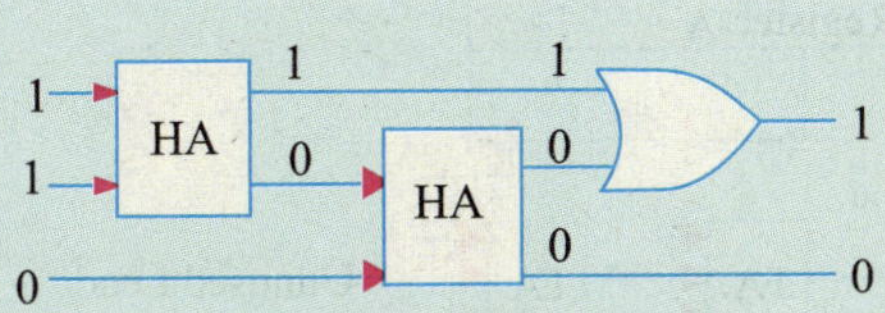

Fig. 70.78

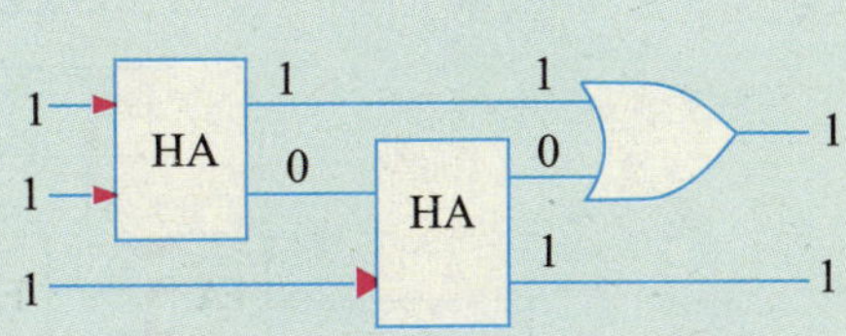

Fig. 70.79

***(i)* A = 1, B = 1, Ci = 0**

The full adder with these three inputs is shown in Fig. 70.78. First half adder gives a sum of 0 and a carry of 1. The second *HA* gives a sum of 0 with a carry of 0. The final output is : *SUM* 0, *CARRY* **1.** As we know from the rules of binary addition, $1 + 1 + 0 = 10_2$ (*i.e* decimal 2).

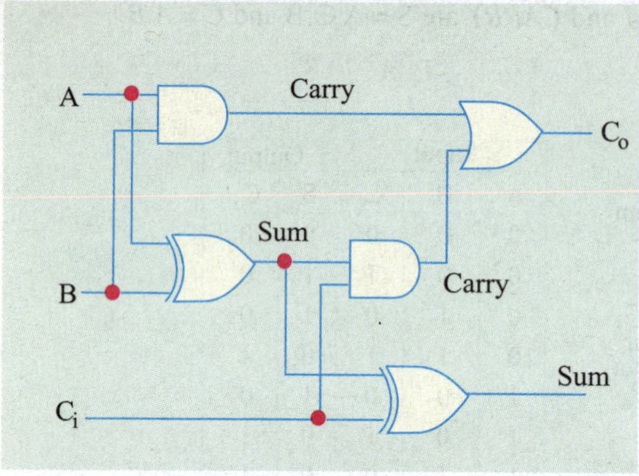

Fig. 70.80

***(ii)* A = 1, B = 1, C = 1**

As detailed in Fig. 70.79, we get a final *SUM* 1 with a *CARRY* 1. The result conforms to the binary addition : $1 + 1 + 1 = 11_2$ (*i.e.* decimal 3)

Detailed Circuit

Fig. 70.80 shows more circuit details of a full adder. The two half adders have been replaced by their *XOR* and *AND* gates. The final carry is given by the *OR* gate and final sum by the *XOR* gate of the second adder.

70.32. Parallel Binary Adder

For adding two 4-bit numbers, we need 4 full adders *connected in parallel* as shown in Fig. 70.81. The two numbers being added are $A_3\ A_2\ A_1\ A_0$ and $B_3\ B_2\ B_1\ B_0$ and their sum is $S_4\ S_3\ S_2\ S_1\ S_0$.

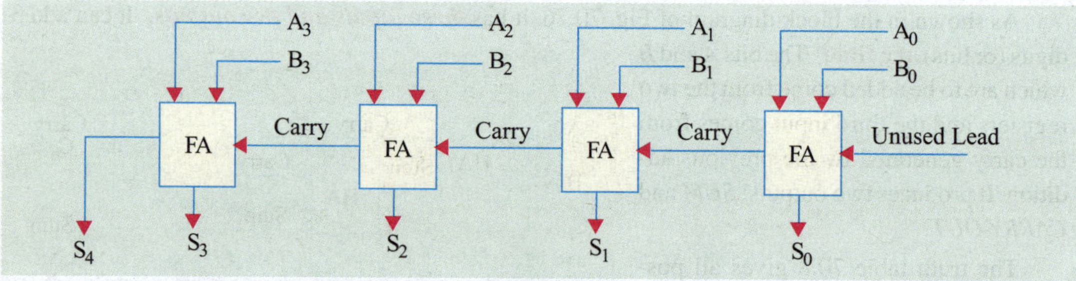

Fig. 70.81

The first adder could be a half adder though we may use a full adder but leave its *CARRY*-IN lead unconnected. As seen, different bits are fed to the four adders from two parallel registers which hold these bits. The final *SUM* appears as a 5-digit display.

Operation

The actual operation may be better understood with the help of the diagram of Fig. 70.82. Suppose, we want to add the following two 4-bit numbers.

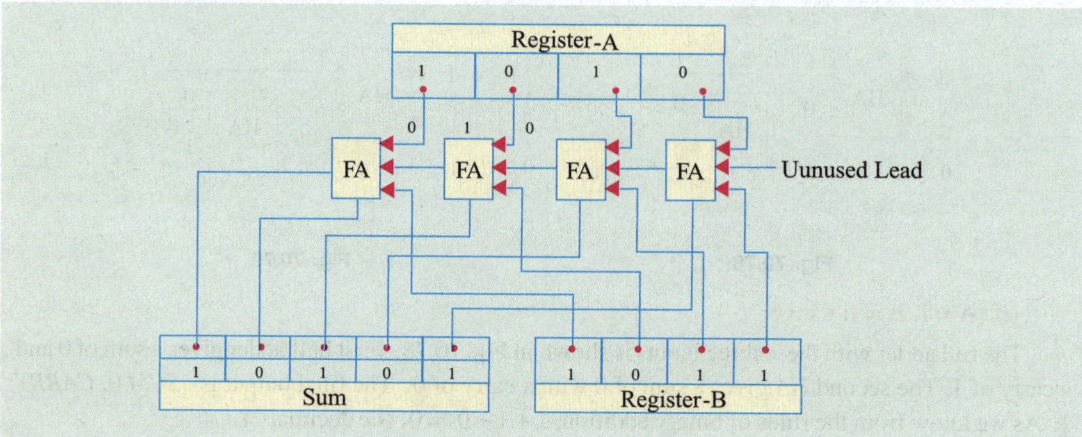

Fig. 70.82

$$\begin{array}{r} 1010 \\ +1011 \\ \hline 10101 \end{array} \qquad\qquad \begin{array}{r} 10 \\ +11 \\ \hline 21 \end{array}$$

The first adder performs 0 + 1 binary addition, giving a sum of 1 and a carry of 0. The two bits 0 and 1 are supplied ***simultaneously*** from the two registers *A* and *B*. The sum 1 appears on the display panel and carry 0 is passed on to the next full adder.

The next adder adds 1 + 1 +0 carry = sum 0 with a carry 1. The third adder performs 0 + 0 + 1 carry = 1 with carry 0. The fourth adder adds 1 + 1 + 0 carry = sum 0 with carry 1 both of which appear on the display unit. Hence, the final addition of the two numbers appears as 10101.

Table 70.10

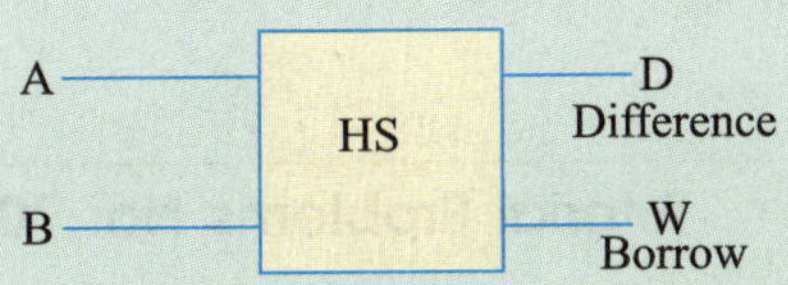

Input		Output	
A	B	W	D
0	0	0	0
0	1	1	1
1	0	0	1
1	1	0	0

Fig. 70.83

It may be noted that the largest binary numbers that can be added by this parallel adder are 1111 and 1111 which give a sum of 11110_2 (decimal 30). To increase its capacity, more full adders may be connected at the left end of Fig. 70.82. For example, for adding 6-bit numbers, we will have to add two more adders thus making a total of **six**.

70.33. Half Subtractor

It can subtract only two binary digits ***at a time*** and produce an output of a difference and a borrow. As shown in the block diagram of Fig. 70.83, it has two inputs and two outputs.

The operation of a half subtractor is based on the rules of binary subtraction illustrated in the truth Table 70.10 for all possible input/output combinations. The difference output in the fourth column has the same logic pattern as when *A* is *XOR*ed with *B* (same was the case for *SUM* in Art. 70.30). Hence, we can use an *XOR* gate to get the difference of two bits. The borrow output in the third column can be obtained by *AND*ing $\overline{A}$ with *B*.

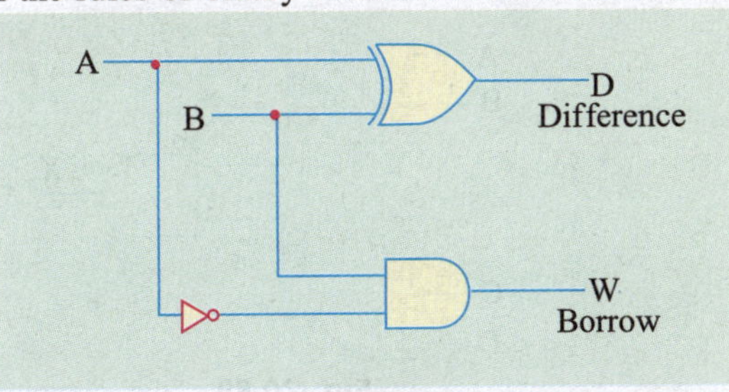

Fig. 70.84

The circuit for a half subtractor is shown in Fig. 70.84.

As mentioned earlier, the logical equations for the difference and borrow are given by

$D = A \oplus B$ and $W = \overline{A}\,B$

70.34. Full Subtractor

As shown in the block diagram of Fig. 70.85, it has three inputs and two outputs. As explained above, half subtractor can handle only 2 bits at a time and can be used for the least significant column of a subtraction problem. A full subtractor can, however, take care of higher-order columns.

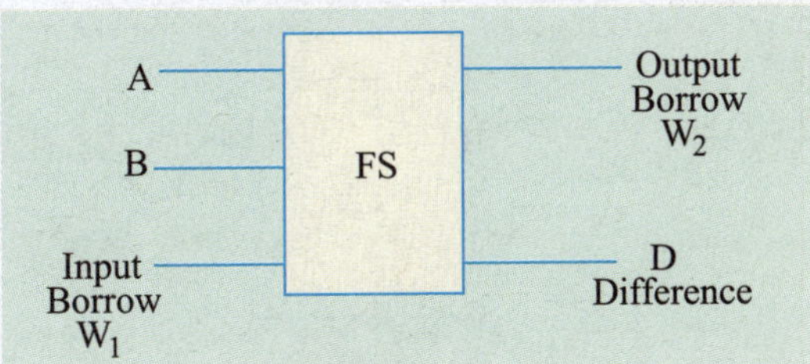

Fig. 70.85

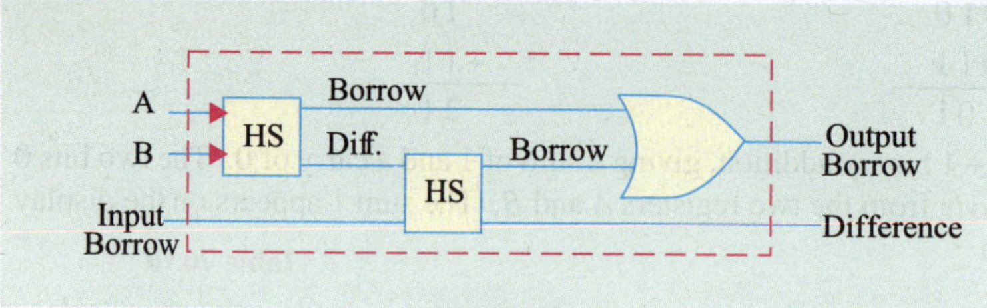

Fig. 70.86

As shown in Fig. 70.86, a full subtractor consists of two half subtractors and one *OR* gate.

It may be remarked here that by cascading 4 full subtractors, we can directly subtract 4-bit numbers *i.e.* we can subtract $B_3\ B_2\ B_1\ B_0$ from $A_3\ A_2\ A_1\ A_0$.

Tutorial Problems No. 70.1

1. Find the Boolean equation for the output of the logic circuit shown in Fig. 70.87. What would be the output if $A = 1$, $B = 0$, $C = 1$, $D = 1$ **[$X = AB + CD$; 1]**

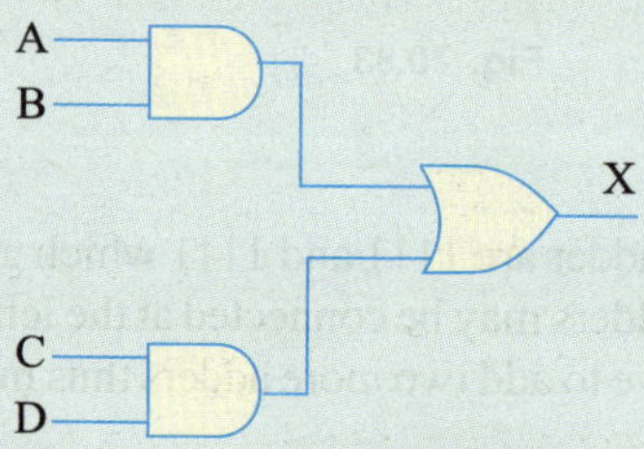

Fig. 70.87

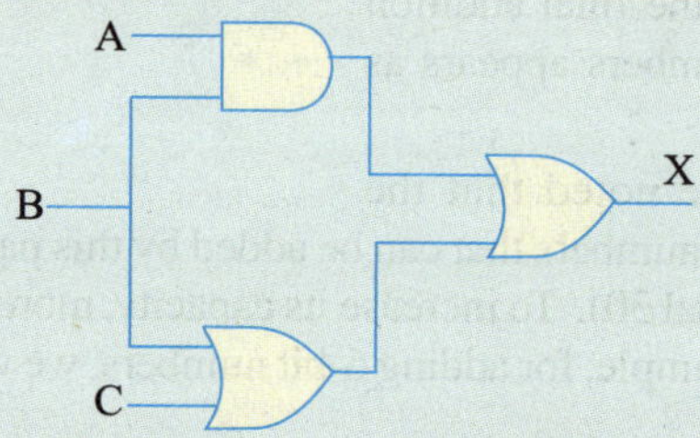

Fig. 70.88

2. What is the output equation of the logic circuit shown in Fig. 70.88 ? Evaluate the output if $A = 1$, $B = 1$, $C = 0$. **[$X = AB + (B + C)$; 1]**
3. After finding the Boolean equation for the circuit shown in Fig. 70.89, compute the output if $A = 1$, $B = 0$, $C = 1$, $D = 0$. **[$X = (A + B)(C + D)$; 1]**

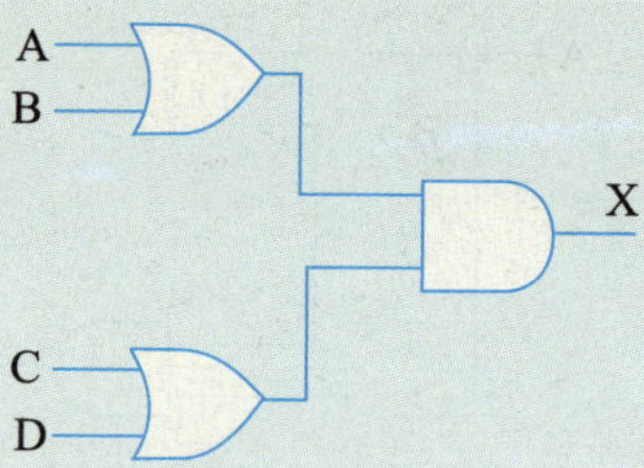

Fig. 70.89

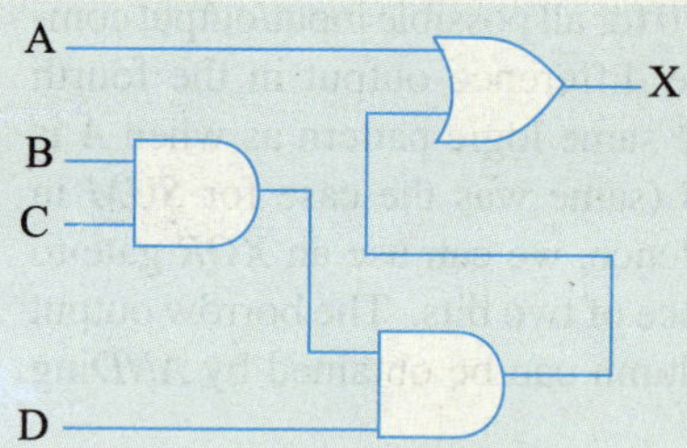

Fig. 70.90

4. Translate the hardware shown in Fig. 70.90 into a Boolean expression. Compute the value of the output if $A = 0$, $B = 1$, $C = 0$, $D = 1$. **[$X = A + BCD$; 0]**
5. What is the Boolean expression for the logic diagram shown in Fig. 70.91? Evaluate its output if $A = 1$, $B = 1$, and $C = 1$. **[$X = AB + \bar{C}$; 1]**

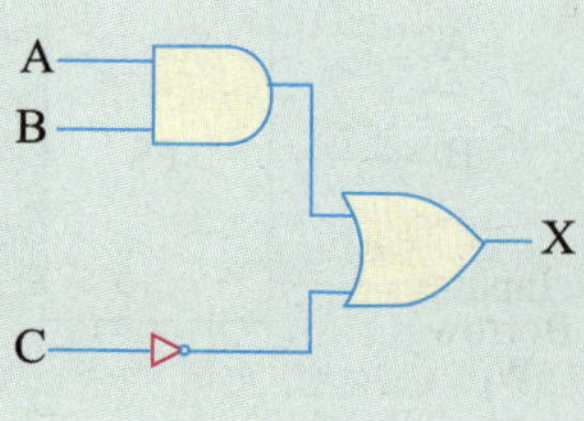

Fig. 70.91

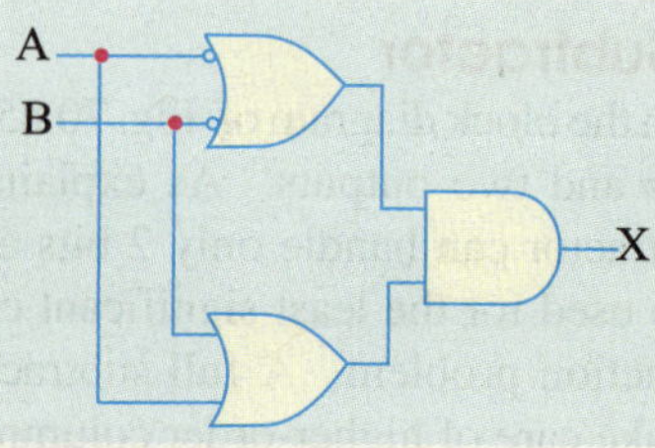

Fig. 70.92

6. Find Boolean expression for the logic circuit of Fig. 70.92. What is the output if $A = 1$, $B = 1$?

[$X = (\overline{A} + \overline{B}).(A + B)$; 0]

7. Give the logic functions performed by the circuits shown in Fig. 70.93 (*a*), (*b*) and (*c*).

[(*a*) X = AB+(C+D) (*b*) X = $(A + B) + \overline{C}$ (*c*) $\overline{(A + B)CD}$]

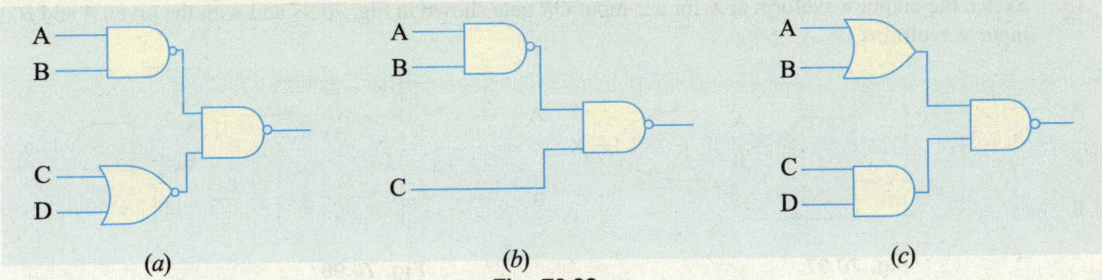

Fig. 70.93

8. State the logic functions of the circuits shown in Fig. 70.94 (*a*), (*b*) and (*c*).

[(*a*) $\overline{(A + B)(CD)}$ (*b*) AB + CD (*c*) (A + B). C.D.E]

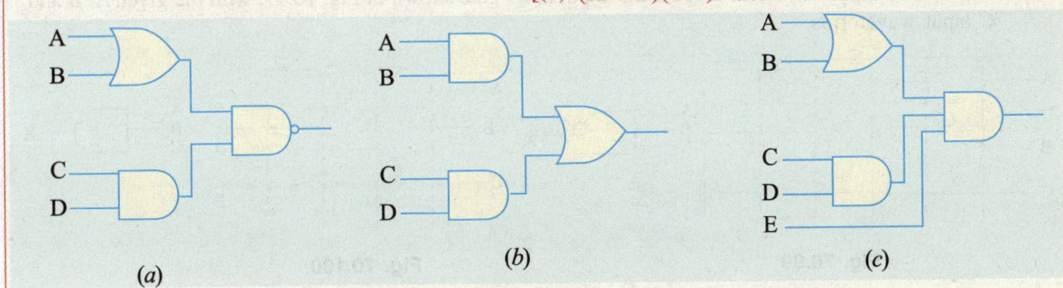

Fig. 70.94

9. State the logic functions performed by the circuits of Fig. 70.95 (*a*), (*b*) and (*c*).

[(*a*) $\overline{AB}.\overline{(C + D)}$ (*b*) AB − $\overline{C}$ (*c*) A B]

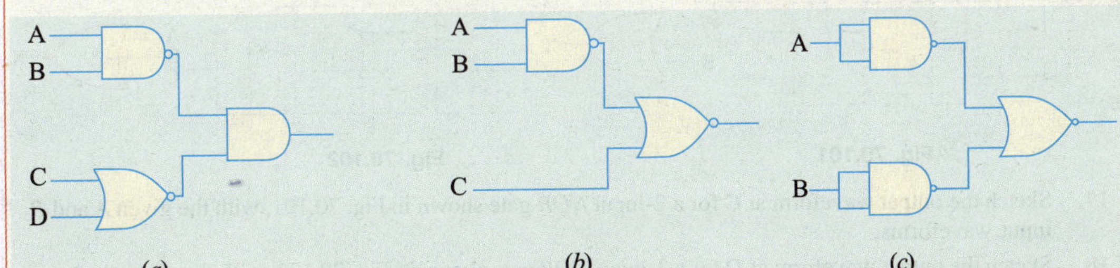

Fig. 70.95

10. Write the logic functions for the circuits shown in Fig. 70.96 (*a*), (*b*) and (*c*).

[(*a*) A(B + C) + CD (*b*) (A + B) + BC (*c*) $A\overline{B} + \overline{C}D$]

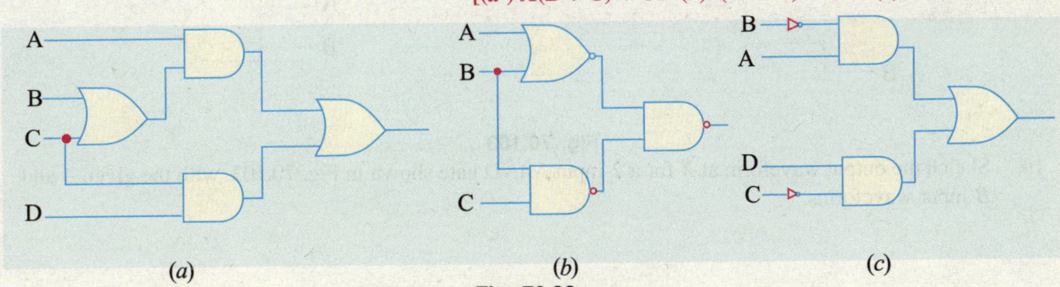

Fig. 70.96

11. What would be the output signal if two signals $A = 101011_2$ and $B = 110101_2$ are applied to the inputs of an *AND* gate? **[100 001]2**

12. What would be the output signal if two input binary signals given by A = 100101 and B = 110110 are applied to (*a*) *OR* gate (*b*) *NAND* gate and (*c*) *XNOR* gate?

[(*a*) $(110111)_2$ (*b*) $(011011)_2$ (*c*) $(101100)_2$]

13. Sketch the output waveform at *X*-for a 2-input *OR* gate shown in Fig. 70.97 and with the given *A* and *B* input waveforms.

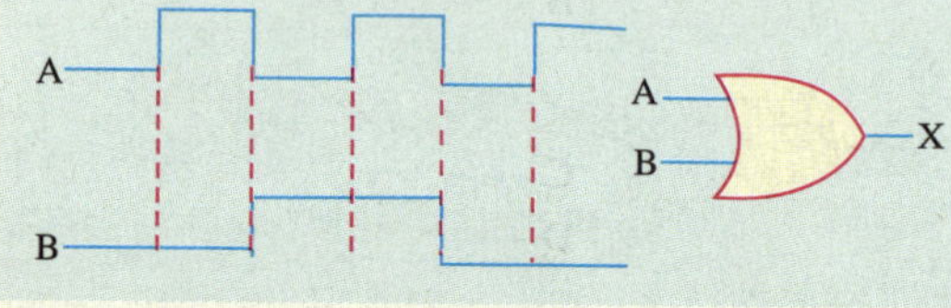

Fig. 70.97

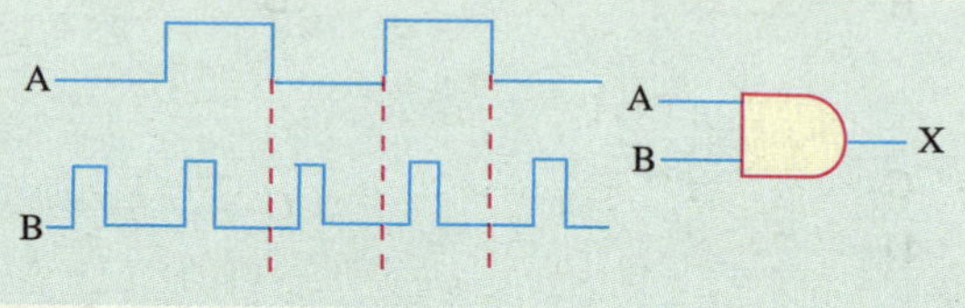

Fig. 70.98

14. Sketch the output waveform of a 2-input *AND* gate shown in Fig. 70.98 with the input waveforms *A* and *B*.

15. Sketch the output waveform at *X* for a 3-input *AND* gate shown in Fig. 70.99, with the given *A*, *B* and *C* input waveforms.

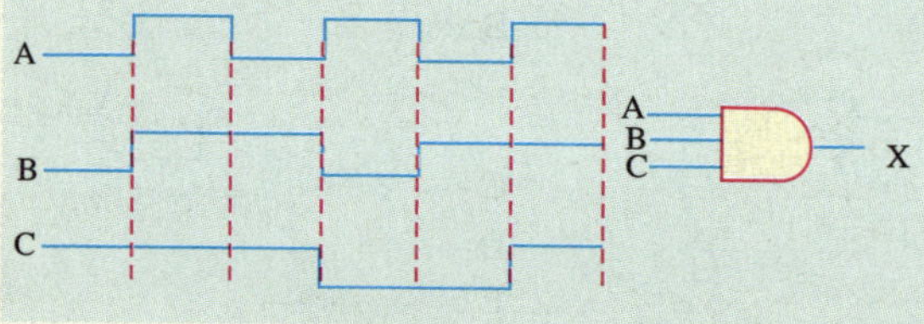

Fig. 70.99

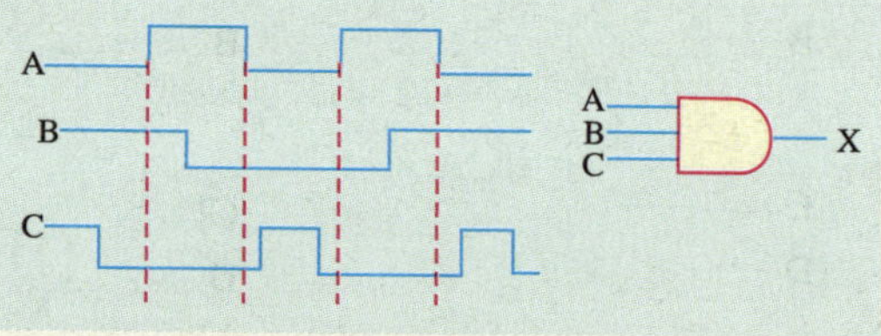

Fig. 70.100

16. Sketch the output waveform at *X* for a 3-input *AND* gate shown in Fig. 70.100, with the given *A*, *B* and *C* input waveforms.

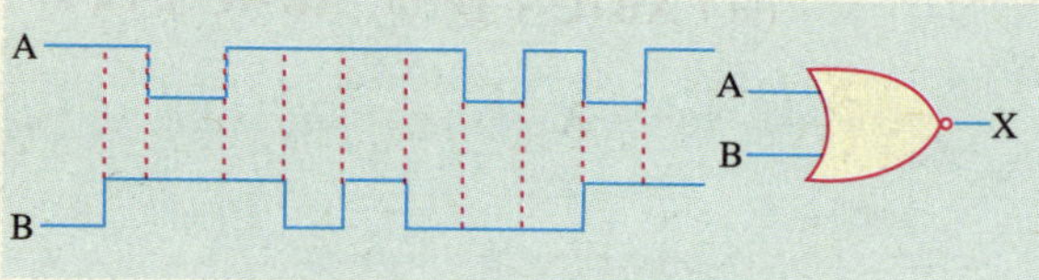

Fig. 70.101

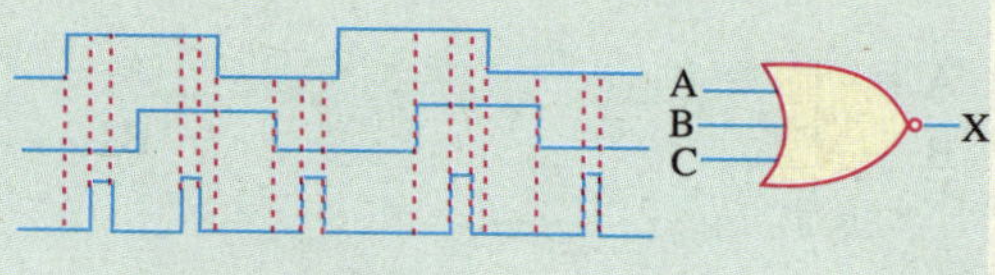

Fig. 70.102

17. Sketch the output waveform at *C* for a 2-input *NOR* gate shown in Fig. 70.101, with the given *A* and *B* input waveforms.

18. Sketch the output waveform at *D* for a 3-input *NOR* gate shown in Fig. 70.102, with the given *A*, *B* and *C* input waveforms.

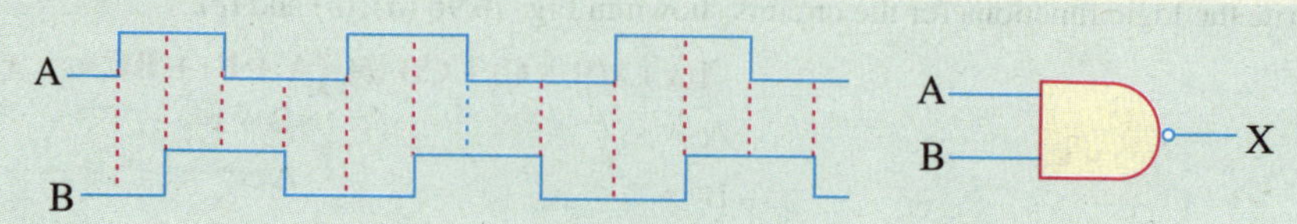

Fig. 70.103

19. Sketch the output waveform at *X* for a 2-input *NAND* gate shown in Fig. 70.103, with the given *A* and *B* input waveforms.

20. Sketch the output waveform at X for a 3-input *NAND* gate shown in Fig. 70.104, with the given A, B and C input waveforms.

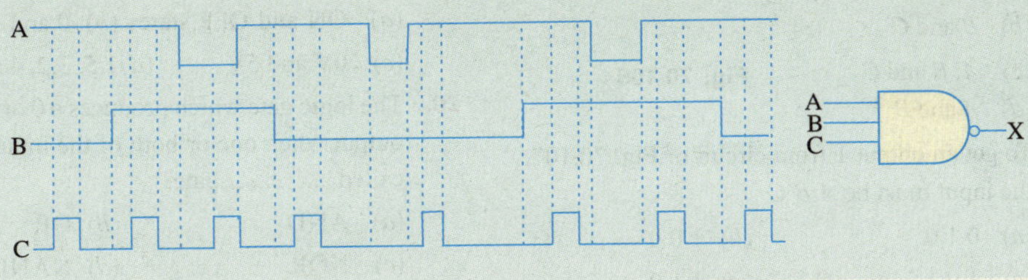

Fig. 70.104

OBJECTIVE TESTS – 70

1. A logic gate is an electronic circuit which
(*a*) makes logic decision
(*b*) allows electron flow only in one direction
(*c*) works on binary algebra
(*d*) alternates between 0 and 1 values.

2. In positive logic, logic state 1 corresponds to
(*a*) positive voltage
(*b*) higher voltage level
(*c*) zero voltage level
(*d*) lower voltage level.

3. In negative logic, the logic state 1 corresponds to
(*a*) negative voltage
(*b*) zero voltage
(*c*) more negative voltage
(*d*) lower voltage level.

4. The voltage levels of a negaive logic system
(*a*) must necessarily be negative
(*b*) may be negative or positive
(*c*) must necessarily be positive
(*d*) must necessarily be 0V and –5V

5. The output of a 2-input *OR* gate is zero only when its
(*a*) both inputs are 0
(*b*) either input is 1
(*c*) both inputs are 1
(*d*) either input is 0.

6. An *XOR* gate produces an output only when its two inputs are
(*a*) high (*b*) low
(*c*) different (*d*) same.

7. An AND gate
(*a*) implements logic addition
(*b*) is equivalent to a series switching circuit
(*c*) is an any-or-all gate
(*d*) is equivalent to a parallel switching circuit.

8. When an input electrical signal A = 10100 is applied to a NOT gate, its output signal is
(*a*) 01011 (*b*) 10101
(*c*) 10100 (*d*) 00101.

9. The only function of a NOT gate is to
(*a*) stop a signal
(*b*) recomplement a signal
(*c*) invert an input signal
(*d*) act as a universal gate.

10. A NOR gate is ON only when all its inputs are
(*a*) ON (*b*) positive
(*c*) high (*d*) OFF.

11. For getting an output from an XNOR gate, its both inputs must be
(*a*) high (*b*) low
(*c*) at the same logic level
(*d*) at the opposite logic levels.

12. In a certain 2-input logic gate, when $A = 0$, $B = 0$, then $C = 1$ and when $A = 0$, $B = 1$, then again $C = 1$. It must be .... gate.
(*a*) XOR (*b*) AND
(*c*) NAND (*d*) NOR

13. The logic symbol shown in Fig. 70.105 represents
(*a*) single-output AND gate
(*b*) NAND gate
(*c*) NAND gate used as NOT gate
(*d*) NOR gate.

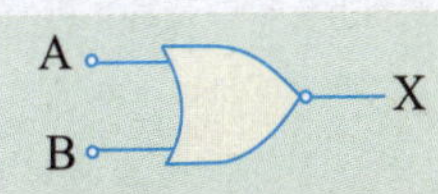

Fig. 70.105

14. The output from the logic gate shown in Fig. 70.106 will be available when inputs ...... are present.

(*a*) *A* and *C*

(*b*) *B* and *C*

(*c*) *A*, *B* and *C*

(*d*) *A* and *B*

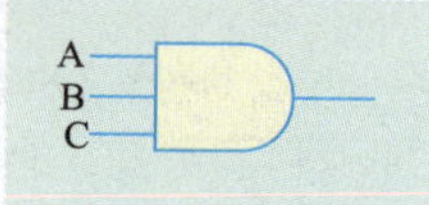

Fig. 70.106

15. To get an output 1 from circuit of Fig. 70.107, the input must be *A B C*

(*a*) 0 1 0 (*b*) 1 0 0

(*c*) 1 0 1 (*d*) 1 1 0

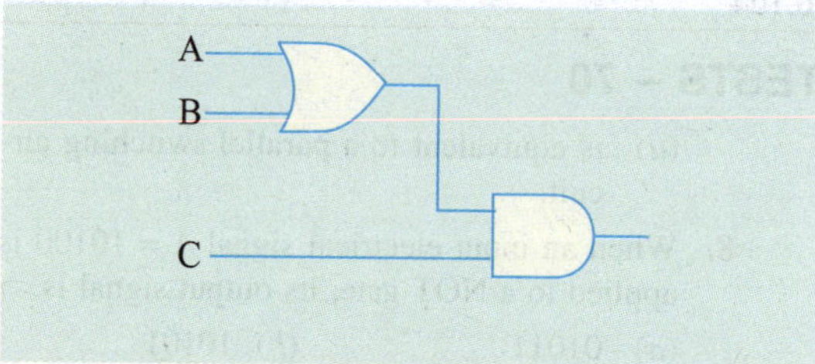

Fig. 70.107

16. Which of the following logic gates in Fig. 70.108 will not have an output of 1 ?

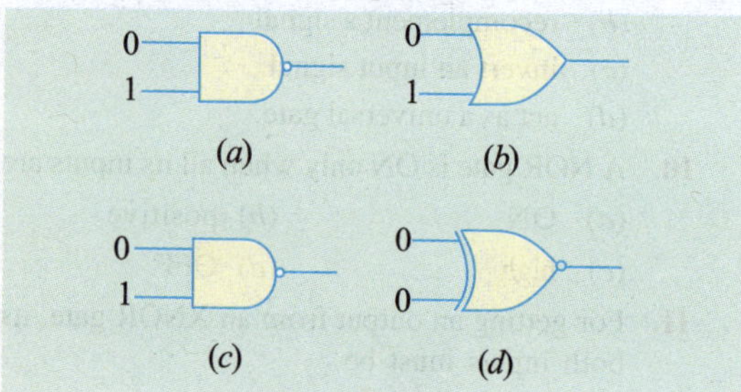

Fig. 70.108

17. A half-adder can be constructed from

(*a*) two XNOR gates only

(*b*) one XOR and one OR gate with their outputs connected in parallel

(*c*) one XOR and one OR gate with their inputs connected in parallel

(*d*) one XOR gate and one AND gate

18. The digital equivalent of an electric series circuit is the .... gate.

(*a*) NOR (*b*) NAND

(*c*) OR (*d*) AND

19. Which of the following represents analog data ?

(*a*) ON and OFF states (*b*) 0 and 1

(*c*) 0V and 5V (*d*) 1.5, 3.2, 4 and 5V

20. The logic gate which produces a 0 or low-level output when one or both of the inputs are 1 is called .... gate.

(*a*) AND (*b*) OR

(*c*) NOR (*d*) NAND

21. The output X of the gated network shown in Fig. 70.109 is

(*a*) $\overline{AB} \cdot \overline{CD} \cdot \overline{EF}$

(*b*) $\overline{AB} + \overline{CD} + \overline{EF}$

(*c*) AB + CD + EF

(*d*) (A + B) (C + D) (E + F)

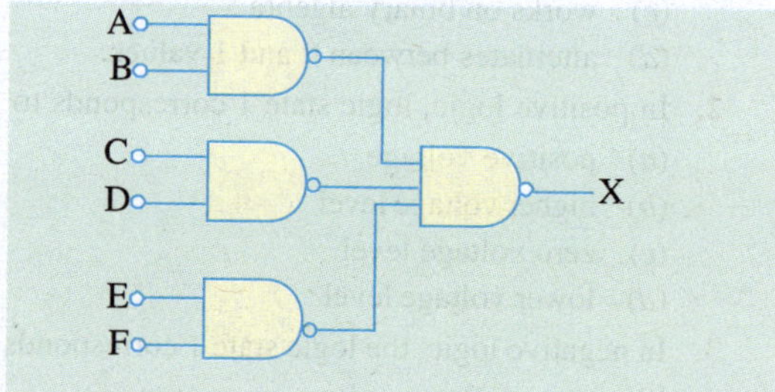

Fig. 70.109

22. The digital circuit shown in Fig 70.110 generates a modified clock pulse at the output. Choose the correct output waveform from the options given below. ***(GATE; 2004)***

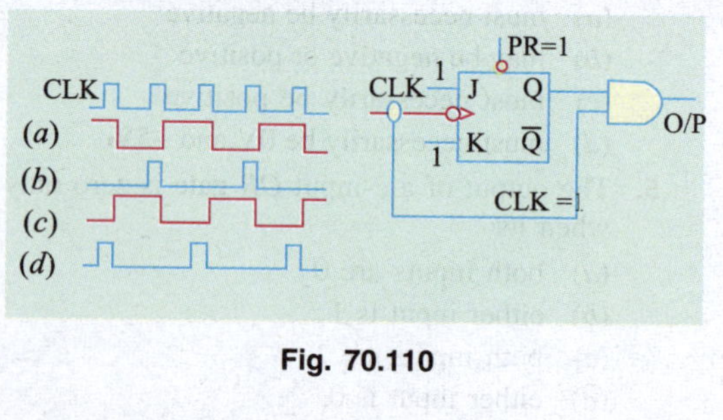

Fig. 70.110

ANSWERS

1. (*a*) **2.** (*b*) **3.** (*c*) **4.** (*b*) **5.** (*a*) **6.** (*c*) **7.** (*b*) **8.** (*a*) **9.** (*c*) **10.** (*d*)
11. (*c*) **12.** (*c*) **13.** (*d*) **14.** (*c*) **15.** (*c*) **16.** (*a*) **17.** (*c*) **18.** (*d*) **19.** (*d*) **20.** (*c*)
21. (*c*) **22.** (*d*)

CHAPTER 71

BOOLEAN ALGEBRA AND LOGIC FAMILIES

Learning Objectives

- Unique Feature of Boolean Algebra
- Laws of Boolean Algebra
- Equivalent Switching Circuits
- DeMorgan's Theorem's
- The Sum-of-Products (SOP) Form
- The Standard SOP Form
- The Standard POS Form
- The Karnaugh Map
- The Four-variable Karnaugh Map
- Square Adjacency in Karnaugh Map
- Mapping Directly on Karnaugh Map from a Truth Table
- "Don't Care" Conditions
- Main Logic Families
- Saturated and Non-saturated Logic Circuits
- DC supply voltage
- Noise Immunity
- Noise Margin-Power Dissipation
- Power Dissipation versus Frequency
- Propagation Delay
- Speed-Power Product
- RTL Circuit
- DTL Circuit
- TTL Circuit
- TTL Sub-families
- ECL Circuit
- MOS family

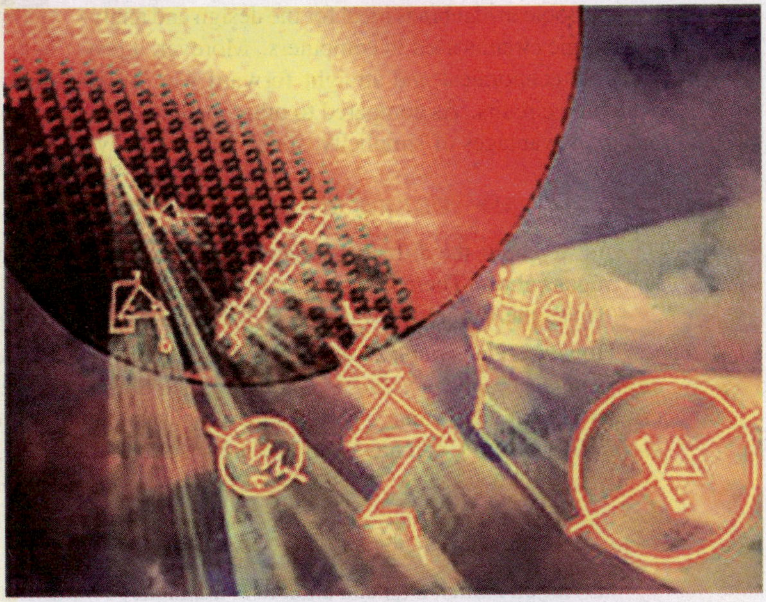

DNA Logic gates have been developed and are the first step towards creating a computer that has a structure of that of electronic PC

71.1. Introduction

Boolean algebra, named after its pioneer George Boole (1815-1864) is the algebra of logic presently applied to the opertion of computer devices. The rules of this algebra are based on human reasoning. It originated from the study of how we reason, what lines of reasoning are valid and what constitutes proof etc.

George Boole (1815–1864)

Starting with his investigation of the laws of thought, Boole developed in 1854 a mathematical system of logic in which he expressed truth functions *as symbols* and then manipulated these symbols to arrive at a conclusion. His new system was not the **ordinary numerical algebra** we know from our high school days but a totally new system called **logic algebra.** For example, in Boolean algebra $A + A = A$ and not $2A$ as is the case in ordinary algebra.

Boolean algebra remained in the realm of philosophy till 1938 when Claude *E.* Shannon used it to solve relay logic problems. As we know, all thinking and logic is concerned with finding answers to binary or two-valued questions like : is it good or bad, right or wrong, true or false etc. This binary nature of logic is exactly like the binary working of relay and switching circuits where relay is either energised or not, light is ON or OFF or pulse is present or not. Because of its very logical nature, Boolean algebra is ideal for the design and analysis of logic circuits used in computers. Morever, it provides an economical and straight forward way of describing computer circuitry and complicated switching circuits. As compared to other mathematical tools of analysis and design, Boolean algebra has the advantages of simplicity, speed and accuracy.

71.2. Unique Feature of Boolean Algebra

As we know, the different variables used in ordinary algebra can have ***any value*** including plus and minus values. There is no restriction on the value they can assume. For example, in the equation $2x + 3y = z$, the variables x, y and z can take on any value available in the entire field of real numbers.

However, the variables used in Boolean algebra have a unique property *i.e.* they can assume ***only one of the two possible values of 0 and 1.*** Each of the variable used in a logical or Boolean equation can assume only the value 0 or 1. For example, in the logical equation $A + B = C$, **each** of the three variables A, B and C can have only the values of either 0 or 1. ***This point must be clearly taken note of by the reader for easy understanding of the laws of Boolean algebra.***

71.3. Laws of Boolean Algebra

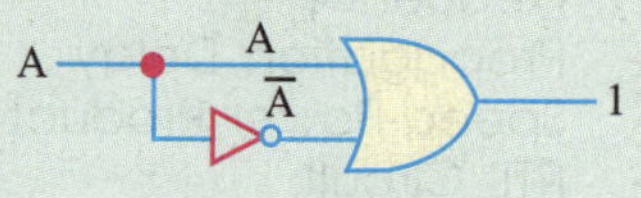

Fig. 71.1

As started earlier, Boolean algebra is a system of Mathematics based on ***logic.*** It has its own set of fundamental laws which are necessary for manipulating different Boolean expressions.

1. OR Laws

These four laws have already been discussed in the previous chapter. These are

Law 1. $A + 0 = A$ Law 3. $A + A = A$

Law 2. $A + 1 = 1$ Law 4. $A + \overline{A} = 1$

The expression given in Law 4 can be understood with the help of Fig. 71.1. Consider the following two possibilities :

(*i*) When $A = 0, \bar{A} = 1$ $\therefore$ $A + \bar{A} = 0 + 1 = 1$

(*ii*) When $A = 1, \bar{A} = 0$ $\therefore$ $A + \bar{A} = 1 + 0 = 1$

2. AND Laws

Law 5. $A . 0 = 0$ Law 7. $A . A = A$

Law 6. $A . 1 = A$ Law 8. $A . \bar{A} = 0$

The expression for Law 8 can be easily understood with the help of the logic circuit of Fig. 71.2. Consider the following two possibilities :

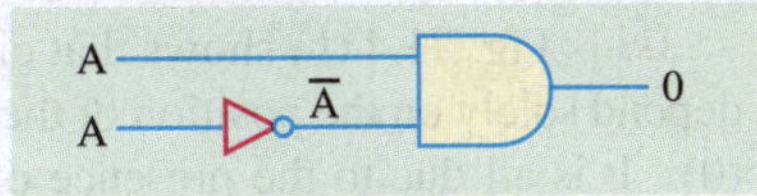

Fig. 71.2

(*i*) When $A = 0, \bar{A} = 1$ $\therefore$ $A . \bar{A} = 0.1 = 0$

(*ii*) When $A = 1, \bar{A} = 0$ $\therefore$ $A . \bar{A} = 1.0 = 0$

3. Laws of Complementation

Law 9. $\bar{0} = 1$ Law 10. $\bar{1} = 0$

Law 11. if $A = 0$, then $\bar{A} = 1$ Law 12. if $A = 1$, then $\bar{A} = 0$

Law 13. $\bar{\bar{A}} = A$

4. Commutative Laws

These laws allow ***change in the position*** of variables in *OR* and *AND* expressions.

Law 14. $A + B = B + A$ Law 15. $A . B = B . A$

These two laws express the fact that the order in which a combination of terms is performed does not affect the final result of the combination.

5. Associative Laws

These laws allow removal of brackets from logical expression and regrouping of variables.

Law 16. $A + (B + C) = (A + B) + C$

Law 17. $(A + B) + (C + D) = A + B + C + D$ Law 18. $A . (B . C) = (A . B) . C$

6. Distributive Laws

These laws permit factoring or multiplying out of an expression.

Law 19. $A (B + C) = AB + AC$ Law 20. $A + BC = (A + B) (A + C)$

Law 21. $A + \bar{A} . B = A + B$

7. Absorptive Laws

These enable us to reduce a complicated logic expression to a simpler form by absorbing some of the terms into existing terms.

Law 22. $A + AB = A$ Law 23. $A . (A + B) = A$

Law 24. $A . (\bar{A} + B) = AB$

The above laws can be used to prove any given Boolean identiry and also for simplifying complicated expressions.

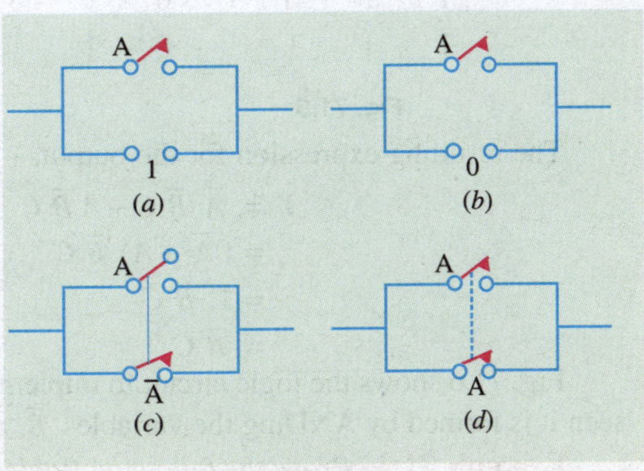

Fig. 71.3

71.4. Equivalent Switching Circuits

The equivalent circuits to illustrate some of the *OR* and *AND* laws given above are shown in Fig. 71.3.

(*i*) Fig. 71.3 (*a*) illustrates $A + 1 = 1$. Here, lower switch is permanently closed representing 1. Hence, value of the *OR* function is 1 (*i.e.* it is *ON*) whatever the value of *A*.

(*ii*) Fig. 71.3 (*b*) represents $A + 0 = A$. Here, function value is determined by A alone.

(*iii*) In Fig. 71.3 (*c*) when $\overline{A}$ opens, $\overline{A}$ closes and *vice versa*. Obviously, whatever the position of A, the circuit would always be *ON* proving that $A + \overline{A} = 1$.

(*iv*) Fig. 71.3 (*d*) proves $A + A = A$. It shows that final result depends on the value of A alone. If $A = 0$, the two switches are open, hence circuit is *OFF*. If $A = 1$, both switches are closed. Hence, circuit is *ON*.

(*v*) In Fig. 71.4 (*a*), the circuit is permanently *OFF* irrespective of the value of A. It is due to ***permanent*** open (0) in the circuit. Hence, it proves $A \,.\, 0 = 0$.

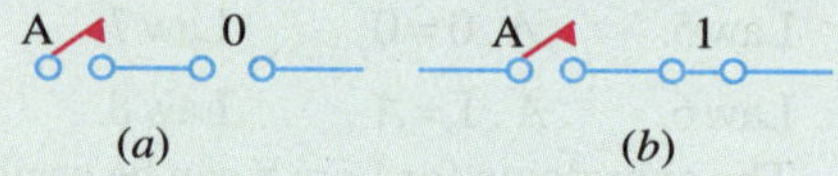

Fig. 71.4

(*vi*) Fig. 71.4 (*b*) shows that circuit conditions will depend solely on the position of the switch. If $A = 1$, circuit is *ON* (1) and when $A = 0$, circuit is *OFF* (0). It is all due to the presence of a ***permanent*** short (1) in the series circuit. Hence, everything depends on A.

Example 71.1. *Prove the following Boolean identity* : $AC + ABC = AC$

Solution. Taking the left hand side expression as y, we get

$$y = AC + ABC = AC\,(1 + B)$$

Now $\quad 1 + B = 1$ —Law 2

$\therefore \quad y = AC \,.\, 1 = AC$ —Law 6

$\therefore \quad AC + ABC = AC$

Example 71.2. *Determine the logic expression for the output Y, from the truth table shown in Fig. 71.5. Simplify and sketch the logic circuit for the simplified expression.*

Solution. There are two 1*s* in the output column of the given truth table. The corresponding binary values are 001 and 101. These values are converted into product terms as follows :

$$001 \rightarrow \overline{A}\;\overline{B}\;C \quad \text{and} \quad 101 \rightarrow A\,\overline{B}\;C.$$

Inputs A	B	C	Output Y	
0	0	0	0	
0	0	1	1	$\rightarrow \overline{A}\;\overline{B}\;C$
0	1	0	0	
0	1	1	0	
1	0	0	0	
1	0	1	1	$\rightarrow A\,\overline{B}\;C$
1	1	0	0	
1	1	1	0	

Fig. 71.5

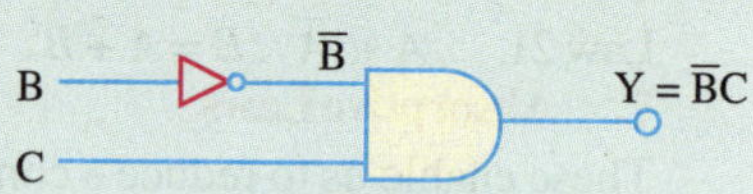

Fig. 71.6

The resulting expression for the output,

$$Y = \overline{A}\;\overline{B}\;C + A\,\overline{B}\;C$$

$$= (\overline{A} + A)\;\overline{B}\;C \qquad \text{...(Law 19)}$$

$$= 1.\;\overline{B}\;C \qquad \text{...(Law 4)}$$

$$= \overline{B}\;C \qquad \text{...(Law 6)}$$

Fig. 71.6 shows the logic circuit to implement the simplified logic expression for the output. As seen it is formed by ANDing the variables $\overline{B}$ and $\overline{C}$. The $\overline{B}$ can be obtained by inverting B.

Example 71.3. *Prove the following Boolean identity* : $(A + B)\,(A + C) = A + BC$

Solution. Putting the left hand side expression equal to y, we get

$$
\begin{aligned}
Y &= (A+B)(A+C) \\
&= AA+AC+AB+BC && \text{—Law 19} \\
&= A+AC+AB+BC && \text{—Law 7} \\
&= A+AB+AC+BC \\
&= A(1+B)+AC+BC && \text{—Law 19} \\
&= A+AC+BC && \text{—Law 2} \\
&= A(1+C)+BC && \text{—Law 19} \\
&= A+BC && \text{—Law 2}
\end{aligned}
$$

$$\therefore \quad (A+B)(A+C) = A+BC$$

Example 71.4. *Prove the following identity* : $A+\bar{A}B = A+B$

Solution. Let the left-hand side expression be put equal to Y.

$$
\begin{aligned}
Y &= A+\bar{A}B = A\,.\,1+\bar{A}B && \text{—Law 6} \\
&= A(1+B)+\bar{A}B && \text{—Law 2} \\
&= A\,.\,1+AB+\bar{A}B && \text{—Law 19} \\
&= A+BA+B\bar{A} && \text{—Law 6 and 15} \\
&= A+B(A+\bar{A}) && \text{—Law 19} \\
&= A+B\,.\,1 && \text{—Law 4} \\
&= A+B && \text{—Law 6}
\end{aligned}
$$

$$\therefore \quad A+\bar{A}B = A+B$$

Example 71.5. *Prove the following Boolean identity* : $(A+B)(A+\bar{B})(\bar{A}+C) = AC$

(Digital Computations, Punjab Univ. 1990)

Solution. Left the left-hand side expressionbe represented by Y.

$$
\begin{aligned}
\therefore \quad Y &= (A+B)(A+\bar{B})(\bar{A}+C) = (AA+A\bar{B}+BA+B\bar{B})(\bar{A}+C) \\
&= (A+AB+A\bar{B})(\bar{A}+C) = [A(1+B)+A\bar{B}](\bar{A}+C) && \because B\bar{B}=0 \\
&= (A+A\bar{B})(\bar{A}+C) = A(1+\bar{B})(\bar{A}+C) && [\because 1+B=1] \\
&= (A\,.\,1)(\bar{A}+C) = A(\bar{A}+C) = A\bar{A}+AC = AC && \because A\bar{A}=0
\end{aligned}
$$

Example 71.6. *Prove the following Boolean identity* :

$$ABC+A\bar{B}C+AB\bar{C} = A+(B+C)$$

(Digital Electronic Systems-I, Kurukshetra Univ. 1991)

Solution. Equating the left-hand side expression to Y, we have

$$
\begin{aligned}
Y &= ABC+A\bar{B}C+AB\bar{C} = AC(B+\bar{B})+AB\bar{C} && \text{—Law 19} \\
&= AC+AB\bar{C} && \text{—Law 4} \\
&= A(C+B\bar{C}) \\
&= A(C+B) && \text{—Law 21} \\
&= A(B+C) && \text{—Law 14}
\end{aligned}
$$

Hence, it proves the given identity.

Example 71.7. *Simplify the following Boolean Expression* :

$$AB\bar{C}+A\bar{B}\bar{C}+\bar{A}BC+ABC+A\bar{B}C$$

(Digital Computations, Punjab Univ. 1992)

Solution. Bringing together those terms which have *two* common letters, we get

$$Y = AB\bar{C}+ABC+A\bar{B}\bar{C}+A\bar{B}C+\bar{A}BC$$

$$= AB(\overline{C} + C) + A\overline{B}(\overline{C} + C) + \overline{A}BC \qquad \text{—Law 19}$$
$$= AB + A\overline{B} + \overline{A}BC \qquad \text{—Law 4}$$
$$= A(B + \overline{B}) + \overline{A}BC \qquad \text{—Law 19}$$
$$= A + \overline{A}BC = A + BC \qquad \text{—Law 4 and Law 21}$$

Example 71.8. *Simplify the following expression and show the minimum gate implementation.*

$$Y = A.B.\overline{C}.\overline{D} + \overline{A}.B\overline{C}.\overline{D} + B.\overline{C}.D$$

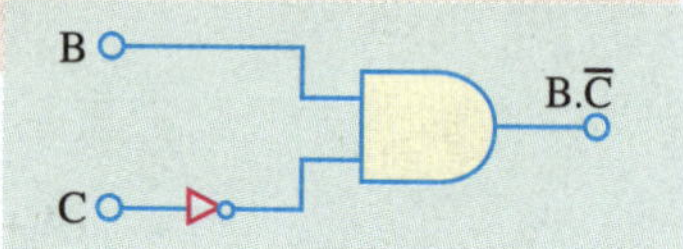

Fig. 71.7

Solution. As seen from *OR* and *AND* laws of Art 71.3, $A + \overline{A} = 1$ and $A.1 = A$

$$\therefore \quad Y = B\overline{C}\overline{D}(A + \overline{A}) + B\overline{C}D = B\overline{C}\overline{D}.1 + B\overline{C}D.$$
$$= B.\overline{C}.\overline{D} + B.\overline{C}.D = B.\overline{C}.(D + \overline{D}) = B.\overline{C}.1 = B.\overline{C}.$$

Minimum gate implementation is shown by the circuit of Fig. 71.7

Example 71.9. *Simplify the following Boolean expression and draw the logic circuits for the simplified expressions.*

(*a*) $Y = \overline{A}BC + A\overline{B}C + ABC + B\overline{C}$ (*b*) $Y = \overline{B}(A + C) + C(\overline{A} + B) + AC$

Solution. (*a*) $Y = \overline{A}BC + A\overline{B}C + ABC + B\overline{C} = BC(A + \overline{A}) + A\overline{B}C + B\overline{C}$

$$= BC + A\overline{B}C + B\overline{C} = B(C + \overline{C}) + A\overline{B}C = A\overline{B}C + B$$

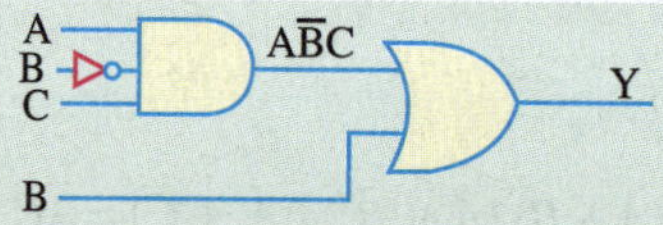

Fig. 71.8

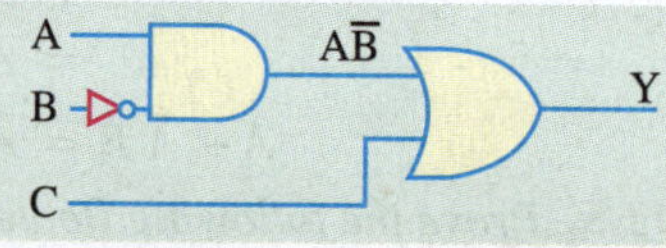

Fig. 71.9

(*b*) $Y = \overline{B}(A + C) + C(\overline{A} + B) + AC = A\overline{B} + \overline{B}C + \overline{A}C + BC + AC$

$$= A\overline{B} + C(\overline{B} + \overline{A} + B + A) = A\overline{B} + C.1 = A\overline{B} + C$$

Example 71.10. *Using truth table, prove that* $A + \overline{A}B = A + B$ *and illustrate the equivalence with the help of a switching circuit.*

Solution. Since there are only two variables *A* and *B*, their number of possible combination is $2^2 = 4$ in terms of 0 and 1.

As seen, $\overline{A}$ is the negative of A. In the fourth column of Table 71.1, $\overline{A}$ has been *AND*ed with *B*. In the fifth column, *A* has been *OR*ed with $\overline{A}B$. The values in last column have been obtained by *OR*ing *A* with *B*. By comparing results of column 5 to 6, the equivalence between the two statements can be proved.

Table No. 71.1

A	B	$\overline{A}$	$\overline{A}B$	$A + \overline{A}B$	$A + B$
0	0	1	0	0	0
0	1	1	1	1	1
1	0	0	0	1	1
1	1	0	0	1	1

Switching circuit of Fig. 71.10 (*a*) represents $(A + \overline{A}B)$. In this circuit, when *A* is open, $\overline{A}$ is closed and *vice versa*. It can be shown that this circuit becomes closed with either *A* or *B* is closed.

(*i*) when *A* is closed, then $\overline{A}$ opens. Circuit is completed *via* the upper branch.

(*ii*) keeping *A* open, when we close *B*, the circuit again becomes closed *via* lower branch because $\overline{A}$ is already closed (due to *A* being open).

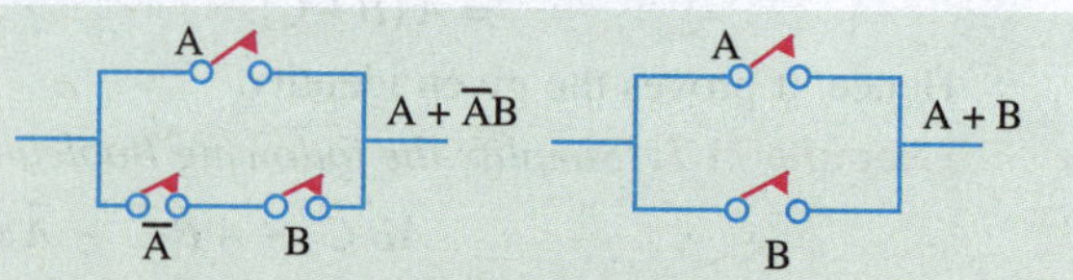

(*a*) (*b*)

Fig. 71.10

Hence, all that we have to do for closing the circuit of Fig. 71.10 (*a*) is to close either switch *A* or *B*. It is exactly what circuit of Fig. 71.10 (*b*) does.

Example 71.11. *Simplify the expression* : $(AB + C)(AB + D)$

Solution. Let

$$
\begin{aligned}
Y &= (AB + C)(AB + D) \\
&= ABAB + ABD + ABC + CD && \text{—Law 19} \\
&= AABB + ABD + ABC + CD \\
&= AB + ABD + ABC + CD && \text{—Law 7} \\
&= AB(1 + D) + ABC + CD = AB + ABC + CD && \text{—Law 2} \\
&= AB(1 + C) + CD = AB + CD && \text{—Law 2}
\end{aligned}
$$

$$\therefore \quad (AB + C)(AB + D) = AB + CD$$

71.5. DE Morgan's Theorem

These two theorems (or rules) are a great help in *simplifying complicated logical expressions.* The theorems can be started as under :

Law 25. $\overline{A + B} = \overline{A} \, . \, \overline{B}$ Law 26. $\overline{A \, . \, B} = \overline{A} + \overline{B}$

The first statement says that ***the complement of a sum equals the product of complements.*** The second statement says that ***the complement of a product equals the sum of the complements.*** In fact, it allows transformation from a sum-of-products from to a product-of-sum from.

As seen from the above two laws, the procedure required for taking out an expression from under a *NOT* sign is as follows :

1. ***complement the given expression i.e., remove the overall NOT sign***
2. ***change all ANDs to ORs and all the ORs to ANDs.***
3. ***complement or negate all individual variables.***

As an illustration, take the following example

$$
\begin{aligned}
\overline{A + BC} &= A + BC && \text{—step 1} \\
&= A(B + C) && \text{—step 2} \\
&= \overline{A}\,(\overline{B} + \overline{C}) && \text{—step 3}
\end{aligned}
$$

Next, consider this example,

$$
\begin{aligned}
\overline{(\overline{A} + B + \overline{C})(\overline{A} + B + C)} &= (\overline{A} + B + \overline{C})(\overline{A} + B + C) && \text{—step 1} \\
&= \overline{A}B\overline{C} + \overline{A}B\overline{C} && \text{—step 2} \\
&= \overline{\overline{A}}\,\overline{B}\,\overline{\overline{C}} + \overline{\overline{A}}\,\overline{B}\,\overline{C} && \text{—step 3} \\
&= A\,\overline{B}\,C + A\,\overline{B}\,\overline{C}
\end{aligned}
$$

This process is called **demorganization.** It should, however, be noted that opposite procedure would be followed in bringing an expression under the *NOT* sign. Let us bring the expression $\overline{A} + \overline{B} + \overline{C}$ under the *NOT* sign.

$$
\begin{aligned}
\overline{A} + \overline{B} + \overline{C} &= \overline{\overline{A}} + \overline{\overline{B}} + \overline{\overline{C}} && \text{—step 3} \\
&= A + B + C \\
&= ABC && \text{—step 2} \\
&= \overline{ABC} && \text{—step 1}
\end{aligned}
$$

Fig. 71.11 shows the circuits to illustrate De Morgan's theorems.

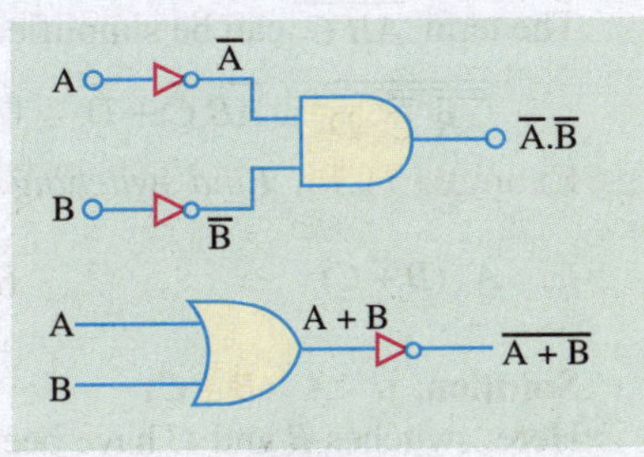

Fig. 71.11

As seen, basic logic function can be either an *OR* gate or an *AND* gate.

Example 71.12. *Demorganize the expression :* $\overline{(A+B)(C+D)}$

De Morgan (1806–1871)

Solution. The procedure is as explained above.

$$\overline{(A+B)(C+D)} = (A+B)(C+D) \quad \text{—step 1}$$

$$= (AB)+(CD) \quad \text{—step 2}$$

$$= \overline{AB} + \overline{CD}$$

$$= (\overline{A}+\overline{B})(\overline{C}+\overline{D})$$

$$= \overline{A}\,\overline{C}+\overline{A}\overline{D}+\overline{B}\overline{C}+\overline{B}\overline{D} \quad \text{—step 3}$$

Example 71.13. *Simplify each of the following expressions using De Morgan's theorems :*

(*a*) $\overline{A\,\overline{(B+\overline{\overline{C}})}D}$

(*b*) $\overline{(M+\overline{N})(\overline{M}+N)}$ (*c*) $\overline{\overline{\overline{AB}\,\overline{C}}\,D}$

Solution. Please note that there is more than one way of simplifying the expressions given in part (*a*), (*b*) and (*c*).

(*a*)

$$\overline{A\,\overline{(B+\overline{\overline{C}})}D} = A\,\overline{(B+\overline{C})}\,D \quad \text{... step 1}$$

$$= A+\overline{\overline{(B+\overline{C})}}+D \quad \text{... step 2}$$

$$= \overline{A}+\overline{\overline{B+\overline{C}}}+\overline{D} \quad \text{... step 3}$$

$$= \overline{A}+B+\overline{C}+\overline{D} \quad \text{...}(\because \overline{\overline{B+\overline{C}}} = B+\overline{C})$$

(*b*)

$$\overline{(M+\overline{N})(\overline{M}+N)} = (M+\overline{N})(\overline{M}+N) \quad \text{... step 1}$$

$$= (M\overline{N})+(\overline{M}N) \quad \text{... step 2}$$

$$= (\overline{M}\,\overline{\overline{N}})+\overline{\overline{M}}\,\overline{N} \quad \text{... step 3}$$

$$= \overline{M}N+M\overline{N} \quad \text{... step 4}$$

(*c*)

$$\overline{\overline{\overline{AB}\,\overline{C}}\,D} = \overline{\overline{AB}\,\overline{C}}\,D \quad \text{... step 1}$$

$$= \overline{\overline{AB}\,\overline{C}}+D \quad \text{... step 2}$$

$$= \overline{\overline{\overline{AB}\,\overline{C}}}+\overline{D} \quad \text{... step 3}$$

$$= \overline{AB}\,\overline{C}+\overline{D} \quad \text{...}(\because \overline{\overline{\overline{AB}\,\overline{C}}} = \overline{AB}\,\overline{C})$$

The term $\overline{AB}\,\overline{C}$ can be simplified further by De Morganising the term $\overline{AB}$ as $\overline{A}+\overline{B}$. Thus

$$\overline{\overline{\overline{AB}\,\overline{C}}\,D} = \overline{AB}\,\overline{C}+\overline{D} = (\overline{A}+\overline{B})\,\overline{C}+\overline{D} = \overline{A}\,\overline{C}+\overline{B}\,\overline{C}+\overline{D}$$

Example 71.14. *Find switching circuits for the following logic expressions :*

(*i*) $A\,.\,(B+C)$ (*ii*) $A\overline{B}+CD$ (*iii*) $(\overline{A}B+AC)\,\overline{C}$

(Industrial Electronics, City & Guilds, London)

Solution. (*i*) $A\,.\,(B+C)$

Here, switches *B* and *C* have been *OR*ed *i.e.* connected in parallel. This parallel circuit is connected in series with switch *A* because $(B+C)$ has been *AND* ed with *A*. hence, the circuit becomes as shown in Fig. 71.12 (*a*). As seen, it is a series-parallel circuit.

(*ii*) $A\overline{B} + CD$

Here, $A\overline{B}$ has been *OR*ed with *CD*. We can easily make out that a 2-brached circuit is needed for this logic expression. One branch contains switch *A* in series with $\overline{B}$ (with one contact point shows raised to indicate negation) and the other branch contains two series-connected switches *C* and *D* as shown in Fig. 71.12 (*b*).

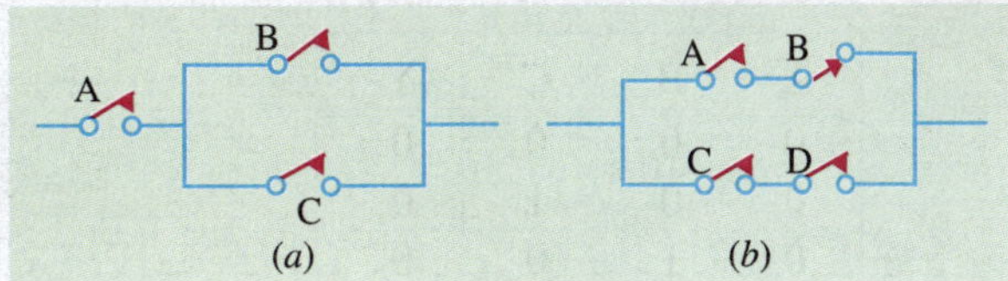

Fig. 71.12

(*iii*) $(\overline{A}B + AC)\,\overline{C}$

From the look of it, we can make out that it consists of a series-parallel circuit as shown in Fig. 71.13. $\overline{A}\,B$ has been *OR*ed with *AC i.e.* $\overline{A}\,B$ and *AC* are in parallel. Of course, $\overline{A}$ and *B* are in series in one branch where as *A* and *C* are in series in the other branch. Both these parallel branches are in series with $\overline{C}$.

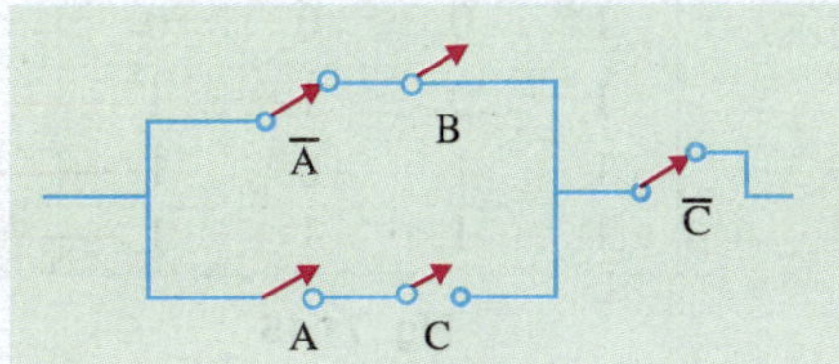

Fig. 71.13

Example 71.15. *Prove that 3-input NAND gate of Fig. 71.14 (a) is equivalent to the bubbled AND gate of Fig. 71.14 (b).*

Solution. The output of *NAND* gate is $\overline{ABC}$ and that of bubbled *OR* gate is $\overline{A} + \overline{B} + \overline{C}$. We have to show that the above two expressions are equivalent.

De Morgan's theorem can be used to prove the above equivalence,

$\overline{ABC} = ABC$ —step 1

$= A + B + C$ —step 2

$= \overline{A} + \overline{B} + \overline{C}$ —step 3

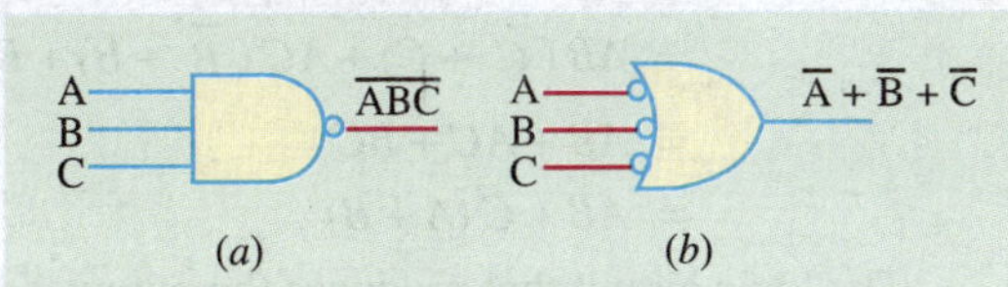

Fig. 71.14

71.6. Duals

Basic duality underlies all Boolean algebra. ***Each expression has its dual which is as true as the original expression.*** For getting the dual of a given Boolean expression, the procedure is to convert

1. all 1*s* to 0*s* and all 0*s* to 1*s*.
2. all *AND*s to *OR*s and all *OR*s to *AND*s.

The dual so obtained ***is also found to be true.***

Some of the Boolean relations and their duals are given in Table 71.2.

Table No. 71.2

Relation		*Dual Relation*	
$A\,.\,0$	$=0$	$A+1$	$=1$
$A\,.\,A$	$=A$	$A+A$	$=A$
$A\,.\,\overline{A}$	$=0$	$A+\overline{A}$	$=1$
$A\,.\,1$	$=A$	$A+0$	$=A$
$A\,.\,(A+B)$	$=A$	$A+AB$	$=1$
$A\,.\,(\overline{A}+B)$	$=AB$	$A+\overline{A}\,B$	$=A+B$

Example 71.16. *Design a logic circuit whose output is HIGH* **only** *when a majority of the inputs A, B and C are HIGH.*

Solution. Since there are three inputs, *A*, *B* and *C*, therefore whenever two or more than two (*i.e.* a majority) inputs are HIGH, the output is HIGH. This situation can be represented in the form of a truth table as shown in Fig. 71.15.

A	B	C	Y	
0	0	0	0	
0	0	1	0	
0	1	0	0	
0	1	1	1	→ $\overline{A}BC$
1	0	0	0	
1	0	1	1	→ $A\overline{B}C$
1	1	0	1	→ $AB\overline{C}$
1	1	1	1	→ ABC

Fig. 71.15

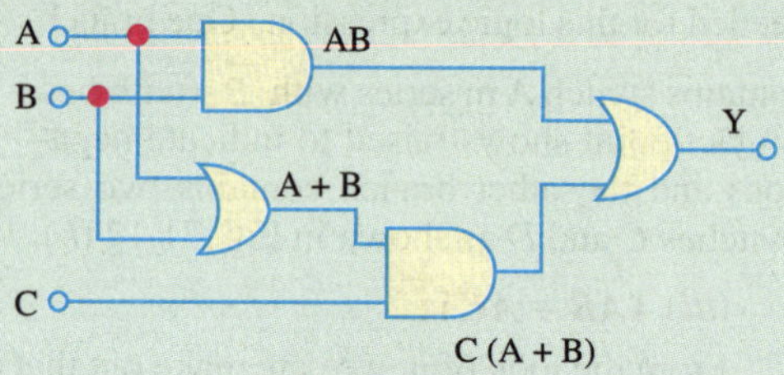

Fig. 71.16

There are four 1*s* in the output column of the truth table. The corresponding binary values are 011, 101, 110 and 111 respectively. Converting these values into product terms and summing up all the terms, we get

$$Y = AB\overline{C} + A\overline{B}C + \overline{A}BC + ABC$$

Adding the term *ABC* two times from our side in the Boolean expression for the output,

$$Y = AB\overline{C} + A\overline{B}C + \overline{A}BC + ABC + ABC + ABC \qquad (\because ABC + ABC + ABC = ABC)$$

Bringing together those terms which have two common letters, we get,

$$\begin{aligned} Y &= AB\overline{C} + ABC + A\overline{B}C + ABC + \overline{A}BC + ABC \\ &= AB(\overline{C} + C) + AC(\overline{B} + B) + BC(\overline{A} + A) \\ &= AB + AC + BC \qquad (\because \overline{C} + C = \overline{B} + B = \overline{A} + A = 1) \\ &= AB + C(A + B) \end{aligned}$$

The logic circuit that produces the output $Y = AB + C(A + B)$ is as shown in Fig. 71.16.

Alternatively :

You can also arrange the equation,

$$Y = AB + AC + BC$$

as

$$Y = A(B + C) + BC \qquad \ldots(i)$$

or

$$Y = (A + C)B + AC \qquad \ldots(ii)$$

If you implement equation (*i*) or (*ii*) using AND and OR logic gates, the number of logic gates is used will still be the same (4).

Example 71.17. *Determine the Boolean expression for the logic circuit shown in Fig. 71.17. Simplify the Boolean expression using Boolean Laws and De Morgan's theorem. Redraw the logic circuit using the simplified Boolean expression.*

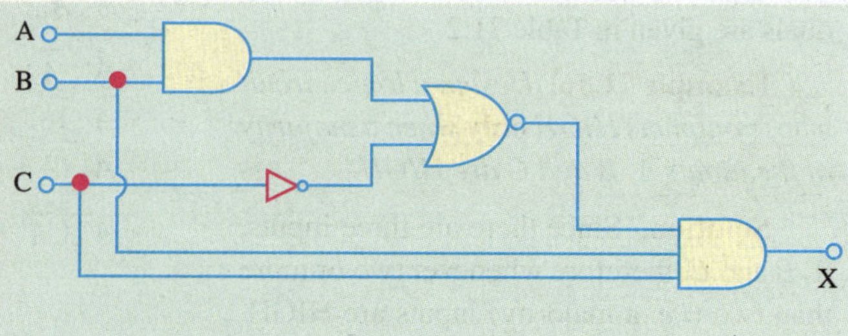

Fig. 71.17

Solution. The output of the given circuit can be obtained by determining the output of each logic gate while working from left to right.

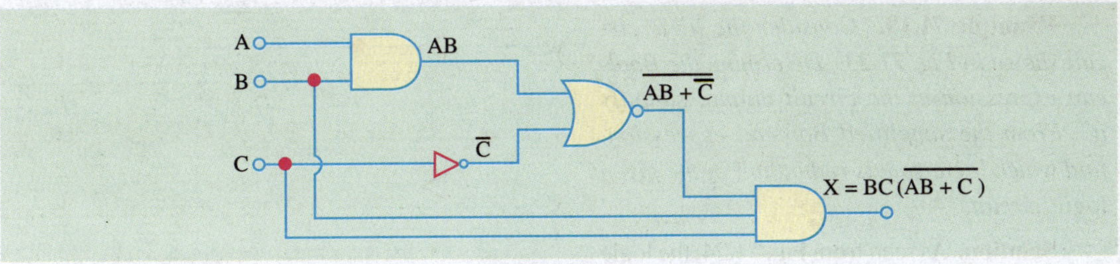

Fig. 71.18

As seen from the logic circuit shown in Fig. 71.18. The output of the circuit is,

$$X = BC\ \overline{(AB + \overline{C})}$$

The output, X can be simplified by De Morganizing the term $\overline{(AB + \overline{C})}$ as shown below.

$$\begin{aligned}
BC\ \overline{(AB + \overline{C})} &= BC\,\overline{(AB + \overline{C}\,)} && \text{... step 1}\\
&= BC\,\overline{(A + B)}\,.\,\overline{\overline{C}} && \text{... step 2}\\
&= BC\ (\overline{A} + \overline{B})\,.\,\overline{\overline{C}} && \text{... step 3}\\
&= BC\ (\overline{A} + \overline{B})\ C && \text{... Law 13}\\
&= BC\ (\overline{A} + \overline{B}) && \text{... Law 7}\\
&= \overline{A}\,BC + BC\ \overline{B} &&\\
&= \overline{A}\,BC + 0 && \text{... Law 8}\\
&= \overline{A}\,BC && \text{... Law 1}
\end{aligned}$$

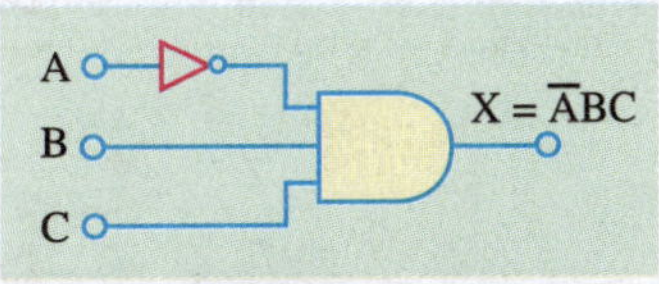

Fig. 71.19

The logic circuit with a simplified Boolean expression $X = \overline{A}\ BC$ is as shown in Fig. 71.19.

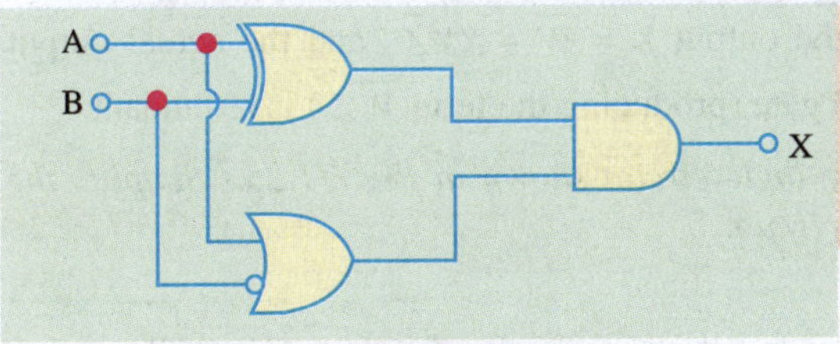

Fig. 71.20

Example 71.18. *Determine the output X of a logic circuit shown in Fig. 71.20. Simplify the output expression using Boolean Laws and theorems. Redraw the logic circuit with the simplified expression.*

Solution. The output of the given logic circuit can be obtained by determining the output of each logic gate while working from left to right.

As seen from Fig. 71.21, the output,

$$\begin{aligned}
X &= (\overline{A}B + A\overline{B})\,(A + \overline{B})\\
&= \overline{A}\,BA + A\,\overline{B}\,A + \overline{A}\,B\,\overline{B} + A\,\overline{B}\ \overline{B}\\
&= 0 + A\,\overline{B}\,A + 0 + A\,\overline{B}\ \overline{B} && \text{... Law 8}\\
&= A\,\overline{B} + A\,\overline{B} && \text{... Law 7 and Law 1}\\
&= A\,\overline{B} && \text{... Law 3}
\end{aligned}$$

Using the simplified Boolean expression, the logic circuit is as shown in Fig. 71.22.

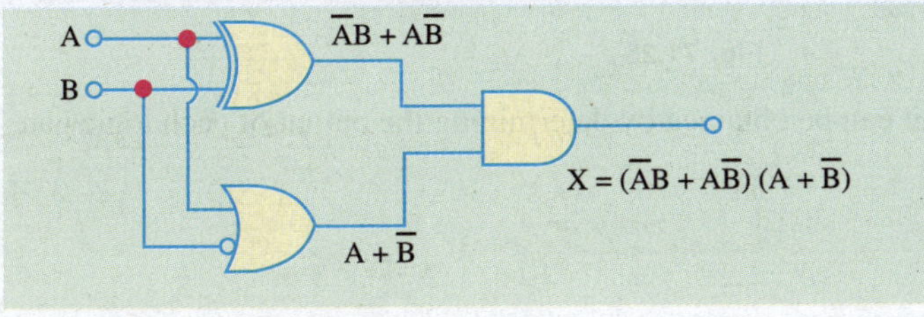

Fig. 71.21

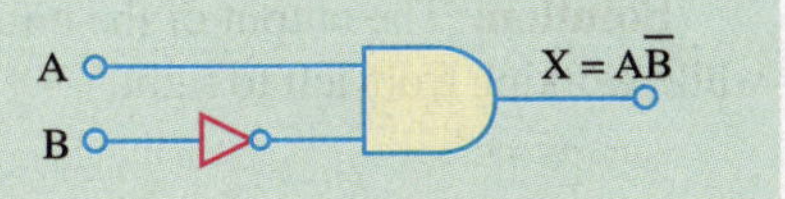

Fig. 71.22

Example 71.19. *Consider the logic circuit shown in Fig. 71.23. Determine the Boolean expression at the circuit output, simplify it. From the simplified Boolean expression, find which logic gate is redundant in the given logic circuit.*

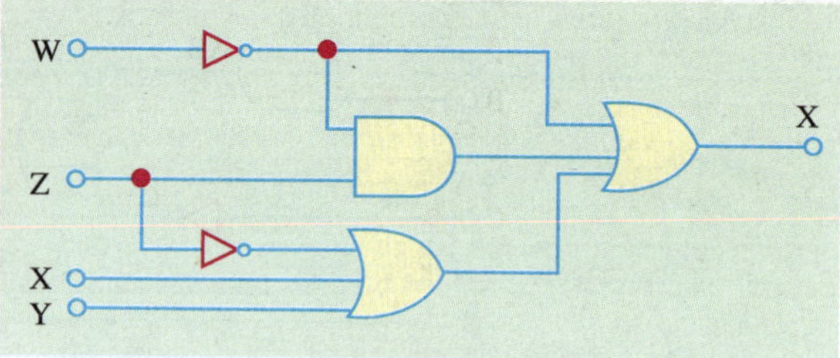

Fig. 71.23

Solution. As seen from Fig. 71.24, the logic circuit output,

$$X = \overline{W} + \overline{W}Z + XY\overline{Z}$$

The expression can be simplified as follows :

$$= \overline{W} + \overline{W}Z + XY\overline{Z}$$

$$= \overline{W}\,(1 + Z) + XY\overline{Z} \qquad \text{—Law 19}$$

$$= \overline{W} + XY\overline{Z} \qquad \text{—Law 2}$$

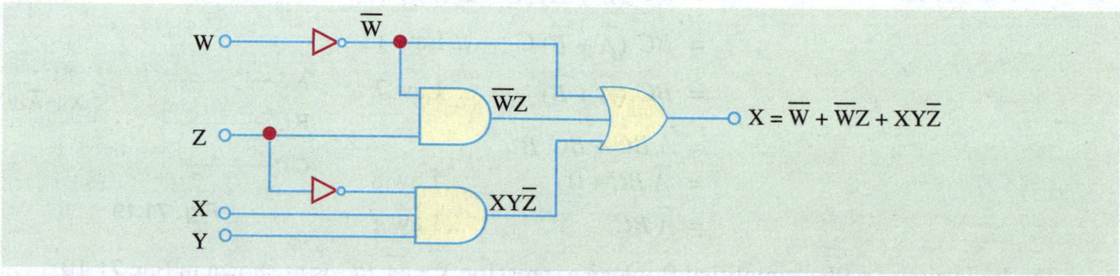

Fig. 71.24

From the simplified Boolean expression of the output $X = \overline{W} + XY\overline{Z}$, and the actual output $\overline{W} + \overline{W}Z + XY\overline{Z}$, we find that the two-input *AND* gate (producing the term $\overline{W}Z$) is redundant.

Example 71.20. *Determine the output of the logic circuit shown in Fig. 71.25. Simplify the output Boolean expression and sketch the logic circuit.*

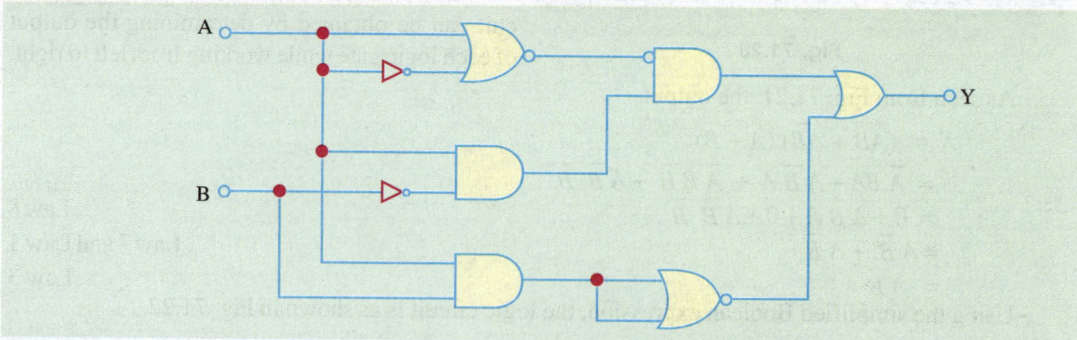

Fig. 71.25

Solution. The output of the circuit can be obtained by determining the output of each logic gate while working from left to right.

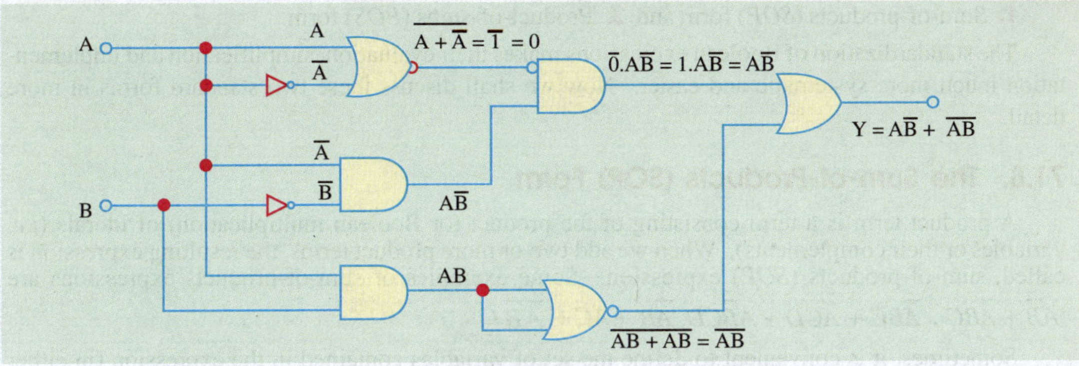

Fig. 71.26

As seen from the circuit shown in Fig. 71.26, we find that the output,

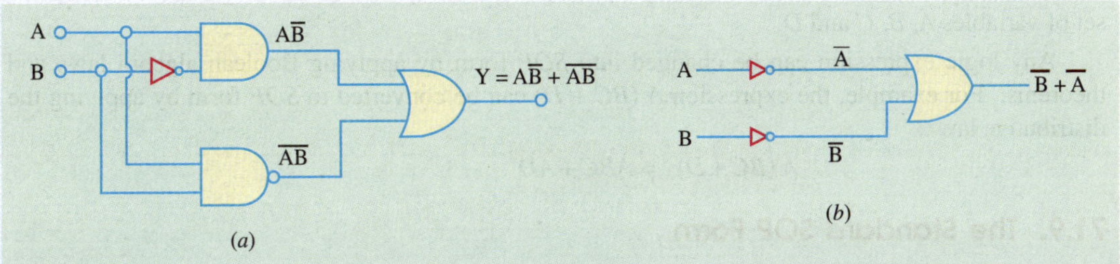

Fig. 71.27

$$y = A\overline{B} + \overline{AB}$$

The sketch of a logic circuit for the simplified Boolean expression is as shown in Fig. 71.27.

Alternatively :

$$\begin{aligned} y &= A\overline{B} + \overline{AB} \\ &= A\overline{B} + (\overline{A} + \overline{B}) \\ &= (A\overline{B} + \overline{B}) + \overline{A} \\ &= (A + 1)\,\overline{B} + \overline{A} \\ &= \overline{B} + \overline{A} \end{aligned}$$

The logic circuit to implement this logic equation is as shown in Fig. 71.27(*b*). Notice the difference in terms of the number and type of logic gates used in the circuits shown in Fig. 71.27. So when you are simplifying and designing logic circuits, it is always possible to have more than one solution.

The circuit shown in Fig. 71.27 shows that it makes use of one inverter (or NOT gate), one AND gate, one NAND gate and one OR gate. In other words, there are four different types of logic gates. However the logic circuit of Fig. 71.27 (*b*) makes use of only three logic gates (two inverters and one OR gate).

71.7. Standard Forms of Boolean Expressions

All Boolean expressions, regardless of their form, can be converted into either of the two following standard forms :

1. Sum-of-products (*SOP*) form and **2.** Product-of-sums (*POS*) form.

The standardization of Boolean expressions makes their evaluation, simplification and implementation much more systematic and easier. Now we shall discuss these two standard forms in more detail.

71.8. The Sum-of-Products (SOP) Form

A product term is a term consisting of the product (or Boolean multiplication) of literals (*i.e.* variables or their complements). When we add two or more product terms, the resulting expression is called, sum-of-products (*SOP*) expression. Some examples of sum-of-products expressions are $A\overline{B} + \overline{A}BC$, $\overline{A}BC + A\overline{C}D + A\overline{B}CD$, $\overline{A}B + AC + \overline{A}\,\overline{B}\,C$

Sometimes, it is convenient to define the set of variables contained in the expression (in either complemented or uncomplemented form) as a ***domain***. For example, domain of the expression $A\overline{B}$ + AB is the set of variables *A* and *B*. Similarly the domain of the expression $\overline{A}\,BC + ABC + A\,\overline{B}\,CD$ is the set of variables *A, B, C* and *D*.

Any logic expression can be changed into *SOP* form by applying Boolean algebra laws and theorems. For example, the expression *A* (*BC* + *D*) can be converted to *SOP* form by applying the distributive law :

$$A\,(BC + D) \;=\; ABC + AD$$

71.9. The Standard SOP Form

So far, we have seen *SOP* (sum-of-products) expressions in which some of the product terms do not contain all the variables in the domain of the expression. For example, the expression, $\overline{A}\,BC$ + $A\,\overline{C}\,D + A\,\overline{B}\,CD$ has a domain made up of the variables *A, B, C* and *D*. But notice that the first two terms contains only three variables, *i.e. D* or $\overline{D}$ is missing from the first term and *B* or $\overline{B}$ is missing from the second term.

A standard *SOP* expression is defined as an expression in which all the variables in the domain apper in each product term. For example $\overline{A}\,BCD + AB\,\overline{C}\,D + A\,\overline{B}\,CD$ is a standard *SOP* expression. Standard *SOP* expressions are important in constructing truth tables and in karnaugh map simplification method.

It is very straightforward to convert non-standard product term to a standard *SOP* using Boolean algebra. Each product term in the *SOP* expression that does not contain all the variables in the domain has to be expanded to standard form to include all the variables in the domain and their complements. As stated below, a nonstandard *SOP* expression is converted into standard form using Boolean algebra law 4 : $A + \overline{A} = 1$ *i.e.* a variable added to its complement equals 1.

Step 1. Multiply each nonstandard product term by a term made up of the sum of a missing variable and its complement. This results in two product terms. It is possible because we know that we can multiply anything by 1 without changing its value.

Step 2. Repeat step 1 until all resulting product terms contain all variables in the domain in either complemented or uncomplemented form. Note that in converting a product term to a standard form, the number of product terms is doubled for each missing variable.

For example, suppose we want to convert the Boolean expression $\overline{A}\,BC + A\,\overline{C}\,D + A\,\overline{B}\,CD$ to a standard *SOP* form. Then following the above procedure we proceed as below :

$$\overline{A}\,BC\,(D + \overline{D}) + A\,(B + \overline{B})\,\overline{C}\,D + A\,\overline{B}\,CD$$

$$= \overline{A}\,BCD + \overline{A}\,BC\,\overline{D} + AB\,\overline{C}\,D + A\,\overline{B}\,\overline{C}\,D + A\,\overline{B}\,CD$$

The expression given above is a standard *SOP* expression.

71.10. The Product-of-sums (POS) Form

The sum term is a term consisting of the sum (or Boolean addition) of literals (*i.e.* variables or their complements). When we multiply two or more sum terms, the resulting expression is called product-of-sums (*POS*). Some examples of *POS* form are $(A + \overline{B})(\overline{A} + B + C)$, $(\overline{A} + B + C)(A + \overline{C} + D)(A + \overline{B} + C + D)$ and $(\overline{A} + \overline{B})(A + C)(\overline{A} + B + C)$.

It may be carefully noted that a *POS* expression, can contain a single-variable term as in $\overline{A}\ (A + \overline{B} + C)(B + \overline{C} + \overline{D})$. In *POS* expression, a single overbar cannot extend over more than one variable, although more than one variable in a term can have an overbar.

71.11. The Standard POS Form

So far, we have seen *POS* (product-of-sums) expressions in which some of the sum terms do not contain all the variables in the domain of the expression. For example, the expression, $(\overline{A} + B + C)$, $(A + \overline{C} + D)(A + \overline{B} + C + D)$ has a domain made up of variables, *A, B, C* and *D*. Notice that the complete set of variables in the domain is not represented in the first two terms of the expression, *i.e.* D or $\overline{D}$ is missing in the first term and B or $\overline{B}$ is missing in the second term.

A standard *POS* expression is defined as an expression in which all the variables in the domain appear in each sum term. For example $(\overline{A} + B + C + D)(A + B + \overline{C} + D)(A + \overline{B} + C + D)$ is a standard *POS* form. Any nonstandard *POS* expression can be converted to a standard form using Boolean algebra.

Each sum term in an *POS* expression that does not contain all the variables in the domain can be expanded to standard form to include all variables in the domain and their complements. As stated below : a nonstandard *POS* expression is converted into a standard form using Boolean algebra Law 8 : $A \cdot \overline{A} = 0$, *i.e.* a variable multiplied by its complemented equals 0.

Step 1. Add to each nonstandard product term a term made up of the ptoduct of a missing variable and its complement. This results in two sum terms. This is possible because we know that we can add 0 to anything without changing its value.

Step 2. Apply law 20, *i.e.* $A + BC = (A + B)(A + C)$

Step 3. Repeat Step 1 until all resulting sum terms contain all variables in the domain in either complemented or uncomplemented form.

For example we want to convert the Boolean expression,

$$(\overline{A} + B + C)(A + \overline{C} + D)(A + \overline{B} + C + D)$$

into a standard *POS* form. Then following the above prodcedure, we proceed as below :

$$(\overline{A} + B + C + D\overline{D})(A + B\overline{B} + \overline{C} + D)(A + \overline{B} + C + D)$$

$$= (\overline{A} + B + C + D)(\overline{A} + B + C + \overline{D})(A + B + \overline{C} + D)(A + \overline{B} + \overline{C} + D)(A + \overline{B} + C + D)$$

The expression given above is a standard *POS* expression

71.12. The Karnaugh Map

The Karnaugh map (or simply a *K*-map) is similar to a truth table because it presents all the possible values of input variables and the resulting output for each value. However, instead of being organised into columns and rows like a truth table, the Karnaugh map is an array of squares (or cells) in which each square represents a binary value of the input variables. The squares are arranged in a way so that simplification of given expression is simply a matter of grouping the squares. Karnaugh maps can be used for expression with two-three, four, and five variable Karnaugh maps to illustrate the principles. Karnaugh map with five-variables is beyond the scope of this book. For higher number of variables, a Quine-McClusky method can be used. This method is also beyond the scope of this book.

The number of squares in a Karnaugh map is equal to the total number of possible input variable combinations (as is the number of rows in a truth table). For two variables, the number of square is $2^2 = 4$, for three variables, the number of squares is $2^3 = 8$ and for four variables, the number of squares is $2^4 = 16$.

71.13. The Two-variable Karnaugh Map

Fig. 71.28 (*a*) shows a two-variable Karnaugh map. As seen, it is an array of four squares. In this cass, *A* and *B* are used for two variables although any other two letters could be used. The binary values of *A* (*i.e.* 0 and 1) are indicated along the left side as $\overline{A}$ and *A* (notice the sequence) and the binary values of *B* are indicated across the top as $\overline{B}$ and *B*. The value of a given square is the value of *A* at the left in the same row combined with the value of *B* at the top in the same column. For example, a square in the upper left corner has a value of $\overline{A}\ \overline{B}$ and a square in the lower right corner has a value of *AB*. Fig. 71.28 (*b*) shows the standard product terms represented by each square in the Karnaugh map.

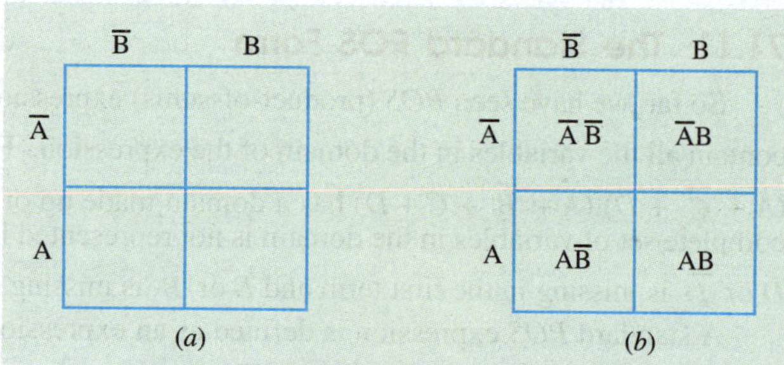

Fig. 71.28

71.14. The Three-variable Karnaugh Map

Fig. 71.29 (*a*) shows a three-variable Karnaugh map. As seen it is an array of eight squares. In this case, *A*, *B* and *C* are used for the variables although any other three letters could be used. The value of *A* and *B* are along the left side (notice the sequence carefully) and the values of *C* are across the top.

The value of a given square is the values of *A* and *B* at the left in the same row combined with the value of *C* at the top in the same column. For example, a square in the upper left corner has a value of $\overline{A}\ \overline{B}\ \overline{C}$ and a square in the bottom right corner has a value of $A\ \overline{B}\ C$. Fig. 71.29 (*b*) shows the product terms that are represented by each square in the Karnaugh map.

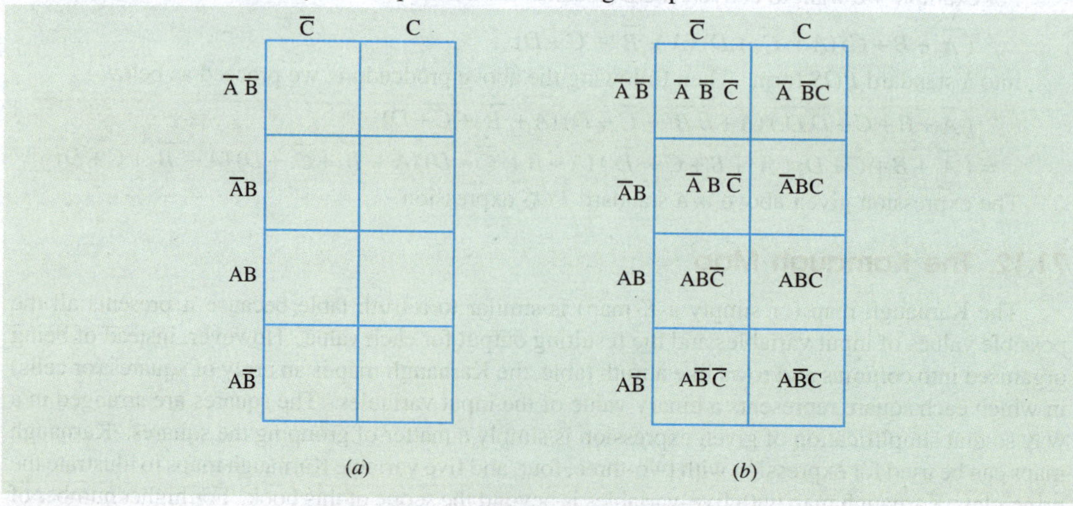

Fig. 71.29

71.15. The Four-variable Karnaugh Map

Fig. 71.30 (*a*) shows a four-variable Karnaugh map. As seen, it is an array of sixteen squares. In this case *A, B, C* and *D* are used for the variables. The values of *A* and *B* are along the left side, and the values of *C* and *D* are across the top. The sequence of the variable values may be noted carefully.

	$\overline{C}\,\overline{D}$	$\overline{C}\,D$	$C\,D$	$C\,\overline{D}$
$\overline{A}\,\overline{B}$				
$\overline{A}\,B$				
AB				
$A\overline{B}$				

(*a*)

	$\overline{C}\,\overline{D}$	$\overline{C}\,D$	$C\,D$	$C\,\overline{D}$
$\overline{A}\,\overline{B}$	$\overline{A}\,\overline{B}\,\overline{C}\,\overline{D}$	$\overline{A}\,\overline{B}\,\overline{C}\,D$	$\overline{A}\,\overline{B}\,C\,D$	$\overline{A}\,\overline{B}\,C\,\overline{D}$
$\overline{A}\,B$	$\overline{A}\,B\,\overline{C}\,\overline{D}$	$\overline{A}\,B\,\overline{C}\,D$	$\overline{A}\,B\,C\,D$	$\overline{A}\,B\,C\,\overline{D}$
AB	$A\,B\,\overline{C}\,\overline{D}$	$A\,B\,\overline{C}\,D$	$A\,B\,C\,D$	$A\,B\,C\,\overline{D}$
$A\overline{B}$	$A\,\overline{B}\,\overline{C}\,\overline{D}$	$A\,\overline{B}\,\overline{C}\,D$	$A\,\overline{B}\,C\,D$	$A\,\overline{B}\,C\,\overline{D}$

(*b*)

Fig. 71.30

The value of a given square is the values of *A* and *B* at the left in the same row combined with the values of *C* and *D* at the top in the same column. For example, a square in the upper right corner has a value $\overline{A}\,\overline{B}\,C\,\overline{D}$ and a square in the lower left corner has a value $A\,\overline{B}\,\overline{C}\,\overline{D}$. Fig. 71.30 (*b*) shows the standard product terms that are represented by each square in the four-variable Karnaugh map.

71.16. Square Adjacency in Karnaugh Map

We have already discussed a two-variable Karnaugh map, a three-variable Karnaugh map and a four-variable Karnaugh map. Now we shall discuss the concept of square adjacency in a Karnaugh map.

It will be interesting to know that the squares in a Karnaugh map are arranged in such a way that there is only a single-variable change between adjacent squares. Adjacency is defined as a single-variable change. It means the squares that differ by only one variable are adjacent. For example, in a three-variable Karnaugh map shown in Fig. 71.29 (*b*), the $\overline{A}\,B\,\overline{C}$ square is adjacent to $\overline{A}\,\overline{B}\,\overline{C}$ square, the $\overline{A}BC$ square and the $AB\overline{C}$ square. It may be carefully noted that square with values that differ by more than one variable are not adjacent. For example, the $\overline{A}\,B\,\overline{C}$ square is not adjacent to the $\overline{A}\,\overline{B}\,C$ square, the *ABC* square, the $A\,\overline{B}\,\overline{C}$ square or the $A\,\overline{B}\,C$ square. In other words, each square is adjacent to the squares that are immediately next to it on any of its four sides. However, a square is not adjacent to the squares that diagonally touch any of its corners.

It may also be noted that squares in the top row are adjacent to the corresponding squares in the bottom row and squares in the outerleft column are adjacent to the corresponding squares in the outer right column. This is called "**wrap around**" adjacency because we can think of the map as wrapping around from top to bottom to form a cylinder or from left to right to form a cylinder. Fig. 71.31 (*a*) and (*b*) shows the square adjacencies with a three-variable and a four-variable Karnaugh maps respectively.

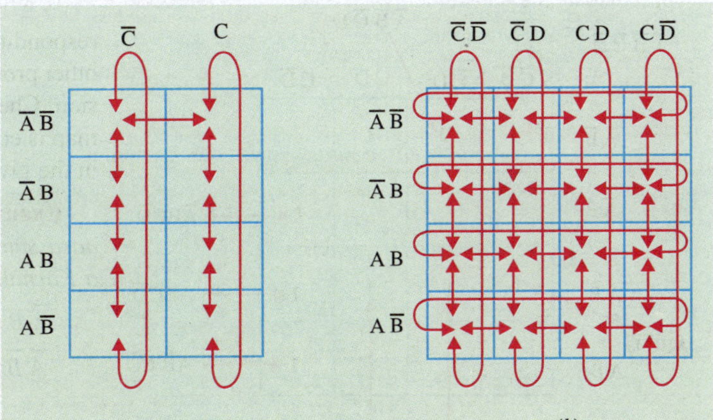

(*a*) (*b*)

Fig. 71.31

Notice the square adjacencies in a four variable Karnaugh map :

Here for example, the square $\overline{A}\,\overline{B}\,\overline{C}\,\overline{D}$ is adjacent to $\overline{A}\,\overline{B}\,\overline{C}\,D$ square, $\overline{A}\,B\,\overline{C}\,D$ square, $A\,\overline{B}\,\overline{C}\,\overline{D}$ square and $\overline{A}\,\overline{B}\,C\,\overline{D}$ square. Similarly $\overline{A}\,B\,\overline{C}\,D$ square is adjacent to $\overline{A}\,\overline{B}\,\overline{C}\,D$ square, $\overline{A}\,B\,\overline{C}\,\overline{D}$ square, $\overline{A}BCD$ square and $AB\overline{C}D$ square.

71.17. Mapping a Standard SOP Expression on the Karnaugh Map

Consider a *SOP* (sum-of-products) expression. $\overline{A}\,\overline{B}\,\overline{C} + \overline{A}\,B\,\overline{C} + A\overline{B}C + A\,\overline{B}\,\overline{C}$. In order to map this expression on the Karnaugh map, we need a three variable Karnaugh-map because the given expression has three variables *A, B,* and *C*. Then select the first product term $\overline{A}\,\overline{B}\,\overline{C}$ and enter 1 in the corresponding square (*i.e.* the first row and the first column) as shown in Fig. 71.32.

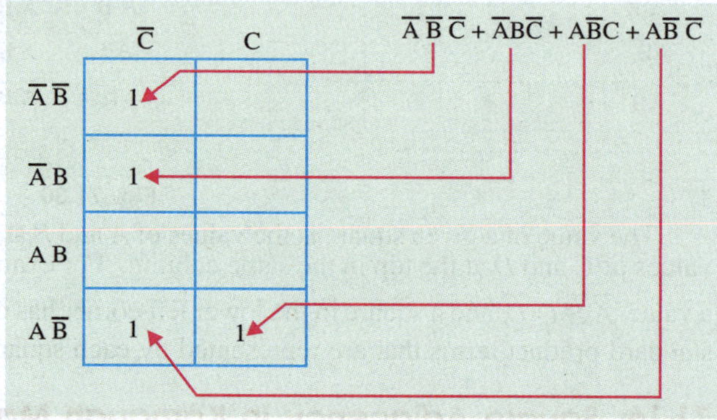

Fig. 71.32

Similarly, for the second product term, $\overline{A}\,B\,\overline{C}$ place a 1 in the second row and first column. Repeat this process for the other two product terms, *i.e.* $A\,\overline{B}\,C$ and $A\,\overline{B}\,\overline{C}$. The squares that do not have 1 are the squares for which the expression is 0. Usually when working with sum-of-products expressions, the 0*s* are left off the map.

Example 71.21. *Map the following SOP expression on the Karnaugh map :* $\overline{A}\,\overline{B}\,C + \overline{A}BC + AB\overline{C} + ABC$.

Solution. Sketch a three variable Karnaugh map as shown in Fig. 71.33. Select the first product term $\overline{A}\,\overline{B}\,C$ and enter 1 in the corresponding square. Similarly enter 1 for the other product terms in the given *SOP* expression. Check that number of 1*s* in the Karnaugh map is equal to the number of product terms in the given *SOP* expression.

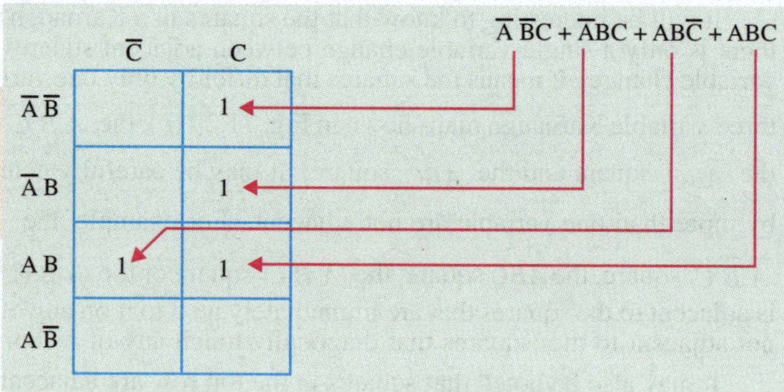

Fig. 71.33

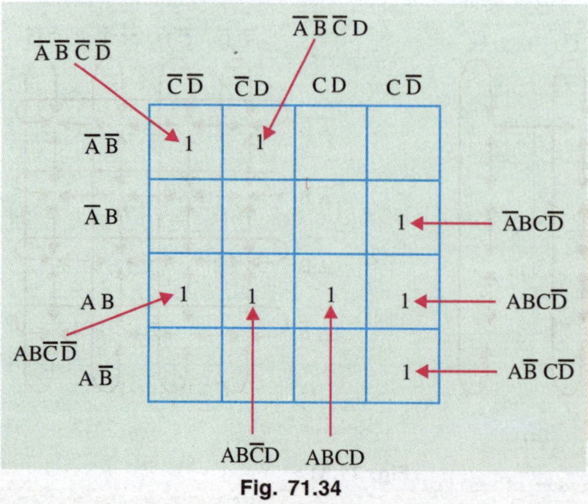

Fig. 71.34

Example 71.22. *Map the following standard sum-of-products (SOP) expression on a Karnaugh map :*

$$\overline{A}\,B\,C\,\overline{D} + A\,B\,C\,\overline{D} + A\,B\,\overline{C}\,\overline{D} + \overline{A}\,\overline{B}\,\overline{C}\,\overline{D} + A\,B\,\overline{C}D + \overline{A}\,\overline{B}\,\overline{C}\,D + A\,\overline{B}\,C\,\overline{D} + ABCD$$

Solution. Sketch a four variable Karnaugh map. Select the first product term $\overline{A}\,B\,C\,\overline{D}$ from the given *SOP* expression and enter 1 in the corresponding square as shown in Fig. 71.34. Similarly enter 1 for the other product terms in the given *SOP* expression. Check that number of 1*s* in the Karnaugh Map is equal to the number of product terms in the given *SOP* expression.

71.18. Mapping a Nonstandard SOP Expression on the Karnaugh Map

A nonstandard sum-of-products (*SOP*) expression is a one that has product terms with one or more missing variables. In such a case, Boolean expression must first be converted to a standard form by a procedure explained in Art. 71.9.

Let us consider an example to illustrate the procedure for mapping a nonstandard *SOP* expression on the Karnaugh map. Suppose we have the *SOP* expression :

$$\overline{A} + A\overline{B} + AB\overline{C}$$

As seen, this expression is obviously not in standard form because each product term does not have three variables. The first term is missing two variables, the second term is missing one variable and the third term is standard. In order to convert the given nonstandard *SOP* expression to a standard form, we multiply the first product term by $B + \overline{B}$ and $C + \overline{C}$, and the second term by $C + \overline{C}$. Expanding the resulting expression, we get,

$$\overline{A}\,(B+\overline{B})\,(C+\overline{C}) + A\overline{B}\,(C+\overline{C}) + AB\overline{C}$$

$$= \overline{A}BC + \overline{A}\,\overline{B}\,C + \overline{A}\,B\,\overline{C} + \overline{A}\,\overline{B}\,\overline{C} + A\overline{B}C + A\,\overline{B}\,\overline{C} + AB\overline{C}$$

Rearranging the expression for our convenience, we get

$$\overline{A}\,\overline{B}\,\overline{C} + \overline{A}\,\overline{B}\,C + \overline{A}\,B\,\overline{C} + \overline{A}BC + AB\overline{C} + A\,\overline{B}\,\overline{C} + A\overline{B}C$$

This expression can be mapped on a three-variable Karnaugh map as shown in Fig. 71.35.

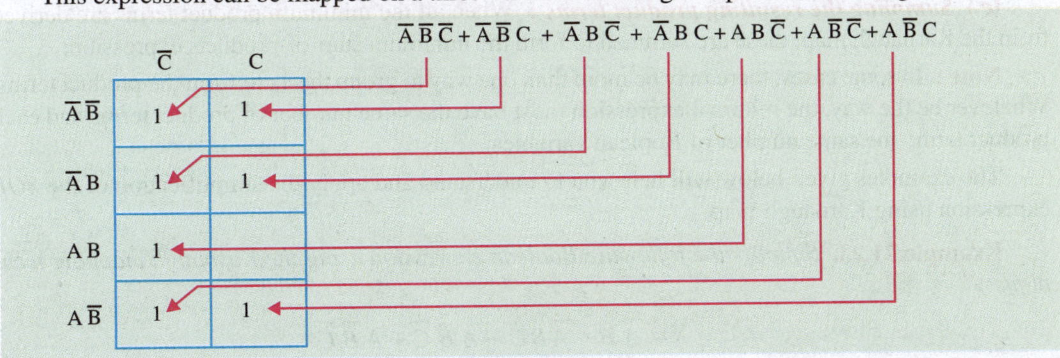

Fig. 71.35

71.19. Simplification of Boolean Expression Using Karnaugh Map

The process that results in an expression containing the minimum number of possible terms with the minimum number of variables is called **simplification** or **minimization.** After the *SOP* (or the Boolean) expression has been mapped on the Karnaugh map, there are three steps in the process of obtaining a minimum *SOP* expression. The three steps are : (*a*) grouping the 1*s*, (*b*) determining the product term for each group and (*c*) summing the resulting product terms.

(*a*) ***Grouping the 1s*** **:** We can group the 1*s* on the Karnaugh map according to the following rules by enclosing those adjacent squares containing 1*s*. The objective is to maximize the size of the groups and to minimize the number of groups.

1. A group must contain either 1, 2, 4, 8 or 16 squares. In the case of two-variable Karnaugh map, 4 squares is the maximum group, for three-variable map, 8 squares are the maximum group and so on.
2. Each square in the group must be adjacent to one or more squares in that same group but all squares in the same group do not have to be adjacent to each other.
3. Always include the largest possible number of 1*s* in a group in accordance with rule 1.
4. Each 1 on the Karnaugh map must be included in at least one group. The 1*s* already in a group can be included in another group as long as the overlapping groups include non-common 1*s*.

(b) Determining the Product Term for each group **:** Following are the rules that are applied to find the minimum product terms and the minimum sum-of-products expression :

1. ***Group the squares that have 1s*** **:** Each group of squares containing 1*s* creates one product term composed of all variables that occur in only one form (either uncomplemented or complemented) within the group. Variables that occur both uncomplemented and complemented within the group are eliminated. These are known as contradictory variables.
2. In order to determine the minimum product term for each group, we need to look at the standard methodology for three-variable and four-variable Karnaugh map respectively.
   - *(i)* ***For a three-variable K-map*** **:** **(1)** for 1-square, group we get a three-variable product term, **(2)** for a 2-square group, we get a two-variable product term, **(3)** for a 4-square product term, we get a one-variable product term.
   - *(ii)* ***For a four-variable K-map*** **:** **(1)** For a 1-square group, we get a four-variable product term, **(2)** for a 2-square group, we get a three-variable product term, **(3)** for a 4-square group, we get a two-variable product term and **(4)** for a 8-square group, we get a one-variable product term.

(c) Summing the resulting product terms **:** When all the minimum product terms are derived from the Karnaugh map, these are summed to form the minimum sum-of-products expression.

Note : In some cases, there may be more than one way to group the 1*s* to form the product terms. Whatever be the way, the minimal expression must have the same number of product terms and each product term, the same number of Boolean variables.

The examples given below will help you to understand and apply the simplification of the *SOP* expression using Karnaugh map.

Example 71.23. *Simplify the following Boolean expression using the Karnaugh mapping technique :*

$$X = \overline{A}\,B + \overline{A}\,\overline{B}\,C + A\,B\,\overline{C} + A\,\overline{B}\,\overline{C}$$

Solution. The first step is to map the given Boolean expression on the Karnaugh map. Notice that there are three variables *A, B* and *C* in the Boolean expression, therefore we need a three-variable Karnaugh map.

The Boolean expression to be mapped is,

$$X = \overline{A}\,B + \overline{A}\,\overline{B}\,C + A\,B\,\overline{C} + A\,\overline{B}\,\overline{C}$$

Note that the given Boolean expression is a nonstandard *SOP* expression because the first product term $\overline{A}\,B$ has the variable *C* missing in it. This can be converted into a standard *SOP* form by modifying the expression as below.

$$\begin{aligned} X &= \overline{A}\,B\,.1 + \overline{A}\,\overline{B}\,C + A\,B\,\overline{C} + A\,\overline{B}\,\overline{C} \\ &= \overline{A}\,B\,(C+\overline{C}) + \overline{A}\,\overline{B}\,C + A\,B\,\overline{C} + A\,\overline{B}\,\overline{C} && \ldots(C+\overline{C}=1) \\ &= \overline{A}\,B\,C + \overline{A}\,B\,\overline{C} + \overline{A}\,\overline{B}\,C + A\,B\,\overline{C} + A\,\overline{B}\,\overline{C} && \ldots(1) \end{aligned}$$

Equation (1) can be mapped on the Karnaugh map as shown in Fig. 71.36. In order to simplify the given expression, the 1's can be grouped together as shown by the loop around the 1's. The two 1's in the first column are grouped together and the term we get is $A\overline{C}$. This is because of the fact that the squares within this group contain both B and $\overline{B}$, so these variables are eliminated. Similarly, the two 1's in the second column are grouped together and the term we get is $\overline{A}C$. This is because of the fact that squares in this group contain both B and $\overline{B}$ which is eliminated. similarly, the two 1's in the second row are grouped together and the term we get is $\overline{A}B$. This is because of the fact that squares in this group conteins both C and $\overline{C}$ which is eliminated. Summing up the three product term, the simplified expression is

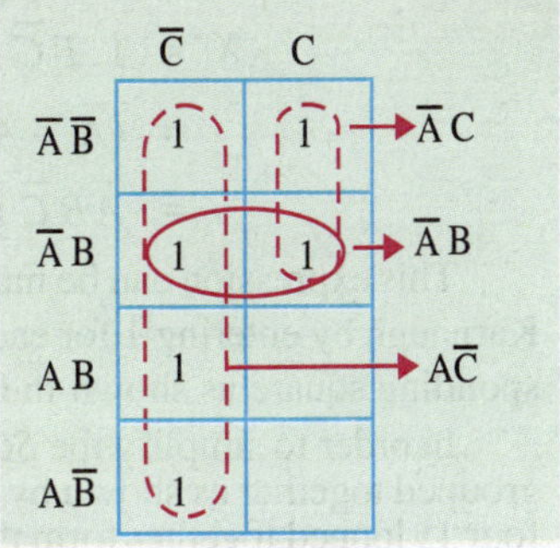

Fig. 71.36

$$X = \overline{A}B + \overline{A}C + A\overline{C}.$$

Example 71.24. *Simplify the following SOP expression using the Karnaugh mapping procedure :*

$$X = \overline{A}B\overline{C}D + A\overline{B}\overline{C}D + \overline{A}\overline{B}\overline{C}D + AB\overline{C}D + AB\overline{C}\overline{D} + ABCD$$

Solution. First of all, notice that the given *SOP* expression is already in the standard from *i.e.* all the product terms in the given expression have all the four variables *A, B, C* and *D*.

Next sketch a four-variable Karnaugh map. Select the first product term $(\overline{A}B\overline{C}D)$ from the given expression and enter 1 in the corresponding square as shown in Fig. 71.37. Similarly enter 1 for the other product terms in the given *SOP* expression to complete the mapping. In order to simplify the given *SOP* expression, the 1*s* can be grouped together as shown by the loop around the 1*s*. The four 1*s* in the second column are grouped together and the product term we get is $\overline{C}D$. This is because of the fact that squares within this group contain both A and $\overline{A}$ and B and $\overline{B}$, so these variables are eliminated.

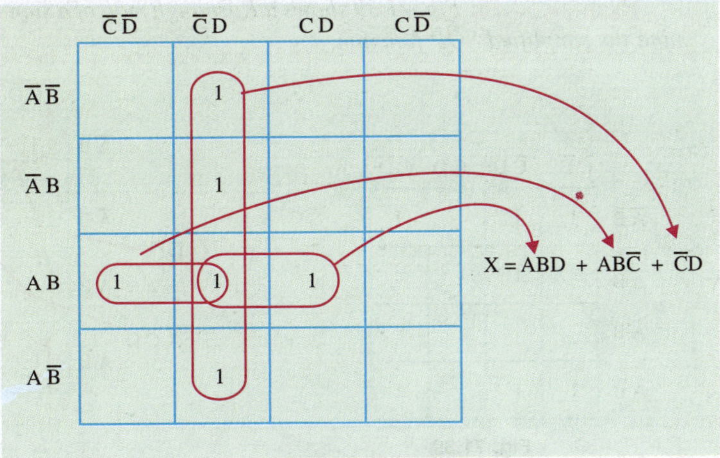

Fig. 71.37

The two 1*s* in the first and second columns can be grouped together. This group contains both D and $\overline{D}$, so this variable is eliminated and the resulting product term is $AB\overline{C}$. Similarly the two 1*s* in the second and third columns can be grouped together. This group contains both C and $\overline{C}$, so this variable is eliminated and the resulting product term is ABD.

The resulting minimal or simplified *SOP* expression is obtained by summing up the three product terms $\overline{C}D$, $AB\overline{C}$ and ABD as shown below :

$$X = ABD + AB\overline{C} + \overline{C}D.$$

Example 71.25. *Simplify the following SOP expression using the Karnaugh mapping technique.*

$$X = B\overline{C}\overline{D} + \overline{A}B\overline{C}D + AB\overline{C}D + \overline{A}BCD + ABCD$$

Solution. First of all, notice that the given *SOP* expression is in the nonstandard *SOP* form because the first product term $(B\overline{C}\overline{D})$ has a variable A or $\overline{A}$ missing in it. Let us convert the given *SOP* expression into a standard *SOP* form as shown below :

$$X = 1.B\overline{C}\,\overline{D} + \overline{A}\,B\,\overline{C}\,D + AB\overline{C}D + \overline{A}BCD\ ABCD$$

$$= (\overline{A} + A)\,B\,\overline{C}\,\overline{D} + \overline{A}\,B\,\overline{C}\,D + AB\overline{C}D + \overline{A}BCD + ABCD \qquad ...(\because \overline{A} + A = 1)$$

$$= \overline{A}B\,\overline{C}\,\overline{D} + A\,B\,\overline{C}\,\overline{D} + \overline{A}B\overline{C}D + AB\overline{C}D + \overline{A}BCD + ABCD$$

This expression can be mapped on to the four-variable Karnaugh by entering 1 for each product term in the corresponding square as shown in Fig. 71.38.

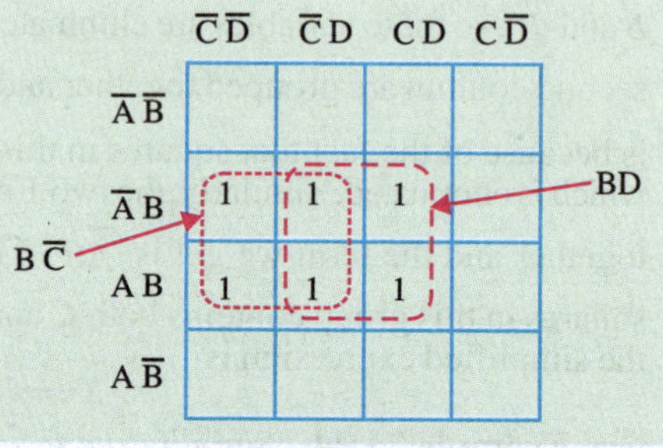

Fig. 71.38

In order to simplify the *SOP* expression, the 1*s* can be grouped together as shown by the loop around the 1*s*. The four 1*s* looped together form the first and second cloumns, contain both A and $\overline{A}$ and D and $\overline{D}$, so these variables are eliminated and the resulting product term is $B\,\overline{C}$. Similarly, the four 1*s* looped together form the second and the third column contain both A and $\overline{A}$ and C and $\overline{C}$, so these variables are eliminated and the resulting product term is BD. The resulting simplified *SOP* expression is the sum of the product terms $B\,\overline{C}$ and BD, *i.e.*,

$$X = B\overline{C} + BD$$

Example 71.26. *Fig. 71.39 shows a Karnaugh map of a sum-of-products (SOP) function. Determine the simplified SOP function.* **(UPSC Civil Services 2000)**

	$\overline{C}\,\overline{D}$	$\overline{C}D$	CD	$C\overline{D}$
$\overline{A}\,\overline{B}$	1	1		1
$\overline{A}B$		1		
AB				
$A\overline{B}$	1	1		1

Fig. 71.39

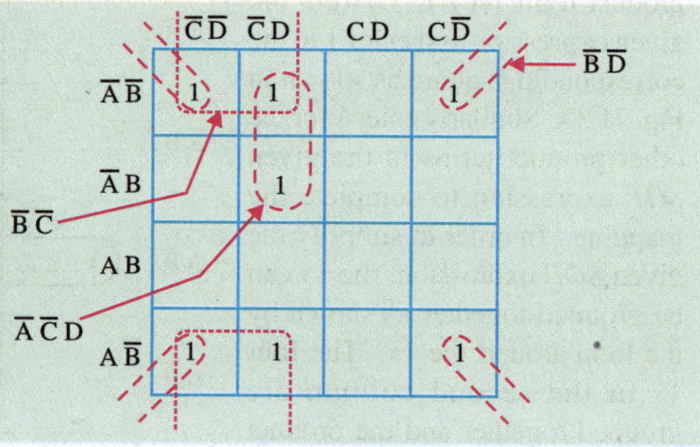

Fig. 71.40

Solution. The grouping of 1*s* is as shown in the Fig. 71.40. Notice the "wrap around" four-square group that includes the 1*s* on four corners of the Karnaugh map. This group produces a product term $\overline{B}\,\overline{D}$. This is determined by observing that the group contains both A and $\overline{A}$ and C and $\overline{C}$, so these variables are eliminated.

Another group of four with "wrap-around" adjacency is formed form the top and the bottom rows of the Karnaugh map. This group overlaps with the previous group and produces a product term $\overline{B}\,\overline{C}$. This is determined by observing that this group contains both A and $\overline{A}$ and D and $\overline{D}$, so these variables are eliminated.

The remaining 1 is absorbed in a overlapping group of two squares. This group produces a three-variable term $\overline{A}\,\overline{C}D$. This is determined by observing that this group contains both B and $\overline{B}$, so this variable is eliminated. This resulting simplified *SOP* function is the sum of the product terms $\overline{B}\,\overline{D}$ + $\overline{B}\,\overline{C}$ and $\overline{A}\,\overline{C}D$, *i.e.*

$$X = \overline{B}\,\overline{D} + \overline{B}\,\overline{C} + \overline{A}\,\overline{C}D$$

71.20. Mapping Directly on Karnaugh Map from a Truth Table

It is possible to map directly on Karnaugh map from a truth table. Recall that a truth table gives the output of a Boolean expression for all possible input variable combinations. Let us illustrate direct mapping through an example of a Boolean expression and its truth table representation.

Let $X = \bar{A}\bar{B}C + \bar{A}BC + A\bar{B}C + AB\bar{C}$. Then its truth table can be indicated as shown in Fig. 71.41 (*a*) and its Karnaugh mapping is shown in Fig. 71.41 (*b*). Notice in the truth table that the output *X* is 1 for four different input variable combinations.

It is evident from the Fig. 71.41 (*a*) and (*b*) that truth table and Karnaugh map are simply different ways to represent a logic function.

Inputs A	B	C	Output X
0	0	0	0
0	0	1	1
0	1	0	0
0	1	1	1
1	0	0	0
1	0	1	1
1	1	0	1
1	1	1	0

(*a*)

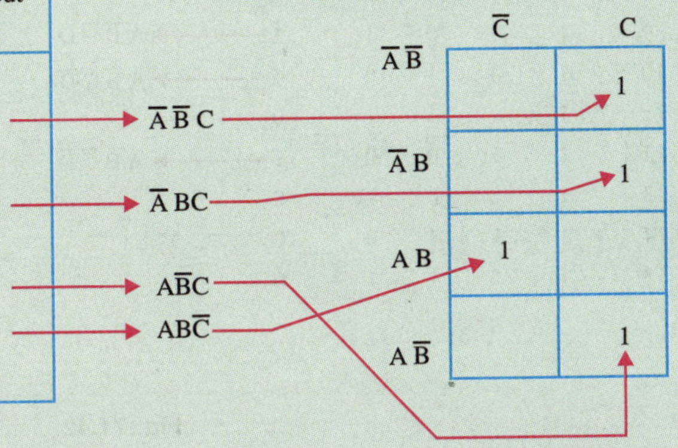

(*b*)

Fig. 71.41

Example 71.27. *Implement the following Boolean expression using minimum number of 3-input NAND gates.*

$$f(A, B, C, D) = \Sigma(1, 2, 3, 4, 7, 9, 10, 12) \qquad \text{(UPSC Engg. Services 1991)}$$

Solution. The given Boolean function indicates that its output is 1 corresponding to the terms indicated within the expression *i.e.*, 1, 2, 3, 4, 7, 9, 10 and 12. This is shown in Fig. 71.42 (*a*). We can map these values directly on to the four-variable Karnaugh map as shown in Fig. 71.42 (*b*). In order to simplify the Boolean expression represented on the Karnaugh map, group the 1*s* as shown in the Fig. 71.42 (*b*). The group of two squares in the first coloum produces a product term $B\bar{C}\bar{D}$. This is determined by observing that the group contains both *A* and $\bar{A}$, so this variable is eliminated. Another group of two squares in the second column produces term $\bar{B}\bar{C}D$. The variable *A* is eliminated because the group contains both *A* and $\bar{A}$.

Another group of two squares in the third column produces the term $\bar{A}\,\bar{C}D$. The variable *B* is eliminated because the group contains both *B* and $\bar{B}$. Still another group of two squares in the fourth column produces the term $\bar{B}C\bar{D}$. The variable *A* is eliminated because the group contains both *A* and $\bar{A}$. Thus the resulting simplified expression,

$$f(A, B, C, D) = B\bar{C}\bar{D} + \bar{B}\bar{C}D + \bar{A}CD + \bar{B}C\bar{D}$$

This can be implemented using 3-input NAND gate as shown in Fig. 71.43.

Decimal Number	Inputs A	B	C	D	Output f	
0	0	0	0	0	0	
1	0	0	0	1	1	→ $\bar{A}\bar{B}\bar{C}D$
2	0	0	1	0	1	→ $\bar{A}\bar{B}C\bar{D}$
3	0	0	1	1	1	→ $\bar{A}\bar{B}CD$
4	0	1	0	0	1	→ $\bar{A}B\bar{C}\bar{D}$
5	0	1	0	1	0	
6	0	1	1	0	0	
7	0	1	1	1	1	→ $\bar{A}BCD$
8	1	0	0	0	0	
9	1	0	0	1	1	→ $A\bar{B}\bar{C}D$
10	1	0	1	0	1	→ $A\bar{B}C\bar{D}$
11	1	0	1	1	0	
12	1	1	0	0	1	→ $AB\bar{C}\bar{D}$
13	1	1	0	1	0	
14	1	1	1	0	0	
15	1	1	1	1	0	

(*a*)

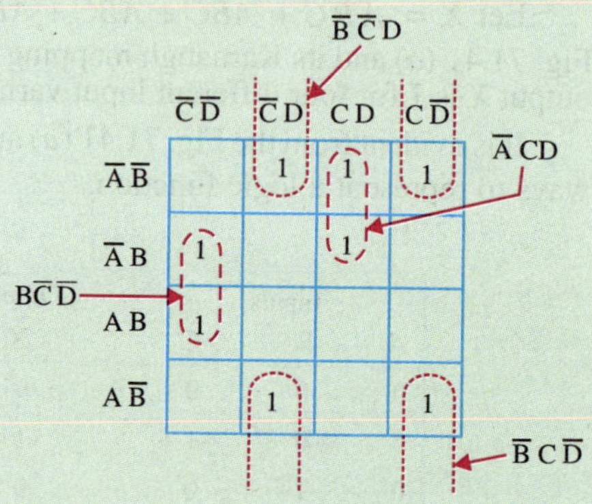

(*b*)

Fig. 71.42

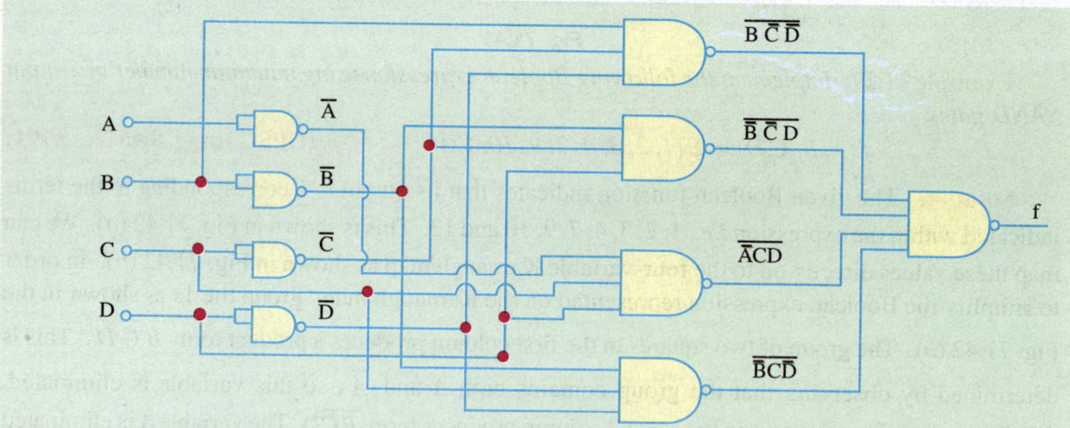

Fig. 71.43

71.21. "Don't Care" conditions

In digital systems design sometimes a situation arises in which some input variable conditions are not allowed. For example in a *BCD* (binary coded decimal) code, there are six invalid combinations : 1010, 1011, 1100, 1101, 1110 and 1111. Since these unallowed states will never occur in an application involving the *BCD* code, they can be treated as "don't care" terms with respect to their effect on the output. That is, for these "don't care" terms either 1 or 0 may be assigned to the output.

Now, we shall discuss as how the "don't care" terms can be used to advantage on the Karnaugh map for simplifying the Logic equations.

Consider for example, a combinational circuit which products a '1' output corresponding to a

BCD input equal and greater than 6. The output is 0 corresponding to a *BCD* input less than 6. The truth table for this situation is as shown in Fig. 71.44 (*a*).

Inputs				Output	
A	*B*	*C*	*D*	*X*	
0	0	0	0	0	
0	0	0	1	0	
0	0	1	0	0	
0	0	1	1	0	
0	1	0	0	0	
0	1	0	1	0	
0	1	1	0	1	→ $\overline{A}BC\overline{D}$
0	1	1	1	1	→ $\overline{A}BCD$
1	0	0	0	1	→ $A\overline{B}\,\overline{C}\,\overline{D}$
1	0	0	1	1	→ $A\overline{B}\,\overline{C}D$
1	0	1	0	*X*	
1	0	1	1	*X*	
1	1	0	0	*X*	
1	1	0	1	*X*	
1	1	1	0	*X*	
1	1	1	1	*X*	

(*a*)

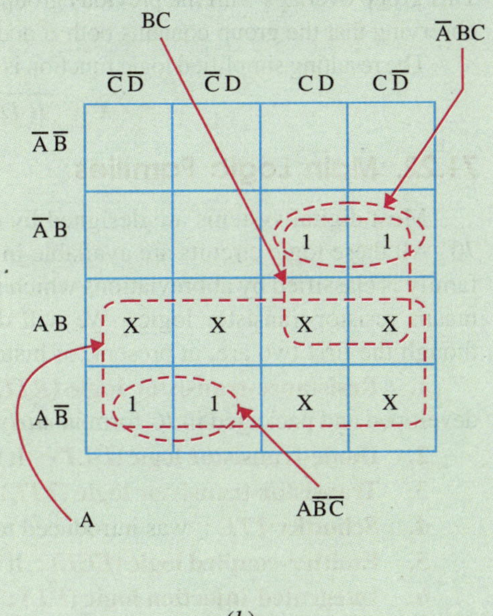

(*b*)

Fig. 71.44

We know that we can place 1*s* directly from the truth table on the Karnaugh map. Similarly we can place *X* for the "don't care" enteries directly on the Karnaugh map as shown in Fig. 71.44 (*b*). When grouping the 1*s*, *Xs* can be treated as 1*s* make a larger grouping or as 0*s*, if they cannot be used to advantage Recall the larger the group of 1*s* the simpler the resulting term will be. Taking advantage of the "don't care" and using them as 1*s*, the resulting expression for the output is $A + BC$. However, if the "don't cares" are not used as 1*s*, the resulting expression is $A\overline{B}\,\overline{C} + \overline{A}BC$. Thus we can see the advantage of using "dont care" terms to get the simplest logic expression.

	$\overline{C}\,\overline{D}$	$\overline{C}D$	CD	$C\overline{D}$
$\overline{A}\,\overline{B}$	1		X	X
$\overline{A}B$		1	X	1
AB		X	1	
$A\overline{B}$	1	X		X

Fig. 71.45

Example 71.28. *Consider the Karnaugh map shown in Fig. 71.45. Determine the logic function represented by the map and simplify it in the minimal form.* **(UPSC Engg. Services 1997)**

Solution. We know that when grouping the 1*s*, X*s* can be treated as 1*s* to make a larger grouping or as 0*s* if they cannot be used to advantage. Recall, the larger the group of 1*s*, the simpler the resulting term will be. Taking advantage of the *Xs* and using them as 1*s*, the grouping of 1*s* and *Xs* is as shown in Fig. 71.46.

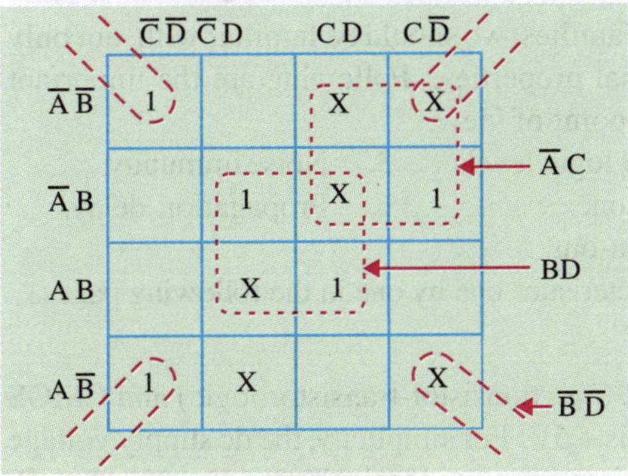

Fig. 71.46

Notice the "wrap-around" four-square group that includes the 1*s* and *Xs* on fours corners of the Karnaugh map. This group produces a product term $\overline{B}\,\overline{D}$. This is determined by observing that the group contains both *A* and $\overline{A}$ and *C* and $\overline{C}$, so these variables are eliminated. Another group of four squares containing 1*s* and *Xs* around the centre of Karnaugh map is formed. This group produces a product term *BD*. This is determined by observing that the group contains both *A* and $\overline{A}$ and *C* and $\overline{C}$, so these variables are eliminated.

Another group of four containing 1*s* and *Xs* is formed near the top right corner of Karnaugh map. This group overlaps with the previous group and produces a product term $\overline{A}\,C$. This is determined by observing that the group contains both *B* and $\overline{B}$ and *D* and $\overline{D}$, so these variables are eliminated.

The resulting simplified logic function is the sum of the three product terms : $\overline{B}\,\overline{D}$, *BD* and $\overline{A}\,C$, *i.e.*,

$$X = \overline{B}\,\overline{D} + BD + \overline{A}\,C$$

71.22. Main Logic Families

Most digital systems are designed by combining various logic functions discussed in Chapter 70. All these logic circuits are available in *IC* modules and are divided into many 'families'. Each family is classified by abbreviations which indicate the type of logic circuit used. For example, *RTL* means resistor-transistor logic. We will discuss the following seven transistor logic families although the first two are, at present, of historic interest only.

1. **Resistance-transistor logic (*RTL*) :** it was the first family group of logic circuits to be developed and packaged in *IC* form in *early* 1960*s*;
2. **Diode-transistor logic (*DLT*) :** It followed *RTL* in *late* 1960*s*;
3. **Transistor-transistor logic (*TTL*) *OR* (T^2L) :** was introduced in the early 1970 *s*;
4. **Schottky *TTL* :** was introduced to improve the speed of *TTL*;
5. **Emitter-coupled logic (*ECL*) :** It is fastest logic line currently available;
6. **Integrated-injection logic (I^2L) :** It is one of the latest of the bipolar types of logic;
7. **Complementary metal-oxide semiconductor (*CMOS*) :** It has the lowest power dissipation of the currently-available logic circuits.

The various logic families discussed above posses different characteristics as detailed below.

71.23. Saturated and Non-saturated Logic Circuits

Those logic circuits in which transistors are driven into saturation are called **saturated** logic circuits or simply **saturated logic.** Those circuits which avoid saturation of their transistors are designed **non-saturated logic.**

The disadvantage of saturated logic is the delay that occurs when the transistors is brought out of saturation. When a transistor is saturated, its base is flooded with carriers. Even when base voltage is switched off, the base remains flooded for some time till all carriers leave it. The time required by the carriers to leave the base is called **saturation delay time** (t_s). Obviously, saturated logic circuits have low switching speeds whereas non-saturated type are much faster. *TTL* is the example of a saturated logic whereas *ECL* represents a non-saturated logic.

71.24. Basic Operating Characteristics and Parameters of Logic Families

When we work with digital ICs from different logic families, we should be familiar with, not only their logical operation but also with the basic operational properties. Following are the important basic operational properties important from the subject point of view.

1. *DC* supply voltage.
2. *TTL* and *CMOS* logic levels
3. Noise immunity
4. Noise margin.
5. Power dissipation.
6. Propagation delay.
7. Speed-power product.
8. Loading and fan-out.

Now we will describe all the above operational characteristics one by one in the following pages.

71.25. DC Supply Voltage

The standard value of the dc supply voltage for *TTL* (*i.e.*, transistor-transistor logic) and CMOS (*i.e.*, complementary metal-oxide semiconductor) device is + 5*V*. For simplicity, the dc supply voltage is usually omitted from the logic circuits. But in practice, it is connected to the V_{CC} or V_{DD} pin of an *IC* package and the ground is connected to the *GND* pin of an *IC* package. Both the voltage and ground are distributed internally to all the logic gates with the package as shown in Fig. 71.47 (*a*). The connections for the single logic gate are as shown in Fig. 71.47 (*b*).

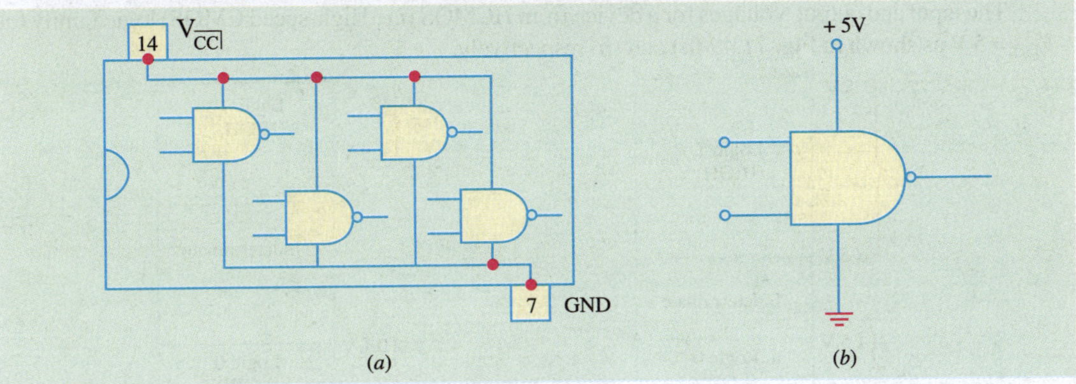

Fig. 71.47

71.26. TTL and CMOS Logic Levels

For TTL circuits, the range of input voltages, that can represent a valid *LOW* (logic 0) and a valid *HIGH* (logic 1) is as shown in Fig. 71.48 (*a*). As seen from this diagram, the range of input voltages that can represent a valid LOW (logic 0) is from 0 to 0.8 V. The LOW input voltage is indicated by the symbol V_{IL}. The lower limit for V_{IL} is represented by $V_{IL(min)}$ and the higher limit for V_{IL}, by $V_{IL(max)}$. The range of input voltages that can represent a valid HIGH (logic 1) is from 2 V to V_{CC} (usually 5 V). The HIGH input voltage is indicated by symbol V_{IH}. The lower limit for V_{IH} is represented by $V_{IH(min)}$ and the higher limit for V_{IH} by $V_{IH(max)}$. Note that the range of values between 0.8 V and 2 V is called ***indeterminate***. This means that when a input voltage is in this range, it can be interpreted as a HIGH or LOW by the logic circuit. Therefore TTL logic gates cannot be operated reliably when input voltages are in this range.

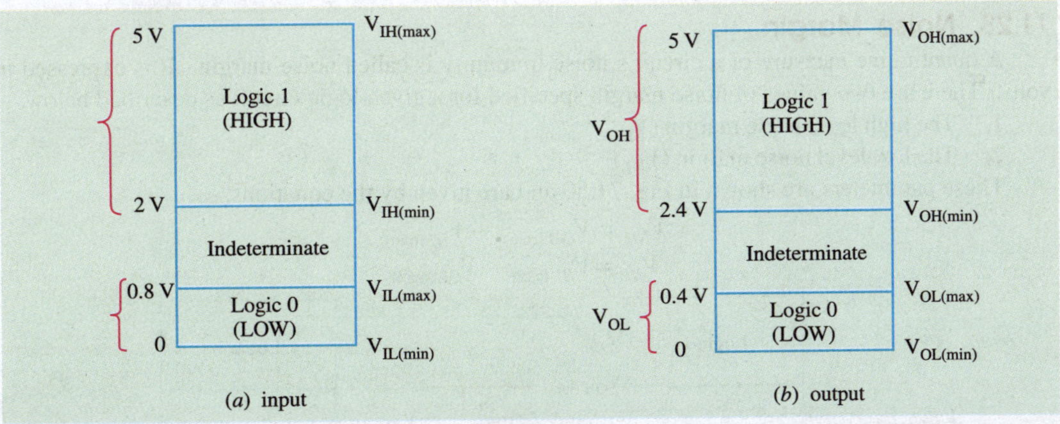

Fig. 71.48

Fig. 71.48 (*b*) shows the range of *TTL* output voltages that can represent valid HIGH (logic 1) and valid *LOW* (logic 0). As seen the range for logic 1 output is from 2.4 V to 5 V and logic 0 output is from 0 to 0.4 V. Note again that the range of values between 0.4 V and 2.4 V is indeterminate. Also note that the output voltage for logic 1 is indicated by the symbol V_{OH}. The lower limit for V_{OH} is represented by $V_{OH(min)}$ and the higher limit for V_{OH} by $V_{OH(max)}$. Similarly the output voltage for logic 0 is indicated by the symbol V_{OL}. The lower limit for V_{OL} is represented by $V_{OL(min)}$ and the higher limit for V_{OL} by $V_{OL(max)}$.

It may be noted from Fig. 71.48 (*a*) and (*b*) that the minimum HIGH output voltage, $V_{OH\,(min)}$ is greater than the minimum HIGH input voltage $V_{IH\,(min)}$. On the other hand, the maximum LOW output voltage, $V_{OL\,(min)}$ is less than the maximum LOW input voltage, $V_{IL(max)}$.

The input and output voltages for a device from HCMOS (*i.e.* High-speed CMOS) logic family for $V_{DD} = 5$ V as shown in Fig. 71.49 (*a*) and (*b*) respectively.

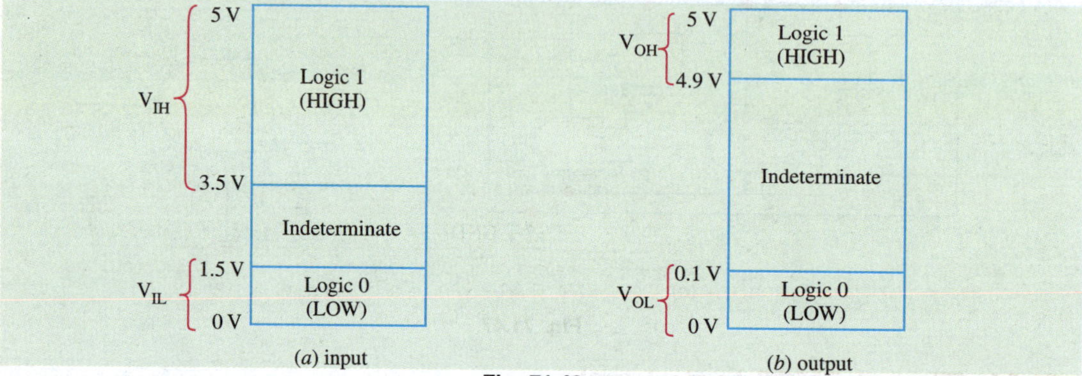

Fig. 71.49

Notice that the input and output voltage values in the HCMOS device are different from that of a TTL device. In a HCMOS device, $V_{IL(max)}$ is 1.5 V (its value is 0.8 V in TTL), $V_{IH\,(min)}$ is 3.5 V (its value is 2 V in TTL), $V_{OL\,(max)}$ is 0.1 V (its value is 0.4 V is TTL), $V_{OH\,(min)}$ is 4.9 V (its value is 2.4 V in TTL).

71.27. Noise Immunity

The noise immunity of a logic circuit refers to the circuit's ability to tolerate noise without causing a false change in its output voltage. The noise voltage is produced by stray electric and magnetic fields on the connecting wires between logic circuits. Sometimes, too much noise voltage cause the voltage at the input of the logic circuit to drop below $V_{IH\,(min)}$ or rise above $V_{IL\,(max)}$. This could produce unpredictable operation in a logic circuit.

71.28. Noise Margin

A quantitative measure of a circuit's noise immunity is called noise margin. It is expressed in volts. There are two values of noise margin specified for a given logic circuit as described below.

1. The high level noise margin (V_{NH})
2. The low level noise margin (V_{NL})

These parameters are shown in Fig. 71.50 and are given by the equations,

$$V_{NH} = V_{OH\,(min)} - V_{IH\,(min)}$$
$$V_{NL} = V_{IL\,(max)} - V_{OL\,(max)}$$

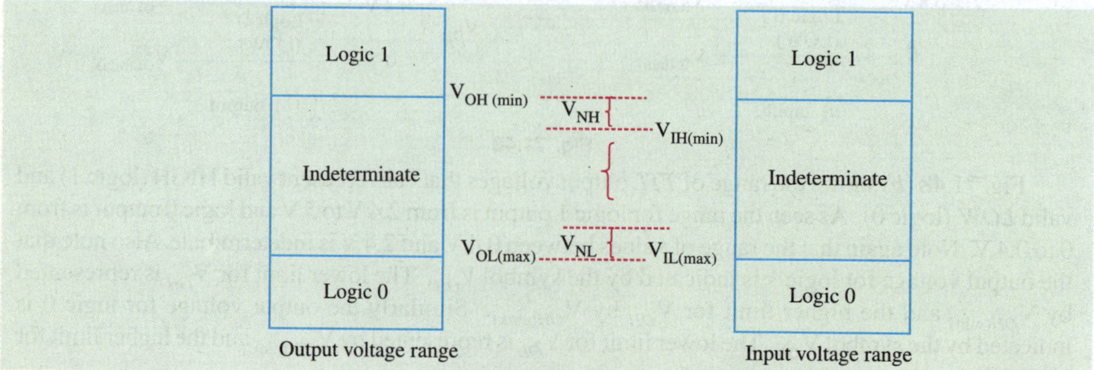

Fig. 71.50

Example 71.29. *Table shows the input/output voltage specifications for the standard TTL family.*

Parameter	*Min (V)*	*Typical (V)*	*Max (V)*
V_{OH}	2.4	3.4	—
V_{OL}	—	0.2	0.4
V_{IH}	2.0	—	—
V_{IL}	—	—	0.8

Using these values, find ***(a)*** *the maximum value of noise spike that can be tolerated when a HIGH output is driving an input,* ***(b)*** *the maximum value of noise spike when a LOW output is driving an input.*

Solution. The maximum value of the noise spike, when driven by a HIGH output,

$$V_{NH} = V_{OH\,(\min)} - V_{IH\,(\min)} = 2.4 - 2.0 = 0.4\text{ V}$$

and the maximum value of the noise spike, when driven by a LOW output,

$$V_{NL} = V_{IL\,(\max)} - V_{OL\,(\max)} = 0.8 - 0.4 = 0.4\text{ V}$$

It is observed that a TTL gate is immunte to 0.4 V of noise for both the HIGH and LOW input states.

71.29. Power Dissipation

As a matter of fact, all logic gates draw current from the dc supply voltage for its normal operation. When the logic gate is in the HIGH output state, it draws an amount of current, I_{CCH}, as shown in Fig. 71.51 (*a*) and when in LOW output state, it draws an amount of current I_{CCL} as shown in Fig. 71.51 (*b*).

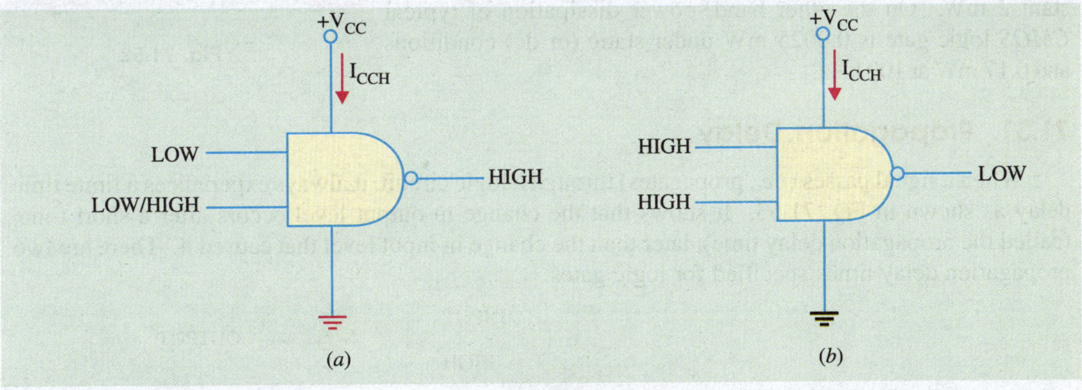

Fig. 71.51

The power dissipation of a logic gate is given by the product of the dc supply votage (V_{CC}) and the amount of current drawn from the suppy (*i.e.,* I_{CCH} or I_{CCL}). Thus power dissipation is given by :

$$P_D = V_{CC} \cdot I_{CCH} \text{ or } V_{CC} \cdot I_{CCL}$$

For example, if I_{CCH} is 2.5 *mA* when V_{CC} *is* 5 V, the power dissipation,

$$P_D = V_{CC} \cdot I_{CCH} = 5\text{ V} \times 2.5\text{ mA} = 12.5\text{ mW.}$$

Usually, the logic gate operates with the inputs, which keep on changing with time (*i.e.,* the input is pulsed). Accordingly the output of a logic gate also switches back and forth between HIGH and LOW. Because of this, the amount of current drawn from the dc supply also varies between I_{CCH} and I_{CCL}. In such a situation, we calculate the average power dissipation. The average power dissipation depends upon the duty cycle and is usually specified for a duty cycle of 50%, the output is HIGH half the time and LOW the other half. Therefore average supply current is :

$$I_{CC} = \frac{I_{CCH} + I_{CCL}}{2}$$

and the average power dissipation is,

$$P_D = V_{CC} . I_{CC}$$

Example 71.30. *A TTL logic gate draws 2 mA when its output is HIGH and 3.5 mA when its output is LOW. Calculate the average power dissipation if the supply voltage is 5 V and the logic gate is operated on 50% duty cycle.*

Solution. The average supply current,

$$I_{CC} = \frac{I_{CCH} + I_{CCL}}{2} = \frac{2\text{mA} + 3.5\text{mA}}{2} = 2.75 \text{ mA}$$

∴ The average power dissipation,

$$P_D = V_{CC} . I_{CC} = 5 \text{ V} \times 2.75 \text{ mA} = \mathbf{13.75 \text{ mW}}$$

71.30. Power Dissipation versus Frequency

Fig. 71.52 shows a graph of power dissipation versus frequency for a *TTL* and a *CMOS* logic gate. As seen from this graph, the power dissipation in a *TTL* circuit is essentially constant over its range of operating frequencies. However, the power dissipation in a *CMOS* circuit is frequency dependent. That is, it is extremely low under zero frequency (or dc) conditions. But the power dissipation increases as the frequency increases. For example, the power dissipation of a typical TTL logic gate is a constant 2 mW. On the other hand, power dissipation of typical *CMOS* logic gate is 0.0025 mW under static (or dc) conditions and 0.17 mW at 100 kHz.

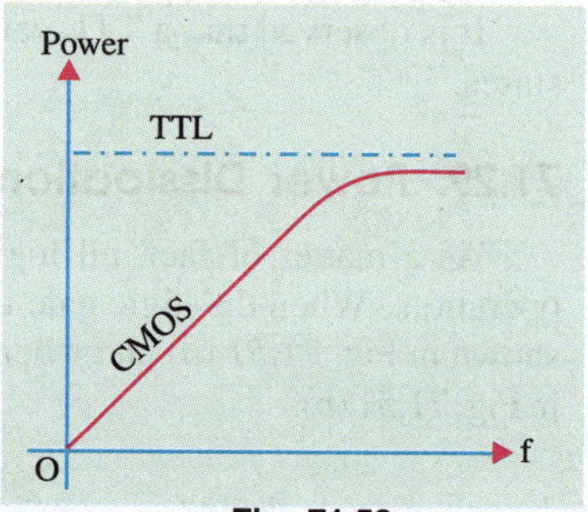

Fig. 71.52

71.31. Propagation Delay

When a signal passes (*i.e.*, propagates) through a logic circuit, it always experiences a finite time delay as shown in Fig. 71.53. It shows that the change in output level occurs after a short time, (called the propagation delay time), later than the change in input level that caused it. There are two propagation delay times specified for logic gates.

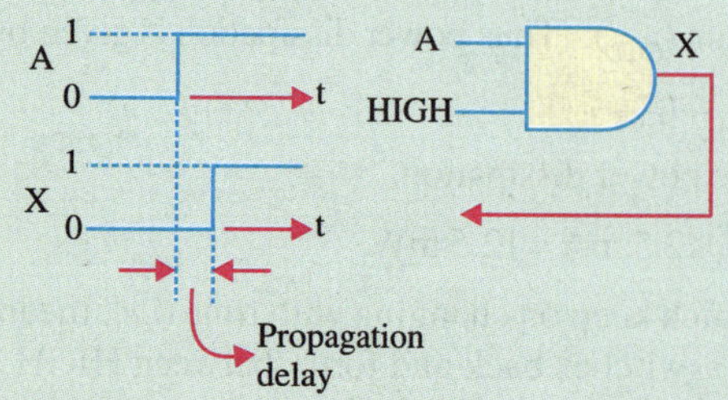

Fig. 71.53

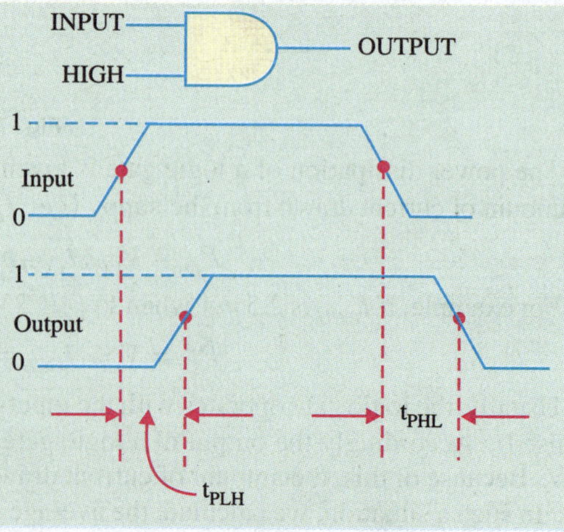

Fig. 71.54

1. t_{PLH} – It is the time interval between a designated point on the input pulse and the corresponding point on the output pulse when the output is changing from LOW to HIGH (or 0 to 1) as shown in Fig. 71.54.

2. t_{PHL} – It is the time interval between a designated point on the input pulse and the corresponding point on the output pulse when the output is changing from HIGH to LOW (or 1 to 0) as shown in Fig. 71.54.

It may be noted that the propagation delay times are indicated in Fig. 71.54 with 50% points on the pulse edges used as references.

The propagation delay time of a logic gate limits its maximum operating frequency. The greater the propagation delay time of a logic gate, the lower is its maximum operating frequency. This means a high speed logic gate is a one that has a small propagation delay time. For example, a logic gate with a delay time of 2 ns is faster than a logic gate that has a delay time of 10 ns.

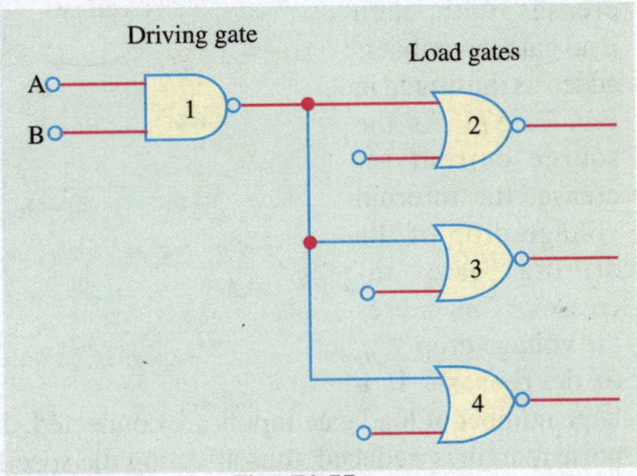

Fig. 71.55

71.32. Speed-Power Product

It provides a basis for comparison of logic circuits when the propagation delay and power dissipation are important considerations in the selection of the type of logic family to be used in certain application. The speed power product is expressed in picojule (pJ). It may be noted carefully that the lower the speed-product, the better is the value. Typically the *CMOS* family has much lower value of speed-power product as compared to the *TTL* logic family. For example, a typical *CMOS* logic family has a speed-power product of 1.5 pJ at 100 Hz while a typical *TTL* has speed-power product of 20 pJ.

71.33. Loading and Fan-out

When the output of any logic gate is connected to one or more inputs of other logic gates, a load on the driving gate is created. This is shown in Fig. 71.55. Here the output of a logic gate (labeled as 1) is connected to the inputs of 3 other logic gates (labeled as 2, 3, and 4). Note that the logic gate labeled 1 is called a ***driving gate*** while the logic gates labeled as 2, 3 and 4 are called ***load gates***.

In any logic family, there is a limit to the number of load gate inputs that a given logic gate can drive. This limit is called the ***fan-out*** of the logic gate. Now we will study the loading and fan-out in *TTL* and *CMOS* logic families in more detail.

Loading and fan-out in TTL family : A *TTL* gate that acts as a driving gate **sources** (*i.e.*, supplies) current to a load gate input in the logic HIGH state and *sinks* (*i.e.*, receives) current from the load gate in the logic LOW state. Fig. 71.56 (*a*) illustrates the current sourcing and (*b*) shows current sinking in the logic gates.

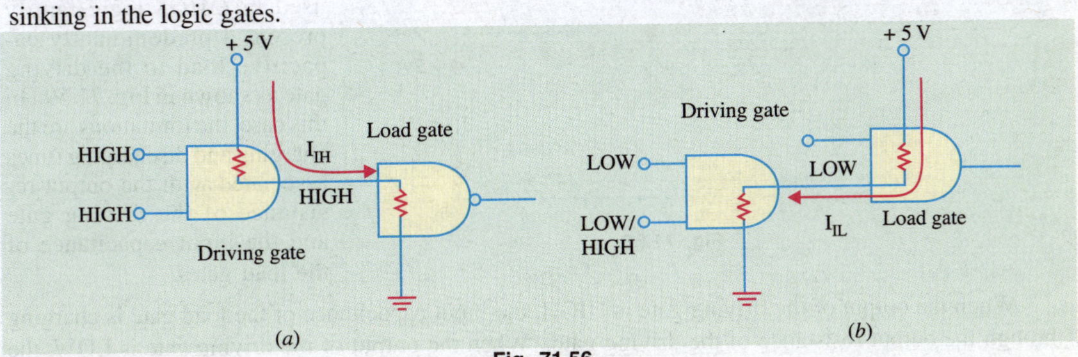

Fig. 71.56

Note that I_{IH} is the current supplied by the driving gate to the load gate when the input of the load gate is HIGH similarly. I_{IL} is the current received by the driving gate from the load gate when the input of the load gate is *LOW*.

As more and more number of load gates are connected to the driving gate, the loading on the driving gate increases. The total source current increases with each load gate input that is added as illustrated in Fig. 71.57. As the source current increase, the internal voltage drop of the driving gate increases. This causes the voltage drop V_{OH} to decreases. If a large number of load gate inputs are connected, drops below $V_{OH\,(min)}$. As a result of this, HIGH level noise margin is reduced, thus affecting the specified operating characteristics of the gate. Moreover, as the total source current increases, the power dissipation of the driving gate increases.

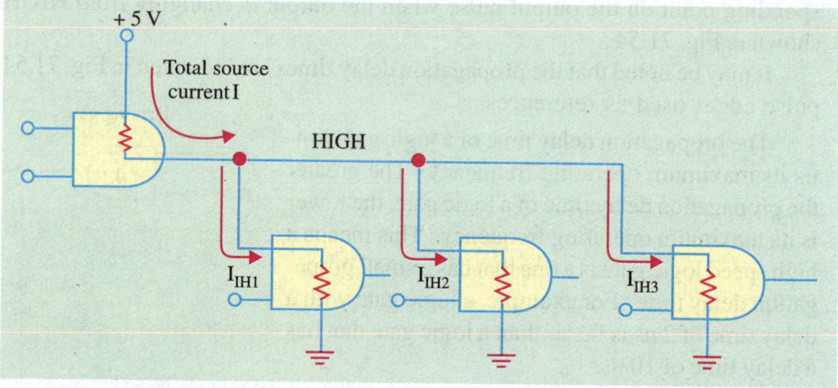

Fig. 71.57

The maximum number of load gate inputs that can be connected without affecting the specified operational characteristics of the driving gate is called ***fan-out.*** Its value is important for designing logic circuits. For example, the standard *TTL* has a fan-out of 10. One input of the same logic family as the driving gate is referred to as a ***unit load.*** This we can also say that a standard *TTL* has a fan-out of 10 unit loads.

The total sink current also increases with each load gate input that is added as shown in Fig. 71.58. As this current increases, the internal voltage drop of the driving gate increases, causing V_{OL} to increase. If a large number of loads are added, V_{OL} exceeds $V_{OL\,(max)}$ and the LOW-level margin is reduced.

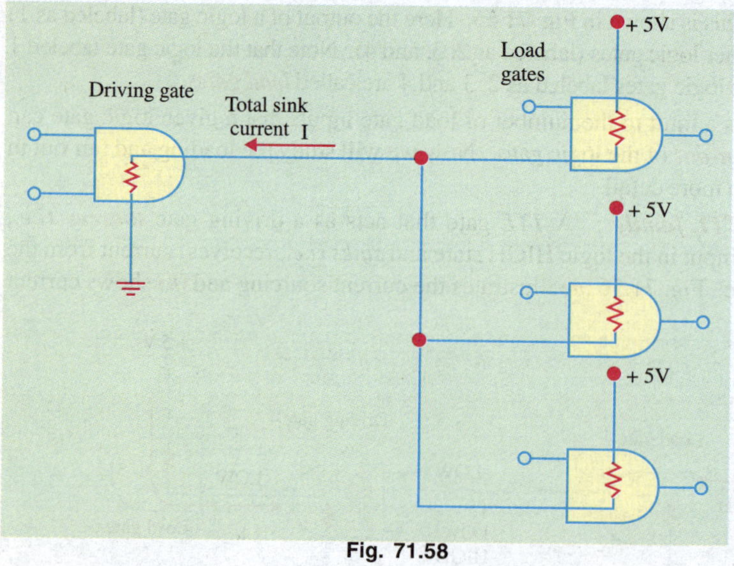

Fig. 71.58

As a matter of fact, in *TTL*, the current sinking capability (in the *LOW* state) is the limiting factor in determining the fan-out.

Loading and fan-out in CMOS: Loading in *CMOS* logic family differs from that in *TTL*. It is because of the fact that the field-effect transistors used in *CMOS* logic family present a predominantly capacitive load to the driving gate as shown in Fig. 71.59. In this case, the limitations are the charging and discharging times associated with the output resistance of the driving gate and the input capacitance of the load gates.

When the output of the driving gate is HIGH, the input capacitance of the load gate is charging through the output resistance of the driving gate. When the output of the driving gate is LOW, the

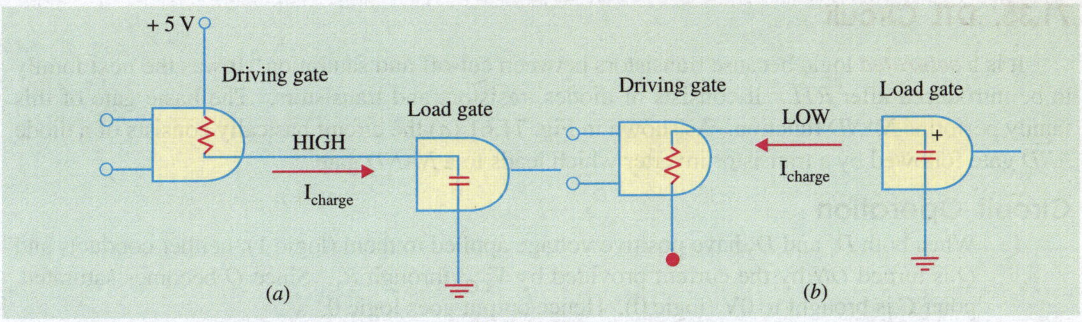

Fig. 71.59

capacitance is discharging. When more and more number of load gate inputs are added to the driving gate output, the total capacitance increases because the input capacitances effectively appear in parallel. This increase in capacitance increases the charging and discharging times. As a result, the maximum operating frequency of the logic gate is reduced. Thus fan-out in a *CMOS* logic family depends upon the operating frequency. The smaller the number of load gate inputs, the greater the maximum operating frequency.

71.34. RTL Circuit

It is a ***saturated* logic.** It uses only transistors and resistors as circuit elements and also resistances in the input to each base. This family is based on the *NOR* circuit shown in Fig. 71.60. All other members of the family are made up of *NOR* cells or variations on them.

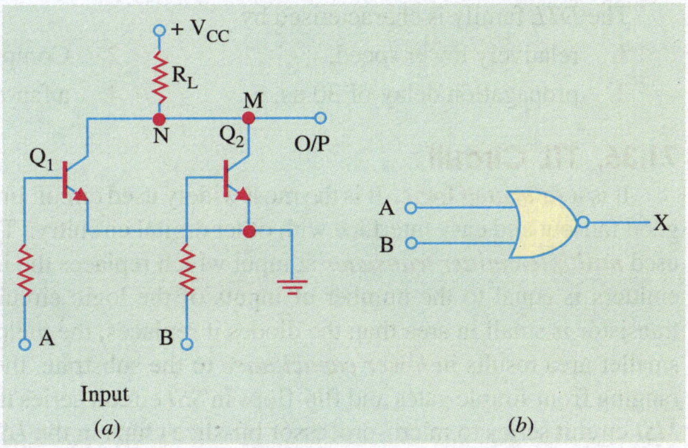

Fig. 71.60

Circuit Operation

We will assume ideal transistors. When both inputs *A* and *B* are 0 V (or logic 0) both transistors are turned *OFF*, hence point *M* goes to + V_{CC} so that output is logic 1.

If either or both input terminals are at + V_{CC} *i.e.* are high (or logic 1), one or both transistors would be fully turned *ON* (*i.e.* saturate) thereby reducing the voltage of point *N* to almost 0 V. Hence, output would be at logic 0.

It is seen that the output is at logic 1 only when ***both inputs are at logic 0*** — the *NOR* logic functions as shown in Fig. 71.60 (*b*).

The *RTL* family has the following characteristics :

1. relatively slow speed,
2. low fan-out of 6 and a fan-in of 4,
3. poor noise immunity,
4. expensive since resistors are required to be fabricated,
5. cannot operate at speed above 4 MHz.

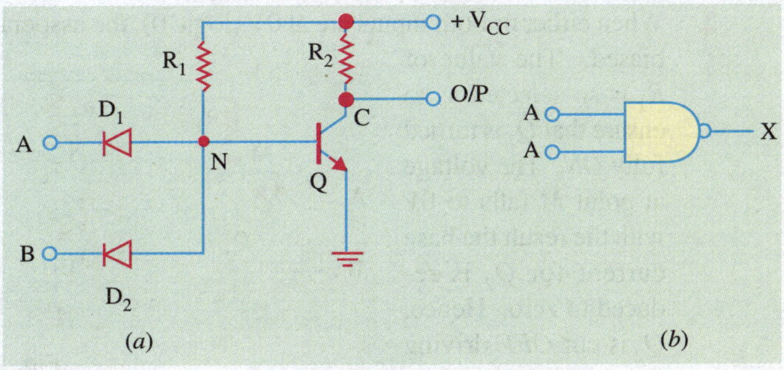

Fig. 71.61

71.35. DTL Circuit

It is a ***saturated*** logic because transistors between cut-off and saturation. It was the next family to be introduced after *RTL*. It consists of diodes, resistors and transistors. The basic gate of this family performs *NAND* function. As shown in Fig. 71.61 (*a*) the circuit basically consists of a diode *AND* gate followed by a transistor inverter which leads to a *NAND* gate.

Circuit Operation

1. When both D_1 and D_2 have positive voltage applied to them (logic 1), neither conducts and Q is turned *ON* by the current provided by V_{CC} through R_1. Since Q becomes saturated, point *C* is brought to 0V (logic 0). Hence output goes logic 0.
2. If either or both inputs are at 0V (logic 0), the associated diode will conduct driving point *N* to ground *i.e.* 0V. Since there is no base voltage for *Q*, it will be cut *OFF* thereby driving point *C* and hence output to V_{CC} *i.e.* logic 1.

It is seen that output is low (a logic 0) only when all inputs are high—the condition for a *NAND* gate.

The *DTL* family is characterised by

1. relatively lower speed,
2. Comparatively better noise immunity,
3. propagation delay of 30 ns,
4. a fan-out of 5.

71.36. TTL Circuit

It is a ***saturated* logic.** It is the most widely used circuit line since early 1970*s* because of its speed, good fan-out and easy interface with other digital circuitry. The unique feature of this circuit is that it used ***multiple-emitter transistor*** at input which replaces the input diodes of the *DTL*. The number of emitters is equal to the number of inputs of the logic circuit (limited to 8). Since a multi-emitter transistor is small in area than the diodes it replaces, the yield from a wafer is increased. Moreover, smaller area results in ***lower capacitance*** to the substrate, thereby wide selection of circuit modules ranging from simple gates and flip-flops in *SSI* circuit series through various registers in computers in *MSI* circuit series to micro-processor bit-slice chips in the *LSI* series.

Basic Circuit

The basic circuit of the *TTL* family is the *NAND* gate cell shown in Fig. 71.62. However, at present *NOR*, *OR* and *AND* gate configurations have also been added to the series.

Circuit Operation

1. If both inptus *A* and *B* are high (logic 1), *E/B* junction of Q_1 is *reverse-biased* so that it has no emitter current. Hence Q_1 is *OFF*. However, its *C/B* junction is ***forward-baised*** supplying base current to Q_2 from V_{CC} *via* R_1. As a result transistor Q_2 is turned fully *ON* (*i.e.* it becomes saturated) driving point *N* to 0*V*. Hence, output is a logic 0.
2. When either or both inputs are at 0*V* (logic 0), the associated *E/B* junction becomes forward-biased. The value of R_1 is so selected as to ensure that Q_1 is turned fully *ON*. The voltage at point *M* falls to 0*V* with the result the base current for Q_2 is reduced to zero. Hence, Q_2 is cut *OFF* driving point *N* and the output to logic 1.

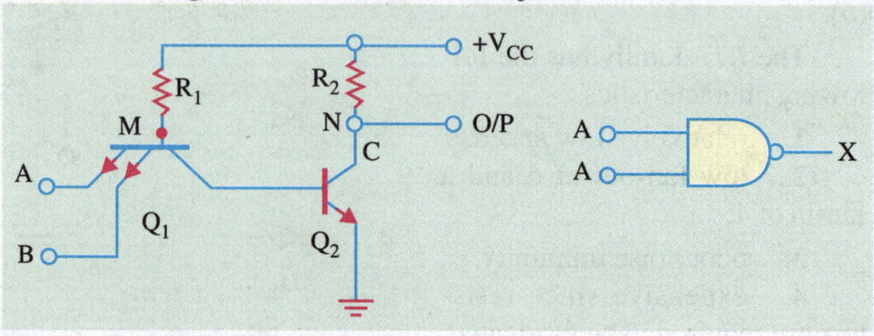

Fig. 71.62

Totem Pole Output

The basic circuited of Fig. 71.62 is never used in practice. Its modified version with an added output stage is in common use (Fig. 71.63). This extra output stage is often known as *totem-pole* stage because the three components Q_3, Q_4 and D are stacked one on top of the other in the manner of a totem-pole. The circuit action is as follows :

1. When Input is High

In this case, the two input terminals have ***positive*** voltage (logic 1). the *E/B* junction is reverse-biased because of which there is no emitter current. Hence, Q_1 is *OFF*. Since *C/B* junction of Q_1 is forward-biased, base current of Q_2 flows from V_{CC} through R_1. Hence, Q is turned *ON*. As a result, potential of point *N* falls so much that Q_3 is turned *OFF*. At the same time Q_4 is turned *ON* by the voltage drop across R_3. Now, when Q_4 is *ON*, its collector potential (*i.e.* potential of point *C*) is nearly that of its emitter. Hence, output is low *i.e.* at logic 0.

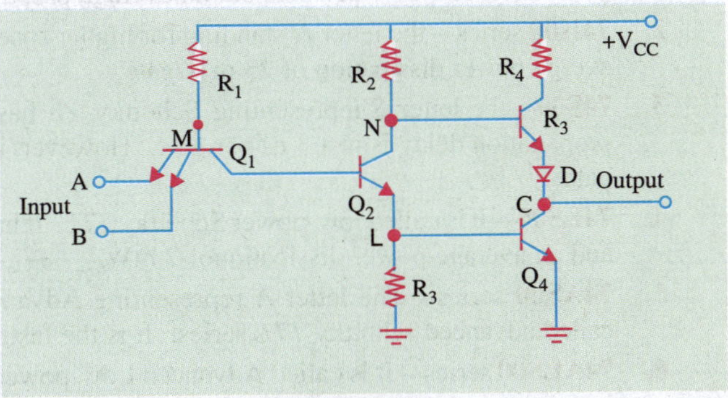

Fig. 71.63

In short, when inputs are at logic 1, Q_1 is *OFF*, Q_2 is *ON*, Q_3 is *OFF* and Q_4 is *ON* because of which output becomes logic 0.

2. When Input is Low

If any of the two inputs or both are low (logic 0), Q_1 turns *ON* and potential of its collector (point *M*) falls. Hence, Q_2 is turned *OFF*, grounding its emitter and the base of Q_4 so that Q_4 is also turned *OFF*.

Since *N* is at V_{CC}, it turns Q_3 *ON*. The potential of point *C* is V_{CC} minus drop in R_4, Q_3 and *D*. Since these drops do not amount to much, output is at logic 1.

It may be noted that when ***even-numbered transistors are ON, the odd-numbered ones are OFF and vice-versa.***

The function of diode *D* in Fig. 71.63 is to prevent both Q_3 and Q_4 from being turned *ON* simultaneously. If both were to be *ON* at the same time, they would offer low immedance to the suppy which will draw excessive current and produce large noise 'spikes' in the output. It may also be noted that the addition of a pàir of totem pole transistor increases the operating speed and output current capability of this circuit. The standard *TTL*-family has

1. greater speed then *DTL*,
2. less noise immunity (0.4 *V*),
3. average propagation delay per gate of 9 *ns*,
4. average power dissipation of 10 mW,
5. a fan-out of 10 meaning one output can drive 10 other *TTL* inputs,

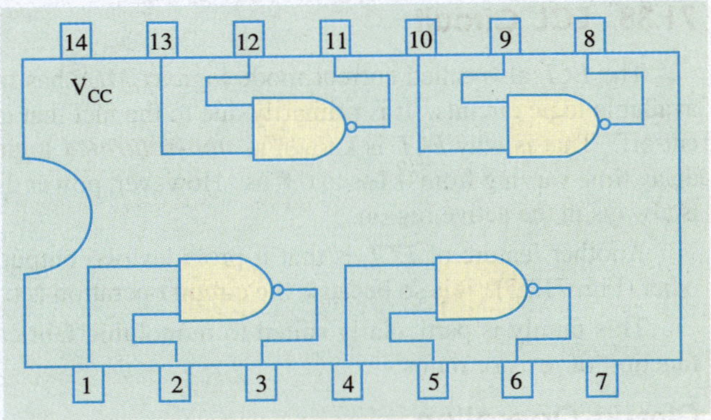

Fig. 71.64

Fig. 71.64 shows a pin-out of a *BEL* 7400 *IC* referred to as 'quad (quadrupole). 2-input *NAND* gate chip'. As seen, there are four separate

2-input *NAND* gates of which any or all may be used at any point in time. It is manufactured by Bharat Electronics LTD. (BEL) Bangalore, India.

71.37. TTL Sub-families

TTL has several ***sub-families*** having different speed and lower dissipation charcteristics as detailed below :

1. **74L00** series—the letter *L* standing for low power consumption. It has an average power dissipation of 1 mW per gate but an average propagation delay of 33 ns.
2. **74H00** series—the letter *H* standing for higher speed. It has a propagation delay of 6 ns but averge power dissipation of 23 mW/gate.
3. **74S00**—the letter *S* representing Schottky. It has the highest speed because its average propagation delay is just 3 ns per gate. However, its average power dissipation is 23 mW/ gate.
4. **74LS00**—It is called low-power Schottky *TTL*. It has an average propagation delay of 9.5 ns and an average power dissipationof 2 mW.
5. **74AS00** series—The letter *A* representing Advanced and *S* stnading for schottky. It is called advanced schottky *TTL* series. It is the fastest *TTL* series.
6. **74ALS00** series—It is called Advanced Low-power Schottky *TTL* series. The 74*ALS* series has the lower speed-power product and the lowest gate power dissipation of all the *TTL* series.
7. **74F00** series —The letter *F* standing for fast. This logic family uses a new *IC* fabrication technique to reduce interdevice capacitances to achieve reduced propagation delays. It has a propagation delay of 3 ns and a power consumption of 6 mW.

Table 71.3 summarizes the important characteristics of the each of the *TTL* sub-families.

Table 71.3

Characteristic	*TTL* sub-family					
	74	74S	74LS	74AS	74ALS	74F
Propagation delay (ns)	9	3	9.5	1.7	4	3
Power Dissipation (mW)	10	20	2	8	1.2	6
Speed-power product (pJ)	90	60	19	13.6	4.8	18
Fan-out	10	20	20	40	20	33

71.38. ECL Circuit

The *ECL* also called current-mode logic (*CML*), has the **highest speed** of any of the currently-available logic circuits. It is primarily due to the fact that transistors never operate fully saturated or cut-off. That is why *ECL* is known as ***non-saturated* logic.** The latest *ECL* series has propagation delay time varying from 0.1ns to 0.8 ns. However, power dissipation is increased since one transistor is always in the active region.

Another feature of *ECL* is that it provides two outputs which are always complement of each other (Fig. 71.65). It is so because the circuit operation is based on a differential amplifier.

This family is particularly suited to monolithic fabrication techniques because logic levels are function of resistor ratios.

Circuit Operation

The basic circuit shown in Fig. 71.65 is combined *OR/NOR* circuit and is operated from a V_{EE} = – 5.2 *V* supply. A built-in constant-current source provides current to the emitters. Strictly speaking, logic 1 is represented by – 0.9 *V* (less negative) and logic 0 by – 1.75 *V* (more negative). Please note that it is a *positive* logic. In negative logic, the functions would be *AND/NAND*. A reference voltage of – 1.29 *V* is applied to the base of Q_3 from a built-in temperature-compensated reference voltage source (*RVS*).

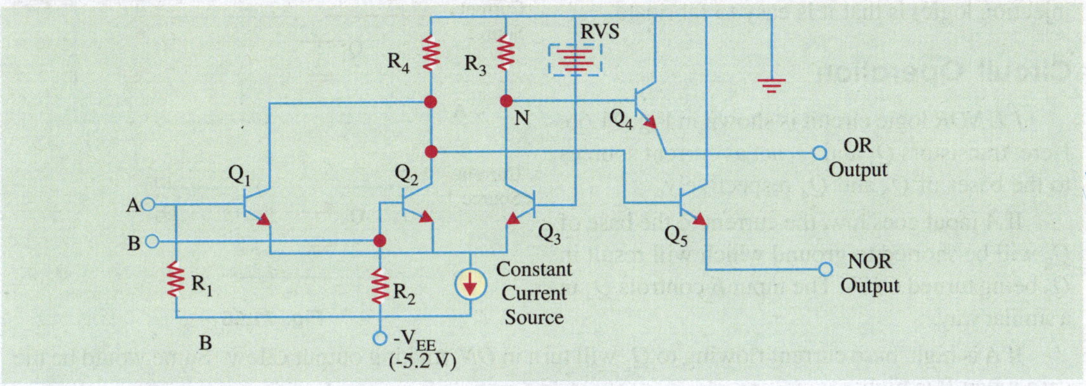

Fig. 71.65

1. When both inptus are logical 0 *i.e.* –1.75 V

In this case, base potential of Q_3 is less negative (or more positive) than the base potential of either Q_1 or Q_2. Hence, Q_3 conducts whilst Q_1 and Q_2 do not. Only enough base current is drawn by Q_3 from RVS so as to remain out of saturation. The collector current of Q_3 develops a voltage of – 1.0 V across R_3 which makes Q_4 to conduct. Transistor Q_4 gives an output voltage at the emitter of about –1.0 – 0.7 = –1.7 V which represents logic 0. Since collector potentials of Q_1 and Q_2 are nearly zero (because they are cut off), the output voltage at the emitter of Q_5 is 0 – 0.7 = –0.7 V which is a logic 1. Obviously, the two outputs are complements of each other.

2. When either input A or B is at logical 1 *i.e.* –0.9 V

In that case, the associated transistor (either Q_1 or Q_2) is turned *ON* while Q_3 is turned *OFF*. The collector potentials of Q_1/Q_2 and are opposite of the previous case. Hence, now Q_5 output is at logical 1 and Q_4 output is at logical 0.

Typical characteristics of an *ECL* family are :

1. propagation delay time per gate of 0.3 ns (meaning extremely fast speed),
2. power dissipation of 25 mW,
3. fan-out of 25 to 50,
4. noise margin from about 0.2 to 0.25 V.

It will be interesting to know that *ECL* family of *ICs* does not include a wide range of general purpose logic devices as do the *TTL* and *CMOS* logic families. *ECL* does include complex, special purpose *ICs* used in applications such as high-speed data transmission, high-speed memories and high-speed arithmetic units. The relatively low noise margins and high power drain of *ECL* are disadvantages compared with *TTL* and *CMOS*. Another drawback is its negative power supply voltage and logic levels which are not compatible with those of other logic families. This makes it difficult to use *ECL* devices in conjunction with *TTL* and/or *CMOS ICs*. Special level-shifting (also called interface) circuits must be connected between *ECL* devices and the *TTL* (or *CMOS*) devices on both input and output.

71.39. I^2L Circuit

It is the latest entry into the bipolar ***saurated*** logic field. It uses no biasing and loading resistors at all! Resistors require lot of power and space on an *IC* chip. Hence, their elimination results in ***higher density circuits operating at much reduce power.*** Because of its **high speed** and **less power dissipa-**

tion, it is used in large computers. Such circuits are also used where high packing density is of prime consideration as in digital wrist watches. I^2L chips are capable of microwatt power dissipation yet can provide high currents when necessary to drive *LED* displays. Another feature of I^2L (integrated injection logic) is that it is easy to fabricate.

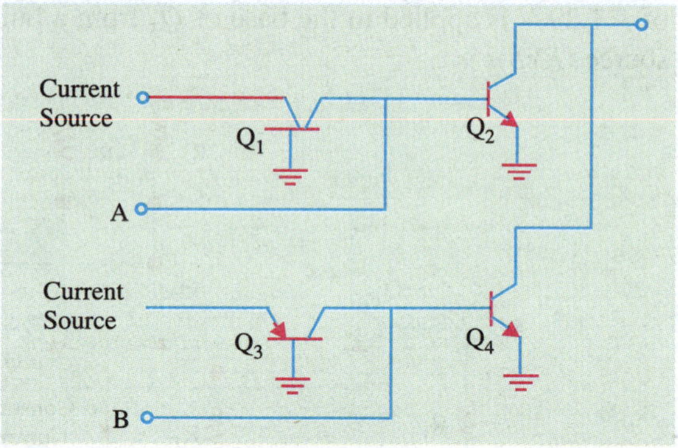

Fig. 71.66

Circuit Operation

I^2L *NOR* logic circuit is shown in Fig. 71.66. Here, transistors Q_1 and Q_2 act as current sources to the bases of Q_3 and Q_4 respectively.

If *A* input goes low, the current to the base of Q_2 will be shorted to ground which will result in Q_2 being turned *OFF*. The input *B* controls Q_4 in a similar way.

If *A* is high, base current flowing to Q_2 will turn in *ON*, making output *C* low. Same would be the case when *B* is high.

It is obvious that output would be high only when ***both inputs*** *A* and *B* are low.

The output would be low when either *A* or *B* or both are high *i.e. NOR* logic function.

71.40. MOS Family

It does not use bipolar transistors but just Enhancement-only *MOSFETs* (Art. 63.14). There are two kinds of digital *MOS* circuits.

(a) one which uses *MOSFETs* of one polarity either all of *N-type* called *NMOS* or all of *P*-type called *PMOS* but not both on the same chip,

(b) the other which employs both *N*-type and *P*-type *MOSFETs* on the same chip. It is called comlementary *MOS* (*CMOS*).

The *MOS* family may be subdivided as under :

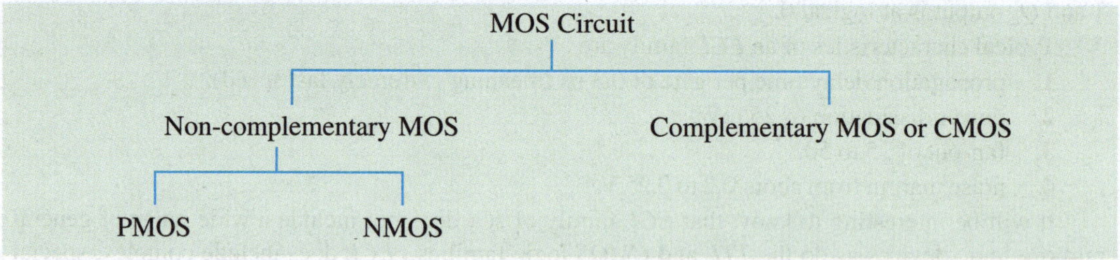

Since a *FET* requires small area, it is possible to fabricate a large number of *MOS* ciruits on a single small chip. Gating arrays with thousands of gates and flip-flops are manufactured in standard containers and are often used in *IC* memories and microprocessors.

Followings are some of the advantages of *MOS ICs* over the bipolar *ICs* (i.e. *TTL*, *ECL* etc.)

1. The *MOS IC* is relatively simple and inexpensive to fabricate.
2. The *MOS* device size is small and it consumers less power. Because of the small size, the *MOS ICs* can accommodate a much larger number of circuit elements on a single chip than bipolar *IC* in the area of large scale integration. This makes them especially well-suited for complex *ICs* such as microprocessor, memory chips etc.
3. *MOS* digital *ICs* normally do not use the *IC* resistor elements that take up so much of the chip area or bipolar *ICs*.

The continuous improvement in *MOS IC* technology has led to the device that are faster than *TTL* devices. Consequently *MOS* devices (especially *CMOS*) have become dominant in the *SSI* (small-scale integration) and *MSI* (medium-scale integration) market.

The major disadvantage of *MOS* devices is that they are susceptible to static-electricity damage. Although we can minimize it by adopting proper handling procedures, yet *TTL* is still more durable for laboratory experimentation. As a result, we are likely to see *TTL* devices used in education as long as they are available.

71.41. PMOS Circuit

A *PMOS NAND* gate is shown in Fig. 71.67 (*a*). As seen, there are no resistors in the circuit. The gate consists of two *E* only *MOSFETs* Q_1 and Q_2 as logic elements and the third one Q_3 as load resistors. When $-V_{DD}$ (say −12 V) is applied, *MOSFETs* will be turned *ON* and when 0 V is applied they would be turned *OFF*. Hence, with positive logic, 0 V would be 1 and − 12 V would be 0 since 1 is assigned to the most positive voltage.

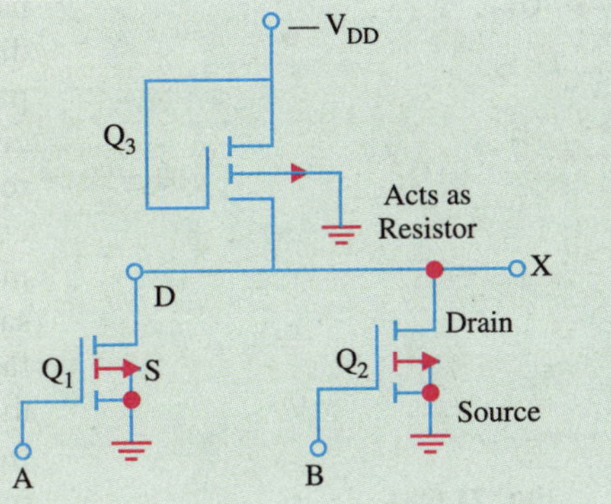

(*a*)

A	B	X
0	0	1
0	1	1
1	0	1
1	1	0

(*b*)

Fig. 71.67

Circuit Operation

1. If any of the two inputs *A* or *B* is at logic 0 (*i.e.* − 12 V), the concerned *MOSFET* would turn *ON* thereby offering a low resistance from drain to source thus causing the output to be nearly 0*V i.e.* a logic 1.
2. The output can be at − 12 V *i.e.* logic 0 only when both inputs *A* and *B* are at 0 V *i.e.* logic 1.

This is the *NAND* function as shown by the truth table of Fig. 71.67 (*b*).

Incidentally, the positive logic *NAND* gate of Fig. 71.67 (*a*) would be the ***negative logic NOR*** gate since the two are identical.

71.42. NMOS Circuit

In Fig. 71.68 (*a*) is shown a two-input *NOR* gate circuit consisting of two *MOSFETs* Q_1 and Q_2 acting as logic elements and Q_3 as a load resistor. For positive logic, 0*V* will be logic 0 and positive voltge (+V_{DD}) will be logic 1.

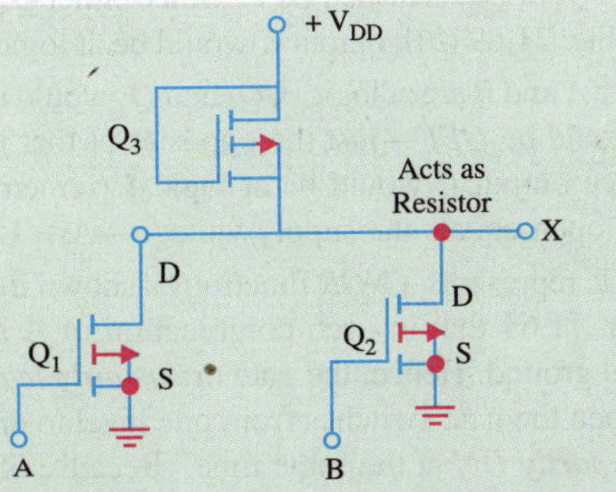

(*a*)

A	B	X
0	0	1
0	1	0
1	0	0
1	1	0

(*b*)

Fig. 71.68

Circuit Operation

1. If any of the inputs *A* or *B* is at logic 1, the corresponding *MOSFET* will conduct causing the output to go low *i.e.* logic 0.
2. If both inputs *A* and *B* are at logic 0, then both *MOSFETs* will be *OFF* driving the output to logic 1.

This is *NOR* function as shown by the truth table of Fig. 71.68 (*b*).

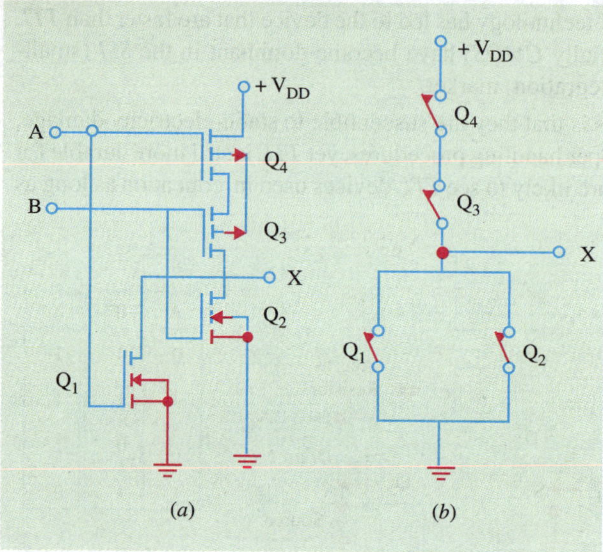

Fig. 71.69

71.43. CMOS Circuit

CMOS logic circuits use both *PMOS* and *NMOS* devices in the same circuit. It gives the advnatage of ***drastic decrease in power dissipation*** (12 nW per gate) and ***increase in speed of operation***. In fact, it has the **lowest power dissipation** amongest different logic families. It has very high packing density *i.e.* larger number of circuits can be placed on a single chip. As a result, it is extensively used in *VLSI* circuits such as on-chip computers and memory systems. The recent silicon-on-saphire *MOS* (*SOSMOS*) is 2 to 4 times faster than the standard *CMOS*. Hence, they are being widely used for everything from electronic watches and calculators to micro processors.

Fig. 71.69 (*a*) shows a *CMOS NOR* circuit which has two *N*-channel *MOSFETs* Q_1 and Q_2 and two *P*-channel *MOSFETs* Q_3 and Q_4. The two inputs *A* and *B* switch between + V_{DD} (logic 1) and ground (logic 0).

Circuit Operation

1. Let *A* = *B* = logic 1 *i.e.* have positive voltage. In that case, Q_1 and Q_2 are *ON* (closed switches) but Q_3 and Q_4 are *OFF* and act as open switches as shown in Fig. 71.69 (*b*). Henc, output *C* is logic 0.
2. If either *A* or *b* is at logic 1, then the associated *N*-channel *MOSFET* (Q_1 or Q_2) is turned *ON* but the associated *P*-channel *MOSFET* (Q_3 or Q_4) is turned *OFF*. Since either Q_3 or Q_4 would be *OFF* [Fig. 71.69 (*b*)], output *C* would be at logic 0.
3. When both *A* and *B* are at logic 0, Q_3 and Q_4 would be *ON* but Q_1 and Q_2 would be *OFF*—just the opposite of that in Fig. 71.69 (*b*). Hence output *C* would be at logic 1 (remember, voltage across an open equals the supply voltage—Art. 1.21.

A	B	C
0	0	1
0	1	0
1	0	0
1	1	0

Fig. 71.70

The above logic represents a *NOR* function as shown in the truth table of Fig. 71.70. It would be observed from Fig. 71.69 that in each combination of *A* and *B*, there is at least ***one open switch*** between + V_{DD} and ground. Hence, the gate draws ***only leakage current from supply for any static state.*** However, when the gate switches from one level to another, some power is consumed becasue two *MOSFETs* are ***partly ON*** at the same time. Because of this reason, power dissipated by *CMOS* circuit *is a **function of input signal frequency.*** Higher the frequency, greater the power dissipation.

71.44. CMOS Sub-families and their characteristics

The *CMOS* family of *ICs* competes directly with *TTL* in the small and medium-scale integration. As *CMOS* technology has produced better and better performance characteristics, it has gradually taken over the field that has been dominated by *TTL* for so long. As a matter of fact, *TTL* devices will be around for a long time, but more and more new equipment is using *CMOS* logic circuits. These days, the *CMOS ICs* provide all the logic-functions that are available in *TTL* Some special-purpose functions provided in *CMOS ICs* are not provided even by *TTL*. Before we look at the various *CMOS* subfamilies, let us define few terms that are used when *ICs* from different families or series are to be used together or as replacements for one another.

(*a*) Pin-to-pin Compatible : Two *ICs* are said to be pin-to-pin compatible when three pin configurations are the same. For example, pin 14 on both *ICs* is V_{CC} supply. Pin 7 on both is ground etc.

(*b*) Functionally Equivalent : Two *ICs* are said to be functionally equivalent when the logic function they perform are exactly the same. For example, both contain four two-input *AND* gates, or two-input *NAND* gates etc.

(*c*) Electrically compatible : Two *ICs* are said to be electricaly compatible when they can be connected directly to each other without taking any special measures to ensure proper operation.

Let us now study the different *CMOS* subfamilies.

1. **5000/14000 series :** The *CMOS* 4000 series or the improved 4000 *B* is the oldest series and was first introduced by *RCA*. This series is functionally equivalent to 14000 series from Motorola. The *CMOS* devices in 4000/14000 series have a very low power dissipation and can operate over a wide range of supply voltage (*i.e.* from 3 to 15 V). These devices are very slow as compared to *TTL* and other *CMOS* subfamilies. The 40H00 series was designed to be faster than 4000 series. It did overcome some of the speed limitations, but it is still much slower than *LSTTL* series. The 4000 series have very low output current capability. The devices in 4000 series are not pin-compatible or electrically compatible with any *TTL* series.
2. **74C00 series :** This is pin-compatible as well as functionally equivalent to *TTL* devices having the same device number. For example, 74C04 is a HEX inverter that has the same pin configuration as the *TTL* 7404 HEX inverter *IC*. The performance characteristics of 74C00 series are the same as those of 4000 series.
3. **74HC/HCT00 series :** The letter *H/HC* is standing for high-speed *CMOS* series. This series has a speed which is 10 times faster than of 74*LS*00 series devices. It has also a higher current capability than that of 74*C*00 series. The 74*HC/HCT00* series *ICs* are pin-compatible with and functionally equivalent to *TTL ICs* with the same device number. This series has become the most widely used series.
4. **74AC/ACT 00 series :** This series is referred as advanced *CMOS* logic. It is functionally equivalent to the *TTL* series but not pin-compatible.
5. **74AHC00 series :** This series is referred to as advanced high-speed *CMOS* logic. It is faster, and has lower-power dissipation. The devices in this series are 3 times faster and can be used as direct replacements for *HC* series devices.
6. **74-BiCMOS series :** These days, the *IC* manufacturers have developed a new logic series-called *BiCMOS* logic. This series combines the best features of bipolar and *CMOS* logic *i.e.* low power characteristics of *CMOS* and high speed characteristics of bipolar circuits. *BiCMOS ICs* are available only in those functions that are used in microprocessor interfacing and memory applications such as latches, buffers, drivers and transceivers.
7. **74-Low Voltage series :** A new series of logic using a nominal supply voltage of 3.3 V has been developed to meet the extremely low power design requirements of battery powered and hand held devices. These *ICs* are being designed into the circuits of notebook computers, mobile radios, hand-held video games, telecommunication equipment, personal digital assistant (*PDA*) and high performance work station computers.

   Table 71.4 shows some of the common Low-voltage families identified by the suffixes as indicated :

Table 71.4

Suffixes	*Low-voltage series*
LV	Low-voltage *HCMOS*
LVC	Low-voltage *CMOS*
LVT	Low-voltage technology
ALVC	Advanced Low-voltage *CMOS*
HLL	High speed Low-power Low-voltage

The power consumption of *CMOS* logic *ICs* decreases approximately with the square of the supply voltage. On the other hand, the propagation delay increases slightly at this reduced voltage. However the speed is restored and even increased by using finer geometry and sub-micron (<1μm) *CMOS* technology that is tailored for low-power and low-voltage applications.

8. **74 AHC/AHCT series :** This is an enchanced version of 74 *HC/HCT*00 series. It provides superior speed and low power consumption. The 74*AHC* series devices have half the static power consumption, one-third the propagation delay, high output drive current, and can operate at V_{CC} of 3.3 to 5 V.

Table 71.5 shows a comparison of the important characteristics of the *CMOS* logic families. It is evident from this table that the devices from 74*AHC*/*AHCT* series has the lowest propagation delay and the smallest value of power dissipation.

Table 71.5

S. No.	*Characteristic*	*CMOS logic family*			
		4000 B	*74 HC/HCT*	*74 BiCMOS*	*74 AHC/AHCT*
1.	Propagation delay (ns)	50	8	2.9	3.7
2.	Power dissipation static (mW)	0.001	0.0025	0.0003 – 7.5	0.000009
3.	Speed-power product (pw-s)	105	15	0.00087 to 22	—

Tutorial Problems No. 71.1

1. Prove the following identities :

   (*i*) $AB + A\overline{B} = A$ (*ii*) $A + A\overline{B} = A + B$ (*iii*) $(A + B)(A + \overline{B}) = A$

2. Simplify the following Boolean expressions :

   (*i*) $A\,\overline{A}\,C$ (*ii*) $ABCD + ABD$ (*iii*) $ABCD + A\,\overline{B}\,C\,D$

   [(*i*) 0, (*ii*) *ABD* (*iii*) *ACD*]

3. Simplify the following Boolean functions :

   (*i*) $A + \overline{A}\,B + AB$ (*ii*) $\overline{A}\,B + \overline{A}\,B + AB + \overline{A}\,B$ (*iii*) $(AB + C)(AB + D)$

   (*iv*) $A + \overline{B}\,C\,(A + B\,\overline{C})$ (*v*) $A\,[B + C\,(\overline{AB + AC})]$ (*vi*) $(A + B)\,.\,(A + C)$

   [(*i*) *A* + *B* (*ii*) B (*iii*) *AB* + *CD* (*iv*) *A* (*v*) *AB* (*vi*) *A* + *B.C*]

4. Simplify the following Boolean expressions :

(*i*) $A\overline{B}C + AB\overline{C} + ABC$ (*ii*) $\overline{A}\,\overline{C} + B\overline{C} + \overline{A}BC + ABC$

(*iii*) $AD + A\overline{B}\,\overline{C} + \overline{B}C\overline{D} + \overline{A}CD + \overline{A}BC\overline{D}$

(*iv*) $A\overline{B}\,\overline{D} + AB\overline{C}\,\overline{D} + ABC\overline{D} + \overline{A}BC\overline{D}$

[(*i*) $A(B + C)$ (*ii*) $B + \overline{A}\,\overline{C}$ (*iii*) $AD + \overline{A}B + A\overline{B}$ (*iv*) $A\overline{B}\,\overline{D} + A\overline{C}\,\overline{D} + BC\overline{D}$]

5. Use truth tables to verify the following identities :

   (*i*) $\overline{A} + B = \overline{A}\,.\,B$ (*ii*) $A + BC = (A + B)(A + C)$ (*iii*) $A(\overline{A} + \overline{B}) = AB$

6. Write the logic equation for the switching circuit of Fig. 71.71. [$A(B + C + D + E)$]

7. Write down the logical expression which describes the working of the circuit of Fig. 71.72. [$AB(C + D)$]

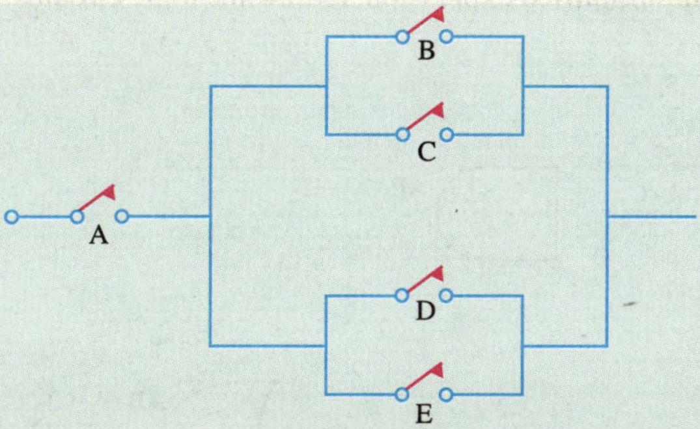

Fig. 71.71

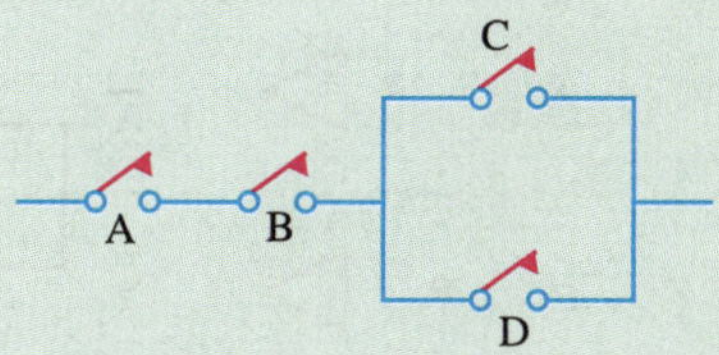

Fig. 71.72

8. Simplify the following Boolean expressions :

   (*i*) $y = AB\overline{C} + A\overline{B}\,\overline{C} + A\overline{B}\,\overline{C} + \overline{A}\,\overline{C}\,\overline{B} + A\overline{B}$

   (*ii*) $y = AB\overline{C}\,\overline{D} + AB\overline{C}D + ABC + ABD + CD$ [(*i*) $A\overline{C} + A\overline{B} + \overline{B}\,\overline{C}$ (*ii*) $AB + CD$]

9. Design a logic circuit whose output is HIGH only when majority of inputs *A*, *B* and *C* are LOW.

10. Determine the Boolean expression for the logic circuit shown in Fig. 71.73. Simplify the Boolean expression using Boolean Laws and De Morgan's theorem. Redraw the logic circuit using the simplified Boolean expression.

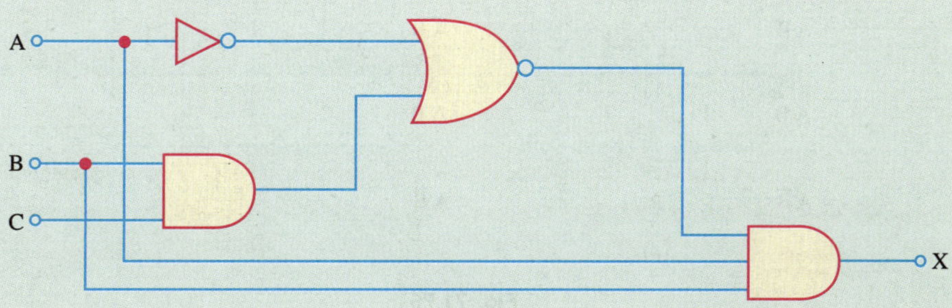

Fig. 71.73

11. Determine the output, *X* of a logic circuit shown in Fig. 71.74. Simplify the output expression using Boolean Laws and theorems. Redraw the logic circuit with the simplified expression. ($\overline{B}\,\overline{C}D + AC$)

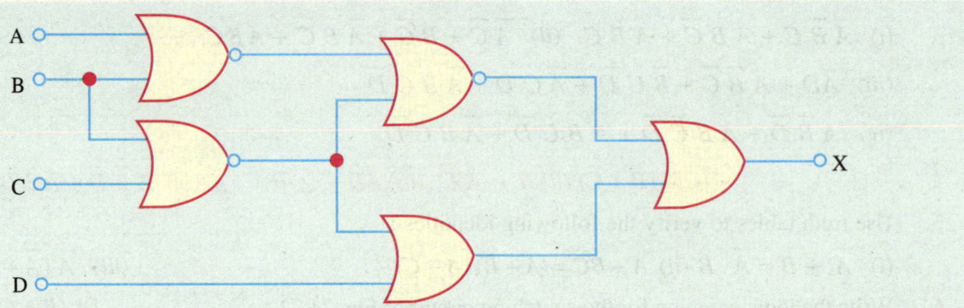

Fig. 71.74

12. Consider the logic circuit shown in Fig. 71.75. Determine the Boolean expression at the circuit output *X*, simplify it. Using the simplified expression, redraw the logic circuit.

$A(\overline{B} + C)$

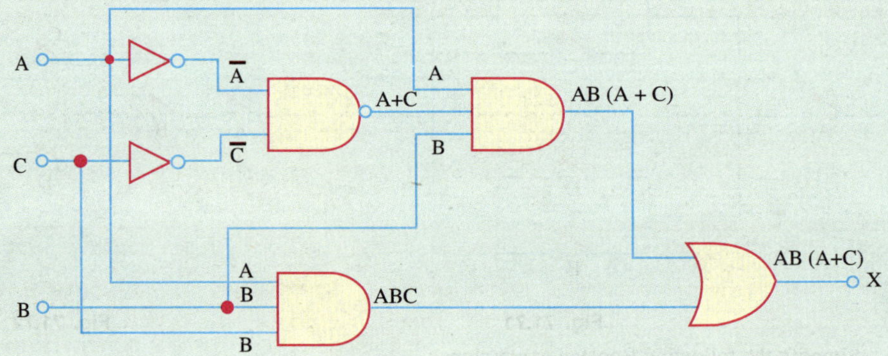

Fig. 71.75

13. Fig. 71.76 (*a*) shows a three-variable Karnaugh map. Group the 1*s* and hence obtain the minimized Boolean expression. $(AB + BC + \overline{A}\,\overline{B}\,\overline{C})$

(*a*)

	$\overline{C}$	C
$\overline{A}\,\overline{B}$	1	
$\overline{A}B$		1
AB	1	1
$A\overline{B}$		

(*b*)

	$\overline{C}\,\overline{D}$	$\overline{C}D$	CD	$C\overline{D}$
$\overline{A}\,\overline{B}$	1	1		
$\overline{A}B$	1	1	1	1
AB				
$A\overline{B}$		1	1	

Fig. 71.76

14. Fig. 71.76 (*b*) shows a four-variable Kamaugh-map. Group the 1*s* and hence obtain the minimized Boolean expression. $(\overline{A}\,\overline{B} + \overline{A}\,C + A\overline{B}D)$

15. A truth table has output 1*s* for these inputs : *ABCD* = 0011, *ABCD* = 0110, *ABCD* = 1000 and *ABCD* = 1100, and 0*s* for the other inputs. Draw the Karnaugh map and find the simplified Boolean equation for the truth table. $(A\overline{C}\,\overline{D} + \overline{A}\,\overline{B}\,CD + \overline{A}BC\overline{D})$

16. Determine the minimized expression for the Karnaugh map shown in Fig. 71.77 (*a*).

$(B\overline{C} + BD + \overline{B}C)$

	$\overline{C}\,\overline{D}$	$\overline{C}D$	CD	$C\overline{D}$
$\overline{A}\,\overline{B}$	0	0	1	1
$\overline{A}B$	1	1	1	0
AB	1	X	X	X
$A\overline{B}$	0	0	X	X

(a)

	$\overline{C}\,\overline{D}$	$\overline{C}D$	CD	$C\overline{D}$
$\overline{A}\,\overline{B}$	1	1		1
$\overline{A}B$	1	1		1
AB			1	1
$A\overline{B}$				

(b)

Fig. 71.77

17. Determine the simplified Boolean expression for the Karnaugh map shown in Fig. 71.77 (*b*).

(UPSC Engg. Services 1995) $(\overline{A}\,\overline{C} + \overline{A}\,\overline{D} + ABC)$

OBJECTIVE TESTS – 71

1. Boolean algebra is essentially based on
(*a*) symbols (*b*) logic
(*c*) truth (*d*) numbers

2. The first person who used Boolean algebra for the design of relay switching circuits was
(*a*) Aristotle (*b*) Boole
(*c*) Shannon (*d*) Ramanujam

3. Different variables used in Boolean algebra can hvae values of
(*a*) 0 or 1 (*b*) low or high
(*c*) true or false (*d*) On or OFF.

4. According to the algebra of logic, $(A + \overline{A})$ equals
(*a*) A (*b*) 1
(*c*) 0 (*d*) $\overline{AA}$.

5. According to the absorptive Laws of Boolean algebra , expressioin $(A + AB)$ equals
(*a*) A (*b*) B
(*c*) AB (*d*) A

6. When we demorganize $\overline{A}\,\overline{B}$, we get
(*a*) $\overline{AB}$ (*b*) $\overline{A} + \overline{B}$
(*c*) $\overline{A + B}$ (*d*) $\overline{AB}$.

7. The dual of the statement $(A + 1) = 1$ is
(*a*) $A \,.\, 1 = A$ (*b*) $A \,.\, 0 = 0$
(*c*) $A + A = A$ (*d*) $A \,.\, A = 1$.

8. The expression $\overline{ABC}$ can be simplified to
(*a*) $\overline{A} \,.\, B \,.\, \overline{C}$ (*b*) $AB + BC + CA$
(*c*) $AB + \overline{C}$ (*d*) $\overline{A} + \overline{B} + \overline{C}$.

9. While obtaining minimal sum-of-products expression,
(*a*) all don't cares are ignored
(*b*) all don't cares are treated as logic 1*s*
(*c*) all don't cares are treated as logic 0*s*
(*d*) only such don't cares that aid minimization are treated as logic 1*s*.

10. In a saturated bipolar logic circuit, transistor operate
(*a*) in deep cut-off (*b*) over active region
(*c*) in saturation
(*d*) just short of saturation

11. Saturated logic circuits have inherently
(*a*) short saturation delay time
(*b*) low switching speed
(*c*) higher power dissipation
(*d*) lower noise immunity.

12. Noise margin is expressed in
(*a*) decibel (*b*) watt
(*c*) volt (*d*) phon

13. DTL family employs
(*a*) resistors and transistors
(*b*) diode and resistor
(*c*) diode and transistors
(*d*) diodes, resistors and transistors.

14. The chief advantage of Schotty *TTL* logic family is its least
(*a*) power dissipation (*b*) propagation delay
(*c*) fan-in (*d*) noise immunity.

15. The main advantage claimed for *ECL* family of logic gates is its
(*a*) very large fan-in
(*b*) use of negative power supply voltage
(*c*) extremely low propagation times
(*d*) least power dissipation.

16. Special feature of an I^2L logic circuit is that it
(*a*) uses ony high-value resistors

(*b*) dissipates negligible power
(*c*) is a bipolar saturated logic
(*d*) is easy to fabricate
(*e*) uses no biasing and loading resistors.

17. A unique advantage feature of *CMOS* logic family is its
(*a*) use of *NMOS* circuits
(*b*) power dissipation in nanowatt range
(*c*) speed
(*d*) dependence on frequency for power dissipation.

18. CMOS circuits are extensively used for on-chip computers mainly because of their extremely
(*a*) low power dissipation
(*b*) large packing density
(*c*) high noise immunity
(*d*) low cost.

19. The main advantage of a CMOS logic family over the TTL family is its
(*a*) much reduced power
(*b*) increased speed of operation
(*d*) extremely low cost
(*d*) series base resistor

20. CMOS logic family uses only
(*a*) MOSFETs and resistors
(*b*) NMOS circuits
(*c*) MOSFETs (*d*) bipolar transistors.

21. Power is drawn by a CMOS circuit only when
(*a*) its output is high (*b*) its output is low
(*c*) it switches logic levels
(*d*) in static state.

22. The most obvious identifying feature of a *TTL* gate is its
(*a*) large fan-out (*b*) high power dissipation
(*c*) interconnected transistors
(*d*) multiemetter input transistor.

23. In digital circuits shottky transistors are preferred over normal transistor because of their
(*a*) lower propagation delay
(*b*) higher propagation delay
(*c*) lower power dissipation
(*d*) higher power dissipation

24. A unique operting feature of *ECL* circuit is its
(*a*) very high speed
(*b*) high power dissipation
(*c*) series base resistor
(*d*) compatibility with other logic series.

25. The fan-in a logic gate refers to the number of
(*a*) input devices that can be connected
(*b*) input terminals (*c*) output terminals
(*d*) circuits output can drive

26. Which of the following statements regarding *ICs* is not correct,
(*a*) *ECL* has the least propagation delay
(*b*) *TTL* has the largest fan out
(*c*) *CMOS* has the biggest noise margin
(*d*) *TTL* has the lowest power consumption.

27. The Boolean equation for *NOR* gate is
(*a*) $\overline{AB} = C$ (*b*) $\overline{A} + \overline{B} = C$
(*c*) $A + B = C$ (*d*) $\overline{A + B} = C.$

28. The Boolean equation for a NAND gate is
(*a*) $\overline{A + B} = C$ (*b*) $\overline{\overline{AB}} = C$
(*c*) $A + B = C$ (*d*) $\overline{A} + \overline{B} = C$

ANSWERS

1. (*b*) **2.** (*c*) **3.** (*a*) **4.** (*b*) **5.** (*a*) **6.** (*c*) **7.** (*b*) **8.** (*d*) **9.** (*d*) **10.** (*c*) **11.** (*b*) **12.** (*c*) **13.** (*d*) **14.** (*b*) **15.** (*c*) **16.** (*e*) **17.** (*b*) **18.** (*b*) **19.** (*a*) **20.** (*c*) **21.** (*c*) **22.** (*d*) **23.** (*a*) **24.** (*a*) **25.** (*b*) **26.** (*d*) **27.** (*d*) **28.** (*d*) **29.** (*b*) **30.** (*a*)

CHAPTER 72

FLIP-FLOPS AND RELATED DEVICES

Learning Objectives

- Flip-Flop (FF)
- Latch
- NAND Gate Latch
- NOR Gate Latch
- Clocked Signals
- Some Main Ideas Common to Clocked Flip-Flops
- Clocked S-C Flip-Flop
- Clocked J-K Flip-Flop
- Clocked D Flip-Flop
- Parallel Transfer of Data Using D-Flip-Flops
- D Latch (Transparent Latch)
- Clocked J-K Flip-Flop with Asynchronous Inputs
- Flip-Flop Timing Parameters
- IC Flip-Flop Timing Values
- Applications of Flip-Flop
- Flip-Flop Synchronization
- Parallel Data Transfer
- Frequency Division
- Counting
- Schmitt Trigger Devices
- Types of one-shots
- Retriggerable One-Shot
- Actual One-Shot Devices
- Clock Generator Circuits
- Schmitt Trigger Oscillator
- 555 Timer Used as an Astable Multivibrator

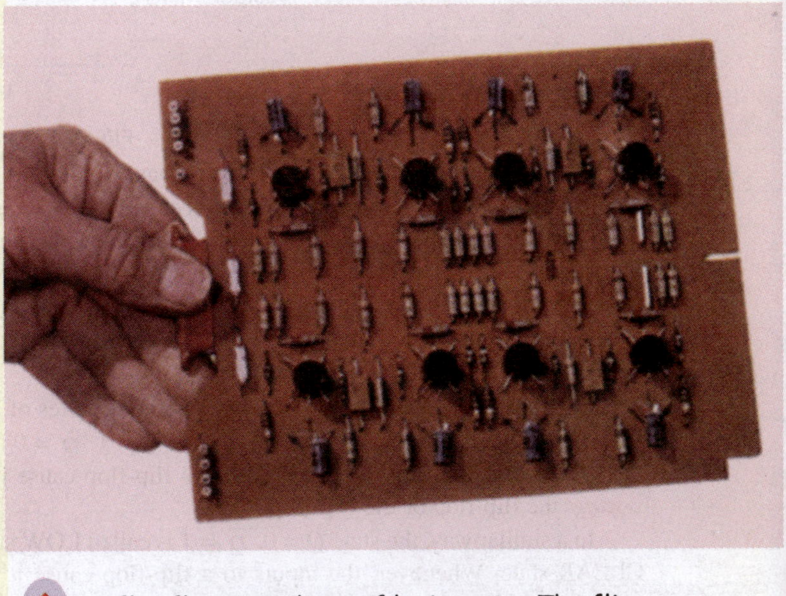

A flip-flop is made up of logic gates. The flip-flpos are extensively used as a memory cell in static random access memory of a computer

72.1. Introduction

The output logic levels of combinational logic circuits at any instant of time are dependent on the logic levels present at the inputs at that time. Any prior input-logic level conditions have no effect on the present outputs. This is due to the fact that combinational logic circuits have no ***memory.*** However, it will be interesting to know that most digital systems are made up of both combinational logic circuits and memory elements.

The most important memory element is the flip-flop (abbreviated as FF). The FF is made up of an assembly of logic gates. It may be noted that even though a logic gate, by itself, has no storage capability, but several such logic gates can be connected together in ways that permit information to be stored. Several different gate arrangements are used to produce these flip-flops.

The flip-flops are used extensively as a memory cell in static random access memory (SRAM) of a computer.

72.2. Flip-Flop (FF)

We have already discussed in the last article that flip-flop (abbreviated as FF) is made up of logic gates and it permits information to be stored in it. Fig. 72.1 (*a*) shows a general type of symbol used for a flip-flop.

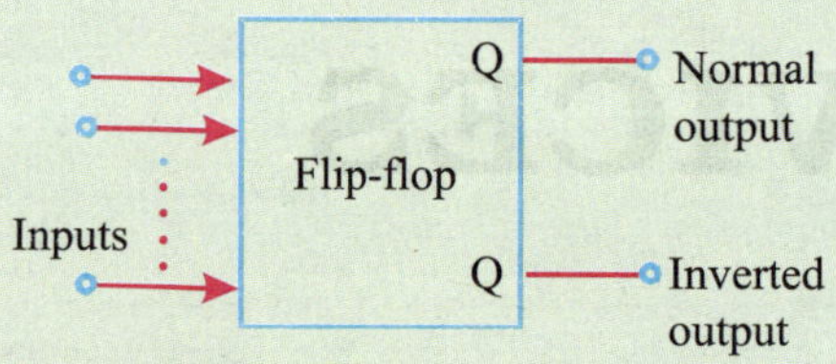

(*a*) Flip-flop symbol

Output states

$Q = 1, \overline{Q} = 0$: It is called HIGH or 1 State; also called SET state

$Q = 0 \ \overline{Q} = 1$: It is called LOW or 0 state; also called CLEAR or RESET state

(*b*) Operating states of Flip-flop

Fig. 72.1

As seen from this diagram, the flip-flop has two outputs labeled as Q and $\overline{Q}$ that are inverse (or complement) of each other. The Q output is called the ***normal*** flip-flop output and $\overline{Q}$ is the ***inverted*** flip-flop output. It may be carefully noted that whenever we refer to the state of a flip-flop, we are referring to the state of its ***normal*** (Q) output. For instance if we say flip-flop is in the HIGH state, we mean that $Q = 1$. On the other hand if we say flip-flop is in LOW state, we mean that $Q = 0$. It is automatically understood that the $\overline{Q}$ state will always be ***inverse*** of Q, *i. e.*, if $Q = 1$, $\overline{Q} = 0$, and if $Q = 0$, $\overline{Q} = 1$.

Fig. 72.1 (*b*) shows the two possible operating states of a flip-flop. One possible state is $Q = 1$, $\overline{Q} = 0$ and the other $Q = 0$, $\overline{Q} = 1$. The state $Q = 1$, $\overline{Q} = 0$ is called HIGH state or 1 state. It is also called *SET* state. Whenever, the inputs to a flip-flop cause it to go to the $Q = 1$ state, we call this ***setting*** the flip-flop or the flip-flop is set.

In a similar way, the state $Q = 0$, $\overline{Q} = 1$ is called LOW state or 0 state. It is also called *RESET* or CLEAR state. Whenever, the inputs to a flip-flop cause it to go to the $Q = 0$ state, we call this ***resetting*** **or** ***clearing*** the flip-flop. In the later part of the chapter we will see that many flip-flops will have a *SET* input and/or RESET (CLEAR) input that is used to drive the flip-flop into a specific output state.

It may be noted from the flip-flop symbol (Fig. 72.1 (*a*)) that a flip-flop can have one or more inputs. These inputs can be used to switch the flip-flop back and forth between its two possible output states.

It will be shown later that most flip-flop inputs need only to be momentarily activated (or pulsed) in order to cause a chage in the output state. The flip-flop output will remain in the new state even after the input pulse is over. This is flip-flop's memory characteristic.

It will be interesting to know that a flip-flop is also known as a latch and a bistable multivibrator. The term ***"latch"*** is used for certain types of flip-flops that will be described later. The term bistable multivibrator is the more technical name but flip-flop is the one used more frequently among the engineers and technologists.

72.3. Latch

A latch is the most basic type of flip-flop circuit. It can be constructed using NAND or NOR gates. Accordingly the latches are of two types :

1. NAND gate latch and
2. NOR gate latch

Both these types of latch are discussed one by one in the following pages.

72.4. NAND Gate Latch

Construction. Fig. 72.2 shows a latch constructed from NAND gates. It is called NAND gate latch or simply NAND latch. As seen from this diagram, the two NAND gates are cross-coupled. This means output of NAND-1 is connected to one of the inputs of NAND-2. Similarly, the output of NAND-2 is connected to one of the inputs of NAND-1. The NAND gate latch outputs are labeled as Q and $\overline{Q}$ respectively. Under normal conditions, these outputs will always be the inverse (or complement) of each other. The latch has two inputs namely SET and CLEAR. The SET input sets Q output of the latch to 1 state while the CLEAR input sets the Q output to the 0 state.

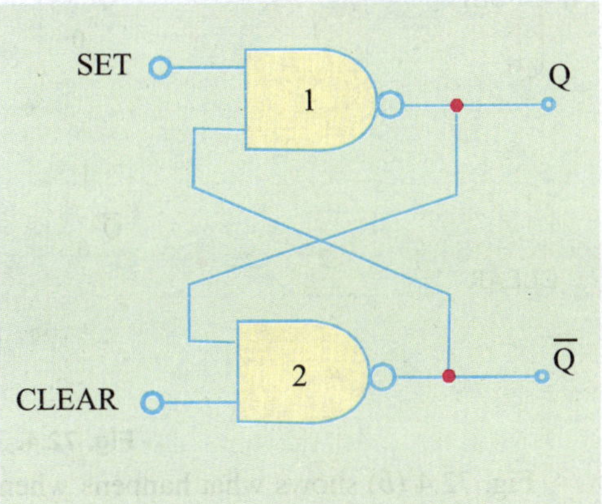

Fig. 72.2. NAND gate latch.

Operation. Let us now understand the operation of NAND gate latch. Normally, the SET and CLEAR inputs of the latch are resting in the HIGH state. Whenever we want to change the latch outputs, one of the two inputs will be pulsed LOW. Let us begin our study by showing that there are two equally likely output states when SET = CLEAR = 1. One possibility is shown in Fig. 72.3 (*a*) where we have $Q = 0$ and $\overline{Q} = 1$. With $Q = 0$, the inputs to the NAND-2 are 0 and 1, which produces $\overline{Q} = 1$. The 1 from $\overline{Q}$ causes NAND-1 to have 1 at both inputs to produce a 0 output at Q. Thus we find that a LOW at the NAND-1 output produces a HIGH at NAND-2 output which in turn keeps the NAND-1 output LOW.

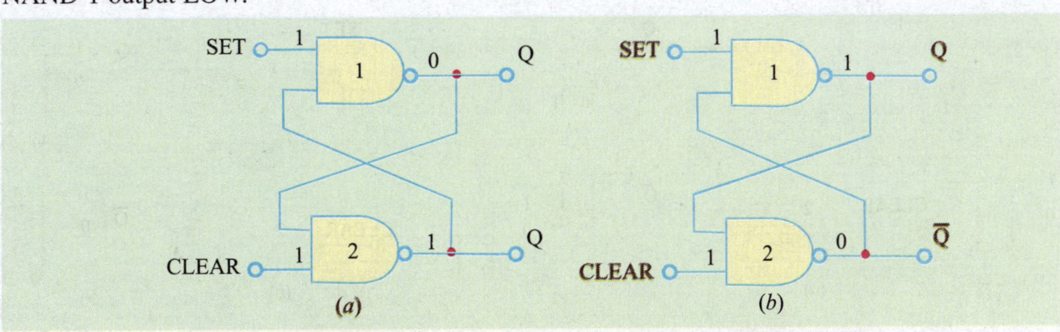

Fig. 72.3. Illustrating the two possible output states of NAND gate latch.

Fig. 72.3 (*b*) shows the second possibility. Here $Q = 1$ and $\overline{Q} = 0$. As seen from this figure, the HIGH from NAND-1 produces a LOW at the NAND-2 output, which in turn keeps the NAND-1 output HIGH. Thus there are two possible output states when SET = CLEAR = 1. However which one of these two states exist will depend on what has occurred previously at the inputs. This is discussed below in more detail where we will take up three situations : (1) setting the latch, (2) clearing the latch and simultaneous setting and clearing the latch.

1. Setting the Latch. We have already mentioned above that if SET = CLEAR = 1, the output Q can be in two possible states *i.e.* $Q = 0$ or $Q = 1$. Now we will study as what happens when the SET input is momentarily pulsed LOW while CLEAR is kept HIGH.

Fig. 72.4 (*a*) shows what happens when $Q = 0$ prior to the occurrence of the pulse. As SET input is pulsed LOW at time 't_0', Q will go HIGH. This will force $\overline{Q}$ to go LOW so that NAND-1 has now two LOW-inputs. Thus when SET returns to 1 state at t_1, the NAND-1 output remains HIGH which in turn keeps the NAND-2 output LOW.

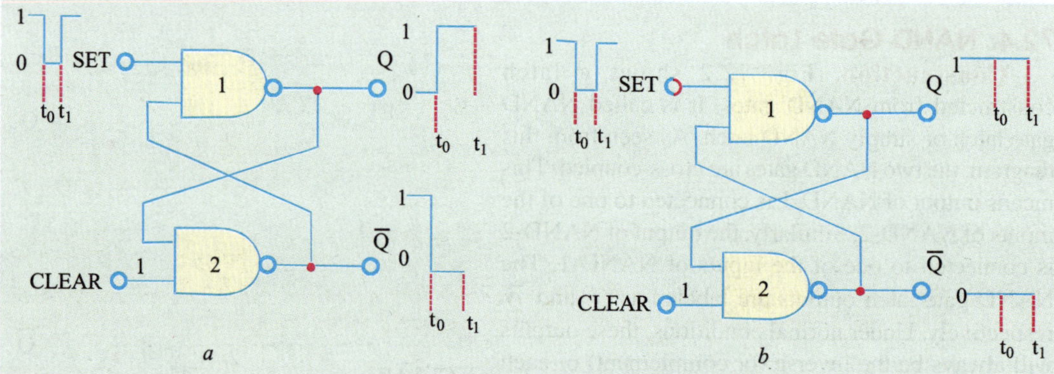

Fig. 72.4. Setting the NAND gate latch

Fig. 72.4 (*b*) shows what happens when $Q = 1$ and $\overline{Q} = 0$. Prior to the application of the SET pulse. Since $\overline{Q} = 0$ is already keeping the NAND-1 output HIGH, the LOW pulse at SET will not change anything. Thus when SET returns HIGH, the latch outputs are still in the $Q = 1$, $\overline{Q} = 0$ state (i.e no change in states).

2. Clearing the latch. Now we will study what happens when CLEAR input is pulsed LOW while SET is kept HIGH. Fig. 72.5 (*a*) shows the situation when $Q = 0$ and $\overline{Q} = 1$ prior to the application of the pulse. Since $Q = 0$ is already keeping the NAND-2 output HIGH, the LOW pulse at CLEAR will not have any effect. When CLEAR returns HIGH, the latch outputs are still $Q = 0$ and $\overline{Q} = 1$.

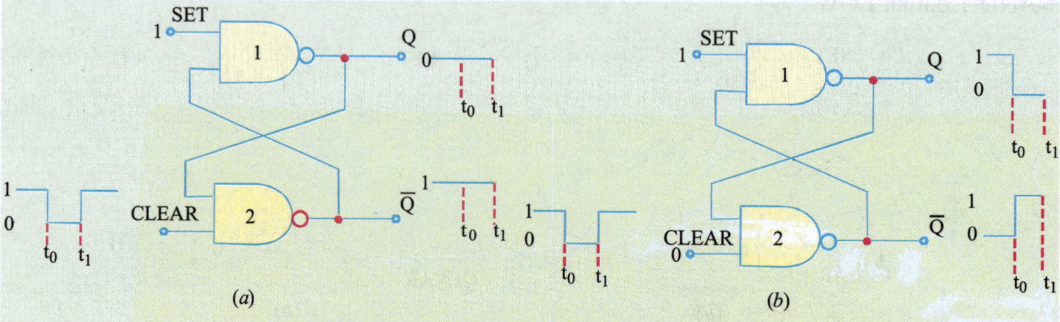

Fig 72.5. Clearing the NAND gate latch.

Fig 72.5 (*b*) shows the situation where $Q = 1$ prior to the occurrence of the CLEAR pulse. As CLEAR is pulsed LOW AT t_0, $\overline{Q}$ will go HIGH and this HIGH forces Q to go LOW because NAND-2 now has two LOW inputs. Thus when CLEAR returns HIGH at t_1, the NAND-2 output remains HIGH. This in turn keeps the NAND-1 output LOW.

3. Simultaneous setting and clearing. If the SET and CLEAR inputs are simultaneously pulsed LOW, this will produce HIGH levels at both NAND outputs so that $Q = \overline{Q} = 1$. This is an undesirable condition as the two outputs are supposed to be inverse of each other. Moreover, when the SET and CLEAR inputs return HIGH, the resulting output state will depend on which input returns HIGH first. Simultaneous transitions back to 1 state will produce unpredictable results. Because of these reasons the SET = CLEAR = 0 condition is normally not used for the NAND latch and is considered as invalid condition.

Inputs		*Output*
SET	*CLEAR*	
1	1	No change
0	1	$Q = 1$
1	0	$Q = 0$
0	0	$Q = \overline{Q} = 1$ (Invalid)

Fig. 72.6. Truth table for NAND gate latch.

The operation of NAND gate latch may be summarized in the form of a truth table as shown in Fig 72.6. Each line of the truth table is described as below :

1. **SET = CLEAR = 1.** This condition is the normal resting state of the latch.The Q and $\overline{Q}$ outputs will remain in the same state in which they were prior to this input condition.
2. **SET = 0, CLEAR = 1.** This condition will always cause the output to go to $Q = 1$ state where it will remain even after SET returns HIGH. This is called setting the latch.
3. **SET = 1, CLEAR = 0.** This condition will always produce $Q = 0$. The output will remain in this state even after CLEAR returns HIGH. This is called clearing or resetting the latch.
4. **SET = CLEAR = 0.** This condition tries to set and clear the latch simultaneously. It produces invalid results and should not be used.

72.5. Alternative Representations of NAND Gate Latch

We have already discussed in the last article about the NAND gate latch operation. It was mentioned that both SET and CLEAR inputs are active-LOW. Further the SET input will set $Q = 1$ when SET goes LOW. On the other hand, the CLEAR input will clear $Q = 0$ when CLEAR goes LOW. Because of this reason NAND gate latch is often drawn using the alternative representation for each NAND gate as shown in Fig 72.7 (*a*). (Recall from Art 70-14 that a NAND gate is equivalent to a bubbled or gate). The bubbles on the inputs of OR gate as well as labelling of the signals as $\overline{\text{SET}}$ and $\overline{\text{CLEAR}}$ indicate the active-LOW status of these inputs.

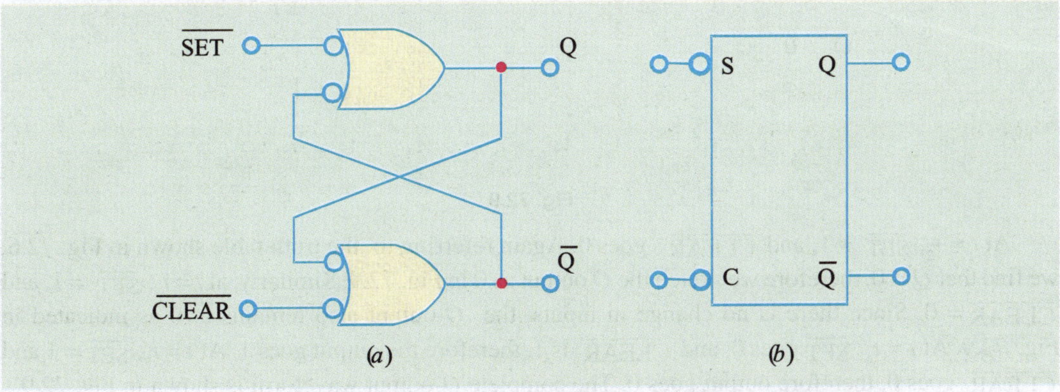

Fig. 72.7. Alternative representations of NAND gate latch.

Fig. 72.7 (*b*) shows a block representation. The *S* and *C* labels represent the SET and CLEAR inputs. While the bubbles at *S* and *C* inputs indicate the active LOW nature of these inputs. So remember whenever we use this block symbol, it represents a NAND gate latch.

Note. The action of ***Clearing*** a flip-flop or a latch is also called ***resetting.*** Both these terms (*i.e.* clearing or resetting) is used interchangeably in the field of digital electronics. Thus a SET-CLEAR latch can also be refered to as SET-RESET latch (or simply *S* - *R* latch).

Example 72.1. *The waveforms shown in Fig* 72.8 (*a*) *are applied to the inputs of the NAND latch shown in Fig.* 72.8 (*b*). *Assume that initially* $Q = 0$ *and determine the* Q – *waveform.*

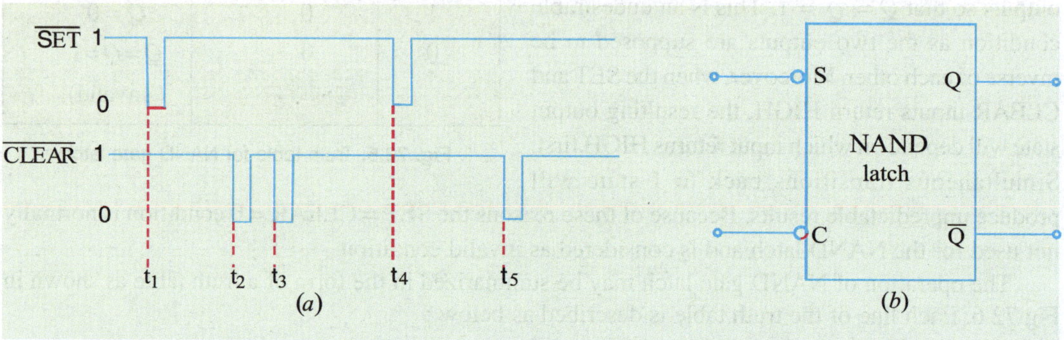

Fig. 72.8

Solution Given :

In order to determine the waveform at *Q*-output, refer to the truth table shown in Fig. 72.6 and the input waveforms. Notice that prior to $t = t_1$, $\overline{\text{SET}} = \overline{\text{CLEAR}} = 1$ and $Q = 0$, At $t = t_1$, $\overline{\text{SET}}$ goes 0 and $\overline{\text{CLEAR}}$ stays at 1. Referring to the truth table in Fig. 72-6, we find that if $\overline{\text{SET}} = 0$, $\overline{\text{CLEAR}} = 1$, $Q = 1$. Therefore the Q – output is indicated as 1 at $t = t_1$ in Fig. 72.9.

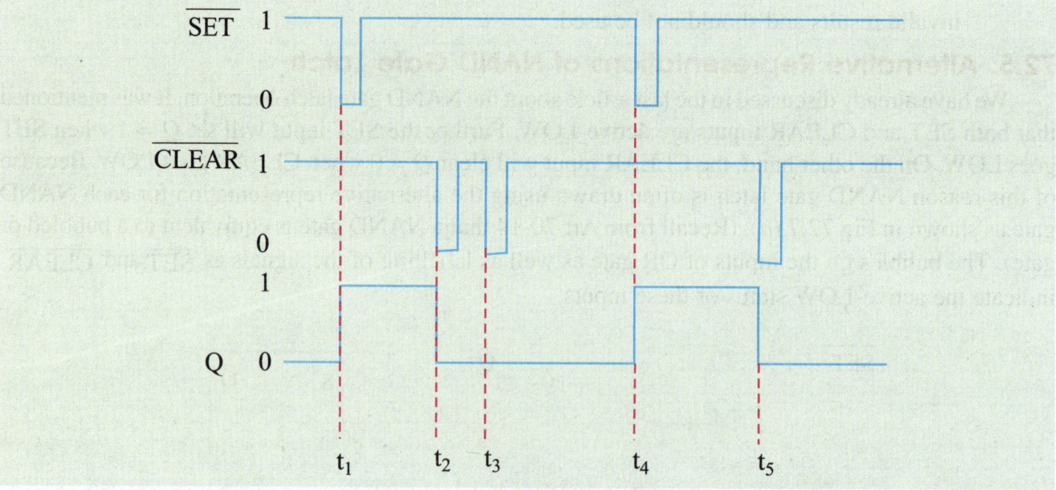

Fig. 72.9

At $t = t_2$, $\overline{\text{SET}} = 1$, and $\overline{\text{CLEAR}}$ goes 0. Again referring to the truth table shown in Fig. 72.6, we find that $Q = 0$, therefore we sketch the Q output as 0 in Fig. 72.9. Similarly, at $t = t_3$, $\overline{\text{SET}} = 1$, and $\overline{\text{CLEAR}} = 0$. Since there is no change in inputs, the *Q*-output also remains at 0 as indicated in Fig. 72.9. At $t = t_4$, $\overline{\text{SET}}$ goes 0, and $\overline{\text{CLEAR}}$ is 1, therefore the output goes 1. At $t = t_5$, $\overline{\text{SET}} = 1$ and $\overline{\text{CLEAR}}$ goes 0, therefore output goes 0. The complete *Q*-output waveform is shown in Fig. 72.9.

Example 72.2. *If the* $\overline{SET}$ *and* $\overline{CLEAR}$ *waveforms shown in Fig 72.10 (a) are applied to the inputs of NAND latch shown in Fig. 72.10 (b) determine the waveform that will be observed on the Q output. Assume that initially Q = 0.*

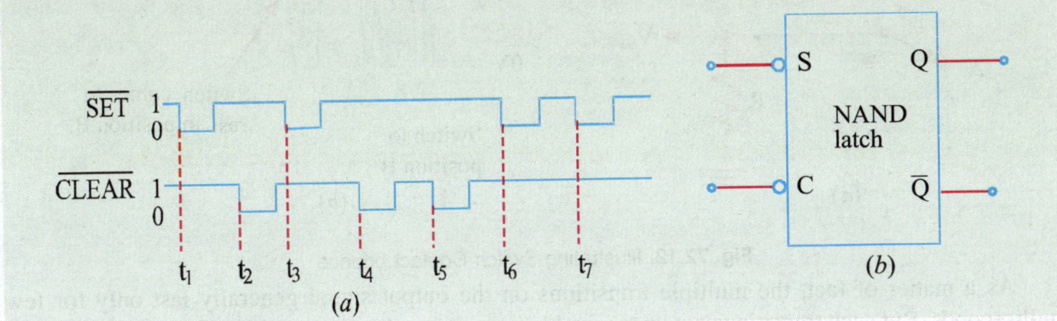

Fig. 72.10

Solution :

In order to determine, the Q-output, let us refer to the truth table shown in Fig. 72 .6 and the input waveforms. At $t = t_1$, $\overline{SET}$ goes from 1 to 0, $\overline{CLEAR}$ = 1, so the output Q goes from 0 to 1 as shown in Fig 72.11. At $t = t_2$, $\overline{SET}$ = 1, $\overline{CLEAR}$ goes from 1 to 0, so does the Q-output (*i.e.* it also goes from 1 to 0). At $t = t_3$, $\overline{SET}$ goes from 1 to 0, $\overline{CLEAR}$ = 1, therefore Q-output goes 1, At $t = t_4$, SET =1, $\overline{CLEAR}$ goes to 0, therefore Q-output goes 0 and so on. The complete waveform at Q-output is shown in Fig. 72.11.

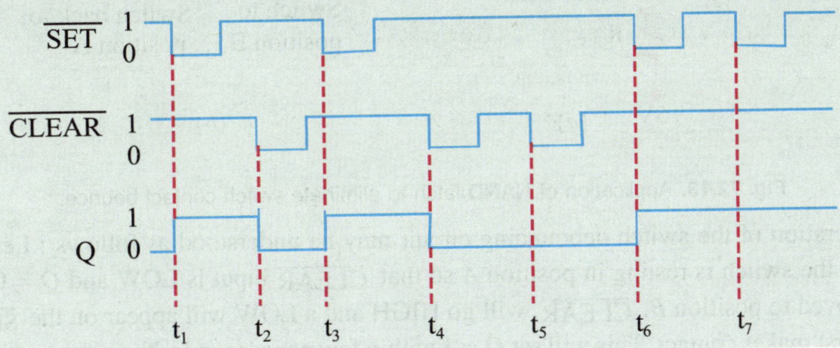

Fig. 72.11

72.6. Application of NAND Latch to Debounce a Mechanical Switch

Consider a mechanical switch with contact points A and B connected to V_{cc} (+ 5V) supply and ground respectively as shown in the Fig. 72.12 (*a*). It has been found experimentally that when the switch moves from contact position A to B, it produces several output voltage transitions as shown in Fig. 72.12 (*b*). It is due to a phenomenon called ***switch bounce*** *i.e.* the switch makes and breaks contact with contact B several times before coming to rest on contact B. In a similar manner, when the switch moves from contact position B to A, it again produces several output voltage transitions before coming to rest on contact A.

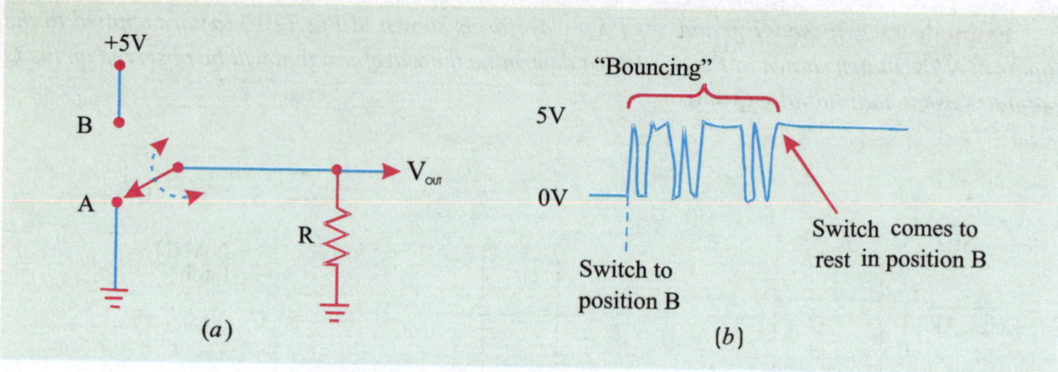

Fig. 72.12. Illustrating Switch Contact bounce

As a matter of fact, the multiple transitions on the output signal generally last only for few milliseconds. But such transitions are unacceptable in many applications. A NAND latch can be used as shown in Fig. 72.13 (*a*) to eliminate the switch bounce.

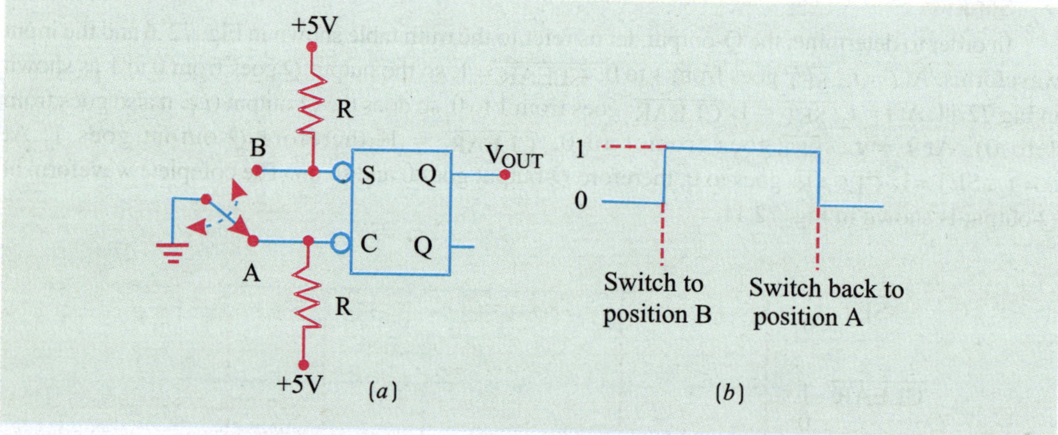

Fig. 72.13. Application of NAND latch to eliminate switch contact bounce.

The operation of the switch debouncing circuit may be understood as follows : Let us assume that initially the switch is resting in position *A* so that $\overline{\text{CLEAR}}$ input is LOW and $Q = 0$. When the switch is moved to position *B*, $\overline{\text{CLEAR}}$ will go HIGH and a LOW will appear on the $\overline{\text{SET}}$ input as the switch first makes contact. This will set $Q = 1$ within few nanoseconds. Now if the switch bounces off contact *B*, $\overline{\text{SET}}$ and $\overline{\text{CLEAR}}$ will both be HIGH and *Q* will not be affected, *i.e.* it will stay HIGH. Thus nothing will happen at *Q* as the switch bounces on and off contact *B* before finally coming to rest in position *B* as shown in Fig. 72.13 (*b*).

In a similar manner, when the switch is moved from position *B* back to position *A*, it will place a LOW on the $\overline{\text{CLEAR}}$ input as it first makes contact. This clears *Q* to the LOW state. It will remain there ever if the switch bounces on and off contact *A* several times before coming to rest.

It is evident from the above discussion that the output at *Q* will consists of single transition each time the switch is moved from one position to the other.

72.7. NOR Gate Latch

Fig. 72.14 (*a*) shows a latch constructed from two cross-coupled NOR gates. It is called NOR gate latch or simply the NOR latch. As seen in the figure, the arrangement is similar to the NAND gate latch (shown in Fig 72.2 on page 2633) except that the Q and $\overline{Q}$ outputs have reversed positions.

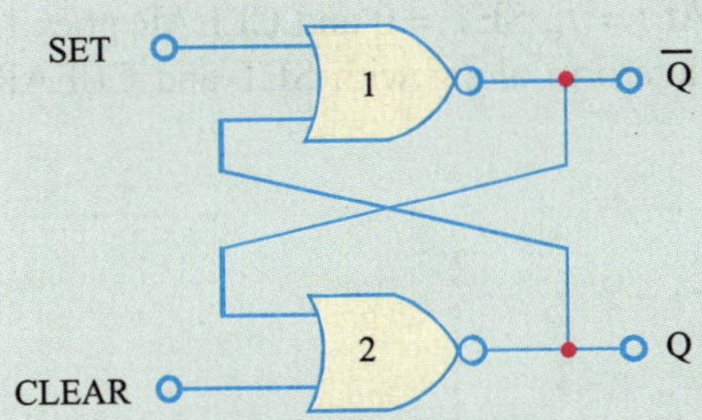

(*a*) Construction

Inputs		*Output*
SET	*CLEAR*	
0	0	No change
1	0	$Q=1$
0	1	$Q=0$
1	1	*Invalid

(*b*) Truth table

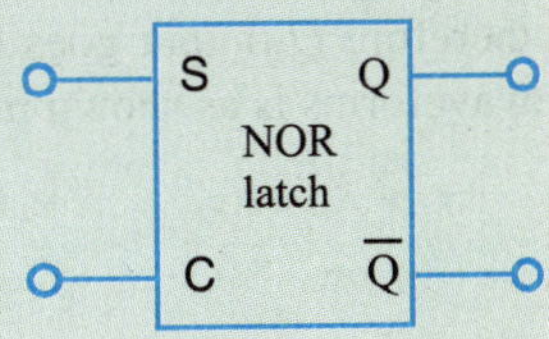

(*c*) Logic symbol

Fig. 72.14. NOR gate latch.

The detailed operation of the NOR latch can be understood exactly in the same manner as for the NAND latch. The results are given in the truth table as shown in Fig 72.14 (*b*) and are summarized as follows :

1. **SET = CLEAR = 0.** This condition is the normal resting state for the NOR latch. It has no effect on the output state. In other words, Q and $\overline{Q}$ will remain in the same state in which they were prior to this input condition.
2. **SET = 1, CLEAR = 0.** This condition will always cause the output to go to Q = 1 state where it will remain even after SET returns to 0.
3. **SET = 0, CLEAR = 1.** This condition will always cause the output to go to Q = 0 state where it will remain even after CLEAR returns to 0.
4. **SET = 1, CLEAR = 1.** This condition tries to set and clear the latch simultaneously. It produces invalid results and should not be used.

It may be carefully noted that the NOR latch operates exactly in the same manner as the NAND latch. However, the SET and CLEAR inputs are active-HIGH rather than active-LOW. Moreover the normal resting state is SET = CLEAR = 0. Further the output Q will be set HIGH by a HIGH pulse on the SET input and it will be cleared LOW by a HIGH pulse on the CLEAR input. Fig 72.14 (*c*) shows the logic symbol for the NOR latch. Notice that there are no bubbles on the S and C inputs (unlike NAND latch) which indicates that these inputs are active-HIGH.

Example 72.3. *The waveforms shown in Fig.* 72.15 (*a*) *are applied to the inputs of the NOR latch shown in Fig.* 72.15 (*b*). *Assume that initially,* Q = 0 *and determine the Q-waveform.*

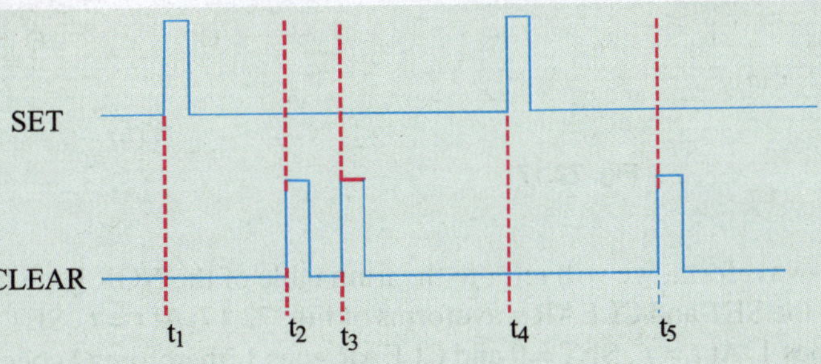

(*a*)

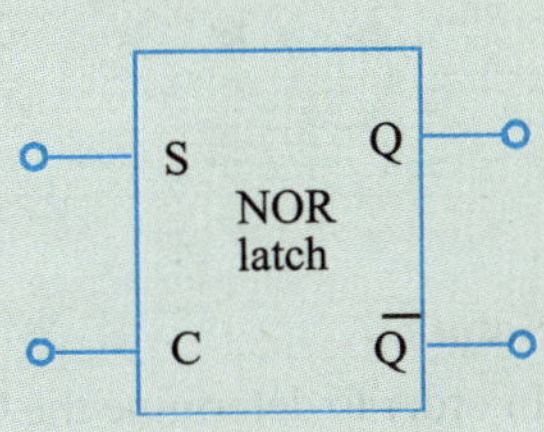

(*b*)

Fig. 72.15

Solution :

In order to determine the Q-waveform, we will refer to the truth table of the NOR latch (shown in

Fig. 72.14 (*b*) page 2639) and the SET and CLEAR waveforms of Fig. 72.15 (*a*). At $t = t_1$, SET goes 1 and CLEAR is 0, therefore *Q*-output also goes 1. AT $t = t_2$, SET = 0 and CLEAR goes 1, therefore the *Q*-output goes 0. At $t = t_3$, SET = 0 and CLEAR goes 1 again, therefore the Q-output remains 0. At $t = t_4$, SET goes 1 and CLEAR is 0, therefore *Q*-output goes 1. At $t = t_5$, SET = 0 and CLEAR goes 1, therefore *Q*-output goes 0. The complete sketch of the *Q*-waveform along with SET and CLEAR waveforms is as shown in Fig. 72.16.

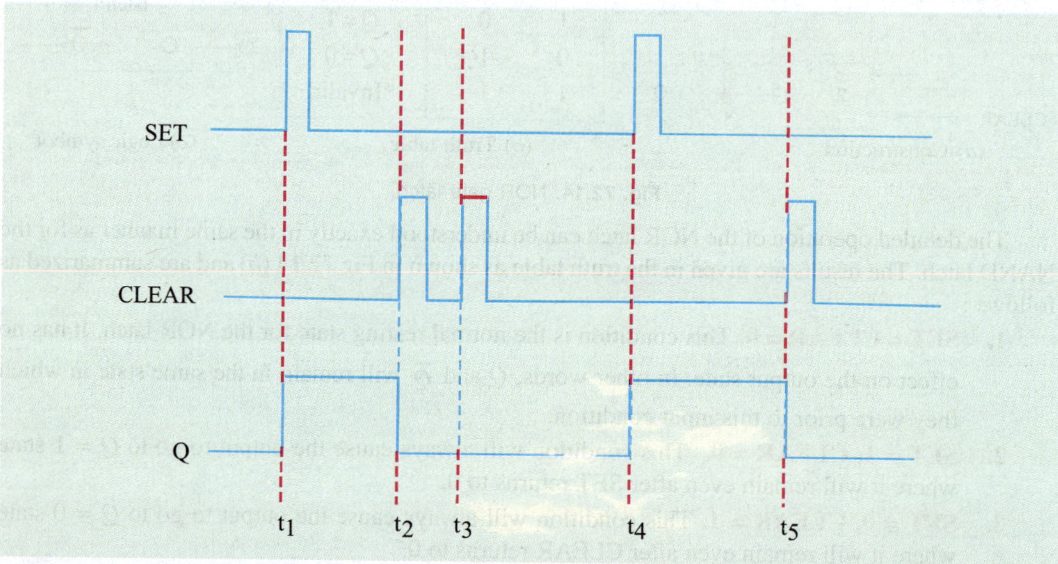

Fig. 72.16

Example 72.4. *If the SET and CLEAR waveforms shown in Fig. 72.17. (a) are applied to the inputs of NOR latch shown in Fig. 72.17. (b), determine the waveform that will be observed on the Q-output. Assume that initially Q = 0.*

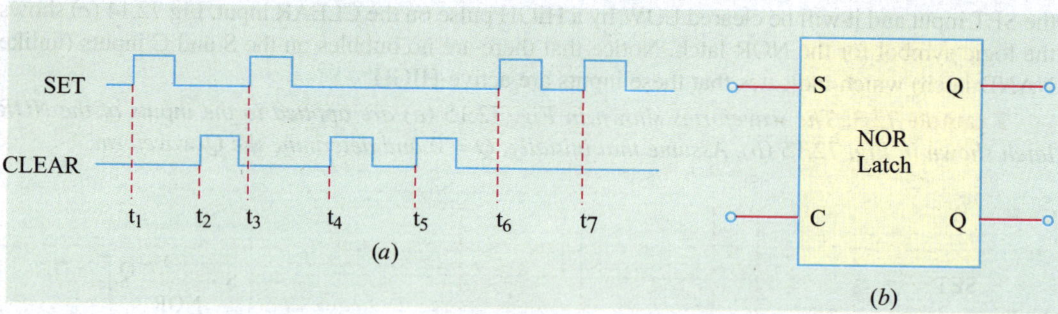

Fig. 72.17

Solution :

In order to determine the Q-waveform, we will refer to the truth table of the NOR latch (shown in Fig. 72.14 (*b*) page 2639) and the SET and CLEAR waveforms of Fig. 72.17. At $t = t_1$, SET goes 1 and CLEAR = 0, therefore *Q* goes 1. At $t = t_2$, SET = 0 and CLEAR goes 1, therefore *Q* goes 0. At $t = t_3$, SET = 1 again, and CLEAR = 0, therefore *Q* goes 1 again. At $t = t_4$, SET = 0. and CLEAR = 1, therefore *Q* goes 0 and so on. A complete sketch of *Q* waveform along with input waveforms is shown in Fig. 72.18.

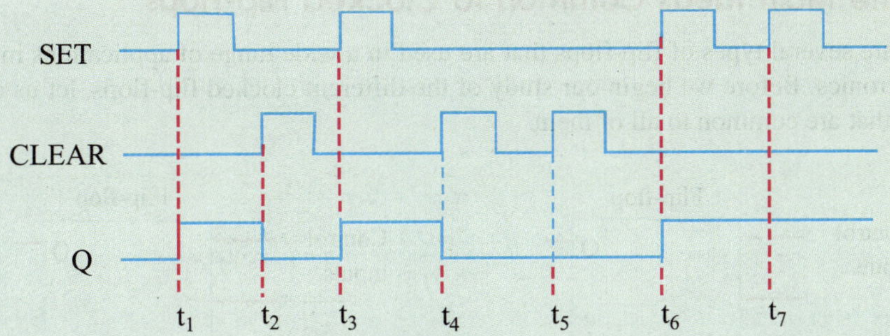

Fig. 72.18

72.8. Clocked Signals

Strictly speaking, the digital systems are of the following two types :

1. Asynchronous systems. In these systems, the outputs of logic circuits can change state any time one or more of the inputs change. Generally, an asynchronous system is more difficult to design and troubleshoot than a synchronous system.

2. Synchronous systems. In these systems, the exact time at which the output can change states are determined by a signal commonly called a ***clock.***

The clock signal is generally a rectangular pulse train as shown in Fig. 72.19 (*a*) or a square wave as shown in Fig. 72.19 (*b*). The clock signal is distributed to all parts of the digital system and most of the system outputs can change state only when the clock makes a transition. The transitions are more commonly referred to as ***edges*** and are pointed out in Fig. 72.19.

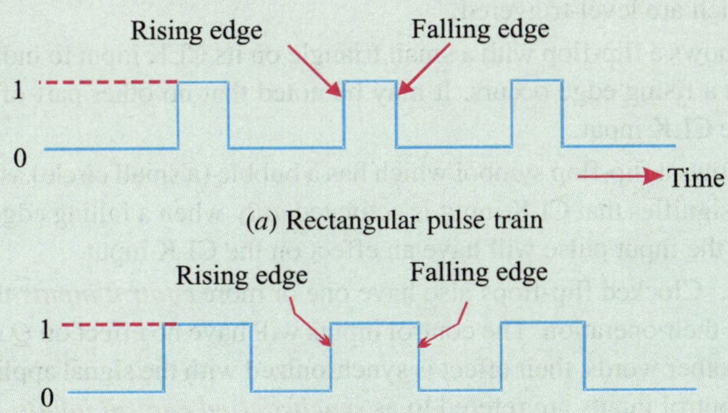

(*a*) Rectangular pulse train

(*b*) Square wave

Fig. 72.19

It may be noted that when the clock changes from 0 to 1, this is called rising edge or ***positive-going transition*** (PGT). On the other hand, when the clock changes from 1 to 0, this is called falling edge or ***negative going transition*** (NGT).

As a matter of fact, most digital systems are principally synchronous (although there are always some asynchronous parts). It is due to the fact that synchronous circuits are easier to design and troubleshoot. The synchronization is accomplished through the use of ***clocked flip-flops*** that are designed to change states on one or the other of the clock's transitions.

72.9. Some Main Ideas Common to Clocked Flip-Flops

There are several types of flip-flops that are used in a wide range of applications in the field of digital electronics. Before we begin our study of the different clocked flip-flops, let us describe the main ideas that are common to all of them.

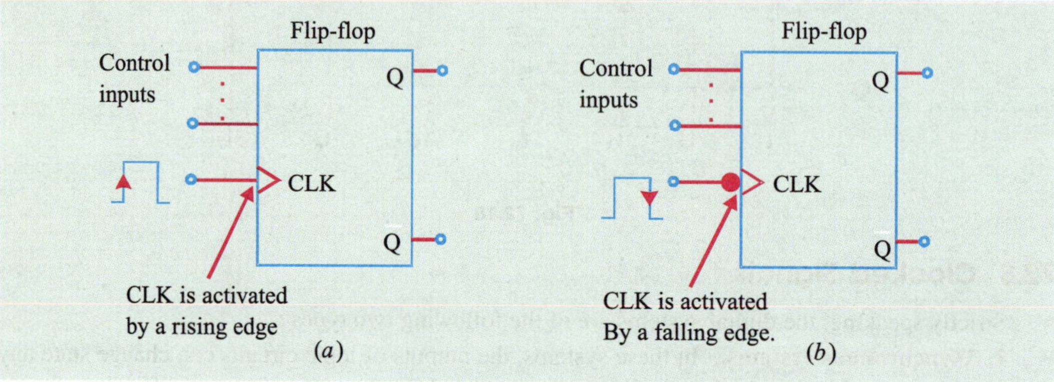

Fig 72.20. Clocked Flip-Flops.

Fig. 72.20. (*a*) and (*b*) shows the logic symbols for a typical clocked flip-flop. As seen, a clocked flip-flop has a clock input and some control inputs. There are described below in more detail.

CLK Input. Clocked flip-flops have a ***clock*** input that is labeled as CLK, CK or CP. However, we will normally use CLK as shown in Fig. 72.20 (*a*) and (*b*). In most clocked flip-flops, the CLK input is ***edge-triggered.*** This means the CLK input is activated by a signal transition. The edge-triggered activation is indicated by the presence of a small triangle on the CLK input. This contrasts with the latches, which are level-triggered.

Fig. 72.20 (*a*) shows a flip-flop with a small triangle on its CLK input to indicate that the input is activated only when a rising edge occurs. It may be noted that no other part of the input pulse will have an effect on the CLK input.

Fig 72.20 (*b*) shows a flip-flop symbol which has a bubble (a small circle) as well as a triangle on its CLK input. This signifies that CLK input is activated ***only*** when a falling edge occurs. Again note that no other part of the input pulse will have an effect on the CLK input.

Control Inputs. Clocked flip-flops also have one or more ***control inputs*** that can have various names depending on their operation. The control inputs will have no effect on Q until the active clock transition occurs. In other words, their effect is synchronized with the signal applied to CLK. Because of this reason, the control inputs are refered to as ***synchronized control inputs.***

72.10. Setup and Hold Times in Clocked Flip-Flops

Strictly speaking, there are two timing requirements that must be met if a clocked flip-flop is to respond relialably to its control inputs when the active CLK edge occurs. These two requirements are : **(1)** set up time, t_S and **(2)** hold time, t_H . Both these requirements are discussed below.

1. Set up time, t_s. It is the time interval immediately preceeding the active edge of the CLK signal during which the control input must be maintained at the proper level. The situation is illustrated in Fig 72.21 (*a*) for a flip-flop that triggers on the rising edge. Usually, the IC manufacturers specify the minimum allowable set up time, t_S (min). If this time requirement is not met, the flip-flop may not respond reliably when the clock edge occurs.

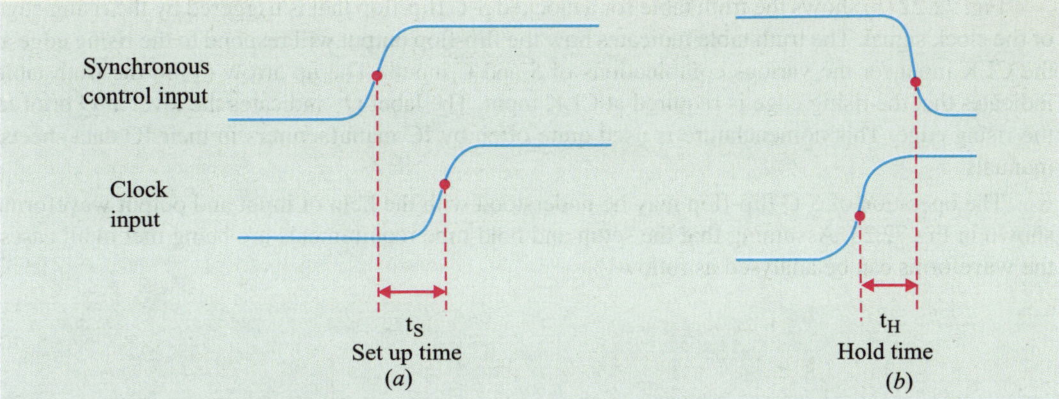

Fig. 72.21. Illustrating the set-up and hold time requirements, for a flip-flop that triggers on the rising edge.

2. Hold time, t_H. It is the time interval immediately following the active edge of the CLK signal during which the synchronous control input must be maintained at the proper level. Refer to Fig. 72.21 (*b*) usually the IC manufacturers specify the minimum acceptable value of hold time, t_H (min). If this requirement is not met, the flip-flop will not trigger reliably.

It is evident from the above discussion that in order to ensure that a clocked flip-flop will respond properly when the active clock edge occurs, the control inputs must be stable (*i.e.* unchanging) for at least a time interval.

1. t_S (min) prior to the active clock edge.
2. t_H (min) after the active clock edge.

A typical value of t_S (min) is in the range of 5 to 50 ns whereas hold times are generally from 0 to 10 ns. It may be carefully noted that the set up and hold times are measured between the 50 per cent points on the edges.

The set up and hold time requirements are extremely important in synchronous systems. This is due to the reason that there will be many situations where the synchronous control inputs to a flip-flop are changing at approximately the same time as the CLK input.

72.11. Clocked *S* - *C* Flip-Flop

Fig. 72.22 (*a*) shows the logic symbol for a clocked *S* - *C* flip-flop that is triggered by the rising edge of the clock signal. In other words this flip-flop can change output states only when a signal applied to its clock input makes a transition from 0 to 1. The *S* and *C* inputs control the state of the flip-flop in the same manner as discussed earlier for the NOR latch. But it may be noted that the flip-flop does not respond to these inputs until the occurrence of the rising edge of the clock signal.

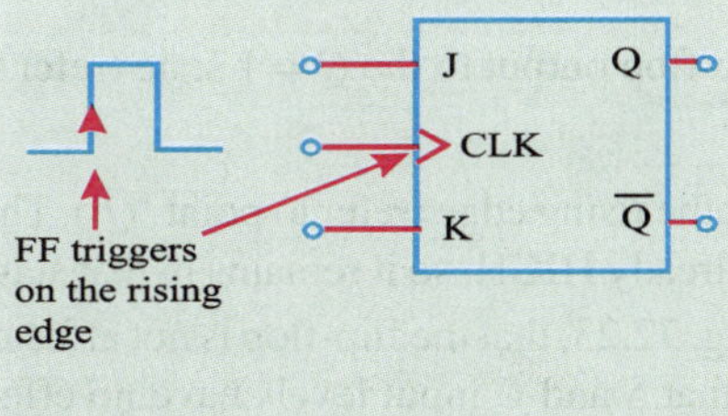

(*a*) Symbol

S (Inputs)	*C*	*CLK*	Output *Q*
0	0	↑	Q_0 (no change)
1	0	↑	1
0	1	↑	0
1	1	↑	Ambiguous

(*b*) Truth table

Fig. 72.22. Clocked *S-C* Flip-Flop.

Fig. 72.22 (*b*) shows the truth table for a clocked *S-C* flip-flop that is triggered by the rising edge of the clock signal. The truth table indicates how the flip-flop output will respond to the rising edge at the CLK input for the various combinations of *S* and *C* inputs. The up arrow (↑) in the truth table indicates that the rising edge is required at CLK input. The label Q_0 indicates the level at *Q* prior to the rising edge. This nomenclature is used quite often by IC manufacturers in their IC data sheets/ manuals.

The operation of *S-C* flip-flop may be understood with the help of input and output waveforms shown in Fig 72.23. Assuming that the setup and hold time requirements are being met in all cases, the waveforms can be analysed as follows :

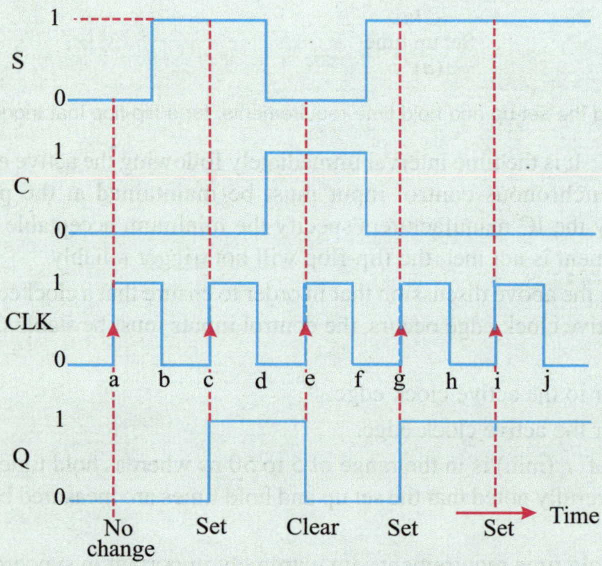

Fig. 72.23. Typical input and output waveforms of a *S-C* flip-flop.

1. Notice that initially all the inputs (S, C and CLK) are 0 and the *Q* output is assumed to be zero, *i.e.* $Q_0 = 0$.
2. When the rising edge of the first clock pulse occurs (refer to point '*a*'), both the *S* and *C* inputs are 0, so the flip-flop output is not affected and remains in the $Q = 0$ state (*i.e.* $Q = Q_0 = 0$).
3. At the occurrence of the rising edge of the second clock pulse (refer to point '*c*'), the *S* input is now HIGH, with *C* input still LOW. This causes the flip-flop output to set to 1 state.
4. At the occurrence of the rising edge of the third clock pulse (refer to point '*e*'), the *S* input is LOW with *C* input HIGH. This causes the flip-flop output to clear to 0 state.
5. The rising edge of the fourth clock pulse sets the flip-flop output to the $Q = 1$ state (refer to point '*g*') because $S = 1$ and $C = 0$.
6. The fifth clock pulse also finds that $S = 1$ and $C = 0$ at the rising edge (refer to point '*Q*'). This situation will produce HIGH output. But since *Q* is already HIGH, so it remains in that state.

It may be carefully noted from the waveforms shown in Fig. 72.23, that the flip-flop is not affected by the falling edge of the clock pulses. It may also be noted that *S* and *C* input levels have no effect on the flip-flop except upon the occurrence of a rising edge of the clock signal.

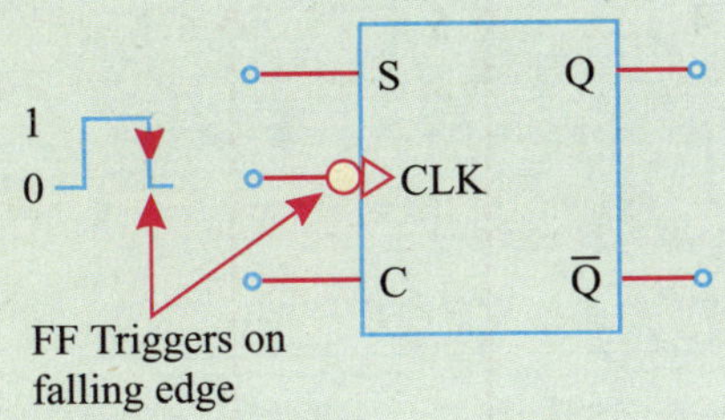

(*a*) Symbol

Inputs			*Output*
S	***C***	***CLK***	***Q***
0	0	↓	Q_0 (no change)
1	0	↓	1
0	1	↓	0
1	1	↓	Ambiguous

(*b*) Truth table

Fig. 72.24. Falling-edge triggered S-C flip-flop.

Fig. 72.24 (*a*) shows the symbol and 72.24 (*b*) the truth table for a clocked *S - C* flip-flop that triggers on the falling edge of the CLK input. Notice the presence of a small circle and a triangle on the CLK input of the flip-flop. These indicate that this flip-flop will trigger only when CLK input goes from 1 to 0. This flip-flop operates in the same way as the rising-edge flip-flop except that the output can change states only on the falling edge of the clock pulses (refer to points *b, d, f, h* and *j* in Fig. 72.23). In actual practice both the rising-edge and falling-edge triggered flip-flops are used in digital systems.

Example 72.5. *Fig. 72.25 (a) shows the SET and CLEAR waveforms applied at the inputs of the clocked S-C flip-flop shown in Fig. 72.25 (b)*

Sketch the waveforms at Q *and* $\overline{Q}$ *outputs of the flip-flop. Assume that initially* $Q = 0$ *and* $\overline{Q} = 1$.

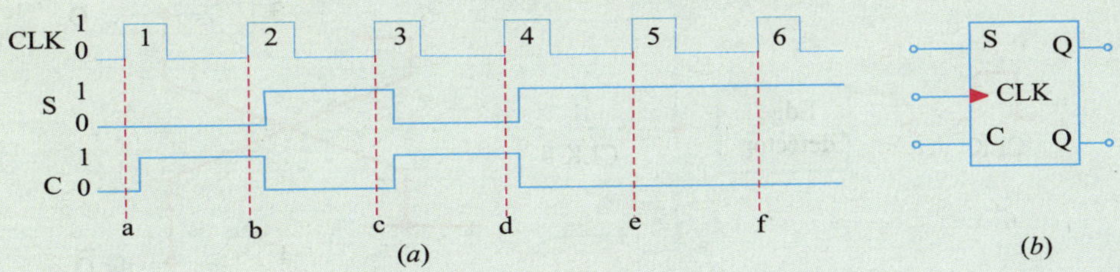

Fig. 72.25

Solution. Given :

In order to determine the *Q*-output, we will refer to the waveforms applied at the flip-flop inputs shown in Fig. 72.25 (*a*) and its truth table shown in Fig. 72.22 (*b*). At point '*a*', $S = C = 0$, therefore $Q_0 = 0$, At point '*b*', $S = 0$ and $C = 1$, therefore Q remains 0. At point '*c*', $S = 1$ and $C = 0$, therefore $Q = 1$. At point '*d*', $S = 0$ and $C = 1$, therefore $Q = 0$. At point '*e*', $S = 1$, $C = 0$, therefore $Q = 1$. At point '*f*', S and C inputs remain the same, therefore Q also remains 1. The sketch of Q-output along with the CLK, S and C waveforms is as shown in Fig. 72.26. The $\overline{Q}$-output waveform is determined by inverting the Q-waveform.

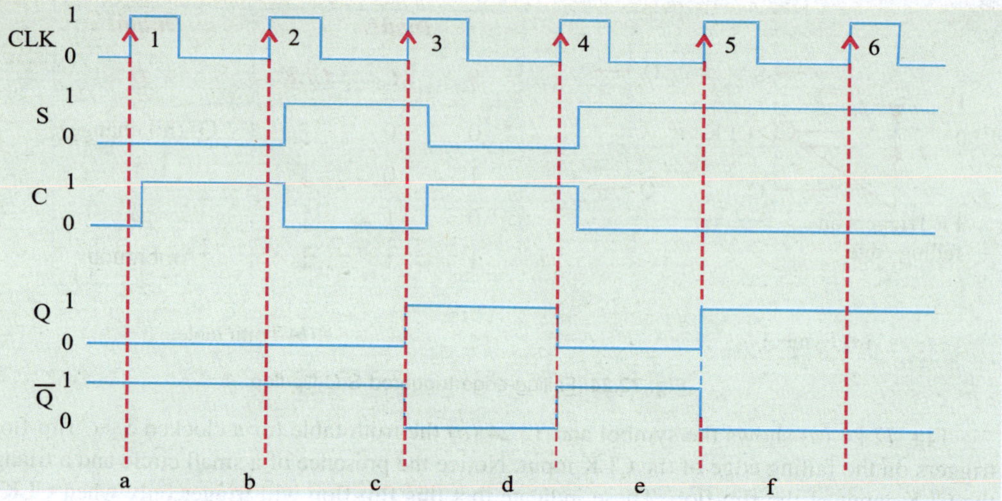

Fig. 72.26

72.12. Internal Circuitry of an Edge-triggered S-C Flip-Flop

Fig . 72.27 shows a simplified version of an internal circuitry of an edge triggered *S-C* flip-flop. Notice that the circuit contains the following three sections :

1. A basic NAND latch
2. A pulse steering circuit and
3. An edge-detector circuit

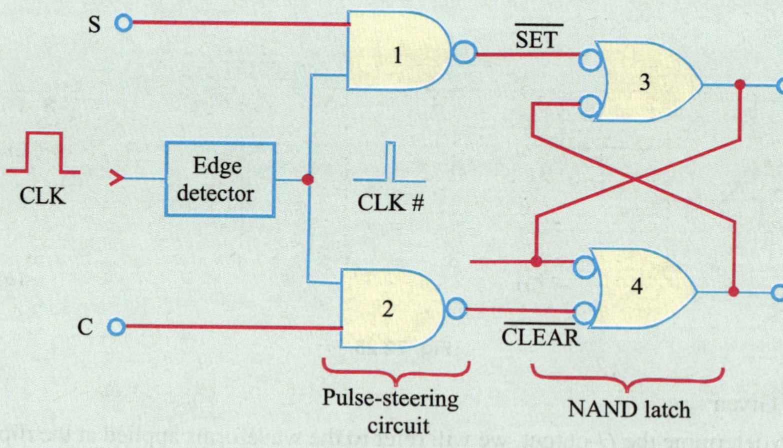

Fig. 72.27. Internal circuitry of an edge-triggered S-C flip-flop.

As seen from the Fig. 72.27, the edge detector produces a narrow positive-going spike (CLK #). The pulse-steering circuit "steers" (or moves) the spike through to the SET and CLEAR input of the latch in accordance with the levels present at *S* and *C* inputs. For example with *S* = 1 and *C* = 0, the spike is inverted and passed through NAND gate-1 to produce a LOW pulse at the SET input of the latch that sets *Q* = 1. With *S* = 0 and *C* = 1, the spike is inverted and passed through NAND gate-2 to produce a LOW pulse at the CLEAR input of the latch that resets *Q* = 0.

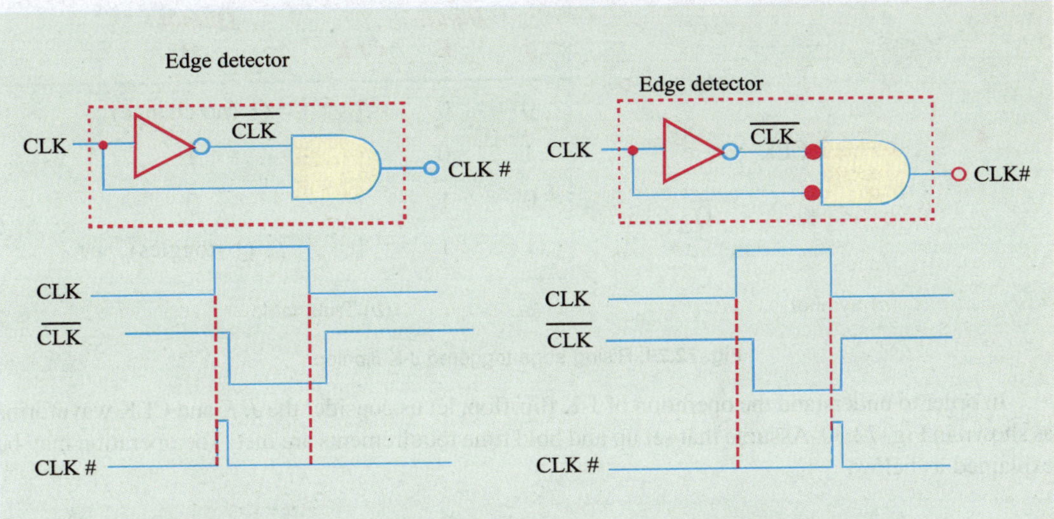

(*a*) Rising edge (*b*) Falling edge

Fig. 72.28. Implementation of edge-detector circuits used in edge triggered flip-flops.

Fig. 72.28 shows the implementation of edge-detector circuits used in edge-triggered flip-flops. Thus Fig. 72.28 (*a*) shows how the spike is generated for edge-triggered flip-flops that trigger on the rising edge of the CLK pulse. As seen from this figure, the INVERTER produces an output with a delay of a few nanoseconds. Because of this the transitions of $\overline{\text{CLK}}$ occur a little bit after that of CLK. Next the AND gate produces an output spike that is HIGH only for the few nanoseconds when both CLK and $\overline{\text{CLK}}$ are HIGH. As a result of this, we get a narrow pulse at edge detector output (CLK #) which occurs on the rising edge of the CLK.

Similarly Fig 72.28(*b*) produces CLK # on the falling edge of CLK for flip-flops that are to trigger on the falling edge.

It may be noted that since the CLK # signal is HIGH for only a few nanoseconds, the *Q* output is affected by the levels at *S* and *C* only for a short time during and after the occurrence of the active edge of CLK. This gives the flip-flop its edge-triggered property.

72.13. Clocked *J-K* Flip-Flop

Fig. 72.29 (*a*) shows the symbol and (*b*) the truth table for a clocked *J-K* flip-flop that is triggered by the rising edge of the clock signal. The *J* and *K* inputs control the state of the flip-flop in the same manner as the *S* and *C* inputs do for the *S-C* flip-flop. However there is one major difference-the $J = K = 1$ condition in *J-K* flip-flop does not result in an ambiguous output unlike $S = C = 1$ in *S-C* flip-flop for, $J = K = 1$ condition, the J-K flip-flop will always go to its opposite state upon the rising edge of the clock signal. This is called ***trigger mode.*** In this mode, if both *J* and *K* are left HIGH, the flip-flop will change states (i.e. toggle) for each rising edge of the clock.

The operation of *J-K* flip-flop for each combination of *J* and *K* is summarized in Fig. 72.29 (*b*) Notice that the truth table is the same except for $J = K = 1$ condition. This condition results in $Q = \overline{Q}_0$ which means that the new value of *Q* will be the inverse of the value it had prior to the rising edge of the clock pulse. It is called toggle operation.

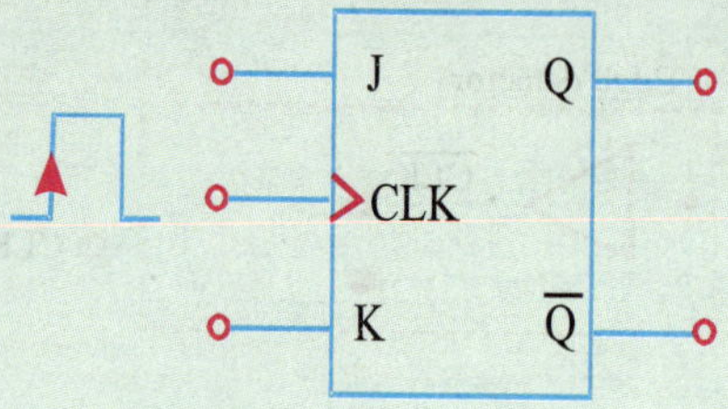

(a) Symbol

Inputs			Output
J	*K*	*CLK*	*Q*
0	0	↑	Q_0 (no change)
1	0	↑	1
0	1	↑	0
1	1	↑	$\overline{Q_0}$ (toggles)

(b) Truth table

Fig. 72.29. Rising edge triggered J-K flip-flop.

In order to understand the operation of J-K flip-flop, let us consider the *J, K* and CLK waveforms as shown in Fig. 72.30. Assume that set up and hold time requirements are met. The operation may be explained as below.

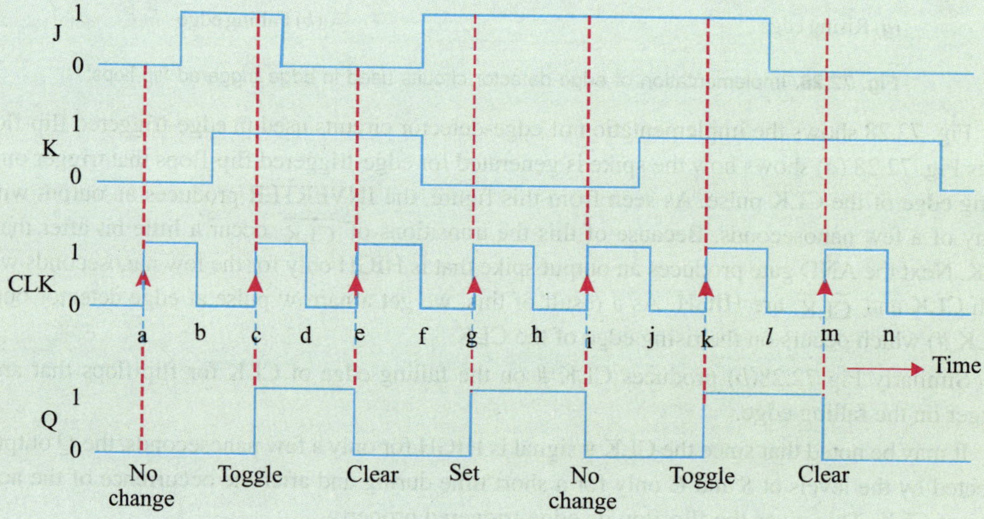

Fig. 72.30

1. Initially all the inputs are 0, and the *Q* output is also assumed to be 0, *i.e.* $Q_0 = 0$.
2. When the rising edge of the first clock pulse occurs (refer to point *a*), the $J = K = 0$ condition exists. Thus the flip-flop does not change its output state, *i.e.* $Q = Q_0 = 0$.
3. When the rising edge of the second clock pulse occurs (refer to point *c*), the $J = K = 1$ condition exists. Thus the flip-flop toggles to its opposite state *i.e.* $Q = \overline{Q}_0 = \overline{0} = 1$.
4. When the rising edge of the third clock pulse occurs (refer to point *e*), $J = 0$ and $K = 1$ condition exists. Thus the flip-flop is cleared to the $Q = 0$ state.
5. When the rising edge of the fourth clock pulse occurs (refer to point *g*), $J = 1$ and $K = 0$ condition exists. This condition sets the output *Q* to 1 state.
6. When the rising edge of fifth clock pulse occurs (refer to point *i*), $J = 1$ and $K = 0$ condition exists. This is the condition that sets the output *Q* to 1 state. However since *Q* is already 1, so it will remain there. Hence no change in the output state.
7. When the rising edge of six clock pulse occurs (refer to point *k*), $J = K = 1$ condition exists. This condition causes the flip-flop to toggle to its opposite state.

8. When the rising edge of seventh clock pulse occurs (refer to point m), $J = 0$ and $K = 1$ condition exists. This condition causes the flop-flop to clear to $Q = 0$ state.

It may be noted from the waveforms that the flip-flop is not affected by the falling edge of the clock pulses. Also the *J* and *K* input levels have no effect except the occurrence of the rising edge of the clock signal. The *J* and *K* inputs by themselves cannot cause the flip-flop to change states.

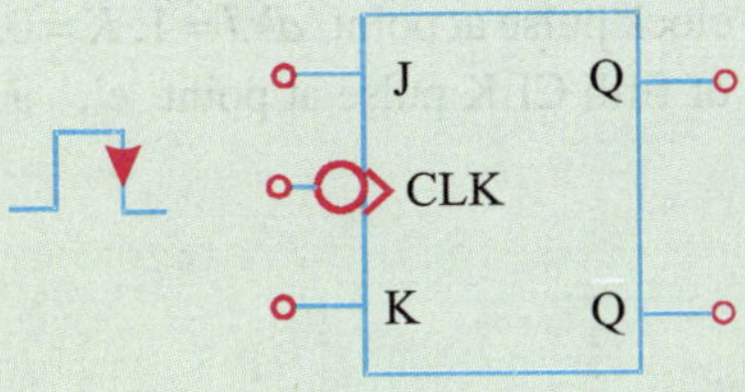

(*a*) Symbol

Inputs			Output
J	*K*	*CLK*	*Q*
0	0	↓	Q_0 (no change)
1	0	↓	1
0	1	↓	0
1	1	↓	Q_0 (toggles)

(*b*) Truth table

Fig. 72.31. Falling edge triggered J-K flip-flop.

Fig. 72.31(*a*) shows the symbol and 72.31 (*b*) the truth table for a clocked *J-K* flip-flop that triggers on the falling edge of the clock pulse. The small circle on the CLK input indicates that the flip-flop will trigger when the CLK input goes from 1 to 0. This flip-flop operates in the same way as the rising edge of the flip-flop of Fig. 72.29 except that the output can change states only on the falling edge of CLK signal (*i.e.* points *b, d, f, h, j, l* and *n*). As a matter of fact, both polarities of edge-triggered *J-K* flip-flops are in common usage in the field of digital electronics.

Strictly speaking, the *J-K* flip-flop is much more versatile than the *S-C* flip-flop because it has no ambiguous states. The $J = K = 1$ condition, which produces the toggling operation, finds extensive use in all types of binary counters.

Example 72.6. Fig 72.32 (*a*) *shows the waveforms applied at J, K and CLK inputs of the clocked J-K flip-flop shown in Fig.* 72.32 (*b*). *Sketch the Q output waveform Assume* $Q = 0$ *initially.*

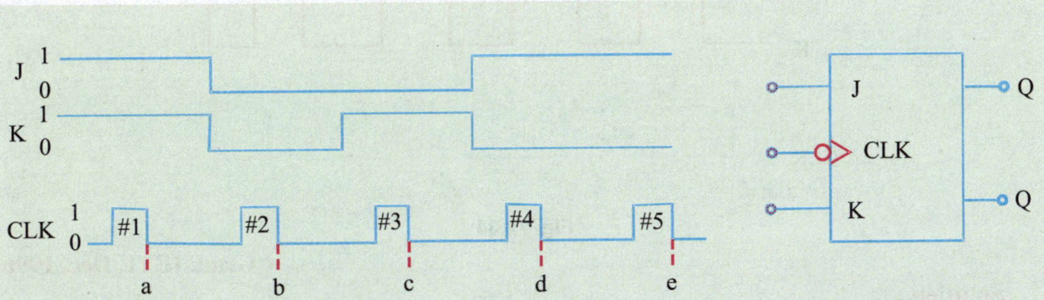

(*a*) (*b*)

Fig. 72.32

Solution :

Notice that flip-flop shown in Fig. 72.32 (*b*) is a clocked *J-K* flip-flop. Also notice the presence of a small circle at the CLK input of the flip-flop. This indicates that the flip-flop will trigger coresponding to the falling edge of the clock pulse. So you need to identify the states of J and K inputs coresponding to the falling edge points of the CLK pulse waveform *i.e.* points *a, b, c, d* and *e* shown in Fig. 72.32.

Using the truth table of clocked *J-K* flip-flop shown in Fig. 72.31, and the given waveforms the *Q*-waveform sketch is shown in Fig. 72.33. Before the arrival of first CLK pulse at point '*a*', all the inputs are 0 and hense $Q = 0 = Q_0$.On the arrival of first CLK pulse at point '*a*', $J = K = l$, the $J - K$ flip-flop will loggle from $Q_0 = 0$. On the arrival of second CLK pulse, at point '*b*', $J = K = 0$ and hence, there is no change in 0/p Q_0 i.e. $Q_0 = 1$. On the arrival of third CLK pulse at point '*c*' J = D, K = 1 and hence the flip-flop is cleared of 0 i.e. $Q_0 = 0$. On the arrival of fourth clock pulse at point '*d*' $J = 1$, $K = 0$, and the flip-flip is set for the output $Q_0 = 1$. Again on the arrival of fifth CLK pulse at point 'e', $J = 1$, $K = 0$, and the output will continue at $Q_0 = 1$

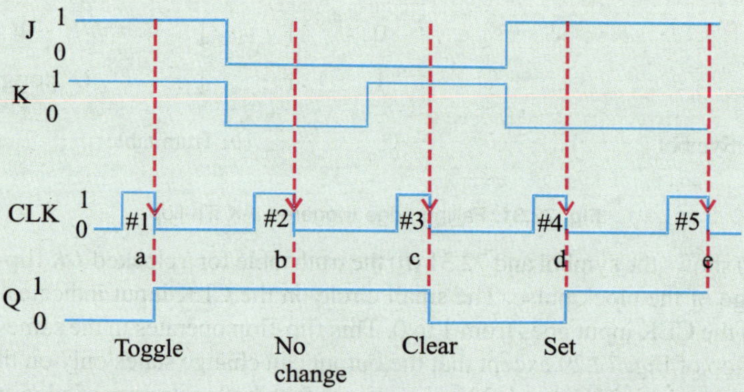

Fig. 72.33

Example 72.7. *What will be the output waveform Q of a J-K flip-flop if the following waveforms are applied at the input? Assume the flip-flop triggers at the falling edge of clock pulse.*

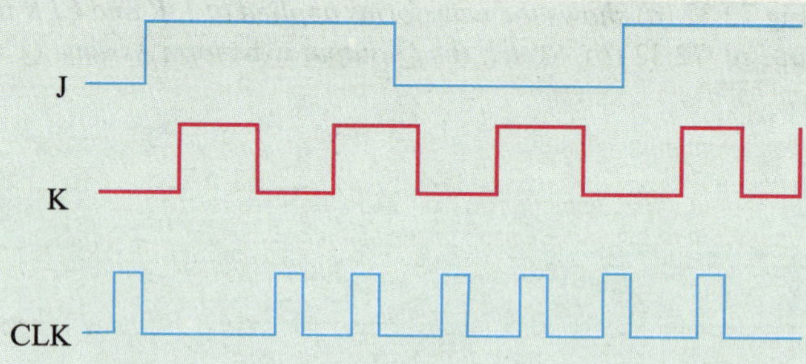

Fig. 72.34

(Grad. IETE Dec. 1996)

Solution :

Recall the truth table of a clocked *J-K* flip-flop, the flip-flop triggers corresponding to the logic levels at the J and K inputs. If $J = K = 1$, the flip-flop output remains in its previous state. If $J = 1$, $K = 0$ the flip-flop output goes 1, if $J = 0$, $K = 1$, the flip-flop goes LOW. However, if $J = K = 1$, the flip-flop output toggles. Keeping it in mind, we can sketch the Q-output waveform as shown in Fig. 72.35.

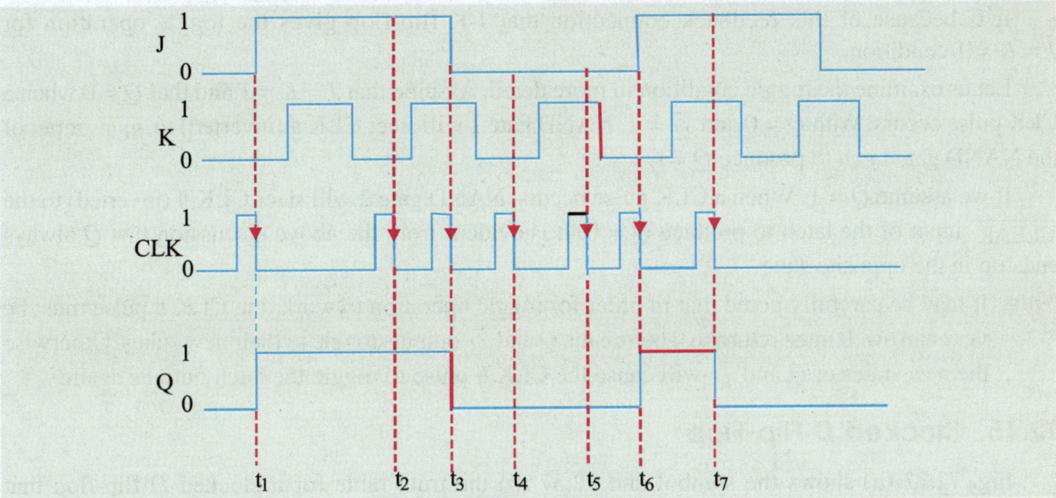

Fig. 72.35

Notice that at t_1, $J = 1$, $K = 0$ therefore the Q–output goes 1. At t_2 , $J = 1$, $K = 0$ so the Q–output remains 1. At t_3, J goes 0, $K = 1$, therefore the Q-output goes 0. At t_4, $J = K = D$. So There is no change in output Hence, $Q_0 = 0$ At t_5 , J is still 0, K is 1, therefore the Q-output stays 0. At t_6 , J goes 1, K is 0, so the Q-output goes 1. At t_7 both J and K are 1, therefore the Q-output toggles and goes 0, as shown in the Fig. 72.35.

72.14 . Internal Circuitry of an Edge-triggered *J-K* Flip-Flop

Fig. 72.36 shows a simplified version of the internal circuitry of an edge-triggered *J-K* flip-flop. As seen, the flip-flop contains the same three sections as the *S-C* triggered flip-flop (refer to Fig 72.27 page 2646), *i.e.*,

1. A basic NAND gate latch
2. A pulse steering circuit and
3. An edge detector circuit

The only difference between the internal circuitry of edge-triggered *J-K*. flip-flop and that of *S-C* flip flop is that in *J-K* flip-flop circuit, Q and $\overline{Q}$ outputs are fed back to the pulse-steering NAND gates.

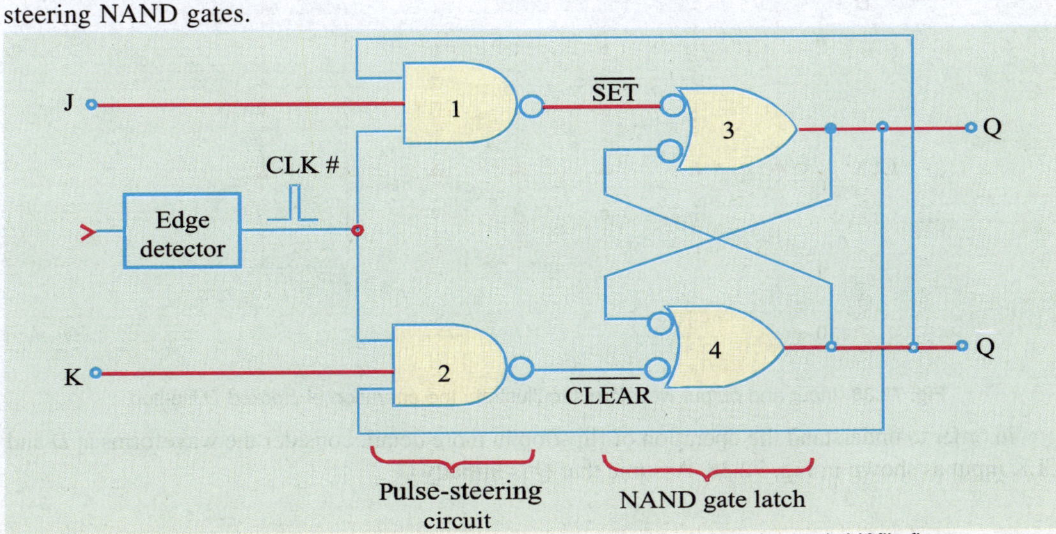

Fig 72.36. A simplified version of an internal circuitry of an edge-triggered *J-K* flip-flop.

It is because of this feedback connection that J-K flip-flop gives the toggle operation for $J = K = 1$ condition.

Let us examine the toggle condition in more detail. Assume that $J = K = 1$ and that $Q = 0$ when a CLK pulse occurs. With $Q = 0$ and $\overline{Q} = 1$, NAND gate 1 will steer CLK # (inverted) to $\overline{\text{SET}}$ input of the NAND gate latch to produce $Q = 1$.

If we assume $Q = 1$. When a CLK pulse occurs, NAND gate 2 will steer CLK # (inverted) to the $\overline{\text{CLEAR}}$ input of the latch to produce $Q = 0$. It is evident from the above discussion that Q always ends up in the opposite state.

Note. It may be carefully noted that in order for toggle operation to work, the CLK # pulse must be very narrow. It must return to 0 before the Q and $\overline{Q}$ outputs toggle to their new states. Otherwise the new states of Q and $\overline{Q}$ will cause the CLK # pulse to toggle the latch outputs again.

72.15. Clocked *D* Flip-Flop

Fig. 72.37 (*a*) shows the symbol and 72.37 (*b*) the truth table for a clocked *D* flip-flop that triggers on the rising edge of the clock pulse. Notice that this flip-flop has only one synchronous control input, *D* which stands for data.

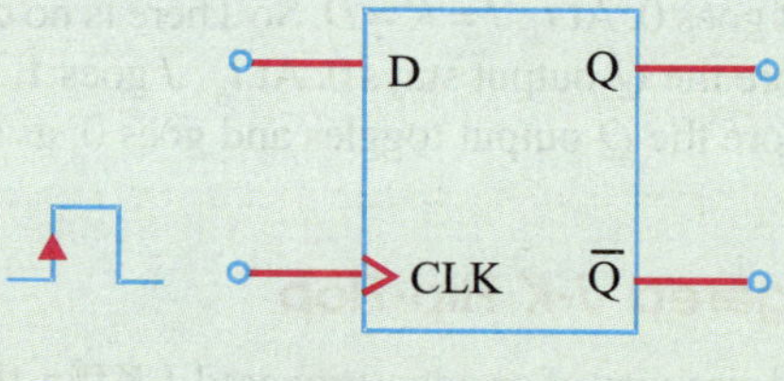

(*a*) Symbol

Inputs		Output
D	***CLK***	***Q***
0	↑	0
1	↑	1

(*b*) Truth table

Fig. 72.37. Clocked D flip-flop.

The operation of the clocked *D* flip-flop is very simple. The output *Q* will go to the same that is present on the D input when the rising edge occurs at the CLK. In other words, the level present at D will be stored in the flip-flop at the instant the rising edge occurs.

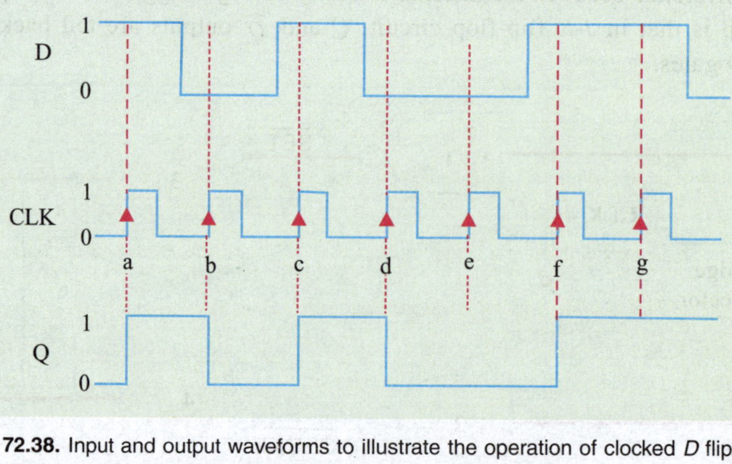

Fig. 72.38. Input and output waveforms to illustrate the operation of clocked *D* flip-flop.

In order to understand the operation of flip-flop in more detail, consider the waveforms at *D* and CLK input as shown in Fig. 72.38. Assume that *Q* is initially 0.

1. When the first rising edge of the CLK pulse occurs (refer to point '*a*'), the *D* input is 1, therefore *Q* will go to 1 state. Even though the *D* input level changes between the points '*a*' and '*b*', it has no effect on *Q*. The output *Q* is storing the 1 that was on *D* at point '*a*'.
2. When the second rising edge of the CLK pulse occurs (refer to point '*b*'), *Q* goes to 0 state since *D* is 0 at that time. The output *Q* stores this 0 value until the rising edge of the third CLK pulse (at point '*c*') causes *Q* to go to 1 since *D* is 1 at that time.
3. In a similar manner, the *Q* output takes on the levels present at *D* when the rising edges occur at points '*d*', '*e*', '*f*' and '*g*'. Notice that *Q* stays 0 at point '*e*' because *D* is still 0.

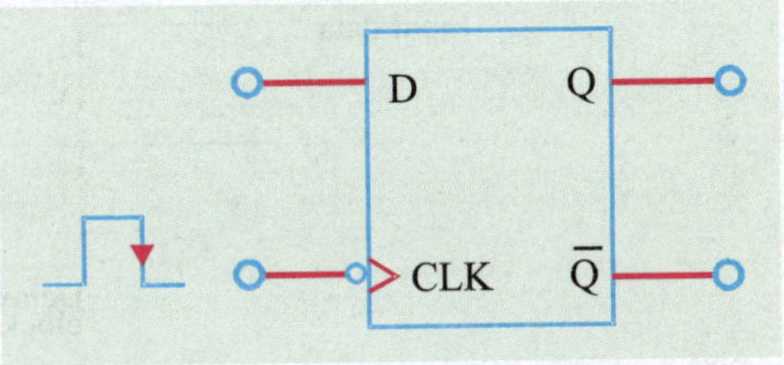

Fig. 72.39. Symbol for a falling-edge triggered *D* flip-flop.

A falling-edge triggered *D* flip-flop operates in the same way as the rising edge triggered *D* flip-flop. However, the difference is that *Q* will take on the value of *D* when a falling edge occurs at the CLK.

A falling-edge triggered *D* flip-flop operates in the same manner as the *D* flip-flop discussed above except that *Q* will take on the value of *D* when a falling edge occurs at the CLK input. The symbol for D flip-flop that triggers on the falling edge has a bubble on the CLK input as shown in Fig. 72.39.

IC 7474 is an example of clocked *D* flip-flop. It contains two rising edge triggered D flip-flops. The pin configuration and some other specification can be found in the data sheet of the device.

72.16. Implementation of *D* Flip-Flop from a *J-K* Flip-Flop

An edge-triggered *D* flip-flop can be obtained easily by adding a single INVERTER to an edge-triggered *J-K* flip-flop as shown in Fig. 72.40. A similar approach can be used to convert a *S-C* flip-flop to a *D* flip-flop.

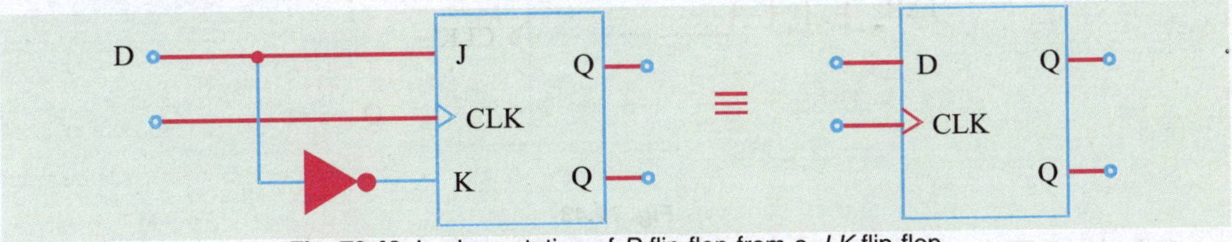

Fig. 72.40. Implementation of *D* flip-flop from a *J-K* flip-flop.

Example 72.8. *Fig 72.41 shows a D flip-flop with the input and clock waveforms applied at their respective inputs. Determine the Q (or output) waveform.*

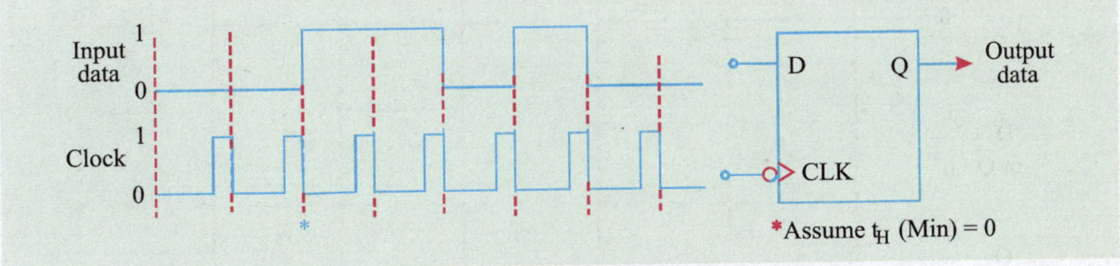

Fig. 72.41

Solution :

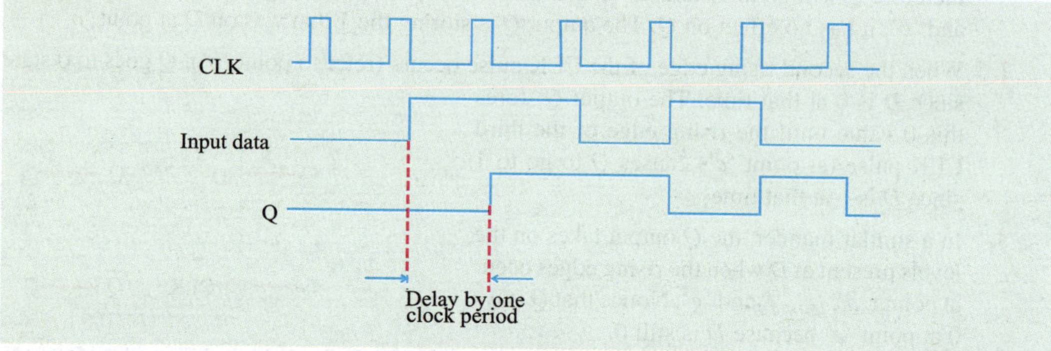

Fig. 72.42

Note. The output (Q) is delayed from the input by one clock period. This is an advantage as a *D*-Flip-flop is used sometimes to delay a binary waveform so that the binary information appears at the output a certain amount of time after it appears at the *D* input.

It is also possible to delay the input by two clock periods. This can be achieved by connecting Q to the D input of a second flip-flop and connect the clock signal to the second flip-flop. The output of the second flip-flop will be delayed by 2 clock periods from the input data.

Example 72.9. *An edge-triggered D Flip-flop can be made to operate in the toggle mode by connecting it as shown in Fig. 72.43. Assume $Q = 0$ initially and determine the Q (output) waveform.*

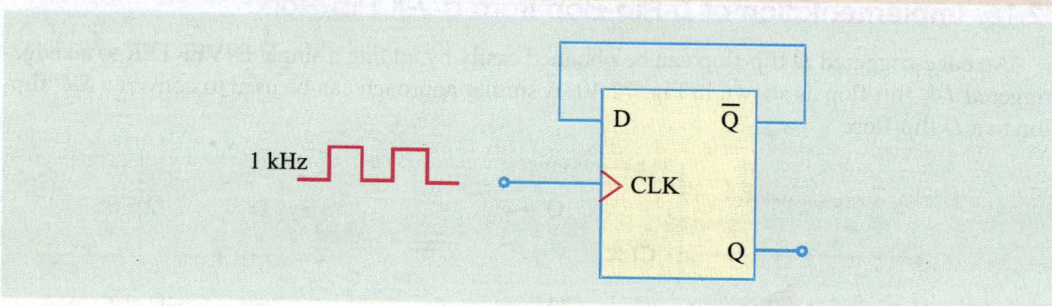

Fig. 72.43

Solution : The clock is 1 kHz waveform. The output waveform can be obtained from the input as shown in Fig. 72.44.

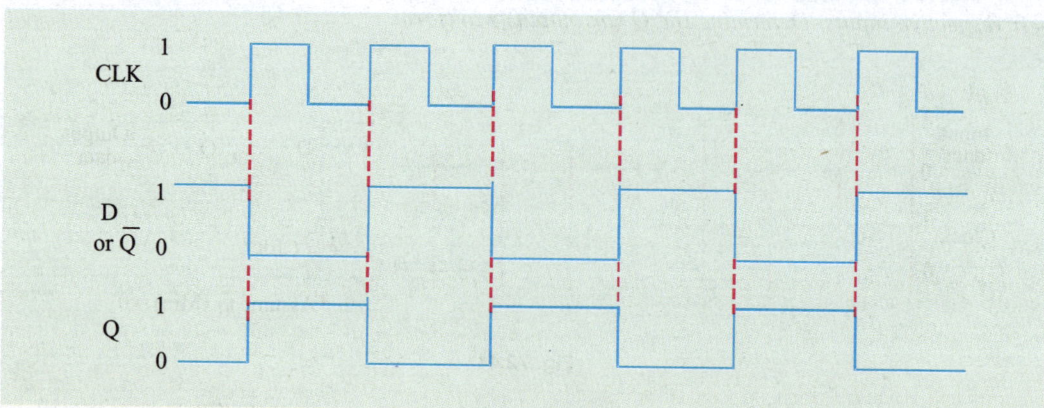

Fig. 72.44

Notice that the output is a square wave of half-the frequency of input *i.e.* 7200 kHz.

Note. A *D*- flip-flop with $\overline{Q}$ tied to its *D*–input as shown in Fig 72.43 is also known as *T* flip-flop where *T* stands for Toggle (or trigger).

72.17. Parallel Transfer of Data Using *D*-Flip-Flops

We have already discussed in Art 72.15 that the *Q*-output of a *D* flip-flop is the same as the *D* input. Let us now study the usefulness of this flip-flop.

Fig. 72.45 shows an application of *D* flip-flop used for parallel transfer of binary data from *X*, *Y*, *Z* – the three outputs of a combinational logic circuit to the outputs Q_1, Q_2 and Q_3 of the *D* flip-flops for storage. The transfer occurs upon application of TRANSFER pulse to the common CLK inputs. The flip-flops can store these values for subsequent processing.

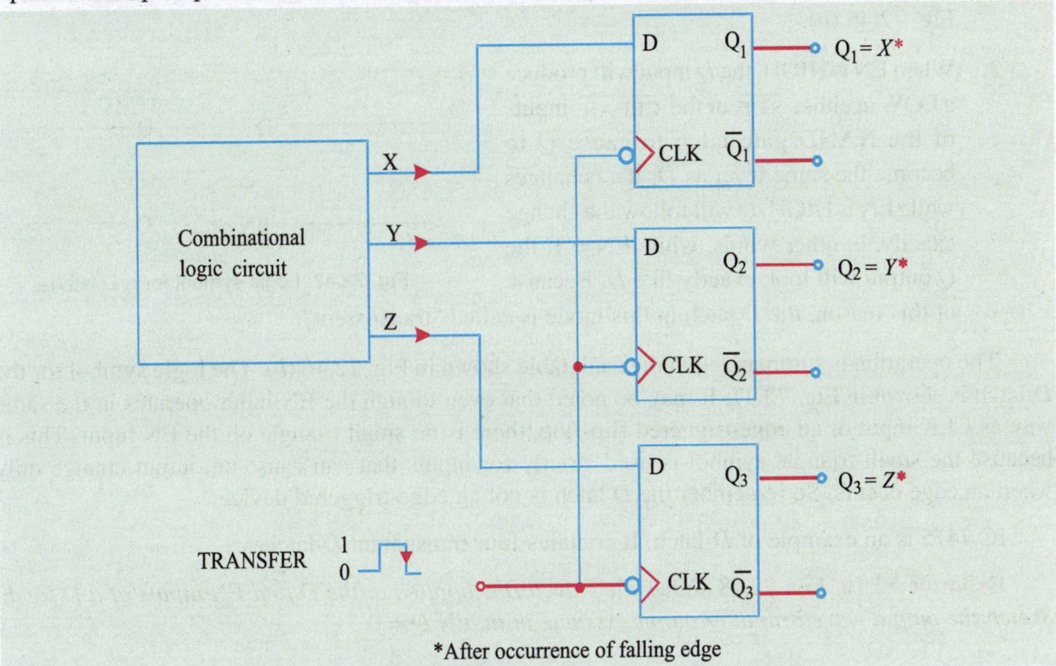

Fig. 72.45. Illustrating parallel transfer of data using *D* flip-flops.

72.18. *D* Latch (Transparent Latch)

We have already discussed in Art. 72.15 about an edge – triggered *D* flip-flop. Such a flip-flop uses an edge-detector circuit to ensure that the output will respond to the *D* input only when the active transition of the clock occurs. If this edge detector circuit is not used, the resultant circuit operates somewhat differently. It is called a *D* latch and has the arrangement shown in Fig 72.46 (*a*).

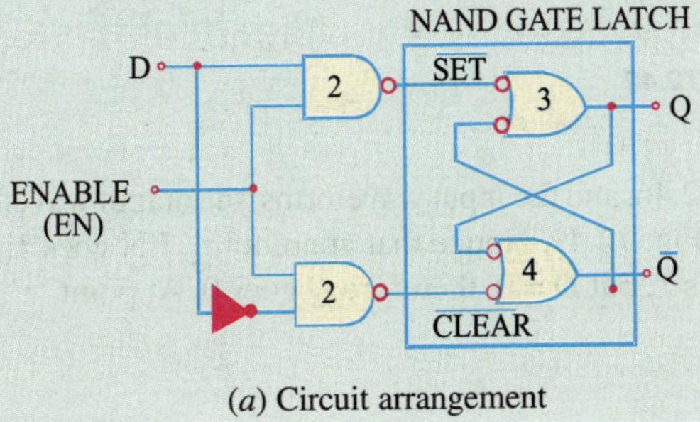

(*a*) Circuit arrangement

Inputs EN	D	Output Q
0	X	Q_0 (no change)
1	0	0
1	1	1

(*b*) Truth table

Fig. 72.46. D latch.

As seen from Fig. 72.46. (*a*), the circuit contains a NAND gate latch, and the steering NAND gates without the edge-detector circuit. The common input to the steering gates is called an enable input (abbreviated as EN) rather than a clock input because its effect on Q and $\overline{Q}$ outputs is not restricted to occurring only on its transitions. The operation of the *D* latch may be explained as follows:

1. When EN is LOW, whatever be the value of *D* input, it is inhibited from affecting the NAND gate latch. It is due to the reason that the outputs of both the steering gates will be held HIGH. Thus the Q and $\overline{Q}$ outputs will stay at whatever level they had just before EN went LOW. In other words, the outputs are "latched" to their current level and cannot change while EN is LOW even if *D* changes. This is represented as Q_0 in the truth table. Refer to Fig. 72.46 (*b*).

2. When EN is HIGH, the *D* input will produce a LOW at either $\overline{\text{SET}}$ or the $\overline{\text{CLEAR}}$ inputs of the NAND gate latch to cause *Q* to become the same level as *D*. If *D* changes while EN is HIGH, *Q* will follow the change exactly. In other words, while EN = 1, the *Q* output will look exactly like *D*. Because of this reason, the *D* latch in this mode is called "transparent".

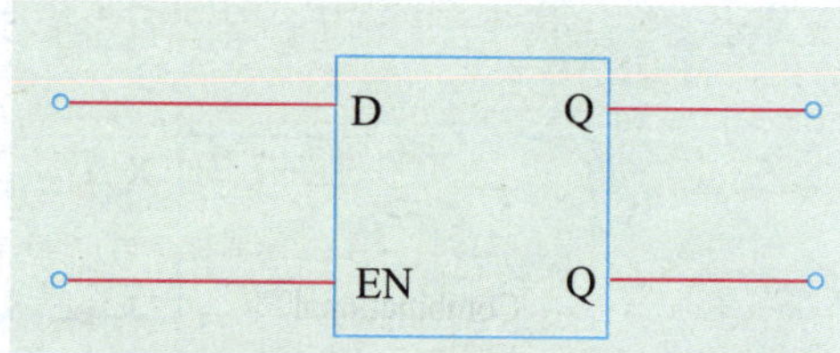

Fig. 72.47. Logic symbol for a D-latchs.

The operation is summarized in the truth table shown in Fig. 72.46 (*b*). The logic symbol for the *D*-latch is shown in Fig. 72.47. It may be noted that even though the EN input operates in the same way as CLK input of an edge-triggered flip-flop, there is no small triangle on the EN input. This is because the small triangle symbol is used strictly for inputs that can cause an output change only when an edge occurs. So remember the *D* latch is not an edge-triggered device.

IC 7475 is an example of *D*-latch. It contains four transparent *D*-latches.

Example 72.10. *Fig. 72.48 shows the waveforms applied at the D and EN inputs of a D latch. Sketch the output waveform at Q-output Assume initially* $Q = 0$

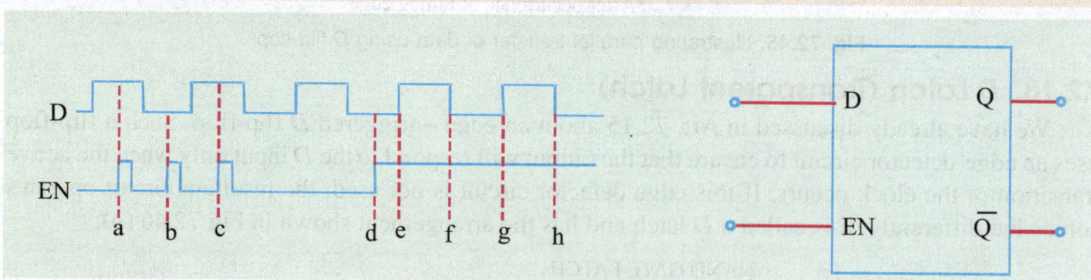

Fig. 72.48

Solution.

Using the truth table of *D*-latch shown in Fig. 72.46, and the input waveforms, the output waveform at *Q* can be obtained and is sketched as shown in Fig. 72.49. Notice that at point '*a*', EN goes 1, and *D* is 1, therefore a output goes 1. At point '*b*' EN is 1, but *D* = 0, therefore *Q* goes 0, At point '*c*', EN is 1, *D* = 1, therefore *Q* goes 1 and so on.

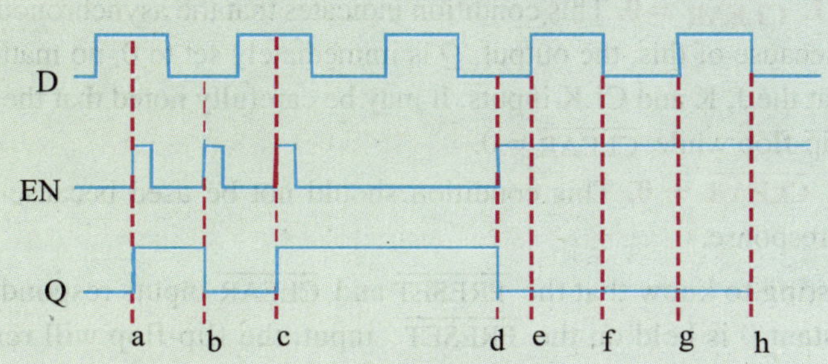

Fig. 72.49

72.19. Clocked *J-K* Flip-Flop with Asynchronous Inputs

Strictly speaking for the clocked flip-flops, the *S, C, J, K* and *D* inputs discussed in the previous articles are refered to as ***control inputs.*** These inputs are also known as ***synchronous inputs.*** These are called synchronous because their effect on the flip-flop output is synchronized with the CLK (*i.e.* clock) input. As discussed earlier, the synchronous control inputs must be used in conjunction with a signal to trigger the flip-flop.

As a matter of fact, most clocked flip-flops also have one-or more ***asynchronous*** inputs. These inputs operate independently of the synchronous inputs and clock input. The asynchronous inputs of a flip-flop can be used to set its output to 1 state or clear to the 0 state at any time regardless of the conditions at the other inputs. In other words, the asynchronous inputs can be used to ***override*** all the other inputs in order to place the flip-flop output in one state or the other. Because of this reason, the asynchronous inputs are also called ***override inputs.***

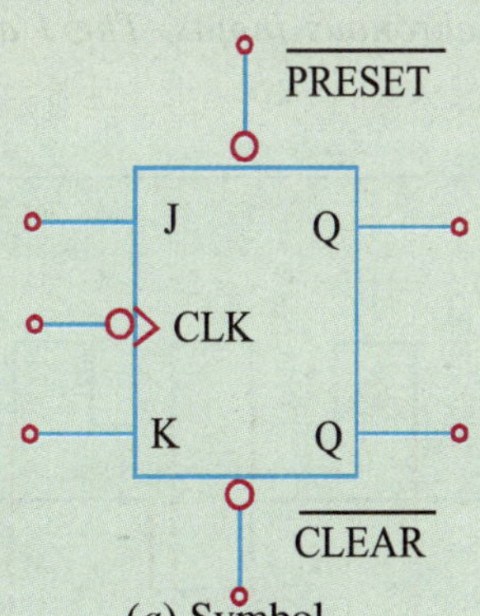

(*a*) Symbol

Inputs		*Output / Flip-flop response*
$\overline{PRESET}$	$\overline{CLEAR}$	
1	1	Clocked operation*
0	1	Q = 1 (regardless of CLK)
1	0	Q = 0 (regardless of CLK)
0	0	Not used

(*b*) Truth table

Fig . 72.50

Fig. 72.50 shows a clocked J-K flip-flop with two asynchronous inputs namely (1) $\overline{\text{PRESET}}$ and (2) $\overline{\text{CLEAR}}$. Both these inputs are active-LOW. This is indicated by the bubbles on the flip-flop symbol. Fig 72.50 (*b*) shows the truth table for the clocked *J-K* flip-flop with asynchronous inputs. Let us study the various cases given in the truth table :

1. **$\overline{\text{PRESET}}$ = $\overline{\text{CLEAR}}$ = 1.** This condition indicates that both the asynchronous inputs are inactive and the flip-flop is free to respond to the *J, K* and CLK inputs. In other words, the flip-flop operates as a normal clocked flip-flop.
2. **$\overline{\text{PRESET}}$ = 0, $\overline{\text{CLEAR}}$ = 1.** This condition indicates that the asynchronous input, $\overline{\text{PRESET}}$ is activated. Because of this, the output, *Q* is immediately set to 1, no matter what conditions are present at the *J, K* and CLK inputs. It may be carefully noted that the CLK input cannot affect the flip-flop while $\overline{\text{PRESET}}$ = 0.

3. $\overline{\text{PRESET}}$ **= 1,** $\overline{\text{CLEAR}}$ **= 0.** This condition indicates that the asynchronous input, $\overline{\text{CLEAR}}$ is activated. Because of this, the output, Q is immediately set to 0, no matter what conditions are present at the J, K and CLK inputs. It may be carefully noted that the CLK input cannot affect the flip-flop while $\overline{\text{CLEAR}}$ = 0.
4. $\overline{\text{PRESET}}$ **=** $\overline{\text{CLEAR}}$ **= 0.** This condition should not be used because it can produce an ambiguous response.

It will be interesting to know that the $\overline{\text{PRESET}}$ and $\overline{\text{CLEAR}}$ inputs respond to dc levels. This means that if a constant 0 is held on the $\overline{\text{PRESET}}$ input, the flip-flop will remain in the $Q = 1$ state regardless of what is occurring at the other inputs. Similarly, a constant 0 at the $\overline{\text{CLEAR}}$ input holds the flip-flop in $Q = 0$ state. Thus the asynchronous inputs can be used to hold the flip-flop in a particular state for any desired interval of time. However, in actual practice, the asynchronous inputs are used to set or clear the flip-flop to the desired state by application of a momentary pulse.

Many clocked flip-flops that are commercially available as ICs, will have both $\overline{\text{PRESET}}$ and $\overline{\text{CLEAR}}$. However some ICs will have only $\overline{\text{CLEAR}}$ input. Some other ICs will have asynchronous inputs that are active-HIGH. For these ICs, the flip-flop symbol would not have a bubble on the asynchronous inputs.

ICs 7476 and 74LS76 are popular *J-K* flip-flops with asynchronous inputs preset and clear. The preset is designated by $\overline{S}_D$ and clear by $\overline{R}_D$. The IC 7476 is a rising edge triggered while 74LS76 is a falling edge triggered device. Each package contains two *J-K* flip-flops.

Example 72.11. Fig. 72.51 (*a*) *shows the logic symbol for a J-K flip-flop that responds to the falling edge on its clock pulse and has active-LOW asynchronous inputs. The J and K inputs are tied HIGH.*

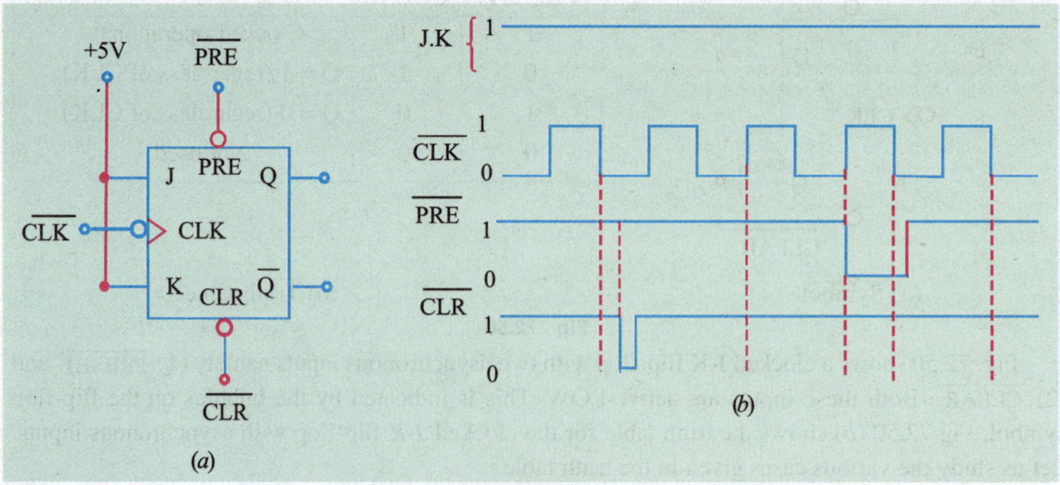

Fig. 72.51

Determine the Q-output in response to the waveforms shown in Fig. 72.51 (*b*). *Assume that initially* $Q = 0$.

Solution. Notice that the *J* and *K* are tied to +5V. The waveforms shown in Fig. 72.51 (*b*) are applied at the inputs of the *J-K* flip flop.

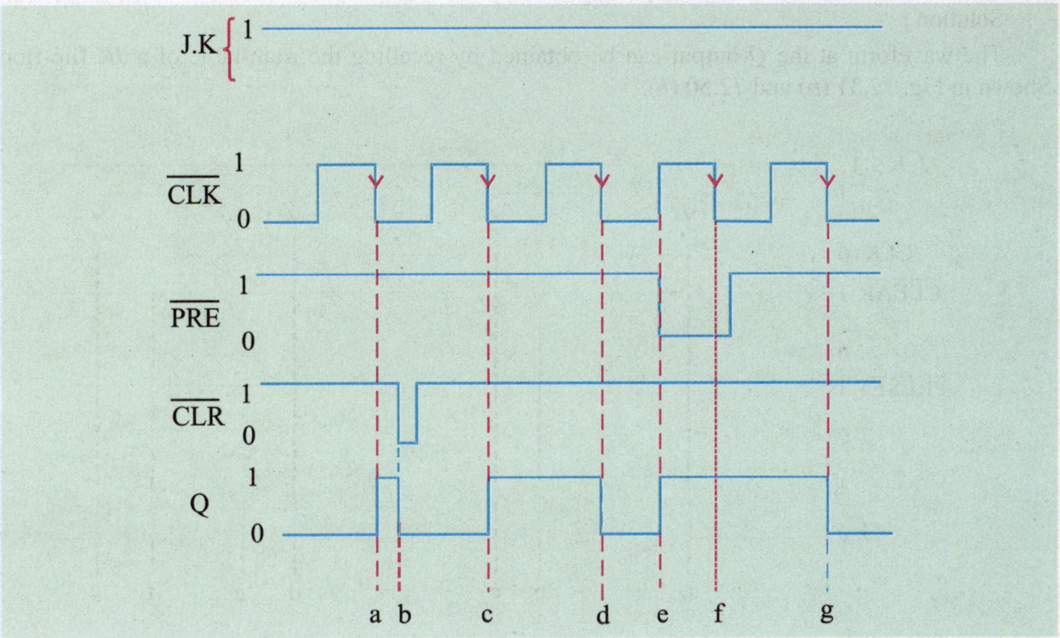

Fig. 72.52

In order to sketch the Q-output waveform, recall the truth table of J-K flip-flop indicated in Fig. 72.31 (*b*) and Fig. 72.50 (*b*). Notice that when $\overline{PRE}=1$ and $\overline{CLR}=1$, the J-K flip-flop will trigger at the falling edge of the clock pulse. Thus at point '*a*', $\overline{PRE}=\overline{CLR}=1$, $J=K=1$, therefore the Q-output toggles *i.e.* it goes from 0 to 1. At point '*b*' $\overline{CLR}$ goes 0, therefore Q goes 0. At point '*c*', $\overline{PRE}=\overline{CLR}=1$, therefore under normal operation J-K flip-flop toggles *i.e.* Q goes from 0 to 1 again. At point '*d*' there is no change in $\overline{PRE}$ or $\overline{CLR}$ inputs. So the Q-output toggles again, *i.e.* Q goes 0. At point '*e*' $\overline{PRE}$ goes 0, therefore Q-output goes 1. At point '*g*', CLK = 1, $\overline{PRE}=\overline{CLR}=1$, and Q toggles to 0. The complete Q-output waveform is shown in Fig. 72.52 along with the input waveforms.

Example 72.12. *Determine the Q-output for the JK flip-flop shown in Fig. 72.53, Assume that Q = 0 initially and remember that the asynchronous inputs override all other inputs.*

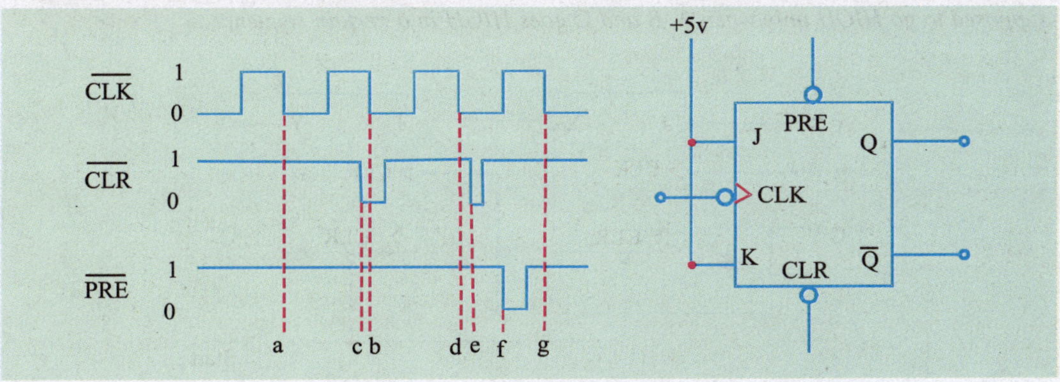

Fig. 72.53

Solution :

The waveform at the Q-output can be obtained by recalling the truth-table of a JK flip-flop. Shown in Fig. 72.31 (*b*) and 72.50 (*b*).

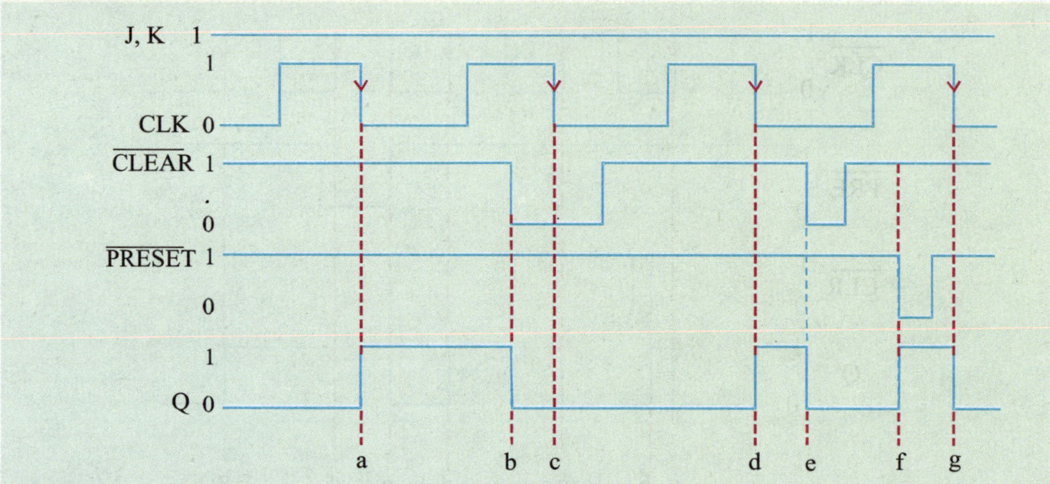

Fig. 72.54

Notice that at point '*a*', $J = K = 1$, $\overline{PRESET} = \overline{CLEAR} = 1$, the falling edge of CLK pulse toggles the J-K flip-flop Q output to 1. At point '*b*' $\overline{CLEAR}$ goes 0, therefore irrespective of the other inputs, Q-output goes 0. At point '*c*' the ouput remains 0 because $\overline{CLEAR}$ is still 0. At point '*d*', $\overline{PRESET} = \overline{CLEAR} = 1$, $J = K = 1$, the J-K flip-flop output toggles on the arrival of falling edge of CLK, *i.e.* Q goes 1. At point '*e*', $\overline{CLEAR}$ goes 0 again, therefore Q-output goes 0. At point '*f*', $\overline{PRESET}$ goes 0, therefore Q-output goes 1. Finally at point '*g*', $\overline{CLEAR} = \overline{PRESET} = 1$, $J = K = 1$, the output toggles on the arrival of falling edge of CLK pulse, *i.e.* Q goes 0.

The complete sketch of Q-output waveform along with other input waveforms is shown in Fig. 72.54.

Example 72.13. *In the circuit of Fig. 72.55, inputs A, B and C are all initially LOW. Output Y is supposed to go HIGH only when A, B and C goes HIGH in a certain sequence.*

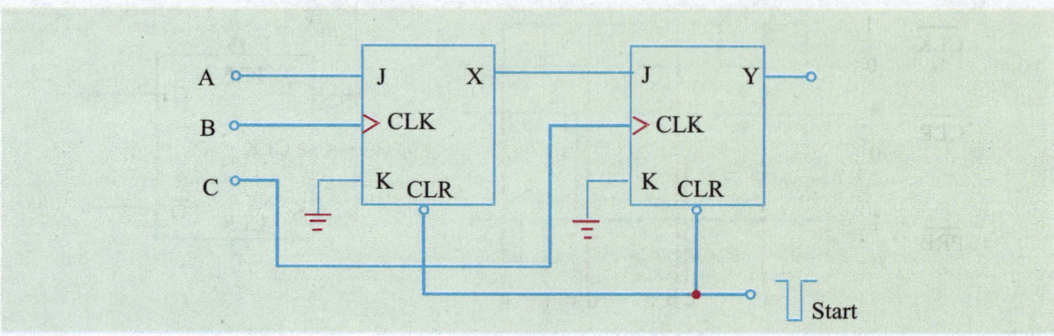

Fig. 72.55

(a) Determine the sequence that will make Y go HIGH.

(b) Explain why the START pulse is needed.

Solution.

(*a*) Sequence at the inputs (A, B and C)

We know that *Y* can go HIGH only when *C* goes HIGH while *X* is already HIGH.

X can go HIGH only if *B* goes HIGH while *A* is HIGH.

Thus the correct sequence that makes Y go HIGH is A, B, C.

(*b*) Need for START pulse

We know that the outputs X and Y need to be cleared to 0 before applying the A, B, C signals. To clear the outputs, we need a negative going START pulse at CLR input. (notice that CLR input of JK flip-flop is active LOW).

72.20. Alternative Designations for Asynchronous Inputs

We have already discussed in the last article about the asynchronous clocked *J-K* flip-flop. These flip-flops are available as ICs and have two asynchronous inputs namely $\overline{\text{PRESET}}$ and $\overline{\text{CLEAR}}$ (or PRESET and CLEAR). IC manufacturers have not agreed on what nomenclature is used for these asynchronous inputs. The most common designations are $\overline{\text{PRE}}$ (short for $\overline{\text{PRESET}}$) and $\overline{\text{CLR}}$ (short for $\overline{\text{CLEAR}}$). The designations $\overline{S}_D$ (direct $\overline{\text{SET}}$) and $\overline{R}_D$ (direct $\overline{\text{RESET}}$) are also used. However we will use the labels $\overline{\text{PRE}}$ and $\overline{\text{CLR}}$ to represent asynchronous inputs. Whenever these asynchronous inputs are active–HIGH, we will not use the overbar to indicate their active-HIGH status *i.e.* PRE and CLR.

72.21. Flip-Flop Timing Parameters

The performance, operating requirements and limitations of flip-flops are specified by several timing parameters found on the data sheet for the device refer to Appendix A. Generally the timing parameters are applicable to all CMOS and TTL flip-flops.

1. Setup and Hold Times. These have already been discussed in Art 72.10 (page 2642). They represent requirements that must be met for reliable flip-flop triggering. The manufacturer's IC data sheet will always specify the minimum values of t_S and t_H. For example, IC 7474 has set up time of 20 ns and hold time of 5 ns. On the other hand 74HC112 has set up time of 25ns and zero hold time.

2. Propagation delay. It is the interval of time required after an input signal has been applied for the resulting output change to occur. Fig. 72.56 (*a*) illustrates the propagation delays that occur in response to a rising edge of the CLK input when *Q*-output changes from 0 to 1. Similarly Fig. 72.56 (*b*) illustrates the propagation delay that occur in response to the rising edge of the CLK input when *Q*-output changes from 1 to 0. It may be noted that these delays are measured between the 50% points (between logic 0 and logic 1 voltage levels) on the input and output waveforms. The same type of delays occur in response to the signals on a flip-flops asynchronous inputs (*i.e.* PRESET and CLEAR). The manufacturer's data sheets usually specify propagation delays in response to all inputs and they usually specify the maximum values for t_{PLH} and t_{PHL}. Notice that t_{PLH} is the delay going from logic LOW to HIGH level whereas t_{PHL} is the delay going from logic HIGH to LOW level.

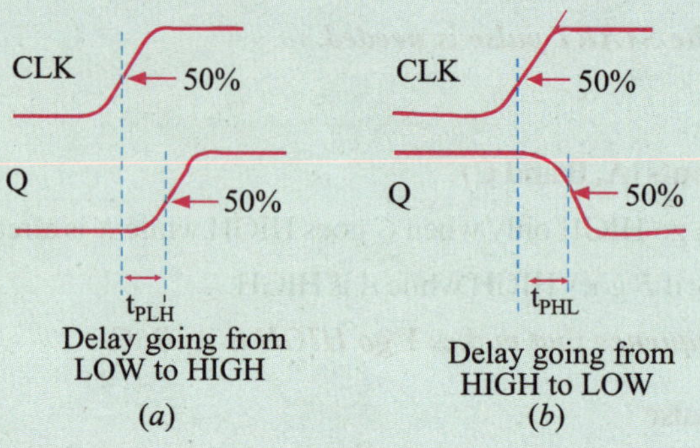

Fig. 72.56

3. Maximum Clock Frequency (f_{max}). This is the highest frequency that may be applied to the CLK input of a flip-flop and still have it triggered reliably. The value of f_{max} limit will vary from flip-flop to flip-flop even with flip-flops having the same device number. For example, f_{max} for IC7474 is 15 MHz, for 74LS112, it is 30 MHz, for 74C74, it is 5 MHz and for 74HC112, it is 20 MHz.

4. Clock Pulse HIGH and LOW Times. The manufacturer will also specify the minimum time duration that the CLK signal must remain LOW before it goes HIGH {represented as $t_w(L)$} and the minimum time that CLK must be kept HIGH before it returns LOW (represented as t_w (H)}. These times are illustrated in Fig. 72.57 (*a*) Failure to meet these minimum time requirements can result in unreliable triggering. It may be noted that these time values are measured between the halfway (50%) points on the signal transitions.

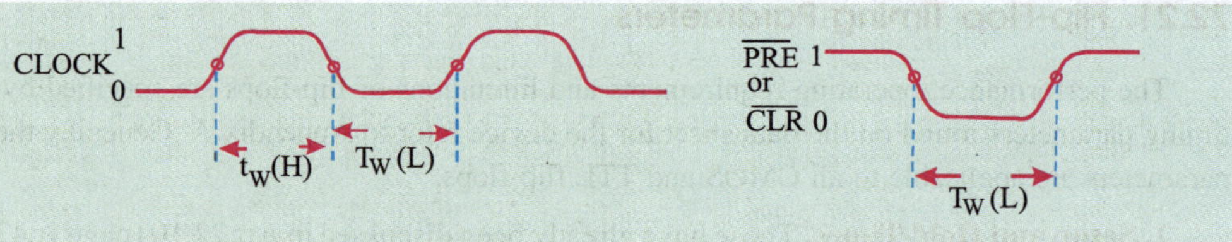

(*a*) Clock Pulse HIGH-LOW times (*b*) Asynchronous active pulse width

Fig. 72.57

5. Asynchronous Active Pulse Width. It is the minimum time duration that a PRESET or CLEAR input must be kept in its active state in order to set or clear the flip-flop reliably. Fig. 72.57 (b) shows t_w (*L*) for active-LOW asynchronous inputs.

6. Clock Transition Times. Strictly speaking, the rise and fall times of a clock waveform should be kept very short for reliable triggering. If the clock signal takes too long to change from one logic level to the other, the flip-flop may trigger erratically or not at all. IC manufacturers usually do not list a maximum transition time requirement for each flip-flop integrated circuit. Instead, it is usually given as a general requirement for all ICs within a given logic family. For example, the transition times should generally be $\leq$ 50 ns for TTL devices and $\leq$ 200 ns for CMOS. These requirements will vary among the different manufacturers and among the various subfamilies within the broad TTL and CMOS logic-families.

72.22. IC Flip-Flop Timing Values

We have already discussed the various flip-flop timing parameters in the last article. Table 72.1. lists all these timing values for the various flip-flops such as 7474, 74 LS112, 74C74 and 74 HC112. Notice that 7474 is a dual edge triggered *D* flip-flop (Standard TTL) device whereas 74LS74 is a dual edge-triggered *J-K* flip-flop (low-power schottky TTL) device. Similarly 74C74 is a dual edge-triggered D flip-flop (metal-gate CMOS) device whereas 74HC112 is a dual edge-triggered *J-K* flip-flop (high-speed CMOS) device.

Table 72.1. IC flip-flop timing values

Parameter	*TTL*		*CMOS*	
(Times in ns)	*7474*	*74LS112*	*74C74*	*74HC112*
t_S	20	20	60	25
t_H	5	0	0	0
t_{PHL} from *CLK* to *Q*	40	24	200	31
t_{PLH} from *CLK* to *Q*	25	16	200	31
t_{PHL} from $\overline{CLR}$ to *Q*	40	24	225	41
t_{PLH} from $\overline{PRE}$ to *Q*	25	16	225	41
t_W(L) *CLK* LOW time	37	15	100	25
t_W(H) *CLK* HIGH time	30	20	100	25
t_W(L) at $\overline{PRE}$ or $\overline{CLR}$	30	15	60	25
f_{Max} in MHz	15	30	5	20

Following are some of the important points which may be noted carefully from Table 72.1.

1. The 74HC series of CMOS devices has timing values that are comparable to those of TTL devices. This can be observed by reading the values below the 7474 column and the values below 74HC112 column.
2. The 74C series devices are much slower than the 74HC series. This can be observed by comparing all the minimum timing requirements below the 74C74 and 74HC112 column.
3. All the flip-flops have nonzero set up time requirement.
4. All of the flip-flops have very low hold-time (t_H) requirements *e.g.* The IC7474 has $t_H = 5$ ns whereas IC74LS112, 74C74 and 74HC112 has $t_H = 0$.

Example 72.14. *The data sheet of a certain flip-flop specifies that the minimum HIGH time $t_W(H)$ for the clock pulse is 30 ns and the minimum LOW time $t_W(L)$ is 37 ns. What is the maximum operating frequency*?

Solution. Given : $t_W(H) = 30$ ns $= 30 \times 10^{-9}$ s and $t_W(L) = 37$ ns $= 37 \times 10^{-9}$ s.

We know that the minimum time period,

$$t_{min} = t_w(H) + t_w(L)$$

$$= \left(30\times10^{-9} + 37\times10^{-9}\right) = 67\times10^{-9}\,\text{s}$$

and the maximum operating frequency,

$$f_{max} = \frac{1}{t_{min}} = \frac{1}{67\times10^{-9}_{ns}} = 14.925\times10^{6}\ Hz$$

$$= \mathbf{14.925\ MHz.}$$

Example 72.15. *Using the flip-flop timing values shown in Table 72.2, determine the followings :*

Table 72.2.

Parameter (time in ns)	*TTL* 7474	74LS112	*CMOS* 74C74	74HC112
t_{PHL} from *CLK* to *Q*	40	24	200	31
t_{PLH} from *CLK* to *Q*	25	16	200	31
t_W(L) at $\overline{PRE}$ or $\overline{CLR}$	30	15	60	25

(*a*) *Assume that initially Q = 0. How long it can take for Q to go HIGH when a rising edge appears at the CLK input of* IC7474?

(*b*) *Assume that initially Q = 1. How long it can take for Q to go LOW in response to the* $\overline{\text{CLR}}$ *input of an* IC 74HC112?

(*c*) *What is the narrowest pulse that should be applied to the* $\overline{\text{CLR}}$ *input of the* IC 74LS112 *flip-flop to clear Q reliably?*

Solution.

(*a*) The rising edge will cause the *Q* output to go from LOW to HIGH. The delay from CLK to *Q* is listed as t_{PLH} = 25 ns for IC7474.

(*b*) For the IC 74HC112, the time required for *Q* to go from HIGH to LOW in response to $\overline{\text{CLR}}$ input is listed as t_{PHL}= 31 ns.

(*c*) For the IC 74HC112, the narrowest pulse at the $\overline{\text{CLR}}$ input is listed as $t_w(L) = 15$ ns.

72.23. Potential Timing Problems in Flip-Flop Circuits

As a matter of fact, in many digital circuits, the output of one flip-flop is connected either directly or through logic gates to the input of another flip-flop and both flip-flops are triggered by the same clock signal. This situation leads to a potential timing problem. In order to understand this problem, consider a situation shown in Fig. 72.58. Here the output of flip-flop (1) Q_1 is connected to the *J* input of flip-flop (2). Notice that both the flip-flops are clocked by the same signal at their CLK inputs.

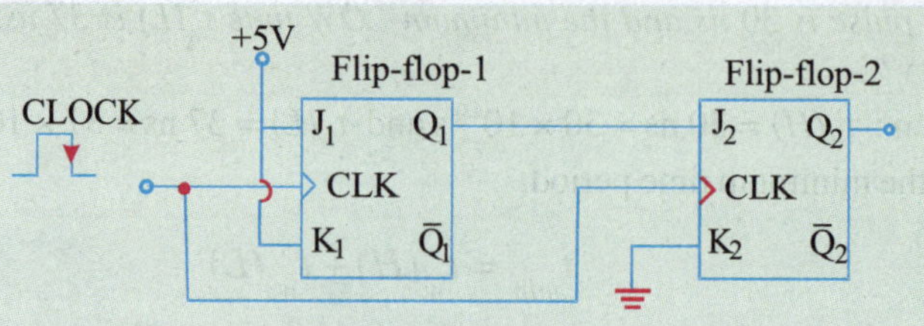

Fig. 72.58

Let us understand the potential timing problem now. Since Q_1 will change on the rising edge of the clock pulse, the J_2 input will be changing as it receives the same rising edge. This could lead to an unpredictable response at Q_2.

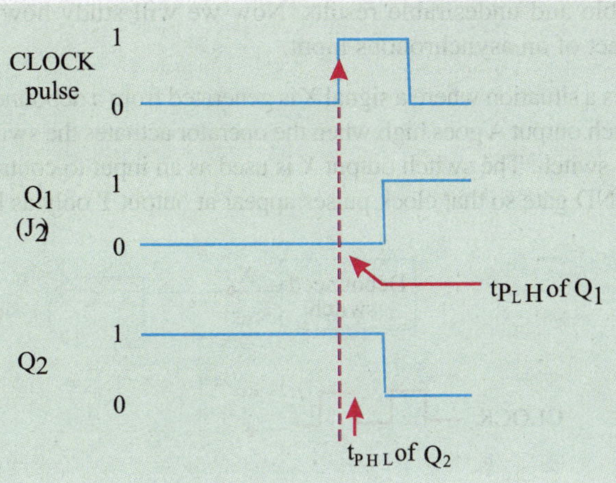

Fig. 72.59

Assume that initially $Q_1 = 0$ and $Q_2 = 1$. Further the flip-flop (1) has $J_1 = K_1 = +5$V (*i.e* HIGH) and flip-flop (2) has $J_2 = Q_1$ and $K_2 = 0$ prior to the occurrence of rising edge of the clock pulse. When the rising edge occurs, Q_1 will toggle to the HIGH state but it will not actually go HIGH until after its propagation delay, t_{PLH} as shown in Fig. 72.59. The same rising edge will reliably clock Q_2 to the LOW state provided that t_{PLH} is greater than Q_2's hold time requirement, t_H as shown in the figure. If this condition is not met, the response of Q_2 will be unpredictable.

Since all edge-triggered flip-flops have hold time (t_H) requirement that is 5 ns or less. Therefore for those flip-flops, situations like that shown in Fig. 72.59 will not be a problem.

Thus in all of the flip-flop circuits that you will study in this book, we will assume that flip-flop's hold time requirement is short enough to respond reliably according to the following rule:

The flip-flop output will go to a state determined by the logic levels present at its synchronous control inputs just prior to the active clock edge.

Applying the above stated rule to the circuit shown in Fig. 72.59, it says that the output of flip-flop (2), Q_2 will go to a state determined by $J_2 = 1$ and $K_2 = 0$ condition that is present just prior to the occurrence of the rising edge of the clock pulse. The fact that J_2 is changing in response to the same rising edge has no effect.

72.24. Applications of Flip-Flop

Although there is a wide variety of applications of edge-triggered (clocked) flip-flops, yet the following are important from the subject point of view :

1. Flip-flop synchronization
2. Data storage and transfer
3. Serial data transfer-shift registers
4. Frequency division
5. Counting

All these applications are discussed one by one in the following pages.

72.25. Flip-Flop Synchronization

Strictly speaking, most digital systems are principally synchronous in their operation. It means, most of the signals will change states in synchronism with the clock transitions. However, in many cases, there will be an external signal that is not synchronised to the clock. In other words such an external signal is an asynchronous signal. Such signals often occur as a result of a human operator actuating an input switch at some random time relative to the clock signal. This randomness can produce unpredictable and undesirable results. Now we will study how a flip-flop can be used to synchronize the effect of an asynchronous input.

Fig. 72.60 shows a situation where a signal *X* is generated from a debounced switch that is actuated by an operator. The switch output A goes high when the operator actuates the switch and goes LOW when the operator releases the switch. The switch output *X* is used as an input to control the passage of the control signal through the AND gate so that clock pulses appear at output *Y* only as long as *X* is HIGH.

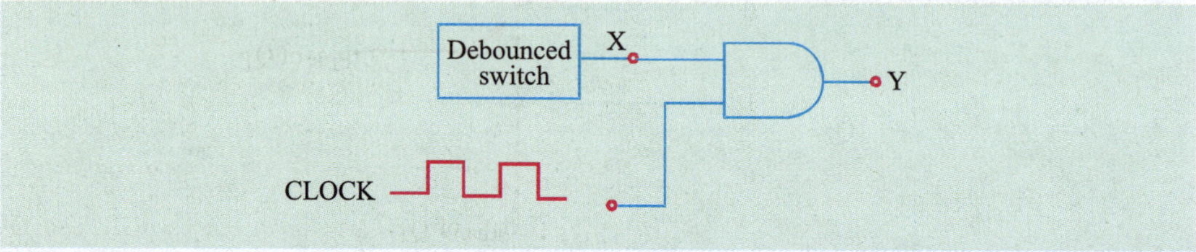

Fig. 72.60

The problem with the circuit shown in Fig. 72.60 is that signal *X* is asynchronous. That is it can change states at any time relative to the clock signal because the exact times when the operator actuates or releases the switch are essentially ***random***. This can produce partial clock pulses at output *Y* if either transition of *X* occurs while the clock signal is HIGH. It is shown in Fig. 72.61. This type of output is often not acceptable. So a method for preventing the appearance of partial pulses at *Y* must be developed. One possible solution is shown in Fig. 72.62.

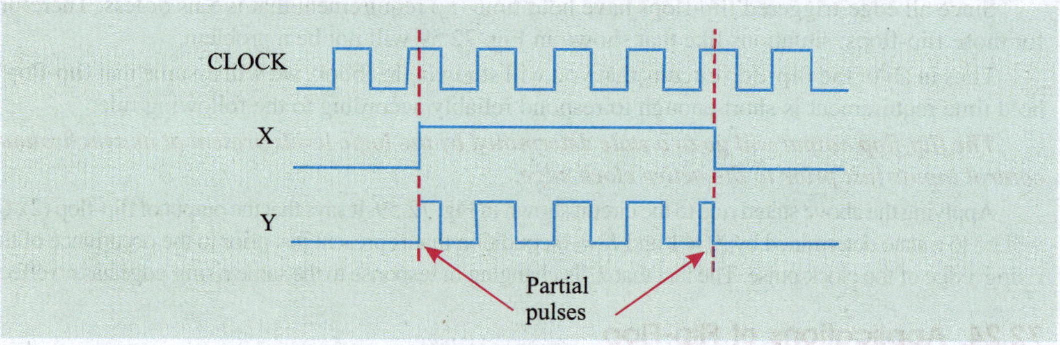

Fig. 72.61

As seen in Fig. 72.62, the *X* signal is connected to the *D* input of a flip-flop. The flip-flop is clocked by the falling edge of the clock signal. Thus when *X* goes HIGH, *Q* will not go HIGH until the next falling edge of the clock at time t_1. This HIGH at *Q* will enable the AND gate to pass subsequent complete clock pulses to *Y* as shown in Fig. 72.63.

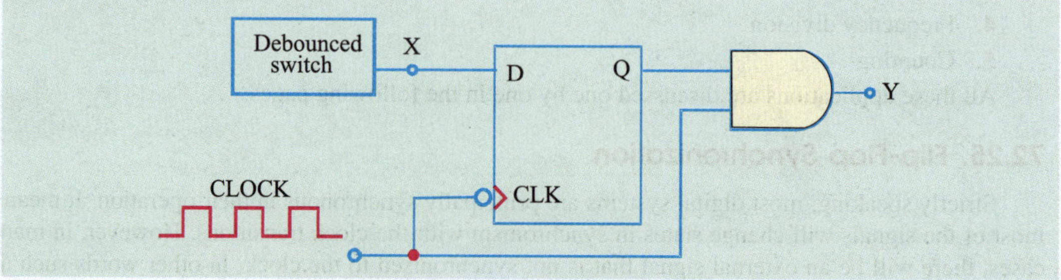

Fig. 72.62

Similarly, when *X* goes LOW, *Q* will not go LOW until the next falling edge of the clock at t_2. Thus the AND gate will not inhibit clock pulses until clock pulse that ends at t_2 has passed through to *Y*. Therefore, output *Y* contains only complete pulses (see Fig. 72.63).

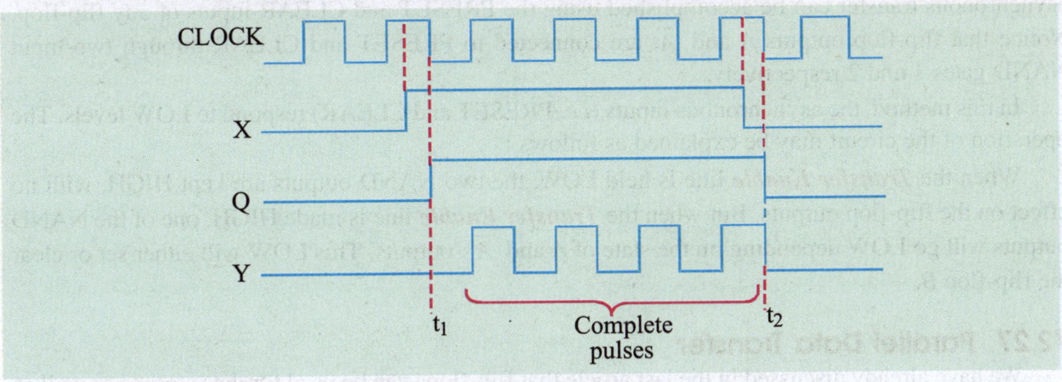

Fig. 72.63

72.26. Data Storage and Transfer

The most common use of flip-flops in the field of digital electronics, is for the storage of data or information. The data may represent numerical values (*e.g.* binary numbers, BCD -binary coded decimal numbers or any data that have been encoded in binary. These data are generally stored in groups of flip-flops called ***registers.*** The operation most often performed on data that are stored in a flip-flop or a register is the ***data transfer*** operation. This operation involves the transfer of data from one flip-flop (or register) to another. The data transfer is of two types: ***(a)*** synchronous and ***(b)*** asynchronous.

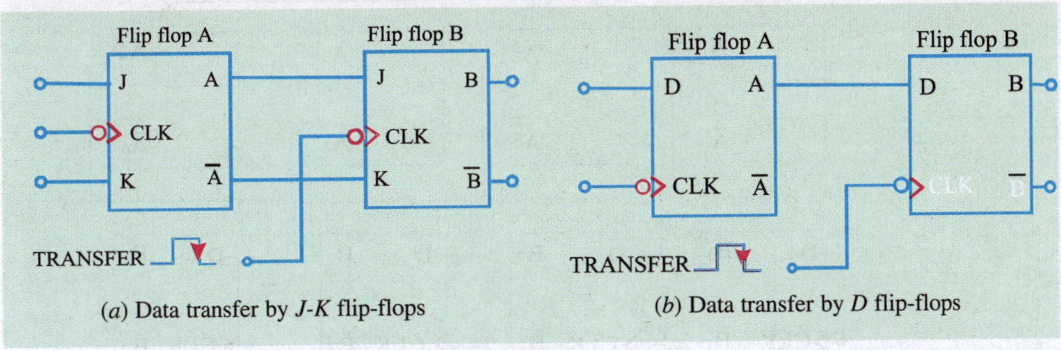

(*a*) Data transfer by *J-K* flip-flops (*b*) Data transfer by *D* flip-flops

Fig. 72.64

Fig. 72.64 (*a*) and (*b*) shows how synchronous data transfer operation can be accomplished between two flip-flops using *J-K* and *D* flip-flops respectively. In each case, the logic value that is currently stored in flip-flop *A* is transferred to flip-flop *B* upon the falling edge of the TRANSFER pulse. Thus after this falling edge, the *B* output will be the same as the *A* output.

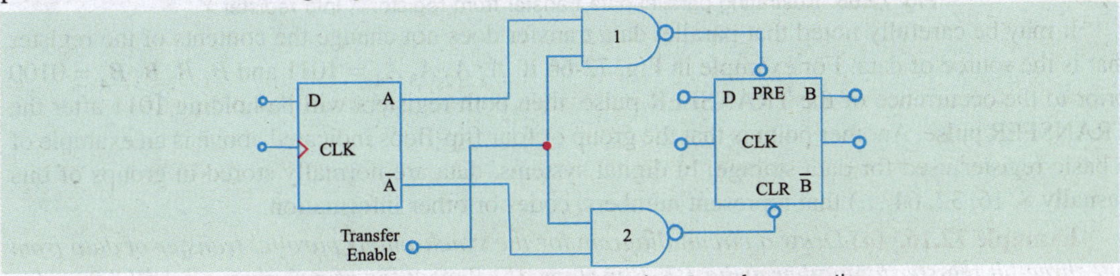

Fig. 72.65. Illustrating asynchronous data transfer operation.

The data transfer illustrated in Fig. 72.64 is called synchronous transfer because the synchronous control and CLK inputs of a flip-flop are used to perform the transfer operation. Fig. 72.65 shows the data transfer that, can be obtained using the asynchronous inputs of a flip-flop. This method called asynchronous transfer can be accomplished using the PRESET and CLEAR inputs of any flip-flop. Notice that flip-flop outputs A and $\overline{A}$ are connected to PRESET and CLEAR through two-input NAND gates 1 and 2 respectively.

In this method, the asynchronous inputs (*i.e.* PRESET and CLEAR) respond to LOW levels. The operation of the circuit may be explained as follows :

When the ***Transfer Enable*** line is held LOW, the two NAND outputs are kept HIGH, with no effect on the flip-flop outputs. But when the ***Transfer Enable*** line is made HIGH, one of the NAND outputs will go LOW depending on the state of A and $\overline{A}$ outputs. This LOW will either set or clear the flip-flop B.

72.27. Parallel Data Transfer

We have already discussed in the last article that flip-flops can be used for data storage as well as for transfer of data from one flip-flop to another. This idea can be further extended for transfer of data from one group of flip-flops (called register A) to another group of flip-flops (called register B). Register A consists of four flip-flops A_1, A_2, A_3 and A_4. Similarly register B also consists of four flip-flops labelled as B_1, B_2, B_3 and B_4 upon application of falling edge of a TRANSFER pulse, the logic level stored in A_1 is transferred to B_1, A_2 is transferred to B_2, A_3 is transferred to B_3 and A_4 is transferred to B_4. It may be noted that the transfer of contents from register A to B occurs at the same time. Because of this reason it is called synchronous transfer or parallel data transfer.

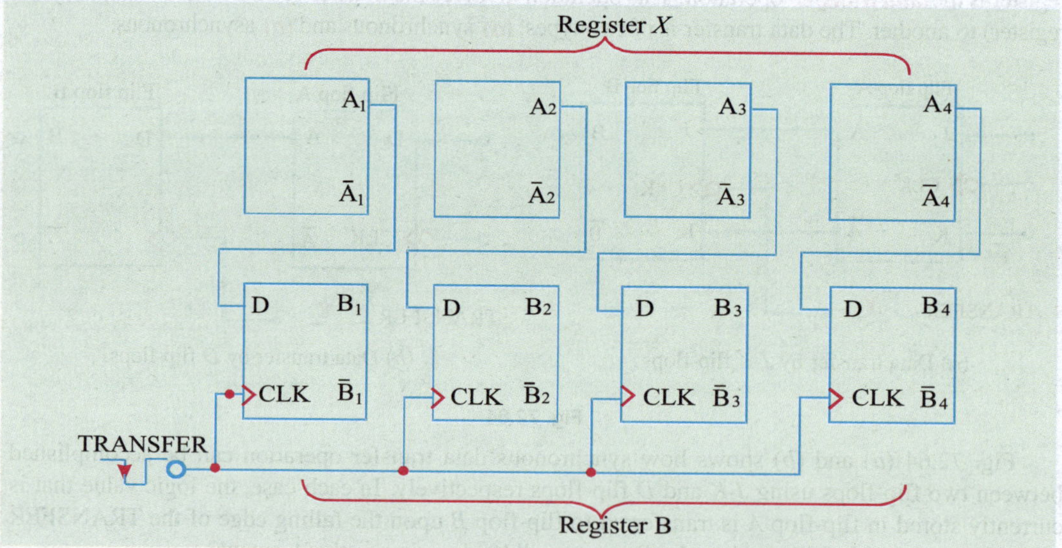

Fig. 72.66. Illustrating parallel data transfer from register X into register Y

It may be carefully noted that parallel data transfer does not change the contents of the register that is the source of data. For example in Fig. 72-66 if $A_1\ A_2\ A_3\ A_4$ = 1011 and $B_1\ B_2\ B_3\ B_4$ = 0100 prior to the occurrence of the TRANSFER pulse, then both registers will be holding 1011 after the TRANSFER pulse. Another point is that the group of four flip-flops indicated above is an example of a basic register used for data storage. In digital systems, data are normally stored in groups of bits (usually 8, 16, 32, 64 ...) that represent numbers, codes or other information.

Example 72.16. (*a*) *Draw a circuit diagram for the synchronous parallel transfer of data from one three-bit register to another using J-K Flip-flops.* (*b*) *Repeat for asynchronous parallel transfer.*

Solution. *(a)* Fig. 72.67 shows the circuit diagram for the synchronous parallel transfer of data from one three-bit register X to another register *Y* using *J-K* flip-flops. Notice that the contents of X_1, X_2, and X_3 are transferred simultaneously into Y_1, Y_2, and Y_3. It may be carefully noted that the circuit allows the parallel transfer of data from the normal outputs as well as the complemented outputs of the flip-flops constituting register *X* .

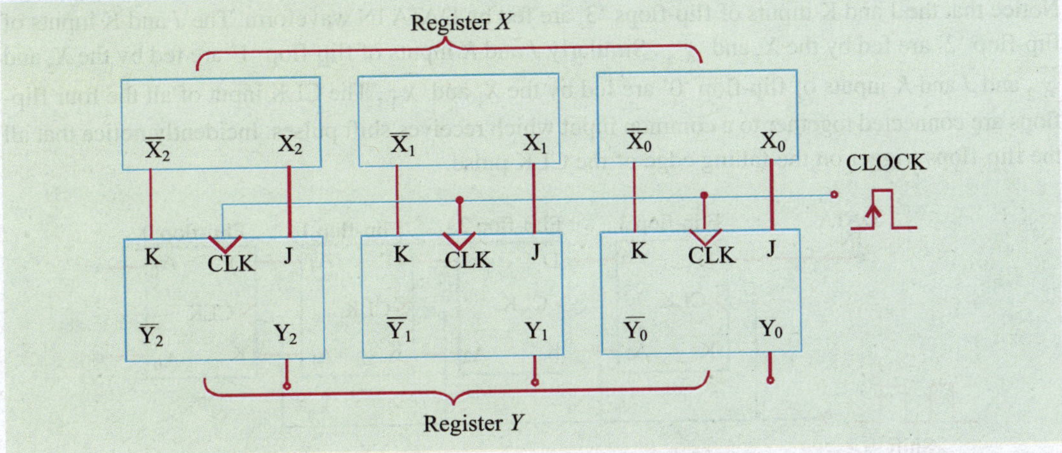

Fig. 72.67. Synchronous parallel transfer of data from one three-bit register to another

(b) Fig. 72-68 shows the circuit diagram for the asynchronous parallel data transfer from one three-bit register to another using *J-K* flip-flops. The circuit allows data transfer from either normal outputs X_0, X_1, and X_2 or the complemented outputs $\overline{X_0}$, $\overline{X_1}$ and $\overline{X_2}$.

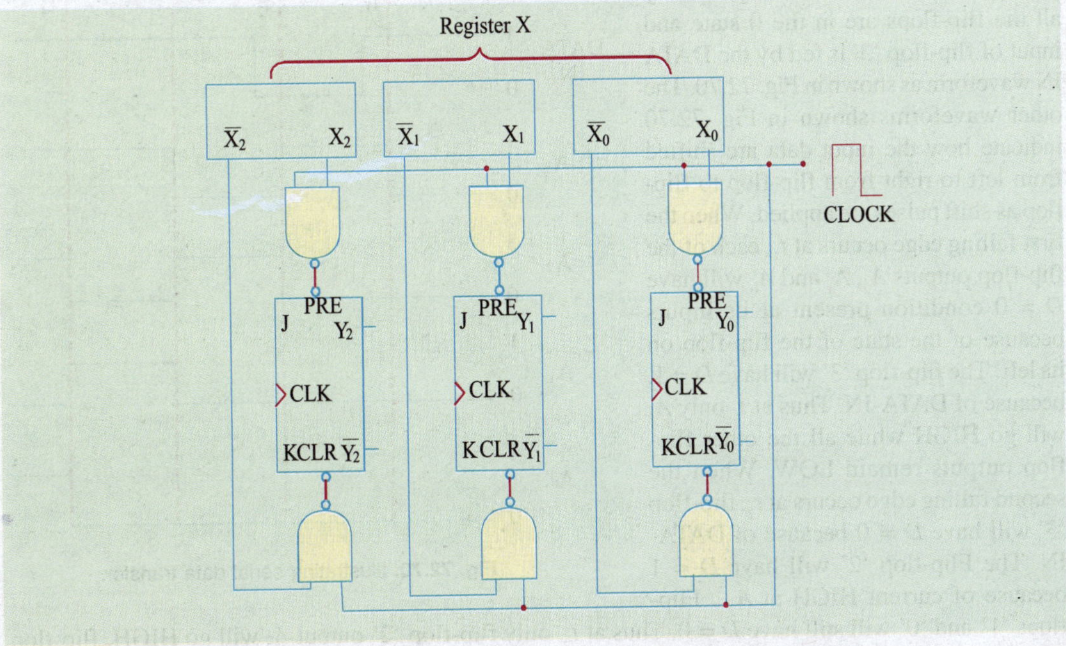

Fig. 72.68. Asynchronous parallel transfer of data from one three-bit register to another.

72.28. Serial Data Transfer : Shift Registers

We have already discussed in the last article that a ***register*** is basically a group of flip-flops to store data. But there is no interconnection between the flip-flops. A ***shift register*** is a group of flip-flops arranged in such a way that the binary numbers stored in the flip-flops are shifted from one flip-flop to the next for every clock pulse. Fig. 72.69 shows a four-bit shift register using *J-K* flip-flops. Notice that the J and K inputs of flip-flops '3' are fed by DATA IN waveform. The *J* and K inputs of flip-flop '2' are fed by the X_3 and $\overline{X}_3$. Similarly *J* and *K* inputs of flip-flop '1' are fed by the X_2 and $\overline{X}_2$ and *J* and *K* inputs of flip-flop '0' are fed by the X_1 and $\overline{X}_1$. The CLK input of all the four flip-flops are connected together to a common input which receives shift pulses. Incidently notice that all the flip-flops trigger on the falling edge of the CLK pulse.

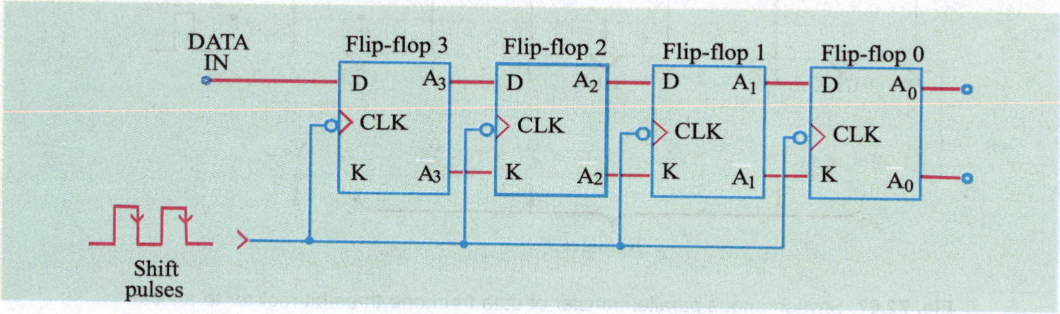

Fig. 72.69. Four-bit shift register.

The operation of the shift register may be understood from the explanation given below. Let us assume that initially all the flip-flops are in the 0 state and input of flip-flop '3' is fed by the DATA IN waveform as shown in Fig. 72.70. The other waveforms shown in Fig. 72.70 indicate how the input data are shifted from left to right from flip-flop to flip-flop as shift pulses are applied. When the first falling edge occurs at t_1, each of the flip-flop outputs A_2, A_1 and A_0 will have $D = 0$ condition present at its inputs because of the state of the flip-flop on its left. The flip-flop '3' will have $D = 1$, because of DATA-IN. Thus at t_1 only A_3 will go HIGH while all the other flip-flop outputs remain LOW. When the second falling edge occurs at t_2, flip-flop '3' will have $D = 0$ because of DATA-IN. The Flip-flop '2' will have $D = 1$ because of current HIGH at A_3. Flip-flops '1' and '0' will still have $D = 0$. Thus at t_2, only flip-flop '2' output A_2 will go HIGH, flip-flop output A_3 will go LOW, and A_2 and A_0 will remain LOW.

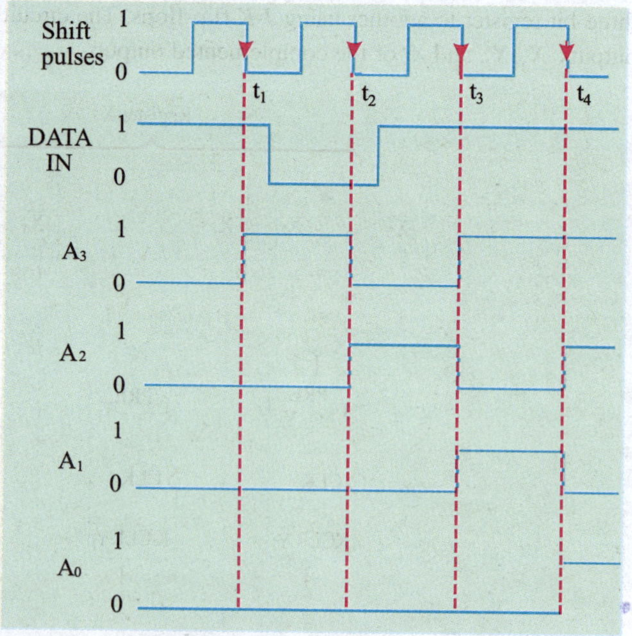

Fig. 72.70. Illustrating serial data transfer.

A similar reasoning can be used to determine how the waveforms change at t_3 and t_4. It may be carefully noted that on each falling edge of the shift pulses, each flip-flop output takes on the level that was present at the output of the flip-flop on its left just *prior* to the falling edge. Of course, flip-flop '3' output X_3 takes on the level that was present at DATA IN just prior to the falling edge.

Note. In the shift register arrangement discussed above, it is necessary that the flip-flops have a very small hold time requirement. This is because of the fact that there are times when the J and K inputs are changing at about the same time as the CLK edge. For example, the flip-flop '3' output, X_3 changes from 1 to 0 in response to the falling edge at t_2, causing the J and K inputs of flip-flop '2' output X_2 to change while its CLK input is changing. Actually because of the propagation delay of flip-flop '3', the J and K inputs of flip-flop '2' won't change for a short time after the falling edge. Because of this reason, a shift register should be implemented using edge-triggered flip-flops that have a t_H value less than one CLK-to-output propagation delay. This later requirement is easily satisfied by most modern edge triggered flip-flops.

72.29. Serial Data Transfer Between Registers

We have already discussed in the last article that a shift register is a group of flip-flops arranged in such a manner that the binary numbers stored in the flip-flops are shifted from one flip-flop to the next after every clock pulse. Now we will study as how data can be serially transferred (or shifted) from one register to the other.

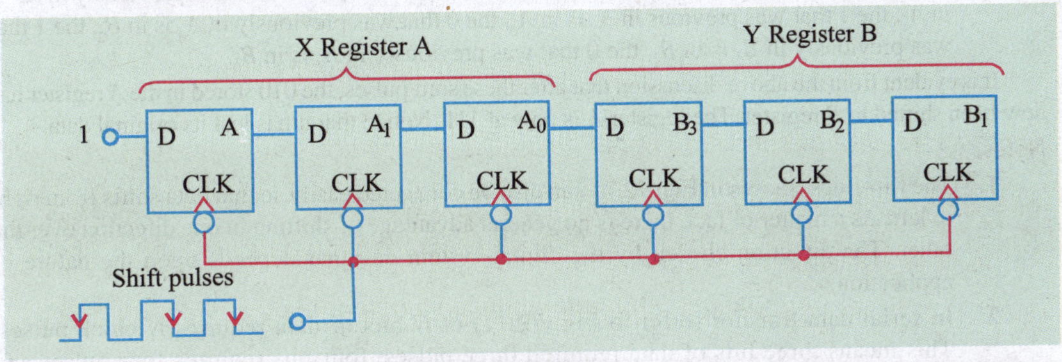

Fig. 72.71

Fig. 72.71 shows three-bit shift registers connected in such a manner that the contents of the A-register will be serially transferred (shifted) into register B. Notice how A_0, output of last flip-flop of register A is connected to the D-input of the first flip-flop of register B. The CLK input of all the flip-flops is connected together at one point where the shift pulses are applied.

Register X			Register Y			
A_2	A_1	A_0	B_2	B_1	B_0	
0	1	0	0	0	0	← Before shift pulses applied
1	0	1	0	0	0	← After first pulse
1	1	0	1	0	0	← After second pulse
1	1	1	0	1	0	← After third pulse

Fig. 72.72

As the shift pulses are applied, the data transfer takes place as follows:

$$A_2 \rightarrow A_1 \rightarrow A_0 \rightarrow B_2 \rightarrow B_1 \rightarrow B_0$$

The flip-flop output A_2 will go to a state determined by its D input. For discussion purpose let us suppose that D is held HIGH, so that A_2 will go HIGH on the first pulse and will remain there. Further let us assume that before any shift pulses are applied, the contents of A register are 010. (*i.e.* $A_2 = 0$, $A_1 = 1$, $A_0 = 0$) and the B register is at 000. As the shift pulse are applied one after the another, the states of each flip-flop output change are indicated in table as shown in Fig. 72.72. Following are some of the points which may be carefully noted from this table.

1. On the falling edge of each clock pulse, each flip-flop takes on the value that was stored in the flip-flop on its left prior to the occurrence of the pulse.
2. After first pulse, A_2 is set to 1, the 0 that was initially in A_2 is in A_1, the 1 that was initially in A_1 is in A_0, the 0 that was initially in A_0 is in B_2, the 0 that was initially in B_2 is in B_1, the 0 that was initially in B_1 is in B_0. The 0 that was initially in Y_0 is lost.
3. After second pulse, A_2 is still at 1 (because D is HIGH), the 1 that was previously in A_2 is in A_1, the 0 that was previously in A_1 is in A_0, the 1 that was previously in A_0 is in B_2, the 0 that was previously in B_2 is in B_1 and 0 that was previously in B_1 is in B_0.
4. After third pulse, A_2 is still at 1 (because D is held HIGH), the 1 that was previously in A_2 is in A_1, the 1 that was previous in A_1 is in A_0, the 0 that was previously in A_0 is in B_2, the 1 that was previously in B_2 is in B_1, the 0 that was previously in B_1 is in B_0.

It is evident from the above discussion that after these shift pulses, the 010 stored in the A register has now been shifted to B register. The register A is now at 111. Notice that it has lost its original data.

Notes.

1. The flip-flops shown in Fig. 72.71 can also be connected easily so that data shifts from right to left. As a matter of fact, there is no general advantage of shifting in one direction over the other. The direction chosen by the digital system designer depends upon the nature of application.
2. In serial data transfer (refer to Fig. 72.71) of N bits of data requires N clock pulses. This means three bits of data required three pulses, four bits requires four pulses and so on. However in parallel transfer, all the data is transferred simultaneously upon the occurrence of a single transfer pulse (refer to Fig. 72.66). In other words it does not matter how many bits are being transferred. It is obvious that parallel transfer is much faster than serial transfer using shift registers. However, parallel transfer requires more interconnections between the sending register (A) and receiving register (B) than does the serial transfer. This is an important consideration when the sending and receiving registers are at a distance from each other. The choice of either parallel or serial transmissions depends on the particular system application and specifications. In actual practice, a combination of two types is used to take advantage of the speed of parallel transfer and economy and simplicity of serial transfer.

72.30. Frequency Division

Many applications require frequency division. For example a quartz (or digital) watch makes use of a quartz crystal to generate a very stable oscillator frequency. This frequency is usually 1MHz or more. In order to advance the "seconds" display once every second, the oscillator frequency is divided by a value that will produce a very stable and accurate 1Hz output frequency. The frequency division can be achieved by using flip-flops.

Fig. 72.73 (*a*) shows a single *J-K* flip-flop connected to toggle (*i.e.* $J = K = 1$). When the clock pulses are applied at its CLK input, the Q output is a square wave with one-half the frequency of the clock input as shown in Fig. 72.73 (*b*). Thus a single flip-flop can be used as a divide-by-2 device. As seen from the figure, the flip-flop changes state on each falling edge of the clock pulse. This results in an output that changes at half the frequency of the clock waveform.

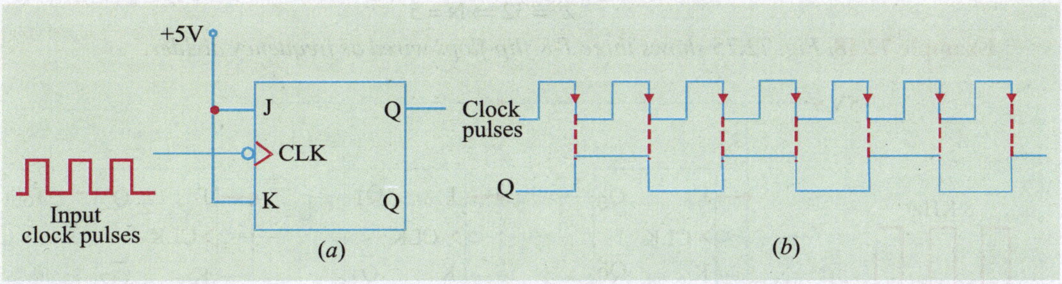

Fig. 72.73

Further division of the clock frequency can be achieved by using the output of one flip-flop as the clock input to the second flip-flop as shown in Fig. 72.74. Notice that the frequency of Q_0 output is divided by 2 by flip-flop 1. Therefore the Q_1 output is one-fourth the frequency of the original clock input. It may be carefully noted that propagation delays are not shown on the timing diagrams.

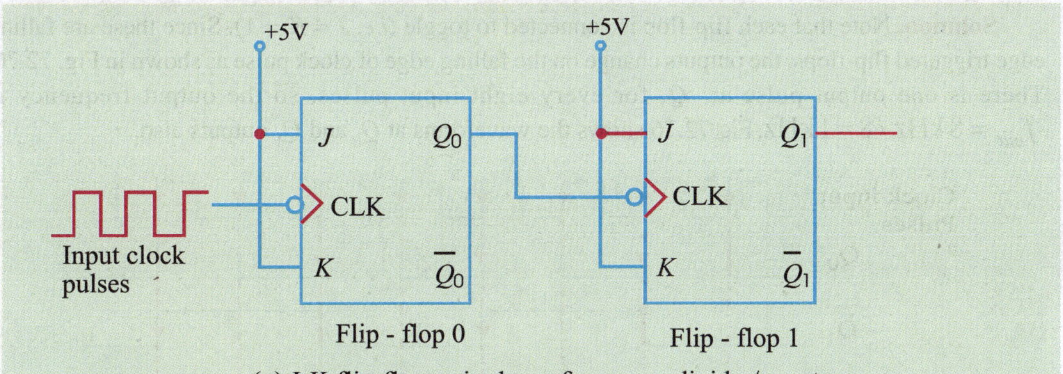

(a) J-K flip-flops wired as a frequence divider/counter (All $\overline{PRE}$ and $\overline{CLR}$ are HIGH)

(b) Timing diagram

Fig. 72.74

By connecting the flip-flops in the way described above, a frequency division of 2^N is achieved, where N is the number of flip-flops. For example, three flip-flops divide the clock frequency by $2^3 = 8$, four flip-flops divide the clock frequency by $2^4 = 16$ and so on.

Example 72.17. *How many flip-flops are required to divide a frequency by thirty-two.*

Solution. We know that a single flip-flop can divide the input frequency by 2. Two flip-flops connected together can divide the frequency by 4. In gereal a frequency division of 2^N can be achieved by using N number of flip-flops. Thus,

$$2^N = 32 \Rightarrow N = 5$$

Example 72.18. *Fig. 72.75 shows three J-K flip-flops wired as frequency divider.*

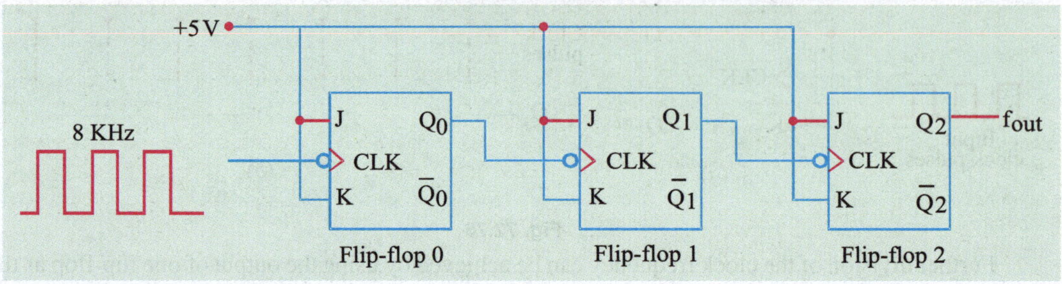

Fig. 72.75

Sketch the frequency waveform at Q_2 output when an 8 kHz square wave input is applied at CLK input of the flip-flop 'D'.

Solution. Note that each flip-flop is connected to toggle (*i.e.* $J = K = 1$). Since these are falling edge triggered flip-flops, the outputs change on the falling edge of clock pulse as shown in Fig. 72.76. There is one output pulse at Q_2 for every eight input pulses, so the output frequency is $f_{out} = 8\text{ kHz}/8 = 1\text{ kHz}$. Fig 72.76 shows the waveforms at Q_0 and Q_1 outputs also.

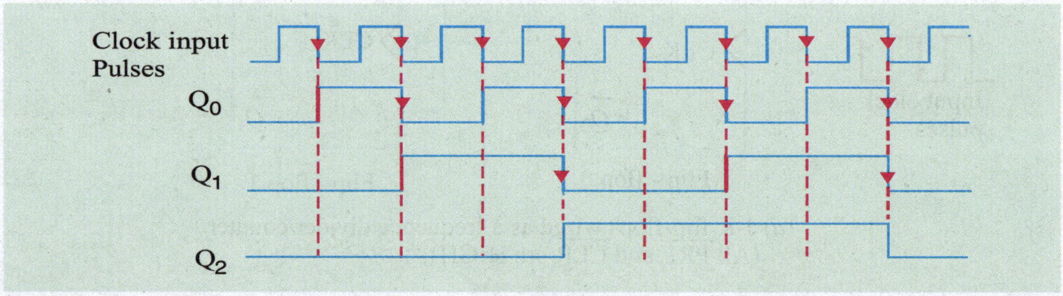

Fig. 72.76

72.31. Counting

Another important application of flip-flops is in digital counters. Fig 72.77 illustrates the concept of a 2-bit counter. Here two *J-K* flip-flops are wired to toggle (*i.e.* $J = K = 1$). Both the flip-flops are initially RESET. Flip-flop 0 toggles on the falling edge of each clock pulse. The *Q*-output of flip-flop 0 clocks flip-flop 1, so that each time Q_0 makes a HIGH-to-LOW transition, flip-flop 1 toggles. The resulting Q_0 and Q_1 waveforms are shown in the figure.

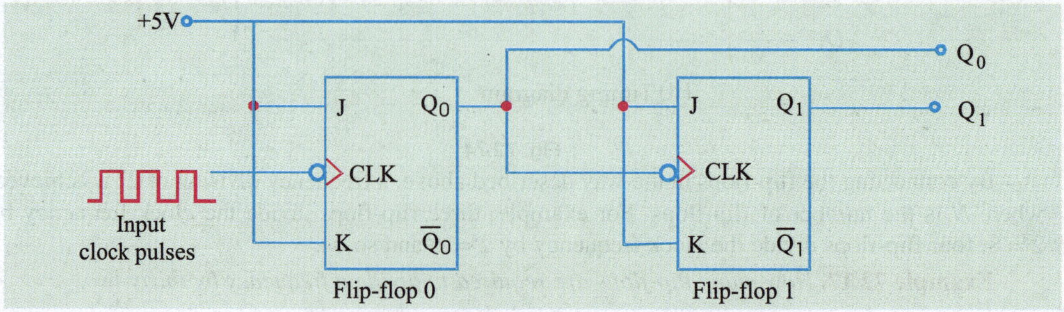

Fig. 72.77. J-K flip-flops wired as a 2-bit counter

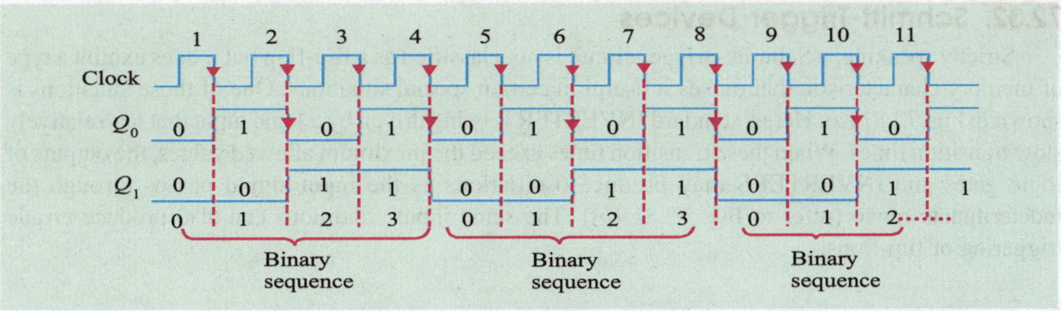

Fig. 72.78

Examine the sequence of Q_0 and Q_1 in Fig. 72.78. Prior to clock pulse 1, $Q_0 = 0$ and $Q_1 = 0$, after clock pulse 1, $Q_0 = 1$ and $Q_1 = 0$, after clock pulse 2, $Q_0 = 0$ and $Q_1 = 1$, and after clock pulse 3, $Q_0 = 1$ and $Q_1 = 1$. If we take Q_0 as the least significant bit, a two bit sequence is produced as the flip-flops are clocked. This binary sequence repeats every four clock pulses as shown in the timing diagram of Fig. 72.78. Thus flip-flops are counting in sequence from 0 to 3 (*i.e.* 00, 01, 10 and 11) and then recycling back to 0 to begin the counting sequence again.

Example 72.19. *Fig. 72.80 shows a J-K flip-flops wired as a 3-bit counter.*

Sketch the output waveforms at Q_0, Q_1 and Q_2. Also show the binary sequence represented by these waveforms.

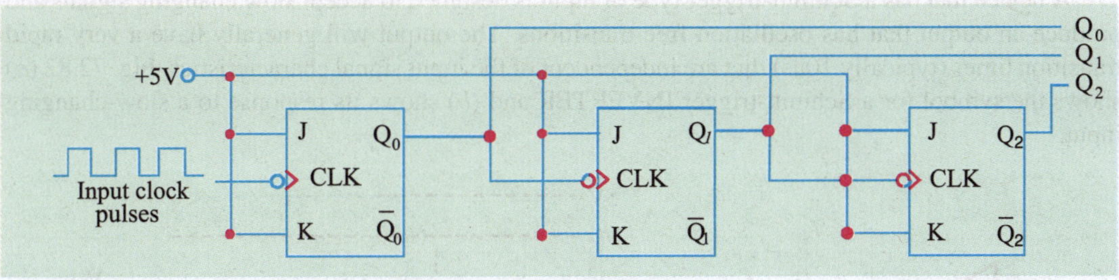

Fig. 72.79

Solution. Fig. 72.80 shows the output waveforms at Q_0, Q_1 and Q_2. Notice that the outputs change on the falling edge of the clock pulses. The outputs go through the binary sequence 000, 001, 010, 011, 100, 101, 110 and 111 as indicated.

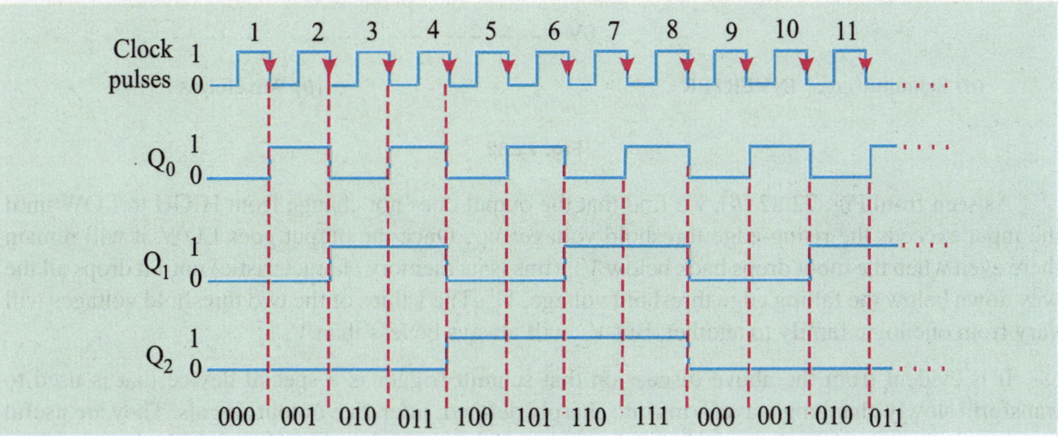

Fig. 72.80

72.32. Schmitt-Trigger Devices

Strictly speaking, a Schmitt –trigger circuit is not classified as a flip-flop but it does exhibit a type of memory characteristic that makes it useful in certain special situations. One of those situations is shown in Fig. 72.81 (*a*). Here a standard INVERTER is being driven by a logic input that has relatively slow transition times. When these transition times exceed the maximum allowed values, the outputs of logic gates and INVERTERS may produce oscillations as the input signal passes through the indeterminate range (refer to Fig. 72.81 (*b*)). The same input conditions can also produce erratic triggering of flip-flops.

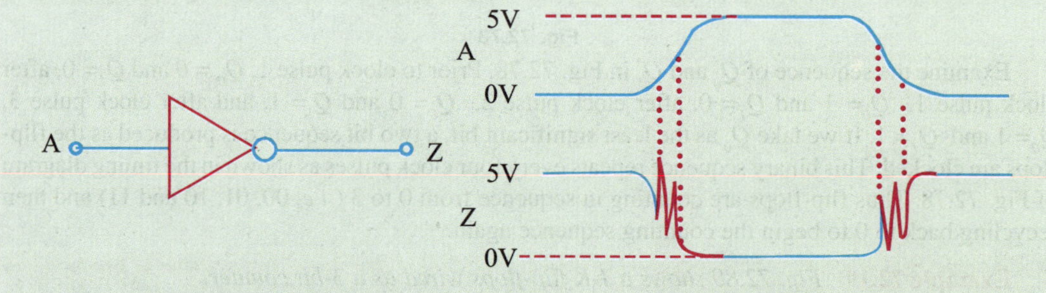

(*a*) Standard INVERTER (*b*) Waveforms

Fig. 72.81

A device that has a Schmitt-trigger type of input is designed to accept slow changing signals and produce an output that has oscillation free transitions. The output will generally have a very rapid transition times (typically 10ns) that are independent of the input signal characteristics. Fig. 72.82 (*a*) shows the symbol for a Schmitt-trigger INVERTER and (*b*) shows its response to a slow-changing input.

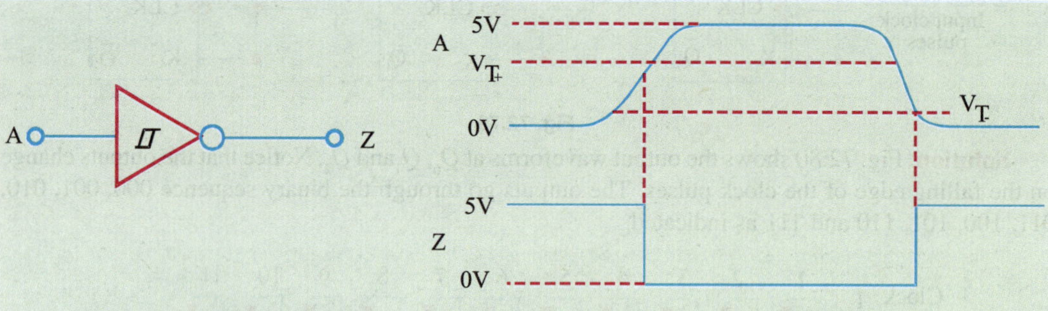

(*a*) Schmitt-trigger INVERTER (*b*) Waveforms

Fig. 72.82

As seen from Fig. 72.82 (*b*), we find that the output does not change from HIGH to LOW until the input exceeds the rising-edge threshold voltage, V_{T+}. Once the output goes LOW, it will remain there even when the input drops back below V_{T+} (this is its memory characteristic) until it drops all the way down below the falling edge threshold voltage, V_{T-}. The values of the two threshold voltages will vary from one logic family to another. But V_{T-} will always be less than V_{T+}.

It is evident from the above discussion that schmitt trigger is a special device that is used to transform slowly changing waveforms into sharply defined, jitter-free output singals. They are useful for changing clock edges that may have slow rise and fall times into straight vertical edges.

There are several ICs with Schmitt-trigger inputs. The 7414, 74LS14 and 74HC14 are hex INVERTER ICs with Schmitt-trigger inputs. On the other hand 7413, 74LS13 and 74HC13 are dual four-input NANDs with Schmitt-trigger inputs.

72.33. One-Shot (Monostable Multivibrator)

A one-shot (abbreviated as *OS*) is also referred to as monostable multivibrator. It is a digital circuit that is somewhat related to a flip flop. For instance, like a flip-flop, the one-shot has two inputs, Q and $\overline{Q}$ which are inverse (or complement) of each other. However, unlike the flip-flop, the one-shot has only one stable output state (normally, $Q = 0$ and $\overline{Q} = 1$). The one-shot remains in this stable state until it is triggered by an input signal. Once triggered, the one-shot outputs switch to the opposite state ($Q = 1, \overline{Q} = 0$). This state is called quasi-stable (i.e. unstable) state. The one-shot remains in this quasi-stable state for a fixed period of time, t_p, which is determined by the RC time constant. The RC time constant in turn is determined by the values of external components connected to one-shot. After a time, t_p, the one-shot outputs return to their resting state until triggered again.

Fig. 72.83 (*a*) shows the logic symbol for one-shot. As seen from this figure, it has one input labeled as trigger input (T), a normal output (Q), and an inverted output ($\overline{Q}$). It needs two external components: resistor (R_T) and a capacitor (C_T) for its operation. The values of R_T and C_T determines the exact value of time period (t_p) for which the one-shot switches to its quasi stable state. The value of t_p can vary from several nanoseconds to several tens of seconds. Fig. 72.83 (*b*) shows the stable state (where $Q = 0$, $\overline{Q} = 1$) and the quasi-stable state (where $Q = 1$ and $\overline{Q} = 0$).

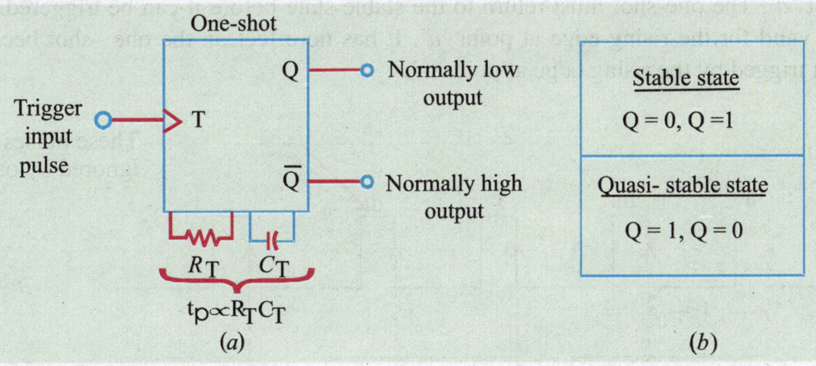

Fig. 72.83. One-shot symbol and Output states.

72.34. Types of One-Shots

Depending upon the operation, the one-shot is of the following two types :

1. Non-retriggerable one-shot and
2. Retriggerable one-shot

Both the type of one-shots are available in the IC form. Now we shall study both these one-shots one by one in the following pages.

72.35. Non-Retriggerable One-Shot

We have already discussed in the last article that one-shot (or a monostable multivibrator) is a device with only one stable state. It is normally in its stable state and will change to its quasi-stable state only when triggered. Once it is triggered, the one-shot remains in its quasi-stable state for a fixed period of time and then automatically returns to its stable state. The time that the device remains in the quasi-stable state determines the pulse width (t_p) of the output.

Fig. 72.84 (*a*) shows the waveforms of a trigger input signal *T* and output of one-shot. Notice that the rising edge of the trigger signal will trigger the one-shot to its quasi-stable state. The one-shot will remain in the quasi-stable state for a time, t_p after which it automatically returns to its stable state.

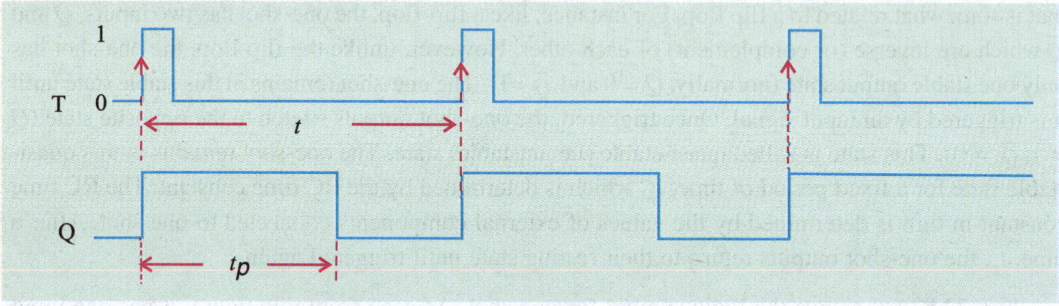

Fig. 72.84

It may be noted from the diagram shown above that one-shot is being triggered at intervals greater than its pulse width (*i.e.* $t > t_p$).

Now we will consider a situation where the one-shot is being triggered at intervals less than its pulse width (*i.e.* $t < t_p$). Refer to Fig. 72.85. Notice that the rising edges of the T signal at points '*b*' has no effect on the one-shot because it has already been trigged to the quasi-stable state by the rising edge at point '*a*'. The one-shot must return to the stable state before it can be triggered. The same argument is valid for the rising edge at point '*d*'. It has no effect on the one –shot because it has already been trigged by the rising edge at point '*c*'.

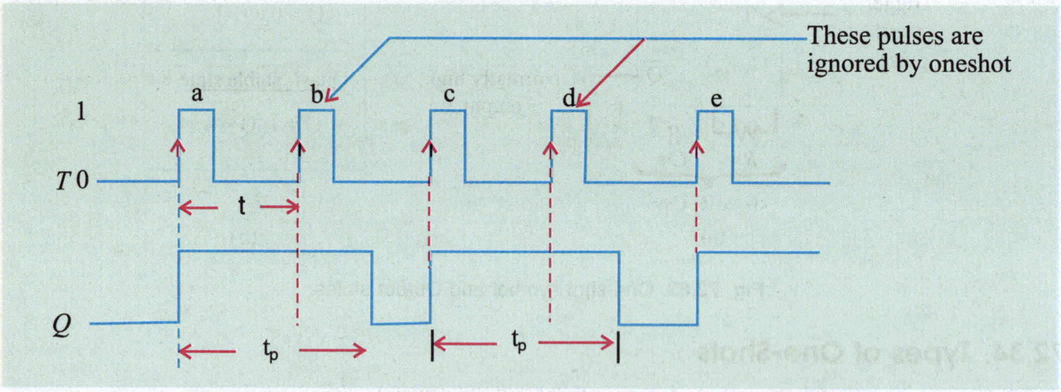

Fig. 72.85

It may be carefully noted that one-shot output pulse duration (t_p) is always the same regardless of the input pulses. The pulse duration t_p, depends only on the values of external components R_T and C_T and the internal one-shot circuitry. In a typical monoshot, the value of t_p is given by the relation.

$$t_p = 0.7\, R_T C_T$$

Example 72.20. *Fig. 72.86 shows three non-retriggerable one-shots connected in a timing chain that produces three sequential output pulses. Note the "1" in front of the pulse on each one-shot symbol to indicate non-retriggerable operation.*

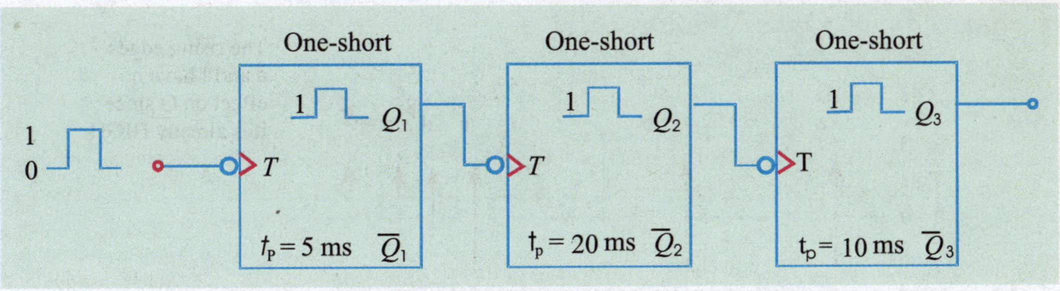

Fig. 72.86

Draw a timing diagram showing the relationship between the input pulse and the three one-shot outputs. Assume an input pulse duration of 10 *ms.*

Solution. Notice that all the one-shots trigger at the falling edge of the input signal. To begin with, the first one-shot triggers at the falling edge of the input signal and the second one-shot triggers at the falling edge of Q_1. The third one-shot triggers at the falling edge of Q_2.

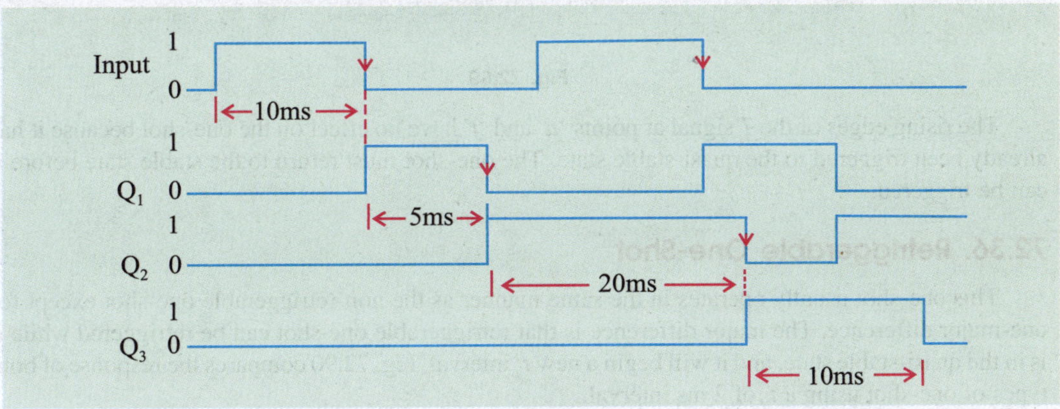

Fig. 72.87

Fig. 72.87 shows a sketch of the timing waveforms at Q_1, Q_2 and Q_3 along with the input trigger waveform.

Example 72.21. *Sketch the waveforms at Q and $\overline{Q}$ outputs of a one-shot which is being triggered by a signal shown in Fig.* 72.88. *Assume that one-shot triggers at the rising edge of the trigger signal and has a t_p =* 1.5 *ms.*

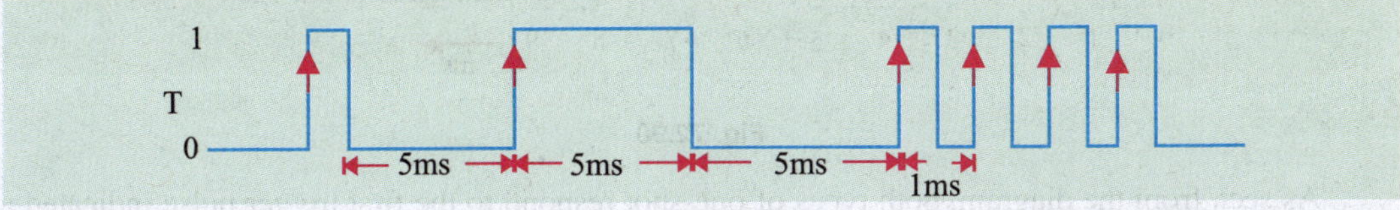

Fig. 72.88

Solution. Fig. 72.89 shows the waveforms at the one-shot outputs Q and $\overline{Q}$ in response to the trigger signal (T). Notice that the rising edges at points *a, b, c* and *e* of the *T* signal, will trigger one-shot to its quasi-stable state. The one-shot will remain in the quasi-stable state for a time, t_p after which it automatically returns to its stable state.

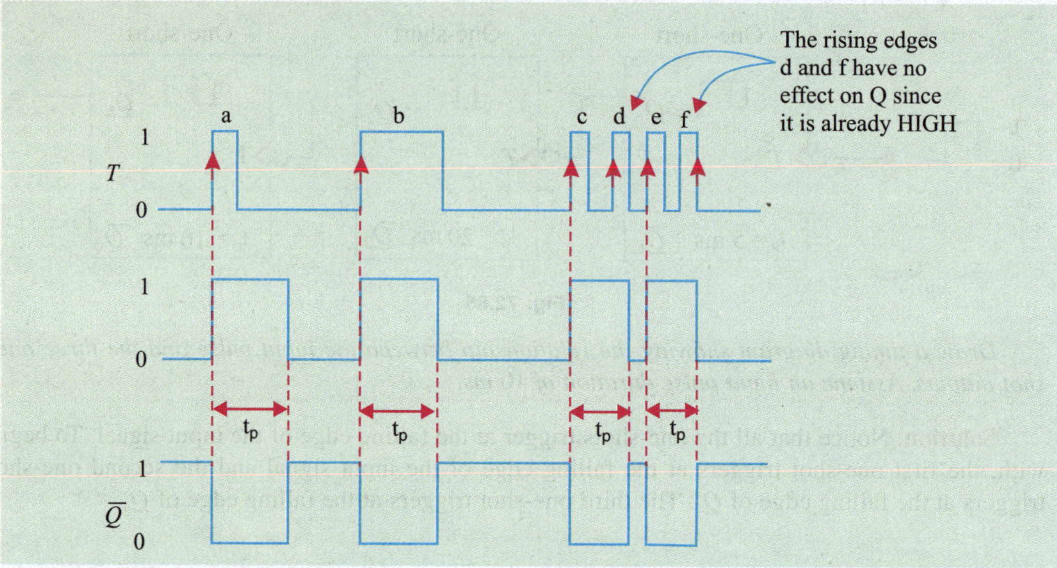

Fig. 72.89

The rising edges of the T signal at points '*d*' and '*f*' have no effect on the one-shot because it has already been triggered to the quasi-stable state. The one-shot must return to the stable state before it can be triggered.

72.36. Retriggerable One-Shot

This one-shot usually operates in the same manner as the non-retriggerable one-shot except for one-major difference. The major difference is that retriggerable one-shot can be retriggered while it is in the quasi-stable state, and it will begin a new t_p interval. Fig. 72.90 compares the response of both types of one-shot using a t_p of 2 ms interval.

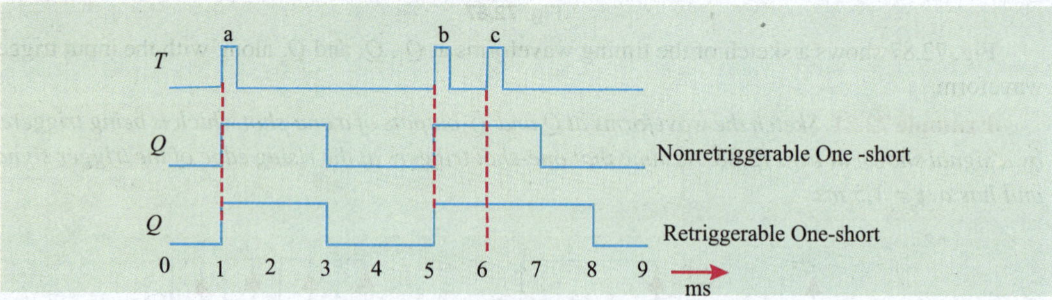

Fig. 72.90

As seen from the diagram, both types of one-shot respond to the first trigger pulse indicated as '*a*' at t = 1 ms by going HIGH for 2 ms and then returning LOW. The second trigger pulse indicated as '*b*' at t = 5 ms triggers both one – shots to the HIGH state. The third trigger pulse indicated as '*c*', at t = 6 ms has no effect on the non-retriggerable one-shot because it is already in the quasi-stable state. However, this trigger pulse will retrigger the retriggerable one-shot to begin a new t_p= 2 ms internal. Thus it will stay HIGH for 2 ms after this third trigger pulse.

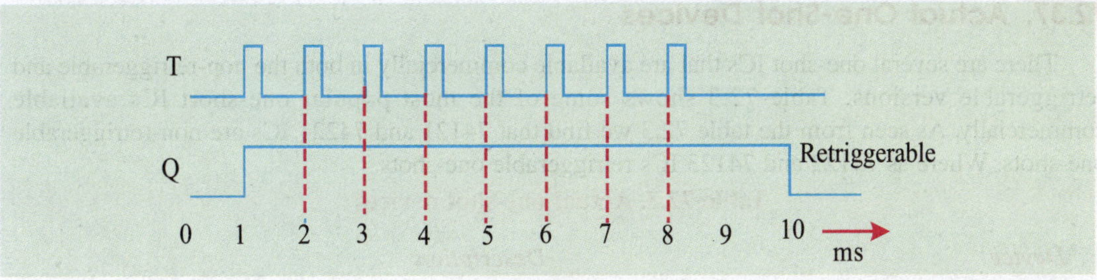

Fig. 72.91

It is evident from the discussion above that a retriggerable one-shot begins a new t_p interval each time a trigger pulse is applied, regardless of the current state of its Q-output. As a matter of fact, trigger pulses can be applied at a rate fast enough that one-shot will always be retriggered before the end of the t_p interval and Q-output will remain HIGH. Such a situation is shown in Fig. 72.91. As seen from this diagram, eight pulses are applied every 1 ms. The Q-output does not return LOW until 2 ms after the last trigger pulse.

Example 72.22. *A retriggerable one-shot can be used as a pulse-frequency detector that detects when the frequency of a pulse input is below a predetermined value. A simple example of this application is shown in Fig.* 72.92. *The operation begins by momentarily closing switch SW*1.

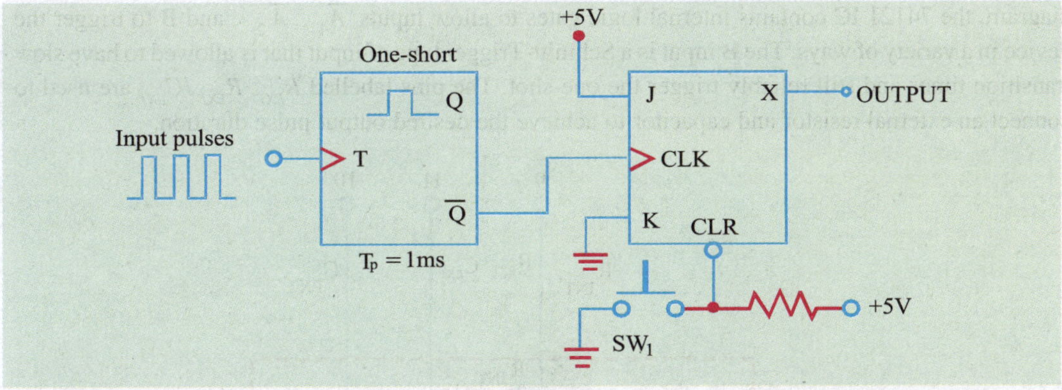

Fig. 72.92

(*a*) *Describe how the circuit responds to input frequencies above* 1*kHz.* (*b*) *Describe how the circuit responds to input frequencies below* 1*kHz.* (*c*) *How would you modify the circuit to detect when the input frequency drops below* 50 kHz?

Solution.

(*a*) Circuit response to input frequencies above 1kHz. Closing the switch SW1, clears X to Zero. Since the one-shot has t_p = 1 ms, the one-shot will be retriggered before the end of the t_p interval for frequencies greater than 1 kHz. Thus $\overline{Q}$ will stay LOW. As a result of this, there is no clock input to *JK* flip-flop and hence X will remain LOW.

(*b*) Circuit response to input frequencies below 1kHz. If the input frequency falls below 1 kHz, the $\overline{Q}$ will return HIGH before the one-shot is triggered again. This positive-going-trigger will clock *JK* flip-flop and X will change to HIGH.

(*c*) Modification of the circuit to detect frequencies below 50 kHz. In order to detect frequencies below 50 kHz, we will have to change t_p of the one-shot to 1/50 kHz = 20 µs.

72.37. Actual One-Shot Devices

There are several one-shot ICs that are available commercially in both the non-retriggerable and retriggerable versions. Table 72.3 shows some of the most popular one-short ICs available commercially. As seen from the table 72.3 we find that 74121 and 74221 ICs are non-retriggerable one-shots. Where as 74122 and 74123 ICs retriggerable one-shots.

Table 72.3. Actual one-shot devices

Device	*Description*
74121	A *single non-retriggerable* one-shot IC in standard TTL series.
74221	A *dual non-retriggerable* one-shot IC in standard TTL series.
74LS221	A *dual non-retriggerable* one-shot IC in Low-power SchottkyTTL series
74HC221	A *dual non-retriggerable* one-shot IC in high-speed CMOS series
74122	A *single retriggerable* one-shot IC in standard TTL series
74LS122	A single retriggerable one-shot in IC low-power Schottky TTL series.
74123	A dual retriggerable one-shot IC in standard TTL series
74LS123	A dual retriggerable one-shot IC in lower-power Schottky TTL series
74HC123	A dual retriggerable one-shot IC in high-speed CMOS series.

Fig. 72.93 shows the logic symbol for the 74121 non-retriggerable one-shot IC. As seen from the diagram, the 74121 IC contains internal logic gates to allow inputs $\overline{A}_1$, $\overline{A}_2$, and B to trigger the device in a variety of ways. The B input is a Schmitt-Trigger type of input that is allowed to have slow transition times and still reliably trigger the one-shot. The pins labelled R_{EXT}, R_{EXT} /C_{EXT} are used to connect an external resistor and capacitor to achieve the desired output pulse duration.

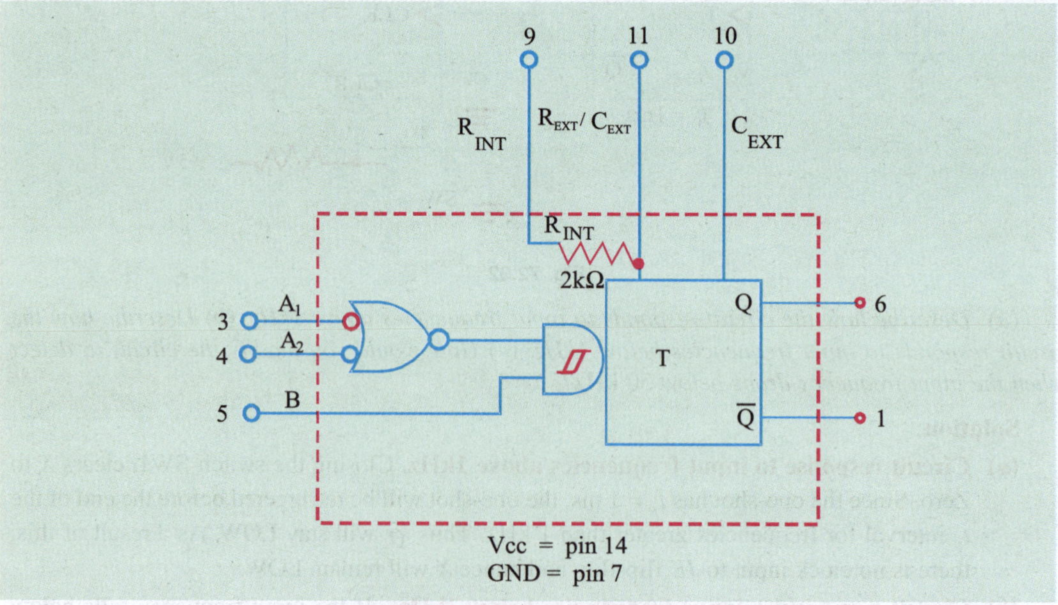

Fig. 72.93. Logic symbol of 74121.

A typical pulse width of 30 ns is produced when no external components are used and the internal timing registor (2 kΩ) is connected to V_{CC} . The pulse width can be set anywhere between 30 ns and 28 ns by the use of external components, (*i.e.* R_{EXT} and C_{EXT}). Mathematically the pulse width.

$$t_p = 0.7\,R\,C_{EXT} \qquad \text{...}(i)$$

where R is either R_{INT} or R_{EXT}. Notice that if R is in kilohms (kΩ) and C_{EXT} is in picofarads (PF), the output pulse width (t_p) is in nanoseconds. Further if R is in kilohms (kΩ) and C_{EXT} is in nanofarads (μF), the output pulse width (t_p) is in microseconds (μS).

Table 72.4 shows the truth table of 74121. Notice that for the first four rows of the table, the Q-output is LOW.

Table 72.4

Inputs			Output	
A_1	A_2	B	Q	$\overline{Q}$
L	X	H	L	L
X	L	H	Ll	Ll
X	X	L	L	L
H	H	X	L	L
H	↓	H	⎍	⊔
↓	H	H	⎍	⊔
↓	↓	H	⎍	⊔
L	X	↑	⎍	⊔
X	L	↑	⎍	⊔

H = High voltage level
L = Low voltage level
X = Don't case
↑ = Rising edge
↓ = Falling edge

It is because of the fact that no trigger pulse is applie at any one of the three inputs. In the next five lines of the truth table, notice that we have a trigger pulse at one of the inputs (A_1 A_2 or B) and the other two inputs are connected LOW or HIGH.

Fig. 72.94 (*a*) shows the logic symbol and 72.94 (*b*) the truth table for 74123 a dual retriggerable one – shot. As seen from the diagram, the 74123 IC contains two retrigger able one-shots. Each retriggerable one-shot contains $\overline{A}_1$, B_1 and R_D to trigger the device in a variety of ways.

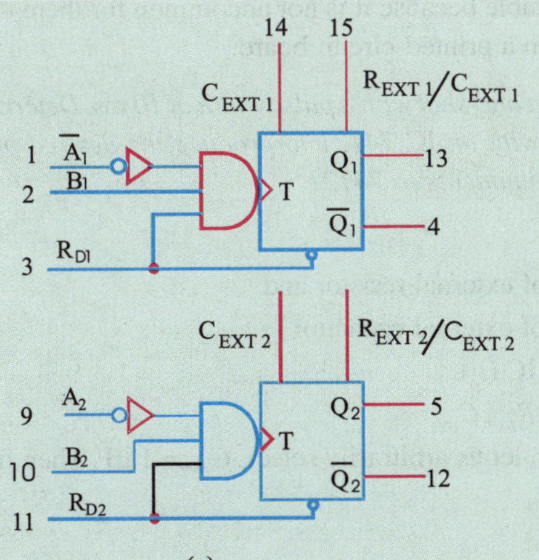

(*a*)

INPUTS			OUTPUTS	
R_D	$\overline{A}$	B	Q	$\overline{Q}$
L	X	X	L	H
X	H	X	L	H
X	X	L	L	H
H	L	↑	⎍	⊔
H	↓	H	⎍	⊔
↑	L	H	⎍	⊔

H = HIGH voltage level
L = LOW voltage X= Don't care
↑ Rising edge
↓ falling edge
⎍ = One HIGH-level pulse
⊔ = one LOW-level pulse

(*b*)

Fig. 72.94

A minimum pulse width of approximately 45 ns is obtained with no external components. Wider pulse widths are achieved by using external components. A general formula for calculating the values of these components for a specified pulse width (t_p), provided C_{EXT} > 1000 pF,

$$t_p = 0.28\,R\,C_{EXT}\left(1+\frac{0.7}{R}\right) \qquad \text{.... }(ii)$$

where 0.28 is a constant determined by a particular type of one-shot, R is either the internal or external resistor, C_{EXT} is in pF and t_p is in ns.

Notes.

1. While selecting the value of R_{EXT}, care should be taken that its value must be in kilohms ($k\Omega$) so that the circuit current is in milli amperes (mA). If the resistor value is chosen in ohms (Ω), the circuit current will be in amperes which the IC will not be able to handle it. On the other hand, if the value of R_{EXT} is in megohms ($M\Omega$), the current will be in micro amperes (μA) which may be too small for internal operation of 74121. Moreover, resistances with values in the range of megohms are susceptible to electrostatic noise.
2. For a desired pulse width t_w, we can determine the values of external components in two ways.
   (*i*) Arbitrarily select the value of capacitor, C_{EXT} and then using equation $t_W = 0.7\ R_{EXT}\ C_{EXT}$, determine the value of $R_{EXT.}$
   (*ii*) Arbitrarily select the value of R_{EXT} and then using the equation, $t_W = 0.7\ R_{EXT}\ C_{EXT}$, determine the value of C_{EXT}. be in microfarads (μF).
   (*iii*) If t_W is required to be in nanoseconds, choose R_{EXT} in kilohms (kΩ) so that C_{EXT} is in picofarads. However if the desired t_W is required to be in milliseconds, still select R_{EXT} is in kilohms ($k\Omega$), but C_{EXT} will now be in microfareds (μF). Care should be taken while selecting the component values in the sense that the components are availably only in certain standard values.
3. While selecting the value of C_{EXT}, care should be taken that it is much larger than any stray capacitance that might be encountered in a typical electronic circuit. Values of capacitance less than 100 pF (0.0001 μF) may be unsuitable because it is not uncommon for there to be 50 pF of stray capacitance between traces in a printed-circuit board.

Example 72.23. *A certain digital circuit requires a one-shot with a pulse width of 10* ms. *Determine the values of external components to be connected with an* IC *74121 to produce the desired pulse width. Also show the connections of the external components to 74121.*

Solution.

Let R_{EXT} = The value of external resistor and

C_{EXT} = The value of external capacitor.

Then we know that the pulse width for a 74121 IC (t_p),

$$10 \times 10^{-3} = 0.7\ R_{EXT}\ C_{EXT} \quad \text{... } (i)$$

Since the desired pulse width is in milli seconds, let us arbitrarily select $C_{EXT} = 1$ μF. Then from equations (*i*).

$$10 \times 10^{-3} = 0.7 \times R_{EXT} \times (1 \times 10^{-6})$$

$$\therefore \quad R_{EXT} = \frac{10\times10^{-3}}{0.7\,(1\times10^{-6})} = 14.28\times10^{-3}\ \Omega = 14.28\ \text{k}\Omega$$

In order to select a resistor value of 14.28 kΩ, we can use a 10 kΩ fixed resistor with a 5 kΩ potentiometer.

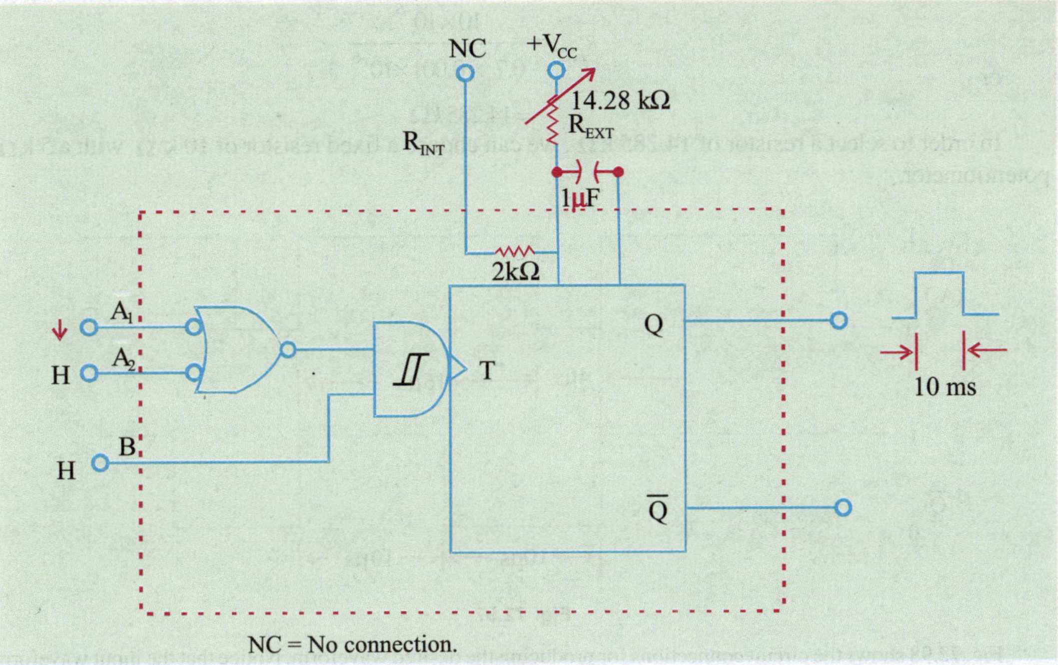

Fig. 72.95

Fig. 72.95 shows a 14.28 k Ω resistor with 1 μF capacitor connected to the 74121 IC for producing a desired pulse width of 10 ms. Notice the settings on $\overline{A}_1$, $\overline{A}_2$ and B inputs of 74121 as well.

Example 72.24. *Design a circuit using a* 74121 *to convert a* 50 *kHz,* 80% *duty cycle square wave to a* 50 *kHz,* 50% *duty cycle square wave.*

Solution.

Fig. 72.96 shows the given square wave. Notice that it has a time period of 1/50 kHz (= 20 μs) and has a pulse width of 80% of 20 μS (= 16 μs).

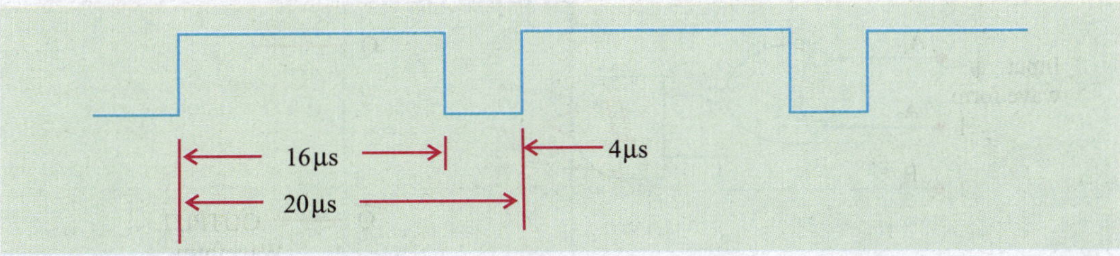

Fig. 72.96

Now we have to stretch the 4 μs negative pulse to 10 μs to make the duty cycle 50 %. If we use the falling edge on the negative pulse to trigger $\overline{A}_1$ input to a 74121 and set the output pulse width (t_p) to 10 μs, we should have the solution as shown in 72.97 (*b*). The output will be taken from $\overline{Q}$ because it provides a negative pulse when triggered.

Let R_{EXT} = The value of external resister and

C_{EXT} = The value of external capacitor,

Then the pulse width (t_p)

$$10 \times 10^{-6} = 0.7\, R_{EXT}\, C_{EXT}$$

Let us arbitrarily select C_{EXT} = 0.001 μ*F* (equal to 1000pf); then

$$10\,\mu s = 0.7 \times R_{EXT} \times 1000 \times 10^{-12}$$

$$R_{EXT} = \frac{10 \times 10^{-6}}{0.7 \times 0.001 \times 10^{-6}}$$

or

$$= 14.285\, k\Omega$$

In order to select a resistor of 14.285 kΩ, we can choose a fixed resistor of 10 k Ω with a 5 kΩ potentiometer.

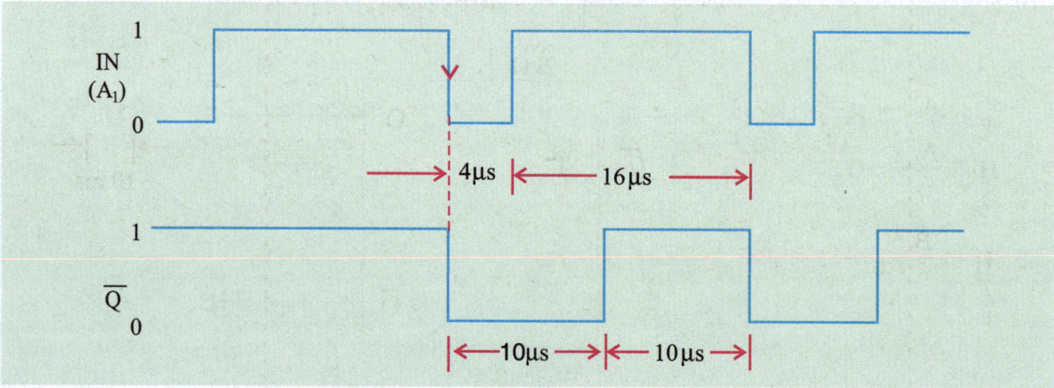

Fig. 72.97

Fig. 72.98 shows the circuit connections for producing the desired waveform. Notice that the input waveform is applied at $\overline{A}_1$ input while $\overline{A}_2$ and B are tied HIGH. The output is obtained from $\overline{Q}$.

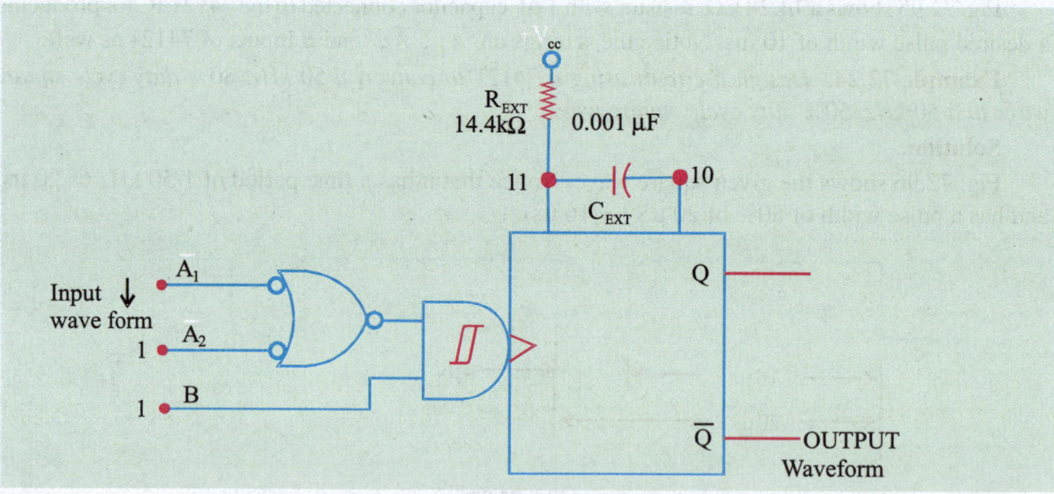

Fig. 72.98

Example 72.25. *Determine the value of R_{EXT} and C_{EXT} that will produce a pulse width of* 50 μS *when connected to a* 74123.

Solution.

Let R_{EXT} = The value of external resistor and

C_{EXT} = The value of external capacitor

Then we know that the pulse width for a 74123 IC (t_p),

$$50 \times 10^{-6} = 0.28\, R_{EXT}\, C_{EXT}\left(1 + \frac{0.7}{R_{EXT}}\right) \quad ...(i)$$

Let us arbitrarily select $C_{EXT} = 0.01\ \mu F$ (recall that C_{EXT} has to be greater than 1000 pF), then, from equation (*i*),

$$50 \times 10^{-6} = 0.28\ R_{EXT}\ (0.01 \times 10^{-6}) \left(1 + \frac{0.7}{R_{EXT}}\right)$$

$$= 0.28\ (R_{EXT} + 0.7)\ (0.01 \times 10^{-6})$$

$$\therefore \qquad R_{EXT} + 0.7 = \frac{50 \times 10^{-6}}{0.28 \times (0.01 \times 10^{-6})} = 17857\ \Omega$$

or $$R_{EXT} = 17857 - 0.7 = 17856.3\ \Omega = 17.9\ \text{k}\ \Omega$$

Thus we can use a fixed resistor of 15 *K* and 5 *K* potentiometer to select $R_{EXT} = 17.9\ K\Omega$.

72.38. Clock Generator Circuits

We have already discussed in Art. 72.2 that flip-flop has two stable states due to which they are known as bistable multivibrators. Further in Art. 72.33 we have discussed that one-shot has one stable state due to which they are referred to as monostable multivibrators. Now we shall discuss a third type of multivibrator which has no stable states. Such a multivibrator is called an astable or free-running multivibrator. An astable multivibrator switches back and forth (*i.e.* oscillates) between two unstable states without any external triggering.

There are several types of astable multivibrators that are in common use but the following three are important from the subject point of view;

1. Schmitt-Trigger oscillator
2. 555 Timer used as astable oscillator.
3. Crystal-controlled clock generator.

Now we shall discuss all the three oscillators one by one in the following pages.

72.39. Schmitt-Trigger Oscillator

Fig. 72.99 shows a Schmitt-trigger INVERTER connected as an oscillator. The signal at V_{OUT} is an approximate square wave with a frequency that depends upon the values of *R* and *C* values. The relationship between the frequency and RC values is shown in Fig. 72.5 for three different Schmitt-trigger INVERTERS.

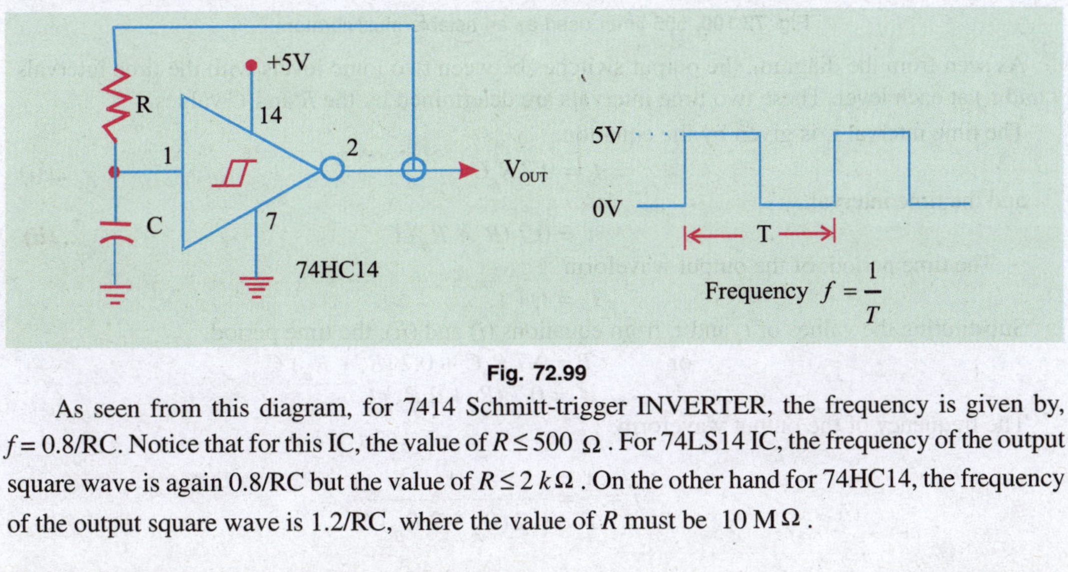

Fig. 72.99

As seen from this diagram, for 7414 Schmitt-trigger INVERTER, the frequency is given by, *f* = 0.8/RC. Notice that for this IC, the value of $R \leq 500\ \Omega$. For 74LS14 IC, the frequency of the output square wave is again 0.8/RC but the value of $R \leq 2\ k\Omega$. On the other hand for 74HC14, the frequency of the output square wave is 1.2/RC, where the value of *R* must be $10\ \text{M}\Omega$.

IC	*Frequency*	*Remarks*
7414	≈ 0.8/RC	(R ≤ 500 Ω)
74LS14	≈ 0.8/RC	(R ≤ 2 kΩ)
74HC14	≈ 1.2/RC	(R ≤ 10 MΩ)

Table 72.5

72.40. 555 Timer Used as an Astable Multivibrator

The 555 IC is a very popular, general purpose timer IC. It can be connected as a one-shot or as an astable multivibrator or as a free running oscillator. Fig. 72.100 shows how external components can be connected to a 555 so that it operates as a free-running oscillator. The output of the oscillator is a repetitive rectangular waveform as shown in Fig. 72.100 (*b*).

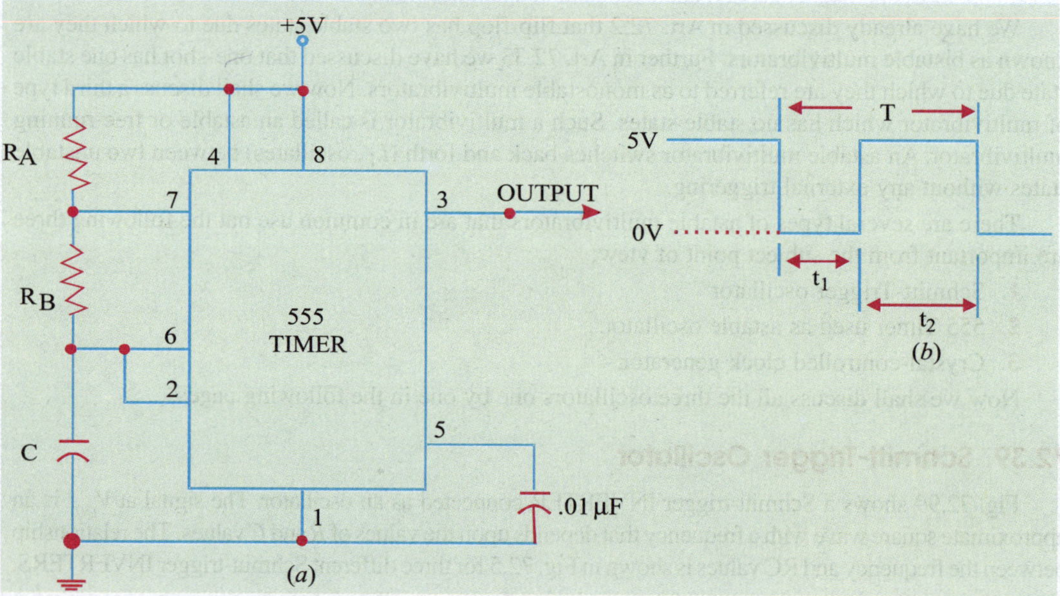

Fig. 72.100. 555 timer used as an astable multivibrator.

As seen from the diagram, the output switches between two logic levels with the time intervals ($t_1$and t_2) at each level. These two time intervals are determined by the *R* and *C* values.

The time interval t_1 is given by the equation,

$$t_1 = 0.7\, R_B C \quad \text{... (i)}$$

and the time interval,

$$t_2 = 0.7\, (R_A + R_B)\, C \quad \text{... (ii)}$$

∴ The time period, of the output waveform

$$T = t_1 + t_2$$

Substituting the values of t_1 and t_2 from equations (*i*) and (*ii*), the time period,

$$\text{or} \quad T = 0.7\, R_B C + 0.7\, (R_A + R_B)\, C$$

$$= 0.7\, (R_A + 2\, R_B)\, C$$

The frequency of the output waveform,

$$f = \frac{1}{T} = \frac{1}{0.7\, (R_A + 2\, R_B)\, C}$$

$$\cong \frac{1.44}{(R_A + 2R_B)C}$$

Further, the duty cycle of the output waveform,

$$\text{duty cycle} = \frac{t_2}{T} \times 100\,\%$$

It may be noted from equations (*i*) & (*ii*) that t_1 and t_2 cannot be equal unless R_A is made zero. This cannot be done without producing excess current through the device. This means, it is impossible to produce an output waveform with a perfect 50% duty cycle. However, it is possible to get very close to 50% duty cycle (*i.e.* $t_1 \approx t_2$) by selecting,

$$R_B \gg R_A$$

It may be carefully noted that the value of R_A must be greater than 1k Ω and the value of *C* must be greater than 500 pF.

Example 72.26. *Fig.* 72.101 *shows the circuit of a* 555 *timer used as an astable multivibrator.*

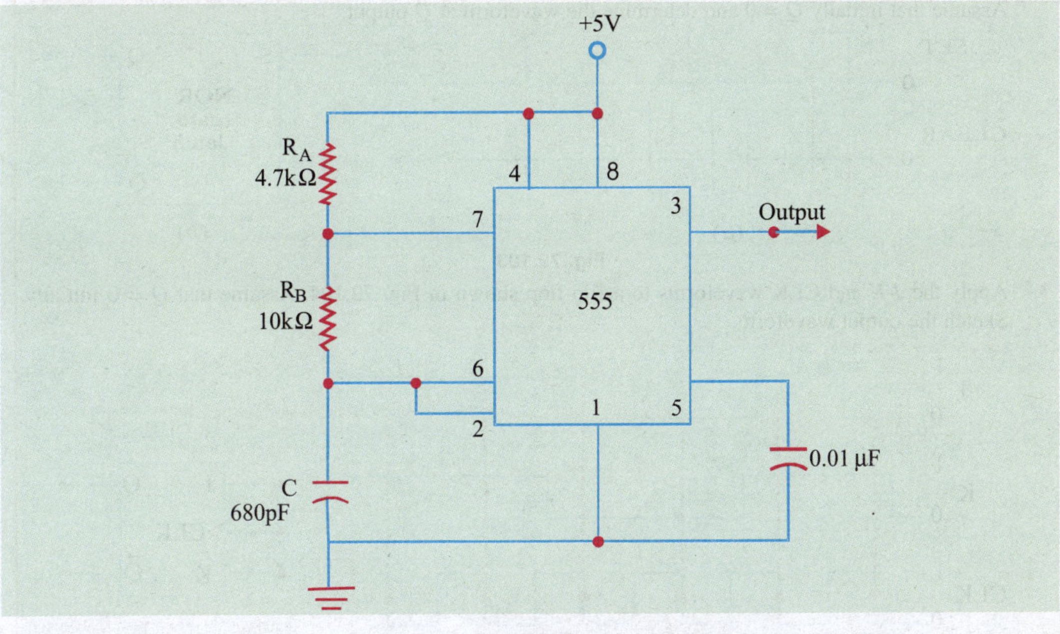

Fig. 72.101

Determine the values t_1, t_2, frequency and duty cycle of the output waveform.

Solution.

$$t_1 = 0.7\ R_B\ C = 0.7 \times (10 \times 10^3) \times (680 \times 10^{-12})$$
$$= 4.76 \times 10^{-6}\ s = 4.76\ \mu s$$
$$t_2 = 0.7\ (R_A + R_B)\ C$$
$$= 0.7 \times (4.7 \times 10^3 + 10 \times 10^3) \times (680 \times 10^{-12})$$
$$= 6.997 \times 10^{-6}\ s \approx 6.997\ \mu s$$
$$T = t_1 + t_2 = 4.76 + 6.997 = 11.757\ \mu s$$
$$f = \frac{1}{T} = \frac{1}{11.757 \times 10^{-6}} = 8.5 \times 10^4 = 85 \text{ kHz}$$

$$\text{Duty cycle} = \frac{t_2}{T} \times 100\% = \frac{6.997}{11.757} = 100\% = 59.5\,\%$$

Tutorial Problems No. 72.1

1. The waveforms of Fig 72.102 (a) are applied to the inputs of a NAND gate latch shown in Fig.72.102. Assume that initially $Q = 0$, and determine the waveform at Q-output.

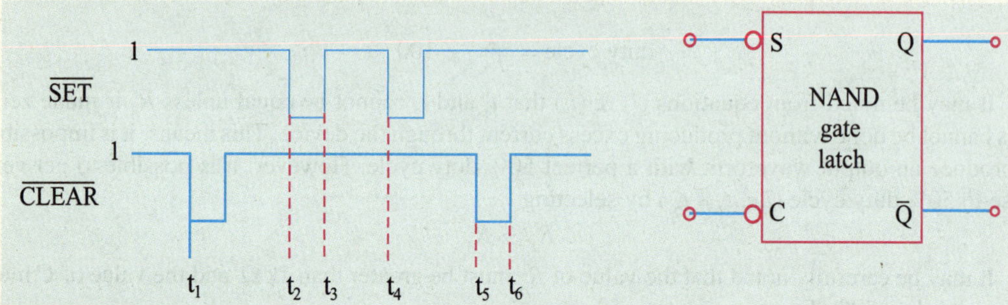

Fig. 72-102

2. The waveforms of Fig. 72.103 (*a*) are applied to the inputs of a NOR gate latch shown in Fig. 72.103 (*b*). Assume that initially $Q = 0$ and determine the waveform at Q output.

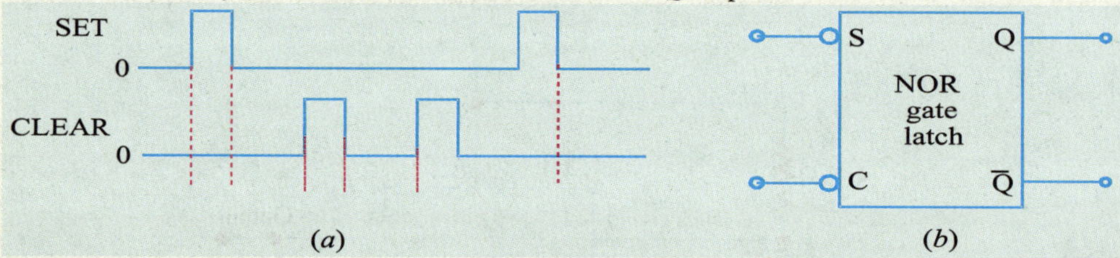

Fig. 72.103

3. Apply the *J-K* and CLK waveforms to a flip-flop shown in Fig. 72.104. Assume that $Q = 0$ initially. Sketch the output waveform.

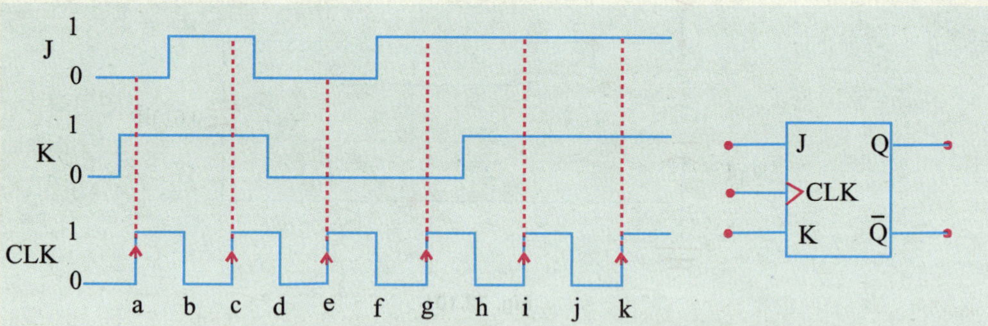

Fig. 72.104

4. Determine the output waveform for a positive-edge triggered *D* flip-flop for the *J, K* and CLK waveforms as shown in Fig. 72.105.

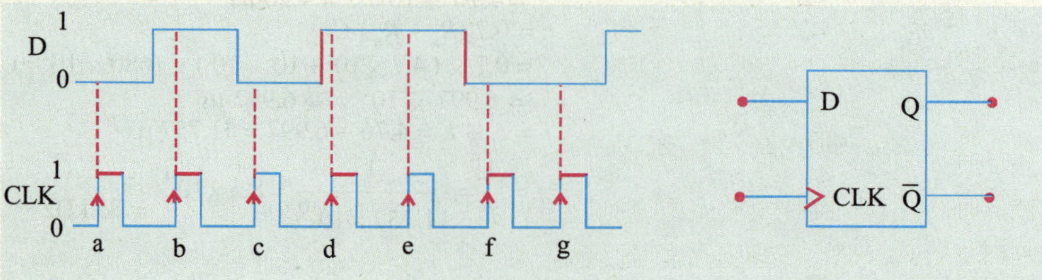

Fig. 72.105

5. Determine the Q waveform for a D latch with ENABLE (EN) and DATA (D) input waveforms shown in Fig. 72.106. Assume $Q = 0$ initially.

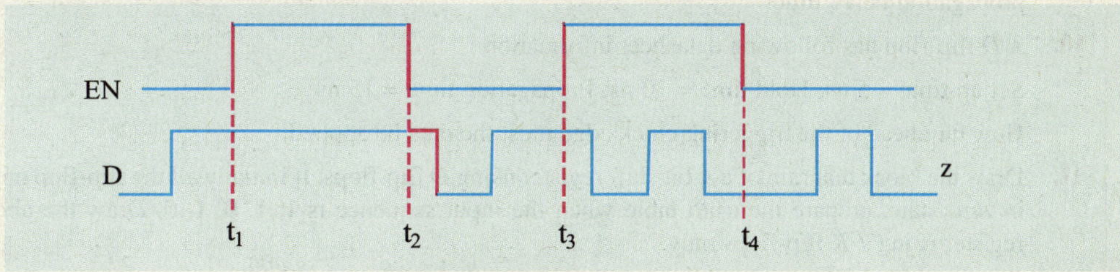

Fig. 72.106

6. Fig. 72.107 (*a*) shows the symbol for a *J-K* flip-flop that responds to a NGT (*i.e.* negative going trigger) on its clock input and has asynchronous inputs. Sketch the output waveform in response to the $\overline{\text{CLK}}$, $\overline{\text{PRE}}$ and $\overline{\text{CLR}}$ waveforms as shown in Fig. 72.107 (*b*).

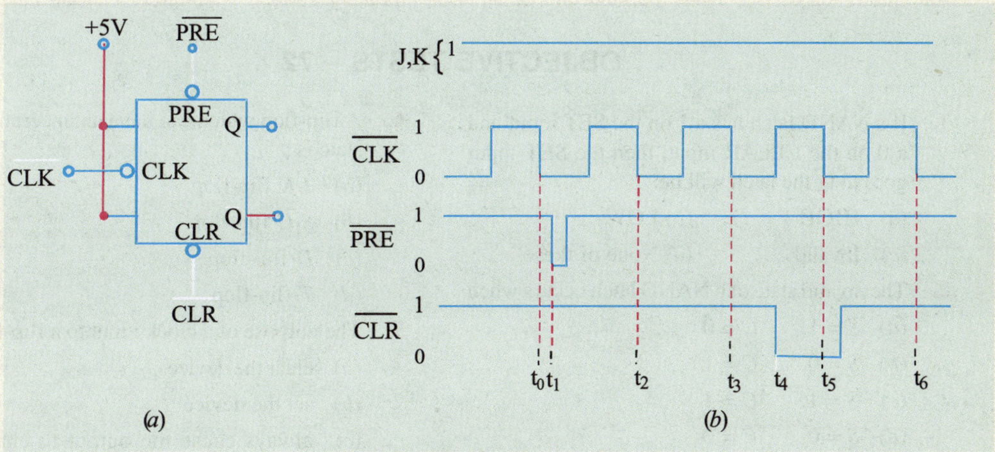

Fig. 72.107

7. Determine the Q output for a negative edge triggered *J-K* flip-flop for the input waveforms shown in Fig. 72.108. Assume $t_H = 0$ and that $Q = 0$ initially.

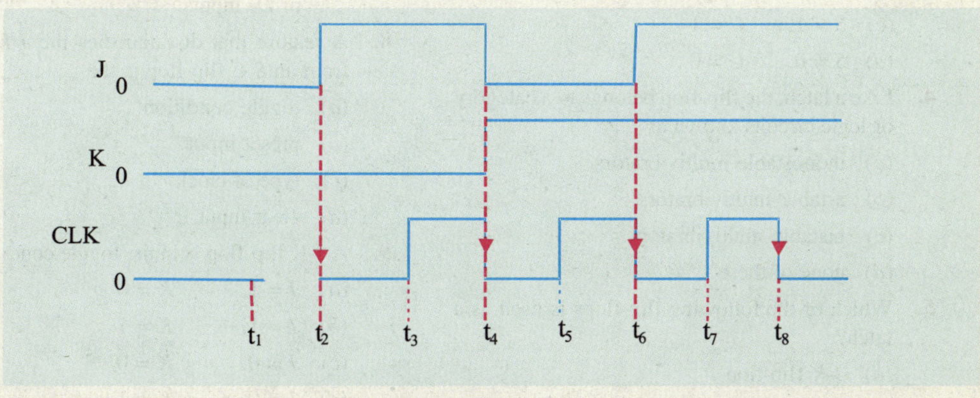

Fig. 72.108

8. A 20-kHz clock signal is applied to a *J-K* flip-flop with $J = K = 1$. Sketch the output waveform and determine its frequency. (**Ans** 10 kHz)

9. In a J-K flip-flop, $J = K = 1$. A 1 MHz square wave is applied at its clock input. It has a propagation delay of 50 ns. Draw the input square wave and the output waveform expected at Q. Show the propagation delay time.

10. A D flip-flop has following datasheet information :

    Set up-time = 5 ns; Hold time = 10 ns; Propagation time = 15 ns

    How far ahead of the triggering clock edge must the data be applied.

11. Draw the block diagram of a 4-bit shift register using D flip-flops. If initially all the flip-flop outputs are in zero state, prepare the truth table when the input sequence is 1, 1, 0, 1, 0. Draw the above shift register using J-K flip-flops only.

12. Consider a 4-bit shift register using J-K flip-flops. Assume that initially, $Q_0 = 1$, $Q_1 = 1$, $Q_2 = 1$ and $Q_3 = 1$. Sketch the output of each flip-flop if an input sequence 101101 is applied to D_0 synchronously with the clock (Q_0 is LSB).

13. Calculate the frequency and the duty cycle of the 555 astable multivibrator output for $C = 0.001\ \mu F$, $R_A = 2.2\ k\Omega$ and $R_B = 100\ k\Omega$. **(Ans** 50.5%)

OBJECTIVE TESTS – 72

1. If a NAND latch has a 1 on the SET input and a 0 on the CLEAR input, then the SET input goes to 0, the latch will be:
   (*a*) HIGH (*b*) LOW
   (*c*) Invalid (*d*) None of these

2. The invalid state of a NAND latch occurs when
   (*a*) $S = 1, \quad C = 0$
   (*b*) $S = 0, \quad C = 1$
   (*c*) $S = 1, \quad C = 1$
   (*d*) $S = 0, \quad C = 0$

3. The invalid state of a NOR latch occurs when
   (*a*) $S = 1, \quad C = 0$
   (*b*) $S = 0, \quad C = 1$
   (*c*) $S = 1, \quad C = 1$
   (*d*) $S = 0, \quad C = 0$

4. Like a latch, the flip-flop belongs to a category of logic circuits known as
   (*a*) monostable multivibrators
   (*b*) astable multivibrators
   (*c*) bistable multivibrators
   (*d*) none of these

5. Which of the following flip-flops is used as a latch?
   (*a*) *J-K* flip-flop
   (*b*) *S-C* flip-flop
   (*c*) *D* flip-flop
   (*d*) *T* flip-flop

6. A flip-flop which can have an uncertain output state is :
   (*a*) *J-K* flip-flop
   (*b*) *S-C* flip-flop
   (*c*) *D* flip-flop
   (*d*) *T* flip-flop

7. The purpose of a clock input to a flip-flop is to
   (*a*) clear the device
   (*b*) set the device
   (*c*) always cause the output to change the states
   (*d*) cause the output to assume a state dependent on the controlling (*S - C, J-K* or *D*) inputs.

8. A feature that distinguishes the *J-K* flip-flop from an *S-C* flip-flop is the
   (*a*) toggle condition
   (*b*) preset input
   (*c*) type of clock
   (*d*) clear input.

9. A J-K flip-flop is in the toggle condition when
   (*a*) $J = 1, \quad K = 0$
   (*b*) $J = 1, \quad K = 1$
   (*c*) $J = 0, \quad K = 0$
   (*d*) $J = 0, \quad K = 1$

10. A *J-K* flip-flop with $J = 1$ and $K = 1$ has a 10 kHz clock input. The Q = output is,
    (*a*) constantly LOW

(b) constantly HIGH

(c) a 5 kHz square wave

(d) a 10 kHz square wave

11. For an edge-triggered *D* flip-flop,

(a) a change in the state of the flip-flop can occur only at a clock pulse edge

(b) the state that flip-flop goes to depends on the *D* input

(c) the output follows the input at each clock pulse

(d) all of these answers

12. The flip-flop shown in Fig. 72.109 logically behaves as

(a) a *D* flip-flop (b) a *J-K* flip-flop

(c) a *T* flip-flop (d) an *R-S* flip-flop

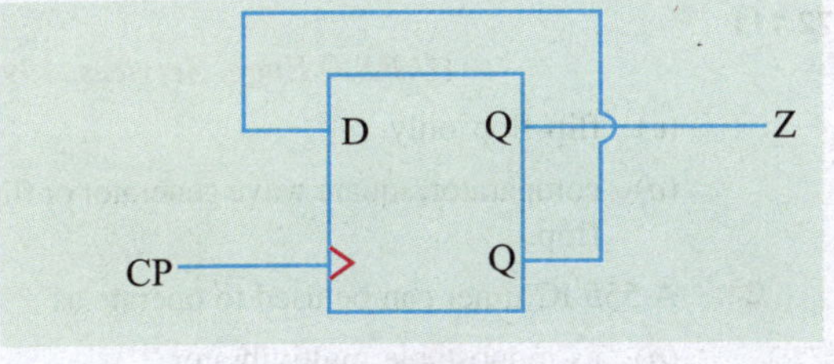

Fig. 72.109

(Grad. I.E.T.E. June, 1997)

13. A *J-K* flip-flop is a device to

(a) divide the frequency by 2

(b) divide the frequency by 4

(c) generate waveform of same frequency as that of the input

(d) cannot be used for frequency division.

14. A 1 μs pulse can be converted into a 1 ms pulse by using

(a) a monostable multivibrator

(b) an astable multivibrator

(c) a bistable multivibrator

(d) a *J-K* flip-flop

(U.P.S.C. Engg. Services, 1999)

15. The following is not a sequential circuit

(a) *J-K* flip-flop (b) counter

(c) full-adder (d) shift register

(Dip. I.E.T.E., Dec. 1996)

16. A one – shot is a type of

(a) monostable multivibrator

(b) astable multivibrator

(c) bistable multivibrator

(d) timer

17. An astable multivibrator

(a) requires a periodic trigger input

(b) has one stable states

(c) is an oscillator

(d) produces a non-periodic pulse output.

18. A 1 msec pulse can be converted to a 10 msec pulse by using:

(a) an astable multivibrator

(b) a monostable multivibrator

(c) a bistable multivibrator

(d) a J-K flip-flop

(U.P.S.C. Engg. Services 1990)

19. A retriggerable one shot is one which

(a) can be triggered only once

(b) has two quassi-stable states

(c) cannot be triggered until full pulse has been outputted.

(d) is capable of being triggered while the output is being generated.

20. The output of the circuit shown in Fig. 72.110 will be

(a) delayed pulses

(b) square waves

(c) trianglular waves

(d) trapezoidal waves.

(U.P.S.C Engg. Services 1997)

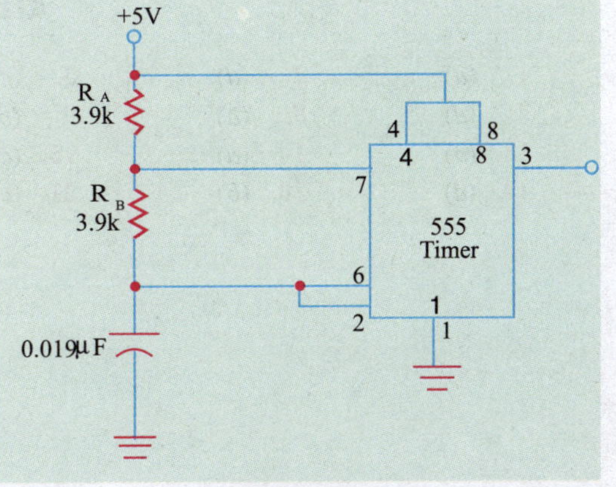

Fig. 72.110

21. The input waveform V_i and the output waveform V_o of a Schmitt NAND are shown in Fig 72.111. The duty cycle of the output waveform is

(*a*) 100 % (*b*) 85.5 %

(*c*) 72.2 % (*d*) 25 %

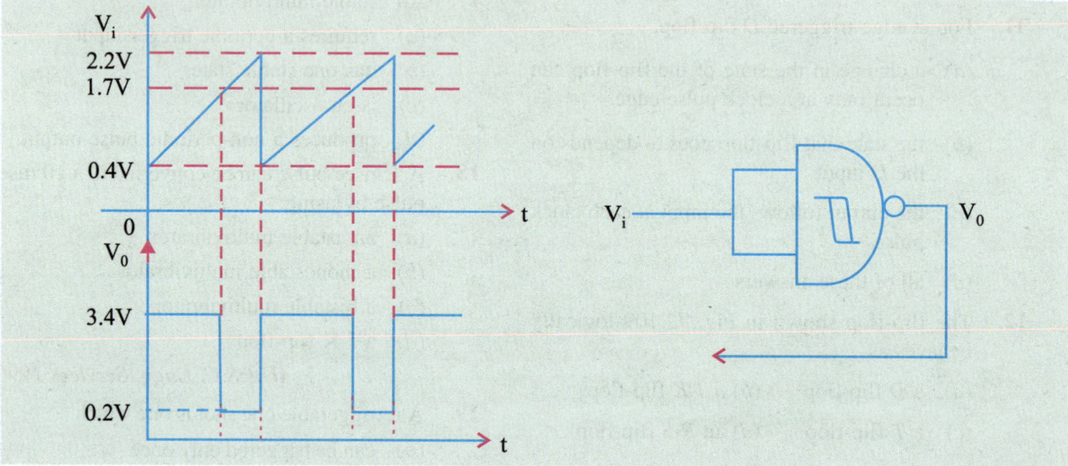

Fig. 72.111

(*U.P.S.C Engg. Services. 1999*)

22. Which of the following circuit is used for production of delays?

(*a*) astable multivibrator

(*b*) bistable multivibrator

(*c*) monostable multivibrator

(*d*) schmitt-trigger

23. A Schmitt-trigger can be used as a

(*a*) comparator only

(*b*) square-wave generator only

(*c*) flip-flop only

(*d*) comparator, square wave generator or flip-flop.

24. A 555 IC timer can be used to operate as

(*a*) as monostable multivibrator

(*b*) an astable multivibrator

(*c*) a voltage controlled oscillator

(*d*) all of the above

ANSWERS

1.	(*a*)	**2.**	(*d*)	**3.**	(*c*)	**4.**	(*c*)	**5.**	(*c*)	**6.**	(*b*)
7.	(*d*)	**8.**	(*a*)	**9.**	(*b*)	**10.**	(*c*)	**11.**	(*d*)	**12.**	(*c*)
13.	(*a*)	**14.**	(*a*)	**15.**	(*c*)	**16.**	(*a*)	**17.**	(*c*)	**18.**	(*b*)
19.	(*d*)	**20.**	(*b*)	**21.**	(*c*)	**22.**	(*c*)	**23.**	(*a*)	**24.**	(*d*)

CHAPTER 73

ELECTRONIC INSTRUMENTS

Learning Objectives

- Introduction
- Analog and Digital Instruments
- Functions of Instruments
- Electronic versus Electrical Instruments
- Essentials of an Electronic Instrument
- The Basic Meter Movement
- Characteristics of Moving Coil Meter Movement
- Variation of Basic Meter Movement
- Converting Basic Meter to DC Ammeter
- Multirange Meter
- Measurement of Current
- Loading Effect of a Voltmeter
- Ohmmeter
- The Multimeter
- Rectifier Type AC Meter
- Electronic Voltmeters
- The Digital Voltmeter (DVM)
- Cathode Ray Tube (CRT)
- Normal Operation of a CRO
- Dual Trace CRO—Dual Beam CRO
- Lissajous Figures
- Applications of a CRO
- The Q Meter
- Logic Analysers
- Signal Generators
- Audio Generators
- Pulse Generators
- RF Generators
- Frequency Synthesizer
- IEEE-488 General Purpose Interface Bus (GPIB) Instruments

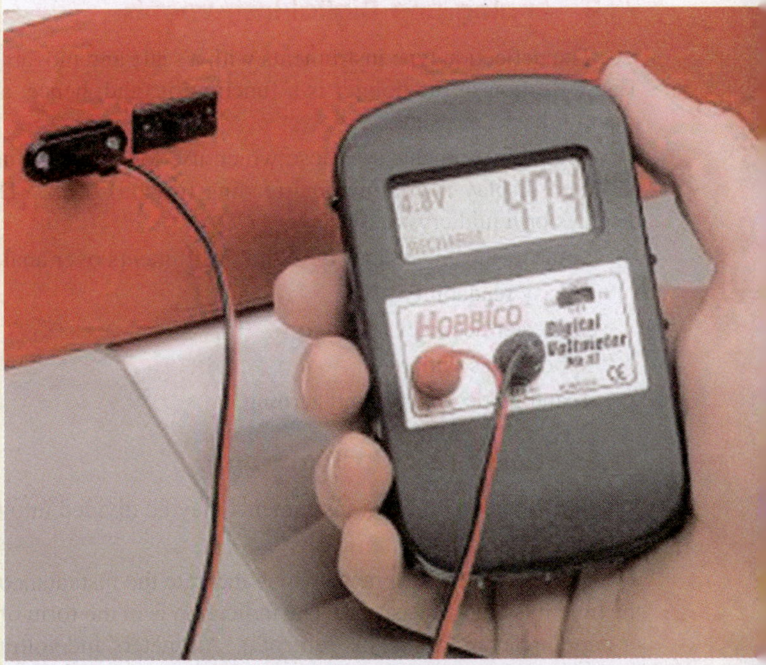

The Digital Voltmeter

73.1. Introduction

Electronic instrumentation is such an interesting field that it combines elements of technologies ranging from the nineteenth to the twenty first centuries. Modern computer-based instrumentation is now evident in every reasonably equipped laboratory and workshop and in catalogs and advertisements of all of the manufacturers. Yet at the root of many space-age instruments is circuitry, such as the wheatstone bridge that is found in nineteenth-century textbooks. Although newer techniques are still in widespread use in new as well as old instruments. In this chapter on electronic instruments you will find both types discussed.

The scientific and technological progress of any nation depends on its ability to measure, calculate and finally, estimate the unknown. Also, the success of an engineer or technician is judged by his ability to measure precisely and to correctly interpret the circuit performance. There are three ways of making such measurements :

(*a*) **by mechanical means**–like measuring gas pressure by Bourdon pressure gauge.

(*b*) **by electrical means**–like measuring potential difference with an electrical voltmeter.

(*c*) **by electronic means**–which is a very sensitive way of detecting the measured quantity because of amplification provided by the active electron device.

The electronic instruments generally have higher sensitivity, faster response and greater flexibility than mechanical or electrical instruments in indicating, recording and, where required, in controlling the measured quantity.

73.2. Analog and Digital Instruments

The deflection type instruments with a scale and movable pointer are called ***analog*** instruments. The deflection of the pointer is a function of (and, hence, analogous to) the value of the electrical quantity being measured.

Digital instruments are those which use logic circuits and techniques to obtain a measurement and then display it in numerical-reading (digital) form. The digital readouts employ either *LED* displays or liquid crystal displays (*LCD*).

Some of the advantages of digital instruments over analog instruments are as under :

1. easy readability
2. greater accuracy
3. better resolution
4. automatic polarity and zeroing

73.3. Functions of Instruments

Functionally, different instruments may be divided into the following three categories :

1. Indicating instruments

These are the instruments which indicate the instantaneous value of quantity being measured, at the time it is being measured. The indication is in the form of pointer deflection (analog instruments) or digital readout (digital instruments). Ammeters and voltmeters are examples of such instruments.

2. Recording instruments

Such instruments provide a graphic record of the variations in the quantity being measured over a selected period of time. Many of these instruments are electromechanical devices which use paper charts and mechanical writing instruments such as an inked pen or stylus.

Electronic recording instruments are of two types :

(*a*) **null type**–which operate on a comparison basis.

(*b*) **galvanometer type**–which operate on deflection type.

3. Controlling instruments

These are widely used in industrial processes. Their function is to control the quantity being measured with the help of information feed back to them by monitoring devices. This class forms the basis of automatic control systems (automation) which are extensively employed in science and industry.

73.4. Electronic Versus Electrical Instruments

Both electrical and electronic instruments measure electrical quantities like voltage and current etc. Purely electrical instruments do not have any built-in amplifying device to increase the amplitude of the quantity being measured. The common dc voltmeter based on moving-coil meter movement is clearly an electrical instrument.

The electronic instruments always include in their make-up some active electron device such as vacuum tube, semiconductor diode or an integrated circuit etc.

As seen, the main distinguishing factor between the two types of instruments is the presence of an electron device in the electronic instruments. Of course, movement of electrons is common to both types, their main difference being that control of electron movement is more effective in electronic instruments than in electrical instruments.

Although electronic instruments are usually more expensive than their electrical counterparts, they offer following advantages for measurements purposes :

1. since electronic instruments can amplify the input signal, they possess very high sensitivity *i.e.* they are capable of measuring extremely small (low-amplitude) signals,
2. because of high sensitivity, their input impedance is increased which means less loading effect when making measurements,
3. they have greater speed *i.e.* faster response and flexibility,
4. they can monitor remote signals.

73.5. Essentials of an Electronic Instrument

As shown Fig. 73.1, an electronic instrument is made up of the following three elements :

1. Transducer

It is the first sensing element and is required only when measuring a non-electrical quantity, say, temperature or pressure. Its function is to convert the non-electrical physical quantity into an electrical signal.

Of course, a transducer is not required if the quantity being measured is already in the electrical form.

2. Signal Modifier

It is the second element and its function is to make the incoming signal suitable for application to the indicating device.

For example, the signal may need amplification before it can be properly displayed. Other types of signal modifiers are : voltage dividers for reducing the amount of signal applied to the indicating device or wave shaping circuits such as filters, rectifiers or chopper etc.

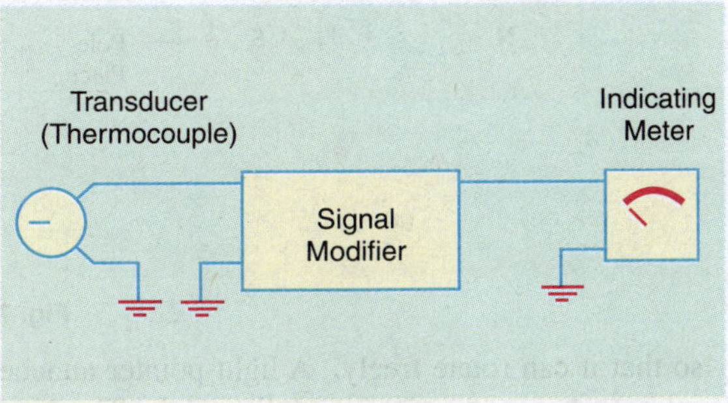

Fig. 73.1

3. Indicating Device

For general purpose instruments like voltmeters, ammeters or ohm meters, the indicating device is usually a deflection type meter as shown in Fig. 73.1. In digital readout instruments, the indicating device is of digitial design.

73.6. Measurement Standards

All instruments, whether electrical or electronic, are calibrated at the time of manufacture against a measurement standard.

1. International Standards

These are defined by international agreement and are maintained at the international Bureau of Weights and Measurements in Paris.

2. Primary Standards

These are maintained at national standards laboratories in each country. They are not available for use outside these laboratories. Their principal function is to calibrate and verify the secondary standards used in industry.

3. Secondary Standards

These are the basic reference standards used by industrial laboratories and are maintained by the particular industry to which they belong. They are periodically sent to national laboratory for calibration and verification against primary standards.

4. Working Standards

These are the main tools of a measurement laboratory and are used to check and calibrate the instrument used in the laboratory.

73.7. The Basic Meter Movement

It is also called D' Arsonval meter movement or a permanent-magnet moving-coil (PMMC) meter movement. Since it is widely used in electronic instruments, it is worthwhile to discuss its construction and principle of operation.

1. Construction

As shown in Fig. 73.2, it consists of a permanent horse-shoe magnet with soft iron pole pieces attached to it. Between the two pole-pieces is situated a cylinder-shaped soft iron core around which moves a coil of fine wire wound on a light metal frame. The metal frame is mounted in jewel bearings

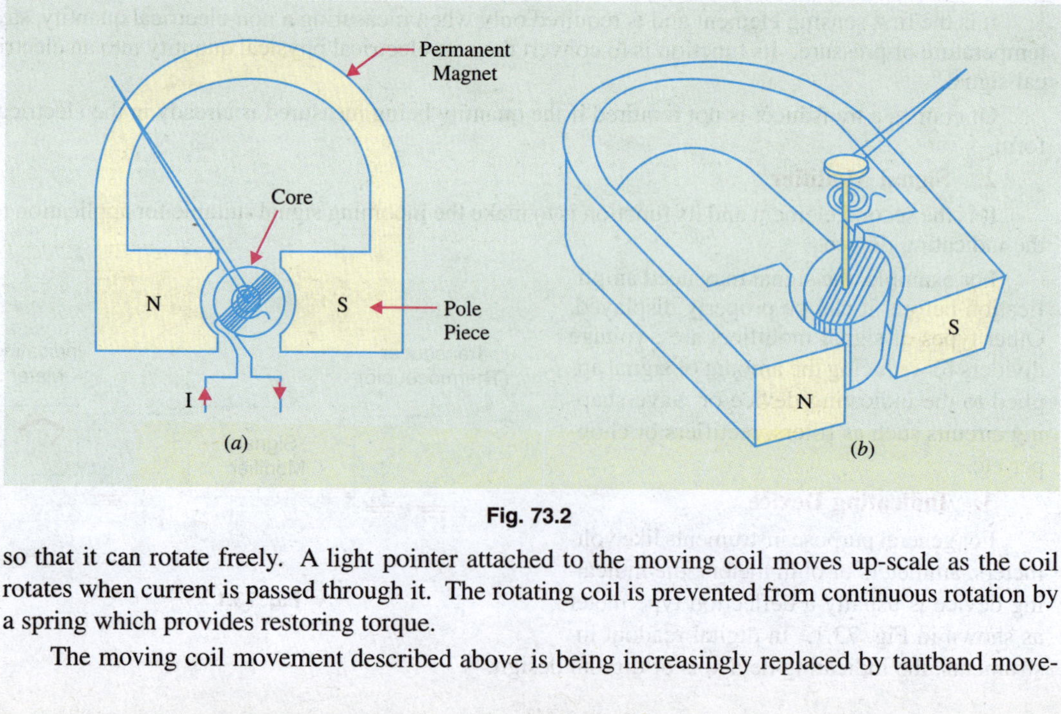

Fig. 73.2

so that it can rotate freely. A light pointer attached to the moving coil moves up-scale as the coil rotates when current is passed through it. The rotating coil is prevented from continuous rotation by a spring which provides restoring torque.

The moving coil movement described above is being increasingly replaced by tautband movement in which the moving coil and the pointer are suspended between bands of spring metal so that the restoring force is torsional. The bands perform two functions (*i*) they support the coil and (*ii*) they provide restoring torque thereby eliminating the pivots and jewels used with coil spring movement.

As compared to pivoted movement, the taut-band has the advantages of

1. greater sensitivity *i.e.* small full-scale deflection current
2. ruggedness,
3. minimal friction,
4. easy to manufacture.

2. Principle of Operation

This meter movement works on the ***motor*** principle and is a current-responding device. The deflection of the pointer is directly proportional to the amount of current passing through the coil.

When direct current flows through the coil, the magnetic field so produced reacts with the field of the permanent magnet. The resultant force turns the coil alongwith its pointer. The amount of deflection is directly proportional to the amount of current in the coil. Hence, their scale is linear. With correct polarity, the pointer reads up-scale to the right whereas incorrect polarity forces the pointer off-scale to the left.

73.8. Characteristics of Moving Coil Meter Movement

We will discuss the following three characteristics :

(*i*) full-scale deflection current (I_m),

(*ii*) internal resistance of the coil (R_m),

(*iii*) sensitivity (*S*).

1. Full-scale Deflection Current (I_m)

It is the current needed to deflect the pointer all the way to the right to the last mark on the calibrated scale. Typical values of I_m for D' Arsonval movement vary from 2 μA to 30 mA.

It should be noted that for smaller currents, the number of turns in the moving coil has to be ***more*** so that the magnetic field produced by the coil is strong enough to react with the field of the permanent magnet for producing reasonable deflection of the pointer. Fine wire has to be used for reducing the weight of the moving coil but it increases its resistance. Heavy currents need thick wire but lesser number of turns so that resistance of the moving coil is comparatively less. The schematic symbol is shown in Fig. 73.3.

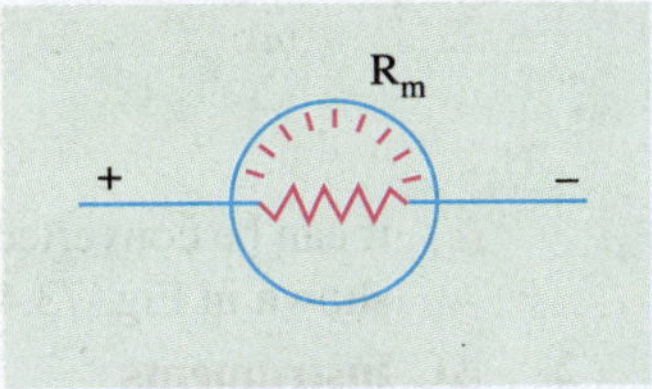

Fig. 73.3

2. Internal Resistance (R_m)

It is the dc ohmic resistance of the wire of the moving coil. A movement with smaller I_m has higher R_m and *vice versa.* Typical values of R_m range from 1.2 Ω for a 30 mA movement to 2 kΩ for a 50 μA movement.

3. Sensitivity (*S*)

It is also known as ***current sensitivity*** or ***sensitivity factor.*** It is given by the reciprocal of full-scale deflection current I_m.

$$\therefore \qquad S = \frac{1}{I_m} \text{ ohm/volt.}$$

For example, the sensitivity of a 50-μA meter movement is

$$S = \frac{1}{50\,\mu A} = \frac{1}{50 \times 10^{-6}}\ \Omega/\text{V} = 20{,}000\ \Omega/\text{V} = 20\ \text{k}\Omega/\text{V}$$

The above figure shows that a full-scale deflection of 50 μA is produced whenever 20,000 Ω of resistance is present in the meter circuit ***for each volt of applied voltage.*** It also represents the ohms-per-volt rating of the meter. The sensitivity of a meter movement depends on the strength of the permanent magnet and number of turns in the coil. Larger the number of turns, smaller the amount of current required to produce full-scale deflection and, hence, higher the sensitivity. A high current sensitivity means a high quality meter movement. It also determines the lowest range that can be covered when the meter movement is modified as an ammeter (Art 73.10) or voltmeter (Art 73.13)

73.9. Variations of Basic Meter Movement

The basic moving-coil system discussed in Art 73.7 can be converted into an instrument to measure dc as well as ac quantities like current, voltage and resistance etc. Without any modification, it can carry a maximum current of I_m can withstand a maximum dc voltage $v = I_m R_m$.

1. DC instruments

(*a*) it can be made into a dc ammeter, milliammeter or microammeter by adding a suitable shunt resistor R_{sh} in parallel with it as shown in Fig. 73.4 (*a*),

(*b*) it can be changed into a dc voltmeter by connecting a multiplier resistor R_{mult} in series with it as shown in Fig. 73.4 (*b*),

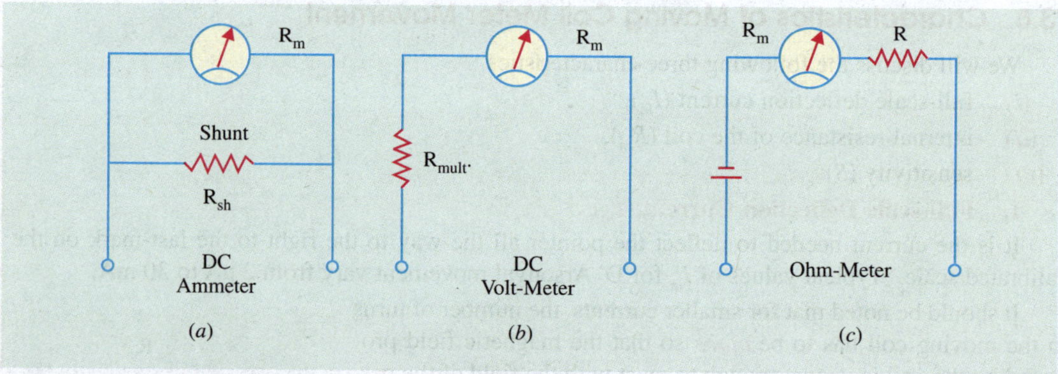

Fig. 73.4

(*c*) it can be converted into an ohmmeter with the help of a battery and series resistor *R* as shown in Fig. 73.4 (*c*).

2. AC Instruments

(*a*) it can be changed into an ac audio-frequency ammeter or voltmeter by simply adding an extra rectifier as shown in Fig. 73.5 (*a*).

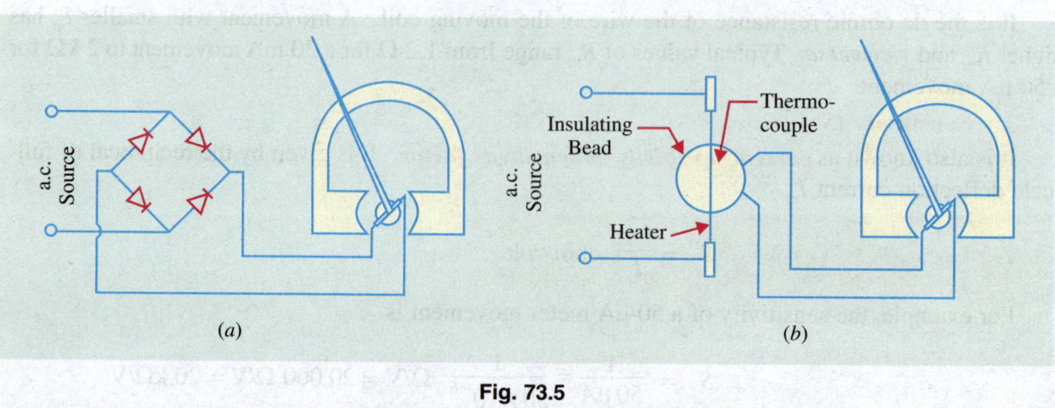

Fig. 73.5

(*b*) it can be converted into a radio frequency ammeter or voltmeter by adding a thermocouple as shown in Fig. 73.5 (*b*).

The above modifications of the basic meter movement have been tabulated below :

- Basic Meter Movement
  - DC Instruments
    - dc ammeter by using a shunt resistor
    - dc voltmeter by using series multiplier resistor
    - ohmmeter by using battery and series resistor
  - AC Instruments
    - audio-frequency ac ammeter or voltmeter by using a rectifier
    - radio-frequency ammeter or voltmeter by using a thermocouple

73.10. Converting Basic Meter to DC Ammeter

As stated earlier and again shown in Fig. 73.6 (*a*), the basic meter movement can carry a maximum current of I_m *i.e.* its full-scale deflection current. However, its current range can be increased (*i.e.* multiplied) to any value by connecting a low resistance (called shunt resistance R_{sh}) in parallel with it as shown in Fig. 73.6 (*b*). The shunted meter works as an ammeter ***with an extended range.***

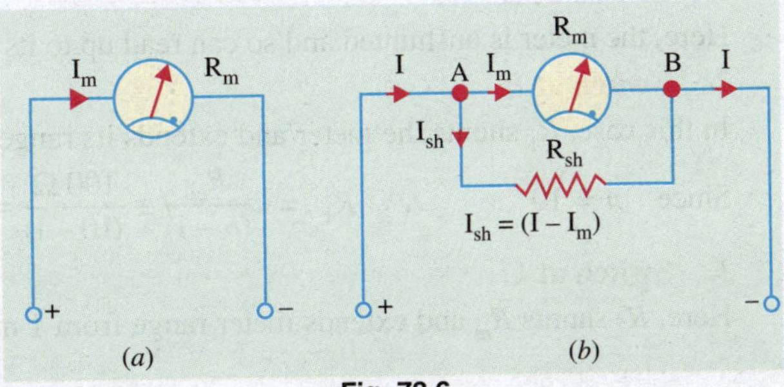

Fig. 73.6

Suppose, we want to measure a line current of I with the help of this meter movement. Obviously, the value of R_{sh} should be such as to shunt or bypass a current of $(I - I_m)$. As seen, range extension is from I_m to I. The ratio $I/I_m = n$ is known as the ***multiplying power or multiplying factor*** of the shunt. It means that a shunt allows the meter to measure current I which is n times larger than I_m.

Value of R_{sh}

In Fig. 73.6 (*b*), voltage across the meter and the shunt is the same because they are joined in parallel.

$$\therefore \quad I_m R_m = I_{sh} \cdot R_{sh} = (I - I_m) R_{sh}$$

$$\therefore \quad R_{sh} = \frac{I_m}{(I - I_m)} \cdot R_m = \frac{1}{(I/I_m - 1)} \cdot R_m \quad \therefore \quad R_{sh} = \frac{R_m}{(n-1)}$$

Here, n is the multiplying factor of the shunt. It is seen that larger the value of n *i.e.* greater the range extension required, smaller the shunt resistance needed. Incidentally, it may be noted that the resistance of the ***shunted*** meter

$$= R_m \parallel R_{sh} = \frac{R_m R_{sh}}{R_m + R_{sh}}$$

It is much less than either R_m or R_{sh}

Example 73.1. *It is required to convert a 5-mA meter with 20 Ω internal resistance into a 5-A ammeter. Calculate*

(*a*) *the value of shunt resistance required*

(*b*) *multiplying factor of the shunt.*

Solution. Here, $I = 5A, I_m = 5\text{ mA} = 0.005\,A, R_m = 20\,\Omega$

(a)
$$R_{sh} = \frac{I_m}{(I - I_m)} . R_m = \frac{0.005}{(5 - 0.005)} \times 20 = \mathbf{0.02\,\Omega\ (approx)}$$

(b)
$$n = \frac{I}{I_m} = \frac{5}{0.005} = \mathbf{1000}$$

Note.
$$R_{sh} = \frac{R_m}{(n-1)} = \frac{20}{(1000-1)} = \frac{20}{999} = 0.02\,\Omega$$
—as found above

Fig. 73.7 shows such an ammeter connected in a load circuit.

73.11. Multirange Meter

The shunt resistance discussed above gives only a single range ammeter. By using universal shunt (also called Ayrton shunt), we can obtain a multirange ammeter as shown in Fig. 73.8.

It is seen that by changing the switch position from *A* to *B* to *C* and finally to *D*, the current range can be extended as desired.

1. Switch at A

Here, the meter is unshunted and so can read up to its full-scale deflection current of 1 mA only.

2. Switch at B

In this case, R_1 shunts the meter and extends its range to 10 mA *i.e.* increases it ten times.

Since $n = 10$ $\quad\therefore\quad R_1 = \dfrac{R_m}{(n-1)} \cong \dfrac{100\,\Omega}{(10-1)} = 11.11\,\Omega$

3. Switch at C

Here, R_2 shunts R_m and extends meter range from 1 mA to 0.1 A *i.e.* to 100 mA. Obviously,

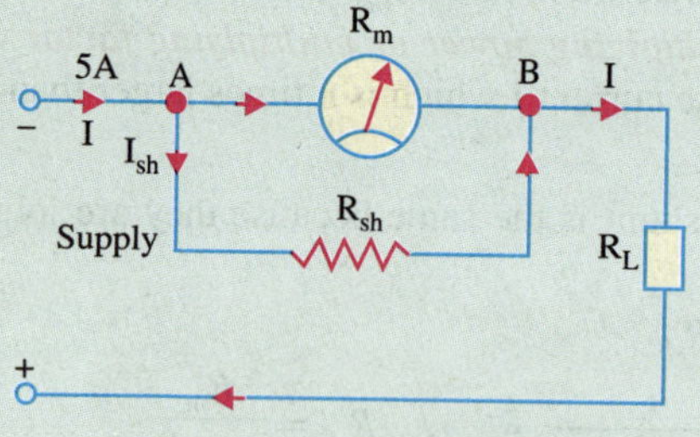

Fig. 73.7

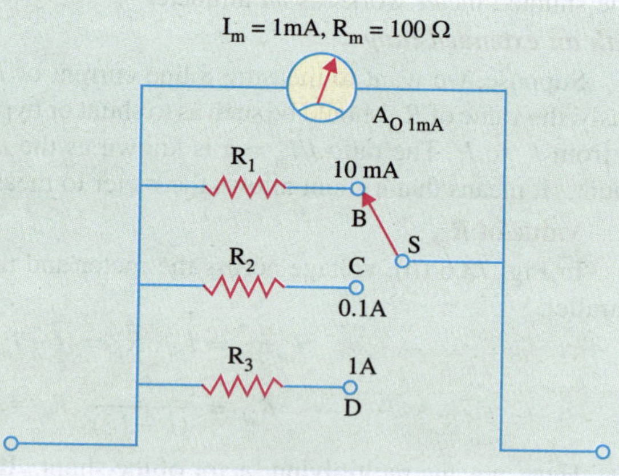

Fig. 73.8

$$n = \frac{100}{1} = 100 \quad \therefore \quad R_2 = \frac{100\,\Omega}{(100-1)} = 1.01\,\Omega$$

4. Switch at D

In this case, R_3 shunts R_m and extends the current range of the meter from 1 mA to 1.0 A *i.e.* 1.0 mA to 1000 mA. Hence $n = 1000/1 = 1000$.

$$\therefore \quad R_3 = \frac{100\,\Omega}{(1000-1)} = 0.1001\,\Omega$$

Incidentally, it may be noted that greater the range extension, smaller the shunt resistance.

Alternative Method

An alternative circuit for range extension is shown in Fig. 73.9. It is called 'add on' method of shunting the meter because resistances can be added one after another for changing the range. Unlike in Fig. 73.8, there is no possibility of the meter being in the circuit without any shunt.

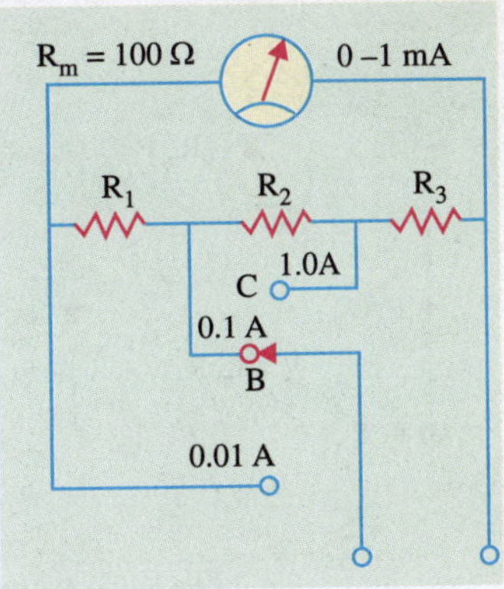

Fig. 73.9

As seen, the universal, shunt consists of three resistances R_1, R_2 and R_3. How they are connected as a shunt is determined by the switch position. When *S* is at position *A*, the combination $(R_1 + R_2 + R_3)$ becomes connected across R_m. When *S* is at position *B*, $(R_2 + R_3)$ become connected in parallel across $(R_1 + R_m)$ and so on.

1. Switch at A

In this case, multiplying factor $n = 10$ mA/1 mA = 10

$$\therefore \qquad (R_1 + R_2 + R_3) = \frac{R_m}{(n-1)} = \frac{100}{9}\ \Omega \qquad ...(i)$$

2. Switch at B

Here, $(R_2 + R_3)$ become in parallel with $(R_1 + R_m)$ or $(R_1 + 100)$. Also, $n = 100/1 = 100$

$$\therefore \qquad R_2 + R_3 = \frac{R_1 + 100}{99} \qquad ...(ii)$$

3. Switch at C

In this position, R_3 is in parallel with $(R_1 + R_2 + 100)$ and $n = 1000/1 = 1000$

$$\therefore \qquad R_3 = \frac{R_1 + R_2 + 100}{999} \qquad ...(iii)$$

Solving for R_1, R_2 and R_3 from Eq. (*i*), (*ii*) and (*iii*) we have

$$R_1 = 10\ \Omega,\ R_2 = 1\ \Omega \text{ and } R_3 = 1/9\ \Omega$$

73.12. Measurement of Current

While measuring current flowing in a circuit, following two points must be kept in mind :

1. The current meter must be connected in series with the circuit where current is to be measured (Fig. 73.7). The full circuit current cannot flow through the meter unless it is made a series component.
2. The dc meter must be connected with the correct polarity for the pointer to read up-scale to the right. Reversed polarity deflects the pointer down-scale to the left forcing it against the stop which can sometime bend the pointer.

73.13. Converting Basic Meter to DC Voltmeter

The basic meter movement can measure a maximum voltage of $I_m R_m$ which is very small [Fig. 73.10 (*a*)]. However, its voltage range can be extended to any value by connecting a large resistance in series with it as shown in Fig. 73.10 (*b*). The series resistance is also called ***multiplier resistance*** because it multiplies the voltage reading capability of the meter many times. It is usually connected inside the voltmeter case.

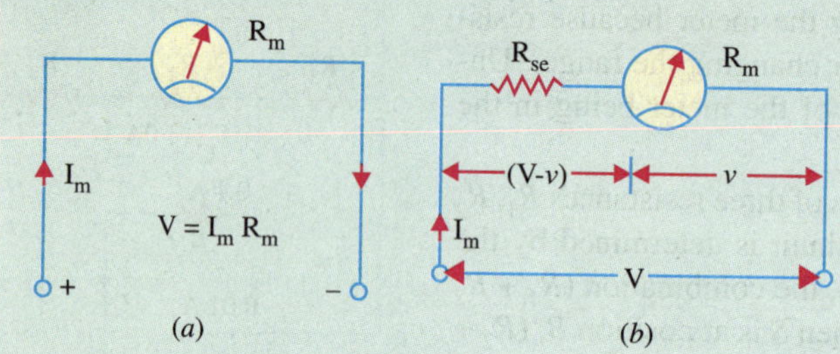

Fig. 73.10

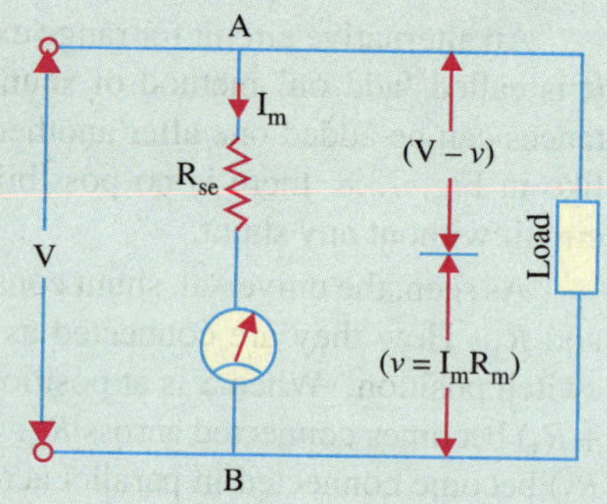

Fig. 73.11

But it should be noted that the voltmeter is connected in parallel with the load across which the voltage is to be measured (Fig. 73.11).

Value of R_{se}

Suppose, it is desired to extend the voltage range of the meter from v to V. The ratio V/v is known as the ***voltage multiplication.*** As seen from Fig. 73.11, drop across R_{se} is $(V - v)$ and current through it is the same as meter current *i.e.* I_m

$$\therefore \quad I_m R_{se} = (V - v) \qquad \text{...}(i)$$

$$\therefore \quad R_{se} = \frac{V - v}{I_m} = \frac{V - I_m R_m}{I_m} = \frac{V}{I_m} - R_m$$

The voltage multiplication (m) can be found from Eq. (i) above,
Dividing both sides by v, we get

$$\frac{I_m R_{se}}{v} = \frac{V}{v} - 1$$

$$\therefore \quad \frac{V}{v} = 1 + \frac{I_m R_{se}}{v} = 1 + \frac{I_m R_{se}}{I_m R_m}$$

$$\therefore \quad \frac{V}{v} = \left(1 + \frac{R_{se}}{R_m}\right) \quad \therefore \quad m = \left(1 + \frac{R_{se}}{R_m}\right)$$

It is seen that for a given meter, higher the series resistance, greater the voltage range extension.

Example 73.2. *A 50-μA meter movement with an internal resistance of 1kΩ is to be used as dc voltmeter of range 50 V. Calculate the*
(a) *multiplier resistance required and* **(b)** *voltage multiplication.*

Solution. (a)

$$R_{se} = \frac{V}{I_m} - R_m$$

$$= \frac{50}{50 \times 10^{-6}} - 1000$$

$$= 10^6 - 1000 = 999{,}000\ \Omega = \mathbf{999\ k\Omega}$$

73.14. Multirange DC Voltmeter

A multirange voltmeter with ranges of 0–5 V, 0–25 V and 0–50 V is shown in Fig. 73.12. Different values of resistors R_1, R_2 and R_3 can be found in the same way as in Art. 73.11. It would be found that for the meter movement shown in figure

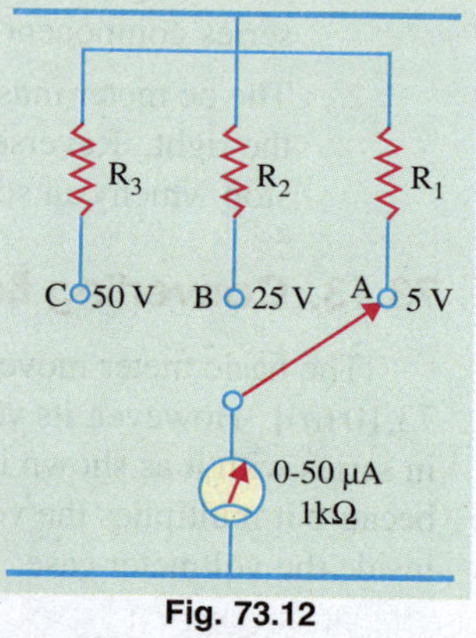

Fig. 73.12

$$R_1 = 99\,\text{K}, R_2 = 499\,\text{K}, R_3 = 999\,\text{K}$$

It is seen that higher the voltage range greater the multiplier resistance required (in almost the same proportion as the ranges).

73.15. Loading Effect of a Voltmeter

When the voltmeter resistance is not high as compared to the resistance of the circuit across which it is connected, the measured voltage becomes less. The decrease in voltage may be negligible or it may be appreciable depending on the sensitivity (ohms-per-volt rating) and input resistance of the voltmeter. It is called voltmeter ***loading effect*** because the voltmeter loads down the circuit across which it is connected. Since input resistance of electronic voltmeter is very high (10 M Ω or more), loading is not a problem in their case.

Consider the circuit shown in Fig. 73.13 in which two 15-K resistors are connected in series across a 100-V dc source. The drop across each is 50V. Now, suppose, that a 30-K voltmeter is connected across R_2 to measure voltage drop across it. Due to loading effect of the voltmeter, the reading is reduced from 50V to 40V as explained below. As seen from Fig. 73.13 (*b*), combined resistance of R_2 and voltmeter is 15 K ‖ 30 K = 10 K.

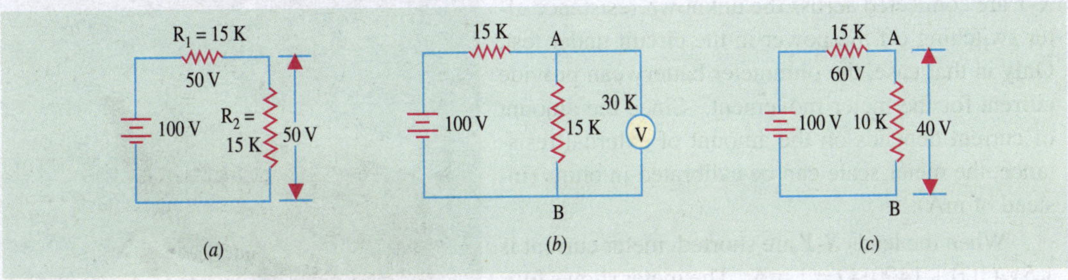

Fig. 73.13

$$\text{drop across 10 K} = \frac{10}{10+15} \times 100 = 40V$$

Loading effect can be minimized by using a voltmeter whose resistance is as high as possible as compared to that of the circuit across which it is connected.

Correction Formula

The loading effect can be neutralized by using the following formula :

$$V_{corr} = V_{means} + \frac{R_1\,R_2}{R_v\,(R_1 + R_2)} . V_{means}$$

where V_{corr} = corrected voltage reading

V_{meas} = measured voltage reading

R_v = voltmeter resistance

R_1, R_2 = voltage dividing resistances in the circuit

In the above case,

$$V_{corr} = 40 + \frac{15 \times 15}{30\,(15+15)} \times 40 = 40 + 10 = \mathbf{50\ V}$$

73.16. Ohmmeter

The basic meter movement can be used to measure resistance. It is combined with a battery and a current-limiting resistance as shown in Fig. 73.14 (*a*). In that case, it is known as an ohmmeter.

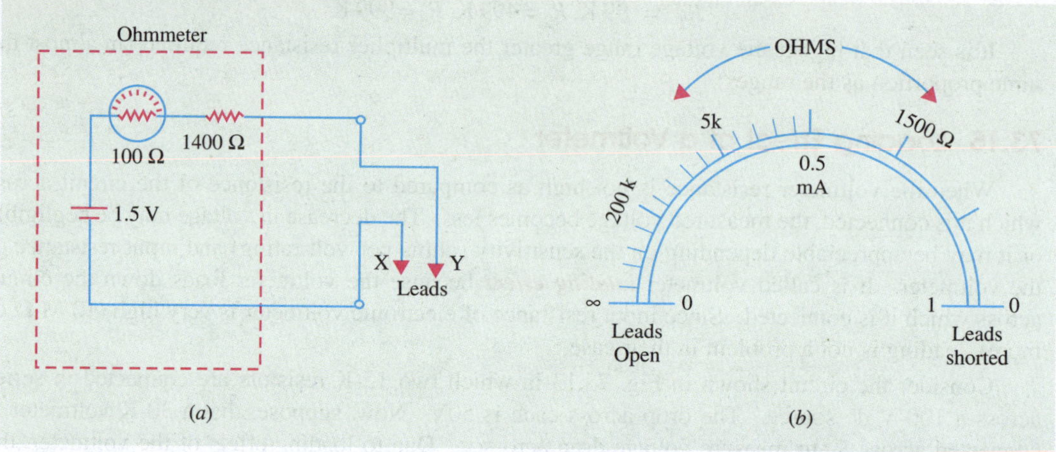

Fig. 73.14

For measuring resistance, the ohm-meter leads *X-Y* are connected across the unknown resistance after switching off the power in the circuit under test. Only in that case, the ohmmeter battery can provide current for the meter movement. Since the amount of current depends on the amount of external resistance, the meter scale can be calibrated in ohms (instead of mA).

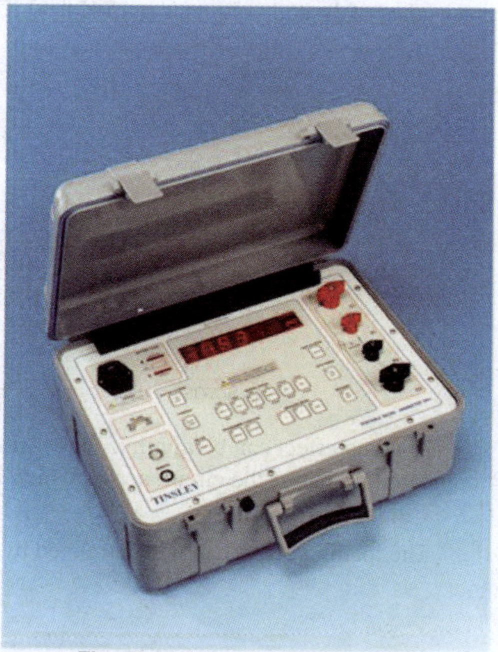

Fig. 73.15. Digital micro-ohmmeter

When the leads *X-Y* are shorted, meter current is $1.5V/(100 + 1400)\ \Omega = 1$ mA. The meter shows full-scale deflection to the right. The ohmmeter reading corresponds to 0 Ω because external resistance is zero. When leads *X-Y* are open *i.e.* do not touch each other, meter current is zero. Hence, it corresponds to infinite resistance on the ohmmeter scale.

Following points about the ohmmeter are worth noting :

1. the resistance scale is non-linear *i.e.* it is expanded at the right near zero ohm and crowded at the left near infinite ohm. This nonlinearity is due to the reciprocal function $I = V/R$.;
2. the ohmmeter reads up-scale regardless of the polarity of the leads because direction of current is determined by the internal battery;
3. at half-scale deflection, external resistance equals the internal resistance of the ohmmeter.
4. the test leads should be shorted and 'ZERO OHMS' control adjusted to bring the pointer to zero on each range.

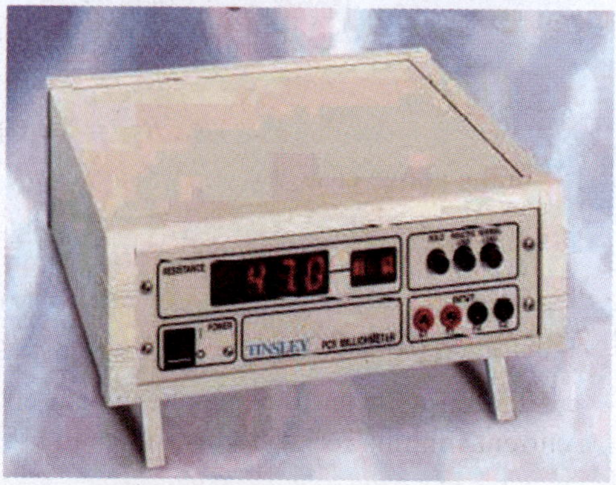

Fig. 73.16. Digital milli-ohmmeter

Fig. 73.15 shows a digital micro-ohmmeter having a range of 1μΩ – 2 kΩ with 3½ digit, 7-segment *LED* display. It has a basic accuracy of ± 0.2% ± 1 digit and is based on a design using *MOS LSI ICs* and glass epoxy *PCB*.

Fig. 73.16 shows a battery-operated portable digital milli-ohmmeter having a measurement range of 200 mΩ – 2 kΩ with an accuracy of ± 0.5% ± 1 digit. It has a 3½ digit 7-segment *LED* display.

73.17. The Multimeter

It is extensively used in cable industry, motor industry, transformer and switchgear industry. It is also called volt-ohm-milliammeter (VOM). It is a general purpose instrument having the necessary circuitry and switching arrangement for measuring ac/dc voltage or ac/dc current or resistance. It is simple, compact and portable because the only power it uses is the battery for the ohm-meter.

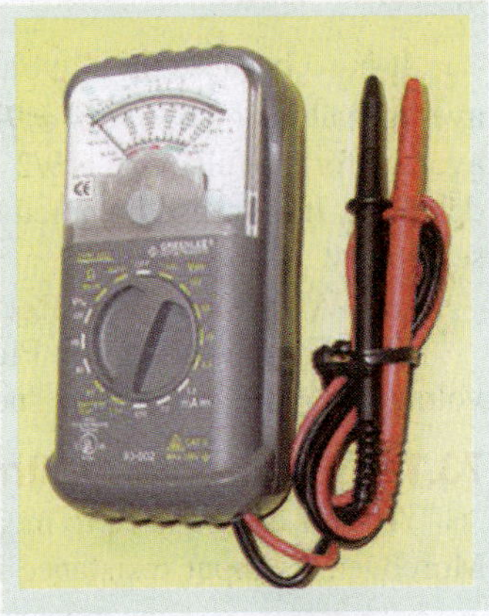

Fig. 73.17. A digital multimeter

Multimeters may be of analog type (Fig. 73.17) or digital type (Fig. 73.18). The analog type is of the pointer and scale type *i.e.* it uses the basic *D*' Arsoval meter movement. However digital multimeters (*DMMs*) are becoming increasingly popular because of their easy readability, numerical display and imporved accuracy.

Fig. 73.18
Courtesy : Fluke Corporation

Fig. 73.17 shows the photograph of an analog multimeter designed primarily for electrical, electronic, radio and TV engineers and technicians. It sells under the brand name of Motwane Multimeter 8 X Mark-III.

It is a 5-function, 30-range meter which measures high ac/dc voltages from 0 to 2.5 kV and ac/dc currents from 0 to 10 A. Its three resistance ranges cover from 0 to 20 MΩ. It is reputed for its excellent reliability, operational simplicity and easy portability.

Fig. 73.18 depicts a digital multimeter which can measure dc voltage upto 1000 V, ac voltages upto 750 V, ac/dc currents from 15 μA to 10 A and resistances from 0 Ω to 100 MΩ. It has a 5 digit multifunction vacuum fluorescent display allowing the user to measure two different parameters of the same signal from one test connection. The user can also view both measurements at the same time.

In Fig. 73.19 is shown a hand-held autoranging, digital multimeter (DMM) having high contrast, 4 digit *LCD* read-out. It has been designed for speed, accuracy and reliability.

Fig. 73.19. Courtesy : Fluke Corporation

73.18. Rectifier Type AC Meter

The D' Arsonval meter movement can be used for measuring alternating quantities provided a rectifier is added to the measuring circuit. A similar rectifier arrangement is found as part of *AC VOLTS* function in multimeters (Art. 73.17). Such as meters are more widely used than either (costly but more accurate) dynamometer type or more delicate thermal and hotwire type.

(a) With Half-Wave Rectifier

The circuit of an ac Voltmeter using half-wave diode recifier is shown in Fig. 73.20. Here, a half-wave rectifier has been combined in series with a dc meter movement.

When used as a dc voltmeter (*i.e.* without rectifier) it would have (say, for example) a range of 10*V*. However, if an ac voltages of rms value 10*V* is applied across input terminals *AB*, it would read 4.5 *V*.

It is so because an *ac* voltage of rms value 10 *V* has a peak value of $10 \times \sqrt{2} = 14$ *V* and an average value of $0.636 \times 14 = 9V$. Since in the half-wave rectified output, one half-cycle is absent, the average for the full cycle is $9/2 = 4.5$ *V*. The meter movement will, therefore read 4.5 *V i.e.* 45% of the dc value. It may also be noted that ac sensitivity of a half-wave ac meter is only 45 percent of the dc sensitivity.

(b) With Full-Wave Rectifier

The circuit is shown in Fig. 73.21. In this case, the meter reading would be 90% of rms input voltage *i.e.* 90% of the dc value.

73.19. Electronic Voltmeters

A *VOM* can be used to measure voltages but it lacks both sensitivity and high input resistances. Moreover, its input resistance is different for each range. The electronic voltmeter (*EVM*), on the other hand, has input resistance ranging from 10 *M* Ω to 100 *M*Ω, thus producing less loading of the circuit under test than the *VOM*. Another advantage of *EVM* is that its input resitance remains constant over all ranges.

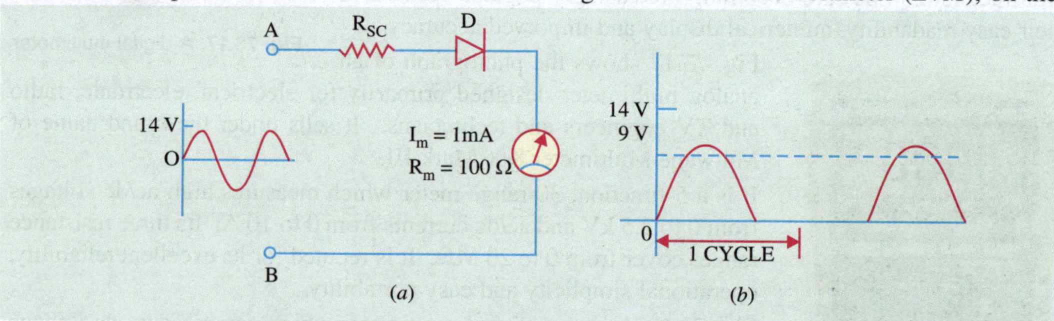

Fig. 73.20

Two types of voltmeters are in use today *(i)* analog and *(ii)* digital. However, a distinction must be made between a digital instrument and an instrument with digital readout. A digital instrument is one which uses internal circuitry of digital design. A digital readout instrument is one whose measuring circuitry is of analog design but the indicating device is of digital design.

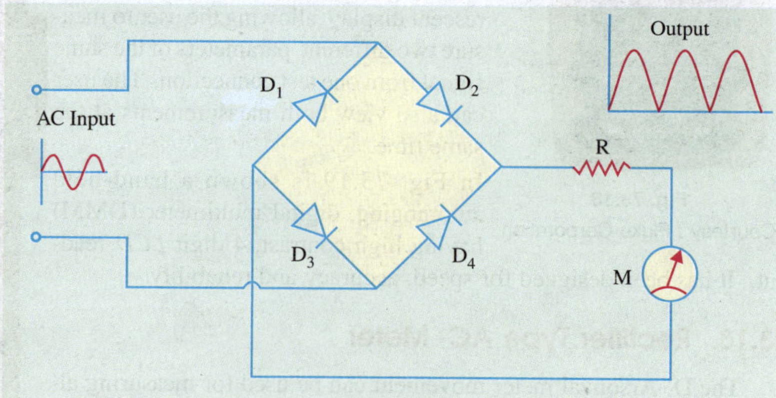

Fig. 73.21

The electronic voltmeters go by a variety of names reflecting the technology used.

(i) vacuum-tube voltmeter (*VTVM*)–it uses vacuum tubes with deflection meter movement,

(ii) digital voltmeters like transistor voltmeter (*TVM*) and *FET* voltmeter (*FETVM*).

73.20. Direct Current *FET VM*

The schematic diagram of a *FET VM* using difference amplifier is shown in Fig. 73.22. The two *FETs* are identical so that increase in the current of one *FET* is offset by corresponding decrease in the source current of the other. The two *FETs* form the lower arms of the balanced bridge circuit whereas the two drain resistors R_D form the upper arms. The meter movement is connected across the drain terminals of the *FETs*.

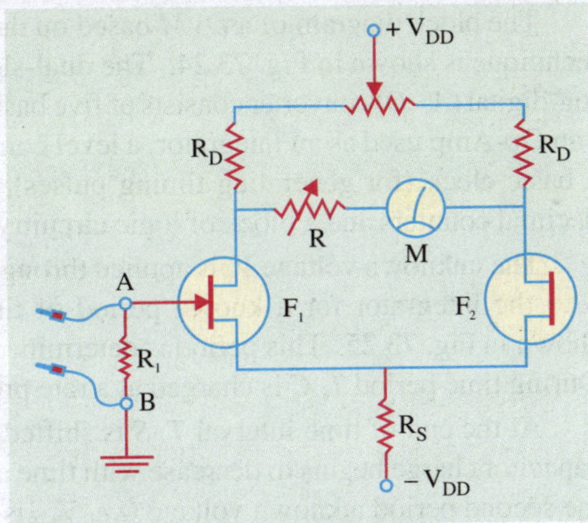

Fig. 73.22

The circuit is balanced under zero-input-voltage condition provided the two *FETs* are identical. In that case, there would be no current through *M*. Zero-Adjust potentiometer is used to get null deflection in case there is a small current through *M* under zero-signal condition.

Full-scale calibration is adjusted with the help of variable resistor *R*.

When positive voltage is applied to the gate of F_1, some current flows through *M*. The magnitude of this current is found to be proportional to the voltage being measured. Hence, meter is calibrated in volts to indicate input voltage.

73.21. Electronic Voltmeter for Alternating Currents

The block diagram of such an *EVM* for ac measurements is shown in Fig. 73.23 where voltage divider allows range selection. The amplifier provides the necessary gain to establish voltmeter sensitivity as well as high input impedance. The negative feed-back ensures stability and accurate overall gain.

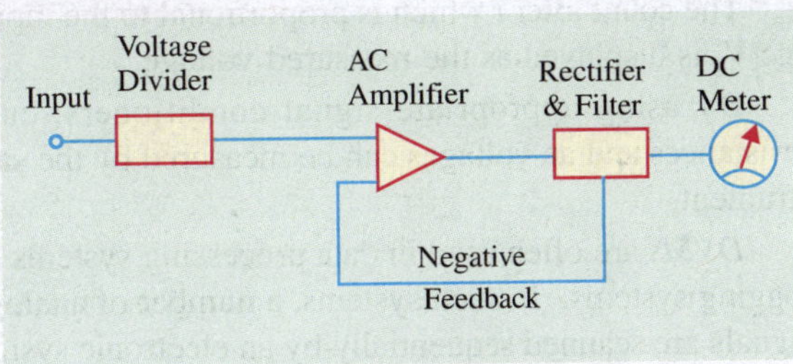

Fig. 73.23

73.22. The Digital Voltmeter (DVM)

Such a voltmeter displays measurements of dc or ac voltages as discrete numerals instead of pointer deflections on a continuous scale as in analog instruments. As compared to other voltmeters, a *DVM* offers the advantages of :

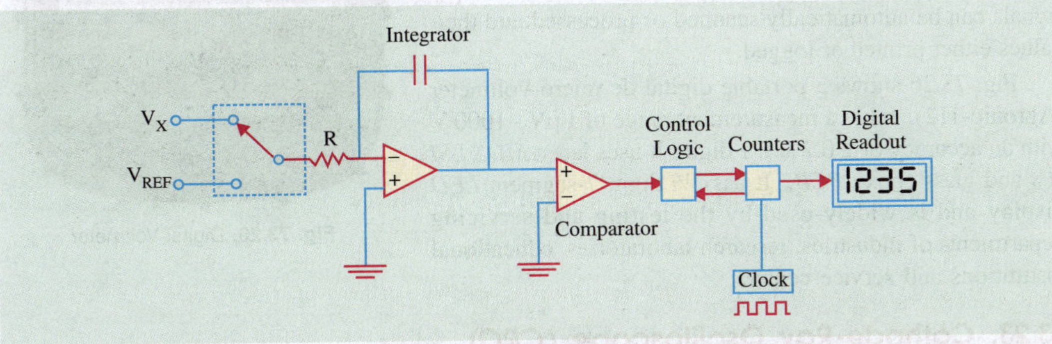

Fig. 73.24

1. greater speed, **2.** higher accuracy and resolution, **3.** no parallax, **4.** reduced human error, **5.** compatibility with other digital equipment for further processing and recording.

With the development and perfection of IC modules, the size and power requirement of *DVMs* have reduced to a level where they can compete with conventional analog instrument both in price and portability.

The block diagram of a *DVM* based on dual-slope technique is shown in Fig. 73.24. The dual-slope analog-digital (*A - D*) converter consists of five basic blocks : an Op-Amp used as an integrator, a level comparator, a basic clock (for generating timing pulses), a set of decimal counters and a block of logic circuitry.

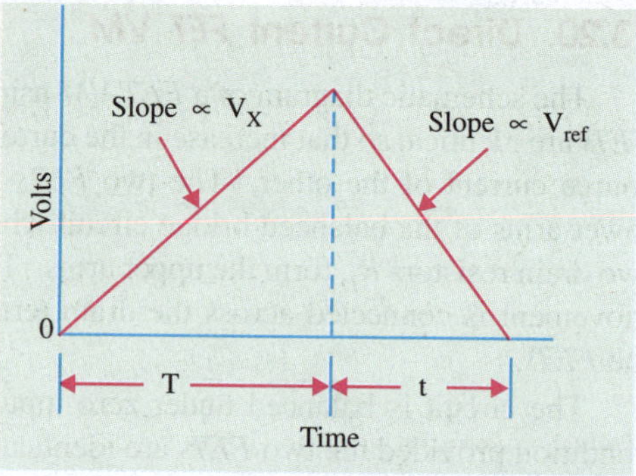

Fig. 73.25

The unknown voltage V_x is applied through switch *S* to the integrator for a known period of time *T* as shown in Fig. 73.25. This period is determined by counting the clock frequency in decimal counters. During time period *T*, *C* is charged at a rate proportional to V_x.

At the end of time interval *T*, *S* is shifted to the reference voltage V_{ref} of ***opposite*** polarity. The capacitor charge begins to decrease with time and results in a down-ward linear ramp voltage. During the second period a known voltage (*i.e.* V_{ref} is observed for an unkown time (*t*). This unknown time *t* is determined by counting timing pulses from the clock until the voltage across the capacitor reaches its basic reference value (reference may be ground or any other basic reference level). From similar triangles of Fig. 73.25.

$$\frac{V_x}{T} = \frac{V_{ref}}{t} \qquad \therefore \quad V_x = \frac{V_{ref}}{t} \times T$$

The count after *t* which is proportional to the input voltage V_x is displayed as the measured voltage.

By using appropriate signal conditioners, currents, resistances and ac voltages can be measured by the same instrument.

DVMs are often used in data processing systems or data logging systems. In such systems, a number of analog input signals are scanned sequentially by an electronic system and then each signal is converted into an equivalent digital value by the *A/D* converter in the *DVM*. The digital value is then transmitted to a printer alongwith the information about the input line from which the signal has been derived. The whole data is then printed out. In this way, a large number of intput signals can be automatically scanned or processed and their values either printed or logged.

Fig. 73.26. Digital Voltmeter

Fig. 73.26 shows a portable digital dc micro-voltmeter (Agronic-112). It has a measurement range of 1 μV - 1000 V with an accuracy of ± 0.2% ± 1 digit. It uses latest *MOS LSI ICs* and glass epoxy *PCB*. It has 3½ digit, 7-segment *LED* display and is widely-used by the testing and servicing departments of industries, research laboratories, educational institutions and service centres.

73.23. Cathode-Ray Oscilloscope (*CRO*)

It is generally referred to as oscilloscope or scope and is the basic tool of an electronic engineer and technician as voltmeter, ammeter and wattmeter are those of an electrical engineer or electrician.

The *CRO* provides a two-dimensional visual display of the signal waveshape on a screen thereby allowing an electronic engineer to 'see' the signal in various parts of the circuit. It, in effect, gives the electronic engineer an eye to 'see' what is happening inside the circuit itself. It is only by 'seeing' the signal waveforms that he/she can correct errors, understand mistakes in the circuit design and thus make suitable adjustments.

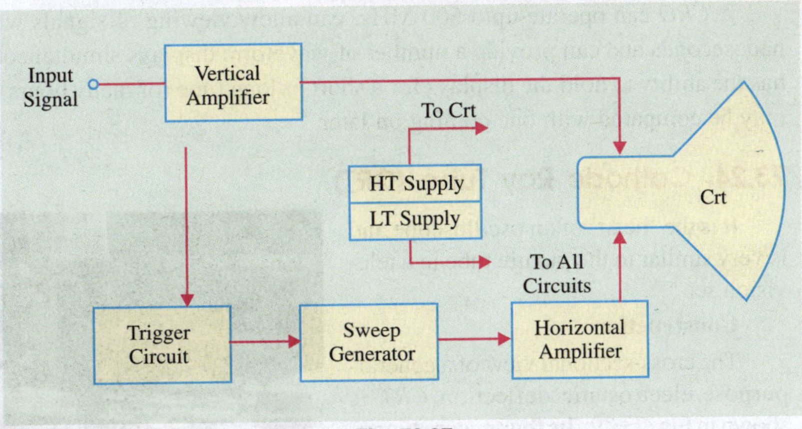

Fig. 73.27

An oscilloscope can display and also measure many electrical quantities like ac/dc voltage, time, phase relationships, frequency and a wide range of waveform characteristics like rise-time, fall-time and overshoot etc. Non-electrical quantities like pressure, strain, temperature and acceleration etc. can also be measured by using different transducers to first convert them into an equivalent voltage.

As seen from the block diagram of an oscilloscope (Fig. 73.27), it consists of the following major sub-systems :

1. **Cathode Ray Tube (*CRT*)**–it displays the quantity being measured.
2. **Vertical amplifier**–it amplifies the signal waveform to be viewed.
3. **Horizontal amplifier**–it is fed with a sawtooth voltage which is then applied to the *X*-plates.
4. **Sweep generator**–produces sawtooth voltage waveform used for horizontal deflection of the electron beam.
5. **Trigger circuit**–produces trigger pulses to start horizontal sweep.
6. **High and low-voltage power supply.**

The operating controls of a basic oscilloscope are shown in Fig. 73.28.

The different **terminals** provide.

1. horizontal amplifier input,
2. vertical amplifier input,
3. sync. input,
4. Z-axis input,
5. external sweep input.

As seen, different **controls** permit adjustment of

1. **Intensity**–for correct brightness of the trace on the screen,
2. **Focus**–for sharp focus of the trace.
3. **Horizontal centering**–for moving the pattern right and left on the screen.
4. **Vertical centering**–for moving the pattern up and down on the screen.
5. **Horizontal gain (also Time/div or Time/cm)**–for adjusting pattern width.
6. **Vertical gain (also volt/div or volt/cm)**–for adjusting pattern height.
7. **Sweep frequency**–for selecting number of cycles in the pattern.
8. **Sync. voltage amplitude**–for locking the pattern.

The different **switches** permit selection of :

1. sweep type,
2. sweep range,
3. sync. type

A *CRO* can operate upto 500 MHz, can allow viewing of signals within a time span of a few nanoseconds and can provide a number of waveform displays simultaneously on the screen. It also has the ability to hold the displays for a short or long time (of many hours) so that the original signal may be compared with one coming on later.

73.24. Cathode Ray Tube (CRT)

It is the 'heart' of an oscilloscope and is very similar to the picture tube in a television set.

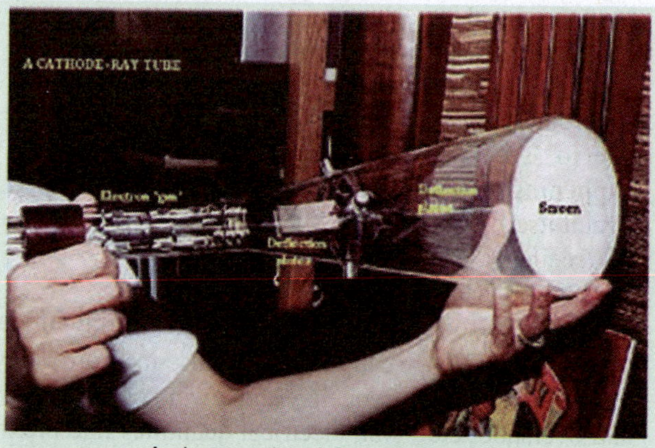

A photography of Cathode ray tube

Construction

The cross-sectional view of a general-purpose electrostatic deflection *CRT* is shown in Fig. 73.29. Its four major components are :

1. an electron gun for producing a stream of electrons,
2. focussing and accelerating anodes-for producing a narrow and sharply-focussed beam of electrons,
3. horizontal and vertical deflecting plates-for controlling the path of the beam,
4. an evacuated glass envelope with a phosphorescent screen which produces bright spot when struck by a high-velocity electron beam.

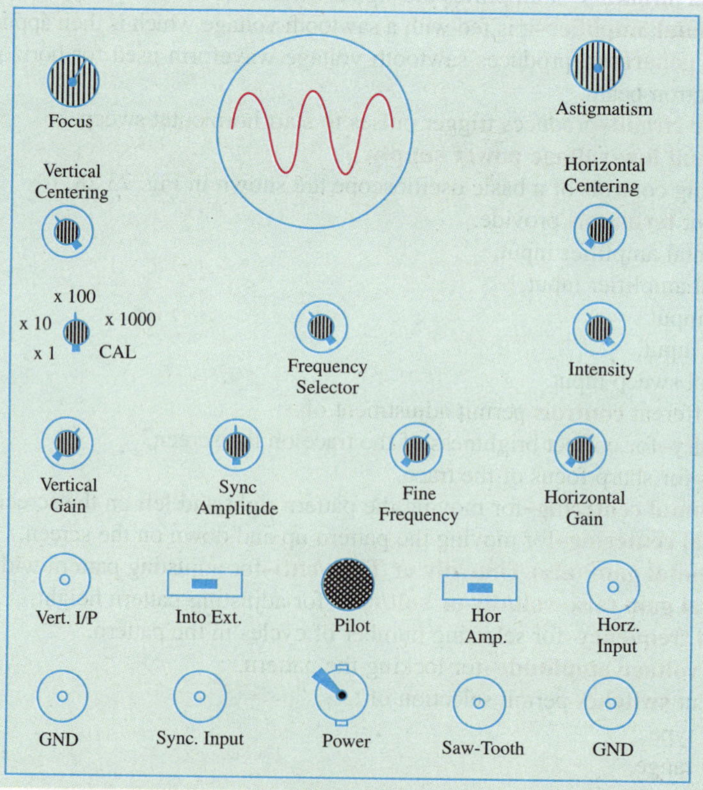

Fig. 73.28

As shown, a *CRT* is a self-contained unit like any electron tube with a base through which leads are brought out for different pins.

1. Electron Gun Assembly

The electron gun assembly consists of an indirectly-heated cathode K, a control grid G, a pre-accelerator anode A_1, focussing anode A_2 and an accelerating anode A_3. The sole function of the electrons gun assembly is to provide a focussed beam of electrons which is accelerated towards the flourescent screen. The electrons are given off by thermionic emission from the cathode. The control grid is a metallic cylinder with a small aperture in line with the cathode and kept at a negative potential with respect to K. The number of electrons allowed to pass through the grid aperture (and, hence, the beam current) depends on the amount of the control grid bias. Since the intensity (or brightness) of the spot S on the screen depends on the strength of beam current, the knob controlling the grid bias is called the ***intensity control.***

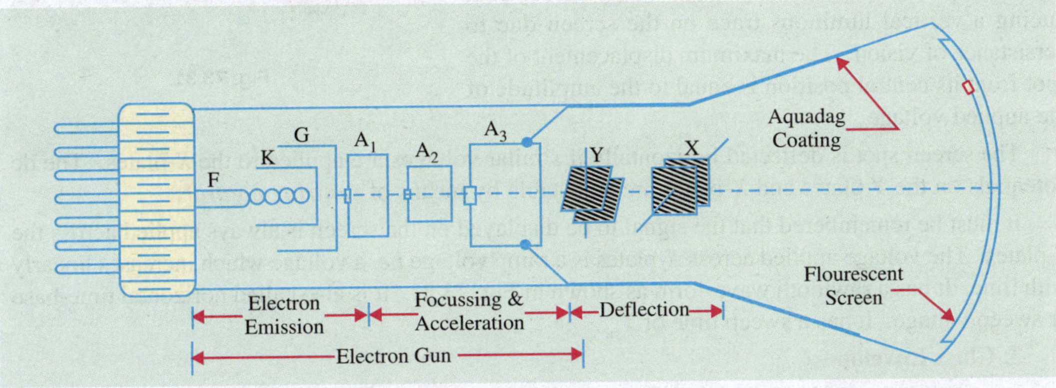

Fig. 73.29

The anodes A_1 and A_3, which are both at positive potential with respect to K, operate to accelerate the electron beam (Fig. 73.30). The cylindrical focussing anode A_2, being at negative potential, repels electrons from all sides and compresses them into a fine beam. The knob controlling the potential of A_2 provides the ***focus control.***

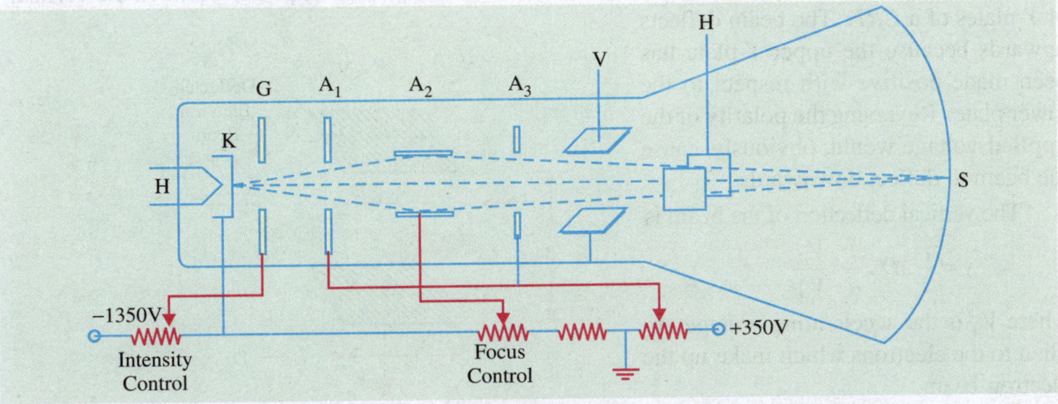

Fig. 73.30

2. Deflecting Plates

Two sets of deflecting plates are used for deflecting the thin pencil-like electronic beam both in the vertical and horizontal directions. The first set marked Y (nearer to the gun) is for vertical deflection and X-set is for horizontal deflection. When no potential is applied across the plates, beam passes between both sets of plates undeflected and produces a bright spot at the centre of the screen.

If upper Y-plate is given a positive potential, the beam is deflected upwards depending on the value of the applied potential. Similarly, the beam (and hence the spot) deflects downwards when lower Y-plate is made positive. However, if an ***alternating*** voltage is applied across the Y-plates, the spot keeps moving up and down thereby producing a vertical luminous trace on the screen due to persistance of vision. The maximum displacement of the spot from its central position is equal to the amplitude of the applied voltage.

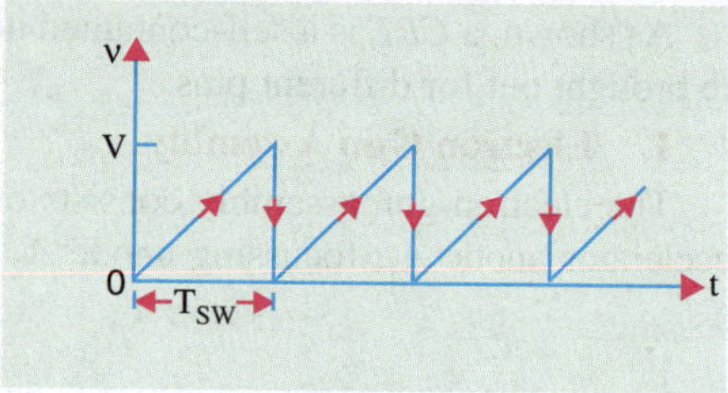

Fig. 73.31

The screen spot is deflected horizontally if similar voltages are applied to the X-plates. The dc potentials on the Y-plates and X-plates are adjustable by means of ***centring controls.***

It must be remembered that the signal to be displayed on the screen is always applied across the Y-plates. The voltage applied across X-plates is a ramp voltage *i.e.* a voltage which increases linearly with time. It has a sawtooth wave-form as shown in Fig. 73.31. It is also called horizontal time-base or sweep voltage. It has a sweep time of T_{sw}.

3. Glass Envelope

It is funnel-shaped having a phosphor-coated screen at its flared end. It is highly-evacuated in order to permit the electron beam to traverse the tube easily. The inside of the flared part of the tube is coated with a conducting graphite layer called Aquadag which is maintained at the same potential as A_3. This layer performs two functions *(i)* it accelerates the electron beam after it passes between the deflecting plates and *(ii)* collects the electrons produces by secondary emission when electron beam strikes the screen. Hence, it prevents the formation of negative charge on the screen.

The screen itself is coated with a thin layer of a flourescent material called phosphor. When struck by high-energy electrons, it glows. In other words, it absorbs the kinetic energy of the electrons and converts it into light-the process being known as ***flourescence.*** That is why the screen is called ***flourescent screen.*** The colour of the emitted light depends on the type of phosphor used.

73.25. Deflection Sensitivity of a *CRT*

Fig. 73.32 shows the upward deflection of an electron beam when it passes between the vertical or Y-plates of a *CRT*. The beam deflects upwards because the upper Y-plate has been made positive with respect to the lower plate. Reversing the polarity of the **applied** voltage would, obviously, cause the beam to deflect downwards.

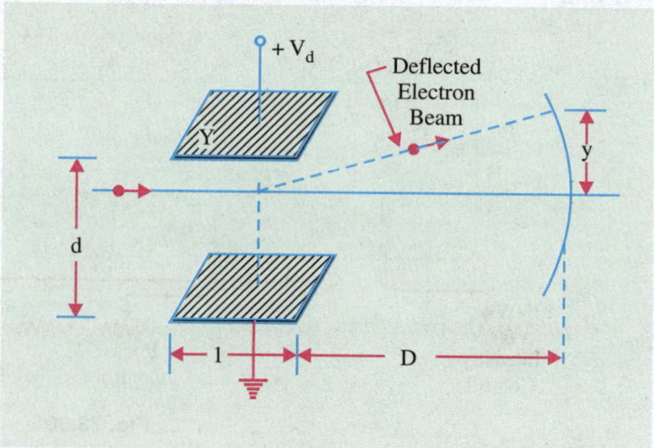

Fig. 73.32

The vertical deflection of the beam is

$$y = \frac{1}{2} . D . \frac{l}{d} . \frac{V_d}{V_A}$$

where V_A is the accelerating voltage applied to the electrons which make up the electron beam.

The deflection sensitivity of a *CRT* is definition as the vertical deflection of the beam on the screen per unit deflecting voltage.

$$\therefore \qquad S = \frac{y}{V_d}$$

Using the above equation, we get $S = \frac{lD}{2dV_A}$

The deflection factor which is defined as the reciprocal of deflection sensitivity is given by $G = 1/S$.

Substituting the value of S from above

$$G = 2.\frac{d}{l}.\frac{V_A}{D} \text{ volt/metre}$$

73.26. Normal Operation of a *CRO*

The signal to be viewed or displayed on the screen is applied across the *Y*-plates of a *CRT*. But to see its waveform or pattern, it is essential to spread it out horizontally from left to right. It is achieved by applying a sawtooth voltage wave (produced by a time base generator) to *X*-plates. Under these conditions, the electron beam would move uniformly from left to right thereby graphic vertical variations of the input signal versus time. Due to repetitive tracing of the viewed waveform, we get a continuous display because of persistence of vision. However, for getting a stable stationary display on the screen, it is essential to synchronize the horizontal sweeping of the beam (sync) with the input signal across *Y*-plates. The signal will be properly synced only when its frequency equals the sweep-generator frequency.

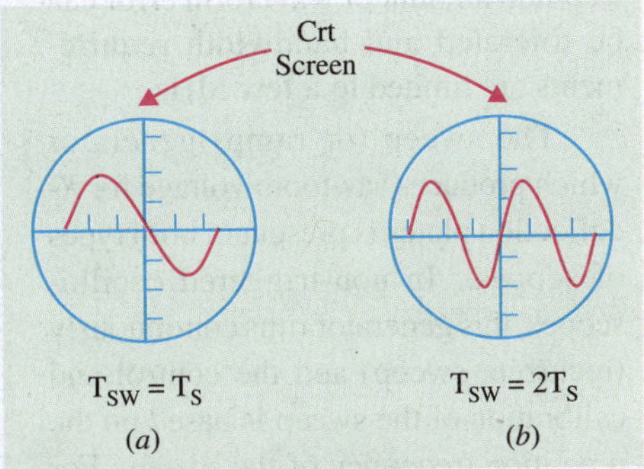

Fig. 73.33

In general, for proper synchronization of time-base with the signal, the condition is

$$T_{sw} = n\,T_s$$

where T_s the time-period of the signal and n is an integer.

If $n = 1$, then $T_{sw} = T_s$ *i.e.* time-periods of the sweep voltage and input signal voltage are equal, then one cycle of the signal would be displayed as shown in Fig. 73.33 (*a*).

On the other hand, if T_{sw} is twice T_s, then two cycles of the signal voltage would be displayed as shown in Fig. 73.33 (*b*). Obviously, three full cycles of the input voltage would be spread out on the screen when $T_{sw} = 3\,T_s$.

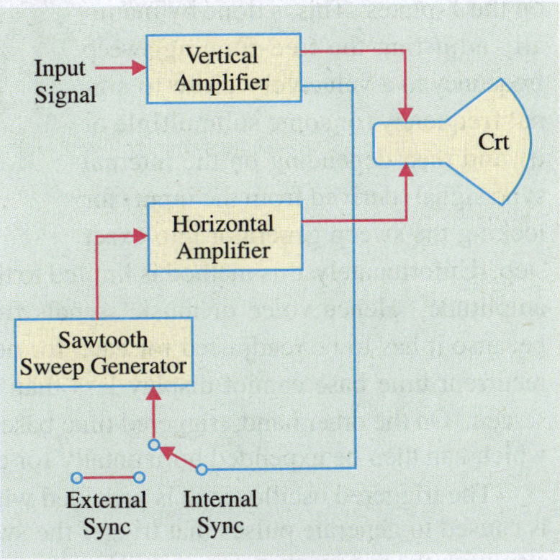

Fig. 73.34

Internal Synchronization

The periodic sawtooth voltage which is applied to *X*-plates for horizontal sweep (or scan) of the beam across the screen is usually provided by the unijunction relaxation oscillator. When the sawtooth voltage falls abruptly to zero, the beam experience no horizontal deflection and hence flies back almost instantly to the original (central) position.

The usual method of synchronizing the input signal is to use a portion of the input signal to trigger the sweep generator so that the frequency of the sweep signal is locked or synchronized to the input signal. It is called internal sync. because the synchronization is obtained by internal wiring connection as shown in the block the diagram of Fig. 73.34.

73.27. Triggered and Non-Triggered Scopes

Oscilloscopes may be classified into two basic types :

1. triggered sweep type.
2. recurrent sweep (free-running) type.

Triggered oscilloscopes, being more sophisticated, are generally used in industrial laboratories and plants, in engineering and technical school laboratories and in all those applications which require study of low- and high-frequency waveforms, for accurate measurement of time and timing relationships etc.

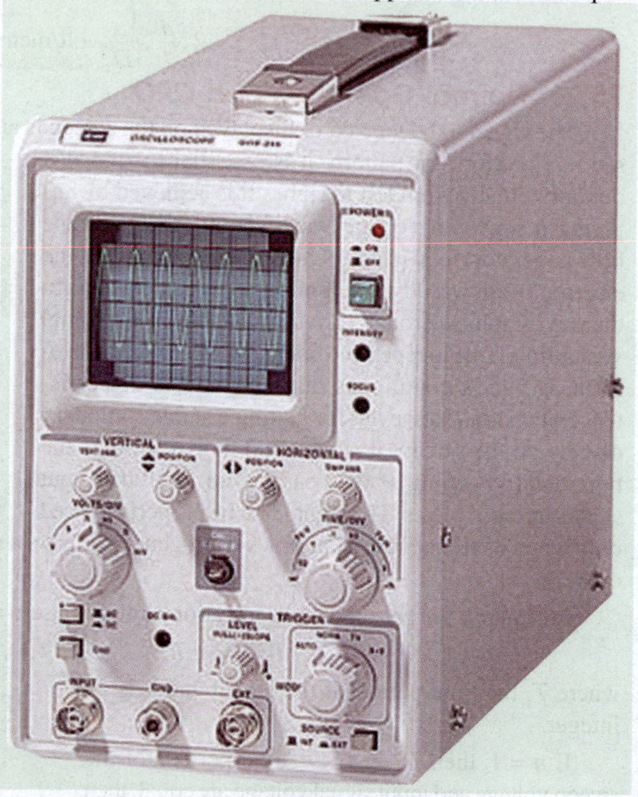

Fig. 73.35. A non-triggered oscilloscope

A non-triggered oscilloscope is generally used in servicing work where a certain amount of waveform error can be tolerated and bandwidth requirements are limited to a few MHz.

The sweep (or ramp) generator which produces sawtooth voltage for *X*-deflection plates is present in both types of scopes. In non-triggered oscilloscopes, this generator runs continuously (recurrent sweep) and the control and calibration of the sweep is based on the repetition freqeuncy of the sweep. For producing a stable stationary display, the sweep frequency has to be forced into synchronization with the input signal on the *Y*-plates. This is done by manually adjusting the free-running sweep freqeuncy to a value very close to signal frequency (or some submultiple of it) and then depending on the internal sync signal (derived from the input) for locking the sweep generator into exact step. Unfortunately, this method is limited to the display of signals which have contant frequency and amplitude. Hence voice or music signals from a microphone cannot be displayed on this scope because it has to be readjusted for each for new change in frequency. Moreover, a free-running or recurrent time base cannot display less than one complete cycle of the input signal on the scope screen. On the other hand, triggered time base can be adjusted to pick out a small part of a waveform which can then be expended horizontally for evaluation of waveform details.

The triggered oscilloscope is provided with a triggered (or driven) sweep. Here the input signal is caused to generate pulses that trigger the sweep thereby ensuring that the sweep is necessarily in step with the trigger that drives it. Hence, screen display remains stable in spite of variations in the frequency or amplitude of the input signal. It means that there is automatic mode of triggering in such scopes. Consequently, input signals of very short duration can be displayed for the simple reason that sweep is initiated by a trigger pulse derived from the waveform under observation.

Fig. 73.35 illustrates a triggered oscilloscope having a bandwidth 0-6 MHz, vertical sensitivity of 10 mV/div and horizontal sweep rate varying from 0.2 μs/div to 0.1 *s*/div.

73.28. Dual Trace CRO

Such oscilloscopes are used extensively by industrial firms and research laboratories. They produce a dual-trace display by means of electronic switching of two separate input signals. As shown in the block diagram of Fig. 73.36, there are two vertical input circuits marked channel *A* and *B* with identical pre-amplifiers. The outputs of the these preamplifiers are fed to an electronic switch which alternately connects them to the main vertical amplifier of the oscilloscope. In this type of scope, there is only one electron beam. The electron switch is also capable of selecting a variety of display modes.

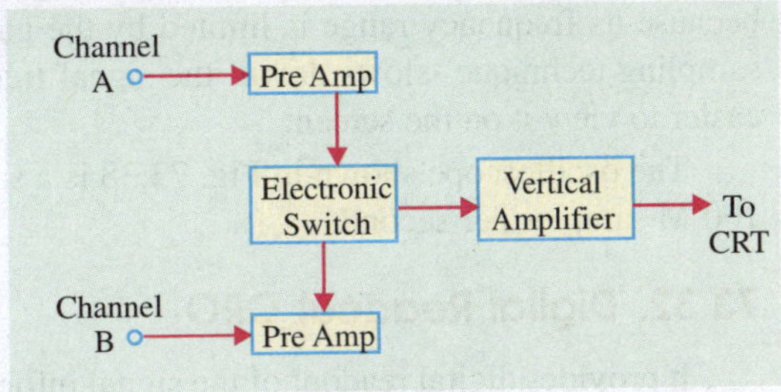

Fig. 73.36

In Fig. 73.37 is shown a dual trace oscilloscope (type *VOS*-26) which has a band-width of 5-15 MHz, vertical sensitivity of 10 mV/cm to 30 V/cm and calibrated sweep speed from 0.3 μs/cm to 10 ms/cm. It is a very sensitive yet simple oscilloscope which assures long life and easy maintenance.

73.29. Dual Beam CRO

Such a *CRO* has two sets of vertical deflection plates and has two electronic beams which produce two separate traces on the scope screen by using the same set of horizontal deflection plates. This scope makes it possible to observe two time-related wave forms at different points *i.e.* the electronic circuit.

Such a scope does not have the same number of display modes as the dual-trace scope yet it is ideally suited for different input signals.

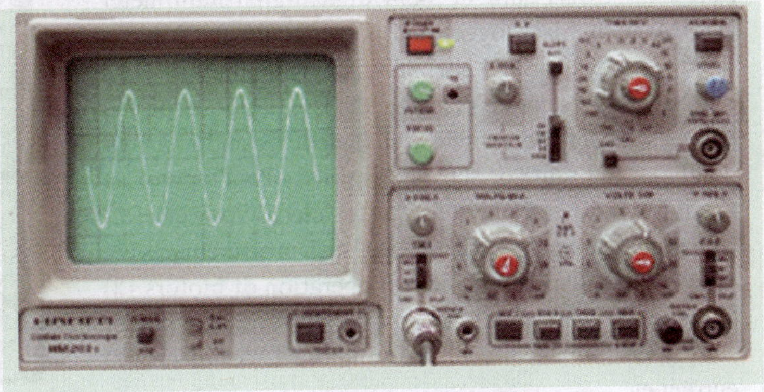

Fig. 73.37. Dual Beam CRO

73.30. Storage Oscilloscope

It can retain a *CRT* display for 10 to 150 hours after the pattern is first produced on the screen. It uses the phenomenon of secondary electron emission to build up and store electrostatic charges on the surface of an insulated target. Such oscilloscopes are especially useful.

1. For real-time observation of events that occur only once.

2. For displaying the waveform of a very low-frequency signal.

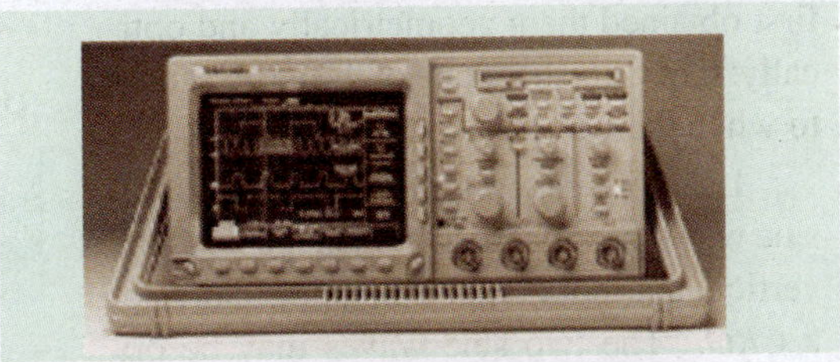

Fig. 73.38. 4 channel storage oscilloscope

Fig. 73.38 shows a 4 channel storage oscilloscope with 400 MHz bandwidth. It has a standard floppy disk drive which makes the saving of screen images or data to a disk, simple. The disk can then be inserted into your personal computer (*PC*) for importing to desk top publishing or spreadsheet programs. The storage oscilloscopes finds their application in biophysics/biomedical research, audio system measurement and analysis, power supply and power-related design, electrophysical and electromechanical system design etc.

73.31. Sampling CRO

It is specifically meant to observe very high frequency repetitive electric signals by using the sampling technique. Such high-frequency signals cannot be viewed by conventional oscilloscopes because its frequency range is limited by the gain-bandwidth product of its vertical amplifier. The sampling technique 'slows down' the signal frequency many thousands of times thereby making it easier to view it on the screen.

The oscilloscope shown in Fig. 73.38 is a sampling cathode ray oscilloscope. Its sample rate is 100 M samples per second.

73.32. Digital Readout CRO

It provides digital readout of the signal information such as voltage or time etc. in addition to the conventional *CRT* display. It consists of a high-speed laboratory *CRO* and an electronic counter, both contained in one cabinet.

73.33. Handheld Battery Operated Oscilloscope

Fig. 73.39 shows a handheld battery operated oscilloscope Model THS 720 P manufactured by Tektronix Corporation. It has built in 3-3/4 digit digital multimeter (DMM) with *data logger and power analyser. It has a bandwidth of over 100 MHz and the sampling rate is as high as 500 M samples per second. The oscilloscope and power analyser can operate simultaneously and independently on the same or separate signals. This type of an oscilloscope is extremely useful for electric/power electronics measurements. Examples of such measurements are : (1) testing and verifying correct operation of motors (2) checking transformer efficiency, (3) verifying power supply performance, (4) measuring the effect of neutral current etc.

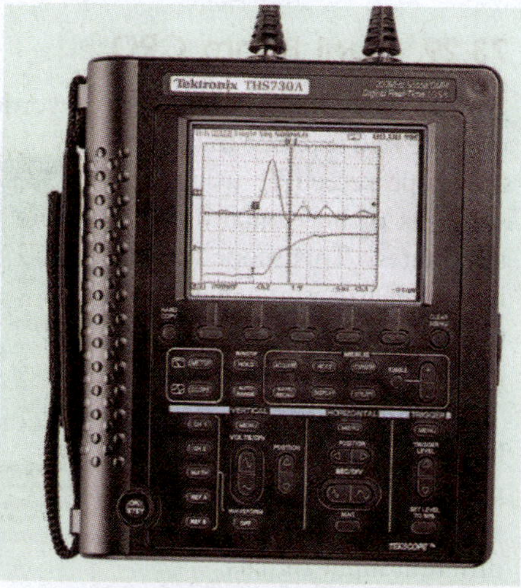

Fig. 73.39. Courtesy : Tektronix Corporation

73.34. Lissajous Figures

Lissajous figures (or patterns) are named in honour of the French scientist who first obtained them geometrically and optically. They illustrate one of the earliest uses to which the *CRO* was put.

Lissajous patterns are formed when two sine waves are applied simultaneously to the vertical and horizontal deflecting plates of a *CRO*. The two sine waves may be obtained from two audio oscillators as shown in Fig. 73.40. Obviously, in this case, ***a sine wave sweeps a sine-wave input signal.*** The shape of the Lissajous pattern depends on

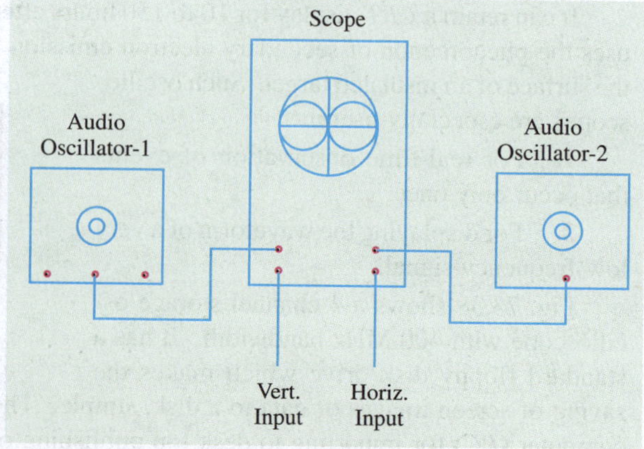

Fig. 73.40

* Data logger is a system which can acquire data and store it in a memory.

the frequency and phase relationship of the two sine waves.

Two sine waves of the *same* frequency and amplitude may produce a straight line, an ellipse or a circle depending on their phase difference (73.41).

In general, the shape of Lissajous figures depends on *(i)* amplitude, *(ii)* phase difference and *(iii)* ratio of frequency of the two waves.

Lissajous figures are used for *(i)* determining an unknown frequency by comparing it with a known frequency *(ii)* checking audio oscillator with a known-frequency signal and *(iii)*checking audio amplifiers and feedback networks for phase shift.

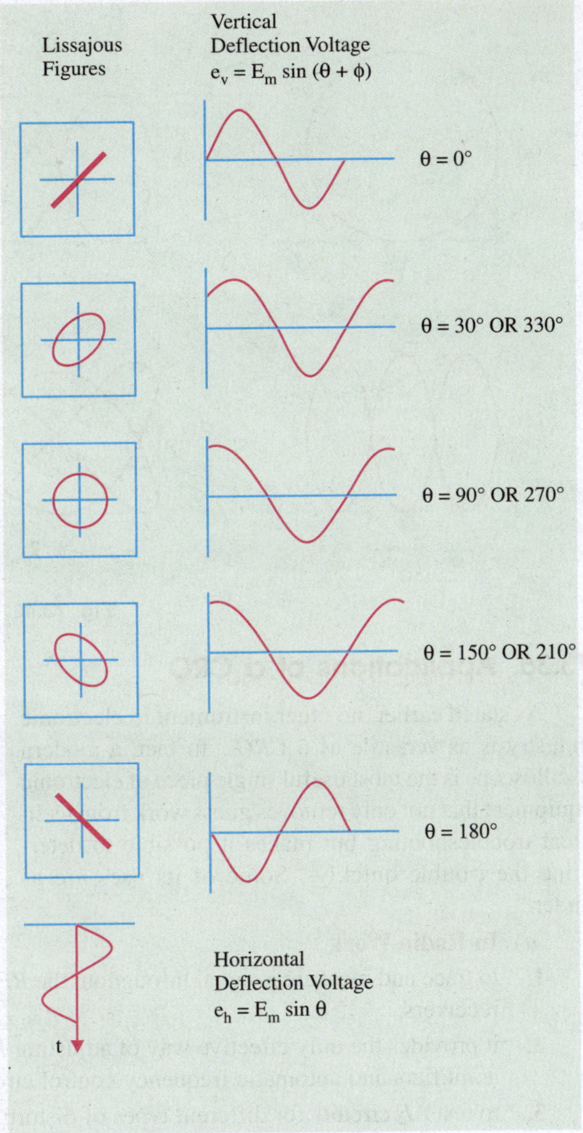

Fig. 73.41

73.35. Frequency Determination with Lissajous Figures

The unknown signal is applied across one set of deflecting plates and a known signal across the other. By studying the resultant Lissajous pattern, unknown frequency can be found.

Depend on the frequency ratio, the various patterns obtained are shown in Fig. 73.42. The ratio of the two frequencies is given by

$$\frac{f_H}{f_v} = \frac{\text{No. of points of horizontal tangency}}{\text{No. of points of vertical tangency}}$$

In Fig. 73.41 (*a*), there is one point of tangency along the horizontal as well as vertical axis. Hence, $f_H = f_v$ *i.e.* the signals have the same frequency. In Fig. 73.42 (*e*) f_H/f_v = 3/2. In other words $f_H = 1.5\, f_v$.

It should be noted in passing that this method of frequency determination has limitations and is being discarded gradually because low-cost digital frequency counters are becoming increasingly available in the market Fig. 73.43. The two main limitatioins of this method are as under :

(i) the numerator and denominator of the frequency ratio must be whole numbers,

(ii) the maximum ratio of frequencies that can be used is 10 : 1. Beyond that, the Lissajous patterns become too complex to analyse.

Fig. 73.43 shows a 10-digit digital frequency counter Model No. PM 6685 manufactured by Fluke Corporation. This can measure frequencies from 10 Hz to 300 MHz. This is an ideal instrument for *R* and *D* laboratories, testing, servicing and even outside the lab environment such as in base station transmitters of large telecommunication networks like *GSM*.

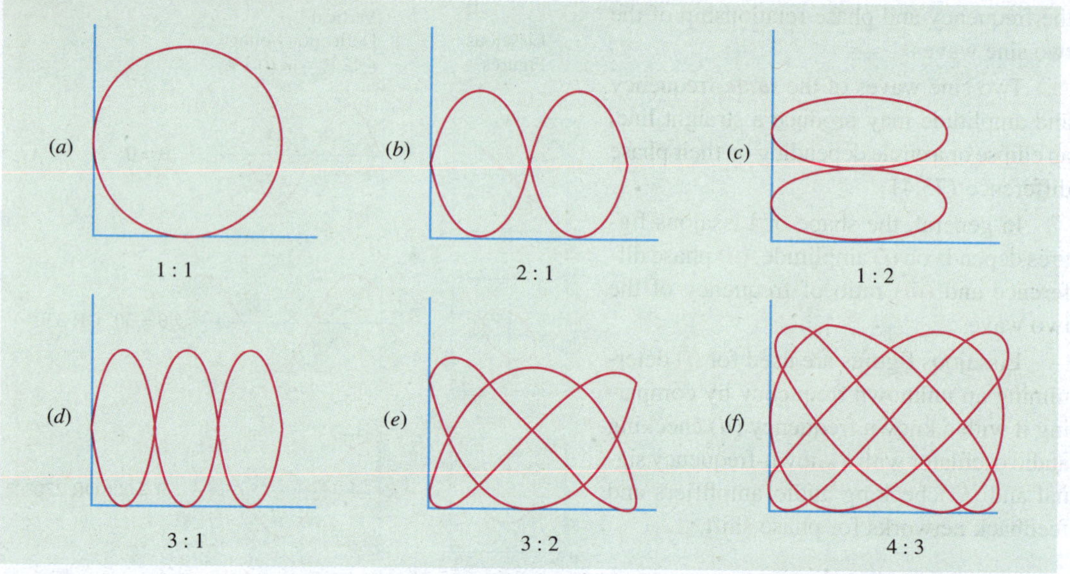

Fig. 73.42

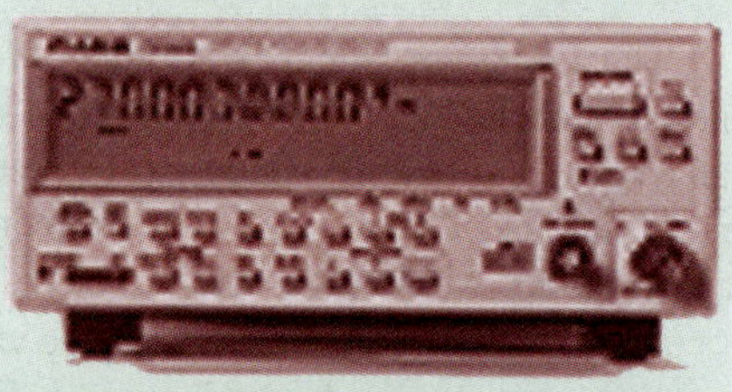

Fig. 73.43
(Courtesy : Fluke Corporation)

73.36. Applications of a CRO

As stated earlier, no other instrument in electronic industry is as versatile as a *CRO*. In fact, a modern oscilloscope is the most useful single piece of electronic equipment that not only removes guess work from technical troubleshooting but makes it possible to determine the trouble quickly. Some of its uses are as under :

(a) In Radio Work

1. to trace and measure a signal throughout the *RF*, *IF* and *AF* channels of radio and television receivers.
2. it provides the only effective way of adjusting *FM* receivers, broadband high-frequency *RF* amplifiers and automatic frequency control circuits;
3. to test *AF* circuits for different types of distortions and other spurious oscillations;
4. to give visual display of waveshapes such as sine waves, square waves and their many different combinations;
5. to trace transistor curves
6. to visually show the composite synchronized TV signal
7. to display the response of tuned circuits etc.

(b) Scientific and Engineering Applications

1. measurement of ac/dc voltages,
2. finding B/H curves for hysteresis loop,
3. for engine pressure analysis,
4. for study of stress, strain, torque, acceleration etc.,
5. frequency and phase determination by using Lissajous figures,
6. radiation patterns of antenna,
7. amplifier gain,

8. modulation percentage,

9. complex waveform as a short-cut for Fourier analysis,

10. standing waves in transmission lines etc.

73.37. The Q Meter

This instrument is designed to measure some of the electrical properties of coils and capacitors by measuring the *Q*-value of an *R-L-C* circuit.

1. Construction

As shown in Fig. 73.44, it essentially consists of

(i) a frequency-calibrated continuously-variable *RF* oscillator,

(ii) a calibrated variable capacitor *C*,

(iii) *VTVM* which is calibrated to read *Q* directly.

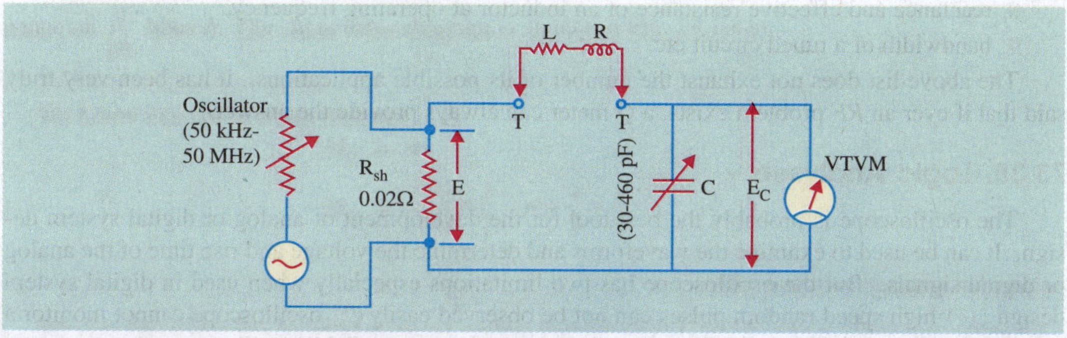

Fig. 73.44

2. Principle of Operation

The basic principle used in *Q*-meter is the resonant rise of the voltage across the capacitor in an *RLC* circuit. The condition for series resonance (Art. 14.10) is

$$X_L = X_C \quad \text{and} \quad E = IR$$

The value of circuit *Q* is

$$Q = \frac{X_L}{R} = \frac{X_C}{R} = \frac{E_C}{E}$$

If the applied voltage *E* is constant, then $Q \propto E_C$. Hence, by measuring voltage drop across *C* under resonant conditions, *Q* can be found. Alternatively, *VTVM* can be calibrated directly in terms of *Q* (rather than voltage).

As seen from Fig. 73.44, the oscillator delivers current to a very small (0.02 Ω) shunt resistance R_{sh} thereby developing a voltage *E* across it. It becomes the applied voltage for the *RLC* circuit and corresponds to the source voltage *E* of Fig. 73.45.

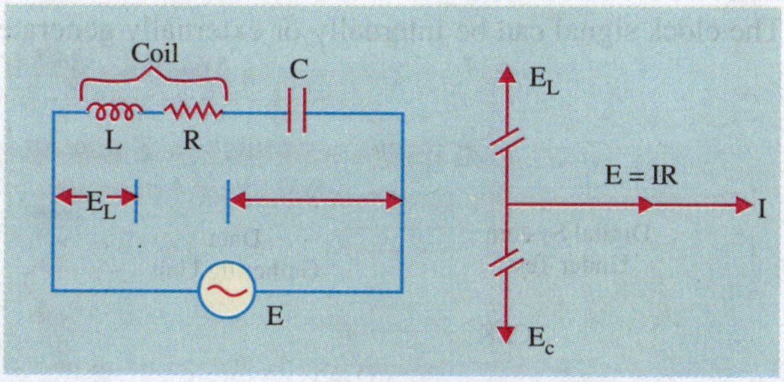

Fig. 73.45

It is measured by a thermocouple meter marked 'Multiply *Q* by'. The voltage E_C across variable *C* is measured by the *VTVM*. The value of *Q*-factor is given by $Q = E_C/E$.

For making measurement, the unknown coil is connected to the test terminals *TT* of the instrument and the circuit is tuned to resonance (Fig. 73.44)

(i) either by setting the oscillator to a given frequency and varying *C*,

(*ii*) or by keeping value of *C* constant and adjusting the oscillator frequency.

The reading on the *VTVM* must be multiplied by the index setting of the 'Multiply *Q* by' meter in order to obtain the actual value of *Q*.

3. Applications

Some of the specialized uses of this instrument are to measure

1. *Q* of a coil,
2. inductance and capacitance,
3. distributed capacitance of a coil,
4. *Q* and *p.f.* of a dielectric material,
5. mutual inductance of coupled circuits,
6. coefficient of coupling,
7. critical coupling,
8. reactance and effective resistance of an inductor at operating frequency,
9. bandwidth of a tuned circuit etc.

The above list does not exhaust the number of its possible applications. It has been very truly said that if ever an *RF* problem exists, a *Q* meter can always provide the answer.

73.38. Logic Analysers

The oscilloscope is probably the best tool for the development of analog or digital system design. It can be used to examine the waveforms and determine the voltage and rise time of the analog or digital signals. But the oscilloscope has two limitations especially when used in digital system design : (1) high speed random pulses can not be observed easily (2) oscilloscope cannot monitor a few signal lines simultaneously. For example, in a commonly available oscilloscope, the maximum number of inputs are four.

For these reasons, the logic analysers has been developed. It operates on a slightly different principle than that of an oscilloscope. Because there are many signal lines in a digital system (such as a microprocessor based system), the data is changing rapidly on each line, a logic analyser must take a *snap shot* of the activities on the lines and store the logic state of each signal in memory for each cycle of the system clock. The conditions under which the snapshot is taken are determined by triggering circuits, which can respond to various combinations of events.

The logic analyser enables the activity of many digital signal points to be recorded simultaneously and then examined in detail. The information is recorded with respect to a clock signal to determine whether they are HIGH or LOW with respect to a defined threshold voltage. This information is stored in memory and is then available for detailed analysis via the logic analyser's display. The clock signal can be internally or externally generated.

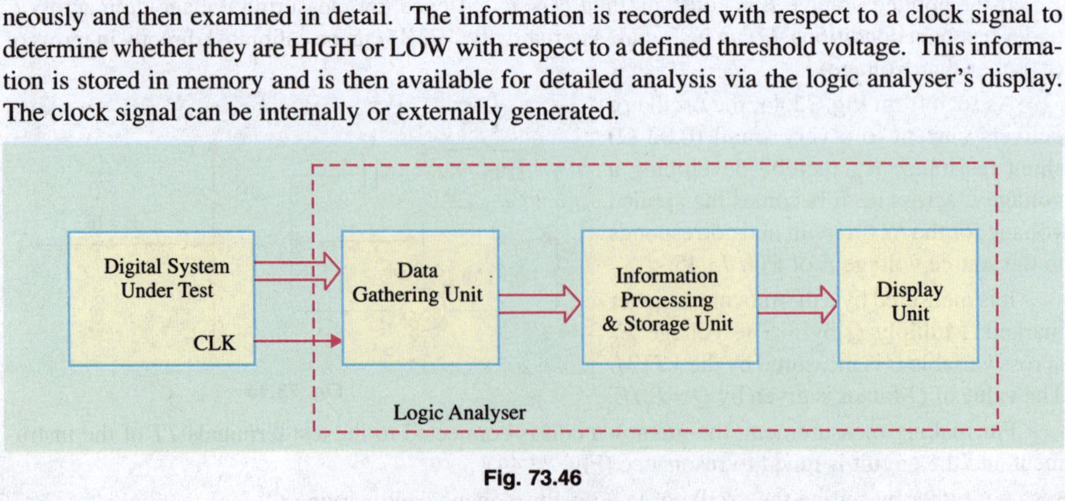

Fig. 73.46

Fig. 73.46 shows a block diagram of a typical logic analyser. It has a data gathering unit, information processing and storage unit and a display unit. The data gathering unit has (1) a pod slots for carrying data from the digital system under test to the logic analyser and (2) a key pad. The key pad is used to enter commands and set up the parameters that the logic anayser will use. The display unit is a cathode ray tube (*CRT*) that displays the command menu for the operator and also displays the output data.

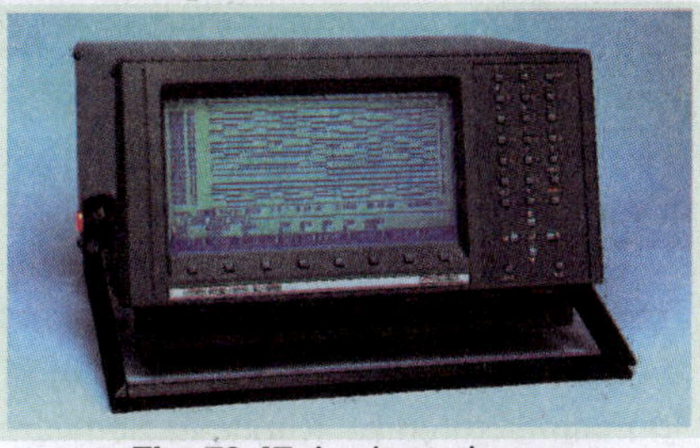

Fig. 73.47. Logic analysers

Applications

- hardware/software debugging
- parametric/mixed signal testing
- hardware simulation and stimulus-response testing
- complex debugging with deep memory.

Fig. 73.47 shows a family of logic analysers, available from Tektronix corporation. Each logic analyser has at least 34 channels, 4-channel digitizing oscilloscope, off-line analysis capability for viewing data and creating setups on a separate PC.

Applications.

Logic analyser is a very powerful tool in the field of microprocessor based system development. Some of its major applications in this area are :

1. Hardware debug and verification.
2. Processor/bus debug and verification.
3. Embedded software integration, debug and verification.

73.39. Spectrum Analysers

The spectrum analyser is an instrument that brings together a superhetrodyne radio receiver with a swept frequency local oscillator and an oscilloscope to present a display of amplitude versus frequency.

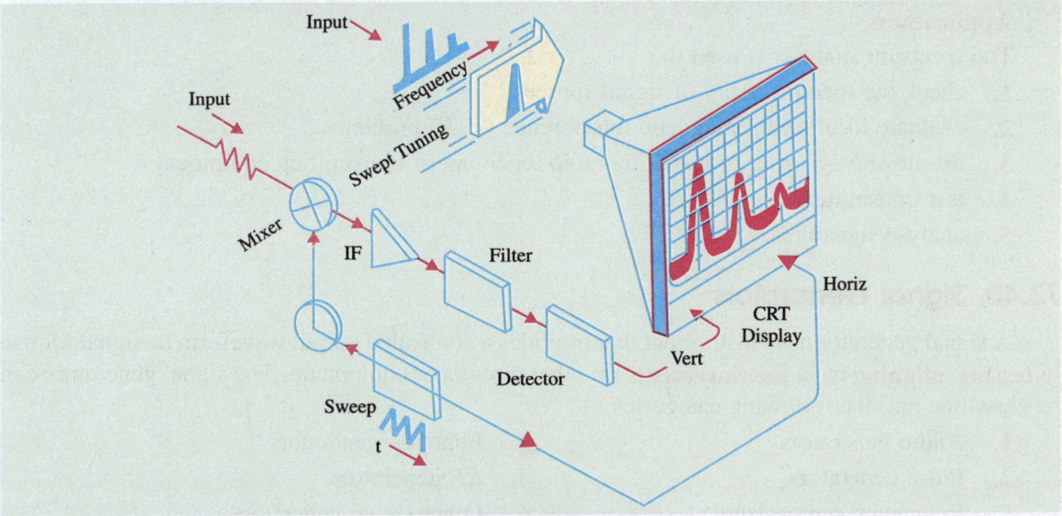

Fig. 73.48

(Courtesy Hewlett Packard)

Fig. 73.48 shows a simple block diagram of a spectrum analyser. As seen, the spectrum analyser is actually a superhetrodyne receiver in which local oscillator is a sweep generator. A low frequency saw-tooth wave is applied to both the sweep oscillator and the horizontal deflection plates of the *CRT*, producing a horizontal deflection that is a function of frequency. The lowest frequency is represented by left side of the trace while the highest frequency is represented by the right side of the trace. The sweep is from left to right.

The input signal is mixed with local oscillator to produce the *IF* (*i.e.* intermediate-frequency or difference) signal. The bandwidth of the *IF* amplifier is relatively narrow. so the output signal at the detector will have a strength that is proportional to the frequency that the local oscillator is converting to the *IF* at that instant. The display will then contain "poles" that represent the amplitudes of the various input frequency components.

There are several spectrum analysers available in the market manufactured by the companies like Rhode & Schwarz (Tektronix), Hewlett Packard (now called Agilent Technologies), and so on. Fig. 73.49 shows two commercially available spectrum analyses.

The spectrum analyser shown in Fig. 73.49 (*a*) is Model *FSE* 30 manufactured by Rhode & Schwarz but marked by Tektronix Corporation.

It has a frequency range from 20 Hz to 76.5 GHz, a bandwidth from 1 Hz to 10 MHz. Another spectrum analyser Model No. 3066 shown in Fig. 73.49 (*b*) is a real-time instrument. It has a frequency range from *DC* to 5 MHz, a bandwidth from *DC* to 3 GHz. The real-time spectrum analyser take a very different approach compared to traditional sweeping spectrum analysers. Rather than acquiring one frequency step at a time, the real time spectrum analyser captures a block of frequencies all at once.

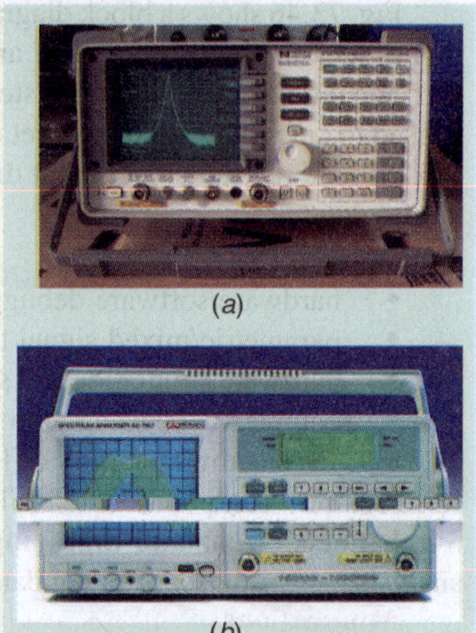

(*a*)

(*b*)

Fig. 73.49

(Courtesy : Rhode & Schwarz and Tektronics Corporation)

It is possible to use computers to do spectrum analysis of the signals. There is a variety of software available over the internet from several companies. Some softwares can be downloaded free of cost from the companies web-sites.

Applications

The spectrum analyser is used to :

1. check the spectral purity of signal sources.
2. evaluate local electromagnetic interference (*EMI*) problems.
3. do site surveys prior to installing radio receiving or transmitting equipment.
4. test transmitters.
5. analyse signatures.

73.40. Signal Generators

A signal generator is an instrument that provides a controlled output waveform or signal for use in **testing, aligning** or in **measurements** on other circuits or equipment. The signal generators can be classified into the following categories :

1. Audio generators
2. Function generators
3. Pulse generators
4. *RF* generators
5. Frequency synthesizers
6. Other signal generators.

73.41. Audio Generators

The audio generators covers the frequency range 20 Hz to 20 KHz, although few models produce signals up to 100 KHz. Audio generators always produce ***pure sine*** waves and most also produce square waves. They uses a 600 Ω output impedance and produce output levels from –40 dB mW to + 4 dB mW.

Two methods of frequency selection are typically used in audio signal generators. ***continuous*** and ***step***. On the continuous type of a dial, we turn a knob to the desired frequency. Many such audio generators have a scale that reads 20 to 200 (or alternatively 2 to 20) and a *range* selector switch determines whether the output frequencies will be 20 to 200 Hz, 200 to 2000 Hz or 2000 to 20,000 Hz. In a step-tuned generator, these controls are replaced by a ***rotary*** or ***pushbutton*** switch bank. As many as four decode switches might be used, although three is a more common number. These will be marked 0 through 100, 0 through 10 and 0.1 through 1.0 in decade steps. A multiplier switch determines whether the actual frequency will be X1, X10, X100 or X1000 the frequency indicated on the selector switches.

Audio signal generator

Fig. 73.50 shows a block diagram of an audio signal generator. The audio oscillator section is usually an *RC* phase-shift oscillator (or a Wien Bridge oscillator) circuit. A power amplifier stage provides buffering between the load and the oscillator and it develops the output signal amplitude. The ac voltmeter at the output is strictly optional, but in some models it is used with a ***level control*** to set precisely the input signal to the attenuator. Not all quality audio signal generators use this feature. So the lack of an ac output meter is not, in itself, indication of quality. In some models, an audio digital frequency counter is used ahead of the attenuator to provide digital display of the output frequency.

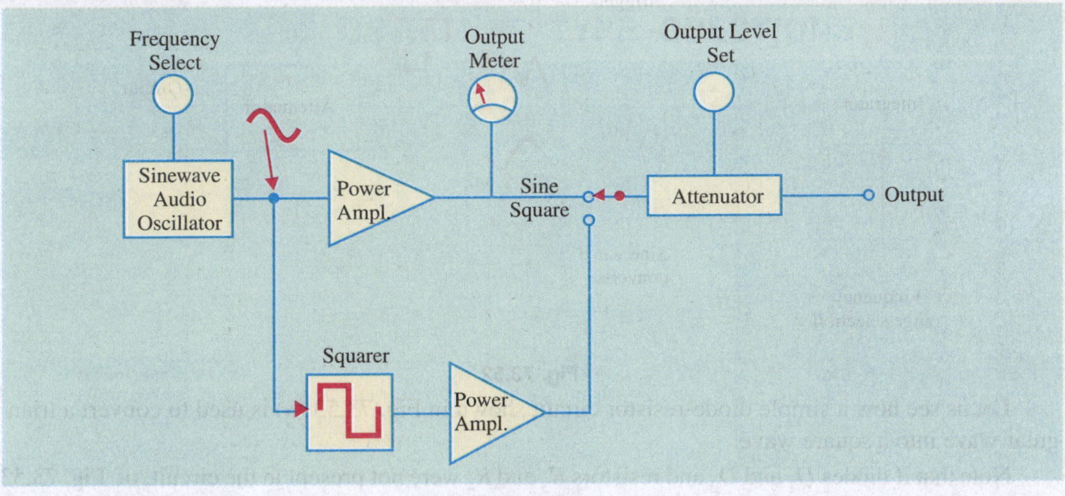

Fig. 73.50

Applications

The audio generators are basically used to test the amplitude and frequency response of audio amplifiers.

73.42. Function Generators

These generators typically, cover at least the same frequency range as audio signal generators (*i.e.* 20 Hz to 20 kHz) but most modern designs have extended frequency ranges. A very common frequency range for function generators is 0.01 Hz to 3 MHz.

The major difference between a function generator and an audio generator is in the number of output waveforms. The audio signal generator produces only sine waves and square waves. While almost all function generators produce these basic waveforms plus triangular waves. Besides this, some function generators also produce sawtooth, pulse and non-symmetrical square waves. Fig. 73.51 shows the controls of a typical function generator.

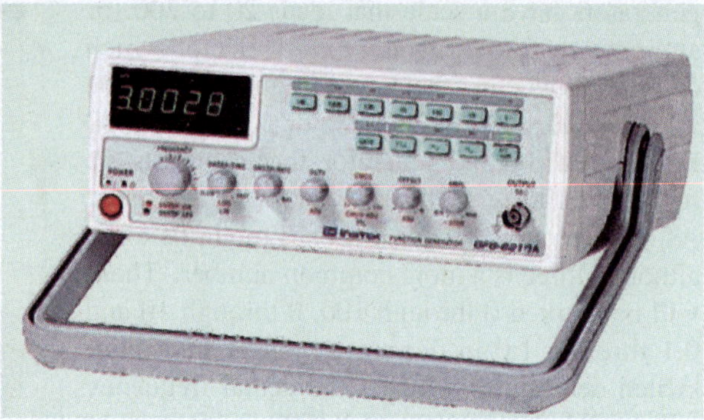

Fig. 73.51

Fig. 73.52 shows a simple block diagram of a function generator. The major parts of a function generator are schmitt trigger, integrator, sine-wave converter and an attenuator. The schmitt trigger converts a slowly varying input signal to a square wave signal. This square wave signal is available at the output as well as it is also connected to the integrator as an input through a potentiometer (R). The potentiometer is used to adjust the frequency of the output signal. The frequency range is adjusted by selecting the appropriate capacitor connected in the integrator circuit.

The sine-wave converter is a six-level (or more) diode-resistor loading circuit.

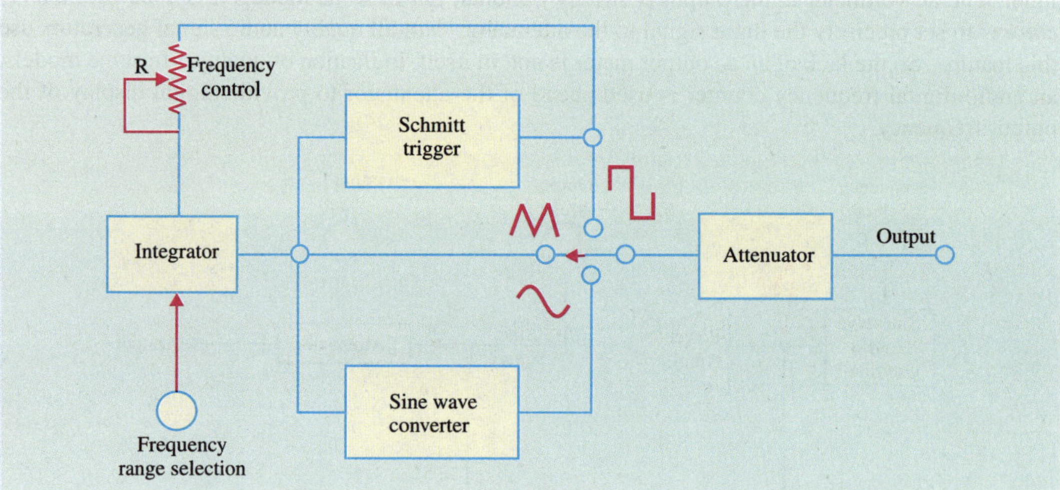

Fig. 73.52

Let us see how a simple diode-resistor circuit shown in Fig. 73.53 (*a*) is used to convert a triangular wave into a square wave.

Note that if diodes D_1 and D_2 and resistors R_3 and R_4 were not present in the circuit, of Fig. 73.53 (*a*), R_1 and R_2 would simply behave as a voltage divider. In this case the output from the circuit would be an attenuated version of the triangular wave :

$$V_0 = V_i \frac{R_2}{R_1 + R_2}$$

With diodes D_1 and R_3 in the circuit, R_1 and R_2 still behave as a simple voltage divider until the voltage drop across R_2, VR_2 exceeds $+ V_1$. At this point D_1 becomes forward biased, and R_3 is effectively in parallel with R_2. Now,

$$V_0 = V_i \frac{R_2 \parallel R_3}{R_1 + R_2 \parallel R_3}$$

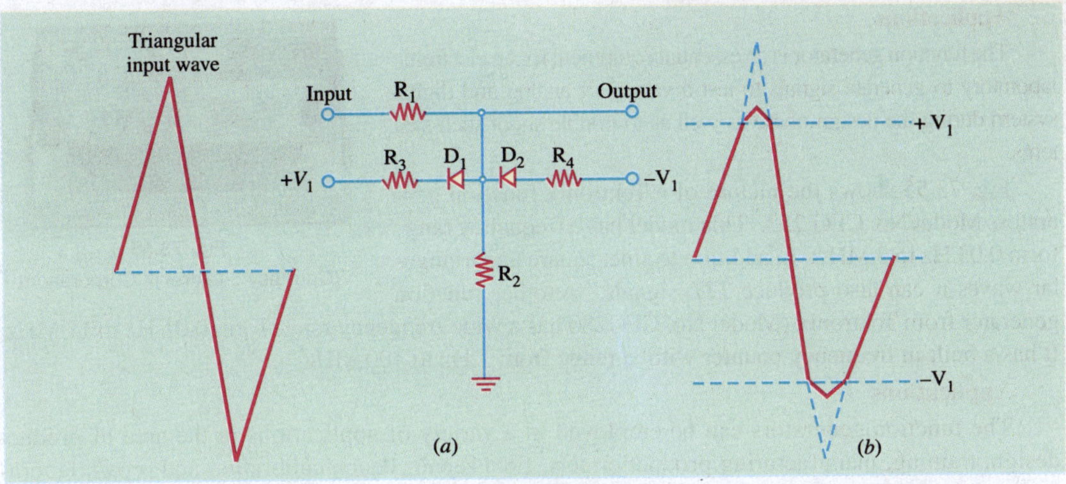

Fig. 73.53

Output voltage levels above + V_1 are attenuated to a greater extent than levels below + V_1. Consequently, the output voltage rises less steeply than without D_1 and R_3 in the circuit (refer to Fig. 73.53 (*b*). When the output falls below + V_1, the diode D_1 is reverse biased. As a result of this, R_3 is no longer in parallel with R_2, and the attenuation is once again $R_2/(R_1 + R_2)$. Similarly during the negative half cycle of the input, the output voltage,

$$V_0 = V_i \frac{R_2}{R_1 + R_2}$$

until V_0 goes below – V_1. Then D_2 becomes forward biased, putting R_4 in parallel with R_2 and making,

$$V_0 = V_i \frac{R_2 \parallel R_4}{R_1 + R_2 \parallel R_4}$$

With $R_3 = R_4$, the negative half-cycle of the output is similar in shape to the positive half-cycle. If we employ six or more diodes, all connected via resistors to different bias voltage levels (refer to Fig. 73.54 (*a*), a good sine-wave approximation can be achieved. With six diodes, the three positive bias voltage levels and three negative bias voltage levels, the slope of the output wave changes three times during each quarter cycle. Assuming correctly selected bias voltages and resistor values, the output wave shape is as shown in Fig. 73.54 (*b*).

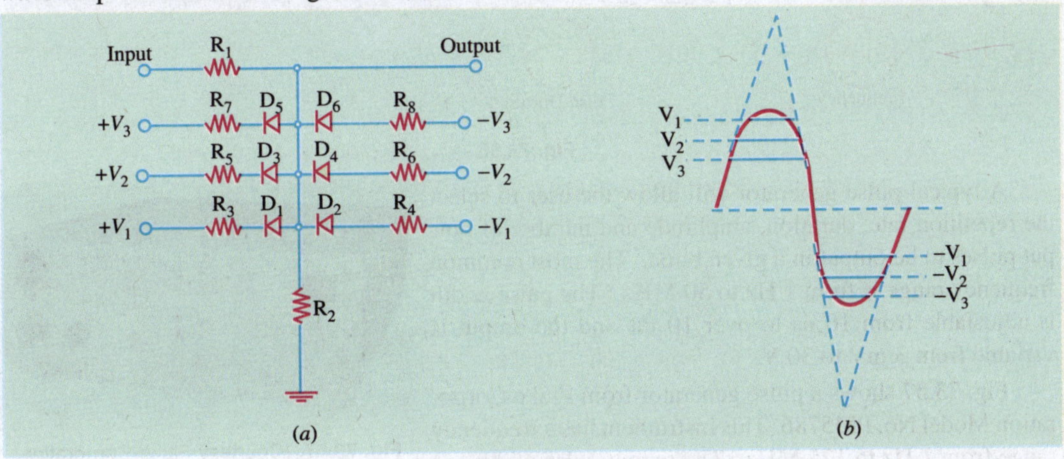

Fig. 73.54

Applications

The function generator is an essential equipment for an electronic laboratory to generate signals to test a variety of analog and digital system during the design phase as well as to trouble shoot such systems.

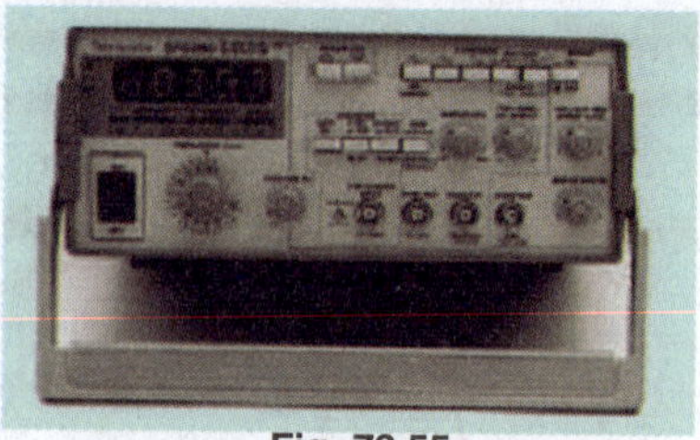

Fig. 73.55
(Courtesy : Tektronix Corporation)

Fig. 73.55 shows the picture of a Tektronix function generators Model No. CFG 253. This model has a frequency range form 0.03 Hz to 3 MHz. In addition to sine, square and triangular waves it can also produce *TTL* signals. Another function generator from Tektronix (Model No. CFG280 has a wide frequency range form 0.01 Hz to 11 MHz. It has a built in frequency counter with a range from 1 Hz to 100 MHz.

Applications

The function generators can be employed in a variety of applications in the area of product design, training, manufacturing production test, field repair, bench calibration and repair, laboratory and research, and education. Mainly they are used for testing amplifiers, filter and digital circuits.

73.43. Pulse Generators

Fig. 73.56 shows the block diagram of a pulse generator. As seen, an astable multivibrator generates square waves. This is used to trigger monostable multivibrator (*i.e.* one-shot). The pulse repetition rate is set by the square-wave frequency. The one-shot triggers on the leading edge of the square-wave and produces one output pulse for each input cycle. The duration of each output pulse is set by the on-time of the one-shot. It may be very short or may approach the period of the square wave. The attenuator facilities output amplitude control and dc level shifting.

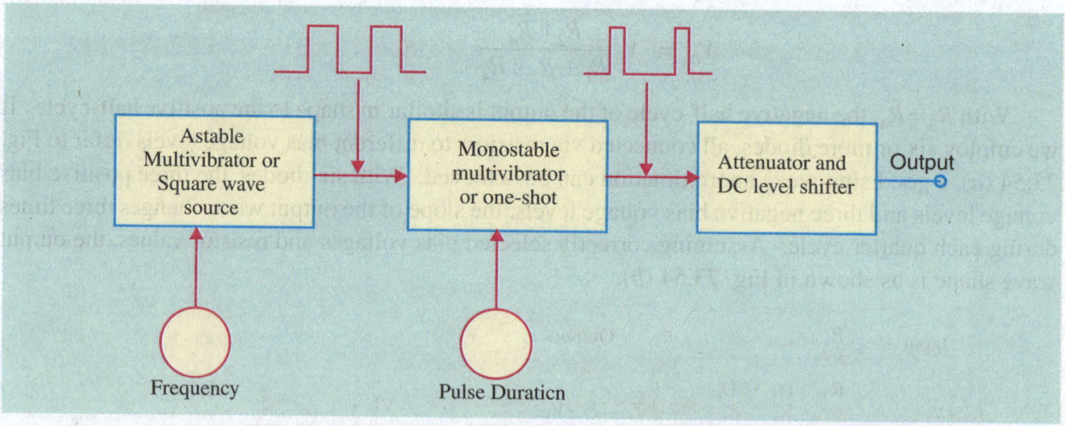

Fig. 73.56

A typical pulse generator will allow the user to select the repetition rate, duration, amplitude and number of output pulses to be output in a given burst. The most common frequency range is from 1 Hz to 50 MHz. The pulse width is adjustable from 10 ns to over 10 ms and the output is variable from 3 mV to 30 V.

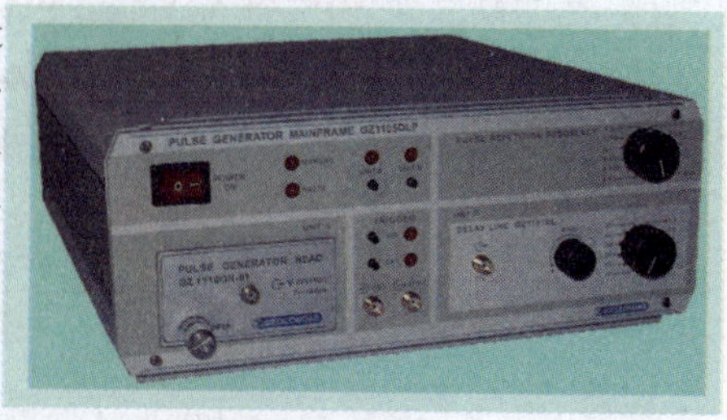

Fig. 73.57. (Courtesy Fluke Corporation)

Fig. 73.57 shows a pulse generator from Fluke Corporation Model No. PM5786. This instrument has a frequency range from 1 Hz to 125 MHz. The output pulse width can be varied from 8 ns to 100 ms. The instrument can also be used to generate pulse bursts. The output voltage level can be adjusted up to 10 V.

Applications

The pulse generators are used extensively to test :

1. Memory circuits
2. Shift registers
3. Counters
4. Other digital components, subsystems and systems.

73.44. RF Generators

A radio frequency (*RF*) signal generator has a sinusoidal output with a frequency somewhere in the range of 100 kHz to 40 GHz region. Fig. 73.58 shows the block diagram of an *RF* generator. As seen, the instrument consists of an *RF* oscillator, an amplifier, a calibrated attenuator and an output level meter.

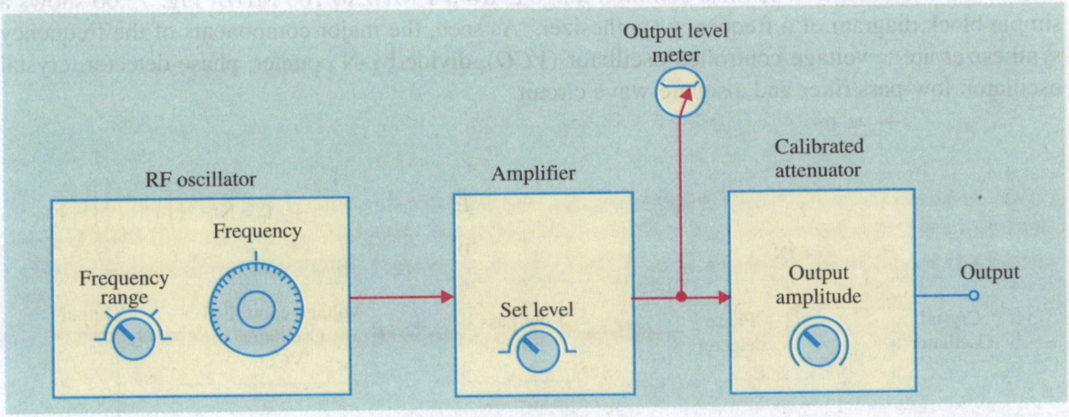

Fig. 73.58

The *RF* oscillator has a continuous frequency control and a frequency range switch, to set the output to any desired frequency. The amplifier includes an output amplitude control. This allows the voltage applied to the attenuator to be set to a calibration point on the output level meter. The output level must always be ***reset*** to this calibration point everytime the frequency is changed. This is necessary to ensure that the output voltage levels are correct, as indicated on the calibrated attenuator.

The oscillator circuit used in an *RF* generator is usually either a Hartley Oscillator or Colpitts oscillator. Most RF signal generators include facilities for amplitude modulation and frequency modulation of the output. Switches are provided on the front panel to allow the user to select no modulation as well as internal or external *AM* or *FM* modulation. It may be noted that each section of the RF generator is ***shielded*** by enclosing it in a metal box. The whole system is then completely shielded. The purpose of shielding is **(1)** to prevent RF interference between the components and **(2)** to prevent the emission of RF energy from any point except the output terminals. As a matter of fact, even the power line is decoupled by means of RF chokes and capacitors to prevent RF emission from it.

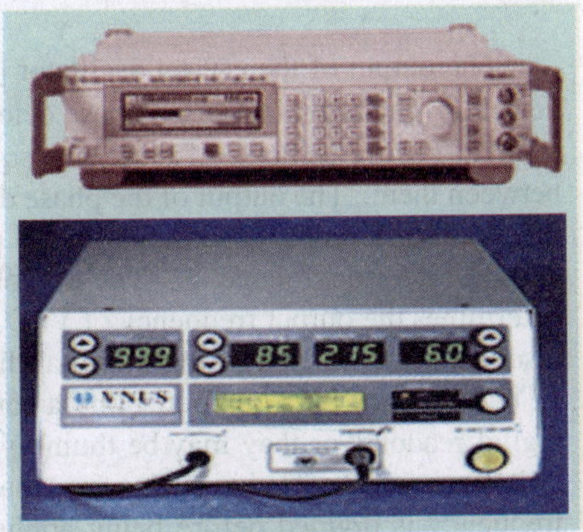

Fig. 73.59
(Courtesy Fluke Corporation)

Fig. 73.59 (*a*) shows an analog RF generator Model No. SML01 manufactured by Tektronics Corporation. This instrument is a general purpose signal generator and is available at low cost. It has a wide frequency range from 9 kHz to 3.3 GHz. Another RF generator Model No. SMP04 shown in Fig. 73.59 (*b*) from Tektronix Corporation is a high precision signal source. It has a wide frequency range from 0.01 GHz to 40 GHz. This instrument can produce, AM–, FM–, phase– and pulse modulated signals as well.

Applications

The RF signal generators are widely used in the area of radar and communication, research and development laboratories, education and training, electromagnetic interference (*EMI*) testing and material testing. Their main applications are :

1. To perform variety of tests on radio transmitters and receivers.
2. To test the amplitude and frequency response of RF amplifiers during the design phase.

73.45. Frequency Synthesizer

It is another type of RF generator that uses ***phase-locked loop*** (*PLL*) to generate output frequencies over a wide range. The most common range is from 1 MHz to 160 MHz. Fig. 73.60 shows a simple block diagram of a frequency synthesizer. As seen, the major components of the frequency synthesizer are : voltage controlled oscillator (*VCO*), divide-by-N counter, phase detector, crystal oscillator, low-pass filter and a square-wave circuit.

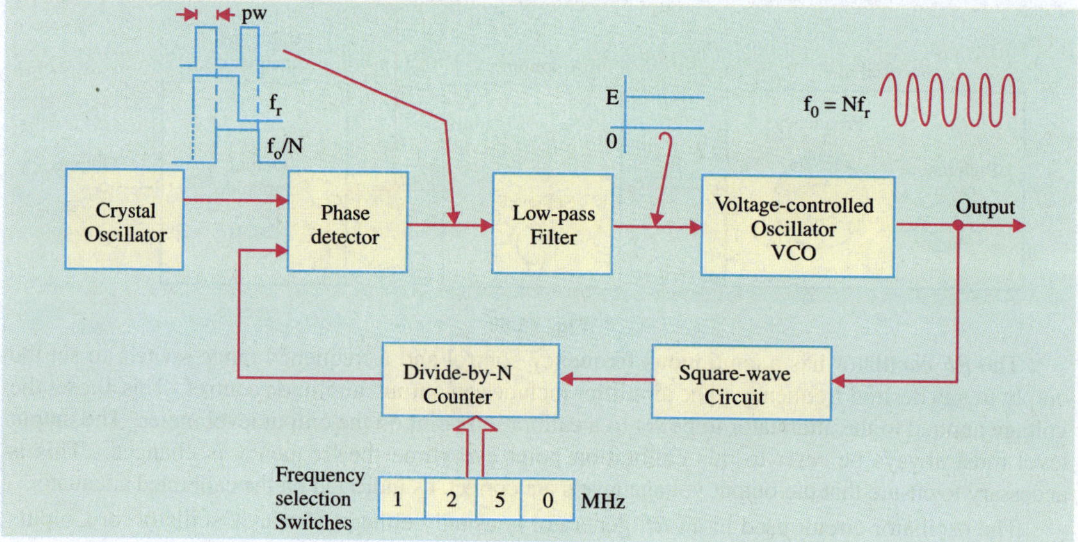

Fig. 73.60

The output of the crystal oscillator (a reference frequency, f_r), is fed into one input of a phase detector. The other input of a phase detector has another square-wave applied to it as shown in the figure. The frequencies of these two square waves is identical but there is a phase difference (ϕ) between them. The output of the phase detector is a pulse waveform with pulse width controlled by the phase difference. The output of the phase detector is applied to the low-pass filter which converts it into a dc voltage, *E*. The dc voltage, *E* is used as the control voltage for the *VCO* and it determines the output frequency of *VCO*. The output of *VCO* is fed to a circuit that converts it into a square wave for triggering a digital divide-by-N counter. The divide-by-N counter divides the *VCO* frequency by a number set by a bank of switches. These switches may be push buttons with digital readouts or they may be thumb-wheel type which indicate their position numerically. The switches are connected in such a way that the displayed number is the factor *N* by which the output frequency is divided before being applied to the phase detector. The switches allow the user to obtain frequency which is any integer multiple of the crystal oscillator frequency.

Applications

The frequency synthesizer is used in almost same areas as the RF signal generators.

73.46. Other Signal Generators

There are some signal generators that do not fit well in various pres-etabilished categories. Some of these signal generators are as discussed below :

1. **RF Markers :** These devices are usually crystal controlled and have a fixed output frequency for use as a reference. These are used to calibrate TV signals.

2. **Digitally Programmable Test Oscillators :** These instruments can have extremely wide frequency range although some versions have much narrow range also. The set frequency can be programmed through the front panel keypad or via a computer interface input such as IEEE-488 general purpose interface bus (commonly known as GPIB).

3. **Arbitrary Waveform Generators :** These instruments allow the user to design and generate virtually any desired waveform. The arbitrary waveform generator is quite useful to perform a variety of tests on communication equipment. For example, a modulated signal that varies over the entire bandwidth and amplitude range of the equipment shown in Fig. 73.61 could be created for testing purpose. Noise could also be superimposed upon the signal and gaps might be introduced between waveform bursts, to investigate the response of the system. Once such a waveform has been designed, it could be stored in the instrument memory and can be recalled repeatedly for production testing.

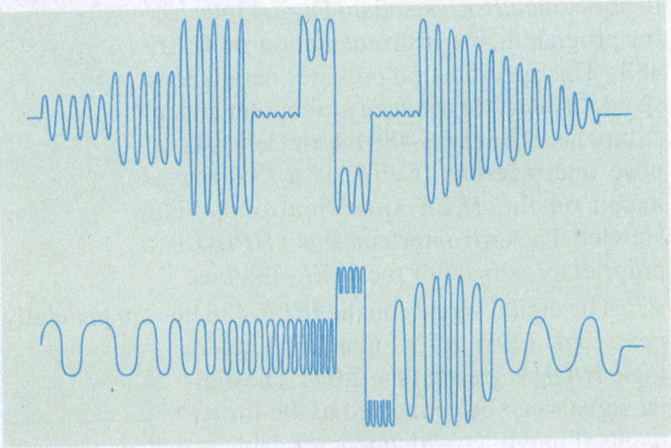

Fig. 73.61

Fig. 73.62 shows an arbitrary waveform generator from Agilent Technologies Model No. 33250A. This instrument can generate real world signals up to 80 MHz. It has a capability to display the waveforms in colour. It can be used as a function generator/pulse generator as well. The instrument has a GPIB/LAN interfaces.

Applications

The arbitary waveform generators are used extensively in the following areas :

1. Communications design and test for producing (*a*) arbitrary IF based signals and (*b*) standard waveforms for communication.
2. Mixed signal design and test.
3. Disk drive read/write design and test.
4. Real word simulations.
5. High-speed low filter data and clock pulse generation.

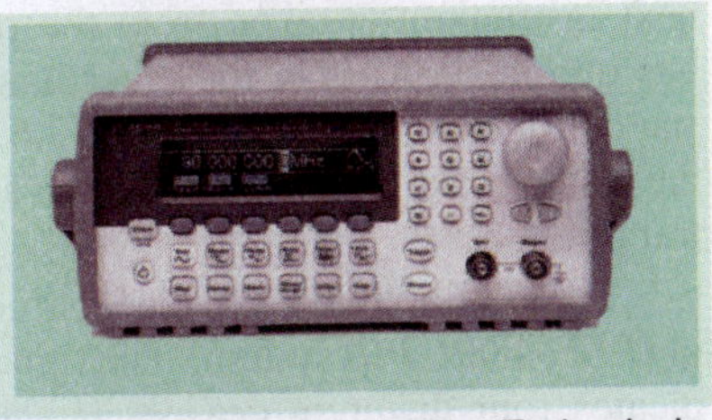

Fig. 73.62 (Courtesy : Agilent Technologies)

73.47. IEEE-488 General Purpose Interface Bus (GPIB) Instruments

These days automatic test equipment (*ATE*) is one of the leading methods for testing electronic equipment in factory production and troubleshooting situations. The basic method is to use a programmable digital computer to control a bank of test instruments. The bank of instruments can be configured for a special purpose or for general use. For example, we could select a particular line-up of equipment needed to test a broadcast audio console and provide a computer program to do a variety of measurements such as gain, frequency response, total harmonic distortion etc. The other possibility could be to do a generalized test set. This method is adopted by number of industries who have many electronic devices or systems to test. There is a main bank of electronic test equipment adapters to make the devices (or systems) under test interconnect with the system and a computer program to do a variety of measurements such as gain, frequency response, total harmonic distortion etc. The other possibility could be to do a generalized test set. This method is adopted by number of industries who have many electronic devices or systems to test. There is a main bank of electronic test equipment adapters to make the devices (or systems) under test interconnect with the system and a computer program for each type of equipment. Such an approach reduces the test equipment cost drastically.

The Institution of Electrical and Electronic Engineers (*IEEE*) has laid out a specification titled *IEEE* standard Digital Interface for programmable instrumentation or *IEEE*-488. This specification provides details for a standard interface between a computer and instruments. The *IEEE*-488 bus or General purpose interface bus (*GPIB*) is a tool that is based on the *IEEE* specifications. The Hewlett-Packard interface bus (*HPIB*) is a proprietary version of the *IEEE*-488 bus.

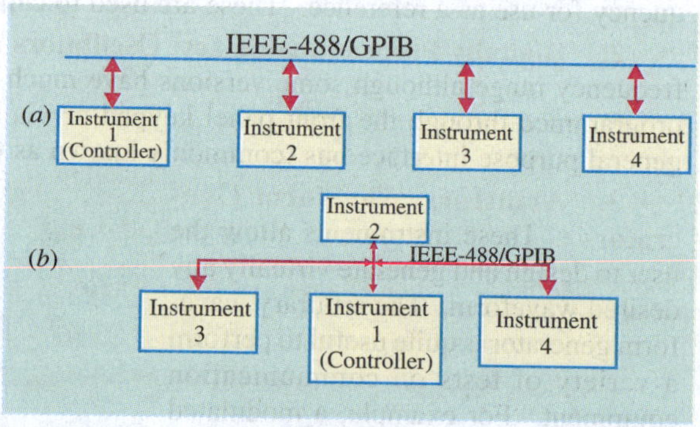

Fig. 73.63

The digital signals on the *IEEE*-488 bus are generally similar to *TTL* (transistor-transistor logic), *i.e.* a logic LOW is less than 0.8 V and a logic *HIGH* is greater than 2.0 V. The digital signals can be connected to the instruments through a multiconductor cable up to 20 metres in length provided that an instrument load is placed every 2 metres. Most *IEEE-488/GPIB* systems operate unrestricted to 250 kilobytes per second or faster with some restrictions.

There are two basic configurations for the *IEEE*-488/*GPIB* system : **(1)** linear and **(2)** star. In the linear configuration shown in Fig. 73.63 (*a*), a tap-off to the next instrument is taken from the previous one in series. On the other hand, in star configuration shown in Fig. 73.63 (*b*), the instruments are connected from a central point.

Fig. 73.64 shows the basic structure of *IEEE*-488/*GPIB* system. The figure indicates four different devices (*i.e.* computer, frequency counter, signal generator and digital multimeter) connected to the bus. The *IEEE*-488/*GPIB* system itself consists of three major buses : **(1)** general interface management (*GIM*) bus. **(2)** data *I/O* (*DIO*) bus and **(3)** data byte transfer (*DBT*) bus.

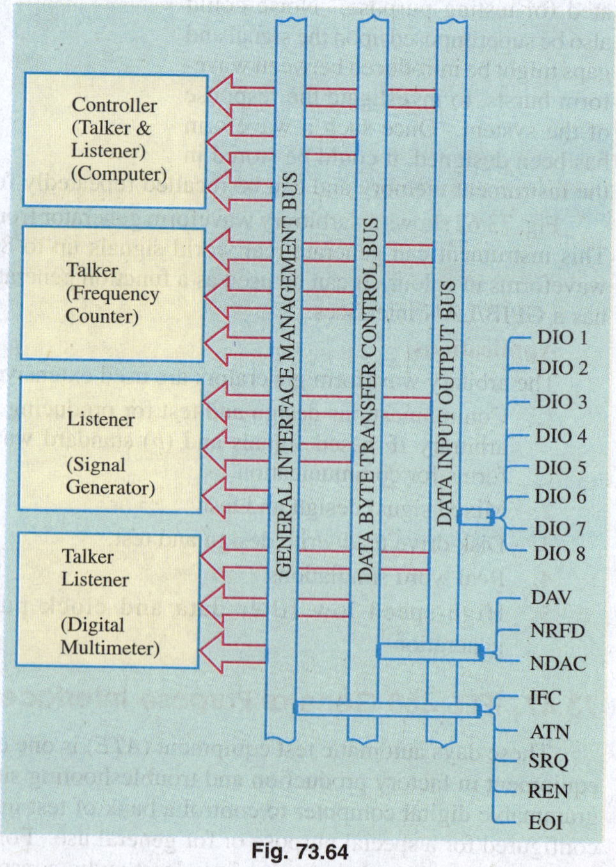

Fig. 73.64

1. General Interface Management (*GIM*) bus : This bus coordinates the whole system and ensures an ordely flow of data over the data input/output (*DIO*) bus. This bus has a number of signals explained below.

***IFC* (Interface Clear Signal) :** This signal is used by the controller to place all devices in a predefined quiescent or standby condition.

***ATN* (Attention Signal) :** This signal is used by the computer/controller to let the system know how data on the *DIO* bus lines are to be interpreted and which device is to respond to the data.

***SRQ* (Service Request Signal) :** This signal is used by a device on the system to ask the controller for attention. It is basically an interupt request signal.

***REN* (Remote Enable Signal) :** This signal is used by the controller to select between two alternate sources of device programming data.

***EOI* (End or Identify Signal) :** This signal is used by talkers for the following two purposes : (1) it will follow the end of a multiple byte sequence of data in order to indicate that the data are now finished. (2) It is also used in conjunction with the *ATN* signal for polling the system.

2. Data *I/O* (*DIO*) bus : This bus is a bidirectional 8-bit data bus that carries data, interface messages and device-dependent messages between the controller, talkers and listeners. This bus sends asynchronously in byte-serial format.

3. Data Byte Transfer (*DBT*) bus: This bus controls the sending of data along the *DIO* bus. There are three signals on this bus as explained below :

***DAV* (Data Valid Signal) :** This signal indicates the availability and validity of data on the line. If the measurement is not finished, for example, the *DAV* signal will be false.

***NRFD* (Not Ready For Data Signal) :** This signal lets the controller know whether or not the specific device addressed is in a condition to receive data.

***NDAC* (Not Data Accepted Signal) :** This signal is used to indicate to the controller whether or not the device accepted the data sent to it over the *DIO* bus.

Each signal line in the bus has a circuit similar to the one shown in Fig. 73.65. As seen from this circuit, each signal line has a pull-up and pull-down resistors, receiver, and driver-circuits. Besides this each signal line has also a shunt protection diode and a stray capacitance.

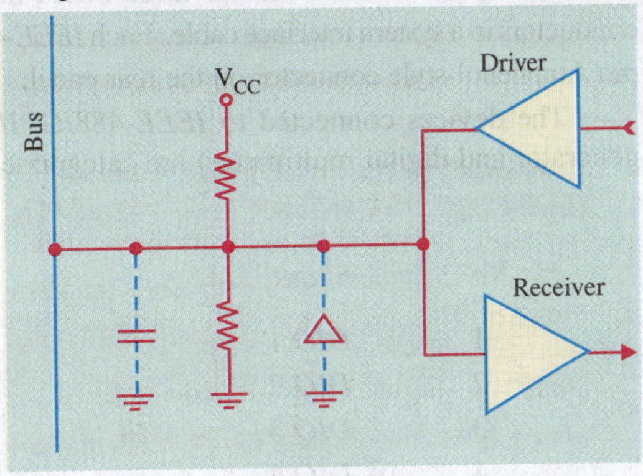

Fig. 73.65

ASCI & GPIB CODE CHART

B7 B6 B5 BITS / B4 B3 B2 B1	ø ø ø (CONTROL)	ø ø 1 (CONTROL)	ø 1 ø (NUMBERS SYMBOLS)	ø 1 1 (NUMBERS SYMBOLS)	1 ø ø (UPPER CASE)	1 ø 1 (UPPER CASE)	1 1 ø (LOWER CASE)	1 1 1 (LOWER CASE)
ø ø ø ø	0 NUL 0 0	20 DLE 10 16	40 0 SP 20 32	60 16 0 30 48	100 0 @ 40 64	120 16 P 50 80	140 0 ` 60 96	160 16 P 70 112
ø ø ø ø	1 GTL SOH 1 1	21 ILO DC1 11 17	41 1 ! 21 33	61 17 1 31 49	101 1 A 41 65	121 17 Q 51 81	141 1 a 61 97	161 17 q 71 113
ø ø ø 1	2 STX 2 2	22 DC2 12 18	42 2 " 22 34	62 18 2 32 50	102 2 B 42 66	122 18 R 52 82	142 2 b 62 98	162 18 r 72 114
ø ø 1 ø	3 ETX 3 3	23 DC3 13 19	43 3 # 23 35	63 19 3 33 51	103 3 C 43 67	123 19 S 52 83	143 3 c 63 99	163 19 s 73 115
ø 1 1 1	4 SDC EOT 4 4	24 DCL DC4 14 20	44 4 $ 24 36	64 20 4 34 52	104 4 D 44 68	124 20 T 54 84	144 7 d 64 100	164 20 t 74 116
ø 1 ø 1	5 PPC ENQ 5 5	25 PPU NAK 15 21	45 5 % 25 37	65 21 5 35 53	105 5 E 45 69	125 21 U 55 85	145 5 e 65 101	165 21 u 75 117
ø 1 1 ø	6 ACK 6 6	26 SYN 16 22	46 6 & 26 38	66 22 6 36 54	106 6 F 46 70	126 22 V 56 86	146 6 f 66 102	166 22 v 76 118
ø 1 1 1	7 BEL 7 7	27 ETB 17 23	47 7 ' 27 39	67 23 7 37 55	107 7 G 47 71	127 23 W 57 87	147 7 g 67 103	167 23 w 77 119
1 ø ø ø	10 GET BS 8 8	30 SPE CAN 18 24	50 8 (28 40	70 24 8 38 56	110 8 H 48 72	130 24 X 58 88	150 8 h 68 104	170 24 x 78 120
1 ø ø 1	11 TCT HT 9 9	31 SPD EM 19 25	51 9) 29 41	71 25 9 39 57	111 9 I 49 73	131 25 Y 53 89	151 9 i 69 105	171 25 y 79 121
1 ø 1 ø	12 LF A 10	32 SUB 1A 26	52 10 * 2A 42	72 26 : 3A 58	112 10 J 4A 74	132 26 Z 5A 90	152 10 j 6A 106	172 26 z 7A 122
1 ø 1 1	13 VT B 11	33 ESC 1B 27	53 11 + 2B 43	73 27 ; 3B 59	113 11 K 4B 75	133 27 [5B 91	153 11 k 6B 107	173 27 { 7B 123
1 1 ø ø	14 FF C 12	34 FS 1C 28	54 12 , 2C 44	74 28 < 3C 60	114 12 L 4C 76	134 28 \ 5C 92	154 12 k 6C 108	174 28 | 7C 124
1 1 ø 1	15 CR D 13	35 GS 1D 29	55 13 — 2D 45	75 29 = 3D 61	115 13 M 4D 77	135 29] 5D 93	155 13 m 6D 109	175 29 } 7D 125
1 1 1 ø	16 SO E 15	36 RS 1E 30	56 14 . 2E 46	76 30 > 3E 62	116 14 N 4E 78	136 30 ^ 5E 94	156 14 n 6E 170	156 30 ~ 7E 126
1 1 1 1	17 SI F 15	37 US 1F 31	57 15 / 2F 47	77 UNL ? 3 F 63	117 15 O 4F 79	137 UNT — 5F 95	157 15 o 6F 117	117 DEL (RUBOUT) 7F 127
	ADDRESSED COMMANDS	UNIVERSAL COMMANDS	LISTEN ADDRESSES		TALK ADDRESSES		SECONDARY ADDRESSES OR COMMANDS	

KEY

25 PPU NAK 15 1	GPIB Code ASCII character decimal

Tektronix ®

Ref : ANSI STD X3. 4-1977

IEEE STD 488-1978, **ISO STD** 646-1973

Fig. 73.66

Fig. 73.66 shows the 7-bit binary signals used in the *IEEE*-488/*GPIB* system for *ASCII* and

GPIB message codes.

The signals defined for the three buses in the *IEEE*-488/*GPIB* systems are implemented as conductors in a system interface cable. Each *IEEE*-488/*GPIB* compatible instrument will have a female 36-pin Amphenol-style connector on the rear panel. The pin-out definitions are given in Table 73.1.

The devices connected to *IEEE*-488/*GPIB* system (*i.e.* computer, frequency counter, signal generator and digital multimeter) are categorised as controller, listener and/or talker.

Table 73.1

Pin No	*Signal Lin*	*Pin No*	*Signal Line*
1	*DIO* 1	13	*DIO* 5
2	*DIO* 2	14	*DIO* 6
3	*DIO* 3	15	*DIO* 7
4	*DIO* 4	16	*DIO* 8
5	*EOI*	17	*REN*
6	*DAV*	18	Ground (6)
7	*NRFD*	19	Ground (7)
8	*NDAC*	20	Ground (8)
9	*IFC*	21	(Ground 9)
10	*SRQ*	22	(Ground 10)
11	*ATN*	23	(Ground 11)
12	Shield	24	Digital Ground

1. **Controller :** Its function is to communicate device addresses and other inface buses to instruments in the system.
2. **Listener :** Its function is to receive commands from other instruments (usually the controller) when the correct address is placed on the bus. The listener acts on the message received but does not send back any data to the controller. The signal generator shown in Fig. 73.22 is an example of a listener.
3. **Talker :** Its function is to respond to the message sent to it by the controller. The frequency counter shown in Fig. 73.62 is an example of a talker.

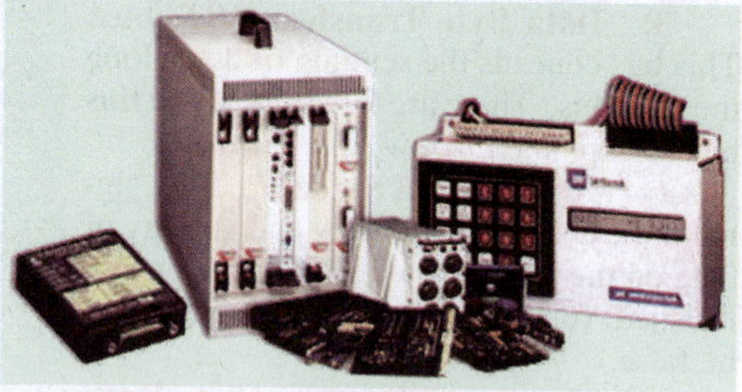

Fig. 73.67

There is also a combination device that accepts commands from the controller to set up ranges, tasks etc. and then returns data back over the *DIO* bus to the controller. The digital multimeter shown in Fig. 73.62 is an example of this category.

The *IEEE*-488 was introduced to the electronic industry in 1977. Since then it has evolved to *IEEE*-488.1 in 1987 and further to *IEEE*-488.2 in 1990. At present the system allows the control upto 14 instruments and it has data transfer rate greater than 1 M bytes/s.

73.48. VXI bus

The *VXI* bus is another fast growing platform for instrumentation systems. It was introduced in 1987 and since then it has experienced tremendous growth and acceptance around the world. *VXI* uses a mainframe chassis with a maximum of 13 slots to hold ***modular instruments on plug-in boards***. Fig. 73.67 shows an example of system using *VXI* instruments. The *VXI* backplane includes the 32-bit *VME* data bus, high performance instrumentation buses for precision timing and synchronization between instrument components, standarized initialization and resource management to ease configuration.

OBJECTIVE TESTS – 73

1. Digital instruments are those which
   (*a*) have numerical readout
   (*b*) use *LED* or *LCD* displays
   (*c*) have a circuitry of digital design
   (*d*) use deflection type meter movement.
2. The main difference between the electronic and electrical instruments is that an electronic instrument contains
   (*a*) an electronic device (*b*) a transducer
   (*c*) a digital readout (*d*) electrons.
3. The essential elements of an electronic instrument are
   (*a*) transducer (*b*) signal conditioner
   (*c*) indicating device (*d*) all of the above.
4. The current sensitivity of a meter is expressed in
   (*a*) ampere (*b*) ohm/ampere
   (*c*) ohm/volt (*d*) ampere/division.
5. The basic meter movement can be converted into an ohmmeter by connecting a .....with it.
   (*a*) high resistance in series
   (*b*) low resistance in parallel
   (*c*) battery in series
   (*d*) battery and a variable resistance in series
6. The D' Arsonval meter movement can be converted into an audio-frequency ac ammeter by adding a ...... to it.
   (*a*) thermocouple (*b*) rectifier
   (*c*) chopper (*d*) transducer.
7. In a linear meter, half-scale deflection occurs when there is ... per cent of the rated current through its coil
   (*a*) 100 (*b*) 25
   (*c*) 50 (*d*) 75
8. A 0-1 mA meter has a sensitivity of
   (*a*) 1 k Ω/V (*b*) 1 mA
   (*c*) 1 k Ω (*d*) 1000 A.
9. A moving coil instrument has a resistance of 10 Ω and takes 40 mA to produce full-scale deflection. The shunt resistance required to convert this instrument for use as an ammeter of range 0 to 2 A is
   (*a*) 0.1021 Ω (*b*) 0.2041 Ω
   (*c*) 0.2561 Ω (*d*) 0.4210 Ω
10. A moving coil ammeter has a fixed shunt of 0.02 ohm resistance. If the coil resistance of the meter is 1000 Ω, a potential difference of 500 mV is required across it for full-scale deflection. Under this condition, the current in the shunt would be
    (*a*) 2.5 A (*b*) 25 A
    (*c*) 0.25 A (*d*) 0.025 A
11. It is desired to convert a 0-1000 μA meter movement, with an internal resistance of 100 Ω into a 0-100 mA meter. The required value of shunt resistance is about
    (*a*) 1 Ω (*b*) 10 Ω
    (*c*) 99 Ω (*d*) 100 Ω
12. Loading effect is principally caused by... instruments
    (*a*) high resistance (*b*) low-sensitivity
    (*c*) high-sensitivity (*d*) high-range
13. A multimeter is used to measure
    (*a*) resistance (*b*) current
    (*c*) voltage (*d*) all of the above
14. A sinusoidal voltage of rms value 10 V is applied to a D' Arsonval movement connected in series with a half-wave rectifier. It will show a reading of... volt
    (*a*) 9 (*b*) 4.5
    (*c*) 10 (*d*) 7.7
15. A *VTVM* produces negligible loading effect on a circuit under test primarily because
    (*a*) it virtually drawn no current from the circuit
    (*b*) of its very high internal resistance
    (*c*) it uses high vacuum tubes
    (*d*) it is a null deflection instrument.
16. In a 3½ digit voltmeter, the largest number that can be read is
    (*a*) 0999 (*b*) 1999
    (*c*) 4999 (*d*) 9999
17. A 3½ digit voltmeter having a resolution of 100 mV can be used to measure maximum voltage of
    (*a*) 100 V (*b*) 200 V
    (*c*) 1000 V (*d*) 5000 V
18. The signal to be observed on the screen of an oscilloscope is applied
    (*a*) across its *X*-plates (*b*) across its *Y*-plates
    (*c*) to the horizontal amplifier
    (*d*) to the trigger circuit.
19. When a 30 V dc is applied to the vertical deflection plates of a cathode ray tube, the bright

spot moves 1 cm away from the centre. If 30 V (rms) ac is applied, then the movement of the spot will be nearly

(*a*) 1 cm (*b*) 1.5 cm
(*c*) 2 cm (*d*) 3 cm

20. Production of a steady stationary display of a signal waveform on the scope screen is due to
(*a*) persistence of vision
(*b*) flourescent material of the screen
(*c*) proper sync. between the signal and the sweep generator
(*d*) electrostatic focussing of the electron beam.

21. Two sinusoidal signals of frequency *f* and 3*f* are applied at *x* and *y* inputs respectively to an oscilloscope. Which one of the following patterns can be observed on the screen ?

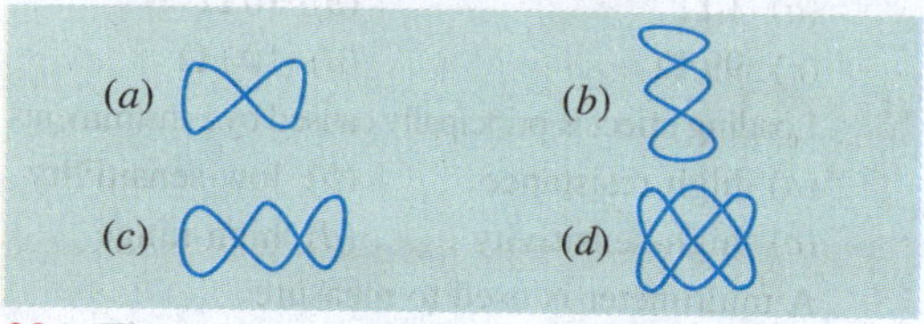

22. The *X*- and *Y*-inputs of a *CRO* are respectively $V \sin \omega t$ and $-V \sin \omega t$. The resulting Lissajous pattern will be
(*a*) a straight line (*b*) a circle
(*c*) an elipse (*d*) a figure of eight

23. The deflection sensitivity of a *CRT* depends inversely on the
(*a*) length of the vertical deflecting plates
(*b*) distance between screen and deflecting plates
(*c*) deflecting voltage
(*d*) separation between *Y*-plates.

24. Two complete signal cycles would be displayed on the screen of a scope when time-period of the sweep generator is ...... the signal time period.
(*a*) half (*b*) twice
(*c*) equal (*d*) thrice

25. A non-triggered oscilloscope is one which
(*a*) has no sweep generator
(*b*) cannot produce a stable stationary screen display
(*c*) has a continuously running time-base generator
(*d*) can display a portion of the input signal wave form.

26. A dual-trace *CRO* has
(*a*) one electron gun (*b*) two electron guns
(*c*) one electron gun and one two-pole switch
(*d*) two electron guns and one two-pole switch

27. The operation a *Q*-meter is based on
(*a*) self-induction (*b*) series resonance
(*c*) mutual induction (*d*) eddy currents.

28. The resolution of a logic analyser is
(*a*) maximum number of input channels
(*b*) the minimum duration of the glitch it can capture
(*c*) its internal clock period
(*d*) the minimum amplitude of the input signal it can display

29. A spectrum analyser can be described as
(*a*) voltage selective frequency meter
(*b*) current selective frequency meter
(*c*) frequency selective voltmeter
(*d*) None of these

ANSWERS

1. (*c*)	**2.** (*a*)	**3.** (*d*)	**4.** (*c*)	**5.** (*d*)	**6.** (*b*)	**7.** (*c*)	**8.** (*a*)	**9.** (*b*)	**10.** (*b*)
11. (*a*)	**12.** (*b*)	**13.** (*d*)	**14.** (*b*)	**15.** (*b*)	**16.** (*a*)	**17.** (*c*)	**18.** (*a*)	**19.** (*c*)	**20.** (*a*)
21. (*b*)	**22.** (*a*)	**23.** (*d*)	**24.** (*b*)	**25.** (*c*)	**26.** (*c*)	**27.** (*b*)	**28.** (*a*)	**29.** (*d*)	

INDEX

A

Absorption laws, 2587
A.C. load line, 2232
A.C./D.C. Analysis (of a transistor), 2213
A-class Amplifier, 2285
Active region (of BJT), 2208
Adder (or summer), 2576
Alpha, cut-off frequency, 2311
Alloy Junction, 2069
Amplifier, 2453, 2466
Amplifier, audio, 2466
Amplifier, buffer, 2453
Amplifier, cascaded, 2310
Amplifier, classification of, 2278
Amplifier, class A, 2285, 2287, 2291
Amplifier, -AB, 2285
Amplifier, -B, 2285
Amplifier, -C, 2285, 2300
Amplifier, differential, 2334, 2282
Amplifier, direct coupled, 2317, 2326
Amplifier, distortion in, 2301
Amplifier, feedback type, 2344
Amplifier, FET., (2383-2385)
Amplifier, frequency response of, 2307, 2308
Amplifier, gain of, 2279
Amplifier, gain bandwidth product, 2311
Amplifier, impedence coupled, 2316
Amplifier, JFET, 2364, 2373
Amplifier, multistage, 2316
Amplifier, negative feedback, (2302-2305)
Amplifier, power distribution, 2295
Amplifier, push-pull, type, (2296-2299)
Amplifier, R-C coupled, (2318-2321)
Amplifier, Transformer coupled, 2317, 2323
Amplifier, tuned, 2300
Amplifier, voltage type, 2277
Amplitude, modulation, 2445
Analog instruments, 2696
AND-gate, 2560
Atom, 2018
Atomic, model (Bohr) 1825
Avalan breakdown, 2066, 2316
Average value, 2155
Azimuthal (or orbital) Quantum Number, (l) 2022
Ayrton Shunt,

B

Band, conduction, 2028
Band, valence, 2028
Band, width, 2028, 2350
B-class Amplifier, 2285
Bakhausen criteria, 2412
Barrier (or Jn. voltage,) 2049
Basic Op.Amp.Series Regulator, 2172
Base bias, 2224, 2225
BCD code, 2546
Beta-cut-off frequency, 2311
Beta-factor, 2142
Beta-rule (or Transfer ratio), 2205
Beta-sensitivity, 2224
Binary system, (2530-2538)
Binary, -addition, 2531
Binary, -division, 2538
Binary, -multiplication, 2536
Binary, -subtraction, 2533
Bipolar Junction Transistor (BJT), 2188
Bi-stable Multi-Vibrator (BMV), 2430
Blocking Oscillator, 2437
Bohr's Atomic Model, 2018, 2019
Boltzman's diode equation, 2062
Boolean Algebra, (2585-2597)
Bonding, 2478
Breakdown, devices 2112
Breakdown-voltage, 2396, 2402
Bubbled gate, 2565

C

Capacitance, diffusion, 2067
Cathode Ray Tube (CRT), 2712
C-B Amplifier, 2278
C-B, Amplifier, configuration, 2191
C-C Amplifier, 2282, 2283
C-C, Amplifier configuration, 2193
C-class Amplifier, 2285
C-E, Amplifier, 2280, 2281
C-E, Amplifier, configuration, 2192
Characteristics, of DE MOSFET, 2378
Characteristics, of DIAC, 2403
Characteristics, of E MOSFET, 2391
Characteristics, of JFET, 2464, 2387
Characteristics, of SCR, 2396
Characteristics, of transistors, 2196
Characteristics, of TRIAC, 2401
Clampers, 2076, 2079
Clippers, 2076, 2162, 2163
Code, alpha-numeric, 2550
Code, BCD, 2546
Code, Excess-3, 2548
Code, five bit, 2549
Code, four bit, 2549
Code, Gray, 2549
Code, hexa-decimal, 2548
Code, Johnson, 2549
Code, octal, 2550
Code, -ring counter, 2550
Code,-seven bit, 2550
Coefficient of, Hall, 2054
Colpitt Oscillator, 2412
Combinational Logic Gate 2575
Comparison: A.M. and F.M. 2467
Comparison, amplifier and oscillator, 2408
Comparison, amplifier configurations, 2284
Comparison, Darlington pair and Emitter Follower, 2332
Comparison,- FETs and Bi-polars, 2376
Comparison, VOM and VTVM, 2708
Complemental subtraction, (of IC's) 2536
Common mode input, 2335
Conduction Bond, 2028
Conductivity, of metals, 2040
Conductivity, of semi-conductors, (2041-2043)
Conversion in between systems, (2528-2544)
Cross-over distortion, 2298
Cut-off point, (2207-2208)

D

D-Latch, 2655
Deflection, sensitivity, 2714
De Morgon's theorem, 2591
DE MOSFET, (2378-2383)
Depletion layer, formation of, 2048
Deviation ratio, 2456
Detection (or Demodulation), (2444-2445)
DIAC, 2403
Die (of an I.C.), 2478
Differential Amplifier, (2334-2338)
Diffusion,-capacitance, 2067
Diffusion, junction, 2070
Diffusion-in solids and gases, 2070
Digital, voltmeter (DVM), 2709
Diodes (2060-2082)
Diode, -AND-gate, 2561
Diode, circuits, 2075
Diode, fabrication, 2069
Diode, GUNN, 2125
Diode, ideal, 2071
Diode, IMPATT, 2125
Diode, LED, 2088
Diode, -OR-GATE, 2556
Diode, photo, 2094
Diode, PIN, 2122
Diode, real, 2074
Diode, Schottly, 2123
Diode, stop recovery, 2324
Diode, tunnel, 2120
Diode, varactor, 2121
Diode, Zener, 2112
Distortion in Amplifiers, 2301
Divide-by-two method (or double dabble) method, 2530
Double side bond suppressed carrier system (DBSC), 2451
Drift velocity (electric), 2045
Duals, 2593

E

EBC-DIC, 2550
Eccles Jordan MV, 2435
Efficiency, Rectifiers, 2132
Electronic Energy Levels (of H-atom), 2021
Electronic, conductivity, (of metals), 2040
Electronic, configuration, 2040
Electronic, mobility, 2040
Electronic, orbits, 2020
Electronic, voltmeter, 2708
Electronic Instruments, 2695
Emergency Lighting systems, 2400
Emitter bias (with two supplies) 2227
Energy, bands, 2026
Energy, in conductors, 2030
Energy, in extrinsic semiconductors, 2037
Energy, in lethium, 2028
Energy, in Insulators, 2030
Energy, in intrinsic semiconductors, 2034
Energy, in solid, 2026
Energy, level, Fermi, 2035
E-only MOSFET, 2382
Epitaxy (of an I.C.) 2479
Epitaxial Junction 2071
Equivalent Switching Circuit, 2587
Equivalent, of Amplifier, 2250
Equivalent, Half-wave, 2135
Equivalent, Transistors 2238
Equivalent T-network, (of CB Circuit), 2254
Etching of an I.C., 2479
Excess-3 code, 2548
Excess Gardy code, 2549
Exclusive, 'OR' gate, 2559
Extrinsic Semiconductor, 2037

F

Factor, noise, 2303
Factor, stability, 2223
F.M. Side-bands, 2457
Fall time, 2427
Feed-back, Amplifiers, 2344, 2362
Feed-back, negative, 2344, 2345
Feed-back, positive, 2344, 2345
Fermi level, 2035
Fermi level, in extrinsic semiconductors, 2036
Fermi level, in intrinsic semiconductors, 2035
Field Effect Transistors, (FET), (2163-2387)
Field Effect Transistors, advantage of, 2377
Field Effect, Transistors, applications of, 2385
Flip-flops, 2435, 2632
Frequency, conversion, 2264
Frequency, deviation, 2455
Frequency, modulation, 2454
Frequency, stability, 2411
Full, Adder (or Summon), 2577
Full, Wave Rectifier (3 phase), 2142
Full-subtractor, 2579
Fundamental of Light, 2088

G

Gain, bandwidth product, 2311
Gain, stability of, 2347
Grady code, 2549

H

Half, Adder (or Summer), 2576
Half-subtractor, 2579
Half-wave Rectifier, 2131
Hall effect, 2054
Harmonics, distortion, 2302
Hertley Oscillator, 2416
Hexi-decimal code, 2548

I

IG-FET, 2114
Immobile ions, 2039
Inter-modulation, distribution, 2302
Integrated circuits, (2471-2501)
Intrinsic semi-conductors, 2034
Intrinsic semi-conductors conductivity of, 2041
Intrinsic, stand-off radio, 2391
Invertor, 2163, 2164

J

Jn-FET, amplifier, 2374
Jn-FET, basic construction, 2364, 2368, 2376
Jn-FET, d.c. biasing of, 2371
Jn-FET, gain, 2374

Jn-FET, gate, 2376
Jn-FET,, static characterstics of, 2367
Jn-FET, source-signal parameters, 2369
Junction, alloy, 2069
Junction, capacitors, 2066
Junction, diffused, 2070
Junction, epitaxial, 2071
Junction, P.N., 2060
Junction, point contact, 2071
Junction, resistance, 2064
Junction, transistor, 2188
Junction, voltages, 2049

K

Karnaugh map, 2599
K(or n = 1) shell, 2024

L

L (or n = 2) shell, 2024
LASCR, 2104
Latch, 2633
Latching devices, 2397
Leakage currents, 2193
Level detector, 2412
Light Emitting Diode (LED), 2088, 2091
Linear, distortion, 2301
Linear Integrated Code (LIC), 2333
Lissajous figures, 2718
Load, lines, 2371
Logic families, 2610
Low current drop-out, 2162

M

M (or n = 3) shell, 2049
Magnetic Spin Quantum Number (ms), 2025
Majority carriers, 2039
Miller effect, 2310
Minority carriers, 2039
Mobile charge carriers, 2039
Modulation, (2441-2470)
Modulation, amplitude, 2445
Modulation, degree of, 2445
Modulation, phase, 2445
Modulation, index, 2456
Modulation, mathematical analysis, 2457
Modulation, method of, 2457
Modulation, frequency, 2454
Mono-stable Multi Vibrator, 2429
MOS-FET, 2378
Multiply by-two method, 2530
Multirange-voltmeters, etc., 2702, 2705
Multi-vibrators, 2429

N

NAND-gate, 2567, 2633
Negative, feed-back, 2344
Negative, logic, 2556
Negative, sequence component,
Negative, scalar, 2342
Noise, immunity from, 2612
Noise, margin, 2612
N-Mos circuit, 2622
Non-inverting Op. Amp. 2358
Non-simusoidal oscillators, 2425
Normally off MOSFET, 2378
NOR-gate, 2566
NOT-gate, 2563
NOR-gate Latch 2638
N-type extrinsic semi-conductor, 2037
Number systems, 2525
Number binary 2226
Number, hexi-decimal, 2526, 2543
Number, octal, 2226, 2540

O

Ohm-meter, 2385
Octal coding, 2526, 2548
Octal system, 2540
One's complement subtraction, 2534
OR-gate, 2556
Orbital Magnetic Quantum Number, 2025
Operating regions, 2208, 2220
Optical Isolators, 2104
Optio-electric devices, (2087-2110)
Oscillators, 2408, 2412
Oscillators, Clapp, 2412
Oscillators, Colpitt, 2412
Oscillators, damped, 2172, 2412

Oscillators, non-sinusoidal, 2425
Oscillators, Pierce (FET), 2422
Oscillators Phase shift type, 2422, 2424
Oscillators, R-C, type 2423, 2424
Oscillators, tuned base (transistor), 2412
Oscillators, tuned base (transistor), 2412
Oscillators, tuned collector (transistor), 2412, 2413
Oscillators, drain (transistor), 2412, 2413
Oscillators, un-damped, 2172
Oscillators, Wien Bridge (A.C.), 2424

P

Parallel Binary Adder, 2578
Passive components, 2472
Paulis' Exclusion principle, 2025
Peak, Clippers 2116
Peak, inverse voltage, (PIV), 2134
Phase, control, 2399
Phase, distortion (delay distortion), 2303
Pinch-off point, 2368
P.N. Junction, (2059-2085)
P.N. applications, 2063
P.N. diode, 2061
P.N. energy band diagram, 2050
P.N. equivalent circuit of, 2068
P.N. equation of static characteristics, 2062
Positive feedback, 2344, 2345
Positive, logic, 2556
P-type extrinsic semi-conductor, 2038
Pulse width, 2427
Push-pull Amplifier, (2296-2299)

Q

Q-factor, 2721
Q-meter, 2721
Q-points, 2222
Quantum Numbers, 2022, 2023, 2024
Quiescent point, 2220

R

R-C coupled Amplifier (2315-2342)
Recombination, (of ions), 2047
Rectifier, single phase (half and full wave), (2129-2136)
Recitifier, six phase, 2143
Rectifier, three phase, (2142-2143)
Rectifiers on controlled, 1948
Regulated Power Supply, (2167-2176)
Resistivity, of conductors, 2039
Resistivity, of extrinsic semi-conductors, 2040
Resistivity, of intrinsic semi-conductor., 2039, 2040
Ripple factor (of various types of Factors), 2133, 2138, 2141, 2149
Rise time, 2427
RTL circuit, 2617

S

Sampling, C.R.O., 2718
Saturation point, 2207, 2208
Saw-tooth Generators, 2426
Schmitt trigger, 2435
S.C.R. (Silicon Controlled Rectifier), (2395-2405)
S.C.R., applications, 2397
S.C.R. biasing of , 2395
S.C.R., controlled circuit, 2395
S.C.R., (*di/dt*)effect, 2398
S.C.R., (*do/dt*) effect, 2399
S.C.R. phase control of, 2160, 2161
S.C.R., light activated 2400
S.C.R., pulse control of, 2159
S.C.R., switch, 2404
S.C.R., two transistor, 2396
Semi-conductors, (2030-2034)
Semi-conductor, diodes, 2030
Side bands, (in Modulation), (2457-2458)
Side bands frequencies, 2457
Silicon Controlled Switch, 2404
Single phase, Shot Multivibrator, 2429
Source, follower, 2373
Special Diodes, (2111-2128)
Stability factor, 2224
Starters, for Induction Motors, (star-delta), 2233
Static characteristics of, diode, 2062
Static characteristic of, JFET, (2363-2387)
Static characteristic of, MOS-FET, 2378
Static characteristic of, SCR, 2395

Static characteristic of, TRIAC, 2401
Storage, CRO, 2710
Storage, Oscilloscope, 2717

T

Theft Alarm, 2400
Thermal, run away, 2195
Thevenin's theorem, 2229
Threshold voltage, 2148
TI (Three Input-OR)-gate, 2559
Time, delay, 2427
T-model, 2253
Transistor, amplifier, 2278
Transistor, basic circuit, 2213
Transistor, biasing rule for, 2189
Transistor, bipolar, 2188
Transistor, capacitance, 2066
Transistor, C.B. configuration, 2191
Transistor, C.C. configuration, 2193
Transistor, C.E. configuration, 2192
Transistor, characteristics of (static), 2196, 2197
Transistor, cut-off in, 2207, 2208
Transistor, equivalent circuits, 2238, 2251
Transistor, fault location in, 2210
Transistor, H-parameters, 2256, 2259
Transistor, hybrid formulae, 2261–2266
Transistor, 'OR'-gate, 2408
Transistor, saturation in, 2207, 2208
Transistor, stability factor, 2223
Transistor, T-model, 2253-2255
Transistor, thermal runaway, 2195
Transistor, TRIAC, 2401, 2402

U

Uni-Junction Transistor (UJT), 2161, 2389, 2400
Uni-junction, Oscillator (or Relarotion), 2157
Un-distorted output (max), 2222

V

Valence Band, 2028
Varacter, 2121
Voltmeter, digital, (DOM), 2709
Voltmeter, electronic type, 2709
Voltmeter, loading effect, 2705
Voltmeter, multimeter (O.C) 2704
VTVM, 2708

W

Wafer (of an IC.) 2479
Wien's Bridge Oscillator, 2424

X

X-'NOR'-gate, 2569

Z

Zener Diode, 2112
Zener Diode, breakdown, 2065
Zener Diode, as peak clipper, 2116
Zener Diode, for meter protection, 2116
Zener divide, as reference elements, 2116
Zener Diode, voltage regulation. 2114
Zener, uses of, 2113
Zero, decibel reference level, 2307

Win Prizes !

Attention: Students

We request you, for your frank assessment, regarding some of the aspects of the book, given as under:

10A 295 A Textbook of Electrical Technology, Vol. IV
B.L. Theraja, A.K. Theraja **Revised Edition 2010**

Please fill up the given space in neat capital letters. Add additional sheet(s) if the space provided is not sufficient, and if so required.

(i) What topic(s) of your syllabus that are important from your examination point of view are not covered in the book ?

....

....

....

....

(ii) What are the chapters and/or topics, wherein the treatment of the subject-matter is not systematic or organised or updated?

....

....

....

....

....

(iii) Have you come across misprints/mistakes/factual inaccuracies in the book? Please specify the chapters, topics and the page numbers.

....

....

....

....

....

(iv) Name top three books on the same subject (in order of your preference - 1, 2, 3) that you have found/heard better than the present book? Please specify in terms of quality (in all aspects).

1 ....

....

2 ....

....

3 ....

....

(v) Further suggestions and comments for the improvement of the book:
....
....
....
....
....

Other Details:

(i) Who recommended you the book? (Please tick in the box near the option relevant to you.)
☐ Teacher ☐ Friends ☐ Bookseller

(ii) Name of the recommending teacher, his designation and address:
....
....
....

(iii) Name and address of the bookseller you purchased the book from:
....
....
....

(iv) Name and address of your institution (Please mention the University or Board, as the case may be).
....
....
....

(v) Your name and complete postal address:
....
....
....

(vi) Write your preferences of our publications (1, 2, 3) you would like to have:
....
....

The best assessment will be awarded half-yearly. The award will be in the form of our publications, as decided by the Editorial Board, amounting to Rs. 300 (total).

Please mail the filled up coupon at your earliest to:

Editorial Department
S. CHAND & COMPANY PVT. LTD.,
Post Box No. 5733, Ram Nagar,
New Delhi 110 055